Advances in Modal Logic

Volume 13

Advances in Modal Logic

Volume 13

Edited by

Nicola Olivetti

Rineke Verbrugge

Sara Negri

and

Gabriel Sandu

ISBN 978-1-84890-341-8

College Publications
Scientific Director: Dov Gabbay
Managing Director: Jane Spurr

http://www.collegepublications.co.uk

Contents

Preface

Advances in Modal Logic (AiML) is an initiative founded in 1995 and aimed at presenting an up-to-date picture of the state of the art in modal logic and its many applications. It consists of a conference series together with volumes based on the conferences. The conference series is the main international forum at which research on all aspects of modal logic is presented. The first installment was held in 1996 in Berlin, Germany, and since then it has been organized biennially, with meetings in 1998 in Uppsala, Sweden; in 2000 in Leipzig, Germany (jointly with ICTL-2000); in 2002 in Toulouse, France; in 2004 in Manchester, UK; in 2006 in Noosa, Australia; in 2008 in Nancy, France; in 2010 in Moscow, Russia; in 2012 in Copenhagen, Denmark; in 2014 in Groningen, the Netherlands; and in 2016 in Budapest, Hungary; in 2018 in Bern, Switzerland. Information about AiML and related events, including conference proceedings, is available at the website `www.aiml.net`.

The thirteenth conference in the AiML series was was organized at the University of Helsinki, Finland, by Sara Negri (University of Genova and University of Helsinki) and Gabriel Sandu (University of Helsinki), with the assistance of Fausto Barbero, Annika Kanckos, Eugenio Orlandelli, and Edi Pavlović. Due to the exceptional situation caused by Covid 19, AiML 2020 was held online on August 24–28, 2020.

The conference website can be found at
`https://www.helsinki.fi/en/conferences/advances-in-modal-logic-2020`.

This volume contains abstracts of invited talks and contributed papers from the conference. The invited talks were given by

- Bahareh Afshari (University of Gothenburg and University of Amsterdam)
- Nick Bezhanishvili (University of Amsterdam)
- Melvin Fitting (City University of New York)
- Nina Gierasimczuk (Technical University of Denmark)

The Programme Committee received 56 regular paper submissions. Of these, 31 were selected for this volume by a reviewing process where every paper received three independent expert reviews.

The volume includes papers on propositional modal logics, their products, predicate modal logics, temporal and epistemic reasoning, modal logic with non-boolean basis, provability and interpretability logics, inquisitive, dynamic, connexive, intuitionistic, substructural, dependence logics and hybrid logics, and other related logics. The topics include history of modal reasoning, decidability and complexity results, proof theory, model theory, interpolation, as well as other related problems in algebraic logic.

In addition, there were 29 submissions for short presentations at the conference, the great majority of them were accepted for presentation.

The members of the Programme Committee for the conference were

- Natasha Alechina (Utrecht University)
- Maria Aloni (University of Amsterdam)
- Philippe Balbiani (CNRS, IRIT Toulouse)
- Guram Bezhanishvili (New Mexico State University)
- Marta Bílková (Charles University Prague)
- Patrick Blackburn (University of Roskilde)
- Agata Ciabattoni (TU Wien)
- Giovanna Corsi (University of Bologna)
- Giovanna D'Agostino (University of Udine)
- Stéphane Demri (CNRS, LSV, ENS Paris-Saclay)
- Hans van Ditmarsch (CNRS, LORIA, University of Lorraine)
- David Fernández-Duque (Ghent University)
- David Gabelaia (TSE Razmadze Mathematical Institute)
- Didier Galmiche (CNRS, LORIA, University of Lorraine)
- Silvio Ghilardi (University of Milan)
- Valentin Goranko (Stockholm University)
- Rajeev Goré (The Australian National University)
- Davide Grossi (University of Groningen)
- Helle Hvid Hansen (Delft University of Technology)
- Wesley Holliday (UC Berkeley)
- Agi Kurucz (King's College London)
- Roman Kuznets (TU Wien)
- Carsten Lutz (University of Bremen)
- George Metcalfe (University of Bern)
- Larry Moss (Indiana University)
- Cláudia Nalon (University of Brasilia)
- Sara Negri (University of Genova and University of Helsinki)
- Eric Pacuit (University of Maryland)
- Xavier Parent (University of Luxembourg)
- Valeria De Paiva (Samsung Research America, Birmingham University)
- Sophie Pinchinat (IRISA, University of Rennes I)
- Mark Reynolds (The University of Western Australia)
- Renate Schmidt (University of Manchester)

- Ilya Shapirovsky (Institute for the Information Transmission Problems)
- Valentin Shehtman (Institute for the Information Transmission Problems)
- Thomas Studer (University of Bern)
- Sara L. Uckelman (Durham University)
- Yde Venema (University of Amsterdam)
- Yanjing Wang (Peking University)
- Michael Zakharyashev (Birbeck University of London)

The Programme Committee was chaired by

- Nicola Olivetti (Aix-Marseille University)
- Rineke Verbrugge (University of Groningen)

The Steering Committee of AiML for 2018–2020 consisted of

- Lev Beklemishev (Steklov Mathematical Institute)
- Guram Bezhanishvili (New Mexico State University)
- Rajeev Goré (Australian National University)
- Giovanna D'Agostino (University of Udine)
- Stéphane Demri (CNRS, France)
- Agi Kurucz (King's College London)
- Sara Negri (University of Genova and University of Helsinki) (local organizer of AiML 2020)
- Nicola Olivetti (Aix-Marseille University)
- Rineke Verbrugge (University of Groningen)

Many other people assisted with the reviewing process, including: Juan Aguilera, Michael Baur, Hugo Bazille, Nick Bezhanishvili, Torben Braüner, Luca Carai, Ivano Ciardelli, Tiziano Dalmonte, Anupam Das, Jeremy Dawson, Sebastian Enqvist, Luis Estrada-González, Saúl Fernández González, Melvin Fitting, Nissim Francez, Peter Fritz, Krisztina Fruzsa, Nikolaos Galatos, Rustam Galimullin, Francesco Antonio Genco, Marianna Girlando, Christopher Hampson, Joost Joosten, Max Kanovich, Stanislav Kikot, Antti Kuusisto, Stepan Kuznetsov, Ori Lahav, Martin Lange, Eveline Lehmann, Bjoern Lellmann, Mateusz Łełyk, Yanjun Li, Churn-Jung Liau, Tim Lyon, Tommaso Moraschini, Massimo Mugnai, Hoang Nga Nguyen, Eugenio Orlandelli, Fedor Pakhomov, Edi Pavlovic, Marcin Przybyłko, Gabriele Pulcini, Vít Punčochář, Ricardo Oscar Rodriguez, Daniel Rogozin, Mikhail Rybakov, Kiana Samadpour Motalebi, Luigi Santocanale, Nenad Savic, Giorgio Sbardolini, Julian Schlöder, François Schwarzentruber, Ian Shillito, Dmitry Shkatov, John Stell, Olim Tuyt, Heinrich Wansing, Xuefeng Wen, Evgeny Zolin. We apologize to anyone whose name was inadvertently left off this list.

We thank the organizers of the conference for their dedication in bringing the conference to life, and to move it online when that became necessary. In particular, we are very grateful to Eugenio Orlandelli for his precious help in the LaTeX processing of the proceedings. We thank the members of the Programme Committee and all other reviewers for the time, professional effort and the expertise that they invested in ensuring the high scientific standards of the conference and its proceedings. We also thank the authors for their excellent contributions. Moreover we thank Jane Spurr for bringing this volume to publication. Special thanks go to Guram Bezhanishvili and Giovanna D'Agostino (Program Chairs of AiML 2018), who supported us with their advice in all phases of the conference.

We would like to thank the Academy of Finland for generously sponsoring the conference within the project *Modalities and conditionals: Systematic and historical studies* (project no. 1308664) and the University of Helsinki for help in the organization and online facilities.

June 24th, 2020
Nicola Olivetti
Rineke Verbrugge
Sara Negri
Gabriel Sandu

Abstracts of Invited Talks

Cyclic Proof Systems for Modal Logics

Bahareh Afshari [1]

University of Amsterdam
Institute for Logic, Language and Computation
Postbus 94242, 1090 GE Amsterdam, The Netherlands

University of Gothenburg
Department of Philosophy, Linguistics and Theory of Science
Box 200, 40530 Göteborg, Sweden

A cyclic proof is a, possibly infinite but, regular derivation tree in which every infinite path satisfies a certain soundness criterion, the form of which depends on the logic under study. Circular and, more generally, non-well-founded derivations are not traditionally regarded as formal proofs but merely as an intermediate machinery in proof-theoretic investigations. They are, however, an important alternative to finitary proofs and in the last decade have helped break some important barriers in the proof theory of logics formalising inductive and co-inductive concepts. Most prominently cyclic proofs have been investigated for: *first-order logic with inductive definitions* [6,8,4], *arithmetic* [18,5,9], *linear logic* [3,10], *modal and dynamic logics* [19,13,17,20,14,2,12,1], *program semantics* [16,11] and *automated theorem proving* [7,15,21].

We focus on cyclic proofs for modal logics, ranging from Gödel-Löb logic to more expressive languages such as the modal μ-calculus, and reflect on how they can contribute to the development of the theory of fixed point modal logic.

References

[1] Afshari, B. and G.E. Leigh. *Lyndon interpolation for the modal mu-calculus*, in: TbiLLC (2019), to appear.

[2] Afshari, B. and G.E. Leigh. *Cut-free completeness for modal mu-calculus*, in: LICS (2017).

[3] Baelde, D., A. Doumane and A. Saurin. *Infinitary proof theory: the multiplicative additive case*, in: CSL (2016).

[4] Berardi, S. and M. Tatsuta. *Classical system of Martin-Löf's inductive definitions is not equivalent to cyclic proof system*, in: LMCS 15(3) (2019).

[5] Berardi, S. and M. Tatsuta. *Equivalence of inductive definitions and cyclic proofs under arithmetic*, in: LICS (2017).

[6] Brotherston, J. *Cyclic proofs for first-order logic with inductive definitions*, in: TABLEAUX (2005), pp. 78–92.

[7] Brotherston, J., N. Gorogiannis and R.L. Petersen. *A generic cyclic theorem prover*, in: APLAS (2012).

[1] b.afshari@uva.nl

[8] Brotherston, J. and A. Simpson. *Sequent calculi for induction and infinite descent*, in: Journal of Logic and Computation 21(6)(2011), pp. 1177–1216.

[9] Das, A. *On the logical complexity of cyclic arithmetic*, in: Logical Methods in Computer Science 16(1) (2019).

[10] De, A. and A. Saurin. *Infinets: The parallel syntax for non-wellfounded proof-theory*, in: TABLEAUX (2019), pp. 297–316.

[11] Docherty S. and R.N.S. Rowe. *A non-wellfounded, labelled proof system for propositional dynamic logic*, in: TABLEAUX (2019), 335–352.

[12] Enqvist, S., H.H. Hansen, C. Kupke, J. Marti and Y. Venema. *Completeness for Game Logic*, in: LICS (2019).

[13] Jungteerapanich N. *A tableau system for the modal μ-calculus*, in: TABLEAUX (2009), pp. 220–234.

[14] Kokkinis, I. and T. Studer. *Cyclic Proofs for Linear Temporal Logic*, in: Ontos Mathematical Logic 6 (2016), 171–192.

[15] Rowe, R.N.S. and J. Brotherston. *Realizability in cyclic proof: Extracting ordering information for infinite descent*, in: TABLEAUX (2017), pp. 295–310.

[16] Santocanale, L. *A calculus of circular proofs and its categorical semantics*, in: FoSSaCS (2002), pp. 357–371.

[17] Shamkanov, D. *Circular proofs for the Gödel-Löb provability logic*, in: Math. Notes 96 (2014), pp. 575–585.

[18] Simpson, A. *Cyclic arithmetic is equivalent to Peano Arithmetic*, in: FOSSACS (2017), pp. 283–300.

[19] Sprenger, C. and M. Dam. *On the structure of inductive reasoning: Circular and tree-shaped proofs in the μ-calculus*, in: FoSSaCS 2003 (2003), pp. 425–440.

[20] Stirling, C. *A tableau proof system with names for modal mu-calculus*, in: EPiCS 42 (2014), pp. 306–318.

[21] Tellez, G. and J. Brotherston. *Automatically verifying temporal properties of pointer programs with cyclic proof*, in: Journal of Automated Reasoning 64(3)(2020), pp. 555–578.

Filtrations, canonical formulas, and axiomatizations of superintuitionistic and modal logics

Nick Bezhanishvili

ILLC, University of Amsterdam
The Netherlands

There are two main tools for establishing the finite model property for modal and superintuitionistic logics: the methods of standard and selective filtrations. For superintutionistic logics standard filtration algebraically corresponds to taking the implication-free reduct of Heyting algebras and selective filtration corresponds to the join-free reduct. The key property that makes these methods work is that these reducts are locally finite. These finite model property proofs can be turned into axiomatization methods using canonical formulas. These formulas were introduced model-theoretically by Zakharyaschev building on the work of Jankov, de Jongh, Fine and Rautenberg. Zakharyaschev's canonical formulas algebraically correspond to join-free reducts of Heyting algebras. Every superintuitionistic logic is axiomatizable by these formulas. Important subclasses of canonical formulas are Jankov-de Jongh, subframe and cofinal subframe formulas giving rise to classes of join-splitting, subframe and cofinal subframe logics, respectively. In this talk, I will also discuss recently introduced stable canonical formulas, which algebraically correspond to implication-free reducts of Heyting algebras. Every superintuionistic logic is also axiomatizable by stable canonical formulas. These formulas give rise to a new class of stable logics.

Modal logic counterpart of Zakharyaschev's canonical formulas axiomatizes all extensions of $\mathsf{K4}$ (this was later generalized to all extensions of the weak transitive logic $\mathsf{wK4}$). This technique is based on the method of selective filtration for transitive modal logics. While selective filtration is very effective in the transitive case, it is less effective for K. This is one of the reasons why canonical formulas do not work well for K. I will discuss how to generalize the technique of stable superintuitionistic canonical formulas to the modal setting. Since the technique of filtration works well for K, we show that this new technique is effective in the non-transitive case as well. However, due to the lack of the master modality in the case of K we need to work with rules as opposed to formulas. I will define stable canonical rules and show that each normal modal logic is axiomatizable by stable canonical rules. For normal extensions of $\mathsf{K4}$ we prove that stable canonical rules can be replaced by stable canonical formulas, thus providing an alternative to Zakharyaschev's axiomatization.

About 'Binding Modalities'

Melvin Fitting [1]

Graduate Center, City University of New York
web page: melvinfitting.org

In classical logic the addition of quantifiers to propositional logic is essentially unique, with some minor variations of course. In modal logic things are not so monolithic. One can quantify over things or over intensions; domains can be the same from possible world to possible world, or shrink, or grow, or follow no pattern, as one moves from a possible world to an accessible one. In 1963 Kripke showed that shrinking or growing domains related to validity of the Barcan and the converse Barcan formulas, but this was a semantic result. Proof theory is trickier. Nested sequents are well behaved, but axiom systems can be unruly. A direct combination of propositional modal axioms and rules with standard quantificational axioms and rules simply proves the converse Barcan formula. It's not easy to get rid of it. Kripke showed how one could do so, but he needed to use a less common axiomatization of the quantifiers. It works, but one has the impression of having a formal proof system with road blocks placed carefully to prevent proofs from veering into the ditch.

Some 40 or more years later, justification logic was created by Artemov, and now there are justification systems that correspond to infinitely many different modal logics. The first justification logic was called LP, for "logic of proofs". It is related to propositional S4. LP was extended to a quantified version by Artemov and Yavorskaya, with a possible world semantics supplied by Fitting. Subsequently Artemov and Yavorskaya transferred their ideas, concerning what they called "binding modalities", back from quantified LP to quantified S4 itself. In the present work we carry their ideas on further to the basic normal modal logic, K, which is not as well-behaved as S4 on these matters. It turns out that this provides a natural intuition for Kripke's non-standard axiomatization from those many years ago. It also relates quite plausibly to the distinction between *de re* and *de dicto*. But now the main work is done through a generalization of the modal operator, instead of through a restriction on allowed quantifier axiomatizations.

[1] melvin.fitting@gmail.com

Learning and Modal Logic: There and Back Again

Nina Gierasimczuk

Technical University of Denmark
Department of Applied Mathematics and Computer Science

Among many interpretations of modal logic the one pertaining to knowledge and belief has been especially buoyant in recent years. The framework of epistemic logic offers a platform for a systematic study of knowledge and belief. Dynamic epistemic logic further extends that way of thinking to cover many kinds of transformations knowledge undergoes in communication, and under other informative events. Such iterated changes can be given a long-term horizon of learning, i.e., they can be seen as ways to acquire a desirable kind of epistemic state. Thus, the question arises: Can modal logic contribute to our understanding of learning processes in general?

The link between dynamic epistemic logic and computational learning theory was introduced in [10,11], where it was shown that exact learning in finite time (also known as finite identification, see [16,17]) can be modelled in dynamic epistemic logic, and that the elimination process of learning by erasing [15] can be seen as iterated upgrade of dynamic doxastic logic. This bridge opened a way to study truth-tracking properties of doxastic upgrade methods on positive, negative, and erroneous input [2,4]. Switching from relational to topological semantics for modal logic allowed characterising favourable conditions for learning in the limit in terms of general topology [3]. This line of research recently culminated in proposing a Dynamic Logic for Learning Theory, which extends Subset Space Logics [7] with dynamic observation modalities and a learning operator [1].

Finite identifiability and its connections with epistemic temporal logic have been further studied in [9]. Learning seen as conclusive epistemic update resulted in designing new types of learners, such as preset learners and fastest learners [14]. Some of those results were later adopted to study learning of action models in dynamic epistemic logic [5,6], and to investigate properties of finite identification from complete data [8]. For an overview of some above contributions one can also consult [12,13].

In my lecture I will overview the modal logic perspective on learnability, drawing from the line of work described above.

References

[1] A. Baltag, N. Gierasimczuk, A. Özgün, A. L. Vargas-Sandoval, and S. Smets. A dynamic logic for learning theory. *Journal of Logical and Algebraic Methods in Programming*, 109:100485, 2019.

[2] A. Baltag, N. Gierasimczuk, and S. Smets. Belief revision as a truth-tracking process. In K. Apt, editor, *TARK'11: Proceedings of the 13th Conference on Theoretical Aspects of Rationality and Knowledge, Groningen, The Netherlands, July 12-14, 2011*, pages 187–190. ACM, New York, NY, USA, 2011.

[3] A. Baltag, N. Gierasimczuk, and S. Smets. On the solvability of inductive problems: A study in epistemic topology. In R. Ramanujam, editor, *Proceedings Fifteenth Conference on Theoretical Aspects of Rationality and Knowledge, TARK 2015, Carnegie Mellon University, Pittsburgh, USA, June 4-6, 2015*, volume 215 of *EPTCS*, pages 81–98, 2015.

[4] A. Baltag, N. Gierasimczuk, and S. Smets. Truth-tracking by belief revision. *Studia Logica*, 107(5):917–947, 2019.

[5] T. Bolander and N. Gierasimczuk. Learning actions models: Qualitative approach. In W. van der Hoek, W. H. Holliday, and W. Wang, editors, *Logic, Rationality, and Interaction - 5th International Workshop, LORI 2015 Taipei, Taiwan, October 28-31, 2015, Proceedings*, volume 9394 of *Lecture Notes in Computer Science*, pages 40–52. Springer, 2015.

[6] T. Bolander and N. Gierasimczuk. Learning to act: qualitative learning of deterministic action models. *Journal of Logic and Computation*, 28(2):337–365, 2018.

[7] A. Dabrowski, L. S. Moss, and R. Parikh. Topological reasoning and the logic of knowledge. *Annals of Pure and Applied Logic*, 78(1):73 – 110, 1996. Papers in honor of the Symposium on Logical Foundations of Computer Science, Logic at St. Petersburg.

[8] D. de Jongh and A. L. Vargas-Sandoval. Finite identification with positive and with complete data. In A. Silva, S. Staton, P. Sutton, and C. Umbach, editors, *Language, Logic, and Computation*, pages 42–63, Springer, Berlin/Heidelberg, 2019. .

[9] C. Dégremont and N. Gierasimczuk. Finite identification from the viewpoint of epistemic update. *Information and Computation*, 209(3):383–396, 2011.

[10] N. Gierasimczuk. Bridging learning theory and dynamic epistemic logic. *Synthese*, 169(2):371–384, 2009.

[11] N. Gierasimczuk. Learning by erasing in dynamic epistemic logic. In A. H. Dediu, A. M. Ionescu, and C. Martin-Vide, editors, *LATA'09: Proceedings of 3rd International Conference on Language and Automata Theory and Applications, Tarragona, Spain, April 2-8, 2009*, volume 5457 of *Lecture Notes in Computer Science*, pages 362–373. Springer, The Netherlands, 2009.

[12] N. Gierasimczuk. *Knowing One's Limits. Logical Analysis of Inductive Inference.* PhD thesis, Universiteit van Amsterdam, The Netherlands, 2010.

[13] N. Gierasimczuk, D. de Jongh, and V. F. Hendricks. Logic and learning. In A. Baltag and S. Smets, editors, *Johan van Benthem on Logical and Informational Dynamics.* Springer, 2014.

[14] N. Gierasimczuk and D. de Jongh. On the complexity of conclusive update. *The Computer Journal*, 56(3):365–377, 2013.

[15] S. Lange, R. Wiehagen, and T. Zeugmann. Learning by erasing. In S. Arikawa and A. Sharma, editors, *ALT*, volume 1160 of *Lecture Notes in Computer Science*, pages 228–241. Springer, 1996.

[16] S. Lange and T. Zeugmann. Types of monotonic language learning and their characterization. In *COLT'92: Proceedings of the 5th Annual ACM Conference on Computational Learning Theory, Pittsburgh, PA, USA, July 27-29, 1992*, pages 377–390. ACM, New York, NY, USA, 1992.

[17] Y. Mukouchi. Characterization of finite identification. In K. Jantke, editor, *AII'92: Proceedings of the International Workshop on Analogical and Inductive Inference, Dagstuhl Castle, Germany, October 5-9, 1992*, volume 642 of *Lecture Notes in Computer Science*, pages 260–267. Springer, Berlin/Heidelberg, 1992.

Contributed Papers

Quantified Reflection Calculus with one modality

Ana de Almeida Borges [1]

University of Barcelona

Joost J. Joosten [2]

University of Barcelona

Abstract

This paper presents the logic $\mathsf{QRC_1}$, which is a strictly positive fragment of quantified modal logic. The intended reading of the diamond modality is that of consistency of a formal theory. Predicate symbols are interpreted as parametrized axiomatizations. We prove arithmetical soundness of the logic $\mathsf{QRC_1}$ with respect to this arithmetical interpretation.

Quantified provability logic is known to be undecidable. However, the undecidability proof cannot be performed in our signature and arithmetical reading. We conjecture the logic $\mathsf{QRC_1}$ to be arithmetically complete. This paper takes the first steps towards arithmetical completeness by providing relational semantics for $\mathsf{QRC_1}$ with a corresponding completeness proof. We further show the finite model property with respect to domains and number of worlds, which implies decidability.

Keywords: Provability logic, strictly positive logics, quantified modal logic, arithmetic interpretations, feasible fragments, decidable logics, finite model property.

1 Introduction

We present a new provability logic $\mathsf{QRC_1}$, standing for Quantified Reflection Calculus with one modality. The best known provability logic is perhaps GL [7]. Recall that GL is a PSPACE decidable propositional modal logic where the modality $\Box$ is used to model formal provability in some base theory such as Peano Arithmetic (PA). Likewise, the dual modality $\Diamond$ is used to model consistency over the base theory. By Solovay's celebrated completeness result [20] we know that, in a sense, the logic GL generates exactly the provable-in-PA structural behavior of formal provability.

Let us make this slightly more precise. By a *realization* $\star$ we mean a map from propositional variables to sentences in the language of Peano Arithmetic.

[1] `anadealmeidagabriel@ub.edu`

[2] `jjoosten@ub.edu`

The realization is extended to all propositional modal formulas by defining $(\varphi \wedge \psi)^\star := \varphi^\star \wedge \psi^\star$ and likewise for other Boolean connectives. Finally, we define $(\Box\varphi)^\star := \Box_{\mathsf{PA}}\varphi^\star$, where $\Box_{\mathsf{PA}}$ is a formula in the language of PA satisfying the Hilbert-Bernays-Löb derivability conditions (see [7]) that arithmetizes formal provability in PA in the sense that $\mathsf{PA} \vdash \chi$ if and only if $\mathbb{N} \models \Box_{\mathsf{PA}}\chi$.[3] We can now paraphrase Solovay's result as $\mathsf{GL} = \{\varphi \mid \forall\star\ \mathsf{PA} \vdash \varphi^\star\}$.

After Solovay's completeness theorem, it was natural to ask whether one could find a logic that generates exactly the provable-in-PA structural behavior of formal provability for (relational) *quantified modal logic.* The main difference with GL is that we now understand a realization $*$ as a map from relation symbols to formulas in the language of Peano Arithmetic such that the free variables match the arity of the relation symbol. Vardanyan showed in [21] that this situation is completely different: now $\{\varphi \mid \forall *\ \mathsf{PA} \vdash \varphi^*\}$ is Π^0_2-complete, a big jump from the PSPACE decidability of GL.

Visser and de Jonge showed that Vardanyan's result can be extended to a wide range of arithmetical theories and called their paper *No escape from Vardanyan's theorem* [22]. Here we shall take some first steps to indeed find an escape to Vardanyan's theorem. We do so by making two adaptations to the standard setting. First, we resort to a very small fragment of relational predicate modal logic called the *strictly positive fragment.* Second, we slightly change the realizations so that we interpret relation symbols not directly as formulas, but as axiomatizations of theories. As such our study follows a recent development of *strictly positive logics* in general (such as [18]) and *reflection calculi* in particular (see [8], [4], and [9]).

Japaridze [17] generalized the logic GL to a polymodal version called GLP, and Beklemishev [2] generalized this further to a transfinite setting yielding GLP_Λ, where for each ordinal $\xi < \Lambda$ there is a provability modality $[\xi]$, and larger ordinals refer to stronger provability notions. The logic GLP_ω has been successfully used in performing a modular ordinal analysis of PA and related systems (see [1], and more recently [6]). A key feature in the ordinal analysis is that consistency operators $\langle n \rangle$ can be interpreted as reflection principles, which are finitely axiomatizable.

However, an interpretation of limit modalities like $\langle \omega \rangle$ would require non finitely axiomatizable reflection schemata. One way to overcome this problem is by resorting to what was coined the *Reflection Calculus* [8], [3] and its transfinite version RC_Λ [11]. Reflection calculi only allow strictly positive formulas, which are based solely on propositional variables, a verum constant, consistency operators, and conjunctions. As such, the arithmetical realizations as above can be taken to be *arithmetical theories* instead of arithmetical formulas.

The logic QRC_1 we present in this paper follows this set-up: we will work with sequents of the form $\varphi \vdash \psi$ where both φ and ψ are strictly positive formulas built up from $\top$, predicate symbols, conjunction, universal quantification and the $\Diamond$ modality. The latter will refer to the usual notion of formal

[3] We refrain from distinguishing a formula φ from its Gödel number $\ulcorner\varphi\urcorner$.

consistency and predicate symbols are interpreted as theories parametrized by the free variables.

Independently of the reflection calculi, other strictly positive modal logics were studied because of their computational desirable properties when compared to their non-strict counterparts (see [18] for an example). In this line, the logic RC can be seen as a PTIME decidable fragment of the PSPACE complete logic GLP (shown in [8]). If indeed the logic QRC_1 we present in this paper turns out to be arithmetically complete, this would yield, in a sense, a shift from undecidability (Π^0_2-completeness) to decidability when resorting to a strictly positive fragment.

2 Quantified Reflection Calculus with one modality

The *Quantified Reflection Calculus with one modality*, or QRC_1, is a sequent logic in a strictly positive predicate modal language.

Towards describing the language of QRC_1, we fix a countable set of variables $x_0, x_1, \ldots$ (also referred to as x, y, z, etc.) and define a signature Σ as a set of constants and a set of relation symbols with corresponding arity (we have no function symbols). We use the letters $c, c_i, \ldots$ to refer to constants and the letters $S, S_i, \ldots$ to refer to relation symbols.

Given a signature, a term t is either a variable or a constant of that signature. Both $\top$ and any n-ary relation symbol applied to n terms are atomic formulas. The set of formulas is the closure of the atomic formulas under the binary connective $\wedge$, the unary modal operator $\Diamond$, and the quantifier $\forall x$, where x is a variable. Formulas are represented by Greek letters such as φ, ψ, χ, etc.

The free variables of a formula are defined as usual. The expression $\varphi[x{\leftarrow}t]$ denotes the formula φ with all free occurrences of the variable x simultaneously replaced by the term t. We say that t is free for x in φ if no occurrence of a free variable in t becomes bound in $\varphi[x{\leftarrow}t]$.

Definition 2.1 Let Σ be a signature and φ, ψ, and χ be any formulas in that language. The axioms and rules of QRC_1 are the following:

(i) $\varphi \vdash \top$ and $\varphi \vdash \varphi$;

(ii) $\varphi \wedge \psi \vdash \varphi$ and $\varphi \wedge \psi \vdash \psi$;

(iii) if $\varphi \vdash \psi$ and $\varphi \vdash \chi$, then $\varphi \vdash \psi \wedge \chi$;

(iv) if $\varphi \vdash \psi$ and $\psi \vdash \chi$, then $\varphi \vdash \chi$;

(v) if $\varphi \vdash \psi$, then $\Diamond\varphi \vdash \Diamond\psi$;

(vi) $\Diamond\Diamond\varphi \vdash \Diamond\varphi$;

(vii) $\Diamond \forall x\, \varphi \vdash \forall x \Diamond\varphi$;

(viii) if $\varphi \vdash \psi$, then $\varphi \vdash \forall x\, \psi$ ($x \notin \mathrm{fv}(\varphi)$);

(ix) if $\varphi[x{\leftarrow}t] \vdash \psi$ then $\forall x\, \varphi \vdash \psi$ (t free for x in φ);

(x) if $\varphi \vdash \psi$, then $\varphi[x{\leftarrow}t] \vdash \psi[x{\leftarrow}t]$ (t free for x in φ and ψ);

(xi) if $\varphi[x{\leftarrow}c] \vdash \psi[x{\leftarrow}c]$, then $\varphi \vdash \psi$ (c not in φ nor ψ).

If $\varphi \vdash \psi$, we say that ψ follows from φ in QRC_1. When the signature is not clear from the context, we write $\varphi \vdash_\Sigma \psi$ instead.

The following easy lemma presents a number of consequences of QRC_1.

Lemma 2.2 *The following are theorems (or derivable rules) of* QRC$_1$*:*

(i) $\forall x\, \forall y\, \varphi \vdash \forall y\, \forall x\, \varphi$;

(ii) $\forall x\, \varphi \vdash \varphi[x{\leftarrow}t]$ *(t free for x in φ);*

(iii) $\forall x\, \varphi \vdash \forall y\, \varphi[x{\leftarrow}y]$ *(y free for x in φ and $y \notin \mathrm{fv}(\varphi)$);*

(iv) *if* $\varphi \vdash \psi$*, then* $\varphi \vdash \psi[x{\leftarrow}t]$ *(x not free in φ and t free for x in ψ);*

(v) *if* $\varphi \vdash \psi[x{\leftarrow}c]$*, then* $\varphi \vdash \forall x\, \psi$ *(x not free in φ and c not in φ nor ψ).*

In order to analyze various aspects of our calculus we define two complexity measures on formulas.

Definition 2.3 Given a formula φ, its *modal depth* $\mathrm{d}_\Diamond(\varphi)$ is defined inductively as follows:

- $\mathrm{d}_\Diamond(\top) := \mathrm{d}_\Diamond(S(x_0, \ldots, x_{n-1})) := 0$;
- $\mathrm{d}_\Diamond(\psi \wedge \chi) := \max\{\mathrm{d}_\Diamond(\psi), \mathrm{d}_\Diamond(\chi)\}$;
- $\mathrm{d}_\Diamond(\forall x\, \psi) := \mathrm{d}_\Diamond(\psi)$;
- $\mathrm{d}_\Diamond(\Diamond\psi) := \mathrm{d}_\Diamond(\psi) + 1$.

Given a finite set of formulas Γ, its modal depth is $\mathrm{d}_\Diamond(\Gamma) := \max_{\varphi\in\Gamma}\{\mathrm{d}_\Diamond(\varphi)\}$.

The definition of quantifier depth $\mathrm{d}_\forall$ is analogous except for:

- $\mathrm{d}_\forall(\forall x\, \psi) = \mathrm{d}_\forall(\psi) + 1$; and
- $\mathrm{d}_\forall(\Diamond\psi) = \mathrm{d}_\forall(\psi)$.

The modal depth provides a necessary condition for derivability, proven by a straightforward induction on $\varphi \vdash \psi$.

Lemma 2.4 *If* $\varphi \vdash \psi$*, then* $\mathrm{d}_\Diamond(\varphi) \geq \mathrm{d}_\Diamond(\psi)$.

In particular, we get irreflexivity for free as stated in the next result. For other calculi this usually requires hard work via either modal or arithmetical semantics, as can be seen in [5], [10], and [12].

Corollary 2.5 *For any formula* φ*, we have* $\varphi \nvdash \Diamond\varphi$.

The following lemma tells us that adding new constants to our signature yields a conservative extension of the calculus.

Lemma 2.6 *Let Σ be a signature and let C be a collection of new constants not yet occurring in Σ. By Σ_C we denote the signature obtained by including these new constants C in Σ. Let φ, ψ be formulas in the language of Σ. Then, if $\varphi \vdash_{\Sigma_C} \psi$, so does $\varphi \vdash_\Sigma \psi$.*

Proof. This is a standard result and a proof for a calculus similar to ours can be found in Section 1.8 of [14]. The idea is to replace every constant from C appearing in the proof of $\varphi \vdash_{\Sigma_C} \psi$ by a fresh variable. It can easily be seen that axioms are mapped to axioms under this replacement, and that the rules are also mapped correctly. The most interesting case is that of the generalization of constants rule, because replacing new constants by variables in the premise

$\varphi[x{\leftarrow}c] \vdash_{\Sigma_C} \psi[x{\leftarrow}c]$ may leave us unable to apply the same rule. Fortunately the term instantiation rule (Rule 2.1.(x)) suffices to complete the proof. □

3 Arithmetical semantics

In this section we look at the intended arithmetical reading of the logic QRC_1. We consider first-order theories in the language $\{0, 1, +, \times, \leq, =\}$ of arithmetic. We refer the reader to [15] for details and definitions. We recall that bounded formulas are those formulas where each quantifier occurs bounded as in $\forall y \leq t$, where y does not occur in t. The Σ_1 formulas are those that arise by existential quantification of bounded formulas. Sets of numbers that can be defined by a Σ_1 formula are called *c.e.* for *computably enumerable.*

The theory $\mathrm{I}\Sigma_1$ contains the defining axioms for our constants and function symbols, say as in Robinson's arithmetic, and moreover allows induction for Σ_1 formulas. It is well-known that $\mathrm{I}\Sigma_1$ proves Σ_1-collection, that is, $\forall\, x{<}z\ \exists\, y\, \varphi(x, y) \to \exists\, y_0\ \forall\, x{<}z\ \exists\, y{<}y_0\, \varphi(x, y)$ where φ is a Σ_1 formula. For the sake of an easy exposition we shall assume that all the theories we work with extend $\mathrm{I}\Sigma_1$. By $\tau(x)$ we denote the elementary formula that presents the standard axiomatization of $\mathrm{I}\Sigma_1$. In particular, $\mathbb{N} \models \tau(n)$ if and only if n is the Gödel number of an axiom of $\mathrm{I}\Sigma_1$.

In the arithmetical interpretation of the propositional logic RC, the propositional variables are mapped to (axiomatizations of) theories, and the conjunction of two theories is interpreted as the union of both theories (corresponding to a disjunction in the sense of either being an axiom of the one or of the other). The arithmetical interpretation of each diamond modality is a consistency notion.

We will fix a provability predicate $\Box_\alpha \varphi$ formalizing the existence of a Hilbert-style proof, which is a sequence of formulas the last of which is φ and such that each element of the sequence is either a logical axiom, an axiom in the sense of α, or the result of applying a rule to earlier elements in the sequence. We denote the dual consistency notion by $\mathrm{Con}_\alpha(\psi)$ and sometimes write Con_α instead of $\mathrm{Con}_\alpha(\top)$.

If we now interpret relation symbols from QRC_1 as theories (parametrized by the free variables), then a universal quantification (which can be conceived of as an infinite conjunction) will be interpreted as an infinite union/disjunction, that is, an existential quantifier. These observations are reflected in Definition 3.1 below.

In this section, we reserve the variables x_i for variables in QRC_1, and the variables y_i, z_i and u are reserved for the arithmetic language with the understanding that the y_i interpret the QRC_1-constants c_i and the z_i interpret the QRC_1-variables x_i. The variable u is reserved for (Gödel numbers of) axioms of the theories that we denote.

Definition 3.1 A realization * takes n-ary predicate symbols in the language of QRC_1 to $(n+1)$-ary Σ_1-formulas in the language of arithmetic, each representing a set of axioms of theories indexed by n parameters. In particular, a realization * is such that $S(\boldsymbol{c}, \boldsymbol{x})^* = \sigma(\boldsymbol{y}, \boldsymbol{z}, u)$ for some Σ_1 formula σ such that for each concrete numerical values for $\boldsymbol{y}, \boldsymbol{z}$ we have that $\mathbb{N} \models \sigma(\boldsymbol{y}, \boldsymbol{z}, u)$ if and only if u is the Gödel number of an axiom of the intended corresponding theory. When we use the vector notation in $S(\boldsymbol{c}, \boldsymbol{x})^* = \sigma(\boldsymbol{y}, \boldsymbol{z}, u)$ we understand that $\boldsymbol{y}$ matches with $\boldsymbol{c}$ and $\boldsymbol{z}$ matches with $\boldsymbol{x}$, and thus if, say, y_i occurs in σ, then c_i occurs in $S(\boldsymbol{c}, \boldsymbol{x})$.

We extend a given realization * to $()^*$ on any formula of $\mathsf{QRC_1}$ as follows:

- $(\top)^* := \tau(u)$;
- $(S(\boldsymbol{c}, \boldsymbol{x}))^* := S(\boldsymbol{c}, \boldsymbol{x})^* \vee \tau(u)$;
- $(\psi(\boldsymbol{c}, \boldsymbol{x}) \wedge \delta(\boldsymbol{c}, \boldsymbol{x}))^* := (\psi(\boldsymbol{c}, \boldsymbol{x}))^* \vee (\delta(\boldsymbol{c}, \boldsymbol{x}))^*$;
- $(\Diamond\psi(\boldsymbol{c}, \boldsymbol{x}))^* := \tau(u) \vee (u = \ulcorner \mathrm{Con}_{(\psi(\boldsymbol{c}, \boldsymbol{x}))^*} \urcorner)$;
- $(\forall x_i\, \psi(\boldsymbol{c}, \boldsymbol{x}))^* := \exists z_i\, (\psi(\boldsymbol{c}, \boldsymbol{x}))^*$.

From now on we omit outer brackets, using the same notation for * and $()^*$. This may lead to confusion for predicate symbols, but the context should tell us which reading to use. We fix the notation $\psi(\boldsymbol{c}, \boldsymbol{x})^* = \psi^*(\boldsymbol{y}, \boldsymbol{z})$ suppressing mention of u when convenient.

Let T be a c.e. theory in the language of arithmetic which extends $\mathrm{I}\Sigma_1$. We define (recall that χ^* will in general depend on $\boldsymbol{y}$ and $\boldsymbol{z}$):

$$\mathcal{QRC}_1(T) := \{\varphi(\boldsymbol{c}, \boldsymbol{x}) \vdash \psi(\boldsymbol{c}, \boldsymbol{x}) \mid \forall\, {}^* T \vdash \forall \theta\, \forall \boldsymbol{y}\, \forall \boldsymbol{z}\, (\Box_{\psi^*}\theta \to \Box_{\varphi^*}\theta)\}.$$

We defer the question of whether $\mathsf{QRC_1} = \mathcal{QRC}_1(T)$ for any sound c.e. T containing $\mathrm{I}\Sigma_1$ to a future paper and prove here only the soundness inclusion.

Theorem 3.2 (Arithmetical soundness) $\mathsf{QRC_1} \subseteq \mathcal{QRC}_1(\mathrm{I}\Sigma_1)$.

Proof. By induction on $\varphi \vdash \psi$. The details can be found in Appendix A. Here we just observe the arithmetical content expressed by of some of the axioms. For example, $\varphi \vdash \top$ is sound because all of our theories extend $\mathrm{I}\Sigma_1$. The $\Diamond\Diamond\varphi \vdash \Diamond\varphi$ axiom reflects provable Σ_1 completeness (see [7]). The $\Diamond \forall x\, \varphi \vdash \forall x \Diamond\varphi$ axiom reflects that if a sum of theories is consistent, then each of the summands is consistent too. □

4 Relational semantics

There have been several proposals for relational semantics for modal predicate logics, from Kripke [19] to many others. Overviews can be found in [16], [13] and [14]. We essentially have first-order models glued together by an accessibility relation. Our interpretation of the universal quantifiers is actualist, which means that $\forall x\, \varphi$ is true at a world w if and only if $\varphi[x{\leftarrow}d]$ is true at w for every d in the domain of w, i.e., for every entity d that exists in that world. It might happen, however, that some other world u has a different domain, and thus that it falsifies $\varphi[x{\leftarrow}e]$ for some specific e.

We proceed by defining frames and relational models.

Definition 4.1 A *frame* $\mathcal{F}$ is a tuple $\langle W, R, \{M_w\}_{w \in W}\rangle$ where:

- W is a non-empty set (the set of worlds, where individual worlds are referred to as w, u, v, etc);
- R is a binary relation on W (the accessibility relation); and
- each M_w is a finite set (the domain of the world w, whose elements are referred to as d, d_0, d_1, etc).

The domain of the frame is $M := \bigcup_{w \in W} M_w$. We say that a frame is *finite* if its set of worlds W is finite. In that case both the relation R and the domain M will be finite as well.

Definition 4.2 A *relational model* $\mathcal{M}$ in a signature Σ is a tuple $\langle \mathcal{F}, \{I_w\}_{w \in W}, \{J_w\}_{w \in W} \rangle$ where:

- $\mathcal{F} = \langle W, R, \{M_w\}_{w \in W} \rangle$ is a frame;
- for each $w \in W$, the interpretation I_w assigns an element of the domain M_w to each constant $c \in \Sigma$, written c^{I_w}; and
- for each $w \in W$, the interpretation J_w assigns a set of tuples $S^{J_w} \subseteq \wp((M_w)^n)$ to each n-ary relation symbol $S \in \Sigma$.

Even though we interpret the universal quantifiers in the actualist way, we cannot allow the domains of each world to be completely unrelated to each other. This is because we want statements such as the axiom $\Diamond \forall x \varphi \vdash \forall x \Diamond \varphi$ to be sound. This axiom forces us to have *inclusive* frames, which means that if w sees a world u, then the domain of w is included or at least embedded in the domain of u. We also require that our frames be transitive, for we want the axiom $\Diamond\Diamond\varphi \vdash \Diamond\varphi$ to be sound. Finally, the interpretation of a constant should indeed be constant throughout (the relevant part of) any useful model. Thus, we introduce the notion of adequate frames and models.

Definition 4.3 A frame $\mathcal{F}$ is *adequate* if the accessibility relation R is:

- inclusive: if wRu, then $M_w \subseteq M_u$; and
- transitive: if wRu and uRv, then wRv.

A model is *adequate* if it is based on an adequate frame and it is:

- concordant: if wRu, then $c^{I_w} = c^{I_u}$ for every constant c.

Note that in an adequate and rooted model the interpretation of the constants is the same at every world.

In order to define truth at a world in a first-order model, we use assignments. A w-assignment g is a function assigning a member of the domain M_w to each variable in the language. In an adequate frame, any w-assignment can be seen as a v-assignment as long as wRv, because $M_w \subseteq M_v$ and hence there is a trivial inclusion (or coercion) $\iota_{w,v} : M_w \to M_v$. If g is such a w-assignment, we represent the corresponding v-assignment $\iota_{w,v} \circ g$ by g^ι when w and v are clear from the context.

Two w-assignments g and h are Γ-*alternative*, denoted by $g \sim_\Gamma h$, if they coincide on all variables other than the ones in Γ. If $\Gamma = \{x\}$, then we write simply x-alternative and $g \sim_x h$.

We extend a given w-assignment g to terms by defining $g(c) := c^{I_w}$ where c is any constant.

Definition 4.4 Let $\mathcal{M} = \langle W, R, \{M_w\}_{w\in W}, \{I_w\}_{w\in W}, \{J_w\}_{w\in W}\rangle$ be a relational model in some signature Σ, and let $w \in W$ be a world, g be a w-assignment, S be an n-ary relation symbol, and φ, ψ be formulas in the language of Σ.

We define $\mathcal{M}, w \Vdash^g \varphi$ (φ is true at w under g) by induction on φ as follows.

- $\mathcal{M}, w \Vdash^g \top$;
- $\mathcal{M}, w \Vdash^g S(t_0, \ldots, t_{n-1})$ iff $\langle g(t_0), \ldots, g(t_{n-1})\rangle \in S^{J_w}$;
- $\mathcal{M}, w \Vdash^g \varphi \wedge \psi$ iff both $\mathcal{M}, w \Vdash^g \varphi$ and $\mathcal{M}, w \Vdash^g \psi$;
- $\mathcal{M}, w \Vdash^g \Diamond\varphi$ iff there is $v \in W$ such that wRv and $\mathcal{M}, v \Vdash^{g^\iota} \varphi$;
- $\mathcal{M}, w \Vdash^g \forall x\, \varphi$ iff for all w-assignments h such that $h \sim_x g$, we have $\mathcal{M}, w \Vdash^h \varphi$.

We now present a number of simple results needed to prove the relational soundness of QRC_1. These are standard observations about either first-order models or Kripke models that we adapted to our case.

Remark 4.5 Let $\mathcal{M}$ be an adequate model, w be any world, g, h be any Γ-alternative w-assignments, and φ be a formula with no free variables in Γ. Then:

$$\mathcal{M}, w \Vdash^g \varphi \iff \mathcal{M}, w \Vdash^h \varphi.$$

Lemma 4.6 (Substitution in formula) *Let $\mathcal{M}$ be an adequate model, w be a world, and $g, \tilde{g}$ be x-alternative w-assignments such that $\tilde{g}(x) = g(t)$. Then for every formula φ with t free for x:*

$$\mathcal{M}, w \Vdash^{\tilde{g}} \varphi \iff \mathcal{M}, w \Vdash^g \varphi[x\leftarrow t].$$

Proof. By induction on φ. We only present the cases of the diamond and of the universal quantifier; the remaining cases are straightforward. We assume without loss of generality that x is a free variable of φ, since otherwise we could use Remark 4.5.

Suppose that φ is $\Diamond\psi$ and assume that $\mathcal{M}, w \Vdash^{\tilde{g}} \Diamond\psi$. Then there is a world v such that wRv and $\mathcal{M}, v \Vdash^{\tilde{g}^\iota} \psi$. Note that $g^\iota \sim_x \tilde{g}^\iota$ and $\tilde{g}^\iota(x) = g^\iota(t)$ (either t is a variable and this is a consequence of $\tilde{g}(x) = g(t)$, or t is a constant and this follows from $t^{I_w} = t^{I_v}$) and thus by the induction hypothesis $\mathcal{M}, v \Vdash^{g^\iota} \psi[x\leftarrow t]$. This gives us $\mathcal{M}, w \Vdash^g \Diamond\psi[x\leftarrow t]$, as desired. The other direction is analogous.

Suppose now that $\varphi = \forall z\, \psi$ and assume that $\mathcal{M}, w \Vdash^{\tilde{g}} \forall z\, \psi$. Note that x and z are different variables, for otherwise x would not be free in φ. Let h be any w-assignment such that $h \sim_z g$. We wish to show $\mathcal{M}, w \Vdash^h \psi[x\leftarrow t]$. Define $\tilde{h}$ such that $\tilde{h} \sim_x h$ and $\tilde{h}(x) := h(t)$. Then by the induction hypothesis we can reduce our goal to $\mathcal{M}, w \Vdash^{\tilde{h}} \psi$. By our assumption, it is enough to check that $\tilde{h} \sim_z \tilde{g}$.

In order to see this, note first that $\tilde{h} \sim_{\{x,z\}} h$ (because $\tilde{h} \sim_x h$). Similarly, $h \sim_{\{x,z\}} g$ and $g \sim_{\{x,z\}} \tilde{g}$. Then $\tilde{h} \sim_{\{x,z\}} \tilde{g}$ by transitivity of $\sim_{\{x,z\}}$. But $\tilde{g}(x) = g(t)$ by assumption; $g(t) = h(t)$ because $g \sim_z h$ (z and t are not the same variable because otherwise t would not be free for x in φ); and $h(t) = \tilde{h}(x)$ by construction of $\tilde{h}$. Thus $\tilde{g}(x) = \tilde{h}(x)$, and $\tilde{h} \sim_z \tilde{g}$.

Towards the other direction, assume that $\mathcal{M}, w \Vdash^g (\forall z\, \psi)[x{\leftarrow}t]$ and that x and z are not the same variable. Let $\tilde{h} \sim_z \tilde{g}$ be a w-assignment. We wish to show $\mathcal{M}, w \Vdash^{\tilde{h}} \psi$. Define $h \sim_x \tilde{h}$ such that $h(x) := g(x)$. Note that $h \sim_z g$ by the transitivity of $\sim_{x,z}$ (using a similar argument to the one above). Thus we know that $\mathcal{M}, w \Vdash^h \psi[x{\leftarrow}t]$ by assumption. It only remains to show that $\tilde{h}(x) = h(t)$, as we can then finally use the induction hypothesis to finish. If t is x there is nothing to show, and t cannot be z, because z is not free for x in $\forall z\, \psi$. Thus, $h(t) = g(t) = \tilde{g}(x) = \tilde{h}(x)$. □

We need some more work to prove the soundness of Rule 2.1.(xi), which we postpone to Appendix B. Otherwise we are ready to prove that QRC_1 is sound with respect to the relational semantics presented above.

Theorem 4.7 (Relational soundness) *If $\varphi \vdash \psi$, then for any adequate model $\mathcal{M}$, for any world $w \in W$, and for any w-assignment g:*

$$\mathcal{M}, w \Vdash^g \varphi \implies \mathcal{M}, w \Vdash^g \psi.$$

Proof. By induction on the proof of $\varphi \vdash \psi$.

In the case of the axioms $\varphi \vdash \top$ and $\varphi \vdash \varphi$, the result is clear, as it is for the conjunction elimination axioms. The conjunction introduction and cut rules follow easily from the definitions.

For the necessitation rule assume the result for $\varphi \vdash \psi$ and further assume that $\mathcal{M}, w \Vdash^g \Diamond\varphi$. Then there is a world v such that wRv and $\mathcal{M}, v \Vdash^{g^\iota} \varphi$. We wish to see $\mathcal{M}, w \Vdash^g \Diamond\psi$. Taking v as a suitable witness, our goal changes to $\mathcal{M}, v \Vdash^{g^\iota} \psi$. Thus the induction hypothesis for v and g^ι finishes the proof.

For the transitivity axiom, $\Diamond\Diamond\varphi \vdash \Diamond\varphi$, assume that $\mathcal{M}, w \Vdash^g \Diamond\Diamond\varphi$. Then there is a world v such that wRv and $\mathcal{M}, v \Vdash^{\iota_{w,v} \circ g} \Diamond\varphi$, and also a subsequent world u such that vRu and $\mathcal{M}, u \Vdash^{\iota_{v,u} \circ (\iota_{w,v} \circ g)} \varphi$. Observing that $\iota_{v,u} \circ (\iota_{w,v} \circ g)$ is the same as $\iota_{w,u} \circ g$, we get $\mathcal{M}, u \Vdash^{\iota_{w,u} \circ g} \psi$, and the transitivity of R provides wRu, which is enough to see $\mathcal{M}, w \Vdash^g \Diamond\varphi$, as desired.

In the case of $\Diamond \forall x\, \varphi \vdash \forall x\, \Diamond\varphi$, assume that $\mathcal{M}, w \Vdash^g \Diamond \forall x\, \varphi$. Then there is $v \in W$ such that wRv and for every v-assignment h with $h \sim_x g^\iota$ we have $\mathcal{M}, v \Vdash^h \varphi$. Let f be any w-assignment such that $f \sim_x g$. Taking v as a suitable world seen by w, we wish to check that $\mathcal{M}, v \Vdash^{f^\iota} \varphi$. By assumption, it is enough to see $f^\iota \sim_x g^\iota$, and this follows from $f \sim_x g$.

For the $\forall$-introduction rule on the right, assume the result for $\varphi \vdash \psi$ towards showing the soundness of $\varphi \vdash \forall x\, \psi$ with $x \notin \mathrm{fv}(\varphi)$. Assume further that $\mathcal{M}, w \Vdash^g \varphi$. Let h be a w-assignment such that $h \sim_x g$. We wish to see that $\mathcal{M}, w \Vdash^h \psi$. Since x is not a free variable in φ, we know that $\mathcal{M}, w \Vdash^h \varphi$ by Remark 4.5. The result follows from the induction hypothesis with w-assignment h.

Consider now the $\forall$-introduction rule on the left. Assume the result for $\varphi[x{\leftarrow}t] \vdash \psi$ with t free for x in φ and assume further that $\mathcal{M}, w \Vdash^g \forall x\, \varphi$. Then for every w-assignment h such that $h \sim_x g$ we have $\mathcal{M}, w \Vdash^h \varphi$. Define $h \sim_x g$ such that $h(x) = g(t)$. We obtain $\mathcal{M}, w \Vdash^g \psi$ by the induction hypothesis and Lemma 4.6.

The term instantiation rule, Rule 2.1.(x), is sound by Lemma 4.6, and the generalization on constants rule, Rule 2.1.(xi), is sound by Lemma B.6. □

5 Relational completeness

We now wish to prove the relational completeness of QRC_1. For every underivable sequent we provide a model that doesn't satisfy it. These models are term models where the worlds are akin to maximal consistent sets. However, since we have no way to express negative formulas, each world is a pair of sets of formulas instead: the set of positive formulas at that world and the set of negative ones.

We start by defining some notions about pairs of formulas, and we write $p, q, \ldots$ to refer to generic pairs that may not have all the necessary properties to be a world in a term model. Given a pair of sets p, the first set is the positive set, or p^+, and the second one is the negative set, or p^-.

Definition 5.1 Given a set of formulas Γ and a formula φ, we say that φ follows from Γ, and write $\Gamma \vdash \varphi$ (overloading the existing notation), if there are formulas $\gamma_0, \ldots, \gamma_n \in \Gamma$ such that $\gamma_0 \wedge \cdots \wedge \gamma_n \vdash \varphi$.

Definition 5.2 Let Φ be a set of formulas.

- A Φ-extension of a pair $p = \langle p^+, p^- \rangle$ is a pair $q = \langle q^+, q^- \rangle$ such that $p^+ \subseteq q^+ \subseteq \Phi$ and $p^- \subseteq q^- \subseteq \Phi$. In that case we write $p \subseteq q \subseteq \Phi$.
- A pair p is consistent if for every $\delta \in p^-$ we have $p^+ \nvdash \delta$.
- A pair $p \subseteq \Phi$ is Φ-maximal consistent if it is consistent and there is no consistent Φ-extension of p.
- A pair p is fully witnessed if for every formula $\forall x\, \varphi \in p^-$ there is a constant c such that $\varphi[x{\leftarrow}c] \in p^-$.
- A pair p is Φ-MCW if it is Φ-maximal consistent and fully witnessed.

Lemma 5.3 *A pair p is Φ-maximal consistent if and only if it is consistent and for every $\varphi \in \Phi$ either $\varphi \in p^+$ or $\varphi \in p^-$.*

Proof. The right-to-left implication is obvious. To check the other one assume that p is Φ-maximal consistent and let $\varphi \in \Phi$. If $p^+ \vdash \varphi$, then $\langle p^+ \cup \{\varphi\}, p^- \rangle$ is still consistent, and thus by maximality it must be that $\varphi \in p^+$. If on the other hand $p^+ \nvdash \varphi$, then $\langle p^+, p^- \cup \{\varphi\} \rangle$ is consistent, and thus we may conclude $\varphi \in p^-$. □

Definition 5.4 Given a set of constants C, the closure of a formula φ under C, written $\mathcal{C\ell}_C(\varphi)$, is defined by induction on the formula as such: $\mathcal{C\ell}_C(\top) := \{\top\}$; $\mathcal{C\ell}_C(S(t_0, \ldots, t_{n-1})) := \{S(t_0, \ldots, t_{n-1})), \top\}$; $\mathcal{C\ell}_C(\varphi \wedge \psi) := \{\varphi \wedge \psi\} \cup \mathcal{C\ell}_C(\varphi) \cup$

$\mathcal{C}\ell_C(\psi)$; $\mathcal{C}\ell_C(\Diamond\varphi) := \{\Diamond\varphi\} \cup \mathcal{C}\ell_C(\varphi)$; and

$$\mathcal{C}\ell_C(\forall x\, \varphi) := \{\forall x\, \varphi\} \cup \bigcup_{c \in C} \mathcal{C}\ell_C(\varphi[x{\leftarrow}c]).$$

The closure under C of a set of formulas Γ is the union of the closures under C of each of the formulas in Γ:

$$\mathcal{C}\ell_C(\Gamma) := \bigcup_{\gamma \in \Gamma} \mathcal{C}\ell_C(\gamma).$$

The closure of a pair p is defined as the closure of $p^+ \cup p^-$.

Note that the closure of a set of closed formulas is itself a set of closed formulas. We often use the concept of closure under a set of constants on an already Φ-maximal pair when we wish to extend the signature of the formulas in Φ with a new set of constants.

Given a consistent pair p, we wish to generate a Φ-maximal consistent and fully witnessed extension of p, for some set of formulas Φ. In the usual Henkin construction this is traditionally accomplished in two steps: first extend the signature to include a constant for each existential statement and add every closed formula of the form $\exists x\varphi \to \varphi[x{\leftarrow}c_\varphi]$ to your set, proving that this didn't break consistency. Then prove a Lindenbaum Lemma to the effect that consistent sets can be extended to maximal consistent sets. The resulting sets will be maximal, consistent, and fully witnessed. However we can not do this because we cannot express implications. Thus if we were to add a witness for every existential formula in our original pair p (read: universal formula in p^-) and then use a Lindenbaum lemma to make it maximal, there could be new existential formulas without witnesses. We might have to iterate the process over and over again, or at least a proof of termination would be non-trivial. Fortunately, this isn't needed. We can manage with a finite set of witnesses, as is shown by the following lemma.

Lemma 5.5 *Given a finite signature Σ with constants C, a finite set of closed formulas Φ in the language of Σ and a consistent pair $p \subseteq \mathcal{C}\ell_C(\Phi)$, there is a finite set of constants $D \supseteq C$ and a pair $q \supseteq p$ in the language of Σ extended by D such that q is $\mathcal{C}\ell_D(\Phi)$-MCW and $\mathrm{d}_\Diamond(q^+) = \mathrm{d}_\Diamond(p^+)$.*

Proof. Let $N := \{c_0, \dots, c_{\mathrm{d}_\forall(\Phi)-1}\}$ and $D := C \cup N$.

Let $q_0 := p$. For every formula φ_i in $\mathcal{C}\ell_D(\Phi)$, if $p^+ \vdash \varphi_i$, define $q_{i+1} = \langle q_i^+ \cup \{\varphi_i\}, q_i^- \rangle$; otherwise define $q_{i+1} = \langle q_i^+, q_i^- \cup \{\varphi_i\}\rangle$. Let $q := q_n$, where n is the size of $\mathcal{C}\ell_D(\Phi)$, i.e., q is what we have at the final iteration of this process.

Now assume by way of contradiction that q is not consistent, and let $\psi \in q^-$ be such that $q^+ \vdash \psi$. Note that for every $\chi \in q^+$ we know that $p^+ \vdash \chi$, because this was the required condition to add χ to q^+ in the first place. Thus, it must be that $p^+ \vdash \psi$. But then the algorithm would have placed ψ in q^+ instead of q^- and we reach a contradiction. We conclude that q is consistent.

Lemma 5.3 tells us that q is $\mathcal{C}\ell_D(\Phi)$-maximal consistent, because every formula of $\mathcal{C}\ell_D(\Phi)$ is either in q^+ or q^-.

On the other hand, we know by Lemma 2.4 that $\mathrm{d}_\Diamond(q^+) \leq \mathrm{d}_\Diamond(p^+)$ because every formula in q^+ is a consequence of p^+. We obtain the equality by observing that $p^+ \subseteq q^+$.

It remains to show that q is fully witnessed. Let $\forall x\,\psi$ be a formula in q^-. We claim that there is $c_i \in N$ such that c_i does not appear in $\forall x\,\psi$. The constants in N are new, so the only way to have a formula $\chi \in \mathcal{C}\ell_D(\Phi)$ with a constant $c_j \in N$ is if the formula $\forall y\,\chi[c_j \leftarrow y]$ is also in $\mathcal{C}\ell_D(\Phi)$, for some variable y that does not appear (free) in χ. Assume then that all the constants in N appear in ψ, and let m be the size of N. Then the formula $\forall y_0 \cdots \forall y_{m-1}\, \forall x\,\psi[c_{s_{m-1}} \leftarrow y_{m-1}] \cdots [c_{s_0} \leftarrow y_0]$ must be in $\mathcal{C}\ell_D(\Phi)$ for some variables y_i and permutation s of the numbers between 0 and $m-1$. But this formula has quantifier depth $m+1$, which is a contradiction because the closure under any set of constants doesn't change the depth of a set of formulas.

Let then $\forall x\,\psi \in q^-$ and $c_i \in N$ be a constant that does not appear in $\forall x\,\psi$. Then we claim that $\psi[x \leftarrow c_i] \in q^-$. Assume it is not the case. Then it must be that $p^+ \vdash \psi[x \leftarrow c_i]$. Note that c_i does not appear in p^+ and that x is not a free variable of p^+ due to it being a set of closed formulas. Then by Lemma 2.2.(v) we obtain that $p^+ \vdash \forall x\,\psi$, which is a contradiction. □

The next step is to link maximal consistent and fully witnessed pairs through a relation that respects the diamond formulas in the pair. To that end we define $\hat{R}$ and prove some properties about it.

Definition 5.6 The relation $\hat{R}$ between pairs is such that $p\hat{R}q$ if and only if both of following hold:

(i) for any formula $\Diamond\varphi \in p^-$ we have $\varphi, \Diamond\varphi \in q^-$; and

(ii) there is some formula $\Diamond\psi \in p^+ \cap q^-$.

Lemma 5.7 *The relation $\hat{R}$ restricted to consistent pairs is transitive and irreflexive.*

Proof. In order to see that $\hat{R}$ is transitive, assume that $p\hat{R}q\hat{R}r$. We wish to see that $p\hat{R}r$. Let $\Diamond\varphi \in p^-$ be arbitrary. Then $\Diamond\varphi \in q^-$ because $p\hat{R}q$, and then $\varphi, \Diamond\varphi \in r^-$ because $q\hat{R}r$. Let now $\Diamond\psi \in p^+ \cap q^-$. Since $q\hat{R}r$ we know that $\Diamond\psi \in r^-$. Then $\Diamond\psi \in p^+ \cap r^-$.

Regarding irreflexivity, suppose that there is a pair p such that $p\hat{R}p$. Then there must be $\Diamond\psi \in p^+ \cap p^-$, which contradicts the consistency of p. □

There is an equivalent formulation of $\hat{R}$ by looking at the positive sets.

Lemma 5.8 *Given a set of formulas Φ, two sets of constants $C \subseteq D$, and pairs p, q such that p is $\mathcal{C}\ell_C(\Phi)$-maximal consistent and q is $\mathcal{C}\ell_D(\Phi)$-maximal consistent, we have that $p\hat{R}q$ if and only if both of the following hold:*

(i) *for every formula $\Diamond\varphi \in \mathcal{C}\ell_C(\Phi)$, if either $\varphi \in q^+$ or $\Diamond\varphi \in q^+$, then $\Diamond\varphi \in p^+$; and*

(ii) *there is some formula $\Diamond\psi \in p^+ \cap q^-$.*

Proof. Assume that $p\hat{R}q$ and let $\Diamond\varphi \in \mathcal{C}\ell_C(\Phi)$ be such that either $\varphi \in q^+$ or $\Diamond\varphi \in q^+$. Assume by contradiction that $\Diamond\varphi \notin p^+$. Then by Lemma 5.3 we know that $\Diamond\varphi \in p^-$. Thus since $p\hat{R}q$, we obtain both $\varphi \in q^-$ and $\Diamond\varphi \in q^-$. But this contradicts the consistency of q. The last condition holds by the definition of $\hat{R}$.

Assume now that these conditions hold, towards checking that $p\hat{R}q$. Only the first condition is in question. Let $\Diamond\varphi \in p^-$ and assume that $\varphi \notin q^-$. By Lemma 5.3, it must be that $\varphi \in q^+$. Then $\Diamond\varphi \in p^+$, which contradicts the consistency of p. Assume now that $\Diamond\varphi \notin q^-$. By the same token, $\Diamond\varphi$ must be in q^+. Then $\Diamond\varphi \in p^+$, reaching a contradiction again. □

The following lemma states that, given a suitable pair p where $\Diamond\varphi$ holds, we can find a second suitable pair q where φ holds and $p\hat{R}q$.

Lemma 5.9 (Pair existence) *Let Σ be a signature with a finite set of constants C, and Φ be a finite set of closed formulas in the language of Σ. If p is a $\mathcal{C}\ell_C(\Phi)$-MCW pair and $\Diamond\varphi \in p^+$, then there is a finite set of constants $D \supseteq C$ and a $\mathcal{C}\ell_D(\Phi)$-MCW pair q such that $p\hat{R}q$, $\varphi \in q^+$, and $\mathrm{d}_\Diamond(q^+) < \mathrm{d}_\Diamond(p^+)$.*

Proof. Consider the pair $r = \langle\{\varphi\}, \{\delta, \Diamond\delta \mid \Diamond\delta \in p^-\} \cup \{\Diamond\varphi\}\rangle$. Assume that r is not consistent, and thus that there is a formula $\psi \in r^-$ such that $\varphi \vdash \psi$. It cannot be that ψ is $\Diamond\varphi$ due to Lemma 2.5. Thus there is $\Diamond\delta \in p^-$ such that either $\varphi \vdash \delta$ or $\varphi \vdash \Diamond\delta$. By Rule 2.1.(v) we get either $\Diamond\varphi \vdash \Diamond\delta$ or $\Diamond\varphi \vdash \Diamond\Diamond\delta$, which also implies $\Diamond\varphi \vdash \Diamond\delta$ by Axiom 2.1.(vi). This contradicts the consistency of p, which leads us to conclude that r is consistent.

We can now use Lemma 5.5 to obtain a finite set of constants $D \supseteq C$ and a $\mathcal{C}\ell_D(\Phi)$-MCW pair $q \supseteq r$ such that $\mathrm{d}_\Diamond(q^+) = \mathrm{d}_\Diamond(r^+) = \mathrm{d}_\Diamond(\varphi) < \mathrm{d}_\Diamond(p^+)$.

It remains to show that $p\hat{R}q$, but this is clear by the definition of r: for every $\Diamond\delta \in p^-$, the formulas δ and $\Diamond\delta$ are in r^- (and hence in q^-), and the formula $\Diamond\varphi$ is both in p^+ and in q^-. □

We are now ready to define an adequate model $\mathcal{M}[p]$ from any given finite and consistent pair p such that $\mathcal{M}[p]$ satisfies the formulas in p^+ and doesn't satisfy the formulas in p^-. The idea is to build a term model where each world w is a $\mathcal{C}\ell_{M_w}(p)$-MCW pair, and the worlds are related by (a sub-relation of) $\hat{R}$. The worlds in this model will be pairs of formulas in different signatures, as we will add new constants every time we create a new world. However, the model is intended to satisfy only formulas in the original signature of p.

Definition 5.10 Given a finite consistent pair p of closed formulas with constants in a finite set C, we define an adequate model $\mathcal{M}[p]$.

We start by defining the underlying frame in an iterative manner. The root is given by Lemma 5.5 applied to C and p, obtaining D and q. Frame $\mathcal{F}^0$ is then defined such that its set of worlds is $W^0 := \{q\}$, its relation R^0 is empty, and the domain of q is $M^0_q := D$.

Assume now that we already have a frame $\mathcal{F}^i$, and we set out to define $\mathcal{F}^{i+1}$ as an extension of $\mathcal{F}^i$. For each leaf w of $\mathcal{F}^i$, i.e., each world such that there is no world $v \in \mathcal{F}^i$ with wR^iv, and for each formula $\Diamond\varphi \in w^+$, use Lemma 5.9 to obtain a finite set $E \supseteq M_w^i$ and a $\mathcal{C}\ell_E(w)$-MCW pair v such that $w\hat{R}v$, $\varphi \in v^+$, and $\mathrm{d}_\diamond(v^+) < \mathrm{d}_\diamond(w^+)$. Now add v to W^{i+1}, add $\langle w, v\rangle$ to R^{i+1}, and define M_v^{i+1} as E.

The process described above terminates because each pair is finite and the modal depth of p^+ (and consequently of $\mathcal{C}\ell_X(p)$ for any set X) is also finite. Thus there is a final frame $\mathcal{F}^{\mathrm{d}_\diamond(p^+)}$. This frame is inclusive by construction, but not transitive. We obtain $\mathcal{F}[p]$ as the transitive closure of $\mathcal{F}^{\mathrm{d}_\diamond(p^+)}$, which can be easily seen to still be inclusive. Thus the frame $\mathcal{F}[p]$ is adequate.

In order to obtain the model $\mathcal{M}[p]$ based on the frame $\mathcal{F}[p]$, let I_q take constants in C to their corresponding version as domain elements and if w is any other world, let I_w coincide with I_q. This is necessary to make sure that the model is concordant, because q sees every other world, and is sufficient to see that $\mathcal{M}[p]$ is adequate. Finally, given an n-ary predicate letter S and a world w, define S^{J_w} as the set of n-tuples $\langle d_0, \ldots, d_{n-1}\rangle \subseteq (M_w)^n$ such that $S(d_0, \ldots, d_{n-1}) \in w^+$.

Lemma 5.11 *Let p be as above. The following are properties of* $\mathcal{F}[p] = \langle W, R, \{M_w\}_{w\in W}\rangle$ *and* $\mathcal{M}[p] = \langle \mathcal{F}[p], \{I_w\}_{w\in W}, \{J_w\}_{w\in W}\rangle$*:*

(i) *For every world w, its domain M_w is finite.*

(ii) *The set of worlds W is finite.*

(iii) *Every world $w \in W$ is $\mathcal{C}\ell_{M_w}(p)$-maximal consistent and fully witnessed.*

(iv) *For every world $w \in W$, we have $\top \in w^+$.*

(v) *For any two worlds $w, u \in W$, if wRu, then $w\hat{R}u$.*

Proof. These are simple consequences of the definition of $\mathcal{M}[p]$. The finiteness of the domains is achieved in Lemma 5.5, while the finiteness of W is proved in Definition 5.10. For the last property, note that R is the transitive closure of $R^{\mathrm{d}_\diamond(p^+)}$. If $wR^{\mathrm{d}_\diamond(p^+)}u$, then $w\hat{R}u$ by construction. The result then follows by the transitivity of $\hat{R}$ (Lemma 5.7). □

We are almost ready to state the truth lemma, which roughly states that provability at a world w of $\mathcal{M}[p]$ is the same as membership in w^+. However, the signatures of the worlds of $\mathcal{M}[p]$ are more expressive than the signature of the formulas we care about. Furthermore, all the formulas in the worlds of $\mathcal{M}[p]$ are closed, while formulas in general may have free variables. In order to deal with this, we replace the free variables of a formula with constants in the appropriate signature first.

Definition 5.12 Given a formula φ in a signature Σ and a function g from the set of variables to a set of constants in some signature $\Sigma' \supseteq \Sigma$, we define the formula φ^g in the signature Σ' as φ with each free variable x simultaneously replaced by $g(x)$.

Given a variable x, the formula $\varphi^{g\setminus x}$ is as above, except that the variable x doesn't get replaced even if it is free in φ.

Lemma 5.13 (Truth lemma) *Let Σ be a signature with a finite set of constants C. For any finite non-empty consistent pair p of closed formulas in the* *language of* Σ, *world* $w \in \mathcal{M}[p]$, w*-assignment* g, *and formula* φ *in the language of* Σ *such that* $\varphi^g \in \mathcal{C}\ell_{M_w}(p)$, *we have that*

$$\mathcal{M}[p], w \Vdash^g \varphi \iff \varphi^g \in w^+.$$

Proof. By induction on φ. The cases of $\top$ and conjunction are straightforward, so we focus on the other ones.

In the case of the relational symbols, we can take $\varphi = S(x,c)$ without loss of generality, where $c \in C$. Note that $\mathcal{M}[p], w \Vdash^g S(x,c)$ if and only if $\langle g(x), c^{I_w} \rangle \in S^{J_w}$, if and only if $S(g(x), c^{I_w}) \in w^+$. Since $c \in C$, we know by the definition of $\mathcal{M}[p]$ that $c^{I_w} = c$. Thus, we conclude that $\mathcal{M}[p], w \Vdash^g S(x,c)$ if an only if $S(g(x),c) \in w^+$, as desired.

Consider now the case of the universal quantifier. For the left to right implication, suppose that $\mathcal{M}[p], w \Vdash^g \forall x\, \varphi$. Then for every w-assignment $h \sim_x g$ we have $\mathcal{M}[p], w \Vdash^h \varphi$. Thus for each such h we know that $\varphi^h \in w^+$ by the induction hypothesis ($\varphi^h \in \mathcal{C}\ell_{M_w}(p)$ because $(\forall x\, \varphi)^g \in \mathcal{C}\ell_{M_w}(p)$). We want to show that $(\forall x\, \varphi)^g \in w^+$, i.e., that $\forall x\, \varphi^{g \backslash x} \in w^+$. Assume by contradiction that this is not the case. Then, since w is $\mathcal{C}\ell_{M_w}(p)$-maximal consistent, it must be that $\forall x\, \varphi^{g\backslash x} \in w^-$. Let $c \in M_w$ be a witness such that $\varphi^{g\backslash x}[x \leftarrow c] \in w^-$, which exists because w is fully witnessed. Let h be the w-assignment that coincides with g everywhere except at x, where $h(x) = c$. Then $g \sim_x h$ and $\varphi^{g\backslash x}[x \leftarrow c] = \varphi^h$. But this contradicts our earlier observation that for every such h the formula φ^h is in w^+.

For the right to left implication, let $\forall x\, \varphi^{g\backslash x} \in w^+$, and let $h \sim_x g$ be any w-assignment. We want to show that $\mathcal{M}[p], w \Vdash^h \varphi$. By the induction hypothesis this is the same as showing that $\varphi^h \in w^+$. But $\varphi^h = \varphi^{g\backslash x}[x \leftarrow h(x)]$, and this is in w^+ by the completeness and consistency of w.

Finally, consider the case of the diamond. For the left to right implication, assume that $\mathcal{M}[p], w \Vdash^g \Diamond\varphi$. Then there is some world u such that wRu and $\mathcal{M}[p], u \Vdash^{g^\iota} \varphi$.[4] By the induction hypothesis we obtain $\varphi^{g^\iota} \in u^+$, and consequently $\varphi^g \in u^+$. Now, since wRu, we also know that $w\hat{R}u$ by Lemma 5.11.(v), and thus by Lemma 5.8 we obtain $\Diamond\varphi^g \in w^+$ as desired.

For the right to left implication, assume that $(\Diamond\varphi)^g \in w^+$. By the construction of $\mathcal{M}[p]$, there is a world u such that $\varphi^g \in u^+$ (and hence $\varphi^{g^\iota} \in u^+$) and wRu, and then $\mathcal{M}[p], u \Vdash^{g^\iota} \varphi$ by the induction hypothesis, from which we finally conclude $\mathcal{M}[p], w \Vdash^g \Diamond\varphi$. □

Theorem 5.14 (Completeness) *If* $\varphi \not\vdash \psi$, *then there are an adequate finite model* $\mathcal{M}$, *a world* $w \in W$, *and a* w*-assignment* g *such that*

$$\mathcal{M}, w \Vdash^g \varphi \quad \textit{and} \quad \mathcal{M}, w \nVdash^g \psi.$$

Proof. Define a set of new constants $C := \{c_{x_i} \mid x_i \in \mathrm{fv}(p)\}$ and let g be a map from the set of variables to C that assigns c_{x_i} to x_i for each i. Define p

[4] Recall that if g is a w-assignment and wRu, we write g^ι to refer to the u-assignment that behaves like g.

as $\langle \varphi^g, \psi^g \rangle$, and assume it is not consistent, i.e., that $\varphi^g \vdash \psi^g$. Then by (a generalization of) Rule 2.1.(xi) and Lemma 2.6 we would get that $\varphi \vdash \psi$. Thus p is consistent. Let $\mathcal{M}[p]$ be the model generated from p as in Definition 5.10 and let w be the root of this model, which is an extension of p. Lemma 5.13 tells us that $\mathcal{M}[p], w \Vdash^g \varphi$ and $\mathcal{M}[p], w \nVdash^g \psi$ because $\varphi^g \in w^+$ and $\psi^g \notin w^+$.□

We observe that $\mathsf{QRC_1}$ has the finite model property with respect to domains and number of worlds (as a consequence of Lemma 5.11 and Theorem 5.14). It is interesting that $\mathsf{QRC_1}$ is this well behaved while, say, predicate intuitionistic logic doesn't enjoy the finite model property with respect to either of these.

Theorem 5.15 $\mathsf{QRC_1}$ *is decidable.*

Proof. Since $\mathsf{QRC_1}$ is recursively axiomatized, has the finite model property, and it is easy to check whether a given finite model is adequate, Post's Theorem allows us to decide whether $\varphi \vdash \psi$. □

Appendix

A Arithmetical soundness

Here we finally prove the soundness of $\mathsf{QRC_1}$ regarding the arithmetical semantics. The following lemma is standard for Σ_1 axiomatizations α and the reader can consult [7] for details.

Lemma A.1 *For any Σ_1 formula α, we have that*

(i) $\mathrm{I}\Sigma_1 \vdash \mathrm{Con}_\alpha(\mathrm{Con}_\alpha) \to \mathrm{Con}_\alpha$;

(ii) $\mathrm{I}\Sigma_1 \vdash \exists z\, \Box_\alpha \varphi \to \Box_\alpha \exists z\, \varphi$.

Recall that we define $\mathcal{QRC}_1(T)$ for a a c.e. theory T in the language of arithmetic which extends $\mathrm{I}\Sigma_1$ as follows.

$$\mathcal{QRC}_1(T) = \{\varphi(\boldsymbol{c}, \boldsymbol{x}) \vdash \psi(\boldsymbol{c}, \boldsymbol{x}) \mid \forall\, {}^*T \vdash \forall \theta\, \forall \boldsymbol{y}\, \forall \boldsymbol{z}\, (\Box_{\psi^*}\theta \to \Box_{\varphi^*}\theta)\}.$$

In the above we assume that all the free variables other than u in $\psi^* \wedge \varphi^*$ are among the $\boldsymbol{y}$ and $\boldsymbol{z}$. The θ are sentences without free variables. Furthermore, we stress that all realizations map to Σ_1 formulas (modulo provable equivalence).

Theorem A.2 (Arithmetical soundness) $\mathsf{QRC_1} \subseteq \mathcal{QRC}_1(\mathrm{I}\Sigma_1)$.

Proof. We proceed by (an external) induction on the proof of $\varphi \vdash \psi$. We shall briefly comment on some of the cases. The case of the axiom $\varphi \vdash \top$ is clear since by an easy induction on φ we van prove that over predicate logic $\varphi^*(\boldsymbol{y}, \boldsymbol{z}, u) \leftrightarrow \tau(u) \vee \varphi'(\boldsymbol{y}, \boldsymbol{z}, u)$ for some formula φ'. The axioms $\varphi \wedge \psi \vdash \varphi$ are easily seen to be sound since $(\varphi \wedge \psi)^* = \varphi^* \vee \psi^*$, that is, the formula that defines the union of two axiom sets.

The rule that if $\varphi \vdash \psi$ and $\psi \vdash \chi$, then $\varphi \vdash \chi$ is straightforward but the rule that if $\varphi \vdash \psi$ and $\varphi \vdash \chi$, then $\varphi \vdash \psi \wedge \chi$ is slightly more tricky. To see the soundness, we fix a particular realization $*$ and reason in $\mathrm{I}\Sigma_1$. Inside $\mathrm{I}\Sigma_1$ we fix arbitrary $\boldsymbol{y}$, $\boldsymbol{z}$ and θ and assume $\Box_{(\psi \wedge \chi)^*(\boldsymbol{y},\boldsymbol{z})}\theta$, that is, $\Box_{\psi^*(\boldsymbol{y},\boldsymbol{z}) \vee \chi^*(\boldsymbol{y},\boldsymbol{z})}\theta$.

Thus, $\Box_{\psi^*(\boldsymbol{y},\boldsymbol{z})}\big(\forall\, i<n\ \xi_i\to\theta\big)$ for some collection of axioms $\{\xi_i\}_{i<n}$ satisfying $\chi^*(\boldsymbol{y},\boldsymbol{z})$. By the induction hypothesis on $\varphi\vdash\psi$ we obtain $\Box_{\varphi^*(\boldsymbol{y},\boldsymbol{z})}\big(\forall\, i<n\ \xi_i\to\theta\big)$, so that $\Box_\tau\big(\forall\, i<n\ \xi_i\to(\forall\, j<m\ \varphi_j\to\theta)\big)$ for some collection of axioms $\{\varphi_j\}_{j<m}$ satisfying $\varphi^*(\boldsymbol{y},\boldsymbol{z})$. Since all the ξ_i satisfy $\chi^*(\boldsymbol{y},\boldsymbol{z})$ we conclude $\Box_{\chi^*(\boldsymbol{y},\boldsymbol{z})}\big(\forall\, j<m\ \varphi_j\to\theta\big)$. Using now the induction hypothesis on $\varphi\vdash\chi$ we conclude $\Box_{\varphi^*(\boldsymbol{y},\boldsymbol{z})}\big(\forall\, j<m\ \varphi_j\to\theta\big)$ whence $\Box_{\varphi^*(\boldsymbol{y},\boldsymbol{z})}\theta$ as was to be shown.

We will now see the soundness of the necessitation rule, that is, if $\varphi\vdash\psi$, then $\Diamond\varphi\vdash\Diamond\psi$. We fix some realization $*$. The induction hypothesis for $\varphi\vdash\psi$ applied to the formula $\bot$ gives us $\mathrm{I}\Sigma_1\vdash\forall\,\boldsymbol{y},\boldsymbol{z}\,\big(\Box_{\psi^*(\boldsymbol{y},\boldsymbol{z})}\bot\to\Box_{\varphi^*(\boldsymbol{y},\boldsymbol{z})}\bot\big)$, whence

$$\mathrm{I}\Sigma_1\vdash\forall\,\boldsymbol{y},\boldsymbol{z}\,\big(\mathrm{Con}_{\varphi^*(\boldsymbol{y},\boldsymbol{z})}\to\mathrm{Con}_{\psi^*(\boldsymbol{y},\boldsymbol{z})}\big).\tag{A.1}$$

Let π be the standard proof of this. We reason in $\mathrm{I}\Sigma_1$, fixing parameters $\boldsymbol{y},\boldsymbol{z},\theta$ and assuming $\Box_{(\Diamond\psi)^*(\boldsymbol{y},\boldsymbol{z})}\theta$. Since $(\Diamond\psi)^*:=\tau(u)\vee(u=\ulcorner\mathrm{Con}_{\psi^*(\boldsymbol{y},\boldsymbol{z})}\urcorner)$, we conclude $\Box_\tau\big(\mathrm{Con}_{\psi^*(\boldsymbol{y},\boldsymbol{z})}\to\theta\big)$. We combine this proof with the proof π of (A.1) to conclude $\Box_\tau\big(\mathrm{Con}_{\varphi^*(\boldsymbol{y},\boldsymbol{z})}\to\theta\big)$, whence $\Box_{(\Diamond\varphi)^*(\boldsymbol{y},\boldsymbol{z})}\theta$.

The soundness of the axiom $\Diamond\Diamond\varphi\vdash\Diamond\varphi$ is similar, now using Lemma A.1.(i) instead of (A.1).

To see the soundness of the axiom $\Diamond\forall\,x_i\,\varphi\vdash\forall\,x_i\,\Diamond\varphi$ we start by proving a First Claim:

$$\mathrm{I}\Sigma_1\vdash\mathrm{Con}_{(\forall\,x_i\,\varphi)^*(\boldsymbol{y},\boldsymbol{z})}\to\forall\,z_i\,\mathrm{Con}_{\varphi^*(\boldsymbol{y},\boldsymbol{z})}.\tag{A.2}$$

To prove this, we reason in $\mathrm{I}\Sigma_1$ and assume $\exists\,z_i\,\Box_{\varphi^*(\boldsymbol{y},\boldsymbol{z})}\bot$, whence for some number ζ we have that $\Box_{\varphi^*(\boldsymbol{y},\boldsymbol{z})[z_i\leftarrow\zeta]}\bot$. Then a slight variation of Lemma A.1.(ii) allows us to see that $\Box_{\exists\,z_i\,\varphi^*(\boldsymbol{y},\boldsymbol{z})}\bot$, and thus $\Box_{(\forall\,x_i\,\varphi)^*(\boldsymbol{y},\boldsymbol{z})}\bot$.

We now prove a Second Claim:

$$\mathrm{I}\Sigma_1\vdash\Box_{(\forall\,x_i\,\Diamond\varphi)^*(\boldsymbol{y},\boldsymbol{z})}\delta\to\Box_{\tau(u)\vee(u=\ulcorner\forall x_i\mathrm{Con}_{\varphi^*(\boldsymbol{y},\boldsymbol{z})}\urcorner)}\,\delta.\tag{A.3}$$

We observe $(\forall\,x_i\,\Diamond\varphi)^*(\boldsymbol{y},\boldsymbol{z})=\exists\,z_i\,(\Diamond\varphi)^*(\boldsymbol{y},\boldsymbol{z})=\exists\,z_i\,\big(\tau(u)\vee u{=}\ulcorner\mathrm{Con}_{\varphi^*(\boldsymbol{y},\boldsymbol{z})}\urcorner\big)$, the latter being provably equivalent to $\tau(u)\vee\exists\,z_i\,\big(u{=}\ulcorner\mathrm{Con}_{\varphi^*(\boldsymbol{y},\boldsymbol{z})}\urcorner\big)$. To prove the Second Claim, we reason in $\mathrm{I}\Sigma_1$ and assume the antecedent $\Box_{(\forall\,x_i\,\Diamond\varphi)^*(\boldsymbol{y},\boldsymbol{z})}\delta$ fixing some $\boldsymbol{y},\boldsymbol{z},\delta$. Thus, we find a collection of numbers $\{\zeta_j\}_{j<m}$ such that

$$\Box_\tau\big(\forall\,j<m\ \mathrm{Con}_{\varphi^*(\boldsymbol{y},\boldsymbol{z})[z_i\leftarrow\zeta_j]}\to\delta\big).$$

Clearly, $\Box_\tau\big(\forall\,z_i\,\mathrm{Con}_{\varphi^*(\boldsymbol{y},\boldsymbol{z})}\to\forall\,j<m\ \mathrm{Con}_{\varphi^*(\boldsymbol{y},\boldsymbol{z})[z_j\leftarrow\zeta_j]}\big)$, which suffices to prove the Second Claim.

Let us now go back the soundness of the axiom $\Diamond\forall\,x_i\,\varphi\vdash\forall\,x_i\,\Diamond\varphi$. We fix $*$, reason in $\mathrm{I}\Sigma_1$, fix $\boldsymbol{y},\boldsymbol{z},\theta$, and assume $\Box_{(\forall\,x_i\,\Diamond\varphi)^*(\boldsymbol{y},\boldsymbol{z})}\theta$. By the Second Claim and the formalized deduction theorem we get $\Box_\tau(\forall\,z_i\,\mathrm{Con}_{\varphi^*(\boldsymbol{y},\boldsymbol{z})}\to\theta)$. The First Claim now gives us $\Box_{(\Diamond\,\forall\,x_i\,\varphi)^*(\boldsymbol{y},\boldsymbol{z})}\theta$ as was to be shown.

The soundness of the $\forall$-introduction rule on the right, that if $\varphi\vdash\psi$, then $\varphi\vdash\forall\,x_i\,\psi$ ($x_i\notin\mathrm{fv}(\varphi)$), is not hard but contains a subtlety. To prove it we fix

$*$, reason in $\mathrm{I}\Sigma_1$, fix $\boldsymbol{y}, \boldsymbol{z}, \theta$ and assume $\Box_{(\forall x_i\psi)^*(\boldsymbol{y},\boldsymbol{z})}\theta$. Since $(\forall x_i\psi)^*(\boldsymbol{y},\boldsymbol{z}) = \exists z_i\psi^*(\boldsymbol{y},\boldsymbol{z})$ we can find numbers $\{\zeta_j\}_{j<m}$ such that

$$\Box_T(\forall j < m\ \psi^*(\boldsymbol{y},\boldsymbol{z})[z_i{\leftarrow}\zeta_j] \to \theta).$$

Now by the induction hypothesis we get

$$\forall j < m\ \Box_{\varphi^*(\boldsymbol{y},\boldsymbol{z})[z_i\leftarrow\zeta_j]}\psi^*(\boldsymbol{y},\boldsymbol{z})[z_i{\leftarrow}\zeta_j].$$

Since $x_i \notin \mathrm{fv}(\varphi)$ we have $\forall j < m\ \Box_{\varphi^*(\boldsymbol{y},\boldsymbol{z})}\psi^*(\boldsymbol{y},\boldsymbol{z})[z_i{\leftarrow}\zeta_j]$. Using Σ_1-collection we obtain $\Box_{\varphi^*(\boldsymbol{y},\boldsymbol{z})}\,\forall j < m\ \psi^*(\boldsymbol{y},\boldsymbol{z})[z_i{\leftarrow}\zeta_j]$, from which the required $\Box_{\varphi^*(\boldsymbol{y},\boldsymbol{z})}\theta$ follows.

The soundness of the remaining axioms and rules is straightforward and boils down to interchanging universal quantifiers. □

B Relational soundness of the generalization on constants rule

The generalization on constants rule, that if $\varphi[x{\leftarrow}c] \vdash \psi[x{\leftarrow}c]$ then also $\varphi \vdash \psi$ as long as c does not appear in φ nor ψ, is indeed sound with respect to Kripke models but the proof needs a couple of extra definitions and results.

We wish to provide counterparts to Remark 4.5 and Lemma 4.6 for when the change happens in the interpretation of a constant instead of a variable. It is straightforward to check that the interpretation of constants not appearing in a formula is not relevant for the truth of that formula:

Remark B.1 Let $\mathcal{M}$ and $\mathcal{M}'$ be adequate models differing only in their constant interpretations $\{I_w\}_{w\in W}$ and $\{I'_w\}_{w\in W}$. Let w be any world, g be any w-assignment, and φ be a formula whose constants are interpreted in the same way by both $\mathcal{M}$ and $\mathcal{M}'$. Then

$$\mathcal{M}, w \Vdash^g \varphi \iff \mathcal{M}', w \Vdash^g \varphi.$$

However, we need a bit of work to be able to state a counterpart of Lemma 4.6 for constants. We want to be able to replace the interpretation of a constant by an element of the domain of some world w, but this element may not exist in the domains of the worlds below w. Thus we need to first get rid of that part of the model and keep only the sub-graph rooted at w.

Definition B.2 Given a frame $\mathcal{F} = \langle W, R, \{M_w\}_{w\in W}\rangle$ and a world $r \in W$, the *frame restricted at* r, written $\mathcal{F}|_r = \langle W|_r, R|_r, \{M_w\}_{w\in W|_r}\rangle$, is defined as the restriction of $\mathcal{F}$ to the world r and all the worlds accessible from r by R. Thus, $W|_r := \{r\} \cup \{w \in W \mid rRw\}$, and the relation $R|_r$ is R restricted to $W|_r$.

If $\mathcal{M} = \langle \mathcal{F}, \{I_w\}_{w\in W}, \{J_w\}_{w\in W}\rangle$ is a model, then $\mathcal{M}|_r$ is defined as $\langle \mathcal{F}|_r, \{I_w\}_{w\in W|_r}, \{J_w\}_{w\in W|_r}\rangle$.

Remark B.3 If $\mathcal{F}$ is an adequate frame, then so is $\mathcal{F}|_r$ for any $r \in W$. Furthermore, if $\mathcal{M}$ is an adequate model, then so is $\mathcal{M}|_r$.

Remark B.4 Given an adequate model $\mathcal{M}$ and a world $r \in W$, we have that for any formula φ, any world $w \in W_r$ and any w-assignment g:

$$\mathcal{M}, w \Vdash^g \varphi \iff \mathcal{M}|_r, w \Vdash^g \varphi.$$

Definition B.5 Given an adequate model $\mathcal{M} = \langle \mathcal{F}, \{I_w\}_{w \in W}, \{J_w\}_{w \in W} \rangle$, a world $r \in W$, a constant c, and an element of the domain $d \in M_r$, we define $\mathcal{M}|_r[c{\leftarrow}d] := \langle \mathcal{F}|_r, \{I'_w\}_{w \in W|_r}, \{J_w\}_{w \in W|_r} \rangle$ such that its frame is $\mathcal{F}$ truncated at r, the relational symbols interpretation and the interpretation of all constants except for c coincides with that of $\mathcal{M}|_r$, and the interpretation $c^{I'_w}$ of the constant c is d for every $w \in W|_r$.

Lemma B.6 *Given a constant c, a formula φ where c does not appear, an adequate model $\mathcal{M}$, a world w, and a w-assignment g, we have:*

$$\mathcal{M}, w \Vdash^g \varphi \iff \mathcal{M}|_w[c{\leftarrow}g(x)], w \Vdash^g \varphi[x{\leftarrow}c].$$

Proof. We proceed by induction on the formula φ. The cases of $\top$, relational symbols, and conjunction are trivial. We assume that x is free in φ, for otherwise we could use Remarks B.1 and B.4.

Consider the diamond case. If $\mathcal{M}, w \Vdash^g \Diamond\psi$, then there is a world v such that wRv and $\mathcal{M}, v \Vdash^{g^\iota} \psi$. By the induction hypothesis we obtain $\mathcal{M}|_v[c{\leftarrow}g^\iota(x)], v \Vdash^{g^\iota} \psi[x{\leftarrow}c]$. Observe that $\mathcal{M}|_v[c{\leftarrow}g^\iota(x)]$ is the same model as $(\mathcal{M}|_w[c{\leftarrow}g(x)])|_v$, since they share the same frame, the same constant interpretation (because $g(x) = g^\iota(x)$) and the same relational symbol interpretation. Then by Remark B.4 we get $\mathcal{M}|_w[c{\leftarrow}g(x)], v \Vdash^{g^\iota} \psi[x{\leftarrow}c]$ and consequently $\mathcal{M}|_w[c{\leftarrow}g(x)] \Vdash^g \Diamond\psi[x{\leftarrow}c]$, as desired. The other implication is analogous.

Finally, let $\varphi = \forall z\,\psi$ and assume that $\mathcal{M}, w \Vdash^g \forall z\,\psi$. Let $h \sim_z g$ be a w-assignment, and set out to prove $\mathcal{M}|_w[c{\leftarrow}g(x)], w \Vdash^h \psi[x{\leftarrow}c]$ (note that z and x are not the same variable for otherwise x would not be free in φ). Since $h \sim_z g$, we know that $g(x) = h(x)$, so by the induction hypothesis it is enough to show $\mathcal{M}, w \Vdash^h \psi$, which follows from our assumption. The other implication is analogous. □

The above lemma is now enough to show the soundness of Rule 2.1.(xi) with respect to Kripke models.

Acknowledgments. We are grateful to the anonymous reviewers for suggesting a number of improvements to the preliminary version of this paper.

References

[1] Beklemishev, L. D., *Provability algebras and proof-theoretic ordinals, I*, Annals of Pure and Applied Logic **128** (2004), pp. 103–124.

[2] Beklemishev, L. D., *Veblen hierarchy in the context of provability algebras*, in: P. Hájek, L. Valdés-Villanueva and D. Westerståhl, editors, *Logic, Methodology and Philosophy of Science, Proceedings of the Twelfth International Congress*, Kings College Publications, 2005 pp. 65–78.

[3] Beklemishev, L. D., *Calibrating provability logic: From modal logic to Reflection Calculus*, in: T. Bolander, T. Braüner, T. S. Ghilardi and L. Moss, editors, *Advances in Modal Logic 9* (2012), pp. 89–94.

[4] Beklemishev, L. D., *Positive provability logic for uniform reflection principles*, Annals of Pure and Applied Logic **165** (2014), pp. 82–105.

[5] Beklemishev, L. D., D. Fernández-Duque and J. J. Joosten, *On provability logics with linearly ordered modalities*, Studia Logica **102** (2014), pp. 541–566.

[6] Beklemishev, L. D. and F. N. Pakhomov, *Reflection algebras and conservation results for theories of iterated truth*, arXiv:1908.10302 [math.LO] (2019).

[7] Boolos, G. S., "The Logic of Provability," Cambridge University Press, Cambridge, 1993.

[8] Dashkov, E. V., *On the Positive Fragment of the Polymodal Provability Logic GLP*, Mathematical Notes **91** (2012), pp. 318–333.

[9] Fernández-Duque, D. and E. Hermo Reyes, *A self-contained provability calculus for* Γ_0, in: R. Iemhoff, M. Moortgat and R. J. G. B. de Queiroz, editors, *Logic, Language, Information, and Computation - 26th International Workshop, WoLLIC 2019, Utrecht, The Netherlands, July 2-5, 2019, Proceedings*, Lecture Notes in Computer Science **11541** (2019), pp. 195–207.

[10] Fernández-Duque, D. and J. J. Joosten, *Models of transfinite provability logics*, Journal of Symbolic Logic **78** (2013), pp. 543–561.

[11] Fernández-Duque, D. and J. J. Joosten, *Well-orders in the transfinite Japaridze algebra*, Logic Journal of the IGPL **22** (2014), pp. 933–963.

[12] Fernández-Duque, D. and J. J. Joosten, *The omega-rule interpretation of transfinite provability logic*, Annals of Pure and Applied Logic **169** (2018), pp. 333–371.

[13] Gabbay, D. M., V. B. Shehtman and D. P. Skvortsov, "Quantification in Nonclassical Logic Vol. 1," Elsevier, 2009.

[14] Goldblatt, R., "Quantifiers, propositions and identity, admissible semantics for quantified modal and substructural logics," Cambridge University Press, 2011.

[15] Hájek, P. and P. Pudlák, "Metamathematics of First Order Arithmetic," Springer-Verlag, Berlin, Heidelberg, New York, 1993.

[16] Hughes, G. E. and M. J. Cresswell, "A New Introduction to Modal Logic," Routledge, 1996.

[17] Japaridze, G., *The polymodal provability logic*, in: *Intensional logics and logical structure of theories: material from the Fourth Soviet-Finnish Symposium on Logic*, Metsniereba, Tibilisi, 1988 pp. 16–48, in Russian.

[18] Kikot, S., A. Kurucz, Y. Tanaka, F. Wolter and M. Zakharyaschev, *Kripke completeness of strictly positive modal logics over meet-semilattices with operators*, Journal of Symbolic Logic **84** (2019), pp. 533–588.

[19] Kripke, S. A., *Semantical considerations on modal logic*, Acta Philosophica Fennica **16** (1963), pp. 83–94.

[20] Solovay, R. M., *Provability interpretations of modal logic*, Israel Journal of Mathematics **28** (1976), pp. 33–71.

[21] Vardanyan, V. A., *Arithmetic complexity of predicate logics of provability and their fragments*, Doklady Akad. Nauk SSSR **288** (1986), pp. 11–14, in Russian. English translation in Soviet Mathematics Doklady 33, 569–572 (1986).

[22] Visser, A. and M. de Jonge, *No escape from Vardanyan's theorem*, Archive for Mathematical Logic **45** (2006), pp. 539–554.

Quantifying over Asynchronous Information Change

Philippe Balbiani

IRIT, CNRS, Université de Toulouse

Hans van Ditmarsch

LORIA, CNRS, Université de Lorraine

Saúl Fernández González

IRIT, CNRS, Université de Toulouse

Abstract

We propose a logic AAA for Arbitrary Asynchronous Announcements. In this logic, the sending and receiving of messages that are announcements are separated and represented by distinct modalities. Additionally, the logic has a modality that represents quantification over information change in the shape of sequences of sending and receiving events, called histories. We present a complete however infinitary axiomatisation, bisimulation invariance, and various results for the logical semantics, wherein we consider both how the logic is different from asynchronous announcement logic AA and how the logic is different from arbitrary public announcement logic APAL. We also address the expressivity and we demonstrate the preservation of an extended fragment of positive formulas (wherein negations do not occur before epistemic modalities). Finally, we present work in progress on the logic AAM of Asynchronous Action Models and the logic AAAM of Arbitrary Asynchronous Action Models.

Keywords: Modal logic, dynamic epistemic logic, asynchrony, quantifying over information change

1 Introduction

We investigate what agents know and learn in distributed systems wherein the sending and receiving of messages are separated. Notions of asynchronous knowledge and common knowledge have been investigated in distributed computing in works such as [7,13,18,19,20] and in temporal epistemic logics in works such as [9,16,21,23]. Our take on such matters is from within so-called Dynamic Epistemic Logic (DEL) [22,6,11], a modal logic of knowledge and change of knowledge (or belief and change of belief), however not in its standard incarnation wherein message sending and receiving is synchronized and instantaneous, but in a recently investigated version by various researchers wherein these are separated [17,4,5]. These approaches are somewhat different from the asynchrony due to partial observation wherein histories (sequences of messages) of different length may have become indistinguishable for an agent, as in [10].

In [4] a logic AA is presented wherein messages that are announcements are still sent by an 'outside observer' or by the environment, but wherein they are individually received by the agents, unlike in public announcement logic [22] wherein all agents receive the announcement simultaneously (synchronised). The logic contains modalities for the announcement of φ, as in $[\varphi]\psi$. This has still as precondition that φ must be true when announced, but it does not have the effect the φ is received by any agent. For that, there are other modalities $[a]\psi$, for 'after the agent a has received the next announcement, ψ is true'. For example, we can now say that $[p][a]B_a p$: after p has been sent and agent a has received it, the agent knows/believes p. Therefore, in this logic AA we cannot obtain common knowledge that the announcement has been received, although we can still approach common knowledge by iterating announcements such as announcement p, all agents received p, announcement that everybody knows p, everybody received that, announcement that everybody knows that everybody knows p, etc. This is as in the concurrent common knowledge of [21]. In [4], a complete axiomatisation for such a logic is given, as well as special results on the class of $\mathcal{S}5$ models (where all relations are equivalence relations).

Subsequently, [5] investigates the wide spectrum from individual reception of messages as in AA, to partial synchronisation of messages by subgroups of all agents, up to synchronised reception of messages by all agents much akin to public announcement logic PAL — they also enrich the epistemic language with common belief modalities.

In the present work we generalize [4] in two ways: to the logic AA of asynchronous announcements we now add a quantifier $[!]\varphi$ for 'after any sequence of events, φ'. It is motivated by a similar quantifier in the logic APAL [2], that stands for 'after any/arbitrary announcement, φ'. Clearly, in the asynchronous setting we cannot have it merely quantifying over unreceived announcements, as this would not affect the beliefs or knowledge of agents. As the order of reception of announcements may vary greatly and may take place much later after an announcement, and possibly many subsequent announcements, have been sent, the natural form of quantification is therefore over arbitrary sequences of such sending and receiving events. We show that the resulting logic AAA has a complete axiomatisation, and varies in crucial respects from the motivating precedent APAL [2]. Such a logic of arbitrary asynchronous announcement may be, we hope, useful for diverse tasks such as: asynchronous epistemic planning, formalising epistemic protocols in distributed computing, and analysing the fine structure of interacting agents independently executing informative and other actions.

One particular further generalisation is also presented in some detail, namely the similar quantification over asynchronous non-public events (in the sense of events that are not known to be eventually received by all agents), such as an agent a privately receiving information on a proposition p, where an agent b also receives the information that a is privately receiving p although not necessarily simultaneous with a. From the works of Hales and collaborators [14,15] it is known that quantification over action models behaves much better than quantification over announcements. It is decidable, and the quantifier can be eliminated from the language: given $\langle ! \rangle \varphi$, meaning 'there is an action model after which φ holds', a specific action model can be synthesised that, if executable, always results in φ. We conjecture similar results for quantifying over asynchronous action models. In particular, asynchronous synthesis seems a highly desirable future goal.

In Section 2 we present Arbitrary Asynchronous Announcement logic AAA, for which we present various semantic results in Section 3. In Section 4 we address the expressivity, and in Section 5 the preservation (after history extension) of the fragment of *positive formulas*. Section 6 presents a complete infinitary axiomatisation. Finally, Section 7 adresses the generalisation of our results to a logic for quantification over asynchronous action models.

2 The logic AAA

Syntax. Let A be a finite set of epistemic agents and P a countable set of propositional variables. We consider the following language $\mathcal{L}$:

$$\varphi ::= p|\top|\neg\varphi|(\varphi \wedge \varphi)|\hat{B}_a\varphi|\langle\varphi\rangle\varphi|\langle a\rangle\varphi|\langle ! \rangle\varphi,$$

where $p \in P, a \in A$. We follow standard rules for omission of the parentheses. The connectives $\bot, \vee, \rightarrow, \leftrightarrow$ are defined by the usual abbreviations.

We define duals $B_a\varphi = \neg\hat{B}_a\neg\varphi$, $[a]\varphi = \neg\langle a\rangle\neg\varphi$, $[\psi]\varphi = \neg\langle\psi\rangle\neg\varphi$, $[!]\varphi = \neg\langle ! \rangle\neg\varphi$.

Let $\mathcal{L}_{-!}$ be the fragment of this language without the $\langle ! \rangle$ modality.

Consider $A \cup \mathcal{L}$ as an alphabet, with agents and formulas as letters. Variables for words in this language are $\alpha, \beta, \ldots$, and ϵ denotes the empty word. Given a word α over $A \cup \mathcal{L}$, $|\alpha|$ is its length, $|\alpha|_a$ is the number of its a's (for each $a \in A$), $|\alpha|_!$ is the number of its formula occurrences, $\alpha\upharpoonright_!$ is the projection of α to $\mathcal{L}$, and $\alpha\upharpoonright_{!a}$ is the restriction of $\alpha\upharpoonright_!$ to the first $|\alpha|_a$ formulas. These notions have obvious inductive definitions.

For each such word, the formula $\langle\alpha\rangle\varphi$ represents an abbreviation of the sequence of announcement and reading modalities corresponding to the formulas and agents which appear in α, defined recursively as follows:

$$\langle\epsilon\rangle\varphi = \varphi;\ \langle\alpha.\psi\rangle\varphi = \langle\alpha\rangle\langle\psi\rangle\varphi;\ \langle\alpha.a\rangle\varphi = \langle\alpha\rangle\langle a\rangle\varphi,$$

where ϵ is the empty word. Every formula in $\mathcal{L}$ is thus of the form $\langle\alpha\rangle\varphi$ for some $\alpha \in (\mathcal{L} \cup A)^*$.

A *prefix* β of α, notation $\beta \subseteq \alpha$, is an initial sequence of α, inductively defined as: $\alpha \subseteq \alpha$, and if $\beta \subseteq \alpha$, then for all $a \in A$ and $\psi \in \mathcal{L}$, $\beta \subseteq \alpha.a$ and $\beta \subseteq \alpha.\psi$.

For a sequence of announcements and readings to be executable, it is necessary that, whenever an agent is doing her n-th reading, there have been at least n formulas announced. Words satisfying this property will be called *histories*.

Histories. A word α in the language $A \cup \mathcal{L}$ is a *history* if for all prefixes $\beta \sqsubseteq \alpha$ and for all $a \in A$, $|\beta|_! \geq |\beta|_a$.

We denote by $\mathcal{H}$ the set of histories. Obviously, if β is a prefix of a history α, then β is a history too.

View relation. Let α, β be histories and $a \in A$. We define: $\alpha \rhd_a \beta$ iff $|\beta|_a = |\alpha|_a$, $\beta\restriction_{!a} = \alpha\restriction_{!a}$ and $|\beta|_! = |\alpha|_a$. (Note that $\alpha \rhd_a \beta$ iff $\alpha\restriction_{!a} = \beta\restriction_{!a} = \beta\restriction_!$.) The set $\mathsf{view}_a(\alpha) := \{\beta \mid \alpha \rhd_a \beta\}$ is the *view* of a given α. Informally, the view of agent a given history α consists of all the different ways in which all agents can receive the announcements that a received in α, in the same order in which they were received.

Semantics. We will use the following well-founded preorder to define our semantics [1]. First, we define $\deg_{\mathrm{B}} \varphi$, $\deg_! \varphi$ and $\|\varphi\|$ recursively: for $k = !, B$,

$$
\begin{array}{ll}
\deg_k p = 0 & \|p\| = 2 \\
\deg_k \top = 0 & \|\top\| = 1 \\
\deg_k(\neg\varphi) = \deg_k \varphi & \|\neg\varphi\| = \|\varphi\| + 1 \\
\deg_k(\varphi \wedge \psi) = \max\{\deg_k \varphi, \deg_k \psi\} & \|\varphi \wedge \psi\| = \|\varphi\| + \|\psi\| \\
\deg_k(\langle a \rangle \varphi) = \deg_k \varphi & \|\langle a \rangle \varphi\| = \|\varphi\| + 2 \\
\deg_k(\langle \varphi \rangle \psi) = \deg_k \varphi + \deg_k \psi & \|\langle \varphi \rangle \psi\| = 2\|\varphi\| + \|\psi\| \\
\deg_{\mathrm{B}}(\hat{B}_a \varphi) = \deg_{\mathrm{B}} \varphi + 1 & \|\hat{B}_a \varphi\| = \|\varphi\| + 1 \\
\deg_!(\hat{B}_a \varphi) = \deg_! \varphi & \\
\deg_B(\langle ! \rangle \varphi) = \deg_B \varphi & \|\langle ! \rangle \varphi\| = \|\varphi\| + 1 \\
\deg_!(\langle ! \rangle \varphi) = \deg_! \varphi + 1 &
\end{array}
$$

For a word α, we set $\deg_k \alpha := \sum\{\deg_k \psi : \psi \text{ occurs in } \alpha\}$ and

$$\|\epsilon\| = 0, \|\alpha.a\| = \|\alpha\| + 1, \|\alpha.\psi\| = \|\alpha\| + \|\psi\|.$$

Finally, for pairs (α, φ) we set: $\deg_k(\alpha, \varphi) = \deg_k \alpha + \deg_k \varphi$, and $\|(\alpha, \varphi)\| = \|\alpha\| + \|\varphi\|$, and we define a well-founded order $\ll$ as a lexicographical ordering on these quantities, i.e. $(\alpha, \varphi) \ll (\beta, \psi)$ iff

$$
\begin{cases}
\deg_!(\alpha, \varphi) < \deg_!(\beta, \psi), \text{ or} \\
\deg_!(\alpha, \varphi) = \deg_!(\beta, \psi) \;\&\; \deg_B(\alpha, \varphi) < \deg_B(\beta, \psi), \text{ or} \\
\deg_!(\alpha, \varphi) = \deg_!(\beta, \psi) \;\&\; \deg_B(\alpha, \varphi) = \deg_B(\beta, \psi) \;\&\; \|(\alpha, \varphi)\| < \|(\beta, \psi)\|.
\end{cases}
$$

We interpret formulas on models (W, R, V), where $R : A \to \mathcal{P}(W^2)$, with respect to pairs (w, α) where $w \in W$ an $\alpha \in \mathcal{H}$. We define the relations "w agrees with α" ($w \bowtie \alpha$) and "(w, α) satisfies φ" ($w, \alpha \models \varphi$) by $\ll$-induction as it appears in Table 1. A formula φ is ϵ-valid, notation $\models^\epsilon \varphi$, iff for all models (W, R, V) and for all $s \in W$, $s, \epsilon \models \varphi$. A formula φ is $*$-valid, notation $\models^* \varphi$, iff for all models (W, R, V), for all $s \in W$ and for all histories α, $s, \epsilon \models [\alpha]\varphi$.

[1] Similar preorders have been used in [1,3] whithin the context of proofs of completeness of APAL.

$w \bowtie \epsilon$	always;
$w \bowtie \alpha.\varphi$	iff $w \bowtie \alpha$ and $w, \alpha \models \varphi$;
$w \bowtie \alpha.a$	iff $w \bowtie \alpha$;
$w, \alpha \models p$	iff $w \in V(p)$;
$w, \alpha \models \top$	always;
$w, \alpha \models \neg\varphi$	iff $w, \alpha \not\models \varphi$;
$w, \alpha \models \varphi_1 \wedge \varphi_2$	iff $w, \alpha \models \varphi_i$, $i = 1, 2$;
$w, \alpha \models \langle a \rangle \varphi$	iff $\lvert\alpha\rvert_a < \lvert\alpha\rvert_!$ and $w, \alpha.a \models \varphi$;
$w, \alpha \models \langle \psi \rangle \varphi$	iff $w, \alpha \models \psi$ and $w, \alpha.\psi \models \varphi$;
$w, \alpha \models \hat{B}_a \varphi$	iff $t, \beta \models \varphi$ for some $(t, \beta) \in W \times \mathcal{H}$ such that $R_a wt, \alpha \triangleright_a \beta, t \bowtie \beta$
$w, \alpha \models \langle ! \rangle \varphi$	iff $w, \alpha \models \langle \beta \rangle \varphi$ for some $\beta \in (\mathcal{L}_{-!} \cup A)^*$.

Table 1
Semantics of AAA

Note that the $\langle ! \rangle$ modality only quantifies over words wherein $\langle ! \rangle$ does not occur. This is to avoid a circular definition. The dual of $\langle ! \rangle$ is read as follows: $w, \alpha \models [!]\varphi$ if and only if, for every possible sequence $\beta \in (\mathcal{L}_{-!} \cup A)^*$, it is the case that $w, \alpha \models [\beta]\varphi$.

Note moreover that the relation $\triangleright_a$ is not reflexive (it is however postreflexive, in the sense that $\alpha \triangleright_a \beta$ implies $\beta \triangleright_a \beta$). For this reason, it is not the case that $w, \alpha \models B_a \varphi$ implies $w, \alpha \models \varphi$. Our modality is not factual and this is one of the reasons we favour a doxastic interpretation of it over an epistemic one.

We make the assumption that an agent forms her beliefs based on announcements she has so far received, ignoring announcements that have already been made but not yet received by that agent, and ignoring possible future announcements. The usual assumption in distributed computing is that agents also consider it possible that other agents may have received more messages than themselves. This results in a notion of asynchronous *knowledge* instead of our notion of asynchronous *belief*. A technical reason for that restriction is that we do not have a well-defined semantics for such knowledge (the presence of dynamic modalities for information change, rather unlike the usual situation in distributed computing, makes this problematic). However, there are also conceptual reasons. In a dynamic epistemic logic without factual (ontic) change, only a very weak notion of asynchronous knowledge would result. In the DEL setting, all we can hope for about the future is that all facts eventually become commonly known. Taking such histories into account, I can never *know* that you are ignorant (as the dual of knowing whether) about the value of an atomic proposition. After all, that value may already have been announced and you may have received that, but not yet me. Whereas in our semantics I can (correctly) believe that you are ignorant, namely if this is even the case if you have received the same information as me (all announcements that I have received you have also received), or possibly less. Because of the assumption to ignore unreceived and possible future announcements, $B_a[a]\bot$ always true: an agent never believes there are unread announcements.

Example 2.1 Consider a model consisting of two states s and t, indistinguishable for two agents a, b, and where p is only true in s. We then have that $s, \epsilon \models \langle p\rangle\langle a\rangle B_a p$, i.e., $s, \epsilon \models \langle p.a\rangle B_a p$: after the announcement of p and a receiving it, agent a (correctly) believes that p. On the other hand, $s, \epsilon \models \langle p.a\rangle\neg B_b p$, because agent b has not yet received the announcement p. Indeed, we have that $s, \epsilon \models \langle p.a\rangle \hat{B}_a \neg B_b p$ as well as $s, \epsilon \models \langle p.a\rangle \hat{B}_a B_b p$. The view of a on history $p.a$ consists of: $p.a$, $p.b.a$ and $p.a.b$. In the former case (the actual history), b is still uncertain about p, whereas in the latter two cases, b would believe p.

For an example of the use of the quantifier, we note that (similarly to APAL) $\langle !\rangle(B_a p \vee B_a \neg p)$ is a validity of AAA. Whatever the value of p, it can be announced and after agent a receiving it she will either believe p or she will believe $\neg p$.

The following lemma, whose proof is straightforward, will be useful:

Lemma 2.2 *Given a model (W, R, V), $w \in W$, $\varphi \in \mathcal{L}$, $\alpha \in \mathcal{H}$ such that $w \bowtie \alpha$, and $\beta \in (\mathcal{L} \cup A)^*$, the following are equivalent:*

i. $w, \alpha \models \langle\beta\rangle\varphi$;

ii. The concatenation $\alpha.\beta$ is a history, $w \bowtie \alpha.\beta$, and $w, \alpha.\beta \models \varphi$.

3 Semantic results for the logic AAA

Bisimulation. The notion of bisimulation in this framework is, perhaps surprisingly, the usual notion of bisimulation between Kripke models: given (W, R, V) and (W', R', V'), a *bisimulation* is a relation $Z \subseteq W \times W'$ such that, if wZw':

i. $w \in V(p)$ iff $w' \in V'(p)$;

ii. if $R_a wv$, then there exists $v' \in W'$ such that $R'_a w'v'$ and vZv';

iii. if $R'_a w'v'$, then there exists $v \in W$ such that $R_a wv$ and vZv'.

As one might expect, we have the following:

Proposition 3.1 *Let Z be a bisimulation such that wZw', and let $(\alpha, \varphi) \in \mathcal{H} \times \mathcal{L}$. We have:*

$$w, \epsilon \models \langle\alpha\rangle\varphi \textit{ iff } w', \epsilon \models \langle\alpha\rangle\varphi.$$

Proof. See Appendix. □

Under certain constraints, if two states satisfy the same formulas, they are bisimilar. Indeed:

Proposition 3.2 *Let (W, R, V) and (W', R', V') be two models such that $R_a[w]$ and $R'_a[w']$ are finite sets for all $a \in A$, $w \in W$, $w' \in W'$. Set wZw' iff, for all $(\alpha, \varphi) \in \mathcal{H} \times \mathcal{L}$, $w, \epsilon \models \langle\alpha\rangle\varphi$ iff $w', \epsilon \models \langle\alpha\rangle\varphi$. Then Z is a bisimulation.*

Proof. See Appendix. □

Properties of belief. As discussed above, while $B_a\varphi \to \varphi$ is ϵ-valid, (as long as the relation R_a is reflexive) it is not $*$-valid. Other properties of our doxastic modality, however, are $*$-valid. Let $\mathcal{S}5$ denote the class of models where the relations R_a are equivalence relations. We have:

Proposition 3.3 *Let* $\varphi \in \mathcal{L}$. *Then:*

i. $\mathcal{S}5 \models^* B_a\varphi \to \neg B_a\neg\varphi$

ii. $\mathcal{S}5 \models^* B_a\varphi \to B_aB_a\varphi$

iii. $\mathcal{S}5 \models^* \neg B_a\varphi \to B_a\neg B_a\varphi$

Proof. See Appendix. □

Belief before and after update. If an agent will believe φ after a certain sequence of events then the agent believes that there is a sequence of events after which φ holds, but not the other way around. Indeed:

Proposition 3.4 *For all* φ, $\models^\epsilon \langle!\rangle\hat{B}_a\varphi \to \hat{B}_a\langle!\rangle\varphi$, *whereas, for some formula* ψ, $\not\models^\epsilon \hat{B}_a\langle!\rangle\psi \to \langle!\rangle\hat{B}_a\psi$.

Proof. See Appendix. □

Church-Rosser and McKinsey Let us see that neither of the formulas

(CR) $\langle!\rangle[!]\varphi \to [!]\langle!\rangle\varphi$ (McK) $[!]\langle!\rangle\varphi \to \langle!\rangle[!]\varphi$

are ϵ-valid. It is known from APAL that these properties are valid for arbitrary announcement on the class of $\mathcal{S}5$ models (where all accessibility relations are equivalence relations) [2]. As we consider arbitrary relations, this is not unexpected. We address the case $\mathcal{S}5$ at the end of this paragraph.

First let us see (McK) is not ϵ-valid. Let $\varphi = [a]\bot$. Then φ will be satisfied at a pair (w, α) if and only if $|\alpha|_a = |\alpha|_!$. For any history β it is the case that $|\beta|_a \leq |\beta|_!$: let $a^\beta = a...a$ be the concatenation of $|\beta|_! - |\beta|_a$ times the letter a. Then, for every β there exists a word a^β such that $w, \epsilon \models [\beta]\langle a^\beta\rangle[a]\bot$. However, $\langle!\rangle[!][a]\bot$ is never satisfied: indeed, for any history β, let $\top^\beta$ be a concatenation of the formula $\top$ enough times so that $|\beta.\top^\beta|_! > |\beta|_a$. Then we have $w, \epsilon \not\models \langle\beta\rangle[\top^\beta][a]\bot$.

Let us now see a counterexample for (CR). Consider the following one-agent model[2]:

Let $W = \{w_1, w_2, w_3\}$, $R_a = \{(w_1, w_2), (w_2, w_2), (w_2, w_3)\}$ and $V(p) = \{w_1, w_2\}$. We have that $w_1, \epsilon \models \langle!\rangle[!]\hat{B}_a\top$. Indeed, consider the history $p.a.[a]\bot.a$. We can easily prove the following by induction on φ:

> If β is a history having $p.a.[a]\bot.a$ as a prefix, then for all φ, $w_1, \beta \models \varphi$ iff $w_2, \beta \models \varphi$.

In particular, any β having $p.a.[a]\bot.a$ as a prefix will be executable at w_1 iff it is executable at w_2. Now, take any sequence γ such that $p.a.[a]\bot.a.\gamma$ is executable at w_1. There exists a β such that $p.a.[a]\bot.a\gamma \triangleright_a \beta$ and β is

---

[2] We thank Louwe Kuijer for this counterexample

executable at w_1. Note that β is necessarily of the form $\beta = p.a.[a]\bot.a.\gamma'$ for some γ'. But this means, by the previous remark, that β is executable at w_2, and thus $w_1, p.a.[a]\bot.a.\gamma \models \hat{B}_a\top$, which means $w_1, \epsilon \models \langle p.a.[a]\bot.a\rangle[!]\hat{B}_a\top$ and thus $w_1, \epsilon \models \langle ! \rangle [!] \hat{B}_a\top$.

However, $w_1, \epsilon \not\models [!]\langle ! \rangle \hat{B}_a\top$: indeed, consider the sequence $B_a p.a$. It is never the case that $w_1, B_a p.a \models \langle \beta \rangle \hat{B}_a\top$ for any announcement, given that, whenever $B_a p.a.\beta \rhd_a \gamma$, γ has $B_a p$ as its first formula, and therefore γ cannot be compatible with w_2, since $w_2, \epsilon \not\models B_a p$.

Also in APAL (CR) is not valid in general (this can be seen via a similar counterexample), but, as said, only with equivalence relations. Whether CR is valid on AAA on the class of $\mathcal{S}5$ models is an open question.

4 Expressivity of AAA

We assume the usual terminology to compare the expressivity of logics or logical languages with respect to a semantics and a class of models. Given two languages $\mathcal{L}_1$ and $\mathcal{L}_2$ interpreted over the same class $\mathcal{C}$ of models, we say that $\mathcal{L}_1$ is at least as expressive as $\mathcal{L}_2$ with respect to $\mathcal{C}$ iff for all formulas $\varphi_2 \in \mathcal{L}_2$, there exists a formula $\varphi_1 \in \mathcal{L}_1$ such that for all models $\mathfrak{M} \in \mathcal{C}$, $\mathfrak{M} \models \varphi_1$ iff $\mathfrak{M} \models \varphi_2$.

If $\mathcal{L}_1$ is at least as expressive as $\mathcal{L}_2$ and $\mathcal{L}_2$ is at least as expressive as $\mathcal{L}_1$ then $\mathcal{L}_1$ and $\mathcal{L}_2$ are *as expressive*. If $\mathcal{L}_1$ is at least as expressive as $\mathcal{L}_2$ and $\mathcal{L}_2$ is not at least as expressive as $\mathcal{L}_1$ then $\mathcal{L}_1$ is *more expressive than* $\mathcal{L}_2$. In this section we show that the language of AAA is more expressive than that of AA, by showing that there is a formula $\varphi \in \mathcal{L}$ to which no formula $\varphi' \in \mathcal{L}_{-!}$ is equivalent. This is shown somewhat similarly to proving that APAL is more expressive than PAL [2, Proposition 3.13].[3]

Proposition 4.1 *$\mathcal{L}$ is more expressive than $\mathcal{L}_{-!}$ for multiple agents, for the class $\mathcal{S}5$ of models wherein each R_a is an equivalence relation.*

Proof. Suppose that AAA is as expressive as AA in $\mathcal{S}5$ for multiple agents. Consider the formula $\langle ! \rangle (B_a p \wedge B_a \neg B_b p)$. Then there must be a formula $\varphi \in \mathcal{L}_{-!}$ that is equivalent to $\langle ! \rangle (B_a p \wedge B_a \neg B_b p)$. Some propositional variable q will not occur in φ. Now consider $\mathcal{S}5$ models M and M' as below (indistinguishable states are linked, and we assume transitivity of access). Of course, the states in M also need a value for atom q, but this is irrelevant for the proof and therefore not depicted (for example, we can assume q to be false in both).

[3] However, with differences that may be considered of interest. In the APAL proof the property used to demonstrate larger expressivity is $\langle ! \rangle (B_a p \wedge \neg B_b B_a p)$. This property uses that in APAL an announcement results in a growth of common knowledge, it uses the synchronous character of PAL announcements. We use another property, $\langle ! \rangle (B_a p \wedge B_a \neg B_b p)$, and on a slightly different model.

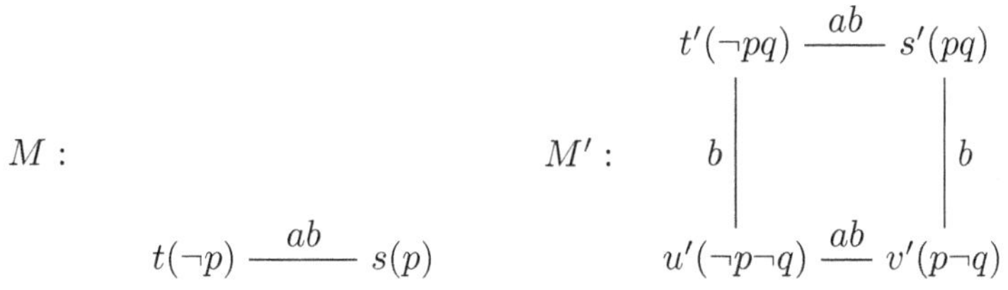

We note that (M, s) is bisimilar to (M', s') if we restrict the clause (i) (for corresponding valuations) to the variable p only. If we now consider formulas $\varphi \in \mathcal{L}_{-!}$ and histories $\alpha \in (\mathcal{L}_{-!} \cup A)^*$ that do not contain the variable q, it can be easily shown by induction on (α, φ) that $s \bowtie \alpha$ iff $s' \bowtie \alpha$ and $s, \alpha \models \varphi$ in M if and only if $s', \alpha \models \varphi$ in M'. As a consequence, for any $\varphi \in \mathcal{L}_{-!}$, we have that $s, \epsilon \models \varphi$ iff $s', \epsilon \models \varphi$.

However, this is not the case in the full language $\mathcal{L}$. Indeed, we have that $s, \epsilon \not\models \langle ! \rangle (B_a p \wedge B_a \neg B_b p)$ in M, whereas $s', \epsilon \models \langle ! \rangle (B_a p \wedge B_a \neg B_b p)$ in M'. The former is because in M, for any history α executable in s, for any announcement in α received by a, a considers it possible that b also received this announcement and thus believes p. The latter is because in M' it holds that $s', (p \vee \neg q).a.b \models B_a p \wedge B_a \neg B_b p$.

This is a contradiction. □

It seems likely, although we did not prove this, that on class $\mathcal{S}5$ for a single agent the $\langle ! \rangle$ modality is definable in AA, such that AAA is then as expressive as AA. However, without any frame properties single-agent AAA is more expressive than AA, again shown similarly to the previous proposition and [2, Prop. 3.14]

Proposition 4.2 *$\mathcal{L}$ is more expressive than $\mathcal{L}_{-!}$ for a single agent.*

Proof. See Appendix. □

A logic is called *compact* if a set of formulas is satisfiable whenever any finite subset is satisfiable.

Proposition 4.3 *The logic* AAA *is not compact.*

Proof. Using the expressivity results, this can be shown by considering the set of formulas

$$\{\langle ! \rangle (B_a p \wedge B_a \neg B_b p)\} \cup \{\neg \langle \beta \rangle (B_a p \wedge B_a \neg B_b p) : \beta \in (\mathcal{L}_{-!} \cup A)^*\}.$$

This set is not satisfiable, but any finite subset is satisfiable, where we use that some variable q must necessarily not occur in such a subset. We then consider M, M' as above. The q-less finite subset will be satisfied at s'. □

5 Positive formulas

In modal logic, the fragment of the language where negations do not bind epistemic modalities is known as the *positive* fragment [8,12,2]. It corresponds to the universal fragment in first-order logic. It has the property that it preserves truth under submodels. In AAA, preservation under submodels is formalised by preservation after history extension. A formula $\varphi \in \mathcal{L}$ is *preserved* iff $\models^* \varphi \to [!]\varphi$. We wish therefore to identify a fragment of the language $\mathcal{L}$ that guarantees preservation.

For AA, it is shown in [4, Prop. 44] that the fragment $\varphi ::= p \mid \neg p \mid \bot \mid (\varphi \wedge \varphi) \mid (\varphi \vee \varphi) \mid B_a\varphi$, that corresponds in a very direct way to the universal fragment, is preserved.

For AAA we wish to expand that frontier, in the direction earlier taken in [12] for synchronous announcements, where a further inductive clause $[\neg\varphi]\varphi$ is added[4], which is further expanded in [2] with an inductive clause $[!]\varphi$ (where $[!]$ is the APAL quantifier over announcements). We will only define a fairly minimal extension and subsequently present some of the difficulties in obtaining a result analogous to those in [12,2], and what the desirable final goal seems to be.

The proof uses a lemma that we therefore present first. Let preorder $\preceq$ on histories be defined as follows: $\alpha \preceq \beta$ iff $\alpha\restriction_! \subseteq \beta\restriction_!$, for all $a \in A$ $|\alpha|_a \leq |\beta|_a$, and for any state s in any model $s \bowtie \beta$ implies $s \bowtie \alpha$. Note that $\alpha \subseteq \beta$ implies $\alpha \preceq \beta$, but not vice versa.

Lemma 5.1 ([4, Lemma 42]) *Let histories α, β and $a \in A$ be given. Suppose $\alpha \preceq \beta$ and $\beta \rhd_a \delta$. Then there is a history γ such that $\gamma \preceq \delta$ and $\alpha \rhd_a \gamma$.*

Consider the following *positive formulas* $\mathcal{L}_+$:

$$\varphi ::= p \mid \neg p \mid \bot \mid (\varphi \wedge \varphi) \mid (\varphi \vee \varphi) \mid B_a\varphi \mid [!]\varphi.$$

We show that positive formulas are preserved.

Proposition 5.2 (Positive implies preserved)
Let $\varphi \in \mathcal{L}_+$. Then $\models_* \varphi \to [!]\varphi$.

Proof. We need to prove the following proposition:

> Let $\varphi \in \mathcal{L}_+$. For all models $M = (W, R, V)$ and $s \in W$, and for all histories α: $s, \epsilon \models [\alpha](\varphi \to [!]\varphi)$.

This is equivalent to

> Let $\varphi \in \mathcal{L}_+$. For all models $M = (W, R, V)$ and $s \in W$, and for all histories α, β such that $\alpha \subseteq \beta$: $s, \epsilon \models [\alpha]\varphi$ implies $s, \epsilon \models [\beta]\varphi$.

A standard inductive proof on the structure of φ fails because in the case $B_a\varphi$ we would need that if $\alpha \subseteq \beta$ and $\beta \rhd_a \delta$, then there is a γ with $\gamma \subseteq \delta$ and $\alpha \rhd_a \gamma$. Such a γ may not exist, namely if many yet unread announcements in δ precede the a in δ that corresponds to the last a in α. However, we can then still find a γ such that $\gamma \preceq \delta$. Therefore, it suffices to show:

---

[4] The possibly strange form of this clause wherein a negation appears has to do with the semantics of public announcement. In PAL, $M, w \models [\neg\varphi]\psi$ iff ($M, w \models \neg\varphi$ implies $M', w \models \psi$) iff ($M, w \models \varphi$ or $M', w \models \psi$), where M' is the model restriction to the states where φ is false. In the disjunctive description, the negation has disappeared.

Lemma 5.3 *Let $\varphi \in \mathcal{L}_+$. For all models $M = (W, R, V)$ and $s \in W$, and for all histories α, β such that $\alpha \preceq \beta$: $s, \epsilon \models [\alpha]\varphi$ implies $s, \epsilon \models [\beta]\varphi$.*

A proof of this Lemma can be found in the Appendix. □

There is no obvious way to expand this fragment of positive formulas with inductive clauses for announcement and reception modalities. The obvious analogue of the $[\neg\varphi]\psi$ from [12] would be $[\neg\varphi.A]\psi$ (and where A represents an arbitrary permutation of all agents in A). But this does not work. For example, consider a model M for one agent a and two variables p, q consisting of four worlds for the four valuations of p and q, and such that these are all indistinguishable for a. Let w be the world where p and q are true. We now have that: $w, \epsilon \models [q.a]B_a q$ whereas $w, p \not\models [q.a]B_a q$, because the a in history $q.a$ reads announcement q in the first case, whereas it reads announcement p in the second case. As long as agent a has not received announcement q, she remains uncertain about the value of q. Differently said, even when $[q.a]B_a q$ is true, $[p.q.a]B_a q$ may be false, so for $\varphi = [q.a]B_a q$ and the quantifier $[!]$ witnessed by $[p]$, $\varphi \rightarrow [!]\varphi$ is false. Further complications might occur when announcements are not immediately received, as in $[p.q.a.a]B_a(p \wedge q)$.

We would like to be able to say that a formula like $[q.a]B_a q$ is positive in the sense that, given that the negation of q is positive (as well as, in this example, q itself), then after q is announced, whenever it is *eventually* received, positive $B_a q$ always remains true afterwards. But we do not have such an eventuality (Kleene-*) modality $\langle a \rangle^*$ in our logical language (yet)! We currently believe that a good candidate for the positive formulas in such a further expanded language is the fragment

$$\varphi ::= p \mid \neg p \mid \bot \mid (\varphi \wedge \varphi) \mid (\varphi \vee \varphi) \mid B_a\varphi \mid [\neg\varphi]\varphi \mid \langle a \rangle^* \varphi \mid [!]\varphi$$

and that such an expanded positive fragment might even syntactically characterize the preserved formulas (with respect to $*$-validity), analogous to van Benthem's result for the (usual) positive fragment [8].

Before moving on, let us point out another property of the positive fragment: when the believed formula φ is positive, and the accessibility relation reflexive, belief becomes factive.

Proposition 5.4 *Let $\varphi \in \mathcal{L}_+$. For any model (W, R, V) such that R_a is reflexive, for all $s \in W$ and α such that $s \bowtie \alpha$, we have $s, \alpha \models B_a\varphi \rightarrow \varphi$. As a consequence, $\mathcal{S}5 \models^* B_a\varphi \rightarrow \varphi$.*

Proof. Suppose $s, \alpha \models B_a\varphi$. Consider $\beta = \delta.\varphi.a^k$, as constructed in the proof of Prop. 3.3 ($\delta.\varphi$ is the prefix of α up until its $|\alpha|_a$-th formula). We have $R_a ss$, $\alpha \rhd_a \beta$, and $s \bowtie \beta$, and thus $s, \beta \models \varphi$. Moreover, since $\delta.\varphi \subseteq \alpha$ and $|\beta|_a = |\alpha|_a$, we have $\beta \preceq \alpha$. By Lemma 5.3, this entails $s, \alpha \models \varphi$. □

6 Axiomatisation of AAA

The axiomatisation of AAA and its completeness proof is based on the axiomatisation of AA [4] and on that of APAL [2] and its completeness uses the method pioneered in [3].

Recall that a formula $\varphi \in \mathcal{L}$ is called *ϵ-valid* if, for every model (W, R, V) and every $w \in W$, it is the case that $w, \epsilon \models \varphi$, and φ is *$*$-valid* if, for every model (W, R, V) and $w \in W$, and for every history α such that $w \bowtie \alpha$, it is the case that $w, \alpha \models \varphi$. In the the present section we provide a complete axiomatization of the logic of ϵ-validities.

Given a symbol # we define a set AF of *admissible forms* as follows:

$$L ::= \# | B_a L | \varphi \to L | \langle \alpha \rangle L,$$

where $\varphi \in \mathcal{L}$, $a \in A$, $\alpha \in \mathcal{H}$. Given $L \in AF$ and $\varphi \in \mathcal{L}$, the formula $L(\varphi)$ is the result of substituting the unique occurrence of # in L by φ.

The following holds:

Lemma 6.1 *Let L be an admissible form. For all $M \in AF$ and for all modal formulas φ, ψ, if $L([!]\varphi) = M([!]\psi)$ then $L = M$ and $\varphi = \psi$.*

Proof. By induction on L. □

The logic AAA consists of the following axioms and rules, for $\alpha \in \mathcal{H}$, $p \in P$, $a \in A$, $L(\#) \in AF$:

(Taut)	All propositional tautologies;
(MP)	If $\vdash \varphi$ and $\vdash \varphi \to \psi$, then $\vdash \psi$;
(Nec_B)	If $\vdash \varphi$, then $\vdash B_a\varphi$;
(K_B)	$B_a(\varphi \to \psi) \to (B_a\varphi \to B_a\psi)$;
($R_{\top 1}$)	$\langle \alpha.a \rangle \top \leftrightarrow \langle \alpha \rangle \top$ if $\lvert\alpha\rvert_a < \lvert\alpha\rvert_!$
($R_{\top 2}$)	$\langle \alpha.a \rangle \top \leftrightarrow \bot$ otherwise;
($R_{\top 3}$)	$\langle \alpha.\varphi \rangle \top \leftrightarrow \langle \alpha \rangle \varphi$;
(R_p)	$\langle \alpha \rangle p \leftrightarrow (\langle \alpha \rangle \top \wedge p)$;
($R_\neg$)	$\langle \alpha \rangle \neg\varphi \leftrightarrow (\langle \alpha \rangle \top \wedge \neg\langle \alpha \rangle \varphi)$;
($R_\wedge$)	$\langle \alpha \rangle (\varphi \wedge \psi) \leftrightarrow (\langle \alpha \rangle \varphi \wedge \langle \alpha \rangle \psi)$;
(R_B)	$\langle \alpha \rangle \hat{B}_a\varphi \leftrightarrow (\langle \alpha \rangle \top \wedge \bigvee_{\alpha \rhd_a \beta} \hat{B}_a \langle \beta \rangle \varphi)$;
([!]-elim)	$L([!]\varphi) \to L([\beta]\varphi)$ (where $\beta \in (\mathcal{L}_{-!} \cup A)^*$);
([!]-int$^\omega$)	If $\vdash L([\beta]\varphi)$ for all $\beta \in (\mathcal{L}_{-!} \cup A)^*$, then $\vdash L([!]\varphi)$

Remark 6.2 *If we remove the last two lines of the above table we obtain the logic* AA*, defined in [4] for the language $\mathcal{L}_{-!}$.*

From now on, AAA will denote both the above axiomatic system and set of all formulas it can derive.

Completeness proof. A *theory* is a set of formulas T such that:

i. AAA $\subseteq T$;

ii. T is closed under (MP): if $\varphi, \varphi \to \psi \in T$, then $\psi \in T$;

iii. T is closed under the rule ([!]$-$int$^\omega$):
If $L([\beta]\varphi) \in T$ for all $\beta \in (\mathcal{L}_{-!} \cup A)^*$, then $L([!]\varphi) \in T$.

A theory is *consistent* if $\bot \notin T$. Note that AAA is the least consistent theory, and $\mathcal{L}$ is the only inconsistent theory. A consistent theory is *maximal* if no proper superset of T is a consistent theory.

The following holds:

Lemma 6.3 *Given a theory T, a formula ψ, and an agent $a \in A$, the sets $T_{B_a} = \{\varphi : B_a\varphi \in T\}$ and $T_\psi = \{\varphi : \psi \to \varphi \in T\}$ are also theories.*

Moreover, $T \subseteq T_\psi$, $\psi \in T_\psi$ and, if $\neg\psi \notin T$, then T_ψ is consistent.

Proof. See Appendix. □

We also have:

Proposition 6.4 (Lindenbaum's Lemma) *A consistent theory can be extended to a maximal consistent theory.*

Proof. See Appendix. □

Now we define a relation between maximal consistent theories as: TR_aS iff, for all φ, $B_a\varphi \in T$ implies $\varphi \in S$ (equivalently, iff $T_{B_a} \subseteq S$).

Proposition 6.5 (Diamond Lemma) *Suppose $\hat{B}_a\varphi \in T$. Then there exists a maximal consistent theory S such that TR_aS and $\varphi \in S$.*

Proof. Consider the theory $(T_{B_a})_\varphi$. First, note that T_{B_a} is a consistent theory, because $\vdash \hat{B}_a\varphi \to \neg B_a\bot$, so $B_a\bot \notin T$ and thus $\bot \notin T_{B_a}$. Moreover, $B_a\neg\varphi \notin T$, thus $\neg\varphi \notin T_{B_a}$. By Lemma 6.3, we thus have that $T_{B_a} \subseteq (T_{B_a})_\varphi$, $\varphi \in (T_{B_a})_\varphi$ and $(T_{B_a})_\varphi$ is consistent. It then suffices to extend $(T_{B_a})_\varphi$ by Lindenbaum's lemma to the desired successor. □

Now we can define our canonical model: let W be the family of maximal consistent theories, let R_a be defined as above and let $V(p) = \{T \in W : p \in T\}$. We have:

Proposition 6.6 (Truth Lemma) *For any history α and formula φ, we have: $T, \epsilon \models \langle\alpha\rangle\varphi$ iff $\langle\alpha\rangle\varphi \in T$.*

Proof. See Appendix. □

We will say that a formula φ is *consistent* if $\nvdash \neg\varphi$ and that a set of formulas Γ is *consistent* if it can be extended to a consistent theory. Note that φ is consistent if and only if the singleton set $\{\varphi\}$ is consistent (for if $\neg\varphi \notin \mathsf{AAA}$, we can extend $\{\varphi\}$ to the consistent theory AAA_φ).

We have:

Theorem 6.7 AAA *is strongly complete with respect to Kripke models.*

Proof. Let Γ be a consistent set of formulas. Then there exists a consistent theory $T_0 \supseteq \Gamma$ and, by Lindenbaum's lemma, a maximal consistent theory $T \supseteq T_0$. We construct the canonical model as above and we have that $T, \epsilon \models \varphi$ for all $\varphi \in \Gamma$. □

7 Asynchronous Action Models

In this final section we shortly present two logics for asynchronous reception of partially observed actions, including quantification over such actions. The reason to present these logics is that they contrast in, we think, interesting ways with the logic AA and with the logic AAA, the main subject of this paper.

7.1 Asynchronous Action Model Logic

Action model logic was proposed by Baltag, Moss and Solecki in [6]. An *action model* is like a relational model but the elements of the domain are called *actions* instead of states, and instead of a valuation a *precondition* is assigned to each domain element. A public announcement corresponds to a singleton action model where the precondition is the announced formula. Under synchronous conditions, executing an action model into a Kripke model means constructing what is known as the *restricted modal product.* This product encodes the new state of information, after action execution. Under asynchronous conditions we do not construct the product model but calculate the belief consequences of actions from the histories, just as for the particular singleton action model that is the public announcement we do not construct model restrictions in AA but instead use the history.

The nature of an asynchronous non-public action is that it is partially observed by the agents, just as in action model logic, but that it is unclear when the different agents partially observe the action, just as in AA. An example of an asynchronous partially observed action when two agents Anne and Bill, who are both ignorant about p, are informed that Anne will receive the truth about some proposition p but not Bill. Suppose that Anne is going to receive the information that p (is true). By the time Bill learns that Anne will be informed in this way, he considers it possible that Anne has already been informed, in which case she now believes p or believes $\neg p$, but he also considers it possible that she has not yet been informed and thus remains igorant about p. Dually, by the time Anne learns that p but Bill has not yet learnt that Anne will be informed about p, Bill incorrectly believes that Anne is ignorant about p.

Action model Formally, an *action model* $\mathcal{E} = (E, S, \mathsf{pre})$ consists of a *domain* E of *actions* $e, f, \ldots$, an *accessibility function* $S : A \to \mathcal{P}(E^2)$, where each S_a is an accessibility relation, and a *precondition function* $\mathsf{pre} : E \to \mathcal{L}$, where $\mathcal{L}$ is a logical language. A pointed action model is a pair $(\mathcal{E}, e)$ where $e \in E$, for which we write $\mathcal{E}_e$. We abuse the language and also call a pointed action model an *action.*

Syntax Similarly to AA we can conceive a modal logical language with $\langle \mathcal{E}_e \rangle \varphi$ as an inductive language construct, for action models $\mathcal{E}$ with *finite domains.* The set of finite pointed action models is called $\mathcal{AM}$.

Histories are words in $(\mathcal{AM} \cup A)^*$. Much like in AA, we will use $\alpha{\restriction_!}$ to refer to the projection of α to $\mathcal{AM}$ and use $\alpha \restriction_{!a}$, $|\alpha|_!$, $|\alpha|_a$ as usual. For such a word to be a history, we again demand that $|\beta|_a \leq |\beta|_!$ for all $\beta \sqsubseteq \alpha$, $a \in A$.

View relation The definition of the $\triangleright_a$ relation in this setting incorporates the partial observablity of action models: given $\alpha \triangleright_a \beta$, we demand that the *action models* appearing in α and β and seen by a are the same. However, for agent a the *actions* in α (points of these action models) may be different from the actions in β. That is, $\alpha \triangleright_a \beta$ iff $|\alpha|_a = |\beta|_a = |\beta|_!$, and for all $i \leq |\alpha|_a$, if $\mathcal{E}_e$ is the i-th action of α and $\mathcal{F}_f$ is the ith action of β, then $\mathcal{E} = \mathcal{F}$ and $S_a ef$. This relation $\triangleright_a$ is post-reflexive, transitive and post-symmetric if we are dealing with $\mathcal{S}5$ action models (wherein all accessibility relations S_a are equivalence relations).

Semantics Define an executability relation $\bowtie$ between states and histories as:

- $w \bowtie \epsilon$,
- $w \bowtie \alpha.a$ iff $w \bowtie \alpha$,
- $w \bowtie \alpha.\mathcal{E}_e$ iff $w \bowtie \alpha$ and $w, \alpha \models \mathsf{pre}(e)$.

With this, the semantics for belief and action model execution are what one might expect, namely:

$w, \alpha \models \langle \mathcal{E}_e \rangle \varphi$ iff $w, \alpha \models \mathsf{pre}(e)$ and $w, \alpha.\mathcal{E}_e \models \varphi$.

$w, \alpha \models \hat{B}_a \varphi$ iff $t, \beta \models \varphi$ for some (t, β) such that $t \bowtie \beta, R_a wt$, and $\alpha \triangleright_a \beta$.

We call this *Asynchronous Action Model Logic* AAM.

Reduction axioms and axiomatisation We recall that the axiomatisation AAA presented in Section 6 consists of the rules and axioms of AA plus an axiom and a rule dedicated to the quantifier (Remark 6.2).

It is straightforward to see that the axiomatisation of AAM is as the axiomatisation of AA where only axiom $R_{\top 3}$ needs to be (analogously) reformulated for action models, whereas the axiom R_B is the same in AA and in AAM, except that, clearly, the relation $\triangleright_a$ used in that axiom now refers to the much more involved view relation for partial observability defined above, where an agent considers all actions possible that are accessible for her given the actual action. These two relevant axioms are:

$$(R'_{\top 3}) \quad \langle \alpha.\mathcal{E}_e \rangle \top \leftrightarrow \langle \alpha \rangle \mathsf{pre}(e);$$
$$(R'_B) \quad \langle \alpha \rangle \hat{B}_a \varphi \leftrightarrow (\langle \alpha \rangle \top \wedge \bigvee_{\alpha \triangleright_a \beta} \hat{B}_a \langle \beta \rangle \varphi).$$

Just as for AA we can show that this axiomatisation is complete with respect to the class of models with empty histories, and that this is again a reduction system, such that every formula in the logical language is equivalent to a formula without dynamic modalities $\langle \mathcal{E}_e \rangle$ for action execution and $\langle a \rangle$ for receiving that information.

To prove that this system is a complete axiomatisation of AAM, we need to define a total preorder $\ll$ from a complexity measure $|.|$ which takes into consideration the precondition formulas present in action models $\mathcal{E}_e$. It therefore seems that this demands that

$$|(\mathcal{E}, e)| = \sum_{e' \in E} |\mathsf{pre}(e')|$$
$$|\alpha| = \sum \{ |(\mathcal{E}, e)| : (\mathcal{E}, e) \text{ occurs in } \alpha \}.$$

We wish to investigate this later and thus show completeness.

7.2 Arbitrary Asynchronous Action Model Logic

A further generalisation is the extension of the logical language with a quantifier $\langle \otimes \rangle$ over action models, such that $\langle \otimes \rangle \varphi$ means that φ is true after the execution of some sequence of finite action models and readings in the current (s, α) pair of the given model.

Let $\mathcal{AM}_{-\otimes}$ be the class of finite pointed action models where $\langle \otimes \rangle$ does not occur in the preconditions. We then get that

$$w, \alpha \models \langle \otimes \rangle \varphi \text{ iff there exists } \beta \in (\mathcal{AM}_{-\otimes} \cup A)^* \text{ such that } w, \alpha \models \langle \beta \rangle \varphi.$$

Let us call the logic with this quantifier AAAM (an extra A, for Arbitrary). Although work on this logic is also very much work in progress, it is illuminating to compare this extension AAAM of AAM with the logic AAA of this submission, wherein we quantify over histories containing announcements. For the synchronous version of arbitrary action model logic, Hales showed in [14] that the restriction to quantifier-free precondition formulas in action models can be relaxed. Hales also showed that we can *synthesise* a multi-pointed action model $\mathcal{E}_F$ (where $F \subseteq \mathcal{D}(\mathcal{E})$) from φ such that $\langle \otimes \rangle \varphi$ is equivalent to $\langle \mathcal{E}_F \rangle \varphi$.

It it were possible to prove similar results for the logic AAAM of arbitrary asynchronous action models, that would be of great interest, as this would then show that AAAM is as expressive as AAM (without quantification), by reducing every formula to one without quantifiers, unlike the larger expressivity of quantifying over asynchronous announcements in AAA; and it would also show decidability of AAAM. Even independent from that, synthesis of asynchronous partially observable actions, and the complexity of such tasks, seems of interest to investigate further.

8 Conclusion

We presented the logic AAA of arbitrary asynchronous announcements, that can be used to reason about agents receiving and sending each other information under asynchronous conditions. We investigated the properties of the arbitrary announcement quantifier, demonstrated bisimulation invariance, the larger expressivity of the logical language with the quantifier, and we showed preservation after history extension of the fragment of the positive formulas. Then, we provided a complete infinitary axiomatisation. Finally, we tentatively described a further generalisation to quantification over action models.

Acknowledgements. Hans van Ditmarsch is also affiliated to IMSc, Chennai, India, as research associate. We thank the AiML reviewers for their comments helping us to improve the paper.

Appendix

Proof of Prop. 3.1. By $\ll$-induction on (α, φ). Trivial for the cases where $(\alpha, \varphi) = (\epsilon, \top)$ and (ϵ, p). For the case where $(\alpha, \varphi) = (\beta.a, \top)$, we note that $w \bowtie \beta.a$ iff $w \bowtie \beta$ and $w, \beta.a \models \top$ iff $w, \beta \models \top$, and thus we can apply induction hypothesis, for $(\beta, \top) \ll (\beta.a, \top)$. For the case $(\alpha, \varphi) = (\beta.\psi, \top)$, we note that $(\beta, \psi) \ll (\beta.\psi, \top)$.

For the cases $(\alpha, \varphi) = (\alpha, \neg\psi)$ and $(\alpha, \psi) = (\alpha, \psi_1 \wedge \psi_2)$, we note that $(\alpha, \psi) \ll (\alpha \neg\psi)$ and $(\alpha, \psi_i) \ll (\alpha, \psi_1 \wedge \psi_2)$.

For the case $(\alpha, \varphi) = (\alpha, \hat{B}_a\psi)$, we have: $w \bowtie \alpha$ iff $w' \bowtie \alpha$ by induction hypothesis applied to $(\alpha, \top)$. If $w, \alpha \models \hat{B}_a\psi$, then there is some $v \in W$ and some history β such that $R_a wv$, $\alpha \triangleright_a \beta$, $v \bowtie \beta$ and $v, \beta \models \psi$. But then there is some $v' \in W'$ with vZv' and $R_a w'v'$ and thus, by induction hypothesis applied to $(\beta, \psi) \ll (\alpha, \hat{B}_a\psi)$, we have $v' \bowtie \beta$, $v', \beta \models \psi$ and thus $w', \alpha \models \hat{B}_a\psi$. The converse is analogous.

For the cases $(\alpha, \psi) = (\alpha, \langle a \rangle \psi)$ and $(\alpha, \psi) = (\alpha, \langle \theta \rangle \psi)$, we note that $(\alpha.a, \psi) \ll (\alpha, \langle a \rangle \psi)$ and $(\alpha.\theta, \psi) \ll (\alpha, \langle \theta \rangle \psi)$.

For the case $(\alpha, \varphi) = (\alpha, \langle ! \rangle \psi)$, we have: on the one hand, $w \bowtie \alpha$ iff $w' \bowtie \alpha$ by induction hypothesis applied to $(\alpha, \top)$. On the other hand, suppose $w, \alpha \models \langle ! \rangle \psi$. Then $w, \alpha \models \langle \beta \rangle \psi$ for some history β which does not contain any occurrences of $\langle ! \rangle$. Therefore $\deg_! \langle \beta \rangle \psi < \deg_! \langle ! \rangle \psi$, and thus by induction hypothesis $w', \alpha \models \langle \beta \rangle \psi$, which entails $w', \alpha \models \langle ! \rangle \psi$. The converse is analogous.□

Proof of Prop. 3.2. It is obvious that condition i. is satisfied. Now, suppose condition ii. fails. That is, for some $v \in W$, we have $R_a wv$ but for all (the finitely many) v' such that $R_a w'v'$ it is not the case that vZv'. Let $R'_a[w'] = \{v'_1, ..., v'_n\}$. For each v'_i there exists some pair (α_i, φ_i) such that either $v, \epsilon \models \langle \alpha_i \rangle \varphi_i$ but $v'_i, \epsilon \not\models \langle \alpha_i \rangle \varphi_i$, or $v, \epsilon \not\models \langle \alpha_i \rangle \varphi_i$ but $v'_i, \epsilon \models \langle \alpha_i \rangle \varphi_i$. Let $\theta_i = \langle \alpha_i \rangle \psi_i$ in the former case and $\theta_i = \neg \langle \alpha_i \rangle \psi_i$ in the latter, and call $\psi = \bigwedge_{i=1}^{n} \theta_i$. Note that $v, \epsilon \models \psi$ and thus $w, \epsilon \models \hat{B}_a\psi$. But then by the definition of Z we have that $w', \epsilon \models \hat{B}_a\psi$, and thus w' has a successor satisfying each formula θ_i: contradiction. Condition iii. is proven similarly. □

Proof of Prop. 3.3. Let $\mathbf{R}_a$ be a relation defined on the set of pairs (s, α) with $s \bowtie \alpha$ as follows:

$$(s, \alpha)\mathbf{R}_a(t, \beta) \quad \text{iff} \quad sR_a t, \alpha \rhd_a \beta, \text{ and } t \bowtie \beta.$$

Note that $s, \alpha \models B_a\varphi$ iff $t, \beta \models \varphi$ for all (t, β) such that $(s, \alpha)\mathbf{R}_a(t, \beta)$. The proof of this result, then consists in showing that $\mathbf{R}_a$ is serial, transitive and Euclidean.

Seriality. Let us see that, for all α, there exists a history β such that $\alpha \rhd_a \beta$ and $s \bowtie \alpha$ implies $s \bowtie \beta$. Let $n := |\alpha|_a \leq |\alpha|_!$ and let φ be the n-th occurrence of a formula in α, so that $\alpha = \delta.\varphi.\gamma$ for some δ, γ. Let $\beta = \delta.\varphi.a^k$, where k is a natural number such that $|\delta.\varphi|_a + k = n$. Then clearly $\alpha \rhd_a \beta$, for β contains n times a and exactly the first n formulas of α, and, if $s \bowtie \alpha$, we have that $s \bowtie \delta.\varphi$, because $\delta.\varphi \subseteq \alpha$, and thus $s \bowtie \delta.\varphi.a^k$. Since R_a is reflexive, this gives that, for any s such that $s \bowtie \alpha$, $(s, \alpha)\mathbf{R}_a(s, \beta)$.

Transitivity. Since R_a and $\rhd_a$ are both transitive, then, clearly, so is $\mathbf{R}_a$.

Euclidicity. Again, since R_a and $\rhd_a$ are Euclidean, so is $\mathbf{R}_a$.

Proof of Prop. 3.4. Let model $\mathfrak{M} = (W, R, V)$ and $s \in W$ be given, and let $\alpha \in (\mathcal{L}_{-!} \cup A)^*$ be such that $s, \epsilon \models \langle \alpha \rangle \hat{B}_a\varphi$. Therefore α is a history, $s \bowtie \alpha$ and $s, \alpha \models \hat{B}_a\varphi$, so that there are t, β such that $R_a st$, $\alpha \rhd_a \beta$, $t \bowtie \beta$, and $t, \beta \models \varphi$. As $t \bowtie \beta$ and $t, \beta \models \varphi$, it follows that $t, \epsilon \models \langle \beta \rangle \varphi$. It therefore follows that $t, \epsilon \models \langle ! \rangle \varphi$. Finally, as $R_a st$, $\epsilon \rhd_a \epsilon$ and $t \bowtie \epsilon$ we conclude $s, \epsilon \models \hat{B}_a \langle ! \rangle \varphi$.

On the other hand, $\hat{B}_a \langle ! \rangle \varphi \to \langle ! \rangle \hat{B}_a \varphi$ is not always ϵ-valid. Consider $\varphi = B_a \neg p$ and the model $M = (W, R, V)$ for a single agent a and atom p and where $W = \{s, t\}$, $R_a = W^2$, and $V(p) = \{s\}$. We then have that $s, \epsilon \models \hat{B}_a \langle ! \rangle B_a \neg p$, because $s, \epsilon \models \hat{B}_a \langle \neg p.a \rangle B_a \neg p$ (because $t, \epsilon \models \langle \neg p.a \rangle B_a \neg p$), whereas clearly

$s, \epsilon \not\models \langle ! \rangle \hat{B}_a B_a \neg p.$ □

Proof of Prop. 4.2. For a single agent we consider the formula $\langle ! \rangle (B_a p \wedge B_a \neg B_a p)$ and proceed as in Prop. 4.1, where in this case we observe that in model N' it holds that $s', (p \vee \neg q).a \models B_a p \wedge B_a \neg B_a p$.

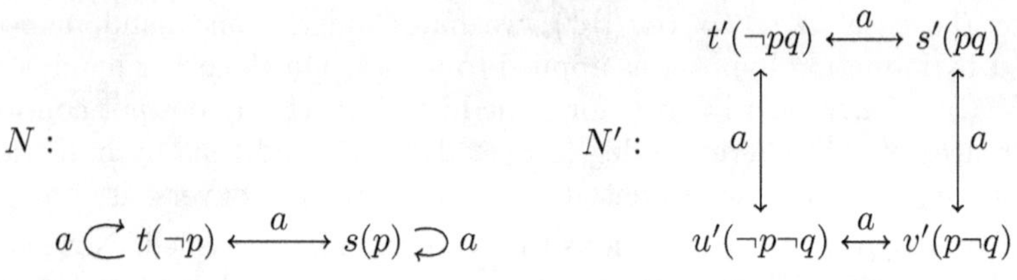

Proof of Lemma 5.3. If $s, \epsilon \models [\alpha]\varphi$, then it is either the case that $s \not\bowtie \alpha$ or $s, \epsilon \models \langle \alpha \rangle \varphi$. In the former case, since $\alpha \preceq \beta$, we get $s \not\bowtie \beta$ and thus trivially $s, \epsilon \models [\beta]\varphi$. For the latter, let us see by induction on the structure of simple positve φ that $s, \epsilon \models \langle \alpha \rangle \varphi$ implies $s, \epsilon \models [\beta]\varphi$.

Case $\bot$. It is never the case that $s, \epsilon \models \langle \alpha \rangle \bot$.

Case atoms. If $s, \epsilon \models \langle \alpha \rangle p$, then $s \in V(p)$, which in turn implies $s, \epsilon \models [\beta]p$. The case for $\varphi = \neg p$ is analogous.

Case conjunction. If $s, \epsilon \models \langle \alpha \rangle (\varphi_1 \wedge \varphi_2)$, then $s, \epsilon \models \langle \alpha \rangle \varphi_i$ for $i = 1, 2$ and thus, by induction hypothesis, $s, \epsilon \models [\beta]\varphi_i$, whence $s, \epsilon \models [\beta](\varphi_1 \wedge \varphi_2)$.

Case disjunction. Analogous.

Case belief. Suppose $s, \epsilon \not\models [\beta] B_a \varphi$. Then $s, \epsilon \models \langle \beta \rangle \hat{B}_a \neg \varphi$, which means there exist t, δ with $R_a st$, $\beta \rhd_a \delta$, $t \bowtie \delta$ and $t, \delta \not\models \varphi$. By Lemma 5.1, there is a γ with $\alpha \rhd_a \gamma$ and $\gamma \preceq \delta$, which gives, by induction hypothesis, $t, \gamma \not\models \varphi$ and thus $s, \alpha \not\models B_a \varphi$.

Case $[!]\varphi$. Suppose $s, \epsilon \not\models [\beta][!]\varphi$. This means that $s, \epsilon \models \langle \beta \rangle \langle ! \rangle \neg \varphi$, i.e. there exists a word $\delta \in (\mathcal{L}_{-!} \cup A)^*$ such that $s \bowtie \beta.\delta$ and $s, \beta.\delta \not\models \varphi$. Since $\alpha \preceq \beta.\delta$, this gives that $s \bowtie \alpha$ and $s, \alpha \not\models \varphi$, and thus $s, \epsilon \not\models \langle \alpha \rangle [!]\varphi$. □

Proof of Lemma 6.3. Checking the first item is easy: if $\varphi \in \mathsf{AAA}$, then $B_a \varphi \in \mathsf{AAA}$ (by necessitation) and $\psi \to \varphi \in \mathsf{AAA}$ (by classical propositional logic). Therefore $B_a \varphi \in T$ and $\psi \to \varphi \in T$, and thus $\varphi \in T_{B_a} \cap T_\psi$.

T_{B_a} is closed under modus ponens because if $\varphi \to \theta \in T_{B_a}$ and $\varphi \in T_{B_a}$, then $B_a(\varphi \to \theta), B_a \varphi \in T$, which by the K axiom plus modus ponens gives $B_a \theta \in T$ and thus $\theta \in T_{B_a}$. For T_ψ, suppose $\varphi \to \theta, \varphi \in T_\psi$. Then $\psi \to (\varphi \to \theta) \in T$ and $\psi \to \varphi \in T$. But note that the former is logically equivalent to $(\psi \to \varphi) \to (\psi \to \theta)$, and, since T is closed under logical equivalence, this means by modus ponens that $\psi \to \theta \in T$ and thus $\theta \in T_\psi$.

For the third condition, suppose $L([\beta]\varphi) \in T_{B_a}$ for all β. Then $B_a L([\beta]\varphi) \in T$ for all β and, since $B_a L(\#)$ is an admissible form, then $B_a L([!]\psi) \in T$, and thus $L([!]\varphi) \in T_{B_a}$. If $L([\beta]\varphi) \in T_\psi$ for all β, then $\psi \to L([\beta]\varphi) \in T$ for all β and, again, since $\psi \to L(\#)$ is an admissible form, this entails $\psi \to L([!]\varphi) \in T$ and therefore $L([!]\varphi) \in T_\psi$.

With respect to the last statement: $\psi \in T_\psi$ because $\psi \to \psi$ is a tautology; if $\neg\psi \notin T$, then $\psi \to \bot \notin T$ thus $\bot \notin T_\psi$, and if $\varphi \in T$, then (since $\varphi \to (\psi \to \varphi)$

is a tautology) $\psi \to \varphi \in T$ and thus $\varphi \in T_\psi$. □

Proof of Prop. 6.4. Let T_0 be a consistent theory. Let $\{\varphi_0, \varphi_1, \varphi_2, ...\}$ be an enumeration of the formulas in $\mathcal{L}$ where each formula appears infinitely many times. For $k \in \omega$ we will construct a consistent theory T_{k+1}, which is a superset of T_k, as follows:

i. If $\neg\varphi_k \notin T_k$, then $T_{k+1} = (T_k)_{\varphi_k}$;

ii. If $\neg\varphi_k \in T_k$ and φ_k is of the form $L([!]\psi)$, then there must exist some $\beta \in (\mathcal{L}_{-!} \cup A)^*$ such that $L([\beta]\psi) \notin T_k$ (for otherwise, by rule iii., we would have that $\varphi_k \in T_k$, in contradiction with the consistency of T_k). We set $T_{k+1} = (T_k)_{\neg L([\beta]\psi)}$.

iii. If $\neg\varphi_k \in T_k$ and φ_k is *not* of the form $L([!]\psi)$, then $T_{k+1} = T_k$.

Each T_k is consistent due to the last statement in the previous Lemma. Then $T = \bigcup_{k\in\omega} T_k$ is consistent. T is trivially closed under modus ponens. For any formula φ_k, either $\neg\varphi_k$ was already in the k-th step of the construction, or φ_k was added to T_{k+1}; therefore T cannot have proper consistent supersets closed under modus ponens. Finally suppose $L([\beta]\psi) \in T$ for all β. If $L([!]\psi) \notin T$, then $\neg L([!]\psi) \in T$ and thus $\neg L([!]\psi) \in T_k$ for some k. Let $m > k$ such that $\varphi_m = L([!]\psi)$. By construction there exists a β such that $\neg L([\beta]\psi) \in T_{m+1} \subseteq T$: contradiction. Therefore T is a maximal consistent theory. □

Proof of Prop. 6.6. By induction on (α, φ).

The case $(\alpha, \varphi) = (\epsilon, \top)$ is trivial. The cases $(\alpha, \varphi) = (\alpha'.\psi, \top)$ and $(\alpha'.a, \top)$ follow from the axioms $R_{\top 1}$, $R_{\top_2}$ and $R_{\top 3}$ and the fact that $(\alpha', \psi) \ll (\alpha'.\psi, \top)$ and $(\alpha', \top) \ll (\alpha'.a, \top)$.

The case (α, p) follows from the definition of $V(p)$ and axiom R_p combined with the fact that $(\alpha, \top) \ll (\alpha, p)$.

The cases $(\alpha, \neg\psi)$ and $(\alpha, \psi_1 \wedge \psi_2)$ follow from $R_\neg$ and $R_\wedge$, respectively, plus the fact that $(\alpha, \psi_i) \ll (\alpha, \psi_1 \wedge \psi_2)$ (for the first case), and $(\alpha, \psi) \ll (\alpha, \neg\psi)$ (for the second case).

Let us see the case $(\alpha, \hat{B}_a\varphi)$: if $T, \epsilon \models \langle\alpha\rangle\hat{B}_a\varphi$, then on the one hand we have that $T \bowtie \alpha$ (i.e., $T, \epsilon \models \langle\alpha\rangle\top$, which by induction hypothesis paired with the fact that $(\alpha, \top) \ll (\alpha, \hat{B}_a\psi)$ gives us that $\langle\alpha\rangle\top \in T$), and on the other hand, $S, \beta \models \varphi$ by some S, β such that R_aTS, $\alpha \rhd_a \beta$ and $S \bowtie \beta$. This means that $S, \epsilon \models \langle\beta\rangle\psi$ and thus (by induction hypothesis due to the fact that $(\beta, \psi) \ll (\alpha, \hat{B}_a\psi)$, we have that $\langle\beta\rangle\psi \in S$. This entails that $\hat{B}_a\langle\beta\rangle\psi \in T$ and thus $\langle\alpha\rangle\top \wedge \bigvee_{\alpha \rhd_a \beta} \hat{B}_a\langle\beta\rangle\psi \in T$, which by R_B gives $\langle\alpha\rangle\hat{B}_a\psi \in T$. For the converse, we use R_B and the Diamond Lemma.

The cases $(\alpha, \langle a\rangle\psi)$ and $(\alpha, \langle\theta\rangle\psi)$ follow directly from the fact that $(\alpha.x, \psi) \ll (\alpha, \langle x\rangle\psi)$ for $x \in \mathcal{L} \cup A$.

Let us see the case $(\alpha, [!]\psi)$. If $T, \epsilon \models \langle\alpha\rangle[!]\psi$, then $T \bowtie \alpha$ and $T, \alpha \models [!]\psi$, which means that, for all $\beta \in (\mathcal{L}_{-!} \cup A)^*$, $T, \epsilon \models \langle\alpha\rangle[\beta]\psi$. By induction hypothesis, noting that $(\alpha, [\beta]\psi) \ll (\alpha, [!]\psi)$ whenever β does not contain occurrences of $[!]$, we have that $\langle\alpha\rangle[\beta]\psi \in T$ for all β and thus $\langle\alpha\rangle[!]\psi \in T$. Conversely, if $\langle\alpha\rangle[!]\psi \in T$, then $\langle\alpha\rangle\top \in T$ (and thus, by IH, $T, \epsilon \models \langle\alpha\rangle\top$, which means

$T \bowtie \alpha$), and, for all $\beta \in (\mathcal{L}_{-!} \cup A)^*$, $\langle\alpha\rangle[\beta]\psi \in T$, which again by induction hypothesis gives $T, \epsilon \models \langle\alpha\rangle[\beta]\psi$ for all β and thus $T, \alpha \models [!]\psi$, whence $T, \epsilon \models \langle\alpha\rangle[!]\psi$. □

References

[1] Balbiani, P., *Putting right the wording and the proof of the Truth Lemma for APAL*, Journal of Applied Non-Classical Logics **25** (2015), pp. 2–19.

[2] Balbiani, P., A. Baltag, H. van Ditmarsch, A. Herzig, T. Hoshi and T. De Lima, *'Knowable' as 'known after an announcement'*, Review of Symbolic Logic **1(3)** (2008), pp. 305–334.

[3] Balbiani, P. and H. van Ditmarsch, *A simple proof of the completeness of APAL*, Studies in Logic **8(1)** (2015), pp. 65–78.

[4] Balbiani, P., H. van Ditmarsch and S. Fernández González, *Asynchronous announcements*, CoRR **abs/1705.03392** (2019).

[5] Balbiani, P., H. van Ditmarsch and S. Fernández González, *From public announcements to asynchronous announcements*, in: *Proc. of 24th ECAI* (2020), to appear.

[6] Baltag, A., L. Moss and S. Solecki, *The logic of public announcements, common knowledge, and private suspicions*, in: *Proc. of 7th TARK* (1998), pp. 43–56.

[7] Ben-Zvi, I. and Y. Moses, *Beyond Lamport's* Happened-before*: On time bounds and the ordering of events in distributed systems*, Journal of the ACM **61** (2014), pp. 13:1–13:26.

[8] van Benthem, J., *One is a lonely number: on the logic of communication*, in: *Logic Colloquium* (2006), pp. 96–129.

[9] Chandy, K. and J. Misra, *How processes learn*, in: *Proc. of the 4th PODC* (1985).

[10] Degremont, C., B. Löwe and A. Witzel, *The synchronicity of dynamic epistemic logic*, in: *Proc. of 13th TARK* (2011), pp. 145–152.

[11] van Ditmarsch, H., W. van der Hoek and B. Kooi, "Dynamic Epistemic Logic", Springer, 2008.

[12] van Ditmarsch, H. and B. Kooi, *The secret of my success*, Synthese **151** (2006), pp. 201–232.

[13] Genest, B., D. Peled and S. Schewe, *Knowledge = Observation + Memory + Computation*, in: *Proc. of 18th FoSSaCS* (2015), pp. 215–229.

[14] Hales, J., *Arbitrary action model logic and action model synthesis*, in: *Proc. of 28th LICS* (2013), pp. 253–262.

[15] Hales, J., "Quantifying over epistemic updates," Ph.D. thesis, School of Computer Science & Software Engineering, University of Western Australia (2016).

[16] Halpern, J. and Y. Moses, *Knowledge and common knowledge in a distributed environment*, Journal of the ACM **37(3)** (1990), pp. 549–587.

[17] Knight, S., B. Maubert and F. Schwarzentruber, *Reasoning about knowledge and messages in asynchronous multi-agent systems*, Mathematical Structures in Computer Science **29** (2019), pp. 127–168.

[18] Kshemkalyani, A. and M. Singhal, "Distributed Computing: Principles, Algorithms, and Systems," Cambridge University Press, 2008.

[19] Lamport, L., *Time, clocks, and the ordering of events in a distributed system*, Communications of the ACM **21** (1978), pp. 558–565.

[20] Mukund, M. and M. Sohoni, *Keeping track of the latest gossip in a distributed system*, Distributed Computing **10** (1997), pp. 137–148.

[21] Panangaden, P. and K. Taylor, *Concurrent common knowledge: Defining agreement for asynchronous systems*, Distributed Computing **6** (1992), pp. 73–93.

[22] Plaza, J., *Logics of public communications*, in: *Proc. of the 4th ISMIS* (1989), pp. 201–216.

[23] Ramanujam, R., *Local knowledge assertions in a changing world*, in: *Proc. of 6th TARK* (1996), pp. 1–14.

Indexed Frames and Hybrid Logics

Philippe Balbiani

IRIT, CNRS, Université de Toulouse

Saúl Fernández González

IRIT, CNRS, Université de Toulouse

Abstract

We define and study the notion of 'indexed frames', i.e., tuples (W_1, W_2, R_1, R_2) where each R_i is a binary relation on $W_1 \times W_2$ such that $R_i(w_1, w_2)(v_1, v_2)$ implies $w_i = v_i$. They generalise, among other things, products of Kripke frames. We show that the logic of indexed frames is the fusion logic $\mathsf{K} \oplus \mathsf{K}$. We show the relation between indexed frames and relativised products and we obtain the different logics of indexed frames when we impose certain constraints on the relations R_1 and R_2. Indexed frames were seemingly first used in [8], whithin a proposal for a broader multi-modal framework called Epistemic Logic of Friendship, allowing for both an epistemic accessibility relation and a 'friendship' relation. The set of agents is encoded in the semantics, and these agents are named using nominal variables (a notion borrowed from hybrid logic) with the novelty that these nominals only refer to the elements of one of the sets. [7] provided an axiomatisation for a fragment of the language. We give a simplified proof of this result and we axiomatise an extension of this fragment.

1 Introduction

This paper is concerned with the very interesting (and, to our knowledge, uncharted) mathematical structure that underlies the framework of *Epistemic Logic of Friendship* introduced by Seligman, Liu and Girard in [8]. (Also studied in [9,10]).

It is not in our scope to study the epistemic and social aspects of EFL. Let us nonetheless briefly recall this framework here: we start off with a bimodal language $\mathcal{L}$, defined as:

$$\phi ::= p|\bot|\neg\phi|(\phi \wedge \phi)|K\phi|F\phi,$$

where $p \in \mathsf{Prop}$, a countable set of propositional variables. K is meant to be read as an epistemic modality ("I know p"), whereas F is a 'frienship' modality ("all my friends p"). We use $\hat{K}$ and $\hat{F}$ as the duals of these operators. Models are of the form $(W, A, \sim, \asymp, V)$, where W and A are nonempty sets ("states" and "agents", respectively), $\sim = \{\sim_a : a \in A\}$ is a family of binary relations on

W indexed by A ($\sim_a \subseteq W^2$ represents agent a's epistemic accessibility), and $\asymp = \{\asymp_w : w \in W\}$ is a family of binary relations on A indexed by W (each representing which agents are friends at world w). $V : \mathsf{Prop} \to 2^{W \times A}$ is a valuation.

We interpret formulas of $\mathcal{L}$ with respect to pairs $(w, a) \in W \times A$, as follows:

$(w, a) \models K\phi$ iff $(v, a) \models \phi$ for all v such that $w \sim_a v$;
$(w, a) \models F\phi$ iff $(w, b) \models \phi$ for all b such that $a \asymp_w b$.

To illustrate this, see the following diagram. It represents a situation with three agents, Alice, Bob and Charlie, wherein at world w Alice has a friend with the property p (represented by the grey nodes) yet she does not know that:

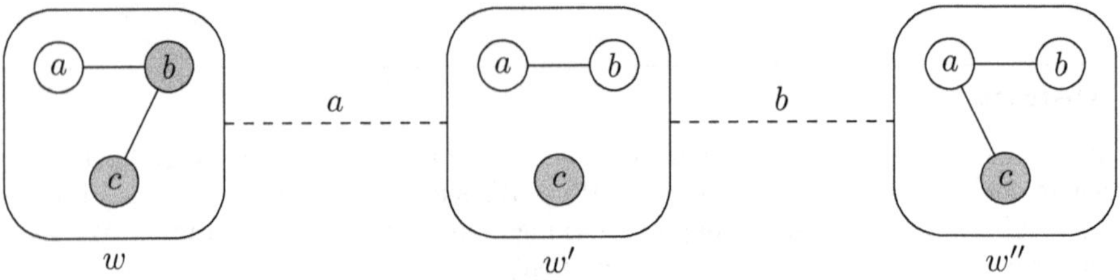

Indeed, it holds that $w, a \models \hat{F}p \wedge \neg K\hat{F}p$. We could also express more complex things such as "Alice does not know Bob and Charlie are friends". In order to do this, we would need to extend the language, as we shall show later. For now, let us focus on this relational structure.

Indexed frames. We have a multi-relational Kripke frame, whose relations are indexed by a set A, in which each state contains a distinct Kripke frame having A as its underlying set.

We shall call these structures *indexed frames.* In Section 2 we study them and provide the complete axiomatisation of the modal logic they give rise to. Note that indexed frames generalise other ways to combine Kripke frames, such as products: recall that, given two Kripke frames $(W_1, R_1), (W_2, R_2)$, their *product* is the birelational Kripke frame $(W_1 \times W_2, R_1^H, R_2^V)$, where $R_1^H(w_1, w_2)(w_1', w_2')$ iff $w_2 = w_2'$ and $R_1 w_1 w_1'$, and $R_2^V(w_1, w_2)(w_1', w_2')$ iff $w_1 = w_1'$ and $R_2 w_2 w_2'$. R_1^H and R_2^V are referred to as the *horizontal* and *vertical* relations, respectively.

One can easily see that a product of two Kripke frames is simply an indexed frame where $\sim_a = \sim_b$ and $\asymp_w = \asymp_v$ for all a, b, w, v. In Subsection 2.2 we show that any subframe of a product of Kripke frames can be turned in a truth-preserving manner into an indexed frame, which will grant us a bunch of extra completeness results.

In Section 3 we show that every formula that is satisfied in an indexed frame can be satisfied in a finite indexed frame.

Naming the agents. Let us go back to the notion "Alice does not know Bob and Charlie are friends". In order to express this in our language, we need to name the agents. This is done in [8] via the introduction of nominal variables and modality $@_n$, directly imported from hybrid logic: see [1,3,5,6]. The language $\mathcal{L}(@)$ extends $\mathcal{L}$ with the atom n and the operator $@_n\phi$, where n belongs to Nom, a countable set of nominal variables. A model for $\mathcal{L}(@)$ is a tuple $(W, A, \sim, \asymp, V)$, as defined above, with the exception that $V : \mathsf{Prop} \cup \mathsf{Nom} \to 2^{W \times A}$ and, for each $n \in \mathsf{Nom}$, $V(n)$ is of the form $W \times \{a\}$ for some $a \in A$. The nominal n can thus be seen as the name of agent a. We now have: $w, a \models n$ iff $V(n) = W \times \{a\}$, and $w, a \models @_n\phi$ iff $w, b \models \phi$, where b is the agent named by n.

A complete axiomatisation of $\mathcal{L}(@)$ was provided for the first time by Sano in [7]. The proof of completeness works (roughly) as follows: first, a cut-free tree sequent calculus is introduced, which is then shown to be sound and complete. Then Sano shows that a formula which is provable in the Hilbert-style system can be converted into a provable tree sequent and, conversely, that from a provable tree sequent one can obtain a formula which is derivable in the Hilbert-style system.

In the conclusion of [7] it is suggested that finding a proof of this result using canonical models is an interesting area of future research. We present such a proof in Section 4 (Subsection 4.1), along with a proof that the logic possesses the finite model property (Subsection 4.2).

Back to friendship logic. For most of this paper we ignore many of the constraints imposed in [8] upon the models in order to make them a realistic framework for a logic of knowledge and friendship, namely: the set of agents A should be finite, the epistemic relations $\sim_a$ should be equivalence relations, the friendship relations $\asymp_w$ should be symmetric and irreflexive, and, optionally, it should be the case that an agent always knows who her friends are (if $w \sim_a v$ and $a \asymp_w b$, then $a \asymp_v b$). We address these properties in Subsection 4.3 and use all the previous results to provide a logic for the exact class of models proposed in [8]. (It is worth noting that, although in Section 4 we stick to the $\sim$ and $\asymp$ symbols to maintain the notation of [8,7], until this moment the reader should not assume they denote equivalence or symmetric relations.)

Another extension. Another operator from hybrid logic is considered in [8]. The operator $\downarrow x.\phi$ allows to name the current agent x, making it possible to refer to it indexically. The resulting extension of $\mathcal{L}(@)$, let us call it $\mathcal{L}(@\downarrow)$, allows to express things like "I have a friend who knows n is friends with me", $\downarrow x.\hat{F}K@_n\hat{F}x$. In Section 5 we provide a sound and complete axiomatization for $\mathcal{L}(@\downarrow)$.

Some proofs have been moved to the Appendix.

2 Indexed Frames

Definition 2.1 An *indexed frame* is a tuple (W, A, R, S) where W and A are nonempty sets, and $R \subseteq A \times W^2$, $S \subseteq W \times A^2$ are ternary relations. We use $R_a ww'$ and $S_w aa'$ to denote, respectively, $(a, w, w') \in R$ and $(w, a, a') \in S$.

We can see R and S as families of binary relations $\{R_a\}_{a\in A}$ and $\{S_w\}_{w\in W}$.

Alternatively, we can see indexed frames as tuples (W, A, R, S) where R and S are binary relations on $W \times A$ such that $R(w,a)(w',a')$ implies $a = a'$ and $S(w,a)(w',a')$ implies $w = w'$.

Let Prop be a countable set of propositional variables. We will consider a language $\mathcal{L}$ as defined in the introduction. We leave aside the epistemic and social considerations and call our modal boxes $\Box_1$ and $\Box_2$ instead of K and F.

Thus our language $\mathcal{L}$ will be $\phi ::= p|\bot|\neg\phi|(\phi \wedge \phi)|\Box_1\phi|\Box_2\phi$, with $p \in \mathsf{Prop}$. We define the other Boolean connectives as usual, the dual modalities $\Diamond_i\phi := \neg\Box_i\neg\phi$ for $i = 1, 2$, and we adopt the standard rules for omission of the parentheses. Given $\phi \in \mathcal{L}$ we define its set of subformulas $\mathsf{subf}\phi$ in the standard way, and its *modal depth*, $\mathsf{md}(\phi)$, recursively as follows:
$\mathsf{md}(p) = \mathsf{md}(\bot) = 0$, $\mathsf{md}(\neg\phi) = \mathsf{md}(\phi)$, $\mathsf{md}(\phi_1 \wedge \phi_2) = \max_{i=1,2} \mathsf{md}(\phi_i)$, $\mathsf{md}(\Box_i\phi) = 1 + \mathsf{md}(\phi)$.

Definition 2.2 An *indexed model* for $\mathcal{L}$ is a tuple $\mathfrak{M} = (W, A, R, S, V)$ where (W, A, R, S) is an indexed frame and $V : \mathsf{Prop} \to 2^{W \times A}$ is a valuation.

We interpret formulas of $\mathcal{L}$ on indexed models with respect to pairs $(w, a) \in W \times A$ as follows:

$(w, a) \models \Box_1\phi$ iff $w', a \models \phi$ for all $w' \in W$ such that $R_a ww'$;

$(w, a) \models \Box_2\phi$ iff $w, a' \models \phi$ for all $a' \in A$ such that $S_w aa'$.

Global truth of formulas in models and validity of formulas in frames are defined as usual.

2.1 The logic of indexed models

Definition 2.3 Given a unimodal logic L, let $\mathsf{Fr}\, L$ be the class of Kripke frames $\mathcal{F}$ such that $\mathcal{F} \models L$. Given unimodal Kripke-complete logics L_1 and L_2 we define $L_1 \circ L_2$ as the logic of indexed frames (W, A, R, S) such that $(W, R_a) \in \mathsf{Fr}\, L_1$ for all $a \in A$ and $(A, S_w) \in \mathsf{Fr}\, L_2$ for all $w \in W$.

Assuming no constraints on the relations R_a and S_w, the logic of indexed models is the fusion logic $\mathsf{K} \oplus \mathsf{K}$, i.e., the least normal modal logic in $\mathcal{L}$ containing the axioms of the minimal modal logic K for each of the $\Box_i$. To express this in terms of the above definition:

Theorem 2.4 $\mathsf{K} \circ \mathsf{K} = \mathsf{K} \oplus \mathsf{K}$.

This result can be proven using a step-by-step construction. For such a proof, see the Appendix. In the next Subsection we shall prove a more general result, and for this we will employ the notion of *relativized products*, studied in [4].

2.2 Indexed frames and relativized products

The following definitions can be found in [4]:

Definition 2.5 Given two families of frames $\mathcal{K}_1$ and $\mathcal{K}_2$, let $\mathcal{K}_1 \times \mathcal{K}_2$ be the family of products of Kripke frames $\mathcal{F}_1 \times \mathcal{F}_2$ such that $\mathcal{F}_i \in \mathcal{K}_i$. Given Kripke-complete unimodal logics L_1, L_2, we define their *(arbitrary) relativized product* as the logic of arbitrary subframes of products of Kripke frames $\mathcal{F}_1 \times \mathcal{F}_2$ such that $\mathcal{F}_i \in \mathsf{Fr}\, L_i$, i.e.,

$$(L_1 \times L_2)^{SF} = \mathsf{Log}\{\mathcal{G} : \mathcal{G} \subseteq \mathcal{F} \text{ for some } \mathcal{F} \in \mathsf{Fr}\, L_1 \times \mathsf{Fr}\, L_2\}.$$

(We say $\mathcal{G} = (W', R'_1, ..., R'_n)$ is a *subframe* of $\mathcal{F} = (W, R_1, ..., R_n)$, denoted $\mathcal{G} \subseteq \mathcal{F}$, whenever $W' \subseteq W$ and each R'_i is the restriction of R_i to W'.)

A logic L is a *subframe logic* if $\mathcal{F} \in \mathsf{Fr}\, L$ and $\mathcal{G} \subseteq \mathcal{F}$ implies $\mathcal{G} \in \mathsf{Fr}\, L$. (Example: S4, because a subframe of a preorder is a preorder; nonexample: the logic of serial frames $\mathsf{K} + \Diamond\top$, because any finite subframe of $(\mathbb{N}, <)$ is not serial.) The following holds:

Proposition 2.6 ([4, Thm. 9.2]) *If L_1, L_2 are subframe logics, $L_1 \oplus L_2 \subseteq (L_1 \times L_2)^{SF}$.*

Moreover, if $L_1, L_2 \in \{\mathsf{K}, \mathsf{T}, \mathsf{K4}, \mathsf{S4}, \mathsf{S5}, \mathsf{S4.3}\}$, then $L_1 \oplus L_2 = (L_1 \times L_2)^{SF}$.

Let us use these results to give a proof of completeness for the logic of indexed frames. Let $\mathcal{F}_i = (W_i, R'_i)$ for $i = 1, 2$. Let $\mathcal{F} = (W, R_1, R_2)$ be a subframe of $\mathcal{F}_1 \times \mathcal{F}_2$. This means that $W \subseteq W_1 \times W_2$, and R_1 and R_2 are the restrictions to W of the horizontal and vertical relations $(R'_1)^H$ and $(R'_2)^V$ respectively.

Consider the indexed frame $\mathcal{G} = (W_1, W_2, R^H, R^V)$, where, for $w_2 \in W_2$,

$$R^H_{w_2} w_1 w'_1 \text{ iff } \begin{cases} (w_1, w_2) \in W \text{ and } (w'_1, w_2) \in W \text{ and } (w_1, w_2) R_1 (w'_1, w_2); & \text{or} \\ (w_1, w_2) \notin W \text{ and } w'_1 = w_1. \end{cases}$$

and, for $w_1 \in W_1$,

$$R^V_{w_1} w_2 w'_2 \text{ iff } \begin{cases} (w_1, w_2) \in W \text{ and } (w_1, w'_2) \in W \text{ and } (w_1, w_2) R_2 (w_1, w'_2); & \text{or} \\ (w_1, w_2) \notin W \text{ and } w'_2 = w_2. \end{cases}$$

Now, let V be a valuation on $\mathcal{F}$ and set $V'(p) = V(p)$ as a valuation on the indexed frame (W_1, W_2, R^H, R^V). The following holds:

Proposition 2.7 *Let ϕ be a formula in the bimodal language, and let $(w_1, w_2) \in W$. Then $\mathcal{F}, V, (w_1, w_2) \models \phi$ iff $\mathcal{G}, V', (w_1, w_2) \models \phi$.*

Proof. By induction on ϕ. Let us see for instance the case $\phi = \Box_1 \psi$.

If $\mathcal{F}, V, (w_1, w_2) \models \Box_1 \psi$, then let w'_1 such that $R^H_{w_2} w_1 w'_1$. Since $(w_1, w_2) \in W$, by definition we have that $(w'_1, w_2) \in W$ and $(w_1, w_2) R_1 (w'_1, w_2)$ in $\mathcal{F}$, which means that $\mathcal{F}, V, (w'_1, w_2) \models \psi$, and, by induction hypothesis $\mathcal{G}, V', (w'_1, w_2) \models \psi$. But since this is true for all w'_1 such that $R^H_{w_2} w_1 w'_1$, we have that $\mathcal{G}, V', (w_1, w_2) \models \Box_1 \psi$. The converse is analogous, noting that $(w_1, w_2) R_1 (w'_1, w_2)$ implies $R^H_{w_2} w_1 w'_1$. □

Moreover, we have the following:

Lemma 2.8 *Suppose $\mathcal{F}$ is a subframe of $(W_1, R'_1) \times (W_2, R'_2)$. Suppose R'_1 (respectively R'_2) has one of the following properties: reflexive; transitive; symmetric; connected; Euclidean. Then, for all w_2, $R^H_{w_2}$ (resp. for all w_1, $R^V_{w_1}$) has the same property.*

Proof. Straightforward by construction of R^H and R^V. □

As a consequence:

Theorem 2.9 *If* $L_1, L_2 \in \{\mathsf{K}, \mathsf{T}, \mathsf{K4}, \mathsf{S4}, \mathsf{S5}, \mathsf{S4.3}\}$, *then* $L_1 \circ L_2 = L_1 \oplus L_2$.

Proof. The inclusion $L_1 \circ L_2 \supseteq L_1 \oplus L_2$ holds by definition of $L_1 \circ L_2$. It suffices to see that $L_1 \circ L_2 \subseteq L_1 \oplus L_2$. If $\phi \notin L_1 \oplus L_2$, then by Proposition 2.6 there exist frames $(W_1, R_1') \in \mathsf{Fr}\, L_1$ and $(W_2, R_2') \in \mathsf{Fr}\, L_2$, a frame $\mathcal{F} = (W, R_1, R_2) \subseteq (W_1, R_1') \times (W_2, R_2')$, a valuation V on $\mathcal{F}$ and a world $(w_1, w_2) \in W$ such that $\mathcal{F}, V, w_1, w_2 \not\models \phi$. But then, the above construction $\mathcal{G} = (W_1, W_2, R^H, R^V)$ satisfies: $(W_1, R^H_{w_2}) \in \mathsf{Fr}\, L_1$ and $(W_2, R^V_{w_1}) \in \mathsf{Fr}\, L_2$ for all w_1, w_2 (Proposition 2.8), and $\mathcal{G}, V', w_1, w_2 \not\models \phi$ (Proposition 2.7); therefore, $\phi \notin L_1 \circ L_2$. □

3 Finite Indexed Model Property

All the logics mentioned so far have the Finite Model Property in the sense that, if a formula is consistent in the logic, there will be a finite model satisfying it [1]. But can we find a finite indexed model satisfying such a formula? The answer is affirmative.

Definition 3.1 A logic L is said to have the *Finite Indexed Model Property (iFMP)* if, given $\phi \notin L$, there exists an indexed model $\mathfrak{M} = (W, A, R, S, V)$ such that W and A are finite, $(W, A, R, S) \models L$, and, for some $(w, a) \in W \times A$, we have $\mathfrak{M}, w, a \not\models \phi$.

Given Kripke-complete unimodal logics L_1 and L_2, let $(L_1 \circ L_2)^f$ be the logic of finite indexed frames of $L_1 \circ L_2$.

Theorem 3.2 $\mathsf{K} \oplus \mathsf{K}$ *has the iFMP, i.e.,* $(\mathsf{K} \circ \mathsf{K})^f = \mathsf{K} \circ \mathsf{K} = \mathsf{K} \oplus \mathsf{K}$.

Proof. This amounts to showing that, if a formula ϕ_0 is satisfied in an indexed model, then there is a finite indexed model that satisfies it. Let $\mathfrak{M} = (W, A, R, S, V)$ and $(w_0, a_0) \in W \times A$ such that $\mathfrak{M}, w_0, a_0 \models \phi_0$.

We define relations $\mathbf{R}$ and $\mathbf{S}$ on $W \times A$ as follows: $(w, a)\mathbf{R}(w', a')$ iff $a = a'$ and $R_a ww'$, and $(w, a)\mathbf{S}(w', a')$ iff $w = w'$ and $S_w aa'$. We will consider chains starting at (w_0, a_0), of the form

$$\alpha = (w_0, a_0)\mathbf{T}_1(w_1, a_1)...\mathbf{T}_k(w_k, a_k),$$

with $k \geq 0$, $\mathbf{T}_i \in \{\mathbf{R}, \mathbf{S}\}$ and $(w_{i-1}, a_{i-1})\mathbf{T}_i(w_i, a_i)$ for $1 \leq i \leq k$. We shall say that such a chain has *length* k (and thus (w_0, a_0) is a chain of length 0). We will call $\mathsf{last}\, \alpha = (w_k, a_k)$.

Fix n to be the modal depth of ϕ_0. We shall construct a finite set of chains of length up to n, in n steps. Let $F_0 = \{(w_0, a_0)\}$. For $0 \leq k \leq n-1$, suppose F_k is a finite set of chains of length k. Let F_{k+1} be a finite set of minimal cardinality satisfying the following property for all $\alpha \in F_k$ and all $\mathbf{T} \in \{\mathbf{R}, \mathbf{S}\}$:

> for any $(w, a) \in W \times A$, if $(\mathsf{last}\, \alpha)\mathbf{T}(w, a)$, then there exists an element $(w', a') \sim_{\phi_0} (w, a)$ such that $\alpha\mathbf{T}(w', a') \in F_{k+1}$,

[1] Indeed, every logic in the set $\{\mathsf{K}, \mathsf{T}, \mathsf{K4}, \mathsf{S4}, \mathsf{S5}, \mathsf{S4.3}\}$ has the FMP and this property is preserved by fusions: see [11].

where $\sim_{\phi_0}$ is the equivalence relation

$(w, a) \sim_{\phi_0} (w', a')$ iff for all $\psi \in \mathsf{subf}\, \phi_0(\mathfrak{M}, w, a \models \psi$ iff $\mathfrak{M}, w', a' \models \psi)$.

It is not hard to see that there is a set of cardinality at most $2 \cdot |F_k| \cdot 2^{|\mathsf{subf}\, \phi_0|}$ satisfying this property. Indeed, for any of the $|F_k|$ choices of α and 2 choices of $\mathbf{T}$, F_{k+1} will contain an element $\alpha\mathbf{T}(w, a)$ for (at most) one representative of each of the (at most) $2^{|\mathsf{subf}\, \phi_0|}$ equivalence classes of $\sim_{\phi_0}$.

Let $F' = F_0 \cup ... \cup F_n$. Let F be the closure of F' under the following property:

if $\alpha \in F$, $length(\alpha) < n$, $\mathbf{T} \in \{\mathbf{R}, \mathbf{S}\}$, $w \in W$ and $a \in A$ occur in F, and $(\mathsf{last}\, \alpha)\mathbf{T}(w, a)$, then $\alpha\mathbf{T}(w, a) \in F$.

Obviously, F' is finite, and so is F.

We construct our finite model $\mathfrak{M}^f = (W^f, A^f, R^f, S^f, V^f)$ where W^f and A^f are the restrictions of W and A to those elements occuring in F, i.e,

$$W^f = \{w \in W : w \text{ occurs in } F\};\ A^f = \{a \in A : a \text{ occurs in } F\};$$

and R^f, S^f and V^f are the corresponding restrictions of R, S, and V. The following holds:

Lemma 3.3 *Let $\alpha \in F$ be a chain of length k, i.e,*

$$\alpha = (w_0, a_0)\mathbf{T}_1(w_1, a_1)...\mathbf{T}_k(w_k, a_k),$$

with $\mathbf{T}_i \in \{\mathbf{R}, \mathbf{S}\}$. Let ϕ be a subformula of ϕ_0 such that $\mathsf{md}(\phi) \leq n - k$. Then, $\mathfrak{M}, w_k, a_k \models \phi$ if and only if $\mathfrak{M}^f, w_k, a_k \models \phi$.

This proves our theorem: it suffices to apply the previous Lemma to the chain (w_0, a_0) of length 0 to obtain $\mathfrak{M}^f, w_0, a_0 \models \phi_0$. □

Remark 3.4 The fact that we are taking a submodel of $\mathfrak{M}$ grants us that we can preserve the universal properties of the relations. This means that, if R is reflexive/ transitive/ symmetric/ connected/ Euclidean, so is R^f. Likewise for S and S^f. This fact, paired with Theorem 2.9, gives us the following result immediately:

Theorem 3.5 *If $L_1, L_2 \in \{\mathsf{K}, \mathsf{T}, \mathsf{K4}, \mathsf{S4}, \mathsf{S5}, \mathsf{S4.3}\}$, then $L_1 \oplus L_2$ has the iFMP. In other words, $(L_1 \circ L_2)^f = L_1 \circ L_2 = L_1 \oplus L_2$.*

4 Epistemic Logic of Friendship

We now consider the framework for an 'epistemic logic of friendship' proposed by [8]. For now, this amounts to adding a set $\mathsf{Nom} = \{n, m, ...\}$ of nominal variables to our language, and extending the language to $\mathcal{L}(@)$, defined as:

$$\phi ::= p | n | \bot | \neg\phi | (\phi \wedge \phi) | K\phi | F\phi | @_n\phi,$$

where $p \in \mathsf{Prop}, n \in \mathsf{Nom}$.

Definition 4.1 Models for $\mathcal{L}(@)$ are of the shape $\mathfrak{M} = (W, A, \sim, \asymp, V)$, where $(W, A, \sim, \asymp)$ is an indexed frame and $V : \mathsf{Prop} \cup \mathsf{Nom} \to 2^{W \times A}$ is a valuation function with the property that, for each $n \in \mathsf{Nom}$, $V(n) = W \times \{a\}$ for some $a \in A$. We refer to this unique a as $a = \underline{n}_V$ (or $a = \underline{n}$ if there is no risk of ambiguity).

A model is *named* whenever, for each $a \in A$, there exists $n \in \mathsf{Nom}$ such that $\underline{n} = a$. (Note that, in a named model, A is at most countable.)

We interpret formulas of $\mathcal{L}(@)$ in named models with respect to pairs $(w, a) \in W \times A$ as follows:

$\mathfrak{M}, w, a \models n$ iff $(w, a) \in V(n)$ (iff $\underline{n} = a$);
$\mathfrak{M}, w, a \models @_n\phi$ iff $\mathfrak{M}, w, \underline{n} \models \phi$.

4.1 Axiomatising $\mathcal{L}(@)$ via canonical models

It is proven in [7], via an argument that employs a tree sequent calculus, that the logic of $\mathcal{L}(@)$ is the system EFL, defined in the table below:

(Taut)	all propositional tautologies	(MP)	from ϕ and $\phi \to \psi$, infer ψ
(K_K)	$K(\phi \to \psi) \to (K\phi \to K\psi)$	(Nec_K)	from ϕ, infer $K\phi$
(K_F)	$F(\phi \to \psi) \to (F\phi \to F\psi)$	(Nec_F)	from ϕ, infer $F\phi$
($\mathrm{K}_@$)	$@_n(\phi \to \psi) \to (@_n\phi \to @_n\psi)$	($\mathrm{Nec}_@$)	from ϕ, infer $@_n\phi$
(Ref)	$@_n n$	(Selfdual)	$\neg @_n\phi \leftrightarrow @_n\neg\phi$
(Elim)	$@_n\phi \to (n \to \phi)$	(Agree)	$@_n @_m\phi \to @_m\phi$
(Back)	$@_n\phi \to F@_n\phi$	(DCom)	$@_n K @_n\phi \leftrightarrow @_n K\phi$
($\mathrm{Rigid}_=$)	$@_n m \to K@_n m$	($\mathrm{Rigid}_\neq$)	$\neg @_n m \to K\neg @_n m$
(Name)	From $@_n\phi$ infer ϕ, where n is fresh in ϕ		
(LBG)	From $L(@_n\hat{F}m \to @_m\phi)$ infer $L(@_nF\phi)$, m fresh in $L(@_nF\phi)$.		

In the last line of the above table, the *necessity forms* $L(\#)$ are defined as:

$$L ::= \# | \phi \to L | @_n KL.$$

In this section we present a novel proof of this result using canonical models. To do this, we consider instead the logic EFL^+, obtained by replacing the rule (LBG) in EFL by the following:

(LBG^+) From $L(@_n\hat{F}m \to @_m\phi)$ for all m fresh in $L(@_nF\phi)$, infer $L(@_nF\phi)$.

The following Lemma can be proven by a straightforward induction on derivations.

Lemma 4.2 EFL *and* EFL^+ *prove the same formulas.*

We thus prove completeness of EFL^+. The following validities will be useful:

Proposition 4.3 *The following are derivable in* EFL*:*

(T1) $\vdash @_m @_n \phi \leftrightarrow @_n \phi$;
(T2) $\vdash n \to (@_n \phi \leftrightarrow \phi)$;
(T3) $\vdash @_n m \to (@_n \phi \leftrightarrow @_m \phi)$;
(T4) $\vdash @_n m \leftrightarrow @_m n$;
(T5) $\vdash @_n (\phi \to \psi) \leftrightarrow (@_n \phi \to @_n \psi)$;
(T6) $\vdash @_n m \to (\phi[k/n] \leftrightarrow \phi[k/m])$, *where* $\phi[k/n]$ *is the formula obtained from* ϕ *by replacing each occurrence of* k *by* n.
(T7) $\vdash @_n m \to @_i K @_n m$, *and* $\vdash @_n \neg m \to @_i K @_n \neg m$;
(T8) $\vdash @_n \hat{F} m \wedge @_m \phi \to @_n \hat{F} \phi$;
(T9) $\vdash @_n F \psi \wedge @_n \hat{F} m \to @_m \psi$;
(R1) *if* $\vdash @_n \hat{F} m \wedge @_m \phi \to \psi$, *then* $\vdash @_n \hat{F} \phi \to \psi$, *with* $m \neq n$ *fresh in* ϕ *and* ψ.

We will say that a formula in $\mathcal{L}(@)$ is a *named formula* whenever it is of the form $@_n \phi$. A *BCN formula* is a Boolean combination of named formulas, and we use BCN to denote the set of such formulas. The following is an immediate consequence of (T1), (T5) and (Selfdual):

Corollary 4.4 *If* $\phi \in BCN$, $n \in \mathsf{Nom}$, *then* $\vdash @_n \phi \leftrightarrow \phi$.

A formula ϕ is *consistent* if $\neg\phi$ is not derivable. The following lemma will be useful later.

Lemma 4.5 *If* n *does not occur in* ϕ, *then* ϕ *is consistent if and only if* $@_n \phi$ *is consistent.*

Proof. If ϕ is inconsistent we have $\vdash \neg\phi$ and thus by (Nec$_@$), $\vdash @_n \neg\phi$, which by (Selfdual) gives that $\vdash \neg @_n \phi$. If $@_n \phi$ is inconsistent then $\vdash \neg @_n \phi$ which by (Selfdual) means $\vdash @_n \neg \phi$ and thus, by (Name), $\vdash \neg\phi$. □

Now we can start our completeness proof. The two above results allow us to focus only on BCN formulas. A *theory* is a set of BCN formulas T such that:

i. $\mathsf{EFL}^+ \cap BCN \subseteq T$;

ii. T is closed under Modus Ponens;

iii. If $L(@_n \hat{F} m \to @_m \phi) \in T$ for all $m \neq n$ not occurring in L or in ϕ, then $L(@_n F \phi) \in T$.

A theory is *consistent* whenever $@_n \bot \notin T$ (for any/all n). It is easy to see that $\mathsf{EFL}^+ \cap BCN$ is the least consistent theory. A consistent theory is *maximal* if no proper superset of it is a consistent theory.

Lemma 4.6 *Given a theory* T, *the set*

$$T_{K_n} = \{\psi \in BCN : \ \vdash \psi \leftrightarrow @_n \phi \text{ for some } @_n K \phi \in T\}$$

is a theory.

Proof. Note the following: for any $\phi \in BCN$, we have that $\phi \in T_{K_n}$ iff $@_n K \phi \in T$. Indeed,if $\phi \in T_{K_n}$, then $\vdash \phi \leftrightarrow @_n \psi$ for some $@_n K \psi \in T$. But then, using (Nec$_K$), (Nec$_@$) and (DCom) in that order we obtain $\vdash @_n K \phi \leftrightarrow$

$@_n K\psi$, and thus $@_n K\phi \in T$. The other direction is trivial and uses that $\vdash @_n\phi \leftrightarrow \phi$. With this:

Rule i. If $\phi \in \mathsf{EFL}^+ \cap BCN$, $m \in \mathsf{Nom}$, $@_n K\phi \in \mathsf{EFL}^+ \cap BCN$ (by applying two Nec rules) and thus $@_n K\phi \in T$, so $\phi \in T_{K_n}$.

Rule ii. If ϕ and $\phi \to \psi \in T_{K_n}$, then $@_n K\phi$, $@_n K(\phi \to \psi) \in T$ and, by applying the K axioms and modus ponens, $@_n K\psi \in T$, and thus $\psi \in T_{K_n}$.

Rule iii. If $L(@_k \hat{F} m \to @_m \phi) \in T_{K_n}$ for all fresh m, then $@_n KL(@_k \hat{F} m \to @_m\phi) \in T$ for all fresh m, and thus, since $@_n KL$ is an admissible form, $@_n KL(@_k F\phi) \in T$, whence $L(@_k F\phi) \in T_{K_n}$.

Lemma 4.7 *Given a theory T and a formula $\phi \in BCN$, the set $T_\phi = \{\psi \in BCN : \phi \to \psi \in T\}$ is a theory containing T and including the formula ϕ, and it is consistent whenever T is consistent and $\neg\phi \notin T$.*

Proof. Rule i. If $\psi \in \mathsf{EFL}^+ \cap BCN$, then $\phi \to \psi \in \mathsf{EFL}^+ \cap BCN$, thus $\psi \in T_\phi$.

Rule ii. Follows from classical propositional logic.

Rule iii. Follows from the fact that, if L is an admissible form, so is $\phi \to L$.

The fact that $\phi \in T_\phi \supseteq T$ is because $\vdash \phi \to \phi$ and $\vdash \psi \to (\phi \to \psi)$. If $\neg\phi \notin T$, then $@_n \neg\phi \notin T$, thus $@_n(\phi \to \bot) \notin T$. Using the K axiom and $\vdash \phi \leftrightarrow @_n\phi$, we obtain $\phi \to @_n\bot \notin T$, and thus $@_n\bot \notin T_\phi$.

Now,

Lemma 4.8 (Lindenbaum's lemma) *A consistent theory can be extended to a maximal consistent theory.*

Proof. Let T_0 be a consistent theory and $(\phi_k)_{k\in\omega}$ be an enumeration of BCN where each formula occurs infinitely many times.

Given a consistent theory T_k, we define a consistent theory T_{k+1} (which extends T_k) as follows:

- If $\neg\phi_k \notin T_k$, then $T_{k+1} = (T_k)_{\phi_k}$.
- If $\neg\phi_k \in T_k$, then:
  - If $\neg\phi_k$ is of the form $\neg L(@_n F\phi)$, then for some fresh m it must be the case that $L(@_n \hat{F} m \to @_m\phi) \notin T_k$, for otherwise we would have by rule iii. that $L(@_n F\phi) \in T_k$, contradicting its consistency. Then we set $T_{k+1} = (T_k)_{\neg L(@_n \hat{F} m \to @_m \phi)}$.
  - Otherwise, $T_{k+1} = T_k$.

Let $T = \bigcup_{k\in\omega} T_k$. Then T is a maximal consistent theory. Consistency is obvious, for each T_k is consistent. Maximality comes from the fact that, for every formula ϕ_k, either $\neg\phi_k$ was already in T_k, or ϕ_k was added to T_{k+1}, therefore it cannot have consistent supersets closed under modus ponens. To see that it is a theory, it suffices to check that Rule iii. is satisfied. And indeed, if $L(@_n F\phi) \notin T$, then $\neg L(@_n F\phi) \in T_k$ for some k. Consider some $k' > k$ such that $\phi_{k'} = \neg L(@_n F\phi)$. Then, by construction, $T_{k'+1}$ must contain $\neg L(@_n \hat{F} m \to @_m\phi)$ for some fresh m, and therefore it is not the case that $L(@_n \hat{F} m \to @_m\phi) \in T$ for all fresh m.

Let MCT denote the set of maximal consistent theories. Given $T, S \in$

MCT, and $n \in \mathsf{Nom}$, we define: $T \sim_n S$ iff $T_{K_n} \subseteq S$.

Lemma 4.9 (Diamond Lemma) *Let $T \in MCT$. We have:*

i. If $@_n\hat{K}\phi \in T$, then there exists $S \in MCT$ such that $T \sim_n S \ni @_n\phi$.

ii. If $@_n\hat{F}\phi \in T$, then there is some $m \neq n$ fresh in ϕ such that $@_n\hat{F}m \wedge @_m\phi \in T$.

Proof. i. Take the consistent theory $(T_{K_n})_{@_n\phi}$ and extend it to the desired successor using Lindenbaum's lemma. Note that T_{K_n} is consistent, for if not, $@_n\bot \in T_{K_n}$, and thus $@_nK@_n\bot \in T$. But, since $@_n\bot$ is equivalent to $\bot$, this means that $@_nK\bot \in T$, contradicting $@_n\hat{K}\phi \in T$. Note moreover that $\neg @_n\phi \notin T_{K_n}$, for if that was the case, $@_nK\neg @_n\phi \in T$, which is equivalent to $\neg @_n\hat{K}\phi \in T$: contradiction. Thus $(T_{K_n})_{@_n\phi}$ is consistent.

ii. If $@_n\hat{F}m \wedge @_m\phi \notin T$ for all fresh m, then $\neg(@_n\hat{F}m) \vee \neg(@_m\phi) \in T$ for all fresh m, and thus, by logical equivalence, $@_n\hat{F}m \to @_m\neg\phi \in T$ for all fresh m, which entails $@_nF\neg\phi \in T$, and therefore $\neg @_n\hat{F}\phi \in T$.

Lemma 4.10 *Let $i \in \mathsf{Nom}$. If $\Gamma \sim_i \Delta$ then, for any $n, m \in \mathsf{Nom}$, we have: $@_nm \in \Gamma$ if and only if $@_nm \in \Delta$.*

Proof. By (T7) of Prop. 4.3: if $@_nm \in \Gamma$, then $@_iK@_nm \in \Gamma$, which entails $@_i@_nm \in \Delta$, and therefore, by the (Agree) axiom, $@_nm \in \Delta$. If $@_nm \notin \Gamma$, by maximal consistency and the (Selfdual) axiom we have that $@_n\neg m \in \Gamma$ and we can proceed similarly to obtain that $@_n\neg m \in \Delta$ and thus $@_nm \notin \Delta$. □

Let ϕ_0 be a consistent formula and let us build a model satisfying it. Take a nominal n_0 not occurring in ϕ_0 and note that $@_{n_0}\phi_0$ is a consistent BCN formula (by Lemma 4.5) and thus the consistent theory $(BCN \cap \mathsf{EFL}^+)_{@_{n_0}\phi_0}$ can be extended (by Lindembaum's lemma) to $\Gamma_0 \in MCT$.

Let W be the set of elements reachable from Γ_0 by the $\sim_n$ relations, i.e.

$$W = \{\Delta \in MCT : \Gamma_0 = \Delta_0 \sim_{n_1} \Delta_1 \sim_{n_2} \dots \sim_{n_k} \Delta_k = \Delta \text{ for some } n_1, \dots, n_k \in \mathsf{Nom}, \Delta_0, \dots, \Delta_k \in MCT\}.$$

Note that this construction guarantees (by Lemma 4.10) that for any $\Gamma \in W$, $@_nm \in \Gamma$ iff $@_nm \in \Gamma_0$. Note moreover that the theorems

$$\vdash @_nn \text{ (Ref)}; \vdash @_nm \leftrightarrow @_mn \text{ (T4)}; \vdash @_nm \wedge @_mi \to @_ni \text{ (conseq. of T3)}$$

guarantee that the binary relation on Nom defined as $n \equiv m$ iff $@_nm \in \Gamma_0$ is an equivalence relation. Let $[n]$ denote the equivalence class of $n \in \mathsf{Nom}$ and let $A = \{[n] :\in \mathsf{Nom}\}$.

For $[n] \in A$, we define $\sim_{[n]} = \sim_n$. Let us see that this is well-defined, which amounts to showing that $\sim_n = \sim_m$ whenever $n \equiv m$. But given $\Gamma, \Delta \in W$, and $n \equiv m$, the fact that $@_nm \in \Gamma \cap \Delta$ paired with (T3) give us that $@_nK\phi \in \Gamma$ iff $@_mK\phi$ in Γ, and $@_n\phi \in \Delta$ iff $@_m\phi \in \Delta$, which entails $\Gamma \sim_n \Delta$ iff $\Gamma \sim_m \Delta$.

For $\Gamma \in W$ we define

$$[n] \preceq_\Gamma [m] \text{ iff } @_n\hat{F}m \in \Gamma.$$

Let us see that this definition does not depend on the choice of representative for the equivalence classes: suppose $@_n\hat{F}m \in \Gamma$ and take $n' \in [n], m' \in [m]$. We have that $@_{n'}\hat{F}m \in \Gamma$, by (T3), and therefore, by (T6), $@_{n'}\hat{F}m' \in \Gamma$.

Finally we define a valuation by setting

$$\begin{aligned} V(p) &= \{(\Gamma, [n]) \in W \times A : @_n p \in \Gamma\}, && p \in \mathsf{Prop}; \\ V(n) &= \{(\Gamma, [n]) : \Gamma \in W\}, && n \in \mathsf{Nom}. \end{aligned}$$

Note that we have defined V so that $\underline{n} = [n]$. We have that

$$\mathfrak{M}^C = (W, A, \sim_{[n]\in A}, \asymp_{\Gamma \in W}, V)$$

is a named model and, moreover:

Lemma 4.11 (Truth Lemma) *For any formula $\phi \in \mathcal{L}(@)$, it is the case that $\mathfrak{M}^C, \Gamma, [n] \models \phi$ if and only if $@_n\phi \in \Gamma$.*

Proof. By induction on ϕ. For the case $\phi = m \in \mathsf{Nom}$ we recall that $\underline{m} = [m]$. For the case $\phi = K\psi$, we use the Diamond Lemma. For the case $\phi = F\psi$, we use the Diamond Lemma for one direction and (T9) for the other. □

With this:

Theorem 4.12 EFL^+ *(and therefore* EFL*) is complete with respect to the class of (not necessarily finite) named indexed models.*

Proof. If ϕ_0 is consistent, so is $@_{n_0}\phi_0$ for n_0 not occurring in ϕ_0, and thus we can construct $\mathfrak{M}^C$ as above and we have that $\mathfrak{M}^C, \Gamma_0, [n_0] \models \phi_0$. □

4.2 Finite models

The following also holds:

Theorem 4.13 EFL *is complete with respect to the class of finite named indexed models.*

This is a consequence of a result very similar to Theorem 3.2: if a formula is satisfied in a model $(W, A, \sim, \asymp, V)$, then there is a finite submodel which satisfies it.

The proof of this result has minimal changes with respect to the proof of Thm. 3.2, and it is sketched in the Appendix.

4.3 Extensions of EFL

In [8] some assumptions are made about the epistemic and social relations in the models. The epistemic relations $\sim_a$ are equivalence relations, whereas the friendship relation $\asymp_w$ is irreflexive and symmetric.

One would expect, for instance, that if the relations $\sim_a$ that give the semantics of the knowledge modality K are reflexive, transitive and symmetric, then this modality should follow the axioms of S5, namely:

$$\vdash K\phi \to \phi; \quad \vdash K\phi \to KK\phi; \quad \vdash \phi \to K\neg K\neg\phi.$$

Similarly, if $\sim_a$ is a preorder, the extra axioms of S4 (i.e. the first two above), should be included to the logic. Let $\mathsf{EFL}+\mathsf{S5}_K$ denote the logic resulting from adding these three axioms to EFL, and let $\mathsf{EFL}+\mathsf{S4}_K$ be the logic resulting from adding the first two. And indeed:

Theorem 4.14 ([7]) $\mathsf{EFL}+\mathsf{S5}_K$ *is sound and complete with respect to the class of models where the* $\sim_a$ *are equivalence relations. Moreover,* $\mathsf{EFL}+\mathsf{S4}_K$ *is sound and complete with respect to the class of models where each* $\sim_a$ *is a preorder.*

The proof of this result in [7] consists in adding corresponding rules to the tree sequent calculus and showing that a provable formula in the Hilbert-style system can be transformed into a provable sequent and vice versa. With the canonical models presented in this text this proof becomes quite straightforward. First, note that thanks to (T5) the following are easily provable in $\mathsf{EFL}+\mathsf{S5}_K$ (and the first two in $\mathsf{EFL}+\mathsf{S4}_K$):

$$\vdash @_n K\phi \to @_n\phi; \quad \vdash @_n K\phi \to @_n KK\phi; \quad \vdash @_n\phi \to @_n K\neg K\neg\phi.$$

With this, the proof of the following lemma is straightforward:

Lemma 4.15 *If the axioms of* S5 *for* K *(resp.* S4*) are present in the logic, each relation* $\sim_n$ *in the canonical model is an equivalence relation (resp. a preorder).*

Remark 4.16 Given that $@_n$ distributes over $\to, \wedge, \vee, \neg$, one can see that there are many examples of formulas ϕ defining a certain frame property from which it is trivial to compute a formula $@_n\psi$ defining the same property in the $\sim_n$ relations of indexed frames. Some obvious questions arise: is this true of any Sahlqvist formula? Can we adapt the notion of Sahlqvist formula to this setting and prove an analogue of the Sahlqvist Completeness Theorem ([2, Thm. 4.42])? We conjecture the answer is affirmative.

Similarly, as pointed out by [7] the following axioms encode irreflexivity and symmetry of the friendship relation $\asymp_w$:

$$\text{(irr)} \quad \neg @_n\hat{F}n \qquad \text{(sym)} \quad @_n\hat{F}m \to @_m\hat{F}n$$

The proof of this lemma is also straightforward:

Lemma 4.17 *If (irr) and (sym) are present in the logic, each relation* $\asymp_\Gamma$ *in the canonical model is irreflexive and symmetric.*

Therefore, and since the rest of the completeness proof proceeds as before, we have a complete axiomatisation of the models proposed by [8]:

Theorem 4.18 $\mathsf{EFL}+\mathsf{S5}_K+(irr)+(sym)$ *is the logic of finite indexed frames* $(W, A, \sim, \asymp)$ *where each* $\sim_a$ *is an equivalence relation and each* $\asymp_w$ *is irreflexive and symmetric.*

Finally, an optional further constraint is that an agent should not doubt who her own friends are. For this one would consider frames with the property:

if $w \sim_a v$, then $a \asymp_w b$ implies $a \asymp_v b$. We will call these *KYF frames* (for "know your friends"). It is again very easy to check that, by adding to the logic the axiom

$$\text{(kyf)} \quad \hat{F}m \rightarrow K\hat{F}m,$$

the resulting canonical model is a KYF frame.

5 Axiomatisation of $\mathcal{L}(@\downarrow)$

In [8] another operator is borrowed from hybrid logic, namely $\downarrow x.\phi$, which names the current agent 'x', allowing to refer to her indexically. We now have, on top of Prop and Nom, a countable set $\mathsf{SVar} = \{x, y, ...\}$ of *state variables.* $\mathcal{L}(@\downarrow)$ is simply $\mathcal{L}(@)$ extended with x and $\downarrow x.\phi$, where $x \in \mathsf{SVar}$. Formulas are read on named indexed models with respect to triples (g, w, a), where $g : \mathsf{SVar} \rightarrow A$ is an assignment function, as follows:

$\mathfrak{M}, g, w, a \models x$ iff $g(x) = a$;
$\mathfrak{M}, g, w, a \models \downarrow x.\phi$ iff $\mathfrak{M}, g^x_a, w, a \models \phi$,
where $g^x_a(y) = g(y)$ for $y \neq x$ and $g^x_a(x) = a$.

Given a formula ϕ and a nominal n, we define $\phi[x/n]$ to be the formula resulting from replacing each *free* occurence of x in ϕ by n. Formally:

Definition 5.1 Given $x \in \mathsf{SVar}$, $n \in \mathsf{Nom}$ and $\phi \in \mathcal{L}(@\downarrow)$:
$\phi[x/n] = \phi$ if $\phi = p \in \mathsf{Prop}, \bot, m \in \mathsf{Nom}$ or $y \in \mathsf{SVar}\backslash\{x\}$; $x[x/n] = n$;
$(\phi \wedge \psi)[x/n] = \phi[x/n] \wedge \psi[x/n]$; $(\downarrow x.\phi)[x/n] = \downarrow x.\phi$;
$(B\phi)[x/n] = B(\phi[x/n])$ if $B = \neg, K, F, @_m$, or $\downarrow y$ $(y \neq x)$;

With this, we can define the logic of the fragment $\mathcal{L}(@\downarrow)$:

Definition 5.2 $\mathsf{EFL}_\downarrow$ is the logic containing the axioms and rules of EFL plus the following axiom and rule:
(DA) $@_n(\downarrow x.\phi \leftrightarrow \phi[x/n])$.
(FV) from $\phi[x/n]$ (with n fresh in ϕ), infer ϕ.

The fact that (DA) is sound can be checked by just unpacking the semantics. The soundness of the (FV) rule is a consequence of the following Lemma, whose proof is an easy induction on ϕ:

Lemma 5.3 *Let $\phi \in \mathcal{L}(@\downarrow)$ and n be fresh in ϕ. Let $\mathfrak{M} = (W, A, \sim, R, V)$ be a model and g an assignment. We define a new valuation in $\mathfrak{M}$ by: $V'(n) = W \times \{g(x)\}$, $V'(m) = V(m)$ for $n \neq m$, $V'(p) = V(p)$ for $p \in$ Prop. Let $\mathfrak{M}' = (W, A, \sim, R, V')$. Then $\mathfrak{M}, w, a, g \models \phi$ iff $\mathfrak{M}', w, a, g \models \phi[x/n]$.*

For completeness we shall use these two lemmas; respectively an application of the (FV) rule, and a straightforward induction on ϕ:

Lemma 5.4 *If ϕ is consistent and n_1,...,n_k are fresh, then $\phi[x_1/n_1]...[x_k/n_k]$ is consistent.*

Lemma 5.5 *Let $\mathfrak{M}$ be a model, ϕ be a formula, g an assignment and $x_1, ..., x_k \in$ SVar. Let $n_1, ..., n_k \in$ Nom such that $\underline{n}_i = g(x_i)$. Then*

$$\mathfrak{M}, w, a, g \models \phi \textit{ iff } \mathfrak{M}, w, a, g \models \phi[x_1/n_1]...[x_k/n_k].$$

Now, we construct our canonical model exactly like before with one caveat: our sets MCT will only contain *BCN formulas without free variables* (i.e. *BCN sentences*). We prove the following variant of the Truth Lemma:

Proposition 5.6 *Let g be an assignment and ϕ a formula whose free variables are $x_1, ..., x_k$. Let $[n_i] = g(x_i)$. Then*

$$\mathfrak{M}, \Gamma, [n], g \models \phi \textit{ iff } @_n\phi[x_1/n_1]...[x_n/n_k] \in \Gamma.$$

With this we can prove completeness:

Theorem 5.7 $\mathsf{EFL}_{\downarrow}$ *is complete with respect to indexed models.*

Proof. Suppose ϕ_0 is a consistent formula. Let $x_1, ..., x_k$ be the free variables of ϕ_0 and $n_0, n_1, ..., n_k$ fresh. Then $\phi_0[x_1/n_1]...[x_k/n_k]$ is a consistent sentence (by Lemma 5.4) and so is

$$@_{n_0}\phi_0[x_1/n_1]...[x_k/n_k]$$

(by Lemma 4.5). We extend this to $\Gamma_0 \in MCT$, we construct the corresponding canonical model and we let g be any assignment such that $g(x_i) = [n_i]$. Then we have by Prop. 5.6 that $\mathfrak{M}, \Gamma_0, [n_0], g \models \phi_0$. □

6 Conclusion

In this paper we have studied several aspects of indexed frames, introduced for the first time (as far as we know) in [8]. We have as well provided axiomatisations for the fragments $\mathcal{L}$ (with several constraints in the relations) and $\mathcal{L}(@\downarrow)$, on top of a novel proof of completeness of EFL for the fragment $\mathcal{L}(@)$.

Some interesting directions for future work include studying the decidability of $\mathcal{L}(@\downarrow)$, resolving the conjecture in Remark 4.16, or otherwise providing a more general version of Thm. 2.9.

But perhaps the most fruitful direction to go from here would be the application of indexed frames to different modal logics wherein some interdependence between the modalities exists. Just as an example, we could think of an epistemic temporal logic where each possible world is a timeline and the set of epistemically accessible worlds changes at every time, modelled using indexed frames.

Acknowledgements. Special acknowledgements are heartily granted to Mina Pedersen and Valentin Shehtman for their valuable suggestions. We also wish to thank the AiML reviewers for their comments helping us to improve the paper.

Appendix

Proof of Theorem 2.4. First we introduce a notion of *indexed pseudo-model.*

Definition .1 An *indexed pseudo-model* is a tuple (W, A, R, S, σ) where (W, A, R, S) is an indexed frame and σ is a function which assigns to every pair $(w, a) \in W \times A$ a $\mathsf{K} \oplus \mathsf{K}$-maximal consistent set, with the following properties:

(C1) If $\Box_1\phi \in \sigma(w, a)$ and $R_a ww'$, then $\phi \in \sigma(w', s)$;

(C2) If $\Box_2\phi \in \sigma(w, a)$ and $S_w aa'$, then $\phi \in \sigma(w', s')$.

The right-to-left direction of C1 and C2 need not hold for certain formulas ϕ and pairs (w, a). We call these situations *defects.* Formally:

Definition .2 A *1-defect* is a tuple (ϕ, w, a) such that $\neg\Box_1\phi \in \sigma(w, a)$ and, for all $w' \in W$ such that $R_a ww'$, $\phi \in \sigma(w', a)$. A *2-defect* is a tuple (ϕ, w, a) such that $\neg\Box_2\phi \in \sigma(w, a)$ and, for all $a' \in A$ such that $S_w aa'$, $\phi \in \sigma(w, a')$.

Given a 1-defect (ϕ, w, a) we can update our pseudo-model into a new pseudo-model without this defect by simply adding a point, as we detail below.

Let $\mathfrak{M} = (W, A, R, S, \sigma)$ be an indexed pseudo-model and (ϕ, w, a) be a 1-defect. That means that $\neg\Box_1\phi \in \sigma(w, a)$ yet $\phi \in \sigma(w', a)$ for all w' such that $R_a ww'$. Note that the set $\{\neg\phi\} \cup \{\psi : \Box_1\psi \in \sigma(w, a)\}$ is consistent,therefore it can be extended by Lindenbaum's lemma to a maximal consistent set Δ. Let $w_0 \notin W$. We define a new pseudo-model in which the defect is not present by $\mathfrak{M}'_1 = (W', A', R', S', \sigma')$, where:

- $W' = W \cup \{w_0\}$; $A' = A$;
- $R' = R \cup \{(a, w, w_0)\}$; $S' = S$;
- for all $a' \in A$, $\sigma'(w_0, a') = \Delta$ and $\sigma'(w', a') = \sigma(w', a')$ for $w' \neq w_0$.

$\mathfrak{M}'_1$ is an indexed pseudo-model. Indeed, suppose $\Box_1\psi \in \sigma'(w', a')$ and $R_{a'}w'w''$. If $w'' \neq w_0$, then $\sigma'(w'', a') = \sigma(w'', a') \ni \psi$. Otherwise, if $w'' = w_0$, then by construction we have that $w' = w$ and $a' = a$. Therefore, since $\Box_1\psi \in \sigma'(w', a') = \sigma(w, a)$ we have by construction that $\psi \in \Delta = \sigma'(w'', a')$. Moreover, we have built $\mathfrak{M}'_1$ such that (ϕ, w, a) is no longer a 1-defect.

In a completely analogous manner, given a 2-defect (ϕ, w, a) we can add an extra point a_0 to A to build a pseudo-model which does not present this defect: $\mathfrak{M}'_2 = (W', A', R', S', \sigma')$ with $W = W'$, $A' = A \cup \{a_0\}$, $R' = R$, $S' = S \cup \{(w, a, a_0)\}$, and $\sigma'(w', a) = \Delta$, for some maximal consistent set Δ containing $\{\psi : \Box_2\psi \in \sigma(w, a)\} \cup \{\neg\phi\}$.

Definition .3 Given an indexed pseudo-model $\mathfrak{M} = (W, A, R, S, \sigma)$ and a 1-defect (resp. a 2-defect) (ϕ, w, a), the $(1, \phi, w, a)$*-update* (resp. $(2, \phi, w, a)$*-update*) of $\mathfrak{M}$ is $\mathfrak{M}'_1$ (resp. $\mathfrak{M}'_2$) as constructed above.

We can now prove that $\mathsf{K} \oplus \mathsf{K}$ is the logic of indexed frames.

Fix a maximal consistent set Σ_0. Let us construct a chain of indexed pseudo-models

$$(\mathfrak{M}^k)_{k\in\omega} = (W^k, A^k, R^k, S^k, \sigma^k)_{k\in\omega}$$

such that, for all k,

i. Σ_0 is in the image of σ^k;

ii. $W^k \subseteq W^{k+1} \subseteq \mathbb{Q}$ and $A^k \subseteq A^{k+1} \subseteq \mathbb{Q}$;

iii. $R^k \subseteq R^{k+1}$ and $S^k \subseteq S^{k+1}$;

iv. $\sigma^{k+1}(w, a) = \sigma^k(w, a)$ if $(w, a) \in W^k \times A^k$.

Initial step: Take $w_0, a_0 \in \mathbb{Q}$ and set $W^0 = \{w_0\}$, $A^0 = \{a_0\}$, $R^0 = S^0 = \varnothing$, and $\sigma^0(w_0, a_0) = \Sigma_0$.

Recursive step. Let $(i_n, \psi_n, w_n, a_n)_{n\in\omega}$ be an enumeration of the set $\{1, 2\} \times \mathcal{L} \times \mathbb{Q} \times \mathbb{Q}$ in which every element appears infinitely many times. Suppose we have constructed $\mathfrak{M}^k = (W^k, A^k, R^k, S^k, \sigma^k)$. Then:

- If $i_k = 1$ and $(w_k, a_k) \in W^k \times A^k$ and (ψ_k, w_k, a_k) is a 1-defect of $\mathfrak{M}^k$, then $\mathfrak{M}^{k+1}$ is the $(1, \psi_k, w_k, a_k)$-update of $\mathfrak{M}^k$;

- If $i_k = 2$ and $(w_k, a_k) \in W^k \times A^k$ and (ψ_k, w_k, a_k) is a 2-defect of $\mathfrak{M}^k$, then $\mathfrak{M}^{k+1}$ is the $(2, \psi_k, w_k, a_k)$-update of $\mathfrak{M}^k$;
- Otherwise, $\mathfrak{M}^{k+1} = \mathfrak{M}^k$.

Finally, let $\mathfrak{M}^\omega = (W^\omega, A^\omega, R^\omega, S^\omega, \sigma^\omega)$, where:

- $W^\omega = \bigcup_{k\in\omega} W^k$; $A^\omega = \bigcup_{k\in\omega} A^k$;
- $R^\omega = \bigcup_{k\in\omega} R^k$; $S^\omega = \bigcup_{k\in\omega} S^k$;
- σ^ω is the unique function such that $\sigma^\omega|_{W^k \times A^k} = \sigma^k$ for all k.

We have:

Lemma .4 *$\mathfrak{M}^\omega$ is an indexed pseudo-model with no defects.*

Proof. The fact that $\mathfrak{M}^\omega$ is an indexed pseudo-model is rather straightforward. Suppose $\Box_1\phi \in \sigma(w,a)$ and $R^\omega_a ww'$ for some $\phi \in \mathcal{L}$, $w, w' \in W^\omega$ and $a \in A^\omega$. Let $k \in \omega$ be the least natural number such that $w, w' \in W^k$ and $a \in A^k$. Then we have that $\Box_1\phi \in \sigma^k(w,a)$ and $R^k_a ww'$, and thus $\phi \in \sigma^k(w',a) = \sigma(w',a)$. Therefore, (C1) is satisfied (and (C2) too via an analogous reasoning).

Let us now see there are no 1-defects (the proof that there are no 2-defects is completely analogous). Suppose that (ϕ, w, a) is a 1-defect of $\mathfrak{M}^\omega$, i.e., $\neg\Box_1\phi \in \sigma^\omega(w,a)$ yet $\phi \in \sigma^\omega(w',a)$ whenever $R^\omega_a ww'$.

Let us consider the least $k \in \omega$ such that $(w,a) \in W^k \times A^k$ and the least $n \geq k$ such that $(1, \phi, w, a) = (i_n, \psi_n, w_n, a_n)$ in the aforementioned enumeration. Then we have that (ϕ, w, a) is a 1-defect in $\mathfrak{M}^n$, and therefore it gets "fixed" in the update $\mathfrak{M}^{n+1}$, i.e., there exists some $w' \in W^{n+1} \setminus W^n$ such that $R^{n+1}_a ww'$ and $\neg\phi \in \sigma^{n+1}(w',a)$. But this means that $R^\omega_a ww'$ and $\neg\phi \in \sigma^\omega(w',a)$: a contradiction. $\Box$

Now,

Lemma .5 (Truth lemma.) *Define a valuation V on $\mathfrak{M}^\omega$ by:*

$$V(p) = \{(w,a) \in W^\omega \times A^\omega : p \in \sigma^\omega(w,a)\}.$$

Then for all $w \in W^\omega$, $a \in A^\omega$ and $\phi \in \mathcal{L}$, $\mathfrak{M}^\omega, w, a \models \phi$ if and only if $\phi \in \sigma^\omega(w,a)$.

Proof. By induction on the structure of ϕ. If $\phi = p$, then the definition of V gives us the result trivially. The induction steps corresponding to $\neg\phi$ and $\phi_1 \wedge \phi_2$ are straightforward.

Now let $\phi = \Box_1\psi$. If $w, a \models \Box_1\psi$, this means that $(w', a) \models \psi$ for every $w' \in W^\omega$ such that $R^\omega_a ww'$. But then by induction hypothesis $\psi \in \sigma^\omega(w',a)$ whenever $R_a ww'$. So, if $\Box_1\psi \notin \sigma^\omega(w,a)$, then $\neg\Box_1\psi \in \sigma^\omega(w,a)$ and thus (ψ, w, a) is a 1-defect, in contradiction with Lemma .4. Thus $\Box_1\psi \in \sigma^\omega(w,a)$. Conversely, suppose $\Box_1\psi \in \sigma^\omega(w,a)$ and $R^\omega_a ww'$. By (C1), this means that $\psi \in \sigma^\omega(w',a)$ which entails, by induction hypothesis, that $(w',a) \models \psi$. Since this is true for all w' with $R_a ww'$, we have that $w, a \models \Box_1\psi$.

The case $\phi = \Box_2\psi$ is analogous. $\Box$

With all this, we can prove the following theorem, from which Thm. 2.4 immediately follows:

Theorem .6 *The fusion logic $\mathsf{K} \oplus \mathsf{K}$ is complete with respect to indexed models.*

Proof. Given a consistent formula ϕ, extend it to a maximal consistent set Σ_0 and construct $\mathfrak{M}^\omega$ by the procedure described above, making sure that Σ_0 is in the image of σ^0. Then we have that there exist $w_0, a_0 \in W^\omega \times A^\omega$ such that $\sigma^\omega(w_0, a_0) = \Sigma_0 \ni \phi$, and therefore by the Truth Lemma $w_0, a_0 \models \phi$. $\Box$

Remark .7 *It is not hard to tweak this proof to show, for instance, that the fusion logic* $\mathsf{S4}_{\Box_1} \oplus \mathsf{K}_{\Box_2}$ *is the logic of indexed models* (W, A, R, S) *where* R_a *is a preorder for all* $a \in A$, *or that* $\mathsf{K}_{\Box_1} \oplus \mathsf{S5}_{\Box_2}$ *is the logic of indexed models wherein the* S_w *are equivalence relations. More generally, this procedure can easily be tweaked in order to provide a proof for every individual instance of Thm. 2.9. However, this proof can help us to go beyond that Theorem and allows us to show, for instance, that the result is true of the logic of serial frames, i.e.,* $(\mathsf{K}+\Diamond\top)\circ(\mathsf{K}+\Diamond\top) = (\mathsf{K}+\Diamond\top)\oplus(\mathsf{K}+\Diamond\top)$.

Proof of Lemma 3.3. By induction on ϕ. The cases for $\phi = p$ and $\phi = \top$ are trivial, and so is the inductive step for $\phi = \neg\psi$.

Case $\phi = \psi_1 \wedge \psi_2$. If $\mathfrak{M}, w_k, a_k \models \psi_1 \wedge \psi_2$, then $\mathfrak{M}, w_k, a_k \models \psi_i$ for $i = 1$ and 2. But then, since $\mathsf{md}\,\psi_i \leq \mathsf{md}\,\psi \leq n - k$, we have by induction hypothesis that $\mathfrak{M}^f, w_k, a_k \models \psi_i$ and thus $\mathfrak{M}^f, w_k, a_k \models \phi$. The converse is analogous.

Case $\phi = \Box_1\psi$. Suppose that $\mathfrak{M}^f, w_k, a_k \models \Box_1\psi$ and take w such that $R_{a_k} w_k w$. Note that $k < n$ because $n - k \geq \mathsf{md}\,\Box_1\psi > 0$, and thus F_{k+1} is defined and contains an element $\alpha\mathbf{R}(w_{k+1}, a_{k+1})$ such that $a_{k+1} = a_k$, $R_{a_k} w_k w_{k+1}$ (and therefore $R^f_{a_k} w_k w_{k+1}$) and $(w_{k+1}, a_{k+1}) \sim_{\phi_0} (w, a_k)$. We have that $\mathfrak{M}^f, w_{k+1}, a_{k+1} \models \psi$ and, since $n - (k+1) = n - k - 1 \geq \mathsf{md}(\Box_1\psi) - 1 = \mathsf{md}\,\psi$, induction hypothesis gives us that $\mathfrak{M}, w_{k+1}, a_{k+1} \models \psi$. By the $\sim_{\phi_0}$ relation, this means that $\mathfrak{M}, w, a_k \models \psi$, and we have thus proven that $\mathfrak{M}, w_k, a_k \models \Box_1\psi$.

Conversely, suppose $\mathfrak{M}, w_k, a_k \models \Box_1\psi$ and $R^f_{a_k} w_k w$. We have that $R_{a_k} w_k w$ and thus $\mathfrak{M}, w, a_k \models \psi$. Since $\alpha\mathbf{R}(w, a_k) \in F$ and its length is $k + 1$, and since $n - (k+1) \geq \mathsf{md}\,\psi$, induction hypothesis applies and we have that $\mathfrak{M}^f, w, a_k \models \psi$. This entails $\mathfrak{M}^f, w_k, a_k \models \Box_1\psi$.

The case $\phi = \Box_2\psi$ is completely analogous. □

Proof of Prop. 4.3.

(T1) to (T6) are proven in Prop. 3 of [7] and Lemma 2 of [3].

(T7) $\vdash @_n m \to @_i K @_n m$.

$\vdash @_n m \to K @_n m$	(Rigid$_=$)
$\vdash @_i @_n m \to @_i K @_n m$	(K$_@$+Nec$_@$)
$\vdash @_n m \to @_i K @_n m$	(T1)

The derivation of $\vdash @_n \neg m \to @_i K @_n \neg m$ is identical but using (Rigid$_{\neq}$ + Selfdual) in the first step.

(T8) $\vdash @_n \hat{F} m \wedge @_m \phi \to @_n \hat{F} \phi$.

$\vdash @_n \hat{F} m \wedge @_m \phi \to @_n \hat{F} m \wedge F @_m \phi$	(Back)
$\vdash @_n \hat{F} m \wedge @_m \phi \to @_n \hat{F} m \wedge @_n F @_m \phi$	(Nec$_@$+K$_@$+T1)
$\vdash @_n \hat{F} m \wedge @_m \phi \to @_n \hat{F}(m \wedge @_m \phi)$	(by modal reasoning: $\Box A \wedge \Diamond B \to \Diamond(A \wedge B)$)
$\vdash @_n \hat{F} m \wedge @_m \phi \to @_n \hat{F} \phi$	(by T2: $\vdash m \wedge @_m \phi \to \phi$)

(T9) $\vdash @_n F \psi \wedge @_n \hat{F} m \to @_m \psi$.

$\vdash @_n F \psi \wedge @_n \hat{F} m \to @_n \hat{F}(m \wedge \psi)$	(modal reasoning: $\Box A \wedge \Diamond B \to \Diamond(A \wedge B)$)
$\vdash m \wedge \psi \to @_m \psi$	(T2)
$\vdash @_n F \psi \wedge @_n \hat{F} m \to @_n \hat{F} @_m \psi$	(two above lines)
$\vdash \hat{F} @_m \psi \to @_m \psi$	(dual of Back)
$\vdash @_n F \psi \wedge @_n \hat{F} m \to @_m \psi$	(two above lines plus (T1))

Before showing (R1), let us show this rule:

(Name')	If $\vdash \phi \to @_m\psi$ and m is fresh, then $\vdash \phi \to \psi$.
$\vdash \phi \to @_m\psi$	(Premise)
$\vdash @_m\phi \to @_m@_m\psi$	(Nec$_@$+K$_@$)
$\vdash @_m\phi \to @_m\psi$	(Agree)
$\vdash @_m(\phi \to \psi)$	(T5)
$\vdash \phi \to \psi$	(Name)

With this:

(R1) If $\vdash @_n\hat{F}m \wedge @_m\phi \to \psi$,
then $\vdash @_n\hat{F}\phi \to \psi$,
with $m \neq n$ fresh in ϕ and ψ.

$\vdash @_n\hat{F}m \wedge @_m\phi \to \psi$	(Premise)
$\vdash @_i@_n\hat{F}m \wedge @_i@_m\phi \to @_i\psi$	(Nec$_@$+K$_@$, i fresh)
$\vdash @_n\hat{F}m \wedge @_m\phi \to @_i\psi$	(T1)
$\vdash @_n\hat{F}m \wedge @_m\phi \to @_m@_i\psi$	(Nec$_@$+K$_@$+T1)
$\vdash @_n\hat{F}m \to @_m(\phi \to @_i\psi)$	(T5)
$\vdash @_nF(\phi \to @_i\psi)$	(BG)
$\vdash @_n\hat{F}\phi \to @_n\hat{F}@_i\psi$	($(\Box(A \to B) \to (\Diamond A \to \Diamond B))$)
$\vdash @_n\hat{F}\phi \to @_n@_i\psi$	(Back)
$\vdash @_n\hat{F}\phi \to @_i\psi$	(T1)
$\vdash @_n\hat{F}\phi \to \psi$	(Name')

Proof sketch of Thm. 4.13. Like Thm. 3.2, this amounts to showing that, given a model satisfying a formula ϕ_0, there is a finite submodel satisfying it.

We define nom_{ϕ_0} to be the (finite) set of nominal variables occuring in ϕ_0, we define $\mathbf{R}, \mathbf{S}$ as in Thm. 3.2 and, for $n \in \mathsf{nom}_{\phi_0}$, we let $(w, a)\mathbf{A}_n(w', a')$ iff $w = w'$ and $a' = \underline{n}$. Given a formula ϕ, we let mod ϕ be the total number of K, F and $@_n$ modalities occurring in ϕ and we let $N =$ mod ϕ_0. We construct a finite set F of chains of length at most N, with the property that, for each relation $\mathbf{T} \in \{\mathbf{R}, \mathbf{S}, \mathbf{A}_n : n \in \mathsf{nom}_{\phi_0}\}$, and each $\alpha \in F$ of length less than N, at least one $\mathbf{T}$-successor of α per equivalence class occurs in F.

Then we consider $\mathfrak{M}^f$ to the the corresponding restriction of $\mathfrak{M}$ to F and we prove that, given a chain α of length $k \leq N$ and a subformula ψ of ψ_0 with mod $\psi \leq N-k$, it is the case that $\mathfrak{M}, \mathsf{last}\,\alpha \models \psi$ iff $\mathfrak{M}^f, \mathsf{last}\,\alpha \models \psi$. This is almost identical to the proof of Lemma 3.3 with the addition of a straightforward induction step for the case $\psi = @_n\theta$. This finishes the proof. $\Box$

Proof of Prop. 5.6. First we note that if a formula has no free variables, the assignment g does not play a role in the semantics (and thus $\mathfrak{M}, \Gamma, [n], g \models \psi$ iff $\mathfrak{M}, \Gamma, [n], g' \models \psi$ for any g, g') and, with this in mind, we first prove:

If ψ is a sentence, then $\mathfrak{M}, \Gamma, [n], g \models \psi$ iff $@_n\psi \in \Gamma$. (*)

This suffices to prove our result: let $x_1, ..., x_k$ be all the free variables of ϕ. Then $\mathfrak{M}, \Gamma, [n], g \models \phi$ if and only if (by Lemma 5.5, noting that $g(x_i) = [n_i] = \underline{n}_i$)

$$\mathfrak{M}, \Gamma, [n], g \models \phi[x_i/n_i]_{i=1}^{k},$$

if and only if (by the result we just proved, noting that $\phi[x_i/n_i]_{i=1}^k$ has no free variables) $@_n\phi[x_i/n_i]_{i=1}^k \in \Gamma$.

We prove (*) by induction on the length of ψ. It is exactly like the proof of Lemma 4.11, with one extra induction step:
$@_n{\downarrow}x.\psi \in \Gamma$ if and only if (by the (DA) axiom) $@_n\psi[x/n] \in \Gamma$, if and only if (by induction hypothesis, since $\psi[x/n]$ has no free variables) $\mathfrak{M}, \Gamma, [n], g \models \psi[x/n]$, if and only if (because the choice of g does not affect the truth value of a sentence) $\mathfrak{M}, \Gamma, [n], g_n^x \models \psi[x/n]$, if and only if (by Lemma 5.5) $\mathfrak{M}, \Gamma, [n], g_n^x \models \psi$, which is the same as $\mathfrak{M}, \Gamma, [n], g \models {\downarrow}x.\psi$.

References

[1] Areces, C. and B. ten Cate, *Hybrid logics*, Handbook of Modal Logic (2006), pp. 821–868.

[2] Blackburn, P., M. de Rijke and Y. Venema, "Modal Logic," Cambridge University Press, 2001.

[3] Blackburn, P. and B. ten Cate, *Pure extensions, proof rules, and hybrid axiomatics*, Studia Logica **84** (2006), pp. 277–322.

[4] Gabbay, D. M., A. Kurucz, F. Wolter and M. Zakharyaschev, "Many-dimensional Modal Logics: Theory and Applications," Elsevier North Holland, 2003.

[5] Gargov, G. and V. Goranko, *Modal logic with names*, Journal of Philosophical Logic **22** (1993), pp. 607–636.

[6] Passy, S. and T. Tinchev, *An essay in combinatory dynamic logic*, Information and Computation **93** (1991), pp. 263–332.

[7] Sano, K., *Axiomatizing epistemic logic of friendship via tree sequent calculus*, in: *International Workshop on Logic, Rationality and Interaction*, Springer, 2017, pp. 224–239.

[8] Seligman, J., F. Liu and P. Girard, *Logic in the community*, in: *Indian Conference on Logic and Its Applications*, Springer, 2011, pp. 178–188.

[9] Seligman, J., F. Liu and P. Girard, *Facebook and the epistemic logic of friendship*, in: *Proceedings of the 14th Conference on Theoretical Aspects of Rationality and Knowledge (TARK 2013)*, 2013, pp. 230–238.

[10] Seligman, J., F. Liu and P. Girard, *Knowledge, friendship and social announcements*, in: *Logic across the university: Foundations and applications: Proceedings of the Tsinghua Logic Conference*, 2013.

[11] Wolter, F., *Fusions of modal logics revisited.*, Advances in modal logic **1** (1996), pp. 361–379.

Counterfactuals and dependencies on causal teams: expressive power and deduction systems

Fausto Barbero [1]

Department of Philosophy, University of Helsinki
PL 24 (Unioninkatu 40), 00014 Helsinki, Finland

Fan Yang [2]

Department of Mathematics and Statistics, University of Helsinki
PL 68 (Pietari Kalmin katu 5), 00014 Helsinki, Finland

Abstract

We analyze the causal-observational languages that were introduced in Barbero and Sandu (2018), which allow discussing interventionist counterfactuals and functional dependencies in a unified framework. In particular, we systematically investigate the expressive power of these languages in causal team semantics, and we provide complete natural deduction calculi for each language. As an intermediate step towards the completeness, we axiomatize the languages over a generalized version of causal team semantics, which turns out to be interesting also in its own right.

Keywords: Interventionist counterfactuals, causal teams, dependence logic, team semantics.

1 Introduction

Counterfactual conditionals express the modality of *irreality*: they describe what *would* or *might* be the case in circumstances which diverge from the actual state of affairs. Pinning down the exact meaning and logic of counterfactual statements has been the subject of a large literature (see e.g. [15]). We are interested here in a special case: the *interventionist* counterfactuals, which emerged from the literature on causal inference ([14,13,11]). Under this reading, a conditional $\mathbf{X} = \mathbf{x} \;\Box\!\!\rightarrow \psi$ states that ψ would hold if we were to intervene on the given system, by subtracting the variables $\mathbf{X}$ to their current causal mechanisms and forcing them to take the values $\mathbf{x}$.

The *logic* of interventionist counterfactuals has been mainly studied in the semantical context of *deterministic causal models* ([7,8,3,19]), which consist

[1] The author was supported by grant 316460 of the Academy of Finland.

[2] The author was supported by Research Funds of the University of Helsinki and grant 308712 of the Academy of Finland.

of an assignment of values to variables together with a system of *structural equations* that describe the causal connections. In [1], causal models were generalized to *causal teams*, in the spirit of *team semantics* ([12,16]), by allowing a *set* of assignments (a "*team*") instead of a single assignment. This opens the possibility of describing e.g. *uncertainty, observations*, and *dependencies.*

One of the main reasons for introducing causal teams was the possibility of comparing the logic of dependencies of causal nature (those definable in terms of interventionist counterfactuals) against that of contingent dependencies (such as those studied in the literature on team semantics, or in database theory) in a unified semantic framework. [1,2] give anecdotal evidence of the interactions between the two kinds of dependence, but offer no general axiomatizations for languages that also involve contingent dependencies. In this paper we fill this gap in the literature by providing complete deduction systems (in natural deduction style) for the languages $\mathcal{COD}$ and $\mathcal{CO}_{\mathbb{V}}$ (from [1]), which enrich the basic counterfactual language $\mathcal{CO}$, respectively, with atoms of functional dependence $=(\mathbf{X}; Y)$ ("Y is functionally determined by $\mathbf{X}$"), or with the intuitionistic disjunction $\mathbb{V}$, in terms of which functional dependence is definable. We also give semantical characterizations, for $\mathcal{COD}$, $\mathcal{CO}_{\mathbb{V}}$ and the basic counterfactual language $\mathcal{CO}$, in terms of definability of classes of causal teams.

The strategy of the completeness proofs is the following. We introduce a *generalized* causal team semantics, which encodes uncertainty over causal models, not only over assignments. (This semantics is used as a tool towards completeness, but also has independent interest.) We then prove completeness results for this semantics, by incorporating techniques developed in [17,5] for the pure (non-causal) team context. Finally, we extend the calculi to completeness over causal teams by adding axioms which capture the property of being a causal team (i.e. encoding *certainty* about the causal connections).

The paper is organized as follows. Section 2 introduces the formal languages and two kinds of semantics. Section 3 deepens the discussion of the functions which describe causal mechanisms, addressing issues of definability and the treatment of dummy arguments. Section 4 characterizes semantically the language $\mathcal{CO}$ and reformulates in natural deduction form the $\mathcal{CO}$ calculi that come from [2]. Section 5 gives semantical characterizations for $\mathcal{COD}$ and $\mathcal{CO}_{\mathbb{V}}$, and complete natural deduction calculi for both kinds of semantics.

2 Syntax and semantics

2.1 Formal languages

Let us start by fixing the syntax. Each of the languages considered in this paper is parametrized by a (finite) **signature** σ, i.e. a pair (Dom, Ran), where Dom is a nonempty finite set of **variables**, and Ran is a function that associates to each variable $X \in$ Dom a nonempty finite set Ran(X) (called the **range** of X) of **constant symbols** or **values.** [3] We reserve the Greek letter σ for signatures.

[3] Note that we do not encode a distinction between exogenous and endogenous variables into the signatures, as done in [8]. Instead, we follow the style of Briggs [3]. Doing so will

We use a boldface capital letter $\mathbf{X}$ to stand for a sequence $\langle X_1, \dots, X_n \rangle$ of variables; similarly a boldface lower case letter $\mathbf{x}$ stands for a sequence $\langle x_1, \dots, x_n \rangle$ of values. We will sometimes abuse notation and treat $\mathbf{X}$ and $\mathbf{x}$ as sets. We write $\mathrm{Ran}(\mathbf{X}) = \mathrm{Ran}(X_1) \times \cdots \times \mathrm{Ran}(X_n)$.

Now fix a signature $\sigma = (Dom, Ran)$. An atomic σ-formula is an equation $X = x$, where $X \in \mathrm{Dom}$ and $x \in \mathrm{Ran}(X)$. The conjunction of n equations $(\dots((X_1 = x_1 \wedge X_2 = x_2) \wedge X_3 = x_3) \wedge \cdots \wedge X_{n-1} = x_{n-1}) \wedge X_n = x_n$ is abbreviated as $X_1 = x_1 \wedge \cdots \wedge X_n = x_n$ or further as $\mathbf{X} = \mathbf{x}$, and is also called an equation. Compound formulas of the basic language $CO[\sigma]$ are formed by the grammar:

$$\alpha ::= X = x \mid \neg\alpha \mid \alpha \wedge \alpha \mid \alpha \vee \alpha \mid \mathbf{X} = \mathbf{x} \mathrel{\Box\!\!\rightarrow} \alpha$$

where $\mathbf{X} \cup \{X\} \subseteq \mathrm{Dom}$, $x \in \mathrm{Ran}(X)$, $\mathbf{x} \in \mathrm{Ran}(\mathbf{X})$. The connective $\Box\!\!\rightarrow$ is used to form *interventionist counterfactuals.* We abbreviate $\neg(X = x)$ as $X \neq x$, and $X = x \wedge X \neq x$ as $\bot$. Throughout the paper, we reserve the first letters of the Greek alphabet, $\alpha, \beta, \dots$ for $CO[\sigma]$-formulas.

Let us compare our language $CO[\sigma]$ with the existing interventionist counterfactual languages in the literature. The original formulation of $CO[\sigma]$ in [1,2] includes in the syntax another conditional $\supset$, called *selective implication*, which can be defined in our setting in terms of negation and disjunction as $\alpha \supset \beta := \neg\alpha \vee \beta$. Our primitive connective negation $\neg$ was treated in [1,2] as a defined connective. The language $CO[\sigma]$ as defined here can also be seen as the fragment of the language considered by Briggs in [3] in which occurrences of $\vee$ and $\neg$ are not allowed in the antecedents of $\Box\!\!\rightarrow$. Differently from the language for counterfactuals defined by Halpern in [9], in our language $CO[\sigma]$ nesting of counterfactuals to the right of $\Box\!\!\rightarrow$ is allowed (i.e., in $\mathbf{X} = \mathbf{x} \mathrel{\Box\!\!\rightarrow} \alpha$, α can still contain counterfactuals), and any type of variables (exogenous or endogenous) can occur in the antecedents of counterfactuals (i.e., in $\mathbf{X} = \mathbf{x}$) as we do not distinguish exogenous and endogenous variables in the signature σ.

We study in this paper also two extensions of $CO[\sigma]$, obtained by adding the *intuitionistic disjunction* $\mathbin{\mathbb{V}}$, or the dependence atoms $=\!(\mathbf{X}; Y)$:

- $CO_{\mathbb{V}}[\sigma]$: $\varphi ::= X = x \mid \neg\alpha \mid \varphi \wedge \varphi \mid \varphi \vee \varphi \mid \varphi \mathbin{\mathbb{V}} \varphi \mid \mathbf{X} = \mathbf{x} \mathrel{\Box\!\!\rightarrow} \varphi$
- $CO\mathcal{D}[\sigma]$: $\varphi ::= X = x \mid =\!(\mathbf{X}; Y) \mid \neg\alpha \mid \varphi \wedge \varphi \mid \varphi \vee \varphi \mid \mathbf{X} = \mathbf{x} \mathrel{\Box\!\!\rightarrow} \varphi$

Note that we only allow the negation $\neg$ to occur in front of $CO[\sigma]$-formulas.

2.2 Causal teams

We now define the team semantics of our logics over causal teams. We first recall the definition of causal teams adapted from [2].

Fix a signature $\sigma = (\mathrm{Dom}, \mathrm{Ran})$. An **assignment** over σ is a mapping $s : \mathrm{Dom} \to \bigcup_{X \in \mathrm{Dom}} \mathrm{Ran}(X)$ such that $s(X) \in \mathrm{Ran}(X)$ for each $X \in \mathrm{Dom}$.[4] Denote by $\mathbb{A}_\sigma$ the set of all assignments over σ. A **team** T over σ is a set of assignments

result in more general completeness results.

[4] We identify syntactical variables and values with their semantical counterpart, following the conventions in most literature on interventionist counterfactuals, e.g. [7,8,3,19]. In this convention distinct symbols (e.g., x, x') denote distinct objects.

over σ, i.e., $T \subseteq \mathbb{A}_\sigma$.

A **system of functions** $\mathcal{F}$ over σ is a function that assigns to each variable V in a domain $\mathrm{En}(\mathcal{F}) \subseteq \mathrm{Dom}$ a set $PA_V^{\mathcal{F}} \subseteq \mathrm{Dom} \setminus \{V\}$ of **parents** of V, and a function $\mathcal{F}_V : \mathrm{Ran}(PA_V^{\mathcal{F}}) \to \mathrm{Ran}(V)$.[5] Variables in the set $\mathrm{En}(\mathcal{F})$ are called **endogenous variables** of $\mathcal{F}$, and variables in $\mathrm{Ex}(\mathcal{F}) = \mathrm{Dom} \setminus \mathrm{En}(\mathcal{F})$ are called **exogenous variables** of $\mathcal{F}$.

Denote by $\mathbb{F}_\sigma$ the set of all systems of functions over σ, which is clearly finite. We say that an assignment $s \in \mathbb{A}_\sigma$ is **compatible** with a system of functions $\mathcal{F} \in \mathbb{F}_\sigma$ if for all endogenous variables $V \in \mathrm{En}(\mathcal{F})$, $s(V) = \mathcal{F}_V(s(PA_V^{\mathcal{F}}))$.

Definition 2.1 *A* ***causal team*** *over a signature σ is a pair $T = (T^-, \mathcal{F})$ consisting of*

- *a team T^- over σ, called the* ***team component*** *of T,*
- *and a system of functions $\mathcal{F}$ over σ, called the* ***function component*** *of T,*

where all assignments $s \in T^-$ are compatible with the function component $\mathcal{F}$.

Any system $\mathcal{F} \in \mathbb{F}_\sigma$ of functions can be naturally associated with a (directed) graph $G_{\mathcal{F}} = (\mathrm{Dom}, E_{\mathcal{F}})$, defined as $(X, Y) \in E_{\mathcal{F}}$ iff $X \in PA_Y^{\mathcal{F}}$. We say that $\mathcal{F}$ is **recursive** if $G_{\mathcal{F}}$ is acyclic, i.e., for all $n \geq 0$, $E_{\mathcal{F}}$ has no subset of the form $\{(X_0, X_1), (X_1, X_2), \ldots, (X_{n-1}, X_n), (X_n, X_0)\}$. The graph of a causal team T, denoted as G_T, is the associated graph of its function component. We call T **recursive** if G_T is acyclic. Throughout this paper, for simplicity we assume that all causal teams that we consider are recursive.

Intuitively, a causal team T may be seen as representing an assumption concerning the causal relationships among the variables in Dom (as encoded in $\mathcal{F}$) together with a range of hypotheses concerning the actual state of the system (as encoded in T^-). We now illustrate this idea in the following example.

Example 2.2 *The following diagram illustrates a causal team $T = (T^-, \mathcal{F})$.*

T^-:

$U \to$	$X \to$	$Y \to$	Z
0	0	1	2
1	1	2	6

$$\begin{cases} \mathcal{F}_X(U) &= U \\ \mathcal{F}_Y(X) &= X + 1 \\ \mathcal{F}_Z(X, Y, U) &= 2 * Y + X + U \end{cases}$$

The table on the left represents a team T^- consisting of two assignments, each of which is tabulated in the obvious way as a row in the table. For instance, the assignment s of the first row is defined as $s(U) = s(X) = 0$, $s(Y) = 1$ and $s(Z) = 2$. The arrows in the upper part of the table represent the graph G_T of the causal team T. For instance, the arrow from U to Z represents the edge (U, Z) in G_T. The graph contains no cycles, thus the causal team T is recursive. The variable U with no incoming arrows is an exogenous variable. The other variables are endogenous variables, namely, $\mathrm{En}(\mathcal{F}) = \{X, Y, Z\}$. The function component is determined by the system of functions on the right of the above diagram. Each equation defines the "law" that generates the values of an endogenous variable.

---

[5] We identify the set $PA_V^{\mathcal{F}}$ with a sequence, in a fixed lexicographical ordering.

Let $S = (S^-, \mathcal{F})$ and $T = (T^-, \mathcal{G})$ be causal teams over the same signature. We call S a **causal subteam** of T, denoted as $S \subseteq T$, if $S^- \subseteq T^-$ and $\mathcal{F} = \mathcal{G}$.

An equation $\mathbf{X} = \mathbf{x}$ is said to be **inconsistent** if it contains two conjuncts $X = x$ and $X = x'$ with distinct values x, x'; otherwise it is said to be **consistent**.

Definition 2.3 (Intervention) *Let $T = (T^-, \mathcal{F})$ be a causal team over some signature $\sigma = (\mathrm{Dom}, \mathrm{Ran})$. Let $\mathbf{X} = \mathbf{x} (= X_1 = x_1 \wedge \cdots \wedge X_n = x_n)$ be a consistent equation over σ. The* ***intervention*** *$do(\mathbf{X} = \mathbf{x})$ on T is the procedure that generates a new causal team $T_{\mathbf{X}=\mathbf{x}} = (T^-_{\mathbf{X}=\mathbf{x}}, \mathcal{F}_{\mathbf{X}=\mathbf{x}})$ over σ defined as follows:*

- *$\mathcal{F}_{\mathbf{X}=\mathbf{x}}$ is the restriction of $\mathcal{F}$ to $\mathrm{En}(\mathcal{F}) \setminus \mathbf{X}$,*
- *$T^-_{\mathbf{X}=\mathbf{x}} = \{s_{\mathbf{X}=\mathbf{x}} \mid s \in T^-\}$, where each $s_{\mathbf{X}=\mathbf{x}}$ is an assignment compatible with $\mathcal{F}_{\mathbf{X}=\mathbf{x}}$ defined (recursively) as*

$$s_{\mathbf{X}=\mathbf{x}}(V) = \begin{cases} x_i & \text{if } V = X_i, \\ s(V) & \text{if } V \notin \mathrm{En}(T) \cup \mathbf{X}, \\ \mathcal{F}_V(s_{\mathbf{X}=\mathbf{x}}(PA_V^{\mathcal{F}})) & \text{if } V \in \mathrm{En}(T) \setminus \mathbf{X} \end{cases}$$

Example 2.4 *Recall the recursive causal team T in Example 2.2. By applying the intervention $do(X = 1)$ to T, we obtain a new causal team $T_{X=1} = (T^-_{X=1}, \mathcal{F}_{X=1})$ as follows. The function component $\mathcal{F}_{X=1}$ is determined by the equations:*

$$\begin{cases} (\mathcal{F}_{X=1})_Y(X) & = X + 1 \\ (\mathcal{F}_{X=1})_Z(X, Y) & = 2 * Y + X + U \end{cases}$$

The endogenous variable X of the original team T becomes exogenous in the new team $T_{X=1}$, and the equation $\mathcal{F}_X(U) = U$ for X is now removed. The new team component $T^-_{X=1}$ is obtained by the rewriting procedure illustrated below:

U	X	Y	Z
0	**1**	...	...
1	**1**	...	...

⇝

U	X	Y	Z
0	1	**2**	...
1	1	**2**	...

⇝

U	X	Y	Z
0	1	2	**5**
1	1	2	**6**

In the first step, rewrite the X-column with value 1. Then, update (recursively) the other columns using the functions from $\mathcal{F}_{X=1}$. In this step, only the columns that correspond to "descendants" of X will be modified, and the order in which these columns should be updated is completely determined by the (acyclic) graph $G_{T_{X=1}}$ of $T_{X=1}$. Since the variable X becomes exogenous after the intervention, all arrows pointing to X have to be removed, e.g., the arrow from U to X. We refer the reader to [2] for more details and justification for this rewriting procedure.

Definition 2.5 *Let φ be a formula of the language $\mathcal{CO}_{\mathbb{V}}[\sigma]$ or $\mathcal{COD}[\sigma]$, and $T = (T^-, \mathcal{F})$ a causal team over σ. We define the satisfaction relation $T \models^c \varphi$ (or simply $T \models \varphi$) over causal teams inductively as follows:*

- *$T \models X = x \iff$ for all $s \in T^-$, $s(X) = x$.*[6]
- *$T \models =\!(\mathbf{X}; Y) \iff$ for all $s, s' \in T^-$, $s(\mathbf{X}) = s'(\mathbf{X})$ implies $s(Y) = s'(Y)$.*

[6] Note once more that the symbol x is used as both a syntactical and a semantical object.

- $T \models \neg\alpha \iff$ *for all* $s \in T^-$, $(\{s\}, \mathcal{F}) \not\models \alpha$.
- $T \models \varphi \wedge \psi \iff T \models \varphi$ *and* $T \models \psi$.
- $T \models \varphi \vee \psi \iff$ *there are two causal subteams* T_1, T_2 *of* T *such that* $T_1^- \cup T_2^- = T^-$, $T_1 \models \varphi$ *and* $T_2 \models \psi$.
- $T \models \varphi \mathbin{\mathbb{V}} \psi \iff T \models \varphi$ *or* $T \models \psi$.
- $T \models \mathbf{X} = \mathbf{x} \mathrel{\Box\!\!\rightarrow} \varphi \iff \mathbf{X} = \mathbf{x}$ *is inconsistent or* $T_{\mathbf{X}=\mathbf{x}} \models \varphi$.

We write a dependence atom $=(;X)$ with an empty first component as $=(X)$. The semantic clause for $=(X)$ reduces to:

- $T \models =(X)$ iff for all $s, s' \in T^-$, $s(X) = s'(X)$.

Intuitively, the atom $=(X)$ states that X has a constant value in the team. It is easy to verify that dependence atoms are definable in $\mathcal{CO}_{\mathbb{V}}[\sigma]$:

$$=(Y) \equiv \mathop{\mathbb{V}}_{y \in \mathrm{Ran}(Y)} Y = y \quad \text{and} \quad =(\mathbf{X}; Y) \equiv \bigvee_{\mathbf{x} \in \mathrm{Ran}(\mathbf{X})} (\mathbf{X} = \mathbf{x} \wedge =(Y)). \tag{1}$$

It is easy to verify that the *selective implication* $\alpha \supset \varphi := \neg\alpha \vee \varphi$, introduced originally in [1], has the same semantic clause as that in [1]:

- $T \models \alpha \supset \varphi \iff T^\alpha \models \varphi$, where T^α is the (unique) causal subteam of T with team component $\{s \in T^- \mid \{s\} \models \alpha\}$.

Example 2.6 *Consider the causal team* T *and the intervention* $do(X = 1)$ *from Examples 2.2 and 2.4. Clearly,* $T_{X=1} \models Y = 2$*, and thus* $T \models X = 1 \mathrel{\Box\!\!\rightarrow} Y = 2$*. We also have that* $T \models =(Y; Z)$*, while* $T_{X=1} \not\models =(Y; Z)$ *(contingent dependencies are not in general preserved by interventions). Observe that* $T \models Y \neq 2 \vee Y = 2$*, while* $T \not\models Y \neq 2 \mathbin{\mathbb{V}} Y = 2$.

2.3 Generalized causal teams

We introduce here a more general semantics, which will be needed as a tool towards the completeness results for $\mathcal{CO}_{\mathbb{V}}$ and $\mathcal{COD}$.

Given a signature σ, write

$$\mathbb{S}\mathrm{em}_\sigma := \{(s, \mathcal{F}) \in \mathbb{A}_\sigma \times \mathbb{F}_\sigma \mid s \text{ is compatible with } \mathcal{F}\}.$$

The pairs $(s, \mathcal{F}) \in \mathbb{S}\mathrm{em}_\sigma$ can be easily identified with the *deterministic causal models* (also known as *deterministic structural equation models*) that are considered in the literature on causal inference ([14],[13], etc.). One can informally identify a causal team $T = (T^-, \mathcal{F})$ with the set

$$T^g = \{(s, \mathcal{F}) \in \mathbb{S}\mathrm{em}_\sigma \mid s \in T^-\}$$

of deterministic causal models with a uniform function component $\mathcal{F}$ throughout the team. In this section, we introduce a more general notion of causal team, called *generalized causal team*, where the function component $\mathcal{F}$ does not have to be constant thoroughout the team.

Definition 2.7 *A* ***generalized causal team*** T *over a signature* σ *is a set of pairs* $(s, \mathcal{F}) \in \mathbb{S}\mathrm{em}_\sigma$*, that is,* $T \subseteq \mathbb{S}\mathrm{em}_\sigma$.

Intuitively, a generalized causal team encodes uncertainty about which causal model governs the variables in Dom - i.e., uncertainty both on the values of the variables and on the laws that determine them. Our interest in such models here is purely technical, but probabilistic variants of them have been used e.g. to define formal notions of blame ([4,9,10]).

Distinct elements $(s,\mathcal{F}),(t,\mathcal{G})$ of the same generalized causal team may also disagree on what is the set of endogenous variables, or on whether the system is recursive or not. A generalized causal team is said to be **recursive** if, for each pair $(s,\mathcal{F})$ in the team, the associated graph $G_{\mathcal{F}}$ is recursive. In this paper we only consider recursive generalized causal teams.

For any generalized causal team T, define the **team component** of T to be the set $T^- := \{s \mid (s,\mathcal{F}) \in T \text{ for some } \mathcal{F}\}$. A **causal subteam** of T is a subset S of T, denoted as $S \subseteq T$. The **union** $S \cup T$ of two generalized causal teams S, T is their set-theoretic union.

A causal team T can be identified with the generalized causal team T^g, which has a constant function component in all its elements. Conversely, if T is a nonempty generalized causal team in which all elements have the same function component $\mathcal{F}$, i.e., $T = \{(s,\mathcal{F}) \mid s \in T^-\}$, we can naturally identify T with the causal team

$$T^c = (T^-,\mathcal{F}).$$

In particular, a singleton generalized causal team $\{(s,\mathcal{F})\}$ corresponds to a singleton causal team $(\{s\},\mathcal{F})$. Applying a (consistent) intervention $do(\mathbf{X} = \mathbf{x})$ on $(\{s\},\mathcal{F})$ generates a causal team $(\{s_{\mathbf{X}=\mathbf{x}}\},\mathcal{F}_{\mathbf{X}=\mathbf{x}})$ as defined in Definition 2.3. We can then define the result of the intervention $do(\mathbf{X} = \mathbf{x})$ on $\{(s,\mathcal{F})\}$ to be the generalized causal team $(\{s_{\mathbf{X}=\mathbf{x}}\},\mathcal{F}_{\mathbf{X}=\mathbf{x}})^g = \{(s_{\mathbf{X}=\mathbf{x}},\mathcal{F}_{\mathbf{X}=\mathbf{x}})\}$. Interventions on arbitrary generalized causal teams are defined as follows.

Definition 2.8 (Intervention over generalized causal teams) *Let T be a (recursive) generalized causal team, and $\mathbf{X} = \mathbf{x}$ a consistent equation over σ. The intervention $do(\mathbf{X} = \mathbf{x})$ on T generates the generalized causal team $T_{\mathbf{X}=\mathbf{x}} := \{(s_{\mathbf{X}=\mathbf{x}},\mathcal{F}_{\mathbf{X}=\mathbf{x}}) \mid (s,\mathcal{F}) \in T\}$.*

Definition 2.9 *Let φ be a formula of the language $\mathcal{CO}_{\mathbb{V}}[\sigma]$ or $\mathcal{COD}[\sigma]$, and T a generalized causal team over σ. The satisfaction relation $T \models^g \varphi$ (or simply $T \models \varphi$) over generalized causal teams is defined in the same way as in Definition 2.5, except for slight differences in the following clauses:*

- *$T \models^g \neg\alpha$ iff for all $(s,\mathcal{F}) \in T$, $\{(s,\mathcal{F})\} \not\models \alpha$.*
- *$T \models^g \varphi \vee \psi$ iff there are two generalized causal subteams T_1, T_2 of T such that $T_1 \cup T_2 = T$, $T_1 \models \varphi$ and $T_2 \models \psi$.*

Example 2.10 *Consider two function components $\mathcal{F},\mathcal{G}$ over the domain $Dom = \{X,Y,Z\}$ with Z the only endogenous variable, and $\mathcal{F}_Z(X) = 2 * X$ and $\mathcal{G}_Z(X,Y) = X + Y$. Clearly, $\mathcal{F} \neq \mathcal{G}$ as, e.g., the graph of $\mathcal{G}$ contains an additional arrow from X to Z. Consider the generalized causal team $T = \{(s,\mathcal{F}),(s,\mathcal{G}),(t,\mathcal{G})\}$, represented in the following left table:*

T:

X	Y	Z	
2	2	4	$\mathcal{F}$
2	2	4	$\mathcal{G}$
1	3	4	$\mathcal{G}$

$T_{Y=1}$:

X	Y	Z	
2	**1**	4	$\mathcal{F}_{Y=1}$
2	**1**	**3**	$\mathcal{G}_{Y=1}$
1	**1**	**2**	$\mathcal{G}_{Y=1}$

Since both $\mathcal{F}$ and $\mathcal{G}$ are recursive, T is recursive. An intervention $do(Y = 1)$ on T updates each row in the above left table according to its own associated function component, returning the right table. Since Y is exogenous both in $\mathcal{F}$ and $\mathcal{G}$, we have $\mathcal{F}_{Y=1} = \mathcal{F}$ and $\mathcal{G}_{Y=1} = \mathcal{G}$. The value of Z in the first row of the above updated table remains unchanged, because Y is not an argument of $\mathcal{F}_Z$.

We list some closure properties for our logics over both causal teams and generalized causal teams in the next theorem, whose proof is left to the reader, or see [2] for the causal team case.

Theorem 2.11 *Let T, S be (generalized) causal teams over some signature σ.*

Empty team property: *If $T^- = \emptyset$, then $T \models \varphi$.*

Downward closure: *If $T \models \varphi$ and $S \subseteq T$, then $S \models \varphi$.*

Flatness of CO-formulas: *If α is a $CO[\sigma]$-formula, then*
$T \models \alpha \iff (\{s\}, \mathcal{F}) \models^c \alpha$ *for all* $s \in T^-$ *(resp.* $\{(s, \mathcal{F})\} \models^g \alpha$ *for all* $(s, \mathcal{F}) \in T$*).*

The team semantics over causal teams and that over generalized causal teams with a constant function component are essentially equivalent, in the sense of the next lemma, whose proof is left to the reader.

Lemma 2.12 *(i) For any causal team T, we have that $T \models^c \varphi \iff T^g \models^g \varphi$.*

(ii) For any nonempty generalized causal team T with a unique function component, we have that $T \models^g \varphi \iff T^c \models^c \varphi$.

Corollary 2.13 *For any set $\Delta \cup \{\alpha\}$ of $CO[\sigma]$-formulas, $\Delta \models^g \alpha$ iff $\Delta \models^c \alpha$.*

Proof. By Lemma 2.12, $\{(s, \mathcal{F})\} \models^g \beta$ iff $(\{s\}, \mathcal{F}) \models^c \beta$ for any $\beta \in \Delta \cup \{\alpha\}$. Thus, the claim follows from the flatness of $CO[\sigma]$-formulas. □

3 Characterizing function components

3.1 Equivalence of function components

Various notions of similarity among causal models have been considered in the literature (see e.g. [6]), which measure the "distance" between two models in terms of their empirical or counterfactual consequences. We consider here a stricter notion of equivalence, which, as we will see in theorem 3.4, characterizes indistinguishability of causal structures by means of our languages.

Consider a binary function f and an $(n{+}2)$-ary function g defined as $f(X, Y) = X + Y$ and $g(X, Y, Z_1, \ldots, Z_n) = X + Y$. Essentially f and g are the same function: $Z_1, \ldots, Z_n$ are dummy arguments of g. We now characterize this idea in the notion of two function components being equivalent up to dummy arguments.

Definition 3.1 *Let $\mathcal{F}, \mathcal{G}$ be two function components over $\sigma = (\mathrm{Dom}, \mathrm{Ran})$.*

- *Let $V \in \mathrm{Dom}$. Two functions $\mathcal{F}_V$ and $\mathcal{G}_V$ are said to be equivalent up to* *dummy arguments, denoted as* $\mathcal{F}_V \sim \mathcal{G}_V$, *if for any* $\mathbf{x} \in \mathrm{Ran}(PA_V^{\mathcal{F}} \cap PA_V^{\mathcal{G}})$, $\mathbf{y} \in \mathrm{Ran}(PA_V^{\mathcal{F}} \setminus PA_V^{\mathcal{G}})$ *and* $\mathbf{z} \in \mathrm{Ran}(PA_V^{\mathcal{G}} \setminus PA_V^{\mathcal{F}})$, *we have that* $\mathcal{F}_V(\mathbf{xy}) = \mathcal{G}_V(\mathbf{xz})$ *(where we assume w.l.o.g. the shown orderings of the arguments of the functions).*

- *Let* $\mathrm{Cn}(\mathcal{F})$ *denote the set of endogenous variables* V *of* $\mathcal{F}$ *for which* $\mathcal{F}_V$ *is a constant function, i.e., for some fixed* $c \in \mathrm{Ran}(V)$, $\mathcal{F}_V(\mathbf{p}) = c$ *for all* $\mathbf{p} \in PA_V^{\mathcal{F}}$. *We say that* $\mathcal{F}$ *and* $\mathcal{G}$ *are* ***equivalent up to dummy arguments****, denoted as* $\mathcal{F} \sim \mathcal{G}$, *if* $\mathrm{En}(\mathcal{F}) \setminus \mathrm{Cn}(\mathcal{F}) = \mathrm{En}(\mathcal{G}) \setminus \mathrm{Cn}(\mathcal{G})$, *and* $\mathcal{F}_V \sim \mathcal{G}_V$ *holds for all* $V \in \mathrm{En}(\mathcal{F}) \setminus \mathrm{Cn}(\mathcal{F})$.

It is easy to see that $\sim$ is an equivalence relation. The next lemma shows that the relation $\sim$ is preserved under interventions.

Lemma 3.2 *For any function components* $\mathcal{F}, \mathcal{G} \in \mathbb{F}_\sigma$ *and consistent equation* $\mathbf{X} = \mathbf{x}$ *over* σ, *we have that* $\mathcal{F} \sim \mathcal{G}$ *implies* $\mathcal{F}_{\mathbf{X}=\mathbf{x}} \sim \mathcal{G}_{\mathbf{X}=\mathbf{x}}$.

Proof. Suppose $\mathcal{F} \sim \mathcal{G}$. Then $\mathrm{En}(\mathcal{F}) \setminus \mathrm{Cn}(\mathcal{F}) = \mathrm{En}(\mathcal{G}) \setminus \mathrm{Cn}(\mathcal{G})$. Observe that $\mathrm{En}(\mathcal{F}_{\mathbf{X}=\mathbf{x}}) = \mathrm{En}(\mathcal{F}) \setminus \mathbf{X}$ and $\mathrm{Cn}(\mathcal{F}_{\mathbf{X}=\mathbf{x}}) = \mathrm{Cn}(\mathcal{F}) \setminus \mathbf{X}$; and similarly for $\mathcal{G}$. It follows that $\mathrm{En}(\mathcal{F}_{\mathbf{X}=\mathbf{x}}) \setminus \mathrm{Cn}(\mathcal{F}_{\mathbf{X}=\mathbf{x}}) = (\mathrm{En}(\mathcal{F}) \setminus \mathrm{Cn}(\mathcal{F})) \setminus \mathbf{X} = (\mathrm{En}(\mathcal{G}) \setminus \mathrm{Cn}(\mathcal{G})) \setminus \mathbf{X} = \mathrm{En}(\mathcal{G}_{\mathbf{X}=\mathbf{x}}) \setminus \mathrm{Cn}(\mathcal{G}_{\mathbf{X}=\mathbf{x}})$. On the other hand, for any $V \in \mathrm{En}(\mathcal{F}_{\mathbf{X}=\mathbf{x}}) \setminus \mathrm{Cn}(\mathcal{F}_{\mathbf{X}=\mathbf{x}}) = (\mathrm{En}(\mathcal{F}) \setminus \mathrm{Cn}(\mathcal{F})) \setminus \mathbf{X}$, by the assumption, $(\mathcal{F}_{\mathbf{X}=\mathbf{x}})_V = \mathcal{F}_V \sim \mathcal{G}_V = (\mathcal{G}_{\mathbf{X}=\mathbf{x}})_V$. □

We now generalize the equivalence relation $\sim$ to the team level. Let us first consider causal teams. Two causal teams $T = (T^-, \mathcal{F})$ and $S = (S^-, \mathcal{G})$ of the same signature σ are said to be **similar**, denoted as $T \sim S$, if $\mathcal{F} \sim \mathcal{G}$. We say that T and S are **equivalent**, denoted as $T \approx S$, if $T \sim S$ and $T^- = S^-$.

Next, we turn to generalized causal teams. We call a generalized causal team T a **uniform team** if $\mathcal{F} \sim \mathcal{G}$ for all $(s, \mathcal{F}), (t, \mathcal{G}) \in T$. By Lemma 3.2, we know that if T is uniform, so is $T_{\mathbf{X}=\mathbf{x}}$, for any consistent equation $\mathbf{X} = \mathbf{x}$. For any generalized causal team T with $(t, \mathcal{F}) \in T$, write $T^{\mathcal{F}} := \{(s, \mathcal{G}) \in T \mid \mathcal{G} \sim \mathcal{F}\}$. Two generalized causal teams S and T are said to be **equivalent**, denoted as $S \approx T$, if $(S^{\mathcal{F}})^- = (T^{\mathcal{F}})^-$ for all $\mathcal{F} \in \mathbb{F}_\sigma$.

Theorem 3.3 (Closure under causal equivalence) *Let* T, S *be two (generalized) causal teams over* σ *such that* $T \approx S$. *We have that* $T \models \varphi \iff S \models \varphi$.

Proof. The theorem is proved by induction on φ. The case $\varphi = \mathbf{X} = \mathbf{x} \;\Box\!\!\rightarrow \psi$ follows from the fact that $T_{\mathbf{X}=\mathbf{x}} \approx S_{\mathbf{X}=\mathbf{x}}$ (Lemma 3.2). The case $\varphi = \psi \vee \chi$ for causal teams follows directly from the induction hypothesis. We now give the proof for this case for generalized causal teams. We only prove the left to right direction (the other direction is symmetric). Suppose $T \models^g \psi \vee \chi$. Then there are $T_0, T_1 \subseteq T$ such that $T = T_0 \cup T_1$, $T_0 \models^g \psi$ and $T_1 \models^g \chi$. Consider $S_i = \{(s, \mathcal{F}) \in S \mid \{(s, \mathcal{F})\} \approx \{(s, \mathcal{G})\}$ for some $(s, \mathcal{G}) \in T_i\}$ $(i = 0, 1)$. It is easy to see that $S_i \approx T_i$ $(i = 0, 1)$ and $S = S_0 \cup S_1$. By induction hypothesis we have that $S_0 \models^g \psi$ and $S_1 \models^g \chi$. Hence $S \models^g \psi \vee \chi$. □

Thus, none of our languages can tell apart causal teams which are equivalent up to dummy arguments. However, one might not be sure, a priori, that a given argument behaves as dummy for a specific function, as this might e.g. be unfeasible to verify if the variables range over large sets. For this reason, we are keeping this distinction in the semantics instead of quotienting it out.

3.2 Characterizing function components

For any function component $\mathcal{F}$ over some signature σ, define a $\mathcal{CO}[\sigma]$-formula

$$\Phi^{\mathcal{F}} := \bigwedge_{V \in \mathrm{En}(\mathcal{F})} \eta_\sigma(V) \wedge \bigwedge_{V \in (\mathrm{Dom} \setminus \mathrm{En}(\mathcal{F})) \cup \mathrm{Cn}(\mathcal{F})} \xi_\sigma(V).$$

where $\eta_\sigma(V) := \bigwedge\{(\mathbf{W} = \mathbf{w} \wedge PA_V^{\mathcal{F}} = \mathbf{p}) \,\Box\!\!\rightarrow V = \mathcal{F}_V(\mathbf{p})$
$\mid \mathbf{W} = \mathrm{Dom} \setminus (PA_V^{\mathcal{F}} \cup \{V\}),\ \mathbf{w} \in \mathrm{Ran}(\mathbf{W}),\ \mathbf{p} \in \mathrm{Ran}(PA_V^{\mathcal{F}})\}$
and $\xi_\sigma(V) := \bigwedge\{V = v \supset (\mathbf{W}_V = \mathbf{w} \,\Box\!\!\rightarrow V = v)$
$\mid v \in \mathrm{Ran}(V), \mathbf{W}_V = \mathrm{Dom} \setminus \{V\}, \mathbf{w} \in \mathrm{Ran}(\mathbf{W}_V)\}$.
Intuitively, for each non-constant endogenous variable V of $\mathcal{F}$, the formula $\eta_\sigma(V)$ specifies that all assignments in the (generalized) causal team T in question behave exactly as required by the function $\mathcal{F}_V$. For each variable V which, according to $\mathcal{F}$, is exogenous or generated by a constant function, the formula $\xi_\sigma(V)$ states that V is not affected by interventions on other variables. If $V \in \mathrm{Cn}(\mathcal{F})$, then V has both an η_σ and a ξ_σ clause. Overall, the formula $\Phi^{\mathcal{F}}$ is satisfied in a team T if and only if every assignment in T has a function component that is $\sim$-equivalent to $\mathcal{F}$: this nontrivial fact is proved in the next theorem. This result is the crucial element for adapting the standard methods of team semantics to the causal context.

Theorem 3.4 *Let σ be a signature, and $\mathcal{F} \in \mathbb{F}_\sigma$.*

(i) For any generalized causal team T over σ, we have that

$$T \models^g \Phi^{\mathcal{F}} \iff \text{for all } (s, \mathcal{G}) \in T : \mathcal{G} \sim \mathcal{F}.$$

(ii) For any nonempty causal team $T = (T^-, \mathcal{G})$ over σ, we have that

$$T \models^c \Phi^{\mathcal{F}} \iff \mathcal{G} \sim \mathcal{F}.$$

Proof. (i). $\Longrightarrow$: Suppose $T \models^g \Phi^{\mathcal{F}}$ and $(s, \mathcal{G}) \in T$. We show $\mathcal{G} \sim \mathcal{F}$. $\mathrm{En}(\mathcal{F}) \setminus \mathrm{Cn}(\mathcal{F}) \subseteq \mathrm{En}(\mathcal{G}) \setminus \mathrm{Cn}(\mathcal{G})$: For any $V \in \mathrm{En}(\mathcal{F}) \setminus \mathrm{Cn}(\mathcal{F})$, there are distinct $\mathbf{p}, \mathbf{p}' \in \mathrm{Ran}(PA_V^{\mathcal{F}})$ such that $\mathcal{F}_V(\mathbf{p}) \neq \mathcal{F}_V(\mathbf{p}')$. Since $T \models \eta_\sigma(V)$, for any $\mathbf{w} \in \mathit{Ran}(\mathbf{W})$, we have that

$$\{(s, \mathcal{G})\} \models (\mathbf{W} = \mathbf{w} \wedge PA_V^{\mathcal{F}} = \mathbf{p}) \,\Box\!\!\rightarrow V = \mathcal{F}_V(\mathbf{p}),$$
$$\{(s, \mathcal{G})\} \models (\mathbf{W} = \mathbf{w} \wedge PA_V^{\mathcal{F}} = \mathbf{p}') \,\Box\!\!\rightarrow V = \mathcal{F}_V(\mathbf{p}').$$

Thus, $s_{\mathbf{W}=\mathbf{w} \wedge PA_V^{\mathcal{F}}=\mathbf{p}}(V) = \mathcal{F}_V(\mathbf{p}) \neq \mathcal{F}_V(\mathbf{p}') = s_{\mathbf{W}=\mathbf{w} \wedge PA_V^{\mathcal{F}}=\mathbf{p}'}(V)$. So, $V \notin \mathrm{Cn}(\mathcal{G})$, and furthermore V is not exogenous (since the value of an exogenous variable is not affected by interventions on different variables). Thus, $V \in \mathrm{En}(G) \setminus \mathrm{Cn}(\mathcal{G})$.

$\mathrm{En}(\mathcal{G}) \setminus \mathrm{Cn}(\mathcal{G}) \subseteq \mathrm{En}(\mathcal{F}) \setminus \mathrm{Cn}(\mathcal{F})$: For any $V \in \mathrm{En}(\mathcal{G}) \setminus \mathrm{Cn}(\mathcal{G})$, there are distinct $\mathbf{p}, \mathbf{p}' \in \mathit{Ran}(PA_V^{\mathcal{G}})$ such that $\mathcal{G}_V(\mathbf{p}) \neq \mathcal{G}_V(\mathbf{p}')$. Now, if $V \notin \mathrm{En}(\mathcal{F}) \setminus \mathrm{Cn}(\mathcal{F})$, then $T \models \xi_\sigma(V)$. Let $v = s(V)$ and $\mathbf{Z} = \mathbf{W}_V \setminus PA_V^{\mathcal{G}}$. Since $\{(s, \mathcal{G})\} \models V = v$ and $V \notin PA_V^{\mathcal{G}}$, for any $\mathbf{z} \in \mathrm{Ran}(\mathbf{Z})$, we have that

$$\{(s, \mathcal{G})\} \models (\mathbf{Z} = \mathbf{z} \wedge PA_V^{\mathcal{G}} = \mathbf{p}) \,\Box\!\!\rightarrow V = v,$$
$$\{(s, \mathcal{G})\} \models (\mathbf{Z} = \mathbf{z} \wedge PA_V^{\mathcal{G}} = \mathbf{p}') \,\Box\!\!\rightarrow V = v.$$

By the definition of intervention, we must have that $v = s_{\mathbf{Z}=\mathbf{z} \wedge PA_V^{\mathcal{G}}=\mathbf{p}}(V) = \mathcal{G}_V(\mathbf{p}) \neq$

$\mathcal{G}_V(\mathbf{p}') = s_{\mathbf{Z}=\mathbf{z}\wedge PA_V^{\mathcal{G}}=\mathbf{p}'}(V) = v$, which is impossible. Hence, $V \in \mathrm{En}(\mathcal{F}) \setminus \mathrm{Cn}(\mathcal{F})$.

$\mathcal{F}_V \sim \mathcal{G}_V$ for any $V \in \mathrm{En}(\mathcal{F}) \setminus \mathrm{Cn}(\mathcal{F})$: For any $\mathbf{x} \in \mathrm{Ran}(PA_V^{\mathcal{F}} \cap PA_V^{\mathcal{G}})$, $\mathbf{y} \in \mathrm{Ran}(PA_V^{\mathcal{F}} \setminus PA_V^{\mathcal{G}})$ and $\mathbf{z} \in \mathrm{Ran}(PA_V^{\mathcal{G}} \setminus PA_V^{\mathcal{F}})$, since $T \models \eta_\sigma(V)$ and $V \notin PA_V^{\mathcal{G}}$, for any $\mathbf{w} \in \mathrm{Ran}(\mathbf{W})$ with $\mathbf{w} \upharpoonright (PA_V^{\mathcal{G}} \setminus PA_V^{\mathcal{F}}) = \mathbf{z}$, we have that

$$\{(s,\mathcal{G})\} \models (\mathbf{W} = \mathbf{w} \wedge PA_V^{\mathcal{F}} = \mathbf{xy}) \;\Box\!\!\rightarrow V = \mathcal{F}_V(\mathbf{xy}).$$

Then $\mathcal{F}_V(\mathbf{xy}) = s_{\mathbf{W}=\mathbf{w}\wedge PA_V^{\mathcal{F}}=\mathbf{xy}}(V) = \mathcal{G}_V(s_{\mathbf{W}=\mathbf{w}\wedge PA_V^{\mathcal{F}}=\mathbf{xy}}(PA_V^{\mathcal{G}})) = \mathcal{G}_V(\mathbf{xz})$, as required.

$\Longleftarrow$: Suppose that $\mathcal{G} \sim \mathcal{F}$ for all $(s,\mathcal{G}) \in T$. Since the formula $\Phi^{\mathcal{F}}$ is flat, it suffices to show that $\{(s,\mathcal{G})\} \models \eta_\sigma(V)$ for all $V \in \mathrm{En}(\mathcal{F})$, and $\{(s,\mathcal{G})\} \models \xi_\sigma(V)$ for all $V \in (\mathrm{Dom} \setminus \mathrm{En}(\mathcal{F})) \cup \mathrm{Cn}(\mathcal{F})$.

For the former, take any $\mathbf{w} \in \mathrm{Ran}(\mathbf{W})$ and $\mathbf{p} \in \mathrm{Ran}(PA_V^{\mathcal{F}})$, and let $\mathbf{Z} = \mathbf{z}$ abbreviate $\mathbf{W} = \mathbf{w} \wedge PA_V^{\mathcal{F}} = \mathbf{p}$. We show that $\{(s_{\mathbf{Z}=\mathbf{z}}, \mathcal{G}_{\mathbf{Z}=\mathbf{z}})\} \models V = \mathcal{F}_V(\mathbf{p})$. Since $\mathcal{G} \sim \mathcal{F}$, by Lemma 3.2 we have that $\mathcal{G}_{\mathbf{Z}=\mathbf{z}} \sim \mathcal{F}_{\mathbf{Z}=\mathbf{z}}$. Thus,

$$\begin{aligned} s_{\mathbf{Z}=\mathbf{z}}(V) = (\mathcal{G}_{\mathbf{Z}=\mathbf{z}})_V(s_{\mathbf{Z}=\mathbf{z}}(PA_V^{\mathcal{G}_{\mathbf{Z}=\mathbf{z}}})) &= (\mathcal{F}_{\mathbf{Z}=\mathbf{z}})_V(s_{\mathbf{Z}=\mathbf{z}}(PA_V^{\mathcal{F}_{\mathbf{Z}=\mathbf{z}}})) && (\text{since } \mathcal{G}_{\mathbf{X}=\mathbf{x}} \sim \mathcal{F}_{\mathbf{X}=\mathbf{x}}) \\ &= \mathcal{F}_V(s_{\mathbf{Z}=\mathbf{z}}(PA_V^{\mathcal{F}})) && (\text{since } V \notin \mathbf{Z}) \\ &= \mathcal{F}_V(\mathbf{p}). \end{aligned}$$

For the latter, take any $v \in \mathrm{Ran}(V)$ and $\mathbf{w} \in \mathrm{Ran}(\mathbf{W}_V)$. Assume that $\{(s,\mathcal{G})\} \models V = v$, i.e., $s(V) = v$. Since $V \notin \mathrm{En}(\mathcal{F}) \setminus \mathrm{Cn}(\mathcal{F})$ and $\mathcal{F} \sim \mathcal{G}$, we know that $V \notin \mathrm{En}(\mathcal{G})$ or $V \in \mathrm{Cn}(\mathcal{G})$. In both cases we have that $\{(s,\mathcal{G})\} \models \mathbf{W}_V = \mathbf{w} \;\Box\!\!\rightarrow V = v$.

(ii). Let T be a nonempty causal team. Consider its associated generalized causal team T^g. The claim then follows from Lemma 2.12 and item (i). □

Corollary 3.5 *For any generalized causal team T over some signature σ,*

$$T \models \bigvee_{\mathcal{F}\in\mathbb{F}_\sigma} \Phi^{\mathcal{F}} \iff T \text{ is uniform.}$$

The intuituionistic disjunction $\mathbin{\mathbb{V}}$ was shown to have the *disjunction property*, i.e., $\models \varphi \mathbin{\mathbb{V}} \psi$ implies $\models \varphi$ or $\models \psi$, in propositional inquisitive logic ([5]) and propositional dependence logic ([17]). It follows immediately from Theorem 3.4 that the disjunction property of $\mathbin{\mathbb{V}}$ fails in the context of causal teams, because $\models^c \bigvee_{\mathcal{F}\in\mathbb{F}_\sigma} \Phi^{\mathcal{F}}$, whereas $\not\models^c \Phi^{\mathcal{F}}$ for any $\mathcal{F} \in \mathbb{F}_\sigma$. Nevertheless, the intuitionistic disjunction does admit the disjunction property over generalized causal teams.

Theorem 3.6 (Disjunction property) *Let Δ be a set of $\mathcal{CO}[\sigma]$-formulas, and φ, ψ be arbitrary formulas over σ. If $\Delta \models^g \varphi \mathbin{\mathbb{V}} \psi$, then $\Delta \models^g \varphi$ or $\Delta \models^g \psi$. In particular, if $\models^g \varphi \mathbin{\mathbb{V}} \psi$, then $\models^g \varphi$ or $\models^g \psi$.*

Proof. Suppose $\Delta \not\models^g \varphi$ and $\Delta \not\models^g \psi$. Then there are two generalized causal teams T_1, T_2 such that $T_1 \models \Delta$, $T_2 \models \Delta$, $T_1 \not\models \varphi$ and $T_2 \not\models \psi$. Let $T := T_1 \cup T_2$. By flatness of Δ, we have that $T \models \Delta$. On the other hand, by downwards closure, we have that $T \not\models \varphi$ and $T \not\models \psi$, and thus $T \not\models \varphi \mathbin{\mathbb{V}} \psi$. □

4 Characterizing *CO*

In this section, we characterize the expressive power of *CO* over causal teams and present a system of natural deduction for *CO* that is sound and complete over both causal teams and generalized causal teams.

4.1 Expressivity

In this subsection, we show that *CO*-formulas capture the flat class of causal teams (up to $\approx$-equivalence). Our result is analogous to known characterizations of flat languages in propositional team semantics ([18]), with a twist, given by the fact that only the unions of similar causal teams are reasonably defined. We define such unions as follows.

Definition 4.1 *Let $S = (S^-, \mathcal{F}), T = (T^-, \mathcal{G})$ be two causal teams over the same signature σ with $S \sim T$. The union of S and T is defined as the causal team $S \cup T = (S^- \cup T^-, \mathcal{H})$ over σ, where*

- $\mathrm{En}(\mathcal{H}) = (\mathrm{En}(\mathcal{F}) \setminus \mathrm{Cn}(F)) \cap (\mathrm{En}(\mathcal{G}) \setminus \mathrm{Cn}(G))$,
- *and for each $V \in \mathrm{En}(\mathcal{H})$, $PA_V^{\mathcal{H}} = PA_V^{\mathcal{F}} \cap PA_V^{\mathcal{G}}$, and $\mathcal{H}_V(\mathbf{p}) = \mathcal{F}_V(\mathbf{px})$ for any $\mathbf{p} \in PA_V^{\mathcal{F}} \cap PA_V^{\mathcal{G}}$ and $\mathbf{x} \in PA_V^{\mathcal{F}} \setminus PA_V^{\mathcal{G}}$.*

Clearly, $\mathcal{H} \sim \mathcal{F} \sim \mathcal{G}$ and thus $S \cup T \sim S \sim T$.

A formula φ over σ determines a class $\mathcal{K}_\varphi$ of causal teams defined as

$$\mathcal{K}_\varphi = \{T \mid T \models \varphi\}.$$

We say that a formula φ **defines** a class $\mathcal{K}$ of causal teams if $\mathcal{K} = \mathcal{K}_\varphi$.

Definition 4.2 *We say that a class $\mathcal{K}$ of causal teams over σ is*

- ***causally downward closed*** *if $T \in \mathcal{K}$ and $S \subseteq T$ imply $S \in \mathcal{K}$;*
- ***closed under causal unions*** *if, whenever $T_1, T_2 \in \mathcal{K}$ and $T_1 \cup T_2$ is defined, $T_1 \cup T_2 \in \mathcal{K}$;*
- ***flat*** *if $(T^-, \mathcal{F}) \in \mathcal{K}$ iff $(\{s\}, \mathcal{F}) \in \mathcal{K}$ for all $s \in T^-$;*
- ***closed under equivalence*** *if $T \in \mathcal{K}$ and $T \approx T'$ imply $T' \in \mathcal{K}$.*

It is easy to verify that $\mathcal{K}$ is flat iff $\mathcal{K}$ is causally downward closed and closed under causal unions. Any nonempty downward closed class $\mathcal{K}$ of causal teams over σ contains all causal teams over σ with empty team component. The class $\mathcal{K}_\varphi$ is always nonempty as the teams with empty team component are always in $\mathcal{K}_\varphi$ (by Theorem 2.11). By Theorems 2.11 and 3.3, if α is a *CO*-formula, then $\mathcal{K}_\alpha$ is flat and closed under equivalence. The main result of this section is the following characterization theorem which gives also the converse direction.

Theorem 4.3 *Let $\mathcal{K}$ be a nonempty (finite) class of causal teams over some signature σ. Then $\mathcal{K}$ is definable by a CO[σ]-formula if and only if $\mathcal{K}$ is flat and closed under equivalence.*

In order to prove the above theorem, we introduce a *CO*-formula Θ^T, inspired by a similar one in [17], that defines the property "having as team component a subset of T^-". For each causal team T over $\sigma = (\mathrm{Dom}, \mathrm{Ran})$, define

$$\Theta^T := \bigvee_{s \in T^-} \bigwedge_{V \in \mathrm{Dom}} V = s(V).$$

Lemma 4.4 *$S \models \Theta^T$ iff $S^- \subseteq T^-$, for any causal teams S, T over σ.*

Proof. "$\Longrightarrow$": Suppose $S \models \Theta^T$ and $S = (S^-, \mathcal{F})$. For any $s \in S^-$, by downward closure, we have that $(\{s\},\mathcal{F}) \models \Theta^T$, which means that for some $t \in T^-$, $(\{s\},\mathcal{F}) \models V = t(V)$ for all $V \in \text{Dom}$. Since $\{s\}$ and $\{t\}$ have the same signature, this implies that $s = t$, thereby $s \in T^-$.

"$\Longleftarrow$": Suppose $S^- \subseteq T^-$. Observe that $S \models \Theta^S$ and $\Theta^T = \Theta^S \vee \Theta^{T\setminus S}$. Thus, we conclude $S \models \Theta^T$ by the empty team property. □

Lemma 4.5 *Let $S = (S^-,\mathcal{G})$ and $T = (T^-,\mathcal{F})$ be causal teams over σ with $S^-,T^- \neq \emptyset$. Then $S \models \Theta^T \wedge \Phi^{\mathcal{F}} \iff S \approx R \subseteq T$ for some R over σ.*

Proof. By Lemma 4.4 and Theorem 3.4, we have that $S \models \Theta^T \wedge \Phi^{\mathcal{F}}$ iff $S^- \subseteq T^-$ and $\mathcal{G} \sim \mathcal{F}$. It then suffices to show that the latter is equivalent to $S \approx R \subseteq T$ for some R. The right to left direction is clear; conversely, if $S^- \subseteq T^-$ and $\mathcal{G} \sim \mathcal{F}$, then we can take $R = (S^-,\mathcal{F})$. □

Consider the quotient set $\mathbb{F}_\sigma/_\approx$. For each equivalence class $[\mathcal{F}] \in \mathbb{F}_\sigma/_\approx$ choose a unique representative $\mathcal{F}_0$. Denote by $\mathbb{F}^0_\sigma$ the set of all such representatives.

Proof of Theorem 4.3. It suffices to prove the direction "$\Longleftarrow$". For each $\mathcal{F} \in \mathbb{F}^0_\sigma$, let $\mathcal{K}^{\mathcal{F}} := \{(T^-,\mathcal{G}) \in \mathcal{K} \mid \mathcal{G} \sim \mathcal{F}\}$. Clearly $\mathcal{K} = \bigcup_{\mathcal{F}\in\mathbb{F}^0_\sigma} \mathcal{K}^{\mathcal{F}}$. Let $T_{\mathcal{F}} = \bigcup \mathcal{K}^{\mathcal{F}}$, which is well-defined as in Definition 4.1. Since $\mathcal{K}$ is closed under causal unions, $T_{\mathcal{F}} \in \mathcal{K}$. We may assume w.l.o.g. that $T_{\mathcal{F}} = (T^-_{\mathcal{F}},\mathcal{F})$. Let

$$\varphi = \bigvee_{\mathcal{F}\in\mathbb{F}^0_\sigma} (\Theta^{T_{\mathcal{F}}} \wedge \Phi^{\mathcal{F}}).$$

It suffices to show that $\mathcal{K}_\varphi = \mathcal{K}$. For any $S = (S^-,\mathcal{G}) \in \mathcal{K}$, there exists $\mathcal{F} \in \mathbb{F}^0_\sigma$ such that $S \in \mathcal{K}^{\mathcal{F}}$. Let $R = (S^-,\mathcal{F})$. Clearly, $S \approx R \subseteq \bigcup \mathcal{K}^{\mathcal{F}} = T_{\mathcal{F}}$, which by Lemma 4.5 implies that $S \models \Theta^{T_{\mathcal{F}}} \wedge \Phi^{\mathcal{F}}$. Hence, $S \models \varphi$, namely $S \in \mathcal{K}_\varphi$.

Conversely, suppose $S = (S^-,\mathcal{G}) \in \mathcal{K}_\varphi$, i.e., $S \models \varphi$. Then for every $\mathcal{F} \in \mathbb{F}^0_\sigma$, there is $S_{\mathcal{F}} \subseteq S$ such that $S = \bigcup_{\mathcal{F}\in\mathbb{F}^0_\sigma} S_{\mathcal{F}}$ and $S_{\mathcal{F}} \models \Theta^{T_{\mathcal{F}}} \wedge \Phi^{\mathcal{F}}$. Thus, by Lemma 4.5, we obtain that $S_{\mathcal{F}} \approx R_{\mathcal{F}} \subseteq T_{\mathcal{F}}$ for some $R_{\mathcal{F}}$. In particular, we have that $S_{\mathcal{F}} = (S^-_{\mathcal{F}},\mathcal{G}) \sim (T^-_{\mathcal{F}},\mathcal{F}) = T_{\mathcal{F}}$, which gives $\mathcal{G} \sim \mathcal{F}$. But since no two distinct elements in $\mathbb{F}^0_\sigma$ are $\sim$-similar to each other, and $S_{\mathcal{F}} \sim S$ for each $\mathcal{F} \in \mathbb{F}^0_\sigma$, this can only happen if $S^-_{\mathcal{F}} = \emptyset$ for all $\mathcal{F} \in \mathbb{F}^0_\sigma$ except one. Denote this unique element of $\mathbb{F}^0_\sigma$ by $\mathcal{H}$. Now, $S = S_{\mathcal{H}} \approx R_{\mathcal{H}} \subseteq T_{\mathcal{F}} \in \mathcal{K}$. Hence we conclude that $S \in \mathcal{K}$, as $\mathcal{K}$ is causally closed downward and closed under equivalence. □

4.2 Deduction system

The logic $\mathcal{CO}[\sigma]$ over (recursive) causal teams was given in [2] a sound and complete axiomatization which incorporated ideas from [7], [8], [3] and [9]. In this section, we present an equivalent system of natural deduction and show it to be sound and complete also over (recursive) generalized causal teams.

Definition 4.6 *The system of natural deduction for $\mathcal{CO}[\sigma]$ consists of the following rules:*

- *(Parameterized) rules for value range assumptions:*

$\dfrac{}{\bigvee_{x\in\text{Ran}(X)} X = x}\ \mathsf{ValDef}$	$\dfrac{X = x}{X \neq x'}\ \mathsf{ValUnq}$

- *Rules for* $\wedge, \vee, \neg$:

$$\dfrac{\varphi \quad \psi}{\varphi \wedge \psi}\wedge\mathsf{I} \qquad \dfrac{\varphi \wedge \psi}{\varphi}\wedge\mathsf{E} \qquad \dfrac{\varphi \wedge \psi}{\psi}\wedge\mathsf{E}$$

$$\dfrac{\varphi}{\varphi \vee \psi}\vee\mathsf{I} \qquad \dfrac{\varphi}{\psi \vee \varphi}\vee\mathsf{I} \qquad \dfrac{\varphi \vee \psi \quad \begin{array}{c}[\varphi]\\ \vdots\\ \alpha\end{array} \quad \begin{array}{c}[\psi]\\ \vdots\\ \alpha\end{array}}{\alpha}\vee\mathsf{E}$$

$$\dfrac{\begin{array}{c}[\alpha]\\ \vdots\\ \bot\end{array}}{\neg\alpha}\neg\mathsf{I} \qquad \dfrac{\alpha \quad \neg\alpha}{\varphi}\neg\mathsf{E} \qquad \dfrac{\begin{array}{c}[\neg\alpha]\\ \vdots\\ \bot\end{array}}{\alpha}\mathsf{RAA}$$

- *Rules for* $\Box\!\!\rightarrow$:

$$\dfrac{}{(\mathbf{X} = \mathbf{x} \wedge Y = y) \Box\!\!\rightarrow Y = y}\Box\!\!\rightarrow\mathsf{Eff} \qquad \dfrac{\mathbf{X} = \mathbf{x} \Box\!\!\rightarrow W = w \quad \mathbf{X} = \mathbf{x} \Box\!\!\rightarrow \gamma}{(\mathbf{X} = \mathbf{x} \wedge W = w) \Box\!\!\rightarrow \gamma}\Box\!\!\rightarrow\mathsf{Cmp}(1)$$

$$\dfrac{\mathbf{X} = \mathbf{x} \Box\!\!\rightarrow \bot}{\varphi}\Box\!\!\rightarrow\bot\mathsf{E} \qquad \dfrac{}{(\mathbf{Y} = \mathbf{y} \wedge X = x \wedge X = x') \Box\!\!\rightarrow \varphi}\bot\Box\!\!\rightarrow\mathsf{E}(2)$$

$$\dfrac{\mathbf{X} = \mathbf{x} \Box\!\!\rightarrow \varphi \quad \begin{array}{c}[\mathbf{X} = \mathbf{x}]\\ \vdots\\ \mathbf{Y} = \mathbf{y}\end{array} \quad \begin{array}{c}[\mathbf{Y} = \mathbf{y}]\\ \vdots\\ \mathbf{X} = \mathbf{x}\end{array}}{\mathbf{Y} = \mathbf{y} \Box\!\!\rightarrow \varphi}\Box\!\!\rightarrow\mathsf{Sub}_A \qquad \dfrac{\mathbf{X} = \mathbf{x} \Box\!\!\rightarrow \varphi \quad \begin{array}{c}[\varphi]\\ \vdots\\ \psi\end{array}}{\mathbf{X} = \mathbf{x} \Box\!\!\rightarrow \psi}\Box\!\!\rightarrow\mathsf{Sub}_C$$

$$\dfrac{\mathbf{X} = \mathbf{x} \Box\!\!\rightarrow \varphi \quad \mathbf{X} = \mathbf{x} \Box\!\!\rightarrow \psi}{\mathbf{X} = \mathbf{x} \Box\!\!\rightarrow \varphi \wedge \psi}\Box\!\!\rightarrow\wedge\mathsf{I} \qquad \dfrac{\mathbf{X} = \mathbf{x} \Box\!\!\rightarrow \varphi \vee \psi}{(\mathbf{X} = \mathbf{x} \Box\!\!\rightarrow \varphi) \vee (\mathbf{X} = \mathbf{x} \Box\!\!\rightarrow \psi)}\Box\!\!\rightarrow\vee\mathsf{Dst}$$

$$\dfrac{\mathbf{X} = \mathbf{x} \Box\!\!\rightarrow (\mathbf{Y} = \mathbf{y} \Box\!\!\rightarrow \varphi)}{(\mathbf{X}' = \mathbf{x}' \wedge \mathbf{Y} = \mathbf{y}) \Box\!\!\rightarrow \varphi}\Box\!\!\rightarrow\mathsf{Extr}\ (3) \qquad \dfrac{(\mathbf{X} = \mathbf{x} \wedge \mathbf{Y} = \mathbf{y}) \Box\!\!\rightarrow \varphi}{\mathbf{X} = \mathbf{x} \Box\!\!\rightarrow (\mathbf{Y} = \mathbf{y} \Box\!\!\rightarrow \varphi)}\Box\!\!\rightarrow\mathsf{Exp}\ (4)$$

$$\dfrac{\neg(\mathbf{X} = \mathbf{x} \Box\!\!\rightarrow \alpha)}{\mathbf{X} = \mathbf{x} \Box\!\!\rightarrow \neg\alpha}\neg\Box\!\!\rightarrow\mathsf{E} \qquad \dfrac{X_1 \rightsquigarrow X_2 \quad \dots \quad \dots \quad X_{k-1} \rightsquigarrow X_k}{\neg(X_k \rightsquigarrow X_1)}\mathsf{Recur}\ (5)$$

(1) γ is $\Box\!\!\rightarrow$-free. (2) $x \neq x'$. (3) $\mathbf{X} = \mathbf{x}$ is consistent, $\mathbf{X}' = \mathbf{X} \setminus \mathbf{Y}$, $\mathbf{x}' = \mathbf{x} \setminus \mathbf{y}$. (4) $\mathbf{X} \cap \mathbf{Y} = \emptyset$.
(5) $X_i \neq X_j$ $(i \neq j)$, and $X \rightsquigarrow Y$ (meaning "X causally affects Y") is defined as:

$$X \rightsquigarrow Y := \bigvee \big\{\mathbf{Z} = \mathbf{z} \Box\!\!\rightarrow ((X = x \Box\!\!\rightarrow Y = y) \wedge (X = x' \Box\!\!\rightarrow Y = y')) \mid \mathbf{Z} \subseteq \mathrm{Dom} \setminus \{X, Y\}, \mathbf{z} \in \mathrm{Ran}(\mathbf{Z}), x, x' \in \mathrm{Ran}(X), y, y' \in \mathrm{Ran}(Y), x \neq x', y \neq y'\big\}.$$

Note that the above system is parametrized with the signature σ, and the rules with double horizontal lines are invertible. We write $\Gamma \vdash_\sigma \varphi$ (or simply $\Gamma \vdash \varphi$ when σ is clear from the context) if the formula φ can be derived from Γ by applying the rules in the above system. It is easy to verify that all rules in our system are sound for recursive (generalized) causal teams. The axioms and rules in the Hilbert system of [2] are either included or derivable in our natural deduction system, as shown in the next proposition. We refer the reader to [2] for a commentary on the rules for $\Box\!\!\rightarrow$ and a discussion of the soundness of the rule $\Box\!\!\rightarrow\vee\mathsf{Dst}$ (i.e. the *Distribution* rule typical of Stalnaker's counterfactuals).

Proposition 4.7 *The following are derivable in the system for $\mathcal{CO}[\sigma]$:*

(i) $\alpha, \neg\alpha \vee \varphi \vdash \varphi$ *(weak modus ponens)*

(ii) $\mathbf{X} = \mathbf{x} \boxright Y = y \vdash \mathbf{X} = \mathbf{x} \boxright Y \neq y'$ *(Uniqueness)*

(iii) $\mathbf{X} = \mathbf{x} \boxright \varphi \wedge \psi \vdash \mathbf{X} = \mathbf{x} \boxright \varphi$ *(Extraction)*

(iv) $\neg(\mathbf{X} = \mathbf{x} \boxright \alpha) \dashv\vdash \mathbf{X} = \mathbf{x} \boxright \neg\alpha$

(v) $\bigvee_{y \in \mathrm{Ran}(Y)}(\mathbf{X} = \mathbf{x} \boxright Y = y)$ *(Definiteness)*

Proof. Item (i) follows from $\neg$E and $\vee$E. Items (ii),(iii) follow from ValUnq, $\wedge$E and $\boxright$Sub$_C$. For item (iv), the left to right direction follows from $\neg\boxright$E. For the other direction, we first derive by applying $\boxright\wedge$I, and $\boxright\bot$E that

$$\mathbf{X} = \mathbf{x} \boxright \neg\alpha, \mathbf{X} = \mathbf{x} \boxright \alpha \vdash \mathbf{X} = \mathbf{x} \boxright \neg\alpha \wedge \alpha \vdash \mathbf{X} = \mathbf{x} \boxright \bot \vdash \bot$$

Then, by $\neg$I we conclude that $\mathbf{X} = \mathbf{x} \boxright \neg\alpha \vdash \neg(\mathbf{X} = \mathbf{x} \boxright \alpha)$.

For item (v), we first derive by $\boxright$Eff that $\vdash \mathbf{X} = \mathbf{x} \boxright X = x$, where $X = x$ is an arbitrary equation from $\mathbf{X} = \mathbf{x}$. By ValDef we also have that $\vdash \bigvee_{y \in \mathrm{Ran}(Y)} Y = y$. Thus, we conclude by applying $\boxright$Sub$_C$ that $\vdash \mathbf{X} = \mathbf{x} \boxright \bigvee_{y \in \mathrm{Ran}(Y)} Y = y$, which then implies that $\vdash \bigvee_{y \in \mathrm{Ran}(Y)}(\mathbf{X} = \mathbf{x} \boxright Y = y)$ by $\boxright\vee$Dst. □

Theorem 4.8 (Completeness) *Let $\Delta \cup \{\alpha\}$ be a set of $\mathcal{CO}[\sigma]$-formulas. Then $\Delta \vdash \alpha \iff \Delta \models^{c/g} \alpha$.*

Proof. Since our system derives all axioms and rules of the Hilbert system of [2], the completeness of our system over causal teams follows from that of [2]. The completeness of the system over generalized causal teams follows from the fact that $\Delta \models^c \alpha$ iff $\Delta \models^g \alpha$, given by Corollary 2.13. □

5 Extensions of $\mathcal{CO}$

5.1 Expressive power of $\mathcal{CO}_{\mathbb{V}}$ and $\mathcal{COD}$

In this section, we characterize the expressive power of $\mathcal{CO}_{\mathbb{V}}$ and $\mathcal{COD}$ over causal teams. We show that both logics characterize all nonempty causally downward closed team properties up to causal equivalence, and the two logics are thus expressively equivalent. An analogous result can be obtained for generalized causal teams, but we omit it due to space limitations.

Theorem 5.1 *Let $\mathcal{K}$ be a nonempty (finite) class of causal teams over some signature σ. Then the following are equivalent:*

(i) $\mathcal{K}$ is causally downward closed and closed under equivalence.

(ii) $\mathcal{K}$ is definable by a $\mathcal{CO}_{\mathbb{V}}[\sigma]$-formula.

(iii) $\mathcal{K}$ is definable by a $\mathcal{COD}[\sigma]$-formula.

By Theorems 2.11 and 3.3, for every $\mathcal{CO}_{\mathbb{V}}[\sigma]$- or $\mathcal{COD}[\sigma]$-formula φ, the set $\mathcal{K}_\varphi$ is nonempty, causally downward closed and closed under causal equivalence. Thus items (ii) and (iii) of the above theorem imply item (i). Since dependence atoms $=(\mathbf{X}; Y)$ are definable in $\mathcal{CO}_{\mathbb{V}}[\sigma]$ (see Equation (1)), item (iii) implies item (ii). It then suffices to show that item (i) implies item (iii). In this proof, we make essential use of a formula Ξ^T that resembles, in the causal setting, a similar formula introduced in [17] in the pure team setting.

Given any causal team $T = (T^-, \mathcal{G})$ over σ, let $\overline{T} = (\mathbb{A}_\sigma \setminus T^-, \mathcal{G})$ and $\mathcal{G}_0 \in \mathbb{F}^0_\sigma$ be such that $[\mathcal{G}_0] = [\mathcal{G}]$. If $T^- \neq \emptyset$ and $|T^-| = k+1$, define a $\mathcal{COD}[\sigma]$-formula

$$\Xi^T := (\chi_k \vee \Theta^{\overline{T}}) \vee \bigvee_{\mathcal{F} \in \mathbb{F}^0_\sigma \setminus \{\mathcal{G}_0\}} \Phi^{\mathcal{F}},$$

where the formula χ_k is defined inductively as

$$\chi_0 = \bot, \ \chi_1 = \bigwedge_{V \in \mathrm{Dom}} {=}(V), \ \text{ and } \chi_k = \underbrace{\chi_1 \vee \cdots \vee \chi_1}_{k \text{ times}} \ (k > 1).$$

Lemma 5.2 *Let S, T be two causal teams over some signature σ with $T^- \neq \emptyset$. Then, $S \models \Xi^T \iff$ for all $R : T \approx R$ implies $R \not\subseteq S$.*

Proof. First, observe that the formula χ_k characterize the cardinality of causal teams S, in the sense that

$$S \models \chi_k \ \text{ iff } \ |S^-| \leq k. \tag{2}$$

Indeed, clearly, $S \models \chi_0$ iff $S^- = \emptyset$, $S \models \chi_1$ iff $|S^-| \leq 1$, and for $k > 1$, $S \models \chi_k$ iff $S = S_1 \cup \cdots \cup S_k$ with each $S_i \models \chi_1$ iff $|S^-| \leq k$.

Now we prove the lemma. Let $S = (S^-, \mathcal{H})$. "$\Longrightarrow$": Suppose $S \models \Xi^T$. If $\mathcal{H} \not\sim \mathcal{G}$, then $T = (T^-, \mathcal{G}) \approx (T^-, \mathcal{G}') = R$ implies $\mathcal{G}' \neq \mathcal{H}$, thereby $R \not\subseteq S$. Now, suppose $\mathcal{H} \sim \mathcal{G} \sim \mathcal{G}_0$. If $S^- = \emptyset$, then since $T^- \neq \emptyset$, the statement holds. If $S^- \neq \emptyset$, then by Lemma 3.4(ii), we know that no nonempty subteam of S satisfies $\bigvee_{\mathcal{F} \in \mathbb{F}^0_\sigma \setminus \{\mathcal{G}_0\}} \Phi^{\mathcal{F}}$. Thus there exist $S_1, S_2 \subseteq S$ such that $S_1^- \cup S_2^- = S^-$,

$$S_1 \models \chi_k \text{ and } S_2 \models \Theta^{\overline{T}}. \tag{3}$$

By (2), the first clause of the above implies that $|S_1^-| \leq k$. Since $|T^-| = k+1 > k$, this means that $T^- \setminus S_1^- \neq \emptyset$. By Lemma 4.5, it follows from the second clause of (3) and the fact that $S_2 \models \Phi^{\mathcal{G}}$ (given again by Lemma 3.4(ii)) that $S_2 \approx R_0 \subseteq \overline{T}$ for some R_0. Thus, $T^- \cap S_2^- = \emptyset$. Altogether, we conclude that $T^- \not\subseteq S^-$. Thus, for any R such that $R \approx T$, we must have that $R^- = T^- \not\subseteq S^-$, thereby $R \not\subseteq S$.

"$\Longleftarrow$": Suppose $T \approx R$ implies $R \not\subseteq S$ for all R. If $\mathcal{H} \not\sim \mathcal{G} \sim \mathcal{G}_0$, then by Lemma 3.4(ii) we have that $S \models \bigvee_{\mathcal{F} \in \mathbb{F}^0_\sigma \setminus \{\mathcal{G}_0\}} \Phi^{\mathcal{F}}$, thereby $S \models \Xi^T$, as required. Now, suppose $\mathcal{H} \sim \mathcal{G}$. The assumption then implies that $T^- \not\subseteq S^-$. Let $S_1 = (S^- \cap T^-, \mathcal{H})$ and $S_2 = (S^- \setminus T^-, \mathcal{H})$. Clearly, $S^- = S_1^- \cup S_2^-$, and it suffices to show that (3) holds. By definition we have that $S_2^- \subseteq (\overline{T})^-$, which implies the second clause of (3) by Lemma 4.4. To prove the first clause of (3), by (2) it suffices to verify that $|S_1^-| \leq k$. Indeed, since $T^- \not\subseteq S^-$, we have that $T^- \supsetneq S^- \cap T^- = S_1^-$. Hence, $|S_1^-| < |T^-| = k+1$, namely, $|S_1^-| \leq k$. □

Now we are in a position to prove the main theorem of the section.

Proof of Theorem 5.1. We prove that item (i) implies item (iii). Let $\mathcal{K}$ be a nonempty finite class of causal teams as described in item (i). Since $\mathcal{K}$ is nonempty and causally downward closed, all causal teams over σ with empty team component belong to $\mathcal{K}$. Thus, every causal team $T \in \mathbb{C}_\sigma \setminus \mathcal{K}$ has a nonempty team component, where $\mathbb{C}_\sigma$ denotes the (finite) set of all causal teams over σ. Now, define $\varphi = \bigwedge_{T \in \mathbb{C}_\sigma \setminus \mathcal{K}} \Xi^T$. We show that $\mathcal{K} = \mathcal{K}_\varphi$.

For any $S \notin \mathcal{K}$, i.e., $S \in \mathbb{C}_\sigma \setminus \mathcal{K}$, since $S \subseteq S$ and $S^- \neq \emptyset$, by Lemma 5.2 we have that $S \not\models \Xi^S$. Thus $S \not\models \varphi$, i.e., $S \notin \mathcal{K}_\varphi$. Conversely, suppose $S \in \mathcal{K}$. Take any $T \in \mathbb{C}_\sigma \setminus \mathcal{K}$. If $T \approx R \subseteq S$ for some R, then since $\mathcal{K}$ is closed under equivalence and causally closed downward, we must conclude that $T \in \mathcal{K}$, which is a contradiction. Thus, by Lemma 5.2, $S \models \Xi^T$. Hence $S \models \varphi$, i.e., $S \in \mathcal{K}_\varphi$. □

5.2 Axiomatizing $CO_{\mathbb{V}}$ over generalized causal teams

In this section, we introduce a sound and complete system of natural deduction for $CO_{\mathbb{V}}[\sigma]$, which extends of the system for $CO[\sigma]$, and can also be seen as a variant of the systems for propositional dependence logics introduced in [17].

Definition 5.3 *The system of natural deduction for $CO_{\mathbb{V}}[\sigma]$ consists of all rules of the system of $CO[\sigma]$ (see Definition 4.6) together with the following rules, where note that in the rules $\vee E$, $\neg I$, $\neg E$, RAA and $\neg\Box\!\!\rightarrow I$ from Definition 4.6 the formula α ranges over $CO[\sigma]$-formulas only:*

- *Additional rules for $\vee$:*

$$\frac{\varphi \vee \psi}{\psi \vee \varphi}\ \vee\mathsf{Com} \qquad \frac{(\varphi \vee \psi) \vee \chi}{\varphi \vee (\psi \vee \chi)}\ \vee\mathsf{Ass} \qquad \frac{\varphi \vee \psi \quad \begin{matrix}[\varphi]\\ \vdots \\ \chi\end{matrix}}{\chi \vee \psi}\ \vee\mathsf{Sub}$$

- *Rules for $\mathbb{V}$:*

$$\frac{\varphi}{\varphi \mathbin{\mathbb{V}} \psi}\ \mathbb{V}\mathsf{I} \qquad \frac{\varphi}{\psi \mathbin{\mathbb{V}} \varphi}\ \mathbb{V}\mathsf{I} \qquad \frac{\varphi \mathbin{\mathbb{V}} \psi \quad \begin{matrix}[\varphi]\\ \vdots \\ \chi\end{matrix} \quad \begin{matrix}[\psi]\\ \vdots \\ \chi\end{matrix}}{\chi}\ \mathbb{V}\mathsf{E}$$

$$\frac{\varphi \vee (\psi \mathbin{\mathbb{V}} \chi)}{(\varphi \vee \psi) \mathbin{\mathbb{V}} (\varphi \vee \chi)}\ \vee\mathbb{V}\mathsf{Dst} \qquad \frac{\mathbf{X} = \mathbf{x} \mathrel{\Box\!\!\rightarrow} \psi \mathbin{\mathbb{V}} \chi}{(\mathbf{X} = \mathbf{x} \mathrel{\Box\!\!\rightarrow} \psi) \mathbin{\mathbb{V}} (\mathbf{X} = \mathbf{x} \mathrel{\Box\!\!\rightarrow} \chi)}\ \Box\!\!\rightarrow\!\mathbb{V}\mathsf{Dst}$$

The rules in our system are clearly sound. We now proceed to prove the completeness theorem. An important lemma for the theorem states that every $CO_{\mathbb{V}}[\sigma]$-formula φ is provably equivalent to the $\mathbb{V}$-disjunction of a (finite) set of $CO[\sigma]$-formulas. Formulas of this type are called *resolutions* of φ in [5].

Definition 5.4 *Let φ be a $CO_{\mathbb{V}}[\sigma]$-formula. Define the set $\mathcal{R}(\varphi)$ of its* ***resolutions*** *inductively as follows:*

- $\mathcal{R}(X = x) = \{X = x\}$,
- $\mathcal{R}(\neg\alpha) = \{\neg\alpha\}$,
- $\mathcal{R}(\psi \wedge \chi) = \{\alpha \wedge \beta \mid \alpha \in \mathcal{R}(\psi), \beta \in \mathcal{R}(\chi)\}$,
- $\mathcal{R}(\psi \vee \chi) = \{\alpha \vee \beta \mid \alpha \in \mathcal{R}(\psi), \beta \in \mathcal{R}(\chi)\}$,
- $\mathcal{R}(\psi \mathbin{\mathbb{V}} \chi) = \mathcal{R}(\psi) \cup \mathcal{R}(\chi)$,

- $\mathcal{R}(\mathbf{X} = \mathbf{x} \,\Box\!\!\rightarrow \varphi) = \{\mathbf{X} = \mathbf{x} \,\Box\!\!\rightarrow \alpha \mid \alpha \in \mathcal{R}(\varphi)\}$.

The set $\mathcal{R}(\varphi)$ is clearly a finite set of $CO[\sigma]$-formulas.

Lemma 5.5 *For any formula $\varphi \in CO_{\mathbb{V}}[\sigma]$, we have that $\varphi \dashv\vdash \bigvee\!\!\!\!\bigvee \mathcal{R}(\varphi)$.*

Proof. We prove the lemma by induction on φ. If φ is $X = x$ or $\neg\alpha$ for some $CO[\sigma]$-formula α, then $\mathcal{R}(\varphi) = \{\varphi\}$, and $\varphi \dashv\vdash \bigvee\!\!\!\!\bigvee \mathcal{R}(\varphi)$ holds trivially.

Now, suppose $\psi \dashv\vdash \bigvee\!\!\!\!\bigvee \mathcal{R}(\psi)$ and $\chi \dashv\vdash \bigvee\!\!\!\!\bigvee \mathcal{R}(\chi)$. If $\varphi = \psi \wedge \chi$, observing that $\theta_0 \wedge (\theta_1 \mathbin{\mathbb{V}} \theta_2) \dashv\vdash (\theta_0 \wedge \theta_1) \mathbin{\mathbb{V}} (\theta_0 \wedge \theta_2)$ (by $\mathbb{V}$E, $\mathbb{V}$I,$\wedge$I,$\wedge$E), we derive by $\mathbb{V}$I, $\mathbb{V}$E that

$$\psi \wedge \chi \dashv\vdash (\bigvee\!\!\!\!\bigvee \mathcal{R}(\psi)) \wedge (\bigvee\!\!\!\!\bigvee \mathcal{R}(\chi)) \dashv\vdash \bigvee\!\!\!\!\bigvee \{\alpha \wedge \beta \mid \alpha \in \mathcal{R}(\psi), \beta \in \mathcal{R}(\chi)\} \dashv\vdash \bigvee\!\!\!\!\bigvee \mathcal{R}(\psi \wedge \chi).$$

If $\varphi = \psi \vee \chi$, we have analogous derivations using the fact that $\theta_0 \vee (\theta_1 \mathbin{\mathbb{V}} \theta_2) \dashv\vdash (\theta_0 \vee \theta_1) \mathbin{\mathbb{V}} (\theta_0 \vee \theta_2)$ (by $\vee\mathbb{V}$Dst, $\mathbb{V}$I, $\mathbb{V}$E and $\vee$Sub) and $\mathbb{V}$I, $\mathbb{V}$E.

If $\varphi = \psi \mathbin{\mathbb{V}} \chi$, then by applying $\mathbb{V}$I and $\mathbb{V}$E, we have that

$$\psi \mathbin{\mathbb{V}} \chi \dashv\vdash (\bigvee\!\!\!\!\bigvee \mathcal{R}(\psi)) \mathbin{\mathbb{V}} (\bigvee\!\!\!\!\bigvee \mathcal{R}(\chi)) \dashv\vdash \bigvee\!\!\!\!\bigvee (\mathcal{R}(\psi) \cup \mathcal{R}(\chi)) \dashv\vdash \bigvee\!\!\!\!\bigvee \mathcal{R}(\psi \mathbin{\mathbb{V}} \chi).$$

If $\varphi = \mathbf{X} = \mathbf{x} \,\Box\!\!\rightarrow \psi$, then

$$\begin{aligned}
\mathbf{X} = \mathbf{x} \,\Box\!\!\rightarrow \psi \;&\dashv\vdash \mathbf{X} = \mathbf{x} \,\Box\!\!\rightarrow \bigvee\!\!\!\!\bigvee \mathcal{R}(\psi) && (\Box\!\!\rightarrow\text{Sub}_C)\\
&\dashv\vdash \bigvee\!\!\!\!\bigvee \{\mathbf{X} = \mathbf{x} \,\Box\!\!\rightarrow \alpha \mid \alpha \in \mathcal{R}(\psi)\} && (\Box\!\!\rightarrow\mathbb{V}\text{Dst, and } \mathbb{V}\text{I}, \Box\!\!\rightarrow\text{Sub}_C, \mathbb{V}\text{E})\\
&\dashv\vdash \bigvee\!\!\!\!\bigvee \mathcal{R}(\mathbf{X} = \mathbf{x} \,\Box\!\!\rightarrow \psi).
\end{aligned}$$

□

Theorem 5.6 (Completeness) *Let $\Gamma \cup \{\psi\}$ be a set of $CO_{\mathbb{V}}[\sigma]$-formulas. Then $\Gamma \vdash \psi \iff \Gamma \models^g \psi$.*

Proof. We prove the "$\Longleftarrow$" direction. Observe that there are only finitely many classes of causal teams of signature σ. Thus, any set of $CO_{\mathbb{V}}[\sigma]$-formulas is equivalent to a single $CO_{\mathbb{V}}[\sigma]$-formula, and it then suffices to prove the statement for $\Gamma = \{\varphi\}$.

Now suppose $\varphi \models \psi$. Then by Lemma 5.5 and soundness we have that $\bigvee\!\!\!\!\bigvee \mathcal{R}(\varphi) \models \bigvee\!\!\!\!\bigvee \mathcal{R}(\psi)$. Thus, for every $\gamma \in \mathcal{R}(\varphi)$, $\gamma \models^g \bigvee\!\!\!\!\bigvee \mathcal{R}(\psi)$, which further implies, by Lemma 3.6, that there is an $\alpha_\gamma \in \mathcal{R}(\psi)$ such that $\gamma \models \alpha_\gamma$. Since γ, α_γ are $CO[\sigma]$-formulas, and the system for $CO_{\mathbb{V}}[\sigma]$ extends that for $CO[\sigma]$, we obtain by the completeness theorem of $CO[\sigma]$ (Theorem 4.8) that $\gamma \vdash \alpha_\gamma$. Applying $\mathbb{V}$I and Lemma 5.5, we obtain $\gamma \vdash \bigvee\!\!\!\!\bigvee \mathcal{R}(\psi) \vdash \psi$ for each $\gamma \in \mathcal{R}(\psi)$. Thus, by Lemma 5.5 and repeated applications of $\mathbb{V}$E, we conclude that $\varphi \vdash \bigvee\!\!\!\!\bigvee \mathcal{R}(\varphi) \vdash \psi$. □

5.3 Axiomatizing $CO_{\mathbb{V}}$ over causal teams

The method for the completeness proof of the previous subsection cannot be used for causal team semantics, as it makes essential use of the disjunction property of $\mathbb{V}$, which fails over causal teams. However, since causal teams can be regarded as a special case of generalized causal teams, all the rules in the system for $CO_{\mathbb{V}}$ over generalized causal teams are also sound over causal teams. We can then axiomatize $CO_{\mathbb{V}}$ over causal teams by extending the system of $CO_{\mathbb{V}}$ for generalized causal teams with an axiom characterizing the property of being uniform, i.e. "indistinguishable" from a causal team.

Definition 5.7 *The system for $CO_{\mathbb{V}}[\sigma]$ over causal teams consists of all rules* *of $\mathcal{CO}_{\mathbb{V}}[\sigma]$ over generalized causal teams (Def. 5.3) plus the following axiom:*

$$\frac{}{\mathbb{V}_{\mathcal{F}\in\mathbb{F}_\sigma}\Phi^{\mathcal{F}}}\ \mathsf{Unf}$$

By Theorem 3.4(ii), the axiom Unf is clearly sound over causal teams.

Lemma 5.8 *For any set $\Gamma\cup\{\psi\}$ of $\mathcal{CO}_{\mathbb{V}}[\sigma]$-formulas, $\Gamma\models^c\psi$ iff $\Gamma, \mathbb{V}_{\mathcal{F}\in\mathbb{F}_\sigma}\Phi^{\mathcal{F}}\models^g\psi$.*

Proof. $\Longleftarrow$: Suppose $T\models^c\Gamma$ for some causal team T. Consider the generalized causal team T^g generated by T. By Lemma 2.12, $T^g\models^g\Gamma$. Since T^g is uniform, Corollary 3.5 gives that $T^g\models^g\mathbb{V}_{\mathcal{F}\in\mathbb{F}_\sigma}\Phi^{\mathcal{F}}$. Then, by assumption, we obtain that $T^g\models^g\psi$, which, by Lemma 2.12 again, implies that $T\models^c\psi$.

$\Longrightarrow$: Suppose $T\models^g\Gamma$ and $T\models^g\mathbb{V}_{\mathcal{F}\in\mathbb{F}_\sigma}\Phi^{\mathcal{F}}$ for some generalized causal team T. By Corollary 3.5 we know that T is uniform. Pick $(t,\mathcal{F})\in T$. Consider the generalized causal team $S=\{(s,\mathcal{F})\mid s\in T^-\}$. Observe that $T\approx S$. Thus, by Theorem 3.3, we have that $S\models^g\Gamma$, which further implies, by Lemma 2.12(ii), that $S^c\models^c\Gamma$. Hence, by the assumption we conclude that $S^c\models^c\psi$. Finally, by applying Lemma 2.12(ii) and Theorem 3.3 again, we obtain $T\models^g\psi$. □

Theorem 5.9 (Completeness) *Let $\Gamma\cup\{\psi\}$ be a set of $\mathcal{CO}_{\mathbb{V}}[\sigma]$-formulas. Then $\Gamma\models^c\psi\iff\Gamma\vdash^c\psi$.*

Proof. Suppose $\Gamma\models^c\psi$. By Lemma 5.8, we have that $\Gamma, \mathbb{V}_{\mathcal{F}\in\mathbb{F}_\sigma}\Phi^{\mathcal{F}}\models^g\psi$, which implies that $\Gamma, \mathbb{V}_{\mathcal{F}\in\mathbb{F}_\sigma}\Phi^{\mathcal{F}}\vdash\psi$, by the completeness theorem (5.6) of the system for $\mathcal{CO}_{\mathbb{V}}[\sigma]$ over generalized causal teams. Thus, $\Gamma\vdash\psi$ by axiom Unf. □

5.4 Axiomatizing $\mathcal{COD}$

We briefly sketch the analogous axiomatization results for the language $\mathcal{COD}$ over both semantics.

Over generalized causal teams, the system for $\mathcal{COD}[\sigma]$ consists of all the rules of the system for $\mathcal{CO}[\sigma]$ (Definition 4.6) together with $\vee$Com, $\vee$Ass, $\vee$Sub (the "additional rules for $\vee$" from Definition 5.3) and the new rules for dependence atoms defined below:

$$\frac{X=x}{=\!(X)}\ \mathsf{DepI_0}$$

$$\frac{\begin{matrix}[=\!(X_1)] & \dots & [=\!(X_n)]\\ & \vdots & \\ & =\!(Y) & \end{matrix}}{=\!(X_1,\dots,X_n;Y)}\ \mathsf{DepI}$$

$$\frac{\varphi \qquad \begin{matrix}\forall x\in Ran(X)\\ [\varphi[X=x/\,=\!(X)]]\\ \vdots\\ \psi\end{matrix}}{\psi}\ \mathsf{Dep_0E}\ (*)$$

$$\frac{=\!(X_1,\dots,X_n;Y)\qquad =\!(X_1)\ \dots\ =\!(X_n)}{=\!(Y)}\ \mathsf{DepE}$$

(*) $\varphi[X=x/\,=\!(X)]$ stands for the formula obtained by replacing *a specific occurrence* of $=\!(X)$ in φ with $X=x$.

These rules for dependence atoms generalize the corresponding rules in the pure team setting as introduced in [17]. The completeness theorem of the system can be proved by generalizing the corresponding arguments in [17]. Analogously to the case for $\mathcal{CO}_{\mathbb{V}}$, in this proof we use the fact that every formula φ is (semantically) equivalent to a formula $\mathbb{V}_{i\in I}\alpha_i$ in disjunctive normal form, where each α_i is a $\mathcal{CO}[\sigma]$-formula obtained from φ by replacing every dependence atom $=\!(\mathbf{X};Y)$ by a formula $\bigvee_{\mathbf{x}\in \mathrm{Ran}(\mathbf{X})}(\mathbf{X}=\mathbf{x}\wedge Y=y)$ with y ranging over all of $\mathrm{Ran}(Y)$. The disjunctive formula $\mathbb{V}_{i\in I}\alpha_i$ is not in the language of $\mathcal{COD}$, but we can prove in the system of $\mathcal{COD}$ (by applying the additional rules in the table above) that $\alpha_i\vdash\varphi$ $(i\in I)$, and that

$$\Gamma,\alpha_i\vdash\psi \text{ for all } i\in I\Longrightarrow\Gamma,\varphi\vdash\psi.$$

These mean *in effect* that "$\varphi\dashv\vdash\mathbb{V}_{i\in I}\alpha_i$". The completeness theorem for $\mathcal{COD}$ is then proved using essentially the same strategy as that for $\mathcal{CO}_{\mathbb{V}}$ (Thm. 5.6).

Over causal teams, using the same method as in the previous section, the complete system for $\mathcal{COD}[\sigma]$ can be defined as an extension of the above generalized causal team system with two additional axioms 1Fun and NoMix:

$$\frac{}{\bigwedge_{V\in \mathrm{Dom}}\Big(\beta_{\mathrm{En}}(V)\supset(\bigwedge_{\mathbf{w}\in\mathbf{W}_V}\mathbf{W}_V=\mathbf{w}\;\Box\!\!\rightarrow\;=\!(V))\Big)}\ \textsf{1Fun}\ (1)$$

$$\frac{}{\bigwedge_{V\in Dom}\bigwedge\{\Xi_*^{\{a,b\}}\mid(a,b)\in\mathbb{S}\mathrm{em}_\sigma^2,\ \{a\}\models\beta_{\mathrm{En}}(V),\ \{b\}\not\models\beta_{\mathrm{En}}(V)\}}\ \textsf{NoMix}\ (2)$$

(1) $\mathbf{W}_V=\mathrm{Dom}\setminus\{V\}$, and $\beta_{\mathrm{En}}(V):=\bigvee_{X\in\mathbf{W}_V}\beta_{\mathrm{DC}}(X,V)$, where each $\beta_{\mathrm{DC}}(X,V)$ is the $\mathcal{CO}[\sigma]$-formula from [2] expressing the property "X is a direct cause of V":

$$\beta_{\mathrm{DC}}(X,V):=\bigvee\{(\mathbf{Z}=\mathbf{z}\wedge X=x)\;\Box\!\!\rightarrow\;V=v,\ (\mathbf{Z}=\mathbf{z}\wedge X=x')\;\Box\!\!\rightarrow\;V=v' \mid x,x'\in\mathrm{Ran}(X),\ v,v'\in\mathrm{Ran}(V),\ \mathbf{Z}=\mathrm{Dom}\setminus\{X,V\},\ \mathbf{z}\in\mathrm{Ran}(\mathbf{Z}),\ x\neq x',v\neq v'\}.$$

(2) $\Xi_*^{\{a,b\}}$ is defined otherwise the same as $\Xi^{\{a,b\}}$ except that χ_1 is redefined as

$$\chi_1:=\bigwedge_{V\in Dom}(=\!(V)\wedge\bigwedge_{\mathbf{w}\in Ran(\mathbf{W}_V)}(\mathbf{W}_V=\mathbf{w}\;\Box\!\!\rightarrow\;=\!(V))).$$

The axiom 1Fun states that the endogenous variables are governed by a unique function; the axiom NoMix guarantees that all members of the generalized causal team agree on what is the set of endogenous variables. Together, these two additional axioms characterize the *uniformity* of the generalized causal team in question (or they are equivalent to the formula Unf in $\mathcal{CO}_{\mathbb{V}}[\sigma]$), thus allow for a completeness proof along the lines of Section 5.3.

6 Conclusion

We have answered the main questions concerning the expressive power and the existence of deduction calculi for the languages that were proposed in [1] and [2], and which involve both (interventionist) counterfactuals and (contingent) dependencies. In the process, we have introduced a generalized causal team semantics, for which we have also provided natural deduction calculi. We point out that our calculi are sound only for *recursive* systems, i.e., when the causal graph is acyclic. The general case (and special cases such as the "Lewisian" systems considered in [19]) will require a separate study. We point out, however, that each of our deduction systems can be adapted to the case of *unique-solution* (possibly generalized) causal teams by replacing the Recur rule with an inference rule that expresses the *Reversibility* axiom from [7].

Our work shows that many methodologies developed in the literature on team semantics can be adapted to the generalized semantics and, to a lesser extent, to causal team semantics. On the other hand, a number of peculiarities emerged that set apart these semantic frameworks from the usual team semantics: for example, the failure of the disjunction property over causal teams. We believe the present work may provide guidelines for the investigation of further notions of dependence and causation in causal team semantics and its variants.

References

[1] Barbero, F. and G. Sandu, *Team semantics for interventionist counterfactuals and causal dependence*, in: *Proceedings CREST@ETAPS 2018*, pp. 16–30.

[2] Barbero, F. and G. Sandu, *Team semantics for interventionist counterfactuals: observations vs. interventions*, Journal of Philosophical Logic (to appear), 2020.

[3] Briggs, R., *Interventionist counterfactuals*, Philosophical Studies: An International Journal for Philosophy in the Analytic Tradition **160** (2012), pp. 139–166.

[4] Chockler, H. and J. Y. Halpern, *Responsibility and blame: A structural-model approach*, Journal of Artificial Intelligence Research 22 (2004): 93-115.

[5] Ciardelli, I., *Questions in logic,* 2016, PhD thesis, Universiteit van Amsterdam.

[6] Eva, B., R. Stern, and S. Hartmann, *The similarity of causal structure*, Philosophy of Science 86.5 (2019): 821-835.

[7] Galles, D. and J. Pearl, *An axiomatic characterization of causal counterfactuals*, Foundations of Science **3** (1998), pp. 151–182.

[8] Halpern, J. Y., *Axiomatizing causal reasoning*, J. Artif. Int. Res. **12** (2000), pp. 317–337.

[9] Halpern, J. (2016). *Actual Causality*, Cambridge, Massachusetts; London, England: The MIT Press.

[10] Halpern, J. Y. and M. Kleiman-Weiner, *Towards Formal Definitions of Blameworthiness, Intention, and Moral Responsibility*, Proceedings of the Thirty-Second AAAI Conference on Artificial Intelligence (AAAI-18), 2018.

[11] Hitchcock, C., *The Intransitivity of Causation Revealed in Equations and Graphs*, The Journal of Philosophy **98** (2001), pp. 273-299.

[12] Hodges, W., *Compositional semantics for a language of imperfect information*, Logic Journal of the IGPL **5** (1997), pp. 539–563.

[13] Pearl, J., *Causality: Models, Reasoning, and Inference*, Cambridge University Press, New York, NY, USA, 2000.

[14] Spirtes, P., C. Glymour and R. N. Scheines, *Causation, Prediction, and Search*, Lecture Notes in Statistics **81**, Springer New York, 1993.

[15] Starr, W., *Counterfactuals*, in: *The Stanford Encyclopedia of Philosophy*, `https://plato.stanford.edu/archives/fall2019/entries/counterfactuals/`.

[16] Väänänen, J., *Dependence Logic: A New Approach to Independence Friendly Logic*, London Mathematical Society Student Texts **70**, Cambridge University Press, 2007.

[17] Yang, F. and J. Väänänen, *Propositional logics of dependence*, Annals of Pure and Applied Logic **167** (2016), pp. 557 – 589.

[18] Yang, F. and J. Väänänen, *Propositional team logics*, Annals of Pure and Applied Logic **168** (2017), pp. 1406 – 1441.

[19] Zhang, J., *A Lewisian logic of causal counterfactuals*, Minds and Machines **23** (2013), pp. 77–93.

the causal graph is acyclic. The general case (no special cases such as the [illegible] systems considered in [illegible]) will require a separate study. We point out, however, that each of our deduction systems can be adapted to the case of [illegible] relation (possibly generalized) causal [illegible] by [illegible] the [illegible] rule with [illegible] that [illegible] the [illegible] from [illegible].

Our work shows that many methodologies developed in the literature on team semantics can be adapted to the generalized setting, and to a lesser extent to causal team semantics. On the other hand, a number of [illegible] have emerged that [illegible] these [illegible] frameworks from [illegible] team semantics, for example, the failure of the [illegible] property over causal teams. We believe the present work may provide guidelines for the investigation of further [illegible] causal [illegible].

References

[illegible]

[illegible] W. [illegible] *[illegible] of [illegible]* [illegible]

[16] Väänänen, J. [illegible] [illegible] *[illegible]* [illegible]

[17] [illegible] and J. Väänänen. [illegible] *[illegible] of [illegible]* [illegible] 101 (2010), [illegible]

[18] Yang, F. and J. Väänänen. *[illegible]* [illegible] 168 (2017), pp. 1406–1441.

[19] [illegible] *[illegible]* [illegible] (2013).

Temporal interpretation of intuitionistic quantifiers

Guram Bezhanishvili and Luca Carai [1]

Department of Mathematical Sciences
New Mexico State University
Las Cruces NM 88003, USA

Abstract

We show that intuitionistic quantifiers admit the following temporal interpretation: $\forall xA$ is true at a world w iff A is true at every object in the domain of every future world, and $\exists xA$ is true at w iff A is true at some object in the domain of some past world. For this purpose we work with a predicate version of the well-known tense propositional logic S4.t. The predicate logic Q°S4.t is obtained by weakening the axioms of the standard predicate extension QS4.t of S4.t along the lines Corsi weakened QK to Q°K. The Gödel translation embeds the predicate intuitionistic logic IQC into QS4 fully and faithfully. We provide a temporal version of the Gödel translation and prove that it embeds IQC into Q°S4.t fully and faithfully; that is, we show that a sentence is provable in IQC iff its translation is provable in Q°S4.t. Faithfulness is proved using syntactic methods, while we prove fullness utilizing the generalized Kripke semantics of Corsi.

Keywords: Intuitionistic quantifiers, temporal interpretation, Gödel translation.

1 Introduction

Unlike classical connectives, intuitionistic connectives lack symmetry. It was noted already by McKinsey and Tarski [17] that Heyting algebras (which are algebraic models of intuitionistic propositional calculus IPC) are not symmetric even in the weak sense, meaning that the order-dual of a Heyting algebra may no longer be a Heyting algebra. In contrast, Boolean algebras (which are algebraic models of classical propositional calculus) are symmetric in the strong sense, meaning that the order-dual of a Boolean algebra A is not only a Boolean algebra, but even isomorphic to A.

This non-symmetry has been addressed by several authors, resulting in the concepts of bi-Heyting algebras and symmetric Heyting algebras. Bi-Heyting

[1] Email addresses: guram@nmsu.edu, lcarai@nmsu.edu
Acknowledgment: We would like to thank the reviewers whose comments have improved the presentation of the paper.

algebras are obtained by adding to the signature of Heyting algebras a binary operation of co-implication, while symmetric Heyting algebras by adding a de Morgan negation (and then co-implication becomes de Morgan dual of implication). The order-dual of a bi-Heyting algebra is again a bi-Heyting algebra, and the order-dual of a symmetric Heyting algebra A is even isomorphic to A. Thus, the class of bi-Heyting algebras is symmetric in the weak sense, while the class of symmetric Heyting algebras is symmetric in the strong sense (hence the name).

The Gödel translation of IPC into S4 extends to a translation of the Heyting-Brouwer calculus HB of Rauszer [18] into the tense extension S4.t of S4, which has the future S4-modality $\Box_F$ and the past S4-modality $\Box_P$. The algebraic models of HB are bi-Heyting algebras, and implication is interpreted using $\Box_F$ and co-implication using $\Box_P$.

This story of non-symmetry also extends to intuitionistic quantifiers. Let IQC be the intuitionistic predicate calculus and QS4 the predicate S4. Not only the intuitionistic quantifiers $\forall x$ and $\exists x$ are not definable from each other (unlike the classical quantifiers), but the Gödel translation $(\ )^t$ of IQC into QS4 is asymmetric in that $(\forall xA)^t = \Box\forall xA^t$ and $(\exists xA)^t = \exists xA^t$. This is manifested in the interpretation of intuitionistic quantifiers in Kripke models. Indeed, a world w of a Kripke model satisfies $\forall xA$ iff A is true at every object of the domain D_v of every world v accessible from w, while w satisfies $\exists xA$ iff A is true at some object in the domain D_w of w. If we think of the worlds of a Kripke model as "states of knowledge," and the order between the states is temporal, then we can interpret the intuitionistic universal quantifier as "for every object in the future," while the existential quantifier as "for some object in the present."

In this article we present a more symmetric interpretation of intuitionistic quantifiers as "for every object in the future" for $\forall x$ and "for some object in the past" for $\exists x$. We show that such interpretation is supported by translating IQC fully and faithfully into a predicate tense logic by an appropriate modification of the Gödel translation. As far as we know, this approach has not been considered in the past. One obvious obstacle is that it is unclear what predicate tense logic to choose for such a translation. Indeed, a natural candidate would be the standard predicate extension QS4.t of S4.t. However, since QS4.t proves the Barcan formula, and hence the Kripke frames validating QS4.t have constant domains, IQC does not translate fully into QS4.t. Instead we work with a weaker logic in which the universal instantiation axiom

$$\forall xA \to A(y/x)$$

is replaced by a weaker version

$$\forall y(\forall xA \to A(y/x)).$$

This approach is along the lines of Kripke [15], Hughes and Cresswell [13], Fitting and Mendelsohn [6], and Corsi [3] who considered modal predicate logics without the Barcan and/or converse Barcan formulas. The generalized Kripke frames considered in this semantics have two domains associated to each world, an inner domain and an outer domain. The inner domains are always contained in the outer domains and are not necessarily increasing. While variables are interpreted in the outer domains, the scope of quantifiers is restricted to the inner domains.

Utilizing this approach, we define a tense predicate logic $\mathsf{Q}^\circ\mathsf{S4.t}$ which is sound with respect to the generalized Kripke semantics with nonempty increasing inner domains and constant outer domains. We modify the Gödel translation to define a temporal translation of IQC into $\mathsf{Q}^\circ\mathsf{S4.t}$ as follows:

$$\begin{aligned}
\bot^t &= \bot \\
P(x_1,\ldots,x_n)^t &= \Box_F P(x_1,\ldots,x_n) \qquad \text{for each n-ary predicate symbol } P \\
(A \wedge B)^t &= A^t \wedge B^t \\
(A \vee B)^t &= A^t \vee B^t \\
(A \to B)^t &= \Box_F(A^t \to B^t) \\
(\forall x A)^t &= \Box_F \forall x A^t \\
(\exists x A)^t &= \Diamond_P \exists x A^t
\end{aligned}$$

Here $\Box_F$ is the $\mathsf{S4}$-modality interpreted as "always in the future" and $\Diamond_P$ is the $\mathsf{S4}$-modality interpreted as "sometime in the past." Thus, the modification of the Gödel translation concerns the clause for $\exists x A$. Our main result states that this translation is full and faithful in the following sense:

Main Theorem.

- For any formula A in the language of IQC, we have

  $$\mathsf{IQC} \vdash A \quad \text{iff} \quad \mathsf{Q}^\circ\mathsf{S4.t} \vdash \forall x_1 \cdots \forall x_n A^t$$

  where $x_1, \ldots, x_n$ are the free variables in A.

- If A is a sentence, then

  $$\mathsf{IQC} \vdash A \quad \text{iff} \quad \mathsf{Q}^\circ\mathsf{S4.t} \vdash A^t.$$

The proof of this surprising result is along the lines of the standard proof of fullness and faithfulness of the Gödel translation of IQC into $\mathsf{QS4}$. We would like to stress that the main challenge is not so much the proof itself, but rather finding the "right" predicate tense modal logic into which to translate IQC. We find it of interest to explore philosophical (as well as practical) consequences of this new temporal point of view on IQC.

The paper is structured as follows. In Section 2 we recall the intuitionistic predicate logic IQC and its Kripke completeness. In Section 3 we briefly review the basics of modal predicate logics and their Kripke semantics, including weaker modal predicate logics. In Section 4 we recall the tense propositional logic $\mathsf{S4.t}$, consider its standard predicate extension $\mathsf{QS4.t}$, and then introduce its weakening Q°S4.t which is our main tense predicate logic of interest. We conclude the section by observing that Q°S4.t is sound with respect to a version of the generalized Kripke semantics studied by Kripke [15], Hughes and Cresswell [13], Fitting and Mendelsohn [6], and Corsi [3]. Our main result, that IQC embeds into Q°S4.t fully and faithfully, is proved in Section 5. We prove faithfulness syntactically, while fullness is proved semantically. We conclude the paper with Section 6 in which we describe some open problems our study has generated. Finally, the Appendix contains the proofs of some technical lemmas used in Sections 4 and 5.

2 The intuitionistic predicate logic

Let IQC be the intuitionistic predicate logic. We recall that the language $\mathcal{L}$ of IQC consists of countably many individual variables $x, y, \ldots$, countably many n-ary predicate symbols $P, Q, \ldots$ (for each $n \geq 0$), the logical connectives $\bot, \wedge, \vee, \to$, and the quantifiers $\forall, \exists$. We do not add any constants to $\mathcal{L}$ since this results in the temporal translation not being faithful (see Remark 5.11).

Formulas are defined as usual by induction and are denoted with upper case letters $A, B, \ldots$. Let x, y be individual variables and A a formula. If x is a free variable of A and does not occur in the scope of $\forall y$ or $\exists y$, then we denote by $A(y/x)$ the formula obtained from A by replacing all the free occurrences of x by y.

The following definition of IQC is taken from [9, Sec 2.6]. We point out that, unlike [9], we prefer to work with axiom schemes, and hence do not need the inference rule of substitution.

Definition 2.1 The intuitionistic predicate logic IQC is the least set of formulas of $\mathcal{L}$ containing all substitution instances of theorems of IPC, the axiom schemes

(i) $\forall x A \to A(y/x)$ Universal instantiation (UI)

(ii) $A(y/x) \to \exists x A$

(iii) $\forall x(A \to B) \to (A \to \forall x B)$ with x not free in A

(iv) $\forall x(A \to B) \to (\exists x A \to B)$ with x not free in B

and closed under the inference rules

$$\frac{A \quad A \to B}{B} \text{ Modus Ponens (MP)} \qquad \frac{A}{\forall x A} \text{ Generalization (Gen)}$$

We next describe Kripke semantics for IQC (see [16,8]).

Definition 2.2 An IQC-*frame* is a triple $\mathfrak{F} = (W, R, D)$ where

- W is a nonempty set whose elements are called the *worlds* of $\mathfrak{F}$.
- R is a partial order on W.

- D is a function that associates to each $w \in W$ a nonempty set D_w such that wRv implies $D_w \subseteq D_v$ for each $w, v \in W$. The set D_w is called the *domain* of w.

Definition 2.3

- An *interpretation* of $\mathcal{L}$ in $\mathfrak{F}$ is a function I associating to each world w and any n-ary predicate symbol P an n-ary relation $I_w(P) \subseteq (D_w)^n$ such that wRv implies $I_w(P) \subseteq I_v(P)$.
- A *model* is a pair $\mathfrak{M} = (\mathfrak{F}, I)$ where $\mathfrak{F}$ is an IQC-frame and I is an interpretation in $\mathfrak{F}$.
- Let w be a world of $\mathfrak{F}$. A *w-assignment* is a function σ associating to each individual variable x an element $\sigma(x)$ of D_w. Note that if wRv, then σ is also a v-assignment.
- Let σ and τ be two w-assignments and x an individual variable. Then τ is said to be an *x-variant* of σ if $\tau(y) = \sigma(y)$ for all $y \neq x$.

We next recall the definition of when a formula A is true in a world w of a model $\mathfrak{M} = (\mathfrak{F}, I)$ under the w-assignment σ, written $\mathfrak{M} \models^\sigma_w A$.

Definition 2.4

$$
\begin{array}{lll}
\mathfrak{M} \models^\sigma_w \bot & & \text{never} \\
\mathfrak{M} \models^\sigma_w P(x_1, \ldots, x_n) & \text{iff} & (\sigma(x_1), \ldots, \sigma(x_n)) \in I_w(P) \\
\mathfrak{M} \models^\sigma_w B \wedge C & \text{iff} & \mathfrak{M} \models^\sigma_w B \text{ and } \mathfrak{M} \models^\sigma_w C \\
\mathfrak{M} \models^\sigma_w B \vee C & \text{iff} & \mathfrak{M} \models^\sigma_w B \text{ or } \mathfrak{M} \models^\sigma_w C \\
\mathfrak{M} \models^\sigma_w B \to C & \text{iff} & \text{for all } v \text{ with } wRv, \text{ if } \mathfrak{M} \models^\sigma_v B, \text{ then } \mathfrak{M} \models^\sigma_v C \\
\mathfrak{M} \models^\sigma_w \forall x B & \text{iff} & \text{for all } v \text{ with } wRv \text{ and each } v\text{-assignment } \tau \\
& & \text{that is an } x\text{-variant of } \sigma,\ \mathfrak{M} \models^\tau_v B \\
\mathfrak{M} \models^\sigma_w \exists x B & \text{iff} & \text{there exists a } w\text{-assignment } \tau \\
& & \text{that is an } x\text{-variant of } \sigma \text{ such that } \mathfrak{M} \models^\tau_w B
\end{array}
$$

Definition 2.5

- We say that A is *true* in a world w of $\mathfrak{M}$, written $\mathfrak{M} \models_w A$, if for all w-assignments σ, we have $\mathfrak{M} \models^\sigma_w A$.
- We say that A is *true* in $\mathfrak{M}$, written $\mathfrak{M} \models A$, if for all worlds $w \in W$, we have $\mathfrak{M} \models_w A$.
- We say that A is *valid* in a frame $\mathfrak{F}$, written $\mathfrak{F} \models A$, if for all models $\mathfrak{M}$ based on $\mathfrak{F}$, we have $\mathfrak{M} \models A$.

We have the following well-known completeness of IQC with respect to Kripke semantics.

Theorem 2.6 ([16]) *The intuitionistic predicate logic* IQC *is sound and complete with respect to Kripke semantics; that is, for each formula A,*

$$\mathsf{IQC} \vdash A \text{ iff } \mathfrak{F} \models A \text{ for each } \mathsf{IQC}\text{-frame } \mathfrak{F}.$$

3 Modal predicate logics

Modal predicate logics were first studied by Barcan [1] and Carnap [2] in 1940s. The semantic study of modal predicate logics was initiated by Kripke [14,15] in late 1950s/early 1960s. Since then many completeness results have been obtained with respect to Kripke semantics, but there is also a large body of incompleteness results, which is one of the reasons that the model theory of modal predicate logics is less advanced than that of modal propositional logics (see, e.g., [9,10] and the references therein).

Let K be the least normal modal propositional logic and let QK be the standard predicate extension of K. The language $\mathcal{L}_\Box$ of QK is the extension of $\mathcal{L}$ with the modality $\Box$. Since the modal logics we consider are based on the classical logic, it is sufficient to only consider the logical connectives $\bot, \to$ and the quantifier $\forall$. The logical connectives $\wedge, \vee, \neg, \leftrightarrow$, the quantifier $\exists$, and the modality $\Diamond$ are treated as usual abbreviations.

We next recall the definition of QK (see, e.g., [9, Sec 2.6], but note, as in Section 2, that we work with axiom schemes instead of having the inference rule of substitution).

Definition 3.1 The modal predicate logic QK is the least set of formulas of $\mathcal{L}_\Box$ containing all substitution instances of theorems of K, the axiom schemes (i) and (iii) of Definition 2.1, and closed under (MP), (Gen), and

$$\frac{A}{\Box A} \quad \text{Necessitation (N)}$$

The definition of QK-*frames* $\mathfrak{F} = (W, R, D)$ is the same as that of IQC-frames (see Definition 2.2) with the only difference that R can be an arbitrary relation. *Models* are also defined the same way, but without the requirement that wRv implies $I_w(P) \subseteq I_v(P)$. The connectives and quantifiers are interpreted at each world in the usual classical way, and

$$\mathfrak{M} \models^\sigma_w \Box A \text{ iff } (\forall v \in W)(wRv \Rightarrow \mathfrak{M} \models^\sigma_v A).$$

Truth and *validity* of formulas are defined as usual.

We next give a brief history of first Kripke completeness results for modal predicate logics. In 1959 Kripke [14] proved Kripke completeness of predicate S5. In late 1960s Cresswell [4,5] (see also Hughes and Cresswell [12]), Schütte [19], and Thomason [20] proved Kripke completeness of predicate T and S4. Kripke completeness of QK was first established by Gabbay [7, Thm. 8.5] [2]:

[2] We would like to thank Ilya Shapirovsky and Valentin Shehtman for useful discussions on the history of Kripke completeness for modal predicate logics.

Theorem 3.2 *The modal predicate logic* QK *is sound and complete with respect to Kripke semantics.*

The following two principles play an important role in the study of modal predicate logics. They were first considered by Barcan [1].

$\Box\forall x A \to \forall x \Box A$	converse Barcan formula	(CBF)
$\forall x \Box A \to \Box\forall x A$	Barcan formula	(BF)

It is easy to see that CBF is a theorem of QK. Indeed, this follows from Theorem 3.2 and the fact that domains of each QK-frame are increasing. On the other hand, a QK-frame validates BF iff it has *constant domains*, meaning that wRv implies $D_w = D_v$, and we have the following well-known theorem (see, e.g., [7, Thm. 9.3]):

Theorem 3.3 *The logic* QK + BF *is sound and complete with respect to the class of* QK*-frames with constant domains.*

A modal predicate logic whose Kripke frames have neither increasing nor decreasing domains was considered already by Kripke [15]. Building on this work, Hughes and Cresswell [13, pp. 304–309] introduced a similar predicate modal logic and proved its completeness with respect to a generalized Kripke semantics. Fitting and Mendelsohn [6, Sec. 6.2] gave an alternate axiomatization of this logic. Building on the work of Fitting and Mendelsohn, Corsi [3] defined the system Q°K whose axiomatization contains a weakening of the universal instantiation axiom.

Definition 3.4 The logic Q°K is the least set of formulas of $\mathcal{L}_\Box$ containing all substitution instances of theorems of K, the axiom schemes

(i) $\forall y(\forall x A \to A(y/x))$ (UI°)

(ii) $\forall x(A \to B) \to (\forall x A \to \forall x B)$

(iii) $\forall x \forall y A \leftrightarrow \forall y \forall x A$

(iv) $A \to \forall x A$ with x not free in A

and closed under (MP), (Gen), and (N).

Remark 3.5 In Definition 3.4, replacing UI° with UI yields an equivalent definition of QK. Therefore, Q°K is contained in QK.

Kripke frames for Q°K generalize Kripke frames for QK by having two domains, inner and outer.

Definition 3.6 A Q°K*-frame* is a quadruple $\mathfrak{F} = (W, R, D, U)$ where

- (W, R) is a K-frame.
- D is a function that associates to each $w \in W$ a set D_w. The set D_w is called the *inner domain* of w.

- U is a nonempty set containing the union of all the D_w. The set U is called the *outer domain* of $\mathfrak{F}$.

Definition 3.6 is a particular case of the frames considered by Corsi [3] where increasing outer domains are allowed. For our purposes, taking a fixed outer domain U is sufficient. We recall from [3] how to interpret $\mathcal{L}_\Box$ in a Q°K-frame $\mathfrak{F} = (W, R, D, U)$.

Definition 3.7

- An *interpretation* of $\mathcal{L}_\Box$ in $\mathfrak{F}$ is a function I associating to each world w and an n-ary predicate symbol P an n-ary relation $I_w(P) \subseteq U^n$.
- A *model* is a pair $\mathfrak{M} = (\mathfrak{F}, I)$ where $\mathfrak{F}$ is a Q°K-frame and I is an interpretation in $\mathfrak{F}$.
- An *assignment* in $\mathfrak{F}$ is a function σ that associates to each individual variable an element of U.
- If σ and τ are two assignments and x is an individual variable, τ is said to be an *x-variant* of σ if $\tau(y) = \sigma(y)$ for all $y \neq x$.
- We say that an assignment σ is *w-inner* for $w \in W$ if $\sigma(x) \in D_w$ for each individual variable x.

We next recall from [3] the definition of when a formula A is true in a world w of a model $\mathfrak{M} = (\mathfrak{F}, I)$ under the assignment σ, written $\mathfrak{M} \models^\sigma_w A$.

Definition 3.8

$\mathfrak{M} \models^\sigma_w \bot$		never
$\mathfrak{M} \models^\sigma_w P(x_1, \ldots, x_n)$	iff	$(\sigma(x_1), \ldots, \sigma(x_n)) \in I_w(P)$
$\mathfrak{M} \models^\sigma_w B \to C$	iff	$\mathfrak{M} \models^\sigma_w B$ implies $\mathfrak{M} \models^\sigma_w C$
$\mathfrak{M} \models^\sigma_w \forall x B$	iff	for all x-variants τ of σ with $\tau(x) \in D_w$, $\mathfrak{M} \models^\tau_w B$
$\mathfrak{M} \models^\sigma_w \Box B$	iff	for all v such that wRv, $\mathfrak{M} \models^\sigma_v B$

Definition 3.9 A formula A is *true* in a model $\mathfrak{M} = (\mathfrak{F}, I)$ at the world $w \in W$ (in symbols $\mathfrak{M} \models_w A$) if for all assignments σ, we have $\mathfrak{M} \models^\sigma_w A$. The definition of *truth* in a model and *validity* in a frame are the same as in Definition 2.5.

We have the following completeness result for Q°K, see [3, Thm. 1.32] and its proof.

Theorem 3.10 Q°K *is sound and complete with respect to the class of* Q°K-*frames.*

Definition 3.11 Let $\mathfrak{F} = (W, R, D, U)$ be a Q°K-frame.

- We say that $\mathfrak{F}$ has *increasing inner domains* if wRv implies $D_w \subseteq D_v$ for each $w, v \in W$.

- We say that $\mathfrak{F}$ has *decreasing inner domains* if wRv implies $D_v \subseteq D_w$ for each $w, v \in W$.
- If $\mathfrak{F}$ has both increasing and decreasing inner domains, we say that it has *constant inner domains.*

The following axiom scheme guarantees nonempty inner domains (hence the abbreviation):

$$\forall x A \to A \text{ with } x \text{ not free in } A \qquad \text{(NID)}$$

The next proposition is not difficult to verify (see, e.g., [6, Sec. 4.9] and [3, pp. 1487–1488]).

Proposition 3.12 *Let* $\mathfrak{F} = (W, R, D, U)$ *be a* Q°K*-frame.*

- $\mathfrak{F}$ *validates* CBF *iff* $\mathfrak{F}$ *has increasing inner domains.*
- $\mathfrak{F}$ *validates* BF *iff* $\mathfrak{F}$ *has decreasing inner domains.*
- $\mathfrak{F}$ *validates* NID *iff* $\mathfrak{F}$ *has nonempty inner domains.*

We have the following completeness results for logics obtained by adding CBF, BF, and NID to Q°K (see [3, Thms. 1.30, 1.32, and Footnote 7]):

Theorem 3.13

- Q°K + CBF *is sound and complete with respect to the class of* Q°K*-frames with increasing inner domains.*
- Q°K+CBF+BF *is sound and complete with respect to the class of* Q°K*-frames with constant inner domains.*
- *Adding* NID *to the above two logics or to* Q°K *yields completeness of the resulting logics with respect to the corresponding classes of frames which have nonempty inner domains.*

On the other hand, completeness of Q°K + BF remains open (see [3, p. 1510]).

4 The logic Q°S4.t

The tense predicate logic we will translate IQC into is based on the well-known tense propositional logic S4.t. We use $\Box_F$ ("always in the future") and $\Box_P$ ("always in the past") as temporal modalities. Then $\Diamond_F$ ("sometime in the future") and $\Diamond_P$ ("sometime in the past") are usual abbreviations $\neg\Box_F\neg$ and $\neg\Box_P\neg$.

Definition 4.1 The logic S4.t is the least set of formulas of the tense propositional language containing all substitution instances of S4-axioms for both $\Box_F$ and $\Box_P$, the axiom schemes

(i) $A \to \Box_P \Diamond_F A$

(ii) $A \to \Box_F \Diamond_P A$

and closed under (MP) and

$$\frac{A}{\Box_F A} \quad \Box_F\text{-Necessitation } (\mathrm{N_F}) \qquad\qquad \frac{A}{\Box_P A} \quad \Box_P\text{-Necessitation } (\mathrm{N_P})$$

Relational semantics of S4.t consists of Kripke frames $\mathfrak{F} = (W, R)$ where R is reflexive and transitive. As usual, propositional letters are interpreted as subsets of W, classical connectives as the corresponding set-theoretic operations on the powerset of W, and for temporal modalities we set:

$$\begin{array}{lll} w \models \Box_F A & \text{iff} & (\forall v \in W)(wRv \Rightarrow v \models A) \\ w \models \Box_P A & \text{iff} & (\forall v \in W)(vRw \Rightarrow v \models A) \end{array}$$

It is well known that S4.t is sound and complete with respect to its relational semantics.

Let $\mathcal{L}_T$ be the bimodal predicate language obtained by extending $\mathcal{L}$ with two modalities $\Box_F$ and $\Box_P$.

Definition 4.2 The logic QS4.t is the least set of formulas of $\mathcal{L}_T$ containing all substitution instances of theorems of S4.t, the axiom schemes (i) and (iii) of Definition 2.1, and closed under (MP), (Gen), ($\mathrm{N_F}$), and ($\mathrm{N_P}$).

The following are temporal versions of CBF and BF:

$\Box_F \forall x A \to \forall x \Box_F A$	converse Barcan formula for $\Box_F$	($\mathsf{CBF_F}$)
$\forall x \Box_F A \to \Box_F \forall x A$	Barcan formula for $\Box_F$	($\mathsf{BF_F}$)
$\Box_P \forall x A \to \forall x \Box_P A$	converse Barcan formula for $\Box_P$	($\mathsf{CBF_P}$)
$\forall x \Box_P A \to \Box_P \forall x A$	Barcan formula for $\Box_P$	($\mathsf{BF_P}$)

The proof that QK $\vdash$ CBF (see, e.g., [15, p. 88]) can be adapted to prove that QS4.t $\vdash$ $\mathsf{CBF_F}$ and QS4.t $\vdash$ $\mathsf{CBF_P}$. It is also well known that $\mathsf{CBF_F}$ and $\mathsf{BF_P}$, as well as $\mathsf{CBF_P}$ and $\mathsf{BF_F}$ are derivable from each other in any tense predicate logic. Therefore, all four are theorems of QS4.t. This is reflected in the fact that QS4.t-frames have constant domains. Indeed, QS4.t is complete with respect to this semantics (see Section 6). But this is problematic for translating IQC fully into QS4.t since IQC-frames with constant domains validate the additional axiom $\forall x(A \vee B) \to (A \vee \forall x B)$, where x is not free in A, which is not a theorem of IQC (see, e.g., [8, p. 53, Cor. 8]).

Consequently, we need to work with a weaker logic than QS4.t. To this end, we introduce the logic Q°S4.t, which weakens QS4.t the same way Q°K weakens QK.

Definition 4.3 The logic Q°S4.t is the least set of formulas of $\mathcal{L}_T$ containing all substitution instances of theorems of S4.t, the axiom schemes (i), (ii), (iii), (iv) of Q°K (see Definition 3.4), NID, $\mathsf{CBF_F}$, and closed under (MP), (Gen), ($\mathrm{N_F}$), and ($\mathrm{N_P}$).

As follows from Proposition A.1 in the Appendix, $\mathsf{BF_P}$ is a theorem of $\mathsf{Q^\circ S4.t}$. In fact, $\mathsf{CBF_F}$ and $\mathsf{BF_P}$ are derivable from each other and the other axioms of $\mathsf{Q^\circ S4.t}$.

Definition 4.4 A $\mathsf{Q^\circ S4.t}$-*frame* is a $\mathsf{Q^\circ K}$-frame $\mathfrak{F} = (W, R, D, U)$ (see Definition 3.6) with nonempty increasing inner domains whose accessibility relation is reflexive and transitive.

Models and assignments are defined as in Definition 3.7. The clauses of when a formula A of $\mathcal{L}_T$ is true in a world w of a $\mathsf{Q^\circ S4.t}$-model $\mathfrak{M} = (\mathfrak{F}, I)$ under the assignment σ, written $\mathfrak{M} \vDash_w^\sigma A$, are defined as in Definition 3.8, but we replace the $\Box$-clause with the following two clauses:

$$\begin{array}{lll} \mathfrak{M} \vDash_w^\sigma \Box_F B & \text{iff} & (\forall v \in W)(wRv \Rightarrow \mathfrak{M} \vDash_v^\sigma B) \\ \mathfrak{M} \vDash_w^\sigma \Box_P B & \text{iff} & (\forall v \in W)(vRw \Rightarrow \mathfrak{M} \vDash_v^\sigma B) \end{array}$$

For formulas of $\mathcal{L}_T$ we define truth in a model and validity in a frame as in Definition 3.9.

Theorem 4.5 $\mathsf{Q^\circ S4.t}$ *is sound with respect to the class of* $\mathsf{Q^\circ S4.t}$-*frames; that is, for each formula A of $\mathcal{L}_T$ and* $\mathsf{Q^\circ S4.t}$-*frame $\mathfrak{F}$, from* $\mathsf{Q^\circ S4.t} \vdash A$ *it follows that $\mathfrak{F} \vDash A$.*

Proof. It is sufficient to show that each axiom scheme is valid in all $\mathsf{Q^\circ S4.t}$-frames and that each rule of inference preserves validity. This can be done by direct verification. We only show that the axiom scheme $\mathsf{CBF_F}$ is valid in all $\mathsf{Q^\circ S4.t}$-frames. Let $\mathfrak{M} = (\mathfrak{F}, I)$ be a $\mathsf{Q^\circ S4.t}$-model, $w \in W$, and σ an assignment. If $\mathfrak{M} \vDash_w^\sigma \Box_F \forall x A$, then for all v with wRv we have $\mathfrak{M} \vDash_v^\sigma \forall x A$. This implies that for each x-variant τ of σ with $\tau(x) \in D_v$ we have $\mathfrak{M} \vDash_v^\tau A$. Since $D_w \subseteq D_v$, this is in particular true for x-variants τ of σ with $\tau(x) \in D_w$. Therefore, for each x-variant τ of σ with $\tau(x) \in D_w$ and for each v with wRv we have $\mathfrak{M} \vDash_v^\tau A$. Thus, for each x-variant τ of σ with $\tau(x) \in D_w$, we have $\mathfrak{M} \vDash_w^\sigma \Box_F A$. Consequently, $\mathfrak{M} \vDash_w^\sigma \forall x \Box_F A$. This shows that $\mathfrak{F} \vDash \Box_F \forall x A \to \forall x \Box_F A$ for each $\mathsf{Q^\circ S4.t}$-frame $\mathfrak{F}$. $\Box$

On the other hand, completeness of $\mathsf{Q^\circ S4.t}$ remains an interesting open problem, which is related to the open problem of completeness of $\mathsf{Q^\circ K} + \mathsf{BF}$ (see Section 6).

5 The translation

In this section we prove our main result that the temporal modification (described in the Introduction) of the Gödel translation embeds IQC into $\mathsf{Q^\circ S4.t}$ fully and faithfully. Our strategy is to prove faithfulness of the translation syntactically, while fullness will be proved by semantical means, utilizing Kripke completeness of IQC.

Our syntactic proof of faithfulness is based on the following technical lemma, the proof of which we give in the Appendix. To keep the notation simple, we denote lists of variables by bold letters. If $\mathbf{x} = x_1, \ldots, x_n$, we write $\forall\mathbf{x}$ for $\forall x_1 \cdots \forall x_n$. We point out that it is a consequence of axioms (ii) and (iii) of $\mathsf{Q^\circ K}$ that from the point of view of provability in $\mathsf{Q^\circ S4.t}$, the order of variables in $\forall\mathbf{x}$ does not matter.

Lemma 5.1

(i) *Let C be an instance of an axiom scheme of* IQC *and* $\mathbf{x}$ *the list of free variables in C. Then* $\mathsf{Q^\circ S4.t} \vdash \forall\mathbf{x}\, C^t$.

(ii) *Let A, B be formulas of $\mathcal{L}$,* $\mathbf{x}$ *the list of variables free in $A \to B$,* $\mathbf{y}$ *the list of variables free in A, and* $\mathbf{z}$ *the list of variables free in B. If* $\mathsf{Q^\circ S4.t} \vdash \forall\mathbf{x}(A \to B)^t$ *and* $\mathsf{Q^\circ S4.t} \vdash \forall\mathbf{y} A^t$*, then* $\mathsf{Q^\circ S4.t} \vdash \forall\mathbf{z} B^t$.

(iii) *Let A be a formula of $\mathcal{L}$, x a variable,* $\mathbf{y}$ *the list of variables free in A, and* $\mathbf{z}$ *the list of variables free in $\forall x A$. If* $\mathsf{Q^\circ S4.t} \vdash \forall\mathbf{y} A^t$*, then* $\mathsf{Q^\circ S4.t} \vdash \forall\mathbf{z}\, (\forall x A)^t$.

Proof. For (i) see the proof of Lemma A.5, for (ii) see the proof of Lemma A.6, and for (iii) see the proof of Lemma A.7. □

Theorem 5.2 *Let A be a formula of $\mathcal{L}$ and $x_1, \ldots, x_n$ the free variables of A. If* $\mathsf{IQC} \vdash A$*, then* $\mathsf{Q^\circ S4.t} \vdash \forall x_1 \cdots \forall x_n A^t$.

Proof. The proof is by induction on the length of the proof of A in IQC. If A is an instance of an axiom of IQC, then the result follows from Lemma 5.1(i). Lemma 5.1(ii) takes care of the case in which the last step of the proof of A is an application of (MP). Finally, if the last step of the proof of A is an application of (Gen) to the variable x, use Lemma 5.1(iii). □

Remark 5.3 We are prefixing the translation of A with $\forall x_1 \cdots \forall x_n$ because it is not true in general that $\mathsf{IQC} \vdash A$ implies $\mathsf{Q^\circ S4.t} \vdash A^t$. For example, if A is an instance of the universal instantiation axiom, which is an axiom of IQC, then A^t is not in general a theorem of $\mathsf{Q^\circ S4.t}$.

Definition 5.4

- For an IQC-frame $\mathfrak{F} = (W, R, D)$ let $\overline{\mathfrak{F}} = (W, R, D, U)$ where $U = \bigcup\{D_w \mid w \in W\}$.
- For an IQC-model $\mathfrak{M} = (\mathfrak{F}, I)$ let $\overline{\mathfrak{M}} = (\overline{\mathfrak{F}}, I)$.

Remark 5.5

- It is obvious that $\overline{\mathfrak{F}}$ is a $\mathsf{Q^\circ S4.t}$-frame.
- If I is an interpretation in $\mathfrak{F}$, then I is also an interpretation in $\overline{\mathfrak{F}}$ because for each n-ary predicate letter P we have $I_w(P) \subseteq D_w^n \subseteq U^n$. Therefore, $\overline{\mathfrak{M}}$ is well defined.

- The w-assignments in $\mathfrak{F}$ are exactly the w-inner assignments in $\overline{\mathfrak{F}}$.

The proof of the following technical lemma is given in the Appendix.

Lemma 5.6 *If A is a formula of $\mathcal{L}$, then* $\mathsf{Q^\circ S4.t} \vdash A^t \to \Box_F A^t$.

Proof. See the proof of Lemma A.2. $\Box$

Lemma 5.7 *Let A be a formula of $\mathcal{L}$, $\mathfrak{M} = (\mathfrak{F}, I)$ a* $\mathsf{Q^\circ S4.t}$*-model, and σ an assignment in $\mathfrak{F}$. If $v, w \in W$ with vRw, then $\mathfrak{M} \vDash_v^\sigma A^t$ implies $\mathfrak{M} \vDash_w^\sigma A^t$.*

Proof. Suppose vRw and $\mathfrak{M} \vDash_v^\sigma A^t$. By Lemma 5.6 and Theorem 4.5, $\mathfrak{M} \vDash_v^\sigma A^t \to \Box_F A^t$. Therefore, $\mathfrak{M} \vDash_v^\sigma \Box_F A^t$, which yields $\mathfrak{M} \vDash_w^\sigma A^t$ because vRw. $\Box$

Proposition 5.8 *Let A be a formula of $\mathcal{L}$, $\mathfrak{M} = (\mathfrak{F}, I)$ an* IQC*-model based on an* IQC*-frame $\mathfrak{F} = (W, R, D)$, and $w \in W$.*

(i) *For each w-assignment σ,*

$$\mathfrak{M} \vDash_w^\sigma A \text{ iff } \overline{\mathfrak{M}} \vDash_w^\sigma A^t.$$

(ii) *If $x_1, \ldots, x_n$ are the free variables of A, then*

$$\mathfrak{M} \vDash_w A \text{ iff } \overline{\mathfrak{M}} \vDash_w \forall x_1 \cdots \forall x_n A^t.$$

Proof. (i). Induction on the complexity of A. Let A be an atomic formula $P(x_1, \ldots, x_n)$. Since wRv implies $I_w(P) \subseteq I_v(P)$ and R is reflexive, we have

$$\begin{aligned}
\mathfrak{M} \vDash_w^\sigma P(x_1, \ldots, x_n) &\text{ iff } (\sigma(x_1), \ldots, \sigma(x_n)) \in I_w(P) \\
&\text{ iff } (\forall v \in W)(wRv \Rightarrow (\sigma(x_1), \ldots, \sigma(x_n)) \in I_v(P)) \\
&\text{ iff } \overline{\mathfrak{M}} \vDash_w^\sigma \Box_F P(x_1, \ldots, x_n) \\
&\text{ iff } \overline{\mathfrak{M}} \vDash_w^\sigma P(x_1, \ldots, x_n)^t
\end{aligned}$$

The cases where $A = \bot$, $A = B \wedge C$, and $A = B \vee C$ are straightforward.

If $A = B \to C$, then using the inductive hypothesis, we have

$$\begin{aligned}
\mathfrak{M} \vDash_w^\sigma B \to C &\text{ iff } (\forall v \in W)(wRv \Rightarrow (\mathfrak{M} \vDash_v^\sigma B \Rightarrow \mathfrak{M} \vDash_v^\sigma C)) \\
&\text{ iff } (\forall v \in W)(wRv \Rightarrow (\overline{\mathfrak{M}} \vDash_v^\sigma B^t \Rightarrow \overline{\mathfrak{M}} \vDash_v^\sigma C^t)) \\
&\text{ iff } \overline{\mathfrak{M}} \vDash_w^\sigma \Box_F(B^t \to C^t) \\
&\text{ iff } \overline{\mathfrak{M}} \vDash_w^\sigma (B \to C)^t
\end{aligned}$$

If $A = \forall x B$, then using the inductive hypothesis, we have

$$\begin{aligned}\mathfrak{M} \models^\sigma_w \forall x B \text{ iff } & (\forall v \in W)(wRv \Rightarrow \text{for each } v\text{-assignment } \tau \text{ that is} \\ & \quad \text{an } x\text{-variant of } \sigma \text{ we have } \mathfrak{M} \models^\tau_v B) \\ \text{iff } & (\forall v \in W)(wRv \Rightarrow \text{for each assignment } \tau \text{ that is} \\ & \quad \text{an } x\text{-variant of } \sigma \text{ with } \tau(x) \in D_v \text{ we have } \overline{\mathfrak{M}} \models^\tau_v B^t) \\ \text{iff } & \overline{\mathfrak{M}} \models^\sigma_w \Box_F \forall x B^t \\ \text{iff } & \overline{\mathfrak{M}} \models^\sigma_w (\forall x B)^t\end{aligned}$$

If $A = \exists x B$, then using the inductive hypothesis, reflexivity of R, Lemma 5.7, and the fact that vRw implies $D_v \subseteq D_w$, we have

$$\begin{aligned}\mathfrak{M} \models^\sigma_w \exists x B \text{ iff } & \text{there is a } w\text{-assignment } \tau \text{ that is an } x\text{-variant of } \sigma \\ & \quad \text{such that } \mathfrak{M} \models^\tau_w B \\ \text{iff } & \text{there is an assignment } \tau \text{ that is an } x\text{-variant of } \sigma \\ & \quad \text{with } \tau(x) \in D_w \text{ such that } \overline{\mathfrak{M}} \models^\tau_w B^t \\ \text{iff } & \text{there is } v \in W \text{ such that } vRw \text{ and an assignment } \rho \text{ that is} \\ & \quad \text{an } x\text{-variant of } \sigma \text{ with } \rho(x) \in D_v \text{ such that } \overline{\mathfrak{M}} \models^\rho_v B^t \\ \text{iff } & \overline{\mathfrak{M}} \models^\sigma_w \Diamond_P \exists x B^t \\ \text{iff } & \overline{\mathfrak{M}} \models^\sigma_w (\exists x B)^t\end{aligned}$$

(ii). By Definition 2.5, $\mathfrak{M} \models_w A$ iff $\mathfrak{M} \models^\sigma_w A$ for each w-assignment σ. As noted in Remark 5.5, w-assignments in $\mathfrak{F}$ are exactly the w-inner assignments in $\overline{\mathfrak{F}}$. Therefore, by (i), $\mathfrak{M} \models_w A$ iff $\overline{\mathfrak{M}} \models^\sigma_w A^t$ for each w-inner assignment σ. It follows from the interpretation of the universal quantifier in $\overline{\mathfrak{M}}$ that $\overline{\mathfrak{M}} \models^\sigma_w A^t$ for each w-inner assignment σ iff $\overline{\mathfrak{M}} \models_w \forall x_1 \cdots \forall x_n A^t$. Thus, $\mathfrak{M} \models_w A$ iff $\overline{\mathfrak{M}} \models_w \forall x_1 \cdots \forall x_n A^t$. □

Theorem 5.9 *Let A be a formula of $\mathcal{L}$ and $x_1, \ldots, x_n$ the free variables of A. If* $\mathsf{Q^\circ S4.t} \vdash \forall x_1 \cdots \forall x_n A^t$, *then* $\mathsf{IQC} \vdash A$.

Proof. Suppose $\mathsf{IQC} \nvdash A$. Theorem 2.6 implies that there is an IQC-model $\mathfrak{M}$ such that $\mathfrak{M} \not\models_w A$ for some world w. By Proposition 5.8(ii), $\overline{\mathfrak{M}} \not\models_w \forall x_1 \cdots \forall x_n A^t$. Thus, $\mathsf{Q^\circ S4.t} \nvdash \forall x_1 \cdots \forall x_n A^t$ by Theorem 4.5. □

By putting Theorems 5.2 and 5.9 together we arrive at the main result of the paper mentioned in the introduction.

Theorem 5.10

- *Let A be a formula of $\mathcal{L}$ and $x_1, \ldots, x_n$ the free variables of A. We have*

$$\mathsf{IQC} \vdash A \textit{ iff } \mathsf{Q^\circ S4.t} \vdash \forall x_1 \cdots \forall x_n A^t.$$

- *If A is a sentence of $\mathcal{L}$, then*

$$\mathsf{IQC} \vdash A \text{ iff } \mathsf{Q^\circ S4.t} \vdash A^t.$$

Remark 5.11 If we allow constants in $\mathcal{L}$, Theorem 5.9 is no longer true in its current form. Indeed, constants in IQC and Q°S4.t behave like free variables and we would have the problem described in Remark 5.3. However, it can be adjusted as follows. Let A be a formula containing free variables $x_1, \ldots, x_n$ and constants $c_1, \ldots c_m$. If $A(y_1/c_1, \ldots, y_m/c_m)$ is the formula obtained by replacing all the constants with fresh variables $y_1, \ldots, y_m$, then IQC $\vdash A$ iff Q°S4.t $\vdash \forall x_1 \cdots \forall x_n \forall y_1 \cdots \forall y_m A^t(y_1/c_1, \ldots, y_m/c_m)$.

6 Open problems

As follows from Theorem 4.5, Q°S4.t is sound with respect to the class of Q°S4.t-frames. However, its completeness remains an interesting open problem. The standard Henkin construction was modified by Hughes and Cresswell [13] and Corsi [3] to obtain completeness of Q°K. If we adapt their technique to Q°S4.t, we obtain two relations R_F and R_P on the canonical model, one coming from $\Box_F$ and the other from $\Box_P$. There does not seem to be an obvious way to select an appropriate submodel in which the restrictions of these two relations are inverses of each other because the outer domains of accessible worlds are forced to increase by the construction. This problem disappears when constructing the canonical model for QS4.t because the presence of $\mathsf{BF_F}$ and $\mathsf{CBF_P}$ in each world allows us to select witnesses without expanding the domains of accessible worlds, thus yielding that QS4.t is sound and complete with respect to the class of QS4.t-frames.

The problem of completeness of Q°S4.t seems to be closely related to the open problem, stated in [3, p. 1510], of whether Q°K + BF is Kripke complete. It appears that answering one of these problems could also provide an answer to the other.

One of the reviewers pointed out that another natural direction is to study the intermediate predicate logics and the corresponding extensions of Q°S4.t for which our temporal translation remains full and faithful. Finally, it is worth investigating whether other tense predicate logics (such as the ones considered in [11]) could be used for translating IQC fully and faithfully. Some such systems admit presheaf semantics which is more general than Kripke semantics.

Appendix

A Additional facts needed in Sections 4 and 5

Proposition A.1 Q°S4.t $\vdash \mathsf{BF_P}$.

Proof. We first show that Q°S4.t $\vdash \Diamond_F \forall x B \to \forall x \Diamond_F B$ for any formula B. We have the proof

1. $\forall x(\forall xB \to B)$
2. $\forall x\Box_F(\forall xB \to B)$
3. $\Box_F(\forall xB \to B) \to (\Diamond_F\forall xB \to \Diamond_F B)$
4. $\forall x\Box_F(\forall xB \to B) \to \forall x(\Diamond_F\forall xB \to \Diamond_F B)$
5. $\forall x(\Diamond_F\forall xB \to \Diamond_F B)$
6. $\forall x\Diamond_F\forall xB \to \forall x\Diamond_F B$
7. $\Diamond_F\forall xB \to \forall x\Diamond_F B$

Here 1 is an instance of UI°; 2 is obtained from 1 by adding $\Box_F$ inside $\forall x$ by applying ($\mathsf{N_F}$), $\mathsf{CBF_F}$, and (MP); 3 is a substitution instance of the K-theorem $\Box_F(C \to D) \to (\Diamond_F C \to \Diamond_F D)$ for $\Box_F$; 4 is obtained from 3 by first adding and then distributing $\forall x$ inside the implication by applying (Gen), axiom (ii) of $\mathsf{Q^\circ K}$, and (MP); 5 follows from 2 and 4 by (MP); 6 is obtained from 5 by distributing $\forall x$; and 7 follows from 6 and axiom (iv) of $\mathsf{Q^\circ K}$.

We now prove $\forall x\Box_P A \to \Box_P\forall xA$.

1. $\forall x\Box_P A \to \Box_P\Diamond_F\forall x\Box_P A$
2. $\Diamond_F\forall x\Box_P A \to \forall x\Diamond_F\Box_P A$
3. $\Box_P\Diamond_F\forall x\Box_P A \to \Box_P\forall x\Diamond_F\Box_P A$
4. $\Diamond_F\Box_P A \to A$
5. $\forall x\Diamond_F\Box_P A \to \forall xA$
6. $\Box_P\forall x\Diamond_F\Box_P A \to \Box_P\forall xA$
7. $\forall x\Box_P A \to \Box_P\forall xA$

Here 1 is an instance of axiom (i) of $\mathsf{S4.t}$; 2 is an instance of $\Diamond_F\forall xB \to \forall x\Diamond_F B$ proved above; 3 and 6 follow from 2 and 5 by adding and distributing $\Box_P$ in the implication; 4 is an instance of the $\mathsf{S4.t}$-theorem $\Diamond_F\Box_P C \to C$; 5 is obtained from 4 by adding and distributing $\forall x$; and 7 follows from 1, 3, and 6. □

Lemma A.2 *If A is a formula of $\mathcal{L}$, then* $\mathsf{Q^\circ S4.t} \vdash A^t \to \Box_F A^t$ *and* $\mathsf{Q^\circ S4.t} \vdash \Diamond_P A^t \to A^t$.

Proof. We only prove that $\mathsf{Q^\circ S4.t} \vdash A^t \to \Box_F A^t$ since it implies that $\mathsf{Q^\circ S4.t} \vdash \Diamond_P A^t \to A^t$. The proof is by induction on the complexity of A. If $A = \bot$, then $A^t = \bot$ and it is clear that $\mathsf{Q^\circ S4.t} \vdash \bot \to \Box_F\bot$.

If A is either an atomic formula $P(x_1, \ldots, x_n)$ or of the form $B \to C$ or $\forall xB$, then A^t is of the form $\Box_F D$. Therefore, the 4-axiom $\Box_F D \to \Box_F\Box_F D$ implies that in all these cases $\mathsf{Q^\circ S4.t} \vdash A^t \to \Box_F A^t$.

If $A = \exists xB$, then $A^t = \Diamond_P\exists xB^t$. So $\Box_F A^t = \Box_F\Diamond_P\exists xB^t$ and $\mathsf{Q^\circ S4.t} \vdash \Diamond_P\exists xB^t \to \Box_F\Diamond_P\exists xB^t$ because it is a substitution instance of the $\mathsf{S4.t}$-theorem $\Diamond_P C \to \Box_F\Diamond_P C$. Finally, if $A = B \wedge C$ or $A = B \vee C$, then we have $A^t = B^t \wedge C^t$ or $A^t = B^t \vee C^t$. By inductive hypothesis, $\mathsf{Q^\circ S4.t} \vdash B^t \to \Box_F B^t$ and $\mathsf{Q^\circ S4.t} \vdash C^t \to \Box_F C^t$. Since $\mathsf{Q^\circ S4.t} \vdash (\Box_F B^t \wedge \Box_F C^t) \to \Box_F(B^t \wedge C^t)$ and $\mathsf{Q^\circ S4.t} \vdash (\Box_F B^t \vee \Box_F C^t) \to \Box_F(B^t \vee C^t)$, we obtain $\mathsf{Q^\circ S4.t} \vdash (B^t \wedge C^t) \to \Box_F(B^t \wedge C^t)$ and $\mathsf{Q^\circ S4.t} \vdash (B^t \vee C^t) \to \Box_F(B^t \vee C^t)$. □

Lemma A.3 *The following are theorems of* Q°S4.t*:*

(i) $\forall y(A(y/x) \to \exists x A)$.

(ii) $\forall x(A \to B) \to (A \to \forall x B)$ *if x is not free in A.*

(iii) $\forall x(A \to B) \to (\exists x A \to B)$ *if x is not free in B.*

Proof. Follows from [3, Lem. 1.3]. □

Lemma A.4 *For formulas A, B of $\mathcal{L}$, the following are theorems of* Q°S4.t.

(i) $\Box_F(\Box_F \forall x A^t \to A^t)$ *if x is not free in A.*

(ii) $\forall y \Box_F(\Box_F \forall x A^t \to A(y/x)^t)$.

(iii) $\Box_F(A^t \to \Diamond_P \exists x A^t)$ *if x is not free in A.*

(iv) $\forall y \Box_F(A(y/x)^t \to \Diamond_P \exists x A^t)$.

(v) $\Box_F(\Box_F \forall x \Box_F(A^t \to B^t) \to \Box_F(A^t \to \Box_F \forall x B^t))$ *if x is not free in A.*

(vi) $\Box_F(\Box_F \forall x \Box_F(A^t \to B^t) \to \Box_F(\Diamond_P \exists x A^t \to B^t))$ *if x is not free in B.*

Proof. Note that x is free in A iff it is free in A^t, and $A(y/x)^t = A^t(y/x)$.

(i). We have the proof

1. $\forall x A^t \to A^t$
2. $\Box_F \forall x A^t \to A^t$
3. $\Box_F(\Box_F \forall x A^t \to A^t)$

where 1 is an instance of NID because x is not free in A^t; 2 is obtained from 1 by applying the T-axiom for $\Box_F$; 3 is obtained from 2 by ($\mathrm{N_F}$).

(ii). We have the proof

1. $\forall y(\forall x A^t \to A^t(y/x))$
2. $\forall y(\Box_F \forall x A^t \to A^t(y/x))$
3. $\forall y \Box_F(\Box_F \forall x A^t \to A^t(y/x))$

where 1 is an instance of UI°; 2 follows from 1 by applying the T-axiom for $\Box_F$ inside $\forall y$; 3 is obtained from 2 by introducing $\Box_F$ inside $\forall y$.

(iii). We have the proof

1. $A^t \to \exists x A^t$
2. $A^t \to \Diamond_P \exists x A^t$
3. $\Box_F(A^t \to \Diamond_P \exists x A^t)$

where 1 is an instance of $C \to \exists x C$, with x not free in C, which is equivalent to NID; 2 follows from 1 by the T-axiom for $\Diamond_P$; 3 is obtained from 2 by ($\mathrm{N_F}$).

(iv). We have the proof

1. $\forall y(A^t(y/x) \to \exists x A^t)$
2. $\forall y(A^t(y/x) \to \Diamond_P \exists x A^t)$
3. $\forall y \Box_F(A^t(y/x) \to \Diamond_P \exists x A^t)$

where 1 follows from Lemma A.3(i); 2 follows from 1 by applying the T-axiom for $\Diamond_P$ inside $\forall y$; 3 is obtained from 2 by introducing $\Box_F$ inside $\forall y$.

(v). We have the proof

1. $\forall x(A^t \to B^t) \to (A^t \to \forall x B^t)$
2. $\forall x \Box_F(A^t \to B^t) \to (A^t \to \forall x B^t)$
3. $\Box_F \forall x \Box_F(A^t \to B^t) \to (\Box_F A^t \to \Box_F \forall x B^t)$
4. $\Box_F \forall x \Box_F(A^t \to B^t) \to (A^t \to \Box_F \forall x B^t)$
5. $\Box_F \forall x \Box_F(A^t \to B^t) \to \Box_F(A^t \to \Box_F \forall x B^t)$
6. $\Box_F(\Box_F \forall x \Box_F(A^t \to B^t) \to \Box_F(A^t \to \Box_F \forall x B^t))$

where 1 follows from Lemma A.3(ii); 2 follows from 1 by applying the T-axiom for $\Box_F$; 3 is obtained from 2 by adding and distributing $\Box_F$; 4 follows from 3 by Lemma A.2; 5 is obtained from 4 by adding and distributing $\Box_F$ and getting rid of one $\Box_F$ in the antecedent using the 4-axiom; 6 follows from 5 by ($\mathrm{N_F}$).

(vi). We have the proof

1. $\forall x(A^t \to B^t) \to (\exists x A^t \to B^t)$
2. $\forall x(A^t \to B^t) \to (\exists x \Diamond_P A^t \to B^t)$
3. $\forall x \Box_F(A^t \to B^t) \to (\exists x \Diamond_P A^t \to B^t)$
4. $\forall x \Box_F(A^t \to B^t) \to (\Diamond_P \exists x A^t \to B^t)$
5. $\Box_F \forall x \Box_F(A^t \to B^t) \to \Box_F(\Diamond_P \exists x A^t \to B^t)$
6. $\Box_F(\Box_F \forall x \Box_F(A^t \to B^t) \to \Box_F(\Diamond_P \exists x A^t \to B^t))$

where 1 follows from Lemma A.3(iii); 2 follows from 1 by Lemma A.2; 3 follows from 2 by applying the T-axiom for $\Box_F$; 4 follows from 3 and the fact that $\mathsf{Q^\circ S4.t} \vdash \Diamond_P \exists x A^t \to \exists x \Diamond_P A^t$ because it is a consequence of $\mathsf{BF_P}$; 5 is obtained from 4 by adding and distributing $\Box_F$; 6 follows from 5 by ($\mathrm{N_F}$). □

Lemma A.5 *If C is an instance of an axiom scheme of* IQC *and* $\mathbf{x}$ *is the list of free variables in C, then* $\mathsf{Q^\circ S4.t} \vdash \forall \mathbf{x}\, C^t$.

Proof. If C is an instance of a theorem of IPC, then it follows from the faithfulness of the Gödel translation in the propositional case that C^t is a theorem of Q°S4.t (since $\Box_F$ is an S4-modality). Applying (Gen) to each free variable of C^t then yields a proof of $\forall \mathbf{x}\, C^t$ in Q°S4.t. Translations of the axiom schemes of Definition 2.1 give:

$$(\forall x A \to A(y/x))^t = \Box_F(\Box_F \forall x A^t \to A(y/x)^t)$$
$$(A(y/x) \to \exists x A)^t = \Box_F(A(y/x)^t \to \Diamond_P \exists x A^t)$$

$$(\forall x(A \to B) \to (A \to \forall x B))^t$$
$$= \Box_F(\Box_F \forall x \Box_F(A^t \to B^t) \to \Box_F(A^t \to \Box_F \forall x B^t))$$
$$(\forall x(A \to B) \to (\exists x A \to B))^t$$
$$= \Box_F(\Box_F \forall x \Box_F(A^t \to B^t) \to \Box_F(\Diamond_P \exists x A^t \to B^t))$$

If C is an instance of one of these axiom schemes, then we obtain a proof of $\forall \mathbf{x}\, C^t$ in Q°S4.t by Lemma A.4 and by applying (Gen) to the free variables of C. More precisely, for the first axiom we use (i) of Lemma A.4 when x is not free in A and (ii) when x is free in A. Similarly, for the second axiom we use (iii) or (iv) of Lemma A.4. Finally, for the third axiom we use (v) and for the fourth axiom we use (vi) of Lemma A.4. □

Lemma A.6 *Let A, B be formulas of $\mathcal{L}$, $\mathbf{x}$ the list of variables free in $A \to B$, $\mathbf{y}$ the list of variables free in A, and $\mathbf{z}$ the list of variables free in B. If* Q°S4.t $\vdash \forall \mathbf{x}(A \to B)^t$ *and* Q°S4.t $\vdash \forall \mathbf{y} A^t$*, then* Q°S4.t $\vdash \forall \mathbf{z} B^t$.

Proof. Let $\mathbf{u}$ be the list of variables free in A but not in B, $\mathbf{v}$ the list of variables free in B but not in A, and $\mathbf{w}$ the list of variables free in both A and B. We then have that $\mathbf{x}$ is the union of $\mathbf{u}$, $\mathbf{v}$, and $\mathbf{w}$; $\mathbf{y}$ is the union of $\mathbf{u}$ and $\mathbf{w}$; and $\mathbf{z}$ is the union of $\mathbf{v}$ and $\mathbf{w}$. Thus, we want to show that if Q°S4.t $\vdash \forall \mathbf{u}\, \forall \mathbf{v}\, \forall \mathbf{w}(A \to B)^t$ and Q°S4.t $\vdash \forall \mathbf{u}\, \forall \mathbf{w} A^t$, then Q°S4.t $\vdash \forall \mathbf{v}\, \forall \mathbf{w} B^t$. We have the proof

1. $\forall \mathbf{u}\, \forall \mathbf{v}\, \forall \mathbf{w}\, \Box_F(A^t \to B^t)$
2. $\forall \mathbf{u}\, \forall \mathbf{w}\, \forall \mathbf{v}\, \Box_F(A^t \to B^t)$
3. $\forall \mathbf{u}\, \forall \mathbf{w}\, \forall \mathbf{v}\, (\Box_F A^t \to \Box_F B^t)$
4. $\forall \mathbf{u}\, \forall \mathbf{w}\, (\Box_F A^t \to \forall \mathbf{v}\, \Box_F B^t)$
5. $\forall \mathbf{u}\, \forall \mathbf{w}\, \Box_F A^t \to \forall \mathbf{u}\, \forall \mathbf{w}\, \forall \mathbf{v}\, \Box_F B^t$
6. $\forall \mathbf{u}\, \forall \mathbf{w} A^t$
7. $\forall \mathbf{u}\, \forall \mathbf{w}\, \Box_F A^t$
8. $\forall \mathbf{u}\, \forall \mathbf{w}\, \forall \mathbf{v}\, \Box_F B^t$
9. $\forall \mathbf{u}\, \forall \mathbf{w}\, \forall \mathbf{v}\, B^t$

10 $\forall \mathbf{w}\, \forall \mathbf{v}\, B^t$

11 $\forall \mathbf{v}\, \forall \mathbf{w}\, B^t$

where 1 and 6 are assumptions; 2 and 11 follow from 1 and 10 by switching the order of quantification; 3 is obtained from 2 by distributing $\Box_F$ inside the universal quantifiers; 4 follows from Lemma A.3(ii) because all the variables in $\mathbf{v}$ are not free in $\Box_F A^t$; 5 is obtained by distributing the universal quantifiers; 7 follows from 6 by introducing $\Box_F$ inside the quantifiers; 8 is obtained by (MP) from 5 and 7; 9 follows from 8 by the T-axiom for $\Box_F$; 10 follows from 9 by NID because no variable in $\mathbf{u}$ is free in B^t. □

Lemma A.7 *Let A be a formula of $\mathcal{L}$, x a variable, $\mathbf{y}$ the list of variables free in A, and $\mathbf{z}$ the list of variables free in $\forall x A$. If* Q°S4.t $\vdash \forall \mathbf{y} A^t$*, then* Q°S4.t $\vdash \forall \mathbf{z}\, (\forall x A)^t$.

Proof. If x is in $\mathbf{y}$, then without loss of generality we may assume that $\mathbf{y}$ is $\mathbf{z}$ concatenated with x. Therefore, by assumption we have $\mathsf{Q^\circ S4.t} \vdash \forall \mathbf{z}\, \forall x A^t$. If x is not in $\mathbf{y}$, then $\mathbf{y} = \mathbf{z}$. Thus, by (Gen) for x and by switching the order of quantifiers, we again obtain $\mathsf{Q^\circ S4.t} \vdash \forall \mathbf{z}\, \forall x A^t$. We can then introduce $\Box_F$ inside the quantifiers to obtain $\mathsf{Q^\circ S4.t} \vdash \forall \mathbf{z}\, \Box_F \forall x A^t$ which means $\mathsf{Q^\circ S4.t} \vdash \forall \mathbf{z}\, (\forall x A)^t$. $\Box$

References

[1] Barcan, R. C., *A functional calculus of first order based on strict implication*, J. Symbolic Logic **11** (1946), pp. 1–16.

[2] Carnap, R., *Modalities and quantification*, J. Symbolic Logic **11** (1946), pp. 33–64.

[3] Corsi, G., *A unified completeness theorem for quantified modal logics*, J. Symbolic Logic **67** (2002), pp. 1483–1510.

[4] Cresswell, M. J., *A Henkin completeness for T*, Notre Dame J. Formal Logic **8** (1967), pp. 186–190.

[5] Cresswell, M. J., *Completeness without the Barcan formula*, Notre Dame J. Formal Logic **9** (1968), pp. 75–80.

[6] Fitting, M. and R. L. Mendelsohn, "First-order modal logic," Synthese Library **277**, Kluwer Academic Publishers Group, Dordrecht, 1998.

[7] Gabbay, D. M., "Investigations in modal and tense logics with applications to problems in philosophy and linguistics," D. Reidel Publishing Co., Dordrecht-Boston, Mass., 1976.

[8] Gabbay, D. M., "Semantical investigations in Heyting's intuitionistic logic," Synthese Library **148**, D. Reidel Publishing Co., Dordrecht-Boston, Mass., 1981.

[9] Gabbay, D. M., V. B. Shehtman and D. P. Skvortsov, "Quantification in nonclassical logic. Vol. 1," Studies in Logic and the Foundations of Mathematics **153**, Elsevier B. V., Amsterdam, 2009.

[10] Garson, J. W., *Quantification in modal logic*, in: *Handbook of philosophical logic, Vol. 3*, Kluwer Acad. Publ., Dordrecht, 2001 pp. 267–323.

[11] Ghilardi, S. and G. C. Meloni, *Modal and tense predicate logic: models in presheaves and categorical conceptualization*, in: *Categorical algebra and its applications (Louvain-La-Neuve, 1987)*, Lecture Notes in Math. **1348**, Springer, Berlin, 1988 pp. 130–142.

[12] Hughes, G. E. and M. J. Cresswell, "An introduction to modal logic," Methuen and Co., Ltd., London, 1968.

[13] Hughes, G. E. and M. J. Cresswell, "A new introduction to modal logic," Routledge, London, 1996.

[14] Kripke, S. A., *A completeness theorem in modal logic*, J. Symbolic Logic **24** (1959), pp. 1–14.

[15] Kripke, S. A., *Semantical considerations on modal logic*, Acta Philos. Fenn. **16** (1963), pp. 83–94.

[16] Kripke, S. A., *Semantical analysis of intuitionistic logic. I*, in: *Formal Systems and Recursive Functions (Proc. Eighth Logic Colloq., Oxford, 1963)* (1965), pp. 92–130.

[17] McKinsey, J. C. C. and A. Tarski, *On closed elements in closure algebras*, Ann. of Math. **47** (1946), pp. 122–162.

[18] Rauszer, C., *Semi-Boolean algebras and their applications to intuitionistic logic with dual operations*, Fund. Math. **83** (1973/74), pp. 219–249.

[19] Schütte, K., "Vollständige Systeme modaler und intuitionistischer Logik," Ergebnisse der Mathematik und ihrer Grenzgebiete. 2. Folge **42**, Springer-Verlag Berlin Heidelberg, 1968.

[20] Thomason, R. H., *Some completeness results for modal predicate calculi*, in: *Philosophical Problems in Logic. Some Recent Developments*, Reidel, Dordrecht, 1970 pp. 56–76.

Model Completeness and Π_2-rules: the case of Contact Algebras

Nick Bezhanishvili

ILLC, Universiteit van Amsterdam
The Netherlands

Silvio Ghilardi

Department of Mathematics, Università degli Studi di Milano
Italy

Lucia Landi

Department of Mathematics, Università degli Studi di Milano
Italy

Abstract

We give a sufficient condition for deciding admissibility of non-standard inference rules inside a modal calculus $\mathcal{S}$ with the universal modality. The condition requires the existence of a model completion for the discriminator variety of algebras which are models of $\mathcal{S}$. We apply the condition to the case of symmetric strict implication calculus, i.e., to the modal calculus axiomatizing contact algebras. Such an application requires a characterization of duals of morphisms which are embeddings (in the model-theoretic sense). We supply also an explicit infinite set of axioms for the class of existentially closed contact algebras. The axioms are obtained via a classification of duals of finite minimal extensions of finite contact algebras.

Keywords: Contact Algebras, Non-Standard Inference Rules, Model Completeness, Existentially Closed Structures.

1 Introduction

The use of non-standard rules has a long tradition in modal logic starting from the pioneering work of Gabbay [18], who introduced a non-standard rule for irreflexivity. Non-standard rules have been employed in temporal logic in the context of branching time logic [7] and for axiomatization problems [19] concerning the logic of the real line in the language with the Since and Until modalities. General completeness results for modal languages that are sufficiently expressive to define the so-called difference modality have been obtained in [32]. For the use of the non-standard density rule in many-valued logics we refer to [27] and [29].

Recently, there has been a renewed interest in non-standard rules in the context of the region-based theories of space [30]. One of the key algebraic structures in these theories is that of *contact algebras.* These algebras form a discriminator variety, see e.g., [4]. Compingent algebras are contact algebras satisfying two $\forall\exists$-sentences (aka Π_2-sentences) [4,15]. De Vries [15] established a duality between complete compingent algebras and compact Hausdorff spaces. This duality led to new logical calculi for compact Hausdorff spaces in [2] for two-sorted modal language and in [4] for a uni-modal language with a strict implication. Key to these approaches is a development of logical calculi corresponding to contact algebras. In [4] such a calculus is called the *strict symmetric implication calculus* and is denoted by $\mathsf{S^2IC}$. The extra Π_2-axoms of compingent algebras then correspond to non-standard Π_2-rules, which turn out to be admissible in $\mathsf{S^2IC}$. This generates a natural question of investigating admissibility of Π_2-rules in $\mathsf{S^2IC}$ and in general in logical calculi corresponding to discriminator varieties of modal algebras. This is the question that we address in this paper. We connect admissibility of non-standard Π_2-rules with the model completion of the first-order theory of the corresponding algebraic structures. Motivated by this connection, we then provide (an infinite) axiomatization of the model completion of the theory of contact algebras. As far as we are aware this is a first systematic study of admissibility in the context of non-standard inference rules.

The definition of Π_2-rules we give below is taken from [4] and is close to that of Balbiani et al. [2].

Definition 1.1 [Π_2-rule] A Π_2-*rule* is a rule of the form

$$(\rho) \qquad \frac{F(\underline{\varphi}/\underline{x},\underline{p}) \to \chi}{G(\underline{\varphi}/\underline{x}) \to \chi}$$

where F, G are formulas, $\underline{\varphi}$ is a tuple of formulas, χ is a formula, and $\underline{p}$ is a tuple of propositional letters which do not occur in $\underline{\varphi}$ and χ.

Little is known about the problem of recognizing *admissibility* for non-standard rules, although this problem was already raised in [32]. An immediate easy computation shows that whenever a system $\mathcal{S}$ admits *local uniform interpolants*, then the above rule (ρ) is admissble iff the formula $G(\underline{x}) \to E_{\underline{p}}F(\underline{x},\underline{p})$ is provable in $\mathcal{S}$, where $E_{\underline{p}}F(\underline{x},\underline{p})$ is the uniform pre-interpolant of $F(\underline{x},\underline{p})$ wrt the variables $\underline{p}$. [1]

Local uniform interpolants rarely exist: among the systems where they are available we list K, GL, $\mathsf{S4.Grz}$, $\mathsf{S5}$ [6,20,22,28,33]. From the structural point of view, *global uniform interpolants* (i.e. uniform interpolants for the global consequence relation) are more informative, due to their relationship to

[1] We consider part of the definition of uniform pre- and post- interpolants, the fact that they are stable under substitution: in other words, substituing $\underline{\varphi}$ for the $\underline{p}$ in $E_{\underline{p}}F(\underline{x},\underline{p})$ must give the same result as computing $E_{\underline{p}}F(\underline{\varphi}/\underline{x},\underline{p})$ after the substitution (see [20] for a careful analysis).

compact congruences and model completions [22,24,31]. However, the above simple argument for recognizing admissibility of non-standard rules *seems not to go through* via global uniform interpolants. There is no direct implication (in both senses) between the existence of local and of global uniform interpolants: global uniform interpolants fail to exists for K [24] whereas local ones exists. Conversely, there are cases where global interpolants exist and local interpolants do not (this is easily seen from the results of [26] for locally tabular S4-logics, where existence of uniform local/global interpolants reduces to existence of ordinary local/global interpolants and hence to superamalgamation/amalgamation properties).

Non-standard rules are usually investigated in a system $\mathcal{S}$ of modal logic with a global modality. The global modality is known to supply a discriminator term for the class of $\mathcal{S}$-algebras [23]. In such contexts, by the results of [22], existence of uniform interpolants imply (actually it is equivalent to) existence of a model completion for the equational class of $\mathcal{S}$-algebras. An easy modification of the arguments in [22], shows also that the existence of global uniform interpolants for $\mathcal{S}$ implies the existence of a model completion $T_{\mathcal{S}}{}^{\star}$ for the theory $T_{\mathcal{S}}$ axiomatizing the universal class of *simple* $\mathcal{S}$-algebras. In this paper, we first show that the latter condition (namely existence of a model completion $T_{\mathcal{S}}{}^{\star}$ for $T_{\mathcal{S}}$) is sufficient to characterize non-standard $\mathcal{S}$-rules. This characterization yields effective recognizability of non-standard rules, if quantifier elimination in $T^{\star}_{\mathcal{S}}$ is effective. The latter is certainly the case when $\mathcal{S}$ is decidable and locally tabular. We apply this general result to the case of contact algebras, where we show that the model completion of the theory of simple algebras exists and provide also an axiomatization for it.

2 Π_2-rules and model completions

A *modal signature* Σ is a finite signature comprising Boolean operators $\wedge, \vee, \rightarrow, \leftrightarrow, \neg$ as well as additional operators of any arity called the *modal* operators. Among modal operators, there is a distinguished unary operator $[\forall]$, called the *global* or *universal* modality. Out of Σ-symbols and out of a countable set of variables $x, y, z, \dots, p, q, r, \dots$ one can build the set of propositional Σ-*formulae*. Σ-formulae might be indicated both with the greek letters $\phi, \psi, \dots$ and the latin capital letters $F, G, \dots$. Notations such as $F(\underline{x})$ mean that the Σ-formula F contains at most the variables from the tuple $\underline{x}$. A *modal system* $\mathcal{S}$ (over the modal signature Σ) is a set of Σ-formulae comprising tautologies, the axioms:

$$[\forall]\phi \rightarrow \phi, \qquad [\forall]\phi \rightarrow [\forall][\forall]\phi,$$

$$\phi \rightarrow [\forall]\neg[\forall]\neg\phi, \qquad [\forall](\phi \rightarrow \psi) \rightarrow ([\forall]\phi \rightarrow [\forall]\psi),$$

$$\bigwedge_i [\forall](\phi_i \leftrightarrow \psi_i) \rightarrow (O(...\phi_i...) \leftrightarrow O(...\psi_i...)) \quad \text{(for all } O \in \Sigma).$$

and closed under the rules of modus ponens (MP) (form ϕ and $\phi \rightarrow \psi$ infer ψ), uniform substitution (US) (from $F(\underline{x})$ infer $F(\underline{\psi}/\underline{x})$), and necessitation (N) (from ϕ infer $[\forall]\phi$). We often write $S \vdash \Phi$ or $\vdash_{\mathcal{S}} \phi$ for $\phi \in S$. *We let a* *modal signature Σ and a modal system $\mathcal{S}$ be fixed for the remaining part of this section.* Formulas in $\mathcal{S}$ will be called *$\mathcal{S}$-axioms.* We say that $\mathcal{S}$ is decidable iff the relation $\mathcal{S} \vdash \phi$ is decidable. We also say that $\mathcal{S}$ is *locally tabular* iff for every finite tuple of propositional variables $\underline{x}$ there are finitely many formulae $\psi(\underline{x}), \ldots, \psi_n(\underline{x})$ such that for every further formula $\phi(\underline{x})$ there is some i in $1, \ldots, n$ such that $\mathcal{S} \vdash \phi \leftrightarrow \psi_i$.

We now consider the effect of the addition of Π_2-rules (see Definition 1.1) to a system $\mathcal{S}$.

Definition 2.1 [Proofs with Π_2-rules] Let Θ be a set of Π_2-rules. For a formula φ, we say that φ is *derivable* in $\mathcal{S}$ using the Π_2-rules in Θ, and write $\vdash_{\mathcal{S}+\Theta} \varphi$, provided there is a proof $\psi_1, \ldots, \psi_n$ such that $\psi_n = \varphi$ and each ψ_i is an instance of an axiom of $\mathcal{S}$, or is obtained either by (MP) or (N) from some previous ψ_j's, or there is $j < i$ such that ψ_i is obtained from ψ_j by an application of one of the Π_2-rules $\rho \in \Theta$. The latter means the following, for ρ like in Definition 1.1: $\psi_j = F(\underline{\xi}/\underline{x}, \underline{p}) \to \chi$ and $\psi_i = G(\underline{\xi}/\underline{x}) \to \chi$, where F, G are formulas, $\underline{\xi}$ is a tuple of formulas, χ is a formula, and $\underline{p}$ is a tuple of propositional letters not occurring in $\underline{\xi}, \chi$.

We are interested in characterizing those Π_2-rules that can be freely used in a system without affecting its deductive power.

Definition 2.2 A rule ρ is *admissible* in the system $\mathcal{S}$ if for each formula φ, from $\vdash_{\mathcal{S}+\rho} \varphi$ it follows that $\vdash_{\mathcal{S}} \varphi$.

We may view our modal signature Σ as a first-order signature and Σ-formulae as terms in such a signature. For a modal system $\mathcal{S}$, an $\mathcal{S}$-algebra is a Boolean algebra with operations (one operation of suitable arity for each $O \in \Sigma$) satisfying $[\forall]\top = \top$ and $\phi = \top$ for every $\mathcal{S}$-axioms ϕ. We call an $\mathcal{S}$-algebra *simple* iff the universal first-order condition $\forall x\, ([\forall]x = \top \vee [\forall]x = \bot)$ holds. This agrees with the standard definition from universal algebra, because it can be shown that congruences in a $\mathcal{S}$-algebra bijectively correspond to $[\forall]$-*filters*, i.e. to filters F satisfying the additional condition that $a \in F$ implies $[\forall]a \in F$. We call $T_{\mathcal{S}}$ the equational first-order theory of *simple non degenerate $\mathcal{S}$-algebras* (an $\mathcal{S}$-algebra is non degenerate iff $\bot \neq \top$). A standard Lindenbaum construction proves the *algebraic completeness theorem*, namely that for every ϕ we have $\mathcal{S} \vdash \phi$ iff the identity $\phi = \top$ holds in all $\mathcal{S}$-algebras (and hence iff $\phi = \top$ holds in all simple $\mathcal{S}$-algebras, because $\mathcal{S}$-algebras are a discriminator variety).

With each Π_2-rule ρ given in Definition 1.1, we can associate the following $\forall\exists$-statement in the *first-order* language of $\mathcal{S}$-algebras:

$$\Pi(\rho) := \forall \underline{x}, z\Big(G(\underline{x}) \not\leq z \Rightarrow \exists \underline{y} : F(\underline{x}, \underline{y}) \not\leq z\Big).$$

Theorem 2.3 *Suppose that the universal theory $T_{\mathcal{S}}$ has a model completion $T^{\star}_{\mathcal{S}}$. Then a Π_2-rule ρ is admissible in $\mathcal{S}$ iff $T^{\star}_{\mathcal{S}} \models \Pi(\rho)$.*

Proof. In our general setting [4, Theorem 6.12] holds, replacing the system SIC mentioned there with our generic system $\mathcal{S}$ (the proof of this generalization is reported in the Appendix below as Theorem A.4 and follows the very same arguments as the analogous result of [4]). Using that theorem, we have to show that $T^\star_{\mathcal{S}} \models \Pi(\rho)$ holds iff every simple $\mathcal{S}$-algebra $\mathcal{B}$ can be embedded into some simple $\mathcal{S}$-algebra $\mathcal{C}$ which satisfies $\Pi(\rho)$. This is shown below using the fact that $\Pi(\rho)$ is a Π_2-sentence. Recall that models of $T^\star_{\mathcal{S}}$ are just the existentially closed simple $\mathcal{S}$-algebras (see [12, Proposition 3.5.15]).

Suppose for the left to right direction that $T^\star_{\mathcal{S}} \models \Pi(\rho)$ holds and let $\mathcal{B}$ be any simple $\mathcal{S}$-algebra. Then $\mathcal{B}$ embeds into an existentially closed simple $\mathcal{S}$-algebra $\mathcal{C}$ (this is a general model-theoretic fact [12]); as mentioned above, since $T_{\mathcal{S}}$ has a model completion $T^\star_{\mathcal{S}}$, the existentially closed simple $\mathcal{S}$-algebras are an elementary class and are precisely the models of $T^\star_{\mathcal{S}}$. Thus $\mathcal{B}$ embeds into $\mathcal{C}$ and $\mathcal{C}$ satisfies $\Pi(\rho)$, because $T^\star_{\mathcal{S}} \models \Pi(\rho)$.

Conversely, suppose that every simple $\mathcal{S}$-algebra $\mathcal{B}$ can be embedded into some simple $\mathcal{S}$-algebra $\mathcal{C}$ which satisfies $\Pi(\rho)$. Pick $\mathcal{B}$ such that $\mathcal{B} \models T^\star_{\mathcal{S}}$ and let $\Pi(\rho)$ be $\forall \underline{x} \exists \underline{y} H(\underline{x}, \underline{y})$, where H is quantifier free. Let $\underline{b}$ be a tuple from the support of $\mathcal{B}$. Then we have $\mathcal{C} \models \exists \underline{y} H(\underline{b}, \underline{y})$ for some extension $\mathcal{C}$ of $\mathcal{B}$. As $\mathcal{B}$ is existentially closed, this immediately entails that $\mathcal{B} \models \exists \underline{y} H(\underline{b}, \underline{y})$. Since the $\underline{b}$ was arbitrary, we conclude that $\mathcal{B} \models \Pi(\rho)$, as required. □

Checking whether a Π_2-rule is admissible or not now amounts to checking whether $T^\star_{\mathcal{S}} \models \Pi(\rho)$ holds or not. The latter can be done via quantifier elimination in $T^\star_{\mathcal{S}}$. We give sufficient conditions for this to be effective.

Corollary 2.4 *Let $\mathcal{S}$ be decidable and locally tabular. Assume also that simple $\mathcal{S}$-algebras enjoy the amalgamation property. Then admissibility of Π_2-rules in $\mathcal{S}$ is effective.*

Proof. Local tabularity of $\mathcal{S}$ implies local finiteness[2] of $T_{\mathcal{S}}$. For universal locally finite theories in a finite language, amalgamability is a necessary and sufficient condition for existence of a model completion [25,34]. Quantifier elimination in $T^\star_{\mathcal{S}}$ is effective because there are only finitely many non-equivalent formulae in a fixed finite number of variables, because of Lemma A.3 from the Appendix and because of the following folklore lemma. □

Lemma 2.5 *The quantifier-free formula $R(\underline{x})$ provably equivalent in $T^\star_{\mathcal{S}}$ to an existential formula $\exists \underline{y} H(\underline{x}, \underline{y})$ is the strongest quantifier free formula $G(\underline{x})$ implied (modulo $T_{\mathcal{S}}$) by $H(\underline{x}, \underline{y})$.*

Proof. Recall that $T_{\mathcal{S}}$ and $T^\star_{\mathcal{S}}$ are co-theories [12], i.e. they prove the same universal formulae. Thus we have the following chain of equivalences:

$$\begin{array}{c} T_{\mathcal{S}} \vdash H(\underline{x}, \underline{y}) \to G(\underline{x}) \\ \hline T^\star_{\mathcal{S}} \vdash H(\underline{x}, \underline{y}) \to G(\underline{x}) \\ \hline T^\star_{\mathcal{S}} \vdash \exists \underline{y} H(\underline{x}, \underline{y}) \to G(\underline{x}) \\ \hline T^\star_{\mathcal{S}} \vdash R(\underline{x}) \to G(\underline{x}) \\ \hline T_{\mathcal{S}} \vdash R(\underline{x}) \to G(\underline{x}) \end{array}$$

[2] Recall that a class of algebras is *locally finite* if every finitely generated algebra in this class if finite, see [11, Section 14.2] for the connection between local finiteness and local tabularity.

yielding the claim. □

We point out that there might be different ways (other than Corollary 2.4) to exploit Theorem 2.3 in order to decide admissibility of Π_2-rules (for instance, as mentioned in the introduction, computability of global interpolants offers a powerful opportunity, given the relationship between model completions and uniform global interpolants [22]). However Corollary 2.4 gives a simple criterion, independent of more sophisticated machinery, which is useful for the application of this paper. In Section 4, we give an example of the application of Corollary 2.4 for recognizing an admissible rule.

The usefulness of Corollary 2.4 lies in the fact that its only real requirement is the amalgamation property, besides local tabularity. Whenever local tabularity holds, finitely presented algebras are finite, thus it is sufficient to establish amalgamability for *finite* algebras (this is easily seen by compactness of first-order logic, because, in the end, amalgamation property can be established by showing the consistency of some joined Robinson diagrams). Whenever a "good" duality is established, amalgamation of finite algebras turns out to be equivalent to dual amalgamation for finite frames, which is usually much easier to check. We will now give a couple of simple (non-)examples.

Example 2.6 If the modal signature contains only the global modality $[\forall]$, we have the locally tabular logic S5. Finite simple non degenerate S5-algebras are dual to finite nonempty sets and onto maps, for which dual amalgamation trivially holds (by standard pullback construction), see, e.g., [11, Thm. 14.23].

Example 2.7 The logic of difference [14,32] has in addition to the global modality a unary operator D subject to the axioms

$$[\forall]\phi \leftrightarrow (\phi \wedge \neg D\neg\phi), \qquad \phi \to D\neg D\neg\phi, \qquad DD\phi \to \phi \vee D\phi.$$

This logic axiomatizes Kripke frames where the accessibility relation is inequality. Local finiteness can be established for instance by the method of irreducible models [21]. Amalgamation however fails. To see this, notice that the simple frames for this logic are sets endowed with a relation E such that $w_1 \neq w_2 \to w_1 E w_2$. Now let $X = \{x_1, \ldots, x_5\}$, $Y = \{y_1, \ldots, y_5\}$ and $Z = \{z_1, z_2\}$. Let $x_i E_X x_j$ iff $i \neq j$ for $1 \leq i, j \leq 5$, $y_i E_Y y_j$ iff $i \neq j$ for $1 \leq i, j \leq 5$ and $z_i E_Z z_j$ for $i, j = 1, 2$. Let also $f : X \to Z$ and $g : Y \to Z$ be such that $f(x_1) = f(x_2) = f(x_3) = g(y_1) = g(y_2) = z_1$ and $f(x_4) = f(x_5) = g(y_3) = g(y_4) = g(y_4) = z_5$. Then it is easy to see that f and g are p-morphisms. If a dual amalgam exists, then there must exist a frame (U, E_U) and onto p-morphims $h : U \to X$ and $j : U \to Y$ such that $f \circ h = g \circ j$. However, an easy argument shows that U should contain more than 5 points. Moreover, for $u, v \in U$ with $u \neq v$ we should have $u E_U v$. But then there will be distinct points in U mapped by f to some x_i, which would entail that x_i is reflexive, which is a contradiction.

3 Symmetric Strict Implication and Contact Algebras

In this section we first review some material from [4]. Let us consider the modal signature comprising, besides the global modality $[\forall]$, a binary operator $\rightsquigarrow$, which we call *strict implication*, subject to the following axioms (we keep the same numeration as in [4] and add axiom (A0) which is seen as a definition of $[\forall]$ in [4]).

(A0) $[\forall]\varphi \leftrightarrow (\top \rightsquigarrow \varphi)$,
(A1) $(\bot \rightsquigarrow \varphi) \wedge (\varphi \rightsquigarrow \top)$,
(A2) $[(\varphi \vee \psi) \rightsquigarrow \chi] \leftrightarrow [(\varphi \rightsquigarrow \chi) \wedge (\psi \rightsquigarrow \chi)]$,
(A3) $[\varphi \rightsquigarrow (\psi \wedge \chi)] \leftrightarrow [(\varphi \rightsquigarrow \psi) \wedge (\varphi \rightsquigarrow \chi)]$,
(A4) $(\varphi \rightsquigarrow \psi) \rightarrow (\varphi \rightarrow \psi)$,
(A5) $(\varphi \rightsquigarrow \psi) \leftrightarrow (\neg\psi \rightsquigarrow \neg\varphi)$,
(A8) $[\forall]\varphi \rightarrow [\forall][\forall]\varphi$,
(A9) $\neg[\forall]\varphi \rightarrow [\forall]\neg[\forall]\varphi$,
(A10) $(\varphi \rightsquigarrow \psi) \leftrightarrow [\forall](\varphi \rightsquigarrow \psi)$,
(A11) $[\forall]\varphi \rightarrow (\neg[\forall]\varphi \rightsquigarrow \bot)$,

Inference rules are modus ponens and necessitation. It can be shown (see [4]) that this system (called *symmetric strict implication calculus* $\mathsf{S^2IC}$) matches our requirements from Section 2. Moreover $\mathsf{S^2IC}$ is locally tabular and simple $\mathsf{S^2IC}$-algebras are those $\mathsf{S^2IC}$-algebras $\mathcal{B}$ where we have that $a \rightsquigarrow b$ is either $\bot$ or $\top$. Thus in a simple non-degenerate $\mathsf{S^2IC}$-algebra, the operation $\rightsquigarrow$ is in fact the characteristic function of a binary relation $\prec$. It can be proved that the characteristic function of a binary relation $\prec$ on a Boolean algebra gives rise to an $\mathsf{S^2IC}$-algebra structure iff it satisfies the following conditions:

(S1) $0 \prec 0$ and $1 \prec 1$;
(S2) $a \prec b, c$ implies $a \prec b \wedge c$;
(S3) $a, b \prec c$ implies $a \vee b \prec c$;
(S4) $a \leq b \prec c \leq d$ implies $a \prec d$;
(S5) $a \prec b$ implies $a \leq b$;
(S6) $a \prec b$ implies $\neg b \prec \neg a$.

Non-degenerate Boolean algebras endowed with a relation $\prec$ satisfying the above conditions (S1)-(S6) are called *contact algebras.* [3] Since the theory of non degenerate simple $\mathsf{S^2IC}$-algebras is essentially the same (in fact, it is a syntactic variant) as the universal theory Con of contact algebras, we shall investigate the latter in order to apply Corollary 2.4. What we have to show in order to check the hypotheses of such a corollary is just that Con is amalgamable.

To prove amalgamability, we need a duality theorem. In [5,10,16] a duality theorem is established for the category of contact algebras and $\prec$*-maps* (a map $\mu : (\mathcal{B}, \prec) \rightarrow (\mathcal{C}, \prec)$ among contact algebras is said to be a $\prec$-map iff it is a Boolean homomorphism such that $a \prec b$ implies $\mu(a) \prec \mu(b)$). We shall make use of that theorem but we shall modify it, because for amalgamation we need

[3] It is more common to use in contact algebras the *contact relation* δ [30] which is given by $a\delta b$ iff $a \not\prec \neg b$. However, we stick with our notation to stay close to our main reference [4].

a duality for contact algebras and *embeddings* in the model theoretic sense (this means that an embedding is an injective map that not only *preserves* but also *reflects* the relation $\prec$). We first recall the duality theorem of [5], giving just the minimum information that is indispensable for our purposes.

We say that a binary relation R on a topological space X is *closed* if R is a closed subset of $X \times X$ in the product topology. Let StR be the category having (i) as objects the pairs (X, R), where X is a (non empty) Stone space and R is a closed, reflexive and symmetric relation on X, and (ii) as arrows the continuous maps $f : (X, R) \to (X', R')$ which are *stable* (i.e. such that xRy implies $f(x)R'f(y)$ for all points x, y in the domain of f). We define a contravariant functor

$$(-)^{\star} : \mathsf{StR}^{op} \to \mathsf{Con}_s$$

into the category Con_s of contact algebras and $\prec$-maps as follows:

- for an object (X, R), the contact algebra $(X, R)^{\star}$ has $\mathsf{Clop}(X)$ the clopens of X as carrier set (with union, intersection and complement as Boolean operations) and its relation $\prec$ is given by $C \prec D$ iff $R[C] \subseteq D$ (here we used the abbreviation $R[C] = \{x \in X \mid sRx \text{ for some } s \in C\}$);
- for a stable continuous map $f : (X, R) \to (X', R')$, the map $f^{\star}$ is the inverse image along f.

Theorem 3.1 ([5,16]) *The functor $(-)^{\star}$ establishes an equivalence of categories.*

We now intend to restrict this equivalence to the category Con_e of contact algebras and embeddings. To this aim we need to identify a suitable subcategory StR_e of StR. Now StR_e has the same objects as StR, however a stable continuous map $f : (X_1, R_1) \to (X_2, R_2)$ is in StR_e iff it satisfies the following additional condition:

$$\forall x, y \in X_2 \; [xR_2y \;\Leftrightarrow\; \exists \tilde{x}, \tilde{y} \in X_1 \text{ s.t. } f(\tilde{x}) = x,\; f(\tilde{y}) = y \;\&\; \tilde{x}R_1\tilde{y}] \qquad (1)$$

Notice that, since R_2 is reflexive, it turns out that a map satisfying (1) must be surjective. We call the stable maps satisfying (1) *regular stable maps*, because it can be shown that these maps are just the regular epimorphisms in the category StR.

Theorem 3.2 *The functor $(-)^{\star}$, suitably restricted in its domain and codomain, establishes an equivalence of categories between* StR_e *and* Con_e.

Proof. We need to show that f satisfies condition (1) above iff $f^{\star}$ is an embedding between contact algebras, i.e. iff it satisfies the condition

$$(R_1[f^{-1}(U)] \subseteq f^{-1}(V) \;\Leftrightarrow\; R_2[U] \subseteq V) \quad \forall\, U, V \in \mathsf{Clop}(X_2) \qquad (2)$$

where $\mathsf{Clop}(X_2)$ is the set of clopens of the Stone space X_2. We tranform condition (2) up to equivalence. First notice that, by the adjunction between direct and inverse image, (2) is equivalent to

$$(f(R_1[f^{-1}(U)]) \subseteq V \;\Leftrightarrow\; R_2[U] \subseteq V) \quad \forall\, U, V \in \mathsf{Clop}(X_2) \qquad (3)$$

Now, in compact Hausdorff spaces, closed relations and continuous functions map closed sets to closed sets, hence $f(R_1[f^{-1}(U)])$ is closed and so, since clopens are a base for closed sets, (3) turns out to be equivalent to

$$(f(R_1[f^{-1}(U)]) = R_2[U]) \ \ \forall\, U \in \mathsf{Clop}(X_2) \tag{4}$$

We now claim that (4) is equivalent to

$$f(R_1[f^{-1}(\{x\})]) = R_2[\{x\}] \ \ \forall x \in X_2 \tag{5}$$

In fact, (5) implies (4) because all operations $f(-), R[-], f^{-1}(-)$ preserve set-theoretic unions. The converse implication holds because of Esakia's lemma below applied to the down-directed system $\{U \in \mathsf{Clop}(X_2) \mid x \in U\}$. Notice that Esakia's lemma applies because $f \circ R_1 \circ f^{op}$ and R_2 are symmetric relations, since R_1 and R_2 are symmetric (here we view f and $f^{-1} = f^{op}$ as relations via their graphs).

Now it is sufficient to observe that (5) is equivalent to the conjunction of (1) and stability. □

We will now prove a version of Esakia's lemma for our spaces. Esakia's lemma normally speaks about the inverse of a relation R, but here we need a version which holds for R-images because our relation is symmetric.

Lemma 3.3 (Esakia, Lemma 3.3.12 in [17]) *Let X be a compact Hausdorff space, and R a point-closed*[4] *symmetric binary relation on X. Then for each downward directed family $\mathcal{C} = \{C_i\}_{i \in I}$ of nonempty closed subsets of X, we have $R[\bigcap_{i\in I} C_i] = \bigcap_{i \in I} R[C_i]$.*

Proof. The inclusion $R[\bigcap_{i\in I} C_i] \subseteq \bigcap_{i\in I} R[C_i]$ is trivial. Now suppose $x \in \bigcap_{i\in I} R[C_i]$. Then $x \in R[C_i]$ for each C_i and, by symmetry, $R[x] \cap C_i$ is nonempty for each $i \in I$. But as C_i-s are downward directed, all the finite intersections $R[x] \cap C_{i_1} \cap ... \cap C_{i_n}$ (with $i_j \in I$ for $j \in \{1, ..., n\}$) are nonempty. By compactness, the infinite intersection (which equals $R[x] \cap \bigcap_{i\in I} C_i$) is nonempty and so, by symmetry, $x \in R[\bigcap_{i\in I} C_i]$. □

Now we are ready to show that Corollary 2.4 applies.

Theorem 3.4 *The universal theory* Con *of contact algebras has the amalgamation property. Therefore, as it is also locally finite,* Con *has a model completion.*

Proof. As we observed in Section 2, it is sufficient to prove amalgamation for finite algebras (by local finiteness and by a compactness argument based on Robinson diagrams). Finite algebras are dual to discrete Stone spaces, hence it is sufficient to show the following.

---

[4] A binary relation R on a topological space X is said to be *point-closed* if $\forall x \in X$ $R[x]$ is closed in X. A closed relation in a compact Hausdorff space maps closed sets to closed sets via $R[-]$, hence it is point-closed.

(+) Given finite nonempty sets X_A, X_B, X_C endwed with reflexive and symmetric relations R_A, R_B, R_C and given regular stable maps $f : (X_B, R_B) \to (X_A, R_A)$, $g : (X_B, R_B) \to (X_A, R_A)$, there exist (X_D, R_D) (with reflexive and symmetric R_D) and regular stable maps $\pi_1 : (X_D, R_D) \to (X_B, R_B)$, $\pi_2 : (X_D, R_D) \to (X_C, R_C)$, such that $f \circ \pi_1 = g \circ \pi_2$.

Statement (+) is easily proved by taking as $(X_D, R_D), \pi_1, \pi_2$ the obvious pullback with the two projections. □

4 A Set of Axioms for Con*

Theorem 3.4 gives the possibility of applying Corollary 2.4 to recognize admissible rules. We give here another algorithm, slightly different from that of Corollary 2.4. We recall that Con* is the theory of existentially closed contact algebras [12]. The following result (given that Con is locally finite) is folklore (a detailed proof of the analogous statement for Brouwverian semilattices is in the ArXiv version of [9] as [8, Proposition 2.16]).

Theorem 4.1 *Let $(\mathcal{B}, \prec)$ be a contact algebra. We have that $(\mathcal{B}, \prec)$ is existentially closed iff for any finite subalgebra $(\mathcal{B}_0, \prec) \subseteq (\mathcal{B}, \prec)$ and for any finite extension $(\mathcal{C}, \prec) \supseteq (\mathcal{B}_0, \prec)$ there exists an embedding $(\mathcal{C}, \prec) \hookrightarrow (\mathcal{B}, \prec)$ such that the following diagram commutes*

$$\begin{array}{ccc} (\mathcal{B}_0, \prec) & \hookrightarrow & (\mathcal{B}, \prec) \\ \downarrow & \nearrow & \\ (\mathcal{C}, \prec) & & \end{array}$$

Example 4.2 Consider the Π_2-rule:

$$(\rho 9) \qquad \frac{(p \rightsquigarrow p) \wedge (\varphi \rightsquigarrow p) \wedge (p \rightsquigarrow \psi) \to \chi}{(\varphi \rightsquigarrow \psi) \to \chi}$$

This rule is admissible in $\mathsf{S^2IC}$ [4, Theorem 6.15]. We will now give an alternative and more automated proof of this result. Translating $\Pi(\rho 9)$ into the equivalent language of contact algebras, we obtain (see statement $(S9)$ from Section 6.3 of [4])

$$x \prec y \Rightarrow \exists z\ (z \prec z \wedge x \prec z \prec y) \tag{6}$$

According to Theorem 2.3, we have to show that (6) is provable in Con*. Note that (6) expresses interesting (order-)topological properties. It is valid on (X, R) iff R is a Priestley quasi-order [5, Lemma 5.2]. Also it is valid on a compact Hausdorff space X iff X is a Stone space [3, Lemma 4.11].

If we follow the procedure of Corollary 2.4 (which is based on Lemma 2.5), we first compute the quantifier-free formula equivalent in Con* to $\exists z\ (z \prec z \wedge x \prec z \prec y)$ by taking the conjunction of the (finitely many) quantifier-free first-order formulae $\phi(x, y)$ which are implied (modulo Con) by $z \prec z \wedge x \prec z \prec y$: this is, up to equivalence, $x \prec y$. Now, in order to show the admissibility of $(\rho 9)$ is sufficient to observe that Con $\models x \prec y \Rightarrow x \prec y$.

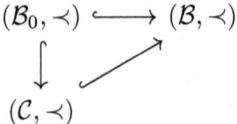

As an alternative, we can rely on Theorem 4.1 and show that (6) is true in every existentially closed contact algebra. To this aim, it is sufficient to enumerate all contact algebras $\mathcal{B}_0$ generated by two elements a, b such that $\mathcal{B}_0 \models a \prec b$ and to show that all such algebras embed in a contact algebra $\mathcal{C}$ generated by three elements a, b, c such that $\mathcal{C} \models c \prec c \wedge a \prec c \prec b$ (this can be done automatically for instance using a model finder tool). Both of the above procedures are heavy and not elegant, but they are nevertheless mechanical and do not require ingenious *ad hoc* constructions (such as e.g., the construction of Lemma 5.4 in [4]).

Theorem 4.1 implicitly supplies an infinite set of axioms for the model completion of the theory of contact algebras. Such an axiomatization is not however very informative, as it comes from generic model-theoretic facts. In this section, we supply a better axiomatization, following the same strategy used in [13] for the case of amalgamable locally finite varieties of Heyting algebras and in [9] for the case of Brouwverian semilattices. The strategy consists of classifying minimal extensions via the so-called 'signatures'.

It is evident that the Theorem 4.1 still holds if we limit its statement to finite *minimal* extensions $(\mathcal{C}, \prec)$ of $(\mathcal{B}_0, \prec)$ (such an extension $(\mathcal{C}, \prec)$ is said to be minimal iff it is proper and every proper extension contains it, up to isomorphism). Using our Duality Theorem 3.2 restricted to the finite discrete case, we can characterize the dual spaces $(X_\mathcal{C}, R_\mathcal{C})$ and $(X_{\mathcal{B}_0}, R_{\mathcal{B}_0})$ and the dual stable map $f : (X_\mathcal{C}, R_\mathcal{C}) \to (X_{\mathcal{B}_0}, R_{\mathcal{B}_0})$ corresponding to such minimal extensions.

Proposition 4.3 *Let $(\mathcal{B}_0, \prec) \hookrightarrow (\mathcal{C}, \prec)$ be an embedding between finite contact algebras, with dual regular stable map $f : (X_\mathcal{C}, R_\mathcal{C}) \to (X_{\mathcal{B}_0}, R_{\mathcal{B}_0})$. The embedding is minimal iff (up to isomorphism) there are a finite set Y, finite subsets $S_1, S_2 \subseteq Y$ and elements $x \in X_{\mathcal{B}_0}, x_1 \in X_\mathcal{C}, x_2 \in X_\mathcal{C}$ such that:*

(i) *$X_{\mathcal{B}_0}$ is the disjoint union $Y \oplus \{x\}$;*

(ii) *$X_\mathcal{C}$ is the disjoint union $Y \oplus \{x_1, x_2\}$;*

(iii) *f restricted to Y is the identity map and $f(x_1) = f(x_2) = x$;*

(iv) *the restrictions of $R_\mathcal{C}$ and of $R_{\mathcal{B}_0}$ to Y coincide;*

(v) *$R_\mathcal{C}[x_1] \setminus \{x_1\} = S_1$ and $R_\mathcal{C}[x_2] \setminus \{x_2\} = S_2$;*

(vi) *$R_{\mathcal{B}_0}[x] \setminus \{x\} = S_1 \cup S_2$.*

Proof. First notice that, as a consequence of (1), if the cardinality of $X_{\mathcal{B}_0}$ and of $X_\mathcal{C}$ are the same, then f is an isomorphism. This is seen as follows: we already observed that condition (1) implies surjectivity and in case of the same finite cardinality surjectivity implies injectivity. Preservation and reflection of the relation follow by stability and (1) again.

In addition, if the cardinality of $X_\mathcal{C}$ is equal to the cardinality of $X_{\mathcal{B}_0}$ plus one (this is precisely the case mentioned in the statement of the proposition), then f cannot be properly factored, hence it is minimal. We show that all minimal maps arise in this way.

In general, if the cardinality of $X_{\mathcal{C}}$ is bigger than the cardinality of $X_{\mathcal{B}_0}$, we can define the following factorization of f. Pick some $x \in X_{\mathcal{B}_0}$ having more than one preimage and split $f^{-1}(\{x\})$ as $T_1 \cup T_2$, where T_1, T_2 are disjoint and non-empty. We have that $X_{\mathcal{C}}$ is the disjoint union $X \oplus T_1 \oplus T_2$ for some set X and $X_{\mathcal{B}_0}$ is the disjoint union $Y \oplus \{x\}$ for some set Y. Define a discrete dual space (Z, R_Z) as follows. Z is the disjoint union $Y \oplus \{x_1, x_2\}$ for new x_1, x_2 and R_Z is the reflexive and symmetric closure of the following sets of pairs: (i) the pairs (z_1, z_2) for $z_1 R_{\mathcal{B}_0} z_2$ and $z_1, z_2 \in Y$; (ii) the pairs (x_i, u) for $u \in f(R_{\mathcal{C}}[T_i])$ $(i = 1, 2)$; (iii) the pair (x_1, x_2), but only in case $T_1 \cap R_{\mathcal{C}}[T_2] \neq \emptyset$. Then it is easily seen that f factorizes as $h \circ \tilde{f}$ in StR_e, where: (I) $\tilde{f}$ maps T_1 to x_1, T_2 to x_2 and acts as f on X; (II) h is the identity on Y and maps both x_1, x_2 to x.

Now h produces the data required by the proposition and $\tilde{f}$ must be an isomorphism if f is minimal. □

Notice that the above conditions (i)-(vi) determine uniquely the finite minimal extension over the contact algebras dual to $(X_{\mathcal{B}_0}, R_{\mathcal{B}_0})$ except for a detail: they do not specify whether we have $x_1 R_{\mathcal{C}} x_2$ or not. So the data x, S_1, S_2 and $Y = X_{\mathcal{B}_0} \setminus \{x\}$ (lying *inside* $X_{\mathcal{B}_0}$) determine in fact *two* minimal expansions of the contact algebra dual to $(X_{\mathcal{B}_0}, R_{\mathcal{B}_0})$.

The next step is to *re-dualize* the data of Proposition 4.3 inside a given finite contact algebra. We first need some notation.

Definition 4.4 Let $(\mathcal{B}_0, \prec_{\mathcal{B}_0})$ be a finite subalgebra of the contact algebra $(\mathcal{B}, \prec_{\mathcal{B}})$. Then, for $b \in \mathcal{B}$, we define $[b]^{\prec_{\mathcal{B}_0}} := \bigcap\{x \in \mathcal{B}_0 \mid b \prec_{\mathcal{B}} x\}$.

The notion of a signature given below dualizes and internalizes the data of Proposition 4.3 (the further bit $\star$ is used to distinguish the two possible minimal extensions).

Definition 4.5 Let $(\mathcal{B}_0, \prec_{\mathcal{B}_0})$ be a finite contact algebra. We call a *signature* in $(\mathcal{B}_0, \prec_{\mathcal{B}_0})$ a tuple $(b, \tilde{c_1}, \tilde{c_2})$, where $b \in \mathcal{B}_0$ is an atom, and $\tilde{c_1}, \tilde{c_2} \in \mathcal{B}_0$ are such that $[b]^{\prec_{\mathcal{B}_0}} \wedge \neg b = \tilde{c_1} \vee \tilde{c_2}$. A *marked signature* in $(\mathcal{B}_0, \prec_{\mathcal{B}_0})$ is a tuple $(b, \tilde{c_1}, \tilde{c_2}, \star)$, where $(b, \tilde{c_1}, \tilde{c_2})$ is a signature and $\star \in \{0, 1\}$.

We are now ready to produce our first-order axiomatization of existentially closed contact algebras.

Theorem 4.6 *A contact algebra $(\mathcal{B}, \prec_{\mathcal{B}})$ is existentially closed if and only if, for any finite subalgebra $(\mathcal{B}_0, \prec_{\mathcal{B}_0}) \subseteq (\mathcal{B}, \prec_{\mathcal{B}})$, the following conditions hold:*

(i) *for every marked signature $(b, \tilde{c_1}, \tilde{c_2}, 1)$ in $(\mathcal{B}_0, \prec_{\mathcal{B}_0})$, there exist $b_1, b_2 \in \mathcal{B} \setminus \{0\}$ such that $b = b_1 \vee b_2$, $b_1 \wedge b_2 = \bot$, $[b_i]^{\prec_{\mathcal{B}_0}} = \tilde{c_i} \vee b$ for $i \in \{1, 2\}$ and $b_1 \not\prec_{\mathcal{B}} \neg b_2$;*

(ii) *for every marked signature $(b, \tilde{c_1}, \tilde{c_2}, 0)$ in $(\mathcal{B}_0, \prec_{\mathcal{B}_0})$, there exist $b_1, b_2 \in \mathcal{B} \setminus \{0\}$ such that $b = b_1 \vee b_2$, $b_1 \wedge b_2 = \bot$, $[b_i]^{\prec_{\mathcal{B}_0}} = \tilde{c_i} \vee b$ for $i \in \{1, 2\}$ and $b_1 \prec_{\mathcal{B}} \neg b_2$.*

Proof. ($\Rightarrow$) Let $(\mathcal{B}_0, \prec_{\mathcal{B}_0}) \hookrightarrow (\mathcal{B}, \prec_{\mathcal{B}})$ be a finite subalgebra, and let $(b, \tilde{c_1}, \tilde{c_2}, \star)$ be a signature in $(\mathcal{B}_0, \prec_{\mathcal{B}_0})$. Let $(\mathcal{B}_0, \prec_{\mathcal{B}_0}) \hookrightarrow (C, \prec_C)$ be the finite minimal extension whose dual satisfies the conditions (i)-(vi) of Proposition 4.3 (and also $x_1 R_\mathcal{C} x_2$ iff $\star = 1$). Then it is clear that there exist b_1, b_2 satisfying (i) (for the case $\star = 1$) or (ii) (for the case $\star = 0$) inside $(C, \prec_\mathcal{C})$. Thanks to Theorem 4.1, we know that there exists an embedding $(C, \prec_\mathcal{C}) \hookrightarrow (B, \prec_\mathcal{B})$ that fixes $(\mathcal{B}_0, \prec_{\mathcal{B}_0})$. Via this embedding, the required b_1, b_2 are moved to $(B, \prec_\mathcal{B})$: they still satisfy the conditions required by (i) and (ii) because such conditions can be expressed as first-order ground conditions with parameters in $(\mathcal{B}_0, \prec_{\mathcal{B}_0})$ [5] and hence they are preserved through embeddings.

($\Leftarrow$) Here we use our Duality Theorem 3.2. We are given a finite contact subalgebra $(\mathcal{B}_0, \prec_{\mathcal{B}_0})$ of $(\mathcal{B}, \prec_\mathcal{B})$ and a finite minimal extension $(\mathcal{C}, \prec_\mathcal{C})$ of its. The situation, in the dual category is the following:

$$(X_{\mathcal{B}_0}, R_{\mathcal{B}_0}) \xleftarrow{\bar{f}} (X_\mathcal{B}, R_\mathcal{B})$$
$$f \uparrow \quad \overset{\tilde{f}}{\dashleftarrow}$$
$$(X_\mathcal{C}, R_\mathcal{C})$$

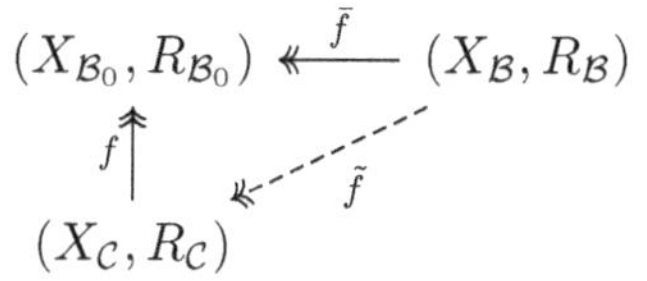

where $\bar{f}$ is dual to the inclusion $(\mathcal{B}_0, \prec_{\mathcal{B}_0}) \hookrightarrow (\mathcal{B}, \prec_\mathcal{B})$ and f satisfies the conditions (i)-(vi) of Proposition 4.3. We need to define $\tilde{f}$ so that the above triangle commutes in the category StR_e. Recall that the two spaces $X_{\mathcal{B}_0}, X_\mathcal{C}$ are discrete, but $X_\mathcal{B}$ is not.

By hypothesis, we know that there exist non empty disjoint clopens $U_{b_1}, U_{b_2} \in \mathsf{Clop}(X_\mathcal{B})$ such that $\bar{f}^{-1}(\{x\}) = U_{b_1} \cup U_{b_2}$. According to Definition 4.4, we have that the clopen defined by $[b_i]^{\prec_{\mathcal{B}_0}}$ is the intersection of the family $\bar{f}^{-1}(T)$ varying T among the subsets of $X_{\mathcal{B}_0}$ such that $R_\mathcal{B}[U_{b_i}] \subseteq \bar{f}^{-1}(T)$, i.e. varying T among the subsets of $X_{\mathcal{B}_0}$ such that $\bar{f}(R_\mathcal{B}[U_{b_i}]) \subseteq T$. Since this intersection is precisely $\bar{f}^{-1}(\bar{f}(R_\mathcal{B}[U_{b_i}]))$, according to our hypothesis, we have $\bar{f}^{-1}(\bar{f}(R_\mathcal{B}[U_{b_i}])) = \bar{f}^{-1}(S_i) \cup \bar{f}^{-1}(\{x\})$. Since $\bar{f}^{-1}$ is injective, we conclude

$$\bar{f}(R_\mathcal{B}[U_{b_1}]) = S_1 \cup \{x\} \quad \text{and} \quad \bar{f}(R_\mathcal{B}[U_{b_2}]) = S_2 \cup \{x\}. \tag{7}$$

In case $x_1 R_\mathcal{C} x_2$ holds, we use hypothesis (i) and in case it does not hold, we use hypothesis (ii). To sum up, recalling that $b_1 \not\prec \neg b_2$ dualizes to $R_\mathcal{B}[U_{b_1}] \cap U_{b_2} \neq \emptyset$, we get

$$x_1 R_\mathcal{C} x_2 \iff R_\mathcal{B}[U_{b_1}] \cap U_{b_2} \neq \emptyset. \tag{8}$$

We define $\tilde{f}$ as follows: $\tilde{f}(z) = \bar{f}(z)$ for $z \notin U_{b_1} \cup U_{b_2}$, $\tilde{f}(z) = x_1$ for $z \in U_{b_1}$, $\tilde{f}(z) = x_2$ for $z \in U_{b_2}$. By Proposition 4.3(iii), it is clear that $\tilde{f} \circ f = \bar{f}$. The continuity of $\tilde{f}$ is also immediate.

Let us check *stability*, namely that for $y_1, y_2 \in X_\mathcal{B}$ such that $y_1 R_\mathcal{B} y_2$, we have $\tilde{f}(y_1) R_\mathcal{C} \tilde{f}(y_2)$. We distinguish three cases:

1. $y_1, y_2 \notin U_{b_1} \cup U_{b_2}$
2. $y_1 \in U_{b_1}$, $y_2 \notin U_{b_1} \cup U_{b_2}$
3. $y_1, y_2 \in U_{b_2} \cup U_{b_2}$

---

[5] If $a_1, \ldots, a_n$ are the elements of $\mathcal{B}_0$, then $[b_i]^{\prec_{\mathcal{B}_0}} = \tilde{c}_i \vee b$ can be written as $\bigwedge_{j=1}^{n} (b_i \prec a_j \leftrightarrow (\tilde{c}_i \vee b \leq a_j))$.

(by the symmetry of $R_{\mathcal{B}}$, this enumeration is exhaustive, up to exchanging the role of U_{b_1} and of U_{b_2}). Case 1 is covered by the stability of $\bar{f}$ and Proposition 4.3(iv). Case 3 is covered by the reflexivity of $R_{\mathcal{C}}$ and (8). In Case 2, we have $y_2 \in R_{\mathcal{B}}[U_{b_1}]$, thus $\tilde{f}(y_2) = \bar{f}(y_2) \in S_1$ by (7) (and by the fact that $\bar{f}(y_2) \neq x$). Thus we conclude $x_1 = \tilde{f}(y_1) R_{\mathcal{C}} \tilde{f}(y_2)$ by Proposition 4.3(v).

It remains to prove that for all $z_1, z_2 \in X_{\mathcal{C}}$ such that $z_1 R_{\mathcal{C}} z_2$ we have

$$\exists y_1, y_2 \in X_{\mathcal{B}} \text{ s.t. } \tilde{f}(y_1) = z_1 \;\&\; \tilde{f}(y_2) = z_2 \;\&\; y_1 R_{\mathcal{B}} y_2 \tag{9}$$

(this is condition (1)). Again we distinguish three cases:

a. $z_1, z_2 \notin \{x_1, x_2\}$

b. $z_1 = x_1$, $z_2 \notin \{x_1, x_2\}$

c. $z_1 = x_1$, $z_2 = x_2$.

Case a is covered by Proposition 4.3(iii) and by the fact that $\bar{f}$ satisfies condition (1). Case c is covered by (8). In Case b, $z_2 \in S_1$ by Proposition 4.3(v) and by (7) there are $y_1 \in U_{b_1}$ and y_2 such that $y_1 R_{\mathcal{B}} y_2$ and $\bar{f}(y_2) = z_2$. Then, $\tilde{f}(y_1) = x_1 = z_1$ and $\tilde{f}(y_2) = \bar{f}(y_2) = z_2$ (we have $\tilde{f}(y_2) = \bar{f}(y_2)$ because $\bar{f}(y_2) = z_2 \neq x$, so that $y_2 \notin U_{b_1} \cup U_{b_2}$). □

Theorem 4.6 gives a first order axiomatization because the reference to finite subalgebras can be replaced by a suitable string of universal quantifiers. However, the above axiomatization is infinite, thus determining whether there exists a finite axiomatization (and supplying one, in case of a positive answer) remains at the moment an open question.

In principle, the axiomatization supplied by Theorem 4.6 should be naturally convertible (using Lemma A.3 below) into a basis for admissible Π_2-rules for the symmetric strict implication calculus, once the notion of a basis for admissible Π_2-rules is suitably defined. We leave this task for future research. Connections with the literature on admissibility of standard inference rules in contact algebras [1] should also be developed: our non-standard rules have the particular shape (ρ) outlined in Definition 1.1 and they trivialize if they are standard (i.e., if p does not occur in the formula F from the premise); however it could be interesting to analyze more general formats for non-standard rules encompassing standard inference rules.

References

[1] Balbiani, P. and Ç. Gencer, *Admissibility and unifiability in contact logics*, in: *TbiLLC 2013, Gudauri, Georgia*, 2013, pp. 44–60.

[2] Balbiani, P., T. Tinchev and D. Vakarelov, *Modal logics for region-based theories of space*, Fund. Inform. **81** (2007), pp. 29–82.

[3] Bezhanishvili, G., *Stone duality and Gleason covers through de Vries duality*, Topology Appl. **157** (2010), pp. 1064–1080.

[4] Bezhanishvili, G., N. Bezhanishvili, T. Santoli and Y. Venema, *A strict implication calculus for compact Hausdorff spaces*, Ann. Pure Appl. Logic **170** (2019), 102714.

[5] Bezhanishvili, G., N. Bezhanishvili, S. Sourabh and Y. Venema, *Irreducible equivalence relations, Gleason spaces, and de Vries duality*, Appl. Categ. Structures **25** (2017), pp. 381–401.

[6] Bílková, M., *Uniform interpolation and propositional quantifiers in modal logics*, Studia Logica **85** (2007), pp. 1–31.

[7] Burgess, J. P., *Decidability for branching time*, Studia Logica **39** (1980), pp. 203–218.

[8] Carai, L. and S. Ghilardi, *Existentially closed brouwerian semilattices* (2017). URL `https://arxiv.org/abs/1702.08352`

[9] Carai, L. and S. Ghilardi, *Existentially closed Brouwerian semilattices*, J. Symb. Log. **84** (2019), pp. 1544–1575.

[10] Celani, S., *Quasi-modal algebras*, Math. Bohem. **126** (2001), pp. 721–736.

[11] Chagrov, A. and M. Zakharyaschev, "Modal logic," Oxford Logic Guides **35**, The Clarendon Press, New York, 1997, xvi+605 pp.

[12] Chang, C.-C. and J. H. Keisler, "Model Theory," North-Holland Publishing Co., Amsterdam-London, 1990, third edition.

[13] Darnière, L. and M. Junker, *Model completion of varieties of co-Heyting algebras*, Houston J. Math. **44** (2018), pp. 49–82.

[14] de Rijke, M., *The modal logic of inequality*, J. Symbolic Logic **57** (1992), pp. 566–584. URL `https://doi.org/10.2307/2275293`

[15] de Vries, H., "Compact spaces and compactifications. An algebraic approach," Ph.D. thesis, University of Amsterdam (1962), available at the ILLC Historical Dissertations Series (HDS-23).

[16] Dimov, G. and D. Vakarelov, *Topological representation of precontact algebras and a connected version of the Stone duality theorem—I*, Topology Appl. **227** (2017), pp. 64–101.

[17] Esakia, L., "Heyting algebras," Trends in Logic—Studia Logica Library **50**, Springer, Cham, 2019, duality theory,.

[18] Gabbay, D., *An irreflexivity lemma with applications to axiomatizations of conditions on tense frames*, in: *Aspects of philosophical logic (Tübingen, 1977)*, Synthese Library **147**, Reidel, Dordrecht-Boston, Mass., 1981 pp. 67–89.

[19] Gabbay, D. and I. Hodkinson, *An axiomatization of the temporal logic with Until and Since over the real numbers*, J. Logic Comput. **1** (1990), pp. 229–259.

[20] Ghilardi, S., *An algebraic theory of normal forms*, Ann. Pure Appl. Logic **71** (1995), pp. 189–245.

[21] Ghilardi, S., *Irreducible models and definable embeddings*, in: *Logic Colloquium '92 (Veszprém, 1992)*, Stud. Logic Lang. Inform., CSLI Publ., Stanford, CA, 1995 pp. 95–113.

[22] Ghilardi, S. and M. Zawadowski, "Sheaves, games, and model completions," Trends in Logic—Studia Logica Library **14**, Kluwer Academic Publishers, Dordrecht, 2002, x+243 pp., a categorical approach to nonclassical propositional logics.

[23] Jipsen, P., *Discriminator varieties of Boolean algebras with residuated operators*, in: C. Rauszer, editor, *Algebraic Methods in Logic and Computer Science*, Banach Center Publications **28**, Polish Academy of Sciences, 1993 pp. 239–252.

[24] Kowalski, T. and G. Metcalfe, *Uniform interpolation and coherence*, Ann. Pure Appl. Logic **170** (2019), pp. 825–841.

[25] Lipparini, P., *Locally finite theories with model companion*, , **72**, Accademia Nazionale dei Lincei, 1982.

[26] Maksimova, L. L., *Interpolation theorems in modal logics and amalgamable varieties of topological Boolean algebras*, Algebra i Logika **18** (1979), pp. 556–586, 632.

[27] Metcalfe, G. and F. Montagna, *Substructural fuzzy logics*, J. Symb. Log. **72** (2007), pp. 834–864.

[28] Shavrukov, V., *Subalgebras of diagonalizable algebras of theories containing arithmetic*, Dissertationes Mathematicae **CCCXXIII** (1993).

[29] Takeuti, G. and S. Titani, *Intuitionistic fuzzy logic and intuitionistic fuzzy set theory*, J. Symb. Log. **49** (1984), pp. 851–866.

[30] Vakarelov, D., *Region-based theory of space: algebras of regions, representation theory, and logics*, in: *Mathematical problems from applied logic. II*, Int. Math. Ser. (N. Y.) **5**, Springer, New York, 2007 pp. 267–348.

[31] van Gool, S. J., G. Metcalfe and C. Tsinakis, *Uniform interpolation and compact congruences*, Ann. Pure Appl. Logic **168** (2017), pp. 1927–1948.

[32] Venema, Y., *Derivation rules as anti-axioms in modal logic*, J. Symbolic Logic **58** (1993), pp. 1003–1034.

[33] Visser, A., *Uniform interpolation and layered bisimulation*, in: P. Hájek, editor, *Gödel 96: Logical foundations on mathematics, computer science and physics – Kurt Gödel's legacy* (1996).

[34] Wheeler, W. H., *Model-companions and definability in existentially complete structures*, Israel J. Math. **25** (1976), pp. 305–330.

Appendix

A An Admissibility Criterion

We report here the statement and the proof of Theorem 6.12 of [4], generalized to a system $\mathcal{S}$ satisfying the conditions of Section 2. We need the same series of results as in [4], starting from Theorem 6.6 of [4] (we need only a slightly simplified version of the last theorem because we do not consider proofs with assumptions):

Theorem A.1 *For every set of Π_2-rules Θ and for every formula ψ, we have that $T_{\mathcal{S}} \cup \{\Pi(\rho) \mid \rho \in \Theta\} \models \psi = \top \iff \vdash_{\mathcal{S}+\Theta} \psi$.*

Proof. The right-to-left direction is a trivial induction on the length of a proof witnessing $\vdash_{\mathcal{S}+\Theta} \psi$. For the other side, we need a modified Lindembaum construction. Suppose that $\nvdash_{\mathcal{S}+\Theta} \psi$. For each rule $\rho_i \in \Theta$, we add a countably infinite set of fresh propositional letters to the set of existing propositional letters. Then we build the Lindenbaum algebra $\mathcal{B}$ over the expanded set of propositional letters, where the elements are the equivalence classes $[\varphi]$ under provable equivalence in $\mathcal{S} + \Theta$. Next we construct a maximal $[\forall]$-filter M of $\mathcal{B}$ such that $\neg[\forall]\psi \in M$ and for every rule $\rho_i \in \Theta$

$$(\rho_i) \qquad \frac{F_i(\underline{\varphi}/\underline{x}, \underline{p}) \to \chi}{G_i(\underline{\varphi}/\underline{x}) \to \chi}$$

and formulas $\underline{\varphi}, \chi$:

(†) if $[G_i(\underline{\varphi}) \to \chi] \notin M$, then there is a tuple $\underline{p}$ such that $[F_i(\underline{\varphi}, \underline{p}) \to \chi] \notin M$.

To construct M, let $\Delta_0 := \{\neg[\forall]\varphi\}$, a consistent set. We enumerate all formulas φ as $(\varphi_k : k \in \mathbb{N})$ and all tuples $(i, \underline{\varphi}, \chi)$ where $i \in \mathbb{N}$ and $\underline{\varphi}, \chi$ are as in the particular rule ρ_i, and we build the sets $\Delta_0 \subseteq \Delta_1 \subseteq \cdots \subseteq \Delta_n \subseteq \ldots$ as follows (notice that, according to the construction below, for all n and $\theta \in \Delta_n$, we have $\vdash_{\mathcal{S}+\Theta} \theta \leftrightarrow [\forall]\theta$).

- For $n = 2k$, if $\nvdash_{\mathcal{S}+\Theta} \bigwedge \Delta_n \to [\forall]\varphi_k$, let $\Delta_{n+1} = \Delta_n \cup \{\neg[\forall]\varphi_k\}$; otherwise let $\Delta_{n+1} = \Delta_n$.
- For $n = 2k+1$, let $(l, \underline{\varphi}, \chi)$ be the k-th tuple. If $\nvdash_{\mathcal{S}+\Theta} \bigwedge \Delta_n \to (G_l(\underline{\varphi}) \to \chi)$, let $\Delta_{n+1} = \Delta_n \cup \{\neg[\forall](F_l(\underline{\varphi}, \underline{p}) \to \chi)\}$, where $\underline{p}$ is a tuple of proposi-

tional letters for ρ_l not occurring in $\underline{\varphi}, \chi$, and any of θ with $\theta \in \Delta_n$ (we can take $\underline{p}$ from the countably infinite additional propositional letters which we have reserved for the rule ρ_l). Otherwise, let $\Delta_{n+1} = \Delta_n$.

Let M be

$$\{ [\theta] \mid \text{ there are } \theta_1, \ldots, \theta_n \in \bigcup_{n\in\mathbb{N}} \Delta_i \text{ such that } \vdash_{\mathcal{S}+\Theta} \theta_1 \wedge \cdots \wedge \theta_n \to \theta\} .$$

It is clear that M is a proper $[\forall]$-filter not containing $[\psi]$.[6] Also, by the even steps of the construction of the sets Δ_n, it contains either $[[\forall]\theta]$ or $[\neg[\forall]\theta]$ for every θ, thus M is a maximal $[\forall]$-filter. Finally, the odd steps of the construction of the sets Δ_n ensure that M satisfies (†): in fact, if $[G_i(\underline{\varphi}) \to \chi] \notin M$, then by step $n = 2k+1$, we have $[\neg[\forall](F_l(\underline{\varphi}, \underline{p}) \to \chi)] \in M$ and if also $[F_i(\underline{\varphi}, \underline{p}) \to \chi] \in M$, then $[[\forall](F_i(\underline{\varphi}, \underline{p}) \to \chi)] \in M$ (because M is a $[\forall]$-filter) and so M would not be proper, a contradiction. Therefore, we can conclude that M satisfies all the desired properties.

By (†), the quotient of $\mathcal{B}$ by M satisfies each $\Pi(\rho_i)$; such a quotient is a simple algebra, because M is maximal as a $[\forall]$-filter. Moreover, since $[\neg[\forall]\psi] \in M$, we have that $[\neg[\forall]\psi]$ maps to $\top$, so $[[\forall]\psi]$ maps to $\bot$ in the quotient. Thus, $[\varphi]$ does not map to $\top$ in the quotient, and hence $T_{\mathcal{S}} \cup \{\Pi(\rho) \mid \rho \in \Theta\} \not\models \psi = \top$. □

Definition A.2 Given a quantifier-free first-order formula $\Phi(\underline{x})$, we associate with it the term (aka the propositional modal formula) $\Phi^*(\underline{x})$ as follows:

$$\begin{aligned} (t(\underline{x}) = u(\underline{x}))^* &= [\forall](t(\underline{x}) \leftrightarrow u(\underline{x})) \\ (\neg\Psi)^*(\underline{x}) &= \neg\Psi^*(\underline{x}) \\ (\Psi_1(\underline{x}) \wedge \Psi_2(\underline{x})^* &= \Psi_1^*(\underline{x}) \wedge \Psi_2^*(\underline{x}). \end{aligned}$$

The following lemma is immediate:

Lemma A.3 *Let $\mathcal{B}$ be a simple $\mathcal{S}$-algebra and let $\Phi(\underline{x})$ be a quantifier-free formula. Then we have $\mathcal{B} \models \Phi(\underline{a}/\underline{x})$ iff $\mathcal{B} \models (\Phi(\underline{a}/\underline{x}))^* = \top$, for every tuple $\underline{a}$ from $\mathcal{B}$.*

Theorem A.4 (Admissibility Criterion) *A Π_2-rule ρ is admissible in $\mathcal{S}$ iff for each simple $\mathcal{S}$-algebra $\mathcal{B}$ there is a simple $\mathcal{S}$-algebra $\mathcal{C}$ such that $\mathcal{B}$ is a substructure of $\mathcal{C}$ and $\mathcal{C} \models \Pi(\rho)$.*

Proof. ($\Rightarrow$) Suppose that the rule ρ

$$(\rho) \quad \frac{F(\underline{\varphi}/\underline{x}, \underline{p}) \to \chi}{G(\underline{\varphi}/\underline{x}) \to \chi}$$

[6] The fact that M is proper comes from the fact that $\not\vdash_{\mathcal{S}+\Theta} \bigwedge \Delta_n \to \bot$. This is clear for even n and for $n = 0$. For odd $n = 2k+1$, suppose that $\vdash_{\mathcal{S}+\Theta} \bigwedge \Delta_k \to [\forall](F_l(\underline{\varphi}, \underline{p}) \to \chi)$ and that $\not\vdash_{\mathcal{S}+\Theta} \bigwedge \Delta_k \to (G_l(\underline{\varphi}) \to \chi)$. Then, by the axiom $[\forall]\phi \to \phi$ from Section 2, we have $\vdash_{\mathcal{S}+\Theta} F_l(\underline{\varphi}, \underline{p}) \to (\bigwedge \Delta_k \to \chi)$ and also (applying the rule ρ_l of the k-th tuple) $\vdash_{\mathcal{S}+\Theta} G_l(\underline{\varphi}) \to (\bigwedge \Delta_k \to \chi)$, yielding a contradiction.

is admissible in $\mathcal{S}$. It is sufficient to show that there exists a model $\mathcal{C}$ of the theory

$$T = T_{\mathcal{S}} \cup \{\Pi(\rho)\} \cup \Delta(\mathcal{B})$$

where $\Delta(\mathcal{B})$ is the diagram of $\mathcal{B}$ [12, p. 68]. Suppose for a contradiction that T has no models, hence is inconsistent. Then, by compactness, there exists a quantifier-free first-order formula $\Psi(\underline{x})$ and a tuple $\underline{x}$ of variables corresponding to some $\underline{a} \in \mathcal{B}$ such that

$$T_{\mathcal{S}} \cup \{\Pi(\rho)\} \models \neg\Psi(\underline{a}/\underline{x}) \text{ and } \mathcal{B} \models \Psi(\underline{a}/\underline{x}).$$

By Theorem A.1, $\mathcal{S} + \rho$ is complete with respect to the simple $\mathcal{S}$-algebras satisfying $\Pi(\rho)$. Therefore, by Lemma A.3, we have $T_{\mathcal{S}} \cup \{\Pi(\rho)\} \models (\neg\Psi(\underline{x}))^* = \top$ and also $\vdash_{\mathcal{S}+\rho} (\neg\Psi(\underline{x}))^*$, where $(-)^*$ is the translation given in Definition A.2. By admissibility, $\vdash_{\mathcal{S}} (\neg\Psi(\underline{x}))^*$. Thus, for the valuation v into $\mathcal{B}$ that maps $\underline{x}$ to $\underline{a}$, we have $v((\neg\Psi(\underline{x}))^*) = 1$, so $v((\Psi(\underline{x}))^*) = 0$. This contradicts the fact that $\mathcal{B} \models \Psi(\underline{a}/\underline{x})$. Consequently, T must be consistent, and hence it has a model.

($\Leftarrow$) Suppose $\vdash_{\mathcal{S}} F(\underline{\varphi}, \underline{p}) \to \chi$ with $\underline{p}$ not occurring in $\underline{\varphi}, \chi$. Let $\mathcal{B}$ be a simple $\mathcal{S}$-algebra and let v be a valuation on $\mathcal{B}$. By assumption, there is a simple $\mathcal{S}$-algebra $\mathcal{C}$ such that $\mathcal{B}$ is a substructure of $\mathcal{C}$ and $\mathcal{C} \models \Pi(\rho)$. Let $i : \mathcal{B} \hookrightarrow \mathcal{C}$ be the inclusion. Then $v' := i \circ v$ is a valuation on $\mathcal{C}$. For any $\underline{c} \in C$, let v'' be the valuation that agrees with v' except for the fact that it maps the $\underline{p}$ into the $\underline{c}$. Since $\vdash_{\mathcal{S}} F(\underline{\varphi}/\underline{x}, \underline{p}) \to \chi$, by the algebraic completeness theorem[7] we have $v''(F(\underline{\varphi}/\underline{x}, \underline{p}) \to \chi) = \top$. This means that for all $\underline{c} \in C$, we have $F(v'(\underline{\varphi}), \underline{c}) \le v'(\chi)$. Therefore, $\mathcal{C} \models \forall \underline{y}\big(F(v'(\underline{\varphi}), \underline{y}) \le v'(\chi)\big)$. Since $\mathcal{C} \models \Pi(\rho)$, we have $\mathcal{C} \models G(v'(\underline{\varphi})) \le v'(\chi)$. Thus, as $G(v'(\underline{\varphi})) \le v'(\chi)$ holds in $\mathcal{C}$, we have that $G(v(\underline{\varphi})) \le v(\chi)$ holds in $\mathcal{B}$. Consequently, $v(G(\underline{\varphi}) \to \chi) = \top$. Applying the algebraic completeness theorem again yields that $\vdash_{\mathcal{S}} G(\underline{\varphi}) \to \chi$ because $\mathcal{B}$ is arbitrary, and hence ρ is admissible in $\mathcal{S}$. □

[7] This is Theorem A.1 for $\Theta = \emptyset$.

Spinozian Model Theory

Justin Bledin and Yitzhak Melamed

Johns Hopkins University

Abstract

This paper is an excerpt from a larger project that aims to open a new pathway into Spinoza's *Ethics* by formally reconstructing an initial fragment of this text. The semantic backbone of the project is a custom-made Spinozian model theory that lays out some of the formal prerequisites for more fine-grained investigations into Spinoza's fundamental ontology and modal metaphysics. We implement Spinoza's theory of attributes using many-sorted models with a rich system of identity that allows us to clarify the puzzling status of such logical principles as the Substitution of Identicals and Transitivity of Identity in Spinoza's thought. The intensional structure of our Spinozian models also captures his proposal that states of affairs can be necessitated or excluded by the essences of particular things, an *essence-relative* modality that should be of interest to philosophers who have sought to rehabilitate the concept of essence in contemporary analytic metaphysics.

Keywords: Spinoza, formal history of philosophy, modal logic, many-sorted model theory, logic of essence

1 Introduction

Spinoza's magnum opus *Ethica More Geometrico Demonstrata* (*Ethics* for short) is not an easy or clear work, to put it mildly.[1] To make matters worse, many commentators have complained of shoddy construction, arguing that Spinoza's logical argumentation breaks down often and early in the text (see for instance [11,10,1]). In this paper, we present part of a broader effort to reevaluate this pessimistic story of "Spinoza, the Logician" by reconstructing an initial fragment of the *Ethics* through E1p15 within the modern framework of quantified modal logic [8].

In his opening definitions and axioms, Spinoza introduces the building blocks of his ontology—substance [*substantia*], attribute [*attributum*], mode

[1] All quotations from Spinoza's works and letters are from Curley's translation [17]. We rely on Gebhardt's critical edition [16] for the Latin text. Passages in the *Ethics* are referred to using the following standard abbreviations: 'a' for axiom, 'c' for corollary, 'e' for explanation, 'p' for proposition, 's' for scholium, and 'd' for a definition when it appears immediately to the right of the part of the book or a demonstration in all other cases (so, for example, E1d1 is the first definition of Part One of the *Ethics*, and E1p15d is the demonstration of the fifteenth proposition of Part One). We use the abbreviation 'KV' for *Short Treatise on God, Man, and His Well-Being* and 'Ep.' for *Letters*.

[*modus*], and God [*Deus*]—and presents several constraints on the ontological, conceptual, and causal relations that obtain between these protagonists. In the propositions themselves, he establishes core properties of substance, such as that it is self-caused (E1p7) and infinite in its own kind (E1p8). By the time he reaches E1p15, Spinoza has already established his *substance monism*: God, a substance with an infinity of attributes (E1d6), exists (E1p11), and is unique (E1p14), and all inheres in God (E1p15). To the extent that Spinoza's demonstrations fail in this crucial early stage of the *Ethics*, this threatens the metaphysical foundations of his entire project.

The semantic backbone of our reconstruction is a custom-made Spinozian model theory, which we develop in the present installment of our work. This theory lays out some of the formal prerequisites for more fine-grained formal investigations into Spinoza's fundamental ontology and modal metaphysics. We implement Spinoza's theory of attributes using many-sorted models with a rich system of identity—our models include no less than three distinct notions of numerical identity—that allows us to clarify the puzzling status of such logical principles as the Substitution of Identicals [2,3] and Transitivity of Identity [9] in Spinoza's thought. The intensional structure of our Spinozian models also captures his proposal that states of affairs can be necessitated or excluded by the essences of particular things, an *essence-relative* modality that should be of interest to philosophers who have sought to rehabilitate the concept of essence in contemporary analytic metaphysics [5,6,7].

Given Spinoza's metaphysical views, bringing in the modern apparatus of possible worlds—and, indeed, allowing for domains consisting of multiple entities—might seem like overkill, or even plain distortion. In addition to his substance monism, we take Spinoza to establish a *necessitarianism* later in Part One according to which every actual state of affairs is necessary—things could not have been otherwise. However, it is important to keep in mind that Spinoza has to *argue* for these doctrines, and some of his main conclusions are drawn only after a lot of careful preliminary work (see E1p33 and its scholia). So, we don't want to build too much of Spinoza's metaphysics directly into our models, which must be capable of representing not only the positions that Spinoza eventually arrives at in the *Ethics* but also alternative metaphysical possibilities that he rules out through his argumentation, such as universes with multiple co-existing substances and non-necessary facts.

That said, to capture some of Spinoza's own idiosyncratic views about the universe, our Spinozian models have a few non-standard twists. We motivate these in sections 2 and 3, where we implement Spinoza's theory of attributes and essence-relative notions of possibility and conceivability—and provide more overview of Spinoza's philosophy in the process. We then present our full Spinozian model theory in §4 and conclude in §5.

2 Modeling the Attributes

At the heart of Spinoza's ontology is the distinction between substance and mode. The hallmark of substance, according to Spinoza, is its *independence*.

In his definition of substance (E1d3), he tells us that substance is "in itself [*in se*]" (i.e., inheres in itself) and "conceived through itself [*per se concipitur*]". Later, in E1p7d, Spinoza proves that substance is independent in a third sense: substance is the cause of itself and is not caused by any other thing.

In contrast, modes are by their nature *dependent* beings. In his definition of mode (E1d5), Spinoza asserts that a mode is an "affection" (roughly, a quality) of substance, and then he spells out how modes are dependent in two senses in which substance is not: a mode is "in another [*in alio*] through which it is also conceived [*per alio concipitur*]"—that is, a mode inheres in and must be understood through something other than itself. In E1p16c1, Spinoza also establishes that modes are causally dependent in that they must be caused by another, namely God.

The other two protagonists of Spinoza's metaphysical system—attributes and God—raise some pressing interpretative puzzles. In E1d4, Spinoza defines an attribute as "what the intellect perceives of substance, as constituting its essence [*id, quod intellectus de substantia percipit, tanquam ejusdem essentiam constituens*]". Spinoza then defines God in E1d6 as "a being absolutely infinite [*ens absolutè infinitum*]", which is spelled out further as "a substance consisting of an infinity of attributes, of which each one expresses an eternal and infinite essence". Unlike Descartes [4], who rules out the possibility of one substance having more than one principal attribute, Spinoza allows substances—or rather the one divine substance, God—to have multiple attributes. Of God's many attributes, we humans have access to only two: Thought and Extension (see E2a5, E2p13, and Ep. 64). But God has infinitely more attributes beyond our epistemic and causal reach (E1d6 and Ep. 56 (IV/261/14)).

A common, though oversimplified, taxonomy divides interpretations of Spinoza's attributes into two camps: subjectivist and objectivist. While the subjective position goes back to Hegel, the locus classicus is Wolfson [19] where Spinoza is taken to claim that attributes are inventions of the finite perceiving mind. Because Wolfson's reading has been subjected to devastating (and to our mind justified) critique, we mention it only to set it aside. Our own working position is objectivist, in at least the sense that we do not take attributes to be inventions of the human mind. Following Garrett [9] (and echoing Melamed [12]), we regard Spinoza as a proponent of a "strong ontological pluralism" according to which one thing can have more than one "fundamental manner or kind of existence, reality, or being". The idea that there can be more than one kind of existence—this is the "ontological pluralism" part—might not strike at least some philosophers as especially peculiar given the distinction between concrete and abstract objects, particulars and universals, or the divine and mundane. On Garrett's interpretation, however, Spinoza takes this further in proposing that a *single* thing can have existence of different kinds—this is the "strong" part. Indeed, God is a substance having infinitely many fundamental kinds of existence, each of which might be regarded as one of God's attributes. Spinoza's "thinking substance" (E2p7s) is God existing as a thinking thing, "extended substance" (E1p15s, E2p7s) is God existing as extended, and likewise for the other unknown attributes shrouded in darkness—they too should be regarded as different kinds of existence (E1p20d: "each of [God's] attributes expresses existence"). The same goes for finite things: on Spinoza's ingenious solution to the Mind-Body Problem, your mind is *you* existing as a thinking thing while your body is *you* existing as an extended thing (E2p21s, E3p2s).[2]

To formally capture Spinoza's idea that one and the same thing can have existence of many different kinds, we adopt a *many-sorted* model theory [14,15,18]. Unlike single-sorted models for modal logic, which include a set of possible worlds $\mathcal{W}$ where each world $w \in \mathcal{W}$ is assigned a single domain $\mathcal{D}(w)$ consisting of the entities that exist in this world, many-sorted models assign a potentially infinite number of domains of quantification to each world. Where $S = \{s_1, s_2, ...\}$ is an index set of *sorts*, a many-sorted model $\mathcal{M}$ for S assigns each $w \in \mathcal{W}$ a domain $\mathcal{D}_{s_1}(w)$ of existents of sort s_1, a domain $\mathcal{D}_{s_2}(w)$ of existents of sort s_2, and so forth (we work with variable domain models). Given a sort $s \in S$, the variables x_s, y_s, ... range over things of this sort and $\forall x_s$, $\exists y_s$, ... quantify over $\mathcal{D}_s(w)$. While entities in different sortal domains of a world are generally regarded as distinct, we repurpose these models to allow for one and the same thing to exist in multiple sortal domains.

For purposes of modeling Spinoza's metaphysics, we work with an infinite set of sorts that includes the sort Th of thinking entities (or rather, entities existing as thinking), the sort Ex of extended entities (or entities existing as extended), and sorts corresponding to all the other attributes. We call these "secondary sorts" because our models also include a "primary sort" $\aleph$ whose domain $\mathcal{D}_{\aleph}(w)$ at a world $w \in \mathcal{W}$ consists of all the entities having any kind of existence in w conceived in their fullness as multifaceted beings:[3]

Spinozian sorts: $\mathcal{S}_{Spinoza} = \{\aleph, \mathtt{Th}, \mathtt{Ex}, ...\}$

The $\aleph$-sort affords a bird's-eye view of a pluralistic ontology. To theorize at this global layer is *not* to substract (or abstract away) the attributes from substance and its modes because on the interpretation we work with, to strip away all the attributes of a thing would be to deny it existence of *any* kind. On the contrary, entities are regarded from the "$\aleph$-perspective" as having *all* their attributes—for instance, Spinoza adopts this all-encompassing perspective in defining God as a being consisting of an infinity of attributes in E1d6.

To help keep track of the identity of entities across different sortal domains,

[2] While Garrett doesn't emphasize the role of the intellect in all this, we take E1d4 to impose a substantive condition on fundamental kinds of existence—these kinds correspond to how substance and its modes can be perceived by the *infinite* intellect (see E2p7s). Adopting one perspective, the infinite intellect perceives God as a thinking thing (E2p1) and its various modes as modes of thought. Adopting another perspective, the infinite intellect perceives God as an extended thing (E2p2) and its modes as modes of extension.

[3] For Spinoza, substance and its modes are *infinitely*-faceted, existing as thinking, extended, and so forth. However, our models can also represent alternative universes with entities existing in only finitely many sortal domains, and perhaps only one. While we often speak of the entities in $\mathcal{D}_{\aleph}(w)$ as "multifaceted", strictly speaking they needn't be.

we assume that Spinozian models come equipped with a family of (partial) *projection functions* $\pi_{\texttt{Th}}$, $\pi_{\texttt{Ex}}$, ... that at each world $w \in \mathcal{W}$ "project" the multifaceted beings dwelling in the primary domain $\mathcal{D}_{\aleph}(w)$ of this world to single-faceted entities in its secondary sortal domains (to project a multifaceted thing is to home in on its having existence of this or that kind):

> **Projection into secondary domains**
> For any world $w \in \mathcal{W}$ and secondary sort $s \in \{\texttt{Th}, \texttt{Ex}, ...\}$, the projection function $\pi_s(w)$ maps entities from $\mathcal{D}_{\aleph}(w)$ into $\mathcal{D}_s(w)$. When defined, $\pi_s(w)(a_{\aleph})$ is $a_{\aleph}$ in w as perceived by the (infinite) intellect as a being of sort s.[4]

For instance, if '$\mathsf{God}_{\aleph}$' denotes Spinoza's absolutely infinite substance at w, then $\pi_{\texttt{Th}}(w)(\mathsf{God}_{\aleph})$ is the thinking substance (i.e., $\mathsf{God}_{\texttt{Th}}$), $\pi_{\texttt{Ex}}(w)(\mathsf{God}_{\aleph})$ is the extended substance (i.e., $\mathsf{God}_{\texttt{Ex}}$), and so forth. Because any single entity is singular under any kind of existence, and the primary domain $\mathcal{D}_{\aleph}(w)$ of a world w includes all the entities with any kind of existence in w, we require that each $\pi_s(w)$ is a one-one injective function, each member of a secondary domain $\mathcal{D}_s(w)$ is the $\pi_s(w)$-projection of some member of $\mathcal{D}_{\aleph}(w)$, and (in the other direction) every multifaceted being $a_{\aleph} \in \mathcal{D}_{\aleph}(w)$ is projected into at least one of the secondary domains of w—i.e., $\pi_s(w)(a_{\aleph})$ is defined for some $s \in \{\texttt{Th}, \texttt{Ex}, ...\}$.

Given the different sortal layers in Spinozian models, we can identify a number of (genuine) identity relations. First, there is the "standard" identity relation $=$ that any element in any domain of a model stands in with respect to itself and to no other element (i.e., $=$ is the diagonal relation):

> **Standard identity**
> $a_s = b_{s'}$ at w iff $a_s, b_{s'}$ are the same element in a model.

Second, there is what we call the "projective" identity of any multifaceted being in the $\aleph$-domain with each of its projections. For $s \in \{\texttt{Th}, \texttt{Ex}, ...\}$,

> **Projective identity**
> $a_{\aleph} =_{\mathrm{P}} b_s$ at w iff $\pi_s(w)(a_{\aleph}) = b_s$.

Third, there is the "cross-attribute" or "trans-attribute" identity of these projections. For $s, s' \in \{\texttt{Th}, \texttt{Ex}, ...\}$,

> **Cross-attribute identity**
> $a_s =_{\mathrm{C}} b_{s'}$ at w iff there is some $c_{\aleph}$ s.t. $c_{\aleph} =_{\mathrm{P}} a_s$ and $c_{\aleph} =_{\mathrm{P}} b_{s'}$ at w.

All three relations are genuine identity relations in the sense that if $a_s = b_{s'}$, $a_s =_{\mathrm{C}} b_{s'}$, or $a_s =_{\mathrm{P}} b_{s'}$, then a_s and $b_{s'}$ are *one and the same thing*, though a_s and $b_{s'}$ might still differ in terms of their kind or kinds of existence. In

[4] Spinoza rephrases his definition of attribute in E2p7s, referring to an attribute as "whatever can be perceived by an *infinite intellect* as constituting the essence of substance" (italics added). So, apparently, Spinoza has God's infinite intellect in mind in E1d4.

monistic ontologies involving only a single kind of existence, there is room for only a single notion of identity. However, in a pluralistic setting where one is referring to and quantifying over entities with more than one fundamental kind of existence, it is useful to have multiple notions of identity in play to capture more fine-grained notions of sameness and difference.

Note that the three identity relations differ with respect to their logical properties. Standard identity is an equivalence relation, as is cross-attribute identity on the secondary sortal domains where it applies. However, projective identity is neither reflexive nor symmetric, and is transitive only in a vacuous sense as we can never have $a_s =_{\mathrm{P}} b_{s'}$ and $b_{s'} =_{\mathrm{P}} c_{s''}$ for any $a_s, b_{s'}, c_{s''}$. That said, we can get failures of transitivity if we consider *combinations* of our identity relations. As discussed above, we can have $\mathsf{God}_{\aleph} =_{\mathrm{P}} \mathsf{God}_{\mathtt{Th}}$ (E2p1: "God is a thinking thing") and $\mathsf{God}_{\aleph} =_{\mathrm{P}} \mathsf{God}_{\mathtt{Ex}}$ (E2p2: "God is an extended thing"), but while $\mathsf{God}_{\mathtt{Th}} =_{\mathrm{C}} \mathsf{God}_{\mathtt{Ex}}$, $\mathsf{God}_{\mathtt{Th}} \neq \mathsf{God}_{\mathtt{Ex}}$. This transitivity failure reveals how standard identity is a stricter notion than cross-attribute identity over the secondary domains where these notions both apply. Cross-attribute identity is the appropriate notion of identity when we are counting substances and modes in the ontology but are not concerned with the distinction between different kinds of existence. From this coarse-grained perspective, $\mathsf{God}_{\mathtt{Th}}$ and $\mathsf{God}_{\mathtt{Ex}}$ are numerically identical because we are talking about one and the same substance. On the other hand, standard identity is the appropriate notion when we are counting things as distinct when they have different kinds of existence. From this more fine-grained perspective, $\mathsf{God}_{\mathtt{Th}}$ and $\mathsf{God}_{\mathtt{Ex}}$ are numerically distinct because the former is God existing as thinking while the latter is God existing as extended. [5]

Our identity relations also differ in terms of their substitutional properties. All the predicates introduced in this paper are referentially transparent contexts with respect to the standard identity relation. For instance, where '$\mathsf{Extended}(t)$' and '$\mathsf{Affection}(t,t')$' formalize *t is extended* and *t is an affection of t′*, the following conditionals hold:

If $\mathsf{Extended}(a_s)$ and $a_s = b_{s'}$, then $\mathsf{Extended}(b_{s'})$.
If $\mathsf{Affection}(a_s, b_{s'})$, $a_s = c_{s''}$, and $b_{s'} = d_{s'''}$, then $\mathsf{Affection}(c_{s''}, d_{s'''})$.

In models that accurately capture Spinoza's philosophy, many predicates will also turn out to be referentially transparent with respect to projective identity in the restricted sense that if they hold with respect to some elements in the primary domain $\mathcal{D}_{\aleph}(w)$, and these elements are all projected into the same secondary sortal domain $\mathcal{D}_s(w)$, then the predicates hold with respect to these projections as well. For instance, where '$\mathsf{Substance}(t)$' and '$t \rightsquigarrow t'$' formalize *t is a substance* and *t causes t′*, the following conditionals hold:

If $\mathsf{Substance}(a_{\aleph})$ and $a_{\aleph} =_{\mathrm{P}} b_s$, then $\mathsf{Substance}(b_s)$.
If $a_{\aleph} \rightsquigarrow b_{\aleph}$, $a_{\aleph} =_{\mathrm{P}} c_s$, and $b_{\aleph} =_{\mathrm{P}} d_s$, then $c_s \rightsquigarrow d_s$.

---

[5] See also [9] for discussion of the status of the transitivity of identity in Spinoza's philosophy.

On the other hand, many predicates will be referentially opaque contexts for projective identity in its full generality—for example, it can be the case that $\mathsf{Affection}(a_{\aleph}, b_{\aleph})$, $a_{\aleph} =_{\mathrm{P}} c_{\mathrm{Ex}}$, $b_{\aleph} =_{\mathrm{P}} d_{\mathrm{Th}}$, but $\neg\mathsf{Affection}(c_{\mathrm{Ex}}, d_{\mathrm{Th}})$. As Della Rocca observes in [2,3], "attribute contexts" like '$\mathsf{Extended}(t)$' and related attribute-sensitive predicates are also referentially opaque with respect to cross-attribute identity (though Della Rocca does not phrase his observation in these terms)—for example, we can have $\mathsf{Extended}(a_{\mathrm{Ex}})$, $a_{\mathrm{Ex}} =_{\mathrm{C}} b_{\mathrm{Th}}$, but $\neg\mathsf{Extended}(b_{\mathrm{Th}})$.

3 Modeling Possibility and Conceivability

While Spinoza's modal metaphysics remains the subject of considerable debate (see [13] for helpful discussion and references), we interpret him as a strict necessitarian. This commitment is *strongly* suggested in various places in the *Ethics*, such as in E1p29, where Spinoza proves that "in nature there is nothing contingent", and in E1p33, where he proves that "Things could have been produced by God in no other way, and in no other order than they have been produced." Even though Spinoza is a necessitarian, there are still good reasons to have multiple worlds available in our models.

First, as mentioned, Spinoza argues for his necessitarianism only in the second half of Part One—the argumentation only really gets going in E1p16, where this paper leaves off—so we don't want to presuppose this doctrine in our model theory, which should be capable of representing rival views. While any model that accurately incorporates Spinoza's modal commitments will be one in which the actual world is the sole metaphysical possibility, we allow for models that include more than one metaphysically possible world in order to represent alternative views that Spinoza rejects.

Furthermore, even in models encoding Spinoza's necessitarianism wherein actuality and metaphysical possibility coincide, there are benefits to having *metaphysically impossible* worlds around. With such worlds, we can capture Spinoza's rich modal metaphysics and more nuanced necessitarianism, which asserts not simply that things could not have been otherwise but that things could not have been otherwise *by virtue of God's essence*—the full natural order flows from the necessity of God's essence (E1p16). In the first scholium immediately following E1p33d, Spinoza goes on to distinguish between two different *sources* or *grounds* of the necessary existence/nonexistence of things: [6]

> A thing is called necessary either by reason of its essence or by reason of its cause. For a thing's existence follows necessarily either from its essence and definition or from a given efficient cause. And a thing is also called impossible from these same causes—namely, either because its essence, *or* definition, involves a contradiction, or because there is no external cause which has been determined to produce such a thing.

For Spinoza, everything that exists necessarily exists and everything that does

---

[6] Spinoza famously endorses a strong version of the Principle of Sufficient Reason according to which there is a cause or reason for the existence or nonexistence of each thing (E1p11d2).

not exist necessarily fails to exist, but there are different reasons that things are ruled into or out of existence. The existence and nonexistence of some things is necessitated by their own essence or nature. In E1p7, for instance, Spinoza proves that "it pertains to the nature of a substance to exist". As for nonexistence, there are square circles and other "Chimeras [*Chymaeram*]" that fail to exist by virtue of their essence (E1p11d2). In contrast, the existence or nonexistence of other things is due to their (external) efficient cause—thus, for example, the existence of a broken window (and the nonexistence of a non-broken window) might be due to the impact of a rock crashing through it. More generally, we can think of the essence or real definition of a thing (or things) as settling certain subject matters while leaving open how things stand with respect to other matters. Spinoza argues that God's essence necessitates the full *ordo naturae*, but the essence of any nonsubstance, taken by itself, must leave many subject matters unsettled, such as the matter of this nonsubstance's own existence (E1p24).

Though we might need as few as one possible world to represent the metaphysically possible, the abundance of worlds in our Spinozian models is helpful for modeling what is necessitated by the essences of things and what is possible relative to these essences, which for Spinoza can outstrip the metaphysically possible. At this point, it is helpful to assume that the domain assignments $\mathcal{D}_{\aleph}$, $\mathcal{D}_{\mathtt{Th}}$, $\mathcal{D}_{\mathtt{Ex}}$, ... map each world $w \in \mathcal{W}$ to sets of existents ("local domains") drawn from "global domains" $\mathfrak{D}_{\aleph}$, $\mathfrak{D}_{\mathtt{Th}}$, $\mathfrak{D}_{\mathtt{Ex}}$, ... of the respective sorts; that is, $\mathcal{D}_s(w) \subseteq \mathfrak{D}_s$ for each $s \in \mathcal{S}_{Spinoza}$. The global domains include all the things we wish to theorize about, whether existent or nonexistent, possible or impossible, conceivable or inconceivable—a global domain can even include Chimeras like square circles, mountains without valleys, and so forth. Let $\mathfrak{D}^*$ be the union of these global domains. We assume that along with the world-internal structure already introduced to implement Spinoza's theory of attributes, a Spinozian model includes an *essence function* $\mathcal{E}$ that assigns to each $a \in \mathfrak{D}^*$ the set of propositions necessitated or forced by its essence:

> **Essence-relative necessity**
> The essence function $\mathcal{E}$ maps each element $a \in \mathfrak{D}^*$ to a set of propositions $\mathcal{E}(a) \subseteq \mathcal{P}(\mathcal{W})$, where $P \in \mathcal{E}(a)$ iff P is true in virtue of a's essence.[7]

The worlds in the intersection $\bigcap \mathcal{E}(a)$, which we call the "essence set" of a, are compatible with every proposition necessitated by a's essence or nature whereas worlds outside this intersection are excluded by a's essence. Spinoza's claim in E1p7 that it pertains to the essence of a substance to exist entails that if a is a substance, then $\mathcal{E}(a)$ includes the proposition that a exists, and therefore every world in $\bigcap \mathcal{E}(a)$ is one in which a exists. In contrast, if a is a nonsubstance (i.e., a is a mode (E1p4d)), and thus, for Spinoza it exists by virtue of its efficient causes (E1p24), then $\mathcal{E}(a)$ cannot include the proposition

[7] To represent what follows from the essences of multiple things taken together, one could define essence functions on the set of nonempty subsets of $\mathfrak{D}^*$, rather than on $\mathfrak{D}^*$ itself.

that a exists, and $\bigcap \mathcal{E}(a)$ can include worlds in which a fails to exist.

As for metaphysical possibility itself, we assume that what is metaphysically possible is dependent on what is possible with respect to essences—specifically, to be metaphysically possible is to be possible relative to the essences of *all* things (in fact, for Spinoza only God's essence need be taken into account):

> **Metaphysical possibility**
> w is a metaphysical possibility iff $w \in \bigcap \mathcal{E}(a)$ for each $a \in \mathfrak{D}^*$.

Introducing the name '@' for the actual world in a model, we require that $@ \in \mathcal{W}$ be possible relative to the essence of any thing, which ensures that the actual world is metaphysically possible:

> **Actual is possible:** $@ \in \bigcap \mathcal{E}(a)$ for each $a \in \mathfrak{D}^*$.

Spinoza's claim that God's essence fixes the order of Nature can be captured by the requirement that $\bigcap \mathcal{E}(\mathsf{God}_{\aleph}) = \{@\}$. This enforces that @ is the *only* metaphysically possible world in the model. But, again, Spinoza has to argue for this position, and so we also allow for models in which God's essence leaves open multiple metaphysical possibilities.

To summarize, there are three kinds of worlds in Spinozian models. First, there are metaphysically possible worlds, such as @, which are compatible with the essences of all things. Second, there are metaphysically impossible worlds which are compatible with the essence of no thing. These worlds will not play an important role in what follows and can be disregarded.[8] Third, there are metaphysically impossible worlds which, though ruled out by the essences of all things when taken together, are nevertheless compatible with the essence of some particular thing (or things) and can therefore be used to capture how this essence leaves various subject matters unsettled. A metaphysically impossible world $w \in \mathcal{W}$ lying in the essence set $\bigcap \mathcal{E}(a)$ of some $a \in \mathfrak{D}^*$ might still be regarded as an open possibility in the restricted sense that the essence of a alone doesn't rule out this world.

This brings us to the notion of *conceivability*, which is intimately related to essence-relative modality in Spinoza's philosophy and appears in several key texts at the beginning of the *Ethics* (see for example E1d1, E1a7, E1p10s, E1p11d1, and E1p14). At least in his early period, Spinoza seems to think that conceivability amounts to the possibility of positing certain ideas in an infinite intellect (see P4 in the KV). However, we want to remain fairly noncommittal about how Spinoza understands conceivability in his later philosophy. So, we hardwire conceivability into our Spinozian models by taking them to include a *conceivability function* $\mathcal{C}$ that assigns to each $a \in \mathfrak{D}^*$ the set of propositions conceivable about it in the "narrow sense"—when attending only to its essence (see [13])—which we call the "conceivability set" of a:

---

[8] These metaphysically impossible worlds might still be regarded as *epistemically possible* (see E1p33s1). The notion of epistemic possibility is crucial to Spinoza's theory of human psychology and the supervening disciplines of ethics and political philosophy.

Conceivability
The conceivability function $\mathcal{C}$ maps each member $a \in \mathfrak{D}^*$ to a set of propositions $\mathcal{C}(a) \subseteq \mathcal{P}(\mathcal{W})$, where $P \in \mathcal{C}(a)$ iff P is conceivable about a when considering only a's essence.

Spinoza's notion of conceivability is a rich topic that requires a great deal more attention than we can offer here (see [13] for further discussion). Particularly important is the connection between what pertains to the essence of a thing and what is conceivable about it given its essence, which can be made precise using the essence and conceivability functions in our models. For instance, E1a7 requires that if the proposition that a does not exist lies in the conceivability set $\mathcal{C}(a)$ ("If a thing can be conceived as not existing..."), then the proposition that a exists is not a member of $\mathcal{E}(a)$, and the essence set $\bigcap \mathcal{E}(a)$ can include worlds in which a fails to exist ("...its essence does not involve existence"). More generally, Spinoza seems to think that if the essence of a thing necessitates certain facts about this thing, then the thing cannot be conceived in ways that conflict with these essentialist facts, and this can be spelled out as the set-theoretic constraint that every proposition in its conceivability set is compatible with its essence set.[9]

4 The Spinozian Language and Model Theory

In this section, we present (most of) the formal Spinozian language used in our project and describe its model theory. First, the language: to represent the logical forms of sentences in the initial fragment of *Ethics* up through E1p15, we adopt a language whose logical symbols include the standard sentential connectives, the actualist quantifiers '$\forall$' and '$\exists$', the possibilist quantifiers 'Π' and 'Σ', and variables indexed to every Spinozian sort. For quantificational purposes, we also help ourselves to unindexed variables 'x', 'y', ... and overlined unindexed variables '$\bar{x}$', '$\bar{y}$', ... for denoting things of any sort in $\mathcal{S}_{Spinoza}$ and of any secondary sort in $\{\mathtt{Th}, \mathtt{Ex}, ...\}$ respectively. Whereas standard first-order languages have only a single symbol for identity, we have three:

Identity symbols: '$=$', '$=_{\mathrm{P}}$', and '$=_{\mathrm{C}}$' for standard, projective, and cross-attribute identity

The Spinozian language also includes the following modal operators:

Necessity-by-essence operators: '$\Box_t$' (read: *It is necessitated by the essence of t that...*) for each term t (constant or variable) in the language

Metaphysical necessity and possibility operators: '$\Box$' (read: *It is metaphysically necessary that...*) and '$\Diamond$' (read: *It is metaphysically possible that...*)

Conceivability operators: '$\blacklozenge_t$' (read: *It is conceivable about t when conceived solely in terms of its essence that...*) for each term t

9 We do not build such correspondences between $\mathcal{E}$ and $\mathcal{C}$ directly into our Spinozian models so that axioms like E1a7 have some work to do.

The remaining logical symbols enable us to talk about sorts (notation: the sortal subscripts or lack thereof on argument positions of predicates indicate whether they can be instantiated by things of any sort (t), multifaceted things only ($t_{\aleph}$), single-faceted things only ($\bar{t}$), or things of some specific secondary sort ($t_{\mathtt{Th}}$, $t_{\mathtt{Ex}}$, ...)):

Sortal-projective predicates

$\mathsf{Same\text{-}sort}(t, t')$:	*t and t' are the same sort of thing*
$\mathsf{All\text{-}sorts}(t_{\aleph})$:	*$t_{\aleph}$ is projected into each of the infinitely many secondary sortal domains*

Turning to the nonlogical symbols of the language, we need a long laundry list of additional predicate for translating the text. Among these are the following predicates, which Spinoza defines in E1d1-E1d8:

$\mathsf{Causa\text{-}sui}(t)$:	*t is a cause of itself*
$\mathsf{Finite\text{-}in\text{-}kind}(\bar{t})$:	*$\bar{t}$ is finite in its own kind*
$\mathsf{Substance}(t)$:	*t is substance*
$\mathsf{Attribute}(\bar{t})$:	*$\bar{t}$ is an attribute*
$\mathsf{Mode}(t)$:	*t is a mode*
$\mathsf{Affection}(t, t')$:	*t is an affection of t'*
$\mathsf{God}(t_{\aleph})$:	*$t_{\aleph}$ is God*
$\mathsf{Abs\text{-}infinite}(t_{\aleph})$:	*$t_{\aleph}$ is absolutely infinite*
$\mathsf{Free}(t)$:	*t is free*
$\mathsf{Eternal}(t)$:	*t is eternal*

In addition to the predicate '$\mathsf{God}(t_{\aleph})$', the language also has the constants '$\mathsf{God}_{\aleph}$', '$\mathsf{God}_{\mathtt{Th}}$', '$\mathsf{God}_{\mathtt{Ex}}$', ... for referring directly to God, both as a multifaceted substance existing in the $\aleph$-domain and as a single-faceted substance (thinking substance, extended substance, and so on) existing in a secondary domain.

Well-formed formulae of the formal Spinozian language are generated from its lexicon through the usual grammar. We interpret these formulae relative to a pointed Spinozian model and to a variable assignment g that maps each variable of the language to some member of the corresponding global sortal domain(s)—where $\mathfrak{D}^*$ is the union of all the global domains and $\bar{\mathfrak{D}}^*$ is the union of only the global secondary domains, $g(x_s) \in \mathfrak{D}_s$, $g(x) \in \mathfrak{D}^*$, and $g(\bar{x}) \in \bar{\mathfrak{D}}^*$. Our official definition of a Spinozian model integrates the intraworld structure from section 2 with the interworld structure from section 3 and adds a function for interpreting the nonlogical symbols in the language:

Spinozian models

A many-sorted Spinozian model $\mathcal{M}$ is an ordered tuple consisting of a nonempty set $\mathcal{W}$ with designated point $@ \in \mathcal{W}$, a domain assignment $\mathcal{D}_s$ for each sort $s \in S_{Spinoza}$ mapping every world $w \in \mathcal{W}$ to a set of entities drawn from a global domain $\mathfrak{D}_s$, a projection function π_s for each secondary sort $s \in \{\mathtt{Ex}, \mathtt{Th}, ...\}$, an essence function $\mathcal{E}$, a conceivability function $\mathcal{C}$, and an interpretation function $\mathcal{I}$ mapping each constant in the language to a member of the corresponding global domain and each n-adic nonlogical predicate symbol and world w to an n-ary relation over the global domains:

a. For each constant c_s, $\mathcal{I}(c_s) \in \mathfrak{D}_s$.
b. For each nonlogical predicate P, $\mathcal{I}(P(t_1, ..., t_n), w) \subseteq \mathfrak{D}^{*n}$.

Note that we interpret constant and predicate symbols over the global domains of the model, and not just over the local domains of worlds. This keeps the model theory flexible: constants can refer to both existing and non-existing things at a world, and predicates can be instantiated by both existents and nonexistents.

The semantics for the non-modal fragment of the language is relatively standard. We first compute the extensions of terms in the usual way:

Term denotations
The denotation $[\![t]\!]_{\mathcal{M},g,w}$ of term t at w with respect to $\mathcal{M}$ and g is defined as follows:

a. $[\![c_s]\!]_{\mathcal{M},g,w} = \mathcal{I}(c_s)$.
b. $[\![x_s]\!]_{\mathcal{M},g,w} = g(x_s)$, $[\![x]\!]_{\mathcal{M},g,w} = g(x)$, and $[\![\bar{x}]\!]_{\mathcal{M},g,w} = g(\bar{x})$.

We then compositionally assign satisfaction conditions to well-formed formulae using these denotations. Starting with atomic formulae, there are three cases to consider: predications, equations, and sortal-projective claims. To evaluate an n-adic nonlogical predicate symbol applied to n terms, we check to see whether the denotations of these terms stand in the n-ary relation expressed by the predicate:

Interpretation of predications

$$\mathcal{M}, g, w \models P(t_1, ..., t_n) \quad \text{iff} \quad \langle [\![t_1]\!]_{\mathcal{M},g,w}, ..., [\![t_n]\!]_{\mathcal{M},g,w} \rangle \in \mathcal{I}(P(t_1, ..., t_n), w)$$

Standard, projective, and cross-attribute identity claims are evaluated by checking whether the denotations of terms on either side of the relevant identity symbol are identical in the senses discussed in Section 2:

Interpretation of identity claims

$$\begin{array}{lll} \mathcal{M}, g, w \models t = t' & \text{iff} & [\![t]\!]_{\mathcal{M},g,w} = [\![t']\!]_{\mathcal{M},g,w} \\ \mathcal{M}, g, w \models t_{\aleph} =_{\mathrm{P}} \bar{t} & \text{iff} & [\![t_{\aleph}]\!]_{\mathcal{M},g,w} =_{\mathrm{P}} [\![\bar{t}]\!]_{\mathcal{M},g,w} \\ \mathcal{M}, g, w \models \bar{t} =_{\mathrm{C}} \bar{t}' & \text{iff} & [\![\bar{t}]\!]_{\mathcal{M},g,w} =_{\mathrm{C}} [\![\bar{t}']\!]_{\mathcal{M},g,w} \end{array}$$

As for the sortal predicates '$\mathsf{Same\text{-}sort}(t, t')$' and '$\mathsf{All\text{-}sorts}(t_{\aleph})$', the former checks whether its arguments are of the same sort while the latter checks whether the multifaceted entity denoted by its argument is projected into each of the infinitely many secondary sortal domains:

Interpretation of sortal-projective claims

$\mathcal{M}, g, w \models \mathsf{Same\text{-}sort}(t, t')$ iff $[\![t]\!]_{\mathcal{M},g,w} \in \mathfrak{D}_s$ iff $[\![t']\!]_{\mathcal{M},g,w} \in \mathfrak{D}_s$ for each sort $s \in \mathcal{S}_{Spinoza}$

$\mathcal{M}, g, w \models \mathsf{All\text{-}sorts}(t_{\aleph})$ iff $\pi_s(w)([\![t_{\aleph}]\!]_{\mathcal{M},g,w})$ is defined for each secondary sort $s \in \{\mathtt{Th}, \mathtt{Ex}, ...\}$

Moving on to the sentential connectives, we assume that they have the classical semantics:

Interpretation of sentential connectives

$\mathcal{M}, g, w \models \neg\varphi$ iff $\mathcal{M}, g, w \not\models \varphi$

$\mathcal{M}, g, w \models \varphi \wedge \psi$ iff $\mathcal{M}, g, w \models \varphi$ and $\mathcal{M}, g, w \models \psi$

$\mathcal{M}, g, w \models \varphi \vee \psi$ iff $\mathcal{M}, g, w \models \varphi$ or $\mathcal{M}, g, w \models \psi$

$\mathcal{M}, g, w \models \varphi \rightarrow \psi$ iff $\mathcal{M}, g, w \not\models \varphi$ or $\mathcal{M}, g, w \models \psi$

$\mathcal{M}, g, w \models \varphi \equiv \psi$ iff $\mathcal{M}, g, w \models \varphi \rightarrow \psi$ and $\mathcal{M}, g, w \models \psi \rightarrow \varphi$

We also give a standard treatment of universal/existential quantification, though we have a range of quantificational options corresponding to the different quantificational domains in our models (reflected in the availability of both actualist and possibilist quantifiers and the range of variable types in the language). Starting with actualist quantification over specific local secondary domains and letting $g_{[x_s \mapsto a_s]}$ be the variant assignment that is exactly like the variable assignment g except it sends the variable x_s to a_s, we evaluate quantified statements of the form '$\forall x_s \varphi$' and '$\exists x_s \varphi$' as follows:

Interpretation of actualist quantifiers

$\mathcal{M}, g, w \models \forall x_s \varphi$ iff for all $a_s \in \mathcal{D}_s(w)$, $\mathcal{M}, g_{[x_s \mapsto a_s]}, w \models \varphi$

$\mathcal{M}, g, w \models \exists x_s \varphi$ iff for some $a_s \in \mathcal{D}_s(w)$, $\mathcal{M}, g_{[x_s \mapsto a_s]}, w \models \varphi$

Actualist quantification with unindexed variables is analogous—where $\mathcal{D}^*(w)$ is the union of all the local domains of w (i.e., the set of all existents of any sort in w) and $\bar{\mathfrak{D}}^*(w)$ is the union of only the local secondary domains (i.e., the set of all existents of any secondary sort in w), we give the following additional semantic entries:

Interpretation of actualist quantifiers (continued)

$\mathcal{M}, g, w \models \forall x \varphi$ iff for all $a \in \mathcal{D}^*(w)$, $\mathcal{M}, g_{[x \mapsto a]}, w \models \varphi$

$\mathcal{M}, g, w \models \exists x \varphi$ iff for some $a \in \mathcal{D}^*(w)$, $\mathcal{M}, g_{[x \mapsto a]}, w \models \varphi$

$\mathcal{M}, g, w \models \forall \bar{x} \varphi$ iff for all $\bar{a} \in \bar{\mathfrak{D}}^*(w)$, $\mathcal{M}, g_{[\bar{x} \mapsto \bar{a}]}, w \models \varphi$

$\mathcal{M}, g, w \models \exists \bar{x} \varphi$ iff for some $\bar{a} \in \bar{\mathfrak{D}}^*(w)$, $\mathcal{M}, g_{[\bar{x} \mapsto \bar{a}]}, w \models \varphi$

We also allow for possibilist quantification over the global domains of a model. Whereas '$\forall$' and '$\exists$' quantify over only the existing things in a world, the possibilist universal and existential quantifiers 'Π' and 'Σ' quantify over all things, whether they exist or not, and whether they are conceivable at the world of evaluation or not. We provide the following clauses for general quantified statements of the form '$\Pi x \varphi$' and '$\Sigma x \varphi$' (the remaining cases are similar):

Interpretation of possibilist quantifiers

$\mathcal{M}, g, w \models \Pi x\varphi$ iff for all $a \in \mathfrak{D}^*$, $\mathcal{M}, g_{[x\mapsto a]}, w \models \varphi$

$\mathcal{M}, g, w \models \Sigma x\varphi$ iff for some $a \in \mathfrak{D}^*$, $\mathcal{M}, g_{[x\mapsto a]}, w \models \varphi$

In some parts of the *Ethics*, Spinoza clearly has actualist quantification in mind. In others, he needs possibilist quantification. In still others it is unclear what he intends to quantify over. In our broader project, we adopt a conservative methodology and try to get by with only actualist quantification as much as possible. We introduce possibilist quantification only when it is absolutely required (starting with our treatment of E1p11d2).

The remaining entries are for the modal operators whose semantics involves the essence and conceivability functions. The essence of a thing necessitates that φ iff the proposition expressed by φ with respect to $\mathcal{M}$ and g (i.e., the set $\{w : \mathcal{M}, g, w \models \varphi\}$) lies in the essence function $\mathcal{E}$ applied to this thing:

Interpretation of necessary-by-essence operators

$\mathcal{M}, g, w \models \Box_t\varphi$ iff $\{w : \mathcal{M}, g, w \models \varphi\} \in \mathcal{E}([\![t]\!]_{\mathcal{M},g,w})$

Metaphysical necessity/possibility is necessity/possibility relative to the essences of all members of the global domains of the model:

Interpretation of metaphysical modality operators

$\mathcal{M}, g, w \models \Box\varphi$ iff for all $v \in \mathcal{W}$ s.t. $v \in \bigcap\mathcal{E}(a)$ for each $a \in \mathfrak{D}^*$, $\mathcal{M}, g, v \models \varphi$

$\mathcal{M}, g, w \models \Diamond\varphi$ iff for some $v \in \mathcal{W}$ s.t. $v \in \bigcap\mathcal{E}(a)$ for each $a \in \mathfrak{D}^*$, $\mathcal{M}, g, v \models \varphi$

Finally, it is conceivable about a thing that φ when attending to only its essence iff the proposition expressed by φ with respect to $\mathcal{M}$ and g is a member of the conceivability set of this thing:

Interpretation of conceivability operators

$\mathcal{M}, g, w \models \blacklozenge_t\varphi$ iff $\{w : \mathcal{M}, g, w \models \varphi\} \in \mathcal{C}([\![t]\!]_{\mathcal{M},g,w})$

Having recursively assigned satisfaction conditions to every formulae of the Spinozian language relative to a pointed model and variable assignment, we can next define truth for its sentences in the usual way, and we identify valid arguments as those that preserve truth in all pointed models.

5 Conclusion

With this model theory in place, one can next get down to work formally reconstructing Spinoza's demonstrations in the *Ethics* (as we have done in the full version of this project). The strict requirement of formal proof provides a powerful diagnostic tool for identifying tacit premises, redundancies, and potential errors in Spinoza's "Geometric manner". Spoiler: pace Leibniz [11], Bennett [1], and others, we find Spinoza to be a skilled logician whose deductive argumentation, for the most part, holds together remarkably well. While many of Spinoza's proofs are enthymemes that require implicit unstated premises to go through, prolonged exegetical gymnastics isn't required to fill in most of the holes, and many of the unstated premises are trivial. Our positive assessment extends to Spinoza's modal reasoning: far from being an incompetent modal logician, Spinoza operates nimbly with complex modal concepts in many of his demonstrations, which is all the more impressive given that he had nothing like the modern technology of modal logic at his disposal.

Making this case on Spinoza's behalf must be left for another occasion. But even from developing the core model theory in this paper, we hope to have already helped to undermine a common perception among philosophers and scholars of the history of philosophy that precise philosophical formalization is inconsistent with historical precision. Precise philosophy and precise history of philosophy needn't come at the expense of one another, and in the current study we strived to achieve both kinds of precision by developing a rigorous formal architecture for theorizing about Spinoza's ontology and modal metaphysics.

References

[1] Bennett, J., "A Study of Spinoza's Ethics," Hackett, Indianapolis, 1984.

[2] Della Rocca, M., *Spinoza's argument for the identity theory*, Philosophical Review **102** (1993), pp. 183–213.

[3] Della Rocca, M., "Representation and the Mind-Body Problem in Spinoza," Oxford University Press, New York, 1996.

[4] Descartes, R., "The Philosophical Writings of Descartes *(3 volumes). Translated by John Cottingham, Robert Stoothoff, and Dugald Murdoch*," Cambridge University Press, Cambridge, 1985.

[5] Fine, K., *Essence and modality*, Philosophical Perspectives **8** (1994), pp. 1–16.

[6] Fine, K., *The logic of essence*, Journal of Philosophical Logic **24** (1995), pp. 241–273.

[7] Fine, K., *Semantics for the logic of essence*, Journal of Philosophical Logic **29** (2000), pp. 543–584.

[8] Fitting, M. and R. L. Mendelsohn, "First-Order Modal Logic," Kluwer Academic Publishers, Dordrecht, 1998.

[9] Garrett, D., *The indiscernibility of identicals and the transitivity of identity in Spinoza's logic of the attributes*, in: Y. Y. Melamed, editor, *Cambridge Critical Guide to Spinoza's Ethics*, Cambridge University Press, Cambridge, 2017 pp. 12–42.

[10] Jarrett, C., *The logical structure of Spinoza's* Ethics*, part 1*, Synthese **37** (1978), pp. 15–65.

[11] Leibniz, G. W., "Philosophical Paper and Letters. *Edited by Leroy E. Loemker*," D. Reidel, Dordrecht, 1969.

[12] Melamed, Y. Y., *Spinoza's deification of existence*, Oxford Studies in Early Modern Philosophy **6** (2012), pp. 75–104.

[13] Newlands, S., "Reconceiving Spinoza," Oxford University Press, Oxford, 2018.

[14] Schmidt, A., *Über deduktive theorien mit mehreren sorten von grunddingen*, Mathematische Annalen **115** (1938), pp. 485–506.

[15] Schmidt, A., *Die zulässigkeit der behandlung mehrsortiger theorien mittels der üblichen einsortigen prädikatenlogik*, Mathematische Annalen **123** (1951), pp. 187–200.

[16] Spinoza, "Opera *(4 volumes). Edited by Carl Gebhardt*," Carl Winter, Heidelberg, 1925.

[17] Spinoza, "The Collected Works of Spinoza *(2 volumes). Edited and translated by Edwin Curley*," Princeton University Press, Princeton, 1985-2016.

[18] Wang, H., *Logic of many-sorted theories*, Journal of Symbolic Logic **17** (1952), pp. 105–116.

[19] Wolfson, H. A., "The Philosophy of Spinoza *(2 volumes)*," Harvard University Press, Cambridge, 1934.

Another Problem in Possible World Semantics

Yifeng Ding [1]

University of California, Berkeley

Wesley H. Holliday [2]

University of California, Berkeley

Abstract

In "A Problem in Possible-World Semantics," David Kaplan presented a consistent and intelligible modal principle that cannot be validated by any possible world frame (in the terminology of modal logic, any neighborhood frame). However, Kaplan's problem is tempered by the fact that his principle is stated in a language with propositional quantification, so possible world semantics for the basic modal language without propositional quantifiers is not directly affected, and the fact that on careful inspection his principle does not target the *world* part of possible world semantics—the *atomicity* of the algebra of propositions—but rather the idea of propositional quantification over a *complete* Boolean algebra of propositions. By contrast, in this paper we present a simple and intelligible modal principle, without propositional quantifiers, that cannot be validated by any possible world frame precisely because of their assumption of atomicity (i.e., the principle also cannot be validated by any atomic Boolean algebra expansion). It follows from a theorem of David Lewis that our logic is as simple as possible in terms of modal nesting depth (two). We prove the consistency of the logic using a generalization of possible world semantics known as *possibility semantics.* We also prove the completeness of the logic (and two other relevant logics) with respect to possibility semantics. Finally, we observe that the logic we identify naturally arises in the study of Peano Arithmetic.

Keywords: modal logic, Kripke incompleteness, Kripke inconsistency, atomic inconsistency, possibility semantics, algebraic semantics, Kaplan's paradox

1 Introduction

In his paper "A Problem in Possible-World Semantics" [17], written for a festschrift for Ruth Barcan Marcus, David Kaplan argued that there is "a problem in the conceptual/mathematical foundation of possible-world semantics (PWS) which threatens its use as a correct basis for doing the model theory of intensional languages" (p. 41). The problem is that certain consistent and

[1] yf.ding@berkeley.edu

[2] wesholliday@berkeley.edu

intelligible modal principles cannot be true in any possible world model. Kaplan's example is the following principle, stating that for any proposition p, it is possible that the property expressed by Q holds of p and only p (up to necessary equivalence of propositions):

$$\forall p \Diamond \forall q (Qq \leftrightarrow \Box(p \leftrightarrow q)). \tag{A}$$

For what sentential operators Q does (A) hold? As Kaplan writes:

> Perhaps, for every proposition, it is possible that it and only it is *Queried* [That is, it is asked whether it is the case that p....]. Or Perhaps not. It shouldn't really matter. There may be no operator expressible in English which satisfies (A). Still, *logic* shouldn't rule it out. (p. 43)

Yet standard possible world semantics rules out (A). For if propositions are in one-to-one correspondence with sets of possible worlds,[3] and $\Diamond\varphi$ (resp. $\Box\varphi$) is true if and only if φ is true at some (resp. all) worlds, then the truth of (A) requires that for every set P of worlds, there is a world w_P where the Q-property holds only of P. In other words, the truth of (A) requires an injective function sending every set of worlds to a world, contradicting Cantor's theorem.

Kaplan's paradox, as it has come to be called, has been much discussed (see, e.g., [19,20,27,1,24]). From our perspective, it has at least two weaknesses as a problem *for possible world semantics*. First, as (A) involves quantification over propositions in the object language, Kaplan's paradox does not pose a direct problem for possible world semantics for modal languages without propositional quantifiers. Second, even if we want propositional quantification, on careful inspection (A) does not in fact target the *world* part of possible world semantics.

To see why not, let us first consider a general algebraic semantics for propositional modal logic with propositional quantifiers as in, e.g., [12,3,4]. We expand a Boolean algebra B with a unary operation f on B. A valuation v assigns to each propositional variable an element of B as its semantic value. Semantic values are then assigned recursively to all formulas of the language with respect to v using operations on B associated with the sentential connectives. Boolean connectives are interpreted using the corresponding Boolean operations in B; the sentential operator Q is interpreted using the operation f; and $\Diamond$ (resp. $\Box$) is interpreted using the operation that sends a to $\top$ if $a \neq \bot$, and otherwise sends a to $\bot$ (resp. the operation that sends a to $\top$ if $a = \top$, and otherwise sends a to $\bot$). Finally, the most natural way to interpret the propositional quantifiers is to assume that B is a *complete* Boolean algebra and then to take the semantic value of $\forall p\varphi$ with respect to v to be the meet in B of all the semantic values of φ with respect to every valuation that differs from v at most in the semantic value it assigns to p.

This algebraic semantics does not make the crucial "world" assumption of

---

[3] *General frame* semantics, in the terminology of modal logic (see, e.g., [2, § 5.5]), is not committed to the view that every set of worlds corresponds to a proposition, so it does not fall under what we call "standard possible world semantics" here.

possible world semantics—that the algebra of propositions is *atomic*—and yet on this algebraic semantics, the semantic value of (A) must still be $\bot$.[4] Thus, (A) targets the idea of propositional quantification over a *complete* Boolean algebra of propositions. Over an incomplete Boolean algebra of propositions, there is a way of interpreting (A) as true—see [14, § 4].

In this paper, we present another problem in possible world semantics, which is not subject to the two criticisms of Kaplan's problem above. After a brief review of possible world semantics in Section 2, in Section 3 we present a simple and intelligible modal principle, without propositional quantifiers, that cannot be validated by any possible world frame precisely because of their assumption of atomicity, i.e., the principle also cannot be validated by any atomic Boolean algebra expansion. It follows from a theorem of David Lewis [18] that our logic is as simple as possible in terms of modal nesting depth (two). Using a generalization of possible world semantics known as *possibility semantics*, reviewed in Section 4, we prove the consistency of the logic in Section 5. We also prove the completeness of the logic (and two other relevant logics) with respect to possibility semantics, via completeness with respect to algebraic semantics in Section 6. In Section 7, we observe that the logic we identify naturally arises in the study of Peano Arithmetic. Finally, we conclude in Section 8 with some open questions for future research.

2 Possible World Semantics

We are interested in semantics for the following bimodal language.

Definition 2.1 Let $\mathcal{L}$ be the language generated by the following grammar:

$$\varphi ::= p \mid \neg\varphi \mid (\varphi \wedge \varphi) \mid \Box\varphi \mid Q\varphi,$$

where p belongs to a countably infinite set Prop of propositional variables. We treat the other connectives $\vee$, $\to$, and $\leftrightarrow$ as abbreviations as usual, and we define $\Diamond\varphi := \neg\Box\neg\varphi$.

According to possible world semantics, propositions (what sentences express) are in one-to-one correspondence with sets of possible worlds, as propositions are in one-to-one corresopndence with truth conditions and truth condi-

[4] For if not, then noting that the semantic value of a formula of the form $\Diamond\varphi$ is either $\bot$ or $\top$, no matter what the valuation of p is the semantic value of $\forall q(Qq \leftrightarrow \Box(p \leftrightarrow q))$ must not be $\bot$. Given that the semantic value of $\Box(p \leftrightarrow q)$ is either $\top$ or $\bot$, and it is the former iff the valuations of p and q are the same, we see that the semantic value of $Qq \leftrightarrow \Box(p \leftrightarrow q)$ is either just the semantic value of Qq in case that p and q have the same value, or it is the complement of the semantic value of Qq in case that p and q have different values. Taking the meet of them as we vary the value of q, if the value of p is $b \in B$, then the value of $\forall q(Qq \leftrightarrow \Box(p \leftrightarrow q))$ is $h(b) := f(b) \wedge \bigwedge_{b' \in B\setminus\{b\}} \neg f(b')$, and as we said, $h(b) > \bot$. However, it is also easy to see that for any $b_1 \neq b_2$ in B, $h(b_1) \wedge h(b_2) \leq f(b_1) \wedge \neg f(b_2) \wedge f(b_2) = \bot$, and given that both $h(b_1)$ and $h(b_2)$ are not $\bot$, $h(b_1) \neq h(b_2)$. Thus, we have found an antichain $C = \{h(b) \mid b \in B\}$ in B, whose cardinality is the same as the cardinality of B. But this is impossible: by the completeness of B, any subset of C has a join in B, and for any two different subsets, they have different joins, rendering the cardinality of B to be $2^{|C|} > |C|$.

tions are satisfied or not satisfied at possible worlds. On this view, *neighborhood models* [22,25,23] give us the most general way to model propositional operators, operators that do not distinguish different syntactic ways of expressing the same proposition. We review the definitions in the current bimodal setting.

Definition 2.2 A *neighborhood frame* is a tuple $\mathfrak{F} = \langle W, N_\Box, N_Q \rangle$ where:

(i) W is a nonempty set,

(ii) $N_\Box : W \to \wp(\wp(W))$ and $N_Q : W \to \wp(\wp(W))$.

A *model* based on $\mathfrak{F}$ is a pair $\mathcal{M} = \langle \mathfrak{F}, V \rangle$ where $V : \mathsf{Prop} \to \wp(W)$.

Definition 2.3 Given a model $\mathcal{M} = \langle \mathfrak{F}, V \rangle$ based on $\mathfrak{F} = \langle W, N_\Box, N_Q \rangle$, $w \in W$, and formula φ, we define $\mathcal{M}, w \vDash \varphi$ as follows:

(i) $\mathcal{M}, w \vDash p$ iff $w \in V(p)$;

(ii) $\mathcal{M}, w \vDash \neg\varphi$ iff $\mathcal{M}, w \nvDash \varphi$;

(iii) $\mathcal{M}, w \vDash (\varphi \wedge \psi)$ iff $\mathcal{M}, w \vDash \varphi$ and $\mathcal{M}, w \vDash \psi$;

(iv) $\mathcal{M}, w \vDash \Box\varphi$ iff $\{v \in W \mid \mathcal{M}, v \vDash \varphi\} \in N_\Box(w)$;

(v) $\mathcal{M}, w \vDash Q\varphi$ iff $\{v \in W \mid \mathcal{M}, v \vDash \varphi\} \in N_Q(w)$.

Moreover, for each formula φ, let $[\![\varphi]\!]^{\mathcal{M}} = \{w \in W \mid \mathcal{M}, w \vDash \varphi\}$.

Definition 2.4 A neighborhood frame $\mathfrak{F} = \langle W, N_\Box, N_Q \rangle$ *validates* a formula φ ($\mathfrak{F} \vDash \varphi$) iff for any model $\mathcal{M}$ based on $\mathfrak{F}$ and $w \in W$, $\mathcal{M}, w \vDash \varphi$.

On the logical side, we start with the definition of congruential modal logics, which can be seen as the broadest class of extensions of classical logic with propositional operators (under the assumption that formulas are logically equivalent iff they express the same proposition). For any frame $\mathfrak{F}$, $\{\varphi \in \mathcal{L} \mid \mathfrak{F} \vDash \varphi\}$ is such a logic.

Definition 2.5 A *congruential modal logic* for $\mathcal{L}$ is a set L of formulas containing all propositional tautologies and closed under modus ponens, uniform substitution, and the congruence rule for each $O \in \{\Box, Q\}$: if $\varphi \leftrightarrow \psi \in \mathsf{L}$, then $O\varphi \leftrightarrow O\psi \in \mathsf{L}$. L is *inconsistent* if $\mathsf{L} = \mathcal{L}$. For any $\Gamma \subseteq \mathcal{L}$, let $\mathsf{Cong}(\Gamma)$ be the smallest congruential modal logic extending Γ. For any congruential modal logic L and $\varphi \in \mathcal{L}$, define $\mathsf{L} + \varphi$ to be $\mathsf{Cong}(\mathsf{L} \cup \{\varphi\})$.

3 The Split Principle

Let S be the smallest congruential modal logic containing $\Box\top$ and

$$p \to \big(\Diamond(p \wedge Qp) \wedge \Diamond(p \wedge \neg Qp)\big). \qquad \text{(SPLIT)}$$

Suppose $\Box$ is the knowledge modality of an agent a. Then intuitively (SPLIT) says that if p is true, then it is compatible with a's knowledge that p is true while property Q holds of p, and it is also compatible with a's knowledge that p is true while property Q does not hold of p. For example, if we interpret Qp as Kaplan suggested, as *p is queried*, then (SPLIT) says that if p is true, then it is compatible with a's knowledge that p is true while p is queried by some agent, and it is also compatible with a's knowledge that p is true while p is not queried by some agent. We do not think that semantics should forbid an epistemic logic for reasoning about a's knowledge in which (SPLIT) is a theorem.[5] (Later we will see an arithmetic interpretation validating (SPLIT) in which $\Diamond$ has an "epistemic" reading as *consistency in Peano Arithmetic*; and before then we will see an interpretation involving future contingents after Theorem 5.2.) And yet, it is forbidden by possible world semantics:

Theorem 3.1 *No neighborhood frame validates* S.

In fact, no atomic Boolean algebra expansion validates S, but for readers more familiar with possible world semantics we first give the proof in terms of neighborhood frames (see Proposition 6.5 for the algebraic analogue).

Proof. Suppose $\mathfrak{F} = \langle W, N_\Box, N_Q \rangle$ validates S. Define a model $\mathcal{M} = \langle \mathfrak{F}, V \rangle$ such that for some $w \in W$, $V(p) = \{w\}$, so $\mathcal{M}, w \vDash p$. Then since $\mathfrak{F}$ validates (SPLIT), we have $\mathcal{M}, w \vDash \Diamond(p \wedge Qp) \wedge \Diamond(p \wedge \neg Qp)$, i.e., $[\![\neg(p \wedge Qp)]\!]^{\mathcal{M}} \notin N_\Box(w)$ and $[\![\neg(p \wedge \neg Qp)]\!]^{\mathcal{M}} \notin N_\Box(w)$. Since $V(p)$ is a singleton set, either $[\![p \wedge Qp]\!]^{\mathcal{M}} = \varnothing$ or $[\![p \wedge \neg Qp]\!]^{\mathcal{M}} = \varnothing$, so $[\![\neg(p \wedge Qp)]\!]^{\mathcal{M}} = W$ or $[\![\neg(p \wedge \neg Qp)]\!]^{\mathcal{M}} = W$. Combining the previous two steps, we have $W \notin N_\Box(w)$, which contradicts the fact that $\mathfrak{F}$ validates $\Box\top$. $\Box$

Syntactically, this logic is inconsistent with some additional principles for Q that are common in the study of specific propositional operators such as necessity and knowledge. However, these principles should not be imposed on arbitrary propositional operators (and they are even dubious for a certain notion of *querying*).

Proposition 3.2

(i) $\mathsf{S} + (Q(p \wedge q) \to Qp)$ *is inconsistent. In other words, the* Q *operator in* S *cannot be monotone.*

(ii) $\mathsf{S} + (Q(p \vee q) \to Qp)$ *is inconsistent. In other words, the* Q *operator in* S *cannot be antitone.*

(iii) *In* S*, the following two rules are derivable:*

$$\frac{\varphi \to Q\varphi}{\neg\varphi}, \qquad \frac{\varphi \to \neg Q\varphi}{\neg\varphi}.$$

(iv) *If we expand the language with propositional quantifiers and consider* $\mathsf{S}\Pi$*, the congruential extension of* S *with the standard axioms and rules for propositional quantifiers (see [3] for the axioms and rules), then* $\exists pQp$ *and hence* $\Box\exists pQp$ *are derivable.*

---

[5] One objection, suggested by a referee, is to consider p being the proposition *nothing is queried*. Then it is not plausible that it is consistent with a's knowledge that $p \wedge Qp$. Indeed, (SPLIT) only makes sense in an epistemic logic for reasoning about the knowledge of an agent a who knows that some proposition is queried (see Proposition 3.2.iv below). Once again, however, semantics should not forbid such an epistemic logic with (SPLIT) as a theorem.

Proof. In $\mathsf{S} + (Q(p \wedge q) \to Qp)$, we have the following derivation:

1 $\vdash Q(p \wedge \neg Qp) \to Qp$ [Monotonicity]

2 $\vdash ((p \wedge \neg Qp) \wedge Q(p \wedge \neg Qp)) \leftrightarrow \bot$ [Boolean reasoning]

3 $\vdash \Diamond((p \wedge \neg Qp) \wedge Q(p \wedge \neg Qp)) \leftrightarrow \bot$ [Congruence and $\Box\top$]

4 $\vdash (p \wedge \neg Qp) \to \Diamond((p \wedge \neg Qp) \wedge Q(p \wedge \neg Qp))$ [(SPLIT), Boolean reasoning]

5 $\vdash (p \wedge \neg Qp) \leftrightarrow \bot$ [From 3 and 4]

6 $\vdash \Diamond(p \wedge \neg Qp) \leftrightarrow \bot$ [Congruence and $\Box\top$]

7 $\vdash p \to \Diamond(p \wedge \neg Qp)$ [(SPLIT) and Boolean reasoning]

8 $\vdash p \leftrightarrow \bot$ [Boolean reasoning]

Clearly, then, $\mathsf{S} + (Q(p \wedge q) \to Qp)$ is inconsistent. For $\mathsf{S} + (Q(p \vee q) \to Qp)$, we have the following derivation:

1 $\vdash p \leftrightarrow ((p \wedge Qp) \vee p)$ [Boolean reasoning]

2 $\vdash Qp \leftrightarrow Q((p \wedge Qp) \vee p)$ [Congruence]

3 $\vdash Qp \to Q(p \wedge Qp)$ [Boolean reasoning and Antitonicity]

4 $\vdash ((p \wedge Qp) \wedge \neg Q(p \wedge Qp)) \leftrightarrow \bot$ [Boolean reasoning]

5 $\vdash \Diamond((p \wedge Qp) \wedge \neg Q(p \wedge Qp)) \leftrightarrow \bot$ [Congruence and $\Box\top$]

6 $\vdash (p \wedge Qp) \to \Diamond((p \wedge Qp) \wedge \neg Q(p \wedge Qp))$ [(SPLIT) and Boolean reasoning]

7 $\vdash (p \wedge Qp) \leftrightarrow \bot$ [From 5 and 6]

8 $\vdash \Diamond(p \wedge Qp) \leftrightarrow \bot$ [Congruence and $\Box\top$]

9 $\vdash p \to \Diamond(p \wedge Qp)$ [(SPLIT) and Boolean reasoning]

10 $\vdash p \leftrightarrow \bot$ [Boolean reasoning]

For the two rules, note that if $\vdash \varphi \to Q\varphi$, then $\vdash (\varphi \wedge \neg Q\varphi) \leftrightarrow \bot$. Then by the congruence of $\Diamond$ and $\Box\top$, $\vdash \Diamond(\varphi \wedge Q\varphi) \leftrightarrow \bot$. Using (SPLIT), we have $\vdash \varphi \to \Diamond(\varphi \wedge Q\varphi)$. Thus $\vdash \neg\varphi$. The derivation for the other rule is similar, using $\vdash \varphi \to \Diamond(\varphi \wedge \neg Q\varphi)$.

Finally, we derive $\exists pQp$:

1 $\vdash \neg\exists pQp \to \Diamond(\neg\exists pQp \wedge Q\neg\exists pQp) \wedge \Diamond(\neg\exists p\neg Qp \wedge \neg Q\neg\exists pQp)$ [(SPLIT)]

2 $\vdash Q\neg\exists pQp \to \exists pQp$ [Π-principles]

3 $\vdash (\neg\exists pQp \wedge Q\neg\exists pQp) \to (\neg\exists pQp \wedge \exists pQp)$ [Boolean reasoning]

4 $\vdash (\neg\exists pQp \wedge Q\neg\exists pQp) \leftrightarrow \bot$ [Boolean reasoning]

5 $\vdash \Diamond(\neg\exists pQp \wedge Q\neg\exists pQp) \leftrightarrow \bot$ [Congruence and $\Box\top$]

6 $\vdash \neg\exists pQp \to \bot$ [From 1 and 5]

7 $\vdash \exists pQp$ [Boolean reasoning]

Since we have the congruence rule and $\Box\top$, we can necessitate $\exists pQp$ and then obtain $\Box\exists pQp$. □

Remark 3.3 A referee informed us of a paper by Hansson and Gärdenfors [10] in which four bimodal axioms are identified that are (i) valid in an atomless Boolean algebra expanded with two operations for interpreting the two modalities but (ii) not valid on any neighborhood frame. The congruential logic axiomatized by these four axioms is strictly stronger than S (but weaker than the logic EST below). We will go beyond Hansson and Gärdenfors by proving the soundness and completeness of our neighborhood-inconsistent logic S—and the logics ES and EST below—with respect to *complete* Boolean algebra expansions, as well as by providing an arithmetic interpretation of EST.

4 Possibility Semantics

Below we will prove that S is consistent using a generalization of possible world semantics known as *possibility semantics* [16,11,13]. A key feature of possibility semantics is that it does not require the algebra of propositions to be atomic. The basic ideas are that (i) formulas are evaluated at partial *possibilities*, ordered by a refinement relation $\sqsubseteq$, so that $x \sqsubseteq y$ ("x refines y") implies that x settles as true (resp. false) every formula that y settles as true (resp. false) and possibly more; (ii) a formula is true (resp. false) at a possibility iff there is no refinement of the possibility that makes the formula false (resp. true); and (iii) a possibility settling a formula as false is equivalent to settling its negation as true (so it suffices to keep track of just the relation $\Vdash$ of settling true), and a possibility settling a conjunction as true is equivalent to settling both conjuncts as true. As for the modal operators, we interpret them using the neighborhood version of possibility semantics from [11, Remark 2.42] and [13] defined below.

Given a partially ordered set $\langle S, \sqsubseteq\rangle$, let $\mathcal{RO}(S, \sqsubseteq)$ be the collection of all $X \subseteq S$ that are *regular downsets* of $\langle S, \sqsubseteq\rangle$:

(i) for every $x \in X$, $\downarrow x := \{x' \in S \mid x' \sqsubseteq x\} \subseteq X$ ("persistence");

(ii) for every $x \notin X$, $\exists x' \sqsubseteq x\ \forall x'' \sqsubseteq x'\ x'' \notin X$ ("refinability").

In possibility semantics, *propositions* are regular downsets in a poset of possibilities. Below we define the analogue of neighborhood frames in possibility semantics, which differ from neighborhood frames in possible world semantics by (i) replacing the set W of worlds with a poset $\langle S, \sqsubseteq\rangle$ or possibilities and (ii) putting conditions on the neighborhood functions such that for any operator O and proposition $X \in \mathcal{RO}(S, \sqsubseteq)$, the set $\{x \in S \mid X \in N_O(x)\}$ of possibilities in which "$O(X)$ is true" is also a proposition in $\mathcal{RO}(S, \sqsubseteq)$.

Definition 4.1 A *neighborhood possibility frame* is a tuple $\mathfrak{F} = \langle S, \sqsubseteq, N_\Box, N_Q\rangle$ where:

(i) $\langle S, \sqsubseteq\rangle$ is a partially ordered set;

(ii) $N_\Box : S \to \wp(\mathcal{RO}(S, \sqsubseteq))$ and $N_Q : S \to \wp(\mathcal{RO}(S, \sqsubseteq))$ are such that for $O \in \{\Box, Q\}$:
   (a) if $X \in N_O(x)$ and $x' \sqsubseteq x$, then $X \in N_O(x')$ ("persistence");
   (b) if $X \notin N_O(x)$, then $\exists x' \sqsubseteq x\ \forall x'' \sqsubseteq x'\ X \notin N_O(x'')$ ("refinability").

A *model* based on $\mathfrak{F}$ is a pair $\mathcal{M} = \langle \mathfrak{F}, V\rangle$ where $V : \mathsf{Prop} \to \mathcal{RO}(S, \sqsubseteq)$.

Definition 4.2 Given a model $\mathcal{M} = \langle \mathfrak{F}, V \rangle$ based on $\mathfrak{F} = \langle S, \sqsubseteq, N_\Box, N_Q \rangle$, $x \in S$, and formula φ, we define $\mathcal{M}, x \Vdash \varphi$ as follows:

(i) $\mathcal{M}, x \Vdash p$ iff $x \in V(p)$;
(ii) $\mathcal{M}, x \Vdash \neg\varphi$ iff for all $x' \sqsubseteq x$, $\mathcal{M}, x' \nVdash \varphi$
(iii) $\mathcal{M}, x \Vdash (\varphi \wedge \psi)$ iff $\mathcal{M}, x \Vdash \varphi$ and $\mathcal{M}, x \Vdash \psi$;
(iv) $\mathcal{M}, x \Vdash \Box\varphi$ iff $\{y \in S \mid \mathcal{M}, y \Vdash \varphi\} \in N_\Box(x)$;
(v) $\mathcal{M}, x \Vdash Q\varphi$ iff $\{y \in S \mid \mathcal{M}, y \Vdash \varphi\} \in N_Q(x)$.

Moreover, for each formula φ, let $[\![\varphi]\!]^{\mathcal{M}} = \{x \in S \mid \mathcal{M}, x \Vdash \varphi\}$.

Lemma 4.3 *For any formula φ and model $\mathcal{M}$ based on a neighborhood possibility frame $\mathfrak{F} = \langle S, \sqsubseteq, N_\Box, N_Q \rangle$, $[\![\varphi]\!]^{\mathcal{M}} \in \mathcal{RO}(S, \sqsubseteq)$.*

Definition 4.4 A neighborhood possibility frame $\mathfrak{F} = \langle S, \sqsubseteq, N_\Box, N_Q \rangle$ *validates* a formula φ iff for any model $\mathcal{M}$ based on $\mathfrak{F}$ and $x \in S$, $\mathcal{M}, x \Vdash \varphi$.

Proposition 4.5 *Given a model $\mathcal{M} = \langle \mathfrak{F}, V \rangle$ based on $\mathfrak{F} = \langle S, \sqsubseteq, N_\Box, N_Q \rangle$, $x \in S$, and formulas φ and ψ:*

- $\mathcal{M}, x \Vdash (\varphi \vee \psi)$ *iff* $\forall x' \sqsubseteq x\ \exists x'' \sqsubseteq x'$: $\mathcal{M}, x'' \Vdash \varphi$ *or* $\mathcal{M}, x'' \Vdash \psi$;
- $\mathcal{M}, x \Vdash (\varphi \to \psi)$ *iff* $\forall x' \sqsubseteq x$, *if* $\mathcal{M}, x' \Vdash \varphi$, *then* $\mathcal{M}, x' \Vdash \psi$;
- $\mathcal{M}, x \Vdash (\varphi \leftrightarrow \psi)$ *iff* $\forall x' \sqsubseteq x$, $\mathcal{M}, x' \Vdash \varphi$ *iff* $\mathcal{M}, x' \Vdash \psi$.

5 Consistency

Our goal in this section is to show that S is consistent by constructing a possibility frame validating it. For this, we first extend S so that we can treat $\Box$ in the simplest way possible and focus on the behaviour of Q.

Definition 5.1 Let ES be the smallest congruential modal logic extending S with the following axioms:

$$\Box p \to p,\ p \to \Box\Diamond p,\ \Box p \to \Box\Box p,\ \Box(p \leftrightarrow q) \to \Box(Qp \leftrightarrow Qq).$$

Let EST be the smallest congruential modal logic extending ES by the T axiom for Q: $Qp \to p$.

Note that in ES, the first three extra axioms make $\Box$ an S5 box. The last extra axiom $\Box(p \leftrightarrow q) \to \Box(Qp \leftrightarrow Qq)$ intuitively says that if two propositions are indistinguishable by $\Box$, then their Q'ed versions are also indistinguishable by $\Box$. The reason we can further add the T axiom for Q and retain consistency [6] is, roughly speaking, that what $Qp \wedge \neg p$ means is not essential to the validity of (SPLIT). More precisely, letting $Q^*\varphi$ abbreviate $(Q\varphi \wedge \varphi)$, note that (SPLIT) is in a congruential modal logic if and only if

$$p \to (\Diamond(p \wedge Q^*p) \wedge \Diamond(p \wedge \neg Q^*p))$$

[6] We make no claim that the T axiom should hold for a particular operator Q such as *Queried*, but the stronger the logic we prove to be consistent, the stronger our result.

is also in the logic, since simply by Boolean reasoning, $p \wedge Q^*p$ is provably equivalent to $p \wedge Qp$, and $p \wedge \neg Q^*p$ is provably equivalent to $p \wedge \neg Qp$. Clearly $Q^*p \to p$ is in any congruential modal logic. Thus, Q^*p is in a sense the essential part of Qp that makes (SPLIT) valid, and $Qp \wedge \neg p$ is not relevant to the splitting of p by Q. Now we show that not only is S consistent, but in fact the stronger logic EST is consistent.

Theorem 5.2 *The logic* EST *is consistent.*

Proof. Consider the full infinite binary tree $2^{<\omega}$:

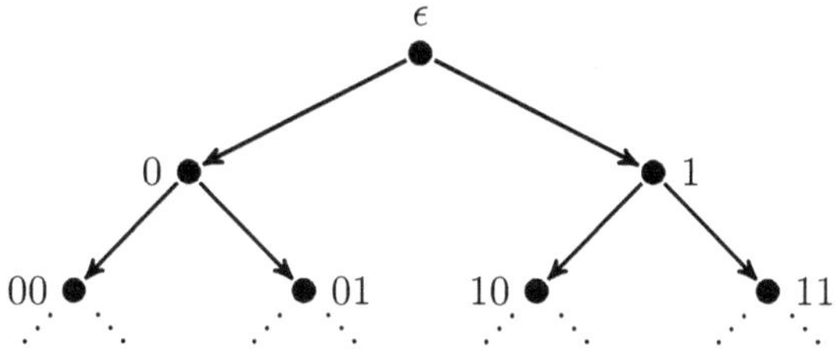

For $x \in 2^{<\omega}$, let $\mathsf{Par}(x)$ be the parent of x in the tree and $x0$ and $x1$ the two extensions of x by 0 and 1, respectively. In general, when y is an initial segment of x, we write $x \sqsubseteq y$ (refinements are lower down). To facilitate the definition of N_Q, for any $P \in \mathcal{RO}(2^{<\omega}) := \mathcal{RO}(2^{<\omega}, \sqsubseteq)$ and any $x \in 2^{<\omega}$, if $x \in P$, let $\mathsf{Firstin}(x, P)$ be the shortest initial segment of x that is in P, and otherwise let it be undefined. Since P is a downset, $\mathsf{Firstin}(x, P)$ is also the only y such that $x \sqsubseteq y$, $y \in P$, and $\mathsf{Par}(y) \notin P$. Moreover, $P = \bigcup_{x \in P} {\downarrow}\mathsf{Firstin}(x, P)$.

Now define N_Q by the following clause: for any $P \in \mathcal{RO}(2^{<\omega})$ and $x \in w^{<\omega}$,

$$P \in N_Q(x) \text{ iff } x \in P \text{ and } x \sqsubseteq \mathsf{Firstin}(x, P)0. \tag{1}$$

We can also define N_Q inductively as follows:

$$\begin{aligned} N_Q(\epsilon) &= \varnothing; \\ N_Q(x0) &= N_Q(x) \cup \{P \in \mathcal{RO}(2^{<\omega}) \mid x \in P, \mathsf{Par}(x) \notin P\}; \\ N_Q(x1) &= N_Q(x); \end{aligned}$$

but this definition is slightly harder to work with. We invite readers to verify that the inductive definition is equivalent to the definition by (1).

To show that this definition will give us a possibility frame, we claim that for any $P \in \mathcal{RO}(2^{<\omega})$, $Q(P) := \{x \in 2^{<\omega} \mid P \in N_Q(x)\} \in \mathcal{RO}(2^{<\omega})$. Pick any $P \in \mathcal{RO}(2^{<\omega})$. Now we show the two requirements for $Q(P) \in \mathcal{RO}(2^{<\omega})$.

- Suppose that $P \in N_Q(x)$ and $x' \sqsubseteq x$. By (1), $x \in P$ and $x \sqsubseteq \mathsf{Firstin}(x, P)0$. Since P is a downset, $x' \in P$. By the definition of $\mathsf{Firstin}$, clearly $\mathsf{Firstin}(x', P) = \mathsf{Firstin}(x, P)$. Hence $x' \sqsubseteq x \sqsubseteq \mathsf{Firstin}(x, P)0 = \mathsf{Firstin}(x', P)0$. Thus, $P \in N_Q(x')$. This shows that $Q(P)$ is a downset.
- Suppose that $x \notin Q(P)$, that is, $P \notin N_Q(x)$. Now we want to find a $x' \sqsubseteq x$

such that $\downarrow x' \cap Q(P) = \varnothing$. If $x \notin P$, then given that $P \in \mathcal{RO}(2^{<\omega})$, pick x' such that $x' \sqsubseteq x$ and $\downarrow x' \cap P = \varnothing$. Clearly, by the first conjunct of (1), $Q(P) \subseteq P$, and so $\downarrow x' \cap Q(P) = \varnothing$. Hence we are left with the case where $x \in P$. In this case, since $P \notin N_Q(x)$, it must be that $x \not\sqsubseteq \mathsf{Firstin}(x, P)0$. But then, for any $x' \sqsubseteq x$, $\mathsf{Firstin}(x', P) = \mathsf{Firstin}(x, P)$, and hence $x' \not\sqsubseteq \mathsf{Firstin}(x', P)0$ (note that we are in a tree here, and there can be only one path from $\mathsf{Firstin}(x, P)$ to x' through x). Thus, every $x' \sqsubseteq x$ fails the second conjunct of (1), and $\downarrow x \cap Q(P) = \varnothing$. This concludes the case where $x \in P$. Note that the above proof establishes the following:

$$\text{whenever } x \in P \text{ yet } x \notin Q(P), \downarrow x \cap Q(P) = \varnothing. \qquad (2)$$

This will be useful when we show that (SPLIT) is valid.

Now define $N_\Box$ such that for every $x \in 2^{<\omega}$, $N_\Box(x) = \{2^{<\omega}\}$. Then it is easy to see that for any $P \in \mathcal{RO}(2^{<\omega})$,

$$\Box(P) := \{x \in 2^{<\omega} \mid P \in N_\Box(x)\} = \begin{cases} 2^{<\omega} & \text{if } P = 2^{<\omega} \\ \varnothing & \text{otherwise.} \end{cases}$$

Clearly, either way, $\Box(P) \in \mathcal{RO}(2^{<\omega})$. Hence, $\mathfrak{T} := \langle 2^{<\omega}, \sqsubseteq, N_\Box, N_Q \rangle$ is a possibility frame. It is routine to verify that $\mathfrak{T}$ validates $\Box p \to p$, $p \to \Box\Diamond p$, and $\Box p \to \Box\Box p$. They all amount to discussing two cases: $V(P) = 2^{<\omega}$ and $V(P) \neq 2^{<\omega}$. It is also not hard to verify $\Box(p \leftrightarrow q) \leftrightarrow \Box(Qp \leftrightarrow Qq)$. The cases to discuss here are $V(p) = V(q)$ and $V(p) \neq V(q)$. In the former case, for all $x \in 2^{<\omega}$, $\langle \mathfrak{T}, V \rangle, x \Vdash p \leftrightarrow q$ and $\langle \mathfrak{T}, V \rangle, x \Vdash Qp \leftrightarrow Qq$. Hence the same goes for $\Box(p \leftrightarrow q)$ and $\Box(Qp \leftrightarrow Qq)$. In the case that $V(p) \neq V(q)$, $\langle \mathfrak{T}, V \rangle, \epsilon \nVdash p \leftrightarrow q$. Then trivially for any $x \in 2^{<\omega}$, $\langle \mathfrak{T}, V \rangle, x \Vdash \Box(p \leftrightarrow q) \to \Box(Qp \leftrightarrow Qq)$.

Now consider (SPLIT) $= p \to (\Diamond(p \wedge Qp) \wedge \Diamond(p \wedge \neg Qp))$. To see that this is valid, first note that $\langle \mathfrak{T}, V \rangle, x \Vdash \Diamond\varphi$ iff there exists $x' \in \mathfrak{T}$ such that $\langle \mathfrak{T}, V \rangle, x' \Vdash \varphi$. Now suppose that $\langle \mathfrak{T}, V \rangle, x \Vdash p$. This means that $x \in V(p)$. Now consider $y = \mathsf{Firstin}(x, V(p))$. Clearly, by definition, $y0 \in Q(V(p))$ and hence $\langle \mathfrak{T}, V \rangle, y0 \Vdash Qp$. Now consider $y1$. Clearly, $y1 \in V(p)$ since $V(p)$ is a downset. But $y1 \notin Q(V(p))$ since $y1 \not\sqsubseteq y0 = \mathsf{Firstin}(y1, V(p))0$. Hence $\downarrow y1 \cap Q(V(p)) = \varnothing$ by (2). Thus, $\langle \mathfrak{T}, V \rangle, y1 \Vdash \neg Qp$. By the semantics of $\Diamond$ and $\wedge$ then, $\langle \mathfrak{T}, V \rangle, x \vDash (\Diamond(p \wedge Qp) \wedge \Diamond(p \wedge \neg Qp))$. Since V and x are arbitrary, we have shown that (SPLIT) is valid on $\mathfrak{T}$. ◻

The possibility frame $\mathfrak{T}$ in the above proof can be given a natural interpretation. The partially ordered set $\langle 2^{<\omega}, \sqsubseteq \rangle$ naturally models the finitary outcomes of an infinite sequence of coin flips (say that 0 represents heads and 1 represents tails), and a crucial property is that every possibility can be further extended into two incompatible possibilities. This matches our intuitive understanding of a world with future contingencies such as random coin flips: at any time, there is at least one more coin to be flipped, and either outcome is possible.

Then our formal definition of N_Q clearly makes $Q\varphi$ express the following: φ is now true, and the first coin flipped after φ became true landed heads up.

We can also avoid temporal talk and instead speak of truth-making: φ, and the coin after the one that (exactly) makes φ true lands heads up. On this reading of Q, (SPLIT) says that if φ is true, then it is possible that φ is true and the coin after the one that makes φ true lands heads up, and it is also possible that φ is true and the coin after the one that makes φ true lands tails up.

In addition to consistency, we will prove the following completeness theorem.

Theorem 5.3 *The logic* EST *(resp.* ES, S*) is the logic of all neighborhood possibility frames that validate* EST *(resp.* ES, S*). In other words,* EST, ES, *and* S *are* possibility complete.

This will be a corollary of the completeness theorem in the next section based on algebraic semantics.

6 Completeness

In this section, we consider algebraic semantics that generalizes possible world semantics and possibility semantics. This will help us understand exactly what it takes to validate S, ES, and EST and show that they are possibility complete.

Definition 6.1 A *Boolean algebra expansion* (BAE) $\mathcal{B}$ is a triple $\langle B, \Box, Q\rangle$ where B is a Boolean algebra and $\Box$, Q are two unary functions on B. We define $\Diamond$ and other derived operations on $\mathcal{B}$ as usual. For convenience, we omit the parentheses for the argument of unary functions as appropriate.

A valuation V on $\mathcal{B}$ is function $V : \mathsf{Prop} \to \mathcal{B}$. Then the semantics for $\mathcal{L}$ is defined by extending V to $\widehat{V} : \mathcal{L} \to \mathcal{B}$ homomorphically:

- $\widehat{V}(p) = V(p)$ for $p \in \mathsf{Prop}$;
- $\widehat{V}(\neg\varphi) = \neg\widehat{V}(\varphi)$; $\widehat{V}(\varphi \wedge \psi) = \widehat{V}(\varphi) \wedge \widehat{V}(\psi)$;
- $\widehat{V}(\Box\varphi) = \Box\widehat{V}(\varphi)$; $\widehat{V}(Q\varphi) = Q\widehat{V}(\varphi)$.

To highlight the algebra whose operations are used when obtaining $\widehat{V}$ from V, especially when V may be regarded as a valuation on two different BAEs, we may write $\widehat{V}^{\mathcal{B}}$. We say that φ is *valid* on $\mathcal{B}$ if for all valuation V on $\mathcal{B}$, $\widehat{V}(\varphi) = \top$, where $\top$ is the top element of $\mathcal{B}$.

Considering the structure of the underlying Boolean algebra, we call a BAE *complete* or a $\mathcal{C}$-BAE (resp. *atomic*, an $\mathcal{A}$-BAE) if its Boolean algebra part is a complete (resp. atomic) Boolean algebra. Then $\mathcal{CA}$-BAEs are complete and atomic BAEs. On the logical side, for $X \in \{\mathcal{C}, \mathcal{A}, \mathcal{CA}\}$, we say a set of formulas Γ is X-consistent iff there is an X-BAE validating Γ, and we say that it is X-complete iff it is the logic of the class of X-BAEs validating it (cf. [21]).

From the algebraic perspective, neighborhood frames correspond to $\mathcal{CA}$-BAEs while neighborhood possibility frames corresponds to $\mathcal{C}$-BAEs. We spell this out for possibility frames, the key fact being that the regular downsets of $\langle S, \sqsubseteq\rangle$—which are just the regular open sets in the topology on S whose opens are downsets of $\langle S, \sqsubseteq\rangle$—form a complete Boolean algebra (see, e.g., [9, § 4]). For proofs and further discussion of the following facts relating neighborhood possibility frames and BAEs, see [13].

Proposition 6.2 *For any possibility frame* $\mathcal{F} = \langle S, \sqsubseteq, N_\Box, N_Q \rangle$, *let* $\mathcal{F}^\mathsf{b} = \langle \mathcal{RO}(S, \sqsubseteq), \Box, Q \rangle$ *where:*

- $\mathcal{RO}(S, \sqsubseteq)$ *is the complete Boolean algebra of regular downsets of* $\langle S, \sqsubseteq \rangle$;
- $O(P) = \{x \in S \mid P \in N_O(x)\}$ *for* $O \in \{\Box, Q\}$.

Then $\mathcal{F}^\mathsf{b}$ *is a* $\mathcal{C}$*-BAE, and any valuation* $V : \mathsf{Prop} \to \mathcal{RO}(S, \sqsubseteq)$ *on* $\mathcal{F}$ *is also a valuation on* $\mathcal{F}^\mathsf{b}$ *and vice versa. Moreover, by a simple induction, for any* φ, $[\![\varphi]\!]^{\langle \mathcal{F}, V \rangle} = \widehat{V}(\varphi)$. *Hence* $\mathcal{F}$ *validates* φ *iff* $\mathcal{F}^\mathsf{b}$ *validates* φ.

Proposition 6.3 *For any complete Boolean algebra* B, *let* $B_\bot$ *be the result of deleting* $\bot$ *from* B *and* $\leq_\bot$ *the result of restricting* $\leq$, *the lattice ordering of* B, *to* $B_\bot$. *Then* $\mathcal{RO}(B_\bot, \leq_\bot)$ *is isomorphic to* B *through the least upper bound* lub *operation. (Note that* $\mathsf{lub}(\varnothing) = \bot$.*)*

Thus, for any $\mathcal{C}$*-BAE* $\mathcal{B} = \langle B, \Box, Q \rangle$, *define* $\mathcal{B}_\mathsf{u} = \langle B_\bot, \leq_\bot, N_\Box, N_Q \rangle$ *where* $N_O(b) = \{P \in \mathcal{RO}(B_\bot, \leq_\bot) \mid b \leq O(\mathsf{lub}(P))\}$ *for* $O \in \{\Box, Q\}$. *Then* $\mathcal{B}_\mathsf{u}$ *is a possibility frame, and* $(\mathcal{B}_\mathsf{u})^\mathsf{b}$ *is isomorphic to* $\mathcal{B}$, *again through* lub. *Hence* $\mathcal{B}$ *validates a formula* φ *iff* $\mathcal{B}_\mathsf{u}$ *validates* φ.

A simple corollary of these two propositions is that a congruential modal logic is possibility complete iff it is $\mathcal{C}$-complete. Hence to show that S, ES, and EST are possibility complete, we show first that they are $\mathcal{C}$-complete. To this end, we begin by translating the two defining axioms of S into their conditions for being valid on BAEs.

Proposition 6.4 *A BAE* $\mathcal{B} = \langle B, \Box, Q \rangle$ *validates* S *iff the following hold:*

(i) $\Diamond\bot = \bot$ *and*

(ii) *for any* $b \in \mathcal{B}$, $\Diamond(b \wedge Qb) \geq b$ *and* $\Diamond(b \wedge \neg Qb) \geq b$.

A simple corollary is the following (cf. the more complicated $\mathcal{A}$-inconsistent *normal* polymodal logic in [28]).

Proposition 6.5 *If a BAE validates* S, *then it is atomless. Hence* S *is* $\mathcal{A}$*-inconsistent.*

Proof. Suppose a BAE $\mathcal{B}$ validates S. Pick any $b \in \mathcal{B}$ such that $b \neq \bot$. Then consider $b_1 = b \wedge Qb$ and $b_2 = b \wedge \neg Qb$. Since $\mathcal{B}$ validates S, by the previous proposition, we have (i) and (ii). By (ii), $\Diamond b_1 \geq b$ and $\Diamond b_2 \geq b$. Hence neither b_1 nor b_2 is $\bot$ since $\Diamond\bot = \bot$ by (i). But clearly $b_1 \vee b_2 = b$. Hence neither of them is b as otherwise the other is $\bot$. Thus, $\bot < b_1 < b$, so b is not an atom.□

Now we are already able to show that S is $\mathcal{C}$-complete.

Theorem 6.6 S *is the logic of the* $\mathcal{C}$*-BAEs validating it. Indeed, letting* $\mathcal{H} = \langle H, \Box, Q \rangle$ *be the Lindenbaum algebra of* S *and* H^+ *the MacNeille completion of the Boolean algebra* H, *there is a way to extend* $\Box$ *and* Q *to* $\Box^+$ *and* Q^+ *on* H^+ *such that* $\mathcal{H}$ *is a subalgebra of* $\mathcal{H}^+ = \langle H^+, \Box^+, Q^+ \rangle$ *(so that* $\mathcal{H}^+$ *refutes all formulas not in* S*) and* $\mathcal{H}^+$ *still validates* S.

Proof. Let $\mathcal{H} = \langle H, \Box, Q \rangle$ be the Lindenbaum algebra of S. By standard algebraic logical theory, $\mathcal{H}$ validates S, and for every $\varphi \notin$ S, there is a valuation

V_φ on $\mathcal{H}$ such that $\widehat{V_\varphi}(\varphi) \neq \top$. Since $\mathcal{H}$ validates S, by Proposition 6.5, H is atomless. Now let H^+ be the MacNeille completion of H, which is the unique (up to isomorphism) complete Boolean algebra with H being a dense subalgebra of it (in the sense that for every $b \in H^+$ such that $\bot < b$, there is a $b' \in H$ such that $\bot < b' \leq b$). (See Chap. 25 of [7] for more.) Clearly then H^+ is also atomless. Now we extend $\Box$ and Q to H^+. First, note that for any $b \in H^+ \setminus H$, there exist $b_1, b_2 \in H^+ \setminus H$ such that $b = b_1 \vee b_2$. To find such b_1 and b_2, first by density pick an $a \in H$ such that $\bot < a < b$ (note that $b \notin H$ and hence $\bot < b$). Then $b' = b \wedge \neg a$ must not be in H since otherwise $b = a \vee b'$ would also be in H. Now that H is atomless, pick $a_1, a_2 \in H \setminus \{\bot\}$ such that $a = a_1 \vee a_2$. Then let $b_1 = a_1 \vee b'$ and $b_2 = a_2 \vee b'$. Clearly $b = b_1 \vee b_2$. To see that $b_1 \notin H$, note that if it is in H, then $b' = b_1 \wedge \neg a_1$ must also be in H, contradicting that $b' \notin H$. The same reasoning applies to b_2. To fix the construction of b_1 and b_2, we can first fix an enumeration of H, which is countable, and then pick a and a_1, a_2 by going through this enumeration.

Now we define $\Box^+$ and Q^+ by the following:

$$\Box^+ b = \begin{cases} \Box b & \text{if } b \in H \\ \bot & \text{if } b \in H^+ \setminus H, \end{cases} \qquad Q^+ b = \begin{cases} Qb & \text{if } b \in H \\ b_1 & \text{if } b \in H^+ \setminus H. \end{cases}$$

Then it is easy to see by the construction of b_1 and b_2 that for every $b \in H^+ \setminus H$, $b \wedge Q^+ b$ (which is just b_1) and $b \wedge \neg Q^+ b$ (which is just b_2) are also in $H^+ \setminus H$. Also, $\Diamond^+ b := \neg \Box^+ \neg b = \top$ for all $b \in H^+ \setminus H$ since $b \notin H$ iff $\neg b \notin H$. Hence for any $b \in H^+ \setminus H$, $\Diamond^+(b \wedge Q^+ b) = \Diamond^+(b \wedge \neg Q^+ b) = \top \geq b$. Thus, by a simple discussion by cases, Proposition 6.4 applies, and $\mathcal{H}^+$ validates S. By construction, $\mathcal{H}$ is a subalgebra of $\mathcal{H}^+$. So $\mathcal{H}^+$ does not validate any formula not in S since $\mathcal{H}$ does not. Therefore, S is the logic of $\mathcal{H}^+$, a $\mathcal{C}$-BAE. □

The above strategy by MacNeille completion applies almost identically to ES and EST except that we need to focus on simple S5 algebras, those BAEs such that the $\Box$ operator essentially tests whether a proposition is $\top$ or not, so that the $\Box^+$ defined in the above proof does not destroy the validity of the $\mathsf{S5}$ axioms. To this end, we first need the following definitions.

Definition 6.7 Let $\mathcal{B} = \langle B, \Box, Q \rangle$ be a BAE. Then:

- $\mathcal{B}$ is *simple S5* if for any $b \in B$, if $b = \bot$ then $\Diamond b = \bot$, and otherwise $\Diamond b = \top$;
- $\mathcal{B}$ is *splitting* if for any $b \in B$, if $b \neq \bot$, then $b \wedge Qb \neq \bot$ and $b \wedge \neg Qb \neq \bot$;
- $\mathcal{B}$ is *deflationary* if $Qb \leq b$ for all $B \in \mathcal{B}$;
- $\mathcal{B}$ is *properly deflationary* if it is both splitting and deflationary; note that this is equivalent to: $\bot < Qb < b$ for all $b \in \mathcal{B} \setminus \{\bot\}$ and $Q\bot = \bot$.

Proposition 6.8 *A simple S5 BAE validates* ES (*resp.* EST) *iff it is also splitting* (*resp. properly deflationary*).

Proof. Let $\mathcal{B}$ be a simple S5 BAE. Then automatically $\mathcal{B}$ validates the $\mathsf{S5}$

axioms for $\Box$ and also the axiom $\Box(p \leftrightarrow q) \to \Box(Qp \leftrightarrow Qq)$, since for any valuation V on $\mathcal{B}$, $\widehat{V}(\Box(p \leftrightarrow q))$ is either $\top$ or $\bot$. If it is $\bot$, the axiom is trivially evaluated to $\top$. If it is $\top$, then $\widehat{V}(p \leftrightarrow q) = \top$, and hence $V(p) = V(q)$. Then $\widehat{V}(Qp) = \widehat{V}(Qq)$ and hence $\widehat{V}(\Box(Qp \leftrightarrow Qq))$ is also $\top$.

For (SPLIT), it is enough to see that $\widehat{V}(\Diamond(p \wedge Qp))$ (resp. $\widehat{V}(\Diamond(p \wedge \neg Qp))$) is either $\top$ or $\bot$, and it is the former iff $V(p) \wedge QV(p) \neq \bot$ (resp. $V(p) \wedge \neg QV(p) \neq \bot$). Then the validity of (SPLIT) translates to the condition that $\mathcal{B}$ is splitting by a simple discussion of whether $V(p) = \bot$.

For the axiom $Qp \to p$, clearly it is valid iff $\mathcal{B}$ is deflationary. $\Box$

Theorem 6.9 ES *is complete with respect to the class of all simple S5 splitting $\mathcal{C}$-BAEs.* EST *is complete with respect to the class of all simple S5 properly deflationary $\mathcal{C}$-BAEs.*

Proof. Let L be either ES or EST. Then take an arbitrary $\delta \notin \mathsf{L}$. We need to find a simple S5 splitting $\mathcal{C}$-BAE that refutes δ, and in the case that $\mathsf{L} = \mathsf{EST}$, the algebra should also be deflationary.

Consider the Lindenbaum algebra $\mathcal{H}$ of L, with $[\cdot]$ the function that sends formulas to their equivalence classes under the provable equivalence relation in L. Since $\delta \notin \mathsf{ES}$, $[\delta] \neq \top_{\mathcal{H}}$. Let $\mathcal{U}$ be an ultrafilter of the Boolean algebra base of $\mathcal{H}$ that does not contain $[\delta]$. Now define $\sim$ on $\mathcal{H}$ by $[\varphi] \sim [\psi]$ iff $\Box(\varphi \leftrightarrow \psi) \in \mathcal{U}$. This is well defined because if both $\varphi \leftrightarrow \varphi'$ and $\psi \leftrightarrow \psi'$ are in $\mathsf{ES} \subseteq \mathsf{L}$, then $\Box(\varphi \leftrightarrow \psi) \leftrightarrow \Box(\varphi' \leftrightarrow \psi')$ is also in $\mathsf{ES} \subseteq \mathsf{L}$. More importantly, $\sim$ is a congruence relation because in L we have the following theorems, with the last being a defining axiom:

- $(\Box(\varphi \leftrightarrow \psi) \wedge \Box(\varphi' \leftrightarrow \psi')) \to \Box((\varphi \wedge \varphi') \leftrightarrow (\psi \wedge \psi'))$;
- $\Box(\varphi \leftrightarrow \psi) \to \Box(\neg\varphi \leftrightarrow \neg\psi)$;
- $\Box(\varphi \leftrightarrow \psi) \to \Box(\Box\varphi \leftrightarrow \Box\psi)$;
- $\Box(\varphi \leftrightarrow \psi) \to \Box(Q\varphi \leftrightarrow Q\psi)$.

Hence we can take the quotient $\mathcal{S} = \mathcal{H}/{\sim}$. Let π be the quotient map for $\sim$, and let V be the composition of π after $[\cdot]$. Now we make three claims:

(i) $\mathcal{S}$ validates L. It is a standard exercise to show that $\mathcal{H}$ validates L. Since $\mathcal{S}$ is a quotient of $\mathcal{H}$, $\mathcal{S}$ also validates L.

(ii) $\mathcal{S}$ is a simple S5 algebra. For this, we just need to show that if $b \in \mathcal{S}$ is not $\top_{\mathcal{S}}$, then $\Box_{\mathcal{S}} b = \bot_{\mathcal{S}}$. This is again standard using the S5 axioms.

(iii) $V|_{\mathsf{Prop}}$ is a valuation on $\mathcal{S}$, $V = \widehat{V|_{\mathsf{Prop}}}$, and $V(\delta) \neq \top_{\mathcal{S}}$.

By Proposition 6.5 and 6.8, we know then that $\mathcal{S}$ is atomless and splitting. Thus $\mathcal{S}$ is a simple S5 splitting algebra that refutes δ by V, and moreover if $\mathsf{L} = \mathsf{EST}$, $\mathcal{S}$ is also deflationary. Thus, all that is left to do is to complete $\mathcal{S}$ while preserving the three properties: being simple S5, splitting, and deflationary (if $\mathcal{S}$ is deflationary). For this, write $\mathcal{S} = \langle S, \Box, Q \rangle$, and let S^+ be the MacNeille completion S. Then pick a function $j : S^+ \to S^+$ such that for every non-bottom $b \in S^+$, $\bot < j(b) < b$. Such a j exists since $\mathcal{S}$ and hence S^+ are atomless. In fact, since S is dense in S^+ by the construction of MacNeille completion, $j(b)$ can be picked in S according to an enumeration of S (note that S is countable). Then define $\mathcal{S}^+ = \langle S^+, \Box^+, Q^+ \rangle$ by

$$\Box^+ b = \begin{cases} \top & \text{if } b = \top \\ \bot & \text{otherwise,} \end{cases} \qquad Q^+ b = \begin{cases} Qb & \text{if } b \in S \\ j(b) & \text{if } b \in S^+ \setminus S. \end{cases}$$

Then clearly:

- $\mathcal{S}$ embeds into $\mathcal{S}^+$ by the identity map, and hence V_{Prop} is also a valuation on $\mathcal{S}^+$ and $V = \widehat{V|_{\mathsf{Prop}}}$;
- $\mathcal{S}^+$ is a simple S5 splitting algebra since $b \wedge j(b), b \wedge \neg j(b) > \bot$;
- if $\mathcal{S}$ is deflationary, meaning that $Qb \leq b$ for all $b \in S$, then $\mathcal{S}^+$ is also deflationary, since $j(b) \leq b$ for all $b \in S^+ \setminus S$ as well;
- $\mathcal{S}^+$ is complete.

Hence δ is refuted by V on $\mathcal{S}^+$, a simple S5 splitting $\mathcal{C}$-BAE that is deflationary if $\mathcal{L} = \mathsf{EST}$. □

An important observation about the two proofs of the $\mathcal{C}$-completeness of S, ES, and EST is that the refuting $\mathcal{C}$-BAEs we constructed are very special: their Boolean reducts are all (isomorphic to) the MacNeille completion of the countable atomless Boolean algebra, since the Lindenbaum algebra of S and the quotients of the Lindenbaum algebra of ES and EST are all countable (since the language we started with is countable) and atomless (since they all validate S). Let us call this special complete Boolean algebra B_{mca}. Then we can say that S, ES, and EST are not just $\mathcal{C}$-complete, but also B_{mca}-complete. A corollary of this is that these three logics are not just possibility complete but also complete with respect to possibility frames based on the full infinite binary tree $2^{<\omega}$.

To see this, we observe that just as $\mathcal{C}$-completeness and possibility completeness are equivalent by Propositions 6.2 and 6.3, B_{mca}-completeness and $2^{<\omega}$-completeness are also equivalent. The following proposition is the core of this new equivalence.

Proposition 6.10 *$\mathcal{RO}(2^{<\omega})$ is (isomorphic to) the MacNeille completion of the countable atomless Boolean algebra.*

Proof. Given the defining property of MacNeille completion, it is enough to see that there is a dense subalgebra of $\mathcal{RO}(2^{<\omega})$ that is countable and atomless. The subalgebra generated by principal downsets (i.e., downsets of the form $\{x \in 2^{<\omega} \mid x \sqsubseteq s\}$ for $s \in 2^{\omega}$) is such a subalgebra. □

With the above proposition, we can state the analogues of Proposition 6.2 and Proposition 6.3.

Proposition 6.11 *For any possibility frame $\mathcal{F} = \langle 2^{<\omega}, N_\Box, N_Q \rangle$ based on $2^{<\omega}$, $\mathcal{F}^{\mathsf{b}}$ is of the form $\langle B_{mca}, \Box, Q \rangle$, a BAE based on B_{mca}.*

Proposition 6.12 *For any BAE $\langle B_{mca}, \Box, Q \rangle$ based on B_{mca}, define neigh-*

borhood functions $N_\Box$ and N_Q on $2^{<\omega}$ by the following clause where σ is any isomorphism from $\mathcal{RO}(2^{<\omega})$ to B_{mca}: for any $O \in \{\Box, Q\}$, $s \in 2^{<\omega}$, and $X \in \mathcal{RO}(2^{<\omega})$, $X \in N_O(s)$ iff $s \in \sigma^{-1}(O(\sigma(X)))$. Then, $(\langle 2^{<\omega}, N_\Box, N_Q\rangle)^{\mathsf{b}} \cong \langle B_{mca}, \Box, Q\rangle$ with σ being the isomorphism.

Thus, a logic is complete with respect to neighborhood possibility frames based on $2^{<\omega}$ iff it is complete with respect to BAEs based on B_{mca}. This completes the proof of the following strengthening of Theorem 5.3.

Theorem 6.13 *The logic* S *(resp.* ES, EST*) is the logic of all neighborhood possibility frames based on $2^{<\omega}$ that validate* S *(resp.* ES, EST*).*

Now that we have seen that S, which is defined by two very simple axioms, is consistent and $\mathcal{C}$-complete yet $\mathcal{A}$-inconsistent, we briefly comment on whether we may have a logic that is also consistent and $\mathcal{A}$-inconsistent but is defined by even simpler axioms. Recall that given a set Γ of formulas, $\mathsf{Cong}(\Gamma)$ is the smallest congruential modal logic containing Γ. Now let $\mathsf{BAE}(\Gamma)$ be the class of BAEs validating Γ. Then the following theorem is due to Lewis [18].

Theorem 6.14 (Lewis) *For every set Γ of formulas of modal depth at most 1,* $\mathsf{Cong}(\Gamma)$ *is complete with respect to all finite BAEs in* $\mathsf{BAE}(\Gamma)$*. Since finite BAEs are all complete and atomic,* $\mathsf{Cong}(\Gamma)$ *is $\mathcal{CA}$-complete.*

Hence, $\{\Box\top, (\textsc{Split})\}$ is optimal in terms of modal depth: depth 2. We can also show that it is optimal in terms of the number of propositional variables used: just 1. For this, let the language $\mathcal{L}$ now include the propositional constant $\top \notin \mathsf{Prop}$ such that for any valuation V, $\widehat{V}(\top) = \top$ on any BAE. Then we have the following simple theorem.

Theorem 6.15 *If $\Gamma \subseteq \mathcal{L}$ contains only formulas that do not use any propositional variable in* Prop*, then* $\mathsf{Cong}(\Gamma)$ *is $\mathcal{CA}$-complete.*

Proof. Let Γ be a set of variable-free formulas and $\mathsf{L} = \mathsf{Cong}(\Gamma)$. L is trivially $\mathcal{CA}$-complete if it is inconsistent. Hence we assume that it is consistent. Consider $\mathcal{H} = \langle H, \langle \nabla_i\rangle_{i\le n}\rangle$, the Lindenbaum algebra of L (here we do not assume that $\mathcal{L}$ has only $\Box$ and Q as modalities). Let $\mathcal{H}^+$ be the BAE where its Boolean base H^+ is the canonical extension of H, and its operations ∇_i^+ are defined by

$$\nabla_i^+(\boldsymbol{a}) = \begin{cases} \nabla_{\mathcal{L}}(\boldsymbol{a}) & \text{if } \boldsymbol{a} \in H^{\text{arity}(\nabla_i)} \\ \top & \text{otherwise.} \end{cases}$$

Then let V_1 be the constantly $\top$ valuation V_1 on $\mathcal{H}^+$, which is also a valuation on $\mathcal{H}$. Since by construction $\mathcal{H}$ is a subalgebra of $\mathcal{H}^+$, $\widehat{V_1}^{\mathcal{H}^+} = \widehat{V_1}^{\mathcal{H}}$. In particular, for any $\varphi \in \Gamma$, $\widehat{V_1}^{\mathcal{H}^+}(\varphi) = \widehat{V_1}^{\mathcal{H}}(\varphi) = \top$ since φ is valid on $\mathcal{H}$. But a variable-free formula is valid iff it is evaluated to $\top$ in any valuation. So all formulas in Σ are still valid on $\mathcal{H}^+$. Since $\mathcal{H}$ is a subalgebra of $\mathcal{H}^+$, formulas that are invalid in $\mathcal{H}$ are still invalid in $\mathcal{H}^+$. Hence the validities of $\mathcal{H}^+$ are precisely $\mathsf{Cong}(\Gamma)$. Since H^+ is a canonical extension, $\mathcal{H}^+$ is a $\mathcal{CA}$-BAE. Hence $\mathsf{Cong}(\Gamma)$ is $\mathcal{CA}$-complete. □

However, $\{\Box\top, (\textsc{Split})\}$ is not optimal in terms of the number of modal operators used. Peter Fritz in his presentation [6] of his paper [5] defined the unimodal logic $\mathsf{Uni3}$, the smallest congruential modal logic containing the following axioms:

$$(\Box\top \wedge p) \leftrightarrow \Box(\Box\top \to (p \wedge \Box(\Box\top \wedge p))) \tag{\textsc{Uni3Ax1}}$$
$$(\Box\top \wedge p) \leftrightarrow \Box(\Box\top \to (p \wedge \neg\Box(\Box\top \wedge p))) \tag{\textsc{Uni3Ax2}}$$
$$\Box\Box\Box\top \tag{\textsc{Uni3Ax3}}$$
$$\neg\Box\bot. \tag{\textsc{Uni3Ax4}}$$

It can be shown that $\mathsf{Uni3}$ is consistent yet $\mathcal{A}$-inconsistent. Hence, an open problem here is whether there is a consistent yet $\mathcal{A}$-inconsistent logic that can be axiomatized by using only 1 modal operator, 1 (or more) propositional variables, and modal depth 2. It is also not known whether $\mathsf{Uni3}$ is $\mathcal{C}$-complete.

7 Split in Peano Arithmetic

In this section, we show how EST arises naturally in the study of Peano Arithmetic and in particular the problem of uniform density [26]. Following [26], define the following sequence of subtheories of PA:

$$\mathsf{AR}_0 = \mathrm{I}\Delta_0 + \mathsf{Exp}, \quad \mathsf{AR}_{n+1} = \mathrm{I}\Sigma_{n+1}.$$

Recall that Exp is the formula stating the totality of the exponential function defined by a Δ_0 formula (see [8, p. 299]), $\mathrm{I}\Delta_0$ is Peano Arithmetic with the induction schema applied only to Δ_0 formulas, and $\mathrm{I}\Sigma_{n+1}$ is Peano Arithemetic with the induction schema only applied to Σ_{n+1} formulas. For each n, AR_{n+1} extends AR_n, with their union being the usual PA. AR_0 is also known as *elementary arithmetic* (EA). These theories are uniformly recursively axiomatized. Hence there is a formula with two free variables, $\mathsf{Prov}(x, y)$, such that $\mathsf{Prov}(n, \ulcorner\varphi\urcorner)$ expresses "φ is provable in AR_n" in PA. For convenience, let $\mathsf{Pr}_x\varphi$ stand for $\mathsf{Prov}(x, \ulcorner\varphi\urcorner)$. Then define $\mathsf{Q}\varphi$ to be

$$\varphi \wedge \forall x(\mathsf{Pr}_x(\varphi \to \mathsf{Pr}_x(\varphi \to \bot)) \to \mathsf{Pr}_x(\varphi \to \bot)).$$

If we write $\mathsf{Con}_n\varphi$ for "φ is consistent in AR_n", then $\mathsf{Q}\varphi$ can be equivalently defined as

$$\varphi \wedge \forall x(\mathsf{Con}_x\varphi \to \mathsf{Con}_x(\varphi \wedge \mathsf{Con}_x\varphi)).$$

For example, $\mathsf{Q}\top$ is equivalent to the formula $\forall x(\mathsf{Con}_x\top \to \mathsf{Con}_x\mathsf{Con}_x\top)$, which intuitively says that for every n, if AR_n is consistent, then it is consistent in AR_n that the system AR_n is consistent. While it sounds trivial to us, PA is not able to prove or disprove this $\mathsf{Q}\top$. The following two lemmas are shown in [26] (note that their notation is C_φ instead of $\mathsf{Q}\varphi$).

Lemma 7.1 ([26], Lemma 3.4) *For any φ and ψ, if $\mathsf{PA} \vdash \varphi \leftrightarrow \psi$, then $\mathsf{PA} \vdash \mathsf{Q}\varphi \leftrightarrow \mathsf{Q}\psi$.*

Lemma 7.2 ([26], Lemma 3.5) *If φ is consistent in* PA*, then both* $\varphi \wedge \mathsf{Q}\varphi$ *and* $\varphi \wedge \neg\mathsf{Q}\varphi$ *are consistent in* PA.

Then it follows immediately from Proposition 6.8 that the logic of this arithmetic Q is at least EST.

Theorem 7.3 *Let H be the Lindenbaum algebra of* PA*. Let $\Box$ be defined on H by $\Box[\varphi] = [\top]$ if $\mathsf{PA} \vdash \varphi$ and $\Box[\varphi] = [\bot]$ otherwise. Define Q on H by $Q[\varphi] = [\mathsf{Q}\varphi]$. Then $\langle H, \Box, Q\rangle$ validates* EST.

Thus, (SPLIT) is not only consistent and intelligible but even has a natural arithmetic interpretation.

8 Conclusion

As with other results showing that certain modal logics are incomplete with respect to possible world semantics but complete with respect to more general semantics (see [15] and references therein), we take the results of this paper to be more positive than negative, as they lead to interesting new questions for the foundations of modal logic. We conclude by mentioning a few questions.

First, on the more philosophical side, we would like to identify more modal operators for which (SPLIT) is intuitively valid. We think that the study of truth-makers or counterfactuals is the most promising path. A related question: as we have shown in Theorem 5.2, monotonicity is inconsistent with S, but what other principles are inconsistent with S? Answering this question will help us narrow down possible interpretations of the $\Box$ and Q operators validating S.

On the more technical side, a first question is whether there are congruential extensions of S (or ES or EST) that are not $\mathcal{C}$-complete. This is essentially a test of how widely applicable our method of MacNeille completion is in proving completeness with respect to $\mathcal{C}$-BAEs. A further question is which extensions of S (or ES or EST) are *tree complete*, that is, complete with respect to a class of possibility frames whose underlying posets of possibilities are trees or even finitely branching trees. We have seen in Theorem 6.13 that the three logics, S, ES, and EST, are all tree complete (indeed, $2^{<\omega}$-complete). But the general picture for congruential logics extending these logics is not clear. It may also be interesting to see how we can axiomatize the logic of the possibility frame based on $2^{<\omega}$ defined in the proof of Theorem 5.2.

Finally, as we mentioned in Section 6, it remains to be seen whether there is a consistent but $\mathcal{A}$-inconsistent congruential modal logic axiomatized using 1 modality, 1 propositional variable, and modal nesting depth 2.

Acknowledgements

A version of this paper was presented as a Philosophy Colloquium talk at UC San Diego on October 19, 2018. We thank the audience for helpful feedback. We also thank the three anonymous referees for AiML for valuable comments.

References

[1] Bacon, A., J. Hawthorne and G. Uzquiano, *Higher-order free logic and the Prior-Kaplan paradox*, Canadian Journal of Philosophy **46** (2016), pp. 493–541.

[2] Blackburn, P., M. de Rijke and Y. Venema, "Modal Logic," Cambridge University Press, New York, 2001.

[3] Ding, Y., *On the logics with propositional quantifiers extending* S5Π, in: G. Bezhanishvili, G. D'Agostino, G. Metcalfe and T. Studer, editors, *Advances in Modal Logic, Vol. 12*, College Publications, London, 2018 pp. 219–235.

[4] Ding, Y., *On the logic of belief and propositional quantification* (2020), UC Berkeley Working Paper in Logic and the Methodology of Science.
URL `https://escholarship.org/uc/item/7476g21w`

[5] Fritz, P., *Post completeness in congruential modal logics*, in: L. D. Beklemishev, S. Demri and A. Maté, editors, *Advances in Modal Logic, Vol. 11*, College Publications, London, 2016 pp. 288–301.

[6] Fritz, P., *Post completeness in congruential modal logics (slides)* (2016).
URL `http://phil.elte.hu/aiml2016/downloads/slides/fritz.pdf`

[7] Givant, S. and P. Halmos, "Introduction to Boolean algebras," Springer Science & Business Media, 2008.

[8] Hájek, P. and P. Pudlák, "Metamathematics of first-order arithmetic," Cambridge University Press, Cambridge, 2017.

[9] Halmos, P. R., "Lectures on Boolean Algebras," D. Van Norstrand Company, Inc., Princeton, 1963.

[10] Hansson, B. and P. Gärdenfors, *A guide to intensional semantics*, in: S. Halldén, editor, *Modality, Morality and Other Problems of Sense and Nonsense: Essays dedicated to Sören Halldén*, CWK Gleerup Bokförlag, Lund, 1973 pp. 151–167.

[11] Holliday, W. H., *Possibility frames and forcing for modal logic (February 2018)* (2018), UC Berkeley Working Paper in Logic and the Methodology of Science.
URL `https://escholarship.org/uc/item/0tm6b30q`

[12] Holliday, W. H., *A note on algebraic semantics for S5 with propositional quantifiers*, Notre Dame Journal of Formal Logic **60** (2019), pp. 311–332.

[13] Holliday, W. H., *Possibility semantics*, in: J. Derakhshan, M. Fitting, D. Gabbay, M. Pourmahdian, A. Rezus and A. S. Daghighi, editors, *Research Trends in Contemproary Logic*, College Publications, London, Forthcoming .

[14] Holliday, W. H. and T. Litak, *One modal logic to rule them all?*, in: G. Bezhanishvili, G. D'Agostino, G. Metcalfe and T. Studer, editors, *Advances in Modal Logic, Vol. 12*, College Publications, London, 2018 pp. 367–386.

[15] Holliday, W. H. and T. Litak, *Complete additivity and modal incompleteness*, The Review of Symbolic Logic **12** (2019), pp. 487–535.

[16] Humberstone, L., *From worlds to possibilities*, Journal of Philosophical Logic **10** (1981), pp. 313–339.

[17] Kaplan, D., *A problem in possible world semantics*, in: W. Sinnott-Armstrong, D. Raffman and N. Asher, editors, *Modality, morality, and belief: essays in honor of Ruth Barcan Marcus*, Cambridge University Press, Cambridge, 1995 pp. 41–52.

[18] Lewis, D., *Intensional logics without iterative axioms*, Journal of Philosophical Logic **3** (1974), pp. 457–466.

[19] Lewis, D., "On the Plurality of Worlds," Basil Blackwell, Oxford, 1986.

[20] Lindström, S., *Possible world semantics and the liar*, in: A. Rojszczak, J. Cachro and G. Kurczewski, editors, *Philosophical Dimensions of Logic and Science*, Springer, Dordrecht, 1999 pp. 297–316.

[21] Litak, T., "An Algebraic Approach to Incompleteness in Modal Logic," Ph.D. thesis, Japan Advanced Institute of Science and Technology (2005).

[22] Montague, R., *Universal Grammar*, Theoria **36** (1970), pp. 373–398.

[23] Pacuit, E., "Neighborhood Semantics for Modal Logic," Springer, Cham, 2017.

[24] Priest, G., *Paradoxical propositions*, Philosophical Issues **28** (2018), pp. 300–307.

[25] Scott, D., *Advice on modal logic*, in: K. Lambert, editor, *Philosophical Problems in Logic: Some Recent Developments*, D. Reidel Publishing Company, Dordrecht, 1970 pp. 143–173.
[26] Shavrukov, V. Y. and A. Visser, *Uniform density in lindenbaum algebras*, Notre Dame Journal of Formal Logic **55** (2014), pp. 569–582.
[27] Uzquiano, G., *Modality and paradox*, Philosophy Compass **10** (2015), pp. 284–300.
[28] Venema, Y., *Atomless varieties*, The Journal of Symbolic Logic **68** (2003), pp. 607–614.

A circular proof system for the hybrid μ-calculus

Sebastian Enqvist [1]

Stockholm University, Department of Philosophy
Universitetsvägen 10 D
Frescati, Stockholm

Abstract

We present a cut-free circular proof system for the hybrid μ-calculus, and prove its soundness and completeness. The system is an adaptation of a circular proof system for the modal μ-calculus due to Stirling, and uses a system of annotations to keep track of fixpoint unfoldings. The language considered here extends the μ-calculus with nominals and satisfaction operators, but not the converse modality. This version of the hybrid μ-calculus is known to have the finite model property, unlike the version that includes converse. The presence of nominals and satisfaction operators causes some non-trivial difficulties to deal with in the completeness proof. In particular we need to be careful about what information attached to nominals to keep and what to discard, and furthermore the structure of traces in a proof-tree becomes more complicated. Still, it turns out that the proof system is complete with the same global condition for validity as Stirling's system. The key tool that we develop for the completeness proof is a proof-search game, in which one of the players attempts to construct a proof in a restricted normal form making use of certain derived rules. We conclude the paper with some tasks for future research, which include proving completeness of a cut-free non-circular sequent calculus, and extending the system developed here to incorporate converse modalities.

Keywords: Hybrid logic, μ-calculus, circular proofs, completeness, automata

1 Introduction

Circular and non-wellfounded proofs are a powerful method for reasoning with fixpoints, and have been considered in a number of contexts [19,6,22,3,4,21]. For the modal μ-calculus, a circular proof system with names for keeping track of fixpoint unfoldings was developed by Stirling [23], building on work by Jungteerapanich [12] and bearing similarities with earlier systems using variables for ordinal approximations [6]. Recently Stirling's system has been simplified and used by Afshari and Leigh to give a cut-free complete sequent system for the

[1] This research was supported by the Swedish Research Council grant 2015-01774.

modal μ-calculus [2]. This provides a novel completeness proof for Kozen's axiomatization [15] that avoids the intricate detour via disjunctive normal forms in Walukiewicz's proof [27]. Building on this work, circular proofs were used in [8] to settle the open problem of completeness of Parikh's logic of games [17].

The present work is intended as a step towards exploring the use of circular proofs to provide complete finitary proof systems for richer *extensions* of the modal μ-calculus. A number of such extensions have been presented in the literature, including the two-way or "full" μ-calculus [26], hybrid μ-calculus [20] and guarded fixpoint logic [11]. In many cases such extensions remain decidable. However, complete proof systems mostly appear to be lacking. Some work in this area does exist: a generic completeness result for coalgebraic versions of the μ-calculus (including extensions like the graded μ-calculus) was presented in [10]. This general result does not cover the hybrid μ-calculus however, since the sort of global conditions that are expressible in hybrid logics are out of scope for the framework in [10]. An infinitary proof system for the two-way μ-calculus was proved complete in [1].

As a proof of concept, we shall develop a cut-free Stirling-style circular proof system for the hybrid μ-calculus. Orignally introduced by Sattler and Vardi in [20], the hybrid μ-calculus features *nominals*, which are used to name points in a model, and *satisfaction operators* that describe what is true at a named point in a model. We shall follow Tamura [24] by not including converse modalities in our language, unlike Sattler and Vardi. Tamura shows that the hybrid μ-calculus without converse has the finite model property, unlike the more expressive version considered by Sattler and Vardi. We also mention a version of the hybrid μ-calculus involving a binder modality, which was investigated in [13] under the name "fully hybrid μ-calculus". This logic is undecidable, and therefore seems out of scope for the kind of methods that we consider here.

The presence of nominals and satisfaction operators already presents some non-trivial challenges for the completeness proof, and addressing these difficulties gives some guidelines on how to deal with proof theory for fixpoint logics that lack the tree model property. In a manner of speaking, we are continuing here along Sattler and Vardi's line of working with logics that lack the tree model property "as if they had the tree model property" [20], but taking the idea in a proof-theoretic direction.

Proofs have been removed or shortened due to page limitations. For a longer version of this paper including detailed proofs, see the preprint availabe online at `https://arxiv.org/abs/2001.04971`.

2 Preliminaries

2.1 The hybrid μ-calculus

The hybrid μ-calculus was initially introduced by Sattler and Vardi in [20]. Their version of the language included a global modality and converse modalities. Here, we shall be considering the weaker version of the hybrid μ-calculus that was studied by Tamura in [24]. For ease of notation we consider the language with only a single box and diamond, but all the results and proofs presented here easily extend to a multi-modal version of the language.

The language $\mathcal{L}$ of the hybrid μ-calculus is given by the following grammar:

$$\varphi := p \mid \neg p \mid \mathsf{i} \mid \neg \mathsf{i} \mid \varphi \vee \varphi \mid \varphi \wedge \varphi \mid \Diamond \varphi \mid \Box \varphi \mid \mathsf{i}{:}\varphi \mid \mu x.\varphi \mid \nu x.\varphi$$

Here, p and x are members of a fixed countably infinite supply Prop of propositional variables, and i comes from a fixed countably infinite supply Nom of nominals. For $\eta x.\varphi$ with $\eta \in \{\mu, \nu\}$, we impose the usual constraint that no occurrence of x in φ is in the scope of a negation, and we also require that each occurrence of x in φ is within the scope of some modality ($\Box$ or $\Diamond$). This latter extra constraint means that we restrict attention to *guarded* formulas. This is a fairly common assumption, and it is well known that removing the constraint of guardedness does not increase the expressive power of the language. It is not an entirely innocent assumption however, since putting a formula in its guarded normal form may cause an exponential blow-up in the size of a formula [5].

Note also that the language is presented in negation normal form. It is routine to verify, given the semantics presented below, that the language is semantically closed under negation, and furthermore there is a simple effective procedure for converting formulas in the extended language with explicit negation of all formulas into formulas in negation normal form.

Free and bound variables of a formula are defined in the usual manner. A *literal* is a formula of the form p or $\neg p$ where $p \in \mathsf{Prop}$, or of the form i or $\neg \mathsf{i}$ where $\mathsf{i} \in \mathsf{Nom}$. We introduce the following abbreviations:

$$\mathsf{i} \approx \mathsf{j} := \mathsf{i}{:}\mathsf{j} \qquad \mathsf{i} \not\approx \mathsf{j} := \mathsf{i}{:}\neg \mathsf{j}$$

These formulas express identity and non-identity, respectively, of the values assigned to the nominals i, j in a model.

Definition 1 *Let φ be any formula in $\mathcal{L}$ and let $x, y \in \mathsf{Prop}$ be bound variables in φ. We say that y is* dependent *on x, written $x <_\varphi y$, if there is a subformula of φ of the form $\eta y.\psi$ in which there is a free occurrence of x. We denote the reflexive closure of $<_\varphi$ by $\leq_\varphi$.*

Definition 2 *We say that a formula φ is* locally well-named *if $<_\varphi$ is irreflexive, no variable occurs both free and bound in φ, and no variable is bound by both μ and ν in φ.*

Note that every formula is equivalent to a locally well-named one up to renaming of bound variables (α-equivalence).

Proposition 2.1 (Afshari & Leigh -17) *If $\eta x.\varphi(x)$ is locally well-named then so is $\varphi(\eta x.\varphi)$.*

Convention 1 *We shall assume throughout the paper that all formulas are locally well-named. Given a locally well-named formula we refer to a bound variable x as a μ-variable if it is bound (only) by μ in φ, and a ν-variable if it is bound (only) by ν.*

Semantics of the hybrid μ-calculus is a simple extension of the usual Kripke semantics for the modal μ-calculus.

Definition 3 *A* Kripke model *is a tuple* $\mathcal{M} = (W, R, V, A)$ *where* W *is a non-empty set members of which will be referred to as* points, $R \subseteq W \times W$ *is the* accessibility relation *over* W, $V : \mathsf{Prop} \to \mathcal{P}(W)$ *is a* valuation *of the propositional variables and* $A : \mathsf{Nom} \to W$ *is an* assignment *of a value in* W *to each nominal.*

Given a Kripke model $\mathcal{M} = (W, R, V, A)$, the interpretation $[\![\varphi]\!]_{\mathcal{M}}$ of a formula φ is defined by the usual recursive clauses for boolean connectives and modalities. Semantics of least fixpoint operators is given according to the Knaster-Tarski Theorem [14,25] as:

$$[\![\mu x.\varphi(x)]\!]_{\mathcal{M}} := \bigcap\{Z \subseteq W \mid [\![\varphi]\!]_{\mathcal{M}[Z/x]} \subseteq Z\},$$

where $\mathcal{M}[Z/x]$ is like $\mathcal{M}$ except that its valuation maps the variable x to Z. For greatest fixpoint operators we have the dual definition:

$$[\![\nu x.\varphi(x)]\!]_{\mathcal{M}} := \bigcup\{Z \subseteq W \mid Z \subseteq [\![\varphi]\!]_{\mathcal{M}[Z/x]}\}.$$

For nominals and satisfaction operators, we have the following clauses: $[\![\mathsf{i}]\!]_{\mathcal{M}} = \{A(\mathsf{i})\}$ and $[\![\mathsf{i} : \varphi]\!]_{\mathcal{M}} = \{w \in W \mid A(\mathsf{i}) \in [\![\varphi]\!]\}$. In other words, $[\![\mathsf{i} : \varphi]\!]_{\mathcal{M}} = W$ if $A(\mathsf{i}) \in [\![\varphi]\!]$, and $[\![\mathsf{i} : \varphi]\!]_{\mathcal{M}} = \emptyset$ otherwise. Given a formula φ and a pointed Kripke model $(\mathcal{M}, w)$ (a model with a distinguished point), we write $\mathcal{M}, w \Vdash \varphi$ to say that $w \in [\![\varphi]\!]_{\mathcal{M}}$.

This semantics may be referred to as the *denotational* semantics of the μ-calculus. The μ-calculus also has an *operational* semantics in the form of a game semantics, which is often easier to work with and neatly captures the intuitive meaning of least and greatest fixpoints (i.e. "finite looping" vs "infinite looping"). In this game semantics it is convenient to work with the (Fischer-Ladner) *closure* $c(\varphi)$ of a formula. The precise definition is a straightforward adaptation of that in [20], with the new clause that $\mathsf{i}:\theta \in c(\varphi)$ imples $\theta \in c(\varphi)$.

Throughout the paper we assume familiarity with basic notions concerning board games and parity games (see [9] for a very brief introduction). Given a Kripke model $\mathcal{M} = (W, R, V, A)$, the *evaluation game* for a formula $\rho \in \mathcal{L}$ in the model $\mathcal{M}$ is a two-player board game between players **Ver**, **Fal**, the set of positions of which is $W \times c(\rho)$, with player assignments and moves defined as follows:

- For a position of the form (w, l) where l is a literal, the set of available moves is $\emptyset$. The position is assigned to **Fal** if $\mathcal{M}, w \Vdash l$ and is assigned to **Ver** otherwise.
- For a position of the form $(w, \varphi O \psi)$ where $O \in \{\wedge, \vee\}$, the available moves are (w, φ) and (w, ψ). The position is assigned to **Ver** if $O = \vee$ and is assigned to **Fal** if $O = \wedge$.
- For a position of the form $(w, O\varphi)$ where $O \in \{\Diamond, \Box\}$, the set of available moves is $\{(v, \varphi) \in W \mid wRv\}$. The position is assigned to **Ver** if $O = \Diamond$ and is assigned to **Fal** if $O = \Box$.

- For a position of the form $(w, \mathsf{i} : \varphi)$, the unique avaliable move is $(A(\mathsf{i}), \varphi)$. The player assignment is arbitrary in this case since there is only one move, but as a convention we assign such positions to player **Ver**.
- For a position of the form $(w, \eta x.\varphi(x))$, the unique available move is $(w, \varphi(\eta x.\varphi))$. By convention we assign such positions to **Ver**.

Partial plays, full plays and strategies for players are defined as usual. Note that if a full play is finite, then the player to which the last position is assigned must be "stuck", i.e. the set of available moves is empty. So the winning condition of finite full plays is defined by declaring the player who got stuck to be the loser of the play. For infinite plays $(w_1, \varphi_1)(w_2, \varphi_2)(w_3, \varphi_3)...$, say that a fixpoint variable x is *unfolded* at the index i if φ_i is of the form $\eta x.\psi(x)$.

Proposition 2.2 *For any (locally well-named) formula ρ and any infinite play π in the evaluation game in $\mathcal{M}$, there is a unique $<_\rho$-minimal variable x that is unfolded infinitely many times on π.*

We shall often refer to the $<_\rho$-minimal variable unfolded infinitely often on π as the *highest ranking* variable that is unfolded infinitely often. We can now define the winning condition of infinite plays: the winner is **Ver** if the highest ranking variable that gets unfolded infinitely often is a ν-variable (relative to ρ), and the winner is **Fal** otherwise.

A strategy is called *positional* if it only depends on the last position of a play, i.e. it can be described as a choice function from positions to available moves. Since the evaluation game is a parity game, and parity games have positional determinacy [7,28], we have:

Proposition 2.3 *The evaluation game of any formula in a model is determinate, and the winning player at any given position has a positional winning strategy.*

As expected the operational semantics agrees with the denotational one:

Proposition 2.4 *Given a pointed Kripke model $(\mathcal{M}, w)$ and a formula ρ, we have $\mathcal{M}, w \Vdash \rho$ if and only if the position (w, ρ) is winning for **Ver** in the evaluation game.*

3 Infinite proofs

In this section we define an infinite sequent-style proof system **Inf** for the hybrid μ-calculus. This proof system will be used as a tool to prove completeness of the finite circular proof system that will be introduced in Section 5.1. The infinite system presented here is essentially dual to an infinite tableau system for the hybrid μ-calculus. An important difference from the tableaux developed by Sattler and Vardi in [20] is that the system is cut-free, which is required since the finitary circular system we shall present later will also be cut-free. Sattler and Vardi's automata-theoretic approach relies on "guessing" all the relevant information about some nominals at the start of the tableau construction. In the dual setting of sequent calculi this amounts to starting the proof construction with a series of cuts.

3.1 The system Inf

We will work with a sequent style proof system, where a sequent is a finite set of formulas interpreted as an implicit disjunction. It will be convenient to require that every formula in a sequent starts with some satisfaction operator, so each sequent has the form:

$$\mathsf{i}_1 : \varphi_1, ..., \mathsf{i}_n : \varphi_n$$

This means that our proof system will only prove formulas of this shape. However, this is not a serious restriction: given a formula φ that is not in the required format, we can always replace it with the formula $\mathsf{i}:\varphi$ where i is some arbitrarily chosen, fresh nominal not appearing in φ. Clearly $\mathsf{i}:\varphi$ is then semantically valid if and only if φ is, and we may regard any proof of $\mathsf{i}:\varphi$ as a proof of φ.

The system has two axioms, which are the law of exluded middle and an identity axiom:

$$\mathsf{i}:p, \mathsf{i}:\neg p \qquad\qquad \mathsf{i} \approx \mathsf{i}$$

Here, p is a nominal or a propositional variable. Rules of inference are given in Figure 1. We remark that, in the modal rule Mod, the nominal j must be fresh, i.e. it cannot appear in any formula in the conclusion of the rule.

$$\frac{\Gamma, \mathsf{i}:\varphi \wedge \psi, \mathsf{i}:\varphi \qquad \Gamma, \mathsf{i}:\varphi \wedge \psi, \mathsf{i}:\psi}{\Gamma, \mathsf{i}:\varphi \wedge \psi}\wedge \qquad\qquad \frac{\Gamma, \mathsf{i}:\varphi \vee \psi, \mathsf{i}:\varphi, \mathsf{i}:\psi}{\Gamma, \mathsf{i}:\varphi \vee \psi}\vee$$

$$\frac{\Gamma, \mathsf{i}:\eta x.\phi(x), \mathsf{i}:\phi(\eta x.\phi(x))}{\Gamma, \mathsf{i}:\eta x.\phi(x)}\eta x$$

$$\frac{\Gamma, \mathsf{i}:\phi, \mathsf{j}:\phi, \mathsf{i} \not\approx \mathsf{j}}{\Gamma, \mathsf{i}:\phi, \mathsf{i} \not\approx \mathsf{j}}\mathsf{Eq} \qquad \frac{\Gamma, \mathsf{i}:(\mathsf{j}:\varphi), \mathsf{j}:\varphi}{\Gamma, \mathsf{i}:(\mathsf{j}:\varphi)}\mathsf{Glob} \qquad \frac{\Gamma, \mathsf{i} \not\approx \mathsf{j}, \mathsf{j} \not\approx \mathsf{i}}{\Gamma, \mathsf{i} \not\approx \mathsf{j}}\mathsf{Com}$$

$$\frac{\Gamma, \mathsf{i}:\Box\varphi, \mathsf{i}:\Diamond\Psi, \mathsf{j}:\varphi, \mathsf{j}:\Psi}{\Gamma, \mathsf{i}:\Box\varphi, \mathsf{i}:\Diamond\Psi}\mathsf{Mod} \qquad\qquad \frac{\Gamma}{\Gamma \cup \Psi}\mathsf{Weak}$$

Fig. 1. Rules of **Inf**

In an application of the modal rule as shown in Figure 1, we refer to $\mathsf{i}:\Box\varphi$ as the principal formula. The expression $\mathsf{i}:\Diamond\Psi$ is short-hand for $\{\mathsf{i}:\Diamond\psi \mid \psi \in \Psi\}$, and likewise $\mathsf{j}:\Psi$ abbreviates $\{\mathsf{j}:\psi \mid \psi \in \Psi\}$. The intuition behind the modal rule is that, if the formulas $\Box\varphi, \Diamond\psi_1, ..., \Diamond\psi_n$ are all false at a point named i, then this must be witnessed by some point that we can give an arbitrary name j, and at which all the formulas $\varphi, \psi_1, ..., \psi_n$ are false. In an application of the rule Eq as shown in the figure, the formula $\mathsf{i}:\phi$ is called the principal formula and $\mathsf{i} \not\approx \mathsf{j}$ the *side formula*. In all other cases where a notion of principal formula makes sense, it should be clear from the form of the rules what the principal formula is. Note that we do allow that the premises and conclusion of a rule application are all the same sequent.

Definition 4 *A rule application is said to be* repeating *if all premises are equal to the conclusion.*

Definition 5 *A* **Inf**-proof, *or proof-tree, is a ranked labelled tree where the label of a node specifies the sequent appearing at the node, the rule application of which the node is the conclusion (if any), and the principal formula (if any), and such that the labels of children of a node are the premises of the specified rule application.*

We shall often abuse terminology slightly by referring to the sequent appearing at a node in a proof as the label of the node. To distinguish *valid* proofs from invalid ones, we need a notion of trace.

Definition 6 *A* partial trace $\boldsymbol{t}$ *(of length $k \leq \omega$) on a branch β of an* ***Inf****-proof Π is a sequence $(u_j, \mathsf{i}_j : \psi_j)_{j<k}$ such that for each j, u_j is a node on β whose label contains $\mathsf{i}_j : \psi_j$, u_{j+1} is the unique successor of u_j in β whenever $j+1 < k$, and one of the following conditions holds if $j+1 < k$:*

(i) $\mathsf{i}_j : \psi_j = \mathsf{i}_{j+1} : \psi_{j+1}$. *We sometimes refer to such parts of traces as "silent steps".*

(ii) $\mathsf{i}_j : \psi_j = \mathsf{i}_j : (\theta_1 \vee \theta_2)$ *is the principal formula in an application of the $\vee$-rule, and $\mathsf{i}_{j+1} : \psi_{j+1} \in \{\mathsf{i}_j : \theta_1, \mathsf{i}_j : \theta_2)\}$.*

(iii) $\mathsf{i}_j : \psi_j = \mathsf{i}_j : (\theta_1 \wedge \theta_2)$ *is the principal formula in an application of the $\wedge$-rule, and $\mathsf{i}_{j+1} : \psi_{j+1} = \mathsf{i}_j : \theta_1$ or $\mathsf{i}_{j+1} : \psi_{j+1} = \mathsf{i}_j : \theta_2$ depending on whether u_{j+1} is the left or right premise of the rule.*

(iv) $\mathsf{i}_j : \psi_j = \mathsf{i}_j : \mathsf{i}' : \theta$ *is the principal formula in an application of the* Glob*-rule, and $\mathsf{i}_{j+1} : \psi_{j+1} = \mathsf{i}' : \theta$.*

(v) $\mathsf{i}_j : \psi_j$ *is the principal formula in an application of the* Eq*-rule with side formula $\mathsf{i}_j \not\approx \mathsf{i}'$, and $\mathsf{i}_{j+1} : \psi_{j+1} = \mathsf{i}' : \psi_j$.*

(vi) $\mathsf{i}_j : \psi_j = \mathsf{i}_j : \eta x.\theta(x)$ *is the principal formula in an application of the η-rule, and $\mathsf{i}_{j+1} : \psi_{j+1} = \mathsf{i}_j : \theta(\eta x.\theta(x))$. In this case we say that an* unfolding *of variable x occurs on the trace $\boldsymbol{t}$ at the index j.*

(vii) u_j *is the conclusion of an application of the* Mod*-rule labelled $\Gamma, \mathsf{i} : \Box\theta, \mathsf{i} : \Diamond\Psi$, the premise is labelled $\Gamma, \mathsf{j} : \theta, \mathsf{j} : \Psi, \mathsf{i} : \Box\theta, \mathsf{i} : \Diamond\Psi$, $\mathsf{i}_j : \psi_j = \mathsf{i} : \Box\theta$, and $\mathsf{i}_{j+1} : \psi_{j+1} = \mathsf{j} : \theta$.*

(viii) u_j *is the conclusion of an application of the* Mod*-rule labelled $\Gamma, \mathsf{i} : \Box\theta, \mathsf{i} : \Diamond\Psi$, the premise is labelled $\Gamma, \mathsf{j} : \theta, \mathsf{j} : \Psi, \mathsf{i} : \Box\theta, \mathsf{i} : \Diamond\Psi$, and for some $\psi \in \Psi$, $\mathsf{i}_j : \psi_j = \mathsf{i} : \Diamond\psi$ and $\mathsf{i}_{j+1} : \psi_{j+1} = \mathsf{j} : \psi$.*

A trace is said to be infinite if it is of length ω. We say that the infinite trace $\boldsymbol{t}$ is trivial *if for some $j < \omega$, $\mathsf{i}_j : \psi_j = \mathsf{i}_m : \psi_m$ for all $j \leq m < \omega$. A non-trivial infinite trace is said to be* good *if the highest ranking fixpoint variable that is unfolded infinitely many times on $\boldsymbol{t}$ is a ν-variable.*

Note that traces move along branches in the direction from conclusions to premises, i.e. traces travel away from the root, and not the other way around. Note also that we do not require traces to start at the root, but adding this constraint would make no substantial difference since every formula appearing in a sequent somewhere in a proof can be connected to a trace starting at the root.

Definition 7 *An **Inf**-proof is said to be* valid *if every infinite branch contains a good trace, and every leaf is labelled by an axiom.*

In order to produce finite circular proofs later on it will be important to carefully apply the weakening rule to discard formulas that are no longer needed and so maintain an upper bound on the size of sequents. The following terminology will play an important role in this regard.

Definition 8 *Given an **Inf**-proof* Π *for some formula* $\mathsf{r}:\rho$*, a nominal* j *appearing in* Π *is said to be* original *if it appears in* $\mathsf{r}:\rho$*. A formula appearing in* Π *is said to be a* ground formula *if it is of the form* $\mathsf{j}:\psi$ *where* j *is an original nominal.*

Definition 9 *An **Inf**-proof is said to be* frugal *if at most finitely many sequents appear in the proof.*

3.2 Derived rules

We shall allow the use of derived rules in proof constructions, as abbreviations of their derivations. These derived rules will be used to formulate a proof search game, which is the main technical tool needed for our completeness proof for **Inf**, and are based on two ideas:

- For all rules except Weak we define what we will call its *narrow* counterpart, which is a derived rule of **Inf**. These rules will be used to automatically discard formulas that will no longer be needed (using Weak), but keep those formulas that might be needed later in the proof construction.
- Two additional derived rules that we call the *deterministic* rule and the *ground rule* will be used to isolate the "essential" choices for the player that tries to construct a proof. These choices will be restricted to two types:

(i) Applications of the Mod-rule to introduce new nominals.
(ii) *Repeating* applications of other rules, which only serve to introduce traces.

Narrow rules We define the narrow rule versions as follows. For the $\wedge$- and $\vee$-rules, the η-rules, the Com-rule, the Glob-rule and the Eq-rule, if the principal formula is a ground formula, then the narrow version of the rule is the same as the standard one. Otherwise, it is defined as follows: we first apply the standard version of the rule, and immediately after we apply the weakening rule to all premises in order to remove the principal formula. For example, if i is a non-original nominal then an instance of the narrow $\wedge$-rule is:

$$\frac{\Gamma, \mathsf{i}:\varphi \qquad \Gamma, \mathsf{i}:\psi}{\Gamma, \mathsf{i}:(\varphi \wedge \psi)}$$

corresponding to the derivation:

$$\dfrac{\mathsf{Weak}\ \dfrac{\Gamma, \mathsf{i}\!:\!\varphi}{\Gamma, \mathsf{i}\!:\!(\varphi \wedge \psi), \mathsf{i}\!:\!\varphi} \qquad \dfrac{\Gamma, \mathsf{i}\!:\!\psi}{\Gamma, \mathsf{i}\!:\!(\varphi \wedge \psi), \mathsf{i}\!:\!\psi}\ \mathsf{Weak}}{\Gamma, \mathsf{i}\!:\!(\varphi \wedge \psi)}\ \wedge$$

The narrow version of the rule Mod is a bit different from the others: if the principal formula $\mathsf{i} : \Box\varphi$ is a ground formula then the rule is the same as Mod. Otherwise, an instance of the narrow rule consists of an application of the modal rule immediately followed by an application of the weakening rule in order to remove *all* formulas of the form $\mathsf{k}\!:\!\theta$ that appear in the premise, and for which k is not an original nominal. For example, if i is a non-original nominal and j is original, then the following is an instance of the narrow Mod-rule:

$$\dfrac{\mathsf{k}\!:\!\varphi, \mathsf{k}\!:\!\psi, \mathsf{j}\!:\!\theta}{\mathsf{i}\!:\!\Box\varphi, \mathsf{i}\!:\!\Diamond\psi, \mathsf{i}\!:\!p, \mathsf{j}\!:\!\theta}$$

If j is non-original then the corresponding instance would be:

$$\dfrac{\mathsf{k}\!:\!\varphi, \mathsf{k}\!:\!\psi}{\mathsf{i}\!:\!\Box\varphi, \mathsf{i}\!:\!\Diamond\psi, \mathsf{i}\!:\!p, \mathsf{j}\!:\!\theta}$$

Note that what counts as an instance of the narrow rules depends on what nominals are considered original, which in turn depends on the root formula of the proof-tree. We therefore emphasize that these rules are not explicitly part of the proof system **Inf**, but only serve as tools for the completeness proof.

Next, we define the deterministic rule and the ground rule. To make these rules precise we need the following:

Convention 2 *Throughout the rest of the paper we fix an arbitrary well-ordering $\prec$ over all formulas (which restricts to a well-ordering over the set of nominals since each nominal is a formula). Furthermore we fix an arbitrary well-ordering over the set of all instances of rules in **Inf**. We overload the notation and denote also this well-ordering by $\prec$.*

The deterministic rule The *deterministic rule* is defined as follows: given a sequent Γ, if there are no applicable instances of the narrow $\wedge$-rule, the narrow $\vee$-rule, the narrow Glob or narrow η-rules *except repeating ones*, then the deterministic rule does not apply. Otherwise, the deterministic rule applies uniquely as follows: we pick the $\prec$-smallest formula in Γ which is the principal formula in an applicable non-repeating instance of one of these rules, we pick the $\prec$-smallest such rule instance for which it is the principal formula, and we apply that rule.

Note that if we repeatedly apply the deterministic rule starting from some sequent Γ until it no longer applies, then this process must eventually terminate. The assumption that all formulas are guarded plays an important role here, without guardedness the process could go on indefinitely via fixpoint unfoldings.

The ground rule The *ground rule* is designed to deterministically apply the Eq-rule and the Com-rule in the same way as the deterministic rule, but also to ensure that original nominals are given special treatment. It is defined as follows: we consider the original nominals appearing in a sequent Γ. If possible, apply the $\prec$-smallest applicable rule instance for which one of the following conditions holds:

(i) it is a non-repeating instance of the narrow Com-rule with principal formula $\mathsf{i} \not\approx \mathsf{j}$, where both i and j are original nominals, or:

(ii) it is a non-repeating instance of the narrow Eq-rule with principal formula $\mathsf{j} : \varphi$ and side formula $\mathsf{j} \not\approx \mathsf{i}$, where i is a $\prec$-minimal original nominal for which such a rule instance applies.

If there are no such rule instances available then the ground rule does not apply. Like the deterministic rule, the process of repeatedly applying the ground rule must eventually terminate.

4 Completeness for Inf

4.1 A game for building Inf-proofs

To prove completeness we shall make use of a proof search game, played between two players **Ver** (the proponent) and **Fal** (the opponent). We fix a root formula $\mathsf{r} : \rho$, so that what counts as a narrow rule is defined relative to this root formula as before, and similarly with the deterministic rule and the ground rule.

Definition 10 *An instance of the weakening rule is called* terminal *if its premise is an axiom.*

Definition 11 *The* **Inf**-game *is a board game, defined as usual by specifying its positions, player assignments and admissible moves for positions and winning conditions on full (finite or infinite) plays.*

Positions: *Game positions are of two types: sequents, which belong to* **Ver**, *and pairs of sequents, which belong to* **Fal**. *We sometimes refer to positions belonging to* **Ver** *as "basic positions".*

Moves for Fal: *Given a position belonging to* **Fal**, *consisting of a pair of sequents, the player simply chooses one of the sequents from the pair.*

Moves for Ver: *Given a position belonging to* **Ver**, *consisting of a sequent Γ, if Γ is an axiom then the game ends and* **Ver** *is declared the winner. Otherwise available moves are defined as follows:*

- *If the deterministic rule is applicable to Γ then this is the only move allowed for* **Ver**.
- *If the deterministic rule is not applicable, but the ground rule is applicable, then this is the only move allowed for* **Ver**.
- *If neither the deterministic rule nor the ground rule are applicable, then the possible moves of* **Ver** *are the narrow modal rule, terminal applications of the weakening rule,* repeating *applications of narrow rules or repeating applications of Weakening.*

If π is a partial play ending with some sequent Γ, then we often refer to Γ as the label *of π. Note that since we allow repeating applications of Weakening,* **Ver** *never gets stuck. So the only full finite plays are those that end in an* *axiom, and are won by* ***Ver****. Thus to finish the construction of the* ***Inf****-game it remains only to decide the winner of an infinite play.* Traces *on a play of the* ***Inf****-game are defined similarly as traces in proof trees, the only difference being that a trace on a play* π *of length* $k \leq \omega$ *is now an object of the form* $(\pi_n, \mathsf{i}_n : \varphi_n)_{n<k}$ *where each* π_n *is an initial segment of the play* π*, and for each* $n+1 < k$ *the initial segment* π_{n+1} *extends* π_n *with a single move. Given an infinite play* π*,* ***Ver*** *is then declared the winner if the play contains a good trace, and otherwise the winner is* ***Fal****.*

We now draw some simple consequences of how the **Inf**-game has been designed.

Proposition 4.1 *In any sequent of the form* $\Gamma, \mathsf{i}:\psi$ *appearing in a play of the* ***Inf****-game,* ψ *contains no non-original nominals.*

Proof. Just observe that all the admissible moves preserve this condition. □

From this proposition a few useful facts follow:

Proposition 4.2 *If a play of the* ***Inf****-game contains any sequent of the form* $\Gamma, \mathsf{i} \not\approx \mathsf{j}$*, then* j *is an original nominal.*

Proof. Special case of Proposition 4.1. □

Proposition 4.3 *For each nominal* i*, and each partial play* π *in the* ***Inf****-game, the label of* π *contains at most* k *formulas of the form* $\mathsf{i}:\psi$*, where* k *is linear in the size of the root formula.*

Proof. Easy using Proposition 4.1. . □

Proposition 4.4 *For any sequent* Γ *appearing in a play of the* ***Inf****-game, at most one non-original nominal appears in* Γ*.*

Proof. The only moves that can introduce new non-original nominals are applications of the narrow modal rule, and by design each instance of this rule erases all occurrences of non-original nominals other than the new nominal that was introduced. □

Like the games for satisfiability checking for the modal μ-calculus introduced in [16], the proof search game is determinate:

Proposition 4.5 *The* ***Inf****-game is determinate, i.e. at every position there is a player who has a winning strategy.*

A crucial part of proving completeness of **Inf** is to show that the standard "good trace" condition on valid proofs, in terms of traces going from the root up along a single branch, is not too strong. At first sight it may seem that we need to consider a more general condition, allowing traces to jump between different occurrences of the same nominal. In this subsection we prove a useful result that deals with this issue.

Definition 12 *Let* S *be a set of plays in the* ***Inf****-game. A* good trace loop *on* S *in the* ***Inf****-game is a sequence of partial traces* $\langle \boldsymbol{t}_1 .... \boldsymbol{t}_n \rangle$ *for which there exist* $\pi_1, ..., \pi_n \in S$ *such that:*

- *Each* $\mathbf{t}_i$ *is a partial trace on* π_i,
- *Each* $\mathbf{t}_i$ *starts and ends with ground formulas,*
- *Each* $\mathbf{t}_{i+1}$ *starts with the last formula of* $\mathbf{t}_i$,
- *The trace* $\mathbf{t}_1$ *starts with the last formula of* $\mathbf{t}_n$, *and*
- *At least one variable is unfolded on some trace* $\mathbf{t}_i$ *and the highest ranking such variable is a* ν*-variable.*

Note that in the following lemma, our focus is on analyzing winning strategies for **Fal** in the proof search game, rather than strategies for **Ver**. The explanation for this is as follows. As winning strategies for **Ver** correspond to proofs, we may think of strategies of **Fal** as providing *refutations*. The completeness proof for **Inf** will build counter-models from such refutations. Rather than building the counter-model from an arbitrary refutation of the root formula, we will start by showing that if such a refutation exists, then there is a refutation of some sequent containing the root formula, with certain properties that make it ideally suited for constructing a counter-model.

Lemma 4.6 *Suppose that* ***Fal*** *has a winning strategy in the* ***Inf****-game for* $\mathsf{i}:\rho$. *Then there exists a sequent* Φ *containing* $\mathsf{i}:\rho$ *and a winning strategy* σ *for* ***Fal*** *in the* ***Inf****-game with starting position* Φ, *such that the following conditions hold:*

(i) *For every sequent appearing in a* σ*-guided play, its ground formulas are the same as the ground formulas in* Φ.

(ii) *The set of* σ*-guided plays does not contain any good trace loops.*

Proof. We first prove the following claim:

Claim 4.7 *There exists some sequent* Φ *such that:*

- $\mathsf{r}:\rho \in \Phi$,
- ***Fal*** *has a winning strategy* σ *in the* ***Inf****-game at the starting position* Φ,
- *For every sequent* Γ *that appears in some* σ*-guided partial play, the ground formulas appearing in* Γ *are the same as the ground formulas in* Φ.

PROOF OF CLAIM Let τ be the winning strategy assumed to exist for **Fal**. First note that the ground formulas appearing in τ-guided partial plays in the **Inf**-game are increasing in the sense that, whenever Γ' appears later than Γ in a partial play, all ground formulas in Γ are also in Γ'. This is because the only admissible rule application that can remove a ground formula is a terminal application of the weakening rule, the premise of which is an axiom. Such applications of weakening never happen in any τ-guided partial play, since such a play would be a loss for **Fal**.

We construct a series of τ-guided partial plays $\pi_0, \pi_1, \pi_2 ...$, where each π_i is an initial segment of π_{i+1}. For each i we let G_i be the set of ground formulas appearing on the last position of π_i. We shall maintain the invariant that, for all proper initial segments π' of π_{i+1}, the ground formulas appearing in the last sequent of π' are contained in G_i. Let π_0 be the start position of the **Inf**-game. Suppose that π_i has been constructed. If there is no τ-guided partial play π' extending π_i in which the last sequent contains ground formulas not in G_i, then we are done: for all τ-guided partial plays extending this play, the ground formulas appearing in all sequents must be equal to G_i, and τ provides a winning strategy for **Fal** in the **Inf**-game for the label of π_i. If there is some τ-guided partial play π' extending π_i in which the last sequent contains ground formulas not in G_i, then just pick π_{i+1} to be its smallest initial segment for which this holds. This procedure must eventually terminate, since otherwise we get an infinite and strictly increasing sequence of sets of ground formulas $G_0 \subset G_1 \subset G_2 ...$, which is impossible since there are only finitely many possible ground formulas that can appear in any play. ◀

Now let Φ and σ be as in the previous claim. Given a σ-guided play π, let $\uparrow \pi$ be the set of partial plays π' such that $\pi \cdot \pi'$ is a σ-guided partial play. Our aim is to find a σ-guided play π such that $\uparrow \pi$ does not contain any good trace loops; we can then simply take the label of π to the sequent claimed to exist in the statement of the Proposition, and we obtain the required winning strategy for **Fal** by assigning the move $\sigma(\pi \cdot \pi')$ to a partial play π'.

Given a good trace loop $\langle \boldsymbol{t}_1, ..., \boldsymbol{t}_n \rangle$, let its *kind* be the set of triples:

$$\{(\mathsf{i}_1 : \varphi_1, X_1, \mathsf{j}_1 : \psi_1), ..., (\mathsf{i}_n : \varphi_n, X_n, \mathsf{j}_n : \psi_n)\}$$

such that for each $m \in \{1, ..., n\}$, the trace $\boldsymbol{t}_m$ begins with $\mathsf{i}_m : \varphi_m$, ends with $\mathsf{j}_m : \psi_m$, and the variables unfolded on $\boldsymbol{t}_m$ are precisely the members of the set X_m. Since each trace in a good trace loop begins and ends with a ground formula, and since there are only finitely many ground formulas, there are finitely many kinds of good trace loops. We shall show how to find a σ-guided play π such that $\uparrow \pi$ does not contain any good trace loops of a given kind. By simply repeating the argument, we can then kill off all the kinds of good trace loop one by one.

So let the kind K be $\{(\mathsf{i}_0 : \varphi_0, X_0, \mathsf{j}_0 : \psi_0), ..., (\mathsf{i}_{n-1} : \varphi_{n-1}, X_{n-1}, \mathsf{j}_{n-1} : \psi_{n-1})\}$. We construct a sequence of partial plays $\pi_0, \pi_1, \pi_2 ...$, where each π_i is an initial segment of π_{i+1}, as follows. If the set of all σ-guided plays does not contain any good trace loops of kind K, we are done. Otherwise, let π_0 be some play on which the part $(\mathsf{i}_0 : \varphi_0, X_0, \mathsf{j}_0 : \psi_0)$ appears, which must exist. Note that we have a partial trace $\boldsymbol{t}_0$ on π_0 leading from $\mathsf{i}_0 : \varphi_0$ to $\mathsf{j}_0 : \psi_0$ on which exactly the variables X_0 were unfolded; since the first formula is a ground formula, and these are the same in all positions in all σ-guided plays, we can simply "pad" the partial trace with silent steps repeating the same formula to extend it to a trace on the whole play π_0. Now we repeat the procedure: if $\uparrow \pi_0$ does not contain any good trace loop, then we are done. Otherwise, we can extend π_0 in the same way to a partial play π_1 containing a trace $\boldsymbol{t}_1$ appearing *after* π_0, such that $\boldsymbol{t}_1$ starts with $\mathsf{i}_1 : \varphi_1$, ends with $\mathsf{j}_1 : \psi_1$ and the variables unfolded are precisely X_1. Then since $\boldsymbol{t}_0$ and $\boldsymbol{t}_1$ end and start respectively with the same ground formula, and ground formulas stay the same, they can be linked together by "padding with silent steps" repeating this formula to form a trace on π_1. It is not hard to see that, if this procedure never terminates, then we end up building an infinite σ-guided play containing a good trace, which is a contradiction since σ was a winning strategy. So the procedure terminates with some π_m, and the proof is finished. □

Note that, since the set of *finite* partial plays in the **Inf**-game is a countable set (being a set of finite sequences over a countable set), we can define a surjective mapping F from the set of nominals to the set of finite σ-guided partial plays, such that $F^{-1}[\pi]$ is infinite for each finite partial play π. We leave the little set theoretic exercise of proving this to the reader. Throughout the rest of this section we fix such a map F. Informally, we think of $F(\mathsf{i})$ as a *tag* attached to the nominal i to remember where it was introduced.

Definition 13 *We say that a full or partial play π of the **Inf**-game is* clean *if, for every initial segment π' of the play ending with the conclusion of an application of the (narrow) modal rule introducing a new nominal j, we have $F(\mathsf{j}) = \pi'$.*

When proving completeness of **Inf** we shall construct a counter-model to the root formula from a winning strategy for **Fal**, and it will then be convenient to restrict attention to clean plays. We are now ready for the main result about the system **Inf**.

Theorem 1 *Let ρ be any formula. The following are equivalent: (a) ρ is valid, (b) **Ver** has a winning strategy in the **Inf**-game for $\mathsf{r}\!:\!\rho$, where r is some fresh nominal, (c) ρ has a valid and frugal **Inf**-proof, (d) ρ has a valid **Inf**-proof.*

Proof. We sketch the proof of the implication (a) $\Rightarrow$ (b), which is the most difficult part of the proof. We prove this by contraposition: suppose there is a winning strategy for **Fal** in the **Inf**-game for $\mathsf{r}\!:\!\rho$. Let Φ be a set of ground formulas containing $\mathsf{r}\!:\!\rho$ and let σ be a winning strategy for **Fal** in the **Inf**-game for premise Φ such that the ground formulas stay the same in every σ-guided play, and the set of σ-guided plays contains no good trace loops. Such Φ and σ exist by Lemma 4.6. We shall construct a countermodel to (the disjunction of) Φ, which gives a countermodel to ρ since $\mathsf{r}\!:\!\rho \in \Phi$.

We construct the model $M = (W, R, A, V)$ using the strategy σ as follows. Let N be the set of nominals i such that i appears in some position in some clean σ-guided play π, and let $\equiv$ be the smallest equivalence relation over N containing all pairs (i, j) for which $\mathsf{i} \not\approx \mathsf{j}$ appears in some position in some clean σ-guided play π.

Claim 4.8 *For each i, the equivalence class $[\mathsf{i}]$ modulo $\equiv$ is either a singleton or contains at least one of the nominals in φ.*

Motivated by this claim, we call a nominal *representative* if its equivalence class is a singleton, or it is the $\prec$-smallest original nominal belonging to its equivalence class. We let W be the set of representative members of N. We set $\mathsf{i}R\mathsf{j}$ iff there is some $\mathsf{j}' \equiv \mathsf{j}$ and a clean σ-guided play in which j' is introduced by an application of the modal rule to the nominal i. Set $A(\mathsf{i})$ to be the representative of $[\mathsf{i}]$. Finally, for a representative i set $\mathsf{i} \in V(p)$ iff $\mathsf{i}\!:\!\neg p$ appears on some clean σ-guided play. We shall show that M is a counter-model to the sequent Φ. The key claims used to prove this are the following:

Claim 4.9 *Suppose that* $\mathsf{i} \equiv \mathsf{j}$ *and that* j *is an original nominal. Then for any basic position* u *appearing in a clean* σ*-guided play, and any* θ*, if* $\mathsf{i}:\theta$ *belongs to* u *then so does* $\mathsf{j}:\theta$.

Claim 4.10 *Suppose that* $\boldsymbol{t}$ *is some partial trace on a clean partial* σ*-guided play* π*, such that the last element of the trace* $\boldsymbol{t}$ *is of the form* $(\pi, \mathsf{k}':\psi)$ *where* $A(\mathsf{k}') = \mathsf{k}$*. If* ψ *is of the form* $\Box\theta$ *or* $\Diamond\theta$*, then there is a clean* σ*-guided play* v *extending* π *and a partial trace on* v *of the form* $(\pi, \mathsf{k}':\psi) \cdot \boldsymbol{u} \cdot (v, \mathsf{k}:\psi)$*, which contains no fixpoint unfoldings.*

We now proceed to show that the sequent Φ is not valid in M. Pick any formula $\mathsf{i} : \varphi \in \Phi$. We shall construct a winning strategy σ' for **Fal** in the evaluation game in M at the starting position $(A(\mathsf{i}), \varphi)$. Inductively, as an invariant we associate with each partial σ'-guided partial play of π' of the form:

$$(\mathsf{j}_1, \psi_1) \cdot \boldsymbol{p} \cdot (\mathsf{j}_n, \psi_n)$$

a sequence of non-empty partial traces $\langle \boldsymbol{t}_1, ..., \boldsymbol{t}_n \rangle$ such that each of these traces $\boldsymbol{t}_k$ belongs to some clean σ-guided partial play π_k, and such that the following conditions hold:

I1: The last element of each trace $\boldsymbol{t}_k$ is of the form $(\pi_k, \mathsf{j}'_k : \psi_k)$ where $A(\mathsf{j}'_k) = \mathsf{j}_k$. Furthermore, if ψ_k is of the form $\Box\theta$ or $\Diamond\theta$, then $\mathsf{j}'_k = \mathsf{j}_k$.

I2: For each $k < n$, if the last element of $\boldsymbol{t}_k$ is $(\pi_k, \mathsf{j}'_k : \psi_k)$ then the first element of $\boldsymbol{t}_{k+1}$ is of the form $(\pi_{k+1}, \mathsf{j}'_k : \psi_k)$. Furthermore, if j_k is not an original nominal then $\pi_k = \pi_{k+1}$.

I3: For $k < n$, a fixpoint unfolding occurs on the trace $\boldsymbol{t}_{k+1}$ iff the same fixpoint is unfolded on $(\mathsf{j}_k, \psi_k) \cdot (\mathsf{j}_{k+1}, \psi_{k+1})$.

Suppose we are given a clean σ'-guided partial play π' of the form $(\mathsf{j}_1, \psi_1) \cdot \boldsymbol{p} \cdot (\mathsf{j}_n, \psi_n)$, and that the associated sequence of partial traces $\langle \boldsymbol{t}_1, ..., \boldsymbol{t}_n \rangle$ has been constructed. Then we can show that, if the last position on π' belongs to **Fal**, then we can define a move for which the invariant (I1) – (I3) can be maintained, and if the last position belongs to **Ver** then the invariant can be maintained for any possible move. This is proved by a case distinction as to the shape of the last position, and uses the claims 4.9, and 4.10. The details of the argument are omitted here.

To finish the proof of (a) $\Rightarrow$ (b), we have given a strategy σ' to **Fal** in the evaluation game such that the invariant (I1) – (I3) is maintained. The strategy σ' ensures that **Fal** never gets stuck, and any lost infinite σ'-guided play produces either an infinite clean σ-guided shadow-play in the **Inf**-game containing a good infinite trace, or a good trace loop on the set of all clean σ-guided plays. In either case we get a contradiction, so σ' is winning for **Fal** and therefore we have found a falsifying model for ρ. $\Box$

5 Finite proofs with names

5.1 The system Saf

In this section we introduce the finitary proof system **Saf**, which is an annotated circular proof system in Stirling's style [23]. We will borrow a more streamlined version of the rules for manipulating annotations from Afshari and Leigh [2]. For each fixpoint variable x we assume that we have a countably infinite supply $\mathsf{x}_0, \mathsf{x}_1, \mathsf{x}_2 ...$ of *names* for that variable. We assume that we have a fixed enumeration of the set of variable names for each variable x, so that we can speak of the n-th variable name for x. The system will be defined taking a strict linear order $<$ over fixpoint variables as a parameter, and in the presentation we assume such an order as given. Given $<$ we write $\mathsf{x} < \mathsf{y}$ for names x, y of variables x, y respectively if $x < y$. Given a word a over the set of variable names and a variable x, we write $\mathsf{a} \leq x$ if there is no variable $y > x$ for which a contains a name of y. Given two words a, b over the set of variable names we write $\mathsf{a} \sqsubseteq \mathsf{b}$ to say that b contains a as a subsequence. For example $\mathsf{xy} \sqsubseteq \mathsf{xzy}$. We write $\mathsf{a} \sqcap \mathsf{b}$ for the longest word c such that $\mathsf{c} \sqsubseteq \mathsf{a}$ and $\mathsf{c} \sqsubseteq \mathsf{b}$ (provided that a longest word with this property exists, otherwise $\mathsf{a} \sqcap \mathsf{b}$ is undefined).

Definition 14 Annotated sequents *will be structures of the form:*

$$\mathsf{a} \vdash \mathsf{i}_1 : \varphi_1^{\mathsf{b}_1}, ..., \mathsf{i}_n : \varphi_n^{\mathsf{b}_n}$$

where $\mathsf{a}, \mathsf{b}_1, ..., \mathsf{b}_n$ *are non-repeating words over the set of variable names (i.e. no variable name appears twice in any of these words), each* b_i *is non-decreasing with respect to the order* $<$*, and* $\mathsf{b}_i \sqsubseteq \mathsf{a}$ *for each* $i \in \{1, ..., n\}$*. The tuple* a *is called the* control *of the sequent.*

A formula ρ will be said to be provable in the system if the sequent $\varepsilon \vdash \mathsf{r} : \rho^{\varepsilon}$ is provable, where the order $<$ on variable names is some arbitrary linearization of $<_\rho$, ε is the empty word, and r is a fresh nominal. We will allow suppressing occurrences of the empty word in our notation, including the control, so that for example the sequent $\varepsilon \vdash \mathsf{r} : \rho^{\varepsilon}$ can be written simply as $\mathsf{r} : \rho$.

Definition 15 *A sequent in the sense of the system* ***Inf*** *will be called a* plain sequent*. Given an annotated sequent* $\Gamma = \mathsf{a} \vdash \mathsf{i}_1 : \varphi_1^{\mathsf{b}_1}, ..., \mathsf{i}_n : \varphi_n^{\mathsf{b}_n}$*, the* underlying plain sequent $\underline{\Gamma}$ *is the plain sequent* $\mathsf{i}_1 : \varphi_1, ..., \mathsf{i}_n : \varphi_n$.

The system has two axioms, which are the law of exluded middle and an identity axiom, which now have the form:

$$\varepsilon \vdash \mathsf{i} : p^{\varepsilon}, \mathsf{i} : \neg p^{\varepsilon} \qquad \varepsilon \vdash \mathsf{i} \approx \mathsf{i}^{\varepsilon}$$

Here, p is a nominal or a propositional variable. Rules of inference are given in Figure 2. Note that Γ, Ψ here denote sets of annotated formulas rather than plain formulas. The rules are subject to the following constraints:

Mod: The nominal j must be fresh.

ηx: $\mathsf{b} \leq x$.

Rec(x): $\mathsf{b} \leq x$, and x is a fresh variable name for x.

Exp: $\mathsf{a} \sqsubseteq \mathsf{a}'$, $\mathsf{b}_i \sqsubseteq \mathsf{b}_i'$ and $\mathsf{b}_i' \sqcap \mathsf{a} \sqsubseteq \mathsf{b}_i$ for each i [2].

Reset(x): The variable x does not appear in any formula in Γ.

$$\frac{\mathsf{a} \vdash \Gamma, \mathsf{i}:\varphi \wedge \psi^{\mathsf{b}}, \mathsf{i}:\varphi^{\mathsf{b}} \qquad \mathsf{a} \vdash \Gamma, \mathsf{i}:\varphi \wedge \psi^{\mathsf{b}}, \mathsf{i}:\psi^{\mathsf{b}}}{\mathsf{a} \vdash \Gamma, \mathsf{i}:\varphi \wedge \psi^{\mathsf{b}}} \wedge$$

$$\frac{\mathsf{a} \vdash \Gamma, \mathsf{i}:\varphi \vee \psi^{\mathsf{b}}, \mathsf{i}:\varphi^{\mathsf{b}}, \mathsf{i}:\psi^{\mathsf{b}}}{\mathsf{a} \vdash \Gamma, \mathsf{i}:\varphi \vee \psi^{\mathsf{b}}} \vee \qquad \frac{\mathsf{a} \vdash \Gamma, \mathsf{i}:\phi^{\mathsf{b}}, \mathsf{j}:\phi^{\mathsf{b}}, \mathsf{i} \not\approx \mathsf{j}^{\mathsf{c}}}{\mathsf{a} \vdash \Gamma, \mathsf{j}:\phi^{\mathsf{b}}, \mathsf{i} \not\approx \mathsf{j}^{\mathsf{c}}} \mathsf{Eq}$$

$$\frac{\mathsf{a} \vdash \Gamma, \mathsf{i}:\Box\varphi^{\mathsf{b}}, \mathsf{i}:\Diamond\Psi, \mathsf{j}:\varphi^{\mathsf{b}}, \mathsf{j}:\Psi}{\mathsf{a} \vdash \Gamma, \mathsf{i}:\Box\varphi^{\mathsf{b}}, \mathsf{i}:\Diamond\Psi} \mathsf{Mod} \qquad \frac{\mathsf{a} \vdash \Gamma, \mathsf{i} \not\approx \mathsf{j}^{\mathsf{b}}, \mathsf{j} \not\approx \mathsf{i}^{\mathsf{b}}}{\mathsf{a} \vdash \Gamma, \mathsf{i} \not\approx \mathsf{j}^{\mathsf{b}}} \mathsf{Com}$$

$$\frac{\mathsf{a} \vdash \Gamma, \mathsf{i}:\eta x.\phi(x)^{\mathsf{b}}, \mathsf{i}:\phi(\eta x.\phi(x))^{\mathsf{b}}}{\mathsf{a} \vdash \Gamma, \mathsf{i}:\eta x.\phi(x)^{\mathsf{b}}} \eta x \qquad \frac{\mathsf{a} \vdash \Gamma, \mathsf{i}:(\mathsf{j}:\varphi)^{\mathsf{b}}, \mathsf{j}:\varphi^{\mathsf{b}}}{\mathsf{a} \vdash \Gamma, \mathsf{i}:(\mathsf{j}:\varphi)^{\mathsf{b}}} \mathsf{Glob}$$

$$\frac{\mathsf{ax} \vdash \Gamma, \mathsf{i}:\nu x.\varphi(x)^{\mathsf{b}}, \mathsf{i}:\varphi(\nu x.\varphi(x))^{\mathsf{bx}}}{\mathsf{a} \vdash \Gamma, \mathsf{i}:\nu x.\varphi(x)^{\mathsf{b}}} \mathsf{Rec(x)} \qquad \frac{\mathsf{a} \vdash \Gamma}{\mathsf{a} \vdash \Gamma \cup \Psi} \mathsf{Weak}$$

$$\frac{\mathsf{a} \vdash \Gamma, \mathsf{i}_1:\varphi_1^{\mathsf{bx}}, ..., \mathsf{i}_n:\varphi_n^{\mathsf{bx}}}{\mathsf{a} \vdash \Gamma, \mathsf{i}_1:\varphi_1^{\mathsf{bxx_1c_1}}, ..., \mathsf{i}_n:\varphi_n^{\mathsf{bxx_nc_n}}} \mathsf{Reset(x)} \qquad \frac{\mathsf{a} \vdash \mathsf{i}_1:\varphi_1^{\mathsf{b}_1}, ..., \mathsf{i}_n:\varphi_n^{\mathsf{b}_n}}{\mathsf{a}' \vdash \mathsf{i}_1:\varphi_1^{\mathsf{b}_1'}, ..., \mathsf{i}_n:\varphi_n^{\mathsf{b}_n'}} \mathsf{Exp}$$

Fig. 2. Rules of **Saf**

A **Saf**-proof is a labelled tree where the labels specify a sequent assigned to a node and the last rule application (for non-leaf nodes), and such that the children of a node are labelled with the premises of the specified rule application. Although valid proofs will always be finite it will be useful to consider infinite **Saf**-proofs as well. We say that the variable name x is *reset* in an instance of the rule Reset(x).

Definition 16 *A* ***Saf****-proof will be considered* valid *if it is a finite proof-tree, and there is a map f from non-axiom leaves to non-leaves, such that:*

[2] Note that $\mathsf{b}_i' \sqcap \mathsf{a}$ is well-defined here: since $\mathsf{b}_i' \sqsubseteq \mathsf{a}'$ and $\mathsf{a} \sqsubseteq \mathsf{a}'$, and since a' is non-repeating, any two variable names occurring in both b_i' and a must appear only once and in the same order in both words. From this follows that the set of words c such that $\mathsf{c} \sqsubseteq \mathsf{b}_i'$ and $\mathsf{c} \sqsubseteq \mathsf{a}$ is a $\sqsubseteq$-directed finite set, so it contains a $\sqsubseteq$-maximal word.

- *$f(l)$ is an ancestor of l, and has the same label.*
- *There is a variable name x that is contained in the control of every node in the path from $f(l)$ to l, and is reset at least once on this path.*

A map f from non-axiom leaves to non-leaves satisfying the first of these conditions is called a back-edge map, *and is* good *if it satisfies the second condition as well. So a finite proof-tree is a valid proof iff it has a good back-edge map.*

We can now state the main result of the paper.

Theorem 2 (Completeness of Saf) *A formula $\mathsf{i}:\varphi$ has a valid **Saf**-proof if and only if it is semantically valid.*

Proof. We only sketch the proof here. For the soundness part, it is a fairly simple exercize to "unfold" a valid **Saf**-proof to an infinite proof-tree in which every infinite branch has a good trace. By forgetting the annotations we can view this as a valid **Inf**-proof, and soundness thus follows from Theorem 1.

For the completeness proof, we reason as follows: first, any valid formula $\mathsf{i}:\varphi$ has a valid and frugal **Inf**-proof Π by Theorem 1. We need to add annotations to the sequents in this proof, possibly inserting some rules of **Saf** for updating annotations, in such a way that we produce an infinite **Saf**-proof for the same conclusion which is still frugal, i.e. contains only finitely many annotated sequents, and satisfies the constraint that on every infinite branch there is some variable name that is reset infinitely many times. The construction of this infinite **Saf**-proof essentially uses annotations to mimick the Safra construction for automata on infinite words, and follows the same reasoning as in [12]. Since the class of proof trees satisfying these criteria is definable in monadic second-order logic, we may apply Rabin's Basis Theorem [18] to find a *regular* infinite **Saf**-proof for the same conclusion, in which every infinite branch has a variable name that is reset infinitely often. Finally, we note that any such regular proof can be "folded back" into a finite proof-tree with a back-edge map that yields a valid finite **Saf**-proof. □

Example 1 *We show a valid **Saf**-proof of the formula $\mathsf{i}:(\Box\neg\mathsf{i} \vee \nu x.\Diamond x)$, which is equivalent to:*

$$\mathsf{i}:(\Diamond\mathsf{i} \to \nu x.\Diamond x)$$

The "†" labels show how the back-edge map connects the leaf to an ancestor.

$$\dfrac{\dfrac{\dfrac{\dfrac{\dfrac{\dfrac{\mathsf{x}_0 \vdash \mathsf{i}:\Box\neg\mathsf{i},\ \mathsf{i}:\Diamond\nu x.\Diamond x^{\mathsf{x}_0}\ \dagger}{\mathsf{x}_0\mathsf{x}_1 \vdash \mathsf{i}:\Box\neg\mathsf{i},\ \mathsf{i}:\Diamond\nu x.\Diamond x^{\mathsf{x}_0\mathsf{x}_1}}\ \mathsf{Reset}(\mathsf{x}_0)}{\mathsf{x}_0 \vdash \mathsf{i}:\Box\neg\mathsf{i},\ \mathsf{i}:\nu x.\Diamond x^{\mathsf{x}_0}}\ \mathsf{Rec}(\mathsf{x}_1)+\mathsf{Weak}}{\mathsf{x}_0 \vdash \mathsf{i}:\Box\neg\mathsf{i},\ \mathsf{j}\not\approx\mathsf{i},\ \mathsf{j}:\nu x.\Diamond x^{\mathsf{x}_0}}\ \mathsf{Eq}+\mathsf{Weak}}{\mathsf{x}_0 \vdash \mathsf{i}:\Box\neg\mathsf{i},\ \mathsf{i}:\Diamond\nu x.\Diamond x^{\mathsf{x}_0}\ \dagger}\ \mathsf{Mod}+\mathsf{Weak}}{\mathsf{i}:\Box\neg\mathsf{i},\ \mathsf{i}:\nu x.\Diamond x}\ \mathsf{Rec}(\mathsf{x}_0)+\mathsf{Weak}}{\mathsf{i}:(\Box\neg\mathsf{i} \vee \nu x.\Diamond x)}\ \vee+\mathsf{Weak}$$

6 Concluding remarks

We conclude with some directions for future work. First of all, with the Stirling-style proof system in place for the hybrid μ-calculus, we should be able to prove cut-free completeness of a sequent system for the hybrid μ-calculus by following the same method of translation between proof systems as in [2]. The proof should not involve any substantial novelties, although the details remain to be checked.

We hope that the methods developed here can be extended to other extended μ-calculi, like guarded fixpoint logic. A first task in this direction is to consider converse modalities, and obtain a Stirling-style circular system for the hybrid μ-calculus including converse modalities. In Vardi's automata-theoretic decision procedure for the two-way μ-calculus, the key component is a finite data structure for encoding generalized traces that can go upwards or downwards along branches in a tableau. This extra component is then removed through a projection operation on automata recognizing valid tableaux. It would be interesting to investigate this construction from a proof-theoretic perspective.

References

[1] Afshari, B., G. Jäger and G. E. Leigh, *An infinitary treatment of full mu-calculus*, in: *International Workshop on Logic, Language, Information, and Computation*, Springer, 2019, pp. 17–34.

[2] Afshari, B. and G. E. Leigh, *Cut-free completeness for modal mu-calculus*, in: *Logic in Computer Science (LICS), 2017 32nd Annual ACM/IEEE Symposium on*, IEEE, 2017, pp. 1–12.

[3] Brotherston, J. and A. Simpson, *Complete sequent calculi for induction and infinite descent*, in: *22nd Annual IEEE Symposium on Logic in Computer Science (LICS 2007)*, IEEE, 2007, pp. 51–62.

[4] Brotherston, J. and A. Simpson, *Sequent calculi for induction and infinite descent*, Journal of Logic and Computation **21** (2010), pp. 1177–1216.

[5] Bruse, F., O. Friedmann and M. Lange, *On guarded transformation in the modal μ-calculus*, Logic Journal of the IGPL **23** (2015), pp. 194–216.

[6] Dam, M. and D. Gurov, *μ-calculus with explicit points and approximations*, Journal of Logic and Computation **12** (2002), pp. 255–269.

[7] Emerson, E. and C. Jutla, *Tree automata, mu-calculus and determinacy (extended abstract)*, in: *Proceedings of the 32nd Symposium on the Foundations of Computer Science* (1991), pp. 368–377.

[8] Enqvist, S., H. H. Hansen, C. Kupke, Y. Venema and J. Marti, *Completeness for game logic*, in: *2019 34th Annual ACM/IEEE Symposium on Logic in computer Science (LICS)*, 2019.

[9] Enqvist, S., F. Seifan and Y. Venema, *Completeness for the modal μ-calculus: separating the combinatorics from the dynamics*, Theoretical Computer Science **727** (2018), pp. 37–100.

[10] Enqvist, S., F. Seifan and Y. Venema, *Completeness for μ-calculi: a coalgebraic approach*, Annals of Pure and Applied Logic **170** (2019), pp. 578–641.

[11] Grädel, E. and I. Walukiewicz, *Guarded fixed point logic*, in: *Proceedings 14th IEEE Symposium on Logic in Computer Science LICS'99*, 1999.

[12] Jungteerapanich, N., *Tableau systems for the modal μ-calculus*, The University of Edinburgh (2010).

[13] Kernberger, D. and M. Lange, *The fully hybrid mu-calculus*, in: *24th International Symposium on Temporal Representation and Reasoning (TIME 2017)*, Schloss Dagstuhl-Leibniz-Zentrum fuer Informatik, 2017.

[14] Knaster, B., *Un théorème sur les fonctions des ensembles*, Annales de la Societé Polonaise de Mathematique **6** (1928), pp. 133–134.

[15] Kozen, D., *Results on the propositional μ-calculus*, Theoretical Computer Science **27** (1983), pp. 333–354.

[16] Niwiński, D. and I. Walukiewicz, *Games for the μ-calculus*, Theoretical Computer Science **163** (1996), pp. 99–116.

[17] Parikh, R., *The logic of games and its applications*, Annals of Discrete Mathematics **24** (1985), pp. 111–139.

[18] Rabin, M., *Automata on infinite objects and church's problem*, American Mathematical Soc. **13** (1972).

[19] Santocanale, L., *A calculus of circular proofs and its categorical semantics*, in: *International Conference on Foundations of Software Science and Computation Structures*, Springer, 2002, pp. 357–371.

[20] Sattler, U. and M. Y. Vardi, *The hybrid μ-calculus*, in: *International Joint Conference on Automated Reasoning*, Springer, 2001, pp. 76–91.

[21] Shamkanov, D. S., *Circular proofs for the Gödel-Löb provability logic*, Mathematical Notes **96** (2014), pp. 575–585.

[22] Sprenger, C. and M. Dam, *On the structure of inductive reasoning: Circular and tree-shaped proofs in the μcalculus*, in: *International Conference on Foundations of Software Science and Computation Structures*, Springer, 2003, pp. 425–440.

[23] Stirling, C., *A tableau proof system with names for modal mu-calculus*, EPiC Series in Computing **42** (2014), pp. 306–318.

[24] Tamura, K., *A small model theorem for the hybrid μ-calculus*, Journal of Logic and Computation **25** (2013), pp. 405–441.

[25] Tarski, A., *A lattice-theoretical fixpoint theorem and its applications*, Pacific Journal of Mathematics **5** (1955), pp. 285–309.

[26] Vardi, M. Y., *Reasoning about the past with two-way automata*, in: *International Colloquium on Automata, Languages, and Programming*, Springer, 1998, pp. 628–641.

[27] Walukiewicz, I., *Completeness of Kozen's axiomatisation of the propositional mu-calculus*, Information and Computation **157** (2000), pp. 142 – 182.
URL `http://www.sciencedirect.com/science/article/pii/S0890540199928365`

[28] Zielonka, W., *Infinite games on finitely coloured graphs with applications to automata on infinite trees*, Theoretical Computer Science **200** (1998), pp. 135–183.

Possibility, consistency, connexivity

Luis Estrada-González [1]

Institute for Philosophical Research
Graduate Program in Philosophy of Science
National Autonomous University of Mexico (UNAM)

Abstract

In this paper I explore how to retain Lewis and Langford's characterization of possibility in terms of consistency and Nelson's idea that all propositions are self-consistent. This would amount to having as logical truths all the formulas of the form $\Diamond A$. I show that in using a very simple three-valued connexive logic to evaluate the Lewis and Langford's definition of modalities, one gets some very interesting results connecting possibilism, the thesis according to which everything is possible, with certain styles of connexivism, especially those with room for contradictory theorems.

Keywords: Aristotle's Theses; connexive logic; possibilism; **LP**.

1 Introduction

In their *Symbolic Logic*, [9], C.I. Lewis and C.H. Langford, building upon previous work of the former (see [8]), famously defined implication in terms of possibility, negation and conjunction. Slightly less famously, Lewis and Langford characterized possibility in terms of consistency, and this led to a definition of implication in terms of consistency, which is a very common idea in the field of connexive logic. Then, there are a number of valid arguments with a connexive flavor in *Symbolic Logic.* All of them include instances of Aristotle's Thesis, $\sim (A \rightarrow \sim A)$, in the premises (or as part of the antecedent, in implicational theorems), and in the system are invalid without such instances of instances of Aristotle's Thesis as premises or antecedents.

Unlike Lewis and Langford, and explicitly reacting against some previous work of the former, Everett J. Nelson held in "Intensional relations" ([16]) that all propositions are self-consistent, so he retained the definition of implication in terms of consistency, but rejected the characterization of possibility in terms of consistency. With Aristotle's Thesis as a logical truth, all the valid arguments

[1] This work was written under the support of the PAPIIT project IN403719 "Intensionality all the way down. A new plan for logical relevance". I presented a previous version at the Ninth Conference Non-Classical Logic. Theory and Applications, held in Toruń. I want to thank Hitoshi Omori, Ricardo Arturo Nicolás-Francisco and Heinrich Wansing for extremely helpful comments on previous drafts, as well as the AiML referees for their comments and advice.

with a connexive flavor in *Symbolic Logic* became fully connexive and remain valid even without including Aristotle's Thesis as a premise (or as an antecedent in the implicational theorems).

In this paper, I probe the prospects of having my cake and eating it too. That is, I want to explore how to retain Lewis and Langford's characterization of possibility in terms of consistency and Nelson's idea that all propositions are self-consistent, which amounts in Lewis and Langford's framework to validate all the formulas of the form $\Diamond A$. And here enters a further connexive twist: I will show that in using a very simple three-valued connexive logic introduced in [20] to evaluate Lewis and Langford's definition of modalities, one gets some very interesting results connecting possibilism, the thesis according to which everything is possible, with certain styles of connexivism, especially those with room for contradictory theorems.

Some provisos are in order here. First, my contribution here is not a brand new logic. I am using instead a logic already in circulation to model both Lewis and Langford's characterization of possibility in terms of consistency and Nelson's idea that all propositions are self-consistent, which results in a connexive logic where possibilism is valid. Second, in showing such a model I am not claiming that connexivity and possibilism are equivalent, because they are not. The idea is rather that certain recent brands of connexivity imply possibilism under some characterizations of possibility proposed in the early 20th century, and that the connection has several ramifications for the study of modalities. Third: truth-functional modal logic, of the likes I will analyze here, has been declared a "dead end" many times now, most notably in [5]. I do not think it is, though. True, there are many objections to be overcome, and I will do so in due time in this paper. Lastly: I am not making any strong claims about the truth, correctness or the like of the views presented here. My main claim is about the existence of the connection already mentioned and the worthiness of studying it.

The structure for the remaining of the paper is as follows. In Section 2, I present Lewis and Langford's conceptualization of the notions mentioned in the title. In Section 3, I present Nelson's reaction towards some of the consequences of the Lewis and Langford's proposal and his arguments to prefer a primitive notion of consistency, not definable in terms of possibility. These applications were not considered when such a logic was first presented. In Section 4, I present a limitative result for an attempt to combine both approaches. In Section 5 I show that Omori's **dLP**, which is a connexive logic built upon the $\{\sim, \wedge, \vee\}$-fragment of González-Asenjo/Priest's Logic of Paradox **LP** enriched with a suitable conditional, can support the combination of Lewis and Langford's characterization of possibility in terms of consistency with Nelson's idea that every proposition is self-consistent. By adding more conceptual tools, one can even get a connexive model of Mortensen's possibilism, where not only everything is possible, but nothing is necessary. In Section 6 I address some concerns about this approach. Finally, in Section 7 I present some sets more of modalities allowed in this framework and then I present some conclusions and suggest some paths for future work in this area.

2 Possibility and connexivity

In Lewis and Langford's *Symbolic Logic*, there is a tight connection between the notions of possibility, consistency and (strict) implication. In their formalization ordering, possibility comes first, and implication is defined in the well-known way. However, at the conceptual level, all of them are equally basic. Thus we read, for example: [2]

> The primitive or undefined ideas assumed are the following: (...)
> 4. Self-consistency or possibility: $\Diamond p$. This may be read "p is self-consistent" or "p is possible" or "It is possible that p be true". As will appear later, $\Diamond p$ is equivalent to "It is false that p implies its own negation," (...). The precise logical significance of $\Diamond p$ will be discussed in Section 4. (...)
> The relation of strict implication can be defined in terms of negation, possibility and product:
> *11.02* $(p \rightarrow q) =_{def.} \sim\Diamond(p \wedge \sim q)$

And then in Section 4 there is, as promised, the discussion of the precise logical significance of possibility. Lewis and Langford say:

> When we speak of two propositions as 'consistent,' we mean that it is not possible, with either of them as premise, to deduce the falsity of the other. Thus if $p \rightarrow q$ has the intended meaning "q is deducible from p," then "p is consistent with q" may be defined as follows:
> *17.01* $(p \circ q) =_{def.} \sim(p \rightarrow \sim q)$ [3]

From this, it easily follows that $(p \circ q) \leftrightarrow \Diamond(p \wedge q)$. Thus, possibility or self-consistency $\Diamond p$ would amount to $(p \circ p)$, which in turn would be equivalent to $\sim(p \rightarrow \sim p)$.

The other usual modalities can be defined then as follows:

18.12 $\sim\Diamond p = \sim(p \circ p) = \sim\sim(p \rightarrow \sim p)$
18.13 $\Diamond\sim p = (\sim p \circ \sim p) = \sim(\sim p \rightarrow \sim\sim p)$
18.14 $\sim\Diamond\sim p = \sim(\sim p \circ \sim p) = \sim\sim(\sim p \rightarrow \sim\sim p)$

Now, Lewis and Langford's characterization of possibility, namely
$\Diamond p = (p \circ p) = \sim(p \rightarrow \sim p)$
might look familiar to a logician acquainted with contra-classical logics, as this is a form of *Aristotle's Thesis*, which is one of the characteristic valid schemas

[2] Throughout the paper, the notation of *Symbolic Logic* will be adjusted. Also, Lewis and Langford take classical logic to be basically correct, and that is why they allow certain logical moves that might be in question for other logicians, especially connexivists. Since in this section I am merely presenting their views, I will leave their classical assumptions untouched.

[3] Note that, unlike its Brazilian sibling introduced several years after by da Costa, Nelson's consistency connective is not intended to control Explosion, $A, \sim A \Vdash B$. Note also that this is, at least typographically, the same definition for a connective named variously 'fusion', 'intensional conjunction' or 'multiplicative conjunction' in the relevance logic literature. I will come back to this issue at the end of Section 5.

of *connexive logics.*[4] And this part of *Symbolic Logic* ([9, 154–178]) is indeed full with theorems with a connexive flavor, among them:
17.52 $((p \to q) \land (p \to \sim q)) \to \sim (p \circ p)$
17.57 $((p \to q) \land \sim (p \circ q)) \to \sim (p \circ p)$
17.58 $((p \circ p) \land \sim (p \circ q)) \to \sim (p \to q)$
17.59 $((p \circ p) \land (p \to q)) \to \sim (p \to \sim q)$
17.591 $(p \circ p) \to \sim ((p \to q) \land (p \to \sim q))$
17.6 $p \to (p \circ p)$

To see this more clearly, consider the contrapositive forms of *17.52* and *17.57*:
$\sim\sim (p \circ p) \to \sim ((p \to q) \land (p \to \sim q))$
$\sim\sim (p \circ p) \to \sim ((p \to q) \land \sim (p \circ q))$
Substituting all the occurrences of the consistency connective by its definition and employing Double Negation Elimination, one gets that the two above are equivalent to
$\sim (p \to \sim p) \to \sim ((p \to q) \land (p \to \sim q))$
which is very close to *17.591*, too. One can read this as expressing that if Aristotle's Thesis holds, Abelard's Principle, $\sim ((p \to q) \land (p \to \sim q))$, holds as well. With Residuation, $((p \land q) \to r) \leftrightarrow (p \to (q \to r))$, and the definition of the consistency connective, *17.59* becomes
$(p \circ p) \to ((p \to q) \to \sim (p \to \sim q))$
that is, if Aristotle's Thesis holds, Boethius' Thesis, $(p \to q) \to \sim (p \to \sim q)$, holds too. Given the equivalence between $p \circ p$, $\sim (p \to \sim p)$ and $\Diamond p$, these seem to be, although probably unintended, among the earliest appearances of "hypothetical", "default" (in the terminology of [23]) or "humble" (in the terminology of [7]) connexive theses.

3 Consistency and connexivity

Lewis and Langford's theory of consistency, possibility and implication has the following well-known consequences:
19.1 $\sim (p \circ p) \to \sim (p \circ q)$
19.11 $(p \circ q) \to (p \circ p)$
19.74 $\sim \Diamond p \to (p \to q)$
19.75 $\sim \Diamond \sim p \to (q \to p)$

The latter two are the infamous paradoxes of strict implication: an impossible proposition implies every other proposition, and a necessary proposition is implied by every other proposition. From *19.1*, given the interdefinibility of

[4] A connexive logic validates

$\sim (A \to \sim A)$	Aristotle's Thesis
$\sim (\sim A \to A)$	Variant of Aristotle's Thesis
$(A \to B) \to \sim (A \to \sim B)$	Boethius' Thesis
$(A \to \sim B) \to \sim (A \to B)$	Variant of Boethius' Thesis

and invalidates

$(A \to B) \to (B \to A)$	Non-symmetry of implication

For good introductions to connexive logics, see [11] or [25].

possibility and consistency, it easily follows that
$\sim\Diamond p \rightarrow\sim(p \circ q)$
whose intuitive phrasing would be "Something impossible is incompatible with everything". From *19.11*, given again the interdefinibility of possibility and consistency, one gets:
$(p \circ q) \rightarrow \Diamond p$
that is, if p is compatible with anything at all, p is possible.

Everett J. Nelson [16] reacted against all these consequences. As the reactions and objections to the paradoxes of strict implication are well-known and have come from different sources, I will focus on Nelson's objections to the unnumbered consequences. His starting point is a notion of consistency different from Lewis and Langford's, because according to Nelson there are pairs of propositions p and q such that each of them is impossible but which are nonetheless mutually consistent, for example "$(2+2) \neq 4$" and "$(3+3) \neq 6$". Nelson not only held that there are impossible propositions that are mutually consistent, but he also held that every proposition is self-consistent, even an impossible one, and that this self-consistency of an impossible does not prevent that it might be inconsistent with some other propositions. For example, "$1 = 0$" is consistent with itself, but it is inconsistent with "$3 \neq 2$". Thus, for Nelson, (in)consistency and (im)possibility come apart, and $(p \circ p)$ is a logical truth, it holds even for "$1 = 0$", but $\Diamond p$ is not, as "$1 = 0$" is not possible. Contradictoriness is a sufficient condition for Nelsonian inconsistency, that is, p and $\sim p$ are inconsistent, but as the example regarding "$(2+2) \neq 4$" and "$(3+3) \neq 6$" shows, $\sim p$ and $\sim q$ might be inconsistent too. Surely Nelson's notion of consistency needs a more precise treatment, and below I will offer a precisification, but this should suffice for now as an exemplification of the differences with his and Lewis and Langford's notions of consistency.

Nelson used the notion of consistency sketched above to give a validity condition for the conditional:
$(p \rightarrow q)$ is true if and only if the antecedent is inconsistent with the negation of the consequent.
Thus, for him $\sim(p \rightarrow\sim p)$ is also a logical truth.[5] Were Nelson right that every proposition is self-consistent, that is, if Aristotle's Thesis was a logical truth, all the theorems with a connexive flavor in *Symbolic Logic* would become fully connexive in the sense that they would be valid even without including Aristotle's Thesis as a premise (or antecedent), because if a logical truth implies a certain proposition p, p itself is a logical truth. Of course, this is not the case with Aristotle's Thesis in Lewis and Langford's theory, but it is in Nelson's (and connexive logics in general).[6]

[5] The intuitive argument for it goes as follows. Suppose that every proposition is either true or false, and that a proposition is false if and only if its negation is true. Now consider the conditional $p \rightarrow\sim p$. The negation of the consequent, $\sim\sim p$, is never inconsistent with the antecedent; hence, the conditional is never true. Then, the negation of the conditional is always true.

[6] For a recent detailed study of Nelson's ideas against the background of Lewis' work, see

Now, both ideas —Lewis and Langford's characterization of possibility in terms of consistency and Nelson's self-consistency of every proposition— have certain independent appealing. The problem is that they together entail possibilism, i.e. that every proposition is possible, which is maybe too much to swallow.

4 A limitative result

In fact, Omori proved in [19] that in any logic **L** satisfying

- $A \to A$
- $((A \to A) \to B) \leftrightarrow B$
- $(A \to B) \to ((C \to A) \to (C \to B))$
- $A \leftrightarrow \sim\sim A$
- $\sim (A \to B) \leftrightarrow (A \to \Diamond \sim B)$
- $\sim (A \to B) \leftrightarrow (A \to \sim B)$

plus the rules

- $A, A \to B \vdash_{\mathbf{L}} B$
- $A \to B \vdash_{\mathbf{L}} \Diamond A \to \Diamond B$
- Uniform Substitution

$\Diamond B \leftrightarrow B$ holds as well. It easily follows then that if the possibilist thesis, $\Diamond B$, is added to **L**, it becomes trivial.

For the sake of the argument, one could leave the rules and the first three items out of the discussion as they are valid in positive logic. Double negation can be granted, too. This reduces the room for disagreement to $\sim (A \to B) \leftrightarrow (A \to \Diamond \sim B)$ and $\sim (A \to B) \leftrightarrow (A \to \sim B)$, the Egré-Politzer's and Wansing's theses, respectively. In order to get more sense of what is going in here, let me describe briefly what is behind each of the theses.

Wansing has employed in several contexts a non-standard falsity condition for the conditional. (See for example [24].) More specifically, he has suggested to take the condition of the form "*If* A is true *then* B is false" rather than the condition of the form "A is true *and* B is false" as the falsity condition for a conditional of the form "If A then B", where truth and falsity are not necessarily exclusive. As a byproduct, the resulting logics turn out to be connexive logics which moreover validate the converses of Boethius' Theses, in particular $\sim(A \to B) \to (A \to \sim B)$.

On the other hand, Paul Egré and Guy Politzer [3] carried out an experiment related to the negation of indicative conditionals and considered *weak* conjunctive and conditional formulas of the forms $A \wedge \Diamond \sim B$ and $A \to \Diamond \sim B$, respectively, besides the more familiar *strong* conjunctive and conditional formulas of the forms $A \wedge \sim B$ and $A \to \sim B$, respectively, as formulas equivalent

[10].

to $\sim(A \rightarrow B)$.

Thus, Omori's result means that, up to non-triviality, it is not possible to have in a single framework, on the one hand, possibilism, and on the other, Wansing's view on negated conditionals and Egré and Politzer's view on negated conditionals. As I have showed, a form of possibilism results from combining Lewis and Langford's view on possibility with Nelson's ideas about consistency. The triviality result means that this mixture cannot be further combined with Wansing's and Egré and Politzer's views on negated conditionals.

The expected casualty is possibilism, i.e. the combination of Lewis and Langford's view on possibility with Nelson's ideas about consistency. Nonetheless, one could also question the assumptions leading to $\Diamond B \leftrightarrow B$. In a sense, this result is more problematic than possibilism itself because, informally, it identifies the possible with the actual. Possibilism may not be that much to swallow, at least in comparison with other options. Fortunately, there are certain formal tools already in circulation that can serve to model this strange mixture. The model is decidedly simple, but it has some notorious features.

In the following section it will be clear that, with good reason, one can blame instead half of the Egré-Politzer Thesis —namely, $(A \rightarrow \Diamond\sim B) \rightarrow\sim(A \rightarrow B)$— and make room for possibilism. In particular, one can show that, according to the model, $\Diamond A$ has to be always false in order to validate the Egré-Politzer Thesis.[7]

5 Connexivity

Consider a language $\mathcal{L}$ with a countable set of propositional variables and with at least the connectives of negation, $\sim$, and conditional, $\rightarrow$. Let $V = \{1, 0\}$ be a set of truth values. Consider a family of interpretations of $\mathcal{L}$, that is, relations $\sigma : \mathcal{L} \longrightarrow V$ —excluding, for any $A \in \mathcal{L}$, that both $1 \notin \sigma(A)$ and $0 \notin \sigma(A)$—, with logical validity defined in the following way (where Γ stands for a collection of formulas of $\mathcal{L}$):

$\Gamma \models A$ if and only if, for every σ, if $1 \in \sigma(B)$ for every $B \in \Gamma$, $1 \in \sigma(A)$

Consider now the following evaluation conditions for the conditional:
$1 \in \sigma(A \rightarrow B)$ if and only if $1 \notin \sigma(A)$ or $1 \in \sigma(B)$
$0 \in \sigma(A \rightarrow B)$ if and only if $1 \notin \sigma(A)$ or $0 \in \sigma(B)$
Such a conditional, together with a relatively uncontroversial evaluation condition for negation, like
$1 \in \sigma(\sim A)$ if and only if $0 \in \sigma(A)$
$0 \in \sigma(\sim A)$ if and only if $1 \in \sigma(A)$
satisfies the core of connexive logics, that is, it validates all of Aristotle's and

[7] Note that one could also decide to save the Egré-Politzer Thesis and blame instead *hyperconnexivity*, i.e. the converse of Boethius' Thesis: $\sim(A \rightarrow B) \rightarrow (A \rightarrow\sim B)$. However, discussing this would require a more complex apparatus; actually, the usual relational semantics for modalities would be more suitable. This is left for the forthcoming second part of this investigation. For a different take on Omori's result, see [17].

Boethius' Theses and their variants, and invalidates the Non-Symmetry of Implication.

What one has got so far is the **LP** negation and the Olkhovikov-Cantwell-Omori (OCO) conditional, first introduced in [18] and then introduced independently in [2] and [20]:

A	B	$\sim A$	$A \to B$
{1}	{1}	{0}	{1}
{1}	{1,0}	{0}	{1,0}
{1}	{0}	{0}	{0}
{1,0}	{1}	{1,0}	{1}
{1,0}	{1,0}	{1,0}	{1,0}
{1,0}	{0}	{1,0}	{0}
{0}	{1}	{1}	{1,0}
{0}	{1,0}	{1}	{1,0}
{0}	{0}	{1}	{1,0}

In what follows, and unless the contrary is stated, the arrow will stand for the OCO conditional.

One nice thing about this choice of connectives is that it allows for double negation elimination, and thus some formulas can be simplified to more manageable and familiar forms, for example
$\sim\sim(A \to \sim A)$ $(\sim\Diamond A)$ can be simplified to $(A \to \sim A)$,
$\sim(\sim A \to \sim\sim A)$ $(\Diamond \sim A)$ can be simplified to $\sim(\sim A \to A)$, and
$\sim\sim(\sim A \to \sim\sim A)$ $(\sim\Diamond\sim A)$ can be simplified to $(\sim A \to A)$.

Below there are the evaluations of the Lewis-Langford modalities according to the OCO conditional and the **LP** negation:

A	$\Diamond_L A$ $\sim(A \to\sim A)$	$\sim\Diamond_L A$ $\sim\sim(A \to\sim A)$	$\Diamond_L \sim A$ $\sim(\sim A \to\sim\sim A)$	$\sim\Diamond_L\sim A$ $\sim\sim(\sim A \to\sim\sim A)$
{1}	{1}	{0}	{1,0}	{1,0}
{1,0}	{1,0}	{1,0}	{1,0}	{1,0}
{0}	{1,0}	{1,0}	{1}	{0}

It is easy to check that Wansing's Thesis is valid according to these valuations, but Egré-Politzer's is not. Consider the case when B is true only and A is at least true, and then the right-to-left direction, $(A \to \Diamond \sim B) \to\sim (A \to B)$, is invalid. This seems correct: From the possibility of the consequent's falsity one cannot infer the actual falsity of the whole conditional, this a way too strong falsity condition.

There are several nice things to say about modalities so defined and evaluated with these connectives.

No modal collapse. A, $\Diamond_L A$ and $\sim \Diamond_L \sim A$ are different propositions because they are not equivalent, they do not have the same values under all interpretations, as can be simply checked by looking at the truth tables, and the same goes for $\sim A$, $\Diamond_L\sim A$ and $\sim\Diamond_L A$. True, given the definition of logical consequence as (forwards) truth-preservation in all interpretations, $A \dashv\vdash\sim \Diamond_L \sim A$

holds, which can be seen as a collapse if '$\sim \Diamond_L \sim A$' is understood as "A is necessary". Nonetheless, equivalence and inter-derivability (and co-implication, I would add) are conceptually different, no matter their simultaneous occurrence in several logics, and one has to be careful on what concept is employing to evaluate claims of collapse. I stick to difference in some interpretations as proof of non-equivalence. Nonetheless, $A \dashv\Vdash \sim \Diamond_L \sim A$ is a modal anomaly that has to be explained in due course.

Dualities between modalities and usual modal axioms. For the rest of the paper I will use '$\Box_L A$' as a shorthand for '$\sim \Diamond_L \sim A$'. This is justified, and this is a second nice thing to say about this framework, because the usual dualities between $\Diamond_L A$ and $\Box_L A$ hold[8]:
For all σ,
$\sigma(\Diamond_L A) = \sigma(\sim\Box_L \sim A)$
$\sigma(\sim\Diamond_L A) = \sigma(\Box_L \sim A)$
$\sigma(\Diamond_L \sim A) = \sigma(\sim\Box_L A)$
$\sigma(\sim\Diamond_L \sim A) = \sigma(\Box_L A)$

Also, all the usual modal axioms
(K) $\Box_L(A \to B) \to (\Box_L A \to \Box_L B)$
(T) $\Box_L A \to A$
(4) $\Box_L A \to \Box_L\Box_L A$
hold, as well as the Necessitation Rule
(NEC) From $\Vdash A$ to infer $\Vdash \Box_L A$

So much for the attractive features of the modalities so defined. In the next section, I will discuss some possible objections to this way of combining possibilism and connexivity, but before that, and to make things more interesting, let me add also conjunction and disjunction as evaluated in **LP**, plus the unary consistency connective '$\circ$' defined as
$1 \in \sigma(\circ A)$ if and only if $1 \in \sigma(A)$ and $0 \notin \sigma(A)$ or $0 \in \sigma(A)$ and $1 \notin \sigma(A)$
$0 \in \sigma(\circ A)$ if and only if $1 \in \sigma(A)$ and $0 \in \sigma(A)$
Then one gets the logic **dLP**.[9] Summarizing, (zeroth-order) **dLP** is characterized by the following truth tables:

A	B	$\sim A$	$\circ A$	$A \wedge B$	$A \vee B$	$A \to B$
{1}	{1}	{0}	{1}	{1}	{1}	{1}
{1}	{1,0}	{0}	{1}	{1,0}	{1}	{1,0}
{1}	{0}	{0}	{1}	{0}	{1}	{0}
{1,0}	{1}	{1,0}	{0}	{1,0}	{1}	{1}
{1,0}	{1,0}	{1,0}	{0}	{1,0}	{1,0}	{1,0}
{1,0}	{0}	{1,0}	{0}	{0}	{1,0}	{0}
{0}	{1}	{1}	{1}	{0}	{1}	{1,0}
{0}	{1,0}	{1}	{1}	{0}	{1,0}	{1,0}
{0}	{0}	{1}	{1}	{0}	{0}	{1,0}

[8] The dualities fail in, for example, Wansing's connexive modal logic **CK**; see [24]. The failure might be a good thing, though; see [26] for an argument to that effect.

[9] First presented, with different primitives though, in [18] and then independently in [20].

which is basically **LP** in the $\{\sim, \wedge, \vee\}$-fragment, augmented with the expressive power allowed by the unary consistency connective and the OCO conditional.

It is easy to check that $\Diamond_L A$ and $\circ A$ are not equivalent, a result that would have pleased Nelson, and this means that $A \circ A$ (self-consistency) and $\circ A$ (consistency simpliciter) are not equivalent, either. This seems in the right track: one thing is that a proposition does not imply its own negation (self-consistency), and another is that a proposition does not imply an arbitrary contradiction (consistency).

When one goes fully to **dLP**, the dualities and the validities are preserved, and even more nice things appear. For example, two "relative consistency" binary connectives can be defined. One of them is more "Lewisian", order-sensitive, non-symmetric, as discussed for example in [21]:
$\sigma(A \circ_L B) = \sigma(B)$ unless $\sigma(A) = \{0\}$, and $\sigma(A \circ_L B) = \{1,0\}$ in that latter case [10]
which is but the OCO condtional. The other is more "Nelsonian":
$1 \in \sigma(A \circ_N B)$
$0 \in \sigma(A \circ_N B)$ if and only if $0 \in \sigma(A)$ or $0 \in \sigma(B)$

While in general $A \circ_L B$ and $A \circ_N B$ are not equivalent, $A \circ_L A$ and $A \circ_N A$ are for any A, so let me write $A \circ_X A$ to express such indistinctness. Thus, the self-consistency $A \circ_X A$ of any proposition A is a theorem of **dLP**; again, a result that would have pleased Nelson. Nonetheless, the (Nelsonian) relative consistency of any two propositions, $A \circ_N B$, also becomes a theorem in **dLP**, something that definitely would not have pleased Nelson, "because some propositions are inconsistent with others." [16, 443] However, Nelson thought that the inconsistency of two distinct propositions cannot be determined by pure logic alone, and that is reflected both in his truth table and the evaluation conditions of $A \circ_N B$. [11]

6 Some concerns

Combining zeroth-order logic, functionality, finite many-valuedness, modalities and highly non-classical theses seems like a recipe for disaster. Let me address three potential worries here. I do not aim at dispelling all air of doubt, that seems nearly impossible in philosophical issues; I only want to show that some objections usually raised to approaches like the one presented above are far from being knock-down.

---

[10] For simplicity, I use the following convention. Let v_j and v_k be our two truth values. Then '$\sigma(X) = \{v_j\}$' means that $v_j \in \sigma(X)$ and $v_k \notin \sigma(X)$, whereas '$\sigma(X) = \{v_j, v_k\}$' means that $v_j \in \sigma(X)$ and $v_k \in \sigma(X)$.

[11] Finally, recall that
$(A \circ B) =_{def.} \sim(A \rightarrow \sim B)$
is the usual definition of fusion (an intensional conjunction) $A \circ B$ in the logic **R**. But $(A \circ_L B)$ can difficultly be regarded as a conjunction, for it is true even when no component is true. This reflects the fact that the OCO conditional is false when both antecedent and consequent are false.

Modal anomalies. It has been a long time since Łukasiewicz offered a many-valued analysis of modalities. Since then, such an approach has lived in discredit because they give rise to *modal anomalies*, that is, highly counterintuitive arguments regarding modalities are validated.

The typical criticism raised against the many-valued approaches to possibility and necessity is of the sort of those found in [5]. The objection is basically that the many-valued notions of possibility and necessity validate several counterintuitive arguments. Dugundji's theorem, that no modal logic between Lewis' **S1** and **S5** can be characterized by a finite many-valued matrix, seems to give more content to the criticisms.

Dugundji's result would be a devastating problem if all and only modalities worth considering lied between **S1** and **S5**, but that is not the case. Consider a modality $\mathcal{J}$ satisfying the following two axiom schemas, where '$\rightarrow$' stands again for a generic conditional:

$\mathcal{J}(A \rightarrow B) \rightarrow (\mathcal{J}A \rightarrow \mathcal{J}B)$

$A \rightarrow \mathcal{J}A$

When $\mathcal{J}A$ is identified with $\Box A$ and the two axioms schemas above are added to classical logic, the resulting logic is simply $\mathbf{K}$+$(A \rightarrow \Box A)$, which is characterized by certain single-element frames. However, $\mathcal{J}A$ cannot be rightly identified with $\Box$ precisely because of $A \rightarrow \mathcal{J}A$; on the other hand, it cannot be rightly identified with $\Diamond$ because of $\mathcal{J}(A \rightarrow B) \rightarrow (\mathcal{J}A \rightarrow \mathcal{J}B)$. Moving to a different logic, intuitionistic logic, for example, allows to study interesting models for these axiom schemas. The resulting modality $\mathcal{J}A$ has an hybrid nature, but still closer to possibility, and has appeared in different contexts, interpreted as variedly as "(at some underlying topological space), it is locally the case that" (as in topos theory, where it first appeared) or "under some family of constraints, the hardware device behaves according to" (as in propositional lax logic); see [6, Section 7.6] for an overview of the different standard incarnations of $\mathcal{J}$. This means that in the presence of different logics, some counterintuitive axiom schemas can make sense for certain modalities. And that was Łukasiewicz's reaction to the modal anomalies in his logic: one could try to make sense of the modalities involved as defined within the logic so as to explain away the unintuitiveness of certain axiom schemas. Whether his personal attempt succeeded or not for his logics is a different issue from the correctness of the methodological advice. The case of $\mathcal{J}A$ proves that the attempts are not a priori doomed to fail.

Let me consider explicitly the schemas that worry Font and Hájek, all of them valid in **dLP**:

FH1. $(\Diamond A \wedge \Diamond B) \rightarrow \Diamond(A \wedge B)$

FH2. $(A \rightarrow B) \rightarrow (\Box A \rightarrow \Box B)$

FH3. $(A \rightarrow B) \rightarrow (\Diamond A \rightarrow \Diamond B)$

FH4. $\Box A \rightarrow (B \leftrightarrow \Box B)$

FH5. $\Box A \rightarrow (\Diamond B \leftrightarrow \Box B)$

What I have said above on logics not between **S1** and **S5** could serve to partially alleviate the concerns by Font and Hájek. Nonetheless, a more substantial reply can be given. FH1 seemed problematic because contradictions were for a long time the archetype of impossibility. The instance $(\Diamond A \wedge \Diamond \sim A) \rightarrow \Diamond(A \wedge \sim A)$ was regarded as a counterexample to the schema: even if A and $\sim A$ were separately possible, jointly they are not. That they could be true at some states of evaluation in certain semantics —the objection continues— does not make them less impossible: those states are *impossible* states after all. That is a moot point: one could argue that possibility is a property of propositions relative to states, not of states themselves, and that being possible (at a state) just means to be true at some accessible state. Nonetheless, $\Diamond(A \wedge \sim A)$ should be expected in a framework where the idea that everything is possible is taken seriously. More than a drawback of the logic, this should be a welcomed result. However, there is more to be said in favor of it, and it will become evident when discussing the next objection about possibilism.

Of the next couple of schemas, FH2 and FH3, Font and Hájek said that, had Łukasiewicz decided to interpret the arrow in the antecedent as a strict implication and not as a material one, the resulting schemas would have been more acceptable. The schemas also hold in the **dLP** setting. Note that the charge of unacceptability due to the material nature of the arrow in the antecedent does not apply here, as $A \rightarrow_{OCO} B$ is neither equivalent nor interderivable with $\sim (A \wedge \sim B)$ nor $\sim A \vee B$. It is not a strict conditional either, but it still encapsulates a sort of intensionality in not making plainly true a conditional whose antecedent is not true (only). One could object that even if $A \rightarrow_{OCO} B$ comes with some intensionality within it, it is not strong enough as to be counted as a sufficient condition for $(\Box A \rightarrow_{OCO} \Box B)$, so $(A \rightarrow_{OCO} B) \rightarrow_{OCO} (\Box A \rightarrow_{OCO} \Box B)$ could still be regarded as genuinely anomalous. Nonetheless, the intensionality within $A \rightarrow_{OCO} B$ seems sufficient to imply $(\Diamond A \rightarrow_{OCO} \Diamond B)$.

Nevertheless, Font and Hájek say that the validity of the last two schemas "is the main reason for [their] claim that as a logic of possibility and necessity, it [Łukasiewicz four-valued modal logic] is a dead end." Informally, $\Box A \rightarrow (B \leftrightarrow \Box B)$ expresses that if something is necessary, any truth simpliciter is also a necessary truth. But just recall what the modalities mean in this context. '$\Diamond_L A$' means that it is false that A implies its own negation, and '$\Box_L A$' means that A is implied by its own negation. So $B \rightarrow_{OCO} \Box_L B$ becomes $B \rightarrow_{OCO} (\sim B \rightarrow_{OCO} B)$. The validity of this does not seem so abhorrent.

Nonetheless, even if B is true, $(\sim B \rightarrow_{OCO} B)$ contradicts Aristotle's Thesis, so it must be false. And it is. This implies that both $\Box A \rightarrow (B \leftrightarrow_{OCO} \Box_L B)$ and $\sim(\Box A \rightarrow_{OCO} (B \leftrightarrow \Box B))$ are valid in **dLP**, and the same holds for $\Box_L A \rightarrow_{OCO} (\Diamond_L B \leftrightarrow_{OCO} \Box_L B)$ and $\sim(\Box_L A \rightarrow_{OCO} (\Diamond_L B \leftrightarrow_{OCO} \Box_L B))$, as can be easily checked. This means that there is inconsistency surrounding certain combinations of truth, possibility and necessity, and notice that they are the modal anomalies: both $(\Diamond_L A \wedge \Diamond_L \sim A) \rightarrow_{OCO} \Diamond_L(A \wedge \sim A)$ and$\sim((\Diamond_L A \wedge \Diamond_L \sim A) \rightarrow_{OCO} \Diamond_L(A \wedge \sim A))$ hold as well in **dLP**. In fact, among all the axiom schemas highlighted by Font and Hájek, $(A \rightarrow B) \rightarrow (\Diamond A \rightarrow \Diamond B)$ is the only one that does not come with its negation validated too in **dLP**, and

I have already argued that this should not count as a so terribly bad anomaly. (Modal anomalies are not so widespread; $\Diamond_L A \rightarrow_{OCO} A$ and $\Diamond_L A \rightarrow_{OCO} \Box_L A$ simply fail, for example.)

There is another route to alleviate the concerns regarding the modal anomalies, namely that in non-classical contexts not all theorems need to be created equal: one could distinguish between different degrees of satisfiability; in particular, between different degrees of theoremhood. This does not mean that one needs to change the notion of logical validity of **dLP** to get a more refined set of logical truths; one can keep the usual definition of logical validity, and regarding as an extra task selecting among the logically true propositions those that meet additional criteria. Given that theorems are limit cases of logically valid arguments, I can borrow some terminology from the variety of notions of logical validity already available in the literature. [12]

Let me call then 'p-theorems' those formulas that are never antidesignated; 'T-theorems' those formulas that are always designated; 'supertheorems' those formulas that are always designated and at least once (just) true; and 'q-theorems' those formulas that are always (just) true. [13]

Unlike **LP**, **dLP** has q-theorems and they might take the following forms:
(q-t i) $\circ\circ A$,
(q-t ii) $A©B$, for © $\in \{\wedge, \vee, \rightarrow\}$ and where both A and B are themselves q-theorems, or
(q-t iii) $\sim A$, where A has the form $\sim B$ and B is a **dLP** q-theorem.

Then, in **dLP**, for any formula A, both $\Diamond_L A$ and $\Box_L \Diamond_L A$ are **dLP** T-theorems, but in general only $\Diamond_L A$ is a **dLP**-supertheorem and sometimes it can be a q-theorem (when A already is one), while $\Box_L \Diamond_L A$ can only be at most a **dLP** T-theorem. Also, $A \rightarrow \Box_L A$ is always just a **dLP** T-theorem, but cannot even be a **dLP**-supertheorem, let alone a q-theorem, because it is never just true.

With such a distinction between theorems, **dLP** can become even closer to Mortensen's possibilism. If all **dLP**-theorems are treated on equal footing, there are some **dLP**-theorems of the form $\Box_L A$, whereas according to logical possibilism there should be no logical truths of such form. But with the further distinctions just drawn, formulas of the form $\Box_L A$ are at most **dLP** T-theorems, whereas all formulas of the form $\Diamond_L A$ are at the very least **dLP**-supertheorems.

Possibilism. Another concern is about a very special "modal anomaly", namely the commitment to *possibilism*, as per the validity of $\Diamond_L A$. For example, Béziau [1] has objected to evaluations of modalities like the ones presented here on the grounds that they make possibility "trivial", in the sense that they make everything possible, and that is not a good result for a theory of the

[12] For a good introduction to the topic and critical discussion of it, see [27].

[13] And I will stop here. If one gives falsity a treatment independent of truth, as it should be done logically, one could obtain even more shades of theoremhood, but those already introduced suffice for my purposes here.

possible.

However, beyond the incredulous stare towards the validity of $\Diamond_L A$, no reasons to reject possibilism have been put forward. Let me consider two potential objections. The first is that possibilism entails trivialism; another is that possibilism is ruled out by the very definitions of logical notions.

The proof that possibilism entails triviality is as follows:

1. $\Diamond\Box A \to \Box A$	Axiom (5)
2. $\Box A \to A$	Axiom (T)
3. $\Diamond A$	Possibilism
4. $\Diamond\Box A$	3, Uniform Substitution
5. $\Box A$	1, 4, Detachment
6. A	2, 5, Detachment

However, notice that the axiom (5) occurring in the first line of the proof is not validated by the tables in Section 5: make A just false. What is validated is its "contraposed" version, $\Diamond_L A \to \Box_L \Diamond_L A$. They are not equivalent, and that is rightly so because they are conceptually distinct. '$\Diamond_L A \to \Box_L \Diamond_L A$' expresses that one can go, so to speak, from the possibility of something to the necessity of its possibility, which is right according to this version of possibilism. '$\Diamond_L \Box_L A \to \Box_L A$' expresses something different, namely that from the possibility of necessity of something, one can go to its necessity, which is wrong according to possibilism because nothing, except possibilities, is necessary. So, neither the argument from possibilism to triviality is valid here, nor one has anything as strong as **S5** modalities in the current setting.[14]

There is also the concern that possibilism is ruled out by the very definitions of logical notions. By this I mean that logical notions as characterized by, say, evaluation conditions, imply the untruth of possibilism. Let me consider first the very notion of possibility. To minimize the risk of begging the question, let me move to more common ground, the usual relational falsity condition for possibility:

- $\Diamond A$ is false at a state i iff A is false at all state j related to i.

Suppose for the sake of the argument that A is in fact false at all state j related to i. Does this mean that there is no j related to i where A is true? If the answer is affirmative, one should ask whether that conclusion comes from the falsity condition alone or whether it comes from additional considerations, for example, certain ideas about the structure of truth values, that they are exclusive maybe.

Of course, more elaborate anti-possibilist arguments can be given. They might involve the characterization of other logical notions, such as conditionals, quantifiers or even logical consequence itself. I cannot go through all those arguments. What I want to highlight is that, in any case, one must wonder whether possibilism is ruled out by evaluation conditions alone or whether other, logic-specific elements —such as the number and structure of truth val-

[14] Note that the "contraposed" versions of (K), (T), and (4) do hold, though.

ues, or the properties of the accessibility relation— are used as well.[15]

Finally, as we have seen, possibilism does not prevent having a sensible theory of modalities that satisfies their mutual distinguishability, their usual interdefinability through negation and without a plethora of non-standard modal theorems accompanying $\Diamond_L A$.[16]

Dialetheism. Finally, there is the concern that **dLP** is not only dialetheist, that is, that it makes room for sentences of the forms A and $\sim A$ that are simultaneously satisfiable, but is also dialetheic (or contradictory or negation-inconsistent, hence the '**d**' in front of '**LP**'!), since it validates contradictory theorems such as

$(A\wedge \sim A) \rightarrow\sim A$ and $\sim((A\wedge \sim A) \rightarrow\sim A)$;

$(A\wedge \sim A) \rightarrow (B\vee \sim B)$ and $\sim((A\wedge \sim A) \rightarrow (B\vee \sim B))$

Furthermore, one can define in **dLP** a new negation $\neg A$ as $\sim A \wedge \circ A$ (or $\sim\circ(A \rightarrow\sim\circ\circ A)$, if conjunction is not available), and then obtain the following **dLP** theorems:

$(A \wedge \neg A) \rightarrow B$ and $\sim((A \wedge \neg A) \rightarrow B)$.

People way more than talented than I have spent up to 40 years trying to convince others that dialetheism is not outrageous, and they are still struggling; see [22] for a book-length defense of dialetheism. In these paragraphs I can only aspire to push further to the already converted: if they have given dialetheism a chance, maybe they can give a contradictory logic a chance too.

One attempt of reassurance might use the terminology from the discussion about modal anomalies: the contradictory theorems of **dLP** are just **dLP** T-theorems. Furthermore, it must also be noticed that **dLP** exhibits contradictions in already expected and significant places for some connexive logicians, namely around Explosion, certain forms of Simplification —more specifically, simplification of contradictories— and the irrelevant Safety, i.e. $(A\wedge \sim A) \rightarrow (B\vee \sim B)$. This reminds me of the situation in faced by Meyer and Martin in investigating Aristotle's syllogistic. Meyer and Martin wanted to provide a logic for Aristotle's syllogistic, which was not reflexive. In their logic **SI∼I**, see [12], $A \rightarrow A$ was treated as a borderline case, both a fallacy and a validity, hence the validity of both $A \rightarrow A$ and $\sim(A \rightarrow A)$. Perhaps the contradictory theorems in **dLP** can be treated similarly as borderline cases: they should be invalid, as many connexivists have said, but also the validity of such schemas is almost necessitated by a truth-functional, truth-preserving logic, with the standard evluations for negation, conjunction and disjunction.

---

[15] Chris Mortensen in [14] defends "(logical) possibilism", by which he means the idea that everything is possible (possibilism *stricto sensu*, I would say) and nothing is necessary (non-necessitarianism). See also [15] to complete his picture about possibilism. I have addressed some lacunae and further consequences elsewhere (see [4]).

[16] Mortensen himself found a sort of possibilism around connexivity when in [13] he proved that the logic **E** plus Aristotle's Thesis implies $\Diamond A$ for every A, with $\Diamond A$ defined as $\sim((\sim A \rightarrow\sim A) \rightarrow\sim A)$, which would amount to A in the present context.

7 More modalities

The presence of another, more classical negation in **dLP** allows further definitons of modalities in **dLP**, for example, as follows:

A	$\Diamond A$ $\neg(A \to \neg A)$	$\neg\Diamond A$ $\neg\neg(A \to \neg A)$	$\Diamond\neg A$ $\neg(\neg A \to \neg\neg A)$	$\neg\Diamond\neg A$ $\neg\neg(\neg A \to \neg\neg A)$
{1}	{1}	{0}	{0}	{1}
{1,0}	{1}	{0}	{0}	{1}
{0}	{0}	{1}	{1}	{0}

Let me write '$\Diamond_{L^C}$' for the possibility defined with Aristotle's Thesis written with such a strong negation. In this case, possibilism is lost and $\Diamond_{L^C} A$ and $\neg\Diamond_{L^C}\neg A$, on the one hand, and $\neg\Diamond_{L^C} A$ and $\Diamond_{L^C}\neg A$, on the other, collapse.

But the two negations can interact in interesting ways. For example, $\neg\ (A\ \to\ \neg\ A)$ defines in the three-valued setting what can be called 'the Béziau possibility', a unary connective © such that $\sigma(©A) = \{0\}$ if and only if $\sigma(A) = \{0\}$, and $\sigma(©A) = \{1\}$ in all other cases, so let me write it as '$\Diamond_B$'. Defining modalities based on the Béziau possibility with $\sim$ instead of $\neg$ —namely, $\sim\Diamond_B A$, $\Diamond_B \sim A$, $\sim\Diamond_B \sim A$— produce modalities different from the L^C modalities as follows:

A	$\Diamond_B A$ $\neg(A \to \neg A)$	$\sim\Diamond_B A$ $\sim\neg(A \to \neg A)$	$\Diamond_B \sim A$ $\neg(\sim A \to \neg \sim A)$	$\sim\Diamond_B \sim A$ $\sim\neg(\sim A \to \neg \sim A)$
{1}	{1}	{0}	{0}	{1}
{1,0}	{1}	{0}	{1}	{0}
{0}	{0}	{1}	{1}	{0}

$\sim\neg(\sim A \to \neg \sim A)$ defines 'Béziau's necessity', a unary conective ©A such that it is true if and only if A is true and is false in every other case.

Dually, defining modalities based on the Lewis-Langford possibility with $\neg$ instead of $\sim$ —namely, $\neg\Diamond_L A$, $\Diamond_L\neg A$, $\neg\Diamond_L\neg A$— produce yet another set of new modalities as follows:

A	$\Diamond_L A$ $\sim(A \to\sim A)$	$\neg\Diamond_L A$ $\neg\sim(A \to\sim A)$	$\Diamond_L\neg A$ $\sim(\neg A \to\sim\neg A)$	$\neg\Diamond_L\neg A$ $\neg\sim(\neg A \to\sim\neg A)$
{1}	{1}	{0}	{1,0}	{0}
{1,0}	{1,0}	{0}	{1,0}	{0}
{0}	{1,0}	{0}	{1}	{0}

Notice that these modalities are even closer to Mortensen's possibilism (possibilism proper and non-necessitarianism) right from the outset, without distinguishing between kinds of theorems: all necessities and impossibilities are just false, and all possibilities are always designated. [17]

---

[17] Incidentally, rewriting '$\sim((\sim A \to\sim A) \to\sim A)$' —the possibility used in [13] to show that **E** plus Aristotle's Thesis is possibilist— as '$\neg((\neg A \to \neg A) \to \neg A)$' makes it equivalent to $\Diamond_B$, not to A as before.

8 Conclusions

In this paper, I explored how to retain Lewis and Langford's characterization of possibility in terms of consistency and Nelson's idea that all propositions are self-consistent. I started by presenting Lewis and Langford's conceptualization of the notions of consistency and possibility, and how certain connexive notions appear there. Then I quickly reconstructed Nelson's reaction towards some of the consequences of Lewis and Langford's proposal and his arguments to prefer a primitive notion of consistency, not definable in terms of possibility. After that, I showed that Omori's logic **dLP** can support the combination of Lewis and Langford's characterization of possibility in terms of consistency with Nelson's idea that every proposition is self-consistent, with the corresponding outcome that all the formulas of the form $\Diamond A$ are theorems. By adding more conceptual tools, I showed that one can even get a connexive model of Mortensen's possibilism, where not only everything is possible, but nothing is necessary. Finally, I discussed some worries about the project, for example, regarding some modal anomalies or the motivations for a logic with contradictory theorems. Discussing one of those concerns led me to pay closer attention to the two negations available in **dLP** and then consider how the modalities behave in the presence of each negation, to find further conceptual insights and new formulations of modalities even closer to Mortensen's possibilism.

A paper would not be as enjoyable if it did not open at least twice the number of questions it tried to address. Let me indicate then some avenues for further exploration. It is well-known that the expressive power of a logic is inversely proportional to its deductive power: the more you can prove, the less distinctions you can draw. As **dLP** is based on **LP**, the obvious choice for a weaker logic is one based on **FDE**. If one adds the OCO conditional to **FDE** then one gets a connexive dialetheic expansion of **FDE**, already discussed in [25] under the name 'material connexive logic'. [18] An open problem is then, investigating the exact shape of the multi-modal features of both **dLP** and **dFDE**, including further interactions between possibilism and connexivity. Going four-valued could also alter the truth conditions given for the relative consistency connective and this in turn could produce a non-Nelsonian split between self-implication and self-consistency, with tremendous consequences for the theories of modalities, connexivity itself and so on.

Further connections between connexivity and possibilism, that is, how starting with versions of one can lead to versions of the other, using frameworks not necessarily in the vicinity of **dLP** and **dFDE**, would be worth exploring too.

[18] A connexive variant of the more general version of Belnap-Dunn logic, including the negation $\neg$, was studied in [20] under the name '**dBD**'.

References

[1] Jean-Yves Béziau. A new four-valued approach to modal logic. *Logique et Analyse*, 54(213):109–121, 2011.

[2] John Cantwell. The logic of conditional negation. *Notre Dame Journal of Formal Logic*, 49(3):245–260, 2008.

[3] Paul Egré and Guy Politzer. On the negation of indicative conditionals. In Maria Aloni, Michael Franke, and Floris Roelofsen, editors, *Proceedings of the 19th Amsterdam Colloquium, AC 2013*, pages 10–18. 2013.

[4] Luis Estrada-González. Models of possibilism and trivialism. *Logic and Logical Philosophy*, 22(4):175–205, 2012.

[5] Josep María Font and Petr Hájek. On Łukasiewicz's four-valued modal logic. *Studia Logica*, 70(2):157–182, 2002.

[6] Robert Goldblatt. Mathematical modal logic: A view of its evolution. *Journal of Applied Logic*, 1(5-6):309–392, 2003.

[7] Andreas Kapsner. Humble connexivity. *Logical and Logical Philosophy*, 28(3):513–536, 2019.

[8] Clarence Irving Lewis. *A Survey of Symbolic Logic.* University of California Press, Berkeley, 1918.

[9] Clarence Irving Lewis and Cooper Harold Langford. *Symbolic Logic.* Dover Publications, New York, 1932.

[10] Ed Mares and Francesco Paoli. C.I. Lewis, E.J. Nelson, and the modern origins of connexive logic. *Organon F*, 26(3):405–426, 2019.

[11] Storrs McCall. A history of connexivity. In Francis Jeffry Pelletier Dov M. Gabbay and John Woods, editors, *Handbook of the History of Logic. Vol. 11. Logic: A History of its Central Concepts*, pages 415–449. Elsevier, Amsterdam, 2012.

[12] Robert Meyer and Errol Martin. S revisited. *Australasian Journal of Logic*, 16(3):49–67, 2019.

[13] Chris Mortensen. Aristotle's Thesis in consistent and inconsistent logics. *Studia Logica*, 43(1-2):107–116, 1984.

[14] Chris Mortensen. Anything is possible. *Erkenntnis*, 30(3):319–337, 1989.

[15] Chris Mortensen. It isn't so, but could it be? *Logique et Analyse*, 48(189–192):351–360, 2005.

[16] Everett J. Nelson. Intensional relations. *Mind*, 39(156):440–453, 1930.

[17] Ricardo Arturo Nicolás-Francisco. A note on three approaches to connexivity. *Felsefe Arkivi—Archives of Philosophy*, (51):129–138, 2019.

[18] Grigory K. Olkhovikov. On a new three-valued paraconsistent logic. In *Logic of Law and Tolerance*, pages 96–113. Ural State University Press, Yekaterinburg, 2001. In Russian.

[19] Hitoshi Omori. Towards a bridge over two approaches in connexive logic. *Logical and Logical Philosophy*, 28(3):553–566.

[20] Hitoshi Omori. From paraconsistent logic to dialetheic logic. In Holger Andreas and Peter Verdée, editors, *Logical Studies of Paraconsistent Reasoning in Science and Mathematics*, pages 111–134. Springer, Berlin, 2016.

[21] Claudio Pizzi. Cotenability and the logic of consequential implication. *Logic Journal of the IGPL*, 12(6):561–579, 2004.

[22] Graham Priest. *In Contradiction: A Study of the Transconsistent.* Oxford University Press, 2006. Second edition.

[23] Matthias Unterhuber. Beyond system P –Hilbert-style convergence results for conditional logics with a connexive twist. *IFCoLog Journal of Logics and their Applications*, 3(3):377–412, 2016.

[24] Heinrich Wansing. Connexive modal logic. In Renate Schmidt, Ian Pratt-Hartmann, Mark Reynolds, and Heinrich Wansing, editors, *Advances in Modal Logic, Volume 5*, pages 367–383. King's College Publications, London, 2005.

[25] Heinrich Wansing. Connexive logic. In Edward N. Zalta, editor, *The Stanford Encyclopedia of Philosophy.* CSLI Stanford, Spring 2020 edition, 2020.

[26] Heinrich Wansing and Sergei Odintsov. Constructive predicate logic and constructive modal logic. formal duality versus semantical duality. In Vincent Hendricks, Fabian Neuhaus, Uwe Scheffler, Stig Andur Pedersen, and Heinrich Wansing, editors, *First-Order Logic Revisited*, pages 269–286. Logos Verlag, Berlin, 2004.

[27] Heinrich Wansing and Yaroslav Shramko. Suszko's Thesis, inferential many-valuedness, and the notion of a logical system. *Studia Logica*, 88(3):405–429, 2008.

Modal Logic for Induction

Giulio Fellin[1,3,4] and Sara Negri[2,3] and Peter Schuster[1]

[1] *Università di Verona, Dipartimento di Informatica*
Strada le Grazie 15, 37134 Verona, Italy
e-mail: {giulio.fellin,peter.schuster}@univr.it

[2] *Università di Genova, Dipartimento di Matematica*
via Dodecaneso 35, 16146 Genova, Italy
e-mail: sara.negri@unige.it

[3] *University of Helsinki, Department of Philosophy*
P.O. Box 24 (Unioninkatu 40 A), 00014 University of Helsinki, Finland
e-mail: {giulio.fellin,sara.negri}@helsinki.fi

[4] *Università di Trento, Dipartimento di Matematica*
Via Sommarive 14, 38123 Povo (Trento), Italy
e-mail: giulio.fellin@unitn.it

Abstract

We use modal logic to obtain syntactical, proof-theoretic versions of transfinite induction as axioms or rules within an appropriate labelled sequent calculus. While transfinite induction proper, also known as Noetherian induction, can be represented by a rule, the variant in which induction is done up to an arbitrary but fixed level happens to correspond to the Gödel–Löb axiom of provability logic. To verify the practicability of our approach in actual practice, we sketch a fairly universal pattern for proof transformation and test its use in several cases. Among other things, we give a direct and elementary syntactical proof of Segerberg's theorem that the Gödel–Löb axiom characterises precisely the (converse) well-founded and transitive Kripke frames.

Keywords: Induction principles, elementary proofs, modal logic, proof theory, Kripke model, sequent calculus.

1 Introduction

At least since Peano formalised what we all know as mathematical induction, induction as a proof principle has been the main tool for tidily unwrapping the potential infinite as generated by an a priori incomplete process. This is well reflected by the ubiquity of definitions and proofs by induction in today's ever more formal sciences.

Transfinite induction is a generalisation of mathematical induction from the natural numbers to less down-to-earth well-founded orders, such as the ordinal numbers. More precisely, if (and only if) any given order is well-founded, then *induction* holds: in the sense that a predicate holds everywhere in the given order provided that the predicate is progressive, i.e. propagates from all predecessors of a given element to the element itself.

As a rule of thumb, instances of induction are applicable more directly, and are better behaved proof-theoretically, than the corresponding instances of well-foundedness, which come as extremum principles or chain conditions (see, e.g., Proposition 4.2 below). Characteristic examples include Aczel's Set Induction [1–3] versus von Neumann and Zermelo's Axiom of Foundation or Regularity, and Raoult's Open Induction [4,11,27] as opposed to Zorn's Lemma.

Awareness of this phenomenon brought us to carry over to the inductive side some occurrences of well-foundedness in the modal logic of provability. Perhaps Segerberg's theorem [35], which stood right at the beginning of an impressive development [9], is the most prominent case: the Gödel–Löb axiom characterises exactly the (converse) well-founded and transitive Kripke frames. [1] The observation that those occurrences are rather about induction prompted the present investigation.

Inasmuch as instances of induction are about predicates or subsets, they typically go beyond the given logical level, and actually have a somewhat semantic flavour [12, 13]. By modal logic [5, 24, 26] we now obtain syntactical, proof-theoretic variants of induction: they are expressed as axioms or rules within an adequate labelled sequent calculus [21, 23]. While induction proper, for which we say Noetherian induction, can be mirrored by a rule (Lemma 3.3), the variant in which induction is done up to an arbitrary but fixed point of the given order, which we dub Gödel–Löb induction, happens to correspond (Lemma 3.1) to the homonymous axiom of provability logic [6,7,19,36,38]. [2] In fact the usual way to define validity in a Kripke model for the modal operator $\Box$ lends itself naturally to capture universal validity up to a point.

To verify the practicability of our approach in proof practice, we give a fairly universal pattern for proof transformation, from rather algebraic inductive proofs to formal proofs with the required rules, and test this in several cases. Among other things, we prove with the corresponding modal rules that induction necessitates the order under consideration to be irreflexive (Lemma 4.1), and that every meet-closed inductive predicate on a poset propagates from the irreducible elements to any element whatsoever (Example 3.5) [28, 31, 32]. As a by product we gain the curiosity that Noetherian induction is tantamount to the corresponding chain condition plus irreflexivity (Proposition 4.2). [3] Last but not least we give a direct and elementary syntactical proof (Theorem 4.3) of Segerberg's aforementioned theorem that the Gödel–Löb axiom holds exactly in the (converse) well-founded and transitive Kripke frames. All this can also be useful in proof practice: while it might be cumbersome to prove directly that an induction principle holds for a given order, it is often easier to check properties such as irreflexivity and transitivity, or even chain conditions.

---

[1] See also, for example, Theorem 3.5 of [37], Example 3.9 of [5] and Teorema 7.2 of [24].

[2] This was also called axiom A3 [37], the Löb formula L [5] and axiom G or axiom W [17,23].

[3] Needless to say, this requires some countable choice.

2 Basic modal logic K

Modal logic is obtained from propositional logic by adding the modal operator $\Box$ to the language of propositional logic. A *Kripke model* [18] (X, R, val) is a set X together with an *accessibility relation* R, i.e. a binary relation between elements of X, and a valuation val, i.e. a function assigning one of the truth values 0 or 1 to an element x of X and an atomic formula P. The usual notation is for $\mathrm{val}(x, P) = 1$ is $x \Vdash P$.

We read "xRy" as "y is *accessible* from x" and we read "$x \Vdash P$" as "x *forces* P". Valuations are extended in a unique way to arbitrary formulae by means of inductive clauses:

$$
\begin{aligned}
&x \nVdash \bot \\
&x \Vdash A \supset B \text{ if and only if } x \Vdash A \Rightarrow x \Vdash B \\
&x \Vdash A \wedge B \text{ if and only if } x \Vdash A \text{ and } x \Vdash B \\
&x \Vdash A \vee B \text{ if and only if } x \Vdash A \text{ or } x \Vdash B \\
&x \Vdash \Box A \text{ if and only if } \forall y (xRy \Rightarrow y \Vdash A)
\end{aligned}
$$

We assume that $x \Vdash P$ is decidable for every $x \in X$ and each atomic formula P, which carries over to arbitrary formulae by the inductive clauses. With the intended applications in mind, in place of R we use the inverse accessibility relation $<$, i.e. we stipulate that $y < x$ if and only if xRy. The pair $(X, <)$ is then dubbed *Kripke frame*.

We adopt the variant $\mathbf{G3K}_<$ (see Table 2) of the calculus **G3K** introduced in [21] for the *basic modal logic* **K** with the additional initial sequents

$$
\begin{aligned}
y < x, \Gamma &\to \Delta, y < x && (\sigma_<) \\
y = x, \Gamma &\to \Delta, y = x && (\sigma_=)
\end{aligned}
$$

and the rules for equality (see Table 2). With $\neg A$ defined as $A \supset \bot$, the rules $L\neg, R\neg$ are special cases of $L\supset, R\supset$, and we do not give them explicitly.

The basic idea of the calculus is the syntactical internalisation of Kripke semantics: the calculus operates on labelled formulae $x{:}\ A$, to be read as "x forces A", and on relational formulae $y < x$. For each connective and for the modality $\Box$ the rules are obtained directly from the inductive forcing clauses for compound formulae.

As is common, we denote by $\mathbf{G3K}^*_<$ the extension of $\mathbf{G3K}_<$ with additional rules corresponding to frame properties $*$, The situation is as as laid out in Table 1, in which we use the common abbreviation $\forall y < x\, A$ for $\forall y(y < x \Rightarrow A)$.

Frame property	Rule
Reflexivity $\forall x(x < x)$	$\dfrac{x < x, \Gamma \to \Delta}{\Gamma \to \Delta}\ \mathit{Ref}$
Irreflexivity $\forall x(x \not< x)$	$\dfrac{}{x < x, \Gamma \to \Delta}\ \mathit{Irref}$
Transitivity $\forall x \forall y < x \forall z < y (z < x)$	$\dfrac{x < z, x < y, y < z, \Gamma \to \Delta}{x < y, y < z, \Gamma \to \Delta}\ \mathit{Trans}$

Table 1
Additional rules for $\mathbf{G3K}^*_<$ and the corresponding frame properties

---

Initial sequents

$x\colon P, \Gamma \to \Delta, x\colon P$

$x\colon \Box A, \Gamma \to \Delta, x\colon \Box A$

$y < x, \Gamma \to \Delta, y < x$

$x = y, \Gamma \to \Delta, x = y$

Propositional rules

$$\frac{x\colon A, x\colon B, \Gamma \to \Delta}{x\colon A \wedge B, \Gamma \to \Delta}\ L\wedge \qquad \frac{\Gamma \to \Delta, x\colon A \quad \Gamma \to \Delta, x\colon B}{\Gamma \to \Delta, x\colon A \wedge B}\ R\wedge$$

$$\frac{x\colon A, \Gamma \to \Delta \quad x\colon B, \Gamma \to \Delta}{x\colon A \vee B, \Gamma \to \Delta}\ L\vee \qquad \frac{\Gamma \to \Delta, x\colon A, x\colon B}{\Gamma \to \Delta, x\colon A \vee B}\ R\vee$$

$$\frac{\Gamma \to \Delta, x\colon A \quad x\colon B, \Gamma \to \Delta}{x\colon A \supset B, \Gamma \to \Delta}\ L\supset \qquad \frac{x\colon A, \Gamma \to \Delta, x\colon B}{\Gamma \to \Delta, x\colon A \supset B}\ R\supset$$

$$\frac{}{x\colon \bot, \Gamma \to \Delta}\ L\bot$$

Modal rules

$$\frac{y\colon A, x\colon \Box A, y < x, \Gamma \to \Delta}{x\colon \Box A, y < x, \Gamma \to \Delta}\ L\Box \qquad \frac{y < x, \Gamma \to \Delta, y\colon A}{\Gamma \to \Delta, x\colon \Box A}\ R\Box \quad (y \text{ fresh})$$

Rules for equality

$$\frac{x = x, \Gamma \to \Delta}{\Gamma \to \Delta}\ \mathit{Eq\text{-}Ref} \qquad \frac{x = z, x = y, y = z, \Gamma \to \Delta}{x = y, y = z, \Gamma \to \Delta}\ \mathit{Eq\text{-}Trans}$$

$$\frac{y < z, x = y, x < z, \Gamma \to \Delta}{x = y, x < z, \Gamma \to \Delta}\ \mathit{Repl}_{<_1} \qquad \frac{x < y, z = y, x < z, \Gamma \to \Delta}{z = y, x < z, \Gamma \to \Delta}\ \mathit{Repl}_{<_2}$$

$$\frac{y\colon P, x = y, x\colon P, \Gamma \to \Delta}{x = y, x\colon P, \Gamma \to \Delta}\ \mathit{Repl}_{At}$$

Table 2
The sequent calculus $\mathbf{G3K}_<$

Derivable sequents

$x\colon A, \Gamma \to \Delta, x\colon A$

$x\colon A \supset B, x\colon A, \Gamma \to \Delta, x\colon B$

$\to x\colon \Box(A \supset B) \supset (\Box A \supset \Box B)$

Admissible rule: Substitution

$$\frac{\Gamma \to \Delta}{\Gamma[y/x] \to \Delta[y/x]}\ \mathit{Subs}$$

Admissible rules: Weakening

$$\frac{\Gamma \to \Delta}{x\colon A, \Gamma \to \Delta}\ \mathit{LW} \qquad \frac{\Gamma \to \Delta}{\Gamma \to \Delta, x\colon A}\ \mathit{RW}$$

$$\frac{\Gamma \to \Delta}{y < x, \Gamma \to \Delta}\ \mathit{LW}_< \qquad \frac{\Gamma \to \Delta}{\Gamma \to \Delta, y < x}\ \mathit{RW}_<$$

Admissible rule: Necessitation

$$\frac{\to x\colon A}{\to x\colon \Box A}\ N$$

Admissible rules: Contraction

$$\frac{x\colon A, x\colon A, \Gamma \to \Delta}{x\colon A, \Gamma \to \Delta}\ \mathit{LC} \qquad \frac{\Gamma \to \Delta, x\colon A, x\colon A}{\Gamma \to \Delta, x\colon A}\ \mathit{RC}$$

$$\frac{y < x, y < x, \Gamma \to \Delta}{y < x, \Gamma \to \Delta}\ \mathit{LC}_< \qquad \frac{\Gamma \to \Delta, y < x, y < x}{\Gamma \to \Delta, y < x}\ \mathit{RC}_<$$

Admissible rule: Replacement

$$\frac{y\colon A, x = y, x\colon A, \Gamma \to \Delta}{x = y, x\colon A, \Gamma \to \Delta}\ \mathit{Repl}$$

Admissible rules: Cut

$$\frac{\Gamma \to \Delta, x\colon A \qquad x\colon A, \Gamma' \to \Delta'}{\Gamma, \Gamma' \to \Delta, \Delta'}\ \mathit{Cut}$$

$$\frac{\Gamma \to \Delta, y < x \qquad y < x, \Gamma' \to \Delta'}{\Gamma, \Gamma' \to \Delta, \Delta'}\ \mathit{Cut}_< \qquad \frac{\Gamma \to \Delta, y = x \qquad y = x, \Gamma' \to \Delta'}{\Gamma, \Gamma' \to \Delta, \Delta'}\ \mathit{Cut}_=$$

Table 3

Structural properties and admissible rules of the sequent calculus $\mathbf{G3K}_<$

The calculus $\mathbf{G3K}_<$ satisfies the following structural properties (for more detail see Table 3, and for a proof see Section 11.4 of [23]):

(i) Sequents of the forms

$$x\colon A, \Gamma \to \Delta, x\colon A$$
$$x\colon A \supset B, x\colon A, \Gamma \to \Delta, x\colon B$$
$$\to x\colon \Box(A \supset B) \supset (\Box A \supset \Box B)$$

are derivable in $\mathbf{G3K}^*_<$ for arbitrary modal formulae A and B.

(ii) The rules of substitution, weakening, contraction and replacement for arbitrary formulae are height-preserving admissible in $\mathbf{G3K}^*_<$.

(iii) The rule of necessitation is admissible in $\mathbf{G3K}^*_<$.

(iv) All the rules of the system $\mathbf{G3K}^*_<$ are height-preserving invertible.

(v) The *Cut* rule is admissible in $\mathbf{G3K}_<$.

Since we add the initial sequents $\sigma_<, \sigma_=$, we also need the following:

Lemma 2.1 *Rules* $Cut_<$ *and* $Cut_=$ *are admissible in* $\mathbf{G3K}_<$.

Proof. The proof is induction as the proof of admissibility of *Cut* (see [23], Theorem 11.9), from which we exclude the cases in which the cut formula is principal as no rule has instances of $=, <$ as principal formulae. All the remaining cases are completely analogous to their counterparts in the proof of admissibility of *Cut.* □

Two important results, to which we will collectively refer as *completeness*, carry over from [22]:

Theorem 2.2 *Let* $\Gamma \to \Delta$ *be a sequent in the language of* $\mathbf{G3K}^*_<$. *Then either the sequent is derivable in* $\mathbf{G3K}^*_<$ *or it has a Kripke countermodel with properties* $*$.

Corollary 2.3 *If a sequent* $\Gamma \to \Delta$ *is valid in every Kripke model with the frame properties* $*$, *then it is derivable in the system* $\mathbf{G3K}^*_<$.

2.1 Connective-like rules for propositional variables

In some of the applications below, we will need to add a propositional variable P to the language of **K** that will have a "connective-like" behavior. For instance, suppose that we want a variable P to behave at x as $Q(x) \supset R(x)$. In order to avoid self-referential definitions, we ask Q and R not to contain P. We then add the following clause to the definition of val:

$$x \Vdash P \text{ if and only if } Q(x) \Rightarrow R(x)$$

Doing so, we further add to $\mathbf{G3K}^*_<$ a pair of rules that mirror the logical rules:

$$\frac{\Gamma \to \Delta, Q(x) \qquad R(x), \Gamma \to \Delta}{x\colon P, \Gamma \to \Delta}\, LP \qquad \frac{Q(x), \Gamma \to \Delta, R(x)}{\Gamma \to \Delta, x\colon P}\, RP$$

Since they have the same behavior as the logical connectives, all proofs given or referred to in the last section can easily be generalised to extensions of $\mathbf{G3K}_<$ by rules of this kind. In particular, LP and RP are invertible and completeness still holds. We just point out that in the proof of admissibility of *Cut*, we have to be careful when considering the case in which the cut formula is principal in both premisses. For instance when we transform

$$\dfrac{\dfrac{Q(x), \Gamma \to \Delta, R(x)}{\Gamma \to \Delta, x\colon P}\,RP \qquad \dfrac{\Gamma' \to \Delta', Q(x) \quad R(x), \Gamma' \to \Delta'}{x\colon P, \Gamma' \to \Delta'}\,LP}{\Gamma, \Gamma' \to \Delta, \Delta'}\,Cut$$

into

$$\dfrac{\dfrac{\Gamma' \to \Delta', Q(x) \qquad \dfrac{Q(x), \Gamma \to \Delta, R(x) \qquad R(x), \Gamma' \to \Delta'}{Q(x), \Gamma, \Gamma' \to \Delta, \Delta'}\,Cut_{(<,=)}}{\Gamma, \Gamma', \Gamma' \to \Delta, \Delta', \Delta'}\,Cut_{(<,=)}}{\Gamma, \Gamma' \to \Delta, \Delta'}\,LC,RC \text{ (multiple times)}$$

we have to take into consideration that $Q(x), R(x)$ may be instances of $<, =$.

3 Induction principles

Induction principles are typically not expressible within a first-order language. We now present them as ordinary rules of labelled sequent calculus. To start with, we recall *Noetherian Induction* and define *Gödel–Löb Induction*:

$$\forall y(\forall z < y\, Ez \Rightarrow Ey) \Rightarrow \forall y\, Ey \qquad (\textit{Noeth-Ind})$$
$$\forall x(\forall y < x(\forall z < y\, Ez \Rightarrow Ey) \Rightarrow \forall y < x\, Ey) \qquad (\textit{GL-Ind})$$

They prompt us to consider two rules and an axiom on top of $\mathbf{G3K}_<$ (rule $R\Box$-GLI is rule $R\Box$-L of [23]):

$$\dfrac{y\colon \Box A, \Gamma \to \Delta, y\colon A}{\Gamma \to \Delta, y\colon A}\,NI \qquad \dfrac{y < x, y\colon \Box A, \Gamma \to \Delta, y\colon A}{\Gamma \to \Delta, x\colon \Box A}\,R\Box\text{-}GLI$$

$$\Box(\Box A \supset A) \supset \Box A \qquad (W)$$

Both rules come with the variable condition that y does not appear in Γ, Δ.

Lemma 3.1 *Let a Kripke frame $(X, <)$ be given. The following are equivalent:*

(i) *Axiom W is valid in X for every formula A.*

(ii) *Axiom W is valid in X for every propositional variable A.*

(iii) *Gödel–Löb Induction holds in X, i.e.*

$$\forall x(\forall y < x(\forall z < y\, Ez \Rightarrow Ey) \Rightarrow \forall y < x\, Ey) \qquad (\textit{GL-Ind})$$

for any given predicate $E(x)$ on X.

Proof. (i)⇒(ii). Trivial.
(ii)⇒(iii). Given $E(x)$, pick a propositional variable A and take a valuation such that $x \Vdash A$ if and only if $E(x)$. Then by expanding the definitions we have the following:

$$\begin{aligned}
& x \Vdash \Box(\Box A \supset A) \supset \Box A \\
\Longrightarrow\, & x \Vdash \Box(\Box A \supset A) \Rightarrow x \Vdash \Box A \\
\Longrightarrow\, & \forall y < x\, y \Vdash \Box A \supset A \Rightarrow \forall y < x\, y \Vdash A \\
\Longrightarrow\, & \forall y < x\, (y \Vdash \Box A \Rightarrow y \Vdash A) \Rightarrow \forall y < x\, y \Vdash A \\
\Longrightarrow\, & \forall y < x\, (\forall z < y\, z \Vdash A \Rightarrow y \Vdash A) \Rightarrow \forall y < x\, y \Vdash A \\
\Longrightarrow\, & \forall y < x\, (\forall z < y\, Ez \Rightarrow Ey) \Rightarrow \forall y < x\, Ey
\end{aligned}$$

(iii)⇒(i). Given a formula A, define $E(x)$ as $x \Vdash A$ and read backwards the proof of (ii)⇒(iii). □

Lemma 3.2 *The following are equivalent over* **G3K**$_<$ *without* $R\Box$ *(including the structural rules):*

(i) *Rule* $R\Box$*-GLI,*

(ii) *Rule* $R\Box$ *plus axiom W.*

Proof. Claim 1: $R\Box$-$GLI \Rightarrow R\Box$.

$$\dfrac{\dfrac{y < x, \Gamma \to \Delta, y \colon A}{y < x, y \colon \Box A, \Gamma \to \Delta, y \colon A}\, {\scriptstyle LW}}{\Gamma \to \Delta, x \colon \Box A}\, {\scriptstyle R\Box\text{-}GLI}$$

Claim 2: $R\Box$-$GLI \Rightarrow W$.

$$\dfrac{\dfrac{\dfrac{y < x, y \colon \Box A \supset A, y \colon \Box A, x \colon \Box(\Box A \supset A) \to y \colon A}{y < x, y \colon \Box A, x \colon \Box(\Box A \supset A) \to y \colon A}\, {\scriptstyle L\Box}}{x \colon \Box(\Box A \supset A) \to x \colon \Box A}\, {\scriptstyle R\Box\text{-}GLI}}{\to x \colon \Box(\Box A \supset A) \supset \Box A}\, {\scriptstyle R\supset}$$

Claim 3: $R\Box + W \Rightarrow R\Box$-$GLI$.

$$\dfrac{\dfrac{\dfrac{y < x, y \colon \Box A, \Gamma \to \Delta, y \colon A}{y < x, \Gamma \to \Delta, y \colon \Box A \supset A}\, {\scriptstyle R\supset}}{\Gamma \to \Delta, x \colon \Box(\Box A \supset A)}\, {\scriptstyle R\Box} \qquad x \colon \Box(\Box A \supset A) \to x \colon \Box A}{\Gamma \to \Delta, x \colon \Box A}\, {\scriptstyle Cut}$$

where $x \colon \Box(\Box A \supset A) \to x \colon \Box A$ is derivable from W by invertibility of $R\supset$. □

Therefore the sequent calculus obtained by replacing $R\Box$ by $R\Box$-GLI is an extension of **G3K**$_<$. If we further add the mathematical rules *Trans* and *Irref*, we get the variant **G3KGL**$_<$ of the calculus **G3KGL** [21] obtained by adding the initial sequents $\sigma_<, \sigma_=$ and removing the mathematical rules *Trans*, *Irref*.

Lemma 3.3 *Let a Kripke frame $(X, <)$ be given. The following are equivalent:*

(i) *Rule NI is sound in X.*

(ii) *For every propositional variable A, in X we have*

$$\forall y\,(y \Vdash \Box A \Rightarrow y \Vdash A) \Rightarrow \forall y\, y \Vdash A$$

for any given valuation $\Vdash$ on X.

(iii) *Noetherian Induction holds in X, i.e.*

$$\forall y(\forall z < y\, Ez \Rightarrow Ey) \Rightarrow \forall y\, Ey \qquad (\textit{Noeth-Ind})$$

for any given predicate $E(x)$ on X.

Proof. (i)⇒(ii). Suppose that, for all y, $y \Vdash \Box A$ implies $y \Vdash A$. It follows that the sequent $y\colon \Box A \to y\colon A$ is valid, hence, by completeness, derivable. Applying rule NI we can therefore derive $\to y\colon A$

$$\frac{y\colon \Box A \to y\colon A}{\to y\colon A}\,NI$$

and by soundness we obtain that for all y, $y \Vdash A$.

(ii)⇒(iii). Given $E(x)$, pick a propositional variable A and take a valuation such that $x \Vdash A$ if and only if $E(x)$. Then:

$$\begin{aligned} &\forall y\,(y \Vdash \Box A \Rightarrow y \Vdash A) \Rightarrow \forall y\, y \Vdash A \\ \Longrightarrow &\forall y\,(\forall z < y\, z \Vdash A \Rightarrow y \Vdash A) \Rightarrow \forall y\, y \Vdash A \\ \Longrightarrow &\forall y\,(\forall z < y\, Ez \Rightarrow Ey) \Rightarrow \forall y\, Ey \end{aligned}$$

(iii)⇒(i). Given a formula A, define $E(x)$ as $x \Vdash A$ and read backwards the proof of (ii)⇒(iii). □

The lemmata proved in this section allow us to transform rather algebraic proofs using induction into tree-like derivations in modal logic, following a certain pattern:

Proof transformation pattern Let X be a set endowed with a binary relation $<$. Suppose that we need to show either

(i) a statement of the form $\forall y\, E(y)$ by way of *Noeth-Ind*, or

(ii) a statement of the form $\forall x \forall y < x\, E(y)$ by way of *GL-Ind*.

We consider $(X, <)$ as a Kripke frame, and build a Kripke model as follows. First, we consider a suitable subformula $U(x)$ of $E(x)$ such that it can be encoded in a sequent $Q(x) \to R(x)$, and fix a propositional variable P. We define a valuation such that val: $(x, P) = 1$ if and only if $U(x)$. This is done by adding (variants of) the following rules to the calculus:

$$\frac{\Gamma \to \Delta, Q(x) \qquad R(x), \Gamma \to \Delta}{x\colon P, \Gamma \to \Delta}\,LP \qquad \frac{Q(x), \Gamma \to \Delta, R(x)}{\Gamma \to \Delta, x\colon P}\,RP$$

By means of P, we find a formula A such that $x \Vdash A$ if and only if $E(x)$. We then proceed as follows:

(i) For *Noeth-Ind*: Derive the sequent $y\colon \Box A \to y\colon A$ by using **G3K**$_<$ plus *RP* and *LP*, then apply rule *NI*:

$$\dfrac{\begin{array}{c}\vdots\\ y\colon \Box A \to y\colon A\end{array}}{\to y\colon A}\;NI$$

(ii) For *GL-Ind*: Derive the sequent $y < x, y\colon \Box A \to y\colon A$ by using **G3K**$_<$ plus *RP* and *LP*, then apply rule *R*$\Box$-*GLI*:

$$\dfrac{\begin{array}{c}\vdots\\ y < x, y\colon \Box A \to y\colon A\end{array}}{\Gamma \to \Delta, x\colon \Box A}\;R\Box\text{-}GLI$$

We point out that this pattern is not fully general, as we do not yet have a universal method to find the subformula $U(x)$ needed to define the valuation.

3.1 Examples

Example 3.4 *GL-Ind* implies that $\forall y < x(y \neq x)$.[4]

Proof. [algebraic] In order to apply *GL-Ind*, we need to show that $\forall y < x(\forall z < y(z \neq x) \Rightarrow y \neq x)$. Fix $y < x$ such that $\forall z < y(z \neq x)$. We need to show that $y \neq x$. Suppose $y = x$. Then $x < x$ and $\forall z < x(z \neq x)$, from which we derive $x \neq x$. Therefore $y \neq x$ and we proved our claim. □

Proof. [modal] Fix x. Pick P such that $y \Vdash P$ if and only if $y = x$. This corresponds to the rules

$$\dfrac{y = x, \Gamma \to \Delta}{y\colon P, \Gamma \to \Delta}\;LP \qquad \dfrac{\Gamma \to \Delta, y = x}{\Gamma \to \Delta, y\colon P}\;RP$$

Then our thesis is equivalent to say that $\to x\colon \Box\neg P$ is derivable in **G3K**$_<$ plus *R*$\Box$-*GLI*, *LP* and *RP*:

$$\dfrac{\dfrac{\dfrac{\dfrac{\dfrac{\dfrac{\dfrac{y = x, y < y, y\colon \Box\neg P \to y\colon \bot, y = x}{y = x, y < y, y\colon \Box\neg P \to y\colon \bot, y\colon P}\;RP}{y\colon \neg P, y = x, y < y, y\colon \Box\neg P \to y\colon \bot}\;L\neg}{y = x, y < y, y\colon \Box\neg P \to y\colon \bot}\;L\Box}{y = x, y < x, y\colon \Box\neg P \to y\colon \bot}\;Repl}{y < x, y\colon \Box\neg P, y\colon P \to y\colon \bot}\;LP}{y < x, y\colon \Box\neg P \to y\colon \neg P}\;R\supset}{\to x\colon \Box\neg P}\;R\Box\text{-}GLI$$

□

[4] If we observe that $\forall y < x(y \neq x)$ is just a variant of irreflexivity $\forall x(x \not< x)$, then this result will be for free once we have proved Lemma 4.1 and Theorem 4.3.

Example 3.5 What follows is a somewhat more general formulation of the fact that by Noetherian induction every meet-closed predicate on a poset propagates from the irreducible elements to any element whatsoever [28, 31, 32].

Consider a ternary predicate $x = y \circ z$. We say that x is $\circ$-*reducible* (for short $R^\circ(x)$) if there are $y < x$ and $z < x$ such that $x = y \circ z$.

Let $E(x)$ be a predicate satisfying

$$\frac{x = y \circ z \quad E(y) \quad E(z)}{E(x)} \tag{$*$}$$

for every y, z. Then *Noeth-Ind* implies $\forall x(R^\circ(x) \vee E(x)) \Rightarrow \forall x\, E(x)$.

Proof. [algebraic] Assume that $\forall x(R^\circ(x) \vee E(x))$. In order to apply induction, we need to show that $\forall x(\forall y < x\, E(y) \Rightarrow E(x))$. Fix x such that $\forall y < x\, E(y)$. It now suffices to show $E(x)$. By assumption, we can distinguish two cases:

- Case $E(x)$: Trivial.
- Case $R^\circ(x)$: Take $y < x$ and $z < x$ such that $x = y \circ z$. By $\forall y < x\, E(y)$ we know that $E(y)$ and $E(z)$. This, by $(*)$ implies $E(x)$.

□

Proof. [modal] Pick a propositional variable P such that $x \Vdash P$ if and only if $E(x)$. The hypothesis $(*)$ can be written as:

$$\frac{x\colon P, y\colon P, z\colon P, x = y \circ z, \Gamma \to \Delta}{y\colon P, z\colon P, x = y \circ z, \Gamma \to \Delta}\,(*)$$

The definition of being $\circ$-reducible can be used in the calculus via the rule

$$\frac{x = y \circ z, y < x, z < x, \Gamma \to \Delta}{R^\circ(x), \Gamma \to \Delta}\,LR^\circ$$

where y, z are fresh, together with the appropriate RR° rule. The thesis becomes that from the sequent $\to R^\circ(x), x\colon P$ we can derive $\to x\colon P$ in $\mathbf{G3K}_<$ using *NI*, $(*)$, LR° and RR°. In fact:

$$\dfrac{\dfrac{\to R^\circ(x), x\colon P \qquad \dfrac{\dfrac{\dfrac{\dfrac{x = y \circ z, y < x, z < x, x\colon P, z\colon P, y\colon P, x\colon \Box P \to x\colon P}{x = y \circ z, y < x, z < x, z\colon P, y\colon P, x\colon \Box P \to x\colon P}\,(*)}{x = y \circ z, y < x, z < x, y\colon P, x\colon \Box P \to x\colon P}\,L\Box}{x = y \circ z, y < x, z < x, x\colon \Box P \to x\colon P}\,L\Box}{R^\circ(x), x\colon \Box P \to x\colon P}\,LR^\circ}{x\colon \Box P \to x\colon P}\,Cut}{\to x\colon P}\,NI$$

□

4 Consequences

In this section we apply the tools that we have just developed, in order to revisit certain common properties of the accessibility relation $<$. In particular, this will lead us to useful characterisations of the induction principles that can simplify the task of controlling that they hold in a given structure. We will further shed some more light on the role of transitivity in the calculus.

4.1 Irreflexivity & Noetherianity

The binary relation $<$ on X is said to be *irreflexive* if $\forall x(x \not< x)$, which corresponds to the following rule

$$\dfrac{}{x < x, \Gamma \to \Delta}\ \textit{Irref}$$

Lemma 4.1 *Noetherian Induction implies irreflexivity.* [5]

Proof. To show this claim, we use the syntactical proof pattern introduced in Section 3. Pick P such that $x \Vdash P$ if and only if $x < x$, i.e. such that

$$\dfrac{x < x, \Gamma \to \Delta}{x \colon P, \Gamma \to \Delta}\ LP \qquad\qquad \dfrac{\Gamma \to \Delta, x < x}{\Gamma \to \Delta, x \colon P}\ RP$$

Then we just need to show $\to x \colon \neg P$ in $\mathbf{G3K}_<$ plus *NI*, *LP* and *RP*:

$$\dfrac{\dfrac{\dfrac{\dfrac{\dfrac{\dfrac{x \colon \Box\neg P, x < x \to x < x}{x < x, x \colon \Box\neg P \to x \colon P}\ RP}{x \colon \neg P, x < x, x \colon \Box\neg P \to}\ L\neg}{x < x, x \colon \Box\neg P \to}\ L\Box}{x \colon P, x \colon \Box\neg P \to}\ LP}{x \colon \Box\neg P \to x \colon \neg P}\ R\neg}{\to x \colon \neg P}\ NI$$

From this we also get admissibility of the rule version of irreflexivity:

$$\dfrac{\to x \colon \neg P \qquad \dfrac{\dfrac{x < x, \Gamma \to \Delta, x < x}{x < x, \Gamma \to \Delta, x \colon \neg P}\ RP}{x \colon \neg P, x < x, \Gamma \to \Delta}\ L\neg}{x < x, \Gamma \to \Delta}\ \textit{Cut}$$

□

As in mathematical practice one often talks about ascending chains, we now occasionally switch back to R. So let $y < x$ if and only if xRy: that is, $<$ and R are converse to each other. Notice that $<$ is irreflexive if and only if so is R.

---

[5] This lemma is a formal direct version of "every well-founded relation is irreflexive", to be compared with "Set Induction implies $\forall x(x \notin x)$" [1–3] as a direct version of "Foundation implies $\forall x(x \notin x)$" in axiomatic set theory.

An *infinite R-sequence* is a sequence $(x_i)_{i \in \mathbb{N}}$ of elements of X such that $x_i R x_{i+1}$ for all $i \in \mathbb{N}$. An infinite R-sequence $(x_i)_{i \in \mathbb{N}}$ is *convergent* if there is $i \in \mathbb{N}$ such that $x_j = x_i$ for all $j > i$. We say that R is *well-founded* if there is no infinite R-sequence; and that R is *Noetherian*—for short, R satisfies *Noeth*—if every infinite R-sequence converges.

While the first and second item of the next lemma are well-known to be equivalent, the occurrence of irreflexivity in the third item is due to the fact that a priori R and $<$ need not possess this feature of an order relation.

Proposition 4.2 *The following are equivalent:*

(i) $<$ *satisfies Noetherian Induction.*

(ii) R *is well-founded.*

(iii) R *is irreflexive and Noetherian.*

Proof. The equivalence of the first and the second item is folklore. See Lemma 4.1 for a formal proof that Noetherian Induction implies irreflexivity. If R is well-founded, i.e. there are no infinite R-sequences at all, then R is trivially Noetherian. As for the converse, if R is irreflexive, then no infinite R-sequence converges; whence if, in addition, R is Noetherian, then R is well-founded. □

Notice in this context that if R is Noetherian, it is not always the case that $<$ satisfies *Noeth-Ind*. In fact, the relation R with the following graph

does not satisfy *Noeth-Ind* because it is not irreflexive, but R is Noetherian because the only infinite R-sequence, which is $xRxRxR...$, converges.

4.2 Transitivity & Induction

The binary relation $<$ on X is said to be *transitive* if $\forall x \forall y < x \forall z < y (z < x)$, which corresponds to the following rule

$$\frac{z < x, z < y, y < x, \Gamma \to \Delta}{z < y, y < x, \Gamma \to \Delta}\ \mathit{Trans}$$

In the light of Proposition 4.2, what we prove next in $\mathbf{G3K}_<$ is a formal version of Segerberg's theorem [35] that the Gödel–Löb axiom describes exactly the (converse) well-founded transitive Kripke frames.

Theorem 4.3 *The following are equivalent:*

(i) *Gödel–Löb Induction,*

(ii) *Noetherian Induction + Transitivity.*

Proof. Claim 1: *GL-Ind* ⇒ *Noeth-Ind*. It suffices to show that rule *NI* is admissible in **G3KGL$_<$**:

$$\cfrac{\cfrac{\cfrac{\cfrac{x\colon \Box A, \Gamma \to \Delta, x\colon A}{y\colon \Box A, \Gamma \to \Delta, y\colon A}\,Subs}{y < x, y\colon \Box A, \Gamma \to \Delta, y\colon A}\,LW}{\Gamma \to \Delta, x\colon \Box A}\,R\Box\text{-}GLI \qquad x\colon \Box A, \Gamma \to \Delta, x\colon A}{\cfrac{\Gamma, \Gamma \to \Delta, \Delta, x\colon A}{\Gamma \to \Delta, x\colon A}\,LC, RC \text{ (multiple times)}}\,Cut$$

Claim 2: *GL-Ind* ⇒ *Trans*. To show this claim, we use the syntactical proof pattern introduced in Section 3. Fix x. Pick P such that $y \Vdash P$ if and only if $y < x$, i.e. such that

$$\cfrac{y < x, \Gamma \to \Delta}{y\colon P, \Gamma \to \Delta}\,LP \qquad\qquad \cfrac{\Gamma \to \Delta, y < x}{\Gamma \to \Delta, y\colon P}\,RP$$

It suffices to show that rule *Trans* is admissible in **G3KGL$_<$** plus *LP* and *RP*:

$$\cfrac{\to x\colon \Box(\Box P \wedge P) \qquad \cfrac{\cfrac{\cfrac{\cfrac{\cfrac{\cfrac{z < x, z < y, y < x, \Gamma \to \Delta}{z < x, y\colon \Box P, y\colon P, x\colon \Box(\Box P \wedge P), z < y, y < x, \Gamma \to \Delta}\,LW}{z\colon P, y\colon \Box P, y\colon P, x\colon \Box(\Box P \wedge P), z < y, y < x, \Gamma \to \Delta}\,LP}{y\colon \Box P, y\colon P, x\colon \Box(\Box P \wedge P), z < y, y < x, \Gamma \to \Delta}\,L\Box}{y\colon \Box P \wedge P, x\colon \Box(\Box P \wedge P), z < y, y < x, \Gamma \to \Delta}\,L\wedge}{x\colon \Box(\Box P \wedge P), z < y, y < x, \Gamma \to \Delta}\,L\Box}{z < y, y < x, \Gamma \to \Delta}\,Cut$$

where $\to x\colon \Box(\Box P \wedge P)$ is derived as follows:[6]

$$\cfrac{\cfrac{y < x, y\colon \Box(\Box P \wedge P) \to y\colon \Box P \qquad \cfrac{y < x, y\colon \Box(\Box P \wedge P) \to y < x}{y < x, y\colon \Box(\Box P \wedge P) \to y\colon P}\,RP}{y < x, y\colon \Box(\Box P \wedge P) \to y\colon \Box P \wedge P}\,R\wedge}{\to x\colon \Box(\Box P \wedge P)}\,R\Box\text{-}GLI$$

where $y < x, y\colon \Box(\Box P \wedge P) \to y\colon \Box P$ is derived as follows:

$$\cfrac{\cfrac{z\colon \Box P, z\colon P, z < y, z\colon \Box P, y < x, y\colon \Box(\Box P \wedge P) \to z\colon P}{z\colon \Box P \wedge P, z < y, z\colon \Box P, y < x, y\colon \Box(\Box P \wedge P) \to z\colon P}\,L\wedge}{y < x, y\colon \Box(\Box P \wedge P) \to y\colon \Box P}\,L\Box$$

---

[6] Notice that the sequent $\to x\colon \Box(\Box P \wedge P)$ corresponds to $\forall x \forall y < x(\forall z < y(z < x) \& y < x)$, which is a redundant version of transitivity as $y < x$ is repeated both in the premises and in the conclusions. The reason why we need this version and not the "standard" one (as, for instance, in the case of *Irref* in Lemma 4.1), will become clear in the next subsection.

Claim 3: *Noeth-Ind* + *Trans* $\Rightarrow$ *GL-Ind*. It suffices to show that Axiom W is derivable in $\mathbf{G3K}_<$ plus *NI* and *Trans*:

$$\dfrac{\dfrac{\dfrac{\dfrac{\dfrac{y\colon A, y<x, x\colon \Box(\Box(\Box A \supset A) \supset \Box A), x\colon \Box(\Box A \supset A) \to y\colon A \quad \mathcal{D}_1}{y\colon \Box A \supset A, y<x, x\colon \Box(\Box(\Box A \supset A) \supset \Box A), x\colon \Box(\Box A \supset A) \to y\colon A}\,L\supset}{y<x, x\colon \Box(\Box(\Box A \supset A) \supset \Box A), x\colon \Box(\Box A \supset A) \to y\colon A}\,L\Box}{x\colon \Box(\Box(\Box A \supset A) \supset \Box A), x\colon \Box(\Box A \supset A) \to x\colon \Box A}\,R\Box}{x\colon \Box(\Box(\Box A \supset A) \supset \Box A) \to x\colon \Box(\Box A \supset A) \supset \Box A}\,R\supset}{\to x\colon \Box(\Box A \supset A) \supset \Box A}\,NI$$

where $\mathcal{D}_1$ is the following derivation:

$$\dfrac{\dfrac{y\colon \Box A, y<x, x\colon \Box(\Box(\Box A \supset A) \supset \Box A), x\colon \Box(\Box A \supset A) \to y\colon A, y\colon \Box A \qquad \mathcal{D}_2}{y\colon \Box(\Box A \supset A) \supset \Box A, y<x, x\colon \Box(\Box(\Box A \supset A) \supset \Box A), x\colon \Box(\Box A \supset A) \to y\colon A, y\colon \Box A}\,L\supset}{y<x, x\colon \Box(\Box(\Box A \supset A) \supset \Box A), x\colon \Box(\Box A \supset A) \to y\colon A, y\colon \Box A}\,L\Box$$

where $\mathcal{D}_2$ is the following derivation:

$$\dfrac{\dfrac{\dfrac{\dfrac{z\colon \Box A \supset A, z<x, z<y, y<x, x\colon \Box(\Box(\Box A \supset A) \supset \Box A), x\colon \Box(\Box A \supset A) \to y\colon A, y\colon \Box A, z\colon \Box A \supset A}{z<x, z<y, y<x, x\colon \Box(\Box(\Box A \supset A) \supset \Box A), x\colon \Box(\Box A \supset A) \to y\colon A, y\colon \Box A, z\colon \Box A \supset A}\,L\Box}{z<y, y<x, x\colon \Box(\Box(\Box A \supset A) \supset \Box A), x\colon \Box(\Box A \supset A) \to y\colon A, y\colon \Box A, z\colon \Box A \supset A}\,Trans}{y<x, x\colon \Box(\Box(\Box A \supset A) \supset \Box A), x\colon \Box(\Box A \supset A) \to y\colon A, y\colon \Box A, y\colon \Box(\Box A \supset A)}\,L\supset}{y<x, x\colon \Box(\Box(\Box A \supset A) \supset \Box A), x\colon \Box(\Box A \supset A) \to y\colon A, y\colon \Box A, y\colon \Box(\Box A \supset A)}\,R\Box$$

$\Box$

Proposition 4.2 and Theorem 4.2 help to see that *Noeth-Ind* $\not\Rightarrow$ *GL-Ind*. In fact, the structure

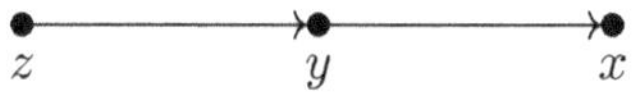

satisfies both *Noeth* and *Irref*, but not *Trans*.

4.3 Transitivity & Cut

The rule *Cut* is known to be admissible in the calculus **G3GL** and thus, by equivalence, in **G3KGL** [23, Theorem 12.20]. As a consequence, *Cut* is also admissible in $\mathbf{G3KGL}_<$ if we add *Trans* and *Irref*. Are these two rules really needed for *Cut* admissibility?

Lemma 4.4 *The following sequents are Cut-free derivable in* $\mathbf{G3KGL}_<$*:*

(i) $x\colon \Box A \to x\colon \Box(A \wedge \Box A)$,[7]

(ii) $x\colon \Box(A \wedge \Box A) \to x\colon \Box\Box A$.

Proof. (i)

$$\dfrac{\dfrac{\dfrac{y\colon A, y<x, y\colon \Box(A \wedge \Box A), x\colon \Box A \to y\colon A}{y<x, y\colon \Box(A \wedge \Box A), x\colon \Box A \to y\colon A}\,L\Box \qquad \mathcal{D}}{y<x, y\colon \Box(A \wedge \Box A), x\colon \Box A \to y\colon A \wedge \Box A}\,R\wedge}{x\colon \Box A \to x\colon \Box(A \wedge \Box A)}\,R\Box\text{-}GLI$$

[7] This is actually the redundant version of transitivity that we had in the proof of Theorem 4.3. Here, the definition of $y \Vdash A$ as $y < x$ is gained by the addition of the premiss $x\colon \Box A$.

where $\mathcal{D}$ is the following derivation:

$$\cfrac{\cfrac{\cfrac{z\colon A, z\colon \Box A, z < y, z\colon \Box A, y < x, y\colon \Box(A \wedge \Box A), x\colon \Box A \to z\colon A}{z\colon A \wedge \Box A, z < y, z\colon \Box A, y < x, y\colon \Box(A \wedge \Box A), x\colon \Box A \to z\colon A}\,L\wedge}{z < y, z\colon \Box A, y < x, y\colon \Box(A \wedge \Box A), x\colon \Box A \to z\colon A}\,L\Box}{y < x, y\colon \Box(A \wedge \Box A), x\colon \Box A \to y\colon \Box A}\,R\Box\text{-}GLI$$

(ii)

$$\cfrac{\cfrac{\cfrac{y\colon A, y\colon \Box A, y < x, y\colon \Box\Box A, x\colon \Box(A \wedge \Box A) \to y\colon \Box A}{y\colon A \wedge \Box A, y < x, y\colon \Box\Box A, x\colon \Box(A \wedge \Box A) \to y\colon \Box A}\,L\wedge}{y < x, y\colon \Box\Box A, x\colon \Box(A \wedge \Box A) \to y\colon \Box A}\,L\Box}{x\colon \Box(A \wedge \Box A) \to x\colon \Box\Box A}\,R\Box\text{-}GLI$$

□

Theorem 4.5 *The Cut rule is not admissible in* $\mathbf{G3KGL}_<$ *without Trans.*

Proof. If *Cut* were admissible, then by Lemma 4.4 the sequent $x\colon \Box A \to x\colon \Box\Box A$ would be *Cut*-free derivable. [8] Let's try to give a *Cut*-free proof:

$$\cfrac{\cfrac{\cfrac{y\colon A, z < y, z\colon \Box A, y < x, y\colon \Box\Box A, x\colon \Box A \to z\colon A}{y\colon A, z < y, z\colon \Box A, y < x, y\colon \Box\Box A, x\colon \Box A \to z\colon A}\,L\Box}{y < x, y\colon \Box\Box A, x\colon \Box A \to y\colon \Box A}\,R\Box\text{-}GLI}{x\colon \Box A \to x\colon \Box\Box A}\,R\Box\text{-}GLI$$

Observe, however, that the upper-most sequent is not derivable in general. In fact, we have a countermodel:

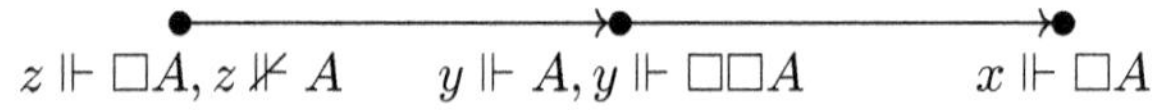

Notice that this is a non-transitive model. □

As a consequence, we get that the assumption of *Trans* is necessary in the aforementioned proof of *Cut*-admissibility in $\mathbf{G3KGL}_<$. [9]

---

[8] The sequent $x\colon \Box A \to x\colon \Box\Box A$ corresponds to transitivity the same way the sequent $x\colon \Box A \to x\colon \Box(A \wedge \Box A)$ corresponds to redundant transitivity from footnote 6. What we are showing is actually that the "standard" version of transitivity can be deduced from the redundant version by using *Cut* and that *Cut* is necessary in any proof of transitivity. This is why we needed the redundant version in the first place.

[9] This may look a bit counterintuitive: a mathematical principle, transitivity, corresponds to a derivable sequent, but is also equivalent, modulo irreflexivity, to a structural rule. However, this is not really astonishing: *Cut* can be viewed as a form of transitivity, as it is a generalisation of the following:

$$\forall C \forall B (B \supset C \Rightarrow \forall A (A \supset B \Rightarrow A \supset C))$$

which is just transitivity of $\supset$ seen as a relation. This is also the reason for which the *Cut* in literature is sometimes called *Trans*, e.g. when dealing with Scott-style entailment relations (cf [34]; for recent work see, e.g., [10, 15, 16, 29, 30, 33, 39]).

5 Future work

The calculus $\mathbf{G3K}_<$ is classical, but the applications studied up to now have a purely constructive proof in their algebraic counterpart. This makes us confident that we can replace $\mathbf{G3K}_<$ by an intuitionistic modal calculus, such as the one presented in [20].

Furthermore, those applications have not yet suggested a general method to find the subformula $U(x)$ required to define the valuation; whence we will next try to pin down such a general method.

Other principles related to induction are worth a closer look. Apart from the notions of Noetherianity discussed in [13, 25], there is Grzegorczyk induction [14], which is a weaker form of induction compatible with reflexivity. Also the principles of transitivity and irreflexivity deserve further investigation, especially in connection with *Cut*-elimination, as well as the variant GH of the Gödel–Löb axiom [8]. There is already some work in progress on relating this approach with Peano Induction, which will likely lead to similar results in Ordinal Induction.

Acknowledgements

The present study was carried out within the projects "A New Dawn of Intuitionism: Mathematical and Philosophical Advances" (ID 60842) funded by the John Templeton Foundation, and "Reducing complexity in algebra, logic, combinatorics - REDCOM" belonging to the programme "Ricerca Scientifica di Eccellenza 2018" of the Fondazione Cariverona. Additional support for this research was granted by the Academy of Finland (research project no. 1308664) and by the Faculty of Arts of the University of Helsinki. All three authors are members of the "Gruppo Nazionale per le Strutture Algebriche, Geometriche e le loro Applicazioni" (GNSAGA) of the Istituto Nazionale di Alta Matematica (INdAM). Last but not least, the authors are grateful to Edi Pavlovic and Daniel Wessel for their interest and suggestions, to Giovanni Sambin for valuable bibliographical indications, and to the anonymous referees for their constructive critique.

References

[1] Aczel, P., *The type theoretic interpretation of constructive set theory*, in: *Logic Colloquium '77 (Proc. Conf., Wrocław, 1977)*, Stud. Logic Foundations Math. **96**, North-Holland, Amsterdam, 1978 pp. 55–66.

[2] Aczel, P. and M. Rathjen, *Notes on constructive set theory*, Technical report, Institut Mittag–Leffler (2000), report No. 40.

[3] Aczel, P. and M. Rathjen, *Constructive set theory* (2010), book draft. URL `https://www1.maths.leeds.ac.uk/~rathjen/book.pdf`

[4] Berger, U., *A computational interpretation of open induction*, in: F. Titsworth, editor, *Proceedings of the Nineteenth Annual IEEE Symposium on Logic in Computer Science* (2004), pp. 326–334.

[5] Blackburn, P., M. Rijke and Y. Venema, "Modal Logic," Cambridge Tracts in Theoretical Computer Science, Cambridge University Press, 2001.

[6] Boolos, G., "The unprovability of consistency. An essay in modal logic." Cambridge University Press, Cambridge etc., 1979.
[7] Boolos, G., "The Logic of Provability," Cambridge University Press, 1993.
[8] Boolos, G. and G. Sambin, *An incomplete system of modal logic*, J. Philos. Logic **14** (1985), pp. 351–358.
[9] Boolos, G. and G. Sambin, *Provability: the emergence of a mathematical modality*, Studia Logica **50** (1991), pp. 1–23.
[10] Cederquist, J. and T. Coquand, *Entailment relations and distributive lattices*, in: S. R. Buss, P. Hájek and P. Pudlák, editors, *Logic Colloquium '98. Proceedings of the Annual European Summer Meeting of the Association for Symbolic Logic, Prague, Czech Republic, August 9–15, 1998*, Lect. Notes Logic **13**, A. K. Peters, Natick, MA, 2000 pp. 127–139.
[11] Coquand, T., *A note on the open induction principle*, Technical report, Göteborg University (1997).
URL `www.cse.chalmers.se/~coquand/open.ps`
[12] Coquand, T. and H. Lombardi, *A logical approach to abstract algebra.*, Math. Structures Comput. Sci. **16** (2006), pp. 885–900.
[13] Crosilla, L. and P. Schuster, *Finite Methods in Mathematical Practice*, in: G. Link, editor, *Formalism and Beyond. On the Nature of Mathematical Discourse*, Logos **23**, Walter de Gruyter, Boston and Berlin, 2014 pp. 351–410.
[14] Dyckhoff, R. and S. Negri, *A cut-free sequent system for Grzegorczyk logic, with an application to the Gödel–McKinsey–Tarski embedding*, J. Logic Comput. **26** (2016), pp. 169–187.
[15] Fellin, G., *The Jacobson Radical: from Algebra to Logic*, Master's thesis. Università di Verona, Dipartimento di Informatica (2018).
[16] Fellin, G., P. Schuster and D. Wessel, *The Jacobson radical of a propositional theory* (2019), submitted.
[17] Gore, R., "Cut-Free Sequent And Tableau Systems For Propositional Normal Modal Logics," Phd thesis, University of Cambridge (1992).
[18] Kripke, S. A., *Semantical analysis of modal logic. I. Normal modal propositional calculi*, Z. Math. Logik Grundlagen Math. **9** (1963), pp. 67–96.
[19] Lindström, P., *Provability logic – a short introduction.*, Theoria **62** (1996), pp. 19–61.
[20] Maffezioli, P., A. Naibo and S. Negri, *The Church–Fitch knowability paradox in the light of structural proof theory*, Synthese **190** (2013), pp. 2677–2716.
[21] Negri, S., *Proof analysis in modal logic*, J. Philos. Log. **34** (2005), pp. 507–544.
[22] Negri, S., *Kripke completeness revisited*, in: G. Primiero, editor, *Acts of Knowledge: History, Philosophy and Logic*, College Publications, 2009 pp. 233–266.
[23] Negri, S. and J. von Plato, "Proof Analysis. A Contribution to Hilbert's Last Problem," Cambridge University Press, Cambridge, 2011.
[24] Orlandelli, E. and G. Corsi, "Corso di logica modale proposizionale," Studi Superiori **1169**, Carrocci editore, 2019, 193 pp.
[25] Perdry, H. and P. Schuster, *Noetherian orders*, Math. Structures Comput. Sci. **21** (2011), pp. 111–124.
[26] Popcorn, S., "First Steps in Modal Logic," Cambridge University Press, Cambridge, 1994.
[27] Raoult, J.-C., *Proving open properties by induction*, Inform. Process. Lett. **29** (1988), pp. 19–23.
[28] Rinaldi, D. and P. Schuster, *A universal Krull–Lindenbaum theorem*, J. Pure Appl. Algebra **220** (2016), pp. 3207–3232.
[29] Rinaldi, D., P. Schuster and D. Wessel, *Eliminating disjunctions by disjunction elimination*, Bull. Symb. Logic **23** (2017), pp. 181–200.
[30] Rinaldi, D., P. Schuster and D. Wessel, *Eliminating disjunctions by disjunction elimination*, Indag. Math. (N.S.) **29** (2018), pp. 226–259, communicated first in *Bull. Symb. Logic* 23 (2017), 181–200.
[31] Schuster, P., *Induction in algebra: a first case study*, in: *2012 27th Annual ACM/IEEE Symposium on Logic in Computer Science* (2012), pp. 581–585, proceedings, LICS 2012, Dubrovnik, Croatia.

[32] Schuster, P., *Induction in algebra: a first case study*, Log. Methods Comput. Sci. **9** (2013), p. 20.

[33] Schuster, P. and D. Wessel, *Resolving finite indeterminacy: A definitive constructive universal prime ideal theorem*, in: *Proceedings of the 35th Annual ACM/IEEE Symposium on Logic in Computer Science*, LICS '20 (2020), p. 820–830.

[34] Scott, D., *Completeness and axiomatizability in many-valued logic*, in: L. Henkin, J. Addison, C. Chang, W. Craig, D. Scott and R. Vaught, editors, *Proceedings of the Tarski Symposium (Proc. Sympos. Pure Math., Vol. XXV, Univ. California, Berkeley, Calif., 1971)*, Amer. Math. Soc., Providence, RI, 1974 pp. 411–435.

[35] Segerberg, K., **13**, 1971.

[36] Smoryński, C., "Self-reference and modal logic," Universitext, Springer-Verlag, New York, 1985.

[37] Solovay, R., *Provability interpretations of modal logic*, Israel Journal of Mathematics **25** (1976), pp. 287–304.

[38] Verbrugge, R., *Provability logic*, in: E. N. Zalta, editor, *The Stanford Encyclopedia of Philosophy*, Metaphysics Research Lab, Stanford University, 2017, fall 2017 edition .

[39] Wessel, D., *Ordering groups constructively*, Comm. Algebra **47** (2019), pp. 4853–4873.

Non-iterative Modal Logics are Coalgebraic

Jonas Forster and Lutz Schröder

Friedrich-Alexander-Universität Erlangen-Nürnberg

Abstract

A modal logic is *non-iterative* if it can be defined by axioms that do not nest modal operators, and *rank-1* if additionally all propositional variables in axioms are in scope of a modal operator. It is known that every syntactically defined rank-1 modal logic can be equipped with a canonical coalgebraic semantics, ensuring soundness and strong completeness. In the present work, we extend this result to non-iterative modal logics, showing that every non-iterative modal logic can be equipped with a canonical coalgebraic semantics defined in terms of a copointed functor, again ensuring soundness and strong completeness via a canonical model construction. Like in the rank-1 case, the coalgebraic semantics is equivalent to a neighbourhood semantics with suitable frame conditions, so that our main result may be phrased as saying that every non-iterative modal logic is complete over its neighbourhood semantics. As an example application of these results, we establish strong completeness of deontic logics with factual detachment, strengthening previous weak completeness results.

Keywords: Coalgebraic logic, neighbourhood semantics, strong completeness, canonical models, deontic logic

1 Introduction

Modal frame axioms are called *non-iterative* if they do not nest modal operators, and *rank-1* if additionally all occurrences of propositional variables are under modal operators; logics are non-iterative or rank-1, respectively, if they can be axiomatized by axioms of the correspondingly restricted shape. Prominent examples include the K-axiom $\Box(a \to b) \to \Box a \to \Box b$, which is rank-1, and the T-axiom $\Box a \to a$, which is non-iterative. Lewis [10] shows that every non-iterative modal logic is *weakly* complete over neighbourhood semantics (i.e. every consistent formula is satisfiable). Without restrictions on the axioms, there are modal logics that are weakly complete but not strongly complete (in the sense that every consistent set of formulae is satisfiable) over neighbourhood semantics [18]. Previous work in coalgebraic logic [17] shows that every rank-1 modal logic is even strongly complete over a canonical coalgebraic semantics that coincides with neighbourhood semantics. In the present paper, we extend this result to non-iterative logics: We show that every (syntactically given) non-iterative modal logic is strongly complete over a canonical coalgebraic semantics, and hence over the equivalent neighbourhood semantics.

Generally, the semantic framework of *coalgebraic logic* [2] supports general proof-theoretic, algorithmic, and meta-theoretic results that can be instantiated to the logic of interest, cutting out much of the repetitive labour associated with the iterative process of designing an application-specific logic. The framework is based on casting state-based models of various types (e.g. relational, probabilistic, neighbourhood-based, or game-based) as coalgebras for a functor, the latter to be thought of as encapsulating the structure of the successors of a state.

It has been shown that the modal logic of the class of *all* coalgebras for a given functor can always be axiomatized in rank 1 [15]. Conversely, as indicated above, every rank-1 logic has a coalgebraic semantics [17]; that is, rank-1 axioms can be absorbed into a functor. The coalgebraic treatment of *non-iterative* axioms thus requires a generalization to copointed functors, to be thought of as incorporating the present state as well as its successors. Indeed it turns out that to obtain strong completeness, it is useful to generalize further to *weakly* copointed functors, in which the present state is virtualized as an ultrafilter, and subsequently restrict to *proper* coalgebras of such weakly copointed functors, in which all these virtual points actually materialize. Our main result thus states more precisely that every non-iterative logic is sound and strongly complete for the class of proper coalgebras of a canonical weakly copointed functor we construct; strong completeness w.r.t. a canonical copointed subfunctor then follows. As indicated above, this result translates back into strong completeness w.r.t. neighbourhood semantics. We complement this statement with an (easier) result showing that the modal logic of a copointed functor can always be equipped with a weakly complete non-iterative axiomatization, justifying the slogan that non-iterative logics are precisely the logics of copointed functors.

We illustrate the use of this result on certain deontic logics that on the one hand avoid the deontic explosion problem (ruling out normality, and hence Kripke semantics) and on the other hand allow for *factual detachment*, embodied in properly non-iterative axioms [20]. The only known semantics for such logics is neighbourhood semantics, and it is only known to be weakly complete (by the mentioned results of Lewis [10], alternatively by a concrete proof given in the online appendix of [20]); our results imply that it is in fact strongly complete.

Organization We recall the syntactic notion of non-iterative modal logic [10] in Section 2. In Section 3, we recall the semantic framework of coalgebraic logic, and discuss copointed and weakly copointed functors. Our main technical tool is the 0-1-step logic of a non-iterative coalgebraic logic, introduced in Section 4. We establish the easier direction of the relationship between non-iterative modal logics and coalgebraic modal logic in Section 5, where we show that the modal logic of coalgebras for a copointed functor is always non-iterative (and has the finite model property). Our main result, which states that conversely, every non-iterative modal logic is strongly complete over a canonical coalgebraic semantics that coincides with neighbourhood semantics, is shown in Sections 6 and 7. In Section 8, we present applications to deontic logics.

Proofs that are omitted or only sketched can be found in the full version [3].

2 Non-iterative Modal Logics

A *(modal) similarity type* Λ is a set of modal operators with associated finite arity. The set $\mathcal{F}(\Lambda)$ of Λ*-formulae* is given by the grammar

$$\phi_1, \dots, \phi_n ::= \bot \mid \neg\phi_1 \mid \phi_1 \wedge \phi_2 \mid L(\phi_1, \dots, \phi_n)$$

where $L \in \Lambda$ has arity n. Additional Boolean operators $\to$, $\leftrightarrow$, $\vee$ and $\top$ can then be defined as usual. We denote by $|\phi|$ the *size* of a formula ϕ, measured as the number of subformulae of ϕ. The grammar does not include propositional atoms as a separate syntactic category; however, these can be cast as nullary modalities. We thus distinguish propositional atoms from propositional variables, which are used to formulate axioms and rules.

Definition 2.1 Let $\mathsf{Prop}(V)$ denote the set of propositional formulae ϕ over a given set V (i.e. $\phi ::= \bot \mid a \mid \neg\phi \mid \phi_1 \wedge \phi_2$, with a ranging over V), and put

$$\Lambda(V) = \{L(a_1, \dots, a_n) \mid L \in \Lambda \text{ n-ary}, a_1, \dots, a_n \in V\}.$$

The elements of V are typically thought of as propositional variables. A *one-step formula* or *rank-1 formula* over V is a formula in $\mathsf{Prop}(\Lambda(\mathsf{Prop}(V)))$, and a *0-1-step formula* or a *non-iterative formula* over V is a formula in $\phi \in \mathsf{Prop}(\Lambda(\mathsf{Prop}(V)) \cup V)$. In words, a formula ϕ over V is non-iterative if it does not contain nested modal operators, and a non-iterative formula ϕ is rank-1 if additionally every variable in ϕ lies under a modal operator. We generally refer to maps of the form $\sigma\colon V \to Z$ that we use to replace entities of type V with entities of type Z in formulae as *Z-substitutions on V*, and write $\phi\sigma$ for the result of applying σ to a formula ϕ over V.

As indicated previously, the axiom $\Box(a \to b) \to \Box a \to \Box b$ (with a, b propositional variables) is rank-1, and $\Box a \to a$ is non-iterative. We define modal logics $\mathcal{L} = (\Lambda, \mathcal{A})$ syntactically by a similarity type Λ and a set $\mathcal{A}$ of *axioms* (in the given similarity type), determining the set of derivable formulae via the usual proof system as recalled below. A logic is *non-iterative* (*rank-1*) if all its axioms are non-iterative (rank-1). Given a logic $\mathcal{L} = (\Lambda, \mathcal{A})$, we say that a Λ-formula ψ is *derivable*, and write $\vdash_{\mathcal{L}} \psi$, if ψ can be derived in finitely many steps via the following rules:

$$(Ax)\ \frac{}{\psi\sigma}(\psi \in \mathcal{A}, \sigma \text{ an } \mathcal{F}(\Lambda)\text{-substitution})$$

$$(P)\ \frac{\phi_1 \quad \dots \quad \phi_n}{\psi}(\{\phi_1, \dots, \phi_n\} \vdash_{PL} \psi) \qquad (C)\ \frac{\phi_1 \leftrightarrow \psi_1 \quad \dots \quad \phi_n \leftrightarrow \psi_n}{L(\phi_1, \dots, \phi_n) \leftrightarrow L(\psi_1, \dots, \psi_n)}$$

where by $\{\phi_1, \dots, \phi_n\} \vdash_{PL} \psi$ we indicate that ψ is derivable from assumptions $\phi_1, \dots, \phi_n$ by propositional reasoning (e.g. propositional tautologies and modus ponens). The last rule is known as the *congruence rule* or *replacement of equivalents.* For a set Φ of Λ-formulae, we write $\Phi \vdash_{\mathcal{L}} \psi$ if $\vdash_{\mathcal{L}} (\phi_1 \wedge \ldots \wedge \phi_n) \to \psi$ for some $\phi_1, \ldots, \phi_n \in \Phi$. We say that Φ is *$\mathcal{L}$-consistent*, or just *consistent*, if $\Phi \nvdash_{\mathcal{L}} \bot$. A formula ϕ is *consistent* if $\{\phi\}$ is consistent.

Remark 2.2 Non-iterative logics can alternatively be presented in terms of proof rules: A *non-iterative rule* ϕ/ψ over V consists of a *premiss* $\phi \in \mathsf{Prop}(V)$ and a *conclusion* $\psi \in \mathsf{Prop}(\Lambda(V) \cup V)$. There are mutual conversions between the two formats, the conversion from axioms to rules being straightforward, and the conversion from rules to axioms being based on Boolean unification: Given a non-iterative rule ϕ/ψ, pick a *projective unifier* [4] of ϕ, i.e. a substitution σ such that $\phi\sigma$ and $\phi \to (a \leftrightarrow \sigma(a))$, for all variables a in ψ, are tautologies, and replace ϕ/ψ with the axiom $\psi\sigma$; further details are as in the rank-1 case [15].

Remark 2.3 As the syntax of the logic itself does not include propositional variables, the above system also does not derive formulae with variables. If desired, propositional variables in formulae can be emulated by introducing fresh propositional atoms (treated as nullary modal operators as indicated above). In particular, if substitution is made to apply to these fresh propositional atoms, then the standard substitution rule $\phi/\phi\sigma$ becomes admissible.

3 Coalgebraic Semantics

We next recall basic definitions in universal coalgebra [14] and coalgebraic logic [2], which will form the underlying semantic framework for our main result. We briefly recall requisite categorical definitions; some familiarity with basic category theory will nevertheless be helpful (e.g. [1]).

The underlying principle of (set-based) universal coalgebra is to encapsulate a type of state-based systems as an endofunctor $T : \mathsf{Set} \to \mathsf{Set}$ (briefly called a *set functor*) where Set is the category of sets and functions. Thus, T assigns to each set X a set TX, and to each map $f\colon X \to Y$ a map $Tf\colon TX \to TY$, preserving identities and composition. We think of TX as a type of structured collections over X. A basic example is the *(covariant) powerset functor* $\mathcal{P}$, which assigns to each set X its powerset $\mathcal{P}X$, and to each map $f\colon X \to Y$ the map $\mathcal{P}f\colon \mathcal{P}X \to \mathcal{P}Y$ that takes direct images, i.e. $\mathcal{P}f(A) = f[A]$ for $A \in \mathcal{P}X$. The most relevant example for our present purposes is the *neighbourhood functor* $\mathcal{N}$, defined as follows. The *contravariant powerset functor* $\mathcal{Q}$ is a functor of type $\mathsf{Set}^{\mathsf{op}} \to \mathsf{Set}$, i.e. reverses the direction of maps; it maps a set X to its powerset $\mathcal{Q}X = \mathcal{P}X$, and a map $f\colon X \to Y$ to the preimage map $\mathcal{Q}f\colon \mathcal{Q}Y \to \mathcal{Q}X$, i.e. $\mathcal{Q}f(B) = f^{-1}[B]$ for $B \in \mathcal{Q}Y$. For any functor F, we indicate by F^{op} the functor that acts like F but on the opposite categories, i.e. with arrows reversed in both domain and codomain. Then, we define $\mathcal{N}$ as the composite

$$\mathcal{N} = \mathcal{Q} \circ \mathcal{Q}^{\mathsf{op}}\colon \mathsf{Set} \to \mathsf{Set}.$$

We think of elements of $\mathcal{N}X$ as neighbourhood systems over X.

Given a functor T, systems are then abstracted as *T-coalgebras* $C = (X, \xi)$ consisting of a set X of *states* and a *transition function* $\xi : X \to TX$. We think of ξ as assigning to each state x a structured collection $\xi(x)$ of successors. E.g.

$\mathcal{P}$-coalgebras are just Kripke frames, assigning as they do to each state a set of successors, and $\mathcal{N}$-coalgebras are neighbourhood frames, where each state receives a collection of neighbourhoods.

Modal operators are semantically interpreted by *predicate liftings* [12,16]:

Definition 3.1 An n-ary *predicate lifting* for a set functor T is a natural transformation $\lambda : \mathcal{Q}^n \to \mathcal{Q} \circ T^{op}$, with $\mathcal{Q}$ being the contravariant powerset functor recalled above. So λ is a family of functions λ_X, indexed over all sets X, such that for all $f : X \to Y$ and $B_i \subseteq Y$, $i = 1, \dots, n$,

$$\lambda_X(f^{-1}[B_1], \dots, f^{-1}[B_n]) = (Tf)^{-1}[\lambda_Y(B_1, \dots, B_n)].$$

A Λ-*structure* $\mathcal{M} = (T, [\![L]\!]_{L\in\Lambda})$ for a signature Λ consists of a functor T and an n-ary predicate lifting $[\![L]\!]$ for every n-ary modal operator $L \in \Lambda$; we say that $\mathcal{M}$ is *based on* T. When there is no danger of confusion, we will occasionally refer to the entire Λ-structure just as T.

Given a Λ-structure $\mathcal{M}$ based on T, we define the satisfaction relation $x \models_C \phi$ between states x in T-coalgebras $C = (X, \xi)$ and Λ-formulae ϕ inductively by

$$\begin{array}{ll} x \not\models_C \bot & \\ x \models_C \neg\phi & \text{iff } x \not\models_C \phi \\ x \models_C \phi \wedge \psi & \text{iff } x \models_C \phi \text{ and } x \models_C \psi \\ x \models_C L(\phi_1, \dots \phi_n) & \text{iff } \xi(x) \in [\![L]\!]([\![\phi_1]\!]_C, \dots, [\![\phi_n]\!]_C) \end{array}$$

where we write $[\![\phi]\!]_C$ (or just $[\![\phi]\!]$) for the *extension* $\{x \in X \mid x \models_C \phi\}$ of ϕ.

Example 3.2 (i) As indicated above, Kripke frames are coalgebras for the powerset functor $\mathcal{P}$. The standard $\Box$ modality is interpreted over $\mathcal{P}$ via the predicate lifting

$$[\![\Box]\!]_X(A) = \{B \in \mathcal{P}X \mid B \subseteq A\},$$

which in combination with the above definition of the satisfaction relation induces precisely the usual semantics of $\Box$.

(ii) *Probabilistic modal logic* [9,6] has unary modal operators L_p indexed over $p \in [0,1] \cap \mathbb{Q}$, with $L_p\phi$ read 'ϕ holds with probability at least p after the next transition step'. It is interpreted over probabilistic transition systems (or Markov chains), which are coalgebras for the *discrete distribution functor* $\mathcal{D}$, given on sets X by taking $\mathcal{D}X$ to be the set of discrete probability distributions on X. The modal operators are then interpreted using the predicate liftings

$$[\![L_p]\!]_X(A) = \{\mu \in \mathcal{D}X \mid \mu(A) \geq p\}.$$

(iii) As seen above, *neighbourhood frames* are coalgebras for the neighbourhood functor $\mathcal{N}$. We capture the usual neighbourhood semantics of the $\Box$ modality by the predicate lifting

$$[\![\Box]\!]_X(A) = \{N \in \mathcal{N}X \mid A \in N\},$$

that is, a state satisfies $\Box\phi$ iff the extension $[\![\phi]\!]$ is a neighbourhood of x. More generally, a Λ-*neighbourhood frame* for a similarity type Λ is a pair $(X, (\nu_L)_{L\in\Lambda})$ consisting of a set X of states and a family of functions $\nu_L \colon X \to \mathcal{P}((\mathcal{P}X)^n)$ for $L \in \Lambda$ n-ary. We refer to subsets of $(\mathcal{P}X)^n$ as *n-ary neighbourhood systems*, and to their elements as *n-ary neighbourhoods*; if $(A_1, \dots, A_n) \in \nu_L(x)$ for n-ary $L \in \Lambda$, then $(A_1, \dots, A_n)$ is an *(n-ary) L-neighbourhood of x*. Satisfaction of modalized formulae by states $x \in X$ is then defined by

$$x \models L(\phi_1, \dots, \phi_n) \text{ iff } ([\![\phi_1]\!], \dots, [\![\phi_n]\!]) \in \nu_L(x);$$

in words, $x \models L(\phi_1, \dots, \phi_n)$ iff $([\![\phi_1]\!], \dots, [\![\phi_n]\!])$ is an L-neighbourhood of x. Λ-neighbourhood frames are coalgebras for the functor $\mathcal{N}_\Lambda$ defined by

$$\mathcal{N}_\Lambda = \textstyle\prod_{L\in\Lambda\ n\text{-ary}} \mathcal{Q} \circ ((\mathcal{Q}^{\mathsf{op}})^n)$$

where product and n-th power $(-)^n$ are pointwise, i.e. $\mathcal{N}_\Lambda X = \prod_{L\in\Lambda\ n\text{-ary}} \mathcal{Q}((\mathcal{Q}X)^n)$. The corresponding predicate liftings are

$$[\![L]\!]_X(A_1, \dots, A_n) = \{(N_L)_{L\in\Lambda} \in \mathcal{N}_\Lambda X \mid (A_1, \dots, A_n) \in N_L\}.$$

Since we work with classical negation, we can reduce all reasoning problems to satisfiability in the usual manner. Given a Λ-structure based on T, a formula ϕ is *valid* if $x \models_C \phi$ for all states x in T-coalgebras C, and a set Φ of formulae is *satisfiable* if there exists a state x in a T-coalgebra C such that $x \models_C \phi$ for all $\phi \in \Phi$. A formula ϕ is satisfiable if $\{\phi\}$ is satisfiable. A logic $\mathcal{L} = (\Lambda, \mathcal{A})$, or just $\mathcal{A}$, is *sound* for $\mathcal{M}$ if all L-derivable formulae are valid over $\mathcal{M}$, *weakly complete* if all consistent formulae are satisfiable (equivalently all valid formulae are derivable), and *strongly complete* if all consistent sets of formulae are satisfiable (which is equivalent to completeness w.r.t. local consequence from possibly infinite sets of assumptions.)

It has been shown that coalgebraic modal logics coincide with rank-1 logics. More precisely, for every Λ-structure $\mathcal{M}$ there exists a rank-1 logic that is weakly complete for $\mathcal{M}$ [15] (strong completeness cannot be expected as coalgebraic modal logics often fail to be compact, e.g. probabilistic modal logic as described in Example 3.2.ii is not compact [15]). Conversely, given a rank-1 logic $\mathcal{L} = (\Lambda, \mathcal{A})$, there is a Λ-structure $\mathcal{M}$ such that $\mathcal{L}$ is sound and strongly complete for $\mathcal{M}$ [17]; this Λ-structure is isomorphic to neighbourhood semantics. Generally speaking, rank-1 axioms can be absorbed into the functor; as a very simple example, the seriality axiom for Kripke frames, $\neg\Box\bot$, can be captured by replacing the powerset functor $\mathcal{P}$ with the non-empty powerset functor $\mathcal{P}^\star$, where $\mathcal{P}^\star X = \{A \in \mathcal{P}X \mid A \neq \emptyset\}$.

To cover non-iterative logics, we therefore need additional structure on the functor that additionally caters for base points: A *copointed functor* (T, ε), or just T when ε is clear from the context, consists of a functor T and a *copoint* ε, i.e. a natural transformation $\varepsilon \colon T \to \mathsf{id}$ where id denotes the identity functor. Coalgebras $C = (X, \xi)$ for a copointed functor are by default required to be *proper*, i.e. $\varepsilon_X \circ \xi = id_X$. Intuitively, a plain functor encapsulates only the possible (structured collections of) successors that can be assigned to a given present state, while a copointed functor additionally retains the information about the present state itself, accessed via the copoint; the properness condition $\epsilon_X \circ \xi$ on coalgebras (X, ξ) of a copointed functor effectively demands that this information is accurate, i.e. applying the copoint to $\xi(x)$ actually returns the present state x.

The main purpose of the information about the present state included in T is to allow imposing relationships between the present point and its collection of successors. Indeed, every functor T can be made copointed by passing to the functor $T \times \mathsf{id}$ (given on sets X by $(T \times \mathsf{id})X = TX \times X$), with $\varepsilon(t, x) = x$; we refer to copointed functors of this shape as *trivially copointed*, as they impose no relationship between the present state and its collection of successors. Copointed functors can absorb non-iterative axioms; e.g. the modal logic **T** is captured by the copointed functor T given by $TX = \{(A, x) \in \mathcal{P}X \times X \mid x \in A\}$ (more details are given in Section 4), which imposes that the present state is among its own successors; that is, proper T-coalgebras are precisely reflexive Kripke frames. This functor T is our first example of a non-trivially copointed functor; note that it is a subfunctor of the trivially copointed functor $\mathcal{P} \times \mathsf{id}$. For purposes of our strong completeness result, we make use of a relaxed notion of copointed functor:

Definition 3.3 A *weakly copointed functor* (T, ε) (or just T when ε is clear from the context) consists of a functor T and a *weak copoint* ε, i.e. a natural transformation $\varepsilon : T \to \mathcal{U}$, where $\mathcal{U}$ denotes the (functor part of) the ultrafilter monad. That is, $\mathcal{U}X$ is the set of ultrafilters on X, and $\mathcal{U}f(\alpha) = \{B \subseteq Y \mid f^{-1}[B] \in \alpha\}$ for $f \colon X \to Y$, $\alpha \in \mathcal{U}X$ (so $\mathcal{U}$ is a subfunctor of the neighbourhood functor $\mathcal{N}$ as in Example 3.2.iii). Then, a T-coalgebra structure $\xi : X \to TX$ is *proper* if $\varepsilon_X \circ \xi = \eta_X$ where $\eta : Id \to \mathcal{U}$ is the unit of the ultrafilter monad, given by $\eta_X(x) = \dot{x} = \{A \in \mathcal{P}X \mid x \in A\}$. Every functor T induces a *trivially weakly copointed functor* $T \times \mathcal{U}$, with second projection as the weak copoint.

Instead of the identity of the present state, a weakly copointed functor contains only a description of the present state, which in general may fail to be realized as an actual state. However, weakly copointed functors relate tightly to copointed functors in the standard sense:

Lemma and Definition 3.4 *Let (T, ε) be a weakly copointed functor. Then*

$$T_c X = \{t \in TX \mid \varepsilon(t) \text{ } principal\}$$

defines a copointed subfunctor of T_c, the copointed part *of T, with copoint ε_c defined by $\varepsilon_c(t) \in \bigcap \varepsilon(t)$. Moreover, every proper T-coalgebra $C = (X, \xi)$ factors through the inclusion $T_c X \hookrightarrow TX$, inducing a coalgebra C_c for the copointed functor T_c. Given a similarity type Λ with assigned predicate liftings for T, we obtain predicate liftings for T_c by restriction; then, a state $x \in X$ satisfies the same Λ-formulae in C as in C_c.*

(Recall that an ultrafilter α is *principal* if $\bigcap \alpha \neq \emptyset$, and then necessarily $|\bigcap \alpha| = 1$.)

Remark 3.5 Indeed, the above lemma implies that weakly copointed functors are not strictly required for our current target results, which are all formulated over proper coalgebras. We nevertheless do involve them in the technical development because of their natural role within coalgebraic logic: The 0-1-step logic, in the sense introduced in the next section, of the canonical Λ-structure, which will be based on a weakly copointed functor, is strongly complete; this would be impossible for any Λ-structure based on a copointed functor. For details, see Remark 6.7.

Remark 3.6 The standard coalgebraic semantics of rank-1 modal logics as recalled above embeds into the copointed setting by converting plain functors T into trivially copointed functors $T \times \mathsf{id}$ or trivially weakly copointed functors $T \times \mathcal{U}$, with modalities interpreted via first projections: Given a Λ-structure based on the functor T, we obtain a Λ-structure based on the trivially copointed functor $T \times \mathsf{id}$ by putting

$$(t, x) \models L(A_1, \ldots, A_n) \quad \text{iff} \quad t \models L(A_1, \ldots, A_n)$$

for $A_1, \ldots, A_n \subseteq X$ and $(t, x) \in TX \times X$, similarly for $T \times \mathcal{U}$. Proper coalgebras for $T \times \mathsf{id}$ and proper coalgebras for $T \times \mathcal{U}$ are both essentially the same as (plain) coalgebras for T, and it is easy to see that the respective modal semantics over T-coalgebras and over proper $T \times \mathsf{id}$- or $T \times \mathcal{U}$-coalgebras are equivalent.

Remark 3.7 The categorical concept of a *comonad* extends the notion of copointed functor by additionally assuming an unfolding operation $\delta : T \to T \circ T$ (the *comultiplication*) satisfying certain equational laws. This amounts to letting T contain information about the entire finite-time future development of the present state: Iterating δ, we can extract evolutions of any depth n, i.e. elements of $T^n X$, from a given element of TX. Comonads can thus be employed to capture iterative frame conditions such as $\Box a \to \Box\Box a$, with the technical caveat that this requires restricting the branching degree of models to avoid set-theoretic existence problems. Since the meta-theory of iterative frame conditions is in general much less well-behaved than that of non-iterative ones (e.g. recall the cited result on failure of strong completeness over neighbourhood semantics [18]), one should manage expectations regarding the perspective of results in comparable generality as the present one.

4 The 0-1-Step Logic

An important driving principle of coalgebraic logic is to reduce metatheoretic properties of a full-blown modal logic with nested modalities, interpreted over coalgebras, to similar properties of a much simpler *one-step logic* where formulae feature precisely one layer of modalities, and are interpreted over structures that essentially model just one transition step (hence the name). To cover non-iterative logics, we need to extend this principle to cover also the current state (besides its successors), arriving at the *0-1-step logic* of the given modal logic. For readability, we *restrict the technical development to unary modalities* from now on; covering higher arities requires no more than additional indexing, and we continue to use higher arities in the examples.

Syntax and derivations In formulae of the 0-1-step logic, we intentionally mix syntax and semantics, replacing propositional variables by their values in a powerset Boolean algebra. That is, given a non-iterative logic $\mathcal{L} = (\Lambda, \mathcal{A})$ and a set X, we take $\mathsf{Prop}(\Lambda(\mathcal{P}X) \cup \mathcal{P}X)$ to be the set of *0-1-step formulae over* $\mathcal{P}X$, referring to elements of $\mathcal{P}X$ as *(interpreted) propositional atoms.* We denote the evaluation of a $\mathsf{Prop}(\mathcal{P}X)$-formula ϕ in the Boolean algebra $\mathcal{P}X$ by $[\![\phi]\!]$, and say that ϕ is *propositionally valid over* $\mathcal{P}X$ if $[\![\phi]\!] = X$. We will identify occurrences of subformulae $\phi \in \mathsf{Prop}(\mathcal{P}X)$ with $[\![\phi]\!]$ when they lie in scope of a modal operator but not otherwise, i.e. on the uppermost level. This evaluation of inner propositional formulae allows us to omit the modal congruence rule. We thus define *0-1-step derivability* $\vdash^{0\text{-}1}_{\mathcal{L}} \psi$ of 0-1-step formulae ψ inductively by the rules

$$\frac{}{\psi\sigma}\ (\psi \in \mathcal{A}, \sigma \text{ a } \mathsf{Prop}(\mathcal{P}X)\text{-substitution})$$

$$\frac{\phi_1, \dots, \phi_n}{\psi}\ (\{\phi_1, \dots, \phi_n\} \vdash_{PL} \psi) \qquad \frac{}{\phi}\ (\phi \in \mathsf{Prop}(\mathcal{P}X), [\![\phi]\!] = X).$$

(Non-iterative rules ϕ/ψ as in Remark 2.2, if present, are also applied in substituted form: if $[\![\phi\sigma]\!] = X$ for a $\mathsf{Prop}(\mathcal{P}X)$-substitution σ, then derive $\psi\sigma$.) That is, $\vdash^{0\text{-}1}_{\mathcal{L}} \psi$ iff ψ is propositionally entailed by

$$\{\psi\sigma \mid \psi \in \mathcal{A}, \sigma \text{ a } \mathsf{Prop}(\mathcal{P}X)\text{-substitution}\} \cup \{\phi \mid \phi \in \mathsf{Prop}(\mathcal{P}X), [\![\phi]\!] = X\}.$$

We write $\Phi \vdash^{0\text{-}1}_{\mathcal{L}} \psi$ if $\vdash^{0\text{-}1}_{\mathcal{L}} (\phi_1 \wedge \ldots \wedge \phi_n) \to \psi$ for some $\phi_1, \dots, \phi_n \in \Phi$. A set Φ of 0-1-step formulae over $\mathcal{P}X$ is *0-1-step consistent* if $\Phi \not\vdash^{0\text{-}1}_{\mathcal{L}} \bot$.

Semantics Fix a weakly copointed functor (T, ε) and a Λ-structure $\mathcal{M}$ based on T. Define the unary predicate lifting ι by $\iota_X(A) = \{t \in TX \mid A \in \varepsilon(t)\}$. The *0-1-step satisfaction* relation $t \models^{0\text{-}1}_X \psi$ between functor elements $t \in TX$ and 0-1-step formulae ψ over $\mathcal{P}X$ is inductively defined by

$$\begin{array}{ll} t \not\models^{0\text{-}1}_X \bot & \\ t \models^{0\text{-}1}_X \neg\phi & \text{iff } t \not\models^{0\text{-}1}_X \phi \\ t \models^{0\text{-}1}_X \phi \wedge \psi & \text{iff } t \models^{0\text{-}1}_X \phi \text{ and } t \models^{0\text{-}1}_X \psi \\ t \models^{0\text{-}1}_X L\phi & \text{iff } t \in [\![L]\!]_X(\phi) \\ t \models^{0\text{-}1}_X B & \text{iff } t \in \iota_X(B) \end{array}$$

where $B \in \mathcal{P}X$ in the last clause, and $[\![\psi]\!]^{0\text{-}1}_X = \{t \in TX \mid t \models^{0\text{-}1}_X \psi\}$. The last clause thus deals with top-level interpreted propositional atoms. Note that in accordance with the above convention, the second to last clause omits interpretation of modal arguments, which are already identified with their interpretation. We say that ψ is *satisfiable* if $[\![\psi]\!]^{0\text{-}1}_X \neq \emptyset$, and we write $TX \models^{0\text{-}1}_X \psi$

if $[\![\psi]\!]_X^{0\text{-}1} = TX$. We generally refer to maps $\tau\colon V \to \mathcal{P}X$ as $\mathcal{P}X$*-valuations*. Given a non-iterative axiom ψ, we write $\psi\tau$ for the 0-1-step formula obtained from ψ by substituting according to τ. Then, ψ is *0-1-step sound* for $\mathcal{M}$ if $TX \models_X^{0\text{-}1} \psi\tau$ for every set X and every $\mathcal{P}X$-valuation τ. Conversely, the logic $\mathcal{L} = (\Lambda, \mathcal{A})$, or just $\mathcal{A}$, is *0-1-step complete* for $\mathcal{M}$ if every 0-1-step formula ψ over $\mathcal{P}X$ such that $TX \models_X^{0\text{-}1} \psi$ is 0-1-step derivable ($\vdash_{\mathcal{L}}^{0\text{-}1} \psi$), equivalently if every 0-1-step consistent formula is satisfiable. The same terminology applies to non-iterative rules (Remark 2.2) (specifically, a non-iterative rule ϕ/ψ is 0-1-step sound if $TX \models_X^{0\text{-}1} \psi\tau$ whenever $[\![\phi\tau]\!] = X$).

To enable an appropriate statement of soundness, we extend the semantics of the logic to allow for *frame conditions*: We refer to a pair (C, π) consisting of a T-coalgebra $C = (X, \xi)$ and a valuation $\pi\colon V \to \mathcal{P}X$ of the propositional variables as a T*-model*. We define satisfaction $x \models_{(C,\pi)} \psi$ of 0-1-step formulae $\psi \in \mathsf{Prop}(\Lambda(\mathsf{Prop}(V) \cup V))$ in states x of T-models (C, π) by the same clauses as for $\models_C$ (Section 3), and additionally

$$x \models_{(C,\pi)} a \quad \text{iff} \quad x \in \pi(a)$$

for $a \in V$. We say that C satisfies the *frame condition* ψ if $x \models_{(C,\pi)} \psi$ for all T-models (C, π). Of course, if C satisfies the frame condition ψ then ψ is *sound* for C, i.e. every state in C satisfies all substitution instances of ψ.

Lemma 4.1 (Soundness) *If a non-iterative axiom ψ over V is 0-1-step sound over a Λ-structure $\mathcal{M}$ based on a weakly copointed functor T, then every proper T-coalgebra satisfies the frame condition ψ; hence, ψ is sound for the class of all proper T-coalgebras.*

We proceed to discuss in more detail how non-iterative axioms are absorbed into (weakly) copointed functors. Given a (weakly) copointed functor T and a set $\mathcal{A}'$ of additional non-iterative axioms, we can pass to the (weakly) copointed subfunctor $T_{\mathcal{A}'}$ of T given by

$$T_{\mathcal{A}'}X = \{t \in T \mid t \models_X^{0\text{-}1} \phi\sigma \text{ for all } \phi \in \mathcal{A}' \text{ and all } \mathcal{P}X\text{-substitutions } \sigma\}$$

and restrict the Λ-structure to $T_{\mathcal{A}'}$ in the evident way. By construction, the axioms in $\mathcal{A}'$ are 0-1-step sound over $T_{\mathcal{A}'}$, and the proper $T_{\mathcal{A}'}$-coalgebras are precisely those proper T-coalgebras that satisfy the axioms in $\mathcal{A}'$ as frame conditions. Moreover, we have

Lemma 4.2 *In the notation introduced above, suppose that the set $\mathcal{A}$ of non-iterative axioms is 0-1 step sound and 0-1 step complete over T. If $\mathcal{A}'$ mentions only finitely many modalities, then $\mathcal{A} \cup \mathcal{A}'$ is 0-1-step complete over $T_{\mathcal{A}'}$.*

Proof (sketch) Observe that if ψ is a 0-1-step formula over $\mathcal{P}X$ such that $T_{\mathcal{A}'}X \models_X^{0\text{-}1} \psi$, with X assumed to be finite w.l.o.g., then $TX \models_X^{0\text{-}1} (\bigwedge \Phi) \to \psi$ where Φ contains representatives up to propositional equivalence of all instances of axioms in $\mathcal{A}'$ under $\mathcal{P}X$-substitutions; the assumptions guarantee that we can take Φ to be finite. □

Example 4.3 (i) We have recalled the coalgebraic view on standard Kripke semantics in Example 3.2.i. The usual axioms of the modal logic K ($\Box\top$ and $\Box(a \to b) \to \Box a \to \Box b$) are 0-1-step complete over the trivially copointed functor $\mathcal{P} \times \mathsf{id}$ induced by the functor $\mathcal{P}$; this is implied by translating the known *one-step* completeness of these axioms over $\mathcal{P}$ [11] into the copointed setting as indicated in Remark 3.6. It follows by Lemma 4.2 that these axioms, together with the T-axiom $\Box a \to a$, are 0-1-step complete for the copointed functor T given by

$$TX = \{(B, x) \in \mathcal{P}X \times X \mid (B, x) \models \Box A \to A \text{ for all } A \in \mathcal{P}X\}.$$

It is easy to see that $TX = \{B, x) \in \mathcal{P}X \times X \mid x \in B\}$, i.e. T coincides with the copointed functor recalled on p. 235, whose proper coalgebras are the reflexive Kripke frames.

(ii) The assumption that the additional axioms only mention finitely many modalities is really needed; without it, the claim fails even in the rank-1 case. For instance, let $\mathcal{S}$ be the *subdistribution functor*, which assigns to a set X the set $\mathcal{S}X$ of discrete subdistributions on X, where a subdistribution is defined like a distribution except that the weight of the whole set is required to be at most 1 rather than equal to 1. We use modalities L_p 'with weight at least p' with the same semantics as in the probabilistic case (Example 3.2.ii). Take the set

$$\mathcal{A}' = \{\neg L_1 \top\} \cup \{L_{1-1/n} \top \mid n \geq 1\}$$

of rank-1 axioms. Then $(\mathcal{S} \times \mathsf{id})_{\mathcal{A}'} X = \emptyset$ for all X, so that $(\mathcal{S} \times \mathsf{id})_{\mathcal{A}'} X \models^{0\text{-}1}_X \bot$, but $\bot$ is not derivable under the given axioms (together with any sound axiomatization of $\mathcal{S}$), as any derivation of $\bot$ could only use a finite subset of $\mathcal{A}'$, and all such finite subsets are clearly consistent.

A key role in the completeness proof will be played by the following subformula property of the 0-1-step logic, which extends [17, Proposition 24] from rank-1 to non-iterative logics.

Proposition 4.4 *Let ψ be a 0-1-step formula over $\mathcal{P}X$ such that $\vdash^{0\text{-}1}_{\mathcal{L}} \psi$. Then ψ is 0-1-step derivable using only $\mathsf{Prop}(\mathfrak{A})$-instances of axioms and $\mathsf{Prop}(\mathfrak{A})$-formulae valid over $\mathcal{P}X$, where $\mathfrak{A} \subseteq \mathcal{P}X$ are the sets occurring in ψ.*

The proof requires some facts about propositional logic.

Lemma 4.5 *Let V and W be disjoint finite sets. For an A-valuation τ on V with $A \subseteq \mathcal{P}X$ and a system of Boolean equations $\phi_i\tau = \psi_i\tau$ for $i = 1, \dots, n$ where $\phi_i, \psi_i \in \mathsf{Prop}(V \cup W)$, if there exists an A-valuation κ for W such that $\phi_i\tau\kappa = \psi_i\tau\kappa$ for $i = 1, \dots, n$, then there exists a $\mathsf{Prop}(V)$-substitution σ on W such that*

(i) *$\phi_i\sigma\tau = \psi_i\sigma\tau$ for $i = 1, \dots, n$*

(ii) *$x\kappa \subseteq [\![x\sigma\tau]\!]$ for $x \in W$ if $|W| = 1$.*

(Claim (i) says effectively that if Boolean equations with coefficients in A are solvable in A, then they are solvable by Boolean combinations of the coefficients that actually occur. Claim (ii) is only needed later.)

Proof. *(i):* This is well-known but we need the construction for Claim (ii). We immediately reduce to a single equation $\phi\tau = \top$ where $\phi = \bigwedge_{i=1}^{n}(\phi_i \leftrightarrow \psi_i)$. We construct σ by induction over $|W|$, with trivial base $|W| = 0$. In the inductive step, we pick $x \in W$ and obtain, by Boolean expansion,

$$\begin{aligned} \phi &\equiv (x \to \phi[\top/x]) \wedge (\neg x \to \phi[\bot/x]) \\ &\equiv (x \to \phi[\top/x]) \wedge (\neg\phi[\bot/x] \to x), \end{aligned}$$

which in turn entails $\neg\phi[\bot/x] \to \phi[\top/x]$, so by assumption the equation $(\neg\phi[\bot/x] \to \phi[\top/x])\tau = \top$ over $W \setminus \{x\}$ is solved by κ, and hence by induction solvable by some $\mathsf{Prop}(V)$-substitution σ'. Thus, the substitution

$$\sigma = [\phi[\top/x]/x]\sigma'$$

for W satisfies $\phi\sigma\tau = \top$.

(ii): Let $W = \{x\}$; we then have constructed $\sigma = [\phi[\top/x]/x]$ in (i). We have to show $\kappa(x) \subseteq [\![\phi[\top/x]\tau]\!]$. Let $y \in \kappa(x)$ and assume w.l.o.g. that ϕ is in CNF, and that x appears in at most one literal in every clause ψ in ϕ. We have to show that $y \in [\![\psi[\top/x]\tau]\!]$. If the literal x appears in ψ, then this holds trivially. Otherwise, ψ must contain some literal not mentioning x whose interpretation contains y, since $\psi\tau\kappa = \top$ by assumption and $y \notin [\![(\neg x)\kappa]\!] = [\![(\neg x)\tau\kappa]\!]$. Therefore $y \in [\![\phi[\top/x]\tau]\!]$ as required. □

Lemma 4.6 *Let $\Phi \subseteq \mathsf{Prop}(V)$, let $\psi \in \mathsf{Prop}(V)$, and let σ be a W-substitution on V and τ a U-substitution on V such that $\tau(a) = \tau(b)$ whenever $\sigma(a) = \sigma(b)$ for all $a, b \in V$, and moreover $\Phi\sigma \vdash_{PL} \psi\sigma$. Then $\Phi\tau \vdash_{PL} \psi\tau$.*

Lemma 4.7 *Let $\Phi \subseteq \mathsf{Prop}(V)$, and let $\psi \in \mathsf{Prop}(V)$. Given a U-substitution σ and a W-substitution τ on V, if $\Phi\sigma \vdash_{PL} \psi\sigma$ then $\Phi\tau \cup \Psi \vdash_{PL} \psi\tau$, where $\Psi = \{\tau(a) \leftrightarrow \tau(b) \mid a, b \in V, \sigma(a) = \sigma(b)\}$.*

Lemma 4.8 *Let V and W be disjoint sets, let $W_0 \subseteq W$, let $\Phi \subseteq \mathsf{Prop}(V)$, let $\psi \in \mathsf{Prop}(W_0)$, and let σ and τ be W-substitutions on V such that $\tau(a) = \tau(b)$ whenever $\sigma(a) = \sigma(b)$ and $\tau(a) = c$ whenever $\sigma(a) = c$ for all $a, b \in V$ and $c \in W_0$, and moreover $\Phi\sigma \vdash_{PL} \psi$. Then $\Phi\tau \vdash_{PL} \psi$.*

Proof. Let σ' and τ' be the W-substitutions on $V \cup W_0$ such that $\sigma'(w) = \tau'(w) = w$ for $w \in W_0$ and $\sigma'(v) = \sigma(v)$, $\tau'(v) = \tau(v)$ for $v \in V$. The claim then follows by Lemma 4.6. □

Proof of Proposition 4.4. Let V be a sufficiently large set of propositional variables. Then there are finite sets Φ_1 of $\mathsf{Prop}(V)$-instances of axioms and $\Phi_2 \subseteq \mathsf{Prop}(V)$ that we can assume to be instantiated by a single $\mathcal{P}X$-valuation σ such that the formulae in $\Phi_2\sigma$ are propositionally valid over $\mathcal{P}X$ and $(\Phi_1 \cup \Phi_2)\sigma \vdash_{\mathrm{PL}} \psi$. By Lemma 4.8, it suffices to show that there is a $\mathsf{Prop}(\mathfrak{A})$-substitution τ that solves the following system of equations:

- For all subformulae $L\rho, L\rho'$ in Φ_1 such that $(L\rho)\sigma = (L\rho')\sigma$ in $\Lambda(\mathcal{P}X)$, we have $(L\rho)\tau = (L\rho')\tau$ in $\Lambda(\mathcal{P}X)$. This amounts to an equation $\rho = \rho'$.
- For all subformula LA in ψ and $L\rho$ in Φ_1 such that $LA = (L\rho)\sigma$ in $\Lambda(\mathcal{P}X)$, we have $LA = (L\rho)\tau$ in $\Lambda(\mathcal{P}X)$. This amounts to an equation $A = \rho$.
- For all subformulae ρ, ρ' in $\Phi_1 \cup \Phi_2$ that do not lie beneath a modal operator and are such that $\rho\sigma = \rho'\sigma$, we have $\rho\tau = \rho'\tau$ in $\mathcal{P}X$. This amounts to an equation $\rho = \rho'$.
- For all subformulae A in ψ and ρ in $\Phi_1 \cup \Phi_2$ that do not lie beneath a modal operator and are such that $\rho\sigma = A$ in $\mathcal{P}X$, we have $\rho\tau = A$ in $\mathcal{P}X$. This amounts to an equation $A = \rho$.

By construction, this system of Boolean equations is solvable by σ, and since only sets from $\mathfrak{A}$ appear in the equations, by Lemma 4.5.(i) it is also solvable by a $\mathsf{Prop}(\mathfrak{A})$-substitution with the required properties. □

5 Copointed Coalgebraic Logics are Non-Iterative

We next establish that weakly copointed functors are indeed characterized by non-iterative axioms; that is, we *fix for this section a Λ-structure $\mathcal{M}$ based on a weakly copointed functor T* and show that there is a set of non-iterative axioms that is sound and weakly complete over the class of all proper T-coalgebras. (We necessarily restrict to weak completeness, since coalgebraic modal logics in general fail to be compact [15]). In more detail, we show that 0-1-step completeness of a non-iterative axiomatization implies its weak completeness over finite models, and we show that the set of all 0-1-step sound non-iterative axioms is 0-1-step complete. The proofs are fairly straightforward generalizations of the rank-1 case [15]. We begin with the latter step:

Theorem 5.1 *The set of all 0-1-step sound 0-1-step axioms is 0-1-step complete.*

Proof. By Remark 2.2, it suffices to show that the set of all 0-1-step sound non-iterative rules is 0-1-step complete. Let $TX \models^{0\text{-}1}_X \psi$ for a 0-1-step formula ψ over $\mathcal{P}X$. Then ψ has the form $\psi = \psi_0\tau$ for $\psi_0 \in \mathsf{Prop}(\Lambda(V_0) \cup V_0)$, with $V_0 \subseteq V$ finite, and a $\mathcal{P}X$-valuation τ. Let ϕ be the conjunction of all clauses χ over V_0 such that $[\![\chi\tau]\!] = X$; then $[\![\phi\tau]\!] = X$. We are thus done once we show that ϕ/ψ_0 is 0-1-step sound. So assume $[\![\phi\sigma]\!] = Y$ for a $\mathcal{P}Y$-valuation σ. We have to show $TY \models^{0\text{-}1}_Y \psi_0\sigma$. For each $y \in Y$ there is $x \in X$ such that for all $a \in V_0$ we have $x \in \tau(a)$ iff $y \in \sigma(a)$ (otherwise there is a clause χ over V_0 such that $X \models \chi\tau$ but $Y \not\models \chi\sigma$, contradicting $Y \models \phi\sigma$). Therefore there is $f : Y \to X$ such that $\sigma(a) = f^{-1}[\tau(a)]$ for all $a \in V_0$. By naturality of predicate liftings (including ι) and commutation of preimage with all Boolean operations, we have $[\![\psi_0\sigma]\!]^{0\text{-}1}_Y = Tf^{-1}[[\![\psi_0\tau]\!]^{0\text{-}1}_X]$, and therefore $TY \models^{0\text{-}1}_Y \psi_0\sigma$ as required. □

We will base all our model constructions on the following central notions:

Definition 5.2 A set Σ of formulae is *closed* if it is closed under subformulae and negations of formulae that are not themselves negations. We write C_Σ

for the set of maximally consistent subsets of Σ. For a Λ-formula ϕ, we write $\hat{\phi} = \{\Phi \in C_\Sigma \mid \phi \in \Phi\}$.

Lemma 5.3 *[15, Lemma 27] Let ϕ be a propositional formula over V, σ a Σ-substitution and $\hat{\sigma}$ a $\mathcal{P}(C_\Sigma)$-valuation with $\hat{\sigma}(a) = \hat{\psi}$ when $\sigma(a) = \psi$. Then $[\![\phi\hat{\sigma}]\!] = C_\Sigma$ iff $\vdash_{\mathcal{L}} \phi\sigma$.*

Definition 5.4 Let Σ be closed. A coalgebra (C_Σ, ξ) is *coherent* if for all $L\psi \in \Sigma$, $\Phi \in C_\Sigma$,

$$\xi(\Phi) \in [\![L]\!]_{C_\Sigma}(\hat{\psi}) \quad \text{iff} \quad L\psi \in \Phi.$$

Lemma 5.5 (Truth lemma [15]) *Let Σ be closed, and let $C = (C_\Sigma, \xi)$ be a coherent T-coalgebra and let $\phi \in \Sigma$. For all $\phi \in \Sigma$ we then have $\Phi \models_C \phi$ iff $\phi \in \Phi$.*

Thus, model constructions reduce to showing the existence of coherent coalgebra structures. The latter requires the following lemma, which for later reuse we prove for possibly infinite Σ:

Lemma 5.6 *Let V_Σ denote the set $\{a_\phi \mid \phi \in \Sigma\}$, and let $\Phi \subseteq \mathsf{Prop}(\Lambda(V_\Sigma) \cup V_\Sigma)$. Let σ be the substitution given by $\sigma(a_\phi) = \phi$, and let $\hat{\sigma}$ be the $\mathcal{P}C_\Sigma$-valuation given by $\hat{\sigma}(a_\phi) = \hat{\phi}$. If $\Phi\sigma$ is consistent, then $\Phi\hat{\sigma}$ is 0-1-step consistent.*

Proof. By contraposition; so assume $\Phi\hat{\sigma} \vdash^{0\text{-}1}_{\Sigma} \bot$. By Proposition 4.4, there is a derivation that uses only $\mathsf{Prop}(\mathfrak{A})$-instances of axioms and $\mathsf{Prop}(\mathfrak{A})$-formulae valid over $\mathcal{P}C_\Sigma$, for $\mathfrak{A} = \{\hat{\phi} \mid \phi \in \mathcal{F}(\Lambda)\}$. We can write the set of these formulae as $\Theta\hat{\sigma}$ for a set $\Theta \subseteq \mathsf{Prop}(\Lambda(V_\Sigma) \cup V_\Sigma)$. By the definition of 0-1-step derivations, it follows that $(\Phi \cup \Theta)\hat{\sigma} \vdash_{PL} \bot$. Now let Ψ denote the set $\{L\rho \leftrightarrow L\rho' \mid \hat{\rho} = \hat{\rho}'\}$. The formulae in Ψ are derivable in $\mathcal{L}$ by Lemma 5.3 and the congruence rule. Similarly, let $\Gamma = \{\phi \leftrightarrow \phi' \mid \hat{\phi} = \hat{\phi}'\}$; the formulae in Γ are $\mathcal{L}$-derivable by Lemma 5.3. By Lemma 4.7, it follows that $(\Phi \cup \Theta)\sigma \cup \Psi \cup \Gamma \vdash_{PL} \bot$ and therefore (again using Lemma 5.3) $\Phi\sigma \vdash_{\mathcal{L}} \bot$. □

Lemma 5.7 (Finite existence lemma) *Let $\mathcal{A}$ be 0-1-step complete, and let Σ be a finite closed set of formulae. Then there exists a coherent proper T-coalgebra structure ξ on C_Σ.*

Proof. Let $\Phi \in C_\Sigma$. We show that the requirements on $\xi(\Phi)$ form a 0-1-step consistent 0-1-step formula, implying existence of $\xi(\Phi)$ by 0-1-step completeness. Take V_Σ, σ and $\hat{\sigma}$ as in Lemma 5.6. Let

$$\chi = \bigwedge_{L\psi \in \Phi} La_\psi \wedge \bigwedge_{\neg L\psi \in \Phi} \neg La_\psi \wedge \bigwedge_{\psi \in \Phi} a_\psi.$$

We need to show that $\chi\hat{\sigma}$ is 0-1-step consistent. By Lemma 5.6, this follows from consistency of $\chi\sigma$, which in turn is implied by consistency of Φ. □

The announced weak completeness result now follows:

Theorem 5.8 (Weak completeness and bounded model property) *Let $\mathcal{A}$ be 0-1-step complete for the Λ-structure $\mathcal{M}$. Then $\mathcal{A}$ is weakly complete over finite proper T-coalgebras; specifically, every consistent formula ϕ is satisfiable in a finite proper T-coalgebra of size at most $2^{|\phi|}$.*

Proof. Let Σ be the smallest closed set containing ϕ. By the finite existence lemma (Lemma 5.7), there is a proper and coherent T-coalgebra ξ on C_Σ; note $|C_\Sigma| \leq 2^{|\phi|}$. Since Σ has only finitely many consistent subsets, the consistent set $\{\phi\}$ is contained in some $\Phi \in C_\Sigma$. By the truth lemma, $\Phi \models_{(C_\Sigma,\xi)} \phi$. □

Remark 5.9 Previous work on the connection between algebraic and coalgebraic semantics [13] has led to results that in particular cover non-iterative frame conditions. The technical setup in the mentioned work features an underlying rank-1 logic, equipped with standard coalgebraic semantics using plain functors, and imposes additional frame conditions as axioms, e.g. non-iterative frame conditions. One of the results obtained [13, Corollary 37] shows that a coalgebraic logic with non-iterative frame conditions is weakly complete over coalgebras satisfying the frame conditions, provided that the frame conditions mention only finitely many modalities. By Remark 3.6 and Lemma 4.2, these assumptions allow combining the given rank-1 logic and the additional frame conditions into a 0-1-step complete logic for the copointed functor defined by the axioms. The weak completeness result therefore follows also from our Theorem 5.8, which moreover applies also to sets of non-iterative frame conditions that mention infinitely many modalities; of course, 0-1-step completeness then needs to be proved without the help of Lemma 4.2. E.g. this will turn out to be possible for the canonical Λ-structure introduced next (Lemma 6.3). All that said, we emphasize again that the results of the present section are obtained by easy extension of previous results in rank-1 coalgebraic logic [15], and included primarily to complete the overall picture; the main technical contribution of the present work lies in the converse implication (to provide a coalgebraic semantics for a given non-iterative logic), tackled next.

6 The Canonical Λ-Structure

We now construct, *for a given non-iterative logic $\mathcal{L} = (\Lambda, \mathcal{A})$ that we fix from now on*, a canonical Λ-structure $\mathcal{M}_\mathcal{L}$ based on a weakly copointed functor $M_\mathcal{L}$ w.r.t. which we show soundness and strong completeness by means of a canonical model construction. As usual, the state space of the canonical model will be the set of maximally consistent sets, denoted $C_\mathcal{L}$ (so $C_\mathcal{L} = C_{\mathcal{F}(\Lambda)}$ in the notation of Section 5).

We construct the functor $M_\mathcal{L}$ as follows. For a set X, $M_\mathcal{L}X$ is the set of maximally 0-1-step consistent subsets of $\mathsf{Prop}(\Lambda(\mathcal{P}X) \cup \mathcal{P}X)$ (i.e. of the set of 0-1-step formulae over $\mathcal{P}X$). For a function $f\colon X \to Y$, we define $M_\mathcal{L}f$ by

$$M_\mathcal{L}f(\Phi) = \{\phi \in \mathsf{Prop}(\Lambda(\mathcal{P}Y) \cup \mathcal{P}Y) \mid \phi\sigma_f \in \Phi\}$$

where σ_f is the $\mathcal{P}X$-substitution on $\mathcal{P}Y$ given by $\sigma_f(A) = f^{-1}[A]$. We define a weak copoint $\varepsilon\colon M_\mathcal{L} \to \mathcal{U}$ by $\varepsilon_X(\Phi) = \Phi \cap \mathcal{P}X$ for $\Phi \in M_\mathcal{L}X$, and interpret $L \in \Lambda$ by

$$[\![L]\!]_X A = \{\Phi \in M_\mathcal{L}X \mid LA \in \Phi\} \qquad \text{for } A \subseteq X.$$

Of course, we intend an element of $M_\mathcal{L}X$ to satisfy precisely the 0-1-step formulae that it contains; indeed, we have

Lemma 6.1 (0-1-step truth lemma) *Let ψ be a 0-1-step formula over $\mathcal{P}X$. Then $\Phi \models_X^{0\text{-}1} \psi$ iff $\psi \in \Phi$, for $\Phi \in M_{\mathcal{L}}X$.*

Since a maximally consistent set in $M_{\mathcal{L}}X$ must in particular contain all $\mathcal{P}X$-instances of the axioms in $\mathcal{A}$, it follows that $\mathcal{A}$ is 0-1-step sound, and hence sound by Lemma 4.1, for $\mathcal{M}_{\mathcal{L}}$.

With a view to proving also 0-1-step completeness, we note a 0-1-step version of the well-known Lindenbaum lemma:

Lemma 6.2 (0-1-step Lindenbaum lemma) *Every 0-1-step consistent set of 0-1-step formulae over $\mathcal{P}X$ is contained in a maximal such set.*

From the 0-1-step truth lemma and the 0-1-step Lindenbaum lemma, 0-1-step completeness is immediate:

Lemma 6.3 *The logic $\mathcal{L}$ is 0-1-step complete for $\mathcal{M}_{\mathcal{L}}$.*

By Theorem 5.8, this implies weak completeness and the finite (in fact, bounded) model property:

Corollary 6.4 *The logic $\mathcal{L}$ is weakly complete over finite proper $M_{\mathcal{L}}$-coalgebras.*

Our main result, established in the next section, will show that $\mathcal{L}$ is in fact *strongly* complete over proper $M_{\mathcal{L}}$-coalgebras (of course, one can then no longer restrict to finite coalgebras). As indicated in the introduction, the canonical Λ-structure is essentially neighbourhood semantics. We proceed to elaborate details.

Recall from Example 3.2.iii that the Λ-neighbourhood functor $\mathcal{N}_\Lambda$ is defined as $\mathcal{N}_\Lambda = \prod_{L\in\Lambda\ n\text{-ary}} \mathcal{Q} \circ ((\mathcal{Q}^{\mathrm{op}})^n)$. Recall that $\mathcal{N}_{\mathcal{L}}$ induces a weakly copointed functor $\mathcal{N}_\Lambda \times \mathcal{U}$. Take $\mathcal{N}_{\mathcal{L}}$ to be the weakly copointed subfunctor of $\mathcal{N}_\Lambda \times \mathcal{U}$ defined by the the axioms $\mathcal{A}$, i.e.

$$\mathcal{N}_{\mathcal{L}} = (\mathcal{N}_\Lambda \times \mathcal{U})_{\mathcal{A}}$$

in notation introduced in Section 4. It is straightforward to see that the proper $\mathcal{N}_{\mathcal{L}}$-coalgebras are precisely the Λ-neighbourhood frames satisfying the frame conditions $\mathcal{A}$. The functors $\mathcal{N}_{\mathcal{L}}$ and $M_{\mathcal{L}}$ are naturally isomorphic via the transformation $\theta\colon M_{\mathcal{L}} \to \mathcal{N}_{\mathcal{L}}$ given by

$$\theta_X(\Phi)_L = (\{A \subseteq X \mid LA \in \Phi\}, \{A \subseteq X \mid A \in \Phi\}),$$

which is also compatible with the predicate liftings. We can thus translate Corollary 6.4 into the language of neighbourhood semantics:

Corollary 6.5 *The logic $\mathcal{L} = (\Lambda, \mathcal{A})$ is weakly complete over the class of finite neighbourhood frames that satisfy the axioms in $\mathcal{A}$ as frame conditions.*

That is, one instance of the coalgebraic weak completeness theorem (Theorem 5.8) is weak completeness of non-iterative modal logics over their neighbourhood semantics as originally proved by Lewis [10]

Remark 6.6 The weak completeness result in the above-mentioned previous work on algebraic-coalgebraic semantics [13, Corollary 37] (see Remark 5.9) similarly puts weak neighbourhood completeness of non-iterative logics in a coalgebraic context: Given a rank-1 logic $\mathcal{L}$, the canonical Λ-structure for the given rank-1 logic satisfies the conditions of [13, Corollary 37], in particular is *one-step complete* (the simpler version of 0-1-step completeness that applies to rank-1 logics) [17], and is isomorphic to the subfunctor of the neighbourhood functor defined by the given rank-1 axioms; [13, Corollary 37] then guarantees that weak completeness is retained in any extension of $\mathcal{L}$ with non-iterative axioms mentioning only finitely many modalities. By comparison, Corollary 6.5 above removes the restriction to finitely many modalities.

Remark 6.7 (Strong 0-1-step completeness) The strong completeness proof for rank-1 canonical structures [17] (which shows that every rank-1 logic is strongly complete over its neighbourhood semantics) can be factored through establishing *strong one-step completeness*, i.e. showing that the one-step logic (the simpler version of the 0-1-step logic that suffices in the rank-1 case) of a canonical structure is strongly complete [17, Remark 55]. Similarly, the 0-1-step logic of the canonical Λ-structure $\mathcal{M}_{\mathcal{L}}$ defined above is strongly complete; that is, for every set X, every consistent set of 0-1-step formulae over $\mathcal{P}X$ is satisfiable over $\mathcal{M}_{\mathcal{L}}$. Indeed, this is immediate from the 0-1-step truth lemma (Lemma 6.1) and the 0-1-step Lindenbaum lemma (Lemma 6.2). On the other hand, the 0-1-step logic of the copointed part of the canonical Λ-structure, or indeed of any copointed functor, clearly fails to be strongly complete: Let α be a non-principal ultrafilter on a set X; then α can be seen as a set of 0-1-step formulae over $\mathcal{P}X$, and as such is consistent; but α is clearly not satisfiable over any copointed functor. Strong completeness of the 0-1-step logic is the moral reason we include weakly copointed functors in the technical development even though, as indicated in Remark 3.5, we could in principle short-circuit them.

7 Strong Completeness

We proceed to prove our main result, strong completeness of non-iterative modal logics over their canonical structure, and hence over their neighbourhood semantics. The centrepiece of the technical development is an existence lemma; we set out to prepare its proof. As usual, one has

Lemma 7.1 (Lindenbaum Lemma) *Every consistent set of Λ-formulae is contained in a maximally consistent set.*

The existence lemma requires us to show 0-1-step consistency of a set of 0-1-step formulae specifying coherence and properness. We start with the following observation, which is fairly immediate by Lemma 5.6:

Lemma 7.2 *Let $\Phi \in C_{\mathcal{L}}$ be a maximally consistent set. Then the set*

$$\{L\hat{\phi} \mid L\phi \in \Phi\} \cup \{\neg L\hat{\phi} \mid \neg L\phi \in \Phi\} \cup \{\hat{\phi} \mid \phi \in \Phi\}$$

of 0-1-step formulae over $\mathcal{P}C_{\mathcal{L}}$ is 0-1-step consistent.

The key step is then to extend the last component of the union above from expressible subsets of $C_{\mathcal{L}}$ to arbitrary subsets:

Lemma 7.3 *Let $\Phi \in C_{\mathcal{L}}$ be a maximally consistent set. Then the set*

$$\{L\hat{\phi} \mid L\phi \in \Phi\} \cup \{\neg L\hat{\phi} \mid \neg L\phi \in \Phi\} \cup \dot{\Phi}$$

of 0-1-step formulae over $\mathcal{P}C_{\mathcal{L}}$ is 0-1-step consistent.

Recall here that $\dot{\Phi} = \{A \subseteq C_{\mathcal{L}} \mid \Phi \in A\}$ is the principal ultrafilter generated by Φ, and note $\dot{\Phi} \supseteq \{\hat{\phi} \mid \phi \in \Phi\}$. The proof makes central use of Lemma 4.5.(i) and (ii) in a step-wise elimination of atoms in $\dot{\Phi} \setminus \{\hat{\phi} \mid \phi \in \Phi\}$ from 0-1-step derivations. With Lemma 7.3 in place, the existence lemma follows straightforwardly:

Lemma 7.4 (Existence lemma) *There exists a coherent proper $M_{\mathcal{L}}$-coalgebra on $C_{\mathcal{L}}$.*

Using the Lindenbaum lemma 7.1 and the truth lemma (Lemma 5.5) in the standard fashion, we then obtain our main result:

Theorem 7.5 *The logic $\mathcal{L}$ is strongly complete over proper $M_{\mathcal{L}}$-coalgebras, and hence over coalgebras for the copointed part (Lemma and Definition 3.4) of $M_{\mathcal{L}}$.*

By the equivalence between the canonical structure and neighbourhood semantics as outlined in Section 6, we thus obtain

Corollary 7.6 *Every non-iterative logic $\mathcal{L} = (\Lambda, \mathcal{A})$ is (sound and) strongly complete over its neighbourhood semantics, i.e. over the class of neighbourhood frames that satisfy the axioms in $\mathcal{A}$ as frame conditions.*

8 Application to Deontic Logic

Deontic logic is concerned with modalities of obligation, such as $O\phi$ 'ϕ is obligatory' and $O(\phi|\psi)$ 'given ψ, ϕ is obligatory' (conditional obligation). It is faced with with specific challenges; e.g., conditional obligations are defeasible, and it is therefore nontrivial to come with principles of *factual detachment*, i.e. of deriving actual from conditional obligations, and moreover one needs to avoid the *deontic explosion* that would be caused by unrestricted normality of the obligation modality: If one had an axiom $(Oa \wedge Ob) \to O(a \wedge b)$, then a single dilemma $(Oa \wedge O\neg a)$ would cause impossible obligations $(O\bot)$, making everything obligatory if additionally monotonicity is imposed. Recent developments in deontic logic often are driven mostly axiomatically, so that the only available semantics is neighbourhood semantics.

As an example, we treat axioms for factual detachment proposed by Straßer [20]. The full logical framework uses principles of adaptive logic to govern the actual factual detachment mechanism; here, we concentrate on the underlying deontic logics called the *base logics* of the framework. The logic distinguishes specific types of obligation respectively called *instrumental* and *proper* (we refer to [20] for their philosophical definition), and has modalities $O(- \mid -)$ (binary conditional obligation), O^i (unary instrumental obligation), O^p (unary proper obligation), and $\bullet^i O(- \mid -)$, $\bullet^p O(- \mid -)$; the latter two binary modalities serve to block factual detachment of instrumental and proper obligations from conditional obligations, respectively. Corresponding dual permission modalities are denoted by replacing O with P. Various axiomatizations are developed as extensions of Goble's logic **CPDM**, which is aimed at avoiding the deontic explosion and is axiomatized in rank 1 [5]. In the online appendix [19] to [20], it is shown that two such logics **CDPM.2d**$^+$ and **CDPM.2e**$^+$ are weakly complete w.r.t. neighbourhood semantics when nesting of modalities is excluded. These logics are non-iterative; they include congruence rules and various rank-1 axioms that we refrain from listing in full, and properly non-iterative axioms

$$(O(a \mid b) \wedge b \wedge \neg \bullet^p O(a \mid b)) \to O^p a \tag{FDp}$$

$$(O(a \mid b) \wedge b \wedge \neg \bullet^i O(a \mid b)) \to O^i a \tag{FDi}$$

$$(O(a \mid b) \wedge \neg a \wedge b) \to \bullet^i O(a \mid b) \tag{fV}$$

$$\begin{aligned} &((P(\neg a \mid b \wedge c) \vee O(\neg a \mid b \wedge c)) \\ &\qquad \wedge b \wedge c \wedge P(b \wedge c \mid b) \wedge O(a \mid b)) \to \bullet^p O(a \mid b) \end{aligned} \tag{Ep}$$

$$\begin{aligned} &((P(\neg a \mid b \wedge c) \vee O(\neg a \mid b \wedge c)) \\ &\qquad \wedge b \wedge c \wedge O(a \mid b)) \to \bullet^i O(a \mid b) \end{aligned} \tag{oV-Ei}$$

where we have converted (Ep) and (oV-Ei) from rules to axioms (Remark 2.2). E.g. (FDp) says that we can detach a proper obligation $O^p a$ from a conditional $O(a \mid b)$ if this is not blocked and b is actually the case, and (fV) say that detaching an instrumental obligation $O^i a$ from a conditional obligation $O(a \mid b)$ is blocked if the obligation is factually violated ($\neg a \wedge b$). By our Corollary 7.6, the fully modal versions (with nested modalities) of both **CDPM.2d**$^+$ and **CDPM.2e**$^+$ are strongly complete w.r.t. neighbourhood semantics.

9 Conclusion and Future Work

We have shown that every non-iterative modal logic is strongly complete over neighbourhood semantics, complementing a classical result by Lewis stating that every such logic is weakly complete and has the finite model property over neighbourhood semantics. Our proof is via coalgebraic semantics, and indeed our main result can be phrased as saying that non-iterative logics are strongly complete over their canonical coalgebraic semantics. A fine point in the coalgebraic semantics is that conceptually, the proof needs to use weakly copointed functors, equipped with a natural transformation into the ultrafilter functor instead of the identity functor like copointed functors, to incorporate non-iterative frame conditions, instead of copointed functors as one would expect. That is, the natural generalization of the construction for the rank-1 case [17], which uses maximally consistent sets in the so-called 0-1-step logic, produces only a weakly copointed functor. Ex post, however, our main result then does imply completeness w.r.t. a copointed subfunctor. We have applied these results to deontic logics allowing factual detachment, obtaining that these logics are strongly complete over neighbourhood semantics, improving on previous weak completeness results [20]. It will be interesting to connect our results to coalgebraic ultrafilter extensions [7] and the coalgebraic Goldblatt-Thomason theorem [8].

References

[1] Awodey, S., "Category Theory," Oxford University Press, 2010.

[2] Cîrstea, C., A. Kurz, D. Pattinson, L. Schröder and Y. Venema, *Modal logics are coalgebraic*, Comput. J. **54** (2011), pp. 31–41.

[3] Forster, J. and L. Schröder, *Non-iterative modal logics are coalgebraic*, arXiv (2020). `http://arxiv.org/abs/2006.05396`

[4] Ghilardi, S., *Unification through projectivity*, J. Log. Comput. **7** (1997), pp. 733–752.

[5] Goble, L., *A proposal for dealing with deontic dilemmas*, in: *Deontic Logic in Computer Science, DEON 2004*, LNAI **3065** (2004), pp. 74–113.

[6] Heifetz, A. and P. Mongin, *Probability logic for type spaces*, Games Econ. Behav. **35** (2001), pp. 31–53.

[7] Kupke, C., A. Kurz and D. Pattinson, *Ultrafilter extensions for coalgebras*, in: *Algebra and Coalgebra in Computer Science, CALCO 2005*, LNCS **3629** (2005), pp. 263–277.

[8] Kurz, A. and J. Rosický, *The Goldblatt-Thomason theorem for coalgebras*, in: *Algebra and Coalgebra in Computer Science, CALCO 2007*, LNCS **4624** (2007), pp. 342–355.

[9] Larsen, K. and A. Skou, *Bisimulation through probabilistic testing*, Inf. Comput. **94** (1991), pp. 1–28.

[10] Lewis, D., *Intensional logics without interative axioms*, J. Philos. Log. **3** (1974), pp. 457–466.

[11] Pattinson, D., *Coalgebraic modal logic: soundness, completeness and decidability of local consequence*, Theor. Comput. Sci. **309** (2003), pp. 177–193.

[12] Pattinson, D., *Expressive logics for coalgebras via terminal sequence induction*, Notre Dame J. Formal Log. **45** (2004), pp. 19–33.

[13] Pattinson, D. and L. Schröder, *Beyond rank 1: Algebraic semantics and finite models for coalgebraic logics*, in: *Foundations of Software Science and Computational Structures, FOSSACS 2008*, LNCS **4962** (2008), pp. 66–80.

[14] Rutten, J., *Universal coalgebra: a theory of systems*, Theor. Comput. Sci. **249** (2000), pp. 3–80.

[15] Schröder, L., *A finite model construction for coalgebraic modal logic*, J. Log. Algebr. Program. **73** (2007), pp. 97–110.

[16] Schröder, L., *Expressivity of coalgebraic modal logic: The limits and beyond*, Theor. Comput. Sci. **390** (2008), pp. 230–247.

[17] Schröder, L. and D. Pattinson, *Rank-1 modal logics are coalgebraic*, J. Log. Comput. **20** (2010), pp. 1113–1147.

[18] Shehtman, V., *On strong neighbourhood completeness of modal and intermediate propositional logic*, in: *JFAK. Essays Dedicated to Johan van Benthem on the Occasion of his 50th Birthday*, Vossiuspers AUP, 1999 .

[19] Straßer, C., *A deontic logic framework allowing for factual detachment – appendix*. `http://www.clps.ugent.be/sites/default/files/publications/DLFFD-appendix.pdf`

[20] Straßer, C., *A deontic logic framework allowing for factual detachment*, J. Appl. Log. **9** (2011), pp. 61–80.

The modal logic of almost sure frame validities in the finite

Valentin Goranko

Department of Philosophy, Stockholm University
Email: valentin.goranko@philosophy.su.se
Department of Mathematics, University of Johannesburg (visiting professorship)

Abstract

A modal formula is almost surely valid in the finite if the probability that it is valid in a randomly chosen finite frame with n states is asymptotically 1 as n grows unboundedly. This paper studies the normal modal logic $\mathbf{ML}^{\mathsf{as}}$ of all modal formulae that are almost surely valid in the finite. Because of the failure of the zero-one law for frame validity in modal logic, the logic $\mathbf{ML}^{\mathsf{as}}$ extends properly the modal logic of the countable random frame $\mathbf{ML}^{\mathsf{r}}$, which was completely axiomatized in a 2003 paper by Goranko and Kapron. The present work studies the logic $\mathbf{ML}^{\mathsf{as}}$, provides a model-theoretic characterisation of its additional validities beyond those in $\mathbf{ML}^{\mathsf{r}}$, and raises some open problems and conjectures regarding the missing additional axioms over $\mathbf{ML}^{\mathsf{r}}$ and the explicit description of the complete axiomatisation of $\mathbf{ML}^{\mathsf{as}}$ which may turn out to hinge on difficult combinatorial-probabilistic arguments and calculations.

Keywords: modal logic, asymptotic probabilities, almost sure frame validities, 0-1 laws, countable random frame, bounded morphisms, axiomatisation

1 Introduction: asymptotic probabilities of logical formulae, 0-1 laws, and almost sure validities

[1] What is the probability that a given modally defined property of Kripke frames holds of a randomly chosen finite Kripke frame? What does it mean for such a property to be 'almost surely valid' in finite Kripke frames? These questions have a good intuitive sense and some potential practical importance (to be discussed briefly further) but, as currently stated, they are imprecise and cannot be answered in general. To make these questions precise, one has to (at least) specify a probability distribution over the class of all finite frames. There is no unique natural such distribution, for at least two reasons:

(i) Finite frames may be considered as *labelled structures* over a concrete finite domain, e.g. a finite set of natural numbers, or as abstract, *unlabelled*

[1] This is a long, but hopefully useful for modal logicians, introduction to the topic.

structures, defined up to isomorphism. These two notions of a structure define two different sample spaces (but, see further).

(ii) In either case, the result is a countably infinite space of structures, which admits uncountably many probability distributions, and none of them can be uniform over all finite structures, because of the countable additivity of the probability measure.

To address the second point we will make some standard assumptions, viz that we first relativise the question above to all finite frames (in either sense) of a fixed size (number of possible worlds) n, all of which we assume to have the same domain $U_n = \{1, ..., n\}$. Then we consider a uniform distribution over all frames with that domain. The latter is equivalent to assuming that a random frame of size n is constructed by assigning with probability $1/2$ an arrow (transition) to each ordered pair of possible worlds in the domain U_n. The difference between the cases of labelled and unlabelled frames is that two randomly constructed frames over U_n that turn out isomorphic are considered the same as unlabelled frames, but not as labelled frames, unless they are identically labelled. Thus, one can define *labelled* and *unlabelled* probability [2] of a given frame property P to hold in a randomly chosen/constructed frame of size n. Then we consider the asymptotic behaviour of these probabilities and their limits as n increases without bound. If these exist, they define the *labelled (resp. unlabelled) (asymptotic) probability in the finite of the property* P. In particular, we define the respective probabilities $\Pr_l(\phi)$ and $\Pr_u(\phi)$ for the frame validity of any modal formula ϕ. It turns out, as shown in [10] (cf. also [18]), that *these probabilities coincide.* The reason for this is that

i) the property of a frame to be *rigid*, i.e. not to have non-trivial automorphisms, has asymptotic probability 1, and

ii) every rigid n-element frame has the same number, viz. $n!$, of non-isomorphic labellings, whence the equality of the asymptotic probabilities.

When $\Pr(P) = 1$ we say that P is *almost surely true in the finite*, while if $\Pr(P) = 0$ we say that P is *almost surely false in the finite.* These apply respectively to first-order (FO) sentences, in terms of the frame properties they define. Since in modal logic we traditionally talk about *validity* and *non-validity* of a modal formula in a given frame, rather than truth and falsity, we say that ϕ is *almost surely valid in the finite* when $\Pr(\phi) = 1$, while ϕ is *almost surely invalid in the finite* if $\Pr(\phi) = 0$. See the precise technical details in Section 2.

It turns out that many natural properties of frames are either almost surely true or almost surely false in the finite. In particular, this is the case for *all first-order definable properties*, which is the celebrated *Zero-one Law for first-order logic (FOL)*, proved first in [13] and independently (and quite differently) in [10]. In the latter, Fagin gave an insightful proof of the 0-1 law for the FO logic of arbitrary relational languages of finite signature, with the case of graphs (i.e. a single binary relation) being representative. Fagin related the almost

[2] Note that computing the labelled probabilities is easy, whereas computing the unlabelled ones is difficult, because it only counts numbers of structures up to isomorphism.

sure truth of FO sentences on finite graphs to the FO theory of the co-called *countable random graph* $\mathfrak{R}$ (aka *Radó graph*) for which Gaifman had proved in [12] that $Th(\mathfrak{R})$ is ω-categorical and axiomatized by an infinite set EXT of *extension axioms*: sentences claiming that every n-tuple of elements in the structure can be extended to an $(n+1)$-tuple in all possible (consistent) ways. The probabilistic aspect of this result is rather surprising: assuming uniform distribution, any randomly constructed countable relational structure is, with probability 1, isomorphic to $\mathfrak{R}$ – thus justifying the term '*the countable random structure*'. That notion and Gaifman's results extend [3] to any finite relational language L. Fagin applied these results, by showing that every extension axiom is almost surely true in the finite. Thus, he provided two purely logical descriptions of the FO sentences σ of any relational language L that are almost surely true in the finite, viz. he showed that the following are equivalent for any FO sentence:

- σ is almost surely true in finite L-structures.
- σ follows from (finitely many) extension axioms in L.
- σ is true in the countable random structure for L.

Consequently, for every FO sentence σ, either it is in $Th(\mathfrak{R})$, hence almost surely true, or its negation is in $Th(\mathfrak{R})$, hence σ is almost surely false; whence the 0-1 law. Fagin's result, which, in particular, states that almost sure truth in the finite is equivalent to logical truth in the respective countable random structure, is often referred to as a *transfer theorem*. That result sparked much interest in the area of finite model theory and further extensive research on 0-1 laws. Such results were proved for several extensions of first-order logic, incl: the extension FOL+LFP of FOL with fixed point operators, in [6]; later subsumed by the 0-1 law for the infinitary logic over bounded number of variables $L^{\omega}_{\infty,\omega}$, in [23]; for some prefix-defined fragments of monadic second-order logic [22], where also strong relations were established between decidability and 0-1 laws of such fragments, etc. Most of these results were proved, like Fagin's result, by a means of suitable versions of the transfer theorem. For a popular and very readable exposition of 0-1 laws in FOL and some extensions see [18], and for such results in fragments of Σ^1_1 see [24].

On the other hand, the 0-1 law easily fails in the presence of a single constant in the language (consider a sentence saying that a given unary predicate is true at the element interpreting that constant) [4]. In second-order logic the 0-1 law

[3] A model-theoretic aside: the random graph $\mathfrak{R}$ is a particular example of a countably infinite homogeneous structure that can be constructed as a Fraïssé limit of a family of sets of finite structures satisfying certain natural closure properties. There are deep model-theoretic connections between homogeneous (more generally, homogenizable) structures, extension properties, asymptotic probabilities and almost sure theories, that generalise Gaifman's and Fagin's results and enable further relativisations and refinements of 0-1 laws (and more generally, limit laws), which go beyond the scope of this paper, so I refer the reader e.g. to [25], [2], and further references therein.

[4] Still, it was proved in [30] that a first-order language with only unary functions does have a *limit law*, in sense that every sentence in that language has an asymptotic probability,

fails badly: think, for instance, of the property of having an odd number of elements. Even its monadic existential fragment $M\Sigma_1^1$ contains sentences with no asymptotic probability, as first proved by Kaufmann (see [27] for a very accessible account of Kaufmann's counterexample, and [28], [29] for stronger such results). While for prefix-defined fragments of $M\Sigma_1^1$ the boundary of 0-1 laws seems to be essentially delineated (see [24] for a survey), little is known in general on that for the full (monadic) second-order logic (M)SOL.

Now, what about modal logic? There are at least two basic relevant notions of modal validity: in Kripke models and in Kripke frames. In the former case the 0-1 law follows immediately from 0-1 law in FOL, since validity of a modal formula in a Kripke model is a FO property [19]. However, for the case of frame validity, which is an essentially universal monadic second-order ($M\Pi_1^1$) property, the 0-1 law cannot be claimed as a consequence of Fagin's theorem. Actually, that 0-1 law was claimed to be proved (by complex combinatorial-probabilistic calculations) in [19]. However, later it was proved in [14] that the respective transfer theorem fails for modal frame validity in the finite, which then cast a doubt on the 0-1 law, too. Indeed, that claim turned out to be wrong, as proved by Le Bars in [29], who provided there a very non-trivial counterexample. Soon thereafter, an erratum [20] was published, pointing out the mistake in [19].

A relatively independent from the 0-1 laws concept, which is in the focus of the present work, is the *almost sure theory* Th_L^{as} of a given logical language L, with respect to the notion of truth or validity under consideration – that is, the set consisting precisely of those sentences of L that are almost surely true (resp. valid) in the finite. Clearly, Th_L^{as} is a well-defined logical theory in a very traditional sense: it contains all valid sentences of the logic and is closed under all finitary rules of inference (as the semantic consequence preserves truth and the asymptotic probability measure is finitely additive). What can one say about the theory Th_L^{as}, in terms of axiomatization and deduction in it, decidability, model-theoretic properties, etc?

The cases of classes of FO structures where 0-1 laws holds by way of transfer theorem are generally easy to analyse thoroughly, because in these cases Th_L^{as} is precisely the (ω-categorical and complete, hence decidable) theory of the countable random structure (resp. universal homogeneous structure. cf. [25], [2], [1]). Curiously, as shown in [19], the respective modal logic of almost sure Kripke model validity, turned out to be already known, viz. *Carnap's modal logic* ([7]), the axioms of which are all modal formulae $\Diamond\phi$ where ϕ is a satisfiable propositional formula. (NB: this is not a normal modal logic, as it is not closed under substitutions.)

However, in cases where 0-1 law fails, or when it holds but not by a suitable transfer from a countable random structure, the question of logical characterisation and, in particular, axiomatization of the respective almost sure theory seems generally quite difficult, and very few such results are known. This ques-

---

though in general not just 0 or 1.

tion arises, in particular, for the modal logic of almost sure frame validity in the finite, hereafter denoted by $\mathbf{ML}^{\mathsf{as}}$, and it is the topic of the present study.

What do we know about $\mathbf{ML}^{\mathsf{as}}$ so far? Both much and little. We know that it is a normal modal logic, extending the normal modal logic $\mathbf{ML}^{\mathsf{r}}$ of the *countable random frame* F^{r} (the frame analogue of the countable random graph $\mathfrak{R}$). The logic $\mathbf{ML}^{\mathsf{r}}$ was studied and axiomatized in [14] where it was also proved to be strictly included in $\mathbf{ML}^{\mathsf{as}}$. (See details in Section 3.) What is not known yet is how the additional axioms, needed to extend the axiomatization of $\mathbf{ML}^{\mathsf{r}}$ to a complete axiomatization of $\mathbf{ML}^{\mathsf{as}}$, look, and even whether $\mathbf{ML}^{\mathsf{as}}$ is recursively axiomatizable. Such questions are inherently difficult, because $\mathbf{ML}^{\mathsf{as}}$ lacks a priori explicit logical semantics in terms of truth and validity in a specific class of models or frames, but rather involves the class of all finite frames as a whole, so it is an *essentially global* concept. In this paper I study the logic $\mathbf{ML}^{\mathsf{as}}$, provide partial answers to these questions, and raise conjectures for their solutions.

Lastly, why should one be interested in $\mathbf{ML}^{\mathsf{as}}$, or in any almost sure theory? Besides being driven by a sheer intellectual curiosity, one can argue that knowing – or being able to identify – the almost sure truths in a given logic may have some practical advantages, e.g. for reducing the *average case* complexity of checking whether a given formula of that logic is valid, by first checking (or, guessing) whether it is almost surely valid. While such argument would probably not make good computational sense for the basic normal modal logic, it may do so for some extensions that are of higher computational complexity, have no finite model property, or are even undecidable. The idea certainly sounds quite reasonable in the case of FOL, which is not only undecidable, but satisfiability of FOL sentences in the finite is not even recursively enumerable (by Trachtenbrot's theorem), whereas the almost sure theory of FOL, being the same as $Th(\mathfrak{R})$, is decidable, and in fact only PSPACE-complete, as proved in [17]. Similar argument might work for other extensions of FOL and (M)SOL, too. As for $\mathbf{ML}^{\mathsf{as}}$, it seems still early to judge whether and what its practical importance may be. One immediate goal of this paper is to at least attract the attention of the modal logic community to this logic.

The paper is organized as follows. After this long introduction and brief technical preliminaries in Section 2, I introduce and compare the logics $\mathbf{ML}^{\mathsf{r}}$ and $\mathbf{ML}^{\mathsf{as}}$ in Section 3. In Section 4, I explore the question of axiomatization of $\mathbf{ML}^{\mathsf{as}}$ and raise some open problems and conjectures. I conclude briefly in Section 5.

2 Preliminaries on modal logic, asymptotic probabilities and almost sure frame validity

Here I provide some technical details on the basic concepts in this paper introduced informally in the introduction. Besides, I assume that the reader has the necessary background in modal logic, including the notions of Kripke model, Kripke frame, truth and validity of modal formulae in these. Familiarity with some basics of the model theory of modal logic, incl. bounded-morphism (aka p-morphism) and characteristic (aka Jankov-Fine) formulae would be helpful, but for the reader's convenience I have included the definitions here.

2.1 Bounded morphisms and characteristic formulae

Definition 2.1 Let $\mathsf{F}_1 = \langle W_1, R_1 \rangle$ and $\mathsf{F}_2 = \langle W_2, R_2 \rangle$ be frames. A mapping $h : W_1 \to W_2$ is a **bounded morphism from** F_1 to F_2 if the following hold:

(i) For all $x, y \in W_1$, if xR_1y then $h(x)R_2h(y)$.

(ii) For all $x \in W_1, t \in W_2$, if $h(x)R_2t$ then xR_1y for some $y \in W_1$ such that $h(y) = t$.

If h is onto, F_2 is called a **bounded-morphic image of** F_1.

Following a commonly used notation, I will often denote by $\mathsf{F}_1 \twoheadrightarrow \mathsf{F}_2$ the claim that F_2 is bounded-morphic image of F_1. An important fact: frame validity of modal formulae is preserved in bounded-morphic images, i.e., if $\mathsf{F}_1 \models \phi$ and $\mathsf{F}_1 \twoheadrightarrow \mathsf{F}_2$, then $\mathsf{F}_2 \models \phi$ (cf. [35], [5], and [15], which are also recommended general references on all other modal logic concepts used here).

The **universal modality** (interpreted by the full Cartesian square of the domain, cf. [16]) will be denoted by $[\mathsf{U}]$, its dual, **existential modality** – by $\langle \mathsf{U} \rangle$, and the basic normal logic **K** extended with $[\mathsf{U}]$ – by $\mathbf{K}_\mathsf{U}$. The language of **K** will be denoted by ML and the extended one with $[\mathsf{U}]$ – by ML_U.

Definition 2.2 ([14]) Let $\mathsf{F} = \langle W, R \rangle$ be any finite frame with $W = \{w_1, \ldots, w_n\}$ and let $\{p_1, \ldots, p_n\}$ be fixed different propositional variables. The **characteristic formula**[5] **of** F **over** $\langle p_1, \ldots, p_n \rangle$ is the formula $\chi_\mathsf{F}(p_1, \ldots, p_n) := \neg[\mathsf{U}]\delta_\mathsf{F}(p_1, \ldots, p_n)$, where δ_F is the 'modal diagram' of F:

$$\delta_\mathsf{F}(p_1, \ldots, p_n) := \bigwedge_{i=1}^{n} \langle \mathsf{U} \rangle p_i \wedge \bigvee_{i=1}^{n} p_i \wedge \bigwedge_{1 \leq i \neq j \leq n} (p_i \to \neg p_j) \wedge$$

$$\bigwedge_{1 \leq i,j \leq n} \{p_i \to \Diamond p_j | w_i R w_j\} \wedge \bigwedge_{1 \leq i,j \leq n} \{p_i \to \neg \Diamond p_j | \neg w_i R w_j\}.$$

When $\{p_1, \ldots, p_n\}$ are fixed or known from the context, I will write simply χ_F.

The following is a variation of a folklore fact (see Remark 2.5). I nevertheless sketch a proof, for the sake of the reader hitherto unfamiliar with it.

Lemma 2.3 ([8], [14]) *For every frame* G *and finite frame* F*:* $\mathsf{G} \twoheadrightarrow \mathsf{F}$ *iff* $\mathsf{G} \not\models \chi_\mathsf{F}$.

Proof. (Sketch) Suppose $\mathsf{G}, V \not\models \chi_\mathsf{F}$ for some valuation V. Then every point $y \in \mathsf{G}$ satisfies exactly one variable $p_{i(y)}$ from $\{p_1, \ldots, p_n\}$. Furthermore, the mapping $f : \mathsf{G} \longrightarrow \mathsf{F}$ defined by $f(y) = w_{i(y)}$ is a surjective bounded morphism. Vice versa, if $f : \mathsf{G} \longrightarrow \mathsf{F}$ is a surjective bounded morphism, then the valuation V on G defined by $V(p_i) = f^{-1}(w_i)$ satisfies $\neg\chi_\mathsf{F}$. □

---

[5] See Remark 2.5.

When both F and G are finite, the lemma above can be strengthened to the claims of the forthcoming Lemma 2.4, where $\mathbf{ML}(\mathsf{F})$ is the normal modal logic of the validities in the frame F and $\mathbf{K}_{\mathsf{U}} + \phi$ is the axiomatic extension of the modal logic $\mathbf{K}_{\mathsf{U}}$ with the axiom scheme ϕ.

Lemma 2.4 ([37]) *For any finite frames* F, G *the following are equivalent:*

(i) $\mathsf{G} \twoheadrightarrow \mathsf{F}$.

(ii) $\mathsf{G} \not\models \chi_{\mathsf{F}}$.

(iii) $\mathbf{ML}(\mathsf{G}) \subseteq \mathbf{ML}(\mathsf{F})$.

(iv) $\mathbf{K}_{\mathsf{U}} + \chi_{\mathsf{F}} \vdash \chi_{\mathsf{G}}$.

(v) *For every modal formula* ϕ*, if* $\mathbf{K}_{\mathsf{U}} + \chi_{\mathsf{G}} \vdash \phi$ *then* $\mathbf{K}_{\mathsf{U}} + \chi_{\mathsf{F}} \vdash \phi$.

Proof. Most of these equivalences are straightforward variations (involving $[\mathsf{U}]$) of widely known and frequently re-discovered facts, that can be found scattered elsewhere (cf. [8] for most of them). One implication is not completely trivial, viz. the implication from (i), (ii), or (iii) to (iv). As noted in [37], the implication from (ii) to (iv) holds in a more general form, viz. for any formula ϕ instead of χ_{F}, with essentially the same proof as for this special case which suffices for our purpose. As [37] has pointed out, the same claim was proved for intuitionistic logic in [33], itself referring to earlier works by Jankov. For further references and more, see the forthcoming Remark 2.5. Nevertheless, I provide here a proof sketch, to make the presentation relatively self-contained for readers not familiar with the more general theory, and also to help them see *why* this result holds, because it is of importance for the logic $\mathbf{ML}^{\mathsf{as}}$, as discussed in Section 4.

(i) $\Rightarrow$ (iv): Suppose $\mathsf{G} \twoheadrightarrow \mathsf{F}$ and fix a bounded morphism $h : \mathsf{G} \to \mathsf{F}$. Let $\mathsf{F} = \langle W_{\mathsf{F}}, R_{\mathsf{F}} \rangle$ with $W_{\mathsf{F}} = \{w_1, ..., w_n\}$ and $\mathsf{G} = \langle W_{\mathsf{G}}, R_{\mathsf{G}} \rangle$ with $W_{\mathsf{G}} = \{u_1, ..., u_m\}$. Suppose $\chi_{\mathsf{F}} = \chi_{\mathsf{F}}(p_1, \ldots, p_n) = \neg[\mathsf{U}]\delta_{\mathsf{F}}(p_1, \ldots, p_n)$ and $\chi_{\mathsf{G}} = \chi_{\mathsf{G}}(q_1, \ldots, q_m) = \neg[\mathsf{U}]\delta_{\mathsf{G}}(q_1, \ldots, q_m)$. Let us define a substitution σ_h on the propositional variables $p_1, ..., p_n$ as follows: for each $i = 1, ..., n$, $\sigma_h(p_i) := \bigvee_{\{j|h(u_j)=w_i\}} q_j$. Intuitively, if we regard $p_1, ..., p_n$ as nominals for $w_1, ..., w_n$ and $q_1, ..., q_m$ respectively as nominals for $u_1, ..., u_m$ then σ_h substitutes each p_i with the syntactic description of the inverse image of w_i in G under h. Now, let us apply σ_h to $\chi_{\mathsf{F}}(p_1, \ldots, p_n)$ and denote the resulting formula by $\xi_{\mathsf{G}\to\mathsf{F}}(q_1, ..., q_m)$. After simple equivalent transformations in $\mathbf{K}_{\mathsf{U}}$, for which there is no space here (but, see them illustrated on an example in the Appendix) $\xi_{\mathsf{G}\to\mathsf{F}}(q_1, ..., q_m)$ is transformed to a formula $\xi'_{\mathsf{G}\to\mathsf{F}}(q_1, ..., q_m)$ of the type $\neg[\mathsf{U}]\delta'_{\mathsf{F}}$, where δ'_{F} is a (long) conjunction with the following property, which can be seen by direct inspection: every conjunct in δ'_{F} is either identical, or follows propositionally (essentially by only applying $A \to B \models A \to (B \vee C)$) from a conjunct in $\delta_{\mathsf{G}}(q_1, \ldots, q_m)$. Thus, $\models \delta_{\mathsf{G}} \to \delta'_{\mathsf{F}}$, hence $\models [\mathsf{U}]\delta_{\mathsf{G}} \to [\mathsf{U}]\delta'_{\mathsf{F}}$, Thus, $\models \neg\chi_{\mathsf{G}} \to \neg\xi'_{\mathsf{G}\to\mathsf{F}}$, hence $\models \xi'_{\mathsf{G}\to\mathsf{F}} \to \chi_{\mathsf{G}}$. Equivalently, $\models \sigma_h(\chi_{\mathsf{F}}) \to \chi_{\mathsf{G}}$. Therefore, $\mathbf{K}_{\mathsf{U}} \vdash \sigma_h(\chi_{\mathsf{F}}) \to \chi_{\mathsf{G}}$, by completeness of $\mathbf{K}_{\mathsf{U}}$, hence $\mathbf{K}_{\mathsf{U}} + \chi_{\mathsf{F}} \vdash \chi_{\mathsf{G}}$.□

Remark 2.5 The characteristic formulae defined here[6] are simplified (due to the availability of the universal modality) variations of the widely called Jankov-Fine formulae, cf. [8, Ch. 9.4]. Such formulae were introduced independently by V.A. Jankov in 1963 [21] (for the intuitionistic logic and finite Heyting algebras) and by D. de Jongh in his 1968 doctoral thesis (for the intuitionistic logic and finite intuitionistic Kripke frames). Their modal logic analogues were invented later, by K. Fine in 1974 [11] for modal logics extending S4 and finite modal algebras, and by W. Rautenberg [31] for modal logics extending K4 and finite Kripke frames. These formulae are at the core of the so called *splitting* techniques and results, initially developed by Jankov (for Heyting algebras), McKenzie (for splitting lattices), Blok, Rautenberg, Kracht, Wolter and others (for splitting lattices of modal logics); see [4] for references. In particular, such formulae were later used by Rautenberg [31] to axiomatize modal logics of finite frames, and generalised and applied further by Kracht [26] and by Zakharyaschev to what he called 'canonical formulae' in [8], used to axiomatize any normal extension of K4. For an algebraic treatment of canonical formulae, see [3].

Thus, the results listed in lemmas 2.3 and 2.4 are essentially not new and apply in a much more general setting[7].

2.2 Asymptotic probabilities and almost sure frame validity in the finite of modal formulae

The class of all finite frames will be denoted by $\mathcal{F}^{\mathsf{fin}}$. Given a modal formula ϕ, the $M\Pi^1_1$-formula expressing the frame condition defined by ϕ (or, any FO sentence equivalent to it, if that frame condition is first-order definable) will be denoted by $FC(\phi)$, and for any class of finite frames $\mathcal{F}$, the subclass of frames in $\mathcal{F}$ where ϕ is valid – by $\mathcal{F}(\phi)$. The set of positive integers is denoted by $\mathbb{N}$.

Given $n \in \mathbb{N}$, let $U_n := \{1, \ldots, n\}$. A **random (labelled) frame of size n** is a frame $\mathsf{F} = (U_n, R)$ obtained by random and independent assignments of truth/falsity of the binary relation R on every pair (x, y) from the set U_n, with probability for truth $p(n)$. The probability space on all n-element frames constructed as above will be denoted by $\mathcal{S}(n,p)$. In this paper I assume $p(n)$ to be the constant 0.5, so the random frame can be obtained by a random assignment of a binary relation on the domain, using uniform distribution. However, the results used and those obtained here hold likewise for any constant probability $p \in (0, 1)$ (cf. [10], [18]).

For any property of frames P, by $\mu_{n,p}(P)$ we denote the classical probability of P in $\mathcal{S}(n,p)$, i.e. the probability that P holds for a randomly constructed n-element frame. In particular, if ϕ is a first-order sentence or a modal formula, $\mu_{n,p}(\phi)$ will denote the probability for ϕ to be true (resp. valid) in a frame from $\mathcal{S}(n,p)$. Note that these are discrete probabilities since $\mathcal{S}(n,p)$ is finite.

[6] I prefer to work with these formulae, rather than with their negations, as defined in [5], for reasons that will become clear in Section 4.

[7] Thanks to Evgeny Zolin [37] for pointing out these links and references.

Now, let us define $\mu_p(\phi) := \lim_{n\to\infty} \mu_{n,p}(\phi)$. If that limit exists, it will be called **the asymptotic probability** [8] **(in the finite)** of ϕ. As the probability is fixed here to $p = 0.5$ we will omit the subscript. We define likewise $\mu_p(P)$ and $\mu(P)$ for any frame property P. A property P is said to be *almost surely true in the finite* if $\mu(P) = 1$ and, respectively, *almost surely false* if $\mu(P) = 0$.

Definition 2.6 A modal formula ϕ is said to be **almost surely valid (in the finite)** if $\mu(\phi) = 1$; respectively, **almost surely invalid** if $\mu(\phi) = 0$.

Note that, by the 0-1 law for FOL, every first-order definable modal formula is either almost surely valid or almost surely invalid in the finite. Moreover [9], this also holds for all modal formulae that define FO property on *finite frames*. For instance, Gödel-Löb formula $\Box(\Box p \to p) \to \Box p$ defines in the finite the class of irreflexive and transitive *finite* frames (cf. [8]); thus it also satisfies the 0-1 law (being almost surely invalid).

Hereafter, for technical convenience we will assume w.l.o.g., that every finite frame of size n that we consider is defined over the set $U_n = \{1, \ldots, n\}$. Thus, the collection of all finite frames $\mathcal{F}^{\mathsf{fin}}$ can be regarded as a proper set. Now, given any set of finite frames $\mathcal{F}$ which contains at least one frame of every (sufficiently large) size n, the probabilities and concepts defined above readily relativise to $\mathcal{F}$, incl. a modal formula being **almost surely valid (resp. invalid) in** $\mathcal{F}$. Further, we say that a set of finite frames $\mathcal{F}$ has an **asymptotic measure 1 (resp. 0)** if the membership to that set has asymptotic probability 1 (resp. 0). An important observation is that for every set with asymptotic measure 1 the absolute and relativised probabilities are equal, hence the absolute and relativised notions of almost sure validity/invalidity coincide.

2.3 The countable random frame F^{r}

The construction of random frames by means of a random pairwise assignment of a binary relation with a given probability p for truth of the relation can be performed on *infinite* sets, too. The outcome of such a random construction on the set $\mathbb{N}$ of natural numbers is called a **countable random frame**. Using combinatorial-probabilistic argument, it was proved in [10] that any countable random relational structure satisfies with probability 1 an infinite sequence EXT of schemes of first-order sentences, called **extension axioms**. For every $n \in \mathbb{N}$, the extension axiom $(\mathsf{EXT})_n$ for frames (directed graphs with loops) is the conjunction of finitely many sentences, each involving a tuple of n distinct variables $\overline{x} = x_1, ..., x_n$ plus another variable y and parameterised by two subsets $I, J \subseteq U_n$, as follows:

$$(\mathsf{EXT})_n = \forall \overline{x} \exists y \left(\bigwedge_{i \neq j} x_i \neq x_j \to \left(\bigwedge_{i \in U_n} x_i \neq y \wedge T(y,y) \wedge \right.\right.$$

[8] Note that this probability measure is not countably additive: $\mu(|\mathsf{F}| = n) = 0$ for every fixed n, while $\mu(\exists n(|\mathsf{F}| = n)) = 1$.

[9] Thanks to Evgeny Zolin [37] for this added remark.

$$\bigwedge_{i\in I} Rx_iy \wedge \bigwedge_{i\in U_n\setminus I} \neg Rx_iy \wedge \bigwedge_{i\in J} Ryx_j \wedge \bigwedge_{j\in U_n\setminus J} \neg Ryx_j \Bigg)\Bigg),$$

where $T(y, y)$ is either Ryy or $\neg Ryy$,

The extension axiom $(\mathsf{EXT})_n$, intuitively says that for every n different points in the frame there is a point which is related to and from each of those, and with itself, in any explicitly prescribed way. Note that if $m < n$ then $(\mathsf{EXT})_n$ implies $(\mathsf{EXT})_m$ on all frames of size at least n. Consequently, every finite set of extension axioms follows almost surely in the finite from a single extension axiom $(\mathsf{EXT})_n$ for a large enough n.

By a result of Gaifman [12] the theory EXT is consistent and ω-categorical, hence complete. The unique countable model F^{r} of EXT is called **the countable random frame**. Using Gaifman's results, Fagin proved (for graphs) in [10] the following **transfer theorem** that for any sentence ψ of FOL:

(i) If $\mathsf{F}^{\mathsf{r}} \models \psi$ then $\mu(\psi) = 1$.

(ii) If $\mathsf{F}^{\mathsf{r}} \not\models \psi$ then $\mu(\psi) = 0$.

This theorem immediately implies the 0-1 law for FOL for frames: every FO sentence is either almost surely true or almost surely false in the finite. Then, by compactness, every almost surely true FO sentence follows from finitely many extension axioms, hence from some instance of $(\mathsf{EXT})_n$. These claims apply likewise to all FO definable (in terms of frame validity) modal formulae.

3 The modal logics of the countable random frame and of almost sure validity

Here we will explore the two normal modal logics in the focus of this study. Most of the content of this section comes from [14], but is included here for the reader's convenience and self-containment of the paper.

Definition 3.1 $\mathbf{ML}^{\mathsf{r}}$ is the modal logic of all formulae valid in F^{r}. $\mathbf{ML}^{\mathsf{as}}$ is the modal logic of all formulae which are almost surely valid in the finite.

Proposition 3.2 ([14])

(i) $\mathbf{ML}^{\mathsf{r}}$ *and* $\mathbf{ML}^{\mathsf{as}}$ *are normal modal logics.*

(ii) *A modal formula* ϕ *is in* $\mathbf{ML}^{\mathsf{r}}$ *iff* $FC(\phi)$ *follows from some extension axiom, hence every such formula is in* $\mathbf{ML}^{\mathsf{as}}$. *Consequently,* $\mathbf{ML}^{\mathsf{r}} \subseteq \mathbf{ML}^{\mathsf{as}}$.

3.1 Complete axiomatization of $\mathbf{ML}^{\mathsf{r}}$

First, we need some basic facts about the countable random frame F^{r}, which easily follow from the extension axioms (cf. [14]):

- F^{r} has a diameter 2: every point can be reached from any point (incl. itself) in 2 R-steps. Indeed, by an instance of the extension axiom scheme $(\mathsf{EXT})_3$: $\mathsf{F}^{\mathsf{r}} \models \forall x \forall y \exists z (Rxz \wedge Rzy)$.
- Every point in F^{r} has infinitely many R-predecessors and infinitely many R-successors and every finite frame is embeddable as a subframe in F^{r}.

For some useful validities and non-obvious non-validities in $\mathbf{ML}^{\mathsf{r}}$, see [14].

Since the extension axiom $(\mathsf{EXT})_3$ is almost surely true in the finite, the subset $\mathcal{F}^{\mathsf{d2}}$ of all finite frames of diameter 2 has asymptotic measure 1. This fact will be of crucial importance further, because it enables us to restrict attention from $\mathcal{F}^{\mathsf{fin}}$ to almost sure validity in $\mathcal{F}^{\mathsf{d2}}$ without extensional change of that notion: every property of finite frames is almost surely true in $\mathcal{F}^{\mathsf{fin}}$ iff it is almost surely true in $\mathcal{F}^{\mathsf{d2}}$. Here is the first important consequence. Note that the universal modality $[\mathsf{U}]$ and the existential modality $\langle\mathsf{U}\rangle$ are simply definable in every frame in $\mathcal{F}^{\mathsf{d2}}$:

$$[\mathsf{U}]p \equiv \Box\Box p, \quad \text{respectively} \quad \langle\mathsf{U}\rangle p \equiv \Diamond\Diamond p.$$

Therefore, these equivalences hold in almost every finite frame, and also in F^{r}. Hereafter, to distinguish the primitives from the definable versions wherever necessary, I use $[\mathsf{U}]$ and $\langle\mathsf{U}\rangle$ as the standard universal/existential modalities, taken as primitives (extending ad hoc the basic modal language) and $[\mathbf{U}]$ and $\langle\mathbf{U}\rangle$ when referring to the operators in the basic modal language defined by the equivalences above. Respectively, $\mathbf{ML}^{\mathsf{r}}_{\mathsf{U}}$ and $\mathbf{ML}^{\mathsf{as}}_{\mathsf{U}}$ will denote the extensions of $\mathbf{ML}^{\mathsf{r}}$ and $\mathbf{ML}^{\mathsf{as}}$ to the language ML_{U}. Note, that every formula ϕ of ML_{U} is trivially translated into a formula ϕ^d of the basic language ML by replacing all occurrences of $[\mathsf{U}]$ and $\langle\mathsf{U}\rangle$ respectively with $[\mathbf{U}]$ and $\langle\mathbf{U}\rangle$. The important property of that translation is that ϕ and ϕ^d are equivalent, hence equally valid, in every frame from $\mathcal{F}^{\mathsf{d2}}$, hence in almost every finite frame, which will suffice for our purposes. In particular, every axiom in $\mathbf{ML}^{\mathsf{as}}_{\mathsf{U}}$ generates its translated axiom in $\mathbf{ML}^{\mathsf{as}}$, which will enable me to state most of the claims and results about $\mathbf{ML}^{\mathsf{as}}$ in the language ML_{U} and for $\mathbf{ML}^{\mathsf{as}}_{\mathsf{U}}$, with the understanding that they apply accordingly to the logic $\mathbf{ML}^{\mathsf{as}}$ of my primary interest.

Theorem 3.3 ([14]) *The following axiomatic system $\boldsymbol{Ax}(\mathbf{ML}^{\mathsf{r}})$ is sound and complete for $\mathbf{ML}^{\mathsf{r}}$ (recall that $[\mathbf{U}]$ and $\langle\mathbf{U}\rangle$ are the defined operators):*

(ML^r_1) K: $\Box(p \to q) \to (\Box p \to \Box q)$.

(ML^r_2) $[\mathbf{U}]p \to p$.

(ML^r_3) $[\mathbf{U}]p \to [\mathbf{U}]\Box p$.

(ML^r_4) $p \to [\mathbf{U}]\langle\mathbf{U}\rangle p$.

(ML^r_5) Scheme MODEXT, consisting of the following axioms for each $n \in \mathbb{N}$:

$$\mathit{MODEXT}_n = \bigwedge_{k=1}^{n} \langle\mathbf{U}\rangle(p_k \wedge \Box q_k) \to \langle\mathbf{U}\rangle \bigwedge_{k=1}^{n} (\Diamond p_k \wedge q_k).$$

The first 4 axiom schemes above come from the axiomatization of $\mathbf{K}_{\mathsf{U}}$ ([16]). It is easy to see that the axiom MODEXT_n is valid in a frame $\mathsf{F} \in \mathcal{F}^{\mathsf{d2}}$ iff for every n points $w_1, \ldots, w_n$ in F there is a point u that is R-reachable from each $w_1, \ldots, w_n$, and each of them is R-reachable from u. This holds for every finite frame, with $[\mathbf{U}]$ and $\langle\mathbf{U}\rangle$ replaced by the primitives $[\mathsf{U}]$, $\langle\mathsf{U}\rangle$. Thus, MODEXT is the modally definable approximation of the extension axioms EXT for FOL.

Proposition 3.4 ([14]) **ML**^{r} *has the finite model property and is decidable, but it is not finitely axiomatizable.*

Thus, **Ax**(**ML**^{r}) derives the 'well-behaved' formulae in **ML**^{as}, viz. those that follow from the extension axioms of FOL. As we will see in Prop.4.1, these include all first-order definable formulae in **ML**^{as}. What about the rest? Maybe, that is all and the logics **ML**^{r}and **ML**^{as}coincide? It turns out, the answer, rather surprisingly on the background of Fagin's Transfer theorem, is 'No', as shown further. To see that, we need to learn more about F^{r} and **ML**^{r}.

3.2 Kernels in finite frames and in F^{r}.

Every bounded morphic image F of a given frame G determines a kernel partition $\mathcal{P}_{\mathsf{F}}$ in G, defined as follows. Given a bounded morphism $h : \mathsf{G} \to \mathsf{F}$, where $\mathsf{F} = \langle W_{\mathsf{F}}, R_{\mathsf{F}} \rangle$ and $\mathsf{G} = \langle W_{\mathsf{G}}, R_{\mathsf{G}} \rangle$, the **kernel partition** $\mathcal{P}_{\mathsf{F}}(\mathsf{G})$ in G consists of the family of clusters $\{h^{-1}(w) \mid w \in W_{\mathsf{F}}\}$. Thus, $\mathcal{P}_{\mathsf{F}}(\mathsf{G})$ is generated by the equivalence relation $\sim_h$ in W_{G}, where $u \sim_h v$ holds iff $h(u) = h(v)$. It satisfies the following properties, determined by F and the definition of bounded morphism. For any two clusters $X = h^{-1}(x)$ and $Y = h^{-1}(y)$ in $\mathcal{P}_{\mathsf{F}}(\mathsf{G})$, either

(i) for each $u \in X$ there is $v \in Y$ such that $uR_{\mathsf{G}}v$, (when $xR_{\mathsf{F}}y$ holds),

or

(ii) for no $u \in X$ there is $v \in Y$ such that $uR_{\mathsf{G}}v$ (when $xR_{\mathsf{F}}y$ does not hold).

Conversely, for every kernel partition in G generated by mapping $h : \mathsf{G} \to \mathsf{F}$ and satisfying the conditions (i) and (ii) above, the mapping h is a bounded morphism from G onto F.

Thus, kernel partitions are an equivalent, and often more visually intuitive way of describing bounded morphisms. Note that existence of a kernel partition with specific FO-definable properties in a frame, like those above, is a $M\Sigma^1_1$-property and, as stated in lemma 2.3, the existence of the kernel partition determined by F in a (randomly selected) frame G is characterised by the non-validity of the respective χ_{F} in that frame. Thus, using existence or non-existence of kernel partitions one can show the non-validity or validity in a given frame of various formulae that are not first-order definable. Here I will give two very simple examples, that will suffice to distinguish **ML**^{r} from **ML**^{as}. Consider the following two frames:

$\mathsf{K}_2 = \langle \{x, y\}, \{(x, x), (x, y), (y, x)\} \rangle$ and

$\mathsf{K}_3 = \langle \{x, y_1, y_2\}, \{(x, x), (x, y_1), (x, y_2), (y_1, x), (y_2, x)\} \rangle$.

(Note that K_2 is a bounded morphic image of K_3.)

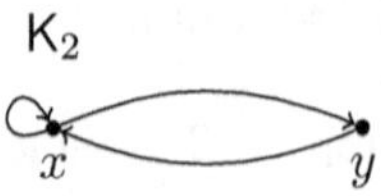

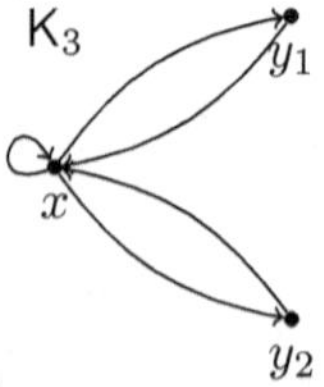

It turns out that the kernel partition $\mathcal{P}_{K_2}$ that K_2 generates in any frame

G for which K_2 is a bounded morphic image corresponds to the well-known notion of a **kernel** in digraphs (cf.[9] but, taking into account that frames are digraphs with loops), whereas the kernel partition that K_3 generates is called **double kernel** in [14] (see details there). Thus, the ML-translated characteristic formula $\chi^d_{K_2}$ (resp. $\chi^d_{K_3}$) is valid precisely in those frames in $\mathcal{F}^{\mathsf{d2}}$ which *do not* have kernels (resp. double kernels). Here are slightly simpler equivalent formulae, where the falsifying valuation of p, (resp. p and q), in any frame with a kernel (resp. double kernel) is that kernel (resp. each of the two sub-kernels).

$\mathsf{NO\text{-}KER} = \langle \mathbf{U} \rangle (p \leftrightarrow \Diamond p)$,

$\mathsf{NO\text{-}DKER} = \langle \mathbf{U} \rangle ((p \vee q) \wedge \Diamond (p \vee q)) \vee \langle \mathbf{U} \rangle (\neg (p \vee q) \wedge (\Box \neg p \vee \Box \neg q))$.

3.3 The finite frames of $\mathbf{ML}^{\mathsf{r}}$.

Even though $\mathbf{ML}^{\mathsf{r}}$ is defined as the logic of the single infinite frame F^{r}, it does have finite frames, as evident from Prop. 3.4. It turns out that they are very simple to describe, as precisely those finite frames that have a 'central point' – a point which is R-related to and from every point (incl. itself). Formally, given a frame $\mathsf{F} = \langle W, R \rangle$, a point $x \in W$ is a **central point** in F if Rxy and Ryx hold for every $y \in W$. The existence of a central point is forced by the axiom scheme MODEXT, and every frame with central point is easily seen to validate MODEXT. Note that both K_2 and K_3 above have central points.

Proposition 3.5 ([14], Lemma 2.4) *For every finite frame* F *the following are equivalent.*

(i) $\mathsf{F} \models \mathbf{ML}^{\mathsf{r}}$

(ii) F *has a central point.*

(iii) F *is a bounded-morphic image of* F^{r}.

(iv) $\mathsf{F}^{\mathsf{r}} \not\models \chi_{\mathsf{F}}$.

(v) F *can be obtained from* F^{r} *by filtration.*

In particular, K_2 and K_3 are bounded morphic images of F^{r}, hence both χ_{K_2} and χ_{K_3} fail in F^{r}, i.e. F^{r} has a kernel and a double kernel.

Corollary 3.6 *For every finite frame* G *without central point:* $\mathsf{F}^{\mathsf{r}} \models \chi_{\mathsf{G}}$, *and hence existence of kernel partition* $\mathcal{P}_{\mathsf{G}}$ *is almost surely false in the finite.*

Thus, Corollary 3.6 provides plenty of (generally) non-first-order definable modal formulae in $\mathbf{ML}^{\mathsf{r}}_{\mathsf{U}}$, respectively in $\mathbf{ML}^{\mathsf{r}}$.

Here is the main technical result in [14], proved by a non-trivial combinatorial-analytic estimation of the expected number of double kernels in a random finite frame from $\mathcal{F}^{\mathsf{d2}}$.

Theorem 3.7 ([14]) *Existence of a double kernel is almost surely false in finite frames. Consequently,* χ_{K_3} *is almost surely valid, hence it is in* $\mathbf{ML}^{\mathsf{as}}$.

Thus, $\chi_{\mathsf{K}_3} \in \mathbf{ML}^{\mathsf{as}}$ but $\chi_{\mathsf{K}_3} \notin \mathbf{ML}^{\mathsf{r}}$, hence the inclusion $\mathbf{ML}^{\mathsf{r}} \subset \mathbf{ML}^{\mathsf{as}}$ is proper. Also, Fagin's transfer theorem fails for frame validity in modal logic.

The technique used in the proof of Theorem 3.7 did not help the authors of [14] to prove the same results for single kernels and χ_{K_2}, and these were left as open questions there. They were proved a little later by Le Bars in [29]. He also proved there that the 0-1 law fails for frame validity in modal logic, by showing that a modified kernel property, defined by the formula

$$\mathsf{MODAL\text{-}KERNEL}: \quad \neg p \wedge q \wedge \Box\Box((p \vee q) \to \neg\Diamond(p \vee q)) \to \Diamond\Box\neg p$$

has no asymptotic probability in the finite.

4 On the axiomatization of the modal logic $\mathbf{ML}^{\mathsf{as}}$

What axioms are needed to add to $\mathbf{Ax}(\mathbf{ML}^{\mathsf{r}})$ in order to axiomatize completely $\mathbf{ML}^{\mathsf{as}}$? We explore this question here, starting with some useful observations. For technical convenience, most of the results will be stated for $\mathbf{ML}^{\mathsf{as}}_{\mathsf{U}}$, instead of $\mathbf{ML}^{\mathsf{as}}$, but they are readily translated to $\mathbf{ML}^{\mathsf{as}}$. I will denote by $\mathbf{Ax}(\mathbf{ML}^{\mathsf{r}}_{\mathsf{U}})$ the axiomatic system $\mathbf{Ax}(\mathbf{ML}^{\mathsf{r}})$ where $[\mathbf{U}]$ and $\langle\mathbf{U}\rangle$ are replaced by the primitives $[\mathsf{U}]$ and $\langle\mathsf{U}\rangle$, with the relevant axioms added (cf. [16]).

4.1 Towards understanding the logic $\mathbf{ML}^{\mathsf{as}}$

Proposition 4.1

(i) *Every first-order definable modal formula which is in* $\mathbf{ML}^{\mathsf{as}}$ *is also in* $\mathbf{ML}^{\mathsf{r}}$.

(ii) *([14]) Every modal formula ϕ in* $\mathbf{ML}^{\mathsf{as}}$ *that defines a purely universal frame condition $FC(\phi)$ is valid.*

Proof. (i) If $\phi \in \mathbf{ML}^{\mathsf{as}}$ and ϕ is first order definable, then $FC(\phi)$ is almost surely true in the finite, hence it follows from an extension axiom. Therefore, $\phi \in \mathbf{ML}^{\mathsf{r}}$, by Proposition 3.2[ii].

(ii) Suppose ϕ is not valid. Then $\neg\phi$ is satisfiable in a finite frame F. The satisfiability of $\neg\phi$ is an existential property, hence preserved in extensions. As F is embeddable in F^{r}, $\neg\phi$ is satisfiable there, too, which contradicts (i). □

So, the missing axioms are neither first-order definable, nor purely universal.

More notation: Given a (possibly infinite) set of frames $\mathcal{F}$, a set of formulae Γ, and a formula ϕ, we denote by $\Gamma \models^{\mathsf{fr}}_{\mathcal{F}} \phi$ the claim that ϕ is valid in every frame from $\mathcal{F}$ in which all formulae of Γ are valid[10]. When $\mathcal{F}$ is the class of all frames I will write simply $\Gamma \models^{\mathsf{fr}} \phi$ and when $\mathcal{F} = \mathcal{F}^{\mathsf{fin}}$ I will write $\Gamma \models^{\mathsf{fr}}_{\mathsf{fin}} \phi$. When $\Gamma = \{\psi\}$, I will write just $\psi \models^{\mathsf{fr}}_{\mathcal{F}} \phi$, respectively $\psi \models^{\mathsf{fr}} \phi$ and $\psi \models^{\mathsf{fr}}_{\mathsf{fin}} \phi$.

I denote by $\mathrm{BM}^{-1}(\mathsf{F})$ the set of finite frames G (over $\mathbb{N}$) such that $\mathsf{G} \twoheadrightarrow \mathsf{F}$.

Note that $\mathbf{ML}^{\mathsf{as}}$ (resp. $\mathbf{ML}^{\mathsf{as}}_{\mathsf{U}}$) is closed under $\models^{\mathsf{fr}}_{\mathsf{fin}}$. Now, what are the finite frames for $\mathbf{ML}^{\mathsf{as}}$ like? (Note that they are the same as those for $\mathbf{ML}^{\mathsf{as}}_{\mathsf{U}}$.) A partial answer follows, that essentially employs for our purpose more general facts listed in Lemma 2.4.

---

[10] This consequence relation is generally not arithmetically definable, hence not recursively axiomatizable, as first shown in [34] by reduction from logical consequence in second-order logic, cf. also discussion in [35]. However, we are only interested here in very special cases of that consequence relation, so no general results can be assumed a priori to hold.

Proposition 4.2 *For every finite frame* F *and a modal formula* $\phi \in \mathrm{ML}_{\mathsf{U}}$*:*

(i) $\mathsf{F} \not\models \phi$ *iff* $\phi \models^{\mathsf{fr}}_{\mathsf{fin}} \chi_{\mathsf{F}}$ *iff* $\phi \models^{\mathsf{fr}} \chi_{\mathsf{F}}$.

(ii) $\mathsf{F} \models \mathbf{ML}^{\mathsf{as}}_{\mathsf{U}}$ *iff* $\chi_{\mathsf{F}} \notin \mathbf{ML}^{\mathsf{as}}_{\mathsf{U}}$.

Proof. (i) Let $\mathsf{F} \not\models \phi$. Then $\mathsf{G} \not\models \phi$ for every frame G such that $\mathsf{G} \twoheadrightarrow \mathsf{F}$. Therefore, for every G such that $\mathsf{G} \models \phi$, it follows that $\mathsf{G} \not\twoheadrightarrow \mathsf{F}$, hence $\mathsf{G} \models \chi_{\mathsf{F}}$. Thus, $\mathsf{F} \not\models \phi$ implies $\phi \models^{\mathsf{fr}} \chi_{\mathsf{F}}$. Further, $\phi \models^{\mathsf{fr}} \chi_{\mathsf{F}}$ obviously implies $\phi \models^{\mathsf{fr}}_{\mathsf{fin}} \chi_{\mathsf{F}}$. Lastly, if $\phi \models^{\mathsf{fr}}_{\mathsf{fin}} \chi_{\mathsf{F}}$ then $\mathsf{F} \not\models \phi$ because $\mathsf{F} \not\models \chi_{\mathsf{F}}$.

(ii) By contraposition, if $\chi_{\mathsf{F}} \in \mathbf{ML}^{\mathsf{as}}_{\mathsf{U}}$ then $\mathsf{F} \not\models \mathbf{ML}^{\mathsf{as}}_{\mathsf{U}}$ because $\mathsf{F} \not\models \chi_{\mathsf{F}}$. Conversely, take $\phi \in \mathbf{ML}^{\mathsf{as}}_{\mathsf{U}}$. If $\mathsf{F} \not\models \phi$ then $\phi \models^{\mathsf{fr}}_{\mathsf{fin}} \chi_{\mathsf{F}}$ by (i), so $\chi_{\mathsf{F}} \in \mathbf{ML}^{\mathsf{as}}_{\mathsf{U}}$. □

From Proposition 4.2 we immediately obtain the following useful fact.

Corollary 4.3 *For any finite frame* F *and* $\phi \in \mathbf{ML}^{\mathsf{as}}_{\mathsf{U}}$*, if* $\mathsf{F} \not\models \phi$ *then* $\chi_{\mathsf{F}} \in \mathbf{ML}^{\mathsf{as}}_{\mathsf{U}}$.

Proposition 4.4 *For any finite frames* F, G*:*

(i) $\mathsf{G} \twoheadrightarrow \mathsf{F}$ *iff* $\chi_{\mathsf{F}} \models^{\mathsf{fr}}_{\mathsf{fin}} \chi_{\mathsf{G}}$.

(ii) *Moreover, if* $\chi_{\mathsf{F}} \models^{\mathsf{fr}}_{\mathsf{fin}} \chi_{\mathsf{G}}$ *then* $\mathbf{K}_{\mathsf{U}} + \chi_{\mathsf{F}} \vdash \chi_{\mathsf{G}}$.

Proof.

(i) Directly from Lemma 2.3 and Proposition 4.2.(i)

(ii) By (i) and Lemma 2.4. Also, χ^d_{G} is derived in the same way in the respectively axiomatized version $\mathbf{K}_{[\mathsf{U}]} + \chi^d_{\mathsf{F}}$ in ML as sketched in Lemma 2.4.□

As noted in the proof of Lemma 2.4, Proposition 4.4(ii) holds likewise for any formula ϕ instead of χ_{F}, but the greater generality seems to be of no use in our case, as all conjectured axioms of $\mathbf{ML}^{\mathsf{as}}_{\mathsf{U}}$ over $\mathbf{ML}^{\mathsf{r}}_{\mathsf{U}}$ are of the type χ_{F}, so the respective conjectured axioms of $\mathbf{ML}^{\mathsf{as}}$ over $\mathbf{ML}^{\mathsf{r}}$ are of the type χ^d_{F}.

4.2 Towards axiomatizing the logics $\mathbf{ML}^{\mathsf{as}}_{\mathsf{U}}$ and $\mathbf{ML}^{\mathsf{as}}$

From the observations made so far we see that natural candidates for additional axioms of $\mathbf{ML}^{\mathsf{as}}_{\mathsf{U}}$ over $\mathbf{Ax}(\mathbf{ML}^{\mathsf{r}}_{\mathsf{U}})$ are the almost surely valid formulae of the type χ_{F} for frames F with central point (recall Corollary 3.6). So, let $\mathcal{C}$ be the set of all finite frames with a central point. Note that $\mathcal{C} \subseteq \mathcal{F}^{\mathsf{d2}}$. Let

$$\Xi^{\mathsf{as}}_{\mathsf{U}} := \{\chi_{\mathsf{F}} \mid \mathsf{F} \in \mathcal{C} \text{ and } \chi_{\mathsf{F}} \in \mathbf{ML}^{\mathsf{as}}_{\mathsf{U}}\}.$$

Then, let Ξ^{as} be the set of translated axioms in ML.

The following conjecture, stated in two equivalent versions, seems natural.

Conjecture 4.5 $\boldsymbol{Ax}(\mathbf{ML}^{\mathsf{r}}_{\mathsf{U}}) \cup \Xi^{\mathsf{as}}_{\mathsf{U}}$ *axiomatizes* $\mathbf{ML}^{\mathsf{as}}_{\mathsf{U}}$.
Respectively, $\boldsymbol{Ax}(\mathbf{ML}^{\mathsf{r}}) \cup \Xi^{\mathsf{as}}$ *axiomatizes* $\mathbf{ML}^{\mathsf{as}}$.

Let us first make an encouraging observation in support of that conjecture. I state the version for $\mathbf{ML}^{\mathsf{as}}_{\mathsf{U}}$; the one for $\mathbf{ML}^{\mathsf{as}}$ is completely analogous.

Proposition 4.6 *For any* $\phi \in \mathbf{ML}^{\mathsf{as}}_{\mathsf{U}}$*:*

(i) $\Xi^{\mathsf{as}}_{\mathsf{U}}(\phi) \models^{\mathsf{fr}}_{\mathcal{C}} \phi$, *where* $\Xi^{\mathsf{as}}_{\mathsf{U}}(\phi) = \Xi^{\mathsf{as}}_{\mathsf{U}} \cap \{\chi_{\mathsf{F}} \mid \phi \models^{\mathsf{fr}}_{\mathsf{fin}} \chi_{\mathsf{F}}\}$.

(ii) $\mathbf{ML}^{\mathsf{r}}_{\mathsf{U}} \cup \Xi^{\mathsf{as}}_{\mathsf{U}} \models^{\mathsf{fr}}_{\mathsf{fin}} \phi$.

Proof. Take $\phi \in \mathbf{ML}^{\mathsf{as}}_{\mathsf{U}}$ and any finite frame F such that $\mathsf{F} \not\models \phi$.
Then, by Corollary 4.3, $\chi_{\mathsf{F}} \in \mathbf{ML}^{\mathsf{as}}_{\mathsf{U}}$. Now:
(i) If $\mathsf{F} \in \mathcal{C}$ then $\mathsf{F} \not\models \Xi^{\mathsf{as}}_{\mathsf{U}}(\phi)$ because $\mathsf{F} \not\models \chi_{\mathsf{F}}$ and $\chi_{\mathsf{F}} \in \Xi^{\mathsf{as}}_{\mathsf{U}}(\phi)$.
(ii) Consider two cases:
– If $\mathsf{F} \in \mathcal{C}$ then $\mathsf{F} \not\models \mathbf{ML}^{\mathsf{r}}_{\mathsf{U}} \cup \Xi^{\mathsf{as}}_{\mathsf{U}}$ by (i).
– if $\mathsf{F} \notin \mathcal{C}$ then $\mathsf{F} \not\models \mathbf{ML}^{\mathsf{r}}_{\mathsf{U}} \cup \Xi^{\mathsf{as}}_{\mathsf{U}}$ because $\mathsf{F} \not\models \mathbf{ML}^{\mathsf{r}}_{\mathsf{U}}$.
Thus, in either case, $\mathsf{F} \not\models \mathbf{ML}^{\mathsf{r}}_{\mathsf{U}} \cup \Xi^{\mathsf{as}}_{\mathsf{U}}$.
By contraposition, if $\mathsf{F} \models \mathbf{ML}^{\mathsf{r}}_{\mathsf{U}} \cup \Xi^{\mathsf{as}}_{\mathsf{U}}$ then $\mathsf{F} \models \phi$.
Hence, $\mathbf{ML}^{\mathsf{r}}_{\mathsf{U}} \cup \Xi^{\mathsf{as}}_{\mathsf{U}} \models^{\mathsf{fr}}_{\mathsf{fin}} \phi$. □

The proposition above provides a model-theoretic characterisation of the additional validities of $\mathbf{ML}^{\mathsf{as}}_{\mathsf{U}}$ (resp. $\mathbf{ML}^{\mathsf{as}}$), beyond those in $\mathbf{ML}^{\mathsf{r}}_{\mathsf{U}}$ (resp. $\mathbf{ML}^{\mathsf{r}}$).

Still, there are two major issues with proving Conjecture 4.5, if true at all:

(i) How to identify the axioms in $\Xi^{\mathsf{as}}_{\mathsf{U}}$?

(ii) How to prove the completeness?

On the first question, let us first make the task a little easier by noting that, due to Corollary 4.3, we only need to identify the axioms χ_{F} for the *minimal* frames $\mathsf{F} \in \mathcal{C}$ such that $\chi_{\mathsf{F}} \in \mathbf{ML}^{\mathsf{as}}_{\mathsf{U}}$, where 'minimal' is in sense (cf. Proposition 4.4) that there is no $\mathsf{F}' \in \mathcal{C}$ such that $\chi_{\mathsf{F}'} \in \mathbf{ML}^{\mathsf{as}}_{\mathsf{U}}$ and $\mathsf{F} \in \mathrm{BM}^{-1}(\mathsf{F}')$, but $\mathsf{F} \not\cong \mathsf{F}'$. Equivalently, we are looking for the maximal sets of frames of the type $\mathrm{BM}^{-1}(\mathsf{F})$ for $\mathsf{F} \in \mathcal{C}$ which have asymptotic measure 0. For that, the membership in $\mathrm{BM}^{-1}(\mathsf{F})$ should almost surely contradict (EXT); equivalently, χ_{F} should follow from (EXT). Being a $M\Pi^1_1$-condition, by compactness χ_{F} should then follow from some $(\mathsf{EXT})_n$, hence some extension axiom η_{F} should fail in all frames in $\mathrm{BM}^{-1}(\mathsf{F})$. To ensure the latter, one should naturally look for η_{F} that fails in F but is preserved in bounded morphic images, so it must fail in all frames from $\mathrm{BM}^{-1}(\mathsf{F})$. A classic result by van Benthem [35, Thm 15.11] characterises the first-order sentences in the language with $=$ and R that are preserved by bounded morphisms as precisely those that are equivalent to ones constructed from atomic formulae, $\top$, and $\bot$ using $\wedge, \vee, \exists, \forall$, and restricted universal quantification $\forall z(Ryz \to \ldots)$ for $z \neq y$. By looking at the syntactic shape of (EXT), one can see that only few of them satisfy the description above. Still, they generate infinitely many axioms, as the next proposition shows.

Proposition 4.7 *There is a subset Φ of infinitely many axioms in $\Xi^{\mathsf{as}}_{\mathsf{U}}$, none of which follows in terms of $\models^{\mathsf{fr}}_{\mathsf{fin}}$ from all others.*

Proof. Consider the sequence of frames $\{\mathsf{F}_n\}_{n \in \mathbb{N}}$ defined as follows.
Let $\mathsf{F}_n = \langle W_n, R_n \rangle$ where $W_n = \{0, 1, ...n\}$, and
$R_n = \{(k, 0) \mid k \in W_n\} \cup \{(0, k) \mid k \in W_n\} \cup \{(k, k+1) \mid k = 1, ..., n-1\}$.
Now, let $\Phi = \{\chi_{\mathsf{F}_n} \mid n \in \mathbb{N}\}$.

Clearly, 0 is a central point, so each F_n is in $\mathcal{C}$. Next, each χ_{F_n} is in $\mathbf{ML}^{\mathsf{as}}_{\mathsf{U}}$. Indeed, note that $\mathrm{BM}^{-1}(\mathsf{F}_n)$ has an asymptotic measure 0 because $\forall x \exists y(Rxy \wedge \neg Ryy)$ is an instance of the extension axiom $(\mathsf{EXT})_1$ that fails in each F_n, hence in every $\mathsf{G} \in \mathrm{BM}^{-1}(\mathsf{F}_n)$, because it is preserved in bounded morphic images. Lastly, each F_n is minimal in the sense above, as it is easy to see that neither of them has proper bounded morphic images (different from F_n and F_0). Let $\Phi^{-n} = \{\chi_{\mathsf{F}_m} \mid 0 < m, m \neq n\}$. Then $\Phi^{-n} \not\models^{\mathsf{fr}}_{\mathsf{fin}} \chi_{\mathsf{F}_n}$ for each $n \in \mathbb{N}$, because $\mathsf{F}_n \models \Phi^{-n}$, while $\mathsf{F}_n \not\models \chi_{\mathsf{F}_n}$. □

The translated set Φ^d provides likewise infinitely many independent axioms in Ξ^{as}. The proposition above makes the following conjecture very likely, but in order to prove it we need either a provably complete infinitary axiomatization of $\mathbf{ML}^{\mathsf{as}}_{\mathsf{U}}$ over $\mathbf{ML}^{\mathsf{r}}_{\mathsf{U}}$ or a proof that $\mathbf{ML}^{\mathsf{as}}_{\mathsf{U}}$ is not recursively axiomatizable.

Conjecture 4.8 *The logic* $\mathbf{ML}^{\mathsf{as}}_{\mathsf{U}}$ *is not finitely axiomatizable over* $\mathbf{ML}^{\mathsf{r}}_{\mathsf{U}}$. *Respectively,* $\mathbf{ML}^{\mathsf{as}}$ *is not finitely axiomatizable over* $\mathbf{ML}^{\mathsf{r}}$.

It is conceivable that additional axioms from $\Xi^{\mathsf{as}}_{\mathsf{U}}$ may be needed to add to Φ for the axiomatization of $\mathbf{ML}^{\mathsf{as}}_{\mathsf{U}}$, and likewise for $\mathbf{ML}^{\mathsf{as}}$. To speculate a little on these, note first that the extension axioms η that fit van Benthem's syntactic description for preservation under bounded morphisms can have at most one universally quantified variable, i.e., be of the type $\forall x \exists y$. Furthermore, for η to fail in some $\mathrm{BM}^{-1}(\mathsf{F})$ such that $\chi_{\mathsf{F}} \in \Xi^{\mathsf{as}}_{\mathsf{U}}$, there must be a negative atom, which can only be $\neg Ryy$. This restricts the syntactic possibilities for η to just a few, that can be easily described. Thereafter, the frames $\mathsf{F} \in \mathcal{C}$ for which such η fails in $\mathrm{BM}^{-1}(\mathsf{F})$, hence the further axioms $\chi_{\mathsf{F}} \in \Xi^{\mathsf{as}}_{\mathsf{U}}$ that are generated by them, are also easily describable. And, now the big unknown is: are these *all* axioms that are missing, or are there more, that are not identifiable in such way? If these are all, then the logic $\mathbf{ML}^{\mathsf{as}}_{\mathsf{U}}$ is recursively (even if not finitely) axiomatizable over $\mathbf{ML}^{\mathsf{r}}_{\mathsf{U}}$ and even stands a chance to be decidable, too, like $\mathbf{ML}^{\mathsf{r}}_{\mathsf{U}}$ is; likewise for $\mathbf{ML}^{\mathsf{as}}$. Otherwise, the problem with the identification of all missing axioms is very likely going beyond logic. Indeed, the question for which frames $\mathsf{F} \in \mathcal{C}$ it holds that $\chi_{\mathsf{F}} \in \mathbf{ML}^{\mathsf{as}}_{\mathsf{U}}$ may then hinge on rather difficult combinatorial-probabilistic calculations, as results in [14] and [29], as well as an empirical study in [32], have indicated.

To sum up: it is currently unknown whether the set $\Xi^{\mathsf{as}}_{\mathsf{U}}$ is even recursively enumerable, though I would conjecture that it is. But even if that is the case, the question whether $\mathbf{ML}^{\mathsf{r}}_{\mathsf{U}} \cup \Xi^{\mathsf{as}}_{\mathsf{U}}$ axiomatizes $\mathbf{ML}^{\mathsf{as}}_{\mathsf{U}}$ remains open. The core of the problem is that we cannot conclude $\mathbf{ML}^{\mathsf{r}}_{\mathsf{U}} \cup \Xi^{\mathsf{as}}_{\mathsf{U}} \vdash \phi$ from $\mathbf{ML}^{\mathsf{r}}_{\mathsf{U}} \cup \Xi^{\mathsf{as}}_{\mathsf{U}} \models^{\mathsf{fr}}_{\mathsf{fin}} \phi$, because we have no recursive axiomatization of $\models^{\mathsf{fr}}_{\mathsf{fin}}$ in ML_{U} (and I currently do not know if one exists). It seems a currently open question whether and when $\Gamma \models^{\mathsf{fr}}_{\mathsf{fin}} \phi$ implies derivability over a suitably recursively axiomatized base logic, beyond the special case established in Proposition 4.4. This is currently unknown to me even for the special case when $\chi_{\mathsf{F}} \models^{\mathsf{fr}}_{\mathsf{fin}} \phi$, where $\chi_{\mathsf{F}} \in \mathbf{ML}^{\mathsf{as}}_{\mathsf{U}}$. Likewise for $\models^{\mathsf{fr}}_{\mathsf{fin}}$ in ML.

An important related question[11] is whether the logic $\mathbf{ML}^{\mathsf{as}}_{\mathsf{U}}$ (resp. $\mathbf{ML}^{\mathsf{as}}$) is *Kripke complete*, i.e. whether it is the modal logic of any class of Kripke frames. If so, it is certainly the modal logic of the class of *all* (not necessarily finite) frames F such that $\mathsf{F} \models \mathbf{ML}^{\mathsf{as}}_{\mathsf{U}}$. Equivalently, the question is whether every non-validity of $\mathbf{ML}^{\mathsf{as}}_{\mathsf{U}}$ is refuted in some (finite, or not) frame F such that

[11] Raised by Evgeny Zolin [37].

$\mathsf{F} \models \mathbf{ML}^{\mathsf{as}}_{\mathsf{U}}$; likewise for $\mathbf{ML}^{\mathsf{as}}$. While this is rather plausible, it does not seem to follow obviously from what is currently known about $\mathbf{ML}^{\mathsf{as}}_{\mathsf{U}}$ (resp. $\mathbf{ML}^{\mathsf{as}}$), so I would add it to the list of currently open problems.

Finally, briefly on the question of proving the completeness of the axiomatization of $\mathbf{ML}^{\mathsf{as}}_{\mathsf{U}}$ and the respective translation for $\mathbf{ML}^{\mathsf{as}}$, if and when it is identified. It is very easy to see that they would be equally complete. This problem seems not less challenging, because – unlike the axioms from the scheme MODEXT – the truly second-order axioms, like those from $\Xi^{\mathsf{as}}_{\mathsf{U}}$, are likely not to be canonical, as the kernel partitions generated in the canonical model by the axioms $\chi_{\mathsf{F}} \in \Xi^{\mathsf{as}}_{\mathsf{U}}$ need not be syntactically definable there. Still, how difficult that problem is can only be assessed when all axioms are explicitly known.

On this note, I leave the question of establishing a provably complete axiomatization of $\mathbf{ML}^{\mathsf{as}}$, while better understood now, still open.

5 Concluding remarks and further challenges

Besides the open questions regarding the axiomatization of $\mathbf{ML}^{\mathsf{as}}$, stated above, many other related problems arise. To mention just one such generic question: given a class $\mathcal{K}$ of Kripke frames, what is the modal logic of almost sure validities of $\mathcal{K}$? The case when the modal logic of $\mathcal{K}$ satisfies the 0-1 law seems to be considerably easier (though, by no means trivial) than the case of $\mathcal{K} = \mathcal{F}^{\mathsf{fin}}$ studied here, as it then boils down to axiomatizing the modal logic of the respective analogue of countable random frame, relativised to the class $\mathcal{K}$, if it exists. Quite promising recent results of that type were announced in [36] for the provability logic and two versions of Grzegorczyk logic.

Further open problems arise when going beyond modal logic, to the full $M\Sigma^1_1$ and $M\Pi^1_1$ on graphs, digraphs, and other important classes of structures. Axiomatizing the almost sure theories of these may very likely lead to quite complicated combinatorial-probabilistic computations proving almost sure existence (resp., non-existence) of kernel partitions. In general, little is known about these so far and the challenge to understand them is wide open.

Acknowledgments

This work was partly supported by research grant 2015-04388 of the Swedish Research Council. I am grateful to the anonymous referees for several corrections and useful comments, and am particularly indebted to Evgeny Zolin, for scrutinising previous versions of the paper and providing numerous corrections, as well as for many valuable comments and references that helped improving the content and simplifying some arguments.

References

[1] Ahlman, O., *Homogenizable structures and model completeness*, Arch. Math. Log. **55** (2016), pp. 977–995.

[2] Ahlman, O., *Simple structures axiomatized by almost sure theories*, Ann. Pure Appl. Logic **167** (2016), pp. 435–456.

[3] Bezhanishvili, G. and N. Bezhanishvili, *An algebraic approach to canonical formulas: Modal case*, Studia Logica **99** (2011), pp. 93–125.
[4] Bezhanishvili, N., *Frame based formulas for intermediate logics*, Studia Logica **90** (2008), pp. 139–159.
[5] Blackburn, P., M. de Rijke and Y. Venema, "Modal Logic," Cambridge UP, 2001.
[6] Blass, A., Y. Gurevich and D. Kozen, *A zero-one law for logic with a fixed-point operator*, Information and Control **67** (1985), pp. 70–90.
[7] Carnap, R., "Logical Foundations of Probability," Univ. of Chicago Press, Chicago, 1950.
[8] Chagrov, A. and M. Zakharyaschev, "Modal Logic," OUP, Oxford, 1997.
[9] de la Vega, W. F., *Kernels in random graphs*, Discrete Math. **82** (1990), pp. 213–217.
[10] Fagin, R., *Probabilities on finite models*, Journal of Symbolic Logic **41** (1976), pp. 50–58.
[11] Fine, K., *Logics containing K4. part I*, J. Symb. Log. **39** (1974), pp. 31–42.
[12] Gaifman, H., *Concerning measures in first-order calculi*, Israel J. of Math. **2** (1964), pp. 1–18.
[13] Glebskii, Y., D. Kogan, M. Liogonki and V. Talanov, *Range and degree of realizability of formulas in the restricted predicate calculus*, Cybernetics **5** (1969), pp. 142–154.
[14] Goranko, V. and B. Kapron, *The modal logic of the countable random frame*, Archive for Mathematical Logic **42** (2003), pp. 221–243.
[15] Goranko, V. and M. Otto, *Model theory of modal logic*, in: P. Blackburn, J. van Benthem and F. Wolter, editors, *Handbook of Modal Logic*, Studies in Logic and Practical Reasoning **3**, Elsevier, 2006 pp. 249–329.
[16] Goranko, V. and S. Passy, *Using the universal modality: Gains and questions*, Journal of Logic and Computation **2** (1992), pp. 5–30.
[17] Grandjean, E., *Complexity of the first–order theory of almost all structures*, Information and Computation **52** (1983), pp. 180–204.
[18] Gurevich, Y., *Zero-one laws: The logic in computer science column*, Bulletin of the European Association for Theoretical Computer Science **46** (1992), pp. 90–106.
[19] Halpern, J. and B. Kapron, *Zero-one laws for modal logic*, Annals of Pure and Applied Logic **69** (1994), pp. 157–193.
[20] Halpern, J. Y. and B. M. Kapron, *Erratum to "zero-one laws for modal logic" [ann. pure appl. logic 69 (1994) 157-193]*, Ann. Pure Appl. Logic **121** (2003), pp. 281–283.
[21] Jankov, V., *On the relation between deducibility in intuitionistic propositional calculus and finite implicative structures*, Dokl. Akad. Nauk SSSR **151** (1963), pp. 1293–1294.
[22] Kolaitis, P. G. and M. Y. Vardi, *0-1 laws and decision problems for fragments of second-order logic*, Inf. Comput. **87** (1990), pp. 301–337.
[23] Kolaitis, P. G. and M. Y. Vardi, *Infinitary logics and 0-1 laws*, Inf. Comput. **98** (1992), pp. 258–294.
[24] Kolaitis, P. G. and M. Y. Vardi, *0-1 laws for fragments of existential second-order logic: A survey*, in: *Proc. of MFCS 2000*, 2000, pp. 84–98.
[25] Koponen, V., *Asymptotic probabilities of extension properties and random l-colourable structures*, Ann. Pure Appl. Logic **163** (2012), pp. 391–438.
[26] Kracht, M., "Tools and Techniques in Modal Logic," Elsevier, 1999.
[27] Le Bars, J., *Counterexamples of the 0-1 law for fragments of existential second-order logic: an overview*, Bulletin of Symbolic Logic **6** (2000), pp. 67–82.
[28] Le Bars, J., *The 0-1 law fails for monadic existential second-order logic on undirected graphs*, Inf. Process. Lett. **77** (2001), pp. 43–48.
[29] Le Bars, J., *Zero-one law fails for frame satisfiability in propositional modal logic*, in: *Proceedings of LICS'2002* (2002), pp. 225–234.
[30] Lynch, J. F., *Probabilities of first-order sentences about unary functions*, Trans. American Mathematical Society **287** (1985), pp. 543–568.
[31] Rautenberg, W., *Splitting lattices of logics*, Archive for Mathematical Logic **20** (1980), pp. 155–159.
[32] Schamm, R., "Zero-one laws and almost sure validities on finite structures," Master's thesis, Rand Afrikaans University (2002).
[33] Skvortsov, D. P., *Remark on a finite axiomatization of finite intermediate propositional logics*, J. Appl. Non Class. Logics **9** (1999), pp. 381–386.

[34] Thomason, S., *Reduction of second-order logic to modal logic*, Zeitschrift für Mathematische Logik und Grundlagen der Mathematik **21** (1975), pp. 107–114.
[35] van Benthem, J. F. A. K., "Modal Logic and Classical Logic," Bibliopolis, 1983.
[36] Verbrugge, R., *Zero-one laws with respect to models of provability logic and two Grzegorczyk logics*, in: *Proc. of Accepted Short Papers of AiML 2018*, pp. 115–119.
[37] Zolin, E., *Personal correspondence* (2020).

Appendix

Example. This example illustrates the details of the derivation sketched in the proof of Lemma 2.4. Consider the frames K_2 and K_3 defined in Section 3.

For convenience, I will rename the points in K_3:
$\mathsf{K}_2 = \langle\{x, y\}, \{(x, x), (x, y), (y, x)\}\rangle$ and
$\mathsf{K}_3 = \langle\{u, v_1, v_2\}, \{(u, u), (u, v_1), (u, v_2), (v_1, u)(v_2, u)\}\rangle$.

The (slightly simplified) characteristic formulae of these are as follows:

$$\chi_{\mathsf{K}_2}(p_x, p_y) := \neg[\mathsf{U}]\big(\langle\mathsf{U}\rangle p_x \wedge \langle\mathsf{U}\rangle p_y \wedge (p_x \vee p_y) \wedge (p_x \to \neg p_y) \wedge \\ (p_x \to \Diamond p_x) \wedge (p_x \to \Diamond p_y) \wedge (p_y \to \Diamond p_x) \wedge (p_y \to \Box\neg p_y)\big).$$

$$\chi_{\mathsf{K}_3}(q_u, q_{v_1}, q_{v_2}) := \neg[\mathsf{U}]\big(\langle\mathsf{U}\rangle q_u \wedge \langle\mathsf{U}\rangle q_{v_1} \wedge \langle\mathsf{U}\rangle q_{v_2} \wedge (q_u \vee q_{v_1} \vee q_{v_2}) \wedge \\ (q_u \to \neg q_{v_1}) \wedge (q_u \to \neg q_{v_2}) \wedge (q_{v_1} \to \neg q_{v_2}) \wedge \\ (q_u \to \Diamond q_u) \wedge (q_u \to \Diamond q_{v_1}) \wedge (q_u \to \Diamond q_{v_2}) \wedge (q_{v_1} \to \Diamond q_u) \wedge (q_{v_2} \to \Diamond q_u) \wedge \\ (q_{v_1} \to \Box\neg q_{v_1}) \wedge (q_{v_1} \to \Box\neg q_{v_2}) \wedge (q_{v_2} \to \Box\neg q_{v_1}) \wedge (q_{v_2} \to \Box\neg q_{v_2})\big).$$

It is easy to check that the mapping $h : \mathsf{K}_3 \to \mathsf{K}_2$ defined by $h(u) = x, h(v_1) = h(v_2) = y$ is a bounded morphism.

The substitution σ_h defined in the proof of Lemma 2.4 acts as follows:

$$\sigma_h(p_x) := q_u, \quad \sigma_h(p_y) := (q_{v_1} \vee q_{v_2}).$$

Respectively,

$$\xi_{\mathsf{K}_3\to\mathsf{K}_2}(q_u, q_{v_1}, q_{v_2}) = \sigma_h(\chi_{\mathsf{K}_2}(p_x, p_y)) = \\ \neg[\mathsf{U}]\big(\langle\mathsf{U}\rangle q_u \wedge \langle\mathsf{U}\rangle(q_{v_1} \vee q_{v_2}) \wedge (q_u \vee (q_{v_1} \vee q_{v_2})) \wedge (q_u \to \neg(q_{v_1} \vee q_{v_2})) \wedge \\ (q_u \to \Diamond q_u) \wedge (q_u \to \Diamond(q_{v_1} \vee q_{v_2})) \wedge ((q_{v_1} \vee q_{v_2}) \to \Diamond q_u) \wedge ((q_{v_1} \vee q_{v_2}) \to \Box\neg(q_{v_1} \vee q_{v_2}))\big).$$

After simple equivalent transformations in $\mathbf{K}_{\mathsf{U}}$, it is transformed to

$$\xi'_{\mathsf{K}_3\to\mathsf{K}_2}(q_u, q_{v_1}, q_{v_2}) = \\ \neg[\mathsf{U}]\big(\langle\mathsf{U}\rangle q_u \wedge (\langle\mathsf{U}\rangle q_{v_1} \vee \langle\mathsf{U}\rangle q_{v_2}) \wedge (q_u \vee q_{v_1} \vee q_{v_2}) \wedge (q_u \to \neg q_{v_1}) \wedge (q_u \to \neg q_{v_2}) \wedge \\ (q_u \to \Diamond q_u) \wedge (q_u \to (\Diamond q_{v_1} \vee \Diamond q_{v_2})) \wedge (q_{v_1} \to \Diamond q_u) \wedge (q_{v_2} \to \Diamond q_u) \wedge \\ (q_{v_1} \to \Box\neg q_{v_1}) \wedge (q_{v_1} \to \Box\neg q_{v_2}) \wedge (q_{v_2} \to \Box\neg q_{v_1}) \wedge (q_{v_2} \to \Box\neg q_{v_2})\big).$$

By a direct inspection, one can see that every conjunct inside the scope of $\neg[\mathsf{U}]$ in $\xi'_{\mathsf{K}_3\to\mathsf{K}_2}(q_u, q_{v_1}, q_{v_2})$ is either identical, or follows propositionally from a conjunct inside the scope of $\neg[\mathsf{U}]$ in $\chi_{\mathsf{K}_3}(q_u, q_{v_1}, q_{v_2})$.

Therefore, $\models \neg\chi_{\mathsf{K}_3}(q_u, q_{v_1}, q_{v_2}) \to \neg\xi'_{\mathsf{K}_3\to\mathsf{K}_2}(q_u, q_{v_1}, q_{v_2})$,
hence $\models \xi'_{\mathsf{K}_3\to\mathsf{K}_2}(q_u, q_{v_1}, q_{v_2}) \to \chi_{\mathsf{K}_3}(q_u, q_{v_1}, q_{v_2})$.
Equivalently, $\models \sigma_h(\chi_{\mathsf{K}_2}(p_x, p_y)) \to \chi_{\mathsf{K}_3}(q_u, q_{v_1}, q_{v_2})$.

Bi-Intuitionistic Logics: a New Instance of an Old Problem

Rajeev Goré [1] Ian Shillito [2]

Research School of Computer Science, Australian National University

Abstract

As anyone who reads the literature on bi-intuitionistic logic will know, the numerous papers by Cecylia Rauszer are foundational but confusing. For example: these papers claim and retract various versions of the deduction theorem for bi-intuitionistic logic; they erroneously claim that the calculus is complete with respect to rooted canonical models; and they erroneously claim the admissibility of cut in her sequent calculus for this logic. Worse, authors such as Crolard, have based some of their own foundational work on these confused and confusing results and proofs.

We trace this confusion to the axiomatic formalism of RBiInt in which Rauszer first characterized bi-intuitionistic logic and show that, as in modal logic, RBiInt can be interpreted as two different consequence relations. We remove this ambiguity by using generalized Hilbert calculi, which are tailored to capture consequence relations.

We show that RBiInt leads to two logics, wBIL and sBIL, with different extensional and meta-level properties, and that they are, respectively, sound and strongly complete with respect to the Kripkean local and global semantic consequence relations of bi-intuitionistic logic. Finally, we explain where they were conflated by Rauszer.

Keywords: Bi-Intuitionistic Logic, Axiomatic Proof Theory, Consequence Relations, Deduction Theorems, Kripke Semantics.

1 Introduction: Confusions

Rauszer's Bi-Intuitionistic logic (RBiInt), introduced in 1974 via an axiomatic calculus [17], is a conservative extension of intuitionistic propositional logic. It adds an extra binary operator $\prec$, dual to the intuitionistic arrow and variously called *exclusion*, *subtraction*, or *co-implication*, and a unary *weak* negation operator $\sim$ definable from $\prec$. In an interdependent series of articles [16,18,19,20,21,22,23], Rauszer studied the algebraic, axiomatic and Kripke-style aspects of this logic. Alas, reviewing the literature on RBiInt can be quite confusing, because, in many places, the status of theorems is unclear if not puzzling. An account of this confusion can be given by three elements.

[1] rajeev.gore@anu.edu.au

[2] ian.shillito@anu.edu.au

First, as is well-known, the usual deduction theorem is: $\Gamma, \varphi \vdash \psi$ iff $\Gamma \vdash \varphi \to \psi$. However the "deduction theorem" is claimed under the following various forms in chronological order: (1) $\Gamma, \varphi \vdash \psi$ iff $\Gamma \vdash \neg{\sim}...\neg{\sim}\varphi \to \psi$ [17]; (2) the usual version above [18]; an explicit retraction of (2) and replacement by (3) $\Gamma, \varphi \vdash \psi$ iff $\Gamma \vdash \neg{\sim}\varphi \to \psi$ [21]; (4) a return to (1) without retracting (3) [23]. Crolard [3] claims that yet another form of the deduction theorem fails to hold.

Second, the Pinto-Uustalu counterexample [14] not only breaks the admissibility of cut in Rauszer's sequent calculus [16] for RBiInt, but also casts doubts on Crolard's work on a formulas-as-types interpretation for RBiInt because of his claim that *"as a by-product of the previous properties [proved by Crolard], we obtain a new proof of this result [Rauszer's cut admissibility]"* [4, p.3].

Third, Rauszer's [20] strong completeness of RBiInt w.r.t. rooted canonical models contradicts Crolard's [3] result that it is not complete for this class.

All this confusion arises from a fundamental problem in the axiomatic proof theory of RBiInt: traditional Hilbert calculi are not designed to treat logics as consequence relations. They lead to an ambiguous notion of derivation from assumptions that can cause us to conflate distinct logics. For example, modal logic as a consequence relation splits into a *strong* and a *weak* version, depending on how the *necessitation rule* is interpreted. Conflating these logics leads to great confusion, notably regarding the deduction theorem [11]. A similar phenomenon, as yet undetected, occurs in RBiInt where traditional Hilbert calculi cannot adequately separate two interpretations of a bi-intuitionistic rule called *DN* (an analogue of the necessitation rule from modal logic).

To pinpoint the confusion, we generalize traditional Hilbert calculi to treat consequence relations rather than just theoremhood. Then, the rules, such as necessitation or *DN*, are expressed in a way that prevents ambiguities about their shape. We use such calculi to explain and fix the fundamental problem of the axiomatic proof theory of RBiInt. Specifically, we give two generalized Hilbert calculi, wBIC and sBIC, for bi-intuitionistic logic that differ only in the shape of the *DN* rule. Unsurprisingly, these systems capture two distinct logics **wBIL** and **sBIL**, which have been conflated in parts of the literature.

Finally, the logics **wBIL** and **sBIL** are shown, respectively, to be sound and strongly complete w.r.t. the Kripkean *local* and *global* semantic consequence relations for bi-intuitionistic logic, mimicking similar results in modal logic via a canonical model construction while using techniques of Sano and Stell [25].

Section 2 contains general definitions of logics as consequence relations and generalized Hilbert calculi. Section 3 contains the problems caused by traditional Hilbert calculi in modal logics, and how generalized Hilbert calculi solve them. Rauszer's traditional Hilbert calculus is in Section 4. Section 5 contains the two generalized Hilbert calculi obtained from Rauszer's axiomatization. In Section 6, we show they define two extensionally distinct logics. Section 7 contains significant theorems distinguishing these logics. Section 8 contains our completeness proofs. In Section 9, we use these results to prove pending claims from Sections 5 and 7. In Section 10, we use the distinctions between our two bi-intuitionistic logics to expose the flaws in Rauszer's results.

2 Preliminaries

In this section we provide the general definitions required to both understand confusions arising in both modal and bi-intuitionistic logic, and avoid them.

We define logics as conventional consequence relations [6,12] where uniform substitution, formal language $\mathcal{L}$ and the set $\mathcal{F}orm_{\mathcal{L}}$ of all formulae of $\mathcal{L}$ is as usual. We use $\varphi, \psi, \chi, ...$ for formulae and $\Gamma, \Delta, ...$ for sets of formulae.

Definition 2.1 Let $\mathcal{L}$ be a formal language. A logic in $\mathcal{L}$ is a set $L \subseteq \{(\Gamma, \varphi) \mid \Gamma \cup \{\varphi\} \subseteq \mathcal{F}orm_{\mathcal{L}}\}$ that satisfies the following properties:

Identity: if $\varphi \in \Gamma$, then $(\Gamma, \varphi) \in L$;

Monotonicity: if $(\Gamma, \varphi) \in L$ and $\Gamma \subseteq \Gamma'$, then $(\Gamma', \varphi) \in L$;

Compositionality: if $(\Gamma, \varphi) \in L$ and $(\Delta, \gamma) \in L$ for all $\gamma \in \Gamma$, then $(\Delta, \varphi) \in L$;

Structurality: if $(\Gamma, \varphi) \in L$, then $(\Gamma^\sigma, \varphi^\sigma) \in L$ for uniform substitution σ.

A logic L is *finitary* if $(\Gamma, \varphi) \in L$ implies there is a finite $\Gamma' \subseteq \Gamma$ with $(\Gamma', \varphi) \in L$.

Thus, technically, a logic is not just a set of theorems but is a consequence relation containing pairs (Γ, φ). Wójcicki [27, pp.xii-xiii, pp.43-51] discusses some interesting aspects of this notion. We then formalize axiomatic systems in a way that generalizes and disambiguates traditional Hilbert calculi. In what follows the notions of formula schema and schema instance are as usual. The letters $A, B, C...$ refer to schemata and $X, Y, Z, ...$ to sets of schemata. We call the axiomatic systems obtained *generalized Hilbert calculi.*

Definition 2.2 Let $\mathcal{L}$ be a language. An *axiom* is a formula schema of $\mathcal{L}$. If $\mathcal{A}$ is a set of axioms, we define $\mathcal{A}^I$ to be the set of instances of axioms of $\mathcal{A}$. An n-ary *rule* $R = (\mathbb{P}, \mathbb{C})$ is a pair where $\mathbb{P} = \{X_1 \vdash B_1, ..., X_n \vdash B_n\}$ is a set of n premises and $\mathbb{C} = (X_{n+1} \vdash B_{n+1})$ is the conclusion, and $\bigcup_{i=1}^{n+1} X_i \cup \{B_i\}$ is a set of schemata of formulae. If R is a rule then we define R^I to be the set of instances of R. A generalized Hilbert calculus in $\mathcal{L}$ is a pair $S = (\mathcal{A}, \mathcal{R})$.

To let a generalized Hilbert calculus define a binary relation we must say which statements of the form $\Gamma \vdash \varphi$ follow from this calculus. To do so, we need to define the notion of derivation in a generalized Hilbert calculus:

Definition 2.3 Let $\mathcal{L}$ be a language. Let $\Gamma \cup \{\varphi\} \in Form_{\mathcal{L}}$ and $S = (\mathcal{A}, \mathcal{R})$ a generalized Hilbert calculus in $\mathcal{L}$. A derivation in S is a tree of expressions, defined inductively as follows:

(Ax): if $\varphi \in \mathcal{A}^I$ then the following is a derivation: $\dfrac{}{\Gamma \vdash \varphi}\ Ax$

(El): if $\varphi \in \Gamma$ then the following is a derivation: $\dfrac{}{\Gamma \vdash \varphi}\ El$

(R): if $\pi_1, \pi_2, ..., \pi_k$ are derivations with respectively $\Gamma_1 \vdash \varphi_1, ..., \Gamma_k \vdash \varphi_k$ as roots and $(\{\Gamma_1 \vdash \varphi_1, ..., \Gamma_k \vdash \varphi_k\}, \Gamma \vdash \varphi) \in R^I$ for some $R \in \mathcal{R}$, then the following is a derivation:

$$\frac{\pi_1 \quad \dots \quad \pi_k}{\Gamma \vdash \varphi}\ R$$

A branch, its length and the length $l(\pi)$ of a derivation π are defined as usual. If there is a derivation in S with $\Gamma \vdash \varphi$ as root, we write $\Gamma \vdash_S \varphi$.

Note that a generalized Hilbert calculus might not define a logic: the relation defined by a system with the lone rule $R = (\emptyset, A \vdash B)$ fails Monotonicity.

3 Theorems and Consequences in Classical Modal Logic

As an example, generalized Hilbert calculi clearly demarcate the existence of two modal logics based on the usual axiomatization $\mathcal{A}_{\mathbf{K}}$ of the basic modal logic **K**. In that setting, the *modus ponens* rule MP is formalized as:

$$\frac{X \vdash A \quad X \vdash A \to B}{X \vdash B}\ MP$$

The *Necessitation* rule, often written as in the middle, can be interpreted either as a weak or strong rule as shown at left and right.

$$\frac{\emptyset \vdash A}{X \vdash \Box A}\ Nec_w \qquad\qquad \frac{A}{\Box A}\ Nec \qquad\qquad \frac{X \vdash A}{X \vdash \Box A}\ Nec_s$$

The calculi wKC $= (\mathcal{A}_{\mathbf{K}}, \{MP, Nec_w\})$ and sKC $= (\mathcal{A}_{\mathbf{K}}, \{MP, Nec_s\})$ respectively define the (*distinct*) logics **wK** and **sK**, corresponding to the extensionally different *local* and *global* Kripkean semantic consequence relations [12].

The most obvious example of their difference, as consequence relations, is that we have $p \vdash_{\mathrm{sKC}} \Box p$ but $p \nvdash_{\mathrm{wKC}} \Box p$. Then, the long-standing debate [11] about the modal deduction theorem is resolved immediately via two simple facts: (1) $p \vdash_{\mathrm{sKC}} \Box p$ but $\nvdash_{\mathrm{sKC}} p \to \Box p$; (2) $p \vdash_{\mathrm{wKC}} \Box p$ iff $\vdash_{\mathrm{wKC}} p \to \Box p$.

Not only does this example show that the two rules added to the same axiomatization do not capture the same logics, as consequence relations, but it also gives sufficient tools to show that these logics differ on their meta-properties. In fact, this partly justifies the fact that the deduction theorem doesn't hold for **sK**, while it is proven to hold for **wK**.

That is, traditional Hilbert calculi allow us to easily confuse the logics **wK** and **sK**. To capture both of them in a traditional Hilbert setting, one has to provide debatable modifications on the notion of derivation. In fact, to capture **sK** one defines the notion of derivation from assumptions as follows [2]:

Definition 3.1 A derivation of φ from assumptions Γ is a list l of formulae ending with φ such that each formula in l is an instance of an axiom of $\mathcal{A}_{\mathbf{K}}$, a member of Γ, or follows via MP or Nec from formulae appearing earlier in l.

While this definition is natural and unproblematic, the notion of derivation from assumptions has to be bent to capture **wK**:

Definition 3.2 A derivation of φ from assumptions Γ is a list of formulae ending with φ, and such that every formula in the list is an instance of an axiom, a member of Γ, follows from formulae appearing before it in the list by MP or *follows from a derivable formula by* Nec.

First, this definition relies on the notion of derivability which really should just be a special case of derivation from assumptions. Second, as it involves the derivability of a formula in the application of Nec, to determine if a list of formulae is a derivation from assumptions or not it is not sufficient to check the list of formulae itself. In other words, the notion defined here is not local as the application of Nec is conditioned on the existence of another derivation. These features bring a lot of confusion on the nature of derivations from assumptions.

A common way to avoid these contortions is to define the notion of derivation from assumptions from the notion of derivation [1,15]:

Definition 3.3 A derivation of φ from assumptions Γ is a derivation of the formula $(\gamma_0 \wedge ... \wedge \gamma_n) \rightarrow \varphi$ for some $n \in \mathbb{N}$ and $\gamma_i \in \Gamma$ for $0 \leq i \leq n$.

Here, some other criticisms can be given. Mainly, it is the striking lack of generality of this definition that we address. More precisely, this definition is not general as there are four types of logics that it cannot capture. First, logics without a conjunction, such as implicational ticket entailment, cannot be captured. Second, the same remark can be made of logics devoid of implication, such as positive modal logic and geometric logic. Third, logics that are not compact are ruled out: it is in their nature to be unable, in some circumstances, to reduce an infinite set of assumptions to a finite one, while this is forced here by the presence of $\gamma_0, ..., \gamma_n$. Finally, no logic for which the deduction theorem fails can be characterized via this definition, as this theorem is built in here.

Generalized Hilbert calculi avoid these issues while easily capturing the logic **wK** by interpreting the necessitation rule as Nec_w. Of course, all of this is well-known for modal logic. We next use generalized Hilbert calculi to show that RBiInt is the victim of a similar confusion: whence our title.

4 Rauszer's Hilbert Calculus for Bi-Intuitionistic Logic

Before showing how bi-intuitionistic logic is captured via generalized Hilbert calculi, we recall Rauszer's traditional Hilbert calculus RBiInt from 1974 [20].

As mentioned above, RBiInt is expressed in the language of intuitionistic logic extended with two operators, i.e. $\prec$ and $\sim$. More formally:

Definition 4.1 Let p, q, r range over a countable set $Prop$ of propositional atoms and let $Log_{BI} = \{\wedge, \vee, \rightarrow, \neg, \prec, \sim\}$ be the set of bi-intuitionistic logical connectives. This pair forms the the language $\mathcal{L}_{BI} := (Log_{BI}, Prop)$ of bi-intuitionistic logic. The formulae $Form_{BI}$ of $\mathcal{L}_{BI}$ are defined as follows:

$$\varphi ::= p \mid \varphi \wedge \varphi \mid \varphi \vee \varphi \mid \varphi \rightarrow \varphi \mid \neg\varphi \mid \varphi \prec \varphi \mid \sim\varphi$$

For convenience, we define $\top := p \rightarrow p$ and $\bot := p \prec p$ for some fixed atomic formula p. The added operators are meant to be the duals of, respectively, $\rightarrow$ and $\neg$. The formula $\varphi \prec \psi$ is usually read as "φ excludes ψ". The formula $\sim \varphi := \top \prec \varphi$, defined dually to $\neg\varphi := \varphi \rightarrow \bot$, is usually called "weak negation". Rauszer's traditional Hilbert calculus RBiInt is defined next [17]:

Definition 4.2 RBiInt consists of the axioms $\mathcal{A}_{BI}$ and rules $\mathcal{R}_{BI}$ below:

RA_1	$(A \to B) \to ((B \to C) \to (A \to C))$	RA_{10}	$(A \to B) \to (\neg B \to \neg A)$
RA_2	$A \to (A \lor B)$	RA_{11}	$A \to (B \lor (A \prec B))$
RA_3	$B \to (A \lor B)$	RA_{12}	$(A \prec B) \to \sim (A \to B)$
RA_4	$(A \to C) \to ((B \to C) \to ((A \lor B) \to C))$		
RA_5	$(A \land B) \to A$	RA_{14}	$\neg(A \prec B) \to (A \to B)$
RA_6	$(A \land B) \to B$	RA_{15}	$(A \to \bot) \to \neg A$
RA_7	$(A \to B) \to ((A \to C) \to (A \to (B \land C)))$	RA_{16}	$\neg A \to (A \to \bot)$
RA_8	$(A \to (B \to C)) \to ((A \land B) \to C)$	RA_{17}	$(\top \prec A) \to \sim A$
RA_9	$((A \land B) \to C) \to (A \to (B \to C))$	RA_{18}	$\sim A \to (\top \prec A)$

$$RA_{13}\ ((A \prec B) \prec C) \to (A \prec (B \lor C)) \qquad \frac{A \quad A \to B}{B}\ MP \qquad \frac{A}{\neg \sim A}\ DN$$

Next, we show that the *Double Negation* rule DN can be interpreted in the context of generalized Hilbert calculi in two main ways, giving different logics.

5 Bi-Intuitionistic Logic As a Consequence Relation

As in the modal case, the traditional Hilbert calculus hides a distinction in the shape of rules. To be more precise, it overlooks the multiple interpretations of DN that are clearly expressed in a generalized Hilbert calculus:

$$\frac{\emptyset \vdash A}{X \vdash \neg \sim A}\ DN_w \qquad\qquad \frac{X \vdash A}{X \vdash \neg \sim A}\ DN_s$$

As we shall see, not only are these rules formally different, but they also have significantly different strength, implying a difference in the consequence relations they define and hence a difference in their logics. To see the difference between the two logics, erroneously identified in Rauszer's work, that emerge from the set of axioms $\mathcal{A}_{BI}$, we define the following generalized Hilbert calculi.

Definition 5.1 We define the generalized Hilbert calculi wBIC $= (\mathcal{A}_{BI}, \mathcal{R}_w)$ and sBIC $= (\mathcal{A}_{BI}, \mathcal{R}_s)$, where $\mathcal{R}_w = \{MP, DN_w\}$ and $\mathcal{R}_s = \{MP, DN_s\}$. We abbreviate $\Gamma \vdash_{wBIC} \varphi$ by $\Gamma \vdash_w \varphi$ and let **wBIL** $= \{(\Gamma, \varphi) \mid \Gamma \vdash_w \varphi\}$ be the consequence relation characterized by wBIC. Similarly we abbreviate $\Gamma \vdash_{sBIC} \varphi$ by $\Gamma \vdash_s \varphi$, and define **sBIL** $= \{(\Gamma, \varphi) \mid \Gamma \vdash_s \varphi\}$.

As there is no guarantee that generalized Hilbert calculi define logics, to assert that **sBIL** and **wBIL** are logics we must show they satisfy Definition 2.1. The single rule derivation of $\Gamma \vdash \varphi$ via (El) shows that **Identity** is satisfied both in **sBIL** and **wBIL**. The other properties need to be proved.

Lemma 5.2 *The following holds for $i \in \{w, s\}$:*

Monotonicity: *if $\Gamma \subseteq \Gamma'$ and $\Gamma \vdash_i \varphi$ then $\Gamma' \vdash_i \varphi$.*

Compositionality: *if $\Gamma \vdash_i \varphi$ and $\Delta \vdash_i \gamma$ for all $\gamma \in \Gamma$, then $\Delta \vdash_i \varphi$*

Structurality: *if $\Gamma \vdash_i \varphi$ then $\Gamma^\sigma \vdash_i \varphi^\sigma$.*

Proof. See the Appendix. □

We are now in position to claim that **sBIL** and **wBIL** are both logics. Furthermore, we can add that they are finitary logics:

Lemma 5.3 *For* $i \in \{s, w\}$*, if* $\Gamma \vdash_i \varphi$*, then* $\Gamma' \vdash_i \varphi$ *for some finite* $\Gamma' \subseteq \Gamma$.

That **sBIL** and **wBIL** are (finitary) logics is all well and good, but we require further work to show that they are *different* logics, as explained next.

6 Extensional Interactions

To prove our claim that **sBIL** and **wBIL** are two logics that were erroneously conflated in the literature we first show they differ on an extensional level.

Claim 6.1 *For* $p \in Prop$*,* $p \vdash_s \neg\sim p$ *and* $p \not\vdash_w \neg\sim p$.

While it is clear that $p \vdash_s \neg\sim p$ holds because DN_s can be applied on $p \vdash p$, we need a semantic argument, that we provide later, to prove that $p \not\vdash_w \neg\sim p$. By accepting this result for now, we can see that the two consequence relations **sBIL** and **wBIL** are extensionally different. However the two consequence relations are closely related. In fact, **sBIL** is an extension of **wBIL**:

Theorem 6.2 *If* $\Gamma \vdash_w \varphi$ *then* $\Gamma \vdash_s \varphi$.

Moreover, they coincide on their sets of *theorems* (derivable from $\emptyset$):

Theorem 6.3 $\emptyset \vdash_s \varphi$ *if and only if* $\emptyset \vdash_w \varphi$.

Traditionally, Theorem 6.3 is an argument against our distinction between **sBIL** and **wBIL** as it identifies the two logics on their sets of theorems. However, as mentioned previously, they are different consequence relations. Given Claim 6.1, **sBIL** and **wBIL** are thus different logics.

7 Deduction and Dual-Deduction Theorems

We proceed to show that **sBIL** and **wBIL** are distinct on a meta-level by proving that both the deduction theorem and its dual hold for **wBIL**, while none hold for **sBIL**. To express these statements we use notions from Sano and Stell [25]. They can be interpreted as an extension of the notion of a logic as a consequence relation of the form (Γ, φ) to the more general form (Γ, Δ).

Definition 7.1 Let $i \in \{w, s\}$ and $\bigvee \Delta$ be the disjunction of all the members of Δ. We define the following:

(i) $\vdash_i [\Gamma \mid \Delta]$ if $\Gamma \vdash_i \bigvee \Delta'$ for some finite $\Delta' \subseteq \Delta$;

(ii) $\not\vdash_i [\Gamma \mid \Delta]$ if it is not the case that $\vdash_i [\Gamma \mid \Delta]$;

(iii) $\bigvee \Delta := \bot$ if $\Delta = \emptyset$;

(iv) $[\Gamma \mid \Delta]$ is complete if $\Gamma \cup \Delta = \mathcal{F}orm_{BI}$.

Pairs of the form $[\Gamma \mid \Delta]$ bring a symmetry, witnessed by the presence of potentially infinite sets of formulae on both sides of the vertical bar, which is not present in expressions such as $\Gamma \vdash \varphi$. Conceptually, this symmetry and the presence of a non-orientated separation symbol $\mid$ suggests a bidirectional reading of a pair $[\Gamma \mid \Delta]$. From left to right such a pair should be read as a *deduction*, while from right to left it should be read as a *refutation*. This interpretation help us understand the duality between $\to$ and $\prec$.

We first require a preliminary result central to the two logics.

Proposition 7.2 *For* $i \in \{w, s\}$*:*

$$\vdash_i [\emptyset \mid (\varphi \prec \psi) \to \chi] \quad \textit{iff} \quad \vdash_i [\emptyset \mid \varphi \to (\psi \vee \chi)]$$

We have not given Rauszer's [17] algebraic semantics for RBiInt, but Proposition 7.2 is an object language analogue of the dual residuation property below:

$$\frac{a \leq b \vee c}{a \prec b \leq c}$$

The deduction theorem is the first theorem to separate the two logics.

Theorem 7.3 (Deduction Theorem) **wBIL** *enjoys the deduction theorem:*

$$\vdash_w [\Gamma, \varphi \mid \psi] \quad \textit{iff} \quad \vdash_w [\Gamma \mid \varphi \to \psi]$$

Next, we give a counter-example for the deduction theorem for **sBIL**.

Proposition 7.4 *We have that* $\vdash_s [p \mid \neg\!\sim p]$ *but* $\nvdash_s [\emptyset \mid p \to \neg\!\sim p]$.

Proof. We prove the first conjunct and postpone the proof of the second to later. Obviously we have $p \vdash_s p$. So we can apply the rule DN_s to obtain $p \vdash_s \neg\!\sim p$, hence $\vdash_s [p \mid \neg\!\sim p]$. □

We leave the following claim as pending:

Claim 7.5 *We have that* $\nvdash_s [\emptyset \mid p \to \neg\!\sim p]$.

This situation is very similar to the modal case: it is well-known that **wK** satisfies the deduction theorem while **sK** does not. However, a variant of this theorem does hold for **sK**: $\Gamma, \varphi \vdash_s \psi$ *iff there exists a* $n \in \mathbb{N}$ *such that* $\Gamma \vdash_s (\varphi \wedge \Box\varphi \wedge ... \wedge \Box^n\varphi) \to \psi$ [2, p.85]. A similar variant of the deduction theorem holds for **sBIL**, but we first need some notation to express it.

Definition 7.6 We define:

(i) for $n \in \mathbb{N}$, let $(\neg\!\sim)^0\varphi := \varphi$ and let $(\neg\!\sim)^{(n+1)}\varphi := \neg\!\sim(\neg\!\sim)^n\varphi$;

(ii) $(\neg\!\sim)^n\Gamma = \{(\neg\!\sim)^n\gamma \mid \gamma \in \Gamma\}$;

(iii) $(\neg\!\sim)^\omega\Gamma = \bigcup_{n\in\mathbb{N}} (\neg\!\sim)^n\Gamma$.

The variant of the deduction theorem below uses the pattern $\neg\!\sim$ as the modal variant uses $\Box$. But it suffices to replace φ by just $(\neg\!\sim)^n\varphi$, without the conjunction of all $(\neg\!\sim)^i\varphi$ for $i \leq n$, as $\neg\!\sim$is a **T** modality satisfying $\neg\!\sim\varphi \to \varphi$. One reviewer noted that Reyes and Zolfaghari [24] show how to interpret this combination as a kind of non-idempotent interior operation on subgraphs.

Theorem 7.7 (Double-Negated Deduction Theorem)

$$\vdash_s [\Gamma, \varphi \mid \psi] \quad \textit{iff} \quad \exists n \in \mathbb{N} \; s.t. \; \vdash_s [\Gamma \mid (\neg\!\sim)^n\varphi \to \psi]$$

Theorem 7.8 (Dual Deduction Theorem) *The following holds:*

$$\vdash_w [\varphi \mid \psi, \Delta] \qquad \textit{iff} \qquad \vdash_w [\varphi \prec \psi \mid \Delta].$$

Proof. Assume that $\vdash_w [\varphi \mid \psi, \Delta]$. By definition we get $\varphi \vdash_w \psi \vee \bigvee \Delta'$ where $\Delta' \subseteq \Delta$ is finite. Using Theorem 7.3 we get $\emptyset \vdash_w \varphi \to (\psi \vee \bigvee \Delta')$. We obtain $\emptyset \vdash_w (\varphi \prec \psi) \to \bigvee \Delta'$ by Proposition 7.2. By Theorem 7.3 again, we obtain $\varphi \prec \psi \vdash_w \bigvee \Delta'$. By definition we get $\vdash_w [\varphi \prec \psi \mid \Delta]$. Note that all the steps used here are based on equivalences. □

Before demonstrating that **sBIL** fails the dual deduction theorem, we remark on the previous theorem. Pairs $[\Gamma \mid \Delta]$ express the duality between $\to$ and $\prec$ on the syntactic level in **wBIL** by showing that $\prec$ plays the same role as $\to$ on the left-hand side of $|$: it internalizes in the object language the relation expressed by our pairs. Just as $\to$ internalizes the deduction relation of expressions such as $\Gamma \vdash \varphi$, dually $\prec$ internalizes the *refutation* relation of expressions such as $\Delta \dashv \varphi$, read "Δ refutes φ" and formalized here as $[\varphi \mid \Delta]$. This interpretation relies on the aforementioned reading of our pairs, from right to left, to express refutations. Fortunately, as we shall show in a separate paper, we can support this interpretation by the fact that **wBIL** can simulate the propositional fragment of Rauszer's refutation system [20, pp.62-63].

The following witnesses the failure of the dual deduction theorem for **sBIL**.

Proposition 7.9 $\vdash_s [p \prec q \mid \neg{\sim}{\sim} q]$ *while* $\nvdash_s [p \mid q, \neg{\sim}{\sim} q]$.

Proof. First, let us prove that $\vdash_s [p \prec q \mid \neg{\sim}{\sim} q]$. By definition, we need to show that $p \prec q \vdash_s \neg \sim\sim q$. We have that $\emptyset \vdash_w q \vee \sim q$, hence $\emptyset \vdash_w p \to (q \vee \sim q)$. By Proposition 7.2 we obtain $\emptyset \vdash_w (p \prec q) \to \sim q$. In turn, by Theorem 7.7 we get $p \prec q \vdash_w \sim q$. Then, by Theorem 6.2 we get that $p \prec q \vdash_s \sim q$. Finally, we can apply the rule DN_s to obtain $p \prec q \vdash_s \neg{\sim}{\sim} q$, hence $\vdash_s [p \prec q \mid \neg{\sim}{\sim} q]$. We leave the following claim as pending:

Claim 7.10 $\nvdash_s [p \mid q, \neg{\sim}{\sim} q]$.

□

While a variant of the deduction theorem exists for **sBIL**, the form or the existence of a variant to the dual deduction theorem is still a mystery to us. For the interested reader: while the deduction theorem fails for **sBIL** because of the rule DN_s, the dual deduction theorem fails for this logic because of the rule MP. It appears that if a variant of the dual deduction theorem exists for **sBIL**, then it must use a "patch" inspired by the structure of MP, as done in the double-negated deduction theorem with DN_s.

On top of the extensional difference between **wBIL** and **sBIL**, the deduction and dual deduction theorems expose their meta-difference. But both differences rely on claims that are still pending. The next section builds on Rauszer's Kripke semantics to resolve these claims.

8 Weak is Local and Strong is Global

wBIL and **sBIL**, proof-theoretically characterized via the generalized Hilbert calculi *wBIC* and *sBIC*, can be captured model-theoretically in a Kripke semantics using well-known notions of semantic consequence: a *local* and a *global* one. This section is devoted to proving these claims.

First we need to define the Kripke semantics [23].

Definition 8.1 A BI-Kripke model $\mathcal{M}$ is a tuple $(W, \leq, I)$, where $(W, \leq)$ is a poset and $I : Prop \to \mathcal{P}(W)$ is an interpretation function obeying persistence: for every $v, w \in W$ with $w \leq v$ and $p \in Prop$, if $w \in I(p)$ then $v \in I(p)$.

The forcing relation of intuitionistic logic is extended to $\prec$ and $\sim$:

Definition 8.2 Given a BI-Kripke model $\mathcal{M} = (W, \leq, I)$, we extend the usual intuitionistic forcing relation between a point $w \in W$ and a formula as follows:

$$\begin{array}{lll} \mathcal{M}, w \Vdash \varphi \prec \psi & \text{iff} & \text{there exists a } v \text{ s.t. } v \leq w,\ \mathcal{M}, v \Vdash \varphi \text{ and } M, v \nVdash \psi \\ \mathcal{M}, w \Vdash \sim\varphi & \text{iff} & \text{there exists a } v \text{ s.t. } v \leq w,\ \mathcal{M}, v \nVdash \varphi \end{array}$$

Let $\Gamma \subseteq \mathcal{F}orm_{BI}$. We write $\mathcal{M}, w \Vdash \Gamma$ if for every $\gamma \in \Gamma$ we have $\mathcal{M}, w \Vdash \gamma$. If $\mathcal{M}, w \Vdash \Gamma$ we say that w is a Γ-point. We write $\mathcal{M} \Vdash \Gamma$ if for every point $w \in W$, $\mathcal{M}, w \Vdash \Gamma$. If $\mathcal{M} \Vdash \Gamma$ we say that $\mathcal{M}$ is a Γ-model.

The main feature of the Kripke semantics for intuitionistic logic is arguably persistence. This property, which we use later, is preserved here:

Lemma 8.3 (Persistence) *Let $\mathcal{M} = (W, \leq, I)$ be a BI-Kripke model and $w \in W$. For all $v \in W$ s.t. $w \leq v$ we have that if $\mathcal{M}, w \Vdash \varphi$ then $\mathcal{M}, v \Vdash \varphi$.*

We are now in position to define the two following notions of semantic consequence in the above-defined Kripke semantics:

Definition 8.4 The local and global consequence relations are as below:

$$\begin{array}{lll} \Gamma \models_l \Delta & \text{iff} & \forall \mathcal{M}. \forall w. (\mathcal{M}, w \Vdash \Gamma \Rightarrow \exists \delta \in \Delta. \mathcal{M}, w \Vdash \delta) \\ \Gamma \models_g \Delta & \text{iff} & \forall \mathcal{M}. (\mathcal{M} \Vdash \Gamma \Rightarrow \forall w \in W. \exists \delta \in \Delta. \mathcal{M}, w \Vdash \delta). \end{array}$$

While the two notions are not generally equivalent in modal logic, they are equivalent in intuitionistic (not bi-intuitionistic) logic. It is easy to see that the local implies the global in full generality. The converse holds for intuitionistic logic for two reasons. First, persistence plays an important role: if a formula is true at a point then it is true at all the successors (the upcone) of that point. Second and more crucially, in an intuitionistic Kripke model, the upcone of a point is bisimilar [2, p.54][13, p.8] in that point with the model itself.

Nonetheless, in the semantics just defined it is not the case that $\Gamma \models_g \Delta$ implies $\Gamma \models_l \Delta$. This can easily be shown by the fact that $p \models_g \neg \sim p$ while $p \not\models_l \neg \sim p$. This fact will help us finally establish the extensional difference between **wBIL** and **sBIL** by proving that local semantic consequence corresponds to **wBIL** and global semantic consequence corresponds to **sBIL**.

We use canonical models on complete pairs $[\Gamma \mid \Delta]$ from Sano and Stell [25]:

Definition 8.5 The canonical model $\mathcal{M}^c = (W^c, \leq^c, I^c)$ is defined in the following way:

(i) $W^c = \{[\Gamma \mid \Delta] : [\Gamma \mid \Delta] \text{ is complete and } \nvdash_w [\Gamma \mid \Delta]\}$;

(ii) $[\Gamma_1 \mid \Delta_1] \leq^c [\Gamma_2 \mid \Delta_2]$ iff $\Gamma_1 \subseteq \Gamma_2$;

(iii) $I^c(p) = \{[\Gamma \mid \Delta] \in W^c : p \in \Gamma\}$.

These pairs are built from unprovable pairs using a bi-intuitionistic version of the Lindenbaum Lemma:

Lemma 8.6 (Lindenbaum Lemma) *If* $\nvdash_w [\Gamma \mid \Delta]$ *then there exist* $\Gamma' \supseteq \Gamma$ *and* $\Delta' \supseteq \Delta$ *such that* $[\Gamma' \mid \Delta']$ *is complete and* $\nvdash_w [\Gamma' \mid \Delta']$.

As usual in canonical model techniques, we prove the crucial Truth Lemma:

Lemma 8.7 (Truth Lemma) *For every* $[\Gamma \mid \Delta] \in W^c$:

$$\psi \in \Gamma \quad \text{iff} \quad \mathcal{M}^c, [\Gamma \mid \Delta] \Vdash \psi.$$

We are now ready to prove the main result of this section:

Theorem 8.8 *The following holds:*

$$\begin{array}{lll} (1) & \vdash_w [\Gamma \mid \Delta] & \text{iff} \quad \Gamma \models_l \Delta \\ (2) & \vdash_s [\Gamma \mid \Delta] & \text{iff} \quad \Gamma \models_g \Delta. \end{array}$$

9 A Semantic Look Back

We use Theorem 8.8, stating that the logics **sBIL** and **wBIL** are respectively sound and complete with respect to the global and local consequence, to fill in the gaps of Sections 6 and 7 by proving the claims left pending there.

First, we can show the extensional difference of the two logics by proving Claim 6.1, which claims that **sBIL** $\nsubseteq$ **wBIL**:

Proof. [of Claim 6.1] On the one hand we obviously have that $\vdash_s [p \mid p]$ hence $\vdash_s [p \mid \neg{\sim}p]$ by DN_s. On the other hand we have that $p \not\models_l \neg{\sim}p$ as shown by the following model $\mathcal{M}_0$ where reflexive arrows are not depicted:

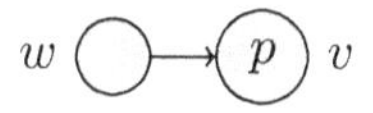

We clearly have $\mathcal{M}_0, v \Vdash p$. We also have $\mathcal{M}_0, v \nVdash \neg{\sim}p$ as $\mathcal{M}_0, v \Vdash {\sim}p$ because $\mathcal{M}_0, w \nVdash p$ and $w \leq v$. By Theorem 8.8 we obtain $\nvdash_w [p \mid \neg{\sim}p]$. □

Second, we resolve Claim 7.5 that **sBIL** fails the deduction theorem.

Proof. [of Claim 7.5] We need to prove that $\nvdash_s [\emptyset \mid p \to \neg{\sim}p]$. Consider the model $\mathcal{M}_0$ above. We have that $\mathcal{M}_0, v \nVdash p \to \neg{\sim}p$, hence $\not\models_g p \to \neg{\sim}p$. By Theorem 8.8 we obtain $\nvdash_s [\emptyset \mid p \to \neg{\sim}p]$. Since applying DN_s to $p \vdash_s p$ gives $\vdash_s [p \mid \neg{\sim}p]$, the deduction theorem does not hold for **sBIL**. □

Lastly, we prove Claim 7.10 that **sBIL** fails the dual deduction theorem:

Proof. [of Claim 7.10] We need to prove that $\nvdash_s [p \mid q \vee \neg{\sim\sim}\, q]$. Consider the following model $\mathcal{M}_1$:

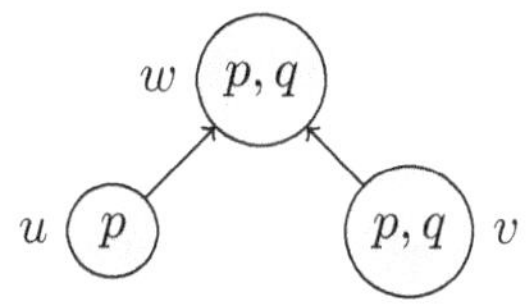

First we have that $\mathcal{M}_1 \Vdash p$. We also have $\mathcal{M}_1, u \nVdash q$ by definition of the valuation. But we also have $\mathcal{M}_1, u \nVdash \neg{\sim}{\sim} q$. In fact $\mathcal{M}_1, w \Vdash {\sim}{\sim} q$ as $v \leq w$ and $\mathcal{M}_1, v \nVdash {\sim} q$: its only predecessor is itself, and it forces q. Consequently we have $p \not\models_g q \vee \neg{\sim}{\sim} q$, which by Theorem 8.8 gives $\nvdash_s [p \mid q \vee \neg{\sim}{\sim} q]$. □

10 Why Rauszer's Proofs are Erroneous

The existence of **sBIL** and **wBIL** justifies our use of the plural bi-intuitionistic logics. We now trace the effect of this bifurcation on Rauszer's works.

As far as we know, neither the existence of **wBIL** and **sBIL**, nor the distinction between them has been highlighted in the literature. While it was certainly not noted in Rauszer's works, it has to be acknowledged that Hiroakira Ono may have suspected something [23, p.7]. Our bifurcation is not important if we only focus on one of the logics and use properties only belonging to it. However if one confuses them by using properties of these logics that are not shared by both of them, then troubles arrive. Unfortunately, such a confusion is made in some of Rauszer's works. As a consequence, various important theorems are asserted with erroneous proofs. The most important of them is the theorem of strong completeness with respect to the Kripke semantics [20]. More precisely there are two proofs for this theorem. The first one [20, Lemma 2.3], is flawed because it ignores restrictions on the use of a lemma proved by Gabbay [7], and this is extraneous to the confusion between the logics **sBIL** and **wBIL**. The second one [20, Theorem 3.5], is a standard completeness proof, involving the construction of a canonical model. However, in this proof, some intermediate lemmas are proved using features which are distinct for these logics. For example the fact that $\vdash_s [\varphi \mid \neg{\sim}\varphi]$ holds is used in the proof of Lemma 3.1 [20], where it is erroneously claimed that a prime filter A is such that if $a \in A$ then $\neg{\sim}a \in A$ and hence ${\sim}a \notin A$. In addition, in the proof of point (3) of Lemma 3.3 [20] the deduction theorem is used implicitly as it relies on a proof provided by Thomason [26] which uses it. Thus, while the proof of Lemma 3.1 [20] suggests the logic used is **sBIL**, the proof of Lemma 3.3 indicates that it must be **wBIL**. Thus the proof of completeness given there, which relies on these two lemmas, is a proof for none of the logics discussed here.

Another strong completeness proof [18] suffers from the same confusion because it relies on the aforementioned completeness proofs [20]. Interestingly, some elements of this paper [18] were corrected [21], but the corrections do not suffice to fix the issue. More precisely, one side of the deduction theorem is changed from $\Gamma \vdash \varphi \to \psi$ to $\Gamma \vdash \neg{\sim}\varphi \to \psi$ [21], but this version also fails for **sBIL** and, in any case, the proofs [20] are not modified to handle the change.

In a nutshell, as the proofs of strong completeness for bi-intuitionistic logic given in Rauszer's PhD thesis [23] are taken from the articles mentioned above, we are left with no actual trace in Rauszer's papers of a correct proof of strong completeness of bi-intuitionistic logic with respect to the Kripke semantics defined. To the best of our knowledge such a proof has only been provided by Sano and Stell [25], but for a different axiomatization. So, our proofs are the first to ensure that Rauszer's axiomatization is strongly complete for the appropriate Kripke semantics in a non-ambiguous way: **sBIL** (**wBIL**) is strongly complete for global (local) semantic consequence in Kripke semantics.

Clearly, providing such a proof is necessary to set the record straight for Rauszer's axiomatization. Furthermore, when compared with the initial proofs, our proofs are useful for avoiding false conclusions hinted at by the former. Most importantly, two proofs of strong completeness [20] involve the construction of a *rooted* canonical model where by "rooted" we understand the following

Definition 10.1 Let $\mathcal{F} = (W, \leq)$ be a BI-Kripke frame. We say that $\mathcal{F}$ is rooted if there is a $w \in W$ such that for every $v \in W$ we have $w \leq^* v$ (but since $\leq$ is reflexive and transitive we can replace $\leq^*$ with $\leq$).

The use of rooted models immediately implies that bi-intuitionistic logic is sound and complete with respect to the class of *rooted* BI-Kripke frames. However, we show that this result fails for both **sBIL** and **wBIL**! Specifically, $\sim p \vee \neg\sim p$ is valid on rooted frames but not valid on the full class of frames.

Lemma 10.2 *Let $\mathcal{F} = (W, \leq)$ be a rooted BI-Kripke frame. For any interpretation function I, we have that $(W, \leq, I) \Vdash \sim p \vee \neg\sim p$.*

Proof. Let r be the root of $\mathcal{F}$ and I an interpretation function, $\mathcal{M} = (W, \leq, I)$ and $w \in W$. As $\mathcal{F}$ is rooted we have that $r \leq w$. If $r \in I(p)$ then persistence and rootedness give $\mathcal{M}, v \Vdash p$ for every $v \in W$, hence $\mathcal{M}, w \Vdash \neg\sim p$. If $r \notin I(p)$ then we get $\mathcal{M}, w \Vdash \sim p$. In each case we obtain $\mathcal{M}, w \Vdash \sim p \vee \neg\sim p$. □

Thus the formula $\sim p \vee \neg\sim p$ is valid on the class of rooted BI-Kripke frames. Now we show that there is a BI-Kripke model $\mathcal{N}$ such that $\mathcal{N} \nVdash \sim p \vee \neg \sim p$. Consider the following model where reflexive arrows are omitted:

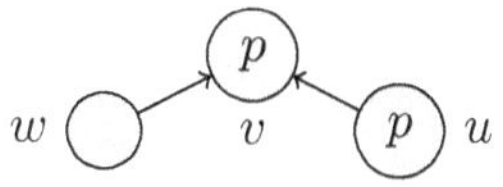

We have that $\mathcal{N}, u \nVdash \sim p$ as the only predecessor of u is itself and $\mathcal{N}, u \Vdash p$. Moreover we have that $\mathcal{N}, w \nVdash p$, hence $\mathcal{N}, v \Vdash \sim p$ which in turn implies $\mathcal{N}, u \nVdash \neg\sim p$. Consequently $\mathcal{N}, u \nVdash \sim p \vee \neg\sim p$.

It can be argued that Crolard [3, p.168] proved that **wBIL** is not complete for the class of rooted frames. But he does not make the distinction between the two logics presented here, nor pinpoint the flaws in Rauszer's proof.

Theorem 8.8 allows us to claim that $\nvdash_i [\emptyset \mid \sim p \vee \neg\sim p]$ for $i \in \{s, w\}$ as $\not\models_j \sim p \vee \neg \sim p$ for $j \in \{g, l\}$. From this, we conclude that neither **sBIL** nor **wBIL** is complete, with their corresponding semantic consequence, for the class of rooted frames: the formula $\sim p \vee \neg\sim p$ is a counterexample to such a claim.

11 Conclusion

Generalized Hilbert calculi effectively provide the tools to clarify the status of rules in axiomatic systems. The distinction between the two logics **sBIL** and **wBIL** can easily be tracked to the obvious difference between the rules DN_w and DN_s in the calculi defining them. Effectively, as in the modal case, different syntactic consequence relations stem from the traditional Hilbert calculus for bi-intuitionistic logic, formalized as generalized Hilbert calculi. The logics **wBIL** and **sBIL** are distinguishable on an extensional level in a similar way to **wK** and **sK**. The similarity with modal logic goes even further as the famous deduction theorem is not a property common to both **sBIL** and **wBIL**. As we have shown, the deduction theorem can be modified to hold in **sBIL**, and the dual deduction theorem holds in **wBIL**, but we have not yet found a modification of the dual deduction theorem for **sBIL**. So, on top of allowing one to clearly detect which logic satisfies the deduction theorem or its dual, generalized Hilbert calculi also prevent the confusions that existed in both the modal [11] and bi-intuitionistic case.

As we have shown, the logics **wBIL** and **sBIL**, respectively, have a local and global semantic counterpart on the class of BI-Kripke frames. Although quite common, this phenomenon finally clarifies the relation between the two logics. It also helps rectify the status of some properties of **sBIL** and **wBIL**, such as the fact that they are not strongly complete with respect to the class of rooted frames.

Finally, the difference between the two logics allows to look at the proof theory of bi-intuitionistic logic from a different angle. We conjecture that the various calculi which have been designed to capture bi-intuitionistic logic [8,9,10,14] are in fact sound and strongly complete for **wBIL**.

There are several directions for further work. First, the diversity of interpretations of the MP rule should be investigated. While we made a case of the multiplicity of interpretations (which we have not exhausted) of the rules DN and Nec, we did not question the shape of the rule MP. We could modify one of the generalized Hilbert calculi defined above to use a modified version of MP where the premisses would be $\emptyset \vdash A$ and $\emptyset \vdash A \to B$. This system would define a logic, but a weird one where $p, p \to q \vdash q$ would not be guaranteed to hold. A second direction, which we are exploring, leads to the algebraic treatment of **wBIL** and **sBIL** as consequence relations [6]. Third, the use of pairs $[\Gamma \mid \Delta]$ suggests a general treatment of logics that would capture both *derivability* and *refutability* calculi in one shot. Finding if such a general framework exists would require further investigations.

Related works: It has to be noted that Sano and Stell's axiomatization [25], when considered in a generalized Hilbert calculus context, also suffers from the same phenomenon as Rauszer's axiomatization. Their rule Mon$-\!\!<$ can be interpreted in the same ways as DN: with a set of assumption in its premise, giving Mon$-\!\!<_s$; or without, giving Mon$-\!\!<_w$. The generalized Hilbert calculus involving the rule Mon$-\!\!<_w$ (Mon$-\!\!<_s$) corresponds to **wBIL** (**sBIL**).

Appendix

Proof. [of Lemma 5.2] **Monotonicity**: Assume $\Gamma \vdash_i \varphi$. Then there is a derivation π of $\Gamma \vdash \varphi$. We prove by induction on $l(\pi)$ that $\Gamma' \vdash_i \varphi$ with $\Gamma \subseteq \Gamma'$. If $l(\pi) = 1$ then either $\varphi \in \Gamma$ or $\varphi \in \mathcal{A}^I$. If $\varphi \in \Gamma$ then $\varphi \in \Gamma'$, hence $\Gamma' \vdash_i \varphi$. If

$\varphi \in \mathcal{A}^I$ then $\Gamma' \vdash_i \varphi$. If $l(\pi) > 1$ then we have to consider the last rule applied. if it is MP then we simply apply the induction hypothesis on the premises and apply the rule to obtain our result. In the case of DN_i we have to distinguish between the case where $i = s$ and $i = w$. If $i = s$ then we can simply use the induction hypothesis on the premises and then apply the rule. If $i = w$ then we simply use the given premise to obtain $\Gamma' \vdash_w \varphi$ as desired.

Compositionality: Assume $\Gamma \vdash_i \varphi$ and that $\Delta \vdash_i \gamma$ for every $\gamma \in \Gamma$. Then we have a derivation π of $\Gamma \vdash \varphi$. We show by induction on the length $l(\pi)$ of π that $\Delta \vdash_i \varphi$. If $l(\pi) = 1$ then either $\varphi \in \Gamma$, or $\varphi \in \mathcal{A}^I$. If $\varphi \in \Gamma$, we have $\Delta \vdash_i \varphi$ by assumption. If $\varphi \in \mathcal{A}^I$, then $\Delta \vdash_i \varphi$. If $l(\pi) > 1$ then consider the last rule applied. If it is MP then we can simply apply the induction hypothesis on the premises and then apply MP to obtain the required conclusion. If it is DN_i, then we must distinguish $i = s$ and $i = w$. If $i = s$, we apply the induction hypothesis on the premise and then the rule. If $i = w$, we apply appropriately the rule, i.e. from $\emptyset \vdash_w \varphi$ to $\Delta \vdash_w \varphi$, to obtain the desired result.

Structurality:Assume $\Gamma \vdash_i \varphi$. Then we have a derivation π of $\Gamma \vdash \varphi$. We will show by induction on $l(\pi)$ that $\Gamma^\sigma \vdash_i \varphi^\sigma$. If $l(\pi) = 1$ then either $\varphi \in \Gamma$ or $\varphi \in \mathcal{A}^I$. If $\varphi \in \Gamma$ then $\varphi^\sigma \in \Gamma^\sigma$, hence $\Gamma^\sigma \vdash \varphi^\sigma$. If $\varphi \in \mathcal{A}^I$ then $\varphi^\sigma \in \mathcal{A}^I$, hence $\Gamma^\sigma \vdash_i \varphi^\sigma$. If $l(\pi) > 1$ then consider the last rule applied. If it is MP then we apply the induction hypothesis on the premises and then apply MP to obtain the required conclusion. If it is DN_i then for both values of i we apply the induction hypothesis on the premise and then the rule. □

Proof. [of Lemma 5.3] Assume $\Gamma \vdash_i \varphi$, giving a derivation π with root $\Gamma \vdash \varphi$. We prove by induction on $l(\pi)$ that there is a finite $\Gamma' \subseteq \Gamma$ such that $\Gamma' \vdash_i \varphi$. If $l(\pi) = 1$ then either $\varphi \in \Gamma$ or $\varphi \in \mathcal{A}^I$. If $\varphi \in \Gamma$, then $\{\varphi\} \subseteq \Gamma$ and $\varphi \vdash_i \varphi$. If $\varphi \in \mathcal{A}^I$, then $\emptyset \subseteq \Gamma$ and $\emptyset \vdash_i \varphi$. If $l(\pi) > 1$ then we consider the last rule applied. If the last rule is MP, then apply the induction hypothesis on the premises to obtain finite $\Gamma', \Gamma'' \subseteq \Gamma$ such that $\Gamma' \vdash_i \psi$ and $\Gamma'' \vdash_i \psi \to \varphi$. Theorem 5.2 delivers $\Gamma' \cup \Gamma'' \vdash_i \psi$ and $\Gamma' \cup \Gamma'' \vdash_i \psi \to \varphi$. Thus MP can be applied to get $\Gamma' \cup \Gamma'' \vdash_i \varphi$, where $\Gamma' \cup \Gamma'' \subseteq \Gamma$ is finite. If the last rule is DN_i, then $i = s$ or $i = w$. If $i = s$, we apply the induction hypothesis on the premise and then the rule. If $i = w$, we apply appropriately the rule to obtain the desired result. □

Proof. [of Proposition 7.2]

($\Rightarrow$) Assume $\vdash_i [\emptyset \mid (\varphi \prec \psi) \to \chi]$, i.e. $\emptyset \vdash_i (\varphi \prec \psi) \to \chi$. From it we can easily obtain $\emptyset \vdash_i ((\varphi \prec \psi) \vee \psi) \to (\chi \vee \psi)$. But as we have $\emptyset \vdash_i \varphi \to ((\varphi \prec \psi) \vee \psi)$ we get $\emptyset \vdash_i \varphi \to (\chi \vee \psi)$, hence $\vdash_i [\emptyset \mid \varphi \to (\chi \vee \psi)]$.

($\Leftarrow$) Assume $\vdash_i [\emptyset \mid \varphi \to (\psi \vee \chi)]$, i.e. $\emptyset \vdash_i \varphi \to (\psi \vee \chi)$. First we have, as an instance of the axiom RA_{11}, $\emptyset \vdash_i (\varphi \prec \psi) \to (\chi \vee ((\varphi \prec \psi) \prec \chi))$. But we have that $\emptyset \vdash_i ((\varphi \prec \psi) \prec \chi) \leftrightarrow (\varphi \prec (\psi \vee \chi))$, so we obtain that $\emptyset \vdash_i (\chi \vee ((\varphi \prec \psi) \prec \chi)) \to (\chi \vee (\varphi \prec (\psi \vee \chi)))$. Thus $\emptyset \vdash_i (\varphi \prec \psi) \to (\chi \vee (\varphi \prec (\psi \vee \chi)))$. However we have that $\emptyset \vdash_i (\varphi \prec (\psi \vee \chi)) \to \sim(\varphi \to (\psi \vee \chi))$. And as we have $\emptyset \vdash_i \varphi \to (\psi \vee \chi)$ by

DN_i we obtain $\emptyset \vdash_i \neg{\sim}(\varphi \to (\psi \vee \chi))$. Consequently we can obtain that $\emptyset \vdash_i (\varphi \prec (\psi \vee \chi)) \to \bot$, hence $\emptyset \vdash_i (\chi \vee (\varphi \prec (\psi \vee \chi))) \to \chi$. This finally implies $\emptyset \vdash_i (\varphi \prec \psi) \to \chi$, hence $\vdash_i [\emptyset \mid (\varphi \prec \psi) \to \chi]$.

□

Proof. [of Theorem 7.3]

($\Leftarrow$) Assume $\vdash_w [\Gamma \mid \varphi \to \psi]$, i.e. $\Gamma \vdash_w \varphi \to \psi$. Then by monotonicity we obtain $\Gamma, \varphi \vdash_w \varphi \to \psi$. Moreover we have that $\Gamma, \varphi \vdash_w \varphi$ as $\varphi \in \Gamma \cup \{\varphi\}$. So by MP we obtain $\Gamma, \varphi \vdash_w \psi$, hence $\vdash_w [\Gamma, \varphi \mid \psi]$.

($\Rightarrow$) Assume $\vdash_w [\Gamma, \varphi \mid \psi]$, i.e. $\Gamma, \varphi \vdash_w \psi$ giving a derivation π of $\Gamma, \varphi \vdash \psi$. We show by induction on the length of π that $\Gamma \vdash_w \varphi \to \psi$. If $l(\pi) = 1$ then either $\psi \in \Gamma \cup \{\varphi\}$ or $\psi \in \mathcal{A}^I$. If $\varphi = \psi$ then we clearly have $\Gamma \vdash_w \varphi \to \psi$. If $\psi \in \Gamma$ then we can deduce $\Gamma \vdash_w \varphi \to \psi$ from the fact that we have $\emptyset \vdash_w p \to (q \to p)$. If $\psi \in \mathcal{A}^I$ then with a similar reasoning $\Gamma \vdash_w \varphi \to \psi$. If $l(\pi) > 1$ then consider the last rule applied. The case of the rule MP is treated as follows. Use the induction hypothesis on the premises of the rule and note that $\emptyset \vdash_w (p \to q) \to ((p \to (q \to r)) \to (p \to r))$: using MP several times one arrives at the establishment of $\Gamma \vdash_w \varphi \to \psi$. If the last rule is DN_w, we have a derivation of $\emptyset \vdash \chi$, so we can apply DN_w to obtain $\emptyset \vdash_w \neg{\sim}\chi$. Then we can use the fact that $\emptyset \vdash_w p \to (q \to p)$ to obtain $\emptyset \vdash_w \varphi \to \neg{\sim}\chi$. By monotonicity we obtain $\Gamma \vdash_w \varphi \to \neg{\sim}\chi$.

□

For the proof of Theorem 7.7 we need the following claim:

Claim .1 $\emptyset \vdash_s \neg(\lambda_1 \prec \lambda_2) \to ({\sim}\lambda_2 \to {\sim}\lambda_1)$

Proof. We have $\emptyset \vdash_s \lambda_1 \to (\lambda_2 \vee (\lambda_1 \prec \lambda_2))$. The rule below is derivable in both systems:

$$\frac{\emptyset \vdash \varphi \to \psi}{\emptyset \vdash (\varphi \prec \chi) \to (\psi \prec \chi)} \prec Mon$$

Indeed, given that $\emptyset \vdash_i \psi \to (\chi \vee (\psi \prec \chi))$ and $\emptyset \vdash_i \varphi \to \psi$, we get $\emptyset \vdash_i \varphi \to (\chi \vee (\psi \prec \chi))$, hence $\emptyset \vdash_i (\varphi \prec \chi) \to (\psi \prec \chi)$ by Proposition 7.2. We can apply $\prec Mon$ to obtain $\emptyset \vdash_s (\top \prec (\lambda_2 \vee (\lambda_1 \prec \lambda_2))) \to {\sim}\lambda_1$. Next we prove that $\emptyset \vdash_s ({\sim}\lambda_2 \wedge \neg(\lambda_1 \prec \lambda_2)) \to {\sim}(\lambda_2 \vee (\lambda_1 \prec \lambda_2))$ to obtain that $\emptyset \vdash_s ({\sim}\lambda_2 \wedge \neg(\lambda_1 \prec \lambda_2)) \to {\sim}\lambda_1$, and hence $\emptyset \vdash_s \neg(\lambda_1 \prec \lambda_2) \to ({\sim}\lambda_2 \to {\sim}\lambda_1)$. First, $\emptyset \vdash_s \top \to ((\lambda_2 \vee (\lambda_1 \prec \lambda_2)) \vee (\top \prec (\lambda_2 \vee (\lambda_1 \prec \lambda_2))))$ is an instance of an axiom. Then by associativity of disjunction we obtain $\emptyset \vdash_s \top \to (\lambda_2 \vee ((\lambda_1 \prec \lambda_2) \vee (\top \prec (\lambda_2 \vee (\lambda_1 \prec \lambda_2)))))$. By Proposition 7.2 we get $\emptyset \vdash_s {\sim}\lambda_2 \to ((\lambda_1 \prec \lambda_2) \vee (\top \prec (\lambda_2 \vee (\lambda_1 \prec \lambda_2))))$. Consequently we easily obtain $\emptyset \vdash_s ({\sim}\lambda_2 \wedge \neg(\lambda_1 \prec \lambda_2)) \to (\top \prec (\lambda_2 \vee (\lambda_1 \prec \lambda_2)))$, i.e. $\emptyset \vdash_s ({\sim}\lambda_2 \wedge \neg(\lambda_1 \prec \lambda_2)) \to {\sim}(\lambda_2 \vee (\lambda_1 \prec \lambda_2))$. □

Proof. [of Theorem 7.7]

($\Rightarrow$) Assume that $\vdash_s [\Gamma, \varphi \mid \psi]$, i.e. that we have a derivation π of $\Gamma, \varphi \vdash \psi$. We reason by induction on the length of π. If $l(\pi) = 1$ then two cases are possible. If the rule applied is Ax then we get $\emptyset \vdash_s \psi$, and as we have that

$\emptyset \vdash_s \psi \to ((\neg\sim)^n\varphi \to \psi)$ for any $n \in \mathbb{N}$ we obtain by MP: $\emptyset \vdash_s (\neg\sim)^n\varphi \to \psi$. By Theorem 5.2 we get $\Gamma \vdash_s (\neg\sim)^n\varphi \to \psi$. If the rule applied is El then either $\psi = \varphi$ and then we get $\emptyset \vdash_s \varphi \to \varphi$ and hence $\Gamma \vdash_s \varphi \to \varphi$, where in the antecedent of the implication $\varphi = (\neg\sim)^0\varphi$. If $l(\pi) \geq 1$ then two cases have to be considered. If the last rule applied is MP then we have by induction hypothesis $\Gamma \vdash_s (\neg\sim)^l\varphi \to \chi$ and $\Gamma \vdash_s (\neg\sim)^m\varphi \to (\chi \to \psi)$ for some χ, $m, l \in \mathbb{N}$. As we have that $\emptyset \vdash_s (\lambda_1 \to \lambda_2) \to ((\lambda_1 \to (\lambda_2 \to \lambda_3)) \to (\lambda_1 \to \lambda_3))$ and $\emptyset \vdash_s \neg\sim\lambda \to \lambda$ we obtain that $\Gamma \vdash_s \neg\sim^n\varphi \to \chi$ for $n = max(m, l)$.

If the last rule applied is DN_s then we get by induction hypothesis that $\Gamma \vdash_s (\neg\sim)^n\varphi \to \chi$. If we prove that $\emptyset \vdash_s \neg\sim(\lambda_1 \to \lambda_2) \to (\neg\sim\lambda_1 \to \neg\sim\lambda_2)$ holds then we can reach our goal by applying DN_s on $\Gamma \vdash_s (\neg\sim)^n\varphi \to \chi$ to obtain $\Gamma \vdash_s \neg\sim((\neg\sim)^n\varphi \to \chi)$ and finally $\Gamma \vdash_s (\neg\sim)^{n+1}\varphi \to \neg\sim\chi$ by MP, hence $\vdash_s [\Gamma \mid (\neg\sim)^{n+1}\varphi \to \neg\sim\chi]$. Let us thus prove $\emptyset \vdash_s \neg\sim(\lambda_1 \to \lambda_2) \to (\neg\sim\lambda_1 \to \neg\sim\lambda_2)$. First note that $\emptyset \vdash_s (\lambda_1 \prec \lambda_2) \to\sim(\lambda_1 \to \lambda_2)$ as it is an instance of the axiom A_{12}. By using A_{10} and MP we obtain $\emptyset \vdash_s \neg\sim(\lambda_1 \to \lambda_2) \to \neg(\lambda_1 \prec \lambda_2)$. Thus, using Claim .1, which can be found just above, we can obtain $\emptyset \vdash_s \neg\sim(\lambda_1 \to \lambda_2) \to (\sim\lambda_2 \to\sim\lambda_1)$. We can instantiate A_{10} again to obtain $\emptyset \vdash_s (\sim\lambda_2 \to\sim\lambda_1) \to (\neg\sim\lambda_1 \to \neg\sim\lambda_2)$, and use this fact with the previous result to finally get $\emptyset \vdash_s \neg\sim(\lambda_1 \to \lambda_2) \to (\neg\sim\lambda_1 \to \neg\sim\lambda_2)$.

($\Leftarrow$) Straightforward use of the rules DN_s and MP with Theorem 5.2.

□

Proof. [of Lemma 8.3] We reason by induction on φ and only show the cases for the added operators:

- $\varphi :=\sim \psi$: $\mathcal{M}, w \Vdash\sim \psi$ then there is a $u \leq w$ such that $\mathcal{M}, u \nVdash \psi$. By transitivity we have $u \leq v$, so there is a $u \leq v$ such that $\mathcal{M}, u \nVdash \psi$. Thus $\mathcal{M}, v \Vdash\sim\psi$.
- $\varphi := \chi \prec \psi$: $\mathcal{M}, w \Vdash \chi \prec \psi$ then there is a $u \leq w$ such that $\mathcal{M}, u \Vdash \chi$ and $\mathcal{M}, u \nVdash \psi$. By transitivity we have $u \leq v$, so there is a $u \leq v$ such that $\mathcal{M}, u \Vdash \chi$ and $\mathcal{M}, u \nVdash \psi$. Thus $\mathcal{M}, v \Vdash \chi \prec \psi$.

□

Proof. [of Lemma 8.6] We start by extending the set Γ to a prime theory Γ' in $\mathcal{L}$ by successive steps. More precisely we create a chain of extensions $\Gamma_0 \subseteq \Gamma_1 \subseteq \Gamma_2 ...$, where $\Gamma_0 = \Gamma$ and $\Gamma' = \bigcup_{k \geq 0} \Gamma_k$. In fact, we take an enumeration of all formulae of $\mathcal{F}orm_{BI}$ and we define Γ_n by induction on $n \in \mathbb{N}$ in the following way:

- $n = 0 : \Gamma_0 = \Gamma$;
- $n \geq 0$: let $\psi_1 \vee \psi_2$ be the first disjunctive sentence of $\mathcal{L}$ that has not yet been treated such that $\vdash_w [\Gamma_n \mid \psi_1 \vee \psi_2]$. Define:

$$\Gamma_{n+1} = \begin{cases} \Gamma_n \cup \{\psi_1\}, & \text{if } \nvdash_w [\Gamma_n, \psi_1 \mid \Delta] \\ \Gamma_n \cup \{\psi_2\}, & \text{otherwise} \end{cases}$$

We first show that $\nvdash_w [\Gamma' \mid \Delta]$. If we show that $\nvdash_w [\Gamma_n \mid \Delta]$ for every $n \in \mathbb{N}$ then we are done. Let us show this statement by induction on n. The base case holds by assumption as $\Gamma_0 = \Gamma$. For the inductive step we have to show that $\nvdash_w [\Gamma_n, \psi_i \mid \Delta]$. But this is obvious as it cannot both be the case that $\vdash_w [\Gamma_n, \psi_1 \mid \Delta]$ and $\vdash_w [\Gamma_n, \psi_2 \mid \Delta]$, otherwise we would have $\vdash_w [\Gamma_n \mid \Delta]$ as $\vdash_w [\Gamma_n \mid \psi_1 \vee \psi_2]$.

Second, we need to show some properties of Γ':

(i) *Consistency*: Γ' is consistent as $\nvdash_w [\Gamma' \mid \Delta]$.

(ii) *Primeness*: Let $\psi_1 \vee \psi_2 \in \Gamma'$ and k the least number such that $\vdash_w [\Gamma_k \mid \psi_1 \vee \psi_2]$. At stage k this $\psi_1 \vee \psi_2$ has not been treated and is treated eventually at a stage $j \geq k$. Then we get that $\psi_1 \in \Gamma_{j+1}$ or $\psi_2 \in \Gamma_{j+1}$, hence $\psi_1 \in \Gamma'$ or $\psi_2 \in \Gamma'$.

(iii) *Closure under deducibility*: Let ψ be a formula such that $\vdash_w [\Gamma' \mid \psi]$. Then $\vdash_w [\Gamma' \mid \psi \vee \psi]$ and as Γ' is prime we get that $\psi \in \Gamma'$.

Third we define $\Delta' = \{\psi \mid \nvdash_w [\Gamma' \mid \psi]\}$. First note that $\Delta \subseteq \Delta'$. Second we obtain that $\mathcal{F}orm_{BI} \setminus \Gamma' = \Delta'$ by definition of derivation and the closure under deducibility of Γ'. So $[\Gamma' \mid \Delta']$ is complete. We obviously obtain that this pair is unprovable: assume otherwise, then there is a finite $\Delta_0 \subseteq \Delta'$ such that $\vdash_w [\Gamma' \mid \bigvee \Delta_0]$, but as Γ' is closed under deducibility and prime we obtain that there is $\psi \in \Delta'$ such that $\psi \in \Gamma'$, which is a contradiction.

So $[\Gamma' \mid \Delta']$ is a complete pair with $\nvdash_w [\Gamma' \mid \Delta']$ and $\Gamma \subseteq \Gamma'$ and $\Delta \subseteq \Delta'$. □

Proof. [of Lemma 8.7] By induction on ψ. We only consider the case for $\prec$:

- $\psi := \psi_1 \prec \psi_2$: ($\Rightarrow$) Assume $\psi_1 \prec \psi_2 \in \Gamma$. We claim that $\nvdash_w [\psi_1 \mid \psi_2, \Delta]$. Suppose it is not the case. Then by definition there is a finite $\Delta_f \subseteq \Delta$ such that $\psi_1 \vdash_w \psi_2 \vee \bigvee \Delta_f$, hence $\vdash_w [\psi_1 \mid \psi_2 \vee \bigvee \Delta_f]$. By Theorem 7.3 we thus obtain $\vdash_w [\emptyset \mid \psi_1 \to (\psi_2 \vee \bigvee \Delta_f)]$. And then by Proposition 7.2 we obtain that $\vdash_w [\emptyset \mid (\psi_1 \prec \psi_2) \to \bigvee \Delta_f]$. But as $\psi_1 \prec \psi_2 \in \Gamma$ and Γ is closed under deducibility we get that $\bigvee \Delta_f \in \Gamma$, which leads to an obvious contradiction. So $\nvdash_w [\psi_1 \mid \psi_2, \Delta]$. Thus by Lemma 8.6 there are $\Gamma' \supseteq \{\psi_1\}$ and $\Delta' \supseteq \Delta \cup \{\psi_2\}$ such that $[\Gamma' \mid \Delta']$ is complete and $\nvdash_w [\Gamma' \mid \Delta']$. Note that $\psi_1 \in \Gamma'$ and $\psi_2 \notin \Gamma'$, hence $\mathcal{M}^c, [\Gamma' \mid \Delta'] \Vdash \psi_1$ and $\mathcal{M}^c, [\Gamma' \mid \Delta'] \nVdash \psi_2$ by induction hypothesis. But we have that $\Delta \subseteq \Delta'$, which implies by completeness that $\Gamma' \subseteq \Gamma$. So $[\Gamma' \mid \Delta'] \leq^c [\Gamma \mid \Delta]$. Consequently $\mathcal{M}^c, [\Gamma \mid \Delta] \Vdash \psi_1 \prec \psi_2$. ($\Leftarrow$) Assume $\mathcal{M}^c, [\Gamma \mid \Delta] \Vdash \psi_1 \prec \psi_2$. Assume for reductio that $\psi_1 \prec \psi_2 \notin \Gamma$. Then $\nvdash_w [\Gamma \mid \psi_1 \prec \psi_2]$. Note that every $\Gamma' \subseteq \Gamma$ is such that $\psi_1 \prec \psi_2 \notin \Gamma'$. And as $\vdash_w [\emptyset \mid \psi_1 \to (\psi_2 \vee (\psi_1 \prec \psi_2))]$ we get for every $\Gamma' \subseteq \Gamma$ such that $[\Gamma' \mid \Delta'] \in W^c$ for some Δ', if $\psi_1 \in \Gamma'$ then $\psi_2 \in \Gamma'$ as Γ' is prime. By induction hypothesis we get that for every such Γ', if $\mathcal{M}^c, [\Gamma' \mid \Delta'] \Vdash \psi_1$ then $\mathcal{M}^c, [\Gamma' \mid \Delta'] \Vdash \psi_2$. This contradicts our assumption $\mathcal{M}^c, [\Gamma \mid \Delta] \Vdash \psi_1 \prec \psi_2$. □

Proof. [of Theorem 8.8] Soundness is straightforward so let us prove (1):

($\Leftarrow$) Here we prove *completeness*. Assume $\nvdash_w [\Gamma \mid \Delta]$. Lemma 8.6 gives us a complete $[\Gamma' \mid \Delta']$ such that $\nvdash_w [\Gamma' \mid \Delta']$, where $\Gamma \subseteq \Gamma'$ and $\Delta \subseteq \Delta'$. Moreover there is no $\delta \in \Delta$ such that $\delta \in \Gamma'$, so by Lemma 8.7 we obtain that in the canonical model of Definition 8.5 the following holds: $\mathcal{M}^c, [\Gamma' \mid \Delta'] \nVdash \delta$ for every $\delta \in \Delta$, while $\mathcal{M}^c, [\Gamma' \mid \Delta'] \Vdash \Gamma$. Consequently, we have that $\Gamma \not\models_l \Delta$.

Then we can prove (2):

($\Leftarrow$) Here we prove *completeness*. Assume $\nvdash_s [\Gamma \mid \Delta]$. We show that $\Gamma \not\models_g \Delta$. Note that $\nvdash_s [(\neg\sim)^\omega\Gamma \mid \Delta]$ from Theorem 7.7. Thus, we get $\nvdash_w [(\neg\sim)^\omega\Gamma \mid \Delta]$ by Theorem 6.2. By the argument used in the strong completeness of **wBIL** we know that there is a pair $[((\neg\sim)^\omega\Gamma)^* \mid \Delta']$ in the canonical model of Definition 8.5 such that $(\neg\sim)^\omega\Gamma \subseteq ((\neg\sim)^\omega\Gamma)^*$ and $\Delta \subseteq \Delta'$. Lemma 8.7 tells us that for all $\delta \in \Delta$ we have $\mathcal{M}^c, [((\neg\sim)^\omega\Gamma)^* \mid \Delta'] \nVdash \delta$ and $\mathcal{M}^c, [((\neg\sim)^\omega\Gamma)^* \mid \Delta'] \Vdash \neg\sim^\omega\Gamma$.

To obtain a proof of $\Gamma \not\models_g \Delta$ we need a Γ-model that has one point that is not a δ-point for all $\delta \in \Delta$. To do so we restrict the canonical model, on the point described above, to obtain a Γ-model. We define $\mathcal{M}^c_\Gamma = (W^c_\Gamma, \leq^c_\Gamma, I^c_\Gamma)$, where $W^c_\Gamma = \{[\Delta_1 \mid \Delta_2] \in W^c \mid$ there is a chain $[((\neg\sim)^\omega\Gamma)^* \mid \Delta']R_1...R_n[\Delta_1 \mid \Delta_2]$, where $R_j \in \{\leq, \geq\}$ for $j \in \mathbb{N}\}$, and I^c_Γ and $\leq^c_\Gamma$ are restrictions of respectively I^c and $\leq^c$ to W^c_Γ. The notion of bisimulation developed by de Groot and Pattinson [5], gives us that $(\mathcal{M}^c, [((\neg\sim)^\omega\Gamma)^* \mid \Delta'])$ and $(\mathcal{M}^c_\Gamma, [((\neg\sim)^\omega\Gamma)^* \mid \Delta'])$ are bisimilar, hence modally equivalent. Thus we have that $\mathcal{M}^c_\Gamma, [((\neg\sim)^\omega\Gamma)^* \mid \Delta] \nVdash \delta$ for every $\delta \in \Delta$, and $\mathcal{M}^c_\Gamma, [((\neg\sim)^\omega\Gamma)^* \mid \Delta] \Vdash (\neg\sim)^\omega\Gamma$. It remains to prove that $\mathcal{M}^c_\Gamma$ is a Γ-model. Let $[\Delta_1 \mid \Delta_2] \in \mathcal{M}^c_\Gamma$. By definition there is a chain $[((\neg\sim)^\omega\Gamma)^* \mid \Delta']R_1...R_n[\Delta_1 \mid \Delta_2]$ such that $R_j \in \{\leq, \geq\}$ for every $j \in \{1, ..., n\}$. We now need the following Claim .2 to conclude that $\mathcal{M}^c_\Gamma, [\Delta_1 \mid \Delta_2] \Vdash (\neg\sim)^\omega\Gamma$. In particular we obtain $\mathcal{M}^c_\Gamma, [\Delta_1 \mid \Delta_2] \Vdash \Gamma$.

Claim .2 *For every chain* $[((\neg\sim)^\omega\Gamma)^* \mid \Delta']R_1...R_n[\Psi_1 \mid \Psi_2]$ *we have that* $\mathcal{M}^c_\Gamma, [\Psi_1 \mid \Psi_2] \Vdash (\neg\sim)^\omega\Gamma$.

Proof. Let $C = [((\neg\sim)^\omega\Gamma)^* \mid \Delta']R_1...R_n[\Psi_1 \mid \Psi_2]$ be a chain. We prove that $\mathcal{M}^c_\Gamma, [\Psi_1 \mid \Psi_2] \Vdash (\neg\sim)^\omega\Gamma$ by induction on the length l of C:

- $l = 0$: then $[\Psi_1 \mid \Psi_2] = [((\neg\sim)^\omega\Gamma)^* \mid \Delta']$ and consequently $\mathcal{M}^c_\Gamma, [((\neg\sim)^\omega\Gamma)^* \mid \Delta'] \Vdash (\neg\sim)^\omega\Gamma$ by Lemma 8.7;
- $l = n+1$: if R_{n+1} is $\leq$ then there is $[\Pi_1 \mid \Pi_2]$ such that $[\Pi_1 \mid \Pi_2] \leq [\Psi_1 \mid \Psi_2]$. By induction hypothesis we get $\mathcal{M}^c_\Gamma, [\Pi_1 \mid \Pi_2] \Vdash (\neg\sim)^\omega\Gamma$ and consequently, by Lemma 8.3 $\mathcal{M}^c_\Gamma, [\Psi_1 \mid \Psi_2] \Vdash (\neg\sim)^\omega\Gamma$. If R_{n+1} is $\geq$ then there is $[\Pi_1 \mid \Pi_2]$ such that $[\Pi_1 \mid \Pi_2] \geq [\Psi_1 \mid \Psi_2]$. By induction hypothesis we get $\mathcal{M}^c_\Gamma, [\Pi_1 \mid \Pi_2] \Vdash (\neg\sim)^\omega\Gamma$. Note that $\neg\sim(\neg\sim)^\omega\Gamma = (\neg\sim)^\omega\Gamma$, so $\mathcal{M}^c_\Gamma, [\Pi_1 \mid \Pi_2] \Vdash \neg\sim(\neg\sim)^\omega\Gamma$. We easily obtain $\mathcal{M}^c_\Gamma, [\Psi_1 \mid \Psi_2] \Vdash (\neg\sim)^\omega\Gamma$.

□

□

References

[1] Blackburn, P., M. de Rijke and Y. Venema, "Modal Logic," Cambridge Tracts in Theoretical Computer Science, Cambridge University Press, 2001.

[2] Chagrov, A. and M. Zakharyaschev, "Modal Logic," Oxford University Press, 1997.

[3] Crolard, T., *Subtractive logic*, TCS **254:1-2** (2001), pp. 151–185.

[4] Crolard, T., *A formulae-as-types interpretation of subtractive logic*, Journal of Logic and Computation **14:4** (2004), pp. 529–570.

[5] de Groot, J. and D. Pattinson, *Hennessy-Milner properties for (modal) bi-intuitionistic logic*, in: *Proc. WoLLIC 2019*, 2019, pp. 161–176.

[6] Font, J. M., "Abstract Algebraic Logic: An Introductory Textbook," Studies in Logic and the Foundations of Mathematics, College Publications, 2016.

[7] Gabbay, D. M., *Applications of trees to intermediate logics*, Journal of Symbolic Logic **37** (1972), pp. 135–138.

[8] Galmiche, D. and D. Méry, *A connection-based characterization of bi-intuitionistic validity*, J. Autom. Reasoning **51** (2013), pp. 3–26.

[9] Goré, R. and L. Postniece, *Combining derivations and refutations for cut-free completeness in bi-intuitionistic logic*, J. Log. Comput. **20** (2010), pp. 233–260.

[10] Goré, R., L. Postniece and A. Tiu, *Cut-elimination and proof-search for bi-intuitionistic logic using nested sequents*, in: *Advances in Modal Logic 7, Nancy, France, 9-12 September 2008*, 2008, pp. 43–66.

[11] Hakli, R. and S. Negri, *Does the deduction theorem fail for modal logic?*, Synthese **187** (2012), pp. 849–867.

[12] Kracht, M., "Tools and Techniques in Modal Logic," Elsevier, 1999.

[13] Patterson, A., *Bisimulation and propositional intuitionistic logic*, in: A. Mazurkiewicz and J. Winkowski, editors, *CONCUR '97: Concurrency Theory* (1997), pp. 347–360.

[14] Pinto, L. and T. Uustalu, *Proof search and counter-model construction for bi-intuitionistic propositional logic with labelled sequents*, in: M. Giese and A. Waaler, editors, *Proc. TABLEAUX* (2009), pp. 295–309.

[15] Popkorn, S., "First Steps in Modal Logic," Cambridge University Press, 1994.

[16] Rauszer, C., *A formalization of the propositional calculus of H-B logic*, Studia Logica: An International Journal for Symbolic Logic **33** (1974), pp. 23–34.

[17] Rauszer, C., *Semi-boolean algebras and their application to intuitionistic logic with dual operations*, Fundamenta Mathematicae LXXXIII (1974), pp. 219–249.

[18] Rauszer, C., *On the strong semantical completeness of any extension of the intuitionistic predicate calculus*, Bulletin de l'Academie Polonaise des Sciences **XXIV** (1976), pp. 81–87.

[19] Rauszer, C., *An algebraic approach to the Heyting-Brouwer predicate calculus*, Fundamenta Mathematicae **96** (1977), pp. 127–135.

[20] Rauszer, C., *Applications of kripke models to Heyting-Brouwer logic*, Studia Logica: An International Journal for Symbolic Logic **36** (1977), pp. 61–71.

[21] Rauszer, C., *Craig interpolation theorem for an extension of intuitionistic logic*, Bulletin de l'Academie Polonaise des Sciences **25** (1977), pp. 127–135.

[22] Rauszer, C., *Model theory for an extension of intuitionistic logic*, Studia Logica **36** (1977), pp. 73–87.

[23] Rauszer, C., "An Algebraic and Kripke-Style Approach to a Certain Extension of Intuitionistic Logic," Dissertationes Mathematicae, 1980.

[24] Reyes, G. E. and H. Zolfaghari, *Bi-heyting algebras, toposes and modalities*, J. Philos. Log. **25** (1996), pp. 25–43.

[25] Sano, K. and J. G. Stell, *Strong completeness and the finite model property for bi-intuitionistic stable tense logics*, in: *Proc. Methods for Modalities 2017*, Electronic Proceedings in Theoretical Computer Science **243** (2017), pp. 105–121.

[26] Thomason, R. H., *On the strong semantical completeness of the intuitionistic predicate calculus*, The Journal of Symbolic Logic **33** (1968), pp. 1–7.

[27] Wójcicki, R., "Theory of logical calculi : basic theory of consequence operations," Kluwer Academic Publishers Dordrecht ; Boston, 1988, xviii, 473 p. ; pp.

Logic-Induced Bisimulations

Jim de Groot

The Australian National University

Helle Hvid Hansen

Delft University of Technology

Alexander Kurz

Chapman University

Abstract

We define a new logic-induced notion of bisimulation (called ρ-bisimulation) for coalgebraic modal logics given by a logical connection, and investigate its properties. We show that it is structural in the sense that it is defined only in terms of the coalgebra structure and the one-step modal semantics and, moreover, can be characterised by a form of relation lifting. Furthermore we compare ρ-bisimulations to several well-known equivalence notions, and we prove that the collection of bisimulations between two models often forms a complete lattice. The main technical result is a Hennessy-Milner type theorem which states that, under certain conditions, logical equivalence implies ρ-bisimilarity. In particular, the latter does *not* rely on a duality between functors T (the type of the coalgebras) and L (which gives the logic), nor on properties of the logical connection ρ.

Keywords: Modal Logic, Coalgebraic Logic, Bisimulation.

1 Introduction

In this paper, we investigate when logical equivalence for a given modal language can be captured by a structural semantic equivalence notion, understood as a form of bisimulation. Our investigation is carried out in the setting of coalgebraic modal logic [21], where semantic structures are given by coalgebras for a functor $\mathsf{T}\colon \mathbf{C} \to \mathbf{C}$ [27]. This allows for a uniform treatment of a wide variety of modal logics [21,25,28]. Coalgebras come with general notions of *behavioural equivalence* and *bisimilarity*, and a logic is said to be *expressive* if logical equivalence implies behavioural equivalence, in which case we have a generalisation of the classic Hennessy-Milner theorem [15].

For **Set**-coalgebras, i.e., when $\mathbf{C} = \mathbf{Set}$, it has been shown that a coalgebraic modal logic is expressive if the language has sufficiently large conjunctions and the set Λ of modalities is *separating*, meaning that they separate points in $\mathsf{T}X$

[24,26,28]. In the more abstract setting of coalgebraic modal logic, where a logic is given by a functor and its semantics by a natural transformation ρ [9,19], a sufficient condition for a logic being expressive is that the so-called mate of ρ is monic [19, Thm. 4.2].

In this line of research, modal logics are often viewed as specification languages for coalgebras. Therefore behavioural equivalence is a given, and the aim is to find expressive logics. However, sometimes the modal language is of primary interest [5] and the relevant modalities need not be separating, see e.g. [12]. This leads us to consider the following question:

Given a possibly non-expressive coalgebraic modal logic, can we characterise logical equivalence by a notion of bisimulation?

Such investigations have been carried out earlier in [4] where a notion of Λ-bisimulation was proposed for **Set**-coalgebras and coalgebraic modal logics with a classical propositional base.

Here we generalise and extend the work of [4] beyond **Set** using the formulation of coalgebraic modal logic via dual adjunctions [9,19,23]. Examples include coalgebras over ordered and topological spaces and modal logics on different propositional bases. After recalling basic definitions of coalgebraic modal logic in Section 2, we define the concept of a *ρ-bisimulation* in Section 3. For **Set**-coalgebras, this is a relation B between coalgebras for which the so-called B-coherent pairs [14,5] give rise to a congruence between complex algebras.

The definition of ρ-bisimulation is structural in the sense that it is defined in terms of the coalgebra structure and the one-step modal semantics ρ. Moreover, it can often be characterised as a greatest fixpoint via relation lifting. We also prove results concerning truth-preservation, composition and lattice structure, and we show that ρ-bisimilarity, like T-bisimilarity for coalgebras. For coalgebras on finite sets, this means that ρ-bisimilarity can be computed by a partition refinement algorithm.

The main technical results are found in Section 4 and concern the distinguishing power of ρ-bisimulations. We first compare ρ-bisimulations with other coalgebraic equivalence notions. Subsequently, we prove a Hennessy-Milner style theorem (Thm. 4.4) in which we give conditions that guarantee that logical equivalence is a ρ-bisimulation. We emphasise that the logic is *not* assumed to be expressive and ρ-bisimilarity will generally differ from bisimilarity for T-coalgebras. Finally, we define a notion of translation between logics and show that if the language of ρ' is a propositional extension of the language of ρ, then ρ-bisimulations are also ρ'-bisimulations (Prop. 4.10).

By instantiating Prop. 4.10, we obtain that for labelled transition systems the ρ-bisimilarity notions for Hennessy-Milner logic [15] and trace logic [19] coincide and are equal to the standard notion of bisimilarity even without assuming image-finiteness. These two logics have the same modalities, which are separating, but trace logic has $\top$ as the only propositional connective.

Due to lack of space, some proofs are omitted here. They will be made available in an extended version of this paper online.

2 Coalgebraic modal logic

We review some background on coalgebraic logic, categorical algebra, and Stone duality. For more see e.g. [27,21,2,3,17]. We write **Set** for the category of sets and functions.

Coalgebraic modal logic generalises modal logic from Kripke frames to coalgebras for a functor T.

Coalgebras can be understood as generalised, state-based systems defined parametrically in the system type T. Formally, we require T to be an endofunctor on a category $\mathbf{C}$. A T-*coalgebra* is then a pair (X, γ) such that $\gamma : X \to \mathsf{T}X$ is a morphism in $\mathbf{C}$. The object X is the state space, and the arrow γ is the coalgebra structure map. A T-*coalgebra morphism* from (X, γ) to (X', γ') is a $\mathbf{C}$-morphism $f : X \to X'$ satisfying $\gamma' \circ f = \mathsf{T}f \circ \gamma$. Together, T-coalgebras and T-coalgebra morphisms form a category which we write as $\mathbf{Coalg}(\mathsf{T})$.

An **algebra** for a functor is the dual notion of a coalgebra. Given an endofunctor $\mathsf{L}\colon \mathbf{A} \to \mathbf{A}$, an L-*algebra* is a pair (A, α) such that $\alpha : \mathsf{L}A \to A$ is a morphism in $\mathbf{A}$. An L-*algebra morphism* from (A, α) to (A', α') is an $\mathbf{A}$-morphism $h : A \to A'$ such that $h \circ \alpha = \alpha' \circ \mathsf{L}h$. We write $\mathbf{Alg}(\mathsf{L})$ for the category of L-algebras and L-algebra morphisms.

Example 2.1 A Kripke frame $(W, R \subseteq X \times X)$ is a coalgebra for the covariant powerset functor $\mathcal{P}\colon \mathbf{Set} \to \mathbf{Set}$ which maps a set to its set of subsets, and a function $f\colon X \to Y$ to the direct image map $f[-]\colon \mathcal{P}X \to \mathcal{P}Y$, by defining $\gamma\colon X \to \mathcal{P}X$ as $\gamma(x) = R[x] = \{y \in X \mid xRy\}$. Similarly, a Kripke model (W, R, V) where V is a valuation of a set P_0 of atomic propositions is a coalgebra for the $\mathbf{Set}$-functor $\mathcal{P}(-) \times \mathcal{P}(P_0)$ (which is constant in its second component) by taking $\gamma(x) = (R[x], V'(x))$ where $V'(x) = \{p \in P_0 \mid x \in V(p)\}$. It can be verified that the ensuing notion of coalgebra morphism coincides with the usual notion of bounded morphism for Kripke frames and Kripke models, respectively.

Example 2.2 Labelled transition systems (LTSs) are coalgebras for the $\mathbf{Set}$-functor $\mathsf{T} = \mathcal{P}(-)^A$ where $\mathcal{P}$ is the covariant powerset functor and A is the set of labels. A coalgebra $\gamma\colon X \to \mathcal{P}(X)^A$ specifies for each state $x \in X$ and label $a \in A$, the set $\gamma(x)(a)$ of a-successors of x. In other words, an LTS is an A-indexed multi-relational Kripke frame, and one verifies that coalgebra morphisms are A-indexed bounded morphisms.

Logical connections To investigate logics for T-coalgebras in this generality, we use the Stone duality approach to modal logic [13,1], but rather than a full duality, here one requires only a dual adjunction $\mathsf{P} : \mathbf{C} \rightleftarrows \mathbf{A} : \mathsf{S}$ (sometimes called a *logical connection*) between a category $\mathbf{C}$ of state spaces and a category $\mathbf{A}$ of algebras that encode a propositional base logic. We emphasise that the functors P and S are contravariant. The classic example is then the instance $\mathcal{Q}_{\mathsf{BA}} : \mathbf{Set} \rightleftarrows \mathbf{BA} : \mathsf{Uf}$ where $\mathcal{Q}_{\mathsf{BA}}$ maps a set to its Boolean algebra of predicates (i.e. subsets), and Uf maps a Boolean algebra to its set of ultrafilters.

We denote the units of a dual adjunction $\mathsf{P} : \mathbf{C} \rightleftarrows \mathbf{A} : \mathsf{S}$ by $\eta^{\mathbf{C}}\colon \mathrm{Id}_{\mathbf{C}} \to \mathsf{SP}$

and $\eta^{\mathbf{A}}\colon \mathrm{Id}_{\mathbf{A}} \to \mathsf{PS}$, and the bijection of Hom-sets $\mathbf{C}(C, \mathsf{S}A) \cong \mathbf{A}(A, \mathsf{P}C)$ in both directions by $(-)^\sharp$. Recall that for $f\colon C \to \mathsf{S}A$, the adjoint transpose of f is $f^\sharp = \mathsf{P}f \circ \eta^{\mathbf{A}}_A$, and for $g\colon A \to \mathsf{P}C$, the adjoint is $g^\sharp = \mathsf{S}g \circ \eta^{\mathbf{C}}_C$.

Coalgebraic Modal Logic Given a dual adjunction $\mathsf{P} : \mathbf{C} \rightleftarrows \mathbf{A} : \mathsf{S}$ and an endofunctor T on $\mathbf{C}$, a *modal logic for* T*-coalgebras* is a pair (L, ρ) consisting of an endofunctor $\mathsf{L}\colon \mathbf{A} \to \mathbf{A}$ (defining modalities) and a natural transformation $\rho : \mathsf{LP} \to \mathsf{PT}$, (defining the *one-step modal semantics*). This data gives rise to a functor $\mathbf{Coalg}(\mathsf{T}) \to \mathbf{Alg}(\mathsf{L})$ which sends a coalgebra (X, γ) to its *complex algebra* $(\mathsf{P}X, \gamma^*)$, where $\gamma^* = \mathsf{P}\gamma \circ \rho_X$. Assuming that $\mathbf{Alg}(\mathsf{L})$ has an initial algebra $\alpha : \mathsf{L}\Phi \to \Phi$, which generalises the Lindenbaum-Tarski algebra, the semantics of (equivalence classes of) formulas is obtained as the unique $\mathbf{Alg}(\mathsf{L})$-morphism $[\![-]\!]_\gamma\colon (\Phi, \alpha) \to (\mathsf{P}X, \gamma^*)$. Viewing the semantics as an $\mathbf{A}$-morphism $[\![-]\!]_\gamma\colon \Phi \to \mathsf{P}X$, its adjoint $\mathrm{th}_\gamma = [\![-]\!]^\sharp_\gamma\colon X \to \mathsf{S}\Phi$, is called the *theory map*, since in the classic case it maps a state in X to the ultrafilter of L-formulas it satisfies. By their definitions, the semantics and the theory map make the following diagrams commute:

$$\begin{array}{ccccc} \mathsf{L}\Phi & & \xrightarrow{\quad\alpha\quad} & & \Phi \\ {\scriptstyle \mathsf{L}[\![\cdot]\!]_\gamma}\downarrow & & & & \downarrow{\scriptstyle [\![\cdot]\!]_\gamma} \\ \mathsf{LP}X & \xrightarrow{\rho_X} & \mathsf{PT}X & \xrightarrow{\mathsf{P}\gamma} & \mathsf{P}X \end{array} \qquad \begin{array}{ccccc} X & & \xrightarrow{\quad\mathrm{th}_\gamma\quad} & & \mathsf{S}\Phi \\ \downarrow{\scriptstyle\gamma} & & & & \downarrow{\scriptstyle \mathsf{S}\alpha} \\ \mathsf{T}X & \xrightarrow{\mathsf{T}\,\mathrm{th}_\gamma} & \mathsf{TS}\Phi & \xrightarrow{\rho^\flat_\Phi} & \mathsf{SL}\Phi \end{array}$$

where $\rho^\flat\colon \mathsf{TS} \to \mathsf{SL}$ is the natural transformation, the so-called mate of ρ, obtained (component-wise) as the adjoint of $\rho\mathsf{S} \circ \mathsf{L}\eta^{\mathbf{A}}$.

Example 2.3 Consider the self-dual adjunction $\mathcal{Q} : \mathbf{Set} \rightleftarrows \mathbf{Set} : \mathcal{Q}$ given in both directions by the contravariant powerset functor $\mathcal{Q}$, which maps a set to its powerset 2^X, and a function $f\colon X \to Y$ to its inverse image map $f^{-1}\colon 2^Y \to 2^X$. In this case, the adjoints are given by transposing. For $f\colon X \to 2^Y$, $f^\sharp\colon Y \to 2^X$ is defined by $f^\sharp(y)(x) = f(x)(y)$.

Considering LTSs as $\mathcal{P}(-)^A$-coalgebras over $\mathbf{Set}$ (cf. Exm. 2.2), we obtain *trace logic for LTSs* [19, Exm. 3.2] by taking $\mathsf{L}^{\mathsf{tr}}\colon \mathbf{Set} \to \mathbf{Set}$ to be the functor $\mathsf{L}^{\mathsf{tr}} = 1 + A \times (-)$ (where $1 = \{*\}$) which encodes a modal signature with a constant modality $\top$ and a unary modality for each $a \in A$. Since $\mathbf{A} = \mathbf{Set}$, trace logic has no other connectives. The initial L^{tr}-algebra consists of finite sequences over A with the empty word as constant, and prefixing with elements from A as the unary operations. That is, L^{tr}-formulas are of the form $\langle a_1 \rangle \cdots \langle a_k \rangle \top$, $k \geq 0$.

We obtain the usual semantics of $\top$ and A-labelled diamonds by defining the modal semantics $\rho^{\mathsf{tr}}\colon 1 + A \times \mathcal{Q}(-) \to \mathcal{Q}(\mathcal{P}(-)^A)$ as $\rho^{\mathsf{tr}}_X(*) = \mathcal{P}(X)^A$ and $\rho^{\mathsf{tr}}_X(a, U) = \{t \in \mathcal{P}(X)^A \mid t(a) \cap U \neq \emptyset\}$. Hence for an LTS (X, γ), $[\![\langle a_1 \rangle \cdots \langle a_k \rangle \top]\!]_\gamma$ is the subset of X consisting of states x that can execute the trace $a_1 \cdots a_k$.

Example 2.4 Consider again LTSs as $\mathcal{P}(-)^A$-coalgebras over $\mathbf{Set}$ (cf. Exm. 2.2), but take now the classic adjunction $\mathcal{Q}_{\mathsf{BA}} : \mathbf{Set} \rightleftarrows \mathbf{BA} : \mathsf{Uf}$. Hennessy-Milner logic [15] (or equivalently, normal multi-modal logic) is here defined as classical propositional logic extended with join-preserving diamonds. This is achieved by defining $\mathsf{L}^{\mathsf{hm}}\colon \mathbf{BA} \to \mathbf{BA}$ as follows. For a Boolean algebra B, $\mathsf{L}^{\mathsf{hm}}B$ is the free Boolean algebra generated by the set $\{\langle a\rangle b \mid b \in B, a \in A\}$ modulo the congruence generated by the usual diamond equations: $\langle a\rangle\bot = \bot$ and $\langle a\rangle(\varphi_1 \vee \varphi_2) = \langle a\rangle\varphi_1 \vee \langle a\rangle\varphi_2$ for all $a \in A$. The modal semantics $\rho^{\mathsf{hm}}\colon \mathsf{L}^{\mathsf{hm}}\mathcal{Q}_{\mathsf{BA}} \to \mathcal{Q}_{\mathsf{BA}}(\mathcal{P}(-)^A)$ is essentially the Boolean extension of ρ^{tr}. In particular, $\rho^{\mathsf{hm}}_X(\langle a\rangle U) = \{t \in \mathcal{P}(X)^A \mid t(a) \cap U \neq \emptyset\}$.

The above description of Hennessy-Milner logic is a special case of a more general approach described in the next example.

Example 2.5 If $\mathbf{A}$ in the dual adjunction is a variety of algebras, we can define a logic (L, ρ) for $\mathsf{T} : \mathbf{C} \to \mathbf{C}$ by *predicate liftings and axioms* as in [20, Def. 4.2] and [23, Thms 4.7 and 8.8]. An *n-ary predicate lifting* is a natural transformation $\lambda : \mathsf{UP}^n \to \mathsf{UPT}$, where $\mathsf{P}^n X$ is the n-fold product of $\mathsf{P}X$ in $\mathbf{A}$ and $\mathsf{U} : \mathbf{A} \to \mathbf{Set}$ is the forgetful functor. Together with a suitable notion of *axioms*, a collection Λ of such predicate liftings yields a functor $\mathsf{L} : \mathbf{A} \to \mathbf{A}$ sending $A \in \mathbf{A}$ to the free algebra generated by $\{\underline{\lambda}(a_1, \ldots, a_n) \mid \lambda \in \Lambda, a_i \in A\}$ modulo (instantiations of) the axioms. Define $\rho : \mathsf{LP} \to \mathsf{PT}$ on generators by $\rho_X(\underline{\lambda}(a_1, \ldots, a_n)) = \lambda_X(a_1, \ldots, a_n) \in \mathsf{PT}X$. If ρ is well-defined then (L, ρ) is a logic for $\mathbf{Coalg}(\mathsf{T})$. All logics in e.g. [4,7,21] are instances hereof.

Next, we interpret positive modal logic [11,10], whose coalgebraic semantics over posets is given in [18, Example 2.4], in topological spaces:

Example 2.6 Consider the dual adjunction $\Omega : \mathbf{Top} \rightleftarrows \mathbf{DL} : \mathsf{pf}$, where Ω takes open subsets of a topological space, viewed as a distributive lattice, and pf takes prime filters of a distributive lattice topologised by the subbase $\{\widetilde{a} \mid a \in A\}$, where $\widetilde{a} = \{p \in \mathsf{pf}A \mid a \in p\}$. The Vietoris functor $\mathsf{V} : \mathbf{Top} \to \mathbf{Top}$ takes $X \in \mathbf{Top}$ to its collection of compact subsets topologised by the subbase consisting of $\boxdot a = \{c \in \mathsf{V}X \mid c \subseteq a\}$ and $\diamonddot a = \{c \in \mathsf{V}X \mid c \cap a \neq \emptyset\}$, where a ranges over the opens of X. For a continuous map $f : X \to X'$ the map $\mathsf{V}f$ takes direct images. The logic functor $\mathsf{N} : \mathbf{DL} \to \mathbf{DL}$ is defined as in [18, Exm. 2.4]. The natural transformation $\rho : \mathsf{N}\Omega \to \Omega\mathsf{V}$ given on generators by $\Box a \mapsto \boxdot a$ and $\Diamond a \mapsto \diamonddot a$ then gives rise to semantics for positive modal logic.

We now recall *linear weighted automata*, see e.g. [8, Section 3.2].

Example 2.7 Let $\mathbb{k}$ be a field and $\mathbf{Vec}_{\mathbb{k}} \rightleftarrows \mathbf{Vec}_{\mathbb{k}}$ the dual adjunction between vector spaces over $\mathbb{k}$ given in both directions by taking dual space via the contravariant functor $(-)^{\vee} = \mathrm{Hom}(-, \mathbb{k}) : \mathbf{Vec}_{\mathbb{k}} \to \mathbf{Vec}_{\mathbb{k}}$. Linear weighted automata for a set A of labels are coalgebras for the endofunctor $\mathsf{W} = \mathbb{k} \times (-)^A$ on $\mathbf{Vec}_{\mathbb{k}}$, where $(-)^A$ is simply the collection of maps $A \to (-)$ with a pointwise vector space structure. We work with the language given by the grammar

$$\varphi ::= 0 \mid p \mid r \cdot \varphi \mid \varphi + \varphi \mid \langle a\rangle\varphi,$$

where $a \in A$, $r \in \mathbb{k}$, and p is a single proposition letter (the termination predicate). Note that, contrary to *loc. cit.*, we also include connectives corresponding to the signature of vector spaces (because we will use vector spaces as algebraic semantics). The interpretation of a formula φ in this (many-valued) setting is a linear map $[\![\varphi]\!] : X \to \Bbbk$. The connectives $0, +$ and r are interpreted via the corresponding operations in vector spaces. For p, we use the nullary predicate lifting $\lambda^p \in \mathsf{U}(\mathsf{W}-)^\vee$, where $\mathsf{U} : \mathbf{Vec}_\Bbbk \to \mathbf{Set}$ is the forgetful functor, given by $\lambda^p_X : \mathsf{W}X \to \Bbbk : (r, t) \mapsto r$. That is $[\![p]\!]_\gamma = \lambda^p_X \circ \gamma : X \to \Bbbk$. As for the diamonds, we use $\lambda^{\langle a \rangle} : \mathsf{U}(-)^\vee \to \mathsf{U}(\mathsf{W}-)^\vee$ defined by

$$\lambda^{\langle a \rangle}_X(m) : \mathsf{W}X \to \Bbbk : (r, t) \mapsto m(t(a)).$$

Concretely, this means that if $[\![p]\!]_\gamma(y) = r \in \Bbbk$ and there is an a-transition $x \xrightarrow{a} y$, then $[\![\langle a \rangle p]\!](x) = r$. Together with the axioms $\langle a \rangle(\varphi + \psi) = \langle a \rangle \varphi + \langle a \rangle \psi$ and $r \cdot \langle a \rangle \varphi = \langle a \rangle (r \cdot \varphi)$ this gives rise to an endofunctor $\mathsf{L} : \mathbf{Vec}_\Bbbk \to \mathbf{Vec}_\Bbbk$, and a logic (L, ρ) for linear weighted automata. One can show that logical equivalence coincides with language semantics if the state-space is finite-dimensional.

Relations as jointly mono spans We are interested in giving certain relations a special status. In **Set**, a binary relation $B \subseteq X \times X$ corresponds to an injective map $B \hookrightarrow X \times X$. This generalises to an arbitrary category (possibly lacking products) via the following notion. A span $X_1 \xleftarrow{\pi_1} B \xrightarrow{\pi_2} X_2$ in a category $\mathbf{C}$ is called *jointly mono* if for all $\mathbf{C}$-arrows h, h' with codomain B it satisfies: if $\pi_1 \circ h = \pi_1 \circ h'$ and $\pi_2 \circ h = \pi_2 \circ h'$ then $h = h'$. We sometimes write the above span as (B, π_1, π_2), leaving codomains implicit. If $\mathbf{C}$ has products, then (B, π_1, π_2) is a jointly mono span if and only if the pairing $\langle \pi_1, \pi_2 \rangle : B \to X_1 \times X_2$ is monic.

The collection of jointly mono spans between two objects $X_1, X_2 \in \mathbf{C}$ can be ordered as follows: $(B, \pi_1, \pi_2) \leq (B', \pi'_1, \pi'_2)$ if there exists a (necessarily monic) map $k : B \to B'$ such that $\pi_i = \pi'_i \circ k$. If $(B, \pi_1, \pi_2) \leq (B', \pi'_1, \pi'_2)$ *and* $(B', \pi'_1, \pi'_2) \leq (B, \pi_1, \pi_2)$, then the two spans must be isomorphic. We write $\mathbf{Rel}(X_1, X_2)$ for the poset of jointly mono spans between X_1 and X_2 up to isomorphism.

Image factorisations and regular epis We will also need a generalisation of image factorisation. A category $\mathbf{C}$ is said to have $(\mathcal{E}, \mathcal{M})$-*factorisations* for some classes $\mathcal{E}$ and $\mathcal{M}$ of $\mathbf{C}$-morphisms, if every morphism $f \in \mathbf{C}$ factorises as $f = m \circ e$ with $e \in \mathcal{E}$ and $m \in \mathcal{M}$. We say that $\mathbf{C}$ has an $(\mathcal{E}, \mathcal{M})$-*factorisation system* [2, Def. 14.1] if moreover both $\mathcal{E}$ and $\mathcal{M}$ are closed under composition, and whenever $g \circ e = m \circ f$, with $e \in \mathcal{E}$ and $m \in \mathcal{M}$, there exists a unique diagonal fill-in d such that $f = d \circ e$ and $g = m \circ d$.

An epi e is *regular* if it is a coequalizer. In a variety, the regular epis are precisely the surjective morphisms. **Set**, **Pos**, **Top**, **Vec**, **Stone** all have a $(\mathcal{R}eg\mathcal{E}pi, \mathcal{M}ono)$-factorisation system.

3 Logic-induced bisimulations

We are now ready to define our logic-induced notion of bisimulation. Throughout this section, we assume we are given a dual adjunction $\mathsf{P} : \mathbf{C} \rightleftarrows \mathbf{A} : \mathsf{S}$, an endofunctor T on $\mathbf{C}$, and a logic (L, ρ) for T-coalgebras. Moreover, we assume that **C** has pullbacks and, in addition, that **A** has pullbacks or **C** has pushouts. Both conditions hold in our examples. In particular, if **A** is variety of algebras then pullbacks exist and are computed as in **Set**.

3.1 Definition and first examples

The basic ingredient for the definition of ρ-bisimulation is the notion of a *dual span*: A jointly mono span $X_1 \xleftarrow{\pi_1} B \xrightarrow{\pi_2} X_2$ in **C** is mapped by P to a cospan $\mathsf{P}X_1 \xrightarrow{\mathsf{P}\pi_1} \mathsf{P}B \xleftarrow{\mathsf{P}\pi_2} \mathsf{P}X_2$ in **A**. Taking its pullback we obtain a jointly mono span in **A**, which we denote by $(\overline{B}, \overline{\pi}_1, \overline{\pi}_2)$ and refer to as the *dual span* of (B, π_1, π_2). In case **C** has pushouts, dual spans exist because dual adjoints send pushouts to pullbacks. In the classic case where $\mathsf{P} = \mathcal{Q}_{\mathsf{BA}} \colon \mathbf{Set} \to \mathbf{BA}$ maps a set to its Boolean algebra of subsets, the dual span $(\overline{B}, \overline{\pi}_1, \overline{\pi}_2)$ consists of *B-coherent pairs (of subsets of X_1 and X_2)*, used in the definition of Λ-bisimulation [4], neighbourhood bisimulation [14], and conditional bisimulation [5]. We proceed to the definition of a ρ-bisimulation.

Definition 3.1 Let $\gamma_1 \colon X_1 \to \mathsf{T}X_1$ and $\gamma_2 \colon X_2 \to \mathsf{T}X_2$ be T-coalgebras. A jointly mono span $X_1 \xleftarrow{\pi_1} B \xrightarrow{\pi_2} X_2$ is a *ρ-bisimulation between γ_1 and γ_2* if

$$\mathsf{P}\pi_1 \circ \gamma_1^* \circ \mathsf{L}\overline{\pi}_1 = \mathsf{P}\pi_2 \circ \gamma_2^* \circ \mathsf{L}\overline{\pi}_2, \tag{1}$$

Definition 3.1 is structural in the sense that it is defined in terms of the coalgebra structure and the one-step modal semantics ρ (via the complex algebras γ_i^*). In particular, it does not refer to the set of all formulas nor to the initial L-algebra. Equation (1) provides a coherence condition that can be checked in concrete settings. We provide examples below. First, we give a more conceptual characterisation in terms of dual spans.

Proposition 3.2 *A jointly mono span (B, π_1, π_2) is a ρ-bisimulation between γ_1 and γ_2 iff its dual span $(\overline{B}, \overline{\pi}_1, \overline{\pi}_2)$ is a congruence between γ_1^* and γ_2^*.*

Proof. ($\Rightarrow$) Equation (1) says that the outer shell of the diagram on the right commutes. The universal property of the pullback $\overline{B}$ then yields a morphism $\beta : \mathsf{L}\overline{B} \to \overline{B}$ such that all squares in (2) commute. ($\Leftarrow$) The existence of such a β implies commutativity of (the outer shell of) the diagram. □

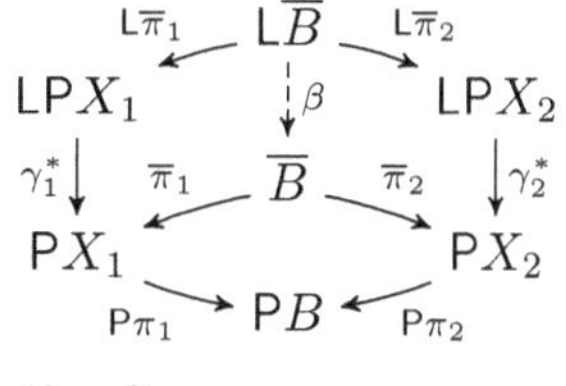

(2)

We instantiate the definition for some of the examples of Section 2.

Example 3.3 Consider the setting of Example 2.5 where **A** is a variety and (L, ρ) is given by predicate liftings and axioms. If **C** is concrete, then (B, π_1, π_2) is a ρ-bisimulation if for all $(x_1, x_2) \in B$, $\lambda \in \Lambda$ and $(a_1, a_2) \in \overline{B} \subseteq \mathsf{P}X_1 \times \mathsf{P}X_2$, we have:

$$\gamma_1(x_1) \in \lambda_{X_1}(a_1) \quad \text{iff} \quad \gamma_2(x_2) \in \lambda_{X_2}(a_2).$$

The notion of a ρ-bisimulation thus generalises that of a Λ-bisimulation from [4,7], where Λ denotes a collection of (open) predicate liftings. Examples 3.4 and 3.5 below are instances hereof.

Example 3.4 In the setting of positive modal logic from Exm. 2.6, a ρ-bisimulation between (X_1, γ_1) and (X_2, γ_2) is a subspace $B \subseteq X_1 \times X_2$ with projections $\pi_i : B \to X_i$ satisfying for all $(x_1, x_2) \in B$ and all B-coherent pairs of opens $(a_1, a_2) \in \Omega X_1 \times \Omega X_2$:

$$\gamma_1(x_1) \subseteq a_1 \text{ iff } \gamma_2(x_2) \subseteq a_2 \quad \text{and} \quad \gamma_1(x_1) \cap a_1 \neq \emptyset \text{ iff } \gamma_2(x_2) \cap a_2 \neq \emptyset.$$

Example 3.5 In the setting of modal vector logic from Exm. 2.7, jointly mono spans are linear subspaces, and the dual span of (B, π_1, π_2) consists of those pairs of $\mathbb{k}$-valued, linear predicates $(h_1, h_2) \in X_1^\vee \times X_2^\vee$ such that $(x_1, x_2) \in B$ implies $h_1(x_1) = h_2(x_2)$. Unravelling the definitions shows that a linear subspace (B, π_1, π_2) of $X_1 \times X_2$ is a ρ-bisimulation between (X_1, γ_1) and (X_2, γ_2), if for all $(x_1, x_2) \in B$, we have $[\![p]\!]_\gamma(x_1) = [\![p]\!]_\gamma(x_2)$, and:

$$\text{if } x_1 \xrightarrow{a} y_1 \text{ and } x_2 \xrightarrow{a} y_2, \text{ then } h_1(y_1) = h_2(y_2) \text{ for all } (h_1, h_2) \in \overline{B}.$$

We say that a span (B, π_1, π_2) between (X_1, γ_1) and (X_2, γ_2) is *truth preserving* if $\mathrm{th}_{\gamma_1} \circ \pi_1 = \mathrm{th}_{\gamma_2} \circ \pi_2$. If $\mathbf{C}$ is concrete, this means that if $(x_1, x_2) \in B$ then x_1 and x_2 have the same theory, i.e., satisfy the same formulas. As desired, ρ-bisimulations are truth-preserving:

Proposition 3.6 *If $X_1 \xleftarrow{\pi_1} B \xrightarrow{\pi_2} X_2$ is a ρ-bisimulation between T-coalgebras (X_1, γ_1) and (X_2, γ_2), then $\mathrm{th}_{\gamma_1} \circ \pi_1 = \mathrm{th}_{\gamma_2} \circ \pi_2$.*

Proof. Let $\beta : \mathsf{L}\overline{B} \to \overline{B}$ be given as in (2), and let $h_\beta : \Phi \to \overline{B}$ be the unique morphism from the initial L-algebra. By construction of β, $\overline{\pi}_i : (\overline{B}, \beta) \to (\mathsf{P}X_i, \gamma_i^*)$ are L-algebra morphisms. By uniqueness of initial morphisms, $[\![-]\!]_{\gamma_i} = \overline{\pi}_i \circ h_\beta$, and hence $\mathsf{S}[\![-]\!]_{\gamma_i} = \mathsf{S}h_\beta \circ \mathsf{S}\overline{\pi}_i$, $i = 1, 2$. Combining this with $\mathsf{S}\overline{\pi}_1 \circ \mathsf{SP}\pi_1 = \mathsf{S}\overline{\pi}_2 \circ \mathsf{SP}\pi_2$ (obtained by applying S to the pullback square of $(\overline{B}, \overline{\pi}_1, \overline{\pi}_2)$), it follows that $\mathsf{S}[\![\cdot]\!]_{\gamma_1} \circ \mathsf{SP}\pi_1 = \mathsf{S}[\![\cdot]\!]_{\gamma_2} \circ \mathsf{SP}\pi_2$. Recall that the theory map is the adjoint of the semantic map, i.e., $\mathrm{th}_{\gamma_i} = \mathsf{S}[\![-]\!]_{\gamma_i} \circ \eta^{\mathbf{C}}_{X_i}$ where $\eta^{\mathbf{C}} : \mathrm{Id}_{\mathbf{C}} \to \mathsf{SP}$ is a unit of the logical connection $\mathsf{P} : \mathbf{C} \rightleftarrows \mathbf{A} : \mathsf{S}$. Together with naturality of $\eta^{\mathbf{C}}$, it now follows that:

$$\begin{aligned} \mathrm{th}_{\gamma_1} \circ \pi_1 = \mathsf{S}[\![\cdot]\!]_{\gamma_1} \circ \eta^{\mathbf{C}}_{X_1} \circ \pi_1 &= \mathsf{S}[\![\cdot]\!]_{\gamma_1} \circ \mathsf{SP}\pi_1 \circ \eta^{\mathbf{C}}_B \\ = \mathsf{S}[\![\cdot]\!]_{\gamma_2} \circ \mathsf{SP}\pi_2 \circ \eta^{\mathbf{C}}_B = \mathsf{S}[\![\cdot]\!]_{\gamma_2} \circ \eta^{\mathbf{C}}_{X_2} \circ \pi_2 &= \mathrm{th}_{\gamma_2} \circ \pi_2 \end{aligned}$$

□

3.2 Lattice structure and composition of ρ-bisimulations

In the remainder of Section 3 we assume that $\mathbf{C}$ is finitely complete and well-powered, hence $\mathbf{Rel}(X_1, X_2)$ is simply the poset of subobjects of $X_1 \times X_2$. Besides, assume that $\mathbf{C}$ has an $(\mathcal{E}, \mathcal{M})$-factorisation system with $\mathcal{M} = \mathcal{M}ono$.

It is well known that bisimulations for $\mathbf{Set}$-based coalgebras are closed under composition if and only if the coalgebra functor preserves weak pullbacks [27]. We know from [4, Exm. 3.3] that Λ-bisimulations do not always compose, even for weak pullback-preserving functors, so the same failure occurs for ρ-bisimulations (cf. Exm. 3.3). However, in special cases we *can* compose.

The composition of two jointly mono spans (B, π_1, π_2) in $\mathbf{Rel}(X_1, X_2)$ and (B', π_2', π_3) in $\mathbf{Rel}(X_2, X_3)$ is given as follows: The pullback (C, c_1, c_3) of π_2 and

π'_2 yields projections $\pi_i \circ c_i : C \to X_i$, and we define $B \circ B'$ via the $(\mathcal{E}, \mathcal{M}ono)$-factorisation of $\langle \pi_1 \circ c_1, \pi_3 \circ c_3 \rangle : C \to X_1 \times X_3$ as $C \twoheadrightarrow B \circ B' \hookrightarrow X_1 \times X_3$.

Call a ρ-bisimulation *full* if both projections are split epi, that is, they have a section. For **Set**-based coalgebras this means that the projections are surjective, i.e., each state in (X_1, γ_1) is ρ-bisimilar to some state in (X_2, γ_2), and vice versa.

Lemma 3.7 *Pullbacks preserve split epimorphisms.*

Lemma 3.8 *Let (f', v) be a pullback of (w, f). If w, f are regular epic and v, f' are split epic then (w, f) is a pushout of (f', v).*

Lemma 3.9 *Let $X_1 \xleftarrow{\zeta_1} S \xrightarrow{\zeta_2} X_2$ and $X_1 \xleftarrow{\pi_1} B \xrightarrow{\pi_2} X_2$ be spans between* T-*coalgebra (X_1, γ_1) and (X_2, γ_2) and suppose $e : S \to B$ is an epi such that $\zeta_i = \pi_i \circ e$. Then (S, ζ_1, ζ_2) satisfies* (1) *if and only if (B, π_1, π_2) does.*

Now we can show that full bisimulations compose.

Proposition 3.10 *Full bisimulations are closed under composition.*

Proof. By Lemma 3.9 it suffices to show that $(C, \pi_1 c_1, \pi_3 c_3)$ satisfies the ρ-bisimulation condition. Since all the π_i are split epic, so are c_1 and c_3 (cf. Lemma 3.7). According to Lemma 3.8 this implies that the square is also a pushout. Therefore (4) below is a pullback, while (1), (2) and (3) are pullbacks by definition. It follows that the outer square is a pullback.

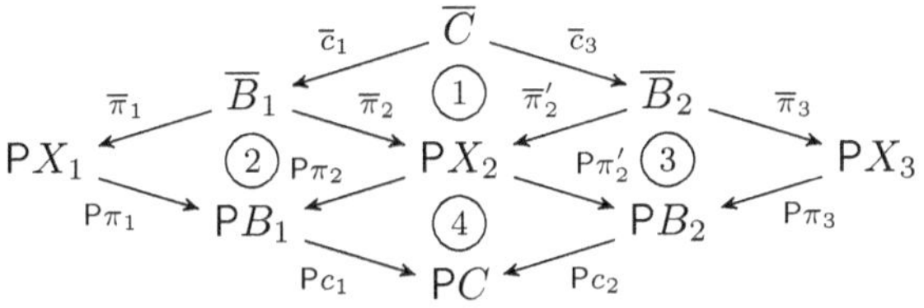

(3)

As a consequence $(\overline{\pi}_1 \overline{c}_1, \overline{\pi}_3 \overline{c}_3)$ is jointly monic. Furthermore, using the fact that B_1 and B_2 are ρ-bisimulations, a straightforward computation shows that $(C, \pi_1 c_1, \pi_3 c_3)$ is a ρ-bisimulation. □

Another well-known result for bisimulations on **Set**-coalgebras is that they form a complete lattice [27]. We now show that, provided **C** has all coproducts, this also holds for ρ-bisimulations. Recall that the empty coproduct $\coprod \emptyset =: \mathbf{0}$ is an initial object, i.e., for all $C \in \mathbf{C}$ there is a unique morphism $!_C : \mathbf{0} \to C$.

Definition 3.11 The *join* of a family $(B_i, \pi_{i,1}, \pi_{i,2})$, $i \in I$, in $\mathbf{Rel}(X_1, X_2)$, is the jointly mono span $\bigcup_{i \in I} B_i$ that arises from the factorisation

$$\coprod_i B_i \twoheadrightarrow \bigcup_i B_i \hookrightarrow X_1 \times X_2 \qquad \left(\text{composite: } \coprod_i \langle \pi_{i,1}, \pi_{i,2} \rangle\right)$$

The *bottom element* (I, ι_1, ι_2) in $\mathbf{Rel}(X_1, X_2)$ is defined by the factorisation of the initial morphism: $\mathbf{0} \twoheadrightarrow I \xhookrightarrow{\langle \iota_1, \iota_2 \rangle} X_1 \times X_2$.

Indeed, $\bigcup_i B_i$ is an upper bound in $\mathbf{Rel}(X_1, X_2)$. Suppose $(B_i, \pi_{i,1}, \pi_{i,2}) \leq (S, s_1, s_2)$ for all i, then there are $t_i : B_i \to S$ such that $\pi_{i,j} = s_j \circ t_i$. From the coproduct we get $t : \coprod_{i \in I} B_i \to S$ and this makes the diagram on the right commute. The factorisation system now gives a diagonal $d : \bigcup_{i \in I} B_i \to S$ witnessing that S is bigger than $\bigcup_{i \in I} B_i$ in $\mathbf{Rel}(X_1, X_2)$.

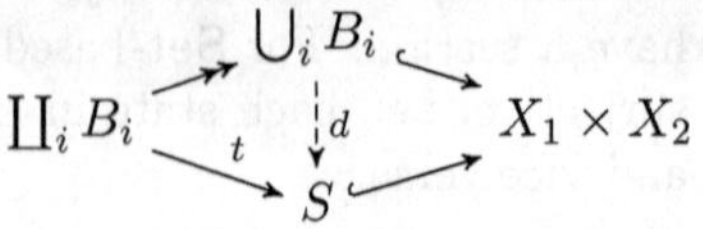

Proposition 3.12 *If $\mathbf{C}$ has an $(\mathcal{E}, \mathcal{M}ono)$-factorisation system, binary products and all coproducts, then the poset $\mathbf{Rel}(X_1, X_2)$ is a complete join-semilattice with join $\bigcup$ and bottom element (I, ι_1, ι_2).*

Proof. Commutativity and associativity of the join follows from the fact that coproducts are commutative and associative. For idempotency note that for every (B, π_1, π_2) in $\mathbf{Rel}(X_1, X_2)$ we have an $(\mathcal{E}, \mathcal{M}ono)$-factorisation $B + B \xrightarrow{\nabla} B \hookrightarrow X_1 \times X_2$, where ∇ is the codiagonal, so $B \cup B = B$.

Next, we show that (I, ι_1, ι_2) is the bottom element in $(\mathbf{Rel}(X_1, X_2), \cup)$. That is, for all (B, π_1, π_2) in $\mathbf{Rel}(X_1, X_2)$, $B \cup I$ is isomorphic to B. By the definition of a coproduct, $B \xrightarrow{\cong} 0 + B \xrightarrow{!_I + \mathrm{id}_B} I + B$ commutes, where i is the inclusion that arises from the coproduct, and because $\mathcal{E}$ is closed under composition, the map $i : B \to I + B$ is in $\mathcal{E}$. By definition of the join, the following commutes:

$$B \hookrightarrow I + B \xrightarrow{\langle \pi_1, \pi_2 \rangle} I \cup B \hookrightarrow X_1 \times X_2$$
$$[\langle \iota_1, \iota_2 \rangle, \langle \pi_1, \pi_2 \rangle]$$

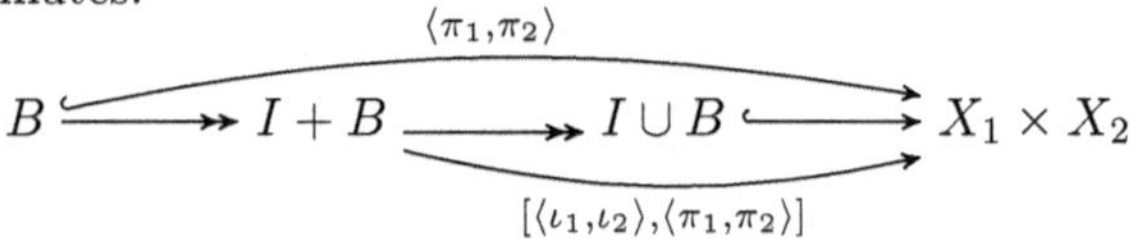

Since factorisation systems are unique up to isomorphism, we get an isomorphism $B \cong B \cup I$. □

We define ρ-*bisimilarity* as the join of all ρ-bisimulations in $\mathbf{Rel}(X_1, X_2)$. The following proposition tells us that ρ-bisimilarity is itself a ρ-bisimulation. Given two T-coalgebras (X_1, γ_1) and (X_2, γ_2), we denote by ρ-$\mathbf{Bis}(\gamma_1, \gamma_2)$ the sub-poset of $\mathbf{Rel}(X_1, X_2)$ of ρ-bisimulations between (X_1, γ_1) and (X_2, γ_2).

Proposition 3.13 *Under the assumptions of Proposition 3.12, ρ-$\mathbf{Bis}(\gamma_1, \gamma_2)$ is closed under joins and bottom element in $\mathbf{Rel}(X_1, X_2)$. Consequently, ρ-$\mathbf{Bis}(\gamma_1, \gamma_2)$ is a complete join semilattice, and hence also a complete lattice.*

While ρ-$\mathbf{Bis}(\gamma_1, \gamma_2)$ is a complete sub-semilattice of $\mathbf{Rel}(X_1, X_2)$, it need not inherit the meets. This resembles the situation for Kripke bisimulations, which are generally not closed under intersections.

Example 3.14 The categories $\mathbf{Set}$, $\mathbf{Top}$ and $\mathbf{Vec}_k$ from Examples 3.4, 3.3 and 3.5 are well-powered, complete and cocomplete, and as mentioned in Section 2 have a $(\mathcal{R}eg\mathcal{E}pi, \mathcal{M}ono)$-factorisation system. Hence ρ-bisimulations for positive modal logic, modal vector logic and coalgebraic geometric logic form complete lattices, and we recover the similar result for Λ-bisimulations in [4].

3.3 Characterisation via relation lifting

Another property of bisimulations for **Set**-coalgebras is that they can be characterised via relation lifting (see e.g. [29, Sec. 2.2]), and that bisimilarity on a coalgebra (X, γ) is a greatest fixpoint of a monotone operator on the lattice of relations $\mathcal{P}(X \times X)$. In this subsection and the following, we show that these results generalise to ρ-bisimulations.

Given X_1, X_2 in **C**, we shall define a monotone map

$$\mathsf{T}^\rho : \mathbf{Rel}(X_1, X_2) \to \mathbf{Rel}(\mathsf{T}X_1, \mathsf{T}X_2)$$

which lifts (B, π_1, π_2) in $\mathbf{Rel}(X_1, X_2)$ to $(\mathsf{T}^\rho B, \mathsf{T}^\rho \pi_1, \mathsf{T}^\rho \pi_2)$ in $\mathbf{Rel}(\mathsf{T}X_1, \mathsf{T}X_2)$. In order to do so, consider the composition, for $i = 1, 2$,

$$\sigma_i : \mathsf{T}X_i \xrightarrow{\eta^\mathbf{C}_{\mathsf{T}X_i}} \mathsf{SPT}X_i \xrightarrow{\mathsf{S}\rho_{X_i}} \mathsf{SLP}X_i \xrightarrow{\mathsf{SL}\overline{\pi}_i} \mathsf{SL}\overline{B} \quad (4)$$

For a concrete example of σ_i, see Example 3.17 below.

Definition 3.15 Given (B, π_1, π_2) in $\mathbf{Rel}(X_1, X_2)$, we define $\mathsf{T}^\rho(B, \pi_1, \pi_2) = (\mathsf{T}^\rho B, \mathsf{T}^\rho \pi_1, \mathsf{T}^\rho \pi_2)$ as the pullback of $\mathsf{T}X_1 \xrightarrow{\sigma_1} \mathsf{SL}\overline{B} \xleftarrow{\sigma_2} \mathsf{T}X_2$.

Observe that $(\mathsf{T}^\rho B, \mathsf{T}^\rho \pi_1, \mathsf{T}^\rho \pi_2)$ is a jointly mono span because it is a pullback. Monotonicity of T^ρ follows from unravelling the definitions. We can now characterise ρ-bisimulations as in [16] using the relation lifting T^ρ.

Theorem 3.16 *A jointly mono span (B, π_1, π_2) between two T-coalgebras (X_1, γ_1) and (X_2, γ_2) is a ρ-bisimulation if and only if there exists a morphism $\delta : B \to \mathsf{T}^\rho B$ in **C** making diagram (5) commute.*

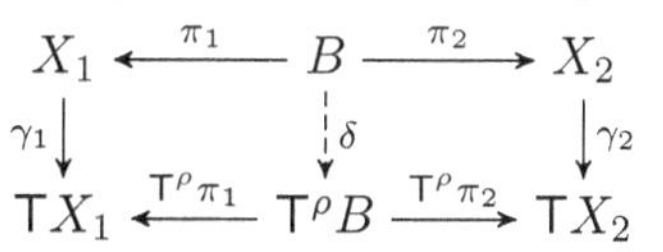

$$\begin{array}{ccccc} X_1 & \xleftarrow{\pi_1} & B & \xrightarrow{\pi_2} & X_2 \\ {\scriptstyle \gamma_1}\downarrow & & {\scriptstyle \delta}\downarrow & & \downarrow{\scriptstyle \gamma_2} \\ \mathsf{T}X_1 & \xleftarrow{\mathsf{T}^\rho \pi_1} & \mathsf{T}^\rho B & \xrightarrow{\mathsf{T}^\rho \pi_2} & \mathsf{T}X_2 \end{array} \quad (5)$$

Proof. *If δ exists, then B is a ρ-bisimulation.* Suppose such a δ exists. In order to show that B is a ρ-bisimulation, we need to show that the outer shell of the left diagram below commutes. Recall that $\eta^\mathbf{C}$ and $\eta^\mathbf{A}$ are the units of the dual adjunction $\mathsf{P} : \mathbf{C} \rightleftarrows \mathbf{A} : \mathsf{S}$.

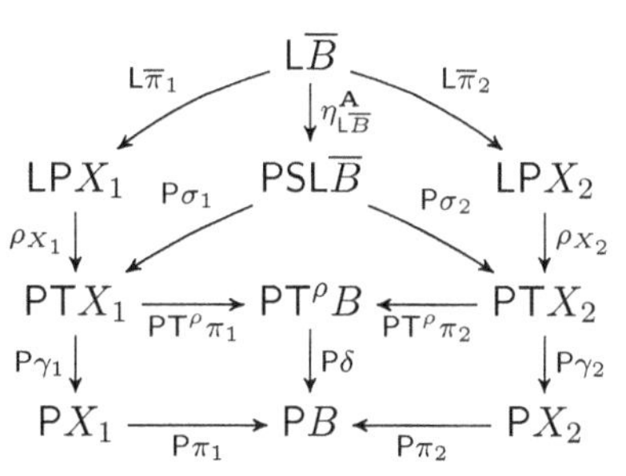

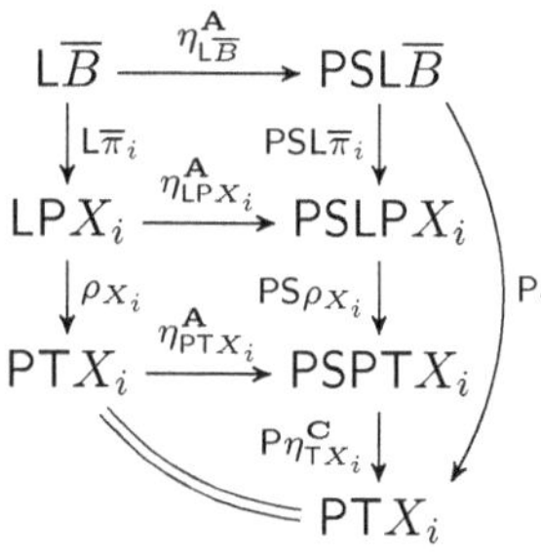

$$(6)$$

Commutativity of the bottom two squares follows from applying P to the diagram in (5). The middle square commutes because of the definition of $\mathsf{T}^\rho B$. The top two squares commute because they are the outer shell of the right diagram in (6). In (6), the right square commutes by definition of σ_i (Eq. 4).

The other two squares commute by naturality of $\eta^{\mathbf{A}}$ and the lower triangle is a triangle identity of the dual adjunction. Therefore the outer shell commutes.

If B is a ρ-bisimulation, then we can find δ. Suppose (B, π_1, π_2) is a ρ-bisimulation. If we can prove that $\sigma_1 \circ \gamma_1 \circ \pi_1 = \sigma_2 \circ \gamma_2 \circ \pi_2$ then we obtain δ as the mediating map induced by the pullback which defines $\mathsf{T}^{\rho}B$, as shown in the diagram (7).

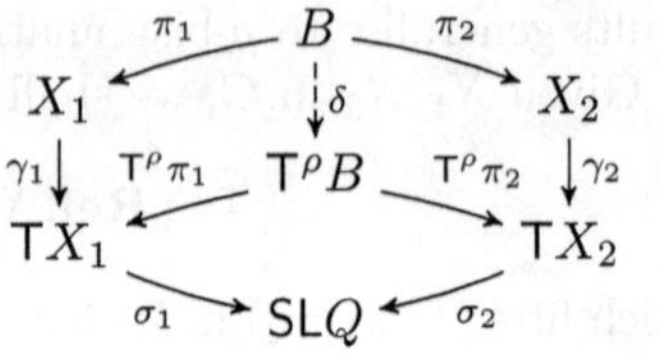

(7)

We claim that the following diagram commutes. Since its outer shell is the same as the outer shell of (7), this proves the proposition. So consider:

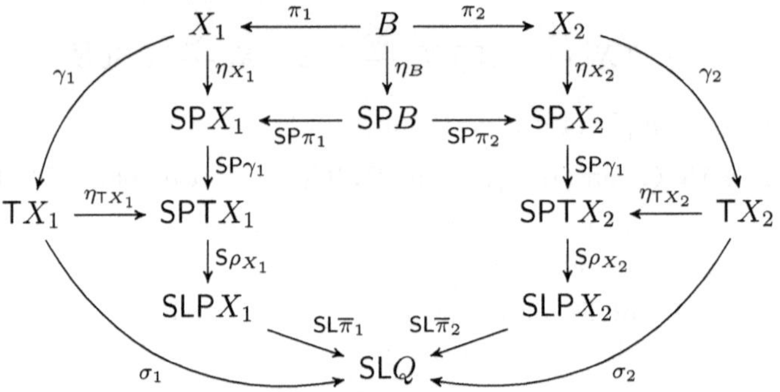

Commutativity of the middle part follows from the fact that B is a ρ-bisimulation. The four top squares commute because η is a natural transformation. The two remaining squares commute by definition of σ_i. □

We work out the explicit description of T^{ρ} in a special case:

Example 3.17 Suppose we work with the classic dual adjunction $\mathcal{Q}_{\mathsf{BA}} : \mathbf{Set} \rightleftarrows \mathbf{BA} : \mathsf{Uf}$, T is an endofunctor on $\mathbf{Set}$, and the logic (L, ρ) is given by predicate liftings and axioms (cf. Example 2.5). Then the type of σ_i is $\mathsf{T}X_i \to \mathsf{UfL}\overline{B}$ and the ultrafilter $\sigma_i(t_i)$ is determined by the elements of the form $\underline{\lambda}(a_1, a_2)$ it contains, where $\lambda \in \Lambda$ and $(a_1, a_2) \in \overline{B}$. Therefore the action of T^{ρ} on (B, π_1, π_2) is given by

$$\mathsf{T}^{\rho}B = \{(t_1, t_2) \in \mathsf{T}X_1 \times \mathsf{T}X_2 \mid \forall \lambda \in \Lambda \text{ and } B\text{-coherent } (a_1, a_2) \\ \text{we have } t_1 \in \lambda_{X_1}(a_1) \Leftrightarrow t_2 \in \lambda_{X_2}(a_2)\}.$$

Informally, these are the pairs in $\mathsf{T}X_1 \times \mathsf{T}X_2$ that cannot be distinguished by lifted B-coherent predicates.

3.4 Characterisation as a (post)fixpoint

As for **Set**-coalgebras, given a relation lifting of T and T-coalgebras (X_1, γ_1), (X_2, γ_2), we can define a map $\mathsf{T}^{\rho}_{\gamma_1,\gamma_2} : \mathbf{Rel}(X_1, X_2) \to \mathbf{Rel}(X_1, X_2)$ by, essentially, taking inverse images under the γ_i. This is a relational version of a predicate transformer on a coalgebra.

Definition 3.18 Given T-coalgebras (X_1, γ_1) and (X_2, γ_2), define $\mathsf{T}^\rho_{\gamma_1,\gamma_2}(B, \pi_1, \pi_1) = (\mathsf{T}^\rho_{\gamma_1,\gamma_2}B, \mathsf{T}^\rho_{\gamma_1,\gamma_2}\pi_1, \mathsf{T}^\rho_{\gamma_1,\gamma_2}\pi_2) \in \mathbf{Rel}(X_1, X_2)$ via the pullback on the right. This is well defined because pullbacks are jointly mono spans.

$$\begin{array}{ccccc}
\mathsf{T}^\rho_{\gamma_1,\gamma_2}B & & \xrightarrow{\mathsf{T}^\rho_{\gamma_1,\gamma_2}\pi_2} & & X_2 \\
 & & & & \downarrow{\scriptstyle \gamma_2} \\
\downarrow{\scriptstyle \mathsf{T}^\rho_{\gamma_1,\gamma_2}\pi_1} & & & & \mathsf{T}X_2 \\
 & & & & \downarrow{\scriptstyle \sigma_2} \\
X_1 & \xrightarrow{\gamma_1} & \mathsf{T}X_1 & \xrightarrow{\sigma_1} & \mathsf{SL}\overline{B}
\end{array}$$

Lemma 3.19 *The map* $\mathsf{T}^\rho_{\gamma_1,\gamma_2} : \mathbf{Rel}(X_1, X_2) \to \mathbf{Rel}(X_1, X_2)$ *is monotone.*

Proof. If $(B, \pi_1, \pi_2) \leq (B', \pi'_1, \pi'_2)$ then there exists an $m : B \to B'$ such that $\pi_i = \pi'_i \circ m$. As a consequence the pullback $\overline{B}'$ is a cone for $\overline{B}$ and we have a mediating map $k : \overline{B}' \to \overline{B}$ satisfying $\overline{\pi}'_i = \overline{\pi}_i \circ k$. Unravelling the definitions reveals that $\mathsf{T}^\rho_{\gamma_1,\gamma_2}B$ with its projections is a cone for $\mathsf{T}^\rho_{\gamma_1,\gamma_2}B'$, hence there is a (unique) map $t : \mathsf{T}^\rho_{\gamma_1,\gamma_2}B \to \mathsf{T}^\rho_{\gamma_1,\gamma_2}B'$ such that $\mathsf{T}^\rho_{\gamma_1,\gamma_2}\pi_i = \mathsf{T}^\rho_{\gamma_1,\gamma_2}\pi'_i \circ t$ which witnesses that $\mathsf{T}^\rho_{\gamma_1,\gamma_2}(B, \pi_1, \pi_2) \leq \mathsf{T}^\rho_{\gamma_1,\gamma_2}(B', \pi'_1, \pi'_2)$. □

As announced, ρ-bisimulations are precisely the post-fixpoints of $\mathsf{T}^\rho_{\gamma_1,\gamma_2}$.

Theorem 3.20 *A relation* $X_1 \xleftarrow{\pi_1} B \xrightarrow{\pi_2} X_2$ *is a* ρ*-bisimulation between* (X_1, γ_1) *and* (X_2, γ_2) *if and only if* $(B, \pi_1, \pi_2) \leq \mathsf{T}^\rho_{\gamma_1,\gamma_2}(B, \pi_1, \pi_2)$.

Proof. If (B, π_1, π_2) is a ρ-bisimulation, then by Theorem 3.16 there is a map $\beta \colon B \to \mathsf{T}^\rho B$ such that diagram (5) commutes. We then get a map $\beta' \colon B \to \mathsf{T}^\rho_{\gamma_1,\gamma_2}B$ from the pullback property of $\mathsf{T}^\rho_{\gamma_1,\gamma_2}B$. Conversely, given $\beta' \colon B \to \mathsf{T}^\rho_{\gamma_1,\gamma_2}B$, we obtain $\beta \colon B \to \mathsf{T}^\rho B$ from the pullback property of $\mathsf{T}^\rho B$. □

Monotonicity of $\mathsf{T}^\rho_{\gamma_1,\gamma_2}$ and the Knaster-Tarski fixpoint theorem imply:

Corollary 3.21 *Under the assumptions of Prop. 3.12,* $\mathsf{T}^\rho_{\gamma_1,\gamma_2}$ *has a greatest fixpoint, and this greatest fixpoint is* ρ*-bisimilarity.*

Example 3.22 We return to the classic setting of Example 3.17. Let (B, π_1, π_2) be a relation between T-coalgebras (X_1, γ) and (X_2, γ_2). Then

$$\mathsf{T}^\rho_{\gamma_1,\gamma_2}B = \{(x_1, x_2) \in X_1 \times X_2 \mid (\gamma_1(x_1), \gamma_2(x_2)) \in \mathsf{T}^\rho B\}.$$

Informally, $\mathsf{T}^\rho_{\gamma_1,\gamma_2}B$ consists of all pairs of worlds whose one-step behaviours are indistiguishable by lifted B-coherent predicates.

4 Distinguishing power

In this section we compare the distinguishing power of ρ-bisimulations with that of other semantic equivalence notions and logical equivalence. We make the same assumptions here as at the start of Section 3. Given a cospan $(X_1, \gamma_1) \to (Y, \delta) \leftarrow (X_2, \gamma_2)$ in $\mathbf{Coalg}(\mathsf{T})$, we call (Y, δ) a *congruence* (of T-coalgebras).

4.1 Comparison with known equivalence notions

We briefly recall three coalgebraic equivalence notions, in descending order of distinguishing power. For more details, see e.g. [4, Def. 3.9].

Definition 4.1 A jointly mono span (B, π_1, π_2) between (X_1, γ_1) and (X_2, γ_2) is a: (i) T-*bisimulation* if there is $t \colon B \to \mathsf{T}B$ such that the π_i become coalgebra morphisms; (ii) *precocongruence* if its pushout $\widehat{\pi}_1 : X_1 \to \widehat{B} \leftarrow X_2 : \widehat{\pi}_1$ can be turned into a congruence between (X_1, γ_1) and (X_2, γ_2), more precisely, if there is $t\colon \widehat{B} \to \mathsf{T}\widehat{B}$ such that $\widehat{\pi}_1$ and $\widehat{\pi}_2$ become coalgebra morphisms; (iii) *behavioural equivalence* if it is a pullback in $\mathbf{C}$ of some cospan $(X_1, \gamma_1) \to (Y, \delta) \leftarrow (X_2, \gamma_2)$.

When T preserves weak pullbacks, all three notions coincide (when considering associated "bisimilarity" notions), but in general, they may differ. In particular, expressive logics can generally only capture behavioural equivalence [14]. The next proposition can be proved in the same way as [4, Prop. 3.10].

Proposition 4.2 *(i) Every* T*-bisimulation is a* ρ*-bisimulation. (ii) Every precocongruence is a* ρ*-bisimulation.*

The converse direction requires additional assumptions.

Proposition 4.3 *If* $\mathbf{C}$ *has pushouts,* P *is faithful, and either (i)* ρ *is pointwise epic or (ii)* $\rho^\flat$ *is pointwise monic and* T *preserves monos, then every* ρ*-bisimulation is a precocongruence. If, in addition,* T *preserves weak pullbacks, then* ρ*-bisimilarity coincides with all three notions in Def. 4.1.*

Proof. Suppose $X_1 \xleftarrow{\pi_1} B \xrightarrow{\pi_2} X_2$ is a ρ-bisimulation with pushout $(\widehat{B}, \widehat{\pi}_1, \widehat{\pi}_2)$ be the pushout. We need to find a coalgebra structure $\zeta : \widehat{B} \to \mathsf{T}\widehat{B}$ which turns $\widehat{\pi}_1$ and $\widehat{\pi}_2$ into coalgebra morphisms. It suffices to show that $\mathsf{T}\widehat{\pi}_1 \circ \gamma_1 \circ \pi_1 = \mathsf{T}\widehat{\pi}_2 \circ \gamma_2 \circ \pi_2$, because then the universal property of the pushout yields the desired ζ. If P is faithful and ρ is pointwise epic, then it suffices to prove that $\mathsf{P}\pi_1 \circ \mathsf{P}\gamma_1 \circ \mathsf{PT}\widehat{\pi}_1 \circ \rho_{\widehat{B}} = \mathsf{P}\pi_2 \circ \mathsf{P}\gamma_2 \circ \mathsf{PT}\widehat{\pi}_2 \circ \rho_{\widehat{B}}$. This follows from the left diagram below, where the outer shell commutes because (B, π_1, π_2) is a ρ-bisimulation and the top two squares commute by naturality of ρ.

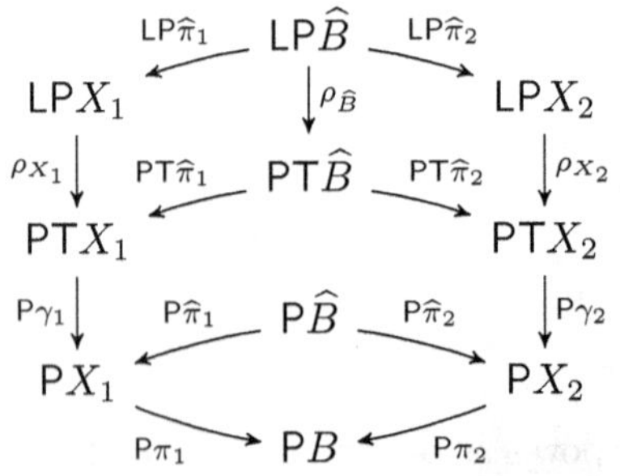

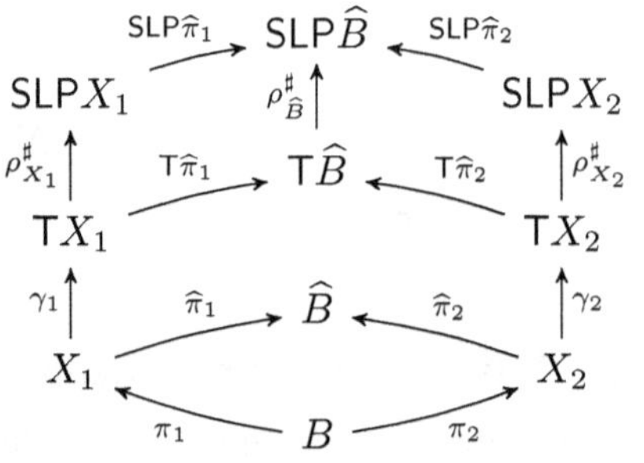

Alternatively, suppose P is faithful (hence $\eta^{\mathbf{C}}\colon \mathrm{Id}_{\mathbf{C}} \to \mathsf{SP}$ is pointwise monic), $\rho^\flat$ is pointwise monic and T preserves monos. Then the transpose $\rho^\sharp_{\widehat{B}} : \mathsf{T}\widehat{B} \to \mathsf{SLP}\widehat{B}$ of $\rho_{\widehat{B}}$ is monic, because $\rho^\sharp_{\widehat{B}} = \mathsf{S}\rho_{\widehat{B}} \circ \eta^{\mathbf{C}}_{\mathsf{T}\widehat{B}} = \rho^\flat_{\mathsf{P}\widehat{B}} \circ \mathsf{T}\eta^{\mathbf{C}}_{\widehat{B}}$, so it suffices to show that $\rho^\sharp_{\widehat{B}} \circ \mathsf{T}\widehat{\pi}_1 \circ \gamma_1 \circ \pi_1 = \rho^\sharp_{\widehat{B}} \circ \mathsf{T}\widehat{\pi}_2 \circ \gamma_2 \circ \pi_2$. But this follows from transposing the left diagram above, which yields the diagram to the right.

When T preserves weak pullbacks, T-bisimilarity coincides with behavioural equivalence [27], and hence also with the largest precocongruence and ρ-bisimilarity. □

We note that condition (ii) in Proposition 4.3 entails that (L, ρ) is expressive [19, Thm. 4.2], i.e., that logical equivalence implies behavioural equivalence. In our abstract setting, *logical equivalence* with respect to (L, ρ) is the kernel pair (B, π, π') of the theory map $\mathrm{th} : X \to \mathsf{S}\Phi$. Hence, (L, ρ) is *expressive* if (B, π, π') is below a behavioural equivalence in $\mathbf{Rel}(X, X)$.

4.2 Hennessy-Milner type theorem

We now prove a partial converse to Proposition 3.6 (truth-preservation). We show that under certain conditions logical equivalence implies ρ-bisimilarity.

Theorem 4.4 *Let $\mathbf{C}' \rightleftarrows \mathbf{A}'$ be the dual equivalence induced by the dual adjunction $\mathbf{C} \rightleftarrows \mathbf{A}$. Suppose that*

- *$\mathbf{C}$ has $(\mathcal{RegEpi}, \mathcal{Mono})$-factorisations for morphisms with domain $\in \mathbf{C}'$;*
- *$\mathbf{C}'$ is closed under regular epimorphic images;*
- *S is faithful and L preserves epis.*

Then for all T-coalgebras (X, γ) with $X \in \mathbf{C}'$, logical equivalence, i.e., the kernel pair (B, π, π') of $\mathrm{th} : X \to \mathsf{S}\Phi$, is a ρ-bisimulation.

Proof. In order to prove that (B, π, π') is a ρ-bisimulation, we need to show that the outer shell of the diagram on the right commutes. From B being the kernel pair of th we have that $(\Phi, [\![\cdot]\!]_\gamma, [\![\cdot]\!]_\gamma)$ is a cone for the pullback $\overline{B}$. Hence we get a morphism $h : \Phi \to \overline{B}$ such that the triangles left and right of h commute, and it is easy to see that all the inner squares and triangles in the diagram on the right commute. Thus, in order to show that the outer shell commutes, it suffices to show that $\mathsf{L}h$ is epic. By the assumption that L preserves epis, it suffices to show that $h : \Phi \to \overline{B}$ is epic. Let $m \circ e$ be the $(\mathcal{RegEpi}, \mathcal{Mono})$-factorisation of th. Then the left diagram in (8) commutes. Since m is monic the upper square is a pullback, and by [2, Proposition 11.33] it is also a pushout. As a consequence, the lower square in the right diagram of (8), obtained from dualising the left one, is a pullback.

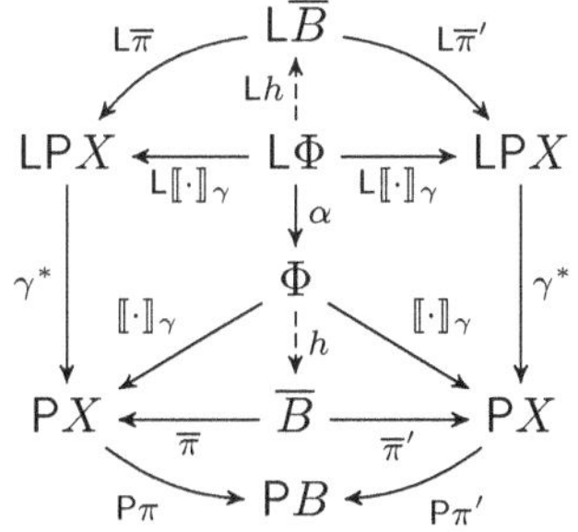

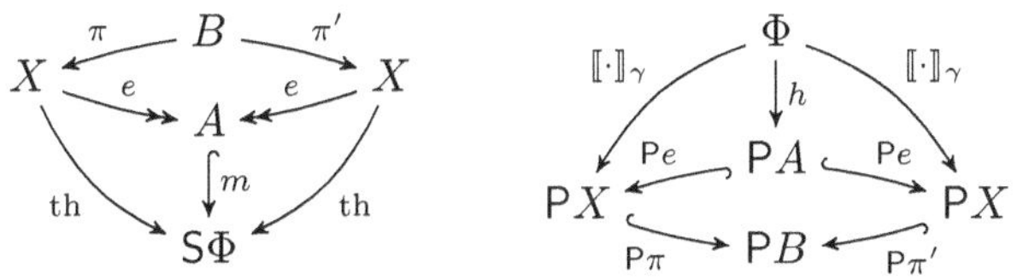

(8)

Here h denotes the adjoint transpose of m. Applying S to h gives the morphism $\mathsf{S}h : \mathsf{SP}A \to \mathsf{S}\Phi$ which by assumption is isomorphic to m (because $A \cong \mathsf{SP}A$). Since S is faithful and m is monic, h and therefore $\mathsf{L}h$ are epic. □

Example 4.5 In the classic case, $\mathbf{Set} \rightleftarrows \mathbf{BA}$ restricts to the full duality between finite sets and finite Boolean algebras. $\mathbf{Set}$ has $(\mathcal{RegEpi}, \mathcal{Mono})$-

factorisations [2, Exm. 14.2(2)]. In **Set** and **BA**, all epis are regular and coincide with surjections [2,6], and finite sets are closed under surjective images. The ultrafilter functor S is faithful. If the logic functor L is given by predicate liftings and relations, then by [23, Remark 4.10] it preserves regular epis, and since all epis are regular, L preserves epis. Applying Theorem 4.4, we recover [4, Theorem 4.5], and thereby all examples given there. In particular, taking (L, ρ) to be Hennessy-Milner logic (Example 2.4), then we recover from Theorem 4.4 that over finite labelled transition systems, logical equivalence implies ρ-bisimilarity for Hennessy-Milner logic.

Remark 4.6 For positive modal logic from Examples 2.6 and 3.4, we have not been able to show that the logic functor $\mathsf{N} : \mathbf{DL} \to \mathbf{DL}$ preserves epis.

Example 4.7 We return to modal vector logic from Examples 2.7 and 3.5. The dual adjunction $\mathbf{Vec}_{\mathbb{k}} \rightleftarrows \mathbf{Vec}_{\mathbb{k}}$ restricts to the well-known self-duality of finite-dimensional vector spaces $\mathbf{FinVec}_{\mathbb{k}}$. The category $\mathbf{Vec}_{\mathbb{k}}$ has $(\mathcal{RegEpi}, \mathcal{Mono})$-factorisations [2, Ex. 14.2] and the regular epis in both $\mathbf{Vec}_{\mathbb{k}}$ and $\mathbf{FinVec}_{\mathbb{k}}$ are the surjections [2, Exm. 7.72]. Moreover, the surjective image of a finite-dimensional vector space is again finite-dimensional, and the functor $(-)^{\vee}$ is faithful. Finally, since L is generated by predicate liftings and axioms it preserves surjections, so we can apply Theorem 4.4 to conclude that logical equivalence and ρ-bisimilarity coincide on W-coalgebras state-spaces in $\mathbf{FinVec}_{\mathbb{k}}$.

Example 4.8 An example where logical equivalence does not imply ρ-bisimilarity is given by trace logic for labelled transitions systems (Example 2.3). The conditions for Theorem 4.4 hold for trace logic, but the induced dual equivalence is in this case trivial, i.e., $\mathbf{C}'$ and $\mathbf{A}'$ are the empty category, hence Theorem 4.4 does not tell us anything.

4.3 Invariance under translations

In this section we assume that $\mathbf{C}$ has pushouts. The example of Hennessy-Milner logic (Exm. 2.4) and trace logic (Exm. 2.3 and 4.8) is a situation where one logic is a reduct of the other. This can be considered a special case of translating a logic into another. We will show under which conditions ρ-bisimilarity is preserved under translations. To make this formal, we first generalise [22, Def 4.1].

Definition 4.9 Assume we are given a "triangle situation" as in diagram (9(a)) such that $\mathsf{P} = \mathsf{U}\mathsf{P}'$, and we have modal semantics $\rho' : \mathsf{L}'\mathsf{P}' \to \mathsf{P}'\mathsf{T}$ and $\rho : \mathsf{L}\mathsf{P} \to \mathsf{P}\mathsf{T}$. A *translation* from (L', ρ') to (L, ρ) is a natural transformation $\tau : \mathsf{L}\mathsf{P} \to \mathsf{U}\mathsf{L}'\mathsf{P}'$ such that $\rho = \mathsf{U}\rho' \circ \tau$, see diagram (9(b)).

$$
\begin{array}{ccc}
\mathsf{T} \circlearrowright \mathbf{C} \begin{array}{l} \xrightarrow{\mathsf{P}'} \mathbf{A}' \circlearrowleft \mathsf{L}' \\ \qquad\quad \mathsf{F}\uparrow \dashv \downarrow\mathsf{U} \\ \xrightarrow{\mathsf{P}} \mathbf{A} \circlearrowleft \mathsf{L} \end{array}
&
\begin{array}{ccc} \mathsf{LP} & \xrightarrow{\tau} & \mathsf{UL'P'} \\ \downarrow{\scriptstyle \rho} & & \downarrow{\scriptstyle \mathsf{U}\rho'} \\ \mathsf{PT} & \xrightarrow{=} & \mathsf{UP'T} \end{array}
&
\begin{array}{ccc} \mathsf{FLP} & \xrightarrow{\tau^{\sharp}} & \mathsf{L'P'} \\ {\scriptstyle \mathsf{F}\rho}\downarrow & & \downarrow{\scriptstyle \rho'} \\ \mathsf{FPT} & \xrightarrow{\varepsilon_{\mathsf{P'T}}} & \mathsf{P'T} \end{array}
\\
\text{(a)} & \text{(b)} & \text{(c)}
\end{array}
\qquad (9)
$$

In (c), ε is the counit of $\mathsf{F} \dashv \mathsf{U}$ (which is adjoint to the identity) because $\mathsf{P} = \mathsf{UP}'$, and $\tau^\sharp$ is the $(\mathsf{F} \dashv \mathsf{U})$-adjoint of τ. In the presence of such a translation, every ρ'-bisimulation is also a ρ-bisimulation. We leave the straightforward proof to the reader. A sufficient condition for the converse is that the transpose $\tau^\sharp$ of τ is epic, see diagram (9(c)). Note that due to the adjunction $\mathsf{F} \dashv \mathsf{U}$, diagram (b) commutes if and only if (c) does. Intuitively, $\tau^\sharp : \mathsf{FLP} \to \mathsf{L'P'}$ being epic formalises that every modality in L' is a propositional combination of a modal formula of L.

Proposition 4.10 *Suppose that $\tau^\sharp$ is pointwise epic. Then every ρ-bisimulation is a ρ'-bisimulation.*

Proof. Commutativity of the outer shell of the following diagram will prove that $X_1 \xleftarrow{\pi_1} B \xrightarrow{\pi_2} X_2$ is a ρ'-bisimulation:

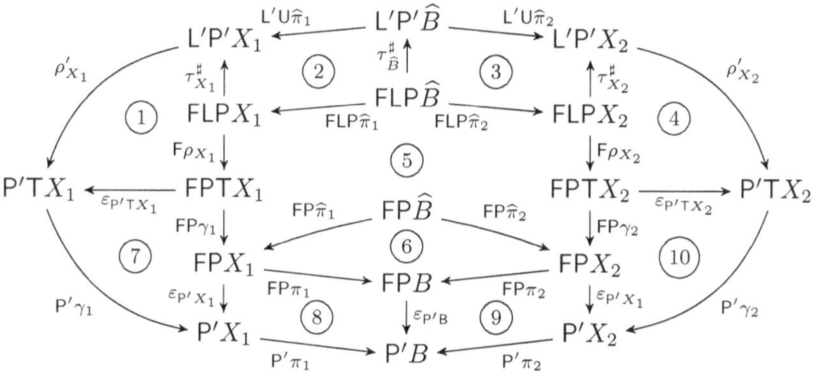

Cells 1 and 4 commute by diagram (c) in (9), and cells 2 and 3 by naturality of $\tau^\sharp$. Commutativity of 5 and 6 together follows from applying F to the diagram witnessing the fact that $X_1 \longleftarrow B \longrightarrow X_2$ is a ρ-bisimulation. Commutativity of the remaining cells follows from the naturality of the counit ε. □

In the setting of Examples 2.4, 2.5 and 3.3, where $\mathbf{A}'$ is a variety of algebras and the logic (L, ρ) is given by predicate liftings and axioms, we can consider the special case of (9) where (L, ρ) is the "modal reduct" of (L', ρ').

Example 4.11 Suppose $\mathbf{A}$ is a variety of algebras with free-forgetful adjunction $\mathsf{F} \dashv \mathsf{U}$. Let (L, ρ) be a logic for T-coalgebras given by a collection Λ of predicate liftings and axioms (Example 2.5). Then we can define $\mathsf{P}_0 = \mathsf{U} \circ \mathsf{P}$, which has dual adjoint $\mathsf{S}_0 = \mathsf{SF}$, where S is the dual adjoint of P. Define the logic functor $\mathsf{L}_0 : \mathbf{Set} \to \mathbf{Set}$ by $\mathsf{L}_0 X = \{\underline{\lambda}_0(a_1, \ldots, a_n) \mid \lambda \in \Lambda a_i \in X\}$ and $\mathsf{L}_0 f(\underline{\lambda}_0(a_1, \ldots, a_n)) = \underline{\lambda}_0(fa_1, \ldots, fa_n)$. Define

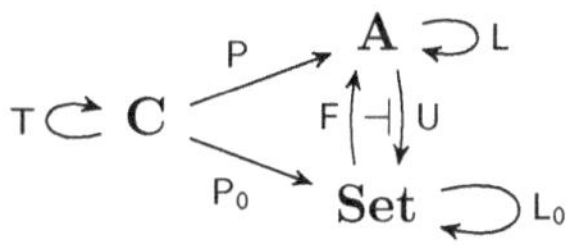

$\tau : \mathsf{L}_0\mathsf{P}_0 \to \mathsf{ULP}$ by $\tau_X(\underline{\lambda}_0(a_1, \ldots, a_n)) = \underline{\lambda}(a_1, \ldots, a_n) \in \mathsf{ULP}X$. The logic (L, ρ) gives rise to the logic (L_0, ρ_0), where $\rho_0 = \mathsf{U}\rho \circ \tau : \mathsf{L}_0\mathsf{P}_0 \to \mathsf{P}_0\mathsf{T}$. Then τ is a translation. One can verify that, in this situation, $\tau^\sharp$ is pointwise epic.

Therefore a jointly mono span $X_1 \longleftarrow B \longrightarrow X_2$ in $\mathbf{C}$ is a ρ-bisimulation if and only if it is a ρ_0-bisimulation. Hence it suffices to look at the underlying sets when verifying whether a jointly mono span is a ρ-bisimulation.

Hennessy-Milner logic and trace logic are a specific instance of Example 4.11.

Example 4.12 Recall trace logic (Exm. 2.3) and Henessy-Milner logic (Exm. 2.4) for LTSs. We gave a concrete definition of the Hennessy-Milner logic functor L^{hm} in Example 2.4. It can be equivalently defined as follows in terms of the free-forgetful adjunction $\mathsf{F} \dashv \mathsf{U}$ of $\mathbf{BA}$ over $\mathbf{Set}$. The join-preservation of the diamond modalities is encoded in L^{hm} by factoring the free-forgetful adjunction $\mathsf{F} \dashv \mathsf{U}$ via the category $\mathbf{JSL}$ of join-semilattices as shown in diagram (10(d)). The logic functor L^{hm} is then defined as $\mathsf{L}^{\mathsf{hm}} = \mathsf{F}\mathsf{L}^{\mathsf{tr}}\mathsf{U} = \mathsf{F}_{\mathsf{BA}}\mathsf{F}_{\mathsf{JSL}}\mathsf{L}^{\mathsf{tr}}\mathsf{U}_{\mathsf{JSL}}\mathsf{U}_{\mathsf{BA}}$. Letting $\mathsf{U} = \mathsf{U}_{\mathsf{JSL}}\mathsf{U}_{\mathsf{BA}}$ and $\mathsf{F} = \mathsf{F}_{\mathsf{BA}}\mathsf{F}_{\mathsf{JSL}}$, we have $\mathcal{Q} = \mathsf{U}\mathcal{Q}_{\mathsf{BA}}$, and in particular $\mathsf{L}^{\mathsf{hm}} = \mathsf{F}\mathsf{L}^{\mathsf{tr}}\mathsf{U}$. The semantics of Hennessy-Milner logic coincides with taking ρ^{hm} to be the adjoint of ρ^{tr}, so trace logic is the modal reduct of Hennessy-Milner logic.

Defining τ as in Exm. 4.11 implies that $\tau^\sharp \colon \mathsf{F}\mathsf{L}^{\mathsf{tr}}\mathcal{Q} \to \mathsf{L}^{\mathsf{hm}}\mathcal{Q}_{\mathsf{BA}}$ is the identity (this follows from $\mathsf{L}^{\mathsf{hm}} = \mathsf{F}\mathsf{L}^{\mathsf{tr}}\mathsf{U}$). Concretely, $\tau^\sharp$ is identity, because formulas of Hennessy-Milner logic are precisely the Boolean combinations of trace logic formulas. Hence, in particular, $\tau^\sharp$ is epic. It now follows from Proposition 4.10 that a ρ^{tr}-bisimulation is a ρ^{hm}-bisimulation (and the converse also holds).

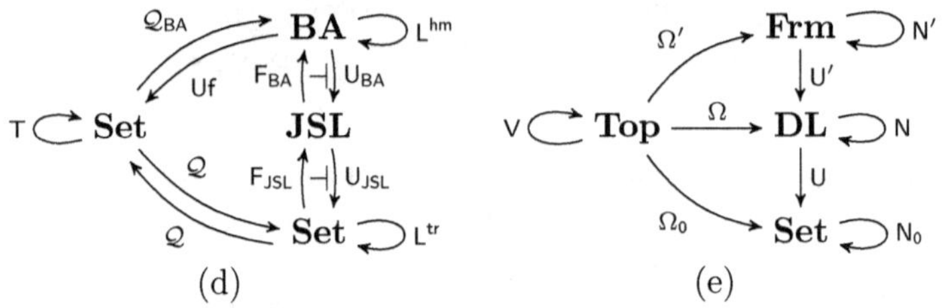

(10)

Example 4.13 We squeeze the topological semantics for positive modal logic from Example 2.6 between two other logics with varying base logics, see diagram (10(e)). Here $\mathbf{Frm}$ is the category of frames and $\Omega' : \mathbf{Top} \to \mathbf{Frm}$ is the functor that sends a topological space to its frame of opens. Let $\mathsf{N}' : \mathbf{Frm} \to \mathbf{Frm}$ be the functor given as in [17, Section III4.3] (known also as the *Vietoris locale*) and define $\rho' : \mathsf{N}'\Omega' \to \Omega'\mathsf{V}$ on generators by $\Box a \mapsto \boxed{\Box} a$ and $\Diamond a \mapsto \boxed{\Diamond} a$. The translation $\tau : \mathsf{N}\Omega \to \mathsf{U}'\mathsf{N}'\Omega'$ given by $\Box a \mapsto \Box a$ and $\Diamond a \mapsto \Diamond a$ is such that $\tau^\sharp$ is epic, thus satisfies the assumptions of Proposition 4.10.

The bottom triangle is an instance of Exm. 4.11. We conclude that a jointly mono span between V-coalgebras (X_1, γ_1) and (X_2, γ_2) is a ρ-bisimulation if and only if it is a ρ'-bisimulation if and only if it is a ρ_0-bisimulation.

5 Conclusion

Our main question was whether we can characterise logical equivalence for (possibly non-expressive) coalgebraic logics by a notion of bisimulation. Towards this goal, we generalised the logic-induced bisimulations in [4] for coalgebraic logics for **Set**-coalgebras to coalgebraic logics parameterised by a dual adjunction. We identified sufficient conditions for when logical equivalence coincides with logic-induced bisimilarity (Thm. 4.4). These are conditions on the categories in the dual adjunction, and *not* on the natural transformation ρ defining (the semantics of) the logic. In particular, we do not require the logic to be expressive.

We found that the distinguishing power of ρ-bisimulations depends on the modalities of the language but not on the propositional connectives. More generally, we showed that certain translations between logics preserve ρ-bisimilarity (Prop. 4.10). Furthermore, as in the expressivity result of [19], ρ-bisimilarity agrees with behavioural equivalence if the mate of ρ is pointwise monic (Prop. 4.3). However, Example 4.12 shows that this is not a necessary condition which raises the question whether one can characterise, purely in terms of ρ, when ρ-bisimilarity coincides with behavioural equivalence.

There are many other avenues for further research. When is a congruence on complex algebras induced by a ρ-bisimulation? Can we drop in Theorem 4.4 the restriction to the subcategory if T is finitary? Can we take quotients with respect to (the largest) ρ-bisimulation on a T-coalgebra?

Moreover, the definition of ρ-bisimulation has a natural generalisation to the order-enriched setting. This gives rise to *ρ-simulations*. Can one prove an ordered Hennessy-Milner theorem where "logical inequality" is recognised by ρ-simulations? Since this question naturally falls into the realm of order-enriched category theory, we will also seek a generalisation to the quantale-enriched setting, accounting for metric versions of simulation.

References

[1] Abramsky, S., *Domain theory in logical form*, Ann. Pure Appl. Logic **51** (1991).

[2] Adámek, J., H. Herrlich and G. E. Strecker, *Abstract and concrete categories* (1990).

[3] Adámek, J., J. Rosický and E. M. Vitale, *Algebraic theories: a categorical introduction to general algebra* (2010).

[4] Bakhtiari, Z. and H. H. Hansen, *Bisimulation for weakly expressive coalgebraic modal logics*, in: *7th Conference on Algebra and Coalgebra in Computer Science, CALCO*, Leibniz International Proceedings in Informatics (LIPIcs), 2017.

[5] Baltag, A. and G. Cinà, *Bisimulation for conditional modalities*, Stud. Log. **106** (2018).

[6] Banaschewski, B., *On the strong amalgamation of Boolean algebras*, Algebra Univers **63** (2010), pp. 235–238.

[7] Bezhanishvili, N., J. de Groot and Y. Venema, *Coalgebraic Geometric Logic*, in: *8th Conference on Algebra and Coalgebra in Computer Science (CALCO 2019)*, Leibniz International Proceedings in Informatics (LIPIcs), 2019.

[8] Bonchi, F., M. M. Bonsangue, M. Boreale, J. Rutten and A. Silva, *A coalgebraic perspective on linear weighted automata*, Information and Computation **211** (2012).

[9] Bonsangue, M. M. and A. Kurz, *Duality for logics of transition systems*, in: *Foundations of Software Science and Computational Structures (FoSSaCS'05)*, LNCS **3441** (2005).

[10] Celani, S. and R. Jansana, *Priestley duality, a Sahlqvist theorem and a Goldblatt-Thomason theorem for positive modal logic*, Logic Journal of the IGPL **7** (1999).

[11] Dunn, J. M., *Positive modal logic*, Studia Logica **55** (1995).

[12] Fan, J., Y. Wang and H. van Ditmarsch, *Almost necessary*, in: *Advances in Modal Logic (AiML'14)* (2014), pp. 178–196.
[13] Goldblatt, R., *Metamathematics of modal logic I*, Rep. on Math. Log. **6** (1976).
[14] Hansen, H. H., C. Kupke and E. Pacuit, *Neighbourhood structures: bisimilarity and basic model theory*, Log. Meth. Comp. Sci. **5** (2009).
[15] Hennessy, M. and R. Milner, *Algebraic laws for nondeterminism and concurrency*, Journal of the ACM (1985).
[16] Hermida, C. and B. Jacobs, *Structural induction and coinduction in a fibrational setting*, Inform. and Comput. **145** (1998).
[17] Johnstone, P. T., "Stone Spaces," Cambridge University Press, 1982.
[18] Kapulkin, K., A. Kurz and J. Velebil, *Expressiveness of positive coalgebraic logic*, Advances in Modal Logic **9** (2014).
[19] Klin, B., *Coalgebraic modal logic beyond sets*, in: *Mathematical Foundations of Programming Semantics (MFPS XXIII)*, ENTCS **173**, 2007, pp. 177 – 201.
[20] Kupke, C., A. Kurz and D. Pattinson, *Algebraic semantics for coalgebraic logics*, in: *Coalgebraic Methods in Computer Science (CMCS'04)*, ENTCS **106**, 2004, pp. 219–241.
[21] Kupke, C. and D. Pattinson, *Coalgebraic semantics of modal logics: An overview*, Theor. Comp. Sci. **412** (2011).
[22] Kurz, A. and R. Leal, *Modalities in the stone age: A comparison of coalgebraic logics*, Theor. Comp. Sci. **430** (2012).
[23] Kurz, A. and J. Rosický, *Strongly complete logics for coalgebras*, Log. Meth. Comp. Sci. **8** (2012).
[24] Moss, L., *Coalgebraic logic*, Annals of Pure and Applied Logic **96** (1999).
[25] Pattinson, D., *Coalgebraic modal logic: soundness, completeness and decidability of local consequence*, Theor. Comp. Sci. **309** (2003).
[26] Pattinson, D., *Expressive logics for coalgebras via terminal sequence induction*, Notre Dame Journal of Formal Logic **45** (2004).
[27] Rutten, J. J. M. M., *Universal coalgebra: a theory of systems*, Theor. Comp. Sci. **249** (2000).
[28] Schröder, L., *Expressivity of coalgebraic modal logic: The limits and beyond*, Theor. Comp. Sci. **390** (2008).
[29] Staton, S., *Relating coalgebraic notions of bisimulation*, Log. Meth. Comp. Sci. **7** (2011).

On the axiomatisation of common knowledge

Andreas Herzig [1]

IRIT, CNRS, Univ. Toulouse 3, France

Elise Perrotin

IRIT, Univ. Toulouse 3, France

Abstract

Standard axiomatisations of the logic of common knowledge contain the greatest fixed-point axiom schema. While such an inductive principle matches our intuitions in the context of temporal logics, it is not immediately obvious in an epistemic context. We propose an axiom schema that we believe to be more intuitive. It says that if it is common knowledge that everybody knows whether φ then it is common knowledge whether φ. Our schema is sound for KT-based common knowledge and moreover complete for S5-based common knowledge. In contrast, it is unsound for logics without the T-axiom. Our axiom schema directly leads to a simple and intuitive axiomatisation of the 'common knowledge whether' operator.

Keywords: Common knowledge, axiomatisation, induction axiom, greatest fixed-point axiom.

1 Introduction

The standard axiomatisations of the logic of common knowledge contain the induction axiom schema, alias greatest fixed-point axiom

$$\texttt{GFP} \quad \mathbf{C}(\varphi \to \mathbf{E}\varphi) \to (\varphi \to \mathbf{C}\varphi),$$

where $\mathbf{C}$ stands for "it is common knowledge that" and $\mathbf{E}$ stands for "everybody knows that" [15,13,9]. An alternative axiomatisation [10,6] has the induction rule

$$\texttt{RGFP} \quad \text{from } \varphi \to \mathbf{E}(\psi \wedge \varphi), \text{ infer } \varphi \to \mathbf{C}\psi.$$

In the proof theory literature there exist sequent system counterparts of these principles, e.g. in [1,11]. Similar axioms and rules were used to axiomatise common belief [3,17].

Such inductive principles are common in temporal logics, where they mirror induction on the natural numbers. There, the reading is obvious and the intuitive meaning is clear. More generally, we can make sense of such principles

[1] http://orcid.org/0000-0003-0833-2782, https://www.irit.fr/~Andreas.Herzig

when interpreted on well-founded orderings. However, at least to the present authors the meaning of the induction axiom schema is less obvious when the modal operator is that of common knowledge, and one might even wonder whether it is a reasonable principle at all. To witness the difficulty to find an intuitive reading to the above principles, consider the reading of RGFP that is given in the introductory chapter of the Handbook of Epistemic Logic:

> "If it is the case that φ is 'self-evident', in the sense that if it is true, then everyone knows it, and, in addition, if φ is true, then everyone knows ψ, we can show by induction that if φ is true, then so is $\mathbf{E}^k(\psi \land \varphi)$ for all k." [22]

The explanations in the standard texts resort to concepts such as 'φ indicates to every agent that ψ is true' [16], 'φ is evident' [18], 'it is public that φ is true' [24], or 'φ is a common basis implying shared belief in ψ' [8]. With these understandings RGFP can be read "if φ is public and indicates ψ to everybody then truth of φ implies that ψ is common knowledge". The formalisation of these supplementary concepts however introduces further complications, see e.g. [5] for a tentative to settle the logic of 'indicates'.

Can the above inductive principles be replaced by principles with more intuitive appeal? We here propose a new axiom schema:

$$\texttt{GFPO} \quad \mathbf{C}(\mathbf{E}\varphi \lor \mathbf{E}\neg\varphi) \to (\mathbf{C}\varphi \lor \mathbf{C}\neg\varphi).$$

Unlike GFP and RGFP, it can be read straightforwardly: "if it is common knowledge that everybody knows whether φ then it is common knowledge whether φ"; or alternatively: "common knowledge that the status of φ is shared knowledge implies that the status of φ is common knowledge". In the present paper we focus on KT- and S5-based common knowledge. We prove the following results:

(i) GFPO is a theorem if the logic of individual knowledge is at least KT;

(ii) GFPO is equivalent to GFP if the logic of individual knowledge is S5;

(iii) GFPO leads to a simple and intuitive axiomatisation of S5-based 'common knowledge whether';

(iv) GFPO is specific to knowledge and fails for belief: contrarily to the status of the standard induction principles, its status differs depending on whether the context is epistemic or doxastic.

Most papers in the literature start by introducing the Kripke semantics and then discuss the axiomatisation of its validities. In contrast, the present paper is semantic-free: all proofs are done syntactically via the axioms and inference rules of the respective systems.

For the sake of simplicity we here only consider shared and common knowledge of the set of all agents. Everything however straightforwardly generalises to common knowledge of arbitrary sets of agents.

The paper is organised as follows. In the next two sections we give the background: two axiom systems for individual knowledge and shared knowledge, KT and S5 (Section 2), and the two standard axiom systems for common

CPC	axiomatics of classical propositional calculus
RN($\mathbf{K}_i$)	from φ, infer $\mathbf{K}_i\varphi$
K($\mathbf{K}_i$)	$\mathbf{K}_i(\varphi \to \psi) \to (\mathbf{K}_i\varphi \to \mathbf{K}_i\psi)$
T($\mathbf{K}_i$)	$\mathbf{K}_i\varphi \to \varphi$
*5($\mathbf{K}_i$)	$\neg\mathbf{K}_i\varphi \to \mathbf{K}_i\neg\mathbf{K}_i\varphi$
Def($\mathbf{E}$)	$\mathbf{E}\varphi \leftrightarrow \bigwedge_{i \in Agt} \mathbf{K}_i\varphi$

Table 1
Axiomatisation of KT (without *5($\mathbf{K}_i$)) and S5 (including *5($\mathbf{K}_i$)) individual knowledge and shared knowledge. The axiom that is not part of the KT axiomatics—i.e., axiom *5($\mathbf{K}_i$)—is starred.

knowledge (Section 3), which we syntactically prove to be equivalent. In Section 4 we prove that the S5-based GFPO axiomatics is equivalent to the standard axiomatics. In Section 5 we axiomatise S5-based 'common knowledge whether'. In Section 6 we discuss how completeness for logics of knowledge that are weaker than S5 could be obtained. In Section 7 we show that our new axiom is unintuitive for logics of belief, understood as logics that do not have the T axiom for individual belief. We conclude in Section 8.

2 Background: individual and shared knowledge

Let *Prop* be a countable set of propositional variables with typical elements $p, q, \ldots$ Let *Agt* be a fixed, finite set of agents with typical elements $i, j, \ldots$ The grammar of formulas is

$$\varphi ::= p \mid \neg\varphi \mid \varphi \wedge \varphi \mid \mathbf{K}_i\varphi \mid \mathbf{E}\varphi \mid \mathbf{C}\varphi,$$

where p ranges over *Prop* and i over *Agt*. The formula $\mathbf{K}_i\varphi$ reads "i knows that φ"; $\mathbf{E}\varphi$ reads "everybody knows that φ", or "it is shared knowledge that φ";[2] finally, $\mathbf{C}\varphi$ reads "it is common knowledge that φ".

A logic of common knowledge is based on a logic of the individual knowledge operators $\mathbf{K}_i$ that is situated between S5 and KT, where the latter is the weakest normal modal logic having the truth axiom $\mathbf{K}_i\varphi \to \varphi$. In this paper we only consider S5 and KT individual knowledge. Table 1 recalls the two axiomatistions as well as the axiom Def($\mathbf{E}$) defining shared knowledge. In order to distinguish the axioms and theorems of S5 from those of KT we adopt the convention that formulas that are not theorems of logic KT are prefixed by "*", such as *5($\mathbf{K}_i$) in Table 1.

The operator $\mathbf{E}$ is a normal modal operator: it obeys the modal schema K and the rule of necessitation RN. Moreover it obeys:

$$\texttt{T}(\mathbf{E}) \quad \mathbf{E}\varphi \to \varphi.$$

2 Many authors use the adjective 'mutual' instead of 'shared'. We opted for the latter because some philosophers such as Stephen Schiffer use the terms 'mutual knowledge' and 'mutual belief' [20] in order to refer to common knowledge and common belief (see e.g. [14]).

It is straightforward to prove that the following holds for logics of individual knowledge from KT on:

Proposition 2.1 *The formula*

$$\texttt{Def2}(\mathbf{Eif}) \quad (\mathbf{E}\varphi \vee \mathbf{E}\neg\varphi) \leftrightarrow \bigwedge_{i \in Agt}(\mathbf{K}_i\varphi \vee \mathbf{K}_i\neg\varphi)$$

is a theorem of the KT axiomatics.

The name of the equivalence anticipates its use in the axiomatics of 'knowing-whether' in Section 5.

Despite the fact that the shared knowledge operator **E** neither obeys positive nor negative introspection, it obeys the B axiom:

Proposition 2.2 *The formula*

$$*\texttt{B}(\mathbf{E}) \quad \neg\varphi \rightarrow \mathbf{E}\neg\mathbf{E}\varphi$$

is a theorem of the S5 axiomatics.

Proof. The proof is simple, but we give it here as we did not find it in the literature:

(i)	$\varphi \rightarrow \mathbf{K}_i\neg\mathbf{K}_i\neg\varphi$	$*\texttt{B}(\mathbf{K}_i)$
(ii)	$\neg\mathbf{K}_i\neg\varphi \rightarrow \neg\mathbf{E}\neg\varphi$	from $\texttt{Def}(\mathbf{E})$
(iii)	$\mathbf{K}_i\neg\mathbf{K}_i\neg\varphi \rightarrow \mathbf{K}_i\neg\mathbf{E}\neg\varphi$	from (ii), $\mathbf{K}_i$ normal
(iv)	$\varphi \rightarrow \mathbf{K}_i\neg\mathbf{E}\neg\varphi$	from (i), (iii)
(v)	$\varphi \rightarrow \mathbf{E}\neg\mathbf{E}\neg\varphi$	from (iv) with $\texttt{Def}(\mathbf{E})$

□

3 Background: two standard axiomatisations of common knowledge

An overview of the different axiomatisations of logics of common knowledge can be found in [17] where the relation between the underlying logic of individual knowledge and the resulting logic of common knowledge is studied in depth. The paper not only considers knowledge, but also belief. As already said above, our new axiom is not appropriate for common belief. Moreover, only two logics of knowledge are in focus in the present section: systems where the logic of $\mathbf{K}_i$ is either KT or S5. (Logics of knowledge between these two are discussed in Section 6.)

In the next two subsections we recall two standard axiomatisations of the logic of common knowledge, one with the induction rule `RGFP` and one with the induction axiom schema `GFP`. We then prove the equivalence of these two axiomatisations.

3.1 With the induction axiom `GFP`

The two axiomatics with the induction axiom schema `GFP` are in Table 2 (left). We distinguish the S5-based from the KT-based axiomatics by starring the supplementary axioms, namely the negative introspection axioms $*\texttt{5}(\mathbf{K}_i)$ and

$*5(\mathbf{C})$. Both axiomatics are due to [9]; others can be found in [15,13]. Such axiomatisations are popular in Dynamic Epistemic Logics [21,23].

It is a standard result in normal modal logics that axiom 4 can be proved from T and 5. In the case of common knowledge, $4(\mathbf{C})$ is already a theorem of the KT-based logic thanks to the induction axiom schema:

Proposition 3.1 *The formula*

$$4(\mathbf{C}) \quad \mathbf{C}\varphi \to \mathbf{CC}\varphi$$

is a theorem of the KT-based GFP *axiomatics.*

Proof.

(i)	$\mathbf{C}(\mathbf{C}\varphi \to \mathbf{EC}\varphi)$	from $\mathtt{FP}'$ and $\mathtt{RN}(\mathbf{C})$
(ii)	$\mathbf{C}(\mathbf{C}\varphi \to \mathbf{EC}\varphi) \to (\mathbf{C}\varphi \to \mathbf{CC}\varphi)$	GFP
(iii)	$\mathbf{C}\varphi \to \mathbf{CC}\varphi$	from (i) and (ii)

□

Proposition 3.2 *Axiom* $*5(\mathbf{C})$ *is redundant in the S5-based* GFP *axiomatics.*

Proof.

(i)	$\neg\mathbf{C}\varphi \to \mathbf{K}_i\neg\mathbf{K}_i\mathbf{C}\varphi$	$*\mathtt{B}(\mathbf{K}_i)$
(ii)	$\mathbf{C}\varphi \to \mathbf{K}_i\mathbf{C}\varphi$	from $\mathtt{FP}'$ and $\mathtt{Def}(\mathbf{E})$
(iii)	$\mathbf{K}_i\neg\mathbf{K}_i\mathbf{C}\varphi \to \mathbf{K}_i\neg\mathbf{C}\varphi$	from (ii), $\mathbf{K}_i$ normal
(iv)	$\neg\mathbf{C}\varphi \to \mathbf{K}_i\neg\mathbf{C}\varphi$	from (i), (iii)
(v)	$\neg\mathbf{C}\varphi \to \mathbf{E}\neg\mathbf{C}\varphi$	from (iv) by $\mathtt{Def}(\mathbf{E})$
(vi)	$\mathbf{C}(\neg\mathbf{C}\varphi \to \mathbf{E}\neg\mathbf{C}\varphi)$	from (v) by $\mathtt{RN}(\mathbf{C})$
(vii)	$\mathbf{C}(\neg\mathbf{C}\varphi \to \mathbf{E}\neg\mathbf{C}\varphi) \to (\neg\mathbf{C}\varphi \to \mathbf{C}\neg\mathbf{C}\varphi)$	GFP
(viii)	$\neg\mathbf{C}\varphi \to \mathbf{C}\neg\mathbf{C}\varphi$	from (vi) and (vii)

□

3.2 With the induction rule RGFP

The two axiomatics with the induction rule RGFP are given in Table 2 (right). They are due to [10,6]; the induction rule can actually be traced back to the analysis of common knowledge in the philosophical literature [24]. Interestingly and contrasting with the GFP axiomatics, the S5 axioms and rules for $\mathbf{C}$ are implicit here:

Proposition 3.3 *The formulas* $\mathtt{K}(\mathbf{C})$, $\mathtt{T}(\mathbf{C})$, $4(\mathbf{C})$, *and* $*5(\mathbf{C})$ *are theorems and the rule* $\mathtt{RN}(\mathbf{C})$ *is derivable in the S5-based* RGFP *axiomatics.*

Proof. The proofs are simple, but we give them here for completeness. $\mathtt{K}(\mathbf{C})$ can be proved by substituting φ by $\mathbf{C}\varphi \wedge \mathbf{C}(\varphi \to \psi)$ in RGFP, using FP and that $\mathbf{E}$ is a normal modal operator. $\mathtt{T}(\mathbf{C})$ can be proved from FP and $\mathtt{T}(\mathbf{E})$. $4(\mathbf{C})$ can be proved by substituting both φ and ψ by $\mathbf{C}\varphi$ in RGFP, using FP and that $\mathbf{E}$ is a normal modal operator. The rule $\mathtt{RN}(\mathbf{C})$ can be derived with RGFP if we

GFP-based axiomatics		RGFP-based axiomatics	
KT	the axiomatics of Table 1	KT	the axiomatics of Table 1
RN(**C**)	from φ, infer $\mathbf{C}\varphi$		
K(**C**)	$\mathbf{C}(\varphi \to \psi) \to (\mathbf{C}\varphi \to \mathbf{C}\psi)$		
T(**C**)	$\mathbf{C}\varphi \to \varphi$		
*5(**C**)	$\neg\mathbf{C}\varphi \to \mathbf{C}\neg\mathbf{C}\varphi$		
FP′	$\mathbf{C}\varphi \to \mathbf{E}\mathbf{C}\varphi$	FP	$\mathbf{C}\varphi \to \mathbf{E}(\varphi \wedge \mathbf{C}\varphi)$
GFP	$\mathbf{C}(\varphi \to \mathbf{E}\varphi) \to (\varphi \to \mathbf{C}\varphi)$	RGFP	from $\varphi \to \mathbf{E}(\psi \wedge \varphi)$, infer $\varphi \to \mathbf{C}\psi$

Table 2
Two axiomatisations of KT-based and S5-based common knowledge: the GFP axiomatics with an induction axiom of [9] (left) and the RGFP axiomatics with an induction rule of [10,6] (right). The principles that are not part of the KT-based axiomatics—i.e., the *5 axioms—are starred.

substitute $\top$ for φ and φ for ψ and use that **E** is a normal modal operator. It is only the proof of *5(**C**) which is a bit longer:

(i) $\varphi \to \mathbf{E}\neg\mathbf{E}\neg\varphi$ — *B(**E**)

(ii) $\mathbf{E}\neg\mathbf{E}\neg\varphi \to \mathbf{E}\neg\mathbf{C}\neg\varphi$ — from FP, **E** normal

(iii) $\mathbf{E}\neg\mathbf{C}\neg\varphi \to \mathbf{E}\mathbf{E}\neg\mathbf{E}\mathbf{C}\neg\varphi$ — from *B(**E**), **E** normal

(iv) $\mathbf{E}\mathbf{E}\neg\mathbf{E}\mathbf{C}\neg\varphi \to \mathbf{E}\mathbf{E}\neg\mathbf{C}\neg\varphi$ — from FP, **E** normal

(v) $\mathbf{E}\neg\mathbf{C}\neg\varphi \to \mathbf{E}(\neg\mathbf{C}\neg\varphi \wedge \mathbf{E}\neg\mathbf{C}\neg\varphi)$ — from (iii), (iv), **E** normal

(vi) $\mathbf{E}\neg\mathbf{C}\neg\varphi \to \mathbf{C}\neg\mathbf{C}\neg\varphi$ — from (v) by RGFP

(vii) $\varphi \to \mathbf{C}\neg\mathbf{C}\neg\varphi$ — from (i), (ii), (vi)

□

3.3 Equivalence of the two axiomatics

The RGFP axiomatics and the GFP axiomatics are both complete for the same semantics (which we do not give here). Therefore all axioms in one system must be derivable in the other, and the inference rules of one system are admissible in the other. We are however not aware of a direct equivalence proof in the respective systems in the literature, so we give it below.[3] We prove the two directions:

(i) in the RGFP axiomatics, K(**C**), T(**C**), *5(**C**), FP′, GFP are theorems and RN(**C**) is derivable;

(ii) in the GFP axiomatics, FP′ is a theorem and RGFP is derivable.

We have already established in Section 3.2 that K(**C**), T(**C**), and *5(**C**) are

[3] The paper by Bucheli et al. [4] establishes that RGFP is derivable from a variant of GFP, $\mathbf{C}(\varphi \to \mathbf{E}\varphi) \to (\mathbf{E}\varphi \to \mathbf{C}\varphi)$ (which they have to choose instead of RGFP because they take K as the logic of individual knowledge). However their proof is indirect, making use of an intermediate system.

	the axiomatics of Table 1
RN(**C**)	from φ, infer $\mathbf{C}\varphi$
K(**C**)	$\mathbf{C}(\varphi \to \psi) \to (\mathbf{C}\varphi \to \mathbf{C}\psi)$
4(**C**)	$\mathbf{C}\varphi \to \mathbf{CC}\varphi$
FP0	$\mathbf{C}\varphi \to \mathbf{E}\varphi$
GFP0	$\mathbf{C}(\mathbf{E}\varphi \lor \mathbf{E}\neg\varphi) \to (\mathbf{C}\varphi \lor \mathbf{C}\neg\varphi)$

Table 3
Alternative axiomatisation of S5 common knowledge: the GFP0 axiomatics.

theorems of the RGFP axiomatics. Second and quite obviously, as **E** is a normal modal operator, we have that FP′ is provable from FP and that, the other way round, FP is provable from FP′ and T(**C**). It remains to prove the equivalence of the induction axiom and the induction rule.

Proposition 3.4 *The induction axiom* GFP *is a theorem of the KT-based* RGFP *axiomatics (and a fortiori of the S5-based* RGFP *axiomatics).*

Proof.

(i) $\mathbf{C}(\varphi \to \mathbf{E}\varphi) \to \mathbf{EC}(\varphi \to \mathbf{E}\varphi)$ — from FP, **E** normal

(ii) $(\mathbf{C}(\varphi \to \mathbf{E}\varphi) \land \varphi) \to \mathbf{E}\varphi$ — from T(**C**)

(iii) $(\mathbf{C}(\varphi \to \mathbf{E}\varphi) \land \varphi) \to (\mathbf{E}\varphi \land \mathbf{EC}(\varphi \to \mathbf{E}\varphi))$ — from (i) and (ii)

(iv) $(\mathbf{C}(\varphi \to \mathbf{E}\varphi) \land \varphi) \to \mathbf{E}((\varphi \land \mathbf{C}(\varphi \to \mathbf{E}\varphi)) \land \varphi)$ — from (iii), **E** normal

(v) $(\mathbf{C}(\varphi \to \mathbf{E}\varphi) \land \varphi) \to \mathbf{C}\varphi$ — from (iv) by RGFP

□

Proposition 3.5 *The induction rule* RGFP *is derivable in the* GFP *axiomatics.*

Proof.

(i) $\varphi \to \mathbf{E}(\psi \land \varphi)$ — hypothesis

(ii) $\mathbf{C}(\psi \land \varphi \to \mathbf{E}(\psi \land \varphi))$ — from (i) by RN(**C**)

(iii) $\mathbf{C}(\psi \land \varphi \to \mathbf{E}(\psi \land \varphi)) \to (\psi \land \varphi \to \mathbf{C}(\psi \land \varphi))$ — GFP

(iv) $\psi \land \varphi \to \mathbf{C}(\psi \land \varphi)$ — from (ii), (iii)

(v) $\varphi \to \psi \land \varphi$ — from (i) by T(**E**)

(vi) $\varphi \to \mathbf{C}\psi$ — from (v), (iv), **C** normal

□

4 An alternative axiomatisation of S5 common knowledge

Table 3 contains a new axiomatics of common knowledge. The main difference w.r.t. the GFP axiomatics is that the induction axiom GFP is replaced by GFP0. A further difference is that our axiomatics explicits 4(**C**), which is a theorem of the GFP and RGFP axiomatics. Finally and thanks to 4(**C**), our version of the fixed-point axiom FP0 is weaker than FP′ (and a fortiori weaker than FP). It is however strong enough to entail T(**C**): $\mathbf{C}\varphi \to \varphi$ (together with Def(**E**) and T($\mathbf{K}_i$)).

Observe that it follows from Proposition 2.1 and the fact that **C** is a normal modal operator that the two axioms

$$\texttt{GFP0}\quad \mathbf{C}(\mathbf{E}\varphi \vee \mathbf{E}\neg\varphi) \to (\mathbf{C}\varphi \vee \mathbf{C}\neg\varphi)$$
$$\texttt{GFP1}\quad \mathbf{C}\bigwedge_{i\in Agt}(\mathbf{K}_i\varphi \vee \mathbf{K}_i\neg\varphi) \to (\mathbf{C}\varphi \vee \mathbf{C}\neg\varphi)$$

are equivalent. The second axiom says that if it is common knowledge that each agent has an epistemic position w.r.t. φ then either φ or $\neg\varphi$ are common knowledge.

4.1 Soundness of the GFP0 axiomatics

We prove soundness w.r.t. the S5-based GFP axiomatics of Table 2. The result holds both for the KT-based and the S5-based version.

The inference rules are the same: RN(**C**) and modus ponens. It remains to show that our axioms of Table 3 are theorems of the S5-based GFP axiomatics. The only ones that are missing there are 4(**C**), FP0, and GFP0. First, 4(**C**) is, by Proposition 3.1, a theorem of the KT-based GFP axiomatics and a fortiori of the S5-based GFP axiomatics. Second, FP0 can be proved from FP′ and T(**C**). Third, here is a proof of GFP0 that relies on T($\mathbf{K}_i$), or rather, its consequence T(**E**):

Proposition 4.1 GFP0 *is a theorem of the KT-based* GFP *axiomatics (and a fortiori of the S5-based* GFP *axiomatics).*

Proof. We distinguish the two cases φ and $\neg\varphi$ and prove that $\mathbf{C}(\mathbf{E}\varphi \vee \mathbf{E}\neg\varphi)$ implies both $\varphi \to \mathbf{C}\varphi$ and $\neg\varphi \to \mathbf{C}\neg\varphi$; from that GFP0 follows by propositional logic reasoning.

(i) $\mathbf{C}(\mathbf{E}\varphi \vee \mathbf{E}\neg\varphi) \to \mathbf{C}(\varphi \to \mathbf{E}\varphi)$ — by T(**E**), RN(**C**), K(**C**)

(ii) $\mathbf{C}(\varphi \to \mathbf{E}\varphi) \to (\varphi \to \mathbf{C}\varphi)$ — GFP

(iii) $\mathbf{C}(\mathbf{E}\varphi \vee \mathbf{E}\neg\varphi) \to (\varphi \to \mathbf{C}\varphi)$ — from (i), (ii)

(iv) $\mathbf{C}(\mathbf{E}\varphi \vee \mathbf{E}\neg\varphi) \to (\neg\varphi \to \mathbf{C}\neg\varphi)$ — from (iii) by uniform subst. of φ by $\neg\varphi$

(v) $\mathbf{C}(\mathbf{E}\varphi \vee \mathbf{E}\neg\varphi) \to (\mathbf{C}\varphi \vee \mathbf{C}\neg\varphi)$ — from (iii), (iv)

□

Therefore all theorems of our new GFP0 axiomatics are also theorems of the GFP axiomatics and, by Proposition 3.5, of the RGFP axiomatics.

4.2 Completeness of the GFP0 axiomatics for S5 knowledge

We prove completeness w.r.t. the S5-based GFP axiomatics. We have already seen in Section 4.1 that the inference rules are the same; it remains to show that the axioms of the S5-based GFP axiomatics of Table 2 that are not in our GFP0 axiomatics are theorems of the latter. These axioms are ∗5(**C**), FP′, and GFP. Proposition 3.2 tells us that ∗5(**C**) can be proved from the rest of the S5-based GFP axiomatics and is therefore redundant: it could be dropped from the GFP axiomatics. Axiom FP′ can be proved from our FP0, 4(**C**), K(**C**), and

$\mathtt{RN}(\mathbf{C})$. It remains to show that GFP is a theorem of our new axiomatics. The next lemma will be instrumental; its proof uses $*\mathtt{B}(\mathbf{E})$ (via Proposition 2.2) and $\mathtt{4}(\mathbf{C})$.

Lemma 4.2 *The schema $\mathbf{C}(\varphi \to \mathbf{E}\varphi) \to \mathbf{C}(\neg\varphi \to \mathbf{E}\neg\varphi)$ is provable from the axiom schemas $\mathtt{K}(\mathbf{C})$, $\mathtt{4}(\mathbf{C})$, $\mathtt{RN}(\mathbf{C})$, FP, $\mathtt{Def}(\mathbf{E})$, and the S5 axioms for $\mathbf{K}_i$.*

Proof. The proof is as follows:

(i)	$\mathbf{C}(\varphi \to \mathbf{E}\varphi) \to \mathbf{E}(\varphi \to \mathbf{E}\varphi)$	by FP, $\mathbf{E}$ normal
(ii)	$\mathbf{E}(\varphi \to \mathbf{E}\varphi) \to (\mathbf{E}\neg\mathbf{E}\varphi \to \mathbf{E}\neg\varphi)$	$\mathbf{E}$ normal
(iii)	$\neg\varphi \to \mathbf{E}\neg\mathbf{E}\varphi$	Proposition 2.2
(iv)	$\mathbf{C}(\varphi \to \mathbf{E}\varphi) \to (\neg\varphi \to \mathbf{E}\neg\varphi)$	from (i), (ii), (iii)
(v)	$\mathbf{CC}(\varphi \to \mathbf{E}\varphi) \to \mathbf{C}(\neg\varphi \to \mathbf{E}\neg\varphi)$	from (iv) by $\mathtt{RN}(\mathbf{C})$ and $\mathtt{K}(\mathbf{C})$
(vi)	$\mathbf{C}(\varphi \to \mathbf{E}\varphi) \to \mathbf{CC}(\varphi \to \mathbf{E}\varphi)$	$\mathtt{4}(\mathbf{C})$
(vii)	$\mathbf{C}(\varphi \to \mathbf{E}\varphi) \to \mathbf{C}(\neg\varphi \to \mathbf{E}\neg\varphi)$	from (v), (vi)

□

Proposition 4.3 *GFP is provable in the GFPO axiomatics.*

Proof. The proof is as follows:

(i)	$\mathbf{C}(\mathbf{E}\varphi \vee \mathbf{E}\neg\varphi) \to (\mathbf{C}\varphi \vee \mathbf{C}\neg\varphi)$	GFPO
(ii)	$\big(\mathbf{C}(\varphi \to \mathbf{E}\varphi) \wedge \mathbf{C}(\neg\varphi \to \mathbf{E}\neg\varphi)\big) \to \mathbf{C}(\mathbf{E}\varphi \vee \mathbf{E}\neg\varphi)$	by $\mathtt{RN}(\mathbf{C})$ and $\mathtt{K}(\mathbf{C})$
(iii)	$\big(\mathbf{C}(\varphi \to \mathbf{E}\varphi) \wedge \mathbf{C}(\neg\varphi \to \mathbf{E}\neg\varphi)\big) \to (\mathbf{C}\varphi \vee \mathbf{C}\neg\varphi)$	from (i) and (ii)
(iv)	$\mathbf{C}(\varphi \to \mathbf{E}\varphi) \to \mathbf{C}(\neg\varphi \to \mathbf{E}\neg\varphi)$	Lemma 4.2
(v)	$\mathbf{C}(\varphi \to \mathbf{E}\varphi) \to (\mathbf{C}\varphi \vee \mathbf{C}\neg\varphi)$	from (iii), (iv)
(vi)	$\mathbf{C}(\varphi \to \mathbf{E}\varphi) \to (\mathbf{C}\varphi \vee \neg\varphi)$	from (v) by $\mathtt{T}(\mathbf{C})$

□

5 Commonly knowing whether

In this section we show that our axiomatics of Table 3 leads to a simple axiomatisation of the S5-based 'common knowledge whether' operator. The axiomatisation of such an operator was left as an open problem in [7], where operators of 'individually knowing whether' were axiomatised.

The first thing we do is to extend our language of "knowing-that" operators $\mathbf{K}_i$, $\mathbf{E}$, and $\mathbf{C}$ by their "knowing-whether" counterparts. We read $\mathbf{Kif}_i\varphi$ as "i knows whether φ"; $\mathbf{Eif}\varphi$ as "it is shared knowledge whether φ"; and $\mathbf{Cif}\varphi$ as "it is common knowledge whether φ". These three epistemic operators are particular modal operators of contingency [19,12,7].

A straightforward possibility is to add to the axiomatics of Table 3 the following axioms:

$\mathtt{Def1}(\mathbf{Kif}_i)$	$\mathbf{Kif}_i\varphi \leftrightarrow (\mathbf{K}_i\varphi \vee \mathbf{K}_i\neg\varphi)$
$\mathtt{Def1}(\mathbf{Eif})$	$\mathbf{Eif}\varphi \leftrightarrow (\mathbf{E}\varphi \vee \mathbf{E}\neg\varphi)$
$\mathtt{Def1}(\mathbf{Cif})$	$\mathbf{Cif}\varphi \leftrightarrow (\mathbf{C}\varphi \vee \mathbf{C}\neg\varphi)$

CPC	axiomatics of classical propositional calculus
Sym($\mathbf{Kif}_i$)	$\mathbf{Kif}_i\varphi \leftrightarrow \mathbf{Kif}_i\neg\varphi$
RE($\mathbf{Kif}_i$)	from $\varphi \leftrightarrow \psi$, infer $\mathbf{Kif}_i\varphi \leftrightarrow \mathbf{Kif}_i\psi$
RN($\mathbf{Kif}_i$)	from φ, infer $\mathbf{Kif}_i\varphi$
Conj($\mathbf{Kif}_i$)	$(\varphi \wedge \psi) \rightarrow (\mathbf{Kif}_i(\varphi \wedge \psi) \leftrightarrow (\mathbf{Kif}_i\varphi \wedge \mathbf{Kif}_i\psi))$
$*45_1$($\mathbf{Kif}_i$)	$\mathbf{Kif}_i\mathbf{Kif}_i\varphi$
$*45_2$($\mathbf{Kif}_i$)	$\mathbf{Kif}_i(\varphi \wedge \mathbf{Kif}_i\varphi)$
Def2($\mathbf{Eif}$)	$\mathbf{Eif}\varphi \leftrightarrow \bigwedge_{i\in Agt} \mathbf{Kif}_i\varphi$
Sym($\mathbf{Cif}$)	$\mathbf{Cif}\varphi \leftrightarrow \mathbf{Cif}\neg\varphi$
RE($\mathbf{Cif}$)	from $\varphi \leftrightarrow \psi$, infer $\mathbf{Cif}\varphi \leftrightarrow \mathbf{Cif}\psi$
RN($\mathbf{Cif}$)	from φ, infer $\mathbf{C}\varphi$
Conj($\mathbf{Cif}$)	$(\varphi \wedge \psi) \rightarrow (\mathbf{Cif}(\varphi \wedge \psi) \leftrightarrow (\mathbf{Cif}\varphi \wedge \mathbf{Cif}\psi))$
$*45_1$($\mathbf{Cif}$)	$\mathbf{Cif}\mathbf{Cif}\varphi$
$*45_2$($\mathbf{Cif}$)	$\mathbf{Cif}(\varphi \wedge \mathbf{Cif}\varphi)$
GFP2	$\mathbf{Cif}\varphi \leftrightarrow (\mathbf{Eif}\varphi \wedge \mathbf{Cif}\mathbf{Eif}\varphi)$
Def2($\mathbf{K}_i$)	$\mathbf{K}_i\varphi \leftrightarrow (\varphi \wedge \mathbf{Kif}_i\varphi)$
Def2($\mathbf{E}$)	$\mathbf{E}\varphi \leftrightarrow (\varphi \wedge \mathbf{Eif}\varphi)$
Def2($\mathbf{C}$)	$\mathbf{C}\varphi \leftrightarrow (\varphi \wedge \mathbf{Cif}\varphi)$

Table 4
Axiomatisation of S5 common knowledge whether: the GFP2 axiomatics.

However, we are going to take another road here, in view of axiomatising the fragment without 'knowing-that' operators. Our axiomatics in Table 4 takes the 'knowing-whether' operators as primitive and defines the 'knowing-that' operators. The first part is proper to $\mathbf{Kif}_i$ and $\mathbf{Eif}$. We might have taken over as well the axiomatics of [7]; the principles Sym($\mathbf{Kif}_i$), RE($\mathbf{Kif}_i$), and RN($\mathbf{Kif}_i$) can also be found there, but we find the rest of our axioms a bit simpler than theirs. Axiom 45_1($\mathbf{Kif}_i$) can be found in [19]. The second part of our axiomatics parallels the first part and moreover has a single greatest fixed-point axiom relating **Eif** and **Cif** (that is perhaps better called a fixed-point axiom *tout court*: its syntactical form is very close to that of a possible fixed-point axiom for common belief $\mathbf{CB}\,\varphi \leftrightarrow (\mathbf{EB}\,\varphi \wedge \mathbf{EB}\;\mathbf{CB}\,\varphi)$). The third part contains the definitions of the 'knowing-that' operators.

We are going to prove soundness and completeness of the axiomatics of Table 4 w.r.t. the S5-based GFP0 axiomatics (more precisely: w.r.t. the extension of the latter by Def1($\mathbf{Kif}_i$), Def1($\mathbf{Eif}$), and Def1($\mathbf{Cif}$)).

Proposition 5.1 *For the S5-based* GFP2 *axiomatics of Table 4, all inference rules are derivable and all axioms are theorems in the S5-based* GFP0 *axiomatics.*

Proof. See the appendix. □

Proposition 5.2 *For the S5-based* GFP0 *axiomatics of Table 3, all inference rules are derivable and all axioms are theorems in the S5-based* GFP2 *axiomatics. Moreover, the equivalences* Def1($\mathbf{K}_i$), Def1($\mathbf{E}$), *and* Def1($\mathbf{C}$) *are theorems in the S5-based* GFP2 *axiomatics.*

Proof. See the appendix. □

It follows from propositions 5.1 and 5.2 that the first two parts of Table 4 provide a sound and complete axiomatisation for the fragment of the language with only 'knowing-whether' operators.

Proposition 5.3 *If formula* φ *has no* $\mathbf{K}_i$, $\mathbf{E}$, $\mathbf{C}$ *operators then* φ *is a theorem of the S5-based* `GFP2` *axiomatics of Table 4 if and only if it is provable without the axioms* `Def2`$(\mathbf{K}_i)$, `Def2`$(\mathbf{E})$, *and* `Def2`$(\mathbf{C})$.

Proof. Suppose no $\mathbf{K}_i$, $\mathbf{E}$, $\mathbf{C}$ occur in φ and suppose φ is a theorem of the S5-based `GFP2` axiomatics. Whenever the proof of φ uses axiom `Def2`$(\mathbf{K}_i)$, `Def2`$(\mathbf{E})$, or `Def2`$(\mathbf{C})$, we can eliminate that axiom by replacing the *definiendum* by the *definiens* everywhere in the proof. □

We end this section by a comment on alternative definitions of 'knowing-whether' group attitudes. As noted in the conclusion of [7], there are more options than those we have considered in this section. We have chosen to define 'shared knowledge whether' as $\mathbf{Eif}\varphi \leftrightarrow (\mathbf{E}\varphi \vee \mathbf{E}\neg\varphi)$. However, instead of requiring that everybody has the same epistemic position about φ one could only require that everybody has *some* epistemic position about φ. This amounts to defining 'weak shared knowledge whether' by $\mathbf{Eif}^w\varphi \leftrightarrow \bigwedge_{i\in Agt}\mathbf{Kif}_i\varphi$. At first glance this is a less demanding notion; however, Proposition 2.1 tells us that $\mathbf{Eif}$ and $\mathbf{Eif}^w$ are equivalent as soon as KT is our basic epistemic logic. Similarly, seemingly weaker definitions of 'common knowledge whether' exist. Instead of requiring that either φ or $\neg\varphi$ is common knowledge, one could only require (a) that it is common knowledge that there is shared knowledge whether φ, or (b) that it is common knowledge that there is weak shared knowledge whether φ. This amounts to replacing $\mathbf{C}\varphi \vee \mathbf{C}\neg\varphi$ in the definition of 'common knowledge whether' either by $\mathbf{C}\,\mathbf{Eif}\varphi$, or by $\mathbf{C}\,\mathbf{Eif}^w\varphi$. Again, these two definitions appear to be weaker than ours, but this fails to be the case. This can be seen from the theorem

$$\texttt{GFP1}\quad \mathbf{C}\bigwedge_{i\in Agt}(\mathbf{K}_i\varphi \vee \mathbf{K}_i\neg\varphi) \to (\mathbf{C}\varphi \vee \mathbf{C}\neg\varphi)$$

of Section 4 (end of the second paragraph) and that can be reformulated as $\mathbf{CEif}^w\varphi \to \mathbf{Cif}\varphi$. We note that both for shared and common knowledge whether, the two options are no longer equivalent for weaker logics, i.e., for logics of belief. We will come back to this in Section 7.

6 Discussion: epistemic logics between KT and S5

We have seen that our new axiom `GFP0` is sound for logics of knowledge, understood as logics where the logic of individual knowledge is at least KT, and that it is complete when the logic of individual knowledge is S5.

We conjecture that the KT-based `GFP0` axiomatics is incomplete. We however do not have a formal proof for the time being. Such a proof would have to delve into semantics: it typically consists in designing a non-standard semantics for which the axiomatics with `GFP0` is complete. We leave this aside for the time being.

Under the hypothesis that the KT-based GFP0 axiomatics is incomplete, one may wonder which axiom is missing to obtain completeness. A tempting avenue is to add the formula $\mathbf{C}(\varphi \to \mathbf{E}\varphi) \to \mathbf{C}(\neg\varphi \to \mathbf{E}\neg\varphi)$ of Lemma 4.2 as an axiom schema to the axiomatics of Table 3. The proof of Proposition 4.3 then gives us completeness because it uses none of the S5 axioms but $\mathtt{T}(\mathbf{K}_i)$. However it can be shown that this amounts to adding $*5(\mathbf{C})$: it can be shown that the formula is equivalent to $*5(\mathbf{C})$ in the presence of $\mathtt{T}(\mathbf{C})$.

Proposition 6.1 *In the GFP-based axiomatics for KT,* $*5(\mathbf{C})$ *and the formula* $\mathbf{C}(\varphi \to \mathbf{E}\varphi) \to \mathbf{C}(\neg\varphi \to \mathbf{E}\neg\varphi)$ *are interderivable.*

Proof. See the appendix. □

Just as common knowledge is necessarily positively introspective even when individual knowledge isn't, it can still be argued that S5 common knowledge can make sense even when individual knowledge is not S5: one can imagine, e.g., that common knowledge is "written on a blackboard", or otherwise easily available to agents such that they are able to immediately verify what is and is not commonly known. We leave further explorations to future work.

7 Discussion: GFP0 is not appropriate for belief

Up to now we have only discussed common knowledge; we now briefly discuss common belief.

Let us write $\mathbf{B}_i\,\varphi$ for "i believes that φ", $\mathbf{EB}\,\varphi$ for "it is shared belief that φ", and $\mathbf{CB}\,\varphi$ for "it is common belief that φ", and let us suppose the logic of the $\mathbf{B}_i$ operators is KD (or, alternatively, any logic without the T axiom).

It is intuitively clear that the belief-version of GFP1,

$$\mathbf{CB} \bigwedge_{i \in Agt} (\mathbf{B}_i\,\varphi \vee \mathbf{B}_i\,\neg\varphi) \to (\mathbf{CB}\,\varphi \vee \mathbf{CB}\,\neg\varphi),$$

should not hold: if there is common belief—and even common knowledge—that everybody has an opinion about φ then it by no means follows that there is common belief about φ.

What about GFP0? The fact that GFP1 is unintuitive need not disqualify GFP0. Indeed, while these two axioms are equivalent in epistemic contexts, they fail to be so in doxastic contexts: in KD45, $\bigwedge_{i \in Agt}(\mathbf{B}_i\,\varphi \vee \mathbf{B}_i\,\neg\varphi)$ does not imply $\mathbf{EB}\,\varphi \vee \mathbf{EB}\,\neg\varphi$, and does not do so a fortiori in KD; and therefore the belief-counterpart of Proposition 2.1 does not hold.

As it turns out, GFP0 is not a reasonable principle of common belief either. This can be highlighted by the following example. Suppose that the set of agents under concern is $Agt = \{1, 2\}$ and that there is a misunderstanding between 1 and 2 about an inform act of a third agent. That third agent is not relevant here, and we suppose that $Agt = \{1, 2\}$. Let us suppose that 1 believes the third agent said p and therefore believes that p is in the common ground ($\mathbf{B}_1\,\mathbf{CB}\,p$), while 2 believes that $\neg p$ is in the common ground ($\mathbf{B}_2\,\mathbf{CB}\,\neg p$). It follows by $4(\mathbf{CB})$ and by the (intuitively still valid) belief-counterpart of FP0 that

$$\mathbf{B}_1\,\mathbf{CB}\,\mathbf{EB}\,p \land \mathbf{B}_2\,\mathbf{CB}\,\mathbf{EB}\,\neg p.$$

As both **CB** and **EB** are normal operators, it follows that

$$\mathbf{B}_1\,\mathbf{CB}\,(\mathbf{EB}\,p \lor \mathbf{EB}\,\neg p) \land \mathbf{B}_2\,\mathbf{CB}\,(\mathbf{EB}\,p \lor \mathbf{EB}\,\neg p),$$

i.e., that $\mathbf{EB}\,\mathbf{CB}\,(\mathbf{EB}\,p \lor \mathbf{EB}\,\neg p)$. The latter is equivalent to $\mathbf{CB}\,(\mathbf{EB}\,p \lor \mathbf{EB}\,\neg p)$ thanks to the belief-version of the fixed-point axiom, which is $\mathbf{CB}\,\varphi \leftrightarrow \mathbf{EB}\,\mathbf{CB}\,\varphi$.

From that the counter-intuitive consequence $\mathbf{CB}\,p \lor \mathbf{CB}\,\neg p$ would follow by the belief-counterpart of `GFP0`.

To sum up, contrarily to the status of the standard induction principles the status of our new versions of the induction axiom differs between knowledge and belief: they are specific to common knowledge and fail for common belef.

8 Conclusion

We have studied the axiomatisation of the logic of common knowledge, coming up with an alternative `GFP0` to the standard induction axiom principles that is intuitively appealing as an axiom for common knowledge. While our proofs are not very difficult, we believe that `GFP0` will lead to presentations of epistemic logic that are intuitively more appealing.

Our investigation may appear somewhat old-fashioned: all our proofs are purely syntactical and we do not use any semantical tools, as was done in 'the syntactic era (1918-1959)' [2, Section 1.7] before Kripke semantics was invented. We nevertheless believe that axiomatic systems provide an important toolbox to understand intuitively what a logical system is able to express and what not. To witness, consider the inference rule `RGFP`: according to the explanations e.g. in [24], the rule says something about φ indicating to everybody that ψ; however and as the equivalence with axiom `GFP` demonstrates, this is not the case: axiom `GFP` of the equivalent `GFP`-based axiomatics has a single schematic variable φ, which shows us that the concept of one proposition indicating another proposition is not accounted for by the Kripke semantics. This is in line with the analysis of [5] where it is argued that this concept cannot be modelled in Kripke semantics and where the authors investigate a different semantical framework.

Acknowledgements

We thank the three reviewers of AiML 2020 for their careful reading and the suggestions they made. Several participants of the Second Workshop on Formal Methods and AI (Rennes, May 2019) provided helpful comments on a previous version of the paper, in particular Hans van Ditmarsch and Yoram Moses.

Appendix

A Proofs of Section 5

A.1 Proof of Proposition 5.1

Proposition 5.1 *For the S5-based* GFP2 *axiomatics of Table 4, all inference rules are derivable and all axioms are theorems in the S5-based* GFP0 *axiomatics.*

We prove each principle of Table 4. We start by the last three definitions so that we can use them in the rest of the proofs.

$$\texttt{Def2}(\mathbf{K}_i) \quad \mathbf{K}_i\varphi \leftrightarrow (\varphi \wedge \mathbf{Kif}_i\varphi)$$

Proof.

(i) $\mathbf{K}_i\varphi \leftrightarrow (\varphi \wedge (\mathbf{K}_i\varphi \vee \mathbf{K}_i\neg\varphi))$ from $\texttt{T}(\mathbf{K}_i)$

(ii) $\mathbf{K}_i\varphi \leftrightarrow (\varphi \wedge \mathbf{Kif}_i\varphi)$ from (i) and $\texttt{Def1}(\mathbf{Kif}_i)$

□

$$\texttt{Def2}(\mathbf{E}) \quad \mathbf{E}\varphi \leftrightarrow (\varphi \wedge \mathbf{Eif}\varphi)$$

Proof. Follow the lines of that of $\texttt{Def2}(\mathbf{K}_i)$: use $\texttt{Def1}(\mathbf{Eif})$ instead of $\texttt{Def1}(\mathbf{Kif}_i)$ and use that $\texttt{T}(\mathbf{E})$ is a theorem. □

$$\texttt{Def2}(\mathbf{C}) \quad \mathbf{C}\varphi \leftrightarrow (\varphi \wedge \mathbf{Cif}\varphi)$$

Proof. Follow the lines of that of $\texttt{Def2}(\mathbf{K}_i)$: use $\texttt{Def1}(\mathbf{Cif})$ instead of $\texttt{Def1}(\mathbf{Kif}_i)$ and $\texttt{T}(\mathbf{C})$ instead of $\texttt{T}(\mathbf{K}_i)$. □

$$\texttt{Sym}(\mathbf{Kif}_i)\text{: } \mathbf{Kif}_i\varphi \leftrightarrow \mathbf{Kif}_i\neg\varphi$$

Proof.

(i) $(\mathbf{K}_i\varphi \vee \mathbf{K}_i\neg\varphi) \leftrightarrow (\mathbf{K}_i\neg\varphi \vee \mathbf{K}_i\neg\neg\varphi)$ $\mathbf{K}_i$ normal

(ii) $\mathbf{Kif}_i\varphi \leftrightarrow \mathbf{Kif}_i\neg\varphi$ from (i) by $\texttt{Def1}(\mathbf{Kif}_i)$

□

$$\texttt{RE}(\mathbf{Kif}_i)\text{: from } \varphi \leftrightarrow \psi\text{, infer } \mathbf{Kif}_i\varphi \leftrightarrow \mathbf{Kif}_i\psi$$

Proof.

(i) $\varphi \leftrightarrow \psi$ hypothesis

(ii) $\mathbf{K}_i\varphi \leftrightarrow \mathbf{K}_i\psi$ from (i), $\mathbf{K}_i$ normal

(iii) $\mathbf{K}_i\neg\varphi \leftrightarrow \mathbf{K}_i\neg\psi$ from (i), $\mathbf{K}_i$ normal

(iv) $(\mathbf{K}_i\varphi \vee \mathbf{K}_i\neg\varphi) \leftrightarrow (\mathbf{K}_i\psi \vee \mathbf{K}_i\neg\psi)$ from (ii), (iii)

(v) $\mathbf{Kif}_i\varphi \leftrightarrow \mathbf{Kif}_i\psi$ from (iv) by $\texttt{Def1}(\mathbf{Kif}_i)$

□

$\texttt{RN}(\mathbf{Kif}_i)$: from φ, infer $\mathbf{Kif}_i\varphi$

Proof.

(i) φ — hypothesis

(ii) $\mathbf{K}_i\varphi$ — from (i), $\mathbf{K}_i$ normal

(iii) $\mathbf{K}_i\varphi \vee \mathbf{K}_i\neg\varphi$ — from (ii)

(iv) $\mathbf{Kif}_i\varphi$ — from (iii) by $\texttt{Def1}(\mathbf{Kif}_i)$

□

$\texttt{Conj}(\mathbf{Kif}_i)$: $(\varphi \wedge \psi) \rightarrow \big(\mathbf{Kif}_i(\varphi \wedge \psi) \leftrightarrow (\mathbf{Kif}_i\varphi \wedge \mathbf{Kif}_i\psi)\big)$

Proof. We prove the two implications $\big(\varphi \wedge \psi \wedge \mathbf{Kif}_i(\varphi \wedge \psi)\big) \rightarrow \mathbf{Kif}_i\varphi$ and $(\varphi \wedge \psi \wedge \mathbf{Kif}_i\varphi \wedge \mathbf{Kif}_i\psi) \rightarrow \mathbf{Kif}_i(\varphi \wedge \psi)$, each time using that we have already proved $\texttt{Def2}(\mathbf{K}_i)$ to be a theorem. For the former:

(i) $\mathbf{K}_i(\varphi \wedge \psi) \rightarrow (\mathbf{K}_i\varphi \vee \mathbf{K}_i\neg\varphi)$ — $\mathbf{K}_i$ normal

(ii) $\big(\varphi \wedge \psi \wedge \mathbf{Kif}_i(\varphi \wedge \psi)\big) \rightarrow \mathbf{Kif}_i\varphi$ — from (i), theorem $\texttt{Def2}(\mathbf{K}_i)$

For the latter:

(i) $(\mathbf{K}_i\varphi \wedge \mathbf{K}_i\psi) \rightarrow \mathbf{K}_i(\varphi \wedge \psi)$ — $\mathbf{K}_i$ normal

(ii) $(\varphi \wedge \mathbf{Kif}_i\varphi \wedge \psi \wedge \mathbf{Kif}_i\psi) \rightarrow \big(\varphi \wedge \psi \wedge \mathbf{Kif}_i(\varphi \wedge \psi)\big)$ from (i), thm. $\texttt{Def2}(\mathbf{K}_i)$

(iii) $(\varphi \wedge \psi \wedge \mathbf{Kif}_i\varphi \wedge \mathbf{Kif}_i\psi) \rightarrow \mathbf{Kif}_i(\varphi \wedge \psi)$ — from (ii)

□

$\texttt{45}_1(\mathbf{Kif}_i)$: $\mathbf{Kif}_i\mathbf{Kif}_i\varphi$

Proof. Similar to the next proof of $\texttt{45}_2(\mathbf{Kif}_i)$. □

$\texttt{45}_2(\mathbf{Kif}_i)$: $\mathbf{Kif}_i(\varphi \wedge \mathbf{Kif}_i\varphi)$

Proof.

(i) $\mathbf{K}_i\varphi \vee \mathbf{K}_i\neg\varphi \vee (\neg\mathbf{K}_i\varphi \wedge \neg\mathbf{K}_i\neg\varphi)$

(ii) $\mathbf{K}_i\varphi \rightarrow \mathbf{K}_i(\varphi \wedge \mathbf{Kif}_i\varphi)$ — from $\texttt{4}(\mathbf{K}_i)$ and thm. $\texttt{Def2}(\mathbf{K}_i)$, $\mathbf{K}_i$ normal

(iii) $\mathbf{K}_i\neg\varphi \rightarrow \mathbf{K}_i\neg(\varphi \wedge \mathbf{Kif}_i\varphi)$ — from $\mathbf{K}_i$ normal

(iv) $(\neg\mathbf{K}_i\varphi \wedge \neg\mathbf{K}_i\neg\varphi) \rightarrow (\mathbf{K}_i\neg\mathbf{K}_i\varphi \wedge \mathbf{K}_i\neg\mathbf{K}_i\neg\varphi)$ — from thm. $\texttt{*5}(\mathbf{K}_i)$

(v) $(\mathbf{K}_i\neg\mathbf{K}_i\varphi \wedge \mathbf{K}_i\neg\mathbf{K}_i\neg\varphi) \rightarrow \mathbf{K}_i\neg\mathbf{Kif}_i\varphi$ — from $\texttt{Def1}(\mathbf{Kif}_i)$, $\mathbf{K}_i$ normal

(vi) $(\neg\mathbf{K}_i\varphi \wedge \neg\mathbf{K}_i\neg\varphi) \rightarrow \mathbf{K}_i\neg(\varphi \wedge \mathbf{Kif}_i\varphi)$ — from (iv), (v), $\mathbf{K}_i$ normal

(vii) $\mathbf{K}_i(\varphi \wedge \mathbf{Kif}_i\varphi) \vee \mathbf{K}_i\neg(\varphi \wedge \mathbf{Kif}_i\varphi)$ — from (i), (ii), (iii), (vi)

(viii) $\mathbf{Kif}_i(\varphi \wedge \mathbf{Kif}_i\varphi)$ — from (vii), $\texttt{Def1}(\mathbf{Kif}_i)$

□

$\texttt{Def2}(\mathbf{Eif})$: $\mathbf{Eif}\varphi \leftrightarrow \bigwedge_{i\in Agt} \mathbf{Kif}_i\varphi$

Proof. This is Proposition 2.1. □

$\texttt{Sym}(\mathbf{Cif})$: $\mathbf{Cif}\varphi \leftrightarrow \mathbf{Cif}\neg\varphi$

Proof. Follow the lines of that of $\texttt{Sym}(\mathbf{Kif}_i)$. □

$\texttt{RE}(\mathbf{Cif})$: from $\varphi \leftrightarrow \psi$, infer $\mathbf{Cif}\varphi \leftrightarrow \mathbf{Cif}\psi$

Proof. Follow the lines of that of $\texttt{RE}(\mathbf{Kif}_i)$. □

$\texttt{RN}(\mathbf{Cif})$ from φ, infer $\mathbf{C}\varphi$

Proof. Follow the lines of that of $\texttt{RN}(\mathbf{Kif}_i)$. □

$\texttt{Conj}(\mathbf{Cif})$ $(\varphi \wedge \psi) \rightarrow \big(\mathbf{Cif}(\varphi \wedge \psi) \leftrightarrow (\mathbf{Cif}\varphi \wedge \mathbf{Cif}\psi)\big)$

Proof. Follow the lines of that of $\texttt{Conj}(\mathbf{Kif}_i)$. □

$*\texttt{45}_1(\mathbf{Cif})$ $\mathbf{Cif}\mathbf{Cif}\varphi$

Proof. Follow the lines of that of $\texttt{45}_1(\mathbf{Kif}_i)$. □

$\texttt{45}_2(\mathbf{Cif})$ $\mathbf{Cif}(\varphi \wedge \mathbf{Cif}\varphi)$

Proof. Follow the lines of that of $\texttt{45}_2(\mathbf{Kif}_i)$. □

$\texttt{GFP2}$ $\mathbf{Cif}\varphi \leftrightarrow (\mathbf{Eif}\varphi \wedge \mathbf{Cif}\mathbf{Eif}\varphi)$

Proof. We prove the three implications $\mathbf{Cif}\varphi \rightarrow \mathbf{Eif}\varphi$, $\mathbf{Cif}\varphi \rightarrow \mathbf{Cif}\mathbf{Eif}\varphi$, and $(\mathbf{Eif}\varphi \wedge \mathbf{Cif}\mathbf{Eif}\varphi) \rightarrow \mathbf{Cif}\varphi$. For the first:

(i) $(\mathbf{C}\varphi \vee \mathbf{C}\neg\varphi) \rightarrow (\mathbf{E}\varphi \vee \mathbf{E}\neg\varphi)$ — from $\texttt{FP0}$

(ii) $\mathbf{Cif}\varphi \rightarrow \mathbf{Eif}\varphi$ — from (i), $\texttt{Def1}(\mathbf{Eif})$, $\texttt{Def1}(\mathbf{Cif})$

For the second:

(i) $\mathbf{C}\varphi \rightarrow \mathbf{CE}\varphi$ — from $\texttt{4}(\mathbf{C})$, $\texttt{FP0}$

(ii) $\mathbf{C}\varphi \rightarrow \mathbf{CEif}\varphi$ — from (i), $\texttt{Def}(\mathbf{Eif})$, normal $\mathbf{C}$

(iii) $\mathbf{C}\neg\varphi \rightarrow \mathbf{CEif}\neg\varphi$ — from (ii) by uniform substitution

(iv) $\mathbf{C}\neg\varphi \rightarrow \mathbf{CEif}\varphi$ — from (iii) by $\texttt{Sym}(\mathbf{K}_i)$, $\texttt{Def}(\mathbf{E})$

(v) $\mathbf{Cif}\varphi \rightarrow \mathbf{CEif}\varphi$ — from (ii), (iv), $\texttt{Def1}(\mathbf{Cif})$

(vi) $\mathbf{Cif}\varphi \rightarrow \mathbf{Cif}\mathbf{Eif}\varphi$ — from (v), $\texttt{Def1}(\mathbf{Cif})$

For the third:

(i) $\mathbf{C}(\mathbf{E}\varphi \vee \mathbf{E}\neg\varphi) \to (\mathbf{C}\varphi \vee \mathbf{C}\neg\varphi)$ GFP0

(ii) $\mathbf{CEif}\varphi \to \mathbf{Cif}\varphi$ from (i), Def1(**Eif**), Def1(**Cif**)

(iii) $(\mathbf{Eif}\varphi \wedge \mathbf{CifEif}\varphi) \to \mathbf{Cif}\varphi$ from (ii), thm. Def2(**C**)

□

A.2 Proof of Proposition 5.2

Proposition 5.2 *For the S5-based* GFP0 *axiomatics of Table 3, all inference rules are derivable and all axioms are theorems in the S5-based* GFP2 *axiomatics. Moreover, the equivalences* Def1($\mathbf{Kif}_i$), Def1(**Eif**), *and* Def1(**Cif**) *are theorems in the S5-based* GFP2 *axiomatics.*

We start by the last three definitions.

Def1($\mathbf{Kif}_i$) $\mathbf{Kif}_i\varphi \leftrightarrow (\mathbf{K}_i\varphi \vee \mathbf{K}_i\neg\varphi)$

Proof.

(i) $(\mathbf{K}_i\varphi \vee \mathbf{K}_i\neg\varphi) \leftrightarrow ((\varphi \wedge \mathbf{Kif}_i\varphi) \vee (\neg\varphi \wedge \mathbf{Kif}_i\neg\varphi))$ from Def2($\mathbf{K}_i$)

(ii) $\mathbf{Kif}_i\neg\varphi \leftrightarrow \mathbf{Kif}_i\varphi$ Sym($\mathbf{Kif}_i$)

(iii) $\mathbf{Kif}_i\varphi \leftrightarrow (\mathbf{K}_i\varphi \vee \mathbf{K}_i\neg\varphi)$ from (i), (ii)

□

Def1(**Eif**) $\mathbf{Eif}\varphi \leftrightarrow (\mathbf{E}\varphi \vee \mathbf{E}\neg\varphi)$

Proof. Follow the lines of that of Def1($\mathbf{Kif}_i$). □

Def1(**Cif**) $\mathbf{Cif}\varphi \leftrightarrow (\mathbf{C}\varphi \vee \mathbf{C}\neg\varphi)$

Proof. Follow the lines of that of Def1($\mathbf{Kif}_i$). □

RN($\mathbf{K}_i$) from φ, infer $\mathbf{K}_i\varphi$

Proof.

(i) φ hypothesis

(ii) $\mathbf{Kif}_i\varphi$ from (i) by RN($\mathbf{Kif}_i$)

(iii) $\varphi \wedge \mathbf{K}_i\varphi$ from Def2($\mathbf{K}_i$)

(iv) $\mathbf{K}_i\varphi$ from (iii)

□

K($\mathbf{K}_i$) $\mathbf{K}_i(\varphi \to \psi) \to (\mathbf{K}_i\varphi \to \mathbf{K}_i\psi)$

Proof.

(i) $(\varphi \wedge \mathbf{Kif}_i\varphi \wedge (\varphi \to \psi) \wedge \mathbf{Kif}_i(\varphi \to \psi)) \to \mathbf{Kif}_i(\varphi \wedge (\varphi \to \psi))$ from $\mathtt{Conj}(\mathbf{Kif}_i)$

(ii) $(\varphi \wedge \mathbf{Kif}_i\varphi \wedge (\varphi \to \psi) \wedge \mathbf{Kif}_i(\varphi \to \psi)) \to \mathbf{Kif}_i(\varphi \wedge \psi)$ from (i) by $\mathtt{RE}(\mathbf{Kif}_i)$

(iii) $(\varphi \wedge \psi \wedge \mathbf{Kif}_i(\varphi \wedge \psi)) \to \mathbf{Kif}_i\psi$ from $\mathtt{Conj}(\mathbf{Kif}_i)$

(iv) $(\varphi \wedge \mathbf{Kif}_i\varphi \wedge (\varphi \to \psi) \wedge \mathbf{Kif}_i(\varphi \to \psi)) \to (\psi \wedge \mathbf{Kif}_i\psi)$ from (ii), (iii)

(v) $\mathbf{K}_i(\varphi \to \psi) \to (\mathbf{K}_i\varphi \to \mathbf{K}_i\psi)$ from (iv) by $\mathtt{Def2}(\mathbf{K}_i)$

□

$\mathtt{T}(\mathbf{K}_i)$ $\quad \mathbf{K}_i\varphi \to \varphi$

Proof.

(i) $(\varphi \wedge \mathbf{Kif}_\varphi) \to \varphi$

(ii) $\mathbf{K}_i\varphi \to \varphi$ from (i) by $\mathtt{Def2}(\mathbf{K}_i)$

□

$*5(\mathbf{K}_i)$ $\quad \neg\mathbf{K}_i\varphi \to \mathbf{K}_i\neg\mathbf{K}_i\varphi$

Proof.

(i) $\mathbf{Kif}_i(\varphi \wedge \mathbf{Kif}_i\varphi)$ $\mathtt{45}_2(\mathbf{Kif}_i)$

(ii) $\mathbf{Kif}_i\mathbf{K}_i\varphi$ from (i) by $\mathtt{Def2}(\mathbf{K}_i)$

(iii) $\mathbf{Kif}_i\neg\mathbf{K}_i\varphi$ from (ii) by $\mathtt{Sym}(\mathbf{Kif}_i)$

(iv) $\neg\mathbf{K}_i\varphi \to (\neg\mathbf{K}_i\varphi \wedge \mathbf{Kif}_i\neg\mathbf{K}_i\varphi)$ from (iii)

(v) $\neg\mathbf{K}_i\varphi \to \mathbf{K}_i\neg\mathbf{K}_i\varphi$ from (iv) by $\mathtt{Def2}(\mathbf{K}_i)$

□

$\mathtt{Def}(\mathbf{E})$ $\quad \mathbf{E}\varphi \leftrightarrow \bigwedge_{i \in Agt} \mathbf{K}_i\varphi$

Proof.

(i) $(\varphi \wedge \mathbf{Eif}\varphi) \leftrightarrow (\varphi \wedge \bigwedge_{i \in Agt} \mathbf{Kif}_i\varphi)$ from $\mathtt{Def2}(\mathbf{Eif})$

(ii) $(\varphi \wedge \mathbf{Eif}\varphi) \leftrightarrow \bigwedge_{i \in Agt}(\varphi \wedge \mathbf{Kif}_i\varphi)$ from (i)

(iii) $\mathbf{E}\varphi \leftrightarrow \bigwedge_{i \in Agt} \mathbf{K}_i\varphi$ from (ii) by $\mathtt{Def2}(\mathbf{E})$, $\mathtt{Def2}(\mathbf{K}_i)$

□

$\mathtt{RN}(\mathbf{C})$ $\quad$ from φ, infer $\mathbf{C}\varphi$

Proof. Follow the lines of that of $\mathtt{RN}(\mathbf{K}_i)$. □

$\mathtt{K}(\mathbf{C})$ $\quad \mathbf{C}(\varphi \to \psi) \to (\mathbf{C}\varphi \to \mathbf{C}\psi)$

Proof. Follow the lines of that of $\mathtt{K}(\mathbf{K}_i)$. □

$\texttt{T}(\mathbf{C})$	$\mathbf{C}\varphi \rightarrow \varphi$

Proof. Follow the lines of that of $\texttt{T}(\mathbf{K}_i)$. □

$\texttt{FP0}$	$\mathbf{C}\varphi \rightarrow \mathbf{E}\varphi$

Proof.

(i) $(\varphi \wedge \mathbf{Cif}\varphi) \rightarrow (\varphi \wedge \mathbf{Eif}\varphi)$ from $\texttt{GFP2}$

(ii) $\mathbf{C}\varphi \rightarrow \mathbf{E}\varphi$ from (i) by $\texttt{Def2}(\mathbf{C})$, $\texttt{Def2}(\mathbf{E})$

□

$\texttt{GFP0}$	$\mathbf{C}(\mathbf{E}\varphi \vee \mathbf{E}\neg\varphi) \rightarrow (\mathbf{C}\varphi \vee \mathbf{C}\neg\varphi)$

Proof.

(i) $(\mathbf{Eif}\varphi \wedge \mathbf{CifEif}\varphi) \rightarrow \mathbf{Cif}\varphi$ from $\texttt{GFP2}$

(ii) $\big((\mathbf{E}\varphi \vee \mathbf{E}\neg\varphi) \wedge \mathbf{Cif}(\mathbf{E}\varphi \vee \mathbf{E}\neg\varphi)\big) \rightarrow \mathbf{Cif}\varphi$
from (i) by thm. $\texttt{Def1}(\mathbf{Eif})$ and $\texttt{RE}(\mathbf{Cif})$

(iii) $\mathbf{C}(\mathbf{E}\varphi \vee \mathbf{E}\neg\varphi) \rightarrow (\mathbf{C}\varphi \vee \mathbf{C}\neg\varphi)$ from (ii) by $\texttt{Def2}(\mathbf{C})$, thm. $\texttt{Def1}(\mathbf{Cif})$

□

B Proof of Proposition 6.1

Proposition 6.1 *In the GFP-based axiomatics for KT,* $*\texttt{5}(\mathbf{C})$ *and the formula* $\mathbf{C}(\varphi \rightarrow \mathbf{E}\varphi) \rightarrow \mathbf{C}(\neg\varphi \rightarrow \mathbf{E}\neg\varphi)$ *are interderivable.*

Proof. From the GFP-based axiomatics for KT and $*\texttt{5}(\mathbf{C})$ (recall that $\texttt{4}(\mathbf{C})$ is derivable from $\texttt{FP}'$, $\texttt{RN}(\mathbf{C})$ and $\texttt{GFP}$):

(i) $\mathbf{CC}(\varphi \rightarrow \mathbf{E}\varphi) \rightarrow \mathbf{C}(\varphi \rightarrow \mathbf{C}\varphi)$ from $\texttt{GFP}$, $\texttt{RN}(\mathbf{C})$, $\texttt{K}(\mathbf{C})$

(ii) $\mathbf{C}(\varphi \rightarrow \mathbf{E}\varphi) \rightarrow \mathbf{CC}(\varphi \rightarrow \mathbf{E}\varphi)$ $\texttt{4}(\mathbf{C})$

(iii) $\mathbf{C}(\varphi \rightarrow \mathbf{E}\varphi) \rightarrow \mathbf{CC}(\neg\mathbf{C}\varphi \rightarrow \neg\varphi)$ from (i), (ii) and $\texttt{4}(\mathbf{C})$

(iv) $\mathbf{C}(\varphi \rightarrow \mathbf{E}\varphi) \rightarrow \mathbf{C}(\mathbf{C}\neg\mathbf{C}\varphi \rightarrow \mathbf{C}\neg\varphi)$ from (iii) and $\texttt{K}(\mathbf{C})$

(v) $\mathbf{C}(\varphi \rightarrow \mathbf{E}\varphi) \rightarrow \mathbf{C}(\neg\mathbf{C}\varphi \rightarrow \mathbf{C}\neg\varphi)$ from (iv) and $*\texttt{5}(\mathbf{C})$

(vi) $\mathbf{C}(\varphi \rightarrow \mathbf{E}\varphi) \rightarrow \mathbf{C}(\neg\varphi \rightarrow \mathbf{E}\neg\varphi)$ from (v), $\texttt{FP}'$ and $\texttt{T}(\mathbf{C})$

From the GFP-based axiomatics for KT and $\mathbf{C}(\varphi \rightarrow \mathbf{E}\varphi) \rightarrow \mathbf{C}(\neg\varphi \rightarrow \mathbf{E}\neg\varphi)$:

(i) $\mathbf{C}(\varphi \rightarrow \mathbf{E}\varphi) \rightarrow \mathbf{C}(\neg\varphi \rightarrow \mathbf{E}\neg\varphi)$ hypothesis

(ii) $\mathbf{C}(\mathbf{C}\varphi \rightarrow \mathbf{EC}\varphi)$ from $\texttt{FP}'$ and $\texttt{RN}(\mathbf{C})$

(iii) $\mathbf{C}(\neg\mathbf{C}\varphi \rightarrow \mathbf{E}\neg\mathbf{C}\varphi)$ from (ii) and (i)

(iv) $\neg\mathbf{C}\varphi \rightarrow \mathbf{C}\neg\mathbf{C}\varphi$ from (iii) and $\texttt{GFP}$

□

References

[1] Alberucci, L. and G. Jäger, *About cut elimination for logics of common knowledge*, Ann. Pure Appl. Log. **133** (2005), pp. 73–99.
URL `https://doi.org/10.1016/j.apal.2004.10.004`

[2] Blackburn, P., M. de Rijke and Y. Venema, "Modal Logic," Cambridge Tracts in Theoretical Computer Science, Cambridge University Press, 2001.

[3] Bonanno, G., *On the logic of common belief*, Math. Log. Q. **42** (1996), pp. 305–311.
URL `https://doi.org/10.1002/malq.19960420126`

[4] Bucheli, S., R. Kuznets and T. Studer, *Two ways to common knowledge*, Electron. Notes Theor. Comput. Sci. **262** (2010), pp. 83–98.
URL `https://doi.org/10.1016/j.entcs.2010.04.007`

[5] Cubitt, R. P. and R. Sugden, *Common knowledge, salience and convention: A reconstruction of David Lewis' game theory*, Economics & Philosophy **19** (2003), pp. 175–210.

[6] Fagin, R., J. Y. Halpern, Y. Moses and M. Y. Vardi, "Reasoning about Knowledge," MIT Press, 1995.

[7] Fan, J., Y. Wang and H. van Ditmarsch, *Contingency and knowing whether*, Rew. Symb. Logic **8** (2015), pp. 75–107.
URL `https://doi.org/10.1017/S1755020314000343`

[8] Fukuda, S., *Formalizing common belief with no underlying assumption on individual beliefs*, Games and Economic Behavior (2020).

[9] Halpern, J. Y. and Y. Moses, *A guide to the modal logics of knowledge and belief: Preliminary draft*, in: *Proceedings of the 9th International Joint Conference on Artificial Intelligence - Volume 1*, IJCAI'85 (1985), pp. 480–490.

[10] Halpern, J. Y. and Y. Moses, *A guide to completeness and complexity for modal logics of knowledge and belief*, Artificial Intelligence **54** (1992), pp. 319–379.

[11] Hill, B. and F. Poggiolesi, *Common knowledge: finite calculus with a syntactic cut-elimination procedure*, Logique et Analyse **58** (2015), pp. 279–306.
URL `https://halshs.archives-ouvertes.fr/halshs-00775822`

[12] Humberstone, I. L., *The logic of non-contingency*, Notre Dame J. Formal Logic **36** (1995), pp. 214–229.
URL `https://doi.org/10.1305/ndjfl/1040248455`

[13] Kraus, S. and D. Lehmann, *Knowledge, belief and time*, TCS **58** (1988), pp. 155–174.
URL `https://doi.org/10.1016/0304-3975(88)90024-2`

[14] Lee, B. P., *Mutual knowledge, background knowledge and shared beliefs: Their roles in establishing common ground*, Journal of Pragmatics **33** (2001), pp. 21–44.

[15] Lehmann, D. J., *Knowledge, common Knowledge and related puzzles (extended summary)*, in: *Proc. PODC 1984*, 1984, pp. 62–67.

[16] Lewis, D. K., "Convention: A Philosophical Study," Harvard University Press, 1969.

[17] Lismont, L. and P. Mongin, *On the logic of common belief and common knowledge*, Theory and Decision **37** (1994), pp. 75–106.
URL `https://doi.org/10.1007/BF01079206`

[18] Monderer, D. and D. Samet, *Approximating common knowledge with common beliefs*, Games and Economic Behavior **1** (1989), pp. 170–190.

[19] Montgomery, H. A. and R. Routley, *Contingency and non-contingency bases for normal modal logics*, Logique et Analyse **9** (1966), p. 318–328.

[20] Schiffer, S., "Meaning," Oxford University Press, 1972.

[21] van Benthem, J., J. van Eijck and B. Kooi, *Logics of communication and change*, Information and Computation **204** (2006), pp. 1620–1662.

[22] van Ditmarsch, H., J. Halpern, W. van der Hoek and B. Kooi, "Handbook of Epistemic Logic," College Publications, 2015.

[23] van Ditmarsch, H., W. van der Hoek and B. Kooi, "Dynamic Epistemic Logic," Springer Publishing Company, Incorporated, 2007, 1st edition.

[24] Vanderschraaf, P. and G. Sillari, *Common knowledge*, in: E. N. Zalta, editor, *The Stanford Encyclopedia of Philosophy*, Stanford University, 2014, spring 2014 edition .

Inquisitive Intuitionistic Logic

Wesley H. Holliday

University of California, Berkeley

Abstract

Inquisitive logic is a research program seeking to expand the purview of logic beyond declarative sentences to include the logic of *questions*. To this end, inquisitive propositional logic extends classical propositional logic for declarative sentences with principles governing a new binary connective of *inquisitive disjunction*, which allows the formation of questions. Recently inquisitive logicians have considered what happens if the logic of declarative sentences is assumed to be intuitionistic rather than classical. In short, what should inquisitive logic be on an intuitionistic base? In this paper, we provide an answer to this question from the perspective of *nuclear semantics*, an approach to classical and intuitionistic semantics pursued in our previous work. In particular, we show how Beth semantics for intuitionistic logic naturally extends to a semantics for inquisitive intuitionistic logic. In addition, we show how an explicit view of inquisitive intuitionistic logic comes via a translation into *propositional lax logic*, whose completeness we prove with respect to Beth semantics.

Keywords: inquisitive logic, intuitionistic logic, Kripke semantics, Beth semantics, algebraic semantics, Heyting algebra, nucleus, lax logic

1 Introduction

Inquisitive logic is a research program seeking to expand the purview of logic beyond declarative sentences to include the logic of *questions* (see, e.g., [7,12,9,8,10]). While classical logic is based on the idea that any *state of the world* that *makes true* certain declarative sentences also makes true certain other declarative sentences, inquisitive logic is based on the idea that any *state of information* that *answers* certain questions (and incorporates the truth of certain declarative sentences) also answers certain other questions (and incorporates the truth of certain other declarative sentences). Thus, one may study a notion of consequence not only between declarative sentences but also between questions, as well as combinations of declaratives and questions.

To formalize this new notion of consequence, the language of inquisitive propositional logic extends that of classical propositional logic for declarative sentences with a new binary connective of *inquisitive disjunction*, $\mathbin{\vee\!\!\!\vee}$, which allows the formation of questions. The formula $p \mathbin{\vee\!\!\!\vee} q$ represents the question of *whether* p *or* q, in contrast to the formula $p \vee q$, which represents the declarative sentence p *or* q. To make this distinction with a formal semantics, classical inquisitive semantics evaluates a formula of the inquisitive language at an information state, understood as a set of classical propositional valuations—the states of the world compatible with the information. An information state *answers the question* $p \vee\!\!\!\vee q$ just in case every valuation in the information state satisfies p or every valuation in the information state satisfies q; by contrast, an information state *supports the declarative* $p \vee q$ just in case every valuation in the information state satisfies p or satisfies q.[1] Thus, while every information state supports the declarative $p \vee \neg p$, not every information state answers the question $p \vee\!\!\!\vee \neg p$. This gives reasoning with inquisitive disjunction an intuitionistic flavor. Yet the logic of declarative sentences (formulas without $\vee\!\!\!\vee$) underlying inquisitive logic is classical.

Recently inquisitive logicians have considered what happens if the logic of declarative sentences is assumed to be intuitionistic rather than classical [24,25,11]. In short, what should inquisitive logic be on an intuitionistic base? This is a natural question not only because of the general interest in intuitionistic logic as a formalization of constructive reasoning with declarative sentences, but also because of the affinity between information-state-based semantics and intuitionistic semantics in the style of Beth [1], Grzegorczyk [17], and Kripke [21]. In fact, the classical inquisitive semantics sketched above may be seen as a special case of intuitionistic Kripke semantics, based on restricting to special Kripke models: the underlying poset of the Kripke model must be the set of all nonempty subsets of a set, ordered by reverse inclusion (a "topless Boolean algebra"), and the valuation of each proposition letter in the Kripke model must be a *regular element* of the Heyting algebra of upsets of the poset, i.e., an upset U such that $U = U^{**}$, where * is the pseudocomplement operation in the Heyting algebra of upsets, which is used in Kripke semantics to interpret the intuitionistic negation connective $\neg$. Restricting the valuation of proposition letters to regular elements, the usual Kripke clauses for $\neg$ and $\wedge$, plus the classical definition of $\vee$ in terms of $\neg$ and $\wedge$, yields classical logic for the declarative fragment of the inquisitive language; then interpreting $\vee\!\!\!\vee$ as the standard Kripke disjunction—as the join (union) in the Heyting algebra of upsets—is responsible for the intuitionistic flavor of $\vee\!\!\!\vee$ noted above.

Given this connection between classical inquisitive semantics and intuitionistic Kripke semantics, how should one modify the semantics to obtain an intuitionistic base logic of declaratives? Ciardelli et al. [11] do so by moving up one level set-theoretically in Kripke models: their semantics evaluates a formula at a subset of a Kripke model, called a *team*. As the points in an intuitionistic Kripke model are traditionally thought of as information states, a team may be thought of as a set of information states—and therefore as a kind of higher-order information state.

In this paper, we pursue a different semantic approach to inquisitive logic on an intuitionistic base. In our semantics, we evaluate formulas of the inquisitive

[1] This is the semantics for proposition letters p and q. For the general recursive clause, see any of the cited references on classical inquisitive logic.

language at individual states in a poset, not at sets of states of a poset. We are able to do so by switching from Kripke semantics on posets to *Beth semantics* on posets. The difference between our approach and that of Ciardelli et al. [11] can be traced to different perspectives on *declarative disjunction* in classical inquisitive semantics. As noted above, in the original approach to inquisitive logic (see, e.g., [8, Def. 2.1.2]), the declarative disjunction $\vee$ is defined in terms of $\neg$ and $\wedge$ using the usual classical definition: $\varphi \vee \psi := \neg(\neg\varphi \wedge \neg\psi)$. Since $\neg$ and $\wedge$ are interpreted as the pseudocomplement and meet (intersection) operations in the Heyting algebra of upsets of a Kripke model, the definition of $\vee$ in terms of $\neg$ and $\wedge$ is equivalent to interpreting $\vee$ as the regularization of the join:

$$V(\varphi \vee \psi) = (V(\varphi) \sqcup V(\psi))^{**}.$$

Thus, we see classical inquisitive semantics as follows:

- the semantic values of formulas are elements of a Heyting algebra of upsets of a special kind of poset (a topless Boolean algebra);
- the semantic values of proposition letters must be regular elements of the Heyting algebra;
- $\neg$ and $\wedge$ are interpreted as pseudocomplement and meet, respectively, in the Heyting algebra;
- the inquisitive disjunction $\mathbin{\vee\!\!\!\vee}$ is interpreted as the join in the Heyting algebra;
- the declarative disjunction $\vee$ is interpreted as the regularization of the join in the Heyting algebra.

The regularization operation $(\cdot)^{**}$ is an example of a *nucleus* on the Heyting algebra of upsets of a poset (see Section 5 for a definition). The fixpoints of this nucleus—the regular elements—form a Boolean algebra, in which the join of two elements is the regularization of their join in the Heyting algebra. This explains why standard inquisitive semantics, which interprets proposition letters as regular elements and interprets declarative disjunction as the regularization of the join in the Heyting algebra of upsets, yields classical logic for the declarative fragment of the inquisitive language. It also explains why inquisitive logic is not closed under uniform substitution of complex formulas for proposition letters, e.g., why $\neg\neg p \to p$ is valid while $\neg\neg(q \mathbin{\vee\!\!\!\vee} r) \to (q \mathbin{\vee\!\!\!\vee} r)$ is not. This happens because while proposition letters are interpreted as regular elements, the join operation in the Heyting algebra can take one out of the algebra of regular elements. To summarize:

- the semantic values of declarative formulas live in the algebra of fixpoints of $(\cdot)^{**}$, while the semantic values of arbitrary formulas may live anywhere in the ambient Heyting algebra.

From this perspective, also adopted in [5], there is a natural way of obtaining semantics for inquisitive logic on an intuitionistic declarative base: we may simply switch from the Boolean nucleus $(\cdot)^{**}$ to a non-Boolean nucleus.

To do so, first note that interpreting declarative disjunction as the regularization of the join in the Heyting algbera of upsets is equivalent to using the following semantic clause:

- a state x in a poset forces $\varphi \vee \psi$ iff for every $x' \geq x$ there is an $x'' \geq x'$ such that x'' forces φ or x'' forces ψ.

In place of this classical interpretation of $\vee$, we give an intuitionistic interpretation of $\vee$ as in Beth semantics:

- a state x in a poset forces $\varphi \vee \psi$ iff every maximal chain [2] through x contains a state that forces φ or a state that forces ψ.

This amounts to interpreting declarative disjunction as the result of applying what we call the *Beth nucleus* to the join in the Heyting algebra of upsets. The Beth nucleus j_b is defined for any upset U of a poset by

$$j_b U = \{x \in X \mid \text{every maximal chain through } x \text{ intersects } U\}.$$

Not only do we interpret declarative disjunction using j_b instead of $(\cdot)^{**}$, but also we require proposition letters to be interpreted as fixpoints of j_b instead of $(\cdot)^{**}$ (i.e., we require that x forces p iff every maximal chain through x contains a state that forces p). Yet the interpretation of inquisitive disjunction $\mathbin{\rotatebox[origin=c]{-90}{$\geqslant$}}$ as join in the Heyting algebra of upsets remains the same. This Beth semantics for the inquisitive language is a special case of a more general algebraic semantics for the inquisitive language based on Heyting algebras equipped with a nucleus, called *nuclear algebras.* Thus, we are extending to the inquisitive setting the *nuclear semantics* for intuitionistic logic studied in our previous work [4].

The starting point on our road to this nuclear approach to inquisitive semantics was the observation that in the classical semantics for inquisitive logic, the nucleus $(\cdot)^{**}$ is used to constrain the valuation of proposition letters and to interpret the declarative disjunction $\vee$ (just as in the *possibility semantics* of [19,18]). By contrast, Ciardelli et al. [11] have a different starting point. They begin by departing from the original definition of declarative disjunction in classical inquisitive logic as $\varphi \vee \psi := \neg(\neg\varphi \wedge \neg\psi)$ and by giving a new semantic clause for classical $\vee$ based on team semantics for dependence logic (see, e.g., [26,27]). In team semantics, disjunction is interpreted as follows:

- an information state T (set of classical valuations) supports $\varphi \vee \psi$ iff there are T', T'' such that $T = T' \cup T''$, T' supports φ, and T'' supports ψ.

Already in the classical setting, this semantics for $\varphi \vee \psi$ is not equivalent to defining $\varphi \vee \psi := \neg(\neg\varphi \wedge \neg\psi)$. For example, with the original definition of $\varphi \vee \psi := \neg(\neg\varphi \wedge \neg\psi)$, the principle

$$((\varphi \vee \varphi) \vee (\varphi \vee \varphi)) \to (\varphi \vee \varphi)$$

is a theorem of inquisitive logic for any φ. Yet with $\vee$ treated as a primitive connective and interpreted using the team semantics above, the principle above

[2] In fact, we will use chains closed under upper bounds, following [4], but this subtlety does not matter here.

is not valid for all φ containing $\mathbin{\rotatebox[origin=c]{-90}{$\geqslant$}}$ (see Example 4.2). Ciardelli et al. [11] extend the team semantics for $\vee$ to the intuitionistic setting by taking T to be a subset of an intuitionistic Kripke model instead of merely a set of classical valuations.

We do not wish to argue that the nuclear approach to inquisitive logic on an intuitionistic base is superior to the team-based approach. Both are natural from different points of view. Starting from the perspective of team semantics for dependence logic, Ciardelli et al. [11] show how to "intuitionize" the semantics, by moving from teams as sets of classical valuations to teams as subsets of a Kripke model. By contrast, starting from the perspective of Beth semantics for intuitionistic logic, we show how to "inquisitivize" the semantics, by adding the Kripke interpretation for $\mathbin{\rotatebox[origin=c]{-90}{$\geqslant$}}$ to Beth semantics. This is why we call our resulting logic "inquisitive intuitionitic logic" in contrast to Ciardelli et al.'s "intuitionistic inquisitive logic."

Our main result is a completeness theorem for inquisitive intuitionistic logic with respect to Beth semantics, which is our answer to the question "What should inquisitive logic be on an intuitionistic base?" But even independently of inquisitive logic, it is a natural question whether one can prove a completeness theorem for the propositional language with two disjunctions $\vee$ and $\mathbin{\rotatebox[origin=c]{-90}{$\geqslant$}}$, with $\vee$ (and proposition letters) interpreted according to Beth semantics, $\mathbin{\rotatebox[origin=c]{-90}{$\geqslant$}}$ interpreted according to Kripke semantics, and $\neg, \to, \wedge$ having their usual interpretations, which are the same in both semantics.

We prove the completeness theorem using a detour through the intuitionistic modal logic of nuclei [16], known as *propositional lax logic* [15]. Propositional lax logic adds to the signature of intuitionistic propositional logic an operator $\bigcirc$, interpreted using the nucleus in a nuclear algebra. A key step in our proof of completeness of inquisitive intuitionistic logic with respect to Beth semantics is a proof of the completeness of propositional lax logic with respect to Beth semantics, i.e., with $\bigcirc$ interpreted as the Beth nucleus j_b on the Heyting algebra of upsets of a poset. Thus, another contribution of the paper is to provide a new semantics for propositional lax logic.

The paper is organized as follows. In Section 2, we recall the standard language of inquisitive logic and present our new semantic proposal: Beth semantics for inquisitive intuitionistic logic. In Section 3, we define for any superintuitionistic logic L its inquisitive version $\mathsf{Inq}(\mathsf{L})$. In this paper, we concentrate on the case where L is the intuitionistic propositional calculus (IPC). Our completeness theorem states that $\mathsf{Inq}(\mathsf{IPC})$ is sound and complete with respect to the class of all Beth frames (posets) according to Beth semantics. Before proving this result, in Section 4 we compare $\mathsf{Inq}(\mathsf{IPC})$ with the system InqI of Ciardelli et al. [11]. We show that the two logics are incomparable in strength. In Section 5, we develop the nuclear perspective on Beth semantics sketched above, which we turn into explicit translations between the language of inquisitive intuitionistic logic and the language of propositional lax logic in Section 6. This lets us transform the problem of proving the completeness of inquisitive intuitionistic logics with respect to Beth semantics into the problem of proving the completeness of propositional lax logics with respect to Beth semantics. We work up to Beth completeness in three stages:

- in Section 6, we obtain completeness with respect to finite nuclear algebras (proved in Appendix A);
- in Section 7, as an intermediate step, we transfer completeness with respect to finite nuclear algebras to completeness with respect to certain finite relational structures, which we call "S-frames," from [3];
- in Section 8, we transfer completeness with respect to finite S-frames to Beth completeness.

2 Beth Semantics for Inquisitive Logic

The *inquisitive intuitionistic language* $\mathcal{L}_{\vee,\mathbin{\vee\!\!\!/}}$ is defined as follows, where p belongs to a countably infinite set Prop of proposition letters:

$$\varphi ::= \bot \mid p \mid (\varphi \wedge \varphi) \mid (\varphi \vee \varphi) \mid (\varphi \to \varphi) \mid (\varphi \mathbin{\vee\!\!\!/} \varphi).$$

As usual, we define $\neg\varphi := \varphi \to \bot$. Let $\mathcal{L}_{\vee}$ be the fragment without $\mathbin{\vee\!\!\!/}$, and let $\mathcal{L}_{\mathbin{\vee\!\!\!/}}$ be the fragment without $\vee$.

Toward introducing our semantics for $\mathcal{L}_{\vee,\mathbin{\vee\!\!\!/}}$, we need the following notions.

Definition 2.1 Given a poset X, we define:

(i) $\mathsf{Up}(X)$ is the set of all upward closed subsets (upsets) of X, i.e., those $U \subseteq X$ such that if $x \in U$ and $x \leq y$, then $y \in U$;

(ii) a *chain* in X is a $C \subseteq X$ such that for all $x, y \in C$, $x \leq y$ or $y \leq x$;

(iii) a *path* in X is a chain C in X that is closed under upper bounds, i.e., if for all $x \in C$, $x \leq y$, then $y \in C$. If $x \in C$, then C is a *path through* x.

Our proposal is to simply extend Beth semantics [1] for intuitionistic logic (following the presentation in [4]) with $\mathbin{\vee\!\!\!/}$ interpreted as in Kripke semantics.

Definition 2.2 For any poset X, $x \in X$, valuation $v : \mathsf{Prop} \to \mathsf{Up}(X)$, and $\varphi \in \mathcal{L}_{\vee,\mathbin{\vee\!\!\!/}}$, we define $X, x \Vdash_v \varphi$ as follows:

(i) $X, x \nVdash_v \bot$; $X, x \Vdash_v p$ iff every path through x intersects $v(p)$;

(ii) $X, x \Vdash_v \varphi \wedge \psi$ iff $X, x \Vdash_v \varphi$ and $X, x \Vdash_v \psi$;

(iii) $X, x \Vdash_v \varphi \vee \psi$ iff every path through x intersects $\{y \in X \mid X, y \Vdash_v \varphi\} \cup \{y \in X \mid X, y \Vdash_v \psi\}$;

(iv) $X, x \Vdash_v \varphi \to \psi$ iff for every $y \geq x$, if $X, y \Vdash_v \varphi$ then $X, y \Vdash_v \psi$;

(v) $X, x \Vdash_v \varphi \mathbin{\vee\!\!\!/} \psi$ iff $X, x \Vdash_v \varphi$ or $X, x \Vdash_v \psi$.

A formula φ is *valid* on X according to inquisitive Beth semantics iff for any valuation $v : \mathsf{Prop} \to \mathsf{Up}(X)$, we have $X, x \Vdash \varphi$ for all $x \in X$ (otherwise φ is *refuted*); φ is valid over a class K of posets iff it is valid on every poset in K.

Example 2.3 Fig. 1 shows a poset (the "Beth comb") with a valuation such that according to Beth semantics, the root node forces $p \vee q$ (as every path through the root contains a node that forces p or a node that forces q) but does not force $p \vee\!\!\!\vee q$ (as the root does not force p and does not force q) and does not force $p \vee \neg p$ (as the path consisting of the nodes along the "spine" of the comb does not contain a node that forces p or a node that forces $\neg p$).

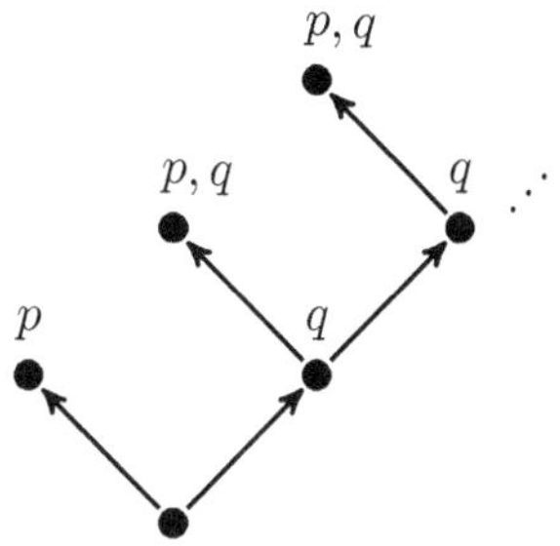

Fig. 1. Beth model for Example 2.3

3 Inquisitive Intuitionistic Logic

We now give a syntactic definition of a family of logical systems, the minimal member of which we will prove complete with respect to Beth semantics.

Definition 3.1 An *inquisitive intuitionistic logic* is a set L of $\mathcal{L}_{\vee,\vee\!\!\!\vee}$ formulas that contains the following formulas and is closed under the following rules, for all $\varphi, \psi, \chi \in \mathcal{L}_{\vee,\vee\!\!\!\vee}$:

- all $\mathcal{L}_{\vee,\vee\!\!\!\vee}$-substitution instances of IPC axioms stated in $\mathcal{L}_{\vee\!\!\!\vee}$;
- $(\alpha \vee \alpha) \to \alpha$ for $\alpha \in \mathcal{L}_{\vee}$;
- $\varphi \to (\varphi \vee \varphi)$; $((\varphi \vee \varphi) \vee (\varphi \vee \varphi)) \to (\varphi \vee \varphi)$; $((\varphi \vee\!\!\!\vee \psi) \vee (\varphi \vee\!\!\!\vee \psi)) \leftrightarrow (\varphi \vee \psi)$;
- $((\varphi \wedge \psi) \vee (\varphi \wedge \psi)) \leftrightarrow ((\varphi \vee \varphi) \wedge (\psi \vee \psi))$;
- rule of modus ponens: if $\varphi \in \mathsf{L}$ and $\varphi \to \psi \in \mathsf{L}$, then $\psi \in \mathsf{L}$;
- rule of replacement of equivalents: if $\varphi \in \mathsf{L}$ and $\psi \leftrightarrow \chi \in \mathsf{L}$, then $\varphi' \in \mathsf{L}$ for any φ' obtained from φ by replacing one or more occurrences of ψ in φ by χ.

The following soundness result is easy to check.

Proposition 3.2 *For any class* K *of posets, the set of* $\mathcal{L}_{\vee,\vee\!\!\!\vee}$*-formulas valid over* K *according to inquisitive Beth semantics is an inquisitive intuitionistic logic.*

One can also consider inquisitive intuitionistic logics based on superintuitionistic logics strictly extending IPC.

Definition 3.3 A *superintuitionistic logic (si-logic) for* $\mathcal{L}_{\vee\!\!\!\vee}$ is a set L of $\mathcal{L}_{\vee\!\!\!\vee}$ formulas that contains the following formulas and is closed under the following rules:

(i) all axioms of IPC stated in $\mathcal{L}_{\vee\!\!\!\vee}$;

(ii) rule of modus ponens: if $\varphi \in \mathsf{L}$ and $\varphi \to \psi \in \mathsf{L}$, then $\psi \in \mathsf{L}$;

(iii) rule of substitution: if $\varphi \in \mathsf{L}$ and φ' is obtained from φ by uniformly substituting formulas for proposition letters in φ, then $\varphi' \in \mathsf{L}$.

Definition 3.4 For any si-logic L for $\mathcal{L}_{\vee\!\!\!\vee}$, let $\mathsf{Inq}(\mathsf{L})$ be the smallest inquisitive intuitionistic logic containing all $\mathcal{L}_{\vee,\vee\!\!\!\vee}$-substitution instances of theorems of L.

In this paper, we concentrate on the smallest inquisitive intuitionistic logic, $\mathsf{Inq}(\mathsf{IPC})$. Our main theorem is the following.

Theorem 3.5 $\mathsf{Inq}(\mathsf{IPC})$ *is sound and complete according to Beth semantics.*

4 Comparison of Inq(IPC) and InqI

Ciardelli et al. [11] syntactically define a system InqI of intuitionistic inquisitive logic. Below we show that InqI and our $\mathsf{Inq}(\mathsf{IPC})$ are incomparable. We refer the reader to [11] for the full definition of InqI and its team semantics, but we will define as much as we need here to distinguish the two logics.

Example 4.1 The axiom

$$(p \vee (q \vee\!\!\!\vee r)) \to ((p \vee q) \vee\!\!\!\vee (p \vee r))$$

of InqI is not valid according to Beth semantics for inquisitive logic. For example, in the poset with the valuation shown in Figure 2, where p is true only at the top left node, q only at the top middle node, and r only at the top right node, the root node satisfies $p \vee (q \vee\!\!\!\vee r)$ but does not satisfy $(p \vee q) \vee\!\!\!\vee (p \vee r)$.

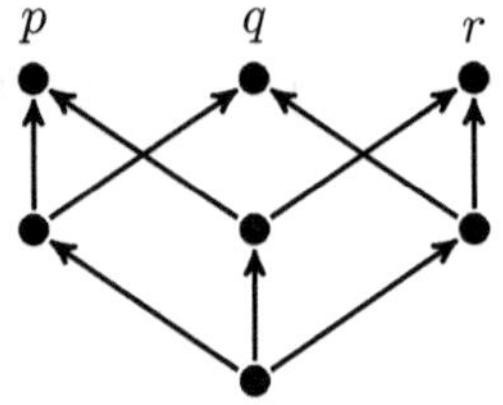

Fig. 2. Beth model for Example 4.1

Example 4.2 The following axiom schemas of $\mathsf{Inq}(\mathsf{IPC})$ have counterexamples according to team semantics for InqI [11]:

(i) $((\varphi \vee \varphi) \vee (\varphi \vee \varphi)) \to (\varphi \vee \varphi)$;

(ii) $((\varphi \wedge \psi) \vee (\varphi \wedge \psi)) \leftrightarrow ((\varphi \vee \varphi) \wedge (\psi \vee \psi))$.

Recall the clauses for $\vee$ and $\vee\!\!\!\vee$ according to team semantics:

- a team T supports $\varphi \vee \psi$ iff there are T', T'' such that $T = T' \cup T''$, T' supports φ, and T'' supports ψ;
- a team T supports $\varphi \vee\!\!\!\vee \psi$ iff T supports φ or T supports ψ.

For (i), take $\varphi := (p_1 \vee\!\!\!\vee p_2) \vee\!\!\!\vee (p_3 \vee\!\!\!\vee p_4)$ and a classical team model (i.e., the Kripke relation R is identity) with $W = \{w_1, w_2, w_3, w_4\}$ and $V(p_i) = \{w_i\}$. Then the team $\{w_1, w_2, w_3, w_4\}$ supports $(\varphi \vee \varphi) \vee (\varphi \vee \varphi)$, because

$\{w_1, w_2, w_3, w_4\} = \{w_1\} \cup \{w_2\} \cup \{w_3\} \cup \{w_4\}$ and each $\{w_i\}$ supports φ. However, $\{w_1, w_2, w_3, w_4\}$ does not support $\varphi \vee \varphi$, because there is no way of writing $\{w_1, w_2, w_3, w_4\}$ as a union of two sets each of which support φ.

For (ii), take $\varphi := p \mathbin{\vee\!\!\!\vee} q$, $\psi := r \mathbin{\vee\!\!\!\vee} s$, and a classical team model with $W = \{w_1, w_2, w_3\}$ such that $V(p) = \{w_1, w_2\}$, $V(q) = \{w_3\}$, $V(r) = \{w_1\}$, and $V(s) = \{w_2, w_3\}$. Then the team $\{w_1, w_2, w_3\}$ supports $\varphi \vee \varphi$, because $\{w_1, w_2, w_3\} = \{w_1, w_2\} \cup \{w_3\}$ and the teams $\{w_1, w_2\}$ and $\{w_3\}$ each support φ; and the team $\{w_1, w_2, w_3\}$ supports $\psi \vee \psi$, because $\{w_1, w_2, w_3\} = \{w_1\} \cup \{w_2, w_3\}$ and the teams $\{w_1\}$ and $\{w_2, w_3\}$ each support ψ. However, the team $\{w_1, w_2, w_3\}$ does not support $(\varphi \wedge \psi) \vee (\varphi \wedge \psi)$, because there is no way of writing $\{w_1, w_2, w_3\}$ as a union of two sets each of which supports $\varphi \wedge \psi$. For the only teams that support $\varphi \wedge \psi$ are the singleton teams.

5 The Nuclear Perspective

The Beth semantics for $\mathcal{L}_{\vee, \mathbin{\vee\!\!\!\vee}}$ in Section 2 can be regarded as a special case of an algebraic semantics based on nuclei.

Definition 5.1 A *nucleus* on a Heyting algebra H is a unary function $j : H \to H$ such that for all $a, b \in H$, $a \leq ja$ (increasing), $jja \leq ja$ (idempotent), and $j(a \wedge b) = ja \wedge jb$ (multiplicative); and j is *dense* if $j0 = 0$.

We denote the meet, join, and implication operations of a Heyting algebra by $\wedge$, $\vee$, and $\to$, trusting that no confusion will arise.

Definition 5.2 A *nuclear algebra* is a pair (H, j) where H is a Heyting algebra and j is a nucleus on H. The algebra is *dense* if j is dense.

The following useful lemma and key theorem are well known.

Lemma 5.3 *For any nuclear algebra (H, j) and $a, b \in H$, we have $j(a \vee b) = j(ja \vee b) = j(a \vee jb) = j(ja \vee jb)$.*

Theorem 5.4 *For any nuclear algebra (H, j), the set $H_j = \{a \in L \mid ja = a\}$ of fixpoints of j is a Heyting algebra, called the* algebra of fixpoints *in (H, j), under the following operations for $a, b \in H_j$: $0_j = j0$, $a \wedge_j b = a \wedge b$, $a \vee_j b = j(a \vee b)$, and $a \to_j b = a \to b$.*

The key idea of the nuclear semantics for $\mathcal{L}_{\vee, \mathbin{\vee\!\!\!\vee}}$ is that the inquisitive disjunction $\mathbin{\vee\!\!\!\vee}$ is interpreted as the join in the nuclear algebra, while the declarative disjunction $\vee$ is interpreted by applying the nucleus to the join.

Definition 5.5 Given a nuclear algebra (H, j) and valuation $v : \mathsf{Prop} \to H_j$, we define $\widehat{v} : \mathcal{L}_{\vee, \mathbin{\vee\!\!\!\vee}} \to H$ by: $\widehat{v}(\bot) = j0$, $\widehat{v}(\varphi \wedge \psi) = \widehat{v}(\varphi) \wedge \widehat{v}(\psi)$, $\widehat{v}(\varphi \vee \psi) = j(\widehat{v}(\varphi) \vee \widehat{v}(\psi))$, $\widehat{v}(\varphi \to \psi) = \widehat{v}(\varphi) \to \widehat{v}(\psi)$, and $\widehat{v}(\varphi \mathbin{\vee\!\!\!\vee} \psi) = \widehat{v}(\varphi) \vee \widehat{v}(\psi)$.

A formula φ is *valid* on (H, j) according to inquisitive nuclear semantics iff for any $v : \mathsf{Prop} \to H_j$, we have $\widehat{v}(\varphi) = 1$ (otherwise φ is *refuted*); and φ is valid over a class K of nuclear algebras iff it is valid on every algebra in K.

The following soundness result is easy to check.

Proposition 5.6 *For any class* K *of nuclear algebras, the set of* $\mathcal{L}_{\vee,\mathbin{\rotatebox[origin=c]{-90}{$\geqslant$}}}$*-formulas valid over* K *according to inquisitive nuclear semantics is an inquisitive intuitionistic logic.*

Beth semantics can be seen as a special case of nuclear semantics using the following intermediate structures from [4].

Definition 5.7 A *nuclear frame* is a triple $(S, \sqsubseteq, j)$ where $(S, \sqsubseteq)$ is a poset and j is a nucleus on the Heyting algebra $\mathsf{Up}(S, \sqsubseteq)$ of upsets of $(S, \sqsubseteq)$.

Example 5.8 The Beth nucleus j_b on $\mathsf{Up}(S, \sqsubseteq)$ is defined by

$$j_b U = \{x \in X \mid \text{every path through } x \text{ intersects } U\}.$$

6 Translation into Lax Logic

The nuclear perspective of the previous section can be made explicit, at the level of the object language, by translating the inquisitive intuitionitic language $\mathcal{L}_{\vee,\mathbin{\rotatebox[origin=c]{-90}{$\geqslant$}}}$ into the language $\mathcal{L}_{\bigcirc}$ of propositional lax logic [15].

Definition 6.1 Let $\mathcal{L}_{\bigcirc}$ be the language defined as follows, where $p \in \mathsf{Prop}$:

$$\varphi ::= \bot \mid p \mid (\varphi \wedge \varphi) \mid (\varphi \vee \varphi) \mid (\varphi \to \varphi) \mid \bigcirc\varphi.$$

Let $\mathcal{L}_{\bigcirc p}$ be the language defined as follows, where $p \in \mathsf{Prop}$:

$$\varphi ::= \bigcirc\bot \mid \bigcirc p \mid (\varphi \wedge \varphi) \mid (\varphi \vee \varphi) \mid (\varphi \to \varphi) \mid \bigcirc\varphi.$$

We first consider the obvious nuclear-algebraic semantics for $\mathcal{L}_{\bigcirc}$, where the sole disjunction $\vee$ of $\mathcal{L}_{\bigcirc}$ is interpreted as the join in the Heyting algebra.

Definition 6.2 Given a nuclear algebra (H, j) and valuation $v : \mathsf{Prop} \to H$, we define $\overline{v} : \mathcal{L}_{\bigcirc} \to H$ by: $\overline{v}(\bot) = 0$, $\overline{v}(\varphi\wedge\psi) = \overline{v}(\varphi)\wedge\overline{v}(\psi)$, $\overline{v}(\varphi\vee\psi) = \overline{v}(\varphi)\vee\overline{v}(\psi)$, $\overline{v}(\varphi \to \psi) = \overline{v}(\varphi) \to \overline{v}(\psi)$, $\overline{v}(\bigcirc\varphi) = j\overline{v}(\varphi)$.

For simplicity, we will drop the overline on v when there is no risk of confusion.

A special case of this nuclear-algebraic semantics for $\mathcal{L}_{\bigcirc}$ is the following *lax Beth semantics* for $\mathcal{L}_{\bigcirc}$.

Definition 6.3 For any poset X, $x \in X$, valuation $v : \mathsf{Prop} \to \mathsf{Up}(X)$, and $\varphi \in \mathcal{L}_{\bigcirc}$, we define $X, x \Vdash_v \varphi$ as follows:

(i) $X, x \nVdash_v \bot$; $X, x \Vdash_v p$ iff $x \in v(p)$;

(ii) $X, x \Vdash_v \varphi \wedge \psi$ iff $X, x \Vdash_v \varphi$ and $X, x \Vdash_v \psi$;

(iii) $X, x \Vdash_v \varphi \vee \psi$ iff $X, x \Vdash_v \varphi$ or $X, x \Vdash_v \psi$;

(iv) $X, x \Vdash_v \varphi \to \psi$ iff for every $y \geq x$, if $X, y \Vdash_v \varphi$ then $X, y \Vdash_v \psi$;

(v) $X, x \Vdash_v \bigcirc\varphi$ iff every path through x intersects $\{y \in X \mid X, y \Vdash_v \varphi\}$.

We now translate $\mathcal{L}_{\vee,\mathbin{\rotatebox[origin=c]{-90}{$\geqslant$}}}$ into $\mathcal{L}_{\bigcirc p}$ as follows.

Definition 6.4 Let ℓ be the translation from $\mathcal{L}_{\vee,\mathbin{\rotatebox[origin=c]{-90}{$\geqslant$}}}$ to $\mathcal{L}_{\bigcirc p}$ defined by: $\ell(\bot) = \bigcirc\bot$, $\ell(p) = \bigcirc p$, $\ell(\varphi \wedge \psi) = \ell(\varphi) \wedge \ell(\psi)$, $\ell(\varphi \vee \psi) = \bigcirc(\ell(\varphi) \vee \ell(\psi))$, $\ell(\varphi \mathbin{\rotatebox[origin=c]{-90}{$\geqslant$}} \psi) = \ell(\varphi) \vee \ell(\psi)$, and $\ell(\varphi \to \psi) = \ell(\varphi) \to \ell(\psi)$.

It is easy to check that the translation is full and faithful in the following sense.

Lemma 6.5 *Let X be a poset and $\varphi \in \mathcal{L}_{\vee,\mathbb{V}}$. Then X validates φ according to inquisitive Beth semantics for $\mathcal{L}_{\vee,\mathbb{V}}$ iff X validates $\ell(\varphi)$ according to lax Beth semantics for $\mathcal{L}_{\bigcirc}$.*

We can also define a translation in the other direction as follows.

Definition 6.6 Let ι be the translation from $\mathcal{L}_{\bigcirc}$ to $\mathcal{L}_{\vee,\mathbb{V}}$ defined by: $\iota(\bot) = \bot$, $\iota(p) = p$, $\iota(\varphi \wedge \psi) = \iota(\varphi) \wedge \iota(\psi)$, $\iota(\varphi \vee \psi) = \iota(\varphi) \mathbb{V} \iota(\psi)$, $\iota(\varphi \to \psi) = \iota(\varphi) \to \iota(\psi)$, and $\iota(\bigcirc\varphi) = \iota(\varphi) \vee \iota(\varphi)$.

For the fragment $\mathcal{L}_{\bigcirc p}$ of $\mathcal{L}_{\bigcirc}$, it is easy to check the following.

Lemma 6.7 *Let X be a poset and $\varphi \in \mathcal{L}_{\bigcirc p}$. Then X validates φ according to lax Beth semantics for $\mathcal{L}_{\bigcirc}$ iff X validates $\iota(\varphi)$ according to inquisitive Beth semantics for $\mathcal{L}_{\vee,\mathbb{V}}$.*

However, the lemma does not extend to all $\varphi \in \mathcal{L}_{\bigcirc}$.

Example 6.8 The formula $\bigcirc p \to p$ is not valid according to lax Beth semantics, but $\iota(\bigcirc p \to p) = (p \vee p) \to p$ is valid according to inquisitive Beth semantics.

Next it is easy to check that composing the translations produces a formula provably equivalent to the original input.

Lemma 6.9 *For any $\varphi \in \mathcal{L}_{\vee,\mathbb{V}}$, the formula $\varphi \leftrightarrow \iota(\ell(\varphi))$ is a theorem of* $\mathsf{Inq(IPC)}$.

Proof. By induction on φ. In the base case, $\bot \leftrightarrow (\bot \vee \bot)$ and $p \leftrightarrow (p \vee p)$ are provable using the axioms of $\mathsf{Inq(IPC)}$. The $\wedge$, $\to$, and $\mathbb{V}$ cases use the inductive hypothesis and replacement of equivalents. For the $\vee$ case, proving

$$(\varphi \vee \psi) \leftrightarrow ((\iota(\ell(\varphi)) \mathbb{V} \iota(\ell(\psi))) \vee (\iota(\ell(\varphi)) \mathbb{V} \iota(\ell(\psi))))$$

uses the inductive hypothesis, replacement of equivalents, and an axiom. □

Though in [15] 'propositional lax logic' (PLL) refers to a single system, we can define a family of lax logics, of which PLL is the smallest.

Definition 6.10 A *propositional lax logic* is a set L of $\mathcal{L}_{\bigcirc}$ formulas that contains the following formulas and is closed under the following rules for all $\varphi, \psi \in \mathcal{L}_{\bigcirc}$:

- all $\mathcal{L}_{\bigcirc}$-substitution instances of IPC axioms stated in $\mathcal{L}_{\vee}$;
- $\varphi \to \bigcirc\varphi$, $\bigcirc\bigcirc\varphi \to \bigcirc\varphi$, and $\bigcirc(\varphi \wedge \psi) \leftrightarrow (\bigcirc\varphi \wedge \bigcirc\psi)$;
- rules of modus ponens and replacement of equivalents.

A *dense* propositional lax logic is a propositional lax logic containing $\bigcirc\bot \to \bot$.

Again soundness is easy to check.

Proposition 6.11 *For any class K of posets, the set of $\mathcal{L}_{\bigcirc}$-formulas valid over K according to lax Beth semantics is a dense propositional lax logic.*

Remark 6.12 To obtain Beth semantics for lax logics without the density axiom, one needs to consider Beth semantics with "strange" or "exploded" worlds that force $\bot$ (see, e.g., [14]), but we leave this for future work.

Here we will consider the (dense) lax logic based on IPC, but the same idea applies to any si-logic.

Definition 6.13 For any si-logic L, let $\mathsf{Lax}(\mathsf{L})$ and $\mathsf{Lax_d}(\mathsf{L})$ be the smallest propositional lax logic and the smallest dense propositional lax logic, respectively, containing all $\mathcal{L}_{\bigcirc}$-substitution instances of L axioms stated in $\mathcal{L}_{\vee}$.

In the Appendix, we prove the following algebraic completeness results.

Theorem 6.14

(i) *If* $\mathsf{Lax}(\mathsf{IPC}) \nvdash \varphi$*, then there is a finite nuclear algebra that refutes* φ*.*

(ii) *If* $\mathsf{Lax_d}(\mathsf{IPC}) \nvdash \varphi$*, then there is a finite dense nuclear algebra that refutes* φ*.*

7 S-Frame Completeness

We now transfer the completeness result of Theorem 6.14 to completeness with respect to certain finite relational structures, which we call "S-frames," from [3]. Proofs of the two lemmas and proposition below can be extracted from [3].

Definition 7.1 An *S-frame* is a triple $\mathfrak{S} = (X, \sqsubseteq, S)$ where $(X, \sqsubseteq)$ is a poset and $S \subseteq X$. $\mathfrak{S}$ is *cofinal* if S is cofinal in $(X, \sqsubseteq)$, i.e., for all $x \in X$ there is a $y \in S$ such that $x \sqsubseteq y$.

S-frames can be constructed from nuclear algebras as follows.

Definition 7.2 Given a nuclear algebra $\mathfrak{A} = (H, j)$, define $\mathfrak{A}_\star := (X, \sqsubseteq, S)$ as follows: X is the set of all prime filters of H; $\sqsubseteq$ is the inclusion order on X; and $S = \{F \in X \mid j^{-1}[F] = F\}$.

Lemma 7.3 *For any nuclear algebra* $\mathfrak{A}$*:*

(i) $\mathfrak{A}_\star$ *is an S-frame;*

(ii) *if* $\mathfrak{A}$ *is dense, then* $\mathfrak{A}_\star$ *is cofinal.*

Conversely, we construct a nuclear algebra from an S-frame as follows.

Definition 7.4 Given an S-frame $\mathfrak{S} = (X, \sqsubseteq, S)$, define the algebra $\mathfrak{S}^\star := (H, j_S)$ as follows: $H = \mathsf{Up}(X)$; for $U \in H$, $j_S U = \{x \in X \mid {\uparrow}x \cap S \subseteq U\}$.

Lemma 7.5 *For any S-frame* $\mathfrak{S}$, $\mathfrak{S}^\star$ *is a nuclear algebra.*

Proposition 7.6 *If* $\mathfrak{A}$ *is a nuclear algebra, then* $\mathfrak{A}$ *embeds into* $(\mathfrak{A}_\star)^\star$*. Moreover, if* $\mathfrak{A}$ *is finite, then the embedding is an isomorphism.*

Say that an S-frame $\mathfrak{S}$ validates/refutes a formula $\varphi \in \mathcal{L}_{\bigcirc}$ just in case $\mathfrak{S}^\star$ validates/refutes φ according to Definition 5.5. Then we obtain the following completeness result from Theorem 6.14, Lemma 7.3, and Proposition 7.6.

Corollary 7.7 *If* $\mathsf{Lax_d}(\mathsf{IPC}) \nvdash \varphi$*, then there is a finite cofinal S-frame that refutes* φ*.*

8 Beth Completeness

In this section, we first prove the Beth completeness of our lax logics and then the Beth completeness of our inquisitive intuitionistic logics.

8.1 Beth Completeness of $\mathsf{Lax_d(IPC)}$

Our strategy is to turn any finite cofinal S-frame refuting a non-theorem φ of $\mathsf{Lax_d(IPC)}$ into a poset refuting φ according to Beth semantics. For this purpose, we use the following key construction.

Definition 8.1 Given an S-frame $(X, \sqsubseteq, S)$, define $(X^\star, \sqsubseteq^\star)$ by:

(i) $X^\star := X \times \mathbb{N}$;

(ii) $\langle x, t\rangle \sqsubseteq^\star \langle x', t'\rangle$ iff one of the following holds:
  (a) $x \sqsubseteq x'$ and $t = t'$;
  (b) $x = x'$, $x \in S$, and $t < t'$;
  (c) $x \sqsubset x'$ and $t < t'$.

Lemma 8.2 *The relation $\sqsubseteq^\star$ is a partial order.*

There are two key properties of $(X^\star, \sqsubseteq^\star)$. First, if $(X, \sqsubseteq, S)$ is finite and cofinal, then every path eventually reaches a pair whose first coordinate is in S.

Lemma 8.3 *Let $(X, \sqsubseteq, S)$ be a finite cofinal S-frame. If C is a path in $(X^\star, \sqsubseteq^\star)$ through $\langle x, t\rangle$, then there is an $\langle x', t'\rangle$ such that $\langle x, t\rangle \sqsubseteq^\star \langle x', t'\rangle \in C$ and $x' \in S$.*

Proof. Let C be a path in $(X^\star, \sqsubseteq^\star)$ through $\langle x, t\rangle$. Thus, $C_\star = \{x' \in X \mid \langle x', t'\rangle \in C\}$ is a chain in $(X, \sqsubseteq)$, which is finite since $(X, \sqsubseteq)$ is finite. Hence $C_\star$ has a maximum, x_{max}. Suppose for contradiction that $C_\star \cap S = \varnothing$, so $x_{max} \notin S$. Since $x_{max} \in C_\star$, there is a t_{max} such that $\langle x_{max}, t_{max}\rangle \in C$. Then $\langle x_{max}, t_{max}\rangle$ is the maximum of C, for if $\langle x', t'\rangle \in C$ and $\langle x_{max}, t_{max}\rangle \sqsubseteq^\star \langle x', t'\rangle$, then since x_{max} is the maximum of $C_\star$, we have $x_{max} = x'$, in which case $x_{max} \notin S$ implies $t_{max} = t'$ by Definition 8.1. Since $(X, \sqsubseteq)$ is finite, $C_\star$ has an upper bound y that is maximal in $(X, \sqsubseteq)$, so $y \in S$ by the cofinality of the S-frame. As $\langle x_{max}, t_{max}\rangle \sqsubseteq^\star \langle y, t_{max}\rangle$ and $\langle x_{max}, t_{max}\rangle$ is the maximum of C, $\langle y, t_{max}\rangle$ is an upper bound of C. Then since C is a path, $\langle y, t_{max}\rangle \in C$, which with $y \in S$ implies $C_\star \cap S \neq \varnothing$, a contradiction. □

Second, we can always create a path in which the first coordinate of the pairs remains forever stuck at some element of S, as follows.

Lemma 8.4 *Let $(X, \sqsubseteq, S)$ be an S-frame, $\langle x, t\rangle \in X^\star$, and $x \sqsubseteq x' \in S$. Then*

$$C := \{\langle x, t\rangle\} \cup \{\langle x', t'\rangle \mid t' > t\}$$

is a path in $(X^\star, \sqsubseteq^\star)$ through $\langle x, t\rangle$.

Proof. Clearly C is a chain in $(X^\star, \sqsubseteq^\star)$. It is also easy to see that C has no upper bound and hence is closed under upper bounds. Thus, C is a path. □

Lemmas 8.3 and 8.4 are the key ingredients for the proposition to follow. First, a *nuclear p-morphism* [4] between nuclear frames $(S, \sqsubseteq, j)$ and $(S', \sqsubseteq', j')$

is a p-morphism f from $(S, \sqsubseteq)$ to $(S', \sqsubseteq')$ such that for $U \in \mathsf{Up}(S', \sqsubseteq')$, $f^{-1}[j'U] = jf^{-1}[U]$; this ensure that f^{-1} is a nucleus-preserving homomorphism from the nuclear algebra $(\mathsf{Up}(S', \sqsubseteq'), j')$ to the nuclear algebra $(\mathsf{Up}(S, \sqsubseteq), j)$; and if f is onto, then f^{-1} is an embedding, which implies that the fixpoint algebra $\mathsf{Up}(S', \sqsubseteq')_{j'}$ embeds into the fixpoint algebra $\mathsf{Up}(S, \sqsubseteq)_j$.

Proposition 8.5 *Let $(X, \sqsubseteq, S)$ be a finite S-frame. The function f defined by $f(\langle x, t\rangle) = x$ is a nuclear p-morphism from the nuclear frame $(X^\star, \sqsubseteq^\star, j_b)$ onto the nuclear frame $(X, \sqsubseteq, j_S)$.*

Proof. First, we check that f is a p-morphism. It is immediate from Definition 8.1 that if $\langle x, t\rangle \sqsubseteq^\star \langle x', t'\rangle$, then $x \sqsubseteq x'$ and hence $f(\langle x, t\rangle) \sqsubseteq f(\langle x', t'\rangle)$. For the back condition, if $f(\langle x, t\rangle) = x \sqsubseteq y$, then we have $\langle x, t\rangle \sqsubseteq^\star \langle y, t\rangle$ and $f(\langle y, t\rangle) = y$. Next we check that for all $U \in \mathsf{Up}(X, \sqsubseteq)$,

$$j_b f^{-1}[U] = f^{-1}[j_S U].$$

Suppose $\langle x, t\rangle \notin j_b f^{-1}[U]$, so there is a path C through $\langle x, t\rangle$ such that $C \cap f^{-1}[U] = \varnothing$. By Lemma 8.3, there is an $\langle x', t'\rangle$ such that $\langle x, t\rangle \sqsubseteq^\star \langle x', t'\rangle \in C$ and $x' \in S$. From $\langle x', t'\rangle \in C$ and $C \cap f^{-1}[U] = \varnothing$, we have $\langle x', t'\rangle \notin f^{-1}[U]$, so $f(\langle x', t'\rangle) = x' \notin U$. From $\langle x, t\rangle \sqsubseteq^\star \langle x', t'\rangle$, we have $x \sqsubseteq x'$. Since $x \sqsubseteq x' \in S \setminus U$, we have $x \notin j_S U$, so $f(\langle x, t\rangle) \notin j_S U$ and hence $\langle x, t\rangle \notin f^{-1}[j_S U]$.

Now suppose $\langle x, t\rangle \notin f^{-1}[j_S U]$, so $f(\langle x, t\rangle) = x \notin j_S U$. Thus, there is an x' such that $x \sqsubseteq x' \in S \setminus U$, which also implies $x \notin U$. By Lemma 8.4, $C := \{\langle x, t\rangle\} \cup \{\langle x', t'\rangle \mid t' > t\}$ is a path through $\langle x, t\rangle$, and it follows from our choice of x' that C does not intersect $f^{-1}[U]$. Hence $\langle x, t\rangle \notin j_b f^{-1}[U]$. □

We now obtain a completeness result for dense lax logic with respect to Beth semantics that is of interest independently of its application to inquisitive logic.

Theorem 8.6 *For any $\varphi \in \mathcal{L}_{\bigcirc}$, if φ is valid on all posets according to lax Beth semantics, then φ is a theorem of* $\mathsf{Lax_d(IPC)}$.

Proof. Suppose φ is not a theorem of $\mathsf{Lax_d(IPC)}$. Then by Corollary 7.7, φ can be refuted according to S-frame semantics on a finite cofinal S-frame $(X, \sqsubseteq, S)$. Then it follows by Lemma 8.5 that φ can be refuted according to lax Beth semantics on a poset. □

8.2 Beth Completeness of $\mathsf{Inq(IPC)}$

In this section, we transfer the completeness result in Theorem 8.6 to Beth completeness for $\mathsf{Inq(IPC)}$. To do so, we use the translation ι of Definition 6.6. Since Lemma 6.7 for ι only applied to the fragment $\mathcal{L}_{\bigcirc p}$ of $\mathcal{L}_{\bigcirc}$, we will also use the following preliminary translation.

Definition 8.7 Let ξ be the translation from $\mathcal{L}_{\bigcirc}$ to $\mathcal{L}_{\bigcirc p}$ defined by: $\xi(\bot) = \bigcirc\bot$; $\xi(p) = \bigcirc p$; $\xi(\varphi \# \psi) = \xi(\varphi) \# \xi(\psi)$ for $\# \in \{\wedge, \vee, \to\}$; $\xi(\bigcirc\varphi) = \bigcirc\xi(\varphi)$.

Definition 8.8 Let $\mathsf{Lax_d(L)}_{\bigcirc p}$ be the logic for $\mathcal{L}_{\bigcirc p}$ whose axioms are all axioms of $\mathsf{Lax_d(L)}$ that belong to $\mathcal{L}_{\bigcirc p}$ and whose rules are modus ponens and replacement of equivalents.

Lemma 8.9 *For any $\varphi \in \mathcal{L}_{\bigcirc p}$, if φ is a theorem of* $\mathsf{Lax_d(L)}$, *then φ is a theorem of* $\mathsf{Lax_d(L)}_{\bigcirc p}$.

Proof. Suppose φ is a theorem of $\mathsf{Lax_d(L)}$, so there exists a proof $\langle \varphi_1, \dots, \varphi_n \rangle$ in $\mathsf{Lax_d(L)}$ with $\varphi_n = \varphi$. Then $\langle \xi(\varphi_1), \dots, \xi(\varphi_n) \rangle$ is a proof in $\mathsf{Lax_d(L)}_{\bigcirc p}$; for if φ_i is an axiom of $\mathsf{Lax_d(L)}$, then $\xi(\varphi_i)$ is an axiom of $\mathsf{Lax_d(L)}_{\bigcirc p}$, and if φ_i is obtained from φ_j and φ_k by modus ponens, then clearly $\xi(\varphi_i)$ is obtained from $\xi(\varphi_j)$ and $\xi(\varphi_k)$ by modus ponens. Let $\bigcirc a_1, \dots, \bigcirc a_k$ be the atomic formulas occurring in φ, so a_i is either a proposition letter or $\bot$. Then $\xi(\varphi_n)$ is obtained from φ_n by replacing each $\bigcirc a_i$ by $\bigcirc\bigcirc a_i$. Thus, by extending $\langle \xi(\varphi_1), \dots, \xi(\varphi_n) \rangle$ with the axioms $\bigcirc\bigcirc a_i \leftrightarrow \bigcirc a_i$ to

$$\langle \xi(\varphi_1), \dots, \xi(\varphi_n), \bigcirc\bigcirc a_1 \leftrightarrow \bigcirc a_1, \dots, \bigcirc\bigcirc a_k \leftrightarrow \bigcirc a_k \rangle$$

and then repeatedly applying replacement of equivalents starting with $\xi(\varphi_n)$, we finally obtain a proof in $\mathsf{Lax_d(L)}_{\bigcirc p}$ of φ_n. □

Lemma 8.10 *For any $\varphi \in \mathcal{L}_{\bigcirc p}$, if φ is a theorem of* $\mathsf{Lax_d(L)}_{\bigcirc p}$, *then $\iota(\varphi)$ is a theorem of* $\mathsf{Inq(L)}$.

Proof. It suffices to show that for any axiom φ of $\mathsf{Lax_d(L)}_{\bigcirc p}$, $\iota(\varphi)$ is a theorem of $\mathsf{Inq(L)}$. The axioms of $\mathsf{Lax_d(L)}_{\bigcirc p}$ are of two kinds: (i) all $\mathcal{L}_{\bigcirc p}$-substitution instances of L-axioms stated in $\mathcal{L}_{\vee}$, and (ii) the axioms for $\bigcirc$. For (i), each $\mathcal{L}_{\bigcirc p}$-substitution instance φ of an L-axiom stated in $\mathcal{L}_{\vee}$ translates to a formula $\iota(\varphi)$ that is also an $\mathcal{L}_{\vee,\mathbb{V}}$-substitution instance of an L-axiom stated in $\mathcal{L}_{\mathbb{V}}$, so $\mathsf{Inq(L)}$ contains $\iota(\varphi)$. For (ii), in each case the ι-translation of an axiom for $\bigcirc$ is an axiom of $\mathsf{Inq(L)}$:

$$\begin{aligned}
\iota(\varphi \to \bigcirc\varphi) &= \iota(\varphi) \to \iota(\bigcirc\varphi) \\
&= \iota(\varphi) \to (\iota(\varphi) \vee \iota(\varphi)), \text{ an axiom of } \mathsf{Inq(L)} \\
\iota(\bigcirc\bigcirc\varphi \to \bigcirc\varphi) &= \iota(\bigcirc\bigcirc\varphi) \to \iota(\bigcirc\varphi) \\
&= (\iota(\bigcirc\varphi) \vee \iota(\bigcirc\varphi)) \to (\iota(\varphi) \vee \iota(\varphi)) \\
&= ((\iota(\varphi) \vee \iota(\varphi)) \vee (\iota(\varphi) \vee \iota(\varphi))) \to (\iota(\varphi) \vee \iota(\varphi)), \\
&\quad \text{an axiom of } \mathsf{Inq(L)} \\
\iota(\bigcirc(\varphi \wedge \psi) \leftrightarrow (\bigcirc\varphi \wedge \bigcirc\psi)) &= \iota(\bigcirc(\varphi \wedge \psi)) \to \iota(\bigcirc\varphi \wedge \bigcirc\psi) \\
&= (\iota(\varphi \wedge \psi) \vee \iota(\varphi \wedge \psi)) \to (\iota(\bigcirc\varphi) \wedge \iota(\bigcirc\psi)) \\
&= ((\iota(\varphi) \wedge \iota(\psi)) \vee (\iota(\varphi) \wedge \iota(\psi))) \to \\
&\quad ((\iota(\varphi) \vee \iota(\varphi)) \wedge (\iota(\psi) \vee \iota(\psi))), \\
&\quad \text{an axiom of } \mathsf{Inq(L)}.
\end{aligned}$$

Finally, for the dense axiom: $\iota(\bigcirc\bot \to \bot) = (\bot \vee \bot) \to \bot$, an axiom of $\mathsf{Inq(L)}$. □

We can now put everything together to prove our main result: completeness of inquisitive intuitionistic logic with respect to Beth semantics.

Theorem 8.11 $\mathsf{Inq}(\mathsf{IPC})$ *is sound and complete with respect to all posets according to Beth semantics.*

Proof. Soundness is easy. For completeness, we have:

φ is valid in Beth semantics for $\mathcal{L}_{\vee,\mathbb{V}}$

$\Rightarrow \ell(\varphi)$ is valid in Beth semantics for $\mathcal{L}_{\bigcirc}$ by Lemma 6.5

$\Rightarrow \ell(\varphi)$ is a theorem of $\mathsf{Lax_d}(\mathsf{IPC})$ by Theorem 8.6

$\Rightarrow \ell(\varphi)$ is a theorem of $\mathsf{Lax_d}(\mathsf{IPC})_{\bigcirc p}$ by Lemma 8.9

$\Rightarrow \iota(\ell(\varphi))$ is a theorem of $\mathsf{Inq}(\mathsf{IPC})$ by Lemma 8.10

$\Rightarrow \varphi$ is a theorem of $\mathsf{Inq}(\mathsf{IPC})$ by Lemma 6.9.

□

9 Conclusion

We have shown the viability of an approach to inquisitive logic on an intuitionistic base using Beth semantics rather than team semantics. As noted, there are two other motivations for this work, independent of inquisitive logic:

- it is natural to consider adding a "Kripke disjunction" $\mathbb{V}$ to Beth semantics and to axiomatize the resulting logic, as we have done with our $\mathsf{Inq}(\mathsf{IPC})$;
- this study unearthed the fact that an old semantics for intuitionistic logic, Beth semantics, can provide a new semantics for (dense) lax logic.

A natural next step, given our general definition of the inquisitive version $\mathsf{Inq}(\mathsf{L})$ of a superintuitionistic logic L, is to investigate the completeness of $\mathsf{Inq}(\mathsf{L})$ for some well-motivated choices of L. One of the axiom schemas of classical inquisitive logic [7] is the schema

$$(\neg\varphi \to (\psi \mathbb{V} \chi)) \to ((\neg\varphi \to \psi) \mathbb{V} (\neg\varphi \to \chi))$$

of the superintuitionistic Kreisel-Putnam logic (KP), which is valid on the special Kripke models used for classical inquisitive logic (recall Section 1). Since we have considered Beth semantics over arbitrary posets, we can refute the KP axiom, but we could also consider restricting to posets satisfying the first-order property corresponding to the KP axiom in Kripke semantics (see, e.g., [6, p. 55]). In fact, in their intuitionistic inquisitive logic, Ciardelli et al. [11] include the schema

$$(\alpha \to (\psi \mathbb{V} \chi)) \to ((\alpha \to \psi) \mathbb{V} (\alpha \to \chi))$$

for α a formula without $\mathbb{V}$, which is equivalent to having the Kreisel-Putnam schema in classical inquisitive logic but not in the intuitionistic setting (see endnote 4 of [11]). We leave the Beth completeness of inquisitive intuitionistic logics with these additional schemas as an open problem.

Acknowledgements

The ideas of this paper were presented at the Workshop on Inquisitive Logic on July 25, 2018 at the University of Amsterdam and later in more developed form at SYSMICS 2019 on January 25, 2019 at the University of Amsterdam. I thank Nick Bezhanishvili, Ivano Ciardelli, Gianluca Grilletti, Vít Punčochář, and other members of the audiences in Amsterdam for helpful feedback, and Yifeng Ding, Guillaume Massas, and the AiML referees for helpful comments on the paper. Special thanks to Guram Bezhanishvili for extensive discussion and for the approach to obtaining the FMP of lax logics in Appendix A, and Sebastian Melzer for spotting a missing axiom in my original version of Definition 3.1.

Appendix

A Proof of Theorem 6.14

In this appendix, we prove the algebraic completeness and finite model property of the lax logics considered in this paper. We first recall the two strategies that have been previously employed to the prove the finite model property for superintuitionistic logics (cf. [2]). The one strategy uses the local finiteness of bounded distributive lattices, while the other that we will build on uses the local finiteness of implicative semilattices, as in the following standard lemmas.

Lemma A.1 *Let $\mathfrak{A}$ be a Heyting algebra, X a finite subset of $\mathfrak{A}$, and $\mathfrak{C}$ the bounded sublattice of $\mathfrak{A}$ generated by X. Then:*

(i) *$\mathfrak{C}$ is a finite Heyting algebra with implication $\to_{\mathfrak{C}}$ given by $a \to_{\mathfrak{C}} b = \bigvee\{x \in \mathfrak{C} \mid x \leq a \to b\}$;*

(ii) *for any $a, b \in \mathfrak{C}$, we have $a \to_{\mathfrak{C}} b \leq a \to b$;*

(iii) *for any $a, b \in \mathfrak{C}$ such that $a \to b \in \mathfrak{C}$, we have $a \to_{\mathfrak{C}} b = a \to b$.*

First proof of FMP of IPC. The approach via Lemma A.1 was utilized by McKinsey and Tarski [22] to prove that IPC has the finite model property. If $\mathsf{IPC} \nvdash \varphi$, then there is a Heyting algebra $\mathfrak{A}$, e.g., the Lindenbaum algebra of IPC, and valuation $v_{\mathfrak{A}}$ on $\mathfrak{A}$ refuting φ. Let $X = \{\overline{v_{\mathfrak{A}}}(\psi) \mid \psi \in Sub(\varphi)\}$, where $Sub(\varphi)$ is the set of subformulas of φ, and generate the finite $\mathfrak{C}$ by X as in Lemma A.1.i. Then $v_{\mathfrak{A}}$ restricts to a valuation $v_{\mathfrak{C}}$ on $\mathfrak{C}$, and for all $\psi \in Sub(\varphi)$, we have $\overline{v_{\mathfrak{A}}}(\psi) = \overline{v_{\mathfrak{C}}}(\psi)$, where the key step of the inductive proof uses the fact about $\to_{\mathfrak{C}}$ in Lemma A.1.iii. Hence $\overline{v_{\mathfrak{A}}}(\varphi) \neq 1$ implies $\overline{v_{\mathfrak{C}}}(\varphi) \neq 1$.

Lemma A.2 *Let $\mathfrak{A}$ be a Heyting algebra, X a finite subset of $\mathfrak{A}$, and $\mathfrak{B}$ the $\{\wedge, \to, 0\}$-subalgebra of $\mathfrak{A}$ generated by X. Then:*

(i) *$\mathfrak{B}$ is a finite Heyting algebra with join $\vee_{\mathfrak{B}}$ given by $a \vee_{\mathfrak{B}} b = \bigwedge\{x \in \mathfrak{B} \mid a \vee b \leq x\}$;*

(ii) *for any $a, b \in \mathfrak{B}$, we have $a \vee b \leq a \vee_{\mathfrak{B}} b$;*

(iii) *for any $a, b \in \mathfrak{B}$ such that $a \vee b \in \mathfrak{B}$, we have $a \vee_{\mathfrak{B}} b = a \vee b$.*

Second proof of FMP of IPC. The approach via Lemma A.2 is due to Diego [13]. One can prove the finite model property of IPC by using the same strategy as above but generating $\mathfrak{B}$ instead of $\mathfrak{C}$ from X. Again one proves that for all $\psi \in Sub(\varphi)$, we have $\overline{v_{\mathfrak{A}}}(\psi) = \overline{v_{\mathfrak{B}}}(\psi)$, but now the key step of the inductive proof uses the fact about $\vee_{\mathfrak{B}}$ in Lemma A.2.iii.

To obtain the finite model property for our lax logics, we need to incorporate nuclei into the above constructions. For this we require the following definition.

Definition A.3

(i) Given posets P and Q, $r : P \to Q$, and $\ell : Q \to P$, we say that (r, ℓ) forms an *adjoint pair* iff for all $p \in P$ and $q \in Q$, $\ell(q) \leq p$ iff $q \leq r(p)$. Then r is the *right adjoint* and ℓ is the *left adjoint.*

(ii) If P and Q are monoids, we say that ℓ is *left exact* if in addition ℓ preserves finite meets.

(iii) A *localization* of a monoid M is a pair (L, ℓ) where L is a submonoid of M and $\ell : M \to L$ is a left exact left adjoint to the inclusion $L \to M$.

The following lemma is well known (see [3, p. 88] and references therein).

Lemma A.4 *There is a one-to-one correspondence between nuclei on a monoid M and its localizations: for any nucleus j on M, we have that (M_j, j) is a localization of M; and for any localization (L, ℓ), we have that ℓ is a nucleus on M such that $M_\ell = L$.*

If the monoid M is in addition a Brouwerian semilattice, then M_j is not only a subalgebra of M but also satisfies the following stronger condition.

Definition A.5 [[23,20]] Let A be a Brouwerian semilattice. A subalgebra T of A is *total* if for any $a \in A$ and $t \in T$, we have that $a \to t \in T$.

The next two lemmas are known, but we include short proofs for the reader's convenience.

Lemma A.6 *Let A be a Brouwerian semilattice and j a nucleus on A.*

(i) *A_j is a total subalgebra of A;*

(ii) *if B is a subalgebra of A, then $A_j \cap B$ is a total subalgebra of B.*

Proof. Part (i) follows from the fact that $j(a \to jb) = a \to jb$, and part (ii) follows from part (i). □

Lemma A.7 *Let A be a Brouwerian semilattice and T a total subalgebra of A. If the inclusion $T \to A$ has a left adjoint ℓ, then (T, ℓ) is a localization of A.*

Proof. Since ℓ is left adjoint to the inclusion, we have $\ell(a) \leq b$ iff $a \leq b$ for all $a \in A$ and $b \in T$. From this it follows that ℓ is order preserving, increasing, and idempotent. To see that it is left exact, let $x, y \in A$. Since ℓ is order preserving, we have $\ell(x \wedge y) \leq \ell(x) \wedge \ell(y)$. For the converse, since ℓ is increasing, we have $x \wedge y \leq \ell(x \wedge y)$, so $x \leq y \to \ell(x \wedge y)$. Since T is a total subalgebra, $y \to \ell(x \wedge y) \in T$, so the adjunction property gives $\ell(x) \leq y \to \ell(x \wedge y)$. Therefore, $\ell(x) \wedge y \leq \ell(x \wedge y)$. From this it follows that $y \leq \ell(x) \to \ell(x \wedge y)$. Since T is a subalgebra, $\ell(x) \to \ell(x \wedge y) \in T$, so applying the adjunction property again yields $\ell(y) \leq \ell(x) \to \ell(x \wedge y)$. Thus, $\ell(x) \wedge \ell(y) \leq \ell(x \wedge y)$. Therefore, (T, ℓ) is a localization of A. □

We now extend Lemma A.2 to the setting of nuclear algebras.

Lemma A.8 *Let $\mathfrak{A}$ be a nuclear algebra, X a finite subset of $\mathfrak{A}$, and $\mathfrak{B}$ the $\{\wedge, \to, 0\}$-subalgebra of $\mathfrak{A}$ generated by X. Then:*

(i) *$\mathfrak{B}$ is a finite nuclear algebra with nucleus $j_{\mathfrak{B}}$ given by $j_{\mathfrak{B}}(a) = \bigwedge\{x \in \mathfrak{A}_j \cap \mathfrak{B} \mid a \leq x\}$;*

(ii) *for any $a \in \mathfrak{B}$, we have $j(a) \leq j_{\mathfrak{B}}(a)$;*

(iii) *for any $a \in \mathfrak{B}$ such that $j(a) \in \mathfrak{B}$, we have $j_{\mathfrak{B}}(a) = j(a)$.*

Proof. For part (i), it follows from Lemma A.2 that $\mathfrak{B}$ is a finite Heyting algebra. By Lemma A.6, $\mathfrak{A}_j \cap \mathfrak{B}$ is a total $\{\wedge, \to, 0\}$-subalgebra of $\mathfrak{B}$. Since $\mathfrak{B}$ is finite, $\mathfrak{A}_j \cap \mathfrak{B}$ is finite, so the inclusion $\mathfrak{A}_j \cap \mathfrak{B} \to \mathfrak{B}$ has a left adjoint ℓ given by $\ell(a) = \bigwedge\{b \in \mathfrak{A}_j \cap \mathfrak{B} \mid a \leq b\}$. Hence by Lemmas A.7 and A.4, ℓ is a nucleus on $\mathfrak{B}$. Thus, $(\mathfrak{B}, \ell)$ is a finite nuclear algebra. In addition, for any $a \in \mathfrak{B}$, we have $j(a) \leq \ell(a)$, and if $j(a) \in \mathfrak{B}$, then $j(a) = \ell(a)$, which yields parts (ii)-(iii). □

We can now prove the FMP for $\mathsf{Lax(IPC)}$ and $\mathsf{Lax_d(IPC)}$.

Theorem A.9

(i) *If $\mathsf{Lax(IPC)} \nvdash \varphi$, then there is a finite nuclear algebra that refutes φ.*

(ii) *If $\mathsf{Lax_d(IPC)} \nvdash \varphi$, then there is a finite dense nuclear algebra that refutes φ.*

Proof. For part (i), suppose $\mathsf{Lax(IPC)} \nvdash \varphi$. Then there is a valuation v on the Lindenbaum algebra $\mathfrak{A}$ of $\mathsf{Lax(IPC)}$ such that $v(\varphi) \neq 1$. Let $S = \{v(\psi) \mid \psi \text{ is a subformula of } \varphi\}$. Let $\mathfrak{B}$ be the $\{\wedge, \to, 0\}$-subalgebra of $\mathfrak{A}$ generated by S. By Lemma A.2, $\mathfrak{B}$ is finite Heyting algebra such that

$$\text{if } a \vee b \in S, \text{ then } a \vee b = a \vee_{\mathfrak{B}} b. \tag{A.1}$$

By Lemma A.8, $(\mathfrak{B}, j_{\mathfrak{B}})$ is a finite nuclear algebra such that

$$\text{if } j(a) \in S, \text{ then } j(a) = j_{\mathfrak{B}}(a). \tag{A.2}$$

Let v' be any valuation on $\mathfrak{B}$ such that for all proposition letters $p \in S$, we have $v'(p) = v(p)$. Then an easy induction using (A.1) and (A.2) shows that for all $\psi \in S$, $v(\psi) = v'(\psi)$, whence $v(\varphi) \neq 1$ in $\mathfrak{A}$ implies $v'(\varphi) \neq 1$ in $\mathfrak{B}$.

For part (ii), observe that if $j(0) = 0$, then $j_{\mathfrak{B}}(0) = 0$. The rest of the proof is the same as for part (i). □

References

[1] Beth, E. W., *Semantic construction of intuitionistic logic*, Mededelingen der Koninklijke Nederlandse Akademie van Wetenschappen **19** (1956), pp. 357–388.

[2] Bezhanishvili, G. and N. Bezhanishvili, *An algebraic approach to filtrations for superintuitionistic logics*, in: J. van Eijck, N. Iemhoff and J. Joosten, editors, *Liber Amicorum Albert Visser*, College Publications, London, 2016 pp. 47–56.

[3] Bezhanishvili, G. and S. Ghilardi, *An algebraic approach to subframe logics. Intuitionistic case*, Annals of Pure and Applied Logic **147** (2007), pp. 84–100.

[4] Bezhanishvili, G. and W. H. Holliday, *A semantic hierarchy for intuitionistic logic*, Indagationes Mathematicae **30** (2019), pp. 403–469.

[5] Bezhanishvili, N., G. Griletti and W. H. Holliday, *Algebraic and topological semantics for inquisitive logic via choice-free duality*, in: *Logic, Language, Information, and Computation. WoLLIC 2019*, LNCS **11541** (2019), pp. 35–52.

[6] Chagrov, A. and M. Zakharyaschev, "Modal logic," Oxford Logic Guides **35**, The Clarendon Press, New York, 1997.

[7] Ciardelli, I., "Inquisitive Semantics and Intermediate Logics," Master's thesis, University of Amsterdam (2009), ILLC Master of Logic Thesis Series MoL-2009-11.

[8] Ciardelli, I., "Questions in logic," Ph.D. thesis, Institute for Logic, Language and Computation, University of Amsterdam (2016).

[9] Ciardelli, I., J. Groenendijk and F. Roelofsen, *Inquisitive semantics: A new notion of meaning*, Language and Linguistics Compass **7** (2013), pp. 459–476.

[10] Ciardelli, I., J. Groenendijk and F. Roelofsen, "Inquisitive Semantics," Oxford University Press, 2018.

[11] Ciardelli, I., R. Iemhoff and F. Yang, *Questions and dependency in intuitionistic logic*, Notre Dame Journal of Formal Logic **61** (2020), pp. 75–115.

[12] Ciardelli, I. and F. Roelofsen, *Inquisitive Logic*, Journal of Philosophical Logic **40** (2011), pp. 55–94.

[13] Diego, A., "Sur les algèbres de Hilbert," Collection de Logique Mathématique, Sér. A, Fasc. XXI, Gauthier-Villars, Paris; E. Nauwelaerts, Louvain, 1966.

[14] Dragalin, A. G., "Mathematical Intuitionism: Introduction to Proof Theory," Translations of Mathematical Monographs **67**, American Mathematical Society, Providence, RI, 1988.

[15] Fairtlough, M. and M. Mendler, *Propositional lax logic*, Information and Computation **137** (1997), pp. 1–33.

[16] Goldblatt, R., *Grothendieck topology as geometric modality*, Zeitschrift für Mathematische Logik und Grundlagen der Mathematik **27** (1981), pp. 495–529.

[17] Grzegorczyk, A., *A philosophically plausible formal interpretation of intuitionistic logic*, Indagationes Mathematicae **26** (1964), pp. 596–601.

[18] Holliday, W. H., *Possibility frames and forcing for modal logic (February 2018)* (2018), UC Berkeley Working Paper in Logic and the Methodology of Science.
URL `https://escholarship.org/uc/item/0tm6b30q`

[19] Humberstone, L., *From worlds to possibilities*, Journal of Philosophical Logic **10** (1981), pp. 313–339.

[20] Köhler, P., *Brouwerian semilattices*, Transactions of the American Mathematical Society **268** (1981), pp. 103–126.

[21] Kripke, S. A., *Semantical analysis of intuitionistic logic I*, in: J. N. Crossley and M. A. E. Dummett, editors, *Formal Systems and Recursive Functions*, North-Holland, Amsterdam, 1965 pp. 92–130.

[22] McKinsey, J. C. C. and A. Tarski, *On closed elements in closure algebras*, Annals of Mathematics **47** (1946), pp. 122–162.

[23] Nemitz, W. C., *Implicative homomorphisms with finite ranges*, Proceedings of the American Mathematical Society **33** (1972), pp. 319–322.

[24] Punčochář, V., *Algebras of information states*, Journal of Logic and Computation **27** (2017), pp. 1643–1675.

[25] Punčochář, V., *Substructural inquisitive logic*, The Review of Symbolic Logic **12** (2019), pp. 296–330.

[26] Väänänen, J., "Dependence Logic: A New Approach to Independence Friendly Logic," Cambridge University Press, Cambridge, 2007.

[27] Väänänen, J., *Modal Dependence Logic*, in: K. R. Apt and R. van Rooij, editors, *New Perspectives on Games and Interaction*, Texts in Logic and Games **4**, Amsterdam University Press, Amsterdam, 2008 pp. 237–254.

Existence, Definedness and Definite Descriptions in Hybrid Modal Logic

Andrzej Indrzejczak [1]

Department of Logic, University of Łódź
Lindleya 3/5, 90–131 Łódź
e-mail:andrzej.indrzejczak@filozof.uni.lodz.pl

Abstract

The paper presents a sequent calculus HFM for first-order hybrid modal logic with lambda operator, existence and definedness predicates. It is particularly useful for dealing with non-rigid and non-designating terms and the apparatus of hybrid logics provides a satisfactory structural proof theory of this logic. Its reduct is shown to be equivalent to Fitting and Mendelsohn's tableau system for first-order modal logic by series of syntactical transformations. Additionally, some account of definite descriptions is formulated in terms of extended calculus HFMD and the whole system satisfies cut elimination theorem.

Keywords: first-order modal logic, hybrid modal logic, definite descriptions, sequent calculus, cut elimination.

1 Introduction

First-Order Modal Logic (FOML) is a field far from commonly accepted solutions. During the last two decades at least three important books due to Fitting and Mendelsohn [10], Garson [11], and Goldblatt [13] provided a detailed treatment of different solutions to philosophical and technical problems connected with FOML. One can also find practically useful deductive systems working even for very sophisticated systems. Prefixed tableaux in [10] and natural deduction (ND) in [11] are good examples, yet, they do not provide well-behaved formulations in the sense of structural proof theory. Recently, the work of Corsi and Orlandelli [25], and Orlandelli [24], provide satisfactory proof-theoretic approach in the framework of labelled sequent calculi (SC). This raises the question if more standard version of SC may be used for that aim. In [16] we provided a standard SC for Garson's version of FOML. In this paper we want to focus on the more demanding approach of Fitting and Mendelsohn (FM) and provide for it a standard version of SC satisfying cut elimination.

[1] The results reported in this paper are supported by the National Science Centre, Poland (grant number: DEC-2017/25/B/HS1/01268).

There are at least two important features which make FM one of the most subtle and expressive logics, providing satisfactory solutions to several philosophical problems involved in FOML. The first feature is connected with the application of predicate abstracts; the second with paying attention to the distinction between existence and definedness. It leads to more flexible treatment of scoping difficulties, of non-rigid and non-denoting terms. Both features are expressible in semantics but the resulting logic is hardly representable in a standard axiomatic way. Therefore, on the level of proof systems (tableaux in this case), additional devices are introduced like double prefixing – of formulae and of terms. This may be seen as the third original feature of their approach although not of the logic but of its tableau presentation. Let us comment on these three features and their significance for our proof-theoretic study.

Predicate abstracts built by means of the lambda operator were introduced to studies on FOML by Thomason and Stalnaker [29] and then the technique was developed by Fitting [9]. In the realm of modal logic this technique is mainly used for taking control over scoping difficulties concerning modal operators but also complex terms like definite descriptions. From the standpoint of proof theory it has additional advantages. In general, introduction of complex terms leads to serious problems connected with unrestricted instantiation of such terms for variables. A freedom of instantiation in quantifier rules usually destroys the subformula property and cut elimination proof. The application of predicate abstracts opens a way to avoid such problems by separation of a direct predication restricted to variables and indirect predication via lambda operators. In consequence, respective rules for quantifiers may be restricted to variables as the only allowed instantiated terms without losing generality.

A distinction between existence and definedness (or denotation) usually goes unnoticed although, as is firmly emphasized in [10] "these are really orthogonal issues. Terms designate; objects exist." It is worth noting that the separation of these notions is also important in studies on constructive mathematics and applications to computer science; see for example Beeson [1] and Fefermann [8]. True, this difference may be easily lost if existence is defined as $\exists x(x = t)$ and definedness as $t = t$ since in the context of negative free logic (NFL) both formulae are provably equivalent. Hence, at least in NFL, this conceptual difference cannot be syntactically represented in a sensible way. However, in FM the application of a richer apparatus enables a syntactical separation of these notions. As a consequence, in the proposed language we can talk about existent and non-existent objects, as well as denoting and non-denoting terms. Suitable predicates are definable in FM; in the system proposed below we introduce them as primitive notions which facilitates syntactic control.

The application of prefixes denoting possible worlds and encoding accessibility relations between them is a well known technique due to Fitting and applied usually to formulae for stating that they hold in respective worlds. In FM prefixes are additionally linked to terms to signify their denotation in worlds denoted by prefixes. In our approach we use this solution not as an auxiliary technical device but as a part of our language. More specifically, we will use a variant of hybrid logic (HL) with twofold application of satisfaction-operators; to formulae and to terms. In contrast to FM we restrict this kind of "rigidification" only to terms which are not variables. Variables in FM are rigid by definition and addition of prefixes is not necessary from this point of view. However, in [10] this technique is applied to variables for enabling control over actualist quantifiers without the need of explicit application of the existence predicate. In our system, the existence predicate is primitive and there is no need to overload variables with unnecessary decoration.

The application of HL to provide well-behaved proof-theoretic representation of FM is not accidental. HL is an interesting generalization of standard modal logic with well established body of general results and extremely rich syntactic resources. The basic language of HL is obtained by the addition of the second sort of propositional atoms called nominals. Informally, they denote propositions true in exactly one world of a model and may serve as names of these worlds. Additionally one can introduce several specific operators; the most important are satisfaction, or "at"-operators, indicating that a formula is satisfied in the world denoted by some nominal. This permits for internalization of the essential part of semantics in the language. What is nice with HL is the fact that changes in the language do not affect seriously the rest of the machinery applied in standard modal logic. In particular, modifications in the relational semantics are minimal. The concept of a frame remains intact, only on the level of models we have some changes. These relatively small modifications of standard modal languages give us many advantages: more expressive language, better behavior in completeness theory, more natural and simpler proof theory. In particular, one may define in HL such frame conditions like irreflexivity, asymmetry, trichotomy and others not expressible in standard modal languages. Proof theory of HL, developed in the framework of tableaux or natural deduction offers even more general approach than application of labels popular in proof theory for standard modal logic.

The aim of the paper is twofold: 1 Extension of HL to obtain fuller expressibility of phenomena so far dealt with only in standard FOML. 2 Providing well-behaved structural proof theory for FM. The original FM is well defined semantically and by means of tableaux which are useful in practice but not fully satisfactory from the theoretical standpoint. Many significant features are introduced as additional technical devices or left implicit (like clauses concerning definedness of terms and existence of objects). What we gain is an SC where all this stuff is introduced explicitly and treated in a uniform fashion by means of well-behaved symmetric and analytic rules satisfying cut elimination.

We start with a brief account of the language and semantics of HL. In section 3 a system HFM (for Hybrid FM) will be presented. Its adequacy is shown indirectly in stages in section 4. First by translation of Fitting's and Mendelsohn's tableau (FM-T) proofs into proofs in some auxiliary calculus HFM1. Then by showing that every proof in HFM1 is simulated in another system HFM2 and vice versa. Finally that HFM2 is equivalent to a reduct of HFM. In section 5 we will extend HFM to cover definite descriptions.

2 Preliminaries

In what follows we assume a standard predicate monomodal language with denumerable sets of predicate symbols $PRED$ (symbolised with P) and function symbols FUN (symbolised with f), both of any arity $n \geq 0$. Incidentally, for representing 0-ary functions we will use c (for individual constant). Individual variables are divided into disjoint sets of bound and free variables (respectively VAR represented by $x, y, z, \ldots$ and PAR, for parameters, represented by $a, b, \ldots$). The set of logical constants comprises boolean and modal connectives: $\neg, \wedge, \vee, \rightarrow, \Box, \Diamond$, (actualist) quantifiers: $\exists, \forall$ and identity predicate $=$. To this basic assortment we add special constants from FM: unary predicates of existence E, nonexistence E^-, definedness D, binary term equality $\approx$, and lambda operator λ for forming predicate abstracts. Categories of terms $TERM$ and formulae FOR are defined in a standard way with an additional clause for the lambda operator:

- if $\varphi \in FOR$ and $t \in TERM$, then $(\lambda x \varphi)t \in FOR$

Hybrid version of this language is obtained by addition of two components: a denumerable set of propositional symbols called nominals $NOM = \{i, j, k, ...\}$ and a denumerable set of unary satisfaction operators (sat-operators) indexed by nominals $@_i$. Following Blackburn and Marx [3] (see also [4], [22], [21]) we will use the latter in two functions; the new clauses are:

- if $\varphi \in FOR$ and $i \in NOM$, then $@_i\varphi \in FOR$
- if $t \in TERM$ and $i \in NOM$, then $@_i t \in TERM$

The first reads "φ is satisfied in a state i". The second – "$@_i t$ names a designatum of t in a state i". For the language of the basic system HFM we restrict the second application of sat-operators to terms other than parameters.

Nominals are introduced for naming states (worlds) of a model domain so in a sense they are terms. However syntactically they are treated as ordinary sentences. In particular, they can be combined by means of boolean and modal connectives. Informally they represent propositions "the name of the actual state is i". On the other hand, they are just names of states when they occur as indices of sat-operators. It is important to note that both nominals and sat-operators are genuine language elements not an extra metalinguistic machinery as in several kinds of labelled systems.

The notion of a frame is defined as for standard FOML and a model is any structure $\mathfrak{M} = \langle \mathcal{W}, \mathcal{R}, \mathcal{D}, d, \mathcal{I}, \mathcal{I}_w \rangle$, where $\mathcal{W}, \mathcal{R}$ is a standard modal frame, $\mathcal{D}$ is a nonempty domain, $d : \mathcal{W} \longrightarrow \mathcal{P}(\mathcal{D})$ is a function which assigns a (nonempty) set of (existent) objects to every world, $\mathcal{I}(i) \in \mathcal{W}$ for every nominal i, and $\mathcal{I}_w$ is a family of world's relative functions of interpretation, defined as follows:

$\mathcal{I}_w(P^n) \subseteq \mathcal{D}^n$, for every n-argument predicate and world;

$\mathcal{I}_w(c) \in \mathcal{D}$, if defined;

$\mathcal{I}_w(f^n) \in \mathcal{D}^{\mathcal{D}^n}$, if defined.

Note that in the last two cases different members of $\mathcal{D}$ and different functions may be selected as designates of c, f in different worlds, so terms (other than variables) are generally non-rigid. Moreover $\mathcal{I}_w$ is partial, i.e. for some w it may be not defined. In case of individual names it means that in some worlds (possibly all) they may be non-denoting. For function symbols it means that corresponding functions are partial, i.e. defined on subsets of $\mathcal{D}^n$.

An assignment v is defined in a standard way as $v : VAR \cup PAR \longrightarrow \mathcal{D}$, hence parameters are rigid terms by definition. An x-variant v' of v is like v with possibly $v'(x) \neq v(x)$; we will use a common notation v^x_0 for x-variant of v with specified value of x. Interpretation $\mathcal{I}^v_w(t)$ of a term t in w under an assignment v is just $v(t)$ for elements of VAR and PAR, $\mathcal{I}_w(t)$ for $t \in FUN$. Hence, $\mathcal{I}^v_w(c) = \mathcal{I}_w(c)$, if it is defined; $\mathcal{I}^v_w(f^n(t_1, ..., t_n)) = \mathcal{I}_w(f^n)\langle\mathcal{I}^v_w(t_1), ..., \mathcal{I}^v_w(t_n)\rangle$, if each $\mathcal{I}^v_w(t_i)$ is defined, and $\langle\mathcal{I}^v_w(t_1), ..., \mathcal{I}^v_w(t_n)\rangle$ is in the domain of $\mathcal{I}_w(f^n)$. In case $t := @_i t'$ it is $\mathcal{I}^v_{\mathcal{I}(i)}(t')$. Hence, attaching the sat-operator $@_i$ to term t' makes it a rigid term, namely an object designed by t' in $\mathcal{I}(i)$ (if there is such an object). That is why it does not make sense to apply sat-operators to parameters; they are already rigid terms. On the other hand in case of complex term $f^n t_1, ..., t_n$ it is not enough to add $@_i$ as a prefix to make it rigid; all arguments must be rigid. In what follows we will use r to denote any rigid term – a parameter or a term with sat-operators attached to all nonparametric components. The clauses for the satisfaction relation are defined as follows:

$\mathfrak{M}, v, w \models P^n(r_1, ..., r_n)$	iff	$\langle\mathcal{I}^v_w(r_1), ..., \mathcal{I}^v_w(r_n)\rangle \in \mathcal{I}_w(P^n)$ and $\mathcal{I}^v_w(r_i) \in \mathcal{D}, i \leq n$
$\mathfrak{M}, v, w \models r_1 = r_2$	iff	$\mathcal{I}^v_w(r_1) = \mathcal{I}^v_w(r_2)$ and $\mathcal{I}^v_w(r_i) \in \mathcal{D}, i \leq 2$
$\mathfrak{M}, v, w \models t_1 \approx t_2$	iff	$\mathcal{I}^v_w(t_1) = \mathcal{I}^v_w(t_2)$ and $\mathcal{I}^v_w(t_i) \in \mathcal{D}, i \leq 2$
$\mathfrak{M}, v, w \models Et$	iff	$\mathcal{I}^v_w(t) \in d(w)$
$\mathfrak{M}, v, w \models Dt$	iff	$\mathcal{I}^v_w(t) \in \mathcal{D}$
$\mathfrak{M}, v, w \models \neg\varphi$	iff	$\mathfrak{M}, v, w \not\models \varphi$
$\mathfrak{M}, v, w \models \varphi \rightarrow \psi$	iff	$\mathfrak{M}, v, w \not\models \varphi$ or $\mathfrak{M}, v, w \models \psi$
$\mathfrak{M}, v, w \models \Box\varphi$	iff	$\mathfrak{M}, v, w' \models \varphi$ for any w' such that $\mathcal{R}ww'$
$\mathfrak{M}, v, w \models \Diamond\varphi$	iff	$\mathfrak{M}, v, w' \models \varphi$ for some w' such that $\mathcal{R}ww'$
$\mathfrak{M}, v, w \models \forall x\varphi$	iff	$\mathfrak{M}, v^x_o, w \models \varphi$ for all $o \in d(w)$
$\mathfrak{M}, v, w \models \exists x\varphi$	iff	$\mathfrak{M}, v^x_o, w \models \varphi$ for some $o \in d(w)$
$\mathfrak{M}, v, w \models (\lambda x\varphi)t$	iff	$\mathcal{I}^v_w(t) \in \mathcal{D}$ and $\mathfrak{M}, v^x_o, w \models \varphi$ for $o = \mathcal{I}^v_w(t)$
$\mathfrak{M}, v, w \models i$	iff	$w = \mathcal{I}(i)$
$\mathfrak{M}, v, w \models @_i\varphi$	iff	$\mathfrak{M}, v, w' \models \varphi$, where $w' = \mathcal{I}(i)$

Note that first two clauses are restricted to rigid terms. Also the only difference between $=$ and $\approx$ is that the latter is defined for all terms. In the original FM semantics [10] atomic formulae and equalities with $=$ are restricted only to variables and have a simpler form:

$\mathfrak{M}, v, w \models P^n(x_1, ..., x_n)$	iff	$\langle v(x_1), ..., v(x_n)\rangle \in \mathcal{I}_w(P^n)$
$\mathfrak{M}, v, w \models x_1 = x_2$	iff	$v(x_1) = v(x_2)$

Rigid terms (other than variables) as arguments of predicates are admissible only as a technical device in FM tableaux. Predicates $E, E^-, D, \approx$ are treated similarly as defined notions. In fact, we could dispense with $\approx$ in FM but it will be necessary later for the treatment of definite descriptions in section 5.

Definitions of truth in a model, satisfiability, validity and entailment are standard. We obtain different normal modal logics by restricting $\mathcal{R}$ suitably.

3 Sequent Calculus HFM

Various proof systems for different hybrid logics were constructed, including tableaux (Blackburn [2], Blackburn and Marx [3], Zawidzki [30]) and natural deduction (ND) (Braüner [6], Indrzejczak [14]). Most of them represent so called sat-calculi where each formula is preceded by the sat-operator. Using sat-calculi instead of calculi working with arbitrary formulae is justified by the fact that φ holds in (any) HL iff $@_i\varphi$ holds, provided i is not present in φ. So a proof of $@_i\varphi$ is equivalent to a proof of φ. One may find several cut-free sat-SC for some HL in different languages independently proposed by Blackburn [2], Braüner [6], Bolander and Braüner [5], Indrzejczak and Zawidzki [19]. In all these cases SC are obtained by translation; either from tableaux or from (normalizable) ND. Hence these systems are cut-free but with no direct syntactical proof for cut elimination. A constructive cut elimination proof for propositional sat-SC was provided by Indrzejczak [15]. In what follows we will use an extended form of this calculus. It consists of the following rules which we divide into several groups for easier reference. Sequents are composed from finite multisets of sat-formulae of the form $@_i\varphi$.

1. Structural rules:

$$(AX)\ \ @_i\varphi \Rightarrow @_i\varphi \quad (C\Rightarrow)\ \frac{@_i\varphi, @_i\varphi, \Gamma \Rightarrow \Delta}{@_i\varphi, \Gamma \Rightarrow \Delta} \quad (\Rightarrow C)\ \frac{\Gamma \Rightarrow \Delta, @_i\varphi, @_i\varphi}{\Gamma \Rightarrow \Delta, @_i\varphi}$$

$$(W\Rightarrow)\ \frac{\Gamma \Rightarrow \Delta}{@_i\varphi, \Gamma \Rightarrow \Delta} \quad (\Rightarrow W)\ \frac{\Gamma \Rightarrow \Delta}{\Gamma \Rightarrow \Delta, @_i\varphi} \quad (Cut)\ \frac{\Gamma \Rightarrow \Delta, @_i\varphi \qquad @_i\varphi, \Pi \Rightarrow \Sigma}{\Gamma, \Pi \Rightarrow \Delta, \Sigma}$$

2. Propositional (Boolean) rules:

$$(\neg\Rightarrow)\ \frac{\Gamma \Rightarrow \Delta, @_i\varphi}{@_i\neg\varphi, \Gamma \Rightarrow \Delta} \qquad (\Rightarrow\neg)\ \frac{@_i\varphi, \Gamma \Rightarrow \Delta}{\Gamma \Rightarrow \Delta, @_i\neg\varphi}$$

$$(\wedge\Rightarrow)\ \frac{@_i\varphi, @_i\psi, \Gamma \Rightarrow \Delta}{@_i(\varphi\wedge\psi), \Gamma \Rightarrow \Delta} \qquad (\Rightarrow\wedge)\ \frac{\Gamma \Rightarrow \Delta, @_i\varphi \qquad \Gamma \Rightarrow \Delta, @_i\psi}{\Gamma \Rightarrow \Delta, @_i(\varphi\wedge\psi)}$$

$$(\Rightarrow\rightarrow)\ \frac{@_i\varphi, \Gamma \Rightarrow \Delta, @_i\psi}{\Gamma \Rightarrow \Delta, @_i(\varphi\rightarrow\psi)} \qquad (\rightarrow\Rightarrow)\ \frac{\Gamma \Rightarrow \Delta, @_i\varphi \qquad @_i\psi, \Gamma \Rightarrow \Delta}{@_i(\varphi\rightarrow\psi), \Gamma \Rightarrow \Delta}$$

3. Modal basic rules:

$$(\Rightarrow \Box)\ \frac{@_i\Diamond j, \Gamma \Rightarrow \Delta, @_j\varphi}{\Gamma \Rightarrow \Delta, @_i\Box\varphi} \qquad (\Box \Rightarrow)\ \frac{\Gamma \Rightarrow \Delta, @_i\Diamond j \quad @_j\varphi, \Gamma \Rightarrow \Delta}{@_i\Box\varphi, \Gamma \Rightarrow \Delta}$$

$$(\Diamond\Rightarrow)\ \frac{@_i\Diamond j,\ @_j\varphi, \Gamma \Rightarrow \Delta}{@_i\Diamond\varphi, \Gamma \Rightarrow \Delta} \qquad (\Rightarrow\Diamond)\ \frac{\Gamma \Rightarrow \Delta, @_i\Diamond j \quad \Gamma \Rightarrow \Delta, @_j\varphi}{\Gamma \Rightarrow \Delta, @_i\Diamond\varphi}$$

where $\varphi \notin NOM$, j does not occur in the conclusion of $(\Rightarrow \Box), (\Diamond \Rightarrow)$.

4. Nominal rules:

These rules are specific for HL and mainly connected with the fact that nominals are formulae of the language, not external devices like prefixes and labels in external proof systems.

$$(@\Rightarrow)\ \frac{@_i\varphi, \Gamma\Rightarrow\Delta}{@_j@_i\varphi,\ \Gamma\Rightarrow\Delta} \qquad (\Rightarrow @)\ \frac{\Gamma\Rightarrow\Delta, @_i\varphi}{\Gamma\Rightarrow\Delta,\ @_j@_i\varphi} \qquad (Ref)\ \frac{@_i i, \Gamma\Rightarrow\Delta}{\Gamma\Rightarrow\Delta}$$

$$(Nom_1)\ \frac{@_i\Diamond k, \Gamma\Rightarrow\Delta}{@_i j, @_j\Diamond k, \Gamma\Rightarrow\Delta} \qquad (Nom_2)\ \frac{@_i\varphi, \Gamma\Rightarrow\Delta}{@_i j, @_j\varphi, \Gamma\Rightarrow\Delta}$$

where φ is atomic or nominal in (Nom_2).

5. Modal frame rules

Rules 1–4 provide an adequate HL formalization of K. In order to cover stronger logics adequate with respect to suitably restricted classes of frames one must add special rules for frame conditions. It may be done in a uniform fashion for many logics by means of standard hybrid translation HT from first-order language into basic hybrid language defined as follows:

$$\begin{array}{lclllcl} HT(Rtt') & = & @_t\Diamond t' & & HT(Pt) & = & @_t p \\ HT(t=t') & = & @_t t' & & HT(\neg\varphi) & = & \neg HT(\varphi) \\ HT(\varphi\vee\psi) & = & HT(\varphi)\vee HT(\psi) & & HT(\exists u\varphi) & = & \exists u HT(\varphi) \end{array}$$

Braüner [6] states that for every basic geometric formula of the form:

$\forall x_1, ..., x_k(\varphi_1 \wedge ... \wedge \varphi_n \rightarrow \exists y_1, ..., y_l(\psi_1 \vee ... \vee \psi_m))$,

where $k \geq 1, l, n, m \geq 0$, each φ_i is an atom and each ψ_i is an atom or finite conjunction of atoms there corresponds a frame rule (FR) of the form:

$$\frac{\Gamma\Rightarrow\Delta, \varphi'_1 \quad ... \quad \Gamma\Rightarrow\Delta, \varphi'_n \qquad \Psi_1, \Gamma\Rightarrow\Delta \quad ... \quad \Psi_m, \Gamma\Rightarrow\Delta}{\Gamma\Rightarrow\Delta}$$

where $k \geq 1, l, n, m \geq 0$, each $\varphi'_i = HT(\varphi_i)$, each Ψ_i is a set of HT-translations of atoms that form conjunction ψ_i and no nominal that corresponds to y_i occurs in $\Gamma_1 - \Gamma_m, \Delta, \varphi'_1 - \varphi'_n$.

6. Specific HFM rules:

$$(\forall E\Rightarrow)\ \frac{\Gamma\Rightarrow\Delta, @_i Eb \quad @_i\varphi[x/b], \Gamma\Rightarrow\Delta}{@_i\forall x\varphi, \Gamma\Rightarrow\Delta} \qquad (\Rightarrow\forall E)\ \frac{@_i Ea, \Gamma\Rightarrow\Delta, @_i\varphi[x/a]}{\Gamma\Rightarrow\Delta, @_i\forall x\varphi}$$

$$(\Rightarrow\exists E)\ \frac{\Gamma\Rightarrow\Delta, @_i Eb \quad \Gamma\Rightarrow\Delta, @_i\varphi[x/b]}{\Gamma\Rightarrow\Delta, @_i\exists x\varphi} \qquad (\exists E\Rightarrow)\ \frac{@_i Ea, @_i\varphi[x/a], \Gamma\Rightarrow\Delta}{@_i\exists x\varphi, \Gamma\Rightarrow\Delta}$$

$$(E-)\ \frac{@_i Ea, \Gamma\Rightarrow\Delta}{\Gamma\Rightarrow\Delta} \qquad (D-)\ \frac{@_i Db, \Gamma\Rightarrow\Delta}{\Gamma\Rightarrow\Delta} \qquad (\approx -)\ \frac{@_i t\approx t, \Gamma\Rightarrow\Delta}{@_i Dt,\ \Gamma\Rightarrow\Delta}$$

$$(=+)\ \frac{@_i\varphi[x/r_1],\Gamma\Rightarrow\Delta}{@_j r_1 = r_2, @_i\varphi[x/r_2],\Gamma\Rightarrow\Delta}\ \text{with } \varphi \text{ atomic formula.}$$

$$(\lambda\Rightarrow)\ \frac{@_iDt, @_i\varphi[x/t^{@_i}],\Gamma\Rightarrow\Delta}{@_i(\lambda x\varphi)t,\Gamma\Rightarrow\Delta} \qquad (\Rightarrow\lambda)\ \frac{\Gamma\Rightarrow\Delta, @_iDt \quad \Gamma\Rightarrow\Delta, @_i\varphi[x/t^{@_i}]}{\Gamma\Rightarrow\Delta, @_i(\lambda x\varphi)t}$$

$$(\Rightarrow E)\ \frac{\Gamma\Rightarrow\Delta, @_iDt \quad \Gamma\Rightarrow\Delta, @_iEb \quad \Gamma\Rightarrow\Delta, @_it^{@_i}=b}{\Gamma\Rightarrow\Delta, @_iEt}$$

$$(E\Rightarrow)\ \frac{@_iDt, @_iEa, @_it^{@_i}=a,\Gamma\Rightarrow\Delta}{@_iEt,\Gamma\Rightarrow\Delta}$$

$$(E^-\Rightarrow)\ \frac{@_iDt,\Gamma\Rightarrow\Delta, @_iEb \quad @_iDt,\Gamma\Rightarrow\Delta, @_it^{@_i}=b}{@_iE^-t,\Gamma\Rightarrow\Delta}$$

$$(\Rightarrow E^-)\ \frac{\Gamma\Rightarrow\Delta, @_iDt \quad @_iEa, @_it^{@_i}=a,\Gamma\Rightarrow\Delta}{\Gamma\Rightarrow\Delta, @_iE^-t}$$

$$(\approx\Rightarrow)\ \frac{@_iDt_1, @_iDt_2, @_it_1^{@_i}=t_2^{@_i},\Gamma\Rightarrow\Delta}{@_it_1\approx t_2,\Gamma\Rightarrow\Delta}$$

$$(\Rightarrow\approx)\ \frac{\Gamma\Rightarrow\Delta, @_iDt_1 \quad \Gamma\Rightarrow\Delta, @_iDt_2 \quad \Gamma\Rightarrow\Delta, @_it_1^{@_i}=t_2^{@_i}}{\Gamma\Rightarrow\Delta, @_it_1\approx t_2}$$

A few conventions were applied in the schemata which will also be used in later sections. a denotes a parameter which is fresh, i.e. occurs only in displayed position and b is any parameter. $t^{@_i} := t$, if t is already rigid; otherwise for $t := c$ it is just $@_ic$ and for $t := f^n t_1, ..., t_n$ it is $@_if^n t_1^{@_i}, ..., t_n^{@_i}$. Moreover, all rules for E and E^- are restricted to $t \notin PAR$. The notion of atomic formula covers formulae of the form $@_iP(r_1...r_n)$ and includes also cases where P is $=, D, E, E^-$; however, in the last two cases with restriction that $r \in PAR$.

The definition of proof is standard, as well as the notions of principal, side and parametric (context) formulae. Also the notion of the height of a proof is standard (the number of nodes of the longest branch). It is important to note that except (Cut) all rules satisfy the generalised subformula property to the effect that in premisses we have only subformulae of the conclusion closed for addition of sat-operator and the following kind of formulae: $@_i\Diamond j, @_ij, @_iEt, @_iDt, @_it=t'$. Moreover one can easily check that arguments of the last three atoms are either (rigidified) terms occuring in the conclusion or fresh parameters. This shows that from the proof-search perspective cut-free HFM is sufficiently analytic. In fact (FM) has also a disadvantage of putting every case in a form of cut-like rule with many premisses composed from formulae of the form $@_i\Diamond j, @_ij$. However, to concrete cases we can apply the rule-generation theorem from [17] which allows us to provide equivalent rules with lower branching factor and active formulae in the conclusion. Since frame expressivity is not our subject here we skip a discussion of these matters.

For cut elimination the key point is that all rules are substitutive and reductive. These notions were introduced by Ciabattoni [7] and applied for general form of cut elimination proof in hypersequent calculi by Metcalfe, Olivetti and Gabbay [23] but can be also applied in the present setting. The former property is connected with the fact that multisets of formulae may be safely substituted for a cut formula which is parametric. It allows for induction on the height of a proof in cases when the cut formula is not principal in at least one premiss of cut. Rules with side conditions concerning fresh parameters or nominals are not fully substitutive but due to the substitution lemma (see Appendix) this problem may be easily overcome. The latter property may be roughly defined as follows: A pair of introduction rules $(\Rightarrow \star)$, $(\star \Rightarrow)$ for a constant $\star$ is reductive if an application of cut on cut formulae introduced by these rules may be replaced by the series of cuts made on less complex formulae. Reductivity permits induction on cut-degree in the course of proving cut elimination. Of course the complexity c of all terms and formulae must be suitably defined:

$c(\alpha) = 0$ for $\alpha \in NOM \cup PAR \cup VAR$; $\quad c(t^{@_i}) = 1$;
$c(@_i\varphi) = c(\neg\varphi) = c(\varphi) + 1$; $c(Ox\varphi) = c(\varphi) + 2$, for $O \in \{\forall, \exists, \lambda\}$;
$c(\varphi \star \psi) = c(\varphi) + c(\psi) + 1$, for $\star$ being a binary connective;
$c(P(r_1, .., r_n)) = max\{c(r_1), ..., c(r_n)\} + 1$ (this clause includes $=$);
$c(Et) = c(E^-t) = c(t) + 2$; $c(Dt) = c(t) + 1$; $c(t_1 \approx t_2) = c(t_1) + c(t_2) + 1$;

For technical reasons we must assume that existence formulae have higher complexity than other atoms with the same arguments, and other rigid terms higher than parameters. One can check by inspection that all rules for compund formulae (including $\approx$) are reductive in pairs. As for $(Nom_1), (Nom_2), (= +)$ introducing nominal and ordinary equalities, they can be principal only in the right premiss of cut; in the left premiss they are always parametric formulae. The same remark applies to Dt which is principal in $(\approx -)$ but can also be principal in $(= +)$. In case of Et and E^-t two situations are possible. With $t \in PAR$ they may be principal also in the right premiss of cut due to $(= +)$, but in the left premiss only parametric. Otherwise in both premisses such formulae are principal due to specific rules for E and E^- which are reductive.

One may easily check that all rules in group 1-4 and 6 are validity-preserving in K models which implies soundness of HFM. The addition of specific rules generated by (FM) extends this to the class of all modal logics axiomatised by geometric formulae. Completeness proof for essentially equivalent propositional SC is provided by Bolander and Braüner [5] (see also [6]). Blackburn and ten Cate [4] provided completeness proof for axiomatic version of HFOML. Fitting and Mendelsohn [10] contain completeness proofs for tableaux formalization of group 6. In what follows instead of proving semantic completeness of HFM we apply a strategy similar to those utilized by Seligman [27] or Blackburn and Marx [3]. We will show that suitably restricted part of HFM is equivalent to FM tableau system by means of a series of purely syntactic transformations.

4 Auxiliary Systems and Transformations

Let us recall briefly FM tableaux (shortly FM-T) for the K-modality. For easier transformations and comparison with SC we will present it in Hintikka-style form, i.e. with sets of (prefixed) formulae as nodes of proof-trees, instead of the original Smullyan-style format defined on single formulae. Prefixes, denoting states in models, are finite lists of integers; for every prefix σ its one-digit extension $\sigma.n$ denotes a state which is accessible from σ (i.e. a state it denotes). Hence FM-T is a kind of external system with a special feature that prefixes are attached not only as prefixes to formulae but also, as subscripts, to terms (including parameters but not bound variables). t_σ syntactically encodes $\mathcal{I}_w(t)$ (where σ denotes w) and is always rigid. t^σ is t_σ if t is a name or a parameter, or just t if it is already rigid (i.e. a term where all components have subscripts). For complex terms, t^σ denotes the result of addition of σ as a subscript to all terms which are not subscripted so far. Rules for connectives are standard so we state only those for $\Box, \forall, =$ and the λ-operator:

$$(\Box)\ \frac{\sigma{:}\Box\varphi, \Gamma}{\sigma.n{:}\varphi, \sigma{:}\Box\varphi, \Gamma} \quad (\neg\Box)\ \frac{\sigma{:}\neg\Box\varphi, \Gamma}{\sigma.n{:}\neg\varphi, \Gamma} \quad (\lambda)\ \frac{\sigma{:}(\lambda x\varphi)t, \Gamma}{\sigma{:}\varphi[x/t^\sigma], \Gamma} \quad (\neg\lambda)\ \frac{\sigma{:}\neg(\lambda x\varphi)t, \Gamma}{\sigma{:}\neg\varphi[x/t^\sigma], \Gamma}$$

$$(\forall)\ \frac{\sigma{:}\forall x\varphi, \Gamma}{\sigma{:}\varphi[x/b_\sigma], \sigma{:}\forall x\varphi, \Gamma} \qquad (\neg\forall)\ \frac{\sigma{:}\neg\forall x\varphi, \Gamma}{\sigma{:}\neg\varphi[x/a_\sigma], \Gamma}$$

$$(=1)\ \frac{\Gamma}{\sigma{:}r = r, \Gamma} \qquad (=2)\ \frac{\sigma{:}r_1 = r_2, \sigma'{:}\varphi[x/r_1], \Gamma}{\sigma{:}r_1 = r_2, \sigma'{:}\varphi[x/r_1], \sigma'{:}\varphi[x/r_2], \Gamma}$$

Side conditions: ($\Box$) $\sigma.n$ occurs in Γ; ($\neg\Box$) $\sigma.n$ is fresh;
($\forall$) b_σ is any parameter with σ (i.e. occuring in Γ, otherwise a new one);
($\neg\forall$) a_σ is fresh; ($\neg\lambda$) t^σ is defined; ($= 2$) φ any formula;
($= 1$) σ occurs in Γ and r is defined (either a parameter or occurs in Γ);

We can get rid of prefixes and use instead the hybrid machinery of nominals with sat-operators obtaining a system HFM-T. It is enough to define a bijective mapping θ from prefixes to nominals in every FM-T proof. We systematically replace all occurences of prefixes in every node with a suitable $@_i$. Moreover, if $\sigma' := \sigma.n$, then we add $@_i\Diamond j$, where $@_i$ and $@_j$ correspond to σ and σ' respectively. As a result, for modals the new rules are obtained of the form:

$$(\Box')\ \frac{@_i\Box\varphi, @_i\Diamond j, \Gamma}{@_j\varphi, @_i\Box\varphi, @_i\Diamond j, \Gamma} \qquad (\neg\Box')\ \frac{@_i\neg\Box\varphi, \Gamma}{@_j\neg\varphi, @_i\Diamond j, \Gamma} \text{ with } j \text{ fresh}$$

This way every node Γ in FM-T proof is transformed into $\Delta, \theta\Gamma$, where $\theta\Gamma$ is a hybrid translation of Γ and Δ is the set of all $@_i\Diamond j$ such that $\sigma, \sigma.n$ occur in Γ, $\theta(\sigma) = i$ and $\theta(\sigma.n) = j$. Since θ can be converted, proofs in both systems are isomorphic and we obtain:

Lemma 4.1 *$1 : \neg\varphi$ has a closed tableau in FM-T iff $@_i\neg\varphi$ (for i not occurring in φ) has a closed tableau in HFM-T*

HFM-T differs from FM-T only in being an internal system. Since we want to work with SC we make two additional, rather cosmetic, changes. Sets are transformed into sequents by moving all negated formulae to succedents (with simultaneous deletion of negations) and finally turning all rules upside-down.

This way we obtain an auxiliary system HFM1 which is quite similar to HFM but for the time being in the language without $D, E, E^-, \approx$. On the other hand, in HFM1 also parameters are prefixed with $@_i$; we call them n-parameters. Moreover, we still need several side conditions, in particular we say that $@_i t$ is defined if t is a parameter or if it appears in the conclusion-sequent. In contrast to HFM, axioms are of the form $\Gamma \Rightarrow \Delta$ with $\Gamma \cap \Delta \neq \varnothing$, and structural rules from group 1 are not required. Propositional part (group 2, 3) is like in HFM but with two slightly different modal rules:

$$(\Box\Rightarrow')\ \frac{@_i\Box\varphi, @_j\varphi, @_i\Diamond j, \Gamma\Rightarrow\Delta}{@_i\Box\varphi, @_i\Diamond j, \Gamma\Rightarrow\Delta} \qquad (\Rightarrow\Diamond')\ \frac{@_i\Diamond j, \Gamma\Rightarrow\Delta, @_i\Diamond\varphi, @_j\varphi}{@_i\Diamond j, \Gamma\Rightarrow\Delta, @_i\Diamond\varphi}$$

The remaining rules look like that:

$$(\forall\Rightarrow)\ \frac{@_i\varphi[x/@_i b], @_i\forall x\varphi, \Gamma\Rightarrow\Delta}{@_i\forall x\varphi, \Gamma\Rightarrow\Delta} \qquad (\Rightarrow\forall)\ \frac{\Gamma\Rightarrow\Delta, @_i\varphi[x/@_i a]}{\Gamma\Rightarrow\Delta, @_i\forall x\varphi}$$

$$(\exists\Rightarrow)\ \frac{@_i\varphi[x/@_i a], \Gamma\Rightarrow\Delta}{@_i\exists x\varphi, \Gamma\Rightarrow\Delta} \qquad (\Rightarrow\exists)\ \frac{\Gamma\Rightarrow\Delta, @_i\exists x\varphi, @_i\varphi[x/@_i b]}{\Gamma\Rightarrow\Delta, @_i\exists x\varphi}$$

$$(=-)\ \frac{@_i r = r, \Gamma\Rightarrow\Delta}{\Gamma\Rightarrow\Delta} \qquad (=+')\ \frac{@_i\varphi[x/r_1], @_j r_1 = r_2, @_i\varphi[x/r_2], \Gamma\Rightarrow\Delta}{@_j r_1 = r_2, @_i\varphi[x/r_2], \Gamma\Rightarrow\Delta}$$

$$(\lambda\Rightarrow')\ \frac{@_i\varphi[x/t^{@_i}], \Gamma\Rightarrow\Delta}{@_i(\lambda x\varphi)t, \Gamma\Rightarrow\Delta} \qquad (\Rightarrow\lambda')\ \frac{\Gamma\Rightarrow\Delta, @_i\varphi[x/t^{@_i}]}{\Gamma\Rightarrow\Delta, @_i(\lambda x\varphi)t}$$

where: $@_i a$ is fresh, whereas $@_i b$ occurs in Γ, Δ, otherwise it is also fresh; r in $(=-)$ and $t^{@_i}$ in $(\Rightarrow\lambda')$ are defined; φ in $(=+')$ is any formula.

Lemma 4.2 *$@_i\neg\varphi$ has closed tableau in HFM-T iff HFM1 $\vdash\ \Rightarrow @_i\varphi$*

The above lemma holds trivially. HFM1 is a quite well-behaved system, however the application of sat-operators to parameters seems to be excessive. After all, parameters are rigid by definition and do not need special indication for that. The problem is that in FM-T the addition of prefixes to parameters plays an additional function; it indicates an existence in a state denoted by a prefix. This of course was transmitted also to HFM1. However, this function can be performed by using the existence predicate, as in the case of free logics. The situation is the same with definedness. In HFM1, exactly as in FM-T, this information is carried out by side conditions added to some rules. Again it is possible to make it explicit by introduction of definedness predicate. In fact, all this, and even more, is present in FM-T [10] in the form of definitions introduced for more compact expression of interesting features of this system:

$$Dt := (\lambda x x = x)t \qquad Et := (\lambda x\exists y(y = x))t$$
$$E^-t := (\lambda x\neg\exists y(y = x))t \qquad t_1 \approx t_2 := (\lambda x(\lambda y x = y)t_2)t_1$$

From this list only the first two are necessary for obtaining the results mentioned above on explicit representation of existence and definedness by means of special formulae instead of side conditions. However, the nonexistence predicate E^- (originally $\bar{E}$) is important for showing the real difference between existence and definedness which is expressed by the thesis $Dt \leftrightarrow Et \vee E^-t$. Term equality $\approx$ is important for definition of suitable rules for definite descriptions in section 5, so we introduce it as well. HFM1 can be enriched in a similar way as FM-T, by introducing definitions for lacking constant predicates. However, it is better to add them as eight additional axioms. For example, for D they have the form:

$$@_iDt, \Gamma \Rightarrow \Delta, @_i(\lambda xx = x)t \quad \text{and} \quad @_i(\lambda xx = x)t, \Gamma \Rightarrow \Delta, @_iDt$$

Moreover, let us notice that we can add to HFM1 three admissible rules: two rules of weakening and cut. Admissibility of weakening can be easilly proven syntactically but with cut it is not so simple. However, cut is validity-preserving and HFM1 is complete, hence admissibility of cut in HFM1 follows. Now we can define another system HFM2 which is based on the same language as HFM, i.e. with $D, E, E^-, \approx$ primitive and without sat-operators attached to parameters. In the propositional basis it is exactly as HFM1, including modal rules; $(= +')$ is also the same. The remaining rules are closer to HFM; in particular $(\Rightarrow \forall E), (\exists E \Rightarrow), (\lambda \Rightarrow)$ are the same, the other ones are:

$$(\forall E{\Rightarrow}') \ \frac{@_i\varphi[x/b], @_i\forall x\varphi, @_iEb, \Gamma \Rightarrow \Delta}{@_i\forall x\varphi, @_iEb, \Gamma \Rightarrow \Delta} \qquad ({\Rightarrow}\lambda'') \ \frac{@_iDt, \Gamma \Rightarrow \Delta, @_i\varphi[x/t^{@_i}]}{@_iDt, \Gamma \Rightarrow \Delta, @_i(\lambda x\varphi)t}$$

$$({\Rightarrow}\exists E') \ \frac{@_iEb, \Gamma \Rightarrow \Delta, @_i\exists x\varphi, @_i\varphi[x/b]}{@_iEb, \Gamma \Rightarrow \Delta, @_i\exists x\varphi} \qquad (\approx -') \ \frac{@_iDt, @_it \approx t, \Gamma \Rightarrow \Delta}{@_iDt, \ \Gamma \Rightarrow \Delta}$$

Note that $(= -)$ is replaced with $(\approx -')$. We add $(E-), (D-)$ as in HFM; these rules are necessary to make explicit what was implicit in HFM1. $(E-)$ is required since in FM semantics world-domains are nonempty. In HFM1 the side condition for $(\forall \Rightarrow)$ and $(\Rightarrow \exists)$ permits introduction of a n-parameter even if no previous application of $(\Rightarrow \forall)$ or $(\exists \Rightarrow)$ provided some. For the present rules an existence formula must be already present in the antecedent, so in case there is no such formula we apply $(E-)$ first (in root first proof-search). Note that if we drop this rule we obtain a variant for logics admitting empty domains. $(D-)$ explicitly shows that (all) parameters are defined. Definedness formulae are also introduced to $(\Rightarrow \lambda'')$ and $(\approx -')$, whereas such a formula added in the premiss-sequent of $(\lambda \Rightarrow)$ plays a similar role for names as in $(D-)$ for parameters. There is one small difference with the original conditions stated by Fitting and Mendelsohn for rules demanding already defined terms. They require that suitable rigid terms should be defined hence if we follow strictly their formulation our definedness formulae in both rules for λ and $(\approx -')$ should be of the form $@_iDt^{@_i}$ instead of $@_iDt$. However, one can easily check that $@_iDt^{@_i}$ and $@_iDt$ are semantically equivalent and for technical reasons using $@_iDt$ in rules is simpler. Instead of definitional axioms we add to HFM2 suitable introduction rules for all additional predicates except D. Rules $(E \Rightarrow)$, $(E^- \Rightarrow)$ and $(\approx \Rightarrow)$ are the same as in HFM; the remaining ones are:

$$(\Rightarrow E')\ \frac{@_iDt, \Gamma \Rightarrow \Delta, @_iEb \quad @_iDt, \Gamma \Rightarrow \Delta, @_it^{@_i} = b}{@_iDt, \Gamma \Rightarrow \Delta, @_iEt}$$

$$(\Rightarrow E^-)\ \frac{@_iDt, @_iEa, @_it^{@_i} = a, \Gamma \Rightarrow \Delta}{@_iDt, \Gamma \Rightarrow \Delta, @_iE^-t}$$

$$(\Rightarrow \approx')\ \frac{@_iDt_1, @_iDt_2, \Gamma \Rightarrow \Delta, @_it_1^{@_i} = t_2^{@_i}}{@_iDt_1, @_iDt_2, \Gamma \Rightarrow \Delta, @_it_1 \approx t_2}$$

As in HFM, in rules for E, E^-, $t \notin PAR$. For D no special rules are needed. Despite several differences concerning rules, and using n-parameters in HFM1 and parameters in HFM2, both systems are equivalent in the sense of provability of the same sequents containing sentences (see Appendix).

Now we are ready to compare HFM2 with HFM. It is straightforward to prove that all rules of HFM2, with the exception of $(= +')$, are derivable in HFM; it is sufficient to apply rule-generation theorem from [17]. $(= +')$ is derivable additionally by induction on the complexity of φ with $(= +)$ in the basis. We cannot in general prove the opposite since already the background propositional hybrid part of HFM contains elements not expressible in FM, like nominal rules (group 4) or rules expressing frame conditions (group 5). This is basically a difference between the expressive powers of external labelled system, as exemplified here by prefixed tableau calculus of Fitting and Mendelsohn, and internal labelled system; the latter have much greater expressive power. However, for the part of HFM restricted to rules from group 1-3 and 6, again rule-generation theorem suffices to demostrate their derivability in HFM2. Therefore HFM2 and HFM (restricted to rules 1-3, 6) are equivalent (see Appendix).

5 Definite Descriptions

We postponed a treatment of definite descriptions since it requires some additional changes. In particular, categories of formulae FOR and terms $TERM$ must be defined simultaneously, and similarly the notion of interpretation of terms and satisfaction of formulae must be treated together. On the other hand, all terms considered so far can be represented as definite descriptions hence we can reduce the category of terms accordingly. We must add the iota-operator $\imath$, for forming definite descriptions out of formulae:

- if $\varphi \in FOR$, then $\imath x\varphi \in TERM$.

Semantically we characterise it by the following clause:

- $\mathcal{I}^v_w(\imath x\varphi) = o$ iff $\mathfrak{M}, v^x_o, w \vDash \varphi$ and no other x-variant of v satisfies φ in w.

Hence definite descriptions are also non-rigid and may be undefined in some (possibly all) worlds. Again addition of $@_i$ to definite description makes it a rigid term; a name of its designatum in $\mathcal{I}(i)$, if it is defined there. Complexity $c(\imath x\varphi) = c(\varphi) + 1$ but for $c(@_i\imath x\varphi) = 1$ which is fixed for all rigid terms.

Syntactically, Fitting and Mendelsohn's approach is based on the form of Hintikka Axiom, of which a "tentative version" takes the form:

$H: t \approx \imath x\varphi \leftrightarrow (\lambda x\varphi)t \wedge \forall y(\varphi[x/y] \rightarrow (\lambda xx = y)t)$, where y is not in φ;

which corresponds directly to the semantic clause. Note however that universal quantifier in H is possibilistic, i.e. it ranges over all elements of the (frame) domain. In FM tableaux its essential content is represented by means of three rules of introduction of implications corresponding to both directions of H. They are weaker in the sense that universal quantifier which introduces unwanted existential commitments is eliminated, and the rule corresponding to $H^{\leftarrow}$, i.e. to the right-left implication, introduces not valid but satisfiable formula containing labelled parameter. Such rules cannot be directly transformed into well-behaved SC rules so, instead of dealing with three FM rules, we introduce three other ones below and directly show that: (1) Hintikka axiom H, as restated in [10], is provable in HFM with these rules; (2) three additional $\imath$-rules are derivable in HFM in the presence of sequent $\Rightarrow H$ as an additional axiom. To realise this aim we must add to our language additional, possibilistic quantifiers symbolised by Tarskian $\bigwedge, \bigvee$. Semantically they are characterised:

- $\mathfrak{M}, v, w \models \bigwedge x\varphi$ iff $\mathfrak{M}, v^x_o, w \models \varphi$ for all $o \in \mathcal{D}$
- $\mathfrak{M}, v, w \models \bigvee x\varphi$ iff $\mathfrak{M}, v^x_o, w \models \varphi$ for some $o \in \mathcal{D}$

Now the system HFMD is HFM with the following additional rules:

$$(\bigwedge\Rightarrow)\ \frac{\Gamma \Rightarrow \Delta, @_i Dt \quad @_i\varphi[x/t^{@_i}], \Gamma \Rightarrow \Delta}{@_i \bigwedge x\varphi, \Gamma \Rightarrow \Delta} \qquad (\Rightarrow\bigwedge)\ \frac{\Gamma \Rightarrow \Delta, @_i\varphi[x/a]}{\Gamma \Rightarrow \Delta, @_i \bigwedge x\varphi}$$

$$(\Rightarrow\bigvee)\ \frac{\Gamma \Rightarrow \Delta, @_i Dt \quad \Gamma \Rightarrow \Delta, @_i\varphi[x/t^{@_i}]}{\Gamma \Rightarrow \Delta, @_i \bigvee x\varphi} \qquad (\bigvee\Rightarrow)\ \frac{@_i\varphi[x/a], \Gamma \Rightarrow \Delta}{@_i \bigvee x\varphi, \Gamma \Rightarrow \Delta}$$

$$(\imath\Rightarrow 1)\ \frac{@_i Dt, @_i\varphi[x/t^{@_i}], \Gamma \Rightarrow \Delta}{@_i t \approx \imath x\varphi, \Gamma \Rightarrow \Delta}$$

$$(\imath\Rightarrow 2)\ \frac{\Gamma \Rightarrow \Delta, @_i\varphi[x/b] \quad @_i Dt, @_i b = t^{@_i}, \Gamma \Rightarrow \Delta}{@_i t \approx \imath x\varphi, \Gamma \Rightarrow \Delta}$$

$$(\Rightarrow\imath)\ \frac{\Gamma \Rightarrow \Delta, @_i Dt \quad \Gamma \Rightarrow \Delta, @_i\varphi[x/t^{@_i}] \quad @_i\varphi[x/a], \Gamma \Rightarrow \Delta, @_i a = t^{@_i}}{\Gamma \Rightarrow \Delta, @_i t \approx \imath x\varphi}$$

and with $(\approx\Rightarrow), (\Rightarrow\approx)$ replaced with:

$$(\approx r1)\ \frac{\Gamma \Rightarrow \Delta, @_i t_1 \approx t_2 \quad @_i t_1^{@_i} = t_2^{@_i}, \Gamma \Rightarrow \Delta}{\Gamma \Rightarrow \Delta}$$

$$(\approx r2)\ \frac{\Gamma \Rightarrow \Delta, @_i t_1 \approx t_2 \quad @_i Dt_i, \Gamma \Rightarrow \Delta}{\Gamma \Rightarrow \Delta}$$

$$(r\approx)\ \frac{\Gamma \Rightarrow \Delta, @_i t_1^{@_i} = t_2^{@_i} \quad @_i t_1 \approx t_2, \Gamma \Rightarrow \Delta}{@_i Dt_1, @_i Dt_2, \Gamma \Rightarrow \Delta}$$

HFMD is adequate since we can prove that rules for $\imath$ are interderivable with $\Rightarrow H$, moreover, cut elimination also holds for it (see Appendix) although this requires some comment. First note that in $(\bigwedge \Rightarrow), (\Rightarrow \bigvee)$ we instantiate variable x with any rigid term but since they have complexity 1, the new rules are also reductive and the proof of cut elimination for HFM is not spoiled. We cannot use restricted form of instantiation, like in case of $(\forall E \Rightarrow), (\Rightarrow \exists E)$, since such system would be incomplete. Consider two formulae: $Er \wedge \forall x\varphi \rightarrow \varphi[x/r]$ and $Dr \wedge \bigwedge x\varphi \rightarrow \varphi[x/r]$. Both are valid and the former is provable in HFM but the second would be unprovable if $(\bigwedge \Rightarrow)$ is restricted to parameters. On the other hand in $(\Rightarrow \bigwedge), (\bigvee \Rightarrow)$ we do not need the additional formula $@_i Da$ in the premiss since parameters are defined everywhere.

Note also that rules of HFMD are defined in such a way that the situation is excluded where some cut formula is principal in both premisses of cut but obtained by means of different kind of rules which are not reductive. In particular, term equality is introduced only by means of rules for $\imath$ and they are reductive. It is the reason why in HFMD we have to replace rules for $\approx$ from HFM with apparently worse equivalents. Since principal formulae of rules for definite descriptions are term equalities there would be a clash with HFM rules. Consider a situation when both cut formulae are term equalities but one is introduced by means of an $\imath$-rule and the other by means of $(\approx\Rightarrow)$ or $(\Rightarrow\approx)$ – an induction on cut-degree fails. However, such situation cannot happen in HFMD where instead of $(\approx\Rightarrow)$ and $(\Rightarrow\approx)$ we have $(\approx r1), (\approx r2), (r \approx)$ which are safe in this respect. Clearly, the new rules are equivalent to $(\approx\Rightarrow), (\Rightarrow\approx)$ (by rule-generation theorem [17] mentioned in section 3) although worse from the proof-search standpoint.

6 Conclusion

We have shown that HL is a sufficiently flexible framework for expressing FM version of FOML. Moreover, in this setting we can formulate a well-behaved SC admitting cut elimination. Although we did not provide a semantic completeness proof it can be carried out either by using a strategy from [4] which is more standard and requires cut, or by means of Hintikka-style saturation technique like in [3] which is possible due to proved cut elimination. Moreover, HFM is formulated in the weak hybrid language as a basis; all additions were taken from FOML of Fitting and Mendelsohn. We can still enrich the language with specifically hybrid constants like nominal quantifiers or $\downarrow$-operator.

On the other hand, our treatment of definite descriptions by means of rules using $\approx$ may be seen as not wholly satisfactory. We obtain a system where $=$ is better characterised than in Indrzejczak [16] but at the cost of some redundancy – two kinds of equality are applied that differ only syntactically but not semantically. The other option would be to characterize definite descriptions by means of special rules for definedness formulae. This is the approach explored by Orlandelli [24] in the framework of labelled SC. In his system definedness predicate is not a part of a language but rather a technical device of the shape $D(t, x, w)$ meaning that t denotes x in w. In our approach this information is divided between $\approx$ and D as unary predicate. The lack of space forbids more extensive comparison of both approaches.

We should add that the treatment of definite descriptions either in terms of rules using some kind of equality or a predicate of definedness still does not provide the characteristics which is separate, in the sense of not exposing other constants in rules except $\imath$-operator. It is worth to explore more general perspective which is connected with using terms on a par with formulae in sequents. This device, introduced by Jaśkowski [20] in his first formulation of ND, was recently succesfully applied in many contexts, for example by Textor [28], Restall [26], Gazzari [12] and Indrzejczak [18]. In hybrid languages a uniform application of sat-operators to formulae and terms seems to offer a particularly interesting and uniform perspective where items are just sat-phrases be it either a formula or a term. We leave this problem for further study.

Appendix

Lemma .1 *If HFM1* $\vdash\ \Gamma \Rightarrow \Delta$, *then HFM2* $\vdash\ \Gamma \Rightarrow \Delta$, *where* Γ, Δ *contain only sentences.*

Proof: Consider a HFM1-proof $\mathcal{D}$. Let $\{p_1, \dots p_n\}$ be the set of all distinct n-parameters occuring in a proof and $\{p'_1, \dots p'_n\}$ the corresponding set of distinct parameters. Going from the root to leaves define for every node $\Gamma \Rightarrow \Delta$ a corresponding sequent $\Pi, \Gamma' \Rightarrow \Delta'$ where Γ', Δ' are obtained from Γ, Δ by simultaneous substitution of every n-parameter p_i with corresponding parameter p'_i. Moreover let $\Pi = \{@_k Ep'_i : p_i := @_k p'_i \in PAR(\Gamma \cup \Delta)\} \cup \{@_i Dt : @_i t \text{ is defined in } \Gamma \cup \Delta\}$. This way we obtain an isomorphic tree $\mathcal{D}'$ of sequents with the same root in the language of HFM2. This tree is not necessarily a HFM2-proof so we must systematically made some adjustments. All leaves of $\mathcal{D}$ which are instances of Ax are trivially axioms of HFM2. However, for the cases of eight definitional axioms we have to provide proofs of their corresponding sequents in HFM2. For two axioms characterising D we have:

$$\dfrac{\dfrac{\dfrac{@_i Dt, \Pi, @_i(@_i t = @_i t), \Gamma \Rightarrow \Delta, @_i(@_i t = @_i t)}{@_i Dt, \Pi, @_i(@_i t = @_i t), \Gamma, \Rightarrow \Delta, @_i(\lambda x x = x)t}(\Rightarrow \lambda'')}{@_i Dt, \Pi, @_i(t \approx t), \Gamma, \Rightarrow \Delta, @_i(\lambda x x = x)t}(\approx\Rightarrow)}{@_i Dt, \Pi, \Gamma \Rightarrow \Delta, @_i(\lambda x x = x)t}(\approx -')$$

and

$$(\lambda \Rightarrow)\ \dfrac{@_i Dt, \Pi, @_i(@_i t = @_i t), \Gamma \Rightarrow \Delta, @_i Dt}{\Pi, @_i(\lambda x x = x)t, \Gamma \Rightarrow \Delta, @_i Dt}$$

For E with $t \notin PAR$:

$$\dfrac{\dfrac{\dfrac{@_i Dt, \Pi, @_i Ea, @_i(a = @_i t), \Gamma \Rightarrow \Delta, @_i(a = @_i t)}{@_i Dt, \Pi, @_i Ea, @_i(a = @_i t), \Gamma \Rightarrow \Delta, @_i \exists y(y = @_i t)}(\Rightarrow \exists E')}{@_i Dt, \Pi, @_i Ea, @_i(a = @_i t), \Gamma \Rightarrow \Delta, @_i(\lambda x \exists y y = x)t}(\Rightarrow \lambda'')}{\Pi, @_i Et, \Gamma \Rightarrow \Delta, @_i(\lambda x \exists y y = x)t}(E \Rightarrow)$$

$$\dfrac{\dfrac{\dfrac{\Sigma, \Pi, \Gamma \Rightarrow \Delta, @_i Ea \qquad \Sigma, \Pi, \Gamma \Rightarrow \Delta, @_i(a = @_i t)}{@_i Dt, @_i Ea, \Pi, @_i(a = @_i t), \Gamma \Rightarrow \Delta, @_i Et}\,(\Rightarrow E')}{@_i Dt, \Pi, @_i \exists y(y = @_i t), \Gamma \Rightarrow \Delta, @_i Et}\,(\exists E \Rightarrow)}{\Pi, @_i(\lambda x \exists yy = x)t, \Gamma \Rightarrow \Delta, @_i Et}\,(\lambda \Rightarrow)$$

where $\Sigma = \{@_i Dt, @_i Ea, @_i(a = @_i t)\}$. If $t \in PAR$ both sequents are provable without $(E \Rightarrow), (\Rightarrow E')$; it justifies our restriction on their application.

For all applications of propositional rules in $\mathcal{D}$ we do not need any changes in $\mathcal{D}'$. For $(\forall \Rightarrow)$ we have by definition of Π that $@_i Eb \in \Pi$ and that $@_i\varphi[x/b] \in \Gamma'$, so in case $@_i b$ was already present in the conclusion of the application of this rule in $\mathcal{D}$ we need no change in $\mathcal{D}'$. In case $@_i b$ was fresh in this application of $(\forall \Rightarrow)$ we must add $@_i Eb$ to Π in the conclusion to secure the correctness of $(\forall E \Rightarrow)$ in HFM2 and apply next $(E-)$ to remove $@_i Eb$. For $(\Rightarrow \forall)$ $@_i a$ occurs only in $@_i\varphi$ but in the corresponding sequent of $\mathcal{D}'$ a occurs also in $@_i Da \in \Pi$, in $@_i Ea \in \Pi$ and in $@_i\varphi \in \Delta'$. Therefore first we delete $@_i Da$ by application of $(D-)$, then the application of $(\Rightarrow \forall E)$ on resulting sequent is correct and yields the desired conclusion. Applications of $(= +'), (\lambda \Rightarrow')$ and $(\Rightarrow \lambda')$ in $\mathcal{D}$ correspond to correct applications of HFM2-versions of respective rules in $\mathcal{D}'$ by definition of Π. For $(= -)$ we apply $(\approx -')$ and $(\approx\Rightarrow)$. □

Before proving the converse let us first demonstrate a derivability of all specific rules of HFM2 in HFM1. Clearly instead of parameters we will use n-parameters here. Quantifier rules $(\forall E \Rightarrow')$ and $(\Rightarrow \exists E')$ are just special versions of HFM1 rules. Derivability of $(\Rightarrow \forall E)$ and $(\exists E \Rightarrow)$ needs a demonstration:

$$\dfrac{\dfrac{\dfrac{\dfrac{\mathcal{D} \qquad @_i(\lambda x \exists yy = x)@_i a \Rightarrow @_i E@_i a}{@_i(@_i a = @_i a) \Rightarrow @_i E@_i a}\,(Cut) \qquad @_i E@_i a, \Gamma \Rightarrow \Delta, @_i\varphi[x/@_i a]}{@_i(@_i a = @_i a), \Gamma \Rightarrow \Delta, @_i\varphi[x/@_i a]}\,(Cut)}{\Gamma \Rightarrow \Delta, @_i\varphi[x/@_i a]}\,(= -)}{\Gamma \Rightarrow \Delta, @_i \forall x\varphi}\,(\Rightarrow \forall)$$

where $\mathcal{D}$ is:

$$\dfrac{\dfrac{@_i(@_i a = @_i a) \Rightarrow @_i \exists yy = @_i a, @_i(@_i a = @_i a)}{@_i(@_i a = @_i a) \Rightarrow @_i \exists yy = @_i a}\,(\Rightarrow \exists)}{@_i(@_i a = @_i a) \Rightarrow @_i(\lambda x \exists yy = x)@_i a}\,(\Rightarrow \lambda')$$

Similarly for $(\exists E \Rightarrow)$. $(\Rightarrow \lambda'')$ needs no justification; for $(\lambda \Rightarrow)$ we have:

$$\dfrac{\dfrac{\dfrac{\dfrac{\dfrac{@_i(t^{@_i} = t^{@_i}) \Rightarrow @_i(t^{@_i} = t^{@_i})}{@_i(t^{@_i} = t^{@_i}) \Rightarrow @_i(\lambda xx = x)t}\,(\Rightarrow \lambda') \qquad @_i(\lambda xx = x)t \Rightarrow @_i Dt}{@_i(t^{@_i} = t^{@_i}) \Rightarrow @_i Dt}\,(Cut) \qquad @_i Dt, @_i\varphi[x/t^{@_i}], \Gamma \Rightarrow \Delta}{@_i(t^{@_i} = t^{@_i}), @_i\varphi[x/t^{@_i}], \Gamma \Rightarrow \Delta}\,(Cut)}{@_i\varphi[x/t^{@_i}], \Gamma \Rightarrow \Delta}\,(= -)}{@_i(\lambda x\varphi)t, \Gamma \Rightarrow \Delta}\,(\lambda \Rightarrow')$$

Derivability of $(\approx -')$:

$$\dfrac{\mathcal{D} \qquad \dfrac{@_i(\lambda x(\lambda yy = x)t)t \Rightarrow @_i(t \approx t) \qquad @_i(t \approx t), @_iDt, \Gamma \Rightarrow \Delta}{@_i(\lambda x(\lambda yy = x)t)t, @_iDt, \Gamma \Rightarrow \Delta}\,(Cut)}{@_iDt, \Gamma \Rightarrow \Delta}\,(Cut)$$

where $\mathcal{D}$ is:

$$\dfrac{@_iDt \Rightarrow @_i(\lambda xx = x)t \qquad \dfrac{\dfrac{\dfrac{@_i(@_it = @_it) \Rightarrow @_i(@_it = @_it)}{@_i(@_it = @_it) \Rightarrow @_i(\lambda yy = @_it)t}\,(\Rightarrow \lambda')}{@_i(@_it = @_it) \Rightarrow @_i(\lambda x(\lambda yy = x)t)t}\,(\Rightarrow \lambda')}{@_i(\lambda xx = x)t \Rightarrow @_i(\lambda x(\lambda yy = x)t)t}\,(\lambda \Rightarrow')}{@_iDt \Rightarrow @_i(\lambda x(\lambda yy = x)t)t}\,(Cut)$$

Derivability of other rules goes similarly, hence we obtain:

Lemma .2 *If HFM2* $\vdash \Gamma \Rightarrow \Delta$, *then HFM1* $\vdash \Gamma \Rightarrow \Delta$

Proof by induction on the height of HFM2-proof. Clearly as a preliminary step we must provide a reverse substitution in all nodes of HFM2-proof, i.e. in Γ, Δ all different parameters must be substituted with different n-parameters in such a way that for any $@_iEb$ or $@_iDb$ in Γ, Δ, b is substituted with $@_ib$. Derivability of all specific rules of HFM2 in HFM1 suffices for the proof. □

Lemma .3 *Provability of H in HFMD*

$$(\Rightarrow \wedge)\,\dfrac{(\imath \Rightarrow 1)\,\dfrac{(\Rightarrow \lambda)\,\dfrac{@_iDt \Rightarrow @_iDt \qquad @_i\varphi[x/t^{@_i}] \Rightarrow @_i\varphi[x/t^{@_i}]}{@_iDt, @_i\varphi[x/t^{@_i}] \Rightarrow @_i(\lambda x\varphi)t}}{@_it \approx \imath x\varphi \Rightarrow @_i(\lambda x\varphi)t} \qquad \mathcal{D}}{@_it \approx \imath x\varphi \Rightarrow @_i(\lambda x\varphi)t \wedge \bigwedge y(\varphi[x/y] \to (\lambda xx = y)t)}$$

where $\mathcal{D}$ is:

$$(\Rightarrow \bigwedge)\,\dfrac{(\Rightarrow\to)\,\dfrac{\dfrac{@_i\varphi[x/a] \Rightarrow @_i\varphi[x/a] \qquad \dfrac{@_iDt \Rightarrow @_iDt \qquad @_it^{@_i} = a \Rightarrow @_it^{@_i} = a}{@_iDt, @_it^{@_i} = a \Rightarrow @_i(\lambda xx = a)t}\,(\Rightarrow \lambda)}{@_it \approx \imath x\varphi, @_i\varphi[x/a] \Rightarrow @_i(\lambda xx = a)t}\,(\imath \Rightarrow 2)}{@_it \approx \imath x\varphi, \Rightarrow @_i\varphi[x/a] \to (\lambda xx = a)t}}{@_it \approx \imath x\varphi \Rightarrow @_i \bigwedge y(\varphi[x/y] \to (\lambda xx = y)t)}$$

Next, the converse:

$$(\wedge \Rightarrow)\,\dfrac{(\lambda \Rightarrow)\,\dfrac{(\Rightarrow \imath)\,\dfrac{@_iDt \Rightarrow @_iDt \qquad @_i\varphi[x/t^{@_i}] \Rightarrow @_i\varphi[x/t^{@_i}] \qquad \mathcal{D}}{@_iDt, @_i\varphi[x/t^{@_i}], @_i \bigwedge y(\varphi[x/y] \to (\lambda xx = y)t) \Rightarrow @_it \approx \imath x\varphi}}{@_i(\lambda x\varphi)t, @_i \bigwedge y(\varphi[x/y] \to (\lambda xx = y)t) \Rightarrow @_it \approx \imath x\varphi}}{@_i(\lambda x\varphi)t \wedge \bigwedge y(\varphi[x/y] \to (\lambda xx = y)t) \Rightarrow @_it \approx \imath x\varphi}$$

where $\mathcal{D}$ is:

$$(D-)\,\dfrac{@_iDa \Rightarrow @_iDa}{\Rightarrow @_iDa} \qquad \dfrac{\dfrac{@_i\varphi[x/a] \Rightarrow @_i\varphi[x/a] \qquad \dfrac{@_iDt, @_it^{@_i} = a \Rightarrow @_it^{@_i} = a}{@_i(\lambda xx = a)t \Rightarrow @_it^{@_i} = a}\,(\lambda \Rightarrow)}{@_i\varphi[x/a] \to (\lambda xx = a)t, @_i\varphi[x/a] \Rightarrow @_it^{@_i} = a}\,(\to\Rightarrow)}{@_i \bigwedge y(\varphi[x/y] \to (\lambda xx = y)t), @_i\varphi[x/a] \Rightarrow @_it^{@_i} = a}\,(\bigwedge \Rightarrow)$$

Lemma .4 *Derivability of HFMD rules (the case of $(\imath \Rightarrow 1), (\imath \Rightarrow 2)$)*

$$\dfrac{\Rightarrow H^{\rightarrow} \qquad \dfrac{@_i t \approx \imath x\varphi \Rightarrow @_i t \approx \imath x\varphi \qquad \dfrac{\dfrac{\dfrac{@_i Dt, @_i\varphi[x/t^{@_i}], \Gamma \Rightarrow \Delta}{@_i(\lambda x\varphi)t, \Gamma \Rightarrow \Delta}\,(\lambda\Rightarrow)}{@_i(\lambda x\varphi)t, @_i \bigwedge y(\varphi[x/y] \rightarrow (\lambda xx = y)t), \Gamma \Rightarrow \Delta}\,(W\Rightarrow)}{@_i(\lambda x\varphi)t \wedge \bigwedge y(\varphi[x/y] \rightarrow (\lambda xx = y)t), \Gamma \Rightarrow \Delta}\,(\wedge\Rightarrow)}{@_i t \approx \imath x\varphi, @_i t \approx \imath x\varphi \rightarrow (\lambda x\varphi)t \wedge \bigwedge y(\varphi[x/y] \rightarrow (\lambda xx = y)t), \Gamma \Rightarrow \Delta}\,(\rightarrow\Rightarrow)}{@_i t \approx \imath x\varphi, \Gamma \Rightarrow \Delta}\,(Cut)$$

$$\dfrac{\dfrac{\dfrac{(D-)\,\dfrac{@_i Db \Rightarrow @_i Db}{\Rightarrow @_i Db} \qquad \dfrac{\Gamma \Rightarrow \Delta, @_i\varphi[x/b] \qquad \dfrac{@_i Dt, @_i t^{@_i} = b, \Gamma \Rightarrow \Delta}{@_i(\lambda xx = b)t, \Gamma \Rightarrow \Delta}\,(\lambda\Rightarrow)}{@_i\varphi[x/b] \rightarrow (\lambda xx = b)t, \Gamma \Rightarrow \Delta}\,(\rightarrow\Rightarrow)}{@_i \bigwedge y(\varphi[x/y] \rightarrow (\lambda xx = y)t), \Gamma \Rightarrow \Delta}\,(\bigwedge\Rightarrow)}{@_i(\lambda x\varphi)t, @_i \bigwedge y(\varphi[x/y] \rightarrow (\lambda xx = y)t), \Gamma \Rightarrow \Delta}\,(W\Rightarrow)}{@_i(\lambda x\varphi)t \wedge \bigwedge y(\varphi[x/y] \rightarrow (\lambda xx = y)t), \Gamma \Rightarrow \Delta}\,(\wedge\Rightarrow)$$

and obtain the conclusion of $(\imath \Rightarrow 2)$ by cut with:

$$\dfrac{\Rightarrow H^{\rightarrow} \qquad \dfrac{@_i t \approx \imath x\varphi \Rightarrow @_i t \approx \imath x\varphi \qquad @_i(\lambda x\varphi)t \wedge \bigwedge y(\varphi[x/y] \rightarrow (\lambda xx = y)t) \Rightarrow \psi}{@_i t \approx \imath x\varphi, @_i t \approx \imath x\varphi \rightarrow (\lambda x\varphi)t \wedge \bigwedge y(\varphi[x/y] \rightarrow (\lambda xx = y)t) \Rightarrow \psi}\,(\rightarrow\Rightarrow)}{@_i t \approx \imath x\varphi \Rightarrow @_i(\lambda x\varphi)t \wedge \bigwedge y(\varphi[x/y] \rightarrow (\lambda xx = y)t)}\,(Cut)$$

where $\psi := @_i(\lambda x\varphi)t \wedge \bigwedge y(\varphi[x/y] \rightarrow (\lambda xx = y)t)$
We prove derivability of $(\Rightarrow \imath)$ in a similar way. □

To prove cut elimination first note that for HFM and HFMD holds:

Lemma .5 (Height-preserving Substitution)
If $\vdash_k \Gamma \Rightarrow \Delta$, *then* $\vdash_k (\Gamma \Rightarrow \Delta)[i/j]$;
If $\vdash_k \Gamma \Rightarrow \Delta$, *then* $\vdash_k (\Gamma \Rightarrow \Delta)[a/r]$.

By lemma 5 every proof may be systematically transformed into regular proof – every fresh parameter and nominal is fresh in the entire proof.

Let cut-degree of cut-formula $@_i\varphi$ be its complexity, i.e. $d@_i\varphi = c(@_i\varphi)$ and proof-degree ($d\mathcal{D}$) be the maximal cut-degree in $\mathcal{D}$.

Technically the proof of cut elimination theorem is an extension of the proof for propositional HL in Indrzejczak [15] (see also [23], [16]) and is based on:

Lemma .6 (Right reduction) *Let* $\mathcal{D}_1 \vdash \Gamma \Rightarrow \Delta, @_i\varphi$ *and* $\mathcal{D}_2 \vdash @_i\varphi^n, \Pi \Rightarrow \Sigma$ *and* $d\mathcal{D}_1, d\mathcal{D}_2 < d@_i\varphi$, *and* $@_i\varphi$ *principal in* $\Gamma \Rightarrow \Delta, @_i\varphi$, *then we can construct a proof* $\mathcal{D}$ *such that* $\mathcal{D} \vdash \Gamma^n, \Pi \Rightarrow \Delta^n, \Sigma$ *and* $d\mathcal{D} < d@_i\varphi$.

Lemma .7 (Left reduction) *Let* $\mathcal{D}_1 \vdash \Gamma \Rightarrow \Delta, @_i\varphi^n$ *and* $\mathcal{D}_2 \vdash @_i\varphi, \Pi \Rightarrow \Sigma$ *and* $d\mathcal{D}_1, d\mathcal{D}_2 < d@_i\varphi$, *then we can construct a proof* $\mathcal{D}$ *such that* $\mathcal{D} \vdash \Gamma, \Pi^n \Rightarrow \Delta, \Sigma^n$ *and* $d\mathcal{D} < d@_i\varphi$.

They hold for SC with substitutive and reductive rules. Lemma 6 makes a reduction on the right, and lemma 7 on the left premiss of cut by induction on the height of respective proofs. The latter in the case of principal cut-formula applies lemma 6. Eventually, lemma 7 yields, by induction on proof-degree:

Theorem .8 *Every proof may be transformed into cut-free proof.* □

References

[1] Beeson, M., "Foundations of Constructive Mathematics," Springer, 1985.

[2] Blackburn, P., *Internalizing labelled deduction*, Journal of Logic and Computation **10/1** (2000), pp. 137–168.

[3] Blackburn, P. and M. Marx, *Tableaux for quantified hybrid logic*, in: *Tableaux 2002, LNAI 2381*, Springer, 2002 pp. 38–52.

[4] Blackburn, P. and B. ten Cate, *Pure extensions, proof rules and hybrid axiomatics*, Studia Logica **84/2** (2006), pp. 277–322.

[5] Bolander, T. and T. Braüner, *Tableau-based decision procedures for hybrid logic*, Journal of Logic and Computation **16/6** (2006), pp. 737–763.

[6] Braüner, T., "Hybrid Logic and its Proof-Theory," Springer, 2011.

[7] Ciabattoni, A., *Automated generation of analytic calculi for logics with linearity*, in: *Proceedings of CSL'04, vol. 3210*, LNCS, 2004 pp. 503–517.

[8] Feferman, S., *Definedness*, Erkenntniss **43** (1995), pp. 295–320.

[9] Fitting, M., *Modal logic should say more than it does*, in: *Computational Logic, Essays in Honor of Alan Robinson*, Cambridge, 1991 pp. 113–135.

[10] Fitting, M. and R. L. Mendelsohn, "First-Order Modal Logic," Kluwer, Dordrecht, 1998.

[11] Garson, J. M., "Modal Logic for Philosophers," Cambridge Univ. Press, 2006.

[12] Gazzari, R., *The calculus of natural calculation*, Studia Logica **submitted** (2020).

[13] Goldblatt, R., "Quantifiers, Propositions and Identity," Cambridge Univ. Press, 2011.

[14] Indrzejczak, A., "Natural Deduction, Hybrid Systems and Modal Logics," Springer, 2010.

[15] Indrzejczak, A., *Simple cut elimination proof for hybrid logic*, Logic and Logical Philosophy **25/2** (2016), pp. 129–141.

[16] Indrzejczak, A., *Cut-free modal theory of definite descriptions*, in: *Advances in Modal Logic 12*, College Publications, 2018 pp. 387–406.

[17] Indrzejczak, A., *Rule-generation theorem and its applications*, Bulletin of the Section of Logic **47/4** (2018), pp. 265–281.

[18] Indrzejczak, A., *A novel approach to equality*, Synthese **to appear** (2020).

[19] Indrzejczak, A. and M. Zawidzki, *Decision procedures for some strong hybrid logics*, Logic and Logical Philosophy **22/4** (2013), pp. 389–409.

[20] Jaśkowski, S., *On the rules of suppositions in formal logic*, Studia Logica **1** (1934), pp. 5–32.

[21] Manzano, M., M. Martins and A. Huertas, *A semantics in equational hybrid propositional type theory*, Bulletin of the Section of Logic **43/3-4** (2014), pp. 121–154.

[22] Manzano, M., M. Martins and A. Huertas, *Completeness in equational hybrid propositional type theory*, Studia Logica **107/6** (2019), pp. 1159–1198.

[23] Metcalfe, G., N. Olivetti and D. Gabbay, "Proof Theory for Fuzzy Logics," Springer, 2008.

[24] Orlandelli, E., *Labelled calculi for quantified modal logics with definite descriptions*, Journal of Logic and Computation **submitted** (2020).

[25] Orlandelli, E. and G. Corsi, *Labelled calculi for quantified modal logics with non-rigid and non-denoting terms*, in: *Proceedings of ARQNL 2018*, CEUR-WS.org, 2018 pp. 64–78.

[26] Restall, G., *Generality and existence 1: Quantification and free logic*, The Review of Symbolic Logic **12/1** (2019), pp. 354–378.

[27] Seligman, J., *Internalization: The case of hybrid logics*, Journal of Logic and Computation **11/5** (2001), pp. 671–689.

[28] Textor, M., *Towards a neo-brentanian theory of existence*, Philosophers' Imprint **17/6** (2017), pp. 1–20.

[29] Thomason, R. and R. Stalnaker, *Modality and reference*, Nous **2** (1968), pp. 359–372.

[30] Zawidzki, M., "Deductive Systems and the Decidability Problem for Hybrid Logics," Univ. of Lodz Press, Jagielonnian Univ. Press, 2013.

Modal Logics with Transitive Closure: Completeness, Decidability, Filtration

Stanislav Kikot [1]

University of Oxford

Ilya Shapirovsky [2]

Steklov Mathematical Institute of Russian Academy of Sciences and Institute for Information Transmission Problems of Russian Academy of Sciences

Evgeny Zolin [3]

National Research University Higher School of Economics, Russian Federation

Abstract

We give a sufficient condition for Kripke completeness of modal logics that have the transitive closure modality. More precisely, we show that if a modal logic admits what we call *definable filtration*, then its enrichment with the transitive closure modality (and the corresponding axioms) is Kripke complete; in addition, the resulting logic has the finite model property and admits definable filtration, too. This argument can be iterated, and as an application we obtain the finite model property for PDL-like expansions of multimodal logics that admit definable filtration.

Keywords: Filtration, decidability, finite model property, transitive closure, PDL.

Introduction

This paper makes a contribution to the study of modal logics enriched by the transitive closure modality.

Modal logics that, in addition to the modal operator $\Box$ for a binary relation R, also contain the operator $\boxplus$ for the transitive closure of R, are quite common [8]. For instance, such are the propositional dynamic logic (PDL) [7] or von Wright's logic $\mathrm{Log}(\mathbb{N}, \mathsf{succ}, <)$ (see [18]). Other examples include logics

[1] staskikotx@gmail.com

[2] ilya.shapirovsky@gmail.com. The work on this paper was supported by the Russian Science Foundation under grant 16-11-10252 and carried out at Steklov Mathematical Institute of Russian Academy of Sciences.

[3] ezolin@gmail.com. The article was prepared within the framework of the HSE University Basic Research Program.

with the operator of common knowledge and 'everyone knows that' in epistemic logics [6] such as the logic Team for collective beliefs and actions [5].

So far, completeness and decidability results for such logics have had bespoke proofs, though many of them are based on Segerberg's [16] and Kozen and Parikh's arguments for PDL [13]. In this paper we present a toolkit for obtaining results on completeness, finite model property, and decidability applicable to a wide range of modal logics with transitive closure.

In Section 2, we recall a (rather general) notion of filtration and come up with a hierarchy of 'admits filtration' notions (including those studied in our earlier work [12]). Section 3 contains our main result (announced in [20]): if the class of models of a logic L *admits definable filtration* (see Definition 2.5), then the axioms of L together with Segerberg's axioms for the transitive closure modality yield a complete axiomatization of the bimodal logic of the class of frames for L augmented by the transitive closure of the accessibility relation. Moreover, the resulting logic has the finite model property (and is decidable, if L was finitely axiomatizable). Section 4 presents examples of logics that satisfy our sufficient condition; we also show how this condition can be 'iterated' for obtaining completeness for 'PDLizations' of a family of logics.

1 Preliminaries

We assume the reader to be familiar with syntax and semantics of multi-modal logic [1,3], so we only briefly recall some notions and fix notation. Let Σ be a (usually finite) alphabet (of indices for modalities). The set $\mathsf{Fm}(\Sigma)$ of *modal formulas* (*over* Σ) is defined from propositional letters $\mathsf{Var} = \{p_0, p_1, \ldots\}$ using Boolean connectives and the modalities $[e]$, for $e \in \Sigma$, according to the syntax:

$$\varphi \ ::= \ \bot \ \mid \ p_i \ \mid \ \varphi \to \psi \ \mid \ [e]\varphi.$$

We use standard abbreviations (e.g., $\top, \wedge$); in particular, $\langle e\rangle\varphi := \neg[e]\neg\varphi$. For a set of formulas Γ, by $\mathsf{Sub}(\Gamma)$ we denote the set of all subformulas of formulas from Γ. We say that Γ is Sub-*closed* if $\mathsf{Sub}(\Gamma) \subseteq \Gamma$.

A (Σ-)*frame* is a pair $F = (W, (R_e)_{e\in\Sigma})$, where $W \neq \varnothing$ and $R_e \subseteq W\times W$ for $e \in \Sigma$. A *model* based on F is a pair $M = (F, V)$, where $V(p) \subseteq W$, for all $p \in \mathsf{Var}$. The *truth relation* $M, x \models \varphi$ is defined in the usual way, e.g.

$$M, x \models [e]\varphi \quad \leftrightharpoons \quad \text{for all } y \in W, \ \text{ if } x\, R_e\, y \ \text{ then } \ M, y \models \varphi.$$

We write $M \models \varphi$ if $M, x \models \varphi$ for all x in M. A formula φ is *valid* on F, notation $F \models \varphi$, if $M \models \varphi$ for all M based on F. For a class of frames $\mathcal{F}$, an $\mathcal{F}$-*model* is a model based on a frame from $\mathcal{F}$.

A (*normal modal*) *logic* (*over* Σ) is a set of formulas L that contains all classical tautologies, the axioms $[e](p \to q) \to ([e]p \to [e]q)$, for each $e \in \Sigma$, and is closed under the rules of modus ponens, substitution, and necessitation (from φ, infer $[e]\varphi$, for each $e \in \Sigma$). An L-*frame* is a frame F such that $F \models L$. *The logic of a class of frames* $\mathcal{F}$ is the set of all formulas that are valid in $\mathcal{F}$. A logic L is (*Kripke*) *complete* if it is the logic of some class of frames. A logic L has the *finite model property* (FMP) if it is the logic of some class of finite frames; or equivalently (see, e.g., [1, Th. 3.28]) if, for every formula $\varphi \notin L$, there is a finite model M such that $M \models L$ and $M \not\models \varphi$. For a logic L, put

$$\begin{aligned}\mathsf{Fr}(L) &= \{F \mid F \text{ is a frame and } F \models L\},\\ \mathsf{Mod}(L) &= \{M \mid M \text{ is a model and } M \models L\}.\end{aligned}$$

Clearly, every $\mathsf{Fr}(L)$-model belongs to $\mathsf{Mod}(L)$. The converse does not hold in general; e.g., the canonical model of a non-canonical logic L is not a $\mathsf{Fr}(L)$-model. But the converse holds in the following special case. A model M is called *differentiated* if any two points in M can be distinguished by a formula.

Lemma 1.1 (See e.g. [10, Ex. 4.9]) *Let $M = (F, V)$ be a finite differentiated model. If all substitution instances of a formula φ are true in M, then $F \models \varphi$.*

In particular, if $M \models L$, where L is a logic, then $F \models L$.

Harrop's theorem. *A finitely axiomatizable logic with the FMP is decidable.*

2 Filtration

The notion of a filtration we introduce below slightly generalizes the standard one (cf. [1, Def. 2.36], [3, Sect. 5.3]) in the following aspect: given a finite set of formulas Γ, we define a filtration as a model obtained by factoring a given model through an equivalence relation that we allow to be *finer* than the one induced by Γ. This modification seems to first appear in [21]; see also [22,23].

Let $M = (W, (R_e)_{e\in\Sigma}, V)$ be a model and Γ a finite Sub-closed set of Σ-formulas. An equivalence relation $\sim$ on W is *of finite index* if the quotient set $W/{\sim}$ is finite. The equivalence relation *induced by* Γ is defined as follows:

$$x \sim_\Gamma y \quad \leftrightharpoons \quad \forall \varphi \in \Gamma\ \big(M, x \models \varphi \ \Leftrightarrow\ M, y \models \varphi\big).$$

Clearly, $\sim_\Gamma$ is of finite index. We say that an equivalence relation $\sim$ *respects* Γ if $\sim\ \subseteq\ \sim_\Gamma$; in other words, if for every class $[x]_\sim \subseteq W$ and every formula $\varphi \in \Gamma$, φ is either true in all points of $[x]_\sim$ or false in all points of $[x]_\sim$.

Definition 2.1 (Filtration) By a *filtration* of a model M that *respects* a set of formulas Γ (or a Γ*-filtration of* M) we mean any model $\widehat{M} = (\widehat{W}, (\widehat{R}_e)_{e\in\Sigma}, \widehat{V})$ that satisfies the following conditions:

- $\widehat{W} = W/{\sim}$, for some equivalence relation of finite index $\sim$ on W;
- the equivalence relation $\sim$ respects Γ, i.e., $x \sim y$ implies $x \sim_\Gamma y$;
- the valuation $\widehat{V}$ is defined on the variables $p \in \Gamma$ canonically: $\widehat{x} \models p \Leftrightarrow x \models p$, for all points $x \in W$, where $\widehat{x} := [x]_\sim$ denotes the $\sim$-class of a point x;
- $R^{\mathsf{min}}_{\sim,e} \subseteq \widehat{R}_e \subseteq R^{\mathsf{max}}_{\Gamma,e}$, for each $e \in \Sigma$. Here $R^{\mathsf{min}}_{\sim,e}$ is the e-th *minimal filtered relation* on $\widehat{W}$, and $R^{\mathsf{max}}_{\Gamma,e}$ is the e-th *maximal filtered relation* on $\widehat{W}$ induced by the set of formulas Γ; they are defined in the usual way:

$$\begin{aligned}\widehat{x}\ R^{\mathsf{min}}_{\sim,e}\ \widehat{y} &\leftrightharpoons \exists x' \sim x\ \exists y' \sim y\colon\ x'\, R_e\, y',\\ \widehat{x}\ R^{\mathsf{max}}_{\Gamma,e}\ \widehat{y} &\leftrightharpoons \text{for every formula } [e]\varphi \in \Gamma\ \big(M, x \models [e]\varphi \ \Rightarrow\ M, y \models \varphi\big).\end{aligned}$$

If $\sim\, =\, \sim_\Phi$ for a finite set of formulas Φ, then we call $\widehat{M}$ a *definable* Γ*-filtration* of M (*through* Φ); we can assume, without loss of generality, that $\Phi \supseteq \Gamma$.

Note that the relations $R^{\min}_{\sim,e}$ and $R^{\max}_{\Gamma,e}$ are well-defined and $R^{\min}_{\sim,e} \subseteq R^{\max}_{\Gamma,e}$. The condition $R^{\min}_{\sim,e} \subseteq \widehat{R}_e$ can be rewritten as $\forall x, y \in W\ (x\, R_e\, y \Rightarrow \widehat{x}\, \widehat{R}_e\, \widehat{y})$. A filtration is always a finite model. The following is the key lemma about filtration (cf. [1, Th. 2.39], [3, Th. 5.23]).

Lemma 2.2 (Filtration lemma) *Suppose that Γ is a finite* Sub*-closed set of formulas and $\widehat{M}$ is a Γ-filtration of a model M. Then, for all points $x \in W$ and all formulas $\varphi \in \Gamma$, we have:* $M, x \models \varphi \;\Leftrightarrow\; \widehat{M}, \widehat{x} \models \varphi$.

2.1 Admissibility of filtration

Definition 2.3 (ADF for classes of frames) We say that a class of frames $\mathcal{F}$ *admits (definable) filtration* if, for any finite Sub-closed set of formulas Γ and an $\mathcal{F}$-model M, there exists an $\mathcal{F}$-model that is a (definable) Γ-filtration of M.

It is well-known that filtration (of the class of all frames) is a method of proving the FMP for complete modal logics; let us state this explicitly.

Lemma 2.4 (AF for frames implies FMP) *If the class of its frames* $\mathsf{Fr}(L)$ *admits filtration and the logic L is Kripke complete, then L has the FMP.*

Proof. If $\varphi \notin L$, then, by completeness of L, there is a frame $F \models L$ with $F \not\models \varphi$. Taking $\Gamma = \mathsf{Sub}(\varphi)$ and Γ-filtrating the model based on F that falsifies the formula φ, we obtain a finite frame $F' \models L$ with $F' \not\models \varphi$. □

Definition 2.5 (ADF for classes of models) We say that a class of models $\mathcal{M}$ *admits (definable) filtration* if, for any finite Sub-closed set of formulas Γ and any $M \in \mathcal{M}$, there is a model in $\mathcal{M}$ that is a (definable) Γ-filtration of M.

The next lemma shows that filtration of the class of all models $\mathsf{Mod}(L)$ is a method of obtaining Kripke (frame!) completeness (and FMP, of course).

Lemma 2.6 (AF for models implies FMP) *If the class of models* $\mathsf{Mod}(L)$ *of a logic L admits filtration, then L has the FMP and hence is Kripke complete.*

Proof. Any normal logic L is model-complete, i.e., $\varphi \in L$ iff $\mathsf{Mod}(L) \models \varphi$; moreover, L is complete w.r.t. a single, *canonical* model M_L. Therefore, if $\varphi \notin L$, then there is a model $M \in \mathsf{Mod}(L)$ such that $M \not\models \varphi$. Let $\Gamma = \mathsf{Sub}(\varphi)$. Take a Γ-filtration $\widehat{M}$ of M such that $\widehat{M} \models L$; here $\widehat{M}$ is a finite model. By taking the filtration of $\widehat{M}$ through the set of all formulas Fm, we obtain a finite differentiated model $\widehat{M}' = (\widehat{F}', \widehat{V}')$ modally equivalent to $\widehat{M}$. So $\widehat{M}' \models L$ and $\widehat{M}' \not\models \varphi$. By Lemma 1.1, $\widehat{F}' \models L$. Thus, every non-theorem of L is falsified in some finite L-frame. So, L is Kripke (frame) complete and even has the FMP.□

So far, we are not aware of any example of a logic whose class of frames (or models) admits filtration, but not definable filtration.

We have two variants of the notion "a logic L admits (definable) filtration":

(I) the class of frames $\mathsf{Fr}(L)$ admits (definable) filtration (Definition 2.3);
(II) the class of models $\mathsf{Mod}(L)$ admits (definable) filtration (Definition 2.5).

In both variants, we filtrate a model $M = (F, V)$ into a model $\widehat{M} = (\widehat{F}, \widehat{V})$. The precondition ($F \models L$) in (I) is stronger than that ($M \models L$) in (II). The postcondition ($\widehat{F} \models L$) in (I) is stronger than ($\widehat{M} \models L$) in (II), too. However, we can always make sure that the finite model $\widehat{M}$ is differentiated. Then $\widehat{M} \models L$ iff $\widehat{F} \models L$. Thus, (II) is stronger than (I). Let us state this explicitly.

Lemma 2.7 (ADF for models implies ADF for frames)
For any logic L, if $\mathsf{Mod}(L)$ admits (definable) filtration, then so does $\mathsf{Fr}(L)$.

Proof. Take any finite Sub-closed set of formulas Γ and a model $M = (F, V)$ with $F \models L$. Then $M \in \mathsf{Mod}(L)$. By assumption, the model M has a (definable) Γ-filtration $\widehat{M} = (\widehat{F}, \widehat{V})$ with $\widehat{M} \models L$. The model $\widehat{M}$ is finite and, without loss of generality, differentiated, by Lemma A.2 (in Appendix). Then $\widehat{F} \models L$, by Lemma 1.1. Thus, $\mathsf{Fr}(L)$ admits (definable) filtration. □

The converse implication in the above lemma does not hold in general. Consider the logic **Ver** of the irreflexive singleton frame; it is axiomatized by $\Box\bot$. One can easily see that the class of its frames $\mathcal{F}$ admits definable filtration. But there are continuum many other logics L with the same class of frames $\mathcal{F}$ (cf. [2], [3, Ex. 10.57]), so that the class *frames* $\mathsf{Fr}(L)$ admits definable filtration, too. However, each of these logics is Kripke incomplete and, by Lemma 2.6, the class of *models* $\mathsf{Mod}(L)$ does not admit filtration.

Next we prove that, for the *canonical* logics, the converse implication holds, and so the notions (I) and (II) coincide, if we consider *definable* filtration. To simplify notation, we work with the unimodal case. Recall that one can build the *canonical* frame $F_T = (W_T, R_T)$ and *canonical* model $M_T = (F_T, V_T)$ not only for a (consistent) normal *logic*, but more generally for a *normal theory* T (which contains all theorems of **K** and is closed under monus ponens and necessitation). Any point $x \in W_T$ is a consistent (never $A, \neg A \in x$) complete (always $A \in x$ or $\neg A \in x$) theory (i.e., closed under modus ponens) containing T.

A logic L is called *canonical* if $F_L \models L$. The following is a well-known fact.

Lemma 2.8 (Canonical generated submodel) *If $T \subseteq T'$ are consistent normal theories, then $M_{T'}$ is a generated submodel of M_T. Similarly for frames.*

Proof. Assume $x \in W_{T'}$, $y \in W_T$, and $x\, R_T y$. To prove that $y \in W_{T'}$, i.e., $T' \subseteq y$, take any formula $A \in T'$. By normality $\Box A \in T'$. Since $T' \subseteq x$, we have $\Box A \in x$. By definition of R_T, we obtain $A \in y$. □

A typical example of a normal theory is the theory of a model $T = \mathsf{Th}(M)$. For a model $M = (W, R, V)$, consider the canonical model M_T of its theory and the *canonical mapping* t from M to M_T defined, for $a \in W$, by

$$t(a) = \mathsf{Th}(M, a) \in W_T.$$

It is monotonic ($a\, R\, b \Rightarrow t(a)\, R_T\, t(b)$), but in general it is neither surjective, nor a p-morphism. The following lemma (proved in Appendix, see Lemma A.3) shows what happens to the canonical mapping if we filtrate both M and M_T through a finite set of formulas Φ.

Lemma 2.9 *Under the above conditions, any finite set of formulas Φ induces a bijection between the quotient sets $W/{\sim_\Phi}$ and $W_T/{\sim_\Phi}$ defined, for $a \in W$, by*

$$f([a]_{\sim_\Phi}) \;:=\; [t(a)]_{\sim_\Phi}.$$

Theorem 2.10 (ADF for frames implies ADF for models) *If L is a canonical logic, then* $\mathsf{Mod}(L)$ *admits definable filtration iff so does* $\mathsf{Fr}(L)$.

Proof. ($\Rightarrow$) By Lemma 2.7. ($\Leftarrow$) IDEA: in order to filtrate a model $M \models L$, we filtrate the canonical model M_T of its theory $T = \mathsf{Th}(M)$ and then use the bijection from Lemma 2.9 to transfer the filtration back to M.

Take a finite Sub-closed set of formulas Γ and a model $M = (W, R, V)$ with $M \models L$. Its theory $T = \mathsf{Th}(M)$ contains L, hence F_T is a generated subframe of F_L, by Lemma 2.8. Since L is canonical, we have $F_L \models L$ and so $F_T \models L$. Thus, M_T is a $\mathsf{Fr}(L)$-model and, by assumption, we can filtrate it.

Therefore, the model M_T has a Γ-filtration $\widehat{M_T} = (\widehat{W_T}, \widehat{R_T}, \widehat{V_T})$ (through some finite set of formulas $\Phi \supseteq \Gamma$) with $\widehat{F_T} \models L$. By Lemma 2.9, there is a bijection f between the finite sets $\widehat{W} = (W/{\sim_\Phi})$ and $\widehat{W_T} = (W_T/{\sim_\Phi})$. Now we build a model $\widehat{M} = (\widehat{W}, \widehat{R}, \widehat{V})$ isomorphic to $\widehat{M_T}$, by putting, for all $a, b \in W$:

$$\widehat{a}\,\widehat{R}\,\widehat{b} \text{ iff } f(\widehat{a})\,\widehat{R_T}\,f(\widehat{b}); \qquad \widehat{a} \models p \text{ iff } f(\widehat{a}) \models p, \text{ for all variables } p \in \Gamma.$$

Since the frames $\widehat{F}$ and $\widehat{F_T}$ are isomorphic and $\widehat{F_T} \models L$, we have $\widehat{F} \models L$. It remains to prove that $\widehat{M}$ is a Γ-filtration (through Φ) of M. Below, we denote $x = t(a) = \mathsf{Th}(M, a)$ and $y = t(b) = \mathsf{Th}(M, b)$, so that $f(\widehat{a}) = \widehat{x}$ and $f(\widehat{b}) = \widehat{y}$.

(var) Let us check that $\widehat{M}, \widehat{a} \models p$ iff $M, a \models p$, for all $p \in \Gamma$. We have:

$$\widehat{M}, \widehat{a} \models p \;\Leftrightarrow\; \widehat{M_T}, \widehat{x} \models p \;\Leftrightarrow\; M_T, x \models p \;\Leftrightarrow\; p \in x \;\Leftrightarrow\; M, a \models p.$$

(min) Let us check that $R^{\mathsf{min}}_{\sim_\Phi} \subseteq \widehat{R}$, i.e., $\forall a, b \in W\ (a\,R\,b \Rightarrow \widehat{a}\,\widehat{R}\,\widehat{b})$.

We use the monotonicity of $t(\cdot)$ and the condition (min) for $\widehat{R_T}$:

$$a\,R\,b \implies t(a)\,R_T\,t(b) \;\Leftrightarrow\; x\,R_T\,y \implies \widehat{x}\,\widehat{R_T}\,\widehat{y} \;\Leftrightarrow\; \widehat{a}\,\widehat{R}\,\widehat{b}.$$

(max) Let us check that $\widehat{R} \subseteq R^{\mathsf{max}}_\Gamma$. Assume $\widehat{a}\,\widehat{R}\,\widehat{b}$. Then $\widehat{x}\,\widehat{R_T}\,\widehat{y}$.

By the condition (max) for $\widehat{R_T}$, we have $\widehat{x}\,((R_T)^{\mathsf{max}}_\Gamma)\,\widehat{y}$.

We need to show that $\widehat{a}\,R^{\mathsf{max}}_\Gamma\,\widehat{b}$. For any formula $\Box A \in \Gamma$, we have:

$$M, a \models \Box A \Leftrightarrow \Box A \in x \Leftrightarrow M_T, x \models \Box A \Rightarrow M_T, y \models A \Leftrightarrow A \in y \Leftrightarrow M, b \models A.$$

This completes the proof of the theorem. □

3 Logics with the transitive closure modality

In this section, $L \subseteq \mathsf{Fm}(\Box)$ is a normal unimodal logic. Let $L^{\boxplus} \subseteq \mathsf{Fm}(\Box, \boxplus)$ be the minimal normal logic that extends L with the following axioms describing the interaction between the modality $\Box$ and the *transitive closure* modality $\boxplus$:

$$(\mathsf{A1})\ \boxplus p \to \Box p, \qquad (\mathsf{A2})\ \boxplus p \to \Box\boxplus p, \qquad (\mathsf{A3})\ \boxplus(p \to \Box p) \to (\Box p \to \boxplus p).$$

Segerberg [16] (see also [17,19]) and later Kozen and Parikh [13] proved that the logic $\mathbf{K}^{\boxplus}$ (and even PDL) is complete and has the FMP; in other words, it is the logic of finite frames of the form (W, R, R^+); hence it is decidable (more exactly, EXPTIME-complete); see also [4] for a constructive variant of completeness theorem. The logic $\mathbf{K}^{\boxplus}$ is known to be not canonical (see Lemma A.5 in Appendix). Thus, even for simple logics we cannot use canonical models as a method of obtaining completeness.

To the best of our knowledge, up to now, there were no general results on the completeness and decidability for the $\boxplus$-companions of logics other than **K**. Here we obtain one such result. We give a condition on L sufficient for the completeness of $L^{\boxplus}$. The condition is strong enough and guarantees not only the completeness, but the FMP of $L^{\boxplus}$; this limits the scope of our approach.

For simplicity, in this section we assume that L is unimodal. The results transfer easily to multi-modal logics. Given a unimodal frame $F = (W, R)$, we denote $F^{\oplus} = (W, R, R^{+})$. Given a class of unimodal frames $\mathcal{F}$, we denote $\mathcal{F}^{\oplus} = \{F^{\oplus} \mid F \in \mathcal{F}\}$. Similarly for a model $M^{\oplus}$ and a class of models $\mathcal{M}^{\oplus}$.

Lemma 3.1 $(W, R, S) \models \{(\mathsf{A1}), (\mathsf{A2}), (\mathsf{A3})\}$ *iff* $R^{+} = S$.

Proof. This is a known fact. Lemma A.4 (in Appendix) gives more details. □

Lemma 3.2 (a) $\mathsf{Mod}(L)^{\oplus} \subseteq \mathsf{Mod}(L^{\boxplus})$. (b) $\mathsf{Fr}(L)^{\oplus} = \mathsf{Fr}(L^{\boxplus})$.

Proof. Any frame of the form (W, R, R^{+}) validates (A1), (A2), and (A3). □

Lemma 3.3 (Conservativity) *For any consistent normal logic L, the logic $L^{\boxplus}$ is a conservative extension of L: if $A \in \mathsf{Fm}(\Box)$ and $L^{\boxplus} \vdash A$, then $L \vdash A$.*

Proof. If $L \nvdash A$, then $M_L \not\models A$ and $M_L^{\oplus} \not\models A$. But $M_L^{\oplus} \models L^{\boxplus}$. So $L^{\boxplus} \nvdash A$. □

3.1 Completeness for logics with the transitive closure modality

In the proof of the main result, we will need to modify a valuation *definably*. By φ^{σ} we denote the application of a substitution $\sigma\colon \mathsf{Var} \to \mathsf{Fm}$ to a formula φ.

Definition 3.4 By a (*modally*) *definable variant* of a model $M = (F, V)$ we mean a model of the form $M^{\sigma} = (F, V^{\sigma})$, for some substitution σ, where the valuation V^{σ} is defined by $V^{\sigma}(p) = V(p^{\sigma})$, for every variable p.

In other words, $M^{\sigma}, x \models p$ iff $M, x \models p^{\sigma}$. By induction one can easily prove:

Lemma 3.5 $M^{\sigma}, x \models \varphi$ *iff* $M, x \models \varphi^{\sigma}$, *for all formulas* φ.

Since a logic is closed under substitutions, we obtain the following fact.

Lemma 3.6 *If L is a logic and $M \models L$, then $M^{\sigma} \models L$, for any substitution σ.*

Recall that the formulas (A1) and (A2) are *canonical*, so they are valid on the canonical frame of $L^{\boxplus}$, for any L. For (A3), this is not the case even for the case $L = \mathbf{K}$: in the canonical frame $F_{\mathbf{K}^{\boxplus}} = (W, R, S)$, only a strict inclusion $R^{+} \subsetneq S$ holds (see lemma A.5 in Appendix).

However, in order to obtain the completeness of $L^{\boxplus}$, we do not necessarily need the converse inclusion $S \subseteq R^{+}$ in the canonical frame of $L^{\boxplus}$. Instead, we do a walk around: given any model $M = (W, R, S, V)$ of $L^{\boxplus}$ (e.g., its canonical model), we remove S, filtrate (W, R, V) into a finite model $(\widehat{W}, \widehat{R}, \widehat{V})$, and then augment it with $(\widehat{R})^{+}$. It only remains to prove that the resulting finite bi-modal model is a definable filtration of the original bi-modal model M; i.e., that $(\widehat{R})^{+}$ is between the minimal and the maximal filtered relations. For maximal, the inclusion follows from the axioms (A1) and (A2) only (see (7) in the proof of Theorem 3.8 below); on the contrary, for the minimal, the required inclusion (see (5) in that proof) holds due to the following remarkable property of (A3).

Let us write $M \models A^*$ if we have $M \models A^\sigma$ for all substitutions σ.

Lemma 3.7 (Induction axiom and minimal filtration)
Let $M = (W, R, S, V) \models (\mathsf{A3})^*$ *and let* $\Phi \subseteq \mathsf{Fm}$ *be finite. Then* $S^{\mathsf{min}}_{\sim_\Phi} \subseteq (R^{\mathsf{min}}_{\sim_\Phi})^+$.

Proof. Denote $r := R^{\mathsf{min}}_{\sim_\Phi}$ and $s := S^{\mathsf{min}}_{\sim_\Phi}$. To prove $s \subseteq r^+$, assume $\widehat{x}\, s\, \widehat{y}$. By definition of the minimal filtered relation $S^{\mathsf{min}}_{\sim_\Phi}$, we can assume, without loss of generality, that $x\, S\, y$. Consider $Y := r^+(\widehat{x}) \subseteq \widehat{W}$. We need to show that $\widehat{y} \in Y$.

Since Φ is finite, every $\sim_\Phi$-equivalence class $\widehat{z} \subseteq W$ is a definable (by some formula) subset of W. Since Y is a *finite* collection of such subsets, their union $\bigcup Y \subseteq W$ is also a definable subset of W. So, there is a formula φ such that, for all $z \in W$, we have: $M, z \models \varphi$ iff $z \in \bigcup Y$ iff $\widehat{z} \in Y$.

Firstly, $M \models \varphi \to \Box\varphi$. Indeed, if $M, a \models \varphi$, $a\, R\, b$, then $\widehat{a} \in Y$, $\widehat{a}\, r\, \widehat{b}$. But Y is closed under r, hence $\widehat{b} \in Y$ and $M, b \models \varphi$. Therefore, $M \models \boxplus(\varphi \to \Box\varphi)$.

Secondly, $M, x \models \Box\varphi$. Indeed, if $x\, R\, z$ then $\widehat{x}\, r\, \widehat{z}$, so $\widehat{z} \in Y$ and $M, z \models \varphi$.

Now we use that $M \models \boxplus(\varphi \to \Box\varphi) \to (\Box\varphi \to \boxplus\varphi)$. Thus, $M, x \models \boxplus\varphi$. Recall that $x\, S\, y$. Then $M, y \models \varphi$, hence $\widehat{y} \in Y$. □

In Appendix (Lemma A.6) we strengthen the above lemma.

Now we come to the main technical tool of our paper.

Theorem 3.8 (Transfer of ADF to logics with transitive closure)
If the class $\mathsf{Mod}(L)$ *admits definable filtration, then so does the class* $\mathsf{Mod}(L^\boxplus)$.

Proof. IDEA:[4] in order to filtrate a model $M = (W, R, S, V) \models L^\boxplus$ for $\Gamma \subseteq \mathsf{Fm}(\Box, \boxplus)$, we build a special set of unimodal formulas Δ and Δ-filtrate the reduct $N = (W, R, V) \models L$ of M into a finite model $\widehat{N} = (\widehat{W}, \widehat{R}, \widehat{V}) \models L$. Then we show that $\widehat{N}^+ = (\widehat{W}, \widehat{R}, (\widehat{R})^+, \widehat{V}) \models L^\boxplus$ is a Γ-filtration of M. More precisely, we first take a modified valuation V^σ and actually filtrate N^σ, not N.

FORMALLY: take a model $M = (W, R, S, V)$ such that $M \models L^\boxplus$ and a finite Sub-closed set of formulas $\Gamma \subseteq \mathsf{Fm}(\Box, \boxplus)$. For each formula $\varphi \in \Gamma$, fix a fresh (not occurring in Γ) variable q_φ. Consider a substitution $\sigma\colon \mathsf{Var} \to \mathsf{Fm}(\Box, \boxplus)$ defined by $\sigma(q_\varphi) = \varphi$ for all $\varphi \in \Gamma$ and $\sigma(p) = p$ for all other variables p. In the definable variant $M^\sigma = (W, R, S, V^\sigma)$ of M we have: $M^\sigma \models q_\varphi \leftrightarrow \varphi$ for all $\varphi \in \Gamma$ (since $\varphi^\sigma = \varphi$), hence $M^\sigma \models \Box q_\varphi \leftrightarrow \Box\varphi$ and even $M^\sigma \models A \leftrightarrow A^\sigma$, for any formula $A \in \mathsf{Fm}(\Box)$. We also have $M^\sigma \models L^\boxplus$ by Lemma 3.6.

Now consider the reduct $N^\sigma = (W, R, V^\sigma)$ of M^σ. Clearly, $N^\sigma \models L$. However, we cannot Γ-filtrate this model, since Γ is a set of *bimodal* formulas.

---

[4] The proof of the main theorem differs from the proof of the corresponding Theorem 2.6 from our paper [12] in the following two aspects. First, in [12] we filtrate a model of the form $M = (W, R, R^+, V)$ such that $(W, R, R^+) \models L^\boxplus$, i.e., $(W, R) \models L$; while here we will filtrate a model of the form $M = (W, R, S, V)$ such that $M \models L^\boxplus$. As a consequence, in the old proof, we had to show that $(R^+)^{\mathsf{min}}_\sim \subseteq (R^{\mathsf{min}}_\sim)^+$, which is quite simple, while here we need to show that $\widehat{S} \subseteq (\widehat{R})^+$, for this we need Lemma 3.7. Secondly, we transform a filtration of (W, R, V) through a set of formulas $\Phi \subseteq \mathsf{Fm}(\Box)$ into a filtration of (W, R, S, V) through some set of formulas $\Phi' \subseteq \mathsf{Fm}(\Box, \boxplus)$, so we need to build Φ' from Φ.

Consider the following finite Sub-closed set of $\Box$-formulas:

$$\Delta \;:=\; \{q_\varphi, \Box q_\varphi \mid \varphi \in \Gamma\} \;\subset\; \mathsf{Fm}(\Box).$$

$\mathsf{Mod}(L)$ admits definable filtration, so there is a Δ-filtration $\widehat{N^\sigma} = (\widehat{W}, \widehat{R}, \widehat{V^\sigma})$ of N^σ through some finite set $\Phi \subseteq \mathsf{Fm}(\Box)$ with $\Delta \subseteq \Phi$ such that $\widehat{N^\sigma} \models L$. Let us change $\widehat{V^\sigma}$ on the variables $p \in \Gamma$ by putting:[5] $\widehat{N^\sigma}, \widehat{x} \models p \leftrightharpoons \widehat{N^\sigma}, \widehat{x} \models q_p$.

Remark. Since we will have several models on the same set of points, we need a more subtle notation. In particular, we have $\widehat{W} = W/{\sim_\Phi^{N^\sigma}}$, this notation shows explicitly in which models we consider the $\sim_\Phi$-equivalence of points.

Now put $\widehat{M} := \widehat{N^\sigma}^{\oplus} = (\widehat{W}, \widehat{R}, (\widehat{R})^+, \widehat{V^\sigma})$. It remains to prove the following.

Claim. *The model $\widehat{M}$ is a Γ-filtration (through Φ^σ) of M.*

(1) We show that $\widehat{W} = W/{\sim_{\Phi^\sigma}^M}$. For any $x \in W$ and $A \in \mathsf{Fm}(\Box)$, we have:

$$N^\sigma, x \models A \quad\Leftrightarrow\quad M^\sigma, x \models A \quad\Leftrightarrow\quad M, x \models A^\sigma.$$

Therefore, for all $x, y \in W$, we have: $(x \sim_\Phi^{N^\sigma} y)$ iff $(x \sim_{\Phi^\sigma}^M y)$.
This allows us to introduce a simpler notation $\sim$ for $\sim_\Phi^{N^\sigma}$ and $\sim_{\Phi^\sigma}^M$.
Since $\widehat{N^\sigma}$ is a Δ-filtration of N^σ, we have: $R_\sim^{\mathsf{min}} \subseteq \widehat{R} \subseteq R_{\sim,\Delta}^{\mathsf{max}}$. $(*)$

(2) The relation $\sim$ respects Γ. Indeed, $\Phi \supseteq \Delta \supseteq \{q_\varphi \mid \varphi \in \Gamma\}$, hence $\Phi^\sigma \supseteq \Gamma$.

(3) We show that $M, x \models p \Leftrightarrow \widehat{M}, \widehat{x} \models p$, for $p \in \Gamma$. Note that $\widehat{M} \models q_p \leftrightarrow p$.

$$\begin{array}{ccccc} M, x \models p & \Leftrightarrow & M^\sigma, x \models q_p & \Leftrightarrow & N^\sigma, x \models q_p \\ & & & & \Updownarrow \\ \widehat{M}, \widehat{x} \models p & \Leftrightarrow & \widehat{M}, \widehat{x} \models q_p & \Leftrightarrow & \widehat{N^\sigma}, \widehat{x} \models q_p \end{array}$$

(4) $R_\sim^{\mathsf{min}} \subseteq \widehat{R}$. This holds by $(*)$.

(5) $S_\sim^{\mathsf{min}} \subseteq \widehat{S}$, where $\widehat{S} := (\widehat{R})^+$. Using (4) and Lemma 3.7, we obtain:

$$S_\sim^{\mathsf{min}} \subseteq (R_\sim^{\mathsf{min}})^+ \subseteq (\widehat{R})^+ = \widehat{S}.$$

(6) $\widehat{R} \subseteq R_{\sim,\Gamma}^{\mathsf{max}}$. Due to $(*)$, it suffices to prove that $R_{\sim,\Delta}^{\mathsf{max}} \subseteq R_{\sim,\Gamma}^{\mathsf{max}}$.
Assume that $\widehat{x}\,(R_{\sim,\Delta}^{\mathsf{max}})\,\widehat{y}$. To show that $\widehat{x}\,(R_{\sim,\Gamma}^{\mathsf{max}})\,\widehat{y}$, take any $\Box\varphi \in \Gamma$. Then:

$$\begin{array}{ccccc} M, x \models \Box\varphi & \overset{(a)}{\Longleftrightarrow} & M^\sigma, x \models \Box q_\varphi & \overset{(b)}{\Longleftrightarrow} & N^\sigma, x \models \Box q_\varphi \\ & & & & \Downarrow (c) \\ M, y \models \varphi & \overset{(a)}{\Longleftrightarrow} & M^\sigma, y \models q_\varphi & \overset{(b)}{\Longleftrightarrow} & N^\sigma, y \models q_\varphi \end{array}$$

We used: (a) Lemma 3.5; (b) $q_\varphi, \Box q_\varphi \in \mathsf{Fm}(\Box)$; (c) $\Box q_\varphi \in \Delta$ and $\widehat{x}\,(R_{\sim,\Delta}^{\mathsf{max}})\,\widehat{y}$.

(7) $\widehat{S} \subseteq S_{\sim,\Gamma}^{\mathsf{max}}$. Due to $(*)$, it suffices to prove that $(R_{\sim,\Delta}^{\mathsf{max}})^+ \subseteq S_{\sim,\Gamma}^{\mathsf{max}}$.
Let us denote $r := R_{\sim,\Delta}^{\mathsf{max}}$ and $s := S_{\sim,\Gamma}^{\mathsf{max}}$. In order to prove that $r^+ \subseteq s$, it suffices to prove two inclusions: $r \subseteq s$ and $r \circ s \subseteq s$.

(7a) Proof of $(r \subseteq s)$. We will use the axiom (A1): $\boxplus p \to \Box p$.
Assume $\widehat{x}\,(R_{\sim,\Delta}^{\mathsf{max}})\,\widehat{y}$. To prove that $\widehat{x}\,(S_{\sim,\Gamma}^{\mathsf{max}})\,\widehat{y}$, take any $\boxplus\varphi \in \Gamma$. Then:

---

[5] We could simply assume that $\widehat{V^\sigma}$ was undefined for the variables $p \in \mathsf{Var}(\Gamma)$ before we defined it here explicitly. This allows us to use the same notation $\widehat{V^\sigma}$ for the amended valuation. Note that this amendment does not change the truth of formulas from Δ in $\widehat{N^\sigma}$.

$$\begin{array}{ccccccc} M,x\models\boxplus\varphi & \overset{(d)}{\Longrightarrow} & M,x\models\Box\varphi & \overset{(a)}{\Longleftrightarrow} & M^\sigma,x\models\Box q_\varphi & \overset{(b)}{\Longleftrightarrow} & N^\sigma,x\models\Box q_\varphi \\ &&&&&& \Downarrow (c) \\ && M,y\models\varphi & \overset{(a)}{\Longleftrightarrow} & M^\sigma,y\models q_\varphi & \overset{(b)}{\Longleftrightarrow} & N^\sigma,y\models q_\varphi \end{array}$$

(d) holds since $M\models\boxplus\varphi\to\Box\varphi$. The explanations of (a,b,c) are the same.

(7b) Proof of $(r\circ s\subseteq s)$. We will use the axiom (A2): $\boxplus p\to\Box\boxplus p$.

Assume $\widehat{x}\,(R^{\mathsf{max}}_{\sim,\Delta})\,\widehat{y}\,(S^{\mathsf{max}}_{\sim,\Gamma})\,\widehat{z}$. To prove $\widehat{x}\,(S^{\mathsf{max}}_{\sim,\Gamma})\,\widehat{z}$, take any $\boxplus\varphi\in\Gamma$. Then:

$$\begin{array}{ccccccc} M,x\models\boxplus\varphi & \overset{(e)}{\Longrightarrow} & M,x\models\Box\boxplus\varphi & \overset{(a)}{\Longleftrightarrow} & M^\sigma,x\models\Box q_{\boxplus\varphi} & \overset{(b)}{\Longleftrightarrow} & N^\sigma,x\models\Box q_{\boxplus\varphi} \\ &&&&&& \Downarrow (c) \\ M,z\models\varphi & \overset{(g)}{\Longleftarrow} & M,y\models\boxplus\varphi & \overset{(a)}{\Longleftrightarrow} & M^\sigma,y\models q_{\boxplus\varphi} & \overset{(b)}{\Longleftrightarrow} & N^\sigma,y\models q_{\boxplus\varphi} \end{array}$$

We used: (e) $M\models\boxplus\varphi\to\Box\boxplus\varphi$; (a) Lemma 3.5; (b) $\Box q_{\boxplus\varphi}\in\mathsf{Fm}(\Box)$; (c) $\Box q_{\boxplus\varphi}\in\Delta$ and $\widehat{x}\,(R^{\mathsf{max}}_{\sim,\Delta})\,\widehat{y}$; (g) $\boxplus\varphi\in\Gamma$ and $\widehat{y}\,(S^{\mathsf{max}}_{\sim,\Gamma})\,\widehat{z}$.

This completes the proof of theorem. □

Note that in (7a) and (7b) we proved inclusions that involve maximal relations, and these inclusions resemble the axioms (A1) and (A2). This is not a coincidence. In Lemma 4.3 of our paper [12], we already made this observation for any *right-linear grammar* axiom and both (A1) and (A2) are right-linear.

Let us summarize the main result on logics with transitive closure. We give two versions. The former theorem uses a rather unusual property (filtration of models). However, its advantage is that one can 'iterate' the application of this theorem (as we do in Section 4), since its premise and conclusion have the same form: "the class of models of a logic admits definable filtration". The latter theorem uses filtration of frames, but additionally requires canonicity.

Theorem 3.9 (Main result, version 1) *Assume that the class of models* $\mathsf{Mod}(L)$ *of a logic* L *admits definable filtration. Then:*

(1) *the class of models* $\mathsf{Mod}(L^{\boxplus})$ *admits definable filtration;*
(2) *hence the logic* $L^{\boxplus}$ *has the finite model property;*
(3) *hence the logic* $L^{\boxplus}$ *is Kripke complete.*

Proof. Assume $\mathsf{Mod}(L)$ admits definable filtration. Then so does $\mathsf{Mod}(L^{\boxplus})$, by Theorem 3.8. By Lemma 2.6, $L^{\boxplus}$ has the FMP and is Kripke complete. □

Theorem 3.10 (Main result, version 2) *Assume that a logic* L *is canonical and the class of its frames* $\mathsf{Fr}(L)$ *admits definable filtration. Then:*

(1) *the class* $\mathsf{Mod}(L^{\boxplus})$ *admits definable filtration;*
(2) *hence the logic* $L^{\boxplus}$ *has the finite model property;*
(3) *hence the logic* $L^{\boxplus}$ *is Kripke complete.*

Proof. If L is canonical and the class $\mathsf{Fr}(L)$ admits definable filtration, then so does the class $\mathsf{Mod}(L)$, by Theorem 2.10. Now apply Theorem 3.9. □

4 PDLization of logics that admit filtration

Now we apply Theorem 3.9 to show that if $\mathsf{Mod}(L)$ admits definable filtration, then the following PDL-like expansions of L have the finite model property.

Definition 4.1 For an alphabet Σ, let $\Sigma^\sharp = \Sigma \cup \{(e \circ f), (e \cup f), e^+ \mid e, f \in \Sigma\}$, assuming that the added symbols are not in Σ. Put $\Sigma^{(0)} = \Sigma$, $\Sigma^{(n+1)} = (\Sigma^{(n)})^\sharp$.

For a frame $F = (W, (R_e)_{e\in\Sigma})$, put $F^\sharp = (W, (R_e)_{e\in\Sigma^\sharp})$, where for $e, c \in \Sigma$,

$$R_{e\circ c} = R_e \circ R_c, \quad R_{e\cup c} = R_e \cup R_c, \quad R_{e^+} = (R_e)^+.$$

Put $F^{(0)} = F$, $F^{(n+1)} = (F^{(n)})^\sharp$.

For a model $M = (F, V)$, we put $M^\sharp = (F^\sharp, V)$ and $M^{(n)} = (F^{(n)}, V)$.

For a logic L over Σ, let $L^\sharp$ be the smallest (normal) logic over $\Sigma^\sharp$ that contains L and the following PDL-like axioms, for all $e, c \in \Sigma$:

$[e \cup c]p \leftrightarrow [e]p \wedge [c]p$,

$[e \circ c]p \leftrightarrow [e][c]p$,

$[e^+]p \to [e]p, \quad [e^+]p \to [e][e^+]p, \quad [e^+](p \to [e]p) \to ([e]p \to [e^+]p)$.

We put $L^{(0)} = L$, $L^{(n+1)} = (L^{(n)})^\sharp$.

The following is a simple analogue of Lemma 3.2.

Lemma 4.2 (a) *$M \models L$ implies $M^\sharp \models L^\sharp$.* (b) *$F \models L$ iff $F^\sharp \models L^\sharp$.*

By an easy induction on n, we obtain

Proposition 4.3 *For a frame F and $n < \omega$, $F \models L$ iff $F^{(n)} \models L^{(n)}$.*

Proposition 4.4 *For a logic L and $n < \omega$, $L^{(n)}$ is conservative over L.*

Proof. As in Lemma 3.3, using ${M_L}^{(n)}$ instead of $M_L^\oplus$ and Lemma 4.2(a). □

Lemma 4.5 *Let L be a logic over Σ, $e, c \in \Sigma$. Let L_1 and L_2 be the logics over $\Sigma \cup \{g\}$, where $g \notin \Sigma$, such that*

L_1 extends L with the axiom $[g]p \leftrightarrow [e]p \wedge [c]p$,
L_2 extends L with the axiom $[g]p \leftrightarrow [e][c]p$.

If $\mathsf{Mod}(L)$ admits definable filtration, then so do $\mathsf{Mod}(L_1)$ and $\mathsf{Mod}(L_2)$.

Proof. Straightforward. Details can be reconstructed from the proof of Lemma 2.3 in [12], which is the analog of our lemma for the classes of frames. □

Theorem 4.6 *Let L be a logic over a finite alphabet Σ. If the class of its models $\mathsf{Mod}(L)$ admits definable filtration, then, for every $n < \omega$, we have:*

(i) *$\mathsf{Mod}(L^{(n)})$ admits definable filtration.*

(ii) *$L^{(n)}$ has the finite model property; a fortiori, $L^{(n)}$ is Kripke complete.*

(iii) *If L is finitely axiomatizable, then $L^{(n)}$ is decidable.*

(iv) *If the class of finite frames of L is decidable, then $L^{(n)}$ is co-recursively enumerable.*

Proof. (i) By Theorem 3.8 and Lemma 4.5, if $\mathsf{Mod}(L)$ admits definable filtration, then so does $\mathsf{Mod}(L^\sharp)$. So, (i) follows by induction on n.

(ii) By Lemma 2.6.

(iii) Note that if L is finitely axiomatizable, then so is $L^{(n)}$. The claim then follows from Harrop's Theorem (see Section 1).

(iv) If the class of finite frames of L is decidable, then the class of finite frames of $L^{(n)}$ is decidable, too. In this case $L^{(n)}$ is co-recursively enumerable, since $L^{(n)}$ is the logic of its finite frames. □

Theorem 4.6 can be generalized for the case when we additionally extend the alphabet with converse modalities. This generalization can be obtained by modifying the proof of Theorem 2.6 in [12].

4.1 Fusions that admit filtration

Here we consider a special kind of definable filtration, called *strict filtration.*

Definition 4.7 If, in terms of Definition 2.1, $\sim = \sim_\Gamma$, then we call the filtration $\widehat{M}$ *strict.* The corresponding notions "a class of frames (or models) *admits strict filtration*" are introduced in the obvious way.

Strict filtration is the most standard variant of filtration; it is well-known that the classes of frames of the logics **K**, **T**, **K4**, **S4**, **S5** admit strict filtration (for the logics **K** and **T**, even the *minimal* strict filtration works; for **K4**, **S4**, **S5**, strict filtration is obtained by taking the transitive closure of the minimal filtered relation [15]).

Let us recall the notion of the *fusion* of logics. Let $L_1, \ldots, L_k$ be logics over finite alphabets $\Sigma_1, \ldots, \Sigma_k$. Without loss of generality we assume that these alphabets are disjoint. The *fusion* $L_1 * \ldots * L_k$ of these logics is the smallest normal logic over the alphabet $\Sigma = \Sigma_1 \cup \ldots \cup \Sigma_k$ that contains $L_1 \cup \ldots \cup L_k$.

It is well-known that the fusion operation preserves Kripke completeness, the finite model property, and decidability [14]. We observe that it also preserves the property "a logic admits strict filtration".

Theorem 4.8 (Fusion and strict filtration) *If classes of frames* $\mathsf{Fr}(L_i)$, $1 \le i \le k$, *admit strict filtration, then* $\mathsf{Fr}\,(L_1 * \ldots * L_k)$ *admits strict filtration.*

Proof. The idea is the same as in the proof of Theorem 3.8. To simplify notation, we consider the case of unimodal logics. Let $L = L_1 * \ldots * L_k$, $M = (F, V)$ be a model on an L-frame $F = (W, R_1, \ldots, R_k)$, $\Gamma \subseteq \mathsf{Fm}(\Box_1, \ldots, \Box_k)$ be finite and Sub-closed. For $\varphi \in \Gamma$, we take fresh variables q_φ, and consider a model $M' = (F, V')$ such that

$$M, x \models \varphi \ \text{ iff } \ M', x \models \varphi \ \text{ iff } \ M', x \models q_\varphi$$

for all x in M. For $1 \le i \le k$, we put:

$$\Gamma_i = \{q_\varphi \mid \varphi \in \Gamma\} \cup \{\Box q_\varphi \mid \Box_i \varphi \in \Gamma\}.$$

Note that $\Gamma_i \subseteq \mathsf{Fm}(\Box)$. Let $\sim_i$ be the equivalence induced by Γ_i in the model $M_i = (W, R_i, V')$, and $\sim_\Gamma$ the equivalence induced by Γ in M. Observe that

$$M_i, x \models \Box q_\varphi \ \text{ iff } \ M, x \models \Box_i \varphi \ \text{ for all } \ \varphi \in \Gamma. \qquad (*)$$

Therefore, one can see that $\sim_i = \sim_\Gamma$ for all i. Put $\widehat{W} = W/{\sim_\Gamma}$. For each i, there exists a filtration $\widehat{M}_i = (\widehat{W}, \widehat{R}_i, \widehat{V}_i)$ of M_i through Γ_i such that $(\widehat{W}, \widehat{R}_i) \models L_i$. The valuations $\widehat{V}_i$ coincide on the variables q_φ. W.l.o.g., they also coincide on other variables (since they do not occur in Γ_i), and that $\widehat{M_i}, \widehat{x} \models p$ iff $M, x \models p$ for each variable $p \in \Gamma$. The resulting valuation on $\widehat{W}$ is denoted by $\widehat{V}$.

Consider the model $\widehat{M} = (\widehat{W}, \widehat{R_1}, \ldots, \widehat{R_k}, \widehat{V})$. Note that its frame validates the fusion L. We claim that $\widehat{M}$ is a filtration of M through Γ. Clearly, $\widehat{R_i}$ contains the i-th minimal filtered relation. To check that $\widehat{R_i}$ is contained in the i-th maximal filtered relation, assume that $\widehat{x}\widehat{R_i}\widehat{y}$, $M, x \models \Box_i\varphi$, and $\Box_i\varphi \in \Gamma$. Then $M_i, x \models \Box q_\varphi$, by (*). Since $\widehat{M_i}$ is a filtration of M_i through Γ_i and $\Box q_\varphi \in \Gamma_i$, we have $\widehat{M_i}, \widehat{y} \models q_\varphi$. By Filtration lemma, $M_i, y \models q_\varphi$. Hence, $M', y \models q_\varphi$ and we conclude that $M, y \models \varphi$, as required. □

Theorem 4.9 *Let $L_1, \ldots, L_k$ be canonical logics and their classes of frames $\mathsf{Fr}(L_i)$, $1 \leq i \leq k$, admit strict filtration. Then, for every $n < \omega$, the logic $(L_1 * \ldots * L_k)^{(n)}$ has the finite model property.*

Proof. The fusion $L = L_1 * \ldots * L_k$ is canonical. By Theorem 4.8, the class $\mathsf{Fr}(L)$ admits strict filtration. Hence $\mathsf{Mod}(L)$ admits definable filtration, by Theorem 2.10. Finally, $(L)^{(n)}$ has the FMP, by Theorem 4.6. □

4.2 A class of formulas that admit strict filtration

We present a collection of modal formulas that admit strict (and so definable) filtration. The obvious candidates are modal formulas whose first-order equivalents belong to a certain FO fragment we call MFP. [6] We define it inductively as the minimal set of FO formulas satisfying the following conditions:

- if x and y are variables, R is a binary relation symbol, then $R(x, y) \in \mathsf{MFP}$ and $x = y \in \mathsf{MFP}$;
- if A and B are in MFP, then $(A \wedge B)$ and $(A \vee B)$ are in MFP;
- if $A \in \mathsf{MFP}$, and v is a variable, then $\forall v\, A$ and $\exists v\, A$ are in MFP;
- if x and y are variables, R is a binary relation symbol, and $A \in \mathsf{MFP}$, then $\forall x \forall y(R(x, y) \to A)$ and $\forall x \forall y(x = y \to A)$ are in MFP.

This definition is the restriction of the fragment $\mathsf{POS} + \forall\mathsf{G}$ from [9] to the first-order language with only binary predicates. Examples of MFP-sentences are reflexivity $\forall x R(x, x)$, symmetry $\forall x \forall y(R(x, y) \to R(y, x))$, and density $\forall x \forall y(R(x, y) \to \exists z\, (R(x, z) \wedge R(z, y)))$, but not transitivity.

FO counterparts of minimal filtrations are strong onto homomorphisms.

Definition 4.10 Given two frames $F = (W, R)$ and $F' = (W', R')$, a map $h \colon W \to W'$ is a *strong onto homomorphism* if h is onto and we have:

- for all $x, y \in W$, if $x\, R\, y$, then $h(x)\, R'\, h(y)$ (*monotonicity*);
- for all $x', y' \in W'$, if $x'\, R'\, y'$, then there exist $x, y \in W$ such that $h(x) = x'$, $h(y) = y'$, and $x\, R\, y$ (*weak lifting*).

Note that a strong homomorphism h from F onto F' induces an equivalence $\sim$ on W defined by $x \sim y$ iff $h(x) = h(y)$, and then F' is isomorphic to the *minimal filtrated frame* $F^{\mathsf{min}}_{\sim} = (W/{\sim}, R^{\mathsf{min}}_{\sim})$. Conversely, if $\widehat{M}$ is a minimal filtration of M, then the map $x \mapsto \widehat{x}$ is a strong homomorphism from F onto $\widehat{F}$.

[6] The abbreviation stems from "*p*reserved under *m*inimal *f*iltration".

Any MFP-formula is preserved under strong onto homomorphisms [9, Prop. 5.2]. Moreover, any FO formula with binary relations that is preserved under strong onto homomorphisms is equivalent to some MFP-formula [11].

Definition 4.11 A modal formula φ is called a *modal* MFP*-formula* if it has a FO equivalent (on frames) in MFP.

Typical examples of modal MFP-formulas are expressions of the form $p \wedge \Diamond q \to \psi$, where ψ is a positive modal formula. Note that these examples are Sahlqvist formulas, and hence canonical.

Theorem 4.12 *For any set Φ of modal* MFP*-formulas over a finite alphabet Σ, the class of frames $\mathsf{Fr}(\mathbf{K}_\Sigma + \Phi)$ admits strict filtration.*

Proof. Denote $\mathcal{F} = \mathsf{Fr}(\mathbf{K}_\Sigma + \Phi)$. Let $M = (F, V)$ be an $\mathcal{F}$-model and Γ a finite Sub-closed set of formulas. Take the minimal filtration $\widehat{M} = (\widehat{F}, \widehat{V})$ of M through Γ; note that this filtration is strict. Then the map $x \mapsto \widehat{x}$ is a strong homomorphism from F onto $\widehat{F}$. Since the set Φ^* of the MFP first-order equivalents of Φ is true in F, it is also true in $\widehat{F}$. Hence $\widehat{M}$ is an $\mathcal{F}$-model. □

From Theorem 4.9, we obtain:

Corollary 4.13 *Let each $L_1, \ldots, L_k$ be any of the logics $\mathbf{K}, \mathbf{T}, \mathbf{K4}, \mathbf{S4}, \mathbf{S5}$, or a logic axiomatized by canonical modal* MFP*-formulas. Then, for any $n < \omega$, the logic $(L_1 * \ldots * L_k)^{(n)}$ has the finite model property.*

5 Conclusions and further research

We proved that if L is a canonical logic, and the class of its frames $\mathsf{Fr}(L)$ admits definable filtration, then the logic $L^{\boxplus}$ is Kripke complete and, moreover, has the FMP (and is decidable, if L was finitely axiomatizable). The first problem we pose is whether we can weaken the pre-conditions and obtain the completeness of $L^{\boxplus}$ without obtaining the FMP.

Problem 1. *If a logic L is canonical, then is the logic $L^{\boxplus}$ Kripke complete?*

Next, can we weaken the 'canonicity' to the 'completeness' in Theorem 3.10?

Problem 2. *If a logic L is complete and the class of its frames $\mathsf{Fr}(L)$ admits definable filtration, then does the same hold for the logic $L^{\boxplus}$?*

The following questions are of more technical character.

Question 1. Is it the case that whenever the class of models (or frames) of a logic L admits filtration, it also admits definable filtration?

Question 2. Let us replace the axiom (A2) $\boxplus p \to \Box\boxplus p$ with (A2′) $\boxplus p \to \boxplus\Box p$ in the logic $\mathbf{K}^{\boxplus}$. Do we obtain the same logic, i.e., does this logic derive (A2)? Note that the frames for it are the same as for $\mathbf{K}^{\boxplus}$, see Lemma A.4(6).

Question 3. Is the logic $\mathbf{K.2}^{\boxplus}$ Kripke complete? (We conjecture: yes.)

Recall that the logic $\mathbf{K.2}$ extends $\mathbf{K}$ with the formula $\Diamond\Box p \to \Box\Diamond p$. It is canonical and hence complete with respect to the class of frames (W, R) that satisfy the first-order *convergence* (or Church–Rosser) condition:

$$\forall x, y, z\,(x\,R\,y\ \wedge x\,R\,z\ \Rightarrow\ \exists w\,(y\,R\,w\ \wedge z\,R\,w)).$$

Our main result is not applicable to this logic, since the class of its frames $\mathsf{Fr}(\mathbf{K.2})$ does not admit filtration, as we established in [12, Theorem 5.4].

If R is convergent, then so is R^+ (easy exercise). Is modal logic able to establish this? That is, can we derive the formula $\diamondplus\boxplus p \to \boxplus\diamondplus p$ in $\mathbf{K.2}^{\boxplus}$? We succeeded in deriving it (see Lemma A.7 in Appendix).

Question 3. In Lemma A.6, the bimodal formula $\boxplus(p \to \Box p) \to (\Box p \to \boxplus p)$ is shown to have the following property crucial for our main result: *if all its substitution instances are true in some model $M = (F, V)$, then this formula is valid on the frame of every definable minimal filtration: if $M \models A^*$ then $F^{\mathsf{min}}_{\sim\Phi} \models A$, for any finite set of formulas Φ.* Are there any other examples of such formulas? How is this property related to the admissibility of filtration, completeness, decidability of a logic axiomatized by such formulas?

Appendix

A.1 On modally differentiated filtration

Any Γ-filtration $\widehat{M}$ of a model M through the same set Γ, i.e., through $\sim_\Gamma$, is always differentiated: indeed, if $[x]_{\sim_\Gamma} \neq [y]_{\sim_\Gamma}$, then the points x and y in M differ by some formula $\varphi \in \Gamma$; by Filtration Lemma 2.2, the truth of all formulas from Γ is preserved, so the points $\widehat{x}$ and $\widehat{y}$ differ by the same formula φ.

On the contrary, a Γ-filtration $\widehat{M}$ of M through some set $\Phi \supseteq \Gamma$ is not necessarily differentiated: in the above argument, x and y will differ by some $\varphi \in \Phi$, and the Filtration Lemma transfers the truth of formulas from Γ only.

Lemma A.2 below resolves this obstacle: by possibly changing the set Φ, a filtration can be made differentiated. We will need the following simple fact.

Proposition A.1 *Let M be a model and $\sim$ an equivalence relation on W of finite index. Then $\sim$ is of the form $\sim_\Phi$, for some finite set of formulas Φ, iff each equivalence class $[x]_\sim$ is defined in M by some formula.*

Proof. If Φ is finite, then every class $[x]_{\sim_\Phi} \subseteq W$ is defined by the formula

$$\bigwedge\big(\{\varphi \mid \varphi \in \Phi \text{ and } M, x \models \varphi\} \cup \{\neg\varphi \mid \varphi \in \Phi \text{ and } M, x \models \neg\varphi\}\big).$$

Conversely, if $\sim$ partitions W into finitely many classes and each class is defined by a formula φ_i, $1 \leq i \leq n$, then clearly $\sim\, = \,\sim_\Phi$ for $\Phi = \{\varphi_1, \ldots, \varphi_n\}$. □

Lemma A.2 *Assume that $\mathsf{Mod}(L)$ admits (definable) filtration. Then for every finite Sub-closed set of formulas Γ and every model $M \in \mathsf{Mod}(L)$, there exists a (definable) Γ-filtration $\widehat{M} \in \mathcal{M}$ of M that is a differentiated model.*

Proof. Idea: first, build a Γ-filtration M_1 of M, then an Fm-filtration M_2 of M_1; finally, build a differentiated Γ-filtration $\widehat{M}$ of M that is isomorphic to M_2.

Formally, let $M = (W, R, V)$, $M \models L$, and let Γ be as stated above.

(1) Since $\mathsf{Mod}(L)$ admits filtration, there is a Γ-filtration $M_1 = (W_1, R_1, V_1)$ of M with $M_1 \models L$. So, $W_1 = W/{\sim}$ for some equivalence relation $\approx$ of finite

index, $\sim$ respects Γ, $R^{\mathsf{min}}_{\approx} \subseteq R_1 \subseteq R^{\mathsf{max}}_{\approx,\Gamma}$, V_1 is defined canonically on $\mathsf{Var}(\Gamma)$.

(2) Let $M_2 = (W_2, R_2, V_2)$ be a filtration of M_1 through the set of all formulas.[7] So, $W_2 = W_1/{\equiv}$, where $\equiv$ is the *modal equivalence* relation; V_2 is canonical on all variables. By the Filtration lemma 2.2, $M_1 \equiv M_2$, so $M_2 \models L$.

(3) Now we build a model $\widehat{M} = (\widehat{W}, \widehat{R}, \widehat{V})$ isomorphic to M_2 as follows. Put $\widehat{W} := W/{\sim}$, where, for all $x, y \in W$, we define an equivalence relation $\sim$ by

$$x \sim y \quad \overset{\text{def}}{\iff} \quad (M_1, [x]_{\approx}) \equiv (M_1, [y]_{\approx}) \quad \iff \quad [[x]_{\approx}]_{\equiv} \;=\; [[y]_{\approx}]_{\equiv}.$$

Claim 1. *The function $h([x]_{\sim}) = [[x]_{\approx}]_{\equiv}$ is a bijection between $\widehat{W}$ and W_2.*

Proof. Easy. This does not rely on the fact that $\approx$ and $\equiv$ are of finite index.

From now on, we denote $\widehat{x} = [x]_{\sim}$.

Claim 2. *The equivalence relation $\sim$ on W respects Γ: if $x \sim y$, then $x \sim_{\Gamma} y$.*

Proof. If $x, y \in W$ and $x \sim y$ then, by the Filtration lemma 2.2, we have:

$$(M, x) \;\sim_{\Gamma}\; (M_1, [x]_{\approx}) \;\sim_{\mathsf{Fm}}\; (M_1, [y]_{\approx}) \;\sim_{\Gamma}\; (M, y).$$

Using the bijection h, we transfer R_2 and V_2 to $\widehat{M}$ in the obvious way:

$$\widehat{x}\,\widehat{R}\,\widehat{y} \;\overset{\text{def}}{\iff}\; h(\widehat{x})\,R_2\,h(\widehat{y}); \qquad \widehat{x} \models q \;\overset{\text{def}}{\iff}\; M_2, h(\widehat{x}) \models q, \text{ for all } q \in \mathsf{Var}.$$

Since the models $\widehat{M}$ and M_2 are isomorphic, we have $\widehat{M} \models L$.

Claim 3. *$\widehat{V}$ is canonical on each $p \in \mathsf{Var}(\Gamma)$: $M, x \models p \;\Leftrightarrow\; \widehat{M}, \widehat{x} \models p$.*

Proof. Indeed: $(M, x) \sim_{\Gamma} (M_1, [x]_{\approx}) \sim_{\mathsf{Fm}} (M_2, [[x]_{\approx}]_{\equiv}) \sim_{\mathsf{Var}} (\widehat{M}, \widehat{x})$.

Claim 4. *The inclusions $R^{\mathsf{min}}_{\sim} \subseteq \widehat{R} \subseteq R^{\mathsf{max}}_{\sim,\Gamma}$ hold.*

Proof. (min) Clearly, $x\,R\,y \Rightarrow [x]_{\approx}\,R_1\,[y]_{\approx} \Rightarrow [[x]_{\approx}]_{\equiv}\,R_2\,[[y]_{\approx}]_{\equiv} \Leftrightarrow \widehat{x}\,\widehat{R}\,\widehat{y}$.
(max) If $\widehat{x}\,\widehat{R}\,\widehat{y}$, then $[[x]_{\approx}]_{\equiv}\,R_2\,[[y]_{\approx}]_{\equiv}$. But $R_2 \subseteq (R_1)^{\mathsf{max}}_{\equiv,\mathsf{Fm}}$. So, for $\Box A \in \Gamma$,
$M, x \models \Box A \Leftrightarrow M_1, [x]_{\approx} \models \Box A \Rightarrow M_1, [y]_{\approx} \models A \Leftrightarrow M, y \models A$.

Claim 5. *If M_1 is a definable filtration of M, then $\widehat{M}$ is definable too.*

Proof. Assume M_1 is a filtration of M through a finite Φ. By Proposition A.1, each $\sim_{\Phi}$-class is defined by some formula φ_i. Each $\sim$-class is the *union* of some $\sim_{\Phi}$-classes (namely, those that are modally equivalent as points in M_1). Hence, each $\sim$-class is defined by the *disjunction* of some formulas φ_i. By Proposition A.1, $\sim = \sim_{\Psi}$, for some set of formulas Ψ, thus $\widehat{M}$ is definable. □

A.2 On filtration of the canonical model of a theory of a model

Lemma A.3 (Filtration and canonical mapping) *Let $M = (W, R, V)$ be a model, $M_T = (W_T, R_T, V_T)$ the canonical model of its theory $T = \mathsf{Th}(M)$, and $t\colon M \to M_T$ the canonical mapping: $t(a) = \mathsf{Th}(M, a) \in W_T$, for $a \in W$.*

Then, for any finite set of formulas Φ, we have a bijection between the (finite) quotient sets $W/{\sim_{\Phi}}$ and $W_T/{\sim_{\Phi}}$ defined, for $a \in W$, by

$$f([a]_{\sim_{\Phi}}) \;:=\; [t(a)]_{\sim_{\Phi}}.$$

Proof. We denote $\widehat{a} := [a]_{\sim_{\Phi}}$. Note that $\widehat{x} = \widehat{y}$ iff $x \cap \Phi = y \cap \Phi$, for all

---

[7] In fact, if a filtration through the set of all formulas is finite, then it is unique, i.e., the minimal and the maximal relations coincide. But here we do not need this fact.

$x, y \in W_T$. Hence, by definition of f, for all $a \in W$ and $x \in W_T$, we have

$$f(\widehat{a}) = \widehat{x} \iff t(a) \cap \Phi = x \cap \Phi.$$

First, let us show that f is well-defined and injective: for all $a, b \in W$:

$$\widehat{a} = \widehat{b} \Leftrightarrow a \sim_\Phi b \Leftrightarrow \mathsf{Th}(M, a) \cap \Phi = \mathsf{Th}(M, b) \cap \Phi \Leftrightarrow [t(a)]_{\sim_\Phi} = [t(b)]_{\sim_\Phi}.$$

To prove that f is surjective, take any $\widehat{x} \in (W_T/{\sim_\Phi})$. Denote $A := \bigwedge(x \cap \Phi')$, where $\Phi' = \Phi \cup \{\neg B \mid B \in \Phi\}$. Clearly, $A \in x$. Now $M \not\models \neg A$, for otherwise $\neg A \in \mathsf{Th}(M) = T \subseteq x$ and x is inconsistent.

Thus, A is satisfiable in M, so $M, a \models A$ for some $a \in W$. We claim that $f(\widehat{a}) = \widehat{x}$, i.e., for all $B \in \Phi$, we have $M, a \models B$ iff $B \in x$. If $B \in x$, then $B \in (x \cap \Phi')$, so $M, a \models B$. If $B \notin x$, then $\neg B \in (x \cap \Phi')$, so $M, a \models \neg B$. □

A.3 On the semantics of Segerberg's axioms

For convenience, let us recall the axioms for the transitive closure modality:

$$(\mathsf{A1})\ \boxplus p \to \Box p, \qquad (\mathsf{A2})\ \boxplus p \to \Box\boxplus p, \qquad (\mathsf{A3})\ \boxplus(p \to \Box p) \to (\Box p \to \boxplus p).$$

We will also consider the following modified axiom: $(\mathsf{A2'})\ \boxplus p \to \boxplus\Box p$.

Lemma A.4 *Let $F = (W, R, S)$ be a bi-modal frame.*

(1) $F \models (\mathsf{A1}) \iff S \supseteq R$.
(2) $F \models (\mathsf{A2}) \iff S \supseteq R \circ S$.
(3) $F \models (\mathsf{A1}) \wedge (\mathsf{A2}) \implies S \supseteq R^+$; *the converse does not hold in general.*
(4) $F \models (\mathsf{A3}) \implies S \subseteq R^+$; *the converse does not hold in general.*
(5) $F \models (\mathsf{A1}) \wedge (\mathsf{A2}) \wedge (\mathsf{A3}) \iff S = R^+$.
(6) $F \models (\mathsf{A1}) \wedge (\mathsf{A2'}) \wedge (\mathsf{A3}) \iff S = R^+$.

Proof. The facts (1) and (2) are well-known. They imply $S \supseteq R^n$, for all $n \geq 1$, and thus (3) follows. Also (3) and (4) imply (5,⇒). So it remains to prove (4), (5,⇐), and (6) and provide counterexamples to (3,⇐) and (4,⇐).

(4,⇒) Assume $F \models (\mathsf{A3})$ and xSy; we need to prove that xR^+y. Denote $P = R^+(x) \subseteq W$; we need to show that $y \in P$. Consider a model $M = (F, V)$ with the valuation $V(p) = P$. Clearly, $M, x \models \Box p$, since $R(x) \subseteq R^+(x) = P$. Next, $M \models p \to \Box p$, since $P \supseteq R(P)$. Hence $M, x \models \boxplus(p \to \Box p)$. But $M, x \models (\mathsf{A3})$. Hence $M, x \models \boxplus p$. Since xSy, we obtain $M, y \models p$ and so $y \in V(p) = P$.

(5,⇐) Suppose $S = R^+$. Clearly, $S \supseteq R$ and $S \supseteq R \circ S$, hence $F \models (\mathsf{A1}) \wedge (\mathsf{A2})$, by (1) and (2). To prove that $F \models (\mathsf{A3})$, take any model $M = (F, V)$ and $x \in W$. Assume that $x \models \boxplus(p \to \Box p)$ and $x \models \Box p$. We need to show that $x \models \boxplus p$, i.e., $y \models p$ for all $y \in S(x)$. Recall that $S = R^+ = \bigcup_{n \geq 1} R^n$. Therefore, it remains to show, for every $n \geq 1$, that $y \models p$ for all $y \in R^n(x)$. We do this by induction. Induction base ($n = 1$) holds since $x \models \Box p$. Induction step: assume $x\, R^{n+1} y$, hence $x\, R^n t\, R\, y$ for some t. By induction hypothesis, $t \models p$. Since $S \supseteq R^+$, we have $S \supseteq R^n$. Thus $x\, S\, t$. Recall that $x \models \boxplus(p \to \Box p)$. Then $t \models p \to \Box p$, whence $t \models \Box p$ and $y \models p$.

(6) Clearly, $F \models (\mathsf{A2'})$ iff $S \supseteq S \circ R$. So, $F \models (\mathsf{A1}) \wedge (\mathsf{A2'})$ implies $S \supseteq R^+$, and $F \models (\mathsf{A3})$ implies $S \subseteq R^+$. Thus (6,⇒) is proved. The implication (6,⇐) is easy, since $S = R^+$ implies $S \supseteq S \circ R$, and so $F \models (\mathsf{A1}) \wedge (\mathsf{A2'}) \wedge (\mathsf{A3})$.

Here is a counterexample $M = (W, R, S, V)$ to (4,$\Leftarrow$): $W = \{a, b, c\}$, $aRbRc$ (R is not transitive), aSc, $V(p) = \{1\}$. Clearly, $S \supset R^+$. But $M, a \not\models$ (A3).

To refute (3,$\Leftarrow$), we show that $S \supseteq R^+$ does not imply $S \supseteq R \circ S$. Take $W = \{a, b\}$, aRb, aSb, bSa. Then $R^+ = R \subseteq S$. But $a(R \circ S)a$ and $\neg(aSa)$. □

A.4 On induction axiom and minimal filtrated frame

Lemma A.5 *The formula* (A3) $\boxplus(p \to \Box p) \to (\Box p \to \boxplus p)$ *is not canonical.*

Proof. Denote $L = \mathbf{K}(\Box, \boxplus) \oplus$ (A3) and its canonical frame $F_L = (W, R, S)$. By Lemma A.4(4), to prove that $F \not\models$ (A3), it suffices to show that $S \not\subseteq R^+$.

The set $\Gamma = \{\neg\boxplus p\} \cup \{\Box^n p \mid n \geq 1\}$ is L-consistent, because every finite set of the form $\{\neg\boxplus p, \Box p, \ldots, \Box^n p\}$ is L-satisfiable (in a chain of length $n+1$). Hence $\Gamma \subseteq x$, for some maximal L-consistent set $x \in W$. Since $\neg\boxplus p \in x$, we have $M_L, x \not\models \boxplus p$ (later, we omit M_L). Hence, for some $y \in W$, we have $x\, S\, y$ and $y \models p$. However, $\neg(x\, R^+\, y)$; indeed, otherwise $x\, R^n\, y$, for some $n \geq 1$, and since $\Box^n p \in x$, we obtain $x \models \Box^n p$ and $y \models p$, a contradiction. □

We could prove the same using variable-free formulas only: put $p := \Diamond\top$.

Let us strengthen Lemma 3.7 (recall that $G \models$ (A3) implies $S \subseteq R^+$). Denote the *minimal filtered* (through Φ) frame by $G^{\mathsf{min}}_{\sim_\Phi} = (W/{\sim_\Phi}, R^{\mathsf{min}}_{\sim_\Phi}, S^{\mathsf{min}}_{\sim_\Phi})$.

Lemma A.6 (Induction axiom and minimal filtrated frame)
Let $M = (W, R, S, V) \models$ (A3)* and let $\Phi \subseteq \mathsf{Fm}$ be finite. Then $G^{\mathsf{min}}_{\sim_\Phi} \models$ (A3).

Proof. The minimal filtration model $\widehat{M} := M^{\mathsf{min}}_{\sim_\Phi} = (G^{\mathsf{min}}_{\sim_\Phi}, \widehat{V})$ is a Φ-filtration of M through Φ, hence it is a finite *differentiated* model (see Section A.1).

Due to Lemma 1.1, in order to prove our lemma, it suffices to show that

$$M \models (\mathsf{A3})^* \quad \text{implies} \quad \widehat{M} := M^{\mathsf{min}}_{\sim_\Phi} \models (\mathsf{A3})^*.$$

Assume $\widehat{M} \not\models$ (A3)$[p := B]$, for some formula B. Then there is $\widehat{x} \in \widehat{W}$ such that **(a)** $\widehat{x} \models \boxplus(B \to \Box B)$, **(b)** $\widehat{x} \models \Box B$, **(c)** $\widehat{x} \not\models \boxplus B$. Hence there is $\widehat{y} \in \widehat{W}$ such that $\widehat{x}\, \widehat{S}\, \widehat{y}$ and **(d)** $\widehat{y} \not\models B$. Since $\widehat{x}\, S^{\mathsf{min}}_{\sim_\Phi}\, \widehat{y}$, without loss of generality, $x\, S\, y$.

Consider $Y := \widehat{V}(B) = \{\widehat{z} \in \widehat{W} \mid \widehat{M}, \widehat{z} \models B\}$. As in Lemma 3.7, Y is a finite collection of definable subsets of W, hence their union $\bigcup Y$ is also a definable subset of W. So, there is a formula φ such that, for all $z \in W$, we have:

$$M, z \models \varphi \;\Leftrightarrow\; z \in \bigcup Y \;\Leftrightarrow\; \widehat{z} \in Y = \widehat{V}(B) \;\Leftrightarrow\; \widehat{M}, \widehat{z} \models B.$$

Now we show that $M, x \not\models$ (A3)$[p := \varphi]$, in contradiction with $M \models$ (A3)*.

(a') $M, x \models \boxplus(\varphi \to \Box\varphi)$. Indeed, take any $a, b \in W$ such that $x\, S\, a\, R\, b$ and $a \models \varphi$. Then $\widehat{x}\, \widehat{S}\, \widehat{a}\, \widehat{R}\, \widehat{b}$ and $\widehat{a} \models B$. Hence $\widehat{b} \models B$ by **(a)**, and so $b \models \varphi$.

(b') $M, x \models \Box\varphi$. Indeed, if $x\, R\, z$, then $\widehat{x}\, \widehat{R}\, \widehat{z}$; hence $\widehat{z} \models B$ by **(b)**, so $z \models \varphi$.

(d') $M, x \not\models \boxplus\varphi$. Indeed, $x\, S\, y$ and $M, y \not\models \varphi$, because $\widehat{y} \not\models B$ by **(d)**. □

A.5 On the logic of convergent frames

For convenience, we repeat the axioms for the transitive closure modality:

(A1) $\boxplus p \to \Box p$, (A2) $\boxplus p \to \Box\boxplus p$, (A3) $\boxplus(p \to \Box p) \to (\Box p \to \boxplus p)$.

Note that in any logic $L^{\boxplus}$, the following inference rule is derivable:

$$\frac{\varphi \to \Box\varphi}{\varphi \to \boxplus\varphi} \qquad (\mathsf{R}\boxplus)$$

Indeed, here is a derivation:
1) $\varphi \to \Box\varphi$. 2) $\boxplus(\varphi \to \Box\varphi)$. 3) $\Box\varphi \to \boxplus\varphi$ by (A3). 4) $\varphi \to \boxplus\varphi$ from 1 and 3.

Furthermore, in any logic $L^{\boxplus}$, the following formula is derivable:

$$\Box p \;\wedge\; \boxplus\Box p \;\to\; \boxplus p \qquad (\mathsf{A}\boxplus)$$

since one of its premises, $\boxplus\Box p$, is stronger then the premise $\boxplus(p \to \Box p)$ in (A3).

Recall that the logic **K.2** extends **K** with the axiom $\Diamond\Box p \to \Box\Diamond p$.

Lemma A.7 (Convergence for transitive closure) $\mathbf{K.2}^{\boxplus} \vdash \diamondplus\boxplus p \to \boxplus\diamondplus p$.

Proof. The proof is in two stages.

(1) We derive $\Diamond\boxplus p \longrightarrow \boxplus\Diamond p$, using $\Diamond\Box\varphi \to \Box\Diamond\varphi$ for $\varphi = p$ and $\varphi = \boxplus p$:

$$\begin{array}{llll}
\Diamond\boxplus p \xrightarrow{(\mathsf{A1})} \Diamond\Box p & \xrightarrow{.2} \Box\Diamond p. & (a) \\
\Diamond\boxplus p \xrightarrow{(\mathsf{A2})} \Diamond\Box\boxplus p & \xrightarrow{.2} \Box\Diamond\boxplus p. & \text{Hence:} \\
\Diamond\boxplus p \xrightarrow{(\mathsf{R}\boxplus)} \boxplus\Diamond\boxplus p & \xrightarrow{(a)} \boxplus\Box\Diamond p. & (\mathrm{b}) \\
\Diamond\boxplus p \xrightarrow{(\mathsf{A}\boxplus)} \boxplus\Diamond p, & \text{obtained from } (a) \text{ and } (b). &
\end{array}$$

(1') We obtain $\diamondplus\Box p \longrightarrow \Box\diamondplus p$ by duality from (1).

(2) Derive $\diamondplus\boxplus p \longrightarrow \boxplus\diamondplus p$ using (1') similarly (replace $\Diamond$ with $\diamondplus$ above). $\Box$

Note that the two stages of the derivation in the above lemma correspond to two inductions needed to prove that R^+ is convergent, assuming that R is convergent. First, by induction on m, one proves:

$$(x\,R^m\,y \text{ and } x\,R\,z) \;\Rightarrow\; \exists t\colon\ (y\,R\,t \text{ and } z\,R^m\,t).$$

Secondly, by induction on n one proves:

$$(x\,R^m\,y \text{ and } x\,R^n\,z) \;\Rightarrow\; \exists t\colon\ (y\,R^n\,t \text{ and } z\,R^m\,t).$$

Now, if $x\,R^+\,y$ and $x\,R^+\,z$, then $x\,R^m\,y$ and $x\,R^n\,z$, for some m, n. Then there is t such that $y\,R^n\,t$ and $z\,R^m\,t$. Hence $y\,R^+\,t$ and $z\,R^+\,t$. So, R^+ is convergent.

This additionally justifies the name 'induction axiom' for the axiom (A3).

Acknowledgements

We are grateful to the three anonymous reviewers for their multiple suggestions and questions on an earlier version of the paper. The work of the second author was supported by the Russian Science Foundation under grant 16-11-10252 and carried out at Steklov Mathematical Institute of Russian Academy of Sciences. The work of the third author was carried out within the framework of the HSE University Basic Research Program.

References

[1] Blackburn, P., M. de Rijke and Y. Venema, "Modal Logic," Cambridge Tracts in Theoretical Computer Science **53**, Cambridge University Press, 2002.

[2] Blok, W., *On the degree of incompleteness of modal logics and the covering relation in the lattice of modal logics*, Technical report, Dept. of Math., Univ. of Amsterdam (1978).

[3] Chagrov, A. and M. Zakharyaschev, "Modal Logic," Oxford Logic Guides **35**, Oxford University Press, 1997.

[4] Doczkal, C. and G. Smolka, *Constructive completeness for modal logic with transitive closure*, in: C. Hawblitzel and D. Miller, editors, *Certified Programs and Proofs – Second International Conference (CPP 2012), Kyoto, Japan, December 13-15, 2012. Proceedings*, Lecture Notes in Computer Science **7679** (2012), pp. 224–239.

[5] Dunin-Keplicz, B. and R. Verbrugge, "Teamwork in multi-agent systems: A formal approach," Wiley Series in Agent Technology **21**, John Wiley & Sons, 2011.

[6] Fagin, R., Y. Moses, J. Y. Halpern and M. Y. Vardi, "Reasoning about knowledge," MIT Press, 2003.

[7] Fischer, M. J. and R. E. Ladner, *Propositional dynamic logic of regular programs*, Journal of computer and system sciences **18** (1979), pp. 194–211.

[8] Gasquet, O., A. Herzig, B. Said and F. Schwarzentruber, *Modal logics with transitive closure*, in: *Kripke's Worlds – An Introduction to Modal Logics via Tableaux*, Studies in Universal Logic, Birkhäuser, 2014 pp. 157–189.

[9] Gheerbrant, A., L. Libkin and C. Sirangelo, *Naïve evaluation of queries over incomplete databases*, ACM Transactions Database Systems **39** (2014), pp. 31:1–31:42.

[10] Goldblatt, R., "Logics of Time and Computation," Number 7 in CSLI Lecture Notes, Center for the Study of Language and Information, 1992, 2nd edition.

[11] Kikot, S., *First-order formulas that are preserved under minimal filtration*, in: *Proceedings of 39th International Workshop of IITP RAS "Information Technologies and Systems 2015"*, 2015, pp. 635–639, (in Russian).

[12] Kikot, S., I. Shapirovsky and E. Zolin, *Filtration safe operations on frames*, in: R. Goré, B. P. Kooi and A. Kurucz, editors, *Advances in Modal Logic*, 10 (2014), pp. 333–352.

[13] Kozen, D. and R. Parikh, *An elementary proof of the completeness of PDL*, Theoretical Computer Science **14** (1981), pp. 113–118.

[14] Kracht, M. and F. Wolter, *Properties of independently axiomatizable bimodal logics*, J. Symbolic Logic **56** (1991), pp. 1469–1485.

[15] Segerberg, K., *Decidability of four modal logics*, Theoria **34** (1968), pp. 21–25.

[16] Segerberg, K., *A completeness theorem in the modal logic of programs*, Notices of the American Mathematical Society **A-522** (1977), pp. 77T–E69.

[17] Segerberg, K., *A completeness theorem in the modal logic of programs*, Banach Center Publications **9** (1982), pp. 31–46.

[18] Segerberg, K., *Von Wright's tense logic*, in: P. A. Schilpp and L. E. Hahn, editors, *The Philosophy of Georg Henrik von Wright*, Open Court, Illinois, 1989 pp. 602–635, appeared 10 years before in Russian: in V.A. Smirnov (ed.), *Logical Deduction*, Nauka Publishers, Moscow, 1979, pp. 173–205.

[19] Segerberg, K., "A Concise Introduction to Propositional Dynamic Logic," 1993.

[20] Shapirovsky, I. and E. Zolin, *On completeness of logics enriched with transitive closure modality*, in: *7th International Conference on Topology, Algebra and Categories in Logic (TACL 2015). Booklet of abstracts.*, 2015, pp. 255–257.

[21] Shehtman, V., *On some two-dimensional modal logics*, in: *8th Congress on Logic Methodology and Philosophy of Science, abstracts*, volume 1 (1987), pp. 326–330.

[22] Shehtman, V., *Filtration via bisimulation*, in: R. Schmidt, I. Pratt-Hartmann, M. Reynolds and H. Wansing, editors, *Advances in Modal Logic*, 5 (2004), pp. 289–308.

[23] Shehtman, V., *Canonical filtrations and local tabularity*, in: R. Goré, B. P. Kooi and A. Kurucz, editors, *Advances in Modal Logic*, 10 (2014), pp. 498–512.

Bisimulational Categoricity

Jędrzej Kołodziejski [1]

Faculty of Mathematics, Informatics and Mechanics
University of Warsaw, Banacha 2
02-097 Warsaw, Poland

Abstract

We introduce and study the notion of *bisimulational categoricity* – the property of having a unique model *up to bisimulation*. We show that: (1) a complete modal theory (i.e. a maximal consistent set of formulae) t has a unique model up to bisimulation iff it has an image-finite model.
We further prove two analogous characterisations: (2) a complete theory t in transitive modal logic (EF-logic) has a unique model up to transitive bisimulation (EF-bisimulation) iff it has a finite model; and (3) a complete theory t in two-way modal logic has a unique model up to two-way bisimulation iff it has a model where every point has finite in- and out-degree.

Keywords: modal logic, model theory, categoricity, bisimulation.

1 Introduction

One of the central notions of classical model theory is that of *categoricity* – a theory is called *categorical* if it has a unique model *up to isomorphism*. In the context of modal logic, bisimilarity seems more appropriate than the isomorphism. One may therefore ask about *bisimulational categoricity*, i.e. the property of having a unique model *up to bisimulation.* [2]

It turns out that the notion of bisimulational categoricity for theories expressed in modal logic is indeed well-behaved and can be characterised in terms of image-finiteness. [3] We show that a complete theory in modal logic has a unique model up to bisimulation iff it has an image-finite model. While the right-to-left implication is (an easy folklore strengthening of) the well-known

[1] j.kolodziejski@mimuw.edu.pl

[2] Somewhat similar idea of finding modal analogues of classical results can be found in Chapter 6 of [12], where the author investigates the number of *non-bisimilar* models of a given modal fixpoint formula – analogically to the result of [9], where the number of *non-isomorphic* models of an MSO formula is considered. Nevertheless, both the result and the involved tools of the mentioned dissertation are rather far from the content of this paper.

[3] Note that, due to the obvious limitations given by the Skolem-Löwenheim Theorem, the classical notion of categoricity of first-order theories is only interesting when models of fixed cardinality are considered. However, unlike with isomorphism, structures of different sizes may still be bisimilar – and so there is no need to relativise *bisimulational* categoricity.

Hennessy-Milner Theorem [6], the left-to-right one requires adaptation of some classical model-theoretic tools and a simple topological argument. As such, our characterization can be thought of both as a completion of the Hennessy-Milner Theorem and as a modal version of the Ryll-Nardzewski Theorem (proven independently by Ryll-Nardzewski [10] Svenonius [11] and Engeler [3]).

Apart from standard modal logic, we provide analogous characterisations for two other interesting logics: transitive modal logic (sometimes known as the *EF-logic* in the context of computer science) and two-way modal logic (i.e. modal logic with both forward and backward modalities). We show that: (i) a complete theory in two-way modal logic has a unique model up to two-way bisimulation iff it has a model where every point has finite in- and out-degree and (ii) a complete theory in the transitive modal logic has a model unique up to transitive bisimulation (also called EF-bisimulation) iff it has a finite model.

In the proof we adapt standard model-theoretic tools to the modal framework and introduce new concepts of *induced modal logics* and *induced bisimulations*, which allow us to uniformly describe a wide range of modal-like logics and their corresponding bisimulations. We also discuss a simple example showing limitations of our method: modal logic enriched with the universal modality fails to have analogous characterisation.

The paper is organised as follows. After this introduction, in Section 2 we recall the basic notions and facts of modal logic and state our first main result, Theorem 2.8. Then, in Section 3 we formally introduce the notion of an inducing assignment, establish some simple related facts and prepare model-theoretic tools for the proof. Finally, in Section 4 we state the other two main theorems – Theorem 4.1 and Theorem 4.2 – and give proofs for all three of them. We conclude with a discussion of limitations of our method.

2 Modal Logic and Bisimulations

We assume the reader to be familiar with basic notions of modal logic ([1] is a good reference). However, for the sake of completeness and to fix notation, we recall the most basic definitions and facts.

Fix a finite set Σ of atomic propositions.

Definition 2.1 A (Kripke) *model* $\mathcal{M}$ for a signature $R = \{R_1, R_2, ..., R_l\}$ of binary relational symbols consists of: a universe M; an interpretation $R_k^{\mathcal{M}} \subseteq M \times M$ for every relation $R_k \in R$; and a valuation $\mathsf{val}^{\mathcal{M}} : \Sigma \to \mathcal{P}(M)$. A *pointed model* is a model with distinguished point – called its *root*. We will usually abuse terminology and call both non-pointed and pointed models just *models* whenever it does not lead to confusion. Moreover, following the notational traditions of modal logic we will skip parentheses and denote pointed models by $\mathcal{M}, p$ instead of $(\mathcal{M}, p)$.

The class of all models over signature R will be denoted $\mathsf{Krip(R)}$. We will typically identify a model with its universe and write $p \in \mathcal{M}$ instead of $p \in M$. Moreover, for the sake of simplicity we write R_k and val, skipping the superscripts whenever the model $\mathcal{M}$ is clear from the context.

Recall the standard syntax and semantics of (poly)modal logic $\mathsf{ML}(R)$ over signature R.

Definition 2.2 The set of formulae of modal logic Φ_R for binary signature R is given by the following grammar:

$$\varphi \mapsto \varphi \vee \varphi \mid \neg\varphi \mid \Diamond_k \varphi \mid a$$

for $a \in \Sigma$ and k such that $R_k \in R$. We use the standard syntactic sugar: $\Box_k\varphi = \neg\Diamond_k\neg\varphi$ and $\varphi \wedge \psi = \neg(\neg\varphi \vee \neg\psi)$. The modal depth of a formula is the maximal nesting of (possibly different) "$\Diamond_k$" operators. In case there is only one operator in R, we skip the subscript and write "$\Diamond$" instead of "$\Diamond_1$".

Definition 2.3 Given a model $\mathcal{M} \in \mathsf{Krip(R)}$, the *semantics map* $[\![\text{-}]\!]^{\mathcal{M}} : \Phi_R \to \mathcal{P}(M)$ is defined inductively as follows:

$$\begin{aligned}
[\![a]\!]^{\mathcal{M}} &= \mathsf{val}^{\mathcal{M}}(a);\\
[\![\varphi \vee \psi]\!]^{\mathcal{M}} &= [\![\varphi]\!]^{\mathcal{M}} \cup [\![\psi]\!]^{\mathcal{M}};\\
[\![\neg\varphi]\!]^{\mathcal{M}} &= M - [\![\varphi]\!]^{\mathcal{M}};\\
[\![\Diamond_k\varphi]\!]^{\mathcal{M}} &= \{p \in M \mid \exists_{q \in [\![\varphi]\!]^{\mathcal{M}}} pR_k^{\mathcal{M}} q\}.
\end{aligned}$$

Definition 2.4 A *bisimulation* between two (not necessarily distinct) models $\mathcal{M}, \mathcal{M}' \in \mathsf{Krip(R)}$ is a relation $Z \subseteq M \times M'$ that satisfies, for every $a \in \Sigma$, $R_k \in R$ and pZp':

- (base condition) $p \in \mathsf{val}(a) \iff p' \in \mathsf{val}(a)$;
- (forth condition) if pR_kq then there exists q' s.t. $p'R_kq'$ and qZq';
- (back condition) if $p'R_kq'$ then there exists q s.t. pR_kq and qZq'.

Pointed models $\mathcal{M}, p$ and $\mathcal{M}', p'$ are said to be *bisimilar* if there exists a bisimulation Z between them s.t. pZp' (notation $\mathcal{M}, p \leftrightarrows \mathcal{M}', p'$). A *functional bisimulation* is a function whose graph is a bisimulation. We will also use the standard characterization of bisimilarity in terms of a bisimulation game between players $\exists$ve and $\forall$dam.

It is widely known that modal logic is invariant under bisimulation, i.e. bisimilar points are always logically indistinguishable. The converse may require an additional assumption of image-finiteness.

Definition 2.5 A model $\mathcal{M} \in \mathsf{Krip(R)}$ is *image-finite* if every point $p \in \mathcal{M}$ has only finitely many R_k-children for every $R_k \in R$.

The classical result of Hennessy and Milner [6] states that, in image-finite models, points that are logically indistinguishable have to be bisimilar. The following example shows that without the assumption of image-finiteness this does not have to be the case.

Example 2.6 The Hedgehogs: $\mathcal{H}, \mathsf{root}_{\mathcal{H}}$ and $\mathcal{H}', \mathsf{root}_{\mathcal{H}'}$ [4]:

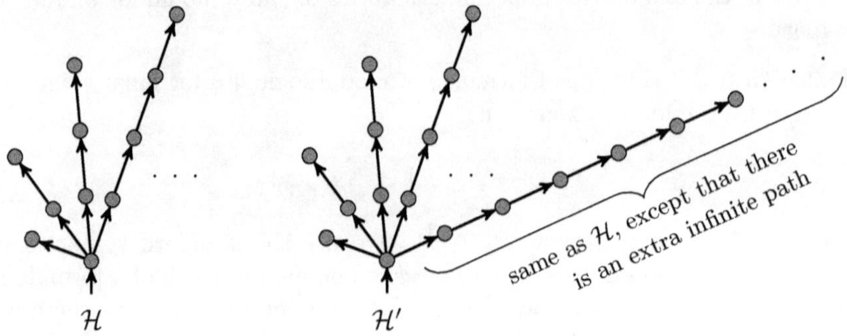

The two models are not bisimilar, as one of them is well-founded but the other is not. However, it is easy to show that they cannot be distinguished by ML formulae. [5]

As it turns out, the above example is an illustration of a general phenomenon, which is that among infinitely many behaviours one can always find a *limit* one that: (i) can be either included or removed from the model but (ii) our local logical means are too weak to tell the difference. This will be the key intuition underlying our characterisation of bisimulational categoricity (i.e. the property of having a unique model up to bisimulation). Roughly, the characterization says that the requirement of image-finiteness – treated up to bisimulation – is not only sufficient but also necessary.

In order to formulate the theorem, we first formally introduce the notion of a *type* – i.e. a maximal consistent set of formulae – analogous to types in first-order model theory (here by *type* we always mean a *complete* one). For the sake of simplicity, let us confine ourselves to the case when the signature consists of a single relation "$\rightarrow$" (the symbol should not be confused with implication: "$\Rightarrow$").

Definition 2.7 Given a point $p \in \mathcal{M} \in \mathsf{Krip}(\{\rightarrow\})$, its *modal type* – denoted $\mathbf{tp}^{\mathcal{M}}(p)$ – is the set $\{\varphi \in \Phi_{\{\rightarrow\}} \mid p \in [\![\varphi]\!]^{\mathcal{M}}\}$ of all modal formulae it satisfies. The set of all modal types will be denoted $\mathbb{T}$.

We are now ready to formulate our first main theorem.

Theorem 2.8 *For every type $t \in \mathbb{T}$, the following are equivalent:*

(1) *t has a unique model up to $\leftrightarroweq$;*

(2) *every model of t is bisimilar to an image-finite model;*

(3) *t has a model which is image-finite.*

---

[4] Here the valuation is not important – for the sake of this example assume $\Sigma = \emptyset$.

[5] In fact, even the full first-order logic cannot distinguish the models, as can be shown using Ehrenfeucht-Fraïssé games.

We will moreover show two analogous characterisations involving two other logics and their corresponding equivalence relations. In order to neatly extract the common part of the structure of the logics we investigate, and because it is interesting in its own right, we formally introduce the notion of induced relations.

3 Induced Relations

Various modal-like logics and bisimilarity-like congruences can be obtained by considering some relation *induced* by the original accessibility relation.

Definition 3.1 Given two binary signatures S, R (*source* and *result*), an *inducing assignment* is an assignment

$$\mathsf{ind} : \mathsf{Krip}(\mathsf{S}) \to \mathsf{Krip}(\mathsf{R})$$

such that every $\mathcal{M} \in \mathsf{Krip}(\mathsf{S})$ has the same universe and valuation as its image $\mathsf{ind}(\mathcal{M})$.

3.1 Induced Logic and Bisimulations

Every inducing assignment gives rise to the induced logic.

Definition 3.2 Given an inducing assignment $\mathsf{ind} : \mathsf{Krip}(\mathsf{S}) \to \mathsf{Krip}(\mathsf{R})$, we define the *induced modal logic* $\mathsf{ML}_{\mathsf{ind}}$ interpreted over $\mathsf{Krip}(\mathsf{S})$. Formulae $\Phi_{\mathsf{ind}} = \Phi_R$ are standard modal formulae over signature R. The semantics map $[\![\text{-}]\!]^{\mathcal{M}}_{\mathsf{ind}} : \Phi_{\mathsf{ind}} \to \mathcal{P}(M)$ is defined with respect to the induced model – on every $\mathcal{M} \in \mathsf{Krip}(\mathsf{S})$ we put:

$$[\![\varphi]\!]^{\mathcal{M}}_{\mathsf{ind}} = [\![\varphi]\!]^{\mathsf{ind}(\mathcal{M})}$$

We say that model $\mathcal{M}, p$ *satisfies* formula φ (notation: $\mathcal{M}, p \models \varphi$) if $\varphi \in [\![\varphi]\!]^{\mathcal{M}}_{\mathsf{ind}}$. Models $\mathcal{M}, p$ and $\mathcal{N}, q$ are *equivalent* (denoted $\mathcal{M}, p \equiv_{\mathsf{ML}_{\mathsf{ind}}} \mathcal{N}, q$) if they satisfy the same $\mathsf{ML}_{\mathsf{ind}}$ formulae.

Similarly to the induced logic, we also define an *induced bisimulation*, where we ignore the original relations and only take the induced ones into account.

Definition 3.3 Given an assignment $\mathsf{ind} : \mathsf{Krip}(\mathsf{S}) \to \mathsf{Krip}(\mathsf{R})$, a relation $Z \subseteq M \times N$ between models $\mathcal{M}, \mathcal{N} \in \mathsf{Krip}(\mathsf{S})$ is an ind-*bisimulation* if it is a bisimulation between $\mathsf{ind}(\mathcal{M})$ and $\mathsf{ind}(\mathcal{N})$. Induced bisimilarity is defined accordingly and denoted $\leftrightarroweq_{\mathsf{ind}}$.

The standard characterization of bisimilarity in terms of a two-player game carries over to the induced setting. Moreover, it follows immediately from invariance of modal logic under bisimulation that for any ind, $\mathsf{ML}_{\mathsf{ind}}$ is invariant under $\leftrightarroweq_{\mathsf{ind}}$:

Proposition 3.4 *For any pair of models $\mathcal{M}, \mathcal{N} \in \mathsf{Krip}(\mathsf{S})$, if $\mathcal{M}, p \leftrightarroweq_{\mathsf{ind}} \mathcal{N}, q$ then $\mathcal{M}, p \equiv_{\mathsf{ML}_{\mathsf{ind}}} \mathcal{N}, q$.*

As it was mentioned, several interesting logics and bisimilarity relations fit well into our induced framework. Let us show a few examples.

Example 3.5 A trivial example is the identity assignment Id. Logic induced by $\mathsf{Id} : \mathsf{Krip}(\mathsf{R}) \to \mathsf{Krip}(\mathsf{R})$ is the same as the original one, i.e. $\mathsf{ML}_{\mathsf{Id}} = \mathsf{ML}(R)$. Likewise, $\leftrightarrows_{\mathsf{Id}}$ equals $\leftrightarrows$.

Example 3.6 Let $\mathsf{ind}_{\leftrightarrows} : \mathsf{Krip}(\{\to\}) \to \mathsf{Krip}(\{\to, \leftarrow\})$ be the assignment that keeps the relation "$\to$" unchanged and additionally introduces its inverse (i.e. a fresh relation "$\leftarrow$" s.t. $p \leftarrow q$ iff $p \to q$ for any two points p and q in the model). Then, $\mathsf{ML}_{\mathsf{ind}_{\leftrightarrows}}$ is the modal logic with *forward* and *backward* (or *future* and *past*) modalities and $\leftrightarrows_{\mathsf{ind}_{\leftrightarrows}}$ is a two-way bisimilarity – where a two-way bisimulation is a relation that is a bisimulation w.r.t. both the accessibility relation and its converse.

Example 3.7 Consider the assignment $\mathsf{ind}_{+} : \mathsf{Krip}(\{\to\}) \to \mathsf{Krip}(\{\to^{+}\})$ that maps a relation to its transitive closure. That way we obtain the transitive modal logic $\mathsf{ML}_{\mathsf{ind}_{+}}$ and transitive bisimilarity $\leftrightarrows_{\mathsf{ind}_{+}}$ – also known as *EF-logic* and *EF-bisimilarity* in the context of computer science (see e.g. [2]).

Example 3.8 Let $\mathsf{ind}_{\forall} : \mathsf{Krip}(\{\to\}) \to \mathsf{Krip}(\{\to, \langle\exists\rangle\})$ be the assignment that keeps "$\to$" and adds a new relation "$\langle\exists\rangle$" which is the full relation on the model's universe. This gives raise to logic $\mathsf{ML}_{\mathsf{ind}_{\forall}}$ being the modal logic with universal modalities and to $\leftrightarrows_{\mathsf{ind}_{\forall}}$ being global bisimilarity.

It is worth to emphasize that the term "logic" as we use it denotes a *set of formulae* together with an appropriate *satisfaction relation* between formulae and models. In particular, it is something different from what is known as *normal modal logic* which is just a *sets of formulae*. For example, *the set of all tautologies* of the transitive modal logic $\mathsf{ML}_{\mathsf{ind}_{+}}$ is precisely *the normal modal logic K4*.

The next example shows that one has to be careful, as in general ind could encode an oracle for arbitrary class of models:

Example 3.9 Let $\mathcal{C}$ be an arbitrary class of pointed models over signature S. The assignment $\mathsf{ind}_{\mathcal{C}} : \mathsf{Krip}(\mathsf{S}) \to \mathsf{Krip}(\mathsf{S} \cup \{\mathsf{R}_{\mathcal{C}}\})$ takes a model $\mathcal{M} \in \mathsf{Krip}(\mathsf{S})$, keeps all the relations from S unchanged and sets $pR_{\mathcal{C}}q$ iff $p = q$ and $\mathcal{M}, p \in \mathcal{C}$ – i.e. $\mathsf{ind}_{\mathcal{C}}$ adds a self-loop labelled by "$\mathcal{C}$" to precisely these points p for which $\mathcal{M}, p \in \mathcal{C}$. Then, the formula $\Diamond_{\mathcal{C}}\top$ is true in $\mathcal{M}, p$ iff $\mathcal{M}, p \in \mathcal{C}$.

3.2 Model Theory – The Space of Types

The notion of modal type can be adapted to the induced setting in a natural way.

Definition 3.10 Given a logic $\mathsf{ML}_{\mathsf{ind}}$, we define an $\mathsf{ML}_{\mathsf{ind}}$-*type* of a point $p \in \mathcal{M} \in \mathsf{Krip}(\mathsf{S})$ – denoted $\mathbf{tp}^{\mathcal{M}}(p)$ – to be the set $\{\varphi \in \mathsf{ML}_{\mathsf{ind}} \mid \mathcal{M}, p \models \varphi\}$. The set of all $\mathsf{ML}_{\mathsf{ind}}$-types will be denoted $\mathbb{T}_{\mathsf{ind}}$.

Along the same lines as in the classical model theory for first-order logic, our types can be equipped with a topology turning it into a Hausdorff space.

Definition 3.11 For any $\varphi \in \mathsf{ML}_{\mathsf{ind}}$, we take the set $\langle\varphi\rangle = \{t \in \mathbb{T}_{\mathsf{ind}} \mid \varphi \in t\}$ of all types containing it. Then, the set $\{\langle\varphi\rangle \mid \varphi \in \mathsf{ML}_{\mathsf{ind}}\}$ is a basis of clopen sets generating a topology on $\mathbb{T}_{\mathsf{ind}}$.

Alternatively, one could obtain the same topology by first picking any enumeration of $\mathsf{ML}_{\mathsf{ind}}$ formulae and then defining a metric $d(t,t') = \frac{1}{n}$ for n being the number of the first formula on which t and t' differ (and 0 if $t = t'$). The underlying intuition is that types which are similar – i.e. hard to distinguish – should be close to each other.

Proposition 3.12 *Analogously to the first-order case, we have that:*

- *the space $\mathbb{T}_{\mathsf{ind}}$ is always Hausdorff;*
- *the logic $\mathsf{ML}_{\mathsf{ind}}$ is compact (i.e. if any finite fragment of a set of formulae t is satisfiable, then so is the entire t) $\iff$ the space $\mathbb{T}_{\mathsf{ind}}$ is compact;*
- *given $T \subseteq \mathbb{T}_{\mathsf{ind}}$, $t \in \mathbb{T}_{\mathsf{ind}}$ is isolated in T $\iff$ there exists a single $\mathsf{ML}_{\mathsf{ind}}$ formula $\varphi \in t$ s.t. $\varphi \notin t'$ for every other $t' \in T$.*

Proof. Observe that by identifying a type with its characteristic function, we can view the space $\mathbb{T}_{\mathsf{ind}}$ as a subspace of $2^{\Phi_{\mathsf{ind}}}$. Since the later is Hausdorff, so is $\mathbb{T}_{\mathsf{ind}}$. Moreover, a subspace of a compact Hausdorff space is compact iff it is closed – and it is easy to check that closedness of $\mathbb{T}_{\mathsf{ind}}$ is the same as logical compactness of $\mathsf{ML}_{\mathsf{ind}}$. The last item follows from the observation that in any topological space, a point is isolated iff it is isolated by a basic open set. □

An important notion that can be generalised to the induced setting is that of modal saturation (also called m-saturation). Our topology on types allows us to capture it in an elegant way.

Definition 3.13 We say that a point p in a model $\mathcal{M} \in \mathsf{Krip(S)}$ is $\mathsf{ML}_{\mathsf{ind}}$-*saturated* if for every $R_k \in R$, the set of types of its R_k-children $\{\mathbf{tp}^{\mathcal{M}}(q) \mid pR_kq\}$ is closed. We call $\mathcal{M}$ $\mathsf{ML}_{\mathsf{ind}}$-*saturated* if all its points are $\mathsf{ML}_{\mathsf{ind}}$-saturated.

In more concrete terms (the way modal saturation is usually defined): $\mathsf{ML}_{\mathsf{ind}}$-saturation means that if any finite fragment of t is realised in some R_k-child of p, then there exists a p's R_k-child realising the entire t. The following is an immediate consequence of an analogous fact for the standard case of $\mathsf{ML}(R)$ and $\leftrightarroweq$:

Theorem 3.14 *Given any two $\mathsf{ML}_{\mathsf{ind}}$-saturated models $\mathcal{M}, \mathcal{M}' \in \mathsf{Krip(S)}$:*

$$\mathcal{M}, p \equiv_{\mathsf{ML}_{\mathsf{ind}}} \mathcal{M}', p' \qquad \underline{\textit{implies}} \qquad \mathcal{M}, p \leftrightarroweq_{\mathsf{ind}} \mathcal{M}', p'$$

for any $p \in \mathcal{M}, p' \in \mathcal{M}'$.

Note that it is immediate that $\mathsf{ML}_{\mathsf{ind}}$-saturation generalises the notion of image-finiteness (w.r.t. the induced relations), as in a Hausdorff space finite sets are always closed.

4 The Main Theorem: Bisimulational Categoricity

After collecting all the necessary notions and tools, we are now ready to state and prove three theorems being the main contribution of this paper (including the already mentioned Theorem 2.8).

Theorem 2.8 *For every type $t \in \mathbb{T}$, the following are equivalent:*

(1) *t has a unique model up to $\leftrightarrows$;*

(2) *every model of t is bisimilar to an image-finite model;*

(3) *t has a model which is image-finite.*

Theorem 4.1 *For every type $t \in \mathbb{T}_{\mathsf{ind}_{\leftrightharpoons}}$, the following are equivalent:*

(1) *t has a unique model up to $\leftrightarrows_{\mathsf{ind}_{\leftrightharpoons}}$;*

(2) *every model of t is $\mathsf{ind}_{\leftrightharpoons}$-bisimilar to a model where every point has finite in- and out-degree;*

(3) *t has a model where every point has finite in- and out-degree.*

Theorem 4.2 *For every type $t \in \mathbb{T}_{\mathsf{ind}_{+}}$, the following are equivalent:*

(1) *t has a unique model up to $\leftrightarrows_{\mathsf{ind}_{+}}$;*

(2) *every model of t is ind_{+}-bisimilar to a finite model;*

(3) *t has a finite model.*

Note that in light of Proposition 4.9, the last theorem implies that when it comes to defining models up to transitive bisimulation, the expressive power of the transitive modal logic does not increase when we move from single formulae to entire theories.

Let us now prove the theorems. Most of the proof is the same in all three cases of Theorems 2.8, 4.1 and 4.2.

4.1 (2) ⇒ (3)

In all the three cases, the implication (2) ⇒ (3) is immediate, as by definition every type has a model.

4.2 (3) ⇒ (1)

Let us now prove a generalisation of the Hennessy-Milner Theorem [6] for $\mathsf{ML}_{\mathsf{ind}}$. It strengthens the standard formulation of Hennessy-Milner-like results in that we only require one of the models to be image-finite (which, in the context of usual modal logic ML, is a well-known folklore strengthening of the original Hennessy-Milner Theorem). It does not require any assumptions on ind and the proof is essentially the same as in the standard case.

Theorem 4.3 (à la Hennessy-Milner) *Assume $\mathcal{M} \in \mathsf{Krip}(\mathsf{S})$ and the induced model $\mathsf{ind}(\mathcal{M})$ is image-finite. Then, for every $\mathcal{M}' \in \mathsf{Krip}(\mathsf{S})$ and every $p \in \mathcal{M}, p' \in \mathcal{M}'$:*

$$\mathcal{M}, p \equiv_{\mathsf{ML}_{\mathsf{ind}}} \mathcal{M}', p' \qquad \underline{\textit{implies}} \qquad \mathcal{M}, p \leftrightarrows_{\mathsf{ind}} \mathcal{M}', p'.$$

Proof. It suffices to show that the relation $\equiv_{\mathsf{ML}} \subseteq M \times M'$ of modal equivalence is itself an ind-bisimulation. The base condition is immediate.

For the back and the forth conditions, let us take $q \equiv_{\mathsf{ML}_{\mathsf{ind}}} q'$, and any $R_k \in R$. By our assumption, q can only have a finite number of R_k-children (in

$\mathsf{ind}(\mathcal{M})$). In particular, they have only a finite number of distinct modal types $t_1, ..., t_n$ – and since $\mathbb{T}_{\mathsf{ind}}$ is a Hausdorff space, we can find pairwise mutually exclusive formulae $\varphi_1, ..., \varphi_n$ s.t. $\varphi_i \in t_i$ but $\varphi_i \notin t_j$ for $i \neq j$. Both q – and by equivalence also q' – satisfy:

$$\Box_k(\bigvee_{i\in\{1,...,n\}} \varphi_i); \qquad \bigwedge_{i\in\{1,...,n\}} \Diamond_k\varphi_i; \qquad \Box_k(\varphi_i \Rightarrow \psi) \text{ for any } \psi \in t_i$$

It follows that the types of R_k-children of q' are exactly $t_1, ..., t_n$. But this implies both the forth and the back conditions, as it means that for every R_k-child of q (or q', respectively) there exists an equivalent R_k-child of q' (resp. q). □

4.3 (1) ⇒ (2)

The last (and hardest to prove) implication is from (1) to (2). Before we proceed, let us recall an elementary topological fact. Since any infinite compact space has to contain a non-isolated point and closed subspaces of a compact space are always compact, it follows that:

Lemma 4.4 *If Y is a closed infinite subset of a compact topological space X, then it contains a point $y \in Y$ that is not isolated in Y.*

As in the classical model theory, we would like to use some good properties based on compactness of the considered logic. However, as shown by Example 3.9, an inducing assignment can encode arbitrary properties and thus in general the logic $\mathsf{ML}_{\mathsf{ind}}$ does not have to be compact. Fortunately, we may overcome this difficulty thanks to additional good properties of the considered assignments.

Lemma 4.5 *Assume that the image of* ind *is axiomatized by a set of sentences A expressed in first-order logic, i.e.:*

$$\mathsf{ind}[\mathsf{Krip}(\mathsf{S})] = \{\mathcal{M} \in \mathsf{Krip}(\mathsf{R}) \mid \mathcal{M} \textit{ satisfies } A\}.$$

Then:

- *the logic $\mathsf{ML}_{\mathsf{ind}}$ is compact;*
- *every $t \in \mathbb{T}_{\mathsf{ind}}$ has an $\mathsf{ML}_{\mathsf{ind}}$-saturated model $\mathcal{M}, r \models t$.*

Proof. For the first item, take any set of formulae $t \subseteq \mathsf{ML}_{\mathsf{ind}}$ and translate it to equivalent set t^{FO} of formulae in first-order logic over the signature $R \cup \{a(x) \mid a \in \Sigma\}$. Observe that t is satisfiable w.r.t. the induced semantics iff $A \cup t^{\mathsf{FO}}$ is satisfiable in $\mathsf{Krip}(\mathsf{R})$ in the standard sense. Hence, compactness of $\mathsf{ML}_{\mathsf{ind}}$ follows from compactness of the first-order logic.

The second item can be proven in a similar way, using the model-theoretic method of elementary saturated extensions. The proof is just a straightforward modification of the standard one (e.g. in [1]) and as such is skipped.[6] □

[6] In fact, if one defines induced first-order logic $\mathsf{FO}_{\mathsf{ind}}$ analogously to the induced modal logic – by interpreting it via ind – the assumption of first-order axiomatizability of $\mathsf{ind}[\mathsf{Krip}(\mathsf{S})]$ allows for a generalisation of van Benthem's theorem saying that $\mathsf{ML}_{\mathsf{ind}}$ is precisely the $\leftrightarroweq_{\mathsf{ind}}$-invariant fragment of $\mathsf{FO}_{\mathsf{ind}}$.

Note that all the assignments $\mathsf{Id}, \mathsf{ind}_{\leftrightarrows}, \mathsf{ind}_{+}$ and $\mathsf{ind}_{\forall}$ satisfy the assumptions of the above lemma.

Before we proceed, let us adapt two basic constructions related to the notion of a bisimulation to our context – generated submodels and bisimulation quotients (also called bisimulation contractions):

Proposition 4.6 (generated submodels) *Let* ind *be either* Id, $\mathsf{ind}_{\leftrightarrows}$ *or* ind_{+}. *Given a model* $\mathcal{M} \in \mathsf{Krip}(\mathsf{S})$ *and a point* $p \in \mathcal{M}$, *the model generated by* p, *denoted* $\langle p \rangle_{\mathcal{M}}$, *is just the submodel of* $\mathcal{M}$ *consisting of points reachable from* p *by a finite path in* $\mathsf{ind}(\mathcal{M})$ *(including* p *itself). Then,* $\mathcal{M}, q \leftrightarroweq_{\mathsf{ind}} \langle p \rangle_{\mathcal{M}}, q$ *for any* $q \in \langle p \rangle_{\mathcal{M}}$.

Proposition 4.7 (quotients) *Let* ind *be either* Id, $\mathsf{ind}_{\leftrightarrows}$ *or* ind_{+}. *For an* ind-*bisimulation* $Z \subseteq \mathcal{M} \times \mathcal{M}$ *being an equivalence relation, there is a model structure on the set of all equivalence classes of* Z *s.t. the projection map* $p \overset{\pi_Z}{\longmapsto} [p]_{/Z}$ *is a functional* ind-*bisimulation. We call that model a quotient of* $\mathcal{M}$ *by* Z *– and denote it* $\mathcal{M}_{/Z}$.[7]

Proof. Both constructions are the same as in the standard case – except for quotients by transitive bisimulations.

Given a model $\mathcal{M}$ and a transitive bisimulation Z being an equivalence relation on M, we can first take the model $\mathcal{M}^{+} = (M, \rightarrow^{\mathcal{M}^{+}}, \mathsf{val}^{\mathcal{M}})$ with $\rightarrow^{\mathcal{M}^{+}}$ being the transitive closure of $\rightarrow^{\mathcal{M}}$.

Observe that $\mathsf{ind}_{+}(\mathcal{M}) = \mathsf{ind}_{+}(\mathcal{M}^{+})$ and hence: (*) the identity map $\mathsf{Id} : M \rightarrow M$ can be seen as a functional transitive bisimulation $\mathsf{Id} : \mathcal{M} \rightarrow \mathcal{M}^{+}$. Moreover, transitivity of $\rightarrow^{\mathcal{M}^{+}}$ implies that: (**) on $\mathcal{M}^{+}$, transitive bisimulations are the same as standard bisimulations.

Since (transitive) bisimulations are closed under compositions, (*) implies that Z is a transitive bisimulation not only on $\mathcal{M}$, but also on $\mathcal{M}^{+}$ – and so by (**) it is also a standard bisimulation on $\mathcal{M}^{+}$. This allows us to quotient (in the standard sense) $\mathcal{M}^{+}$ by Z obtaining $(\mathcal{M}^{+})_{/Z}$. Since the natural projection $\pi_Z : \mathcal{M}^{+} \rightarrow (\mathcal{M}^{+})_{/Z}$ is a functional bisimulation and bisimulations are always instances of transitive bisimulations, the graph of the function π_Z – and therefore by (*) also $\pi_Z \circ \mathsf{Id} : \mathcal{M} \rightarrow (\mathcal{M}^{+})_{/Z}$ – is a transitive bisimulation. □

We are now ready for the proof.

Case 1: $\mathsf{ind} = \mathsf{Id}$

Let us take a model $\mathcal{M}, r$ that is not bisimilar to any image-finite model – we will construct another model that is equivalent, but non-bisimilar to it. We may combine: (i) Lemma 4.5 to obtain an equivalent model which is $\mathsf{ML}(\{\rightarrow\})$-saturated, (ii) Proposition 4.7 to take its quotient by $\leftrightarroweq$ where (by Proposition 3.14) no two points satisfy the same formulae and finally (iii) apply Proposition 4.6 to take a submodel accessible from the root. If such model is not bisimilar

[7] Note that in the case of ind_{+} such quotient does not have to be unique. Nevertheless, it is unique *up to* $\leftrightarroweq_{\mathsf{ind}}$.

to $\mathcal{M}, r$, we are done – so the remaining case is when $\mathcal{M}, r$ has all the properties listed above.

Since by our assumption $\mathcal{M}, r$ is not image-finite, there must exist a point p reachable from r by a finite path and having infinitely many children. The set $T = \{\mathbf{tp}^{\mathcal{M}}(q) \mid p \to q\}$ is an infinite closed subset of a compact space and so by Lemma 4.4 it contains a non-isolated limit type t^{lim} realised in some p's child p^{lim}.

Now, in order to construct another model for t we simply remove the arrow leading from p to p^{lim}:

$$\mathcal{N} = (M, \to^{\mathcal{M}} - \{(p, p^{\mathrm{lim}})\}, \mathsf{val}^{\mathcal{M}})$$

We prove by induction on n that any point $q \in \mathcal{M}$ satisfies exactly the same formulae of modal depth n in both $\mathcal{M}$ and $\mathcal{N}$ (and thus in particular $\mathcal{N}, r \models t$). The base case is obvious. For the induction step, the only interesting case is for p, as prima facie it could satisfy less sentences of the form $\Diamond\varphi$. However, since t^{lim} is not isolated in T, for any $\varphi \in t^{\mathrm{lim}}$ there must be $t' \in T$ s.t. $\varphi \in t'$. By definition of T this means that there is a sibling s of p^{lim} s.t. $\mathcal{M}, s \models t'$ – and so in particular $\mathcal{M}, s \models \varphi$. But modal depth of φ is smaller than that of $\Diamond\varphi$ – so we know by induction hypothesis that $\mathcal{N}, s \models \varphi$, and hence $\mathcal{N}, p \models \Diamond\varphi$.

On the other hand, we will show that $\mathcal{M}, r \not\rightleftharpoons \mathcal{N}, r$, as ∀dam has the following winning strategy in the bisimulation game: (i) First follow the path to the point p in $\mathcal{M}$. If after that ∃ve responds with a point $q \in \mathcal{N}$ other than p, we know that $\mathcal{M}, p \not\equiv_{\mathsf{ML}} \mathcal{N}, q$ (as no two different points are equivalent in $\mathcal{N}$) and so $\mathcal{M}, p \not\rightleftharpoons \mathcal{N}, q$ – which means that ∀dam can now win the game. (ii) If ∃ve responded with the same point $p \in \mathcal{N}$, ∀dam moves to p^{lim} in $\mathcal{M}$. Now ∃ve has to respond with some point $q \in \mathcal{N}$ – but by definition of $\mathcal{N}$ we know that she cannot choose p^{lim}, and so again $\mathcal{M}, p^{\mathrm{lim}} \not\equiv_{\mathsf{ML}} \mathcal{N}, q$, meaning that ∀dam can win the game from that point.

<u>Case 2: $\mathsf{ind} = \mathsf{ind}_{\leftrightarrows}$</u>

In this case, we need a slight modification of the previous construction due to the fact that we deal with two-way modalities and removing an arrow $q \to q'$ changes both sets: q's successors and q''s predecessors.

As in the previous case, we take an $\mathsf{ML}_{\mathsf{ind}_{\leftrightarrows}}$-saturated model of $t \in \mathbb{T}_{\mathsf{ind}_{\leftrightarrows}}$ where any two different points have different types and any point is accessible by a finite path (possibly using forward and backward moves) from the root – s.t. some point $p \in \mathcal{M}$ has infinitely many successors (the case with infinitely many *predecessors* is entirely symmetric). We take the limit t^{lim} of $T = \{\mathbf{tp}^{\mathcal{M}}(q) \mid p \to q\}$ realised by some p^{lim}.

We define $\mathcal{N}$ as follows. First take the disjoint union $\mathcal{N}' = \mathcal{M}_1 + \mathcal{M}_2 + \mathcal{M}_3$, where each $\mathcal{M}_i$ is a copy of $\mathcal{M}$. We will denote the element of $\mathcal{M}_i$ corresponding to $q \in \mathcal{M}$ by q_i. Let us also pick any child $s \in \mathcal{M}$ of p different than p^{lim}. Then, our model $\mathcal{N}$ is just $\mathcal{N}'$ without the arrow $p_2 \to p_2^{\lim}$ and with two additional arrows $p_2 \to s_1$ and $p_3 \to p_2^{\lim}$:

$$\mathcal{N} = (N', \to^{\mathcal{N}'} - \{(p_2, p_2^{\lim})\} \cup \{(p_2, s_1), (p_3, p_2^{\lim})\}, \mathsf{val}^{\mathcal{N}'})$$

A picture of $\mathcal{M}, r$ and $\mathcal{N}, r_1$:

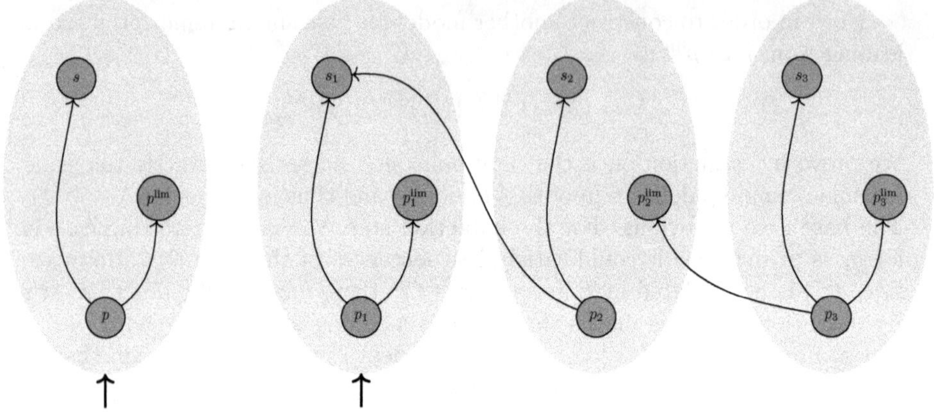

The rest of the proof is analogous to the previous case. We first prove by induction on n that for every $q \in \mathcal{M}$, $\mathcal{M}, q$ and $\mathcal{N}, q_i$ satisfy the same $\mathsf{ML}_{\mathsf{ind}_{\leftrightharpoons}}$-formulae of modal depth n. This boils down to checking several straightforward cases (the one in which we use the fact that $t_{\lim}$ was not isolated is that with p_2's successors).

The winning strategy for ∀dam witnessing $\mathcal{M}, r \not\leftrightarrows_{\mathsf{ind}_{\leftrightharpoons}} \mathcal{N}, r_1$ is as follows: (i) First follow the path from r_1 to $p_2 \in \mathcal{N}$.[8]. In order not to loose, ∃ve has to respond in $\mathcal{M}$ with the only point that is equivalent to p_2, namely p. (ii) Then, ∀dam moves to $p^{\lim}$ in $\mathcal{M}$ and ∃ve has to respond in $\mathcal{N}$ with a point non-equivalent with it – thus loosing the game.

<u>Case 3: $\mathsf{ind} = \mathsf{ind}_+$</u>

This is the most involved case. The key difficulty is that it does not suffice to simply remove arrows from the model to remove them from its *transitive closure*. Consider the following example.

[8] Note that since in this context *accessibility* means *two-way accessibility*, after removing the arrow $p_2 \to p_2^{\lim}$, p_2 does not have to be accessible from r_2. Indeed, it could actually happen that $\mathcal{M}, r \leftrightarrows_{\mathsf{ind}_{\leftrightharpoons}} \mathcal{N}, r_2$. However, we know that s_1 is accessible from r_1 and from there we can move backwards to p_2.

Example 4.8 In the model below, the rightmost black point has a copy of ω (with the reverse order as the accessibility relation) as its children.

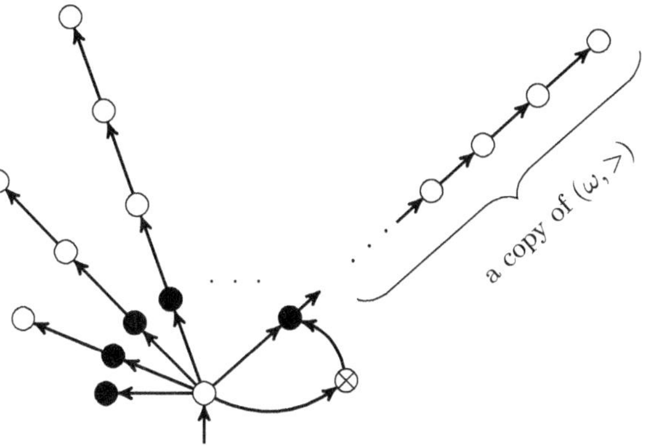

One can check that the type t_{lim} of the rightmost black point *is not isolated* among the types of its black siblings. However, it *is isolated* from the perspective of the crossed point – which in turn is isolated from the perspective of the root. Basing on that observation, it is not hard to show that any model $\mathsf{ML}_{\mathsf{ind}_+}$-equivalent to the one above must realise t_{lim} in a *descendant* (not necessarily a *child*) of its root. In particular, this demonstrates that not every isolated type can be omitted. Nevertheless, we will show that in the presence of a non-isolated type it is always possible to find *some* (possibly different) type that can be omitted.

Let us start with recalling the following well-known fact:

Proposition 4.9 *If $\mathcal{M}, r$ is a finite model, then it is definable in $\mathsf{ML}_{\mathsf{ind}_+}$ up to $\leftrightarrows_{\mathsf{ind}_+}$, i.e. there is an $\mathsf{ML}_{\mathsf{ind}_+}$-formula s.t. every its model is ind_+-bisimilar to $\mathcal{M}, r$. In particular, finite models only realize types isolated in $\mathbb{T}_{\mathsf{ind}^+}$.*

Proof. Since $M = q_1, ..., q_n$ is finite, it realises only finitely many types $t_1, ..., t_n$ (w.l.o.g. all distinct, as otherwise we may quotient the model). Since $\mathbb{T}_{\mathsf{ind}_+}$ is a Hausdorff space, there are mutually exclusive sentences $\varphi_i \in t_i$ for every i. First, define ψ_i to be the formula that describes which atomic propositions belong to t_i and which other types it sees:

$$\begin{gathered}\bigwedge\{a \in \Sigma \mid a \in t_i\} \wedge \\ \Box(\bigvee\{\varphi_j \mid q_i \to^+ q_j\}) \wedge \\ \bigwedge\{\Diamond\varphi_j \mid q_i \to^+ q_j\}\end{gathered}$$

Then, we put:

$$\theta_i = \psi_i \wedge \Box(\bigwedge_{j\in\{1,...,n\}} \{\varphi_j \Rightarrow \psi_j\})$$

It is straightforward that $\theta_i \in t_i$. On the other hand, if $\mathcal{N}, q \models \theta_i$, then already $\mathcal{N}, q \leftrightarrows_{\mathsf{ind}_+} \mathcal{M}, q_i$. Indeed, w.l.o.g. we may assume that such $\mathcal{N}$ is reachable from q and then it is easy to check that: (i) the types of all the points of $\mathcal{N}$ are precisely $\{t_1, ..., t_n\}$, (ii) the map $f : N \to M$ sending a point with type t_i to q_i is a functional bisimulation. It then follows that each type t_i is isolated by its basic neighbourhood $\langle \theta_i \rangle$. □

As in both previous cases, let us take a model $\mathcal{M}, r$ that is infinite, $\mathsf{ML}_{\mathsf{ind}_+}$-saturated, reachable and no two points realise different types – but the model is not bisimilar to a finite one. It follows that the root has infinitely many descendants. We will need the following fact:

Lemma 4.10 *There exists a point $p_\infty \in \mathcal{M}$ s.t. $p_\infty \to^+ p_\infty$ and its type t_∞ is a non-isolated element of $\{\mathbf{tp}^{\mathcal{M}}(q) \mid p_\infty \to^+ q\}$.*

Proof. We will inductively construct a sequence of (not necessarily distinct) points, indexed by countable ordinals $(p_\alpha)_{\alpha<\omega_1} \subseteq \mathcal{M}$ with the property that for any $\alpha < \beta$: (i) $p_\alpha \to^+ p_\beta$ and (ii) $\mathbf{tp}^{\mathcal{M}}(p_\beta)$ is not isolated in $\{\mathbf{tp}^{\mathcal{M}}(q) \mid p_\alpha \to^+ q\}$.

For the induction base, we simply take the root $p_0 = r$.

For $\alpha + 1$, we know by induction assumption that $\mathbf{tp}^{\mathcal{M}}(p_\alpha)$ is not isolated, so by Lemma 4.9 we know that the model generated by p_α has to be infinite (except for the case $\alpha = 0$ where we just know that r has infinitely many descendants). Now we look at the infinite set $T_\alpha = \{\mathbf{tp}^{\mathcal{M}}(q) | \ p_\alpha \to^+ q\}$ and pick some its limit – a non-isolated type $t_{\alpha+1} \in \mathbb{T}_{\mathsf{ind}_+}$ which, by $\mathsf{ML}_{\mathsf{ind}_+}$-saturation, is realised in some descendant $p_{\alpha+1}$ of p_α.

For a limit ordinal α, we fix a subsequence $(\alpha_i)_{i\in\omega} \subseteq \alpha$ of shape ω which is cofinal with α (which exists as α is countable). Take any limit t_α of the set $T_\alpha = \{\mathbf{tp}^{\mathcal{M}}(p_{\alpha_i}) \mid i \in \omega\}$. Since t_α is not isolated and $\mathbb{T}_{\mathsf{ind}_+}$ is Hausdorff, every $\varphi \in t_\alpha$ must belong to infinitely many types from T_α. It follows that there are arbitrary big i s.t. $\varphi \in t_{\alpha_i}$, so every p_{α_j} – and hence by cofinality also every p_β – has a descendant satisfying φ. Hence, by $\mathsf{ML}_{\mathsf{ind}_+}$-saturation, each p_β has a descendant realising t_α – and by our assumptions on $\mathcal{M}$ this point p_α is unique.

Now we claim that $p_\alpha = p_\beta$ for some $\alpha \neq \beta$. Indeed, observe that if $p \to^+ q$, then q cannot satisfy more formulae of the form $\Diamond\varphi$ than p. Since there are only countably many formulae, for sufficiently large α all $\mathbf{tp}^{\mathcal{M}}(p_\alpha)$ may only differ on formulae equivalent to boolean combinations of Σ. But $\mathcal{P}(\Sigma)$ is finite, so $p_\alpha = p_\beta$ for some $\alpha < \beta$ and thus we put $p_\infty = p_\alpha$. It then follows from (i) that $p_\infty \to^+ p_\infty$. Finally, (ii) implies that the type t_∞ is not isolated in $\{\mathbf{tp}^{\mathcal{M}}(q) \mid p_\infty \to^+ q\}$, as desired. □

Now, we can define a new model by removing all the arrows leading to p_∞:

$$\mathcal{N} = (M, \to -\{(q, p_\infty) \mid q \in \mathcal{M}\}, \mathsf{val}^{\mathcal{M}}).$$

Observe that t_∞ is not isolated in $\{\mathbf{tp}^{\mathcal{M}}(q) \mid p \to^+ q\}$ for any ancestor p of p_∞. This allows us, as in the two previous cases, to prove by induction on modal depth that $\mathcal{M}, q \equiv_{\mathsf{ML}_{\mathsf{ind}_+}} \mathcal{N}, q$ for every $q \in \mathcal{M}$. On the other hand, p_∞ is reachable from the root in $\mathcal{M}$ but not in $\mathcal{N}$ – which gives a winning strategy for $\forall$dam in the bisimulation game. Q.E.D.

4.4 Limitations

We end with two examples illustrating limitations of our method. First of all, let us emphasize that our proofs rely on compactness of the logic under consideration – and it is not hard to come up with an example of a non-compact logic which fails to have analogous characterisation. For instance, consider the mix of ML and $\mathsf{ML}_{\mathsf{ind}_+}$ – i.e. the logic having *both* the standard and the transitive modalities. Such logic is not compact and can describe the infinitely branching Hedgehog (Example 2.6) up to bisimulation – by extending its ML-type with an additional sentence: $\Box(\Box\bot \vee \Diamond^+\Box\bot)$ (i.e. "every child of the root either has no children or has a descendant with no children").

Since non-compact logics seem out of our reach, a natural question is if compactness is *sufficient* for analogous characterisation. Unfortunately, this is not the case. The second example shows that even the stronger assumption of first-order axiomatizability of $\mathsf{ind}[\mathsf{Krip}(\mathsf{S})]$ (which implies compactness of $\mathsf{ML}_{\mathsf{ind}}$ by Lemma 4.5) is not sufficient to generalise our characterization to $\mathsf{ML}_{\mathsf{ind}}$. Recall the universal modality induced by $\mathsf{ind}_\forall$ (Example 3.8). The class $\mathsf{ind}_\forall[\mathsf{Krip}(\{\to\})]$ is definable by a single first-order sentence: $\forall_{x,y} x\langle\exists\rangle y$. However, consider the following model $\mathcal{M} \in \mathsf{Krip}(\{\to\})$:

Example 4.11 $M = \omega + 1 = \{0, 1, ..., \omega\}$; $p \to^{\mathcal{M}} q$ iff $p = q + 1$ or $p = q = \omega$. As in Example 2.6 (The Hedgehogs), we assume $\Sigma = \emptyset$.

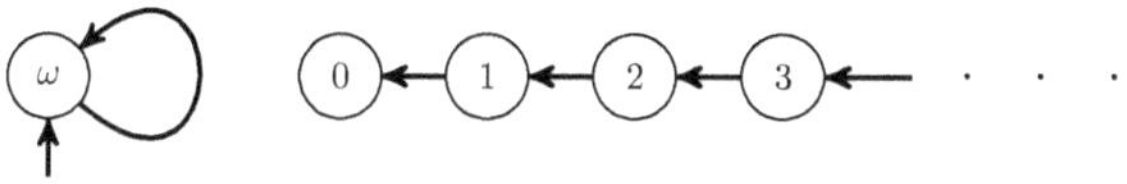

Observe that $\mathcal{M}, p \not\equiv_{\mathsf{ML}_{\mathsf{ind}_\forall}} \mathcal{M}, q$ for all $p \neq q$ – and so every point has infinitely many pairwise non-equivalent $\langle\exists\rangle$-children. However, it is not hard to show that any model equivalent to $\mathcal{M}, \omega$ must be $\mathsf{ind}_\forall$-bisimilar to it. The thing is that although the topological part of our reasoning still works and we may find a limit of the types realised in $\mathcal{M}$ (in fact, in this situation there is precisely one such limit type – the type of ω) – it is not possible to omit that limit type.

Acknowledgement

I would like to thank Mateusz Łełyk and Bartosz Wcisło who supervised my master thesis on (the standard modal logic version of) bisimulational categoricity for their support and many valuable remarks. I am also grateful to Mikołaj Bojańczyk and Grzegorz Lisowski for a number of suggestions concerning style and readability. Last but not least, I would also like to thank the anonymous referees who helped to improve the final version of this paper.

References

[1] Blackburn, P., M. de Rijke and Y. Venema, "Modal Logic," Cambridge University Press, 2002.

[2] Bojańczyk, M. and T. Idziaszek, *Algebra for infinite forests with an application to the temporal logic EF*, in: *CONCUR*, Lecture Notes in Computer Science **5710** (2009), pp. 131–145.

[3] Engeler, E., *A characterization of theories with isomorphic denumerable models*, Notices of the American Mathematical Society **6** (1959).

[4] Goranko, V. and S. Passy, *Using the universal modality: Gains and questions*, Journal of Logic and Computation (1992), pp. 5–30.

[5] Grzegorczyk, A., *On the concept of categoricity*, Studia Logica **13** (1962), pp. 39–66.

[6] Hennessy, M. and R. Milner, *Algebraic laws for non-determinism and concurrency*, Journal of the ACM **32** (1985), pp. 137–161.

[7] Marker, D., "Model Theory: An Introduction," Graduate Texts in Mathematics, Springer, 2013.

[8] Morley, M., *Categoricity in power*, Transactions of the American Mathematical Society **114** (1965), pp. 514–538.

[9] Niwinski, D., *On the cardinality of sets of infinite trees recognizable by finite automata*, in: *MFCS*, 1991.

[10] Ryll-Nardzewski, C., *On the categoricity in power* $\aleph_0$, Bulletin de l'Académie Polonaise des Sciences, Série des Sciences Mathématiques, Astronomiques Et Physiques **7** (1959), pp. 545–548.

[11] Svenonius, L., $\aleph_0$*-categoricity in first-order predicate calculus*, Theoria (Lund) **25** (1959), pp. 82–94.

[12] Wang, Y., "Epistemic Modelling and Protocol Dynamics," Ph.D. thesis, University of Amsterdam (2010).

Reduction of Modal Logic and Realization in Justification Logic

Hirohiko Kushida [1]

Computer Science Program, the Graduate Center, the City University of New York
365 Fifth Avenue, New York, NY, 10016, USA

Abstract

In this paper, we first offer basic results regarding modal logic: (1) a wide range of modal systems can be syntactically reduced to the modal logic K in terms of theoremhood and (2) we can restrict the forms of modal axioms without changing their deductive power in that range of modal logics. Then, based on these results, we offer a new, simple, uniform, and modular proof-theoretical proof of the realization of a wide range of modal logics with possible combinations of modal axioms $T, D, 4, 5$ (including $\mathsf{S5}$) in Justification Logic. We do not use a generalization of sequent calculus, such as hypersequent and nested sequent calculi. We simply utilize the standard cut-free sequent calculus for K and then show, in the realized proof in Justification Logic (corresponding to K), how to recover the realizations of the modal axioms by rewriting terms in the proof.

Keywords: Modal Logic, Justification Logic, Proof Theory, Realization Theorem.

1 Introduction

One of the most common interpretations of modal logic is the epistemic logical interpretation: reading a modal formula $\Box A$ as "A is known." However, the machinery of epistemic logic does not refer to how the knowledge A is attained. Justification Logic offers a tool to refer to a reason or justification for a proposition; a modal formula is of the form $s : A$ with a term s, which is read as "s is a reason or justification for A." Moreover, Justification Logic is equipped with operators on terms: $+$, $\cdot$, $!$ and $?$. The first two are binary and express the concatenation and an application of modus ponens, respectively; the latter two are unary and express positive and negative introspections, respectively. Then, for example, the logical omniscience problem could be avoided, in a sense; it could be viewed as a problem of term complexity. As we deduce a more complicated formula, we have a more complicated term with the formula at the same time. Cf. [6]. We refer to [3], [4], and [25] for a general introduction to the family of systems called Justification Logic.

[1] proof(underscore)hkushida@yahoo.co.jp

One of the fundamental results concerning Justification Logic is the realization theorem, which acts as a bridge to the modal logics. There have been many studies on the realization theorem for major modal logics. The realization theorem for the modal logic S4 was provided in Artemov [1], [2] with the Logic of Proofs, LP, which is the first system of Justification Logic. It makes the following claim: for some realization, that is, some assignment of terms to modality $\Box$, a formula is provable in S4 if and only if the realization of the formula is provable in LP. This result intended to give an arithmetical meaning to S4; a realized modal formula $s : A$ reads "s is a proof of A."

The original proof of the realization theorem in [1], [2] was a proof-theoretical one, using a standard cut-free sequent calculus of S4. Fitting [12] proposed a possible-world semantics for LP and proved the realization theorem using this semantics. The semantics has been studied well and extended for various systems of Justification Logic. It is called Fitting semantics today. Another semantical proof was offered for realization for LP in Fitting [15]. Substructural variants of LP were introduced, and the realization theorems were proved for some modal substructural logics by a proof-theoretic method in Kurokawa and Kushida [19].

Systems with the negative introspection operator were proposed by several authors pursuing epistemological interpretation of LP. Those systems correspond to the modal logic S5. Such a system was first introduced in Artemov et al. [5] and Kazakov [18], and the realization theorem for S5 was proved by a proof-theoretic method.

The negative introspection operator "?" that has been the subject of recent studies is characterized by the formula $\neg s : A \rightarrow ?s : \neg s : A$. It was proposed independently by Pacuit [27] and Rubtsova [28], [29]. The realization theorem was proved for S5 via Fitting semantics in [28], [29].

Fitting [14] offered an elegant proof-theoretical proof of the realization for S5 with the operator "?". Kurokawa and Kushida [20] offered an S5 variant of Linear Logic and proved the realization theorem for it with the corresponding substructural justification logic using a proof-theoretical method.

Nested sequent calculus is an apparatus used to execute an inference rule inside formulas. Although it is not clear if it is a natural expression of logical reasoning, it has been a useful tool to handle some logical systems that are not well-behaved proof-theoretically, such as S5. Motohashi [26] showed that the Intuitionistic Logic can be faithfully embedded in the classical predicate logic via a composition of Gödel's embedding and the standard translation (converting modality to quantifier). This result of [26] is one of the precursors of the method of nested sequent calculus, although it would be difficult to specify the first to have invented any similar kind of apparatus. In [21], the method was applied to a wide range of major modal logics between K and S5 (including the two) in a uniform way; it was shown that those modal logics can be faithfully embedded in the classical predicate logic by Motohashi's method. Later, we applied the method to the realization problem in light of Justification Logic in [22]; it was shown that the modal logic GL can be realized in a variant of LP with free variables using Motohashi's method.

While the realization of subsystems of S4 was proved in Brezhnev [8] proof-theoretically, a proof for modal logics including S5 was offered in a uniform way in Brünnler, Goetschi, and Kuznets [10]; Goetschi and Kuznets [17]; and Borg and Kuznets [7]. They utilized nested sequent systems to prove the realization for a wide range of modal logics between K and S5 (including the two). In particular, the proof in [7] was modular as well as uniform.

In this paper, we offer a new, simple, uniform, and modular proof of the realization of major modal logics extended by additional axioms: what we call $D, T, 4, 5$. These systems are modal logic correspondents to Justification Logic with the above-mentioned operators: $\cdot, +, !, ?$. [2] Our proof is a proof-theoretical one, but we do not use a generalization of sequent calculus, such as nested or hypersequent calculus; rather, we will simply use the standard cut-free sequent calculus for the modal logic K. We will present a reduction theorem of all of those extended modal logics over K. This result is concerned with the research problem treated in Fitting [11] and will be of independent interest apart from Justification Logic and the realization problem. Moreover, we will point out that the form of axioms $D, T, 4, 5$ can be restricted to a kind of normal form without changing their deductive power. This is a basic fact of the nature of modal logic, which seems not to have been published so far. We present the second reduction theorem using this normal form.

Then, we will make a realization for K to a basic system of Justification Logic called J. Then, to obtain realization for the other logics, we will show how to convert some realized formulas to the form of the axioms of Justification Logic by rewriting terms in the proof of a realized formula in J. It will be seen that a circular argument can be avoided in the rewriting algorithm, thanks to the second reduction theorem.

This paper is organized as follows. In §2, we define the modal logics treated in this study. Then, we offer two reduction theorems. It is also pointed out that the well-known modal axioms can be restricted to a sort of normal form. In §3, we define the systems of Justification Logic corresponding to those modal logics and prove the internalization theorem for a basic system of Justification Logic. In §4, we present our proof-theoretic proof of the realization theorem for all the systems defined in a uniform and modular way.

2 Modal Logics and Reduction Theorem

Let us begin with a review of axiomatic systems of the modal logic K and its normal extensions which we are going to handle in this paper. We adopt the propositional connectives: $\rightarrow, \neg$. The other ones are defined in terms of the two, which will be also used below. The unary modal operator $\Box$ is added. The other

[2] We do not handle the modal axiom called "B". We restrict our attention to terms with these operators in Justification Logic, while a new operator is needed to realize systems including "B", as was shown in [17], [7]. However, it is possible to apply our method to prove the realization for those systems including "B". We will touch on this point later in a footnote.

operator $\Diamond$ can be defined in terms of $\Box$, which is not considered in this paper. We use the symbols $\bot$ for the propositional constant and $P, Q, \ldots, P_1, P_2, \ldots$ for propositional variables. The formulas are constructed from atomic formulas in the usual way.

The modal logic K is an axiomatic system for the propositional logic augmented with the axiom $\Box(A \to B) \to (\Box A \to \Box B)$ and the inference rule $A/\Box A$ (Necessitation). We consider axioms called $D, T, 4$ and 5.

D	$\neg\Box\bot$
T	$\Box A \to A$
4	$\Box A \to \Box\Box A$
5	$\neg\Box A \to \Box\neg\Box A$

Then we obtain from K the system $\mathsf{KS}_1 \cdots \mathsf{S}_n$ extended with $S_1 \cdots S_n$ from the schemas $D, T, 4, 5$. As usual, we follow the custom to call the systems KD, KT, KT4, KT5 as D, T, S4, S5, respectively. (KT45 is equivalent to KT5.) By the notation $\mathsf{KS}_1 \cdots \mathsf{S}_n$, we can cover ten systems: D, T, K4, K5, K45, KD4, KD5, KD45, S4, S5. Let L denote any system from these systems.

Now, we show that L can be syntactically reduced to the modal logic K with respect to theoremhood. For L, a finite set of modal formulas α and a natural number n, we define the special formula $X(\alpha, n, \mathsf{L})$ as follows.

L	$X(\alpha, n, \mathsf{L})$
D	$\bigwedge_{0 \le i \le n} \Box^i \neg\Box\bot$
T	$\bigwedge_{\Box B \in \alpha} \bigwedge_{0 \le i \le n} \Box^i(\Box B \to B)$
K4	$\bigwedge_{\Box B \in \alpha} \bigwedge_{0 \le i \le n} \Box^i(\Box B \to \Box\Box B)$
K5	$\bigwedge_{\Box B \in \alpha} \bigwedge_{0 \le i \le n} \Box^i(\neg\Box B \to \Box\neg\Box B)$
K45	$X(\alpha, n, \mathsf{K4}) \wedge X(\alpha, n, \mathsf{K5})$
KD4	$X(\alpha, n, \mathsf{D}) \wedge X(\alpha, n, \mathsf{K4})$
KD5	$X(\alpha, n, \mathsf{D}) \wedge X(\alpha, n, \mathsf{K5})$
KD45	$X(\alpha, n, \mathsf{KD4}) \wedge X(\alpha, n, \mathsf{K5})$
S4	$X(\alpha, n, \mathsf{T}) \wedge X(\alpha, n, \mathsf{K4})$
S5	$X(\alpha, n, \mathsf{T}) \wedge X(\alpha, n, \mathsf{K5})$

Here "$\Box^n$" denotes "$\overbrace{\Box \cdots \Box}^{n-many}$".

Lemma 2.1 *Let α, β be any finite set of modal formulas and n, m be any natural numbers. Then we have the following.*

(1) $\vdash_K X(\alpha \cup \beta, max(n, m), L) \to X(\alpha, n, L) \wedge X(\beta, m, L)$;

(2) $\vdash_K X(\alpha, n+1, L) \to \Box X(\alpha, n, L)$.

Proof. For (1). Suppose $n \ge m$. For any formula C, we have the following derivation by propositional calculus.

$$\begin{aligned} \bigwedge_{\Box B \in \alpha \cup \beta} \bigwedge_{0 \le i \le n} \Box^i C \quad &\to \quad \bigwedge_{\Box B \in \alpha} \bigwedge_{0 \le i \le n} \Box^i C \wedge \bigwedge_{\Box B \in \beta} \bigwedge_{0 \le i \le n} \Box^i C \\ &\to \quad \bigwedge_{\Box B \in \alpha} \bigwedge_{0 \le i \le n} \Box^i C \wedge \bigwedge_{\Box B \in \beta} \bigwedge_{0 \le i \le m} \Box^i C \end{aligned}$$

Thus, we have proven the cases when L is D, T, $\mathsf{K4}$ or $\mathsf{K5}$. By using these results, we can prove the other cases; we handle the case L is $\mathsf{KD5}$. (Other cases are similar.) We have the following derivation by propositional calculus.

$$\begin{array}{rcc} X(\alpha \cup \beta, n, \mathsf{KD5}) & = & X(\alpha \cup \beta, n, \mathsf{D}) \wedge X(\alpha \cup \beta, n, \mathsf{K5}) \\ & \to & X(\alpha, n, \mathsf{D}) \wedge X(\beta, m, \mathsf{D}) \wedge X(\alpha, n, \mathsf{K5}) \wedge X(\beta, m, \mathsf{K5}) \\ & \to & X(\alpha, n, \mathsf{KD5}) \wedge X(\beta, m, \mathsf{KD5}) \end{array}$$

Thus, (1) holds for this case.

For (2). When L is D, T, $\mathsf{K4}$ or $\mathsf{K5}$. For any formula C, we have the following derivation in K.

$$\begin{array}{rcl} \bigwedge_{\Box B \in \alpha} \bigwedge_{0 \le i \le n+1} \Box^i C & \to & \bigwedge_{\Box B \in \alpha} \bigwedge_{1 \le i \le n+1} \Box^i C \\ & \to & \Box \bigwedge_{\Box B \in \alpha} \bigwedge_{0 \le i \le n} \Box^i C \end{array}$$

Thus, (2) holds for these cases.

The other cases can be established by using these results; again, we take the case $\mathsf{L} = \mathsf{KD5}$ only, as the remaining cases are similarly proved. We have the following derivation in K.

$$\begin{array}{rcc} X(\alpha, n+1, \mathsf{KD5}) & = & X(\alpha, n+1, \mathsf{D}) \wedge X(\alpha, n+1, \mathsf{K5}) \\ & \to & \Box X(\alpha, n, \mathsf{D}) \wedge \Box X(\alpha, n, \mathsf{K5}) \\ & \to & \Box [X(\alpha, n, \mathsf{D}) \wedge X(\alpha, n, \mathsf{K5})] \\ & = & \Box X(\alpha, n, \mathsf{KD5}) \end{array}$$

Thus, this case has been proven for (2). □

We call '$\Box A$' in the above definition of $D, T, 4, 5$ *the core* of them. E.g., $\Box(\Box P \wedge \neg P)$ is the core of an axiom T: $\Box(\Box P \wedge \neg P) \to (\Box P \wedge \neg P)$. For a given proof in L, we define $\mathcal{AS}$ (*axiom specification*) to be the set $\{\Box A : \Box A$ is the core of an axiom $D, T, 4$ or 5 used in the proof$\}$.

Lemma 2.2 *For any formula A of modal logic,*
if $\vdash_{\mathsf{L}} A$ *with some* $\mathcal{AS}$, *then* $\vdash_{\mathsf{K}} X(\mathcal{AS}, n, \mathsf{L}) \to A$, *for some n.*

Proof. We proceed by induction on the length of a proof of A in L with $\mathcal{AS}$. When the proof is an axiom of K, $X(\mathcal{AS}, n, \mathsf{L}) = \emptyset$. When the proof is an axiom of D, T, 4 or 5, $\vdash_{\mathsf{K}} X(\mathcal{AS}, 0, \mathsf{L}) \to A$.

• For modus ponens, suppose that A is derived from $B \to A$ and B. By the induction hypothesis, for some $\mathcal{AS}_1, \mathcal{AS}_2, n$ and m, we have $\vdash_{\mathsf{K}} X(\mathcal{AS}_1, n, \mathsf{L}) \to (B \to A)$ and $\vdash_{\mathsf{K}} X(\mathcal{AS}_2, m, \mathsf{L}) \to B$. Then, we obtain $\vdash_{\mathsf{K}} X(\mathcal{AS}_1, n, \mathsf{L}) \wedge X(\mathcal{AS}_2, m, \mathsf{L}) \to A$. By (1) of Lemma 2.1, $\vdash_{\mathsf{K}} X(\mathcal{AS}_1 \cup \mathcal{AS}_2, max(n, m), \mathsf{L}) \to A$.

• For necessitation, suppose that $A = \Box B$ is derived from B. By the induction hypothesis, for some $\mathcal{AS}$ and n, we have $\vdash_{\mathsf{K}} X(\mathcal{AS}, n, \mathsf{L}) \to B$. By necessitation and normality of '□', we obtain $\vdash_{\mathsf{K}} \Box X(\mathcal{AS}, n, \mathsf{L}) \to \Box B$. By (2) of Lemma 2.1, $\vdash_{\mathsf{K}} X(\mathcal{AS}, n+1, \mathsf{L}) \to \Box B$. □

Theorem 2.3 *(the first Reduction Theorem) For any formula A of modal logic, the following two are equivalent.*

(1) $\vdash_{\mathsf{L}} A$ *with* $\mathcal{AS}$*;*
(2) $\vdash_{\mathsf{K}} X(\mathcal{AS}, n, \mathsf{L}) \rightarrow A$*, for some n.* [3]

Proof. It is easily seen that, for any α and any n, $X(\alpha, n, \mathsf{L})$ is provable in L. Then, (2) obviously implies (1). The converse direction immediately follows from Lemma 2.2. □

We take an example to sketch a reduction of proof in KD5 to that in K in Appendix I.

2.1 Restriction of Modal Axioms

Here, we show that modal logics under consideration have the same deductive power if we restrict the form of the axioms in a certain way. We define a normal form of formulas of modal logic as follows.

Definition 2.4 *The normal form of formulas is defined as follows.*

1. $P_1 \wedge \cdots \wedge P_n \rightarrow Q_1 \vee \cdots \vee Q_p$ *is in normal form.*
2. *When* $B_1, \ldots, B_m, C_1, \ldots, C_q$ *are in normal form, so is the following:* $P_1 \wedge \cdots \wedge P_n \wedge \Box B_1 \wedge \cdots \wedge \Box B_m \rightarrow Q_1 \vee \cdots \vee Q_p \vee \Box C_1 \vee \cdots \vee \Box C_q$
3. *If* B *is equivalent (in propositional logic) to a formula in normal form,* B *is also in normal form.*

Theorem 2.5 *(Normal Form Theorem) For any formula* A *of modal logic,* A *is equivalent in* K *to a conjunction of formulas in normal form.*

Proof. We define the *degree* of A, $d(A)$, as follows. $d(P) = 0; d(A \rightarrow B) = d(A) + d(B); d(\neg A) = d(A); d(\Box A) = d(A) + 1$. We proceed by induction on $d(A)$. At first, by propositional logic, A can be transformed into a conjunction of the forms:

$$(\natural) \quad P_1 \wedge \cdots \wedge P_n \wedge \Box B_1 \wedge \cdots \wedge \Box B_m \rightarrow Q_1 \vee \cdots \vee Q_p \vee \Box C_1 \vee \cdots \vee \Box C_q.$$

Here, by this propositional transformation, the formulas (each B_i and each C_j) inside the outmost occurrences of $\Box$ are untouched.

Now, in the base case, A is a conjunction of the form $P_1 \wedge \cdots \wedge P_n \rightarrow Q_1 \vee \cdots \vee Q_p$ and is in normal form. In the induction step, let D denote any B_i or any C_j. By the induction hypothesis, D can be equivalently in K transformed into the form $E_1 \wedge \cdots \wedge E_r$ with each E_i in normal form. Hence, $\vdash_{\mathsf{K}} \Box D \leftrightarrow \Box(E_1 \wedge \cdots \wedge E_r) \leftrightarrow \Box E_1 \wedge \cdots \wedge \Box E_r$. So, we may assume that each B_i in ($\natural$) is already in normal form. As to C_j in ($\natural$), assume that $C_1 = E_1 \wedge \cdots \wedge E_r$ where each E_i is in normal form. Then, ($\natural$) is equivalent to the following.

$$\bigwedge_{1 \leq i \leq r} [P_1 \wedge \cdots \wedge P_n \wedge \Box B_1 \wedge \cdots \wedge \Box B_m \rightarrow Q_1 \vee \cdots \vee Q_p \vee \Box E_i \vee \Box C_2 \vee \cdots \vee \Box C_q]$$

After all, A is equivalent in K to a conjunction of the forms of ($\natural$) where each B_i and each C_j are in normal form. □

[3] We could restrict the set of modal formulas $\mathcal{AS}$ so that the elements come from subformulas of A rather than axioms of a proof in L of A. This direction of research is found in [11]. Here, we cannot make such a restriction because our axiomatic systems do not enjoy the subformula property. Anyway, our concern here lies in the realization of modal logics and constructing $\mathcal{AS}$ this way is enough for our purpose.

We show that the restriction of the core of modal axioms to normal form does not change the deductive power of systems.

Theorem 2.6 *If $\vdash_{\mathsf{L}} A$, then $\vdash_{\mathsf{L}} A$ with $\mathcal{AS}$ consisting of formulas in normal form.*

Proof. It suffices to show that a general form of axiom of T, 4 and 5, respectively, is derivable in K from a restricted form of T, 4 and 5 with the core in normal form, respectively. By the Normal Form Theorem, a formula B is equivalent to $E_1 \wedge \cdots \wedge E_r$ with each E_i in normal form. Note that E_i is in normal form if and only if $\Box E_i$ is in normal form.

On T axiom: we have $\vdash_{\mathsf{K}} \Box B \to B = \Box(E_1 \wedge \cdots \wedge E_r) \to (E_1 \wedge \cdots \wedge E_r). \leftrightarrow (\Box E_1 \wedge \cdots \wedge \Box E_r) \to (E_1 \wedge \cdots \wedge E_r)$. Also, we have $\vdash_{\mathsf{K}} [(\Box E_1 \to E_1) \wedge \cdots \wedge (\Box E_r \to E_r)] \to .(\Box E_1 \wedge \cdots \wedge \Box E_r) \to (E_1 \wedge \cdots \wedge E_r)$. Therefore, $\vdash_{\mathsf{K}} [(\Box E_1 \to E_1) \wedge \cdots \wedge (\Box E_r \to E_r)] \to .\Box B \to B$.

On 4 axiom, it is similar to the case of T axiom.

On 5 axiom, we have $\vdash_{\mathsf{K}} \neg\Box B \to \Box\neg\Box B = \neg\Box(E_1 \wedge \cdots \wedge E_r) \to \Box\neg\Box(E_1 \wedge \cdots \wedge E_r). \leftrightarrow (\neg\Box E_1 \vee \cdots \vee \neg\Box E_r) \to \Box(\neg\Box E_1 \vee \cdots \vee \neg\Box E_r)$. On the other hand, $\vdash_{\mathsf{K}} [(\neg\Box E_1 \to \Box\neg\Box E_1) \wedge \cdots \wedge (\neg\Box E_r \to \Box\neg\Box E_r)] \to .(\neg\Box E_1 \vee \cdots \vee \neg\Box E_r) \to (\Box\neg\Box E_1 \vee \cdots \vee \Box\neg\Box E_r)$. As $\Box F \vee \Box G$ implies $\Box(F \vee G)$ in K for any F and G, we obtain $\vdash_{\mathsf{K}} [(\neg\Box E_1 \to \Box\neg\Box E_1) \wedge \cdots \wedge (\neg\Box E_r \to \Box\neg\Box E_r)] \to .\neg\Box B \to \Box\neg\Box B$.□

Now, we can sharpen the Reduction Theorem.

Theorem 2.7 *(the second Reduction Theorem) For any formula A of modal logic, the following two are equivalent.*

(1) $\vdash_{\mathsf{L}} A$;

(2) $\vdash_{\mathsf{K}} X(\alpha, n, \mathsf{L}) \to A$, for some α and n such that α consists of formulas in normal form.

Proof. Derived by Theorems 2.3 and 2.6. □

Each of Theorems 2.3, 2.5, 2.6, 2.7 is a simple but general observation and would belong to basics of modal logic, although it seems not commonly known. Theorem 2.7 will be useful to give a uniform proof of realization theorem in the following sections and could be thought of to reveal a hidden nature of modal logics together with the realization. [4]

3 Justification Logics and Internalization

Next, we review the corresponding systems of Justification Logic. The formulas of Justification Logic are defined in the same way as modal logic except that

---

[4] As we remarked in the Introduction, we do not handle the axiom "B" of the form $\neg A \to \Box\neg\Box A$. Anyway, the whole argument in this section holds for "B" and the systems with it, and the realization for systems with "B" can be proved by our method in the following sections. However, unfortunately, the modal logics GL and GLS do not satisfy Theorems 2.6 or 2.7, while they do Theorem 2.3 where we have the definitions: $X(\alpha, n, \mathsf{GL}) = \bigwedge_{\Box B \in \alpha} \bigwedge_{0 \le i \le n} \Box^i(\Box(\Box B \to B) \to \Box B)$ and $X(\alpha, n, \mathsf{GLS}) = X(\alpha, n, \mathsf{GL}) \wedge \bigwedge_{\Box B \in \alpha}(\Box B \to B)$. See [23] for a recent development of the study of GLS.

$\Box A$ is replaced with $s : A$, where s is a justification term, or simply, term and defined inductively as follows.

1. Constants $c, d, e, \ldots, c_1, c_2, \ldots$ are justification terms.
2. Variables $x, y, z, \ldots, x_1, x_2, \ldots$ are justification terms.
3. If s and t are justification terms, then so are $s \cdot t, s + t, !s$, and $?s$.

For $\boldsymbol{t} = (t_1, \ldots, t_n)$, by $\cdot(\boldsymbol{t})$ we mean any concatenation of all terms of $(t_1, \ldots, t_n)$ via the operator $\cdot$ in arbitrary order. The term $+(\boldsymbol{t})$ is similarly defined with $+$ in place of $\cdot$. The basic system J is defined by the following axioms and inference rules.

Axioms:
A1. Axioms of classical propositional logic
A2. $s : (A \to B) \to .t : A \to (s \cdot t) : B$
A3. $s : A \to (s + t) : A; \quad t : A \to (s + t) : A$

Rules of Inference:
R1. Modus Ponens: $A, A \to B / B$
R2. Iterated Axiom Necessitation: $A / c_1 : c_2 : \cdots : c_n : A$,
where each c_i $(1 \leq i \leq n)$ is a constant and A is an axiom.

The *constant specification*, CS, in a proof is defined to be the set of formulas introduced by R2 in the proof. We introduce the axioms named $D^j, T^j, 4^j, 5^j$ as follows.

D^j	$\neg s : \bot$
T^j	$s : A \to A$
4^j	$s : A \to !s : s : A$
5^j	$\neg s : A \to ?s : \neg s : A$

For modal logic L= $\mathsf{KS}_1 \cdots \mathsf{S}_n$, the system JL is provided by J augmented with the axioms: $S_1^j, \ldots, S_n^j$.

Let us prove the internalization theorem for J, which is a fundamental property of Justification Logics. Below, for any term s and formula A, by $at(s)$ and $at(A)$ we mean a set of atomic terms (that is, constants and variables) appearing in s and A, respectively.

Theorem 3.1 *(Internalization for J) For any formula A of J, $\vdash_J A$ implies $\vdash_J \cdot(\mathbf{c}) : A$, for some term of the form $\cdot(\mathbf{c})$ such that $at(\cdot(\mathbf{c})) \cap at(A) = \emptyset$.*

Proof. We proceed by induction on the length of a proof of A in J. When the proof is an axiom itself, we can take any fresh constant c so that $c : A$ is provable in J by R2. In the induction step, for the case of R1, by the induction hypothesis, we have terms $\cdot(\mathbf{c})$ and $\cdot(\boldsymbol{d})$ such that the following hold.

$$\vdash_J \cdot(\mathbf{c}) : A \qquad \vdash_J \cdot(\boldsymbol{d}) : (A \to B)$$
$$at(\cdot(\mathbf{c})) \cap at(A) = \emptyset \qquad at(\cdot(\boldsymbol{d})) \cap at(A \to B) = \emptyset$$

If $at(\cdot(\boldsymbol{c})) \cap at(B) \neq \emptyset$, we can substitute fresh constants for some of $\boldsymbol{c}$ to make it empty. (This is possible because any constants of $\boldsymbol{c}$ are introduced in R2 and we can choose any constants in applying R2.) By using A2 and R1, we have $\vdash_J (\cdot(\boldsymbol{c}) \cdot \cdot(\boldsymbol{d})) : B$ and $at(\cdot(\boldsymbol{c}) \cdot \cdot(\boldsymbol{d})) \cap at(B) = \emptyset$.

For the case of R2, when we have $c_n : c_{n-1} : \cdots : c_1 : A$ from an axiom A, we can also have $c_{n+1} : c_n : c_{n-1} : \cdots : c_1 : A$ from A by R2. Here, c_{n+1} is a fresh constant and the desired term. □

We note that Theorem 3.1 is a refinement of the standard form of the Internalization Theorem, which just claims that provability of A implies provability of $s : A$ for some term s.

The following corollary follows straightforwardly; we put the proof to Appendix II due to the lack of space.

Corollary 3.2 *For any formulas $B_1, B_2, \ldots, B_n, A$ and any terms $t_1, t_2, \ldots, t_n$ of* J*,*
$\vdash_J B_1 \wedge B_2 \wedge \cdots \wedge B_n. \rightarrow A$ *implies* $\vdash_J t_1 : B_1 \wedge t_2 : B_2 \wedge \cdots \wedge t_n : B_n. \rightarrow [\cdot(\boldsymbol{c}) \cdot t_1 \cdot t_2 \cdot \cdots \cdot t_n] : A$*, for some justification term $\cdot(\boldsymbol{c})$ such that $at(\cdot(\boldsymbol{c})) \cap at(A) = \emptyset$.* [5]

4 Realization of modal logics

A realization of a formula of modal logic is a replacement of each occurrence of □ in the formula with a justification term. Such a realization is denoted by r (possibly with integer subscripts) and the result of realization r for a formula A is denoted by A^r. Our aim is to prove the following realization theorem for L.

Theorem 4.1 *For any formula A of modal logic,*
$\vdash_L A$ *iff, for some r,* $\vdash_{JL} A^r$.

We are going to prove Theorem 4.1 by reducing it to the following Theorem 4.2 (the realization of K).

Theorem 4.2 *For any formula A of modal logic,*
$\vdash_K A$ *iff, for some r,* $\vdash_J A^r$.

Theorem 4.2 was first proved in Brezhnev [8] by utilizing sequent calculus method initiated by Artemov [2]. We modify the method slightly and naturally; the operator $+$ will be used when two positive occurrences of □ merge in a proof in K.

We make use of the standard sequent calculus for K. A sequent is of the form $\Gamma \Longrightarrow \Delta$. [6] The sequent calculus for K, which we also call K, is defined to be the extension of the sequent calculus for classical propositional logic LK with the following rule. (See, for example, [30] for the full description of LK.)

---

[5] Here, we follow the notation of "association to the left" in restoring brackets of the form $s_1 \cdot s_2 \cdot \cdots \cdot s_m$. That is, $s_1 \cdot s_2 \cdot \cdots \cdot s_m$ is read as $(\cdots((s_1 \cdot s_2) \cdot s_3) \cdot \cdots \cdot s_m)$. On the other hand, $\cdot(\boldsymbol{c})$ is read according to our previous definition of this notation; it can be any term consisting of constants $\boldsymbol{c}$ and the operator $\cdot$.

[6] As usual, by greek capital letters, we mean finite sequences of formulas.

$$\frac{\Gamma \Longrightarrow A}{\Box\Gamma \Longrightarrow \Box A}\ \Box$$

We assume the well-known facts: (i) this sequent calculus is equivalent to the axiomatic system K with respect to theoremhood and (ii) it enjoys the cut-elimination theorem.

For a sequent $S = \Gamma \Longrightarrow \Delta$, its *formula image*, $fi(S)$, is defined to be $\bigwedge\Gamma \to \bigvee\Delta$.

Proof of Theorem 4.2. The 'if' part is proved by using what is called the forgetful projection, say f: for any formula B of modal logic and any realization r, $(B^r)^f = B$. It is easily shown that $(A^r)^f = A$ is provable in K by induction on the length of a proof of A^r in J.

Now we handle the other part. Let us recall the 'normality' of realization introduced in [2]. A normal realization of a formula is one that assigns a variable to each negative occurrence of $\Box$. A realization of a sequent $S = \Gamma \Longrightarrow \Delta$ is defined by: $S^r = (fi(S))^r$. S^r can be also expressed by $\Gamma^r \Longrightarrow \Delta^r$. r for S is normal if S^r is normal.

Let P be a cut-free proof of S in K. In P, we restrict the initial sequent $A \Longrightarrow A$ to the case when A is an atomic formula. For an application of inference rule, an occurrence of $\Box$ in a upper sequent has the (obviously) related occurrence of $\Box$ in the lower sequent. Thus, all occurrences of $\Box$ form a 'forest' in P, where occurrences of $\Box$ are nodes and in-between inference rules are edges. Each occurrence of $\Box$ in the end-sequent is the 'root' of a 'tree'. All of the occurrences of $\Box$ belonging to a specific 'tree' have the same polarity. We call a tree which has positive occurrences of $\Box$ a *positive tree* and one which has negative occurrences of $\Box$ a *negative tree* in P.

We present the Realization Algorithm which assigns a term to each occurrence of $\Box$ in P so that each realized sequent is provable in J.

Realization Algorithm

(Step 1) Assign distinct variables to each negative tree in P, and replace all the nodes $\Box$ in a tree with the assigned variable.

(Step 2) Fix a positive tree in P. We proceed from top to bottom.

2.1. Assign a distinct variable for each leaf of the tree which is introduced by the rule '$\Box$'. Also, assign a uniform variable to all leaves of the tree which are introduced otherwise.

2.2. Keep on assigning the same term in each branch until another branch meets with it or the root is reached.

2.2.1. When two branches of the tree merge by 'c' (contraction) or logical rules, connect the two obtained terms by the operator $+$ and assign the new term to the next node. We take an example of the case when 'c' is involved.

$$\frac{B(\Box C), B(\Box C), \Gamma \Longrightarrow \Delta}{B(\Box C), \Gamma \Longrightarrow \Delta}\ c$$

Here, $\Box C$ occurs negatively in B and positively in the whole sequent. Suppose that one indicated occurrence of $\Box$ of $\Box C$ is replaced with $+(\boldsymbol{x})$ and the other is replaced with $+(\boldsymbol{y})$. Then, replace the related occurrence of $\Box$ in the lower sequent with $[+(\boldsymbol{x})] + [+(\boldsymbol{y})]$.

(Step 3) Update r by replacing variables x used in (Step 2) for the leaves of positive trees introduced by $\Box$ rule as follows.

$$\frac{B_1, \ldots, B_n \Longrightarrow C}{\Box B_1, \ldots, \Box B_n \Longrightarrow \Box C}\ \Box$$

Suppose that $(\Box B_1, \ldots, \Box B_n \Longrightarrow \Box C)^r$ has become $y_1 : B_1^r \wedge \cdots \wedge y_n : B_n^r \to x : C^r$ in (Step 1, 2). By Corollary 3.2, there is some $\cdot(\boldsymbol{d})$ such that if $B_1^r \wedge \cdots \wedge B_n^r. \to C^r$ is provable in J then so is $y_1 : B_1^r \wedge \cdots \wedge y_n : B_n^r \to (\cdot(\boldsymbol{d}) \cdot y_1 \cdot \cdots \cdot y_n) : C^r$. Update r so that $\cdot(\boldsymbol{d}) \cdot y_1 \cdot \cdots \cdot y_n$ is substituted for x.

(The end of the Realization Algorithm)

It is easily seen that this algorithm halts eventually. Also, it surely works correctly. We put the argument for the correctness in Appendix III.

In this way, we can obtain a normal realization of a formula provable in K such that the resulting formula is provable in J. This completes the proof of Theorem 4.2. $\Box$

We note the following point on the normal realization we have constructed from a cut-free proof in K in the proof of Theorem 4.2.

Note. We can take fresh constants for $\cdot(\boldsymbol{d})$ in (Step 3) for each application of rule $\Box$, because those constants are introduced from the rule R2 and we can choose any constant in applying R2. Thus, each leaf in a positive tree is realized to a term which does not share variables or constants with other leaves, and they can merge with the operator $+$ in 'c' or logical inferences, while in the original algorithm in [2], all the nodes in a positive tree have the same term.

Proof of Theorem 4.1. For the 'if' part, it is similarly proved to Theorem 4.2. For the 'only if' part. Suppose that A is provable in L. In light of the second Reduction Theorem (Theorem 2.7), there is a cut-free proof P in K of $X_1, \ldots, X_p \Longrightarrow A$. Here, $X(\alpha, n, \mathsf{L}) = X_1 \wedge \cdots \wedge X_p$ for some n and some set α composed of formulas in normal forms; each X_i is $X(\alpha, n, \mathsf{D})$, $X(\alpha, n, \mathsf{T})$, $X(\alpha, n, \mathsf{K4})$ or $X(\alpha, n, \mathsf{K5})$.

Fix any X_a. We impose the following condition.

($\natural\natural$) There is no application of $c : l$ (contraction on the left hand side) in P on any subformula of X_a.

We can transform P so that ($\natural\natural$) is satisfied; any such application of $c : l$ can be permuted with the following inference so that X_a may be duplicated in the end-sequent. We show this. Proceed from top to bottom. We distinguish cases by the inference below such an application of $c : l$, among which we pick up two cases: $\to : r$ and $\Box$. When it is $\to : r$, we can move the application of

$c : l$ to the right by permuting them, as follows.

$$\dfrac{\dfrac{A, A, \Gamma \Rightarrow \Delta, B}{A, \Gamma \Rightarrow \Delta, B}\, c : l}{\Gamma \Rightarrow \Delta, A \to B}\, \to: r \quad \rhd \quad \dfrac{\dfrac{\dfrac{\dfrac{A, A, \Gamma \Rightarrow \Delta, B}{A, \Gamma \Rightarrow \Delta, A \to B}\, \to: r}{A, \Gamma \Rightarrow \Delta, A \to B, B}\, w}{\Gamma \Rightarrow \Delta, A \to B, A \to B}\, \to: r}{\Gamma \Rightarrow \Delta, A \to B}\, c : r$$

When it is the rule $\Box$, we can exchange it with the application of $c : l$, as follows.

$$\dfrac{\dfrac{A, A, \Gamma \Rightarrow B}{A, \Gamma \Rightarrow B}\, c : l}{\Box A, \Box\Gamma \Rightarrow \Box B}\, \Box \quad \rhd \quad \dfrac{\dfrac{A, A, \Gamma \Rightarrow B}{\Box A, \Box A, \Box\Gamma \Rightarrow \Box B}\, \Box}{\Box A, \Box\Gamma \Rightarrow \Box B}\, c : l$$

After all, we may suppose that (♮♮) holds. Now we apply the Realization Algorithm to the proof P to obtain a proof, say P^*, in J of $X_1^r, \ldots, X_p^r \Longrightarrow A^r$ with some normal realization r. Fix any X_a. X_a has one of the forms:

$\Box^i(\Box B \to B)$;
$\Box^i(\Box B \to \Box\Box B)$;
$\Box^i(\neg\Box B \to \Box\neg\Box B)$.

Here, B is in normal form and can be $\bot$. Each form has two occurrences of a formula B such that the corresponding occurrences of $\Box$ have the opposite polarities inside B. Thus, the normal realization of them can be different: B^{r1} and B^{r2}. The realization of X_a is one of the forms:

$x_1 : \cdots x_i : (u : B^{r2} \to B^{r1})$;
$x_1 : \cdots x_i : (u : B^{r2} \to y_1 : y_2 : B^{r1})$;
$x_1 : \cdots x_i : (\neg y_1 : B^{r1} \to y_2 : \neg u : B^{r2})$.

Our first task is to reconcile B^{r1} and B^{r2} by rewriting terms in P^* and so updating the realization. If B contains no $\Box$, there is nothing to do here. Also, propositional variables are unimportant for the task. Thus, we may suppose that B is of the form:

$$\Box C_1 \wedge \cdots \wedge \Box C_n \to \Box D_1 \vee \cdots \vee \Box D_m$$

Here, each C_i and D_j are in normal form. Suppose that B^{r1} and B^{r2} are of the following forms.

$$B^{r1} = s_1 : C_1^r \wedge \cdots \wedge s_n : C_n^r \to z_1 : D_1^r \vee \cdots \vee z_m : D_m^r$$

$$B^{r2} = w_1 : C_1^r \wedge \cdots \wedge w_n : C_n^r \to t_1 : D_1^r \vee \cdots \vee t_m : D_m^r$$

By induction on $deg(B)$, we show that the realization can be so updated that (i) B^{r1} and B^{r2} are identical, and (ii) the realization of other parts of $X(\alpha, n, \mathsf{L}) \to A$ than X_a can change in such a way that only positive occurrences of a variable are replaced with a term.

As a result, the realization will be no longer normal. By the induction hypothesis, we assume that the realizations of each C_i and D_j in B^{r1} and B^{r2} are identical. We apply the following algorithm.

Rewriting Algorithm

(Step 1) For all $1 \le i \le n$, replace w_i in P^* with s_i. For all $1 \le j \le m$, let t_j^+ be a term obtained from t_j by this replacement.

(Step 2) For all $1 \le j \le m$, replace z_j in P^* with t_j^+.

(The end of Rewriting Algorithm)

Clearly, this algorithm halts eventually, as occurrences to be replaced are finite in each step and the number of those occurrences is reduced after each replacement. Also, the algorithm works correctly; we put the detailed argument in Appendix IV. Here, we note that (Step 1) and (Step 2) essentially reconcile the antecedent of B^{r1} and B^{r2} and the succedent of B^{r1} and B^{r2}, respectively, and (Step 2) does not change that antecedent anymore (as no s_i contains any z_j), thanks to the second Reduction Theorem and the property (♮♮). This is why we could avoid a circular argument in reconciling B^{r1} and B^{r2}. [7]

Also, note that (Step 1, 2) both take the form: for negative occurrences of variables, replace all (negative and positive) occurrences of them with some term. So, even if some preceding application of this Rewriting Algorithm to another conjunct altered some variables which occur only positively in the conjunct X_a^r under consideration it does not lose the applicability of the Rewriting Algorithm to X_a^r.

We have updated the realization r, which is not normal now, so that $X(\alpha, n, \mathsf{L})^r \to A^r$ is provable in J where each conjunct X_a^r of $X(\alpha, n, \mathsf{L})^r$ is of the following form.

$$x_1 : \cdots x_i : (u : B^r \to B^r);$$
$$x_1 : \cdots x_i : (u : B^r \to y_1 : y_2 : B^r);$$
$$x_1 : \cdots x_i : (\neg y_1 : B^r \to y_2 : \neg u : B^r).$$

Our remaining task is to make these forms provable in JL. First, for each conjunct X_a, there are outermost realized modalities, $x_1, \ldots, x_i$. Take fresh constants $c_1, \ldots, c_i$. For each $1 \le a \le i$, replace x_a in P^* with c_a. Then, we distinguish cases according to L. $X(\alpha, n, \mathsf{L}) - X(\alpha, n, \mathsf{L}_0)$ is a formula obtained from $X(\alpha, n, \mathsf{L})$ by removing $X(\alpha, n, \mathsf{L}_0)$. For systems L_0 and L_1, we write $\mathsf{L}_0 \subseteq \mathsf{L}_1$ to mean the latter extends the former.

(Case 1) When $\mathsf{D} \subseteq \mathsf{L}$, $\neg u : \bot$ is an axiom in JL. By R2, $c_1 : \cdots c_i : (\neg u : \bot)$ is provable in JL. Therefore, $\vdash_{\mathsf{JL}} X(\alpha, n, \mathsf{L})^r - X(\alpha, n, \mathsf{D})^r. \to A^r$.

(Case 2) When $\mathsf{T} \subseteq \mathsf{L}$, $u : B^r \to B^r$ is an axiom in JL, and, by R2, $\vdash_{\mathsf{JL}} c_1 : \cdots c_i : (u : B^r \to B^r)$. Therefore, $\vdash_{\mathsf{JL}} X(\alpha, n, \mathsf{L})^r - X(\alpha, n, \mathsf{T})^r. \to A^r$.

(Case 3) When $\mathsf{K4} \subseteq \mathsf{L}$, first replace y_1 and y_2 in P^* with u and $!u$, respectively. Then, $u : B^r \to !u : u : B^r$ is an axiom in JL and, by R2, $c_1 : \cdots c_i : (u : B^r \to !u : u : B^r)$ is provable in JL. Hence, $\vdash_{\mathsf{JL}} X(\alpha, n, \mathsf{L})^r - X(\alpha, n, \mathsf{K4})^r. \to A^r$.

(Case 4) When $\mathsf{K5} \subseteq \mathsf{L}$, first replace y_1 and y_2 in P^* with u and $?u$, respectively. Then, $\neg u : B^r \to ?u : \neg u : B^r$ is an axiom in JL. By applying R2, $\vdash_{\mathsf{JL}} c_1 :$

[7] The second Reduction Theorem and the property (♮♮) are actually based on the same idea: we can rule out positive occurrences of $\wedge$ in modal axioms without changing deductive power. (They are negative in the whole sequent.)

$\cdots c_i : (\neg u : B^r \rightarrow ?u : \neg u : B^r)$. Hence, $\vdash_{\mathsf{JL}} X(\alpha, n, \mathsf{L})^r - X(\alpha, n, \mathsf{K5})^r . \rightarrow A^r$.

The obtained figure is surely a proof in J, because all replacement we executed is so that variables are converted to terms. Each conjunct of $X(\alpha, n, \mathsf{L})^r$ is now of the following form.

$c_1 : \cdots c_i : (u : B^r \rightarrow B^r)$;
$c_1 : \cdots c_i : (u : B^r \rightarrow !u : u : B^r)$;
$c_1 : \cdots c_i : (\neg u : B^r \rightarrow ?u : \neg u : B^r)$.

In this way, we can eliminate every conjunct of $E(\alpha, n, \mathsf{L})^r$ in JL and we obtain the result of provability of A^r in JL. This finishes the proof of Theorem 4.1. □

The realization which we finally constructed is not normal. However, it is obtained from the normal realization we obtained through the Realization Algorithm by assigning terms to variables. Thus, positive occurrences of terms are still composed of negative occurrences of terms. In this sense, the final realization would keep a flavor of normality.

5 Conclusion Remark

In this paper, we offered the reduction theorems of modal logics to the system K, and we proved a basic fact that modal axioms can be restricted to a sort of normal form without changing their deductive power. Then, based on these results, we presented a uniform and modular proof of the realization of major modal logics in Justification Logics using a proof-theoretical method.

As further research problems, it would be intriguing to clarify a semantical meaning of the reduction theorems and normal form theorem (in terms of both possible-world and algebraic semantics.) Also it would be interesting to investgate the extension of the theorems to second-order modal logics. Moreover, it would be intriguing to ask how far our method can be generalized, as recently it turned out in Fitting [16] that there exist infinitely many modal logics that have counterparts in Justification Logic. [8]

Appendix

I. *Example of reduction of proof.*

Here we sketch an example of reduction (as in Theorem 2.3) of proof in KD5 to that in K. Let us consider the formula $\Box(\neg P \vee \neg\Box Q) \rightarrow \Box\Box\neg\Box(P \wedge Q)$ provable in KD5. We permit (i) putting hypotheses in a proof, where of course we cannot apply Necessitation to a formula depending on hypotheses and (ii) applying an inference rule introducing '$\rightarrow$' discharging some hypotheses, a conjunction of which occur as the antecedent of the introduced '$\rightarrow$'. This relaxation is justifiable in the standard axiomatic system for propositional logic and, therefore, our system K. Here is a sketch of its proof in KD5.

[8] The question concerning algebraic model was suggested to me by an anonymous referee.

$$
\begin{array}{ll}
1. & \Box(\neg P \vee \neg\Box Q) \quad \textit{Hypothesis} \\
 & \vdots \\
n_1. & \neg\Box Q \to \Box\neg\Box Q \quad \textit{Axiom 5} \\
 & \vdots \\
n_2. & \Box(\neg P \vee \Box\neg\Box Q) \\
 & \vdots \\
n_3. & \neg\Box\neg\Box Q \to \Box\neg\Box\neg\Box Q \quad \textit{Axiom 5} \\
 & \vdots \\
n_4. & \Box(\Box\neg P \vee \Box\neg\Box Q) \\
 & \vdots \\
n_5. & \neg\Box\bot \quad \textit{Axiom D} \\
 & \vdots \\
n_6. & \Box(\neg\Box P \vee \Box\neg\Box Q) \\
 & \vdots \\
n_7. & \neg\Box P \to \Box\neg\Box P \quad \textit{Axiom 5} \\
 & \vdots \\
n_8. & \Box(\Box\neg\Box P \vee \Box\neg\Box Q) \\
 & \vdots \\
n_9. & \Box\Box\neg\Box(P \wedge Q) \\
n_{10}. & \Box(\neg P \vee \neg\Box Q) \to \Box\Box\neg\Box(P \wedge Q) \quad 1
\end{array}
$$

Then, we can convert this proof to the following proof in K.

$$
\begin{array}{ll}
1. & [\Box(\neg P \vee \neg\Box Q)] \quad \textit{Hypothesis} \\
 & \vdots \\
n_1. & \Box(\neg\Box Q \to \Box\neg\Box Q) \quad \textit{Hypothesis} \\
 & \vdots \\
n_2. & \Box(\neg P \vee \Box\neg\Box Q) \\
 & \vdots \\
n_3. & \Box(\neg\Box\neg\Box Q \to \Box\neg\Box\neg\Box Q) \quad \textit{Hypothesis} \\
 & \vdots \\
n_4. & \Box(\Box\neg P \vee \Box\neg\Box Q) \\
 & \vdots \\
n_5. & \Box\neg\Box\bot \quad \textit{Hypothesis} \\
 & \vdots \\
n_6. & \Box(\neg\Box P \vee \Box\neg\Box Q) \\
 & \vdots \\
n_7. & \Box(\neg\Box P \to \Box\neg\Box P) \quad \textit{Hypothesis} \\
 & \vdots \\
n_8. & \Box(\Box\neg\Box P \vee \Box\neg\Box Q) \\
 & \vdots \\
n_9. & \Box\Box\neg\Box(P \wedge Q) \\
n_{10}. & \Box(\neg P \vee \neg\Box Q) \to \Box\Box\neg\Box(P \wedge Q) \quad 1 \\
n_{11}. & X(\alpha, 1, \mathsf{KD5})^- \to .\Box(\neg P \vee \neg\Box Q) \to \Box\Box\neg\Box(P \wedge Q) \quad n_1, n_3, n_5, n_7 \\
 & \vdots \\
n_{12}. & X(\alpha, 1, \mathsf{KD5}) \to .\Box(\neg P \vee \neg\Box Q) \to \Box\Box\neg\Box(P \wedge Q)
\end{array}
$$

Here, $X(\alpha, 1, \mathsf{KD5})^-$ is a conjunction of the formulas n_1, n_3, n_5, n_7 and

$\alpha = \{\Box Q, \Box\neg\Box Q, \Box\bot, \Box P\}$. Note that $X(\alpha, 1, \mathsf{KD5}) = X(\alpha, 0, \mathsf{KD5}) \wedge X(\alpha, 1, \mathsf{KD5})^{-}$.

II. *Proof of Corollary 3.2.*

Suppose $\vdash_{\mathsf{J}} B_1 \wedge B_2 \wedge \cdots \wedge B_n. \to A$. Then, $\vdash_{\mathsf{J}} B_1 \to (B_2 \to \cdots (B_n \to A)\cdots)$. By Theorem 3.1, for some $\cdot(\mathbf{c})$, $\vdash_{\mathsf{J}} \cdot(\mathbf{c}) : [B_1 \to (B_2 \to \cdots (B_n \to A)\cdots)]$ such that $at(\cdot(\mathbf{c})) \cap at(A) = \emptyset$. We work in J and by induction on n. Suppose that we obtain:

$$t_1 : B_1 \to .t_2 : B_2 \to . \cdots t_i : B_i \to \\ (\cdot(\mathbf{c}) \cdot t_1 \cdot t_2 \cdot \cdots \cdot t_i) : [B_{i+1} \to (\cdots (B_n \to A)\cdots)].$$

The following is an axiom from A2.

$$(\cdot(\mathbf{c}) \cdot t_1 \cdot t_2 \cdot \cdots \cdot t_i) : [B_{i+1} \to (\cdots (B_n \to A)\cdots)] \to . \\ t_{i+1} : B_{i+1} \to (\cdot(\mathbf{c}) \cdot t_1 \cdot t_2 \cdot \cdots \cdot t_i \cdot t_{i+1}) : [B_{i+2} \to (\cdots (B_n \to A)\cdots)]$$

Then, by propositional calculus with the last two formulas, we obtain:

$$t_1 : B_1 \to .t_2 : B_2 \to . \cdots t_i : B_i \to .t_{i+1} : B_{i+1} \to \\ (\cdot(\mathbf{c}) \cdot t_1 \cdot t_2 \cdot \cdots \cdot t_i \cdot t_{i+1}) : [B_{i+2} \to (\cdots (B_n \to A)\cdots)].$$

Thus, we have:

$$t_1 : B_1 \to .t_2 : B_2 \to . \cdots t_n : B_n \to (\cdot(\mathbf{c}) \cdot t_1 \cdot t_2 \cdot \cdots \cdot t_n) : A.$$

and, by propositional calculus,

$$t_1 : B_1 \wedge t_2 : B_2 \wedge \cdots \wedge t_n : B_n. \to (\cdot(\mathbf{c}) \cdot t_1 \cdot t_2 \cdot \cdots \cdot t_n) : A.$$

Here, the desired property on terms is preserved.

III. *Argument for the correctness of the Realization Algorithm.*

We verify the correctness of the algorithm: every realized sequent obtained in there is provable in J. We proceed from top to bottom in P. For an initial sequent S of the form $A \Longrightarrow A$ or $\bot, \Gamma \Rightarrow \Delta$, S^r is an axiom of J. It is easily checked that for every application of a rule, if the realizations of the upper sequents are provable in J then so is that of the lower sequent, except the case when two branches of a positive tree merge via 'c' or logical inferences. These cases are similarly treated. We handle the case of $\wedge : r$ here.

$$\frac{B(\Box C), \Gamma \Longrightarrow \Delta, D \quad B(\Box C), \Gamma \Longrightarrow \Delta, E}{B(\Box C), \Gamma \Longrightarrow \Delta, D \wedge E} \wedge : r$$

Here, $\Box C$ occurs negatively in B and positively in the whole sequent. Suppose that the upper sequent is realized as follows.

$$B^r(s : C^r), \Gamma^r \Longrightarrow \Delta^r, D^r \qquad B^r(t : C^r), \Gamma^r \Longrightarrow \Delta^r, E^r$$

It is easily seen that we can replace the related occurrences of s (the indicated term of $s : C^r$), corresponding to some branches of the positive tree in P, with $s + t$, keeping all the inferences in J; and we can do the same for the related occurrences of t (the indicated term of $t : C^r$), corresponding to other branches of the positive tree in P. Then we obtain a proof of the realized sequent

$$B^r((s+t):C^r), \Gamma^r \Longrightarrow \Delta^r, D^r \qquad B^r((s+t):C^r), \Gamma^r \Longrightarrow \Delta^r, E^r$$

Thus, we can know the realization of the lower sequent of $\wedge : r$ is provable by propositional inferences (corresponding to $\wedge : r$) in J.

IV. *Argument for correctness of the Rewriting Algorithm.*

We consider only the case of the conjunct $X_a = \Box^i(\Box B_{(2)} \to B_{(1)})$. The other cases can be treated similarly. Let $B_{(1)}$ and $B_{(2)}$ be occurrences of B which have become B^{r1} and B^{r2} by the realization, respectively. In the proof P in K, there can be some applications of $\to: l$ to introduce $\Box B_{(2)} \to B_{(1)}$.

$$\begin{array}{cc} P_1 & P_2 \\ \vdots & \vdots \end{array}$$

$$\dfrac{\Gamma \Longrightarrow \Delta, \Box B_{(2)} \quad B_{(1)}, \Gamma \Longrightarrow \Delta}{\Box B_{(2)} \to B_{(1)}, \Gamma \Longrightarrow \Delta} \to: l$$

$$\vdots$$

$$\Box^i(\Box B_{(2)} \to B_{(1)}), X^-(\alpha, n, \mathsf{L}) \Longrightarrow A$$

In the subproof P_1, there can be applications of $\to: r$ introducing $B_{(2)}$.

$$\vdots$$

$$\dfrac{\Box C_1 \wedge \cdots \wedge \Box C_n, \Sigma \Longrightarrow \Theta, \Box D_1 \vee \cdots \vee \Box D_m}{\Sigma \Longrightarrow \Theta, \Box C_1 \wedge \cdots \wedge \Box C_n \to \Box D_1 \vee \cdots \vee \Box D_m} \to: r$$

$$\vdots$$

$$\Gamma \Longrightarrow \Delta, \Box B_{(2)}$$

By the Realization Algorithm, the principal formula of such an application of $\to: r$ becomes of the form:

$$w_1 : C_1^r \wedge \cdots \wedge w_n : C_n^r \to t'_1 : D_1^r \vee \cdots \vee t'_m : D_m^r$$

Here, each t'_i is a subterm of the corresponding t_i in B^{r2}; they become identical when more $+$ are added.

Fix any w_i $(1 \leq i \leq n)$. We show w_i does not appear in any C_j^r or D_k^r.

There can be several occurrences of $\Box C_i$ the $\Box$ of which belongs to the same negative tree and is realized to w_i. Since they must be contracted eventually in P, it cannot appear in any C_j or D_k.

Also, there can be occurrences of $\Box E$ which is the right principal formula of applications of $\Box : r$ having a left principal formula $\Box C_i$ and is realized to w_i. $\Box E$ may be some $\Box D_k$. Two formulas (C_j and $\Box E$) ($j = i$ or $j \neq i$) and (D_k and $\Box E$) are never contracted in cut-free P. This is because to contract such two, C_j of $\Box C_j$ or D_k of $\Box D_k$ must have $\Box E$ as a subformula and $\Box E$ should have one more $\Box$ outside itself. However, since the rule $\Box$ increases the number of $\Box$ by one for each auxiliary formula, $\Box C_j$ or $\Box D_k$ would have also one more $\Box$ outside. Thus, such two formulas are never contracted. [9]

Hence, firstly, the realization inside each C_i and D_k does not use any variable w_i $(1 \leq i \leq n)$, and (Step 1) does not change the realization of any C_i or D_k. [10]

Next, the root of the negative tree which is associated with w_i appears inside X_a and is obviously the only negative occurrence of w_i in the end-sequent, while the roots of the positive trees associated with terms containing w_i can appear inside or outside $X_a = \Box^i(\Box B_{(2)} \to B_{(1)})$ in the end-sequent. So, secondly, for other part of $X(\alpha, n, \mathsf{L}) \to A$ than X_a, (Step 1) can replace only the positive occurrences of variables of $w_i (1 \leq i \leq n)$.

Thus, the execution of (Step 1) guarantees that the antecedents of B^{r1} and B^{r2} become identical and satisfies the desired property (ii).

Concerning (Step 2), we turn to look at P_2. In the subproof P_2, there can be applications of $\to: l$ introducing $B_{(1)}$.

$$\vdots$$

$$\dfrac{\Sigma \Longrightarrow \Theta, \Box C_1 \wedge \cdots \wedge \Box C_n \qquad \Box D_1 \vee \cdots \vee \Box D_m, \Sigma \Longrightarrow \Theta}{\Box C_1 \wedge \cdots \wedge \Box C_n \to \Box D_1 \vee \cdots \vee \Box D_m, \Sigma \Longrightarrow \Theta} \to: l$$

$$\vdots$$

$$B_{(1)}, \Gamma \Longrightarrow \Delta$$

By the Realization Algorithm, the principal formula of such an application of $\to: l$ becomes of the form:

$$s'_1 : C^r_1 \wedge \cdots \wedge s'_n : C^r_n \to z_1 : D^r_1 \vee \cdots \vee z_m : D^r_m$$

---

[9] This is formally proved by induction on the number of applications of rules between each $\Box$ rule which introduce some $\Box C_j$ and the end-sequent.

[10] In other words, we do not have a self-referential realization on any $\Box C_i$ and $\Box D_k$. Generally, this kind of self-reference phenomenon can be avoided in realization of the modal logic K and D, which was shown in Kuznets [24]. Here, we proved the possibility to avoid self-referentiality for a specific form of formulas in a cut-free proof in K.

Here, each s'_i is a subterm of the corresponding one in B^{r1}; they become identical when more $+$ are added.

Fix any z_i $(1 \leq i \leq m)$. There can be applications of $\Box$-rule which have $\Box D_i$ as a left principal formula. Let $\Box E$ be the right principal formula of any such application of $\Box$-rule. By a similar argument to P_1, $\Box E$ cannot merge with any C_j or D_k. So, the realization does not use any z_i there, and (Step 2) does not change any C_j^r or D_k^r.

Moreover, in the subproof above the left upper sequent of the application of $\rightarrow$: l, there is no such application of $\Box$-rule. Because: if there is, $\Box D_i$ appears in the lower sequent of it, it must be contracted below the $\rightarrow$: l with the occurrence of $\Box D_i$ in the right upper sequent of the $\rightarrow$: l, but it contradicts ($\natural\natural$). Therefore, $\Box E$ never merges with any $\Box C_i$. (So, what cannot merge with $\Box E$ is not only C_i but $\Box C_i$.) Hence, no term of s_j $(1 \leq j \leq n)$ contains any z_i, and (Step 2) does not change any s_j. This guarantees that (Step 2) does not change the outcome of (Step 1), and we obtain non circularity of the Rewriting Algorithm. [11]

Finally, by a similar argument to P_1, for other part of $X(\alpha, n, \mathsf{L}) \rightarrow A$ than X_a, (Step 2) can replace only the positive occurrences of variables of $z_i (1 \leq i \leq n)$.

Thus, the execution of (Step 2) guarantees that the succedents of B^{r1} and B^{r2} become identical, the antecedents remain untouched, and satisfies the desired property (ii). Note that after each step of the rewriting process, the obtained figure is surely a proof in J.

References

[1] Artemov S., Operational modal logic, Technical Report MSI 95-29, Cornell University, 1995.

[2] Artemov S., Explicit provability and constructive semantics, The Bulletin of Symbolic Logic, 7(1), 2001, pp.1-36.

[3] Artemov, S. and Fitting, M.: Justification Logic, The Stanford Encyclopedia of Philosophy, 2012.

[4] Artemov, S. and Fitting, M.: Justification Logic: Reasoning with Reasons, Cambridge University Press, 2019.

[5] Artemov, S., E. Kazakov, and D. Shapiro. On logic of knowledge with justifications. Technical Report CFIS 99-12, Cornell University, 1999.

[6] Artemov, S., and R. Kuznets. "Logical omniscience as infeasibility." Annals of pure and applied logic 165, no. 1 (2014): 6-25.

[11] We can adopt another way to avoid a circular argument, instead of further transforming proof to satisfy ($\natural\natural$); we introduce a new operation on terms, say '$*$', and axioms: $s * t : A \rightarrow s : A$ and $s * t : A \rightarrow t : A$. ('$s * t : A$' has a natural meaning, 'both s and t are a justification for A'.) Then, in the Realization Algorithm, we can make the operation $*$ play the same role in negative trees as $+$ does in positive trees. More concretely, for a negative tree, put a distinct variable to each leaf, and use $*$ where two branches meet to concatenate the two terms. Then, no s_i does not contain any z_j because z_j never occurs in the subproof above the application of $\Box$-rule whose right principal formula is realized with s_i. Finally, we can eliminate all occurrences of the operation $*$ by assigning a fresh variable to each negative tree.

[7] Borg A., Kuznets R. (2015) Realization Theorems for Justification Logics: Full Modularity. In: De Nivelle H. (eds) Automated Reasoning with Analytic Tableaux and Related Methods. TABLEAUX 2015. Lecture Notes in Computer Science, vol 9323. Springer, Cham

[8] Brezhnev V., On explicit counterparts of modal logics, Technical Report CFIS 2000-05, Cornell University, 2000.

[9] Brezhnev, V. and R. Kuznets, Making knowledge explicit: How hard it is, Theoretical Computer Science, 357, pp. 23-34, 2006.

[10] Brünnler, K., R. Goetschi, R. Kuznets, A syntactic realization theorem for justification logics, in: L.D. Beklemishev, V. Goranko, V. Shehtman (Eds.), Advances in Modal Logic, vol. 8, College Publications, 2010, pp. 39–58.

[11] Fitting, M., Subformula results in some propositional modal logics. Studia Logica 37, 387-391 (1978) doi: 10.1007/BF02176170

[12] Fitting, M., The logic of proofs, semantically, Ann. Pure Appl. Logic 132 (1) (2005) 1-25.

[13] Fitting, M., Justification logics, logics of knowledge, and conservativity. Ann Math Artif Intell 53, 153-167 (2008)

[14] Fitting M. (2011) The Realization Theorem for S5: A Simple, Constructive Proof. In: van Benthem J., Gupta A., Pacuit E. (eds) Games, Norms and Reasons. Synthese Library (Studies in Epistemology, Logic, Methodology, and Philosophy of Science), vol 353. Springer, Dordrecht.

[15] Fitting M., Realization using the model existence theorem, Journal of Logic and Computation, Volume 26, Issue 1, February 2016, Pages 213–234.

[16] Fitting M., Modal logics, justification logics, and realization, Annals of Pure and Applied Logic, Volume 167, Issue 8, August 2016, Pages 615-648.

[17] Goetschi, R., Kuznets, R., Realization for justification logics via nested sequents: Modularity through embedding. Annals of Pure and Applied Logic 163(9), pp. 1271-1298, 2012.

[18] Kazakov, E. L., Logics of Proofs for S5, M.Phil. thesis, Lomonosov Moscow State University, Faculty of Mechanics and Mathematics, in Russian, 1999.

[19] Kurokawa, H. and H. Kushida, Substructural Logic of Proofs. In: Libkin L., Kohlenbach U., de Queiroz R. (eds) Logic, Language, Information, and Computation. WoLLIC 2013. Lecture Notes in Computer Science, vol 8071. Springer, Berlin, Heidelberg, 2013.

[20] Kurokawa, H. and H. Kushida, Resource Sharing Linear Logic, Journal of Logic and Computation, vol.30(1), pp 281-294, 2020.

[21] Kushida H., Applicability of Motohashi's Method to Modal Logics, Bulletin of the Section of Logic, Volume 34/3, 2005, pp. 121–134.

[22] Kushida, H., On the realization of the provability of logic, manuscript, 2007.

[23] Kushida, H., A Proof Theory for the Logic of Provability in True Arithmetic, to appear in: Studia Logica, 2019.

[24] Kuznets R., Self-Referential Justifications in Epistemic Logic, Journal of Philosophical Logic, 39, pp. 577-590, 2010.

[25] Kuznets, R. and T. Studer, Logics of Proofs and Justifications, College Publications, 2019

[26] Motohashi, N., A faithful interpretation of intuitionistic predicate logic in classical predicate logic, Commentarii Mathematici Universitatis Sancti Pauli 21 (1972), pp. 11–23.

[27] Pacuit E., A note on some explicit modal logics. In *Proceedings of the 5th Panhellenic Logic Symposium*, Athens, 2005.

[28] Rubtsova, N., Evidence reconstruction of epistemic modal logic S5. Lecture Notes in Computer Science, vol. 3967, pp. 313–321, 2006.

[29] Rubtsova, N., On Realization of graphic-modality by Evidence Terms, Journal of Logic and Computation, Volume 16, Issue 5, October 2006, Pages 671–684,

[30] Takeuti G., *Proof Theory,* Second edition, North-Holland, Amsterdam, 1987.

The 'Long Rule' in the Lambek Calculus with Iteration: Undecidability without Meets and Joins

Stepan Kuznetsov [1]

Steklov Mathematical Institute of RAS
8 Gubkina St., Moscow 119991, Russia

Abstract

We consider the Lambek calculus extended with positive iteration as a unary connective. The choice of positive iteration, not Kleene star, is dictated by Lambek's antecedent non-emptiness restriction. Usually iteration is axiomatized either by an inductive schema or by an ω-rule. We consider an intermediate system with a rule which we call the 'long rule,' which reduces iteration of A to explicit treatment of powers of A up to the k-th one, and reusing iteration in the form $A^k \cdot A^+$. In the presence of additive disjunction (union), the 'long rule' is easily derivable. For the 'pure' Lambek calculus without additives this is not the case. For the system with the 'long rule' we prove undecidability. We also investigate connections of this system with the standard inductive-style one.

Keywords: Lambek calculus, iteration, undecidability

1 Introduction

Iteration, or Kleene star, is one of the most basic and at the same time one of the most intriguing algebraic operations appearing in theoretical computer science. Following the line of work by Pratt [23] and Kozen [12], we consider substructural (algebraic) non-commutative logics with two implications (divisions) and iteration as a modality (cf. [24, § 9.5]). The idea of division operations we consider throughout this paper goes back to Krull [14]. From the logical point of view divisions were introduced in the Lambek calculus [19]. The Lambek calculus is a non-commutative intuitionistic variant of Girard's linear logic [7], in the multiplicative-only language (see Abrusci [1]). Thus, the system we are going to consider is *the Lambek calculus* (or non-commutative intuitonistic multiplicative-only linear logic) *extended with iteration.*

[1] sk@mi-ras.ru
This work was performed at the Steklov International Mathematical Center and supported by the Ministry of Science and Higher Education of the Russian Federation (agreement no. 075-15-2019-1614).

Action logic, denoted by **ACT** and introduced by Pratt and Kozen, includes, besides divisions and iteration, also lattice operations: join and meet. Thus, action logic can be viewed as an extension of the multiplicative-additive ('full') Lambek calculus. Following the standard definition of Kleene star as a fixed point, Pratt axiomatizes it using an induction axiom ('pure induction'). In contrast, later works of Buszkowski and Palka [3,21,5] feature a stronger system called *infinitary action logic,* $\mathbf{ACT}_\omega$, with an ω-rule for Kleene star. Buszkowski and Palka show that $\mathbf{ACT}_\omega$ is Π_1^0-complete. Thus, it is undecidable and strictly stronger than action logic with induction axioms/rules of any kind. (As noticed by the author in [16], there exist variations of induction rules which yield systems which are strictly between **ACT** and $\mathbf{ACT}_\omega$.) The question of decidability for **ACT**, posed by Kozen in 1994, was recently solved, by the author of this paper, negatively [18]. This undecidability result applies to the whole range of systems between **ACT** and $\mathbf{ACT}_\omega$. Moreover, its modification [17] gives Σ_1^0-completeness for any logic in this range, provided it is recursively enumerable.

This paper continues the line of [18] and [17]. We now focus on the extension of the Lambek calculus with iteration, but *without* join and meet. Another distinctive feature of the system considered here is the so-called *Lambek's non-emptiness restiction.* Algebraically it means that we allow models without the unit. Lambek's restriction was originally motivated by linguistic applications of the Lambek calculus (see [20, § 2.5]). Here it will help to simplify some of the technicalities in the proofs. We conjecture that our results will also be valid without Lambek's restriction. However, we do not yet claim this, since some technicalities depend on the non-emptiness restriction.

In the presence of Lambek's restriction, we cannot introduce Kleene star itself: one of the axioms for Kleene star, $\mathbf{1} \vdash A^*$, includes the unit (empty antecedent). Instead, we introduce *positive iteration,* A^+. Interestingly, in his pioneering work [10], Kleene himself also avoided using the unit ('empty event') and introduced a binary iteration operation $A * B$, which means $A^* \cdot B$ ("several times A, then B"). In Kleene's notation, A^+ is $A * A$.

2 Preliminaries

Let us formally introduce the Lambek calculus with positive iteration, denoted by $\mathbf{L}^+$. Formulae of $\mathbf{L}^+$ are built from variables using three binary connectives: $\cdot$ (product), $\backslash$ (left division), $/$ (right division), and one unary connective: $^+$ (positive iteration). We formulate $\mathbf{L}^+$ as a sequent calculus, though cut is, unfortunately, not going to be eliminable. Sequents of $\mathbf{L}^+$ are expressions of the form $A_1, \ldots, A_n \vdash B$, where $A_1, \ldots, A_n, B$ are formulae, $n \geq 1$ (empty antecedents are disallowed). Formulae are denoted by capital Latin letters; capital Greek letters stand for sequences of formulae, possibly empty.

The core of $\mathbf{L}^+$ is the Lambek calculus $\mathbf{L}$, with the following axioms and rules of inference:

$$\overline{A \vdash A}$$

$$\frac{\Pi \vdash A \quad \Gamma, B, \Delta \vdash C}{\Gamma, \Pi, A \setminus B, \Delta \vdash C} \qquad \frac{A, \Pi \vdash B}{\Pi \vdash A \setminus B}, \text{ where } \Pi \text{ is non-empty}$$

$$\frac{\Pi \vdash A \quad \Gamma, B, \Delta \vdash C}{\Gamma, B \,/\, A, \Pi, \Delta \vdash C} \qquad \frac{\Pi, A \vdash B}{\Pi \vdash B \,/\, A}, \text{ where } \Pi \text{ is non-empty}$$

$$\frac{\Gamma, A, B, \Delta \vdash C}{\Gamma, A \cdot B, \Delta \vdash C} \qquad \frac{\Pi \vdash A \quad \Delta \vdash B}{\Pi, \Delta \vdash A \cdot B} \qquad \frac{\Pi \vdash A \quad \Gamma, A, \Delta \vdash C}{\Gamma, \Pi, \Delta \vdash C}\ (cut)$$

Axioms and rules for iteration reflect the idea that, algebraically, a^+ should be the least (that is, the strongest) b such that $a \vdash b$ and $a \cdot b \vdash b$:

$$\frac{}{A \vdash A^+} \qquad \frac{}{A, A^+ \vdash A^+} \qquad \frac{A \vdash B \quad A, B \vdash B}{A^+ \vdash B}$$

As one can see, iteration here is axiomatized in a non-sequential style; thus, cut is not eliminable in $\mathbf{L}^+$. Unfortunately, no cut-free sequential version for the inductive axiomatization of iteration is known, in the presence of divisions. Unsuccessful attempts were taken by Jipsen [9] and Pentus [22]. For the logic of Kleene algebras, without division (but with join), a cut-free circular hypersequential system was constructed by Das and Pous [6]. This became possible, because for Kleene algebras the inductively axiomatized logic is complete, that is, admits the ω-rule. For systems with division operations, this is not the case due to complexity reasons (a recursively enumerable set of sequents could not coincide with a Π_1^0-hard one).

As shown by Pratt [23], in the presence of division operations left iteration is also right. This means that the following axiom and rule are derivable in $\mathbf{L}^+$:

$$\frac{}{A^+, A \vdash A^+} \qquad \frac{A \vdash B \quad B, A \vdash B}{A^+ \vdash B}$$

A stronger version of $\mathbf{L}^+$ is obtained by introducing the ω-rule for iteration:

$$\frac{\Gamma, A, \Delta \vdash B \quad \Gamma, A, A, \Delta \vdash B \quad \Gamma, A, A, A, \Delta \vdash B \quad \ldots}{\Gamma, A^+, \Delta \vdash B}$$

Axioms for iteration can also be reformulated in a sequential style:

$$\frac{\Gamma_1 \vdash A \quad \ldots \quad \Gamma_n \vdash A}{\Gamma_1, \ldots, \Gamma_n \vdash A^+}\ (n \geq 1)$$

and in this infinitary system, denoted by $\mathbf{L}^+_\omega$, cut is eliminable. This is essentially due to Palka [21], with necessary modifications connected with Lambek's restriction.

Adding join ($\vee$) and meet ($\wedge$) with the following rules:

$$\frac{\Gamma, A_1, \Delta \vdash C \quad \Gamma, A_2, \Delta \vdash C}{\Gamma, A_1 \vee A_2, \Delta \vdash C} \qquad \frac{\Pi \vdash A_i}{\Pi \vdash A_1 \vee A_2}\ (i = 1, 2)$$

$$\frac{\Gamma, A_i, \Delta \vdash C}{\Gamma, A_1 \wedge A_2, \Delta \vdash C}\ (i = 1, 2) \qquad \frac{\Pi \vdash A_1 \quad \Pi \vdash A_2}{\Pi \vdash A_1 \wedge A_2}$$

to $\mathbf{L}^+$ and $\mathbf{L}^+_\omega$ yields $\mathbf{ACT}^+$ and $\mathbf{ACT}^+_\omega$ respectively. These are positive variants of ordinary and infinitary action logic. Complexity results for $\mathbf{ACT}^+$ and $\mathbf{ACT}^+_\omega$ can be proved by slight modifications of the proofs for systems without Lambek's restriction and with Kleene star instead of positive iteration. Thus, due to Buszkowski [3] and Palka [21] $\mathbf{ACT}^+_\omega$ is Π^0_1-complete; $\mathbf{ACT}^+$ is undecidable [18] (more precisely, Σ^0_1-complete [17]).

In the infinitary case, Buszkowski's Π^0_1-hardness result can be strengthened: $\mathbf{L}^+_\omega$, the system without join and meet, is already Π^0_1-hard [15]. In this paper, we investigate the possibility of performing a similar strengthening of the undecidability result for $\mathbf{ACT}^+$ [18] to $\mathbf{L}^+$. Namely, we prove undecidability not for $\mathbf{L}^+$ itself, but for a system very closely related to $\mathbf{L}^+$.

An important component of the undecidability proof for $\mathbf{ACT}^+$ is the so-called 'long rule' [18], formulated as follows:

$$\frac{A \vdash B \quad A, A \vdash B \quad \ldots \quad A^k \vdash B \quad A^k, A^+ \vdash B}{A^+ \vdash B}$$

Actually, this is a series of rules parametrized by k. In the presence of $\vee$, this rule can be easily derived, for any k, using cut with $A^+ \vdash A \vee A^2 \vee \ldots \vee A^k \vee (A^k \cdot A)$. This can be also performed without $\vee$, but with $\wedge$ and division operations [17]. Notice that the 'long rule' itself includes neither $\vee$, nor $\wedge$, but its derivation in $\mathbf{ACT}^+$ requires one of these connectives.

By $\mathbf{L}^+_\ell$ we denote $\mathbf{L}^+$ with the 'long rule' added as a rule of inference. More precisely, we include instances of the 'long rule' for each k.

The rest of this paper is organized as follows. In Section 3, we prove undecidability of $\mathbf{L}^+_\ell$. In Section 4, we show that, unlike $\mathbf{ACT}^+$ and $\mathbf{L}^+_\omega$, in $\mathbf{L}^+$ the 'long rule' is not *derivable*. The question whether a weaker property, *admissibility* of the 'long rule' in $\mathbf{L}^+$, holds is left open. Section 5 includes some concluding remarks and speculations.

We conclude this section by showing that the 'long rule' is derivable in $\mathbf{L}^+_\omega$ and presenting a contextified (sequent-style) version of the 'long rule.'

Lemma 2.1 *The 'long rule' is derivable in* $\mathbf{L}^+_\omega$.

Proof. In $\mathbf{L}^+_\omega$, one can easily derive $A^n \vdash A^+$ for any $n \geq 1$ (just use the right rule for iteration with $\Gamma_1 = \ldots = \Gamma_n = A$).

Now, given the premises of the 'long rule,' let us establish $A^m \vdash B$ for any $m \geq 1$. Indeed, if $m \leq k$, this sequent is explicitly given. If $m > k$, then we use cut:

$$\frac{A^{m-k} \vdash A^+ \quad A^k, A^+ \vdash B}{A^m \vdash B}\ (cut)$$

Now $A^+ \vdash B$ is derived by the ω-rule. □

Lemma 2.2 *The following 'sequential version' of the 'long rule' is derivable in* $\mathbf{L}_\ell^+$*:*

$$\frac{\Gamma, A, \Delta \vdash B \quad \Gamma, A, A, \Delta \vdash B \quad \dots \quad \Gamma, A^k, \Delta \vdash B \quad \Gamma, A^k, A^+, \Delta \vdash B}{\Gamma, A^+, \Delta \vdash B}$$

Proof. If $\Gamma = G_1, \dots, G_s$, let $\bullet\Gamma = G_1 \cdot \ldots \cdot G_s$; similarly for $\bullet\Delta$. Now $\Gamma, A^+, \Delta \vdash B$ is derived by cut from $A^+ \vdash \bullet\Gamma \setminus B / \bullet\Delta$ and $\Gamma, \bullet\Gamma \setminus B / \bullet\Delta, \Delta \vdash B$. The latter is derivable in $\mathbf{L}$; the derivation for the former is by the 'long rule':

$$\frac{\dfrac{\Gamma, A, \Delta \vdash B}{A \vdash \bullet\Gamma \setminus B / \bullet\Delta} \quad \dots \quad \dfrac{\Gamma, A^k, \Delta \vdash B}{A^k \vdash \bullet\Gamma \setminus B / \bullet\Delta} \quad \dfrac{\Gamma, A^k, A^+, \Delta \vdash B}{A^k, A^+ \vdash \bullet\Gamma \setminus B / \bullet\Delta}}{A^+ \vdash \bullet\Gamma \setminus B / \bullet\Delta}$$

□

3 Undecidability of $\mathbf{L}_\ell^+$

Theorem 3.1 *The derivability problem in* $\mathbf{L}_\ell^+$ *is undecidable.*

The proof of Theorem 3.1 combines ideas of the undecidability proof for **ACT** from [18] and the Π_1^0-hardness proof for $\mathbf{L}_\omega^+$ from [15].

First we encode several kinds of Turing machine behaviour via totality-like properties of context-free grammars. Then we follow the idea of Buszkowski [3] and embed these grammars into the Lambek environment. However, instead of the standard embedding (which goes back to Gaifman [2]) we use Safiullin's [25] construction of Lambek grammars with unique type assignment.

We consider only deterministic Turing machines, and suppose that each Turing machine has a designated *cycling state* q_c in which it gets stuck. (Rules for q_c are as follows: $\langle q_c, a\rangle \to \langle q_c, a, N\rangle$ for any letter a of the inner alphabet; N stands for "no move.") The cycling state can be added to any Turing machine, even if it is not necessary: in this case it can be just made unreachable.

Following the standard way (see [13, Lect. 35]), we encode a configuration of our Turing machine as $b_1 \dots b_{i-1} q b_i b_{i+1} \dots b_n$, if the machine is in state q, observing the i-th letter of the word $b_1 \dots b_n$ in its memory. Protocols are sequences of configurations separated by a special character #, also beginning and ending with #.[2] Let Σ be the alphabet for protocols (including the inner alphabet, the set of states, and #). A protocol is *correct,* if each configuration, starting from the second one, is the successor of the previous configuration. A protocol is a *halting* one, if the last configuration has no successor (the machine cannot proceed one more step forward).

Given a Turing machine $\mathfrak{M}$ and an input word x, one can effectively construct (see [13, Lect. 35], for example) a context-free grammar $\mathcal{G}_{\mathfrak{M},x}$ which

[2] In some other presentations of this construction in textbooks, the code of every second configuration is inverted. For our purposes, this is irrelevant.

generates all words over Σ, except the correct halting protocol of $\mathfrak{M}$ on x (if it exists). This construction gives a reduction of the *non-halting* problem for Turing machines to the totality problem for context-free grammars, and thus establishes Π_1^0-hardness of the latter.

We suppose that $\mathcal{G}_{\mathfrak{M},x}$ is in Greibach normal form [8] and extend it by extra rules for capturing the easy case of non-halting—getting stuck in q_c:

$$\begin{array}{ll} S \Rightarrow \#CU & \\ U \Rightarrow aU & \text{for any } a \in \Sigma \\ U \Rightarrow a & \text{for any } a \in \Sigma \\ C \Rightarrow aC & \text{for any } a \in \Sigma \\ C \Rightarrow q_c U & \\ C \Rightarrow q_c & \end{array}$$

In these rules, non-terminal U generates all non-empty words and C generates all words including q_c. Thus, the rule $S \Rightarrow \#CU$ captures the idea that any word including q_c could not be a correct halting protocol.

We also suppose that $\mathcal{G}_{\mathfrak{M},x}$ has a subgrammar starting with a non-terminal E which generates all words which are incorrect protocols and cannot be fixed by extending to the right. Due to greibachization, the leading # gets removed. For example, such a "bad" protocol could include a configuration which is followed by another configuration which is not its successor. For more details, see [18,17]. We express the idea that such a "bad" protocol cannot be fixed, by adding the following rules:

$$S \Rightarrow \#EU, \qquad S \Rightarrow aU \text{ for } a \neq \#.$$

(The second rule states that a good protocol should always start with #.)

We denote the extended grammar by $\mathcal{G}'_{\mathfrak{M},x}$.

Next, in order to use reasoning in the style of [15], we restrict ourselves to a two-letter alphabet $\{e, f\}$. Let $\Sigma = \{a_1, a_2, \dots, a_N\}$ and define a homomorphism $h \colon \Sigma^+ \to \{e, f\}^+$ on letters as follows:

$$h(a_i) = ef^i = e\underbrace{f \dots f}_{i \text{ times}}.$$

(Then h is uniquely propagated to words as a homomorphism.)

By $h(\mathcal{G}'_{\mathfrak{M},x})$ we denote the image of $\mathcal{G}'_{\mathfrak{M},x}$ under homomorphism h. In order to maintain it in Greibach normal form, for each old rule of the form $A \Rightarrow a_i BC$ we introduce a series of rules

$$A \Rightarrow eX_1, \quad X_1 \Rightarrow fX_2, \quad \dots, \quad X_{i-1} \Rightarrow fX_i, \quad X_i \Rightarrow fBC,$$

where $X_1, \dots, X_i$ are new non-terminal symbols (different for each rule of the original grammar). Translations for rules of the forms $A \Rightarrow a_i B$ and $A \Rightarrow a_i$ is similar.

Next, let us construct the grammar $\widetilde{\mathcal{G}}_{\mathfrak{M},x}$. We extend $h(\mathcal{G}'_{\mathfrak{M},x})$ with rules generating words with subwords of the form ef^m, where $m > N = |\Sigma|$ (these words are not in the image of h):

$S \Rightarrow eF_{\geq N}W$	$F \Rightarrow fF$
$S \Rightarrow eF_{\geq N}$	$F \Rightarrow f$
$S \Rightarrow eFS'$	$F_{\geq N} \Rightarrow fF_{\geq N-1}$
$S' \Rightarrow eFS'$	$F_{\geq N-1} \Rightarrow fF_{\geq N-2}$
$S' \Rightarrow eF_{\geq N}W$	$\dots$
$S' \Rightarrow eF_{\geq N}$	$F_{\geq 3} \Rightarrow fF_{\geq 2}$
$W \Rightarrow eFW$	$F_{\geq 2} \Rightarrow fF$
$W \Rightarrow eF$	

Here S', W, F, and $F_{\geq m}$ $(m = 2, \dots, N)$ are new non-terminal symbols.

Finally, we replace U with W in the 'old' part of the grammar. This will not alter the language, since any word derived from W is either also derived from U, or includes a subword of the form ef^m with $m > N$, which is of course not an h-image of a correct protocol.

This finishes the construction of $\widetilde{\mathcal{G}}_{\mathfrak{M},x}$. From this construction, one can easily see the following property:

Lemma 3.2 *The grammar $\widetilde{\mathcal{G}}_{\mathfrak{M},x}$ generates all words of the language generated by the regular expression $(ef^+)^+$ if and only if $\mathfrak{M}$ does not halt on x. If $\mathfrak{M}$ does halt on x, then $\widetilde{\mathcal{G}}_{\mathfrak{M},x}$ generates all words of this language, except $h(\pi)$, where π is the halting protocol of $\mathfrak{M}$ on x.*

The next step uses Safiullin's construction of Lambek grammar with unique type assignment. This result was published by Safiullin as a short note [25] without detailed proofs. A complete exposition is presented in the Appendix of [15]. We shall need Safiullin's result for grammars over a two-letter alphabet in the following form.

Theorem 3.3 (Safiullin) *Let $\widetilde{\mathcal{G}}$ be a context-free grammar over alphabet $\{e, f\}$ in Greibach normal form. Then there exist formulae E, F, and H_A for each non-terminal A, such that the following holds:*

(i) *a non-empty word w is generated by $\widetilde{\mathcal{G}}$ if and only if the sequent $\Gamma_w \vdash H_S$ is derivable in* **L**, *where Γ_w is a sequence of formulae obtained from w by replacing e with E and f with F (e.g., for $w = effee$ we have $\Gamma_w = E, F, F, E, E$);*

(ii) *for each rule of $\widetilde{\mathcal{G}}$ we have the following sequents derivable in* **L**:

Rule	**Sequent**
$A \Rightarrow eBC$	$E, H_B, H_C \vdash H_A$
$A \Rightarrow fBC$	$F, H_B, H_C \vdash H_A$
$A \Rightarrow eB$	$E, H_B \vdash H_A$
$A \Rightarrow fB$	$F, H_B \vdash H_A$
$A \Rightarrow e$	$E \vdash H_A$
$A \Rightarrow f$	$F \vdash H_A$

In this theorem, the first statement is essentially the result on transforming a context-free grammar into a Lambek grammar with unique type assignment (E is the type for e, F for f, and H_S is the goal type). The second statement is actually a technical lemma (induction step) for proving the "only if" direction in the first statement. However, we shall need the second statement explicitly. Further details of Safiullin's construction are irrelevant for us, we use it as a black box.

Using induction and statement (ii), one can easily prove a strengthening of the "only if" part of statement (i). Namely,

(iii) *if a word α in the alphabet of both terminal and non-terminal symbols is derivable in $\mathcal{G}$ from a non-terminal A (notation: $A \Rightarrow^* \alpha$), then the sequent $\Gamma_\alpha \vdash H_A$ is derivable in* $\mathbf{L}$.

Here Γ_α is obtained from α by replacing e with E, f with F, and each non-terminal B by the corresponding H_B.

Consider the sequent

$$(E \cdot F^+)^+ \vdash H_S,$$

where E, F and H_S are obtained from $\widetilde{\mathcal{G}}_{\mathfrak{M},x}$ by the construction from Theorem 3.3. Now we proceed as in [18], proving one direction for $\mathbf{L}^+_\omega$ and non-halting of $\mathfrak{M}$ on x and the other direction for $\mathbf{L}^+_\ell$ and $\mathfrak{M}$ getting stuck in q_c while running on x.

Lemma 3.4 *The sequent $(E \cdot F^+)^+ \vdash H_S$ is derivable in $\mathbf{L}^+_\omega$ if and only if $\mathfrak{M}$ does not halt on x.*

Proof. The ω-rule is invertible, by cut with $A, \ldots, A \to A^+$. Thus, $(E \cdot F^+)^+ \vdash H_S$ is derivable in $\mathbf{L}^+_\omega$ if and only if so is $\Gamma_w \vdash H_S$ for any word w from the language of the regular expression $(ef^+)^+$. This sequent does not include the iteration modality, so its derivability in $\mathbf{L}^+_\omega$ is equivalent to its derivability in $\mathbf{L}$. By Theorem 3.3, derivability of all these sequents is equivalent to the fact that $\mathcal{G}_{\mathfrak{M},x}$ generates all words satisfying the regular expression $(ef^+)^+$. By Lemma 3.2, this is equivalent to non-halting of $\mathfrak{M}$ on x. □

Lemma 3.5 *If $\mathfrak{M}$ gets stuck in q_c when running on x, then $(E \cdot F^+)^+ \vdash H_S$ is derivable in $\mathbf{L}^+_\ell$.*

Proof. Here the 'long rule' finally comes into play. Let n be the length (in symbols, not in steps) of the protocol of $\mathfrak{M}$ running on x until it reaches q_c.

Using the 'long rule,' we derive $(E \cdot F^+)^+ \vdash H_S$ from the following sequents:

$$(E, F^+)^k \vdash H_S \qquad k \leq n$$
$$(E, F^+)^n, (E \cdot F^+)^+ \vdash H_S$$

The first series of sequents, $(E, F^+)^k \vdash H_S$, is also derived by exhaustive application of the 'long rule,' in its form with sequential contexts (Lemma 2.2), up to $N = |\Sigma|$. The sequents we now have to derive are of the form $\Pi_1, \ldots, \Pi_k \vdash H_S$, where $k \leq n$ and each Π_i is either E, F^s, where $s \leq N$, or E, F^N, F^+.

If all Π_i's are of the form E, F^s, then $\Pi_1, \ldots, \Pi_k \vdash H_S$ does not include $^+$ and is derivable in $\mathbf{L}$ by applying Lemma 3.4 and inverting the ω-rule.

The more interesting case is when our sequent includes E, F^N, F^+. Let Π_{i_0} be the first Π_i of this form. First we notice that $F^+ \vdash H_F$ is derivable in $\mathbf{L}_\ell^+$:

$$\frac{F \vdash H_F \quad F, H_F \vdash H_F}{F^+ \vdash H_F}$$

Here the premises are derivable by Theorem 3.3(ii), due to the rules $F \Rightarrow f$ and $F \Rightarrow fF$. Thus, by cut, we can replace E, F^N, F^+ by E, F^N, H_F.

Moreover, since $F_{\geq N} \Rightarrow^* f^N F$, we can apply cut with $F^N, H_F \vdash H_{F_{\geq N}}$ and replace Π_{i_0} with $E, H_{F_{\geq N}}$. For $i \neq i_0$ we similarly replace Π_i with E, H_F, using either $F \Rightarrow^* f^N F$ or $F \Rightarrow^* f^k$. Thus, the whole sequent is now of the form

$$E, H_F, \ldots, E, H_F, E, H_{F_{\geq N}}, E, H_F, \ldots, E, H_F \vdash H_S,$$

which is derivable due to the following derivation in $\widetilde{G}_{\mathfrak{M},x}$:

$$S \Rightarrow eFS' \Rightarrow^* eF \ldots eFS' \Rightarrow eF \ldots eFeF_{\geq N}W \Rightarrow^* eF \ldots eFeF_{\geq N}eF \ldots eF$$

for $i_0 \neq 1, k$ and similarly (but using different rules of $\widetilde{G}_{\mathfrak{M},x}$) for $i_0 = 1$ and $i_0 = k$.

Finally, the second sequent, $(E, F^+)^n, (E \cdot F^+)^+ \vdash H_S$, is derived in a similar fashion. We applying the 'long rule' with N exhaustively to the instances of F^+ in $(E, F^+)^n$ and consider two cases for premises. If at least one of the instances of E, F^+ becomes E, F^N, F^+, then we again reduce to

$$E, H_F, \ldots, E, H_F, E, H_{F_{\geq N}}, E, H_F, \ldots, E, H_F, (E \cdot F^+)^+ \vdash H_S$$

Next, we notice derivability of $(E \cdot F^+)^+ \vdash H_W$:

$$\dfrac{\dfrac{\dfrac{F^+ \vdash H_F \quad E, H_F \vdash H_W}{E, F^+ \vdash H_W}\,(cut)}{E \cdot F^+ \vdash H_W} \quad \dfrac{\dfrac{F^+ \vdash H_F \quad E, H_F, H_W \vdash H_W}{E, F^+, H_W \vdash H_W}\,(cut)}{E \cdot F^+, H_W \vdash H_W}}{(E \cdot F^+)^+ \vdash H_W}$$

The premises are derivable by Theorem 3.3(ii) due to $W \Rightarrow eF$ and $W \Rightarrow eFW$; $F^+ \vdash H_F$ was established above.

Thus, we reduce to

$$E, H_F, \ldots, E, H_F, E, H_{F_{\geq N}}, E, H_F, \ldots, E, H_F, H_W \vdash H_S,$$

which is derivable by statement (iii) below Theorem 3.3 due to

$$S \Rightarrow^* eF \ldots eFeF_{\geq N}eF \ldots eFW.$$

The second, more interesting case is when each instance of F^+ becomes F^{s_i} for some i:

$$E, F^{s_1}, \ldots, E, F^{s_n}, (E \cdot F^+)^+ \vdash H_S.$$

Recalling $(E \cdot F^+)^+ \vdash H_W$ (see above), we reduce to

$$E, F^{s_1}, \ldots, E, F^{s_n}, H_W \vdash H_S.$$

Next, this sequent can be rewritten in the form

$$\Gamma_{h(w)}, H_W \vdash H_S,$$

where $w = a_{s_1} \ldots a_{s_n}$. Since n is the number of letters in the protocol sufficient for $\mathfrak{M}$ on x to reach the cycling state q_c, the word w either includes q_c, or is an incorrect ("bad") protocol, or does not start with $\#$.

In the first case, we have $w = \#w'$ and $C \Rightarrow^* w'$ in $\mathcal{G}'_{\mathfrak{M},x}$. Thus, we get $C \Rightarrow^* h(w')$ in $\widetilde{\mathcal{G}}_{\mathfrak{M},x}$, and by statement (iii) derive

$$\Gamma_{h(w')} \vdash H_C.$$

Gathering things together and cutting, we get

$$\Gamma_{h(\#)}, H_C, H_W \vdash H_S,$$

which is derivable via statement (iii) and $S \Rightarrow^* h(\#)CW$.

The case where w is a "bad" protocol is similar, using $S \Rightarrow^* h(\#)EW$. Finally, if w starts with $a_{s_1} \neq \#$ we have

$$\Gamma_{h(a_{s_1})}, \Gamma_{h(w')}, H_W \vdash H_S,$$

which is derivable by cut from $\Gamma_{h(w')}, H_W \vdash H_W$ and $\Gamma_{h(a_{s_1})}, H_W \vdash H_S$. These are derivable by statement (iii), using $W \Rightarrow^* h(w')W$ and $S \Rightarrow^* a_{s_1}W$.

This finishes the proof of our key lemma. □

Now we proceed exactly as in [18]. Let

$\mathcal{H} = \{\langle \mathfrak{M}, x\rangle \mid \mathfrak{M}$ halts on $x\}$
$\overline{\mathcal{H}} = \{\langle \mathfrak{M}, x\rangle \mid \mathfrak{M}$ does not halt on $x\}$
$\mathcal{C} = \{\langle \mathfrak{M}, x\rangle \mid \mathfrak{M}$ gets stuck in q_c while running on $x\}$
$\mathcal{K} = \{\langle \mathfrak{M}, x\rangle \mid (E \cdot F^+)^+ \vdash H_S$, where E, F, and H_S come from $\widetilde{\mathcal{G}}_{\mathfrak{M},x}$, is derivable in $\mathbf{L}^+_\ell\}$

By Lemma 3.4 $\mathcal{K} \subseteq \overline{\mathcal{H}}$ (recall that $\mathbf{L}^+_\ell$ is a subsystem of $\mathbf{L}^+_\omega$ by Lemma 2.1); by Lemma 3.2 $\mathcal{C} \subseteq \mathcal{K}$. Since $\mathcal{C}$ and $\mathcal{H}$ are recursively inseparable, $\mathcal{K}$ is undecidable, thus so is the derivability problem for $\mathbf{L}^+_\ell$. Theorem 3.1 proved.

Following the reasoning with effective inseparability of $\mathcal{C}$ and $\mathcal{H}$, presented in [17], we can show Σ_1-completeness of $\mathbf{L}^+_\ell$ and, moreover, any recursively enumerable logic in the range between $\mathbf{L}^+_\ell$ and $\mathbf{L}^+_\omega$. This is performed exactly as for action logic with meet and join.

4 Non-derivability of the 'long rule' in $\mathbf{L}^+$

As one can see from the previous section, the 'long rule' is a crucial component of the undecidability proof. If we could derive this rule in $\mathbf{L}^+$, as it can be done for **ACT** [18], we would get undecidability for $\mathbf{L}^+$.

Unfortunately, as we show in this section, this is not the case: the 'long rule' is not derivable in $\mathbf{L}^+$.

Before proceeding further, let us notice a subtle difference between *derivability* and a weaker notion of *admissibility* of a new rule in a calculus. A rule is called *derivable,* if there exists a derivation of the conclusion of this rule with its premises as hypotheses. This derivation is allowed to use cut. On the other hand, a rule (rule scheme) is *admissible,* if, for any substitution of concrete formulae for meta-variables A, B, C, $\ldots$, derivability of its premises implies derivability of its conclusion.

Clearly, any derivable rule is admissible. The converse implication, however, does not hold. For example, the rule $\frac{A \vdash A \cdot A}{B \vdash C}$ is admissible, but not derivable in $\mathbf{L}^+$. The reason is that $A \vdash A \cdot A$ cannot be derivable for any A. This can be proved by interpretation on language models, see [4]. Indeed, consider cofinite languages over an alphabet. Product (pairwise concatenation) and divisions (defined according to the rules of the Lambek calculus) of cofinite languages yield again cofinite languages. Thus, if we interpret all variables as cofinite languages, then the interpretation of A will be also cofinite, thus, non-empty. But then the shortest word of A does not belong to $A \cdot A$ (the empty word is not allowed due to Lambek's non-emptiness condition). Thus, the rule in question is admissible *ex falso.* On the other hand, it is clearly non-derivable, since $B \vdash C$ is absolutely foreign to $A \vdash A \cdot A$. Unfortunately, the author is not aware of more interesting examples of admissible non-derivable rules—that is, in which there exist derivable instances of the premises.

We claim only non-derivability of the 'long rule.' Its admissibility in $\mathbf{L}^+$ is left as an open question.

Theorem 4.1 *The special case of the 'long rule' for $k = 1$,*[3]

$$\frac{A \vdash B \quad A, A^+ \vdash B}{A^+ \vdash B}$$

is not derivable in $\mathbf{L}^+$.

[3] We could call it 'short rule.'

Proof. We prove non-derivability of this rule by presenting an algebraic counter-model. The appropriate class of algebraic models for $\mathbf{L}^+$ is formed by *residuated semigroups with iteration (RSGI)*, defined as follows.

An RSGI is a partially ordered algebraic structure $(S, \preceq, \cdot, \backslash, /, ^+)$, such that:

(i) $\preceq$ is a partial order on S;

(ii) $(S, \cdot)$ is a semigroup;

(iii) $\backslash$ and $/$ are residuals of $\cdot$ w.r.t. $\preceq$:

$$x \backslash y = \max_{\preceq}\{z \mid x \cdot z \preceq y\}, \qquad y / x = \max_{\preceq}\{z \mid z \cdot x \preceq y\};$$

(iv) for each $x \in S$, $x^+ = \min_{\preceq}\{y \mid x \preceq y \text{ and } x \cdot y \preceq y\}$.

An interpretation function v is just a mapping of variables to elements of S; then it is propagated to formulae. A sequent $A_1, \ldots, A_n \vdash B$ is true under v, if $v(A_1) \cdot \ldots \cdot v(A_n) \preceq v(B)$.

Clearly, the following strong form of soundness holds for $\mathbf{L}^+$ w.r.t. RSGI: if a sequent is derivable from a set of hypotheses, and under a given v all these hypotheses are true, then so is the goal sequent. (The proof of soundness involves using monotonicity of $\cdot$ w.r.t. $\preceq$, which is due to Lambek [19]. Completeness also holds, by a Lindenbaum – Tarski argument, but we shall not need it.)

We shall present an RSGI and its two elements $a, b \in S$, such that $a \preceq b$, $a \cdot a^+ \preceq b$, but $a^+ \not\preceq b$. This will do the job, since if the rule in question were derivable, then, in particular, one could derive $p^+ \vdash q$ from $p \vdash q$ and $p, p^+ \vdash q$ (p and q are variables). This conflicts soundness, by taking $v(p) = a$, $v(q) = b$.

Let us start with a standard example of RSGI, which reflects Lambek's original linguistic motivations,—the algebra of formal languages. For us, it is sufficient to consider languages without the empty word over a one-letter alphabet $\Sigma = \{s\}$. Such languages are in one-to-one correspondence with sets of non-zero natural numbers (the word $\underbrace{s \ldots s}_{n}$ is represented by n). We denote the set of all such sets by $\mathcal{P}(\mathbb{N}_+)$. The elements $\varnothing$ and $\mathbb{N}_+$ of $\mathcal{P}(\mathbb{N}_+)$ (the empty and the total language) will play special rôles in our construction. The set of all other languages is $\mathcal{P}_0(\mathbb{N}_+) = \mathcal{P}(\mathbb{N}_+) - \{\varnothing, \mathbb{N}_+\}$.

Our RSGI will be $\mathcal{P}(\mathbb{N}_+)$ extended by two extra elements:

$$S = \mathcal{P}(\mathbb{N}_+) \cup \{\xi, \eta\} = \mathcal{P}_0(\mathbb{N}_+) \cup \{\varnothing, \mathbb{N}_+, \xi, \eta\}.$$

The partial order $\preceq$ on S is defined as follows:

- on $\mathcal{P}(\mathbb{N}_+)$, the partial order is the subset relation;
- for any $x \in \mathcal{P}_0(\mathbb{N}_+) \cup \{\varnothing\}$, we have $x \prec \xi$; ξ and $\mathbb{N}_+$ are incomparable;
- η is the maximal element: for any $x \in \mathcal{P}(\mathbb{N}_+) \cup \{\xi\}$, we have $x \prec \eta$.

The product operation on S is commutative and defined as follows:

- for $x, y \in \mathcal{P}(\mathbb{N}_+)$, product is defined as pairwise addition:

$$x \cdot y = \{n + m \mid n \in x, m \in y\};$$

- $\varnothing \cdot \xi = \varnothing \cdot \eta = \varnothing$;
- $\xi \cdot x = \eta$ for any $x \neq \varnothing$;
- $\eta \cdot x = \eta$ for any $x \neq \varnothing$.

Associativity of product, $(x \cdot y) \cdot z = x \cdot (y \cdot z)$, is proved as follows. The interesting case is when at least one of x, y, z is ξ or η: otherwise we refer to associativity of formal language multiplication. If one of x, y, z is ξ or η and another one is $\varnothing$, then $(x \cdot y) \cdot z = x \cdot (y \cdot z) = \varnothing$; otherwise $(x \cdot y) \cdot z = x \cdot (y \cdot z) = \eta$.

Now let us define residuals, that is, prove existence of the corresponding maxima. Since our semigroup is commutative, we shall always have $x \setminus y = y \,/\, x$, so it is sufficient to prove existence of $x \setminus y$.

- For $x, y \in \mathcal{P}(\mathbb{N}_+)$, if $x \neq \varnothing$, we have

  $$x \setminus y = \{n \in \mathbb{N}_+ \mid (\forall m \in x)\, n + m \in y\},$$

  as in the algebra of formal languages. Indeed, inside $\mathcal{P}_0(\mathbb{N}_+)$ this is the maximal z such that $x \cdot z \preceq y$. As for ξ and η, we have (since $x \neq \varnothing$) $x \cdot \xi = x \cdot \eta = \eta \not\preceq y$.
- For any y we have $\varnothing \setminus y = \eta$. Indeed, $\varnothing \cdot z \preceq y$ holds for any z (since $\varnothing \cdot z = \varnothing$), so we just take the maximum of the whole S.
- For any x, we have $x \setminus \eta = \eta$. Indeed, $x \cdot z \preceq \eta$ holds for any z (since η is the maximum).
- For any $x \in \mathcal{P}_0(\mathbb{N}_+)$, we have $x \setminus \xi = \mathbb{N}_+$. Indeed, $x \cdot \mathbb{N}_+$ belongs to $\mathcal{P}_0(\mathbb{N}_+)$ and therefore is below ξ in the sense of $\preceq$. On the other hand, the only two elements, which are not below $\mathbb{N}_+$, are ξ and η. For them we have $x \cdot \xi = x \cdot \eta = \eta \not\preceq \xi$.
- We also have $\mathbb{N}_+ \setminus \xi = \mathbb{N}_+$. This happens because of the lack of the empty word (zero in $\mathbb{N}_+$): $\mathbb{N}_+ \cdot \mathbb{N}_+ = \{n \mid n \geq 2\} \preceq \xi$. For ξ and η we have, again, $\mathbb{N}_+ \cdot \xi = \mathbb{N}_+ \cdot \eta = \eta \not\preceq \xi$.
- For any $y \neq \eta$, we have $\eta \setminus y = \varnothing$, since $\eta \cdot \varnothing = \varnothing \prec y$ and $\eta \cdot z = \eta \not\preceq y$ for any $z \neq \varnothing$. (As shown above, $\eta \setminus \eta = \eta$.)
- Similarly, $\xi \setminus \eta = \eta$ (shown above), and for any $y \neq \eta$ we have $\xi \setminus y = \varnothing$ (in particular, $\xi \setminus \xi = \varnothing$). Indeed, $\xi \cdot \varnothing = \varnothing \prec y$ and $\xi \cdot z = \eta \not\preceq y$ for any $z \neq \varnothing$.

Finally, let us define iteration, that is, prove that for any x there exists $x^+ = \min_{\preceq}\{y \mid x \preceq y \text{ and } x \cdot y \preceq y\}$.

- For $x \in \mathcal{P}(\mathbb{N}_+)$, its iteration x^+ is defined traditionally: $x^+ = \{n_1 + \ldots + n_k \mid k \geq 1, n_i \in x\}$. If $x^+ \neq \mathbb{N}_+$, then it is indeed the necessary minimum: it is the minimum in $\mathbb{N}_+$, and two other candidates, ξ and η, are above x^+. The case of $x^+ = \mathbb{N}_+$ is a bit more interesting. Again, $\eta \succ x^+$, so it is not a rival; but ξ is incomparable with $x^+ = \mathbb{N}_+$. Fortunately, ξ fails to satisfy the second condition on y to be considered as a candidate for x^+. If $x \neq \varnothing$, then $x \cdot \xi = \eta \not\preceq \xi$ (if $x = \varnothing$, then $x^+ = \varnothing \neq \mathbb{N}_+$).

- $\xi^+ = \eta$. Indeed, ξ^+ should be ξ or η, and ξ does not suffice, since $\xi \cdot \xi = \eta \not\preceq \xi$. For η, everything is all right: $\xi \preceq \eta$ and $\xi \cdot \eta = \eta \preceq \eta$.
- $\eta^+ = \eta$. Indeed, $\eta \preceq \eta$ and $\eta \cdot \eta = \eta \preceq \eta$. Smaller y's are out of the game, since $\eta \not\preceq y$.

Having defined our specific RSGI S, now let $a = \{1\}$ and $b = \xi$. We have: $a \preceq b$; $a^+ = \mathbb{N}_+$, so $a \cdot a^+ = \{n \mid n \geq 2\} \preceq b$; but $a^+ \not\preceq b$ ($\mathbb{N}_+$ and ξ are incomparable). This finishes our proof. □

An important observation on our RSGI S is that its partial order does not form a lattice structure. Namely, $\mathbb{N}_+$ and ξ have no meet: any element of $\mathcal{P}_0(\mathbb{N}_+)$ is below both, and among them there is no maximal one. Dually, $a = \{1\}$ and $a \cdot a^+ = \{n \mid n \geq 2\}$ have no join: ξ and $\mathbb{N}_+$ are above both and are incomparable. This is by design: once we have a lattice, or at least we have a join of a and $a \cdot a^+$, we can apply the derivation of the 'long rule' in $\mathbf{ACT}^+$.

We also notice that in S iteration a^+ is defined as a fixed point, not as a supremum (that is, S is not *-continuous). Indeed, for $a = \{1\}$ its iteration $a^+ = \mathbb{N}_+$ is the smallest y such that $a \preceq y$ and $a \cdot y \preceq y$. However, a^+ is not $\sup_{\preceq}\{a^n \mid n \geq 1\}$. Indeed, there are two incomparable upper bounds for $a^n = \{n\}$, namely, $\mathbb{N}_+$ and ξ. The latter is a 'fake' iteration, since it is not a fixpoint: $a \cdot \xi = \eta \not\preceq \xi$. The non-*-continuity of S is also for a good reason: otherwise, S would model $\mathbf{L}^+_\omega$, and in this system the 'long rule' is derivable (Lemma 2.1).

5 Concluding Remarks

We have proved undecidability (and Σ_1-completeness) of the Lambek calculus with an inductively axiomatized positive iteration modality, extended with the so-called 'long rule' of the form

$$\frac{A \vdash B \quad A, A \vdash B \quad A^k \vdash B \quad A^k, A^+ \vdash B}{A^+ \vdash B}$$

This result refines the undecidability result for action logic [18], since now we obtain undecidability for a system without additive connectives, meet and join ($\wedge$ and $\vee$).

Another distinctive feature of this paper is the Lambek's non-emptiness restriction imposed on the calculus. We conjecture that the same results hold without this restriction. However, this is left as an open question for further research, since some technicalities, namely, Safiullin's Theorem 3.3 and the counter-model construction in Theorem 4.1, in their current state, depend on Lambek's restriction.

In action logic with meet and join, the 'long rule' is derivable; for the multiplicative-only system $\mathbf{L}^+$ studied in this paper, this is not the case (Theorem 4.1). The question of whether the 'long rule' is admissible in $\mathbf{L}^+$ is still open. If the answer happens to be positive, we shall immediately get undecidability of $\mathbf{L}^+$ (since in this case $\mathbf{L}^+$ and $\mathbf{L}^+_\ell$ derive the same set of sequents). If the answer is negative, then $\mathbf{L}^+_\ell$ is strictly stronger than $\mathbf{L}^+$, and complexity of the latter remains a separate open problem.

Moreover, non-derivability and potential non-admissibility of the 'long rule' brings some light upon the old question on constructing a cut-free calculus for action logic with inductive axiomatizations for iteration. As noticed in the Preliminaries, for systems with inductive-style rules for iteration no cut-free sequential calculi are known. The issues with the 'long rule' discussed in this paper are actually conservativity issues. Since the 'long rule' is not derivable in $\mathbf{L}^+$, this calculus is not a *strongly conservative* fragment of $\mathbf{ACT}^+$. Namely, consider three sequents $p \vdash q$, $p, p^+ \vdash q$, and $p^+ \vdash q$ (premises and conclusion of the 'long rule'). These sequents are formulated in the language of $\mathbf{L}^+$, without $\vee$ and $\wedge$. Actually, they use only one connective, $^+$. However, one can derive the third one from the first and the second ones only in $\mathbf{ACT}^+$ (via a detour through $\vee$), not in $\mathbf{L}^+$. If the 'long rule' happens to be non-admissible, ordinary conservativity would also fail. In this case, in particular, it would be an open question which sequents without $\vee$ and $\wedge$ are derivable in $\mathbf{ACT}^+$—are these sequents exactly theorems of $\mathbf{L}^+_\ell$, or do they form a larger set?

However, if $\mathbf{ACT}^+$ were axiomatized by a sequent calculus (even with a non-standard notion of proof, like a circular one), it would enjoy conservativity. Thus, in view of the issues with the 'long rule,' it looks reasonable to extend our approaches for axiomatizing $\mathbf{ACT}^+$ and search for hypersequential formalisms where $\vee$ or $\wedge$ is incorporated into the meta-syntax (cf. Kozak's system for distributive full Lambek calculus [11]). Notice that the sequents appearing in the 'long rule' do not include division operations (only product and iteration). Thus, the same conservativity issues could potentially appear in the logics of Kleene algebras and lattices without residuals.

These considerations are quite coherent with the complete cut-free circular proof system for Kleene algebras presented by Das and Pous [6]. Their calculus is hypersequential, introducing join ($\vee$) on the meta-syntactic level to the right-hand sides of sequents. The counter-example for cut-free cyclic provability in a system with traditional sequents given by Das and Pous is $A \cdot A^* \vdash A^* \cdot A$, which is quite close to our 'short rule' in Theorem 4.1.

Acknowledgements The author is grateful to Lev Beklemishev, Anupam Das, Max Kanovich, Fedor Pakhomov, Andre Scedrov, Daniyar Shamkanov, and Stanislav Speranski for fruitful discussions. Being a Young Russian Mathematics award winner, the author thanks its jury and sponsors for this high honour. (The work on this particular paper was funded from another source, as mentioned on the first page.)

References

[1] Abrusci, V. M., *A comparison between lambek syntactic calculus and intuitionistic linear logic*, Zeitschrift für mathematische Logik und Grundlagen der Mathematik **36** (1990),

pp. 11–15.

[2] Bar-Hillel, Y., C. Gaifman and E. Shamir, *On the categorial and phrase-structure grammars*, Bulletin of the Research Council of Israel **9F** (1960), pp. 1–16.

[3] Buszkowski, W., *On action logic: equational theories of action algebras*, Journal of Logic and Computation **17** (2007), pp. 199–217.

[4] Buszkowski, W., *Lambek calculus and substructural logics*, Linguistic Analysis **36** (2010), pp. 15–48.

[5] Buszkowski, W. and E. Palka, *Infinitary action logic: complexity, models and grammars*, Studia Logica **89** (2008), pp. 1–18.

[6] Das, A. and D. Pous, *A cut-free cyclic proof system for Kleene algebra*, in: R. Schmidt and C. Nalon, editors, *Automated Reasoning with Analytic Tableaux and Related Methods. TABLEAUX 2017*, Lecture Notes in Computer Science **10501** (2017), pp. 261–277.

[7] Girard, J.-Y., *Linear logic*, Theoretical Computer Science **50** (1987), pp. 1–102.

[8] Greibach, S. A., *A new normal-form theorem for context-free phrase structure grammars*, Journal of the ACM **12** (1965), pp. 42–52.

[9] Jipsen, P., *From residuated semirings to Kleene algebras*, Studia Logica **76** (2004), pp. 291–303.

[10] Kleene, S. C., *Representation of events in nerve nets and finite automata*, in: C. E. Shannon and J. McCarthy, editors, *Automata Studies*, Princeton University Press, 1956 pp. 3–41.

[11] Kozak, M., *Distributive full Lambek calculus has the finite model property*, Studia Logica **91** (2009), p. 201–216.

[12] Kozen, D., *On action algebras*, in: J. van Eijck and A. Visser, editors, *Logic and Information Flow*, MIT Press, 1994 pp. 78–88.

[13] Kozen, D., "Automata and Complexity," Springer-Verlag, New York, 1997.

[14] Krull, W., *Axiomatische Begründung der algemeinen Idealtheorie*, Sitzungsberichte der physikalischmedizinischen Societät zu Erlangen **56** (1924), pp. 47–63.

[15] Kuznetsov, S., *The Lambek calculus with iteration: two variants*, in: J. Kennedy and R. de Queiroz, editors, *Logic, Language, Information, and Computation. WoLLIC 2017*, Lecture Notes in Computer Science **10388**, 2017, pp. 182–198.

[16] Kuznetsov, S., **-continuity vs. induction: divide and conquer*, in: G. Bezhanishvili, G. D'Agostino, G. Metcalfe and T. Studer, editors, *Proceedings of AiML '18*, Advances in Modal Logic **12** (2018), pp. 493–510.

[17] Kuznetsov, S., *Action logic is undecidable*, arXiv preprint 1912.11273 (2019).

[18] Kuznetsov, S., *The logic of action lattices is undecidable*, in: *2019 34th Annual ACM/IEEE Symposium on Logic in Computer Science (LICS)* (2019), pp. 1–9.

[19] Lambek, J., *The mathematics of sentence structure*, American Mathematical Monthly **65** (1958), pp. 154–170.

[20] Moot, R. and C. Retoré, "The logic of categorial grammars: a deductive account of natural language syntax and semantics," Lecture Notes in Computer Science **6850**, Springer, 2012.

[21] Palka, E., *An infinitary sequent system for the equational theory of *-continuous action lattices*, Fundamenta Informaticae **78** (2007), pp. 295–309.

[22] Pentus, M., *Residuated monoids with Kleene star* (2010), unpublished manuscript.

[23] Pratt, V., *Action logic and pure induction*, in: J. van Eijck, editor, *JELIA 1990: Logics in AI*, Lecture Notes in Artificial Intelligence **478** (1991), pp. 97–120.

[24] Restall, G., "An introduction to substructural logics," Routledge, 2000.

[25] Safiullin, A. N., *Derivability of admissible rules with simple premises in the Lambek calculus*, Moscow University Mathematics Bulletin **62** (2007), pp. 168–171.

A Monadic Logic of Ordered Abelian Groups

George Metcalfe [1] Olim Tuyt

Mathematical Institute
University of Bern, Switzerland
{george.metcalfe, olim.tuyt}@math.unibe.ch

Abstract

A many-valued modal logic with connectives interpreted in the ordered additive group of real numbers is introduced as a modal counterpart of the one-variable fragment of a (monadic) first-order real-valued logic. It is shown that the logic is decidable and admits an interpretation of the one-variable fragment of first-order Łukasiewicz logic. Completeness of an axiom system for the modal-multiplicative fragment is established via a Herbrand theorem for its first-order counterpart. A functional representation theorem is then proved for a class of monadic lattice-ordered abelian groups and used to establish completeness of an axiom system for the full logic.

Keywords: Modal Logic, Ordered Groups, Łukasiewicz Logic, Monadic Fragments.

1 Introduction

Many-valued modal logics with connectives interpreted in the ordered additive group of real numbers have been studied in a wide range of different settings. For example, modal logics based on the semantics of Łukasiewicz logic with truth values in the real unit interval have been considered as the basis for fuzzy description logics (see, e.g., [2, 21, 29]), logics for reasoning about belief and probabilities (see, e.g., [16–18, 20]), a Łukasiewicz mu-calculus [25], and as a fragment of continuous logic [4]. Such logics have also been studied from a purely algebraic perspective (see, e.g., [9, 11, 14, 23]) and appear in the guise of lattice-ordered groups (ℓ-groups, for short) with a (co-)nucleus in the study of semantics for substructural logics (see, e.g., [19, 26]). The appeal of these logics is clear: they make use of familiar arithmetical operations on the real numbers and well-studied computational methods (e.g., linear programming), and they relate to groups, arguably the most fundamental structures of classical algebra.

In [15], a minimal real-valued modal logic K(A) was defined as an extension of Abelian logic, the logic of abelian ℓ-groups, introduced independently in [24] as a relevant logic and [8] as a comparative logic. Among the advantages of focussing on modal extensions of Abelian logic are that the language is rich enough to interpret other logics (e.g., modal extensions of Łukasiewicz logic),

[1] This research was supported by Swiss National Science Foundation grant 200021_184693.

$$
\begin{array}{ll}
(\mathrm{B}) & (\varphi \to \psi) \to ((\psi \to \chi) \to (\varphi \to \chi)) \\
(\mathrm{C}) & (\varphi \to (\psi \to \chi)) \to (\psi \to (\varphi \to \chi)) \\
(\mathrm{I}) & \varphi \to \varphi \\
(\mathrm{A}) & ((\varphi \to \psi) \to \psi) \to \varphi \\
(+1) & \varphi \to (\psi \to (\varphi + \psi)) \\
(+2) & (\varphi \to (\psi \to \chi)) \to ((\varphi + \psi) \to \chi) \\
(\overline{0}1) & \overline{0} \\
(\overline{0}2) & \varphi \to (\overline{0} \to \varphi) \\
(\wedge 1) & (\varphi \wedge \psi) \to \varphi \\
(\wedge 2) & (\varphi \wedge \psi) \to \psi \\
(\wedge 3) & ((\varphi \to \psi) \wedge (\varphi \to \chi)) \to (\varphi \to (\psi \wedge \chi)) \\
(\vee 1) & \varphi \to (\varphi \vee \psi) \\
(\vee 2) & \psi \to (\varphi \vee \psi) \\
(\vee 3) & ((\varphi \to \chi) \wedge (\psi \to \chi)) \to ((\varphi \vee \psi) \to \chi)
\end{array}
$$

$$
\frac{\varphi \quad \varphi \to \psi}{\psi}\ (\mathrm{mp}) \qquad \frac{\varphi \quad \psi}{\varphi \wedge \psi}\ (\mathrm{adj})
$$

Fig. 1. An Axiom System for Abelian Logic

the semantics are based directly on structures studied intensively in algebra and computer science, and there exists a natural separation between the group (multiplicative) and lattice (additive) fragments of the logics. Indeed, in [15], a sequent calculus, an axiom system, and a complexity result were obtained for the modal-multiplicative fragment of K(A) as first steps towards addressing the corresponding (much more challenging) problems for the full logic.

In this paper, we introduce a real-valued modal logic S5(A) as the modal counterpart of the one-variable fragment of (monadic) first-order Abelian logic. It is easily proved that S5(A) is decidable and admits an interpretation of the one-variable fragment of first-order Łukasiewicz logic axiomatized in [28]. The main contribution of the paper lies rather with the two distinct methods used to establish completeness results. First, we make use of a Herbrand theorem for the first-order counterpart of S5(A) and basic facts from linear programming to give a syntactic completeness proof for an axiomatization of the modal-multiplicative fragment. For an axiomatization of the full logic, we give an algebraic completeness proof using monadic abelian ℓ-groups, which, similarly to monadic Heyting algebras (see [5,7]) and MV-algebras (see [9,11,14]), may be viewed as abelian ℓ-groups with certain "relatively complete" subalgebras. We adapt a method used in [9] to prove a functional representation theorem for monadic MV-algebras to obtain a similar theorem for monadic abelian ℓ-groups, and then establish completeness with respect to the real-valued semantics via a partial embedding lemma for linearly ordered abelian groups.

2 A Real-Valued Monadic Logic

In this section, we introduce a many-valued modal logic defined over the ordered abelian group $\mathbf{R} = \langle \mathbb{R}, \min, \max, +, -, 0 \rangle$ and prove a Herbrand theorem for the corresponding one-variable fragment of a (monadic) first-order Abelian logic.

Let $\mathcal{L}_{\mathrm{A}}$ be a propositional language with binary connectives $+$, $\to$, $\wedge$, and $\vee$, and a constant $\overline{0}$, where $\neg\varphi := \varphi \to \overline{0}$, $\varphi \leftrightarrow \psi := (\varphi \to \psi) \wedge (\psi \to \varphi)$, $0\varphi := \overline{0}$, and $(n+1)\varphi := \varphi + (n\varphi)$ $(n \in \mathbb{N})$. Let us also denote by $\mathrm{Fm}(\mathcal{L})$ the set of formulas for any propositional language $\mathcal{L}$ over a countably infinite set of variables $\{p_i \mid i \in \mathbb{N}\}$. An axiomatization of Abelian logic — a single-constant version of multiplicative additive intuitionistic linear logic extended with the axiom schema (A) — is presented in Fig. 1 that is complete with respect to both the logical matrix $\langle \mathbf{R}, \mathbb{R}^{\geq 0} \rangle$ and the variety of abelian ℓ-groups (defined in Section 5).

Now let $\mathcal{L}_{\mathrm{A}}^{\Box}$ be $\mathcal{L}_{\mathrm{A}}$ extended with a unary connective $\Box$, where $\Diamond\varphi := \neg\Box\neg\varphi$. An S5(A)-*model* is an ordered pair $\mathfrak{M} = \langle W, V \rangle$ consisting of a non-empty set W and a function $V \colon \{p_i \mid i \in \mathbb{N}\} \times W \to \mathbb{R}$ such that for each $i \in \mathbb{N}$ the function $V_i \colon W \to \mathbb{R};\ w \mapsto V(p_i, w)$ is bounded[2]; V is then extended to the function $V \colon \mathrm{Fm}(\mathcal{L}_{\mathrm{A}}^{\Box}) \times W \to \mathbb{R}$ as follows:

$$\begin{aligned} V(\overline{0}, w) &= 0 & V(\varphi \wedge \psi, w) &= \min(V(\varphi, w), V(\psi, w)) \\ V(\varphi + \psi, w) &= V(\varphi, w) + V(\psi, w) & V(\varphi \vee \psi, w) &= \max(V(\varphi, w), V(\psi, w)) \\ V(\varphi \to \psi, w) &= V(\psi, w) - V(\varphi, w) & V(\Box\varphi, w) &= \bigwedge\{V(\varphi, u) \mid u \in W\}. \end{aligned}$$

By calculation, also

$$V(\neg\varphi, w) = -V(\varphi, w) \quad \text{and} \quad V(\Diamond\varphi, w) = \bigvee\{V(\varphi, u) \mid u \in W\}.$$

A formula $\varphi \in \mathrm{Fm}(\mathcal{L}_{\mathrm{A}}^{\Box})$ is said to be *valid* in $\mathfrak{M}$ if $V(\varphi, w) \geq 0$ for all $w \in W$. If φ is valid in all S5(A)-models, it is called S5(A)-*valid*, written $\models_{\mathrm{S5(A)}} \varphi$.

The logic S5(A) corresponds (as expected) to the one-variable fragment of a (monadic) first-order logic. Consider a first-order language with unary predicate symbols $P_0, P_1, \ldots$ and constants $c_0, c_1, \ldots$. We denote by Fm the set of first-order formulas for this language defined using the propositional connectives of $\mathcal{L}_{\mathrm{A}}$ and the universal quantifier $\forall$ over a countably infinite set of object variables, defining $(\exists x)\alpha := \neg(\forall x)\neg\alpha$. For convenience, we also often write $\bar{c}$ or $\bar{x}$ to denote an n-tuple of constants or variables, and, given $\bar{c} = c_1, \ldots, c_n$ and $\bar{d} = d_1, \ldots, d_m$, let $\bar{d} \subseteq \bar{c}$ stand for $\{d_1, \ldots, d_m\} \subseteq \{c_1, \ldots, c_n\}$.

A $\forall$A-*interpretation* $\mathcal{I} = \langle D_{\mathcal{I}}, v_{\mathcal{I}} \rangle$ consists of a non-empty set $D_{\mathcal{I}}$ and a function $v_{\mathcal{I}}$ that maps terms (constants and variables) to elements of $D_{\mathcal{I}}$, and each P_i $(i \in \mathbb{N})$ to a bounded function from $D_{\mathcal{I}}$ to $\mathbb{R}$. The function $v_{\mathcal{I}}$ is then extended to Fm by defining $v_{\mathcal{I}}(P_i(t)) = v_{\mathcal{I}}(P_i)(v_{\mathcal{I}}(t))$ for each $i \in \mathbb{N}$ and term t, and then inductively, where $v_{\mathcal{I}}[x \mapsto a]$ denotes the map that sends x to

[2] A function $f \colon A \to \mathbb{R}$ is *bounded* if there exists $r \in \mathbb{R}$ such that $|f(a)| \leq r$ for all $a \in A$.

a and coincides elsewhere with $v_{\mathcal{I}}$,

$$v_{\mathcal{I}}(\overline{0}) = 0 \qquad v_{\mathcal{I}}(\alpha \wedge \beta) = \min(v_{\mathcal{I}}(\alpha), v_{\mathcal{I}}(\beta))$$
$$v_{\mathcal{I}}(\alpha + \beta) = v_{\mathcal{I}}(\alpha) + v_{\mathcal{I}}(\beta) \qquad v_{\mathcal{I}}(\alpha \vee \beta) = \max(v_{\mathcal{I}}(\alpha), v_{\mathcal{I}}(\beta))$$
$$v_{\mathcal{I}}(\alpha \to \beta) = v_{\mathcal{I}}(\beta) - v_{\mathcal{I}}(\alpha)) \qquad v_{\mathcal{I}}((\forall x)\alpha) = \bigwedge\{v_{\mathcal{I}}[x \mapsto a](\alpha) \mid a \in D\}.$$

We say that $\mathcal{I}$ *satisfies* $\alpha \in Fm$ if $v_{\mathcal{I}}(\alpha) \geq 0$ and that α is $\forall$A-*valid*, written $\models_{\forall\mathrm{A}} \alpha$, if it is satisfied by all $\forall$A-interpretations.

Now let Fm_1 denote the set of formulas in Fm that contain at most one object variable x and no constants. For each $\varphi \in \mathrm{Fm}(\mathcal{L}_{\mathrm{A}}^{\Box})$, let $\alpha_\varphi \in Fm_1$ be the result of replacing occurrences of $\Box$ by $(\forall x)$ and occurrences of p_i ($i \in \mathbb{N}$) by $P_i(x)$, and, conversely, for any $\alpha \in Fm_1$, let $\varphi_\alpha \in \mathrm{Fm}(\mathcal{L}_{\mathrm{A}}^{\Box})$ be the result of replacing occurrences of $(\forall x)$ by $\Box$ and occurrences of $P_i(x)$ ($i \in \mathbb{N}$) by p_i. Equivalences between S5(A)-validity and $\forall$A-validity then follow directly from the preceding definitions.

Proposition 2.1 *For any* $\varphi \in \mathrm{Fm}(\mathcal{L}_{\mathrm{A}}^{\Box})$ *and* $\alpha \in Fm_1$,

$$\models_{\mathrm{S5(A)}} \varphi \iff \models_{\forall\mathrm{A}} \alpha_\varphi \quad \textit{and} \quad \models_{\forall\mathrm{A}} \alpha \iff \models_{\mathrm{S5(A)}} \varphi_\alpha.$$

It is not hard to check that $\forall$A-validity is preserved by all quantifier-shifts; that is, for any $\alpha, \beta \in Fm$, variable x that does not occur in β, and $\star \in \{+, \wedge, \vee\}$,

$$\models_{\forall\mathrm{A}} (\forall x)(\alpha \star \beta) \leftrightarrow ((\forall x)\alpha \star \beta) \qquad \models_{\forall\mathrm{A}} (\exists x)(\alpha \star \beta) \leftrightarrow ((\exists x)\alpha \star \beta)$$
$$\models_{\forall\mathrm{A}} (\forall x)(\alpha \to \beta) \leftrightarrow ((\exists x)\alpha \to \beta) \qquad \models_{\forall\mathrm{A}} (\exists x)(\alpha \to \beta) \leftrightarrow ((\forall x)\alpha \to \beta)$$
$$\models_{\forall\mathrm{A}} (\forall x)(\beta \to \alpha) \leftrightarrow (\beta \to (\forall x)\alpha) \qquad \models_{\forall\mathrm{A}} (\exists x)(\beta \to \alpha) \leftrightarrow (\beta \to (\exists x)\alpha).$$

Hence for any $\alpha \in Fm$, there exists a prenex $\beta \in Fm$ such that $\models_{\forall\mathrm{A}} \alpha \leftrightarrow \beta$. Moreover, the following Herbrand theorem holds for existential sentences.[3]

Theorem 2.2 *For any quantifier-free formula* $\alpha \in Fm$ *with free variables in* $\bar{x} = x_1, \ldots, x_m$ *and constants in* $\bar{c} = c_1, \ldots, c_n$,

$$\models_{\forall\mathrm{A}} (\exists \bar{x})\alpha \iff \models_{\forall\mathrm{A}} \bigvee\{\alpha(\bar{d}) \mid \bar{d} \subseteq \bar{c}\}.$$

Proof. The right-to-left direction follows using the easily-verified fact that $\models_{\forall\mathrm{A}} \beta(c) \to (\exists y)\beta(y)$ for any $\beta \in Fm$ and constant c. For the converse, we suppose contrapositively that $\not\models_{\forall\mathrm{A}} \bigvee\{\alpha(\bar{d}) \mid \bar{d} \subseteq \bar{c}\}$. Then there exists a $\forall$A-interpretation $\langle D_{\mathcal{I}}, v_{\mathcal{I}}\rangle$ such that $v_{\mathcal{I}}(\alpha(\bar{d})) < 0$ for all $\bar{d} \subseteq \bar{c}$. Consider now the $\forall$A-interpretation $\langle D'_{\mathcal{I}}, v'_{\mathcal{I}}\rangle$ such that $D'_{\mathcal{I}} = \{v_{\mathcal{I}}(c_1), \ldots, v_{\mathcal{I}}(c_n)\}$ and $v'_{\mathcal{I}}$ coincides on $c_1, \ldots, c_n$ with $v_{\mathcal{I}}$ and maps each P_i ($i \in \mathbb{N}$) to the restriction of $v_{\mathcal{I}}(P_i)$ to $D'_{\mathcal{I}}$. Then $v'_{\mathcal{I}}((\exists \bar{x})\alpha) = \bigvee\{v_{\mathcal{I}}(\alpha(\bar{d})) \mid \bar{d} \subseteq \bar{c}\} < 0$. So $\not\models_{\forall\mathrm{A}} (\exists \bar{x})\alpha$. $\Box$

[3] Note that if the logic $\forall$A is extended to allow non-constant function symbols and predicate symbols of arbitrary arity, it will admit Skolemization. However, the logic will then, as in the case of first-order Łukasiewicz logic (see [3, 12] for details), admit only an "approximate Herbrand theorem".

For any $\alpha \in Fm_1$, replacing any free variable x in α with a new constant, then iteratively replacing each positive occurrence of a subformula $(\forall x)\alpha'(x)$ with $\alpha'(c)$ for a new constant c, and finally shifting quantifiers, yields an existential sentence $\beta \in Fm$ such that $\models_{\forall \mathrm{A}} \alpha \iff \models_{\forall \mathrm{A}} \beta$. Theorem 2.2 now tells us that α is $\forall$A-valid if and only if a certain quantifier-free sentence is $\forall$A-valid. But validity of quantifier-free sentences can be checked in the ordered additive group **R** and is decidable [30], so we obtain the following result.

Corollary 2.3 S5(A)*-validity is decidable.*

3 The One-Variable Fragment of Łukasiewicz Logic

In this section, we prove that the one-variable fragment of first-order Łukasiewicz logic axiomatized as a many-valued modal logic by Rutledge in [28] (see also [9, 11, 14, 22]) may be viewed as a fragment of the logic S5(A).

Let $\mathcal{L}_{Ł}^{\Box}$ be a propositional language with a binary connective $\supset$ and unary connectives $\sim$ and $\Box$. An S5(Ł)*-model* is an ordered pair $\mathfrak{M} = \langle W, V \rangle$ consisting of a non-empty set W and a function $V\colon \{p_i \mid i \in \mathbb{N}\} \times W \to [0,1]$ that is extended to a function $V\colon \mathrm{Fm}(\mathcal{L}_{Ł}^{\Box}) \times W \to [0,1]$ by

$$V(\sim\varphi, w) = 1 - V(\varphi, w)$$
$$V(\varphi \supset \psi, x) = \min(1, 1 - V(\varphi, w) + V(\psi, w))$$
$$V(\Box\varphi, w) = \bigwedge\{V(\varphi, u) \mid u \in W\}.$$

An $\mathcal{L}_{Ł}^{\Box}$-formula φ is said to be *valid* in $\mathfrak{M}$ if $V(\varphi, w) = 1$ for all $w \in W$, and called S5(Ł)*-valid*, written $\models_{\mathrm{S5(Ł)}} \varphi$, if it is valid in all S5(Ł)-models. As in the case of S5(A) considered in Section 2, it is straightforward to prove that S5(Ł)-validity corresponds to validity in the one-variable fragment of first-order Łukasiewicz logic (see [28] for details).

Let us fix $\bot := \Box p_0 \wedge \neg\Box p_0$, noting that this constant is interpreted as the same nonpositive real number in all worlds of an S5(A)-model. We define the following map from the set $\mathrm{Fm}_0(\mathcal{L}_{Ł}^{\Box})$ of $\mathcal{L}_{Ł}^{\Box}$-formulas defined over $\{p_i \mid i \in \mathbb{N}^+\}$ to $\mathrm{Fm}(\mathcal{L}_{\mathrm{A}}^{\Box})$:

$$p_i^* = (p_i \wedge \overline{0}) \vee \bot \qquad (\sim\varphi)^* = \varphi^* \to \bot$$
$$(\varphi \supset \psi)^* = (\varphi^* \to \psi^*) \wedge \overline{0} \qquad (\Box\varphi)^* = \Box\varphi^*.$$

We show that $*$ preserves validity between S5(Ł) and S5(A) by identifying the value of $\varphi \in \mathrm{Fm}_0(\mathcal{L}_{Ł}^{\Box})$ in $[0,1]$ with the value of $\varphi^* \in \mathrm{Fm}(\mathcal{L}_{\mathrm{A}}^{\Box})$ in $[\bot, 0]$.

Theorem 3.1 *Let* $\varphi \in \mathrm{Fm}_0(\mathcal{L}_{Ł}^{\Box})$. *Then* $\models_{\mathrm{S5(Ł)}} \varphi$ *if and only if* $\models_{\mathrm{S5(A)}} \varphi^*$.

Proof. Suppose first that φ is not valid in an S5(Ł)-model $\mathfrak{M} = \langle W, V \rangle$. Then $V(\varphi, x_0) < 1$ for some $x_0 \in W$. We consider the S5(A)-model $\mathfrak{M}' = \langle W, V' \rangle$ where $V'(p_0, x) = -1$ and $V'(p_i, x) = V(p_i, x) - 1$ ($i \in \mathbb{N}^+$) for all $x \in W$, noting that $V'(\bot, x) = V'(\Box p_0 \wedge \neg\Box p_0, x) = -1$ for all $x \in W$. It suffices to prove that $V'(\psi^*, x) = V(\psi, x) - 1$ for all $x \in W$ and $\psi \in \mathrm{Fm}_0(\mathcal{L}_{Ł}^{\Box})$, since then $V'(\varphi^*, x_0) = V(\varphi, x_0) - 1 < 0$ and $\not\models_{\mathrm{S5(A)}} \varphi^*$. We proceed by induction on the size (number of symbols) of ψ. For the base case, we have $V'(p_i^*, x) = V'((p_i \wedge \overline{0}) \vee \bot) = V(p_i, x) - 1$ for each $i \in \mathbb{N}^+$. For the inductive step we obtain, using the induction hypothesis,

$$\begin{aligned} V'((\psi_1 \supset \psi_2)^*, x) &= V'((\psi_1^* \to \psi_2^*) \wedge \overline{0}, x) \\ &= \min(V'(\psi_2^*, x) - V'(\psi_1^*, x), 0) \\ &= \min((V(\psi_2, x) - 1) - (V(\psi_1, x) - 1), 0) \\ &= \min(V(\psi_2, x) - V(\psi_1, x), 0) \\ &= \min(1 - V(\psi_1, x) + V(\psi_2, x), 1) - 1 \\ &= V(\psi_1 \supset \psi_2, x) - 1, \end{aligned}$$

and, the case where ψ is $\sim\psi_1$ being very similar, for the modal case,

$$\begin{aligned} V'((\Box\psi_1)^*, x) &= V'(\Box\psi_1^*, x) \\ &= \bigwedge\{V'(\psi_1^*, y) \mid y \in W\} \\ &= \bigwedge\{V(\psi_1, y) - 1 \mid y \in W\} \\ &= \bigwedge\{V(\psi_1, y) \mid y \in W\} - 1 \\ &= V(\Box\psi_1, x) - 1. \end{aligned}$$

Suppose now conversely that φ^* is not valid in an S5(A)-model $\mathfrak{M} = \langle W, V\rangle$. That is, $V(\varphi^*, x_0) < 0$ for some $x_0 \in W$. Observe first that if $V(\Box p_0, x_0) = 0$, then, by a simple induction on the size of $\psi \in \mathrm{Fm}_0(\mathcal{L}_{Ł}^{\Box})$, we obtain $V(\psi^*, x) = 0$ for all $\psi \in \mathrm{Fm}_0(\mathcal{L}_{Ł}^{\Box})$ and $x \in W$, a contradiction. Hence $V(\Box p_0, x_0) \neq 0$. Moreover, by scaling (dividing $V(p_i, y)$ by $|V(\Box p_0, x_0)|$ for each $i \in \mathbb{N}^+$ and $x \in W$), we may assume that $V(\bot, x) = -1$ for all $x \in W$. We consider the S5(Ł)-model $\mathfrak{M}' = \langle W, V'\rangle$ where $V'(p_i, x) = \max(\min(V(p_i, x) + 1, 1), 0)$ for each $x \in W$ and $i \in \mathbb{N}^+$. It then suffices to prove that $V'(\psi, x) = V(\psi^*, x) + 1$ for all $\psi \in \mathrm{Fm}_0(\mathcal{L}_{Ł}^{\Box})$ and $x \in W$ by an easy induction on the size of ψ. $\Box$

The above proof can be extended to obtain an interpretation of the full first-order Łukasiewicz logic into a first-order Abelian logic. In particular, monadic first-order Łukasiewicz logic can be viewed as a fragment of the monadic logic ∀A defined in Section 2. Since the former has been shown by Bou in unpublished work to be undecidable, this is also the case for the latter.

4 The Modal-Multiplicative Fragment

In this section, we use the Herbrand theorem obtained in Section 2 to establish the completeness of an axiom system for the modal-multiplicative fragment of S5(A).[4] Let us consider first the axiom system $\mathcal{A}_{\mathrm{m}}$ defined over the language

[4] Note that we follow here standard terminology from the linear and substructural logic literature in referring to the *multiplicative* fragment of Abelian logic, even though the group multiplication for the real numbers is in fact addition.

$$\begin{array}{ll}
\text{(K)} & \Box(\varphi \to \psi) \to (\Box\varphi \to \Box\psi) \\
\text{(T)} & \Box\varphi \to \varphi \\
\text{(5)} & \Diamond\varphi \to \Box\Diamond\varphi \\
\text{(M)} & \Box(\varphi + \varphi) \to (\Box\varphi + \Box\varphi)
\end{array}$$

$$\frac{\varphi}{\Box\varphi}\ \text{(nec)}$$

Fig. 2. Modal Axiom and Rule Schema

$\mathcal{L}_{\mathrm{m}}$ with connectives $+$, $\to$, and $\overline{0}$ by removing the axiom and rule schema for $\wedge$ and $\vee$ from those presented in Fig. 1 and adding

$$\frac{n\varphi}{\varphi}\ (\mathrm{con}_n) \quad (n \geq 2).$$

It is not hard to show (see, e.g., [8, 24]) that $\mathcal{A}_{\mathrm{m}}$ is complete with respect to the multiplicative fragment of Abelian logic defined by the logical matrix $\langle\langle\mathbb{R}, +, -, 0\rangle, \mathbb{R}^{\geq 0}\rangle$.

Now let $\mathcal{L}_{\mathrm{m}}^{\Box}$ be the language extending $\mathcal{L}_{\mathrm{m}}$ with $\Box$ and let $\mathcal{S}5(\mathcal{A}_{\mathrm{m}})$ be the axiom system defined over $\mathcal{L}_{\mathrm{m}}^{\Box}$ by extending $\mathcal{A}_{\mathrm{m}}$ with the modal axiom and rule schema presented in Fig. 2. Soundness for this system is proved as usual by checking that the axioms are S5(A)-valid and the rules preserve S5(A)-validity.

Lemma 4.1 *Let* $\varphi \in \mathrm{Fm}(\mathcal{L}_{\mathrm{m}}^{\Box})$. *If* $\vdash_{\mathcal{S}5(\mathcal{A}_{\mathrm{m}})} \varphi$, *then* $\models_{\mathrm{S5(A)}} \varphi$.

To prove completeness, we will make use of the fact that occurrences of $\Box$ can be shifted inwards and hence that every formula is provably equivalent in $\mathcal{S}5(\mathcal{A}_{\mathrm{m}})$ to a formula of modal depth at most one. For $\varphi, \psi \in \mathrm{Fm}(\mathcal{L}_{\mathrm{m}}^{\Box})$, let us write $\vdash_{\mathcal{S}5(\mathcal{A}_{\mathrm{m}})} \varphi \equiv \psi$ to denote that $\vdash_{\mathcal{S}5(\mathcal{A}_{\mathrm{m}})} \varphi \to \psi$ and $\vdash_{\mathcal{S}5(\mathcal{A}_{\mathrm{m}})} \psi \to \varphi$.

Lemma 4.2 *For any* $\varphi, \psi \in \mathrm{Fm}(\mathcal{L}_{\mathrm{m}}^{\Box})$,

(i) $\vdash_{\mathcal{S}5(\mathcal{A}_{\mathrm{m}})} \Box(\varphi + \Box\psi) \equiv \Box\varphi + \Box\psi$

(ii) $\vdash_{\mathcal{S}5(\mathcal{A}_{\mathrm{m}})} \Box(\varphi + \neg\Box\psi) \equiv \Box\varphi + \neg\Box\psi$

(iii) $\vdash_{\mathcal{S}5(\mathcal{A}_{\mathrm{m}})} \Box\Box\varphi \equiv \Box\varphi$

(iv) $\vdash_{\mathcal{S}5(\mathcal{A}_{\mathrm{m}})} \Box\neg\Box\varphi \equiv \neg\Box\varphi$

(v) $\vdash_{\mathcal{S}5(\mathcal{A}_{\mathrm{m}})} \Box n\varphi \equiv n\Box\varphi$ *for all* $n \in \mathbb{N}$.

Proof. Derivations for (i)-(iv) are obtained, similarly to other "S5" logics, using the modal axiom schema (K), (T), and (5), and are omitted here. For (v), we note first that $n\Box\varphi \to \Box n\varphi$ is derivable in $\mathcal{S}5(\mathcal{A}_{\mathrm{m}})$ for $n \in \mathbb{N}$ using (nec) and (K) together with the axioms of $\mathcal{A}_{\mathrm{m}}$. For the converse, observe that $\Box(2^k)\varphi \to (2^k)\Box\varphi$ is derivable in $\mathcal{S}5(\mathcal{A}_{\mathrm{m}})$ for $k \in \mathbb{N}$ using repeated applications of (M), (mp), and the $\mathcal{A}_{\mathrm{m}}$-derivable formula $\psi_1 \to (\psi_2 \to (\psi_1 + \psi_2))$. But then also for any $n \geq 1$, we can choose $k \in \mathbb{N}$ such that $2^k \geq n$ and observe that $(\Box n\varphi + (2^k - n)\Box\varphi) \to \Box(2^k)\varphi$ and hence $(\Box n\varphi + (2^k - n)\Box\varphi) \to (2^k)\Box\varphi$ are derivable in $\mathcal{S}5(\mathcal{A}_{\mathrm{m}})$. Since $(((2^k - n)\Box\varphi) \to ((2^k - n)\Box\varphi)) \to \overline{0}$ is derivable in

$\mathcal{S}5(\mathcal{A}_\mathrm{m})$, also $\Box n\varphi \to n\Box\varphi$ is derivable in $\mathcal{S}5(\mathcal{A}_\mathrm{m})$ as required. Finally, for the case $n = 0$ just note that $\Box\overline{0} \to \overline{0}$ is an instance of (T). □

Let us write $\sum_{i=1}^{n} \varphi_i$ to denote $\varphi_1 + \ldots + \varphi_n$ for any $\varphi_1, \ldots, \varphi_n \in \mathrm{Fm}(\mathcal{L}_\mathrm{A}^\Box)$. An easy induction on modal depth using Lemma 4.2 (i)-(iv) yields the following normal form property for modal-multiplicative formulas.[5]

Lemma 4.3 *For any modal-multiplicative formula $\varphi \in \mathrm{Fm}(\mathcal{L}_\mathrm{m}^\Box)$, there exist multiplicative formulas $\varphi_0, \varphi_1, \ldots, \varphi_n, \psi_1, \ldots, \psi_m \in \mathrm{Fm}(\mathcal{L}_\mathrm{m})$ such that*

$$\vdash_{\mathcal{S}5(\mathcal{A}_\mathrm{m})} \varphi \equiv \varphi_0 + \sum_{i=1}^{n} \Box\varphi_i + \sum_{j=1}^{m} \neg\Box\psi_j.$$

Let Fm_m denote the set of first-order formulas in Fm not containing $\wedge$ or $\vee$. The following lemma is a consequence of a well-known duality principle for linear programming stating that either one or another linear system has a solution, but not both (see, e.g., [13]): more precisely, for any $M \in \mathbb{Z}^{m\times n}$, either $y^T M < \mathbf{0}$ for some $y \in \mathbb{R}^m$ or $Mx = \mathbf{0}$ for some $x \in \mathbb{N}^n \backslash \{\mathbf{0}\}$.

Lemma 4.4 *For any quantifier-free and variable-free $\alpha_1, \ldots, \alpha_n \in Fm_m$,*

$$\models_{\forall\mathrm{A}} \alpha_1 \vee \ldots \vee \alpha_n \iff \models_{\forall\mathrm{A}} \sum_{i=1}^{n} \lambda_i\alpha_i \text{ for some } \lambda_1, \ldots, \lambda_n \in \mathbb{N} \text{ not all } 0.$$

Proof. Let $\beta_1, \ldots, \beta_m$ denote the m ground atoms $P_i(c_j)$ that occur in $\alpha_1, \ldots, \alpha_n$. We may assume without loss of generality that $\alpha_j = \sum_{i=1}^{m} m_{ij}\beta_j$ for each $j \in \{1, \ldots, n\}$, where $M = (m_{ij}) \in \mathbb{Z}^{m\times n}$. Then $\models_{\forall\mathrm{A}} \alpha_1 \vee \ldots \vee \alpha_n$ if and only if there does not exist any $y \in \mathbb{R}^m$ such that $y^T M < \mathbf{0}$. Hence, by the duality principle mentioned above, $\models_{\forall\mathrm{A}} \alpha_1 \vee \ldots \vee \alpha_n$ if and only if $Mx = \mathbf{0}$ for some $x \in \mathbb{N}^n \backslash \{\mathbf{0}\}$, which is equivalent to the statement that $\models_{\forall\mathrm{A}} \sum_{i=1}^{n} \lambda_i\alpha_i$ for some $\lambda_1, \ldots, \lambda_n \in \mathbb{N}$ not all zero. □

We now have the tools required to prove our completeness theorem for $\mathcal{S}5(\mathcal{A}_\mathrm{m})$.

Theorem 4.5 *Let $\varphi \in \mathrm{Fm}(\mathcal{L}_\mathrm{m}^\Box)$. Then $\vdash_{\mathcal{S}5(\mathcal{A}_\mathrm{m})} \varphi$ if and only if $\models_{\mathrm{S5(A)}} \varphi$.*

Proof. The left-to-right-direction is Lemma 4.1. For the converse, suppose that φ is S5(A)-valid. By Lemma 4.3, there exist $\varphi_0, \varphi_1, \ldots, \varphi_n, \psi_1, \ldots, \psi_m \in \mathrm{Fm}(\mathcal{L}_\mathrm{m})$ such that $\vdash_{\mathcal{S}5(\mathcal{A}_\mathrm{m})} \varphi \equiv \psi$, where

$$\psi = \varphi_0 + \sum_{i=1}^{n} \Box\varphi_i + \sum_{j=1}^{m} \neg\Box\psi_j.$$

By Lemma 4.1, also ψ is S5(A)-valid and it suffices to prove that $\vdash_{\mathcal{S}5(\mathcal{A}_\mathrm{m})} \psi$. Consider now the $\forall$A-valid (by Proposition 2.1) first-order formula

$$\alpha_\psi = \alpha_{\varphi_0}(x) + \sum_{i=1}^{n} (\forall x)\alpha_{\varphi_i}(x) + \sum_{j=1}^{m} \neg(\forall x)\alpha_{\psi_j}(x).$$

[5] It is not possible to obtain a similar normal form property for all $\varphi \in \mathrm{Fm}(\mathcal{L}_\mathrm{A}^\Box)$ simply by shifting boxes; e.g., $\Box(p \vee (q + \Box r))$ is not equivalent to any formula of modal depth ≤ 1.

Using generalization, renaming of variables, and quantifier shifts,

$$\models_{\forall \mathrm{A}} (\forall y_0)(\forall y_1)\ldots(\forall y_n)(\exists x_1)\ldots(\exists x_m)\big(\sum_{i=0}^{n} \alpha_{\varphi_i}(y_i) + \sum_{j=1}^{m} \neg\alpha_{\psi_j}(x_j)\big).$$

Hence also for constants $\bar{c} = c_0, c_1, \ldots, c_n$,

$$\models_{\forall \mathrm{A}} (\exists x_1)\ldots(\exists x_m)\big(\sum_{i=0}^{n} \alpha_{\varphi_i}(c_i) + \sum_{j=1}^{m} \neg\alpha_{\psi_j}(x_j)\big).$$

An application of Theorem 2.2 then yields, writing $\bar{d}$ for $d_1, \ldots, d_m$,

$$\models_{\forall \mathrm{A}} \bigvee \{\sum_{i=0}^{n} \alpha_{\varphi_i}(c_i) + \sum_{j=1}^{m} \neg\alpha_{\psi_j}(d_j) \mid \bar{d} \subseteq \bar{c}\}.$$

But then by Lemma 4.4, there exist $\lambda_{\bar{d}} \in \mathbb{N}$ for each $\bar{d} \subseteq \bar{c}$ not all 0 satisfying

$$\models_{\forall \mathrm{A}} \sum_{\bar{d} \subseteq \bar{c}} \lambda_{\bar{d}}\big(\sum_{i=0}^{n} \alpha_{\varphi_i}(c_i) + \sum_{j=1}^{m} \neg\alpha_{\psi_j}(d_j)\big).$$

Hence also, letting $\mu = \sum_{\bar{d} \subseteq \bar{c}} \lambda_{\bar{d}}$,

$$\models_{\forall \mathrm{A}} \sum_{i=0}^{n} \mu\alpha_{\varphi_i}(c_i) + \sum_{\bar{d} \subseteq \bar{c}} \lambda_{\bar{d}} \sum_{j=1}^{m} \neg\alpha_{\psi_j}(d_j).$$

Now let us rewrite the second part of this $\forall$A-valid formula to obtain

$$\models_{\forall \mathrm{A}} \sum_{i=0}^{n} \mu\alpha_{\varphi_i}(c_i) + \sum_{i=0}^{n}\sum_{j=1}^{m} \lambda_{ij} \neg\alpha_{\psi_j}(c_i),$$

for some λ_{ij} $(0 \leq i \leq n,\ 1 \leq j \leq m)$ such that $\sum_{i=0}^{n}\sum_{j=1}^{m} \lambda_{ij} = \sum_{\bar{d} \subseteq \bar{c}} \lambda_{\bar{d}} = \mu$.

Then for each $i \in \{0, 1, \ldots, n\}$, we must have

$$\models_{\forall \mathrm{A}} \mu\alpha_{\varphi_i}(c_i) + \sum_{j=1}^{m} \lambda_{ij} \neg\alpha_{\psi_j}(c_i).$$

So also, by Proposition 2.1,

$$\models_{\mathrm{S5(A)}} \mu\varphi_i + \sum_{j=1}^{m} \lambda_{ij} \neg\psi_j.$$

By the completeness of $\mathcal{A}_{\mathrm{m}}$ with respect to $\langle\langle \mathbb{R}, +, -, 0\rangle, \mathbb{R}^{\geq 0}\rangle$, it follows that for each $i \in \{0, 1, \ldots, n\}$,

$$\vdash_{\mathcal{S}5(\mathcal{A}_{\mathrm{m}})} \mu\varphi_i + \sum_{j=1}^{m} \lambda_{ij} \neg\psi_j.$$

But then for each $i \in \{1, \ldots, n\}$, using (nec), (K), (T), and Lemma 4.2,

$$\vdash_{S5(\mathcal{A}_\mathrm{m})} \mu\Box\varphi_i + \sum_{j=1}^{m} \lambda_{ij}\neg\Box\psi_j \quad \text{and also} \quad \vdash_{S5(\mathcal{A}_\mathrm{m})} \mu\varphi_0 + \sum_{j=1}^{m} \lambda_{0j}\neg\Box\psi_j,$$

and using (mp) and the $\mathcal{A}_\mathrm{m}$-axiom (+1),

$$\vdash_{S5(\mathcal{A}_\mathrm{m})} \mu\varphi_0 + \sum_{i=1}^{n} \mu\Box\varphi_i + \mu\sum_{j=1}^{m} \neg\Box\psi_j.$$

Finally, an application of (con_μ) yields $\vdash_{S5(\mathcal{A}_\mathrm{m})} \psi$ as required. □

In principle, this proof strategy can also be used to prove completeness for an axiom system for the full logic S5(A). New variables can be introduced to obtain a depth-one formula as in Lemma 4.3, and Theorem 2.2 can then be applied to the resulting existential sentence to obtain an S5(A)-valid disjunction of quantifier-free sentences. However, the presence of $\wedge$ and $\vee$ requires repeated applications of Lemma 4.4 and currently we are able only to prove completeness using this method for a system with a family of combinatorially defined axioms.

Let us remark finally that, as in the classical setting, the monadic logic $\forall$A restricted to Fm_m coincides (up to equivalence of sentences) with its one-variable fragment. Let $\alpha \in Fm_m$ be any sentence. Repeated applications of the quantifier-shift $\models_{\forall\mathrm{A}} (\forall x)(\alpha_1 + \alpha_2) \leftrightarrow ((\forall x)\alpha_1 + \alpha_2)$, where x is not free in α_2 yield a sentence $\beta \in Fm_m$ such that $\models_{\forall\mathrm{A}} \alpha \leftrightarrow \beta$ and no subformula $(\forall x)\beta'$ of β contains a free variable different to x. Hence we can rename all the bound variables in β to obtain a one-variable sentence $\chi \in Fm_m$ such that $\models_{\forall\mathrm{A}} \alpha \leftrightarrow \chi$. Since S5(A) is decidable (Corollary 2.3), first-order multiplicative Abelian logic provides a first interesting example (as far as we know) of a first-order infinite-valued logic that has a decidable monadic fragment.

5 Monadic Abelian ℓ-Groups

In this section, we introduce abelian ℓ-groups supplemented with a monadic operator as an algebraic semantics for S5(A). Following similar results for monadic Heyting algebras [5] and monadic MV-algebras [14], we describe a correspondence between these algebras and lattice-ordered abelian groups equipped with certain "relatively complete" subalgebras. We then use this correspondence to give a characterization of the "ideals" of these algebras.

An *abelian ℓ-group* is an algebraic structure $\mathbf{G} = \langle G, \wedge, \vee, +, -, 0\rangle$ such that $\langle G, \wedge, \vee\rangle$ is a lattice, $\langle G, +, -, 0\rangle$ is an abelian group, and $+$ is compatible with the lattice order, i.e., $a \leq b$ implies $a + c \leq b + c$ for all $a, b, c \in G$. We call $\mathbf{G}$ an *abelian o-group* if the lattice order $\leq$ is linear. A non-empty subset $H \subseteq G$ that is closed under the operations of $\mathbf{G}$ forms an *ℓ-subgroup* $\mathbf{H}$ of $\mathbf{G}$, where $\mathbf{H}$ is called an *ℓ-ideal* of $\mathbf{G}$ if it is also convex, i.e., if $a, b \in H$, $c \in G$, and $a \leq c \leq b$, then $c \in H$. For any ℓ-ideal $\mathbf{H}$ of $\mathbf{G}$, the set of right cosets of $\mathbf{H}$ in $\mathbf{G}$ forms an abelian ℓ-group $\mathbf{G}/\mathbf{H}$ with lattice order $H + a \leq H + b :\Leftrightarrow a \leq b + c$ for some $c \in H$. We refer to [1] for further details.

Example 5.1 The ordered additive group $\mathbf{R}$ encountered in Section 2 is an abelian o-group. Also important for our purposes are abelian ℓ-groups obtained as sets of functions from a set W to an abelian ℓ-group $\mathbf{G}$ with operations defined pointwise, denoted by $\mathbf{G}^W$. In particular, we will consider the case where $\mathbf{G}$ is an abelian o-group and the bounded functions from W to G form an ℓ-subgroup $\mathbf{B}(W, \mathbf{G})$ of $\mathbf{G}^W$.

A *monadic abelian* ℓ*-group* is an ordered pair $\langle \mathbf{G}, \Box \rangle$ consisting of an abelian ℓ-group $\mathbf{G}$ and a unary operator $\Box$ on G, with defined operator $\Diamond a := -\Box -a$, that satisfies for all $a, b \in G$,

$$\begin{array}{llll} \text{(M1)} & \Box(a+b) \leq \Box a + \Diamond b & \text{(M4)} & \Box(a \wedge b) = \Box a \wedge \Box b \\ \text{(M2)} & \Box a \leq a & \text{(M5)} & \Diamond(a \wedge \Diamond b) = \Diamond a \wedge \Diamond b \\ \text{(M3)} & \Diamond a = \Box \Diamond a & \text{(M6)} & \Box(a+a) = \Box a + \Box a. \end{array}$$

A non-empty subset $H \subseteq G$ forms a *monadic* ℓ*-subgroup* $\langle \mathbf{H}, \Box \rangle$ of $\langle \mathbf{G}, \Box \rangle$ if $\mathbf{H}$ is an ℓ-subgroup of $\mathbf{G}$ that is closed under $\Box$.

Let $\mathsf{M\ell G}$ denote the variety of monadic abelian ℓ-groups. We call $\langle \mathbf{G}, \Box \rangle \in \mathsf{M\ell G}$ *functional* if $\mathbf{G}$ is an ℓ-subgroup of $\mathbf{B}(W, \mathbf{H})$ for a set W and abelian o-group $\mathbf{H}$, and for all $f \in G$, $x \in W$,

$$\Box f(x) = \bigwedge \{ f(y) \mid y \in W \}.$$

If $\Box f(x) = \min\{ f(y) \mid y \in W \}$ for all $f \in G$, $x \in W$, we call $\langle \mathbf{G}, \Box \rangle$ *witnessed*, and in the case where $\mathbf{H}$ is $\mathbf{R}$, we call $\langle \mathbf{G}, \Box \rangle$ *standard*.

Observe now that for any $\langle \mathbf{G}, \Box \rangle \in \mathsf{M\ell G}$, the set $\Box G := \{ \Box a \mid a \in G \} = \{ \Diamond a \mid a \in G \}$ forms an ℓ-subgroup $\Box \mathbf{G}$ of $\mathbf{G}$ satisfying for all $a \in G$,

$$\Box a = \bigvee \{ b \in \Box G \mid b \leq a \}.$$

More generally, an ℓ-subgroup $\mathbf{G}_0$ of an abelian ℓ-group $\mathbf{G}$ is *relatively complete* if $\bigvee \{ b \in \Box G \mid b \leq a \}$ exists for all $a \in G$, or, equivalently, the inclusion map of G_0 in G has a right adjoint $\Box_0 \colon G \to G_0$, i.e., for all $a \in G_0$ and $b \in G$,

$$a \leq \Box_0 b \iff a \leq b.$$

In this case, we obtain an algebraic structure $\langle \mathbf{G}, \Box_0 \rangle$ that satisfies conditions (M1)-(M4) in the definition of a monadic abelian ℓ-group. To ensure, however, that (M5) and (M6) are satisfied, we require also that for all $a, b \in G$,

$$\Box_0(a+a) = \Box_0 a + \Box_0 a \quad \text{and} \quad \Diamond_0(a \wedge \Diamond_0 b) = \Diamond_0 a \wedge \Diamond_0 b,$$

in which case $\langle \mathbf{G}, \Box_0 \rangle \in \mathsf{M\ell G}$ with $\Box_0 G = G_0$, and we call $\mathbf{G}_0$ *m-relatively complete*. Hence we obtain the following result.

Proposition 5.2 *There exists a one-to-one correspondence between monadic abelian* ℓ*-groups* $\langle \mathbf{G}, \Box \rangle$ *and ordered pairs* $\langle \mathbf{G}, \mathbf{G}_0 \rangle$ *of abelian* ℓ*-groups such that* $\mathbf{G}_0$ *is an m-relatively complete* ℓ*-subgroup of* $\mathbf{G}$.

Example 5.3 The universe of any non-trivial relatively complete ℓ-subgroup of an abelian ℓ-group $\mathbf{B}(W, \mathbf{R})$ for some set W is a set of constant functions $\{f\colon W \to \{r\} \mid r \in H\}$, where $\mathbf{H}$ will be $\mathbf{R}$ if the ℓ-subgroup is m-relatively complete, and a one-generated ℓ-subgroup of $\mathbf{R}$ otherwise.

Given a monadic abelian ℓ-group $\langle \mathbf{G}, \Box \rangle$, we say that $\mathbf{K}$ is a *monadic ℓ-ideal* of $\langle \mathbf{G}, \Box \rangle$ if $\mathbf{K}$ is an ℓ-ideal of $\mathbf{G}$ and $a \in K$ implies $\Box a \in K$. It is straightforward to check that in this case, $\langle \mathbf{G}, \Box \rangle / \mathbf{K} := \langle \mathbf{G}/\mathbf{K}, \Box_K \rangle$ with $\Box_K(K+a) := K + \Box a$ is a monadic abelian ℓ-group.

Proposition 5.4 *The monadic ℓ-ideals of a monadic abelian ℓ-group $\langle \mathbf{G}, \Box \rangle$ and the ℓ-ideals of $\Box\mathbf{G}$ are in a one-to-one correspondence implemented by the maps $J \mapsto J \cap \Box G$ and $K \mapsto K^{\Box\Diamond} := \{a \in G \mid \Box a \in K \text{ and } \Diamond a \in K\}$.*

Proof. First consider any ℓ-ideal $\mathbf{K}$ of $\Box\mathbf{G}$. We show that $\mathbf{K}^{\Box\Diamond}$ is a monadic ℓ-ideal of $\mathbf{G}$. For closure under $-$, observe that if $a \in K^{\Box\Diamond}$ (i.e., $\Box a, \Diamond a \in K$), since $\mathbf{K}$ is an ℓ-ideal, $-\Box a = \Diamond - a \in K$ and $-\Diamond a = \Box - a \in K$, so $-a \in K^{\Box\Diamond}$. For closure under $+$, observe that if $a, b \in K^{\Box\Diamond}$ (i.e., $\Box a, \Box b, \Diamond a, \Diamond b \in K$), using property (M1) of monadic abelian ℓ-groups,

$$K \ni \Box a + \Box b \leq \Box(a+b) \leq \Diamond a + \Box b \in K$$
$$K \ni \Box a + \Diamond b \leq \Diamond(a+b) \leq \Diamond a + \Diamond b \in K,$$

so by convexity $\Box(a+b), \Diamond(a+b) \in K$ and hence $a + b \in K^{\Box\Diamond}$. Moreover, using properties (M4) and (M2),

$$K \ni \Box a \wedge \Box b = \Box(a \wedge b) \leq \Diamond(a \wedge b) \leq \Diamond a \wedge \Diamond b \in K,$$

so by convexity again, $a \wedge b \in K^{\Box\Diamond}$. Closure under $\Box$ is clear and convexity is a consequence of the monotonicity of $\Box$ and $\Diamond$. So $\mathbf{K}^{\Box\Diamond}$ is a monadic ℓ-ideal and, since $a = \Box a = \Diamond a$ for any $a \in K$, also $K = K^{\Box\Diamond} \cap \Box G$.

Now consider any monadic ℓ-ideal $\mathbf{J}$ of $\langle \mathbf{G}, \Box \rangle$. Since $\Box\mathbf{G}$ is an ℓ-subgroup of $\mathbf{G}$, it follows easily that $J \cap \Box G$ is the universe of an ℓ-ideal of $\Box\mathbf{G}$. Moreover, $\Box J \subseteq J \cap \Box G$, so $J \subseteq (J \cap \Box G)^{\Box\Diamond}$. Conversely, if $a \in (J \cap \Box G)^{\Box\Diamond}$, then $\Box a, \Diamond a \in J$ and, since $\Box a \leq a \leq \Diamond a$, by convexity, $a \in J$. So $J = (J \cap \Box G)^{\Box\Diamond}$ and we have shown that the maps implement a one-to-one correspondence. □

6 A Completeness Theorem

In this section, we prove the completeness with respect to S5(A)-validity of an axiom system $\mathcal{S}5(\mathcal{A})$ consisting of the axiom and rule schema for Abelian logic in Fig. 1, the modal axiom and rule schema in Fig. 2, and the axiom schema

$$(\wedge\Box)\ \ (\Box\varphi \wedge \Box\psi) \to \Box(\varphi \wedge \psi) \qquad (\wedge\Diamond)\ \ (\Diamond\varphi \wedge \Diamond\psi) \to \Diamond(\varphi \wedge \Diamond\psi).$$

First, a standard Lindenbaum-Tarski argument can be used to prove that $\mathcal{S}5(\mathcal{A})$ is complete with respect to the variety $\mathsf{M\ell G}$ of monadic abelian ℓ-groups.

Lemma 6.1 *Let $\varphi \in \mathrm{Fm}(\mathcal{L}^{\Box}_{\mathrm{A}})$. Then $\vdash_{\mathcal{S}5(\mathcal{A})} \varphi$ if and only if $\mathsf{M\ell G} \models \overline{0} \leq \varphi$.*

The remainder of this section is dedicated to proving the completeness of $\mathcal{S}5(\mathcal{A})$ with respect to first the functional members and then the standard members of MℓG. As a first step towards these results, we show that it suffices to consider monadic abelian ℓ-groups $\langle \mathbf{G}, \Box \rangle$ such that $\Box\mathbf{G}$ is linearly ordered, which, for convenience, we call *chain-monadic* abelian ℓ-groups.

Recall (see e.g. [6]) that a monadic abelian ℓ-group $\langle \mathbf{G}, \Box \rangle$ is a *subdirect product* of a family of monadic abelian ℓ-groups $(\langle \mathbf{H}_j, \Box_j \rangle)_{j\in J}$ if it is a monadic ℓ-subgroup of the direct product $\prod_{j\in J} \langle \mathbf{H}_j, \Box_j \rangle$ such that each projection map $\pi_j \colon \prod_{k\in J} \langle \mathbf{H}_k, \Box_k \rangle \to \langle \mathbf{H}_j, \Box_j \rangle$; $(a_k)_{k\in J} \mapsto a_j$ is surjective. Crucially, if an equation fails in $\langle \mathbf{G}, \Box \rangle$, then it fails in some $\langle \mathbf{H}_j, \Box_j \rangle$. Let us also recall that an ℓ-ideal $\mathbf{K}$ of an abelian ℓ-group $\mathbf{G}$ is called *prime* if $\mathbf{G}/\mathbf{K}$ is linearly ordered.

Lemma 6.2 *Each monadic abelian ℓ-group is isomorphic to a subdirect product of chain-monadic abelian ℓ-groups.*

Proof. Let $\langle \mathbf{G}, \Box \rangle$ be a monadic abelian ℓ-group and let S be the set of all prime ℓ-ideals $\mathbf{P}$ of $\Box\mathbf{G}$. Then $\Box\mathbf{G}/\mathbf{P}$ is linearly ordered for each $\mathbf{P} \in S$ and $\bigcap\{P \mid \mathbf{P} \in S\} = \{0\}$ (see, e.g., [1, Proposition 1.2.9]). By Proposition 5.4, each $\mathbf{P} \in S$ corresponds to a monadic ℓ-ideal $\mathbf{P}^{\Box\Diamond}$ of $\langle \mathbf{G}, \Box \rangle$ such that $\Box\mathbf{G}/\mathbf{P}^{\Box\Diamond}$ is linearly ordered. Moreover, since $\Box a = \Diamond a = 0$ implies $a = 0$ for all $a \in G$, it follows that $\bigcap\{P^{\Box\Diamond} \mid \mathbf{P} \in S\} = \{0\}$ and the map $\sigma \colon \langle \mathbf{G}, \Box \rangle \to \prod_{\mathbf{P}\in S} \langle \mathbf{G}, \Box \rangle / \mathbf{P}^{\Box\Diamond}$; $a \mapsto (P^{\Box\Diamond} + a)_{\mathbf{P}\in S}$ is an embedding between monadic abelian ℓ-groups. Hence, $\langle \mathbf{G}, \Box \rangle$ is isomorphic to a subdirect product of the family of chain-monadic abelian ℓ-groups $(\langle \mathbf{G}, \Box \rangle / \mathbf{P}^{\Box\Diamond})_{\mathbf{P}\in S}$. □

Following a method used in [9] to characterize subdirectly irreducible monadic MV-algebras, we now show that each chain-monadic abelian ℓ-group $\langle \mathbf{G}, \Box \rangle$ admits a functional representation.

Lemma 6.3 *Let $\langle \mathbf{G}, \Box \rangle$ be a chain-monadic abelian ℓ-group and $a \in G$. Then there exists a prime ℓ-ideal $\mathbf{P}$ of $\mathbf{G}$ such that $P + a = P + \Box a$ and $P \cap \Box G = \{0\}$.*

Proof. Let $\langle \mathbf{G}, \Box \rangle$ be a chain-monadic abelian ℓ-group and $a \in G$. We apply Zorn's Lemma to the set $\mathcal{K}$ of all ℓ-ideals $\mathbf{K}$ of $\mathbf{G}$ such that $K \cap \Box G = \{0\}$ and $a - \Box a \in K$, ordered by inclusion. First, we check that $\mathcal{K}$ is non-empty. We show that the ℓ-ideal $\mathbf{K}(a - \Box a)$ of $\mathbf{G}$ generated by the element $a - \Box a$ is in $\mathcal{K}$. By, e.g., [1, Proposition 1.2.3], recalling that $|x| := x \vee -x$ for any $x \in G$,

$$K(a - \Box a) = \{b \in G \mid |b| \leq n|a - \Box a| \text{ for some } n \in \mathbb{N}\}.$$

Let $b \in K(a - \Box a) \cap \Box G$. Then for some $n \in \mathbb{N}$,

$$\begin{aligned} |b| = \Box|b| &\leq \Box(2^n|a - \Box a|) && \text{since } |b| \in \Box G,\ b \in K(a - \Box a) \\ &= 2^n\Box|a - \Box a| && \text{using (M6)} \\ &= 2^n\Box(a - \Box a) && \text{using (M2)} \\ &= 2^n(\Box a - \Box a) && \text{using (M1), (M2), and (M3)} \\ &= 0. \end{aligned}$$

So $b = 0$ and $\mathcal{K} \neq \emptyset$. Moreover, it is easy to see that $\mathcal{K}$ is closed under taking unions of chains, so Zorn's Lemma yields a maximal element $\mathbf{P} \in \mathcal{K}$.

Suppose for a contradiction that $\mathbf{P}$ is not prime. Then there exist $b, c \in G$ with $b \wedge c = 0$ but $b, c \notin P$ (see, e.g., [1, Theorem 1.2.10]). By the maximality of $\mathbf{P}$, there exist $r \in (P(b) \cap \Box G)\backslash\{0\}$ and $s \in (P(c) \cap \Box G)\backslash\{0\}$, where $\mathbf{P}(b)$ and $\mathbf{P}(c)$ are the ℓ-ideals generated by $P \cup \{b\}$ and $P \cup \{c\}$, respectively. Since $\Box\mathbf{G}$ is linearly ordered, we can assume without loss of generality that $|r| \leq |s|$. Convexity then implies that also $r \in P(c) \cap \Box G$. Hence $r \in P(b) \cap P(c) = P(b \wedge c) = P(0) = P$. But $P \cap \Box G = \{0\}$, so $r = 0$, a contradiction. That is, $\mathbf{P}$ is prime. Finally, note that since $a - \Box a \in P$, also $P + a = P + \Box a$. □

Lemma 6.4 *Let $\langle \mathbf{G}, \Box \rangle$ be a chain-monadic abelian ℓ-group and $a \in G\backslash\{0\}$. Then there exists a prime ℓ-ideal $\mathbf{P}$ of $\mathbf{G}$ such that $a \notin P$ and $P \cap \Box G = \{0\}$.*

Proof. Let $\langle \mathbf{G}, \Box \rangle$ be a chain-monadic abelian ℓ-group and $a \in G\backslash\{0\}$. We apply Zorn's Lemma to the set $\mathcal{K}$ of all proper ℓ-ideals $\mathbf{K}$ of $\mathbf{G}$ such that for all $r \in \Box G\backslash\{0\}$, $|a| \wedge |r| \notin K$, ordered by inclusion. To show that $\{0\} \in \mathcal{K}$, it suffices to show that for $a \in G$, $r \in \Box G$, $a \wedge r = 0$ implies that $a = 0$ or $r = 0$. If $a \wedge r = 0$, then also $\Box(a \wedge r) = \Box a \wedge r = 0$ and $\Diamond(a \wedge r) = \Diamond a \wedge r = 0$ using conditions (M4) and (M5), respectively. Since $\Box\mathbf{G}$ is linearly ordered, either $r = 0$ or $\Box a = \Diamond a = 0$, i.e. $r = 0$ or $a = 0$. Moreover, $\bigcup \mathcal{C} \in \mathcal{K}$ for any chain $\mathcal{C} \subseteq \mathcal{K}$, therefore $\mathcal{K}$ contains a maximal element $\mathbf{P}$.

We show next that $\mathbf{P}$ is prime. Consider $b, c \in G$ such that $b \wedge c = 0$ and suppose for a contradiction that $b, c \notin P$. By the maximality of $\mathbf{P}$, neither $\mathbf{P}(b)$ nor $\mathbf{P}(c)$ belongs to $\mathcal{K}$ and so there exist $p, q \in \Box G\backslash\{0\}$ such that $|a| \wedge |p| \in P(b)$ and $|a| \wedge |q| \in P(c)$. Since $\Box\mathbf{G}$ is linearly ordered, we can assume without loss of generality that $|p| \leq |q|$. Then $0 \leq |a| \wedge |p| \leq |a| \wedge |q|$, so by convexity, $|a| \wedge |p| \in P(b) \cap P(c) = P(b \wedge c) = P$, contradicting $\mathbf{P} \in \mathcal{K}$.

Lastly note that $\mathbf{P}$ satisfies the required properties. For, if $a \in P$, then $|a| \in P$ and so by convexity, $|a| \wedge |r| \in P$ for all $r \in \Box G$, contradicting $\mathbf{P} \in \mathcal{K}$. It follows similarly that $P \cap \Box G = \{0\}$. □

Theorem 6.5 *Any chain-monadic abelian ℓ-group $\langle \mathbf{G}, \Box \rangle$ is isomorphic to a witnessed functional monadic abelian ℓ-group.*

Proof. Let $\langle \mathbf{G}, \Box \rangle$ be a chain-monadic abelian ℓ-group, and let $\{\mathbf{P}_i\}_{i \in I}$ be the family of all prime ℓ-ideals $\mathbf{P}$ of $\mathbf{G}$ such that $P \cap \Box G = \{0\}$. It follows from Lemma 6.4 that $\bigcap\{P_i \mid i \in I\} = \{0\}$ and hence that $\sigma\colon \mathbf{G} \to \prod_{i \in I} \mathbf{G}/\mathbf{P}_i$; $a \mapsto (a + P_i)_{i \in I}$ is an embedding between abelian ℓ-groups. Moreover, for each $i \in I$, since $P_i \cap \Box G = \{0\}$, the map $\pi_i \circ \sigma|_{\Box G}$ is an ℓ-embedding, where π_i is the ith projection map.

We make use of a *generalized amalgamation property* for abelian o-groups: that is, for any abelian o-group $\mathbf{H}_0$, family of abelian o-groups $\{\mathbf{H}_j\}_{j \in J}$, and family of ℓ-embeddings $\{\gamma_j\colon \mathbf{H}_0 \to \mathbf{H}_j\}_{j \in J}$, there exists an abelian o-group $\mathbf{H}$ (called the *amalgam*) and family of ℓ-embeddings $\{\sigma_j\colon \mathbf{H}_j \to \mathbf{H}\}_{j \in J}$ such that $\sigma_{j_1} \circ \gamma_{j_1} = \sigma_{j_2} \circ \gamma_{j_2}$ for all $j_1, j_2 \in J$. This property was established by Pierce [27] for families of size 2 and extended to the generalized version in [9].

For the abelian o-group $\Box\mathbf{G}$, family of abelian o-groups $\{\mathbf{G}/\mathbf{P}_i\}_{i\in I}$ and family of ℓ-embeddings $\{\pi_i \circ \sigma|_{\Box G}\colon \Box\mathbf{G} \to \mathbf{G}/\mathbf{P}_i\}_{i\in I}$, we therefore obtain an amalgam $\mathbf{H}$ with ℓ-embeddings $\gamma_i\colon \mathbf{G}/\mathbf{P}_i \to \mathbf{H}$ for each $i \in I$. Defining $\gamma := \prod_{i\in I}\gamma_i\colon \prod_{i\in I} G/P_i \to H^I$ yields an ℓ-embedding $\rho := \gamma\circ\sigma\colon \mathbf{G} \to \mathbf{H}^I$. Observe now that for all $r \in \Box G$ and $i, j \in I$,

$$\rho(r)(i) = \gamma_i(\sigma(r)(i)) = \gamma_i(\pi_i(\sigma(r))) = \gamma_j(\pi_j(\sigma(r))) = \gamma_j(\sigma(r)(j)) = \rho(r)(j).$$

That is, $\rho(r)$ is a constant function. Moreover, for each $a \in G$, there exists, by Lemma 6.3, an $i \in I$ such that $P_i + a = P_i + \Box a$ and hence $\rho(\Box a)(i) = \rho(a)(i)$. So for any $a \in G$ and $i \in I$, we obtain $\rho(\Box a)(i) = \min\{\rho(a)(j) \mid j \in I\}$. □

To prove the promised completeness result for $\mathcal{S}5(\mathcal{A})$, we make use of the following folklore result from the theory of abelian ℓ-groups.

Lemma 6.6 (cf. [10]) *Let* $\mathbf{G}$ *be an abelian o-group. For each finite subset* S *of* G*, there exists a function* $h\colon S \to \mathbb{R}$ *satisfying for all* $a, b, c \in S$*,*

(i) $a \leq b$ *if and only if* $h(a) \leq h(b)$*;*

(ii) *if* $0 \in S$*, then* $h(0) = 0$*;*

(iii) $a + b = c$ *if and only if* $h(a) + h(b) = h(c)$*;*

(iv) $b = -a$ *if and only if* $h(b) = -h(a)$*.*

Theorem 6.7 *Let* $\varphi \in \mathrm{Fm}(\mathcal{L}_{\mathrm{A}}^{\Box})$*. Then* $\vdash_{\mathcal{S}5(\mathcal{A})} \varphi$ *if and only if* $\models_{\mathrm{S5(A)}} \varphi$*.*

Proof. For the left-to-right direction, it is easily checked that the axioms are S5(A)-valid and the rules preserve S5(A)-validity. For the converse, suppose that $\nvdash_{\mathcal{S}5(\mathcal{A})} \varphi$. By Lemmas 6.1 and 6.2, there exist a chain-monadic abelian ℓ-group $\langle \mathbf{G}, \Box\rangle$ and a valuation $e\colon \mathrm{Fm}(\mathcal{L}_{\mathrm{A}}^{\Box}) \to \langle \mathbf{G}, \Box\rangle$ such that $0 \not\leq e(\varphi)$. By Theorem 6.5, we may assume that $\mathbf{G}$ is a witnessed ℓ-subgroup of $\mathbf{B}(W, \mathbf{H})$ for some non-empty set W and abelian o-group $\mathbf{H}$. Hence there exists $x_0 \in W$ such that $e(\varphi)(x_0) < 0$. Let Σ be the set of subformulas of φ. For each $\Box\psi \in \Sigma$, we choose $x_{\Box\psi} \in W$ such that

$$e(\Box\psi)(x_{\Box\psi}) = e(\psi)(x_{\Box\psi}).$$

Let $W' := \{x_{\Box\psi} \in W \mid \Box\psi \in \Sigma\} \cup \{x_0\}$ and define

$$S := \{e(\psi)(x) \mid x \in W',\ \psi \in \Sigma\} \cup \{-e(\psi)(x) \mid x \in W',\ \psi \in \Sigma\} \cup \{0\}.$$

Since both W' and Σ are finite, so is S. Using Lemma 6.6, we obtain a function $h\colon S \to \mathbb{R}$ satisfying the properties (i)-(iv). We consider the standard monadic abelian ℓ-group $\langle \mathbf{B}(W', \mathbf{R}), \Box\rangle$ and any valuation $e'\colon \mathrm{Fm}(\mathcal{L}_{\mathrm{A}}^{\Box}) \to \langle \mathbf{B}(W', \mathbf{R}), \Box\rangle$ such that for each $p \in \Sigma \cap \mathrm{Var}$ and $x \in W'$,

$$e'(p)(x) := h(e(p)(x)).$$

A simple induction on formulas shows that $e'(\psi)(x) = h(e(\psi)(x))$ for all $\psi \in \Sigma$ and $x \in W'$, and in particular,

$$e'(\varphi)(x_0) = h(e(\varphi)(x_0)) < h(0) = 0.$$

Finally, consider the S5(A)-model $\langle W', V \rangle$ where $V(p,x) := e'(p)(x)$ for each $x \in W'$ and observe that $V(\varphi, x_0) = e'(\varphi)(x_0) < 0$. Hence $\not\models_{\mathrm{S5(A)}} \varphi$. $\square$

References

[1] M.E. Anderson and T.H. Feil, *Lattice-Ordered Groups: An Introduction*, Springer, 1988.

[2] F. Baader, S. Borgwardt, and R. Peñaloza, *Decidability and complexity of fuzzy description logics*, KI **31** (2017), no. 1, 85–90.

[3] M. Baaz and G. Metcalfe, *Herbrand's theorem, Skolemization, and proof systems for first-order Łukasiewicz logic*, J. Logic Comput. **20** (2010), no. 1, 35–54.

[4] S. Baratella, *Continuous propositional modal logic*, J. Appl. Non-Classical Logics **28** (2018), no. 4, 297–312.

[5] G. Bezhanishvili, *Varieties of monadic Heyting algebras - part I*, Studia Logica **61** (1998), no. 3, 367–402.

[6] S. Burris and H. P. Sankappanavar, *A course in universal algebra*, Springer, 1981.

[7] X. Caicedo, G. Metcalfe, R. Rodríguez, and O. Tuyt, *The one-variable fragment of Corsi logic*, Proc. WoLLIC, 2019, pp. 70–83.

[8] E. Casari, *Comparative logics and abelian ℓ-groups*, Logic colloquium '88, 1989, pp. 161–190.

[9] D. Castaño, C. Cimadamore, J.P.D. Varela, and L. Rueda, *Completeness for monadic fuzzy logics via functional algebras*, Fuzzy Sets and Systems (2020). In press.

[10] R. Cignoli and D. Mundici, *Partial isomorphisms on totally ordered abelian groups and Hájek's completeness theorem for basic logic*, Multiple-Valued Logic **6** (2001), 89–94.

[11] C. Cimadamore and J.P.D. Varela, *Monadic MV-algebras are equivalent to monadic ℓ-groups with strong unit*, Studia Logica **98** (2011), no. 1–2, 175–201.

[12] P. Cintula, D. Diaconescu, and G. Metcalfe, *Skolemization and Herbrand theorems for lattice-valued logics*, Theor. Comput. Sci. **768** (2019), 54–75.

[13] G.B. Dantzig, *Linear programming and extensions*, Princeton Univ. Press, 1963.

[14] A. di Nola and R. Grigolia, *On monadic* MV-*algebras*, Ann. Pure Appl. Logic **128** (2004), no. 1-3, 125–139.

[15] D. Diaconescu, G. Metcalfe, and L. Schnüriger, *A real-valued modal logic*, Log. Methods Comput. Sci. **14** (2018), no. 1, 1–27.

[16] T. Flaminio, L. Godo, and E. Marchioni, *Logics for belief functions on MV-algebras*, Int. J. Approx. Reason. **54** (2013), no. 4, 491–512.

[17] T. Flaminio and F. Montagna, *MV-algebras with internal states and probabilistic fuzzy logics*, Internat. J. Approx. Reason. **50** (2009), 138–152.

[18] R. Furber, R. Mardare, and M. Mio, *Probabilistic logics based on Riesz spaces*, Log. Methods Comput. Sci. **16** (2020).

[19] N. Galatos and C. Tsinakis, *Generalized MV-algebras*, J. Algebra **283** (2005), 254–291.

[20] L. Godo, P. Hájek, and F. Esteva, *A fuzzy modal logic for belief functions*, Fundam. Inform. **57** (2003), 127–146.

[21] P. Hájek, *Making fuzzy description logic more general*, Fuzzy Sets and Systems **154** (2005), no. 1, 1–15.

[22] P. Hájek, *On fuzzy modal logics S5(C)*, Fuzzy Sets and Systems **161** (2010), 2389–2396.

[23] G. Hansoul and B. Teheux, *Extending Łukasiewicz logics with a modality: Algebraic approach to relational semantics*, Studia Logica **101** (2013), no. 3, 505–545.

[24] R.K. Meyer and J.K. Slaney, *Abelian logic from A to Z*, Paraconsistent logic: Essays on the inconsistent, 1989, pp. 245–288.

[25] M. Mio and A. Simpson, *Łukasiewicz μ-calculus*, Fundam. Inform. **150** (2017), 317–346.

[26] F. Montagna and C. Tsinakis, *Ordered groups with a conucleus*, J. Pure Appl. Algebra **214** (2010), no. 1, 71–88.

[27] K.R. Pierce, *Amalgamations of lattice ordered groups*, Trans. Amer. Math. Soc. **172** (1972), 249–260.

[28] J.D. Rutledge, *A preliminary investigation of the infinitely many-valued predicate calculus*, Ph.D. Thesis, 1959.

[29] U. Straccia, *Reasoning within fuzzy description logics*, J. Artificial Intelligence Res. **14** (2001), 137–166.

[30] V. Weispfenning, *Model theory of abelian ℓ-groups*, Lattice-ordered groups, 1989, pp. 41–79.

Actuality in Intuitionistic Logic

Satoru Niki [1]

School of Information Science, Japan Advanced Institute of Science and Technology
1-1 Asahidai, Nomi, 923-1292, Ishikawa, Japan

Hitoshi Omori [2]

Depaertment of Philosophy I, Ruhr University Bochum
Universitätsstraße 150, 44780, Bochum, Germany

Abstract

In "Empirical Negation", Michael De takes up the challenge of extending intuitionism from mathematical discourse to empirical discourse, and to this end, he introduced an expansion of intuitionistic propositional logic obtained by adding a unary connective called empirical negation. The intuitive reading of empirical negation of A is: it is not the case that there is sufficient evidence at present that A. From a model-theoretic perspective, cashed out in terms of pointed Kripke models for intuitionistic logic, empirical negation of A is forced at a point iff A is not forced at the base point. Then, a simple calculation reveals that double empirical negation of A is forced at a point iff A is forced at the base point. In other words, double empirical negation can be seen as an actuality operator explored by John N. Crossley, Lloyd Humberstone, Martin Davies and more. Based on these, we introduce an expansion of intuitionistic propositional logic obtained by adding actuality. Our main results include sound and strongly complete axiomatization as well as comparisons to closely related systems such as Global Intuitionistic Logic of Satoko Titani as well as LGP of Matthias Baaz.

Keywords: Actuality, Empirical Negation, Global Intuitionistic Logic, Intuitionistic Modal Logic, Completeness, Sequent Calculus.

1 Introduction

In the literature, there are various expansions of intuitionistic logic, based on a number of different motivations. One of the motivations that seems to be popular is to extend intuitionism from mathematical discourse to empirical discourse. To this end, the role played by proof within the mathematical discourse will be played by warrant/evidence/verification/etc. within the empirical discourse.

---

[1] We are grateful to the referees for their helpful comments. Email: satoruniki@jaist.ac.jp

[2] This research was supported by a Sofja Kovalevskaja Award of the Alexander von Humboldt-Foundation, funded by the German Ministry for Education and Research. Email: Hitoshi.Omori@rub.de

The main background of this paper, namely "Empirical Negation" by Michael De, is a contribution within the above motivation, following the discussions led by Michael Dummett and Neil Tennant.[3] De's focus in [16] was on negation, and expanded the language of intuitionistic propositional logic by adding empirical negation. The intuitive reading of empirical negation of A is that "it is not the case that there is sufficient evidence at present that A". Model theoretically, this is formulated with the help of pointed Kripke models for intuitionistic logic. More specifically, empirical negation of A is forced at a point iff A is not forced at the base point. Following De's paper, a Hilbert-style axiomatization was given for the expansion of intuitionistic logic in [17], and in [18], a comparison of empirical negation and classical negation was carried out over subintuitionistic logic, introduced and explored by Greg Restall in [38].

Now, a simple calculation reveals that double empirical negation of A is forced at a point iff A is forced at the base point. In other words, double empirical negation can be seen as an actuality operator explored by John N. Crossley, Lloyd Humberstone, Martin Davies and more. This then gives rise to a natural question of exploring an expansion of intuitionistic logic enriched by actuality operator.[4] The aim of this paper is twofold, and the first aim is to address this question. Although the notion of actuality has been discussed in classical settings (see our brief overview below), no attempts are known, to the best of authors' knowledge, to discuss the notion of actuality based on intuitionistic logic.[5] However, it is of significant interest how we can incorporate the notion along the philosophical foundation of Dummett-Tennant-De. The second aim is to draw some connections to closely related systems. This enables us to uncover links with other logical concepts, such as empirical negation and globality. For this purpose, we shall adopt a language that includes absurdity and therefore negation. Nonetheless we shall also observe how the notion of actuality is independent of that of negation, which is an advantage over an approach that defines actuality in terms of empirical negation. Before moving further, let us briefly review some of the developments in the literature related to our aim.

Actuality The notion of actuality has been studied in modal logic for a long time, and various conceptualizations have been introduced. Even at an early period, Crossley, Humberstone and Davies [14,15] already introduced two different actuality operators, A and $\mathscr{F}$ (read *fixedly*). Each model $\mathcal{M}$ has a distinguished world w^*, and $\mathsf{A}\varphi$ is true at w iff φ is true at w^*. On the other hand, $\mathscr{F}\varphi$ is true at w iff for every model $\mathcal{M}'$, φ is true at $\mathcal{M}'$'s distinguished world w'. These two operators represent different intuitions about whether 'the actual world' is necessarily so or not.

[3] Another interesting direction following Dummett is to discuss not only verification, but also falsification. This path is explored by Andreas Kapsner in great detail in [33].

[4] HO would like to thank Patrick Blackburn for pointing this out and encouraging him to pursue this direction at AiML 2016 in person.

[5] Note that there is a recent work on the notion of actuality based on relevant logics by Shawn Standefer in [42].

Another example for flexible actuality is that of Dominic Gregory [29], whose semantics includes a mapping @, which maps a world w to *its* actual world $@(w)$ in the same model, with a couple of conditions on @. This in particular allows there being more than one actual worlds in a model.[6]

Baaz' LGP and Titatni's GI Recall that Gödel-Dummett logic, introduced in [24] by Dummett, is an extension of intuitionistic logic with the linearity axiom:

$$(A \to B) \vee (B \to A). \tag{Lin}$$

Semantically, this logic is characterised by linear Kripke frames, which enables us to see it as a fuzzy logic in intuitionistic setting.

Then, in [2], Matthias Baaz expanded Gödel-Dummett logic by an additional operator, $\triangle$, which he called a *projection* modality, also later known as Baaz' Delta. The resulting logic is named **LGP**. Semantically, a formula of the form $\triangle A$ attains either the value 1 or 0, and it attains the value 1 iff A has the value 1.[7] In other words, $\triangle A$ is true iff A is valid in the model. Baaz in the same paper also mentions an operator equivalent to empirical negation in the setting of Gödel-Dummett logic (cf. [2, p.33]).

A logic closely related to **LGP** of Baaz is Satoko Titani's *global intuitionistic logic* **GI**, introduced in [46]. This logic, formulated as a sequent calculus, is defined by adding to intuitionistic logic an operator $\Box$ of *globalization*. From a semantic perspective, in terms of algebraic semantics, $\Box$ has the same interpretation as $\triangle$. There is also a fuzzy extension of **GI** called *fuzzy intuitionistic logic with globalization* **GIF** proposed by Gaisi Takeuti and Satoko Titani in [45], whose propositional fragment is equivalent to **LGP** (cf. [13, Remark 3]).

Note here that global intuitionistic logic can be regarded as an instance of intuitionistic modal logics which are equipped with at least two accessibility relations, intuitionistic $\leq$ and modal R. This is studied since 1948 by Frederic B. Fitch in [25], followed by Arthur N. Prior's [37] and R. A. Bull's papers [11,12], and later major developments include [7,8,21,36,39,40,41,48]. Some close connections of global intuitionistic logic to intuitionistic modal logics are studied by Hiroshi Aoyama in [1].

Based on these, this paper is structured as follows. We first introduce intuitionistic logic with actuality operator, called $\mathbf{IPC}^{@}$, both in terms of semantics and proof system, in §2. Then, in §3, we establish the soundness and strong completeness of $\mathbf{IPC}^{@}$. This is followed by a comparison of $\mathbf{IPC}^{@}$ with related systems in §4 and §5. More specifically, $\mathbf{IPC}^{@}$ is compared with intuitionistic logic with empirical negation as well as logic of actuality of Crossley and Humberstone in §4. We then turn to compare $\mathbf{IPC}^{@}$ with **LGP** of Baaz and **GI** of Titani in §5. The paper concludes with a brief summary of our main results and some directions for future research in §6.

---

[6] For more discussions on actuality, see, for instance, [26,32,43].

[7] This condition is closely related to the framework of *simple monadic Heyting algebra* which is explored in detail in [6] by Guram Bezhanishvili. We would like to thank one of the referees for directing our attention to this paper.

2 Semantics and Proof system

After setting up the language, we first present the semantics, and then turn to the proof system.

Definition 2.1 The language $\mathcal{L}_{\perp}^{@}$ consists of a finite set $\{@, \perp, \wedge, \vee, \rightarrow\}$ of propositional connectives and a countable set Prop of propositional variables which we denote by p, q, etc. Furthermore, we denote by Form the set of formulas defined as usual in $\mathcal{L}_{\perp}^{@}$. We denote a formula of $\mathcal{L}_{\perp}^{@}$ by A, B, C, etc. and a set of formulas of $\mathcal{L}_{\perp}^{@}$ by Γ, Δ, Σ, etc.

2.1 Semantics

Definition 2.2 A model for the language $\mathcal{L}_{\perp}^{@}$ is a quadruple $\langle W, g, \leq, V\rangle$, where W is a non-empty set (of states); $g \in W$ (the base state); $\leq$ is a partial order on W with g being the least element; and $V : W \times \mathsf{Prop} \rightarrow \{0, 1\}$ an assignment of truth values to state-variable pairs with the condition that $V(w_1, p) = 1$ and $w_1 \leq w_2$ only if $V(w_2, p) = 1$ for all $p \in \mathsf{Prop}$ and all $w_1, w_2 \in W$. Valuations V are then extended to interpretations I to state-formula pairs by the following conditions:

- $I(w, p) = V(w, p)$;
- $I(w, \perp) = 0$;
- $I(w, @A) = 1$ iff $I(g, A) = 1$;
- $I(w, A \wedge B) = 1$ iff $I(w, A) = 1$ and $I(w, B) = 1$;
- $I(w, A \vee B) = 1$ iff $I(w, A) = 1$ or $I(w, B) = 1$;
- $I(w, A \rightarrow B) = 1$ iff for all $x \in W$: if $w \leq x$ and $I(x, A) = 1$ then $I(x, B) = 1$.

Semantic consequence is now defined in terms of truth preservation at g: $\Gamma \models A$ iff for all models $\langle W, g, \leq, I\rangle$, $I(g, A) = 1$ if $I(g, B) = 1$ for all $B \in \Gamma$.

2.2 Proof System

Definition 2.3 The system $\mathbf{IPC}^{@}$ consists of the following axiom schemata and rules of inference:

$$\perp\rightarrow A \quad \text{(Ax0)}$$
$$A\rightarrow(B\rightarrow A) \quad \text{(Ax1)}$$
$$(A\rightarrow(B\rightarrow C))\rightarrow((A\rightarrow B)\rightarrow(A\rightarrow C)) \quad \text{(Ax2)}$$
$$(A \wedge B)\rightarrow A \quad \text{(Ax3)}$$
$$(A \wedge B)\rightarrow B \quad \text{(Ax4)}$$
$$(C\rightarrow A)\rightarrow((C\rightarrow B)\rightarrow(C\rightarrow(A \wedge B))) \quad \text{(Ax5)}$$
$$A\rightarrow(A \vee B) \quad \text{(Ax6)}$$
$$B\rightarrow(A \vee B) \quad \text{(Ax7)}$$
$$(A\rightarrow C)\rightarrow((B\rightarrow C)\rightarrow((A \vee B)\rightarrow C)) \quad \text{(Ax8)}$$
$$@(A\rightarrow B)\rightarrow(@A\rightarrow @B) \quad \text{(Ax9)}$$
$$@A\rightarrow A \quad \text{(Ax10)}$$
$$@A\rightarrow @@A \quad \text{(Ax11)}$$
$$@A \vee (@A\rightarrow B) \quad \text{(Ax12)}$$
$$@(A\vee B)\rightarrow(@A\vee @B) \quad \text{(Ax13)}$$
$$\frac{A}{@A} \quad \text{(RN)}$$
$$\frac{A \quad A\rightarrow B}{B} \quad \text{(MP)}$$

Finally, we write $\Gamma \vdash A$ if there is a sequence of formulas $B_1, \ldots, B_n, A$, $n \geq 0$, such that every formula in the sequence $B_1, \ldots, B_n, A$ either (i) belongs to Γ; (ii) is an axiom of $\mathbf{IPC}^{@}$; (iii) is obtained by (MP) or (RN) from formulas preceding it in sequence.

Remark 2.4 We will refer to the subsystem of $\mathbf{IPC}^{@}$ which consists of axiom schemata (Ax1)–(Ax8) and a rule of inference (MP) as $\mathbf{IPC}^{+}$.

Note that the deduction theorem does not hold with respect to $\to$ in $\mathbf{IPC}^{@}$. However, we do have a deduction theorem in a slightly different form, and our goal now is to prove this. For this purpose, we begin with some preparations.

Fact 2.5 *The following formulas are provable in* $\mathbf{IPC}^{+}$ *and thus in* $\mathbf{IPC}^{@}$.

$$A \to A \quad (1)$$
$$(A \vee B) \to (B \vee A) \quad (2)$$
$$(A \to (B \to C)) \to (B \to (A \to C)) \quad (3)$$
$$(A \vee B) \to ((B \to C) \to (A \vee C)) \quad (4)$$
$$(A \to (B \to C)) \to ((A \wedge B) \to C) \quad (5)$$

Now, we can prove one direction of the deduction theorem.

Proposition 2.6 *For all* $\Gamma \cup \{A, B\} \subseteq \mathsf{Form}$, *if* $\Gamma, A \vdash B$ *then* $\Gamma \vdash @A \to B$.

Proof. By the induction on the length n of the proof of $\Gamma, A \vdash B$. If $n = 1$, then we have the following three cases.

- If B is one of the axioms of $\mathbf{IPC}^{@}$, then we have $\vdash B$. Therefore, by (Ax1), we obtain $\vdash @A \to B$ which implies the desired result.
- If $B \in \Gamma$, we have $\Gamma \vdash B$, and thus we obtain the desired result by (Ax1).
- If $B = A$, then by (Ax10), we have $@A \to B$ which implies the desired result.

For $n > 1$, then there are two additional cases to be considered.

- If B is obtained by applying (MP), then we will have $\Gamma, A \vdash C$ and $\Gamma, A \vdash C \to B$ lengths of the proof of which are less than n. Thus, by induction hypothesis, we have $\Gamma \vdash @A \to C$ and $\Gamma \vdash @A \to (C \to B)$, and by (Ax2) and (MP), we obtain $\Gamma \vdash @A \to B$ as desired.
- If B is obtained by applying (RN), then $B = @C$ and we will have $\Gamma, A \vdash C$ length of the proof of which is less than n. Thus, by induction hypothesis, we have $\Gamma \vdash @A \to C$. By (Ax9) and (RN), we have $\Gamma \vdash @@A \to @C$. Another application of (Ax9) gives us $\Gamma \vdash @A \to @C$, i.e. $\Gamma \vdash @A \to B$ as desired.

This completes the proof. □

Proposition 2.7 *For all* $\Gamma \cup \{A, B\} \subseteq \mathsf{Form}$, *if* $\Gamma \vdash @A \to B$ *then* $\Gamma, A \vdash B$.

Proof. By the assumption $\Gamma \vdash @A \to B$. Moreover, we have $\Gamma, A \vdash @A$ by (RN). Thus, we obtain the desired result by (MP). □

By combining Propositions 2.6 and 2.7, we obtain the following theorem.

Theorem 2.8 *For all* $\Gamma \cup \{A, B\} \subseteq \mathsf{Form}$, $\Gamma, A \vdash B$ *iff* $\Gamma \vdash @A \to B$.

Let us mention a corollary of the deduction theorem which shall prove vital for the completeness theorem.

Corollary 2.9 *If* $A \vdash C$ *and* $B \vdash C$, *then* $A \vee B \vdash C$.

Proof. If $A \vdash C$ and $B \vdash C$, then by deduction theorem $\vdash @A \to C$ and $\vdash @B \to C$. Thus $\vdash (@A \vee @B) \to C$; now use (Ax13) to deduce $\vdash @(A \vee B) \to C$. By deduction theorem again, we conclude $A \vee B \vdash C$. □

3 Soundness and Completeness

We now turn to prove the soundness and the strong completeness. The proofs are in large part analogous to those of [17,18] which build on [38].

3.1 Soundness

Theorem 3.1 *For* $\Gamma \cup \{A\} \subseteq \mathsf{Form}$*, if* $\Gamma \vdash A$ *then* $\Gamma \models A$.

Proof. By induction on the length of the proof. □

3.2 Key notions for completeness

In below we introduce some concepts used in the argument for completeness.

(i) $\Sigma \vdash_{\pi} A$ iff $\Sigma \cup \Pi \vdash A$.
(ii) Σ is a Π-*theory* iff:
  (a) if $A, B \in \Sigma$ then $A \wedge B \in \Sigma$.
  (b) if $\vdash_{\pi} A \to B$ then (if $A \in \Sigma$ then $B \in \Sigma$).
(iii) Σ is *prime* iff (if $A \vee B \in \Sigma$ then $A \in \Sigma$ or $B \in \Sigma$).
(iv) $\Sigma \vdash_{\pi} \Delta$ iff for some $D_1, \ldots, D_n \in \Delta$, $\Sigma \vdash_{\pi} D_1, \ldots, D_n$.
(v) $\vdash_{\pi} \Sigma \to \Delta$ iff for some $C_1, \ldots, C_n \in \Sigma$ and $D_1, \ldots, D_m \in \Delta$:

$$\vdash_{\pi} C_1 \wedge \cdots \wedge C_n \to D_1 \vee \cdots \vee D_m.$$

(vi) Σ is Π-*deductively closed* iff (if $\Sigma \vdash_{\pi} A$ then $A \in \Sigma$).
(vii) $\langle \Sigma, \Delta \rangle$ is a Π-*partition* iff:
  (a) $\Sigma \cup \Delta = \mathsf{Form}$
  (b) $\nvdash_{\pi} \Sigma \to \Delta$
(viii) Σ is *non-trivial* iff $A \notin \Sigma$ for some formula A.

Lemma 3.2 *If* Γ *is a non-empty* Π*-theory, then* $\Pi \subseteq \Gamma$.

Proof. Take $A \in \Pi$. Then, we have $\Pi \vdash A$. Now since Γ is non-empty, take any $C \in \Gamma$. Then, by (Ax1), we obtain $\Pi \vdash C \to A$, i.e. $\vdash_{\pi} C \to A$. Thus, combining this together with $C \in \Gamma$ and the assumption that Γ is Π-theory, we conclude that $A \in \Gamma$. □

3.3 Extension lemmas

We now introduce a number of lemmas concerning extensions of sets with various properties. For the proofs, cf. [17, §2] which are based on [38].

Lemma 3.3 *If* $\langle \Sigma, \Delta \rangle$ *is a* Π*-partition then* Σ *is a prime* Π*-theory.*

Lemma 3.4 *If* $\nvdash_{\pi} \Sigma \to \Delta$ *then there are* $\Sigma' \supseteq \Sigma$ *and* $\Delta' \supseteq \Delta$ *such that* $\langle \Sigma', \Delta' \rangle$ *is a* Π*-partition.*

Corollary 3.5 *Let* Σ *be a non-empty* Π*-theory,* Δ *be closed under disjunction, and* $\Sigma \cap \Delta = \emptyset$*. Then there is* $\Sigma' \supseteq \Sigma$ *such that* $\Sigma' \cap \Delta = \emptyset$ *and* Σ' *is a prime* Π*-theory.*

Lemma 3.6 *If* $\Sigma \nvdash \Delta$ *then there are* $\Sigma' \supseteq \Sigma$ *and* $\Delta' \supseteq \Delta$ *such that* $\langle \Sigma', \Delta' \rangle$ *is a partition, and* Σ' *is deductively closed.*

We shall mention that the proof of this lemma relies on Corollary 2.9, and consequently on (Ax13). Hence the same argument cannot be directly imitated by a logic lacking this axiom, such as **GIPC** in §5.

Corollary 3.7 *If* $\Sigma \nvdash A$ *then there are* $\Pi \supseteq \Sigma$ *such that* $A \notin \Pi$, Π *is a prime* Π*-theory and is* Π*-deductively closed.*

3.4 Counter-example lemma

Lemma 3.8 *If* Δ *is a* Π*-theory and* $A \to B \notin \Delta$*, then there is a prime* Π*-theory* Γ*, such that* $A \in \Gamma$ *and* $B \notin \Gamma$.

Proof. Let $\Sigma = \{C : A \to C \in \Delta\}$. We check that Σ is a Π-theory. First, if $C_1, C_2 \in \Sigma$ then $A \to C_1, A \to C_2 \in \Delta$. Since $\vdash (A \to C_1 \wedge A \to C_2) \to (A \to (C_1 \wedge C_2))$ and Δ a Π-theory, we have $A \to (C_1 \wedge C_2) \in \Delta$. Thus $C_1 \wedge C_2 \in \Sigma$. Now suppose that $\vdash_\pi C \to D$ and $C \in \Sigma$. Then $\vdash_\pi (A \to C) \to (A \to D)$ and $A \to C \in \Delta$; so $A \to D \in \Delta$ and hence $D \in \Sigma$.

Clearly $A {\in} \Sigma$ and $B \vee \cdots \vee B {\notin} \Sigma$. Based on this, let Δ' be the closure of $\{B\}$ under disjunction. Then $\Sigma \cap \Delta' {=} \emptyset$, and the result follows from Corollary 3.5.□

Note that, since Σ is non-trivial, the obtained Γ is non-trivial as well.

3.5 Completeness

We are now ready to prove the completeness.

Theorem 3.9 *For all* $\Gamma \cup \{A\} \subseteq \mathsf{Form}$*, if* $\Gamma \models A$ *then* $\Gamma \vdash A$.

Proof. We prove the contrapositive. Suppose that $\Gamma \nvdash A$. Then, by Corollary 3.7, there is a $\Pi \supseteq \Gamma$ such that Π is a prime Π-theory, Π-deductively closed and $A \notin \Pi$. Define the interpretation $\mathfrak{A} = \langle X, \Pi, \leq, I \rangle$, where $X = \{\Delta : \Delta$ is a non-trivial prime Π-theory$\}$, $\Delta \leq \Sigma$ iff $\Delta \subseteq \Sigma$ and I is defined thus. For every state Σ and propositional parameter p:

$$I(\Sigma, p) = 1 \text{ iff } p \in \Sigma$$

We show by induction on B that $I(\Sigma, B) = 1$ iff $B \in \Sigma$. We concentrate on the cases where B has the form $@C$ and $C \to D$.

When $B \equiv @C$, if $I(\Sigma, @C) = 1$ then by definition $I(\Pi, C) = 1$. By IH this is equivalent to $C \in \Pi$. Then $C \in \Sigma$ as $\Pi \subseteq \Sigma$ and also $\vdash_\pi @C$ by (RN). Hence $\vdash_\pi C \to @C$ by (Ax1). Now as Σ is a Π-theory, $C \in \Sigma$ implies $@C \in \Sigma$. For the other direction, it suffices to show $@C \in \Sigma$ implies $C \in \Pi$. First note $@C \vee @C \to D \in \Pi$ for all D because Π is Π-deductively closed. Then as Π is a prime theory, for each D either $@C \in \Pi$ or $@C \to D \in \Pi$. That is, either $@C \in \Pi$ or for all D, $@C \to D \in \Pi$. But if the latter, because Σ is a Π-theory, that $\Pi \subseteq \Sigma$ and $\vdash (@C \wedge (@C \to D)) \to D$ imply $D \in \Sigma$ for all D. This contradicts the non-triviality of Σ, so it must be that $@C \in \Pi$. But then $C \in \Pi$ by (Ax10) and Π being a Π-theory.

When $B \equiv C \to D$, by IH $I(\Sigma, C \to D) = 1$ iff for all Δ s.t. $\Sigma \subseteq \Delta$, if $C \in \Delta$ then $D \in \Delta$. Hence it suffices to show that this latter condition is equivalent to $C \to D \in \Sigma$. For the forward direction, we argue by contraposition; so assume $C \to D \notin \Sigma$. Then by Lemma 3.8 we can find find a non-trivial prime

Π-theory Σ' such that $C \in \Sigma'$ but $D \notin \Sigma'$. For the backward direction, assume $C \to D \in \Sigma$ and $C \in \Delta$ for any Δ s.t. $\Sigma \subseteq \Delta$. Then $C \to D \in \Delta$ as well, and so $D \in \Delta$ since Δ is a Π-theory.

It now suffices to observe that $B \in \Pi$ for all $B \in \Gamma$ and $A \notin \Pi$, which in view of the above means $\Gamma \not\models A$. This completes the proof. □

4 Comparison (I)

In this section, we give some comparisons of **IPC**$^{@}$ with **IPC**$^{\sim}$, as given in [16,17], and **S5A** of Crossley and Humberstone, as given in [14].

4.1 Empirical negation and actuality

IPC$^{\sim}$ employs the language $\mathcal{L}^{\sim} = \{\sim, \wedge, \vee, \to\}$, and is axiomatized as follows.

Definition 4.1 The system **IPC**$^{\sim}$ consists of (Ax1)-(Ax8), (MP) and the following axiom schemata and a rule of inference:

$$A \vee {\sim}A \qquad \text{(N1)}$$
$${\sim}A \to ({\sim}{\sim}A \to B) \qquad \text{(N2)}$$
$$\frac{A \vee B}{{\sim}A \to B} \quad \text{(RP)}$$

We shall denote the deducibility in **IPC**$^{\sim}$ by $\vdash_{\sim}$. The deduction theorem holds in the form $\Gamma, A \vdash_{\sim} B$ iff $\Gamma \vdash_{\sim} {\sim}{\sim}A \to B$ (cf. [17, Theorem 2.1]). The corresponding semantics for **IPC**$^{\sim}$ is almost identical to that of **IPC**$^{@}$, except for the valuation of formulas of the form ${\sim}A$, which is given by:

$$I(w, {\sim}A) = 1 \text{ iff } I(g, A) = 0.$$

Remark 4.2 Note that Kosta Došen, in papers [20,22,23], considered negative modalities in models with two relations between worlds, like the models for intuitionistic modal logics, and one of them has the following condition:

$$w \Vdash {\sim}A \text{ iff for some } w' \in W, wRw' \text{ and } w' \nVdash A.$$

Although the modal relation R is absorbed by the intuitionistic relation $\leq$, empirical negation can be seen as having this type of valuation. Interestingly, Došen considered this sort of absorption is a necessary condition for a negative modality to be deemed a 'negation' (cf. [23, p.85]). For a recent discussion on negation understood as negative modality, see [4,5,19]. See also [31] for an up-to-date survey on negation, as well as negative modalities, in general.

Remark 4.3 There are two more things to note with this valuation. First, intuitionistic $\bot$ and consequently the intuitionistic negation $\neg$ is definable in **IPC**$^{\sim}$ by setting $\bot := {\sim}(A \to A)$. Second, since $I(w, {\sim}{\sim}A)$=1 iff $I(g, A)$=1, we see @ is also definable in **IPC**$^{\sim}$ by $@A := {\sim}{\sim}A$.

A natural question then would be whether we can go the opposite direction, namely, is $\sim$ definable in **IPC**$^{@}$? It turns out that this also holds. Since we have $\bot$ in $\mathcal{L}^{@}_{\bot}$, we readily see: $I(w, \neg @A)$=1 iff $I(g, A)$=0. The situation changes once we drop $\bot$ from the language. Let **IPC**$^{@+}$ be defined in the language $\mathcal{L}^{@} = \{@, \wedge, \vee, \to\}$ with (Ax1)-(Ax13), (RN) and (MP). The completeness for **IPC**$^{@+}$ with respect to Kripke models with the base state is readily obtainable by an analogous means to that of **IPC**$^{@}$.

Proposition 4.4 *$\sim$ is not definable in* $\mathbf{IPC}^{@+}$.

Proof. If $\sim$ is definable in $\mathbf{IPC}^{@+}$, then as we have seen $\perp$ is also definable as $\sim(A{\to}A)$. Let F be such a formula. Now choose a model such that $V(w,p){=}1$ for all p and $w{\in}W$. Then by induction on formula we can establish $I(w,A){=}1$ for all A and $w{\in}W$. So in particular, $I(w,F){=}1$ for all $w{\in}W$, a contradiction.□

Therefore $\mathbf{IPC}^{@+}$ may be seen as an intuitionistic system with actuality operator that is independent of negation. This system consequently has an advantage over $\mathbf{IPC}^{@}$ and $\mathbf{IPC}^{\sim}$ when a non-standard notion of negation is espoused. Moreover it offers a suitable starting point for combining intuitionism in empirical discourse and the school of intuitionism which eschews negation altogether, as a result of scepticism towards unrealised concepts (cf. [30]).

4.2 Classical actuality and constructive actuality

We now turn to compare $\mathbf{IPC}^{@}$ to **S5A** of Crossley and Humberstone. To this end, we first review the basics of **S5A**, with a slightly difference in the notation to replace A, for actuality, by @. Then the system is described by the language $\mathcal{L}_m^{@} = \{@, \Box, \perp, \wedge, \vee, \to\}$.

Definition 4.5 [Crossley & Humberstone] An **S5A**-model for the language $\mathcal{L}_m^{@}$ is a triple $\langle W, g, V\rangle$, where W is a non-empty set (of states); $g \in W$ (the base state); and $V : W \times \mathsf{Prop} \to \{0,1\}$ an assignment of truth values to state-variable pairs. Valuations V are then extended to interpretations I to state-formula pairs by the following conditions:

- $I(w,p) = V(w,p)$;
- $I(w,\perp) = 0$;
- $I(w,\Box A) = 1$ iff for all $w \in W$, $I(w,A) = 1$;
- $I(w,@A) = 1$ iff $I(g,A) = 1$;
- $I(w,A \wedge B) = 1$ iff $I(w,A) = 1$ and $I(w,B) = 1$;
- $I(w,A \vee B) = 1$ iff $I(w,A) = 1$ or $I(w,B) = 1$;
- $I(w,A \to B) = 1$ iff $I(w,A) \neq 1$ or $I(w,B) = 1$.

Then, **S5A**-validity is defined in terms of truth at all $w \in W$: $\models_{\mathbf{S5A}} A$ iff for all **S5A**-models $\langle W, g, I\rangle$, $I(w,A) = 1$ for all $w \in W$.

Definition 4.6 [Crossley and Humberstone] The axiomatic proof system for **S5A** consists of the following axioms in addition to any axiomatization of **S5**:

$@(@A \to A)$	(A1)	$@A \leftrightarrow \neg@\neg A$	(A3)	$@A \to \Box@A$ (A5)
$@(A \to B) \to (@A \to @B)$	(A2)	$\Box A \to @A$	(A4)	

We refer to the derivability in **S5A** as $\vdash_{\mathbf{S5A}}$.

Based on these, Crossley and Humberstone established the following result.

Theorem 4.7 (Crossley and Humberstone) *For all $A \in \mathsf{Form}_m^{@}$, $\models_{\mathbf{S5A}} A$ iff $\vdash_{\mathbf{S5A}} A$.*

The above axiomatization seen in view of $\mathbf{IPC}^{@}$ is problematic since the right-to-left direction of (A3) is *not* valid/derivable. However, a slightly different axiomatization will allow us to compare **S5A** and $\mathbf{IPC}^{@}$ more easily.

Proposition 4.8 *Let* $\vdash_{\mathbf{S5A}'}$ *be the derivability in a system obtained from the axiomatic proof system for* **S5A** *by replacing* (A3) *by the following two axioms:*

$$@A \to \neg @\neg A \qquad \text{(A3.1)} \qquad @(A \vee B) \to (@A \vee @B) \qquad \text{(A3.2)}$$

Then, for all $A \in \mathsf{Form}_m^@$, $\vdash_{\mathbf{S5A}'} A$ *iff* $\vdash_{\mathbf{S5A}} A$.

Proof. For the left-to-right direction, it suffices to check that (A3.2) is derivable in **S5A**. In view of (A3), (A3.2) is derivable iff $\vdash_{\mathbf{S5A}} (@\neg A \wedge @\neg B) \to @(\neg A \wedge \neg B)$. But this is obvious since @ is an extension of **K**-modality.

For the other way around, it suffices to prove $\vdash_{\mathbf{S5A}'} @A \vee @\neg A$. Since we have classical tautologies, we have $\vdash_{\mathbf{S5A}'} A \vee \neg A$, and by the rule of necessitation, we have $\vdash_{\mathbf{S5A}'} \Box(A \vee \neg A)$. This implies $\vdash_{\mathbf{S5A}'} @(A \vee \neg A)$ in view of (A4), and finally we make use of (A3.2) to obtain the desired result. □

Remark 4.9 Note first that even though we do not have the necessity operator in $\mathbf{IPC}^@$, the actuality operator also enjoys the following condition:

$$I(w, @A) = 1 \text{ iff for all } w \in W, I(w, A) = 1$$

This is because the base point is the root. Thus, if we regard □ as @ in the above axiomatization of **S5A**, then we can see that all the axiom schemata and rules of inference related to □ and @ in **S5A** are derivable in $\mathbf{IPC}^@$.

Therefore, there is a sense in which $\mathbf{IPC}^@$ is a generalization of **S5A**. But there is also a sense in which this generalization is not simple. More specifically, we obtain the following result.

Proposition 4.10 $\mathbf{IPC}^@$ *plus Peirce's law collapses into* **Triv** *based on* **CL**.

Proof. In view of (Ax10), it suffices to prove $A{\to}@A$ in the extension. Note first that $A{\vee}(A{\to}B)$ is still derivable from an instance of Peirce's law, namely $(((A \vee (A{\to}B)){\to}A){\to}(A{\vee}(A{\to}B))){\to}(A{\vee}(A{\to}B))$. Then as before we obtain $@A{\vee}@(A{\to}B)$, which entails $(@A{\to}@B){\to}@(A{\to}B)$. Take $B \equiv @A$ and we have $(@A{\to}@@A){\to}@(A{\to}@A)$. By (Ax11) and (Ax10), we obtain $A \to @A$.□

Remark 4.11 The above proof does not rely on the existence of $\bot$ in the language, and thus also applies to $\mathbf{IPC}^{@+}$.

5 Comparison (II)

In this section, we offer further comparisons of $\mathbf{IPC}^@$ with **LGP** of Baaz, as given in [2], and **GIPC** of Titani, as given in [46].

5.1 Baaz Delta and actuality

As we mentioned in the introduction, Baaz' logic **LGP** is Gödel-Dummet logic equipped with a projection modality $\triangle$. Let us first look at the precise formulation in [2]. (For the sake of simplicity, we shall hereafter use $\mathcal{L}_\bot^@$ to describe the system, so @ will be used instead of $\triangle$.)

Definition 5.1 [Baaz] Let $V \subseteq [0, 1]$ be a set of *truth values* containing 0 and 1. A *valuation* $\mathfrak{V}$ based on V assigns a truth value in V to each propositional variable. $\mathfrak{V}$ is extended to all propositions by the clauses:

- $\mathfrak{V}(\bot) = 0$
- $\mathfrak{V}(A{\wedge}B) = min(\mathfrak{V}(A), \mathfrak{V}(B))$
- $\mathfrak{V}(A{\vee}B) = max(\mathfrak{V}(A), \mathfrak{V}(B))$
- $\mathfrak{V}(A{\to}B) = \begin{cases} \mathfrak{V}(B) & \text{if } \mathfrak{V}(A) > \mathfrak{V}(B) \\ 1 & \text{if } \mathfrak{V}(A) \leq \mathfrak{V}(B) \end{cases}$
- $\mathfrak{V}(@A) = \begin{cases} 1 & \text{if } \mathfrak{V}(A) = 1 \\ 0 & \text{if } \mathfrak{V}(A) \neq 1 \end{cases}$

Then $\mathbf{GP}(V) := \{A : \mathfrak{V}(A) = 1 \text{ for every } \mathfrak{V} \text{ based on } V\}$.

Definition 5.2 **LGP** is axiomatized by adding the following axiom to $\mathbf{IPC}^{@}$.

$$(A \to B) \vee (B \to A) \tag{Lin}$$

Let V be infinite. Baaz showed the following weak completeness for **LGP**.

Theorem 5.3 (Baaz) *For all $A \in \mathsf{Form}$, $\mathbf{LGP} \vdash A$ iff $A \in \mathbf{GP}(V)$.*

As is well-known (e.g. [27, Theorem 19, Chapter 4]), Kripke-semantically (Lin) corresponds to lineally ordered Kripke frames. Thus as an improvement, we obtain a *strong* completeness proof for **LGP**, in view of Theorem 3.9. More specifically, let us denote $\vdash_l$ and $\models_l$ for the derivability in **LGP** and semantic consequence with respect to the class of linearly ordered models, respectively.

Proposition 5.4 *For all $\Gamma \cup \{A\} \subseteq \mathsf{Form}$, $\Gamma \vdash_l A$ iff $\Gamma \models_l A$.*

Proof. For soundness, we have to check that (Lin) holds in any linearly ordered model. Given a linearly ordered model $\langle W, g, \leq, I\rangle$ and formulas A and B, let us denote $V(A) = \{w : I(w, A) = 1\}$ and $V(B) = \{w : I(w, B) = 1\}$. Then we have $V(A) \subseteq V(B)$ or $V(B) \subseteq V(A)$. Hence $I(g, A \to B \vee B \to A) = 1$.

For completeness, we have to check that the counter-model construction of Theorem 3.9 creates a linearly ordered model. Suppose otherwise. Then there are states Σ_1 and Σ_2 such that neither $\Sigma_1 \subseteq \Sigma_2$ nor $\Sigma_2 \subseteq \Sigma_1$. Then we can find a formula A_1 in Σ_1 not in Σ_2, and A_2 in Σ_2 not in Σ_1. Now as the base state Π is a prime Π-theory, $A_1 \to A_2 \vee A_2 \to A_1 \in \Pi$, and so $A_1 \to A_2 \in \Pi$ or $A_2 \to A_1 \in \Pi$. Without loss of generality, assume the former. Then because Σ_1 is a Π-theory, $A_1 \wedge (A_1 \to A_2) \in \Sigma_1$; thus $A_2 \in \Sigma_1$, a contradiction. Therefore the counter-model has to be linearly ordered. This completes the proof. □

Remark 5.5 The above result clarifies that $\mathbf{IPC}^{@}$ is a generalization of **LGP** to include non-linearly ordered models. To give a further comparison, for **LGP** it is observed in [2] that $\neg\neg A$ is a dual projection operator of $@A$, attaining 1 if $A \neq 0$ and 0 otherwise. In the setting of $\mathbf{IPC}^{@}$, this true-if-not-false type of operator is perhaps better captured by $\neg@\neg A$ (i.e. $\sim\neg A$). $I(w, \neg@\neg A){=}1$ iff for some $u{\in}W, I(u, A){=}1$; so while $\neg\neg A{\to}\neg@\neg A$ holds in general, $\neg@\neg A{\to}\neg\neg A$ does not. One may readily check that this latter implication is equivalent to *the weak excluded middle* $\neg A \vee \neg\neg A$ as an axiom; in particular $\neg@\neg A$ and $\neg\neg A$ becomes equivalent in **LGP**, because (Lin) implies the weak excluded middle.

5.2 A reformulation of global intuitionistic logic

Next we shall consider propositional global intuitionistic logic (to be called **GIPC**). Let us first look at the formulation of the logic in sequent calculus as given in [46,1]. The system will be described in the language $\mathcal{L}^{@}_{\bot}$. Originally, however, $\Box$ was used in place of @, and $\neg$ was taken as primitive, rather than $\bot$. We shall call the calculus **LGJ** and the derivability by $\vdash_{gGI}$.

Definition 5.6 [Titani & Aoyama] The rule of the calculus **LGJ** are as follows.

$$A \Rightarrow A \text{ [Ax]} \qquad\qquad \bot \Rightarrow \text{ [L}\bot\text{]}$$

$$\frac{\Gamma \Rightarrow \Delta}{A, \Gamma \Rightarrow \Delta}\text{[LW]} \qquad \frac{\Gamma \Rightarrow \Delta}{\Gamma \Rightarrow \Delta, A}\text{[RW]}$$

$$\frac{A, A, \Gamma \Rightarrow \Delta}{A, \Gamma \Rightarrow \Delta}\text{[LC]} \qquad \frac{\Gamma \Rightarrow \Delta, A, A}{\Gamma \Rightarrow \Delta, A}\text{[RC]}$$

$$\frac{\Gamma, A, B, \Pi \Rightarrow \Delta}{\Gamma, B, A, \Pi \Rightarrow \Delta}\text{[LE]} \qquad \frac{\Gamma \Rightarrow \Delta, A, B, \Lambda}{\Gamma \Rightarrow \Delta, B, A, \Lambda}\text{[RE]}$$

$$\frac{\Gamma \Rightarrow \Delta, A \qquad A, \Pi \Rightarrow \Lambda}{\Gamma, \Pi \Rightarrow \Delta, \Lambda}\text{[Cut]}$$

$$\frac{A_i, \Gamma \Rightarrow \Delta}{A_1 \wedge A_2, \Gamma \Rightarrow \Delta}\text{[L}\wedge\text{]} \qquad \frac{\Gamma \Rightarrow \Delta, A \qquad \Gamma \Rightarrow \Delta, B}{\Gamma \Rightarrow \Delta, A \wedge B}\text{[R}\wedge\text{]}$$

$$\frac{A, \Gamma \Rightarrow \Delta \qquad B, \Gamma \Rightarrow \Delta}{A \vee B, \Gamma \Rightarrow \Delta}\text{[L}\vee\text{]} \qquad \frac{\Gamma \Rightarrow \Delta, A_i}{\Gamma \Rightarrow \Delta, A_1 \vee A_2}\text{[R}\vee\text{]}$$

$$\frac{\Gamma \Rightarrow \Delta, A \qquad B, \Pi \Rightarrow \Lambda}{A \to B, \Gamma, \Pi \Rightarrow \Delta, \Lambda}\text{[L}\to\text{]} \qquad \frac{A, \Gamma \Rightarrow \bar{\Delta}, B}{\Gamma \Rightarrow \bar{\Delta}, A \to B}\text{[R}\to\text{]}$$

$$\frac{A, \Gamma \Rightarrow \Delta}{@A, \Gamma \Rightarrow \Delta}\text{[L@]} \qquad \frac{\bar{\Gamma} \Rightarrow \bar{\Delta}, A}{\bar{\Gamma} \Rightarrow \bar{\Delta}, @A}\text{[R@]}$$

In the above, $i \in \{1, 2\}$ and $\bar{\Gamma}$ and $\bar{\Delta}$ are finite sequences of @-*closed formulas*, which are formulas built from $\bot$ and formulas of the form $@A$, by the connectives $\wedge, \vee, \to$. For example, $@@A, @A \wedge @(\bot \to C), \neg@(\neg A \vee B)$ are all @-closed formulas. We shall denote @-closed formulas by $\bar{A}, \bar{B}$ and so on.

We wish to compare **GIPC** with **IPC**$^{@}$. For this purpose it is preferable to have at hand a Hilbert-style axiomatization. This we claim to be the following.

Definition 5.7 The system **GIPC** consists of (Ax0)-(Ax12), (MP),(RN) and the following axiom scheme:

$$(@A \to @B) \to @(@A \to B) \tag{Ax14}$$

The derivability in **GIPC** will be denoted by $\vdash_{GI}$.

Remark 5.8 Note that the deduction theorem, in the form of Theorem 2.8, holds for **GIPC** as well, by the same argument.

We now show a lemma before proving that **LGJ** and **GIPC** are equivalent.

Lemma 5.9 *Let $\bar{A}$ be @-closed. Then,* (i) $\vdash_{GI} \bar{A} \vee \bar{A} \to B$, *and* (ii) $\vdash_{GI} \bar{A} \to @\bar{A}$.

Proof. For (i), we argue by induction on the complexity of A.

- If $\bar{A} \equiv \bot$, then $\vdash_{GI} \bot \vee \bot \to B$.
- If $\bar{A} \equiv @A$, then $@A \vee @A \to B$ is an instance of (Ax12).
- If $\bar{A} \equiv \bar{C} \wedge \bar{D}$, then by IH $\vdash_{GI} \bar{C} \vee \bar{C} \to B$ and $\vdash_{GI} \bar{D} \vee \bar{D} \to B$. So $\vdash_{GI} (\bar{C} \wedge \bar{D}) \vee (\bar{C} \wedge \bar{D}) \to B$.
- If $\bar{A} \equiv \bar{C} \vee \bar{D}$, similarly $\vdash_{GI} (\bar{C} \vee \bar{D}) \vee (\bar{C} \vee \bar{D}) \to B$.
- If $\bar{A} \equiv \bar{C} \to \bar{D}$, by IH $\vdash_{GI} \bar{C} \vee \bar{C} \to \bar{D}$ and $\vdash_{GI} \bar{D} \vee \bar{D} \to B$. So $\vdash_{GI} (\bar{C} \to \bar{D}) \vee (\bar{C} \to \bar{D}) \to B$.

For (ii), we similarly argue by induction on A.

- If $\bar{A} \equiv \bot$, then $\bot \to @\bot$ is an instance of (Ax0).
- If $\bar{A} \equiv @A$, then $@A \to @@A$ is an instance of (Ax11).
- If $\bar{A} \equiv \bar{B} \wedge \bar{C}$, then by IH $\vdash_{GI} \bar{B} \to @\bar{B}$ and $\vdash_{GI} \bar{C} \to @\bar{C}$. Thus $\vdash_{GI} \bar{B} \wedge \bar{C} \to @\bar{B} \wedge @\bar{C}$. Now it is easy to check via the deduction theorem that $\vdash_{GI} @\bar{B} \wedge @\bar{C} \to @(\bar{B} \wedge \bar{C})$. Hence $\vdash_{GI} \bar{B} \wedge \bar{C} \to @(\bar{B} \wedge \bar{C})$.
- If $\bar{A} \equiv \bar{B} \vee \bar{C}$, then using the same IH as above, we see $\vdash_{GI} \bar{B} \vee \bar{C} \to @\bar{B} \vee @\bar{C}$. Again it is an easy consequence of the deduction theorem that $\vdash_{GI} @\bar{B} \to @(\bar{B} \vee \bar{C})$ and $\vdash_{GI} @\bar{C} \to @(\bar{B} \vee \bar{C})$. Hence $\vdash_{GI} \bar{B} \vee \bar{C} \to @(\bar{B} \vee \bar{C})$.
- If $\bar{A} \equiv \bar{B} \to \bar{C}$, then using (Ax10) and the IH that $\vdash_{GI} \bar{C} \to @\bar{C}$ we infer $\vdash_{GI} (\bar{B} \to \bar{C}) \to (@\bar{B} \to @\bar{C})$. Thus by (Ax14) $\vdash_{GI} (\bar{B} \to \bar{C}) \to @(@\bar{B} \to \bar{C})$. Also by the IH that $\vdash_{GI} \bar{B} \to @\bar{B}$ we have $\vdash_{GI} (@\bar{B} \to \bar{C}) \to (\bar{B} \to \bar{C})$. So by (RN) and (Ax9), $\vdash_{GI} @(@\bar{B} \to \bar{C}) \to @(\bar{B} \to \bar{C})$. Combining the above two observations, we conclude $\vdash_{GI} (\bar{B} \to \bar{C}) \to @(\bar{B} \to \bar{C})$.

This completes the proof. □

Proposition 5.10 *The following equivalence hold between* **LGP** *and* **GIPC**.
(i) *For all* $A \in \mathsf{Form}$, *if* $\vdash_{GI} A$ *then* $\vdash_{gGI} \Rightarrow A$.
(ii) *For all* $\Gamma, \Delta \subseteq \mathsf{Form}$, *if* $\vdash_{gGI} \Gamma \Rightarrow \Delta$ *then* $\vdash_{GI} \bigwedge \Gamma \to \bigvee \Delta$.

Proof. For (i), given the correspondence in intuitionistic logic, it suffices to consider axioms involving @ and (RN). Here we show cases for (Ax12) and (Ax14), which are stated but not shown in [1, Proposition 2.1]; other cases are immediate.

<u>Ax12</u>

$$\dfrac{\dfrac{\dfrac{@A \Rightarrow @A}{@A \Rightarrow @A, B}\,[\mathrm{RW}]}{\Rightarrow @A, @A \to B}\,[\mathrm{R}\!\to]}{\Rightarrow @A \vee @A \to B}\,[\mathrm{R}\vee],[\mathrm{RC}]$$

<u>Ax14</u>

$$\dfrac{\dfrac{\dfrac{\dfrac{@A \Rightarrow @A \qquad \dfrac{B \Rightarrow B}{@B \Rightarrow B}\,[\mathrm{L}@]}{@A \to @B, @A \Rightarrow B}\,[\mathrm{L}\!\to]}{@A \to @B \Rightarrow @A \to B}\,[\mathrm{R}\!\to]}{@A \to @B \Rightarrow @(@A \to B)}\,[\mathrm{R}@]}{\Rightarrow (@A \to @B) \to @(@A \to B)}\,[\mathrm{R}\!\to]$$

For (ii), we treat here the cases for [R→], [L@] and [R@].

- For [R→], by IH $\vdash_{GI} (\bigwedge \Gamma \wedge A) \to (\bigvee \bar{\Delta} \vee B)$. So $\vdash_{GI} \bigwedge \Gamma \to (A \to (\bigvee \bar{\Delta} \vee B))$. Now by Lemma 5.9 (i), $\vdash_{GI} \bigvee \bar{\Delta} \vee \bigvee \bar{\Delta} \to B$. Thus $\vdash_{GI} \bigwedge \Gamma \to (\bigvee \bar{\Delta} \vee A \to B)$.
- For [L@], by IH $\vdash_{GI} (A \wedge \bigwedge \Gamma) \to \bigvee \Delta$. Then $\vdash_{GI} A \to (\bigwedge \Gamma \to \bigvee \Delta)$. So by

(Ax10) $\vdash_{GI} @A \to (\bigwedge \Gamma \to \bigvee \Delta)$. Hence $\vdash_{GI} (@A \wedge \bigwedge \Gamma) \to \bigvee \Delta$.

- For [R@], by IH $\vdash_{GI} \bigwedge \bar{\Gamma} \to (\bigvee \bar{\Delta} \vee A)$. Then $\vdash_{GI} (\bigwedge \bar{\Gamma} \wedge (\bigvee \bar{\Delta} \to @A)) \to A$. Thus by (RN) and (Ax9), $\vdash_{GI} @(\bigwedge \bar{\Gamma} \wedge (\bigvee \bar{\Delta} \to @A)) \to @A$. Here we note $@(\bigwedge \bar{\Gamma} \wedge (\bigvee \bar{\Delta} \to @A))$ is @-closed. So by Lemma 5.9 (ii), $\vdash_{GI} (\bigwedge \bar{\Gamma} \wedge (\bigvee \bar{\Delta} \to @A)) \to @A$. Also by Lemma 5.9 (i), $\vdash_{GI} \bigvee \bar{\Delta} \vee \bigvee \bar{\Delta} \to @A$. From these we deduce $\vdash_{GI} \bigwedge \bar{\Gamma} \to (\bigvee \bar{\Delta} \vee @A)$.

This completes the proof. □

5.3 Globalization and actuality

We are now ready to compare $\mathbf{IPC}^@$ and **GIPC**. We first observe that the former logic contains the latter.

Proposition 5.11 $\mathbf{IPC}^@ \supseteq \mathbf{GIPC}$.

Proof. It suffices to observe that (Ax14) is derivable in $\mathbf{IPC}^@$. Applying (RN) and (Ax13) to (Ax12), we obtain $\vdash @A \vee @(@A \to B)$. Then on one hand, since $\vdash @A \to ((@A \to @B) \to @B)$ and $\vdash @B \to @(@A \to B)$ (the latter by (Ax1), (RN) and (Ax9)), we have $\vdash @A \to ((@A \to @B) \to @(@A \to B))$. On the other hand, it is immediate that $\vdash @(@A \to B) \to ((@A \to @B) \to @(@A \to B))$. Therefore $\vdash (@A \to @B) \to @(@A \to B)$. □

Remark 5.12 Baaz, in [2], states sequent rules for $\triangle$ of **LGP**. It turns out that the same rules can be used to formulate a calculus for $\mathbf{IPC}^@$. It is obtained from **LGJ** by relaxing [R@] to

$$\frac{\bar{\Gamma} \Rightarrow \Delta, A}{\bar{\Gamma} \Rightarrow \Delta, @A}\ [\mathrm{R}@]$$

By Proposition 5.11, we can use Lemma 5.9 for $\mathbf{IPC}^@$ as well. Then we can argue analogously to Proposition 5.10; the treatments of cases for the new [R@] and (Ax13) are straightforward.

To show that the inclusion of the above proposition is strict, we shall turn to a closely related logic called $\mathbf{TCC}_\omega$. This is a subsystem of $\mathbf{IPC}^\sim$ introduced by A. B. Gordienko in [28] as an extension of Richard Sylvan's logic $\mathbf{CC}_\omega$ (cf. [44]). Its axiomatization is that of $\mathbf{IPC}^\sim$, except (RP) is replaced with

$$\frac{A \to B}{{\sim}B \to {\sim}A}. \qquad \text{(RC)}$$

The deducibility in $\mathbf{TCC}_\omega$ will be denoted $\vdash_t$. It is easy to check that formulas and rules derivable in $\mathbf{IPC}^\sim$ listed in [17, Lemma 2.6, Lemma 2.8] are also derivable in $\mathbf{TCC}_\omega$. In particular, the following formulas and rule are derivable.

$$({\sim}A \to A) \to A \qquad \text{(t1)} \qquad\qquad {\sim}{\sim}A \to A \qquad \text{(t3)}$$

$${\sim}(A \to A) \to B \qquad \text{(t2)} \qquad\qquad \frac{A}{{\sim}{\sim}A} \qquad \text{(t4)}$$

Moreover, the same form of the deduction theorem as $\mathbf{IPC}^\sim$ holds in $\mathbf{TCC}_\omega$.

Quite similarly to the situation with $\mathbf{IPC}^@$ and $\mathbf{IPC}^\sim$, we have the following translations between **GIPC** and $\mathbf{TCC}_\omega$.

Definition 5.13 Let $()^{\sim}$ and $()^{@}$ be translations between $\mathcal{L}^{@}_{\bot}$ and $\mathcal{L}^{\sim}$ such that:

$$\begin{array}{rlrl} p^{\sim} &= p & p^{@} &= p \\ (A \circ B)^{\sim} &= A^{\sim} \circ B^{\sim} & (A \circ B)^{@} &= A^{@} \circ B^{@} \\ (@A)^{\sim} &= {\sim}{\sim}A^{\sim} & ({\sim}A)^{@} &= \neg @A^{@} \\ \bot^{\sim} &= {\sim}(p_0 \to p_0) & & \end{array}$$

where p_0 is a fixed propositional variable, and $\circ \in \{\wedge, \vee \to\}$.

Lemma 5.14 *For all $A \in \mathsf{Form}$, $\vdash_{GI} A \leftrightarrow (A^{\sim})^{@}$ and for all $A \in \mathsf{Form}^{\sim}$, $\vdash_t A \leftrightarrow (A^{@})^{\sim}$.*

Proof. By induction on A. Here we look at the cases $A \equiv @B$ and $A \equiv {\sim}B$.

For the former, we need to show $\vdash_{GI} @B \leftrightarrow \neg @ \neg @(B^{\sim})^{@}$. By IH $\vdash_{GI} B \leftrightarrow (B^{\sim})^{@}$, so its suffices to show $\vdash_{GI} @B \leftrightarrow \neg @ \neg @B$. We first note $\neg @B$ is @-closed, thus $\vdash_{GI} \neg @ \neg @B \leftrightarrow \neg\neg @B$. Also $\vdash_{GI} \neg\neg @B \leftrightarrow @B$ from (Ax12). Therefore we conclude $\vdash_{GI} @B \leftrightarrow \neg @ \neg @B$ as desired.

For the latter, we need $\vdash_t {\sim}B \leftrightarrow ({\sim}{\sim}(B^{@})^{\sim} \to {\sim}(p_0 \to p_0))$. Again by IH $\vdash_t B \leftrightarrow (B^{@})^{\sim}$. Then the equivalence follows by (N2), (t1) and (t2). □

Proposition 5.15 *We have that* (i) *for all $A \in \mathsf{Form}$, $\vdash_{GI} A$ iff $\vdash_t A^{\sim}$, and* (ii) *for all $A \in \mathsf{Form}^{\sim}$, $\vdash_t A$ iff $\vdash_{GI} A^{@}$.*

Proof. By Lemma 5.14, it suffices to show the left-to-right direction.

For (i), we need to check the translations of (Ax9)-(Ax12), (Ax14) and (RN) hold in $\mathbf{TCC}_{\omega}$.

- (Ax9) is translated as ${\sim}{\sim}(A^{\sim} \to B^{\sim}) \to ({\sim}{\sim}A^{\sim} \to {\sim}{\sim}B^{\sim})$, the derivability of which is immediate from the deduction theorem and (RC).
- (Ax10) is translated as ${\sim}{\sim}A^{\sim} \to A^{\sim}$, which is an instance of (t3).
- (Ax11) is translated as ${\sim}{\sim}A^{\sim} \to {\sim}{\sim}{\sim}{\sim}A^{\sim}$. This follows from (N2) and (t1), which imply ${\sim}{\sim}{\sim}A^{\sim} \to {\sim}A^{\sim}$; then use (RC).
- (Ax12) becomes ${\sim}{\sim}A^{\sim} \vee {\sim}{\sim}A^{\sim} \to B^{\sim}$, a consequence of (N1) and (N2).
- For (Ax14), we need to show $\vdash_t ({\sim}{\sim}A^{\sim} \to {\sim}{\sim}B^{\sim}) \to {\sim}{\sim}({\sim}{\sim}A^{\sim} \to B^{\sim})$. First $\vdash_t {\sim}A^{\sim} \vee {\sim}{\sim}A^{\sim}$ from (N1) and ${\sim}{\sim}{\sim}A^{\sim} \to {\sim}A^{\sim}$ as seen above. So $\vdash_t ({\sim}{\sim}A^{\sim} \to {\sim}{\sim}B^{\sim}) \to ({\sim}A^{\sim} \vee {\sim}{\sim}B^{\sim})$. We shall show $\vdash_t ({\sim}A^{\sim} \vee {\sim}{\sim}B^{\sim}) \to {\sim}{\sim}({\sim}{\sim}A^{\sim} \to B^{\sim})$. On one hand, $\vdash_t {\sim}A^{\sim} \to {\sim}{\sim}({\sim}{\sim}A^{\sim} \to B^{\sim})$ from (N2), (t3) and (RC). On the other hand, $\vdash_t {\sim}{\sim}B^{\sim} \to {\sim}{\sim}({\sim}{\sim}A^{\sim} \to B^{\sim})$ from (Ax1) and (RC). Thus $\vdash_t ({\sim}A^{\sim} \vee {\sim}{\sim}B^{\sim}) \to {\sim}{\sim}({\sim}{\sim}A^{\sim} \to B^{\sim})$ as required.
- Finally, (RN) is replicable by (t4).

For (ii), we need to check (N1),(N2) and (RC).

- (N1) is translated into $A^{@} \vee \neg @A^{@}$, which is an instance of (Ax12).
- (N2) is translated into $\neg @A^{@} \to (\neg @ \neg @A^{@} \to B^{@})$. As we observed in Lemma 5.14, $\neg @ \neg @A^{@}$ is equivalent to $\neg\neg @A^{@}$; so it follows from (Ax0).
- For (RC), we need to derive $\neg @B \to \neg @A$ from $A \to B$. This is possible with (RN),(Ax9) and by contraposition.

This completes the proof. □

The translation allows us to use the Kripke semantics for $\mathbf{TCC}_\omega$.

Definition 5.16 [Gordienko] A $\mathbf{TCC}_\omega$-model for $\mathcal{L}^\sim$ is a triple $\langle W, \leq, V\rangle$ with each component as in $\mathbf{IPC}^@$. V is extended to interpretation I analogously, except for the interpretation of $\sim A$, which is given by:

$$I(w, \sim A) = 1 \text{ iff } I(w', A) = 0 \text{ for some } w' \in W.$$

We shall use $\models_t$ for the semantic consequence, defined as follows: $\models_t A$ iff for all $\mathbf{TCC}_\omega$-models $\langle W, \leq, V\rangle$, $I(w, A) = 1$ for all $w \in W$.

Remark 5.17 Note in particular that a model of $\mathbf{TCC}_\omega$ does not necessarily have a base state. If it does, then the interpretation coincides with that of $\mathbf{IPC}^@$.

We are now ready to separate the two systems.

Theorem 5.18 (Gordienko) *For all $A \in \mathsf{Form}^\sim$, $\vdash_t A$ iff $\models_t A$.*

Corollary 5.19 $\mathbf{IPC}^@ \supsetneq \mathbf{GIPC}$.

Proof. First, observe that we have the following valuation for $\sim\sim A$.

$$I(w, \sim\sim A) = 1 \text{ iff } I(w', A) = 1 \text{ for all } w'.$$

Now, if **GIPC** proves (Ax13), then by Proposition 5.15 $\sim\sim(p\vee q) \rightarrow \sim\sim p \vee \sim\sim q$ is provable in $\mathbf{TCC}_\omega$. On the other hand, if we consider a model where $W = \{w, w'\}$, $\leq = \{(w,w), (w', w')\}$, $V(p)=\{w\}$ and $V(q)=\{w'\}$, then $I(w, \sim\sim(p \vee q)) = 1$, but $I(w, \sim\sim p) = I(w, \sim\sim q) = 0$. Hence this is a countermodel for $\sim\sim(p \vee q) \rightarrow \sim\sim p \vee \sim\sim q$. So by the previous theorem, $\not\vdash_t \sim\sim(p \vee q) \rightarrow \sim\sim p \vee \sim\sim q$. A contradiction. Therefore **GIPC** does not prove (Ax13). □

Remark 5.20 Note that given a model of $\mathbf{TCC}_\omega$, we can define a model for $\mathcal{L}^@_\perp$ with the interpretation I such that

$$I(w, @A) = 1 \text{ iff } I(w', A) = 1 \text{ for all } w'.$$

Then, it is not difficult to see that each such model corresponds to the original model similarly to Lemma 5.14 and Proposition 5.15. Therefore, it is an immediate consequence of Theorem 5.18 that this gives a sound and weakly complete Kripke semantics for **GIPC**. (This semantics can be also obtained from Ono's semantics via Gordienko's technique; see below.)

We offer a few more words about **GIPC**. In [36], Hiroakira Ono extensively discussed intutitionistic modal systems which are defined by axioms that classically define **S5** when added to **S4**. Aoyama [1] compared some of these systems with **GIPC**, [8] but he did not compare with the strongest of Ono's systems, $\mathbf{L}_4$. It is defined by (Ax0)-(Ax11), $@A \vee @\neg @A$, (MP) and (RP). The Kripke semantics for $\mathbf{L}_4$ in [36] is characterised by modal relation R that is an equivalence relation; this corresponds to the original semantics of $\mathbf{TCC}_\omega$, from which Gordienko derived [28, Lemma 4.4] the semantics of Definition 5.16. This observation and Proposition 5.15 suggest a close relationship between **GIPC** and $\mathbf{L}_4$. In fact, the two systems turn out to coincide.

[8] Some of the comparisons offered in [1] are also observed by Hidenori Kurokawa in [34].

Proposition 5.21 **GIPC** $=$ $\mathbf{L}_4$

Proof. On one hand, $\neg@A$ is @-closed, so by Lemma 5.9 (ii) $\neg@A \rightarrow @\neg@A$ is derivable in **GIPC**. Thus with (Ax12), $@A \vee @\neg@A$ is derivable in **GIPC**. Consequently **GIPC** contains $\mathbf{L}_4$. On the other hand, $@A \vee @\neg@A$ implies (Ax12) with (Ax0) and (Ax10). Moreover, $(@A \rightarrow @B) \rightarrow @(@A \rightarrow @B)$ is known to be derivable in $\mathbf{L}_4$ (cf. [36, Figure 2.1]), and it is a consequence of (Ax10), (RN) and (Ax9) that $@(@A \rightarrow @B) \rightarrow @(@A \rightarrow B)$ holds, so (Ax14) is also derivable in $\mathbf{L}_4$. Thus $\mathbf{L}_4$ contains **GIPC** as well. □

5.4 Sequent calculi for $\mathbf{TCC}_\omega$ and $\mathbf{IPC}^\sim$

Finally, we shall use the results obtained so far to formulate sequent calculi for $\mathbf{TCC}_\omega$ and $\mathbf{IPC}^\sim$. We begin with introducing an analogue of @-closed for formulas in $\mathcal{L}^\sim$.

Definition 5.22 We define the class of *$\sim$-closed formulas* by the next clauses.
(i) $\bot$, $\sim A$ are $\sim$-closed.
(ii) If $\bar{B}$ and $\bar{C}$ are $\sim$-closed, then $\bar{B} \circ \bar{C}$ is $\sim$-closed, where $\circ \in \{\wedge, \vee, \rightarrow\}$.

It is straightforward to check that if $\bar{A}$ is $\sim$-closed, then $\bar{A}^@$ is @-closed.

Lemma 5.23 *For all* $A \in \mathsf{Form}^\sim$, $\vdash_t \bar{A} \rightarrow \sim\sim\bar{A}$.

Proof. By the above observation and Lemma 5.9 (ii), we have $\vdash_{GI} \bar{A}^@ \rightarrow @\bar{A}^@$. Thus by Proposition 5.15 (i) and Lemma 5.14, $\vdash_t \bar{A} \rightarrow \sim\sim\bar{A}$. □

The sequent rules for $\sim$ corresponding to $\mathbf{TCC}_\omega$ is obtained by the following

$$\frac{\bar{\Gamma} \Rightarrow \bar{\Delta}, A}{\sim A, \bar{\Gamma} \Rightarrow \bar{\Delta}}\,[\mathrm{L}\sim] \qquad \frac{A, \Gamma \Rightarrow \Delta}{\Gamma \Rightarrow \Delta, \sim A}\,[\mathrm{R}\sim]$$

where $\bar{\Gamma}, \bar{\Delta}$ are $\sim$-closed. The sequent calculus **LT** for $\mathbf{TCC}_\omega$ is obtained by adding the above rules to the positive and non-modal fragment of **LGJ** (derivability denoted by $\vdash_{gT}$).

Theorem 5.24 *For all* $\Gamma, \Delta \subseteq \mathsf{Form}^\sim$, $\vdash_{gT} \Gamma \Rightarrow \Delta$ *iff* $\vdash_t \bigwedge\Gamma \rightarrow \bigvee\Delta$.

Proof. For the right-to-left direction, we need to check the cases for (N1),(N2) and (RC). Each case is straightforward. For the right-to-left direction, we must check the cases for [L$\sim$] and [R$\sim$]. The latter case is simple; for the former case, $\vdash_t \bar{\Gamma} \rightarrow (\bigvee\bar{\Delta} \vee A)$ by IH. Then by (MP) and Lemma 5.23, $\bar{\Gamma} \vdash_t \sim\sim\bigvee\bar{\Delta} \vee A$. So by (N2), (RC) and (t3), we obtain $\bar{\Gamma} \vdash_t \sim A \rightarrow \bigvee\bar{\Delta}$. Hence by deduction theorem and Lemma 5.23 again, we conclude $\vdash_t (\bar{\Gamma} \wedge \sim A) \rightarrow \bigvee\bar{\Delta}$. □

A sequent calculus for $\mathbf{IPC}^\sim$ has not been considered before. We can now obtain one by removing the condition that $\bar{\Delta}$ is $\sim$-closed in [L$\sim$]. The correspondence with the Hilbert-style system is straightforwardly demonstrable.

6 Concluding remarks

In this article, we introduced $\mathbf{IPC}^@$, an expansion of **IPC**, obtained by adding actuality operator, and compared with systems including **LGP** of Baaz, **GIPC** of Titani and $\mathbf{IPC}^{\sim}$ of De, obtained by adding projection operator, globalization operator and empirical negation respectively. What emerged is the following hierarchy of systems in $\mathcal{L}^{@}_{\bot}$, each corresponding to a system in $\mathcal{L}^{\sim}$.

(In $\mathcal{L}^{@}_{\bot}$)		(In $\mathcal{L}^{\sim}$)
LGP	$\Longleftrightarrow$	$\mathbf{IPC}^{\sim}$ + (Lin)
$\mathbf{IPC}^{@}$	$\Longleftrightarrow$	$\mathbf{IPC}^{\sim}$
GIPC = $\mathbf{L}_4$	$\Longleftrightarrow$	$\mathbf{TCC}_{\omega}$

With respect to these systems, we make some additional observations and mention a few future directions.

Hybrid logic Since there are clear connections between hybrid logics and logics with actuality operator, and in particular there are some results on hybrid logics based on intuitionistic logic (cf. [9,10]), a comparison of $\mathbf{IPC}^{@}$ to these systems will be of great interest.

Kripke semantics vs. Beth semantics We observed that @ in $\mathbf{IPC}^{@}$ and $\sim$ in $\mathbf{IPC}^{\sim}$ are inter-definable (in the presence of $\bot$ in the language), and similarly for **GIPC** and $\mathbf{TCC}_{\omega}$. As we have noted, a crucial difference between the semantics of $\mathbf{IPC}^{\sim}$ and $\mathbf{TCC}_{\omega}$ (hence the interpretation of @) is that models in the former always has a base state, while the latter in general does not. As a result, Kripke-semantically, even though both @ can be understood as a globalization operator (i.e. true iff true everywhere), only the former can be interpreted as an actuality operator. Yet one may wonder whether one could view @ in **GIPC** as a sort of actuality operator.

Beth semantics offers a possibility for this alternative interpretation. It is a semantics similar to Kripke semantics, but crucially different in that (i) all models have a base state, and (ii) the valuation of disjunction does not require one of the disjuncts to hold in the same world.[9] If we define the clause for $\sim$ as in the Kripke semantics of $\mathbf{IPC}^{\sim}$, we obtain Beth semantics with empirical negation (cf. Appendix). One of the present authors have shown elsewhere in [35] that (RP) is not valid, but $\mathbf{TCC}_{\omega}$ is sound and complete with respect to this semantics. This means that @ in **GIPC** can be understood as actuality operator *with respect to Beth semantics.* Thus there are two types of actuality operator/empirical negation in intuitionistic logic, Kripke-type and Beth-type.

With this kind of perspective, we can connect results related to **GIPC** with empirical negation. For instance, Titani's global intutionistic set theory can be seen as a mathematical theory with Beth-type empirical negation, by reading $\neg\Box$ as $\sim$. This could then encourage the investigation of intuitionistic set theory with Kripke-type empirical negation, as a possible future direction.

Quantifiers Global intuitionistic logic was originally formulated in a first-order language. Moreover, quantification for **LGP** has been investigated in

---

[9] For more information, cf. [47, Chapter 13].

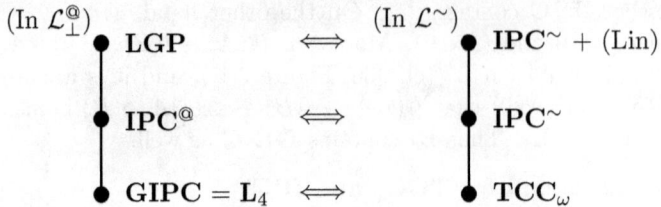

[2,3]. From this perspective, it seems to be a natural direction to consider first-order systems for **IPC**$^{@}$. This can be particularly interesting because like disjunction, existential quantifier has differing interpretations in Kripke and Beth models. Therefore we might be able to find an interesting interaction between quantifiers and modal operators. Moreover, for the purpose of comparing **IPC**$^{@}$ to **S5A** of Crossley and Humberstone, we also need quantifiers, and this will be yet another motivation for adding quantifiers.

Hypersequent calculi The sequent calculus for global intuitionistic logic **GI** defined by Titani and Aoyama is not cut-eliminatable, as observed by Agata Ciabattoni in [13, p.437]. She instead formulated a cut-free hypersequent calculus for **GI** and for **GIF**. We may then expect a similar approach to be quite beneficial in pursuing cut-free sequent calculi for the systems we have considered, namely **IPC**$^{@}$, **IPC**$^{\sim}$ and **TCC**$_{\omega}$.

Appendix

Beth semantics for TCC$_{\omega}$ We shall employ the following notations for sequences and related notions.

- $\alpha, \beta, \ldots$: infinite sequences of the form $\langle \alpha_1, \alpha_2, \ldots \rangle$ of natural numbers.
- $\langle\rangle$: the empty sequence.
- $b, b', \ldots$: finite sequences of the form $\langle b_1, \ldots, b_n \rangle$ of natural numbers.
- $b * b'$: b concatenated with b'.
- $lh(b)$: the length of b.
- $b \preceq b'$: $b * b'' = b'$ for some b''.
- $b \prec b'$: $b \preceq b'$ and $b \neq b'$.
- $\bar{\alpha}n$: α's initial segment up to the nth element.
- $\alpha \in b$: b is α's initial segment.

We define a *tree* to be a set T of finite sequences of natural number such that $\langle\rangle \in T$, $b \in T \vee b \notin T$ and $b \in T \wedge b' \prec b \rightarrow b' \in T$. We call each finite sequence in T as a *node* and $\langle\rangle$ as the *root*. A *successor* of a node b is a node of the form $b * \langle x \rangle$. By *leaves* of T, we mean the nodes of T which do not have a successor, i.e. nodes b such that $\neg\exists x(b * \langle x \rangle) \in T$. A *spread* then is a tree whose nodes always have a successor, i.e. $\forall b \in T \exists x(b * \langle x \rangle \in T)$.

A *Beth model* then is a triple $(W, \preceq, V)$, where $(W, \preceq)$ defines a spread and $V : W \times \mathsf{Prop} \rightarrow \{0,1\}$ an assignment of truth values to state-variable pairs with the condition that:

$V(b,p) = 1$ iff for all $\alpha \in b$ there is m such that $(V(\bar{\alpha}m, p) = 1)$. [covering]

An interpretation I for Beth model is defined by the following clauses.

- $I(b,p) = V(b,p)$;
- $I(b, A \wedge B)=1$ iff $I(b,A)=1$ and $I(b,B)=1$;
- $I(b, A \vee B)=1$ iff for all $\alpha \in b$ there is n such that $I(\bar{\alpha}n, A)=1$ or $I(\bar{\alpha}n, B)=1$;
- $I(b, A \rightarrow B) = 1$ iff for all $b \in W$: if $b \preceq b'$ and $I(b', A)=1$ then $I(b', B)=1$;
- $I(b, {\sim}A) = 1$ iff $I(\langle\rangle, A) = 0$.

The semantic consequence is then defined as in Kripke semantics.

References

[1] Aoyama, H., *The semantic completeness of a global intuitionistic logic*, Mathematical Logic Quarterly **44** (1998), pp. 167–175.

[2] Baaz, M., *Infinite-valued Gödel logics with 0-1-projections and relativizations*, in: P. Hájek, editor, *Gödel '96*, Springer, 1996 pp. 23–33.

[3] Baaz, M., N. Preining and R. Zach, *Completeness of a hypersequent calculus for some first-order godel logics with delta*, in: *36th International Symposium on Multiple-Valued Logic (ISMVL'06)*, IEEE, 2006, pp. 9–9.

[4] Berto, F., *A modality called 'negation'*, Mind **124** (2015), pp. 761–793.

[5] Berto, F. and G. Restall, *Negation on the Australian Plan*, Journal of Philosophical Logic **48** (2019), pp. 1119–1144.

[6] Bezhanishvili, G., *Varieties of monadic Heyting algebras. Part I*, Studia Logica **61** (1998), pp. 367–402.

[7] Bierman, G. M. and V. C. de Paiva, *On an intuitionistic modal logic*, Studia Logica **65** (2000), pp. 383–416.

[8] Božić, M. and K. Došen, *Models for normal intuitionistic modal logics*, Studia Logica **43** (1984), pp. 217–245.

[9] Braüner, T., *Axioms for classical, intuitionistic, and paraconsistent hybrid logic*, Journal of Logic, Language and Information **15** (2006), pp. 179–194.

[10] Braüner, T., *Intuitionistic hybrid logic: Introduction and survey*, Information and Computation **209** (2011), pp. 1437–1446.

[11] Bull, R. A., *A modal extension of intuitionist logic.*, Notre Dame Journal of Formal Logic **6** (1965), pp. 142–146.

[12] Bull, R. A., *MIPC as the formalisation of an intuitionist concept of modality*, The Journal of Symbolic Logic **31** (1966), pp. 609–616.

[13] Ciabattoni, A., *A proof-theoretical investigation of global intuitionistic (fuzzy) logic*, Archive for Mathematical Logic **44** (2005), pp. 435–457.

[14] Crossley, J. N. and L. Humberstone, *The logic of "actually"*, Reports on Mathematical Logic **8** (1977), pp. 1–29.

[15] Davies, M. and L. Humberstone, *Two notions of necessity*, Philosophical Studies: An International Journal for Philosophy in the Analytic Tradition **38** (1980), pp. 1–30.

[16] De, M., *Empirical Negation*, Acta Analytica **28** (2013), pp. 49–69.

[17] De, M. and H. Omori, *More on empirical negation.*, in: *Advances in modal logic*, 2014, pp. 114–133.

[18] De, M. and H. Omori, *Classical and empirical negation in subintuitionistic logic.*, in: *Advances in Modal Logic*, 2016, pp. 217–235.

[19] De, M. and H. Omori, *There is more to negation than modality*, Journal of Philosophical logic **47** (2018), pp. 281–299.

[20] Došen, K., *Negative modal operators in intuitionistic logic*, Publication de l'Instutute Mathematique, Nouv. Ser **35** (1984), pp. 3–14.

[21] Došen, K., *Models for stronger normal intuitionistic modal logics*, Studia Logica **44** (1985), pp. 39–70.

[22] Došen, K., *Negation as a modal operator*, Reports on Mathematical Logic **20** (1986), pp. 15–27.

[23] Došen, K., *Negation in the light of modal logic*, in: *What is Negation?*, Springer, 1999 pp. 77–86.

[24] Dummett, M., *A propositional calculus with denumerable matrix*, The Journal of Symbolic Logic **24** (1959), pp. 97–106.

[25] Fitch, F. B., *Intuitionistic modal logic with quantifiers*, Portugaliae mathematica **7** (1948), pp. 113–118.

[26] Fritz, P., *What is the correct logic of necessity, actuality and apriority?*, The Review of Symbolic Logic **7** (2014), pp. 385–414.

[27] Gabbay, D. M., "Semantical investigations in Heyting's intuitionistic logic," Synthese Library **148**, Springer, 1981.

[28] Gordienko, A. B., *A Paraconsistent Extension of Sylvan's Logic*, Algebra and Logic **46** (2007), pp. 289–296.

[29] Gregory, D., *Completeness and decidability results for some propositional modal logics containing "actually" operators*, Journal of Philosophical Logic **30** (2001), pp. 57–78.

[30] Heyting, A., *G.F.C. Griss and his negationless intuitionistic mathematics*, Synthese (1953), pp. 91–96.

[31] Horn, L. and H. Wansing, *Negation*, in: E. N. Zalta, editor, *The Stanford Encyclopedia of Philosophy*, 2020, Spring 2020 edition .

[32] Humberstone, L., *Two-dimensional adventures*, Philosophical Studies **118** (2004), pp. 17–65.

[33] Kapsner, A., "Logics and falsifications," Trends in logic **40**, Springer, 2014.

[34] Kurokawa, H., *Hypersequent calculi for intuitionistic logic with classical atoms*, Annals of Pure and Applied Logic **161** (2009), pp. 427–446.

[35] Niki, S., *Empirical negation, co-negation and the contraposition rule I: Semantical investigations*, submitted.

[36] Ono, H., *On some intuitionistic modal logics*, Publications of the Research Institute for Mathematical Sciences **13** (1977), pp. 687–722.

[37] Prior, A. N., "Time and modality," Oxford University Press, 1957.

[38] Restall, G., *Subintuitionistic Logics*, Notre Dame Journal of Formal Logic **35** (1994), pp. 116–126.

[39] Servi, G. F., *On modal logic with an intuitionistic base*, Studia Logica **36** (1977), pp. 141–149.

[40] Simpson, A. K., "The proof theory and semantics of intuitionistic modal logic," Ph.D. thesis, University of Edinburgh (1994).

[41] Sotirov, V. H., *Modal theories with intuitionistic logic*, in: *Proceedings of the Conference on Mathematical Logic, Sofia*, 1980, pp. 139–171.

[42] Standefer, S., *Actual issues for relevant logics*, Ergo (forthcoming).

[43] Stephanou, Y., *The meaning of 'actually'*, dialectica **64** (2010), pp. 153–185.

[44] Sylvan, R., *Variations on da Costa C Systems and Dual-Intuitionistic Logics I. Analyses of C_ω and CC_ω*, Studia Logica **49** (1990), pp. 47–65.

[45] Takeuti, G. and S. Titani, *Global intuitionistic fuzzy set theory*, in: P. Hájek, editor, *The Mathematics of Fuzzy Systems*, TUV-Verlag, 1986 pp. 291–301.

[46] Titani, S., *Completeness of global intuitionistic set theory*, The Journal of Symbolic Logic **62** (1997), pp. 506–528.

[47] Troelstra, A. S. and D. van Dalen, "Constructivism in Mathematics: An introduction, volume II," Elsevier, 1988.

[48] Ursini, A., *A modal calculus analogous to K4W, based on intuitionistic propositional logic, I^o*, Studia Logica **38** (1979), pp. 297–311.

A Semantics for a Failed Axiomatization of K

Hitoshi Omori [1] Daniel Skurt [2]

Department of Philosophy I, Ruhr University Bochum
Universitätsstraße 150, 44780, Bochum, Germany

Abstract

In "Yet another "choice of primitives" warning: Normal modal logics", Lloyd Humberstone discussed a failed axiomatization for the normal modal logic K with $\Diamond$ as the only primitive modal operator. More specifically, Humberstone observed that a simple translation of the standard axiomatization for K, where all occurrences of the necessity operator $\Box$ are replaced by $\neg\Diamond\neg$, will not be a complete axiomatization, since $\Diamond p \to \Diamond\neg\neg p$ is not derivable. As a result, the emerging proof system resists the standard Kripke semantics. However, to the best of the authors' knowledge, no semantics for the failed axiomatization of K is known in the literature. The aim of this article is to offer the first sound and complete semantics for the failed axiomatization of K by making use of a semantical framework suggested by John Kearns. In short, Kearns' semantics is a combination of non-deterministic semantics together with an additional hierarchy of valuations. We will also discuss a small question left open by Humberstone in the same paper. In view of the results presented in this article, we hope to establish part of the versatility of Kearns' semantics.

Keywords: Non-deterministic Semantics, Primitive Connectives, Normal Modal Logics.

1 Introduction

Both in classical and nonclassical logics, there is a freedom in choosing the set of primitive connectives. For example, in classical logic, one may take negation and the conditional as primitive connectives, or take all, negation, conjunction, disjunction and conditional as primitive. Or even one single connective known as Sheffer's stroke.

[1] The work reported in this paper was initially supported by JSPS KAKENHI Grant Number JP18K12183, and later by a Sofja Kovalevskaja Award of the Alexander von Humboldt-Foundation, funded by the German Ministry for Education and Research. Some part of the results were presented at *Oberseminar Logik und Sprachtheorie* in Tübingen in May 2019. We would also like to thank the three referees for their careful reading and helpful comments, suggestions and corrections that improved the paper. Email: Hitoshi.Omori@rub.de

[2] Some of the observations of this article were presented at *Trends in Logic XVIII* in Milan in September 2018, at the *Kolloquium Philosophie & Linguistik* in Göttingen in November 2018, at the *Philosophisches Kolloquium* in Leipzig in December 2018. DS would like to thank the audience for their helpful comments and encouragements. Email: Daniel.Skurt@rub.de

We also know, however, that sometimes some additional care is required.[3] For example, if we take negation and disjunction as primitive connectives, then the following set of axioms and the rule of inference, due to Hilbert and Ackermann, are complete with respect to the usual two-valued semantics, where $A \to B$ abbreviates $\neg A \vee B$.

$$(A \vee A) \to A \qquad (A \vee B) \to (B \vee A)$$
$$A \to (A \vee B) \qquad (A \to B) \to ((C \vee A) \to (C \vee B))$$
$$\text{From } A \text{ and } A \to B\text{, infer } B$$

Now, consider negation and conjunction as primitive connectives, and if we simply translate the above set of axioms and the rule of inference with the usual definitions $A \vee B =_{\text{def.}} \neg(\neg A \wedge \neg B)$ and $A \to B =_{\text{def.}} \neg(A \wedge \neg B)$, then we obtain the following:

$$\neg((\neg(\neg A \wedge \neg A)) \wedge \neg A) \qquad \neg(A \wedge \neg\neg(\neg A \wedge \neg B)$$
$$\neg(\neg(\neg A \wedge \neg B) \wedge \neg\neg(\neg B \wedge \neg A))$$
$$\neg((\neg(A \wedge \neg B)) \wedge \neg\neg(\neg(\neg C \wedge \neg A) \wedge \neg\neg(\neg C \wedge \neg B)))$$
$$\text{From } A \text{ and } \neg(A \wedge \neg B)\text{, infer } B$$

However, as observed by Henryk Hiż in [8], the latter system is *not* a complete axiomatization since we may observe that $\neg(\neg p \wedge p)$ is not derivable.

As for modal logics, it was shown by David Makinson in [14] that "*the decision whether to treat the zero-ary falsum operator as primitive or as defined, affects the general structure of the lattice of all modal logics.*" Moreover, in [9], Lloyd Humberstone observed, among other things, that a simple translation of the axiomatization for the modal logic K with the necessity (or "box") as the primitive connective, obtained by replacing the occurrences of "box" by "not diamond not" will *not* be a complete axiomatization, since we may observe that $\Diamond p \to \Diamond\neg\neg p$ is *not* derivable.[4]

Note, however, that a sound and complete semantics for the failed axiomatization of K, which we refer to as $\mathsf{K_f}$, is not yet available in the literature, at least to the best of the authors' knowledge.[5]

Based on these, the aim of this article is to fill in this gap and as a byproduct show the versatility of John Kearns' semantics, devised in [12].[6] More specifically, we will first introduce a sixteen-valued non-deterministic semantics (cf. [1] for a survey) with an additional hierarchy on the set of all valuations for

[3] For an interesting discussion related to this point, but from a wider perspective, see [7] and references therein.

[4] This is also reported by Richmond Thomason in a recent note [21] without any reference to Humberstone's observation. We will not discuss Thomason's note since the eight-valued matrix he introduces seems to be not fully articulated. Note that, as pointed out by Humberstone, there is a four-valued matrix that will establish one of Thomason's results (see Remark 2.6 below).

[5] This is not to say that there are no sound and complete Kripke semantics for the modal logic K with a primitive possibility operator, see for example [2]. In brief, this axiomatization makes use of one more axiom than just the K-axiom. We will return to this point later.

[6] Kearns' semantics was later applied to a larger family of normal modal logics in [16].

$\mathsf{K_f}$ following Kearns. Then we will prove that $\mathsf{K_f}$ is sound and complete with respect to our Kearns' style semantics, now involving sixteen values, instead of four values.[7] Once these are established, we will extend $\mathsf{K_f}$ with additional axioms in order to give a sound and complete semantics for a system we call $\mathsf{S5_f}$. With this semantics we will deal with a problem left open by Humberstone, namely showing the independence of $\Diamond\neg\neg p \to \Diamond p$ from the failed axiomatization.

2 Semantics and proof system

Our language $\mathcal{L}$ consists of the set $\{\neg, \Diamond, \to\}$ of propositional connectives and a countable set Prop of propositional parameters. Furthermore, we denote by Form the set of formulas defined as usual in $\mathcal{L}$. We denote formulas of $\mathcal{L}$ by A, B, C, etc. and sets of formulas of $\mathcal{L}$ by Γ, Δ, Σ, etc.

2.1 Proof system

We first introduce the target system of this article, namely the system $\mathsf{K_f}$. We also define a subsystem that will be sound and complete with respect to the non-deterministic semantics *without* the hierarchy.

Definition 2.1 First, the system $\mathsf{K_f}$ consists of the following axiom schemata and rules of inference:[8]

$$A \to (B \to A) \quad \text{(Ax1)} \qquad (A\to(B\to C))\to((A\to B)\to(A\to C)) \quad \text{(Ax2)}$$

$$(\neg B\to\neg A)\to(A\to B) \quad \text{(Ax3)} \qquad \neg\Diamond\neg(A\to B)\to(\neg\Diamond\neg A\to\neg\Diamond\neg B) \quad \text{(LK)}$$

$$\frac{A \quad A\to B}{B}\ \text{(MP)} \qquad \frac{A}{\neg\Diamond\neg A}\ \text{(RN)}$$

We write $\vdash_{\mathsf{K_f}} A$ for, there is a *proof for* A in K_f if there is a sequence of formulas $B_1, \ldots, B_n, A$ $(n \geq 0)$, such that every formula in the sequence either (i) is an axiom of $\mathsf{K_f}$; or (ii) is obtained by (MP) or (RN) from formulas preceding it in the sequence. Moreover, we define $\Gamma \vdash_{\mathsf{K_f}} A$ iff for a finite subset Γ' of Γ, $\vdash_{\mathsf{K_f}} C_1 \to (C_2 \to (\cdots (C_n \to A)\cdots))$ where $C_i \in \Gamma' (1 \leq i \leq n)$.

Second, we define a subsystem of $\mathsf{K_f}$, referred to as $\mathsf{k_f}$, which is obtained by eliminating (RN) and adding the following schemata:[9]

$$\Diamond\neg\neg(A\to B)\to(\neg\Diamond\neg A\to\Diamond\neg\neg B) \quad (\mathrm{Ak_f1}) \qquad \neg\Diamond\neg\neg(A\to B)\to\neg\Diamond\neg A \quad (\mathrm{Ak_f2})$$

$$\neg\Diamond\neg\neg(A\to B)\to\neg\Diamond\neg\neg B \quad (\mathrm{Ak_f3}) \qquad \Diamond\neg(A\to B)\to\Diamond\neg B \quad (\mathrm{Ak_f4})$$

$$\Diamond\neg\neg\neg A\to\Diamond\neg A \quad (\mathrm{Ak_f5}) \qquad \Diamond\neg A\to\Diamond\neg\neg\neg A \quad (\mathrm{Ak_f6})$$

We define $\Gamma \vdash_{\mathsf{k_f}} A$ (A *can be derived from* Γ) iff there is a sequence of formulas $B_1, \ldots, B_n, A$ $(n \geq 0)$, such that every formula in the sequence either (i) is an

[7] Sixteen values may remind us of the system SIXTEEN$_3$ of Yaroslav Shramko and Heinrich Wansing (cf. [18,19]). However, we could not establish any relation between their semantics and our semantics.

[8] Where (Ax1), (Ax2), (Ax3) and (MP) are a well-known axiomatization of classical propositional logic (cf. [20]) and, (LK) and (RN) are the K-axiom and rule of necessitation expressed in terms of $\neg$ and $\Diamond$.

[9] We would like to thank one of the anonymous reviewers for pointing out the missing axioms.

element of Γ (ii) is an axiom of $\mathsf{k_f}$; or (iii) is obtained by (MP) from formulas preceding it in the sequence.

Remark 2.2 It is rather easy to see that $\mathsf{k_f}$ is a subsystem of $\mathsf{K_f}$. Indeed, note first that in $\mathsf{K_f}$, we have the following rules in view of (LK), (MP) and (RN):

$$\frac{A\to B}{\neg\Diamond\neg A\to\neg\Diamond\neg B}\ ,\quad \frac{A\to(B\to C)}{\neg\Diamond\neg A\to(\neg\Diamond\neg B\to\neg\Diamond\neg C)}.$$

Then, in order to see that (Ak$_\mathsf{f}$1) is derivable in $\mathsf{K_f}$, apply the second rule to $A\to(\neg B\to\neg(A\to B))$. For the rest, apply the first rule to $\neg(A\to B)\to A$, $\neg(A\to B)\to\neg B$, $B\to(A\to B)$, $A\to\neg\neg A$ and $\neg\neg A\to A$, respectively.

2.2 A detour: counter-model of Humberstone

Here, we will review the counter-model used by Humberstone, in [9], to establish that $\Diamond A\to\Diamond\neg\neg A$ is *not* derivable in $\mathsf{K_f}$.

Definition 2.3 [Humberstone] A model for $\mathcal{L}$ is a triple $\langle W,N,V\rangle$ in which W is a set with $\emptyset\neq N\subseteq W$ and V is a function assigning to each propositional variable a subset of W. Given a model $\mathcal{M}=\langle W,N,V\rangle$ we define truth of a formula A at a point $u\in W$ ($\mathcal{M}\models_u A$) as follows:

- $\mathcal{M}\models_u p$ iff $u\in V(p)$, if $p\in\mathsf{Prop}$;
- $\mathcal{M}\models_u B\to C$ iff $\mathcal{M}\not\models_u B$ or $\mathcal{M}\models_u C$;
- $\mathcal{M}\models_u \neg B$ iff $u\in N$ and $\mathcal{M}\not\models_u B$;
- $\mathcal{M}\models_u \Diamond B$ iff for some $v\in W:\mathcal{M}\models_v B$.

A formula A is true in the model $\mathcal{M}=\langle W,N,V\rangle$, (notation: $\mathcal{M}\models A$), just in case for all $u\in N$, we have $\mathcal{M}\models_u A$, and valid (notation: $\models_H A$, where H stands for Humberstone) if it is true in every model.

Fact 2.4 (Theorem 2.1 in [9]) *For all* $A\in\mathsf{Form}$, *if* $\vdash_{\mathsf{K_f}} A$ *then* $\models_H A$.

Fact 2.5 (Corollary 2.2 in [9]) $\not\models_H \Diamond p\to\Diamond\neg\neg p$.[10]

Proof. Consider a two-element model $\mathcal{M}_0=\langle W_0,N_0,V_0\rangle$, with $W_0=\{u,v\}$ and $u\neq v$, $N_0=\{u\}$ and $V_0(p)=\{v\}$. Now, we have $\mathcal{M}_0\models_u\Diamond p$, but $\mathcal{M}_0\not\models_u\Diamond\neg\neg p$. The latter follows since there is no element in W_0, such that $\neg\neg p$ is true. Indeed, for u, we have $u\in N$ but $\mathcal{M}_0\not\models_u p$, and for v, we have $\mathcal{M}_0\not\models_v\neg p$, but $v\notin N$. Therefore, $\mathcal{M}_0\not\models_u\Diamond p\to\Diamond\neg\neg p$. as desired. □

Remark 2.6 Note that the above model $\mathcal{M}_0$ can be seen as a four-valued matrix with its four elements being $1=\{u,v\}, 2=\{u\}, 3=\{v\}$ and $4=\emptyset$, and designated values 1 and 2. Truth tables for the connectives are as follows.

A	$\neg A$	$\Diamond A$
1	4	1
2	4	1
3	2	1
4	2	4

$A\to B$	1	2	3	4
1	1	2	3	4
2	1	1	3	3
3	1	2	1	2
4	1	1	1	1

[10] Thus, $\mathsf{K_f}$ does not enjoy the replacement property, also known as self-extensionality. So, if this property is crucial for modal logics (cf. [15]), then $\mathsf{K_f}$ is not a modal logic.

Then, if we assign the value 3 to p, then $\Diamond p \to \Diamond\neg\neg p$ receives the value 4, as desired. Note, however, that $\Diamond\neg\neg p \to \Diamond p$ will be verified in this model (note that the above matrix can be found in [9, Figure 1]).

2.3 Semantics

We now turn to present the semantics for $\mathsf{K_f}$. To this end, we first introduce the basic Nmatrix which requires sixteen truth values.

Definition 2.7 A $\mathsf{K_f}$-*Nmatrix* for $\mathcal{L}$ is a tuple $M = \langle \mathcal{V}, \mathcal{T}, \mathcal{O}\rangle$, where:

(a) $\mathcal{V} = \{\mathbf{T_1}, \mathbf{T_2}, \mathbf{T_3}, \mathbf{T_4}, \mathbf{t_1}, \mathbf{t_2}, \mathbf{t_3}, \mathbf{t_4}, \mathbf{f_1}, \mathbf{f_2}, \mathbf{f_3}, \mathbf{f_4}, \mathbf{F_1}, \mathbf{F_2}, \mathbf{F_3}, \mathbf{F_4}\}$,

(b) $\mathcal{T} = \{\mathbf{T_1}, \mathbf{T_2}, \mathbf{T_3}, \mathbf{T_4}, \mathbf{t_1}, \mathbf{t_2}, \mathbf{t_3}, \mathbf{t_4}\}$,

(c) For every n-ary connective $*$ of $\mathcal{L}$, $\mathcal{O}$ includes a corresponding n-ary function $\tilde{*}$ from $\mathcal{V}^n$ to $2^{\mathcal{V}} \setminus \{\emptyset\}$ as follows (we omit the brackets for sets):

A	$\tilde{\neg}A$	$\Diamond A$	A	$\tilde{\neg}A$	$\Diamond A$	A	$\tilde{\neg}A$	$\Diamond A$	A	$\tilde{\neg}A$	$\Diamond A$
$\mathbf{T_1}$	$\mathbf{F_4}$	$\mathcal{T}$	$\mathbf{t_1}$	$\mathbf{F_1}$	$\mathcal{T}$	$\mathbf{f_1}$	$\mathbf{t_4}$	$\mathcal{T}$	$\mathbf{F_1}$	$\mathbf{t_1}$	$\mathcal{T}$
$\mathbf{T_2}$	$\mathbf{f_4}$	$\mathcal{T}$	$\mathbf{t_2}$	$\mathbf{f_1}$	$\mathcal{T}$	$\mathbf{f_2}$	$\mathbf{T_4}$	$\mathcal{T}$	$\mathbf{F_2}$	$\mathbf{T_1}$	$\mathcal{T}$
$\mathbf{T_3}$	$\mathbf{F_4}$	$\mathcal{F}$	$\mathbf{t_3}$	$\mathbf{F_1}$	$\mathcal{F}$	$\mathbf{f_3}$	$\mathbf{t_4}$	$\mathcal{F}$	$\mathbf{F_3}$	$\mathbf{t_1}$	$\mathcal{F}$
$\mathbf{T_4}$	$\mathbf{f_4}$	$\mathcal{F}$	$\mathbf{t_4}$	$\mathbf{f_1}$	$\mathcal{F}$	$\mathbf{f_4}$	$\mathbf{T_4}$	$\mathcal{F}$	$\mathbf{F_4}$	$\mathbf{T_1}$	$\mathcal{F}$

$A \tilde{\to} B$	$\mathcal{T}_{13}$	$\mathcal{T}_{24}$	t_{13}	t_{24}	f_{13}	f_{24}	$\mathcal{F}_{13}$	$\mathcal{F}_{24}$
$\mathcal{T}_{13}, \mathcal{T}_{24}$	$\mathcal{T}_{13}$	$\mathcal{T}_{24}$	t_{13}	t_{24}	f_{13}	f_{24}	$\mathcal{F}_{13}$	$\mathcal{F}_{24}$
t_{13}, t_{24}	$\mathcal{T}_{13}$	$\mathcal{T}_{13}$	$\mathcal{T}_{13}, t_{13}$	$\mathcal{T}_{13}, t_{13}$	f_{13}	f_{13}	$f_{13}, \mathcal{F}_{13}$	$f_{13}, \mathcal{F}_{13}$
f_{13}, f_{24}	$\mathcal{T}_{13}$	$\mathcal{T}_{24}$	t_{13}	t_{24}	$\mathcal{T}_{13}$	$\mathcal{T}_{24}$	t_{13}	t_{24}
$\mathcal{F}_{13}, \mathcal{F}_{24}$	$\mathcal{T}_{13}$	$\mathcal{T}_{13}$	$\mathcal{T}_{13}, t_{13}$	$\mathcal{T}_{13}, t_{13}$	$\mathcal{T}_{13}$	$\mathcal{T}_{13}$	$\mathcal{T}_{13}, t_{13}$	$\mathcal{T}_{13}, t_{13}$

where

- $\mathcal{T}_{13} = \{\mathbf{T_1}, \mathbf{T_3}\}$, $\mathcal{T}_{24} = \{\mathbf{T_2}, \mathbf{T_4}\}$, $t_{13} = \{\mathbf{t_1}, \mathbf{t_3}\}$, $t_{24} = \{\mathbf{t_2}, \mathbf{t_4}\}$,
- $\mathcal{F}_{13} = \{\mathbf{F_1}, \mathbf{F_3}\}$, $\mathcal{F}_{24} = \{\mathbf{F_2}, \mathbf{F_4}\}$, $f_{13} = \{\mathbf{f_1}, \mathbf{f_3}\}$, $f_{24} = \{\mathbf{f_2}, \mathbf{f_4}\}$,
- $\mathcal{F} = \{\mathbf{f_1}, \mathbf{f_2}, \mathbf{f_3}, \mathbf{f_4}, \mathbf{F_1}, \mathbf{F_2}, \mathbf{F_3}, \mathbf{F_4}\}$.

A $\mathsf{k_f}$-*valuation* in a $\mathsf{K_f}$-Nmatrix M is a function $v : \mathsf{Form} \to \mathcal{V}$ that satisfies the following condition for every n-ary connective $*$ of $\mathcal{L}$ and $A_1, \ldots, A_n \in \mathsf{Form}$: $v(*(A_1, \ldots, A_n)) \in \tilde{*}(v(A_1), \ldots, v(A_n))$. [11]

Remark 2.8 Note that the above truth table for $\tilde{\to}$ is making use of an unusual abbreviation. The full version is available in the Appendix.

Remark 2.9 This truth-table for implication can be seen as a generalization of the truth-table for implication of the system K, presented in [16, Definition 43]. The similarities will become explicit in the definition of the canonical model (cf. Definition 3.8 below).

Definition 2.10 A is a $\mathsf{k_f}$-*consequence* of Γ ($\Gamma \models_{\mathsf{k_f}} A$) iff for all $\mathsf{k_f}$-valuation v, if $v(B) \in \mathcal{T}$ for all $B \in \Gamma$ then $v(A) \in \mathcal{T}$. In particular, A is a $\mathsf{k_f}$-*tautology* iff $v(A) \in \mathcal{T}$ for all $\mathsf{k_f}$-valuations v.

[11] Note that the definition of $\mathsf{k_f}$-valuations is in the terminology of [16] called legal valuation, which in turn is also called dynamic valuation in [1].

Remark 2.11 Note that with Yuri V. Ivlev another logician considered non-deterministic semantics for a language with modality (cf. [10,11]). He is, however, not dealing with normal modal logics but only fragments without the rule of necessitation. Our system $\mathsf{k_f}$ can therefore also be understood as a system of modality in the sense of Ivlev. [12]

Definition 2.12 Let v be a function $v : \mathsf{Form} \to \mathcal{V}$. Then,

- v is a *0th-level* $\mathsf{K_f}$*-valuation* if v is a $\mathsf{k_f}$-valuation.
- v is a $m+1$*st-level* $\mathsf{K_f}$*-valuation* iff v is an mth-level $\mathsf{K_f}$-valuation and for every sentence A, $v(A) \in \{\mathbf{T_1}, \mathbf{T_2}, \mathbf{T_3}, \mathbf{T_4}\}$ holds if $v'(A) \in \mathcal{T}$ for every mth-level $\mathsf{K_f}$-valuation v'.

Based on these, we define v to be a $\mathsf{K_f}$*-valuation* iff v is an mth-level $\mathsf{K_f}$-valuation for every $m \geq 0$.

Definition 2.13 A is a $\mathsf{K_f}$*-tautology* ($\models_{\mathsf{K_f}} A$) iff $v(A) \in \{\mathbf{T_1}, \mathbf{T_2}, \mathbf{T_3}, \mathbf{T_4}\}$ for every $\mathsf{K_f}$-valuation v.

Remark 2.14 The definition of $\mathsf{K_f}$-valuations involves some complicated construction. Hence, for the sake of simplicity, we will only focus on tautologies, but not consequence relations with possibly non-empty premises for $\mathsf{K_f}$, and similarly for its extensions.

3 Soundness and completeness

We first prove the soundness, and then turn to the completeness result for both $\mathsf{k_f}$ and $\mathsf{K_f}$. The proofs are variants of those in [16].

3.1 Soundness

The soundness for the $\mathsf{k_f}$ consequence relation is rather straightforward.

Proposition 3.1 *For all* $\Gamma \cup \{A\} \subseteq \mathsf{Form}$*, if* $\Gamma \vdash_{\mathsf{k_f}} A$ *then* $\Gamma \models_{\mathsf{k_f}} A$*.*

Proof. It suffices to check that all axioms are $\mathsf{k_f}$-tautologies, and that the rule of inference (MP) preserves the designated values. The details are spelled out in the Appendix. □

For the soundness of $\mathsf{K_f}$, we need the following lemma.

Lemma 3.2 *Assume that* $\vdash_{\mathsf{K_f}} A$ *and that the length of the proof for* A *is* m*. Then, for every* m*th-level* $\mathsf{K_f}$*-valuation* v*,* $v(A) \in \mathcal{T}$*.*

Proof. By induction on the length m of the proof for $\vdash_{\mathsf{K_f}} A$. For the base, case in which $m = 1$, A is one of the axioms. Since axioms are $\mathsf{k_f}$-tautologies, as shown above, $v(A) \in \mathcal{T}$ for every 1st-level $\mathsf{K_f}$-valuation. (Note that by definition, if a sentence is designated for every mth-level $\mathsf{K_f}$-valuation, then it is also designated for every $m+1$st-level $\mathsf{K_f}$-valuation.) For the induction step, assume that the result holds for proofs of the length m, and let $B_1, \ldots, B_m, B_{m+1} (= A)$ be the proof for A. Then, there are the following three cases:

---

[12] For continuations of Ivlev's approach, see for example [3,4,5]. For a little problem with Ivlev's original paper, see [16, §3.3].

- If A is an axiom, then A is designated for every $\mathsf{k_f}$-valuation, and thus for every $m+1$st-level $\mathsf{K_f}$-valuation as well.
- If A is obtained by applying (MP) to B_i and $B_j(=B_i \to A)$, then by induction hypothesis, B_i and B_j are designated for every $\max\{i,j\}$th-level $\mathsf{K_f}$-valuation, and thus for every mth-level $\mathsf{K_f}$-valuation. By the truth table for $\to$, we obtain that A is also designated for every mth-level $\mathsf{K_f}$-valuation. Therefore, A is designated for every $m+1$st-level $\mathsf{K_f}$-valuation as well.
- If A is obtained by applying (RN) to B_i, then by induction hypothesis, B_i is designated for every ith-level $\mathsf{K_f}$-valuation. So, for every $i+1$st-level $\mathsf{K_f}$-valuation, $\neg\Diamond\neg B_i$, i.e. A is designated. Thus, A is designated for every $m+1$st-level $\mathsf{K_f}$-valuation.

This completes the proof. □

Once we have the lemma, soundness for $\mathsf{K_f}$ follows immediately.

Proposition 3.3 *For all* $A \in \mathsf{Form}$, *if* $\vdash_{\mathsf{K_f}} A$ *then* $\models_{\mathsf{K_f}} A$.

Proof. Let the length of the proof for A be m. Then, by the above lemma, A is designated for every mth-level $\mathsf{K_f}$-valuation. Therefore, $v(A) \in \{\mathbf{T_1}, \mathbf{T_2}, \mathbf{T_3}, \mathbf{T_4}\}$ for every $m+1$st-level $\mathsf{K_f}$-valuation v. Since $\mathsf{K_f}$-valuations are also $m+1$st-level $\mathsf{K_f}$-valuations, we obtain that A takes one of the values $\mathbf{T_1}, \mathbf{T_2}, \mathbf{T_3}, \mathbf{T_4}$ for every $\mathsf{K_f}$-valuation, as desired. □

3.2 Completeness

We now turn to prove the completeness. First, we list some provable formulas, without proofs, that will be used in the following proofs.

Proposition 3.4 *The following formulas are provable in* $\mathsf{k_f}$*:*

$$\neg\Diamond\neg A \to (\Diamond\neg B \to \Diamond\neg(A \to B)) \quad (1) \qquad (A \to B) \to ((\neg A \to B) \to B) \quad (3)$$
$$A \to (\neg B \to \neg(A \to B)) \quad (2) \qquad A \to (\neg A \to B) \quad (4)$$

Second, we introduce some standard notions that will be used in the proofs. In what follows, we let L be $\mathsf{k_f}$ or $\mathsf{K_f}$, or their extensions we consider in later sections.

Definition 3.5 For $\Gamma \subseteq \mathsf{Form}$, Γ is an L-*consistent* set iff $\Gamma \nvdash A$ or $\Gamma \nvdash \neg A$ for all $A \in \mathsf{Form}$. Γ is L-*inconsistent* otherwise.

Definition 3.6 For $\Gamma \subseteq \mathsf{Form}$, Γ is *maximal* L-*consistent* set iff Γ is L-consistent and any set of formulas properly containing Γ is L-inconsistent. If Γ is maximal L-consistent set, then we say that Γ is a L-`mcs`.

We then obtain the following well-known lemma. As the proof is standard, we will leave it to the reader.

Lemma 3.7 *For any* $\Sigma \cup \{A\} \subseteq \mathsf{Form}$, *suppose that* $\Sigma \nvdash_{\mathsf{L}} A$. *Then, there is a* $\Pi \supseteq \Sigma$ *such that* Π *is a* L-`mcs`.

We next define the canonical valuation. This will also give us an intuitive reading of the sixteen values.

Definition 3.8 For any $\Sigma \subsetneq \mathsf{Form}$, we define a function $v_\Sigma : \mathsf{Form} \to \mathcal{V}$ as follows.

$$v_\Sigma(B) := \begin{cases} \mathbf{T_1} & \text{if } \Sigma \vdash \neg\Diamond\neg B \text{ and } \Sigma \vdash B \text{ and } \Sigma \vdash \Diamond B \text{ and } \Sigma \vdash \Diamond\neg\neg B \\ \mathbf{T_2} & \text{if } \Sigma \vdash \neg\Diamond\neg B \text{ and } \Sigma \vdash B \text{ and } \Sigma \vdash \Diamond B \text{ and } \Sigma \nvdash \Diamond\neg\neg B \\ \mathbf{T_3} & \text{if } \Sigma \vdash \neg\Diamond\neg B \text{ and } \Sigma \vdash B \text{ and } \Sigma \nvdash \Diamond B \text{ and } \Sigma \vdash \Diamond\neg\neg B \\ \mathbf{T_4} & \text{if } \Sigma \vdash \neg\Diamond\neg B \text{ and } \Sigma \vdash B \text{ and } \Sigma \nvdash \Diamond B \text{ and } \Sigma \nvdash \Diamond\neg\neg B \\ \mathbf{t_1} & \text{if } \Sigma \nvdash \neg\Diamond\neg B \text{ and } \Sigma \vdash B \text{ and } \Sigma \vdash \Diamond B \text{ and } \Sigma \vdash \Diamond\neg\neg B \\ \mathbf{t_2} & \text{if } \Sigma \nvdash \neg\Diamond\neg B \text{ and } \Sigma \vdash B \text{ and } \Sigma \vdash \Diamond B \text{ and } \Sigma \nvdash \Diamond\neg\neg B \\ \mathbf{t_3} & \text{if } \Sigma \nvdash \neg\Diamond\neg B \text{ and } \Sigma \vdash B \text{ and } \Sigma \nvdash \Diamond B \text{ and } \Sigma \vdash \Diamond\neg\neg B \\ \mathbf{t_4} & \text{if } \Sigma \nvdash \neg\Diamond\neg B \text{ and } \Sigma \vdash B \text{ and } \Sigma \nvdash \Diamond B \text{ and } \Sigma \nvdash \Diamond\neg\neg B \\ \mathbf{f_1} & \text{if } \Sigma \vdash \neg\Diamond\neg B \text{ and } \Sigma \nvdash B \text{ and } \Sigma \vdash \Diamond B \text{ and } \Sigma \vdash \Diamond\neg\neg B \\ \mathbf{f_2} & \text{if } \Sigma \vdash \neg\Diamond\neg B \text{ and } \Sigma \nvdash B \text{ and } \Sigma \vdash \Diamond B \text{ and } \Sigma \nvdash \Diamond\neg\neg B \\ \mathbf{f_3} & \text{if } \Sigma \vdash \neg\Diamond\neg B \text{ and } \Sigma \nvdash B \text{ and } \Sigma \nvdash \Diamond B \text{ and } \Sigma \vdash \Diamond\neg\neg B \\ \mathbf{f_4} & \text{if } \Sigma \vdash \neg\Diamond\neg B \text{ and } \Sigma \nvdash B \text{ and } \Sigma \nvdash \Diamond B \text{ and } \Sigma \nvdash \Diamond\neg\neg B \\ \mathbf{F_1} & \text{if } \Sigma \nvdash \neg\Diamond\neg B \text{ and } \Sigma \nvdash B \text{ and } \Sigma \vdash \Diamond B \text{ and } \Sigma \vdash \Diamond\neg\neg B \\ \mathbf{F_2} & \text{if } \Sigma \nvdash \neg\Diamond\neg B \text{ and } \Sigma \nvdash B \text{ and } \Sigma \vdash \Diamond B \text{ and } \Sigma \nvdash \Diamond\neg\neg B \\ \mathbf{F_3} & \text{if } \Sigma \nvdash \neg\Diamond\neg B \text{ and } \Sigma \nvdash B \text{ and } \Sigma \nvdash \Diamond B \text{ and } \Sigma \vdash \Diamond\neg\neg B \\ \mathbf{F_4} & \text{if } \Sigma \nvdash \neg\Diamond\neg B \text{ and } \Sigma \nvdash B \text{ and } \Sigma \nvdash \Diamond B \text{ and } \Sigma \nvdash \Diamond\neg\neg B \end{cases}$$

Remark 3.9 Compared to the definition of the canonical valuation for K in [16, Lemma 52], the number of truth values has doubled, since we are treating $\Diamond B$ and $\Diamond\neg\neg B$ separately.

Lemma 3.10 *If Σ is a* $\mathsf{k_f}$-**mcs**, *then v_Σ is a well-defined* $\mathsf{k_f}$*-valuation.*

Proof. The details are spelled out in the Appendix. □

Remark 3.11 By a careful examination, we also obtain that if Σ is a $\mathsf{K_f}$-mcs, then v_Σ is a well-defined $\mathsf{k_f}$-valuation.

Based on these, we are now ready to prove the completeness of $\mathsf{k_f}$.

Theorem 3.12 *For all $\Gamma \cup \{A\} \subseteq \mathsf{Form}$, if $\Gamma \models_{\mathsf{k_f}} A$ then $\Gamma \vdash_{\mathsf{k_f}} A$.*

Proof. Suppose that $\Gamma \nvdash_{\mathsf{k_f}} A$. Then by Lemma 3.7, we can construct a $\mathsf{k_f}$-mcs Σ_0 such that $\Gamma \subseteq \Sigma_0$. In view of Lemma 3.10, we can define a $\mathsf{k_f}$-valuation v_{Σ_0} such that $v_{\Sigma_0}(B) \in \mathcal{T}$ for every $B \in \Gamma$ and $v_{\Sigma_0}(A) \notin \mathcal{T}$. Thus we have $\Gamma \not\models_{\mathsf{k_f}} A$, as desired. □

For the completeness of $\mathsf{K_f}$, we need one more lemma.

Lemma 3.13 *Let Γ be a* $\mathsf{K_f}$-mcs. *If v_Γ is a* $\mathsf{k_f}$*-valuation, then v_Γ is also an mth-level* $\mathsf{K_f}$*-valuation for every $m \geq 1$, and thus a* $\mathsf{K_f}$*-valuation.*

Proof. By induction on m. For the base case, we prove that v_Γ is 1st-level $\mathsf{K_f}$-valuation. Let A be a sentence that is designated for every $\mathsf{k_f}$-valuation. Assume, for reductio, that $\nvdash_{\mathsf{K_f}} A$. Then, in view of Remark 2.2, $\nvdash_{\mathsf{k_f}} A$. Now, by Lemma 3.7, there is a $\mathsf{k_f}$-mcs Σ such that $\Sigma \nvdash_{\mathsf{k_f}} A$. Now let v_Σ be the $\mathsf{k_f}$-valuation generated by Σ. By the definition of v_Σ, we have that $\Sigma \nvdash_{\mathsf{k_f}} A$, i.e. $v(A) \notin \mathcal{T}$. But this contradicts our assumption that A is designated for every $\mathsf{k_f}$-valuation. Therefore, we have proved that $\vdash_{\mathsf{K_f}} A$. Then by (RN), we obtain $\vdash_{\mathsf{K_f}} \neg\Diamond\neg A$. And by the definition of v_Γ, we obtain that $v_\Gamma(A) \in \{\mathbf{T_1}, \mathbf{T_2}, \mathbf{T_3}, \mathbf{T_4}\}$, as desired.

For the induction step, assume that v_Γ is an mth-level $\mathsf{K_f}$-valuation, and let A be a sentence that is designated for every mth-level $\mathsf{K_f}$-valuation. Assume, for contradiction, that $\nvdash_{\mathsf{K_f}} A$. Again, in view of Remark 2.2, we obtain $\nvdash_{\mathsf{k_f}} A$. Now by Lemma 3.7, there is a $\mathsf{k_f}$-`mcs` Δ such that $\Delta \nvdash_{\mathsf{k_f}} A$. Now let v_Δ be the $\mathsf{k_f}$-valuation generated by Δ. By induction hypothesis, we have that v_Δ is an mth-level $\mathsf{K_f}$-valuation. Moreover, by the definition of v_Δ, we have that $\Delta \nvdash_{\mathsf{k_f}} A$, i.e. $v_\Delta(A) \notin \mathcal{T}$. But this contradicts our assumption that A is designated for every mth-level $\mathsf{K_f}$-valuation. Therefore, we have proved that $\vdash_{\mathsf{K_f}} A$. Then by (RN), we obtain $\vdash_{\mathsf{K_f}} \neg\Diamond\neg A$. And by the definition of v_Γ, we obtain that $v_\Gamma(A) \in \{\mathbf{T_1}, \mathbf{T_2}, \mathbf{T_3}, \mathbf{T_4}\}$, as desired. □

We are now ready to prove completeness for $\mathsf{K_f}$.

Theorem 3.14 *For all $A \in$ Form, if $\models_{\mathsf{K_f}} A$ then $\vdash_{\mathsf{K_f}} A$.*

Proof. Suppose that $\nvdash_{\mathsf{K_f}} A$. Then by Lemma 3.7, we have an $\mathsf{K_f}$-`mcs` Σ_0 such that $\Sigma_0 \nvdash_{\mathsf{K_f}} A$. In view of Remark 3.11, we can define a $\mathsf{k_f}$-valuation v_{Σ_0}, and by Lemma 3.13, this v_{Σ_0} is also a $\mathsf{K_f}$-valuation. Since we have $v_{\Sigma_0}(A)\notin\mathcal{T}$, it is also the case that $v_{\Sigma_0}(A)\notin\{\mathbf{T_1}, \mathbf{T_2}, \mathbf{T_3}, \mathbf{T_4}\}$ (since $v_{\Sigma_0}(A)\in\{\mathbf{T_1}, \mathbf{T_2}, \mathbf{T_3}, \mathbf{T_4}\}$ implies that $v_{\Sigma_0}(A)\in\mathcal{T}$). Thus we obtain $\not\models_{\mathsf{K_f}} A$. □

4 On extensions of failed K

We now have a semantics for $\mathsf{K_f}$, and with this semantics at hand, we can turn to discuss an open problem left by Humberstone in [9]. Let us first explain the problem, and then outline our approach to the problem.

In [9, p.401], Humberstone pointed out that he was not successful in finding an argument, possibly a variation of the above counter-model we reviewed in Definition 2.3, establishing the unprovability of $\Diamond\neg\neg p \to \Diamond p$. Hence this problem was left open (see also [9, p.402]).

Note here that we can easily check that the Humberstone's four-valued semantics verifies both $A{\to}\Diamond A$ and $\Diamond A{\to}\neg\Diamond\neg\Diamond A$, namely axioms for S5. This implies that the unprovability of $\Diamond\neg\neg p \to \Diamond p$ is *not* due to the weakness of $\mathsf{K_f}$. In other words, the warning of choice of primitives carries over to extensions of $\mathsf{K_f}$, as well.

Based on this observation, we will mainly focus on extensions of $\mathsf{K_f}$ obtained by adding axioms for S5, and analyse the open problem of Humberstone in some detail. More specifically, we not only establish the unprovability of $\Diamond\neg\neg p \to \Diamond p$, but also offer sound and complete semantics for extensions obtained by adding one or both of $\Diamond A \to \Diamond\neg\neg A$ and $\Diamond\neg\neg A \to \Diamond A$. In order to show how the number of truth values will be reduced, we will also introduce an extension of $\mathsf{K_f}$ obtained by adding an axiom for T.

4.1 From failed K to failed T

First, the extensions of $\mathsf{K_f}$ and $\mathsf{k_f}$ are obtained as follows.

Definition 4.1 The systems $\mathsf{T_f}$ and $\mathsf{t_f}$ are obtained by adding $A \to \Diamond A$ to $\mathsf{K_f}$ and $\mathsf{k_f}$, respectively. The consequence relations are defined as in Definition 2.1.

For the semantics, we introduce the following Nmatrix obtained by eliminating truth-values of the $\mathsf{K_f}$-Nmatrix.

Definition 4.2 A $\mathsf{T_f}$-*Nmatrix* for $\mathcal{L}$ is a tuple $M = \langle \mathcal{V}, \mathcal{T}, \mathcal{O} \rangle$, where:

(a) $\mathcal{V} = \{\mathbf{T_1}, \mathbf{t_1}, \mathbf{F_1}, \mathbf{F_2}, \mathbf{F_3}, \mathbf{F_4}\}$,

(b) $\mathcal{T} = \{\mathbf{T_1}, \mathbf{t_1}\}$,

(c) For every n-ary connective $*$ of $\mathcal{L}$, $\mathcal{O}$ includes a corresponding n-ary function $\tilde{*}$ from $\mathcal{V}^n$ to $2^{\mathcal{V}} \setminus \{\emptyset\}$ as follows (we omit the brackets for sets):

A	$\tilde{\neg} A$	$\Diamond A$
$\mathbf{T_1}$	$\mathbf{F_4}$	$\mathcal{T}$
$\mathbf{t_1}$	$\mathbf{F_1}$	$\mathcal{T}$
$\mathbf{F_1}$	$\mathbf{t_1}$	$\mathcal{T}$
$\mathbf{F_2}$	$\mathbf{T_1}$	$\mathcal{T}$
$\mathbf{F_3}$	$\mathbf{t_1}$	$\mathcal{F}$
$\mathbf{F_4}$	$\mathbf{T_1}$	$\mathcal{F}$

$A \tilde{\to} B$	$\mathbf{T_1}$	$\mathbf{t_1}$	$\mathbf{F_1}$	$\mathbf{F_2}$	$\mathbf{F_3}$	$\mathbf{F_4}$
$\mathbf{T_1}$	$\mathbf{T_1}$	$\mathbf{t_1}$	$\mathcal{F}_{13}$	$\mathcal{F}_{24}$	$\mathcal{F}_{13}$	$\mathcal{F}_{24}$
$\mathbf{t_1}$	$\mathbf{T_1}$	$\mathcal{T}$	$\mathcal{F}_{13}$	$\mathcal{F}_{13}$	$\mathcal{F}_{13}$	$\mathcal{F}_{13}$
$\mathbf{F_1}$	$\mathbf{T_1}$	$\mathcal{T}$	$\mathcal{T}$	$\mathcal{T}$	$\mathcal{T}$	$\mathcal{T}$
$\mathbf{F_2}$	$\mathbf{T_1}$	$\mathcal{T}$	$\mathcal{T}$	$\mathcal{T}$	$\mathcal{T}$	$\mathcal{T}$
$\mathbf{F_3}$	$\mathbf{T_1}$	$\mathcal{T}$	$\mathcal{T}$	$\mathcal{T}$	$\mathcal{T}$	$\mathcal{T}$
$\mathbf{F_4}$	$\mathbf{T_1}$	$\mathcal{T}$	$\mathcal{T}$	$\mathcal{T}$	$\mathcal{T}$	$\mathcal{T}$

where $\mathcal{F}_{13} = \{\mathbf{F_1}, \mathbf{F_3}\}$, $\mathcal{F}_{24} = \{\mathbf{F_2}, \mathbf{F_4}\}$ and $\mathcal{F} = \{\mathbf{F_1}, \mathbf{F_2}, \mathbf{F_3}, \mathbf{F_4}\}$.

Remark 4.3 The definitions of $\mathsf{t_f}$-valuations, mth-level valuations and consequence relations are exactly as in Definitions 2.7, 2.10, 2.12 and 2.13, with the difference that only the value $\mathbf{T_1}$ is preserved in the hierarchy, respectively.

Proposition 4.4 (Soundness) *For all* $\Gamma \cup \{A\} \subseteq \mathsf{Form}$, *(i) if* $\Gamma \vdash_{\mathsf{t_f}} A$ *then* $\Gamma \models_{\mathsf{t_f}} A$, *and (ii) if* $\vdash_{\mathsf{T_f}} A$ *then* $\models_{\mathsf{T_f}} A$.

Proof. The proof is similar to the proof for Proposition 3.3. □

For the completeness, we need the following definition and lemma as before.

Definition 4.5 For any $\Sigma \subseteq \mathsf{Form}$, we define a function $v_\Sigma : \mathsf{Form} \to \mathcal{V}$ as follows.

$$v_\Sigma(B) := \begin{cases} \mathbf{T_1} & \text{if } \Sigma \vdash \neg\Diamond\neg B \text{ and } \Sigma \vdash B \text{ and } \Sigma \vdash \Diamond B \text{ and } \Sigma \vdash \Diamond\neg\neg B \\ \mathbf{t_1} & \text{if } \Sigma \nvdash \neg\Diamond\neg B \text{ and } \Sigma \vdash B \text{ and } \Sigma \vdash \Diamond B \text{ and } \Sigma \vdash \Diamond\neg\neg B \\ \mathbf{F_1} & \text{if } \Sigma \nvdash \neg\Diamond\neg B \text{ and } \Sigma \nvdash B \text{ and } \Sigma \vdash \Diamond B \text{ and } \Sigma \vdash \Diamond\neg\neg B \\ \mathbf{F_2} & \text{if } \Sigma \nvdash \neg\Diamond\neg B \text{ and } \Sigma \nvdash B \text{ and } \Sigma \vdash \Diamond B \text{ and } \Sigma \nvdash \Diamond\neg\neg B \\ \mathbf{F_3} & \text{if } \Sigma \nvdash \neg\Diamond\neg B \text{ and } \Sigma \nvdash B \text{ and } \Sigma \nvdash \Diamond B \text{ and } \Sigma \vdash \Diamond\neg\neg B \\ \mathbf{F_4} & \text{if } \Sigma \nvdash \neg\Diamond\neg B \text{ and } \Sigma \nvdash B \text{ and } \Sigma \nvdash \Diamond B \text{ and } \Sigma \nvdash \Diamond\neg\neg B \end{cases}$$

Lemma 4.6 *If* Σ *is a* $\mathsf{t_f}$-$\mathtt{mcs}$, *then,* v_Σ *is a well-defined* $\mathsf{t_f}$*-valuation.*

Proof. The details of the proof are exactly as in Lemma 3.10, except that we eliminate the values $\mathbf{T_2}$, $\mathbf{T_3}$, $\mathbf{T_4}$, $\mathbf{t_2}$, $\mathbf{t_3}$, $\mathbf{t_4}$, $\mathbf{f_1}$, $\mathbf{f_2}$, $\mathbf{f_3}$ and $\mathbf{f_4}$. □

Now we can prove the completeness.

Theorem 4.7 (Completeness) *For all* $\Gamma \cup \{A\} \subseteq \mathsf{Form}$, *(i) if* $\Gamma \models_{\mathsf{t_f}} A$ *then* $\Gamma \vdash_{\mathsf{t_f}} A$, *and (ii) if* $\models_{\mathsf{T_f}} A$ *then* $\vdash_{\mathsf{T_f}} A$.

Proof. Similar to the proofs of Theorems 3.12 and 3.14, by making use of Lemma 4.6 instead of Lemma 3.10. We leave the details to the reader. □

4.2 From failed T to failed S5

We now turn to the failed S5 which will serve as the basic system in analyzing Humberstone's open problem. For the proof system, we add three more axioms.

Definition 4.8 The systems $\mathsf{S5_f}$ and $\mathsf{s5_f}$ are obtained by adding $\Diamond\Diamond A{\to}\Diamond A$, $\Diamond\neg\neg\Diamond A{\to}\Diamond A$ and $\Diamond A{\to}\neg\Diamond\neg\Diamond A$ to $\mathsf{T_f}$ and $\mathsf{t_f}$, respectively. The consequence relations $\vdash_{\mathsf{S5_f}}$ and $\vdash_{\mathsf{s5_f}}$ are defined as in Definition 2.1.

For the semantics, the number of truth values will remain the same, but we eliminate non-determinacy for the truth-tables of $\Diamond$.

Definition 4.9 An $\mathsf{S5_f}$-*Nmatrix* for $\mathcal{L}$ is a tuple $M = \langle \mathcal{V}, \mathcal{T}, \mathcal{O} \rangle$, where:

(a) $\mathcal{V} = \{\mathbf{T}_1, \mathbf{t}_1, \mathbf{F}_1, \mathbf{F}_2, \mathbf{F}_3, \mathbf{F}_4\}$,

(b) $\mathcal{T} = \{\mathbf{T}_1, \mathbf{t}_1\}$,

(c) For every n-ary connective $*$ of $\mathcal{L}$, $\mathcal{O}$ includes a corresponding n-ary function $\tilde{*}$ from $\mathcal{V}^n$ to $2^{\mathcal{V}} \setminus \{\emptyset\}$ as follows (we omit the brackets for sets):

A	$\tilde{\neg}A$	$\Diamond A$	$A\tilde{\to}B$	$\mathbf{T}_1$	$\mathbf{t}_1$	$\mathbf{F}_1$	$\mathbf{F}_2$	$\mathbf{F}_3$	$\mathbf{F}_4$
$\mathbf{T}_1$	$\mathbf{F}_4$	$\mathbf{T}_1$	$\mathbf{T}_1$	$\mathbf{T}_1$	$\mathbf{t}_1$	$\mathcal{F}_{13}$	$\mathcal{F}_{24}$	$\mathcal{F}_{13}$	$\mathcal{F}_{24}$
$\mathbf{t}_1$	$\mathbf{F}_1$	$\mathbf{T}_1$	$\mathbf{t}_1$	$\mathbf{T}_1$	$\mathcal{T}$	$\mathcal{F}_{13}$	$\mathcal{F}_{13}$	$\mathcal{F}_{13}$	$\mathcal{F}_{13}$
$\mathbf{F}_1$	$\mathbf{t}_1$	$\mathbf{T}_1$	$\mathbf{F}_1$	$\mathbf{T}_1$	$\mathcal{T}$	$\mathcal{T}$	$\mathcal{T}$	$\mathcal{T}$	$\mathcal{T}$
$\mathbf{F}_2$	$\mathbf{T}_1$	$\mathbf{T}_1$	$\mathbf{F}_2$	$\mathbf{T}_1$	$\mathcal{T}$	$\mathcal{T}$	$\mathcal{T}$	$\mathcal{T}$	$\mathcal{T}$
$\mathbf{F}_3$	$\mathbf{t}_1$	$\mathbf{F}_4$	$\mathbf{F}_3$	$\mathbf{T}_1$	$\mathcal{T}$	$\mathcal{T}$	$\mathcal{T}$	$\mathcal{T}$	$\mathcal{T}$
$\mathbf{F}_4$	$\mathbf{T}_1$	$\mathbf{F}_4$	$\mathbf{F}_4$	$\mathbf{T}_1$	$\mathcal{T}$	$\mathcal{T}$	$\mathcal{T}$	$\mathcal{T}$	$\mathcal{T}$

where $\mathcal{F}_{13} = \{\mathbf{F}_1, \mathbf{F}_3\}$ and $\mathcal{F}_{24} = \{\mathbf{F}_2, \mathbf{F}_4\}$.

Remark 4.10 The definitions of $\mathsf{s5_f}$ valuations, mth-level valuations and consequence relations are exactly as in the Definitions 2.7, 2.10, 2.12 and 2.13.

There is, however, a significant property of $\mathsf{S5_f}$ which distinguishes it from the other systems introduced in this article so far. Indeed, we do not need a whole hierarchy of mth-level valuations, but only two levels are sufficient. We leave the details to the interested reader and refer to [16, §4.4] in which this was observed for S4 and S5.

Proposition 4.11 (Soundness) *For all* $\Gamma \cup \{A\} \subseteq \mathsf{Form}$, *(i) if* $\Gamma \vdash_{\mathsf{s5_f}} A$ *then* $\Gamma \models_{\mathsf{s5_f}} A$, *and (ii) if* $\vdash_{\mathsf{S5_f}} A$ *then* $\models_{\mathsf{S5_f}} A$.

Proof. We only note that the additional axioms are valid in the $\mathsf{S5_f}$-Nmatrix in which all non-determinacies for $\Diamond$ are eliminated. The rest is exactly as in Proposition 4.4. □

Theorem 4.12 (Completeness) *For all* $\Gamma \cup \{A\} \subseteq \mathsf{Form}$, *(i) if* $\Gamma \models_{\mathsf{s5_f}} A$ *then* $\Gamma \vdash_{\mathsf{s5_f}} A$, *and (ii) if* $\models_{\mathsf{S5_f}} A$ *then* $\vdash_{\mathsf{S5_f}} A$.

Proof. We only note that the additional axioms allow us to conclude that if Σ is a $\mathsf{s5_f}$-`mcs`, then v_Σ is a well-defined $\mathsf{s5_f}$-valuation. The proof is by induction, and we only check the case when B is of the form $\Diamond C$.

- If $v_\Sigma(C) = \mathbf{T_1}$, then by IH, we obtain that $\Sigma \vdash \neg\Diamond\neg C$ and $\Sigma \vdash C$ and $\Sigma \vdash \Diamond C$ and $\Sigma \vdash \Diamond\neg\neg C$. Then, by the third conjunct and $\Diamond A \to \neg\Diamond\neg\Diamond A$, we have $\Sigma \vdash \neg\Diamond\neg\Diamond C$, and this also implies $\Sigma \vdash \Diamond C$, $\Sigma \vdash \Diamond\Diamond C$ and $\Sigma \vdash \Diamond\neg\neg\Diamond C$. Then, by the definition of v_Σ, this means $v_\Sigma(\Diamond C) = \mathbf{T_1}$, as desired. Other cases with $v_\Sigma(C)=\mathbf{t_1}$, $v_\Sigma(C)=\mathbf{F_1}$ and $v_\Sigma(C)=\mathbf{F_2}$ are the same.
- If $v_\Sigma(C) = \mathbf{F_3}$, then by IH, we obtain that $\Sigma \nvdash \neg\Diamond\neg C$ and $\Sigma \nvdash C$ and $\Sigma \nvdash \Diamond C$ and $\Sigma \vdash \Diamond\neg\neg C$. Then, by the third conjunct together with $\Diamond\Diamond A \to \Diamond A$ and $\Diamond\neg\neg\Diamond A \to \Diamond A$, we have $\Sigma \nvdash \Diamond\Diamond C$ and $\Sigma \vdash \Diamond\neg\neg\Diamond C$, and we also have $\Sigma \nvdash \neg\Diamond\neg\Diamond C$ and $\Sigma \nvdash \Diamond C$. Then, by the definition of v_Σ, this means $v_\Sigma(\Diamond C) = \mathbf{F_4}$, as desired. The other case with $v_\Sigma(C)=\mathbf{F_4}$ is the same.

The rest of the proof is exactly as in Theorem 4.7. □

4.3 On the open problem of Humberstone

We are now in the position to shift our interest to the problem left open by Humberstone in [9]. The counter-model given by Humberstone, and described in §2.2, invalidates one direction of the equivalence $\Diamond A \leftrightarrow \Diamond\neg\neg A$, namely $\Diamond A \to \Diamond\neg\neg A$, while validating the other. We will now show that the above semantics for $\mathsf{S5_f}$ in the style of Kearns, with one more adjustment, helps us to establish the unprovability of both $\Diamond p \to \Diamond\neg\neg p$ and $\Diamond\neg\neg p \to \Diamond p$.

The adjustment we need to make is, to close the non-determinacies to obtain a six-valued (deterministic) matrix that will do the job for our present purposes. One may have expected that the above semantics will directly give us the counter-model. Unfortunately, this is not the case, at least at the moment, due to the problem of analyticity in Kearns' semantics (cf. [16, Remark 42]).[13] Still, the definition of the canonical valuation strongly suggests that we should be able to give a counter-model, and by following Schiller Joe Scroggs who explored the many-valued extensions of S5 in [17], we may aim at a deterministic extension of our semantics for $\mathsf{S5_f}$.

Definition 4.13 A $\mathsf{dS5_f}$-*matrix* for $\mathcal{L}$ is a tuple $M = \langle \mathcal{V}, \mathcal{T}, \mathcal{O} \rangle$, where:

(a) $\mathcal{V} = \{\mathbf{T_1}, \mathbf{t_1}, \mathbf{F_1}, \mathbf{F_2}, \mathbf{F_3}, \mathbf{F_4}\}$,

(b) $\mathcal{T} = \{\mathbf{T_1}\}$,

(c) For every n-ary connective $*$ of $\mathcal{L}$, $\mathcal{O}$ includes a corresponding n-ary function $\tilde{*}$ from $\mathcal{V}^n$ to $\mathcal{V}$ as follows:

A	$\tilde{\neg}A$	$\Diamond A$	$A \tilde{\to} B$	$\mathbf{T_1}$	$\mathbf{t_1}$	$\mathbf{F_1}$	$\mathbf{F_2}$	$\mathbf{F_3}$	$\mathbf{F_4}$
$\mathbf{T_1}$	$\mathbf{F_4}$	$\mathbf{T_1}$	$\mathbf{T_1}$	$\mathbf{T_1}$	$\mathbf{t_1}$	$\mathbf{F_1}$	$\mathbf{F_4}$	$\mathbf{F_1}$	$\mathbf{F_4}$
$\mathbf{t_1}$	$\mathbf{F_1}$	$\mathbf{T_1}$	$\mathbf{t_1}$	$\mathbf{T_1}$	$\mathbf{T_1}$	$\mathbf{F_1}$	$\mathbf{F_1}$	$\mathbf{F_1}$	$\mathbf{F_1}$
$\mathbf{F_1}$	$\mathbf{t_1}$	$\mathbf{T_1}$	$\mathbf{F_1}$	$\mathbf{T_1}$	$\mathbf{t_1}$	$\mathbf{T_1}$	$\mathbf{t_1}$	$\mathbf{T_1}$	$\mathbf{t_1}$
$\mathbf{F_2}$	$\mathbf{T_1}$	$\mathbf{T_1}$	$\mathbf{F_2}$	$\mathbf{T_1}$	$\mathbf{T_1}$	$\mathbf{T_1}$	$\mathbf{T_1}$	$\mathbf{T_1}$	$\mathbf{T_1}$
$\mathbf{F_3}$	$\mathbf{t_1}$	$\mathbf{F_4}$	$\mathbf{F_3}$	$\mathbf{T_1}$	$\mathbf{t_1}$	$\mathbf{T_1}$	$\mathbf{t_1}$	$\mathbf{T_1}$	$\mathbf{t_1}$
$\mathbf{F_4}$	$\mathbf{T_1}$	$\mathbf{F_4}$	$\mathbf{F_4}$	$\mathbf{T_1}$	$\mathbf{T_1}$	$\mathbf{T_1}$	$\mathbf{T_1}$	$\mathbf{T_1}$	$\mathbf{T_1}$

[13] It is not yet proven that in Kearns' semantics a partial valuation that falsifies a formula can be extended to a full valuation that necessarily falsifies the formula, as well. See also [1] for a discussion on analyticity related to non-deterministic semantics in general.

We refer to the semantic consequence relation defined in terms of preservation of the designated value with the above matrix as $\models_{\mathsf{dS5}}$.

Then, it is tedious but routine to check the following.

Lemma 4.14 *For all* $A \in \mathsf{Form}$, *if* $\vdash_{\mathsf{S5_f}} A$ *then* $\models_{\mathsf{dS5}} A$.

We are now ready to answer Humberstone's open problem.

Theorem 4.15 $\nvdash_{\mathsf{S5_f}} \Diamond p \to \Diamond \neg\neg p$ *and* $\nvdash_{\mathsf{S5_f}} \Diamond \neg\neg p \to \Diamond p$.

Proof. In view of the above lemma, it suffices to check that $\not\models_{\mathsf{dS5}} \Diamond p \to \Diamond \neg\neg p$ and $\not\models_{\mathsf{dS5}} \Diamond \neg\neg p \to \Diamond p$. For the first item, assign $\mathbf{F_2}$ to A of the $\mathsf{dS5_f}$-matrix. Then, $\Diamond p$ receives the value $\mathbf{T_1}$ and $\Diamond \neg\neg p$ receives the value $\mathbf{F_4}$. Therefore, $\Diamond p \to \Diamond \neg\neg p$ receives the non-designated value $\mathbf{F_4}$, as desired. For the second item, assign $\mathbf{F_3}$ to p of the $\mathsf{dS5_f}$-matrix. Then, $\Diamond \neg\neg p$ receives the value $\mathbf{T_1}$ and $\Diamond p$ receives the value $\mathbf{F_4}$. Therefore, $\Diamond \neg\neg p \to \Diamond p$ receives the non-designated value $\mathbf{F_4}$, as desired. □

Remark 4.16 In view of the definition of the canonical valuation, this result is of course something expected. The emphasis here should be that some technical devices are available to make the canonical valuation work as designed, thanks to the semantic framework introduced by Kearns. This is, in turn, giving us some new insight on the problem left open by Humberstone.

Let us now continue by introducing further extensions of $\mathsf{S5_f}$ and $\mathsf{s5_f}$ obtained by adding one of the two formulas $\Diamond A \to \Diamond \neg\neg A$ and $\Diamond \neg\neg A \to \Diamond A$.

Definition 4.17 Let $\mathsf{S5_{fc}}$ and $\mathsf{s5_{fc}}$ be the systems obtained by adding $\Diamond A \to \Diamond \neg\neg A$ to $\mathsf{S5_f}$ and $\mathsf{s5_f}$, respectively. Moreover, let $\mathsf{S5_{fa}}$ and $\mathsf{s5_{fa}}$ be the systems obtained by adding $\Diamond \neg\neg A \to \Diamond A$ to $\mathsf{S5_f}$ and $\mathsf{s5_f}$, respectively.[14]

For the semantics, we need to eliminate one value each that was used to invalidate the additional axiom.

Definition 4.18 The Nmatrices for $\mathsf{S5_{fc}}$ and $\mathsf{S5_{fa}}$ are obtained from the Nmatrix for $\mathsf{S5_f}$, by eliminating the values $\mathbf{F_2}$ and $\mathbf{F_3}$, respectively. More specifically, an $\mathsf{S5_{fc}}$-*Nmatrix* for $\mathcal{L}$ is a tuple $M = \langle \mathcal{V}, \mathcal{T}, \mathcal{O} \rangle$, where:

(a) $\mathcal{V} = \{\mathbf{T_1}, \mathbf{t_1}, \mathbf{F_1}, \mathbf{F_3}, \mathbf{F_4}\}$,

(b) $\mathcal{T} = \{\mathbf{T_1}, \mathbf{t_1}\}$,

(c) For every n-ary connective $*$ of $\mathcal{L}$, $\mathcal{O}$ includes a corresponding n-ary function $\tilde{*}$ from $\mathcal{V}^n$ to $2^{\mathcal{V}} \setminus \{\emptyset\}$ as follows (we omit the brackets for sets):

A	$\tilde{\neg} A$	$\Diamond A$
$\mathbf{T_1}$	$\mathbf{F_4}$	$\mathbf{T_1}$
$\mathbf{t_1}$	$\mathbf{F_1}$	$\mathbf{T_1}$
$\mathbf{F_1}$	$\mathbf{t_1}$	$\mathbf{T_1}$
$\mathbf{F_3}$	$\mathbf{t_1}$	$\mathbf{F_4}$
$\mathbf{F_4}$	$\mathbf{T_1}$	$\mathbf{F_4}$

$A \tilde{\to} B$	$\mathbf{T_1}$	$\mathbf{t_1}$	$\mathbf{F_1}$	$\mathbf{F_3}$	$\mathbf{F_4}$
$\mathbf{T_1}$	$\mathbf{T_1}$	$\mathbf{t_1}$	$\mathcal{F}_{13}$	$\mathcal{F}_{13}$	$\mathbf{F_4}$
$\mathbf{t_1}$	$\mathbf{T_1}$	$\mathcal{T}$	$\mathcal{F}_{13}$	$\mathcal{F}_{13}$	$\mathcal{F}_{13}$
$\mathbf{F_1}$	$\mathbf{T_1}$	$\mathcal{T}$	$\mathcal{T}$	$\mathcal{T}$	$\mathcal{T}$
$\mathbf{F_3}$	$\mathbf{T_1}$	$\mathcal{T}$	$\mathcal{T}$	$\mathcal{T}$	$\mathcal{T}$
$\mathbf{F_4}$	$\mathbf{T_1}$	$\mathcal{T}$	$\mathcal{T}$	$\mathcal{T}$	$\mathcal{T}$

[14] Note that additional subscripts c and a are for consequent and antecedent respectively, indicating the position of the double negation in the additional axiom.

where $\mathcal{F}_{13} = \{\mathbf{F_1}, \mathbf{F_3}\}$.

Moreover, an $\mathsf{S5_{fa}}$-*Nmatrix* for $\mathcal{L}$ is a tuple $M = \langle \mathcal{V}, \mathcal{T}, \mathcal{O}\rangle$, where:

(a) $\mathcal{V} = \{\mathbf{T_1}, \mathbf{t_1}, \mathbf{F_1}, \mathbf{F_2}, \mathbf{F_4}\}$,

(b) $\mathcal{T} = \{\mathbf{T_1}, \mathbf{t_1}\}$,

(c) For every n-ary connective $*$ of $\mathcal{L}$, $\mathcal{O}$ includes a corresponding n-ary function $\tilde{*}$ from $\mathcal{V}^n$ to $2^{\mathcal{V}} \setminus \{\emptyset\}$ as follows (we omit the brackets for sets):

A	$\tilde{\neg} A$	$\tilde{\Diamond} A$	$A \tilde{\to} B$	$\mathbf{T_1}$	$\mathbf{t_1}$	$\mathbf{F_1}$	$\mathbf{F_2}$	$\mathbf{F_4}$
$\mathbf{T_1}$	$\mathbf{F_4}$	$\mathbf{T_1}$	$\mathbf{T_1}$	$\mathbf{T_1}$	$\mathbf{t_1}$	$\mathbf{F_1}$	$\mathcal{F}_{24}$	$\mathcal{F}_{24}$
$\mathbf{t_1}$	$\mathbf{F_1}$	$\mathbf{T_1}$	$\mathbf{t_1}$	$\mathbf{T_1}$	$\mathcal{T}$	$\mathbf{F_1}$	$\mathbf{F_1}$	$\mathbf{F_1}$
$\mathbf{F_1}$	$\mathbf{t_1}$	$\mathbf{T_1}$	$\mathbf{F_1}$	$\mathbf{T_1}$	$\mathcal{T}$	$\mathcal{T}$	$\mathcal{T}$	$\mathcal{T}$
$\mathbf{F_2}$	$\mathbf{T_1}$	$\mathbf{T_1}$	$\mathbf{F_2}$	$\mathbf{T_1}$	$\mathcal{T}$	$\mathcal{T}$	$\mathcal{T}$	$\mathcal{T}$
$\mathbf{F_4}$	$\mathbf{T_1}$	$\mathbf{F_4}$	$\mathbf{F_4}$	$\mathbf{T_1}$	$\mathcal{T}$	$\mathcal{T}$	$\mathcal{T}$	$\mathcal{T}$

where $\mathcal{F}_{24} = \{\mathbf{F_2}, \mathbf{F_4}\}$.

We can then establish soundness and completeness for the four new systems introduced in this subsection. Indeed, all definitions, propositions and theorems are exactly as in §4.1 and §4.2. More specifically, all proofs can be obtained by slightly modifying the proofs of the mentioned subsections, by removing the values $\mathbf{F_2}$ and $\mathbf{F_3}$, respectively. We safely leave the details for the interested reader.

Remark 4.19 Note that we can strengthen Theorem 4.15 as follows: $\not\vdash_{\mathsf{S5_{fa}}} \Diamond p \to \Diamond\neg\neg p$ and $\not\vdash_{\mathsf{S5_{fc}}} \Diamond\neg\neg p \to \Diamond p$. This is precisely because we can make use of the submatrices of the six-valued $\mathsf{dS5_f}$-matrix introduced in Definition 4.13.

4.4 From failed S5 to full S5

As noted by Humberstone in [9, p.401], we obtain the standard normal modal logic K if we extend $\mathsf{K_f}$ by adding $\Diamond A \leftrightarrow \Diamond\neg\neg A$ since this gives us the equivalence $\neg\neg\Diamond\neg\neg A \leftrightarrow \Diamond A$ which corresponds to the $\neg\Box\neg A \leftrightarrow \Diamond A$ used in [2, p.34]. Since this observation also carries over to extensions of K, it is rather natural to introduce the common extension of $\mathsf{S5_{fa}}$ and $\mathsf{S5_{fc}}$ obtained by adding the missing direction.

Definition 4.20 The systems $\mathsf{S5}\Diamond$ and $\mathsf{s5}\Diamond$ are obtained by adding the axiom scheme $\Diamond A \leftrightarrow \Diamond\neg\neg A$ to $\mathsf{S5_f}$ and $\mathsf{s5_f}$, respectively.

For the semantics, seen from the $\mathsf{S5_f}$-Nmatrix, we need to eliminate both values that were used to invalidate the additional axioms. Equivalently, we obtain the same Nmatrix from the $\mathsf{S5_{fa}}$-Nmatrix and the $\mathsf{S5_{fc}}$-Nmatrix by eliminating the values that we used to invalidate the missing direction.

Definition 4.21 An $\mathsf{S5}\Diamond$-*Nmatrix* for $\mathcal{L}$ is a tuple $M = \langle \mathcal{V}, \mathcal{T}, \mathcal{O}\rangle$, where:

(a) $\mathcal{V} = \{\mathbf{T_1}, \mathbf{t_1}, \mathbf{F_1}, \mathbf{F_4}\}$,

(b) $\mathcal{T} = \{\mathbf{T_1}, \mathbf{t_1}\}$,

(c) For every n-ary connective $*$ of $\mathcal{L}$, $\mathcal{O}$ includes a corresponding n-ary function $\tilde{*}$ from $\mathcal{V}^n$ to $2^{\mathcal{V}} \setminus \{\emptyset\}$ as follows (we omit the brackets for sets):

A	$\tilde{\neg}A$	$\Diamond A$
$\mathbf{T}_1$	$\mathbf{F}_4$	$\mathbf{T}_1$
$\mathbf{t}_1$	$\mathbf{F}_1$	$\mathbf{T}_1$
$\mathbf{F}_1$	$\mathbf{t}_1$	$\mathbf{T}_1$
$\mathbf{F}_4$	$\mathbf{T}_1$	$\mathbf{F}_4$

$A\tilde{\to}B$	$\mathbf{T}_1$	$\mathbf{t}_1$	$\mathbf{F}_1$	$\mathbf{F}_4$
$\mathbf{T}_1$	$\mathbf{T}_1$	$\mathbf{t}_1$	$\mathbf{F}_1$	$\mathbf{F}_4$
$\mathbf{t}_1$	$\mathbf{T}_1$	$\mathbf{T}_1, \mathbf{t}_1$	$\mathbf{F}_1$	$\mathbf{F}_1$
$\mathbf{F}_1$	$\mathbf{T}_1$	$\mathbf{T}_1, \mathbf{t}_1$	$\mathbf{T}_1, \mathbf{t}_1$	$\mathbf{T}_1, \mathbf{t}_1$
$\mathbf{F}_4$	$\mathbf{T}_1$	$\mathbf{T}_1, \mathbf{t}_1$	$\mathbf{T}_1, \mathbf{t}_1$	$\mathbf{T}_1, \mathbf{t}_1$

The rest of the details towards soundness and completeness results will be as before. Indeed, all definitions, propositions and theorems are exactly as in §4.1 and §4.2, and all proofs can be obtained by slightly modifying the proofs of the mentioned subsections, by removing both of the values $\mathbf{F}_2$ and $\mathbf{F}_3$.

Remark 4.22 A closer look at the definitions reveals that the Nmatrix of S5◇ and its definition of the canonical valuation are very similar to the ones for S5 given in [16]. In fact, the definitions of the canonical valuations are equivalent since in view of the additional axiom, the distinction between $\Diamond B$ and $\Diamond\neg\neg B$ is redundant. For the Nmatrix, however, this is slightly different since the one in [16], taken from [12], has the following truth-table for $\tilde{\to}$.

$A\tilde{\to}B$	$\mathbf{T}_1$	$\mathbf{t}_1$	$\mathbf{F}_1$	$\mathbf{F}_4$
$\mathbf{T}_1$	$\mathbf{T}_1$	$\mathbf{t}_1$	$\mathbf{F}_1$	$\mathbf{F}_4$
$\mathbf{t}_1$	$\mathbf{T}_1$	$\mathbf{T}_1, \mathbf{t}_1$	$\mathbf{F}_1$	$\mathbf{F}_1$
$\mathbf{F}_1$	$\mathbf{T}_1$	$\mathbf{T}_1, \mathbf{t}_1$	$\mathbf{T}_1, \mathbf{t}_1$	$\mathbf{t}_1$
$\mathbf{F}_4$	$\mathbf{T}_1$	$\mathbf{T}_1$	$\mathbf{T}_1$	$\mathbf{T}_1$

Indeed, there are some additional non-determiniacies in our S5◇-Nmatrix:

- $\mathbf{F}_1\tilde{\to}\mathbf{F}_4$ will be $\mathbf{t}_1$ by having $\neg\Diamond\neg(A \to B) \to (\Diamond A \to \Diamond B)$ as derivable;
- $\mathbf{F}_4\tilde{\to}x$ for all $x\in\mathcal{V}$ will be always $\mathbf{T}_1$ by having $\Diamond\neg(A\to B)\to\Diamond A$ as derivable.

However, at the level of S5◇-valuations, where some valuations will be ruled out, the formulas above will be validated, and thus the two semantics, the one for S5◇ given above, and the one for S5, introduced in [12] are indeed equivalent.

5 Concluding remarks

Let us now briefly summarize our main results of this paper, and then discuss a few items for future directions.

Main results Building on the observation of Humberstone about the choice of primitives in [9], we aimed at offering a sound and complete semantics for the failed axiomatization of the modal logic K_f, a variant of the modal logic K with the possibility operator as the only primitive modal operator, but without $\neg\neg\Diamond\neg\neg A\leftrightarrow\Diamond A$, the key axiom to obtain an axiomatization based on ◇. The resulting semantics is based on a sixteen-valued non-deterministic semantics which can be seen as a variant of the semantics devised by Kearns in [12].

We also discussed an open problem left by Humberstone in [9], showing the independence of not only $\Diamond A \to \Diamond\neg\neg A$ but also $\Diamond\neg\neg A \to \Diamond A$ from $\mathsf{S5}_\mathsf{f}$ (failed axiomatization of S5), therefore also from K_f. To this end, we devised a semantics for $\mathsf{S5}_\mathsf{f}$ based on a six-valued non-deterministic semantics, and used a deterministic extension to establish the unprovability of both $\Diamond p \to \Diamond\neg\neg p$

and $\Diamond\neg\neg p \rightarrow \Diamond p$ in one single matrix.

The extensions of $\mathsf{K_f}$ we discussed in this article can be ordered, from left to right by deductive strength, in the following way:

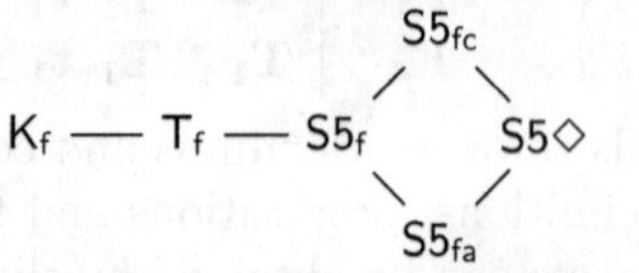

Note that Humberstone also specifically asked for a counter-model for $\Diamond\neg\neg A \rightarrow \Diamond A$ that is closer to his own counter-model we revisited in Definition 2.3. There is a question of how to precisify the notion of closeness to Humberstone's counter-model, but there is one possible answer due to Xuefeng Wen presented in [23, p.71], independently of Humberstone's question.[15] More specifically, Wen's model modifies the standard model $\mathcal{M} = \langle W, R, V\rangle$ for the modal logic K by changing the usual truth condition for $\Diamond$ as follows.

$$\mathcal{M}, x \Vdash \Diamond A \text{ iff } A{=}\neg B \text{ and for some } y \in W, xRy \text{ and } \mathcal{M}, y \nVdash B.$$

Then, we may easily observe that $\Diamond\neg\neg p \rightarrow \Diamond p$ fails in a model with $W = \{w\}$, $R = \{(w, w)\}$ and $V(p) = \{w\}$. Note, however, that $\Diamond A \rightarrow \Diamond\neg\neg A$ is valid in the model suggested by Wen. We therefore leave the task to explore the exact relations between Humberstone's counter-model, Wen's counter-model and our model, for interested readers.

Future directions (I): a systematic study on extensions of failed K Since our motivation came from Humberstone's interesting observations reported in [9], we only focused on extensions of $\mathsf{K_f}$ that were crucial and helpful for our observations. However, this does not exclude a more systematic study of extensions of $\mathsf{K_f}$. We will here offer a sketch of some of the facts that seem to suggest that the landscape of the extensions of $\mathsf{K_f}$ may look quite different from the extensions of K.

First, it is well known that in considering extensions of K, there are two equivalent formulations for many cases. For example, additions of $\Box A \rightarrow A$ and $A \rightarrow \Diamond A$ will both give rise to the modal logic T. This will no longer be the case for extensions of $\mathsf{K_f}$. Indeed, as we observed in §4.1, the extension of $\mathsf{K_f}$ by $A \rightarrow \Diamond A$ was characterized by a semantics obtained by eliminating ten values from the semantics for $\mathsf{K_f}$. However, if we extend $\mathsf{K_f}$ by $\neg\Diamond\neg A \rightarrow A$, then we can only eliminate eight values.

Something similar happens to D-like systems when we add $\neg\Diamond\neg A \rightarrow \Diamond A$ and $\neg\Diamond\neg A \rightarrow \Diamond\neg\neg A$ to $\mathsf{K_f}$. More specifically, the former requires elimination of six values whereas the latter only requires to eliminate four values. Moreover, there will be a deviation from the usual picture in the sense that $\mathsf{K_f}$ plus $\neg\Diamond\neg A \rightarrow A$, a T-like system, and $\mathsf{K_f}$ plus $\neg\Diamond\neg A \rightarrow \Diamond A$, D-like system, are incomparable. Indeed, we may establish that $\neg\Diamond\neg A \rightarrow \Diamond A$ is unprovable in the first system and $\neg\Diamond\neg A \rightarrow A$ is unprovable in the second system in a similar

[15] Our sincere thanks go to one of the anonymous referees for pointing this out.

manner as we did in Theorem 4.15. We can then again order the extensions of $\mathsf{K_f}$, up to $\mathsf{T_f}$, from left to right by deductive strength, in the following way:

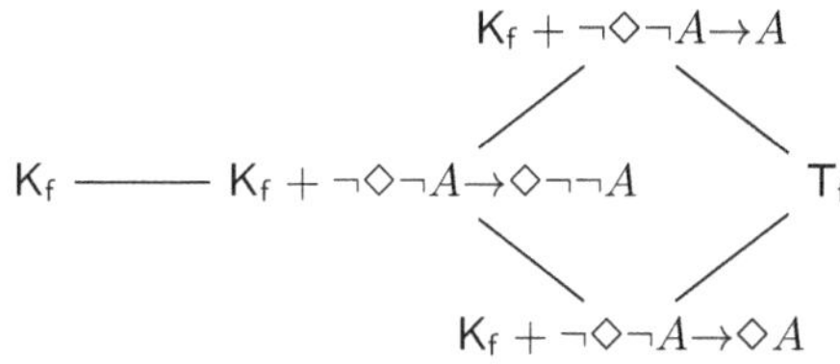

Therefore, it can be safely said, the class of extensions for $\mathsf{K_f}$ looks quite different than usually for normal modal logics.

Future directions (II): Failed axiomatizations of tense logics One of the *examples from life* for failed axiomatizations, as Humberstone put it in [9], is that of tense logics. In particular, he discussed the system $\mathsf{K_t}$, given by S. K. Thomason in [22], introduced as a bimodal logic with two primitive possibility operators. It was shown by Humberstone that the axiomatization for $\mathsf{K_t}$ fails to be complete with respect to Kripke semantics, by a similar argument building on a variation of his own counter-model.

In light of the results of this article, we believe that it is possible to construct sound and complete Kearns' semantics for $\mathsf{K_t}$ or even its extensions. We have not spelled out the details, but probably an Nmatrix with 128 truth-values suffices to prove the desired result for $\mathsf{K_t}$, and for certain extensions of $\mathsf{K_t}$, the number of truth-values should be reduced significantly. The key idea for constructing such semantics, lies in the canonical model construction, where we would need to add conditions for the interaction of the two modal operators.

Future directions (III): Some critical reflections on Kearns' semantics Finally, we left out an important question raised by readers of earlier versions of this article, the question whether this semantics is of any philosophical value. At the moment we are far away from claiming any philosophical significance , unless we follow Kearns' discussion (cf. [13]). In the end, we fully agree with B. J. Copeland in [6, p. 400], when he writes:

> "Philosophically significant semantical arguments can be yielded only by philosophically significant semantics, not by merely formal model theory."

Thus, it remains as a (huge) challenge to explore if we can turn Kearns' semantics into a philosophically significant semantics.

Appendix

In this appendix, we spell out the details left open in the text and give the full truth-table of the $\mathsf{K_f}$ conditional from Definition 2.7.

Details of Proposition 3.1 We will only prove the case for (LK), since the proofs for the other modal axioms as well as the classical axioms and (MP) are similar. Now, suppose that for a $\mathsf{k_f}$-valuation $v(\neg\Diamond\neg(A\to B)\to(\neg\Diamond\neg A\to\neg\Diamond\neg B)) \notin \mathcal{T}$. Then, this implies (1) $v(\neg\Diamond\neg(A \to B)) \in \mathcal{T}$, (2) $v(\neg\Diamond\neg A) \in \mathcal{T}$ and (3) $v(\neg\Diamond\neg B) \notin \mathcal{T}$. Now, we can see that three conditions imply the following, respectively:

(1) then $v(\Diamond\neg(A \to B)) \notin \mathcal{T}$
then $v(\neg(A \to B)) \in \{\mathbf{T_3}, \mathbf{t_3}, \mathbf{T_4}, \mathbf{t_4}, \mathbf{F_3}, \mathbf{f_3}, \mathbf{F_4}, \mathbf{f_4}\}$
then $v(A \to B) \in \{\mathbf{T_1}, \mathbf{T_2}, \mathbf{T_3}, \mathbf{T_4}, \mathbf{f_1}, \mathbf{f_2}, \mathbf{f_3}, \mathbf{f_4}\}$

(2) then $v(\Diamond\neg A) \notin \mathcal{T}$
then $v(\neg A) \in \{\mathbf{T_3}, \mathbf{t_3}, \mathbf{T_4}, \mathbf{t_4}, \mathbf{F_3}, \mathbf{f_3}, \mathbf{F_4}, \mathbf{f_4}\}$
then $v(A) \in \{\mathbf{T_1}, \mathbf{T_2}, \mathbf{T_3}, \mathbf{T_4}, \mathbf{f_1}, \mathbf{f_2}, \mathbf{f_3}, \mathbf{f_4}\}$

(3) then $v(\Diamond\neg B) \in \mathcal{T}$
then $v(\neg B) \in \{\mathbf{T_1}, \mathbf{t_1}, \mathbf{T_2}, \mathbf{t_2}, \mathbf{F_1}, \mathbf{f_1}, \mathbf{F_2}, \mathbf{f_2}\}$
then $v(B) \in \{\mathbf{t_1}, \mathbf{t_2}, \mathbf{t_3}, \mathbf{t_4}, \mathbf{F_1}, \mathbf{F_2}, \mathbf{F_3}, \mathbf{F_4}\}$

By looking at the truth tables, this is not possible.

Proof of Lemma 3.10 Note first that v_Σ is well-defined. The desired result can be proved by induction on the number n of connectives.
(Base): For atomic formulas, this is obvious in view of the definition of v_Σ.
(Induction step): We split the cases based on the connectives.
Case 1. If $B = \neg C$, then we have sixteen cases of which we will prove four.

- If $v_\Sigma(C) \in \{\mathbf{T_2}, \mathbf{T_4}\}$, then by IH, we obtain that $\Sigma \vdash \neg\Diamond\neg C$ and $\Sigma \vdash C$ and $\Sigma \nvdash \Diamond\neg\neg C$. From this, we immediately get $\Sigma \vdash \neg\Diamond\neg\neg C$ and $\Sigma \nvdash \neg C$ and $\Sigma \nvdash \Diamond\neg C$ and $\Sigma \nvdash \Diamond\neg\neg\neg C$. Then, by the definition of v_Σ, this means $v_\Sigma(\neg C) = \mathbf{f_4}$, as desired.
- If $v_\Sigma(C) \in \{\mathbf{f_1}, \mathbf{f_3}\}$, then by IH, we obtain that $\Sigma \vdash \neg\Diamond\neg C$ and $\Sigma \nvdash C$ and $\Sigma \vdash \Diamond\neg\neg C$. From this, we immediately get $\Sigma \nvdash \neg\Diamond\neg\neg C$ and $\Sigma \vdash \neg C$ and $\Sigma \nvdash \Diamond\neg C$ and $\Sigma \nvdash \Diamond\neg\neg\neg C$. Then, by the definition of v_Σ, this means $v_\Sigma(\neg C) = \mathbf{t_4}$, as desired.

The other cases are similar and left to the reader.

Case 2. If $B = \Diamond C$, then we can deal with sixteen cases by splitting into the following two cases.

- If $v_\Sigma(C) \in \{\mathbf{T_i}, \mathbf{t_i}, \mathbf{f_i}, \mathbf{F_i}\}$ with $i \in \{1, 2\}$, then by IH, we obtain that $\Sigma \vdash \Diamond C$. By the definition of v_Σ, we obtain $v_\Sigma(\Diamond C) \in \mathcal{T}$, as desired.
- If $v_\Sigma(C) \in \{\mathbf{T_i}, \mathbf{t_i}, \mathbf{f_i}, \mathbf{F_i}\}$ with $i \in \{3, 4\}$, then by IH, we obtain that $\Sigma \nvdash \Diamond C$. By the definition of v_Σ, we obtain $v_\Sigma(\Diamond C) \notin \mathcal{T}$, i.e. $v_\Sigma(\Diamond C) \in \mathcal{F}$, as desired.

Case 3. If $B = C \to D$, then we have 256 different cases, which can be reduced to eighteen of which will prove four.

- If $v_\Sigma(D) \in \{\mathbf{T_1}, \mathbf{T_3}\}$, then by IH, we obtain that $\Sigma \vdash \neg\Diamond\neg D$ and $\Sigma \vdash D$ and $\Sigma \vdash \Diamond\neg\neg D$. From the first conjunct and (Ak$_f$4) we get $\Sigma \vdash \neg\Diamond\neg(C \to D)$. The second conjunct and (Ax1) gives us $\Sigma \vdash C \to D$ and by the third conjunct together with (Ak$_f$3) we have $\Sigma \vdash \Diamond\neg\neg(C \to D)$. Then, by the definition of v_Σ, this means $v_\Sigma(C \to D) \in \{\mathbf{T_1}, \mathbf{T_3}\}$, as desired.
- If $v_\Sigma(C) \in \{\mathbf{F_1}, \mathbf{F_2}, \mathbf{F_3}, \mathbf{F_4}\}$ and $v_\Sigma(D) \in \{\mathbf{t_1}, \mathbf{t_2}, \mathbf{t_3}, \mathbf{t_4}, \mathbf{F_1}, \mathbf{F_2}, \mathbf{F_3}, \mathbf{F_4}\}$, then by IH, we obtain that $\Sigma \nvdash \neg\Diamond\neg C$ and $\Sigma \nvdash C$ and $\Sigma \nvdash \neg\Diamond\neg D$. From the second conjunct together with (Ax1) and (Ax3) we infer $\Sigma \vdash C \to D$. And

the first conjunct together with (Ak$_{\mathsf{f}}$2) gives us $\Sigma \vdash \Diamond\neg\neg(C \rightarrow D)$. Then, by the definition of v_Σ, this means $v_\Sigma(C \rightarrow D) \in \{\mathbf{T_1}, \mathbf{T_3}, \mathbf{t_1}, \mathbf{t_3}\}$, as desired.

- If $v_\Sigma(C) \in \{\mathbf{t_1}, \mathbf{t_2}, \mathbf{t_3}, \mathbf{t_4}\}$ and $v_\Sigma(D) \in \{\mathbf{f_1}, \mathbf{f_2}, \mathbf{f_3}, \mathbf{f_4}\}$, then by IH, we obtain that $\Sigma \nvdash \neg\Diamond\neg C$ and $\Sigma \vdash C$ and $\Sigma \vdash \neg\Diamond\neg D$ and $\Sigma \nvdash D$. The third conjunct together with (Ak$_{\mathsf{f}}$4) gives us $\Sigma \vdash \neg\Diamond\neg(C \rightarrow D)$, while the first conjunct together with (Ak$_{\mathsf{f}}$2) gives us $\Sigma \vdash \Diamond\neg\neg(C \rightarrow D)$. We also obtain $\Sigma \nvdash C \rightarrow D$ from the second and fourth conjunct together with (2) from Proposition 3.4. Then, by the definition of v_Σ, this means $v_\Sigma(C \rightarrow D) \in \{\mathbf{f_1}, \mathbf{f_3}\}$, as desired.
- If $v_\Sigma(C) \in \{\mathbf{T_1}, \mathbf{T_2}, \mathbf{T_3}, \mathbf{T_4}\}$ and $v_\Sigma(D) \in \{\mathbf{f_2}, \mathbf{f_4}\}$, then by IH, we obtain that $\Sigma \vdash \neg\Diamond\neg C$ and $\Sigma \vdash C$ and $\Sigma \vdash \neg\Diamond\neg D$ and $\Sigma \nvdash D$ and $\Sigma \nvdash \Diamond\neg\neg D$. The third conjunct together (Ak$_{\mathsf{f}}$4) gives us $\Sigma \vdash \neg\Diamond\neg(C \rightarrow D)$. From the second and forth conjunct together with (2) from Proposition 3.4 we obtain $\Sigma \nvdash (C \rightarrow D)$. And (Ak$_{\mathsf{f}}$1), together with the first and the last conjuncts, gives us $\Sigma \nvdash \Diamond\neg\neg(C \rightarrow D)$. Then, by the definition of v_Σ, this means $v_\Sigma(C \rightarrow D) \in \{\mathbf{f_2}, \mathbf{f_4}\}$, as desired.

Truth-table for the conditional in Definition 2.7

$A \rightarrow B$	$\mathbf{T_1}$	$\mathbf{T_2}$	$\mathbf{T_3}$	$\mathbf{T_4}$	$\mathbf{t_1}$	$\mathbf{t_2}$	$\mathbf{t_3}$	$\mathbf{t_4}$	$\mathbf{f_1}$	$\mathbf{f_2}$	$\mathbf{f_3}$	$\mathbf{f_4}$	$\mathbf{F_1}$	$\mathbf{F_2}$	$\mathbf{F_3}$	$\mathbf{F_4}$
$\mathbf{T_1}$	$\mathcal{T}_{13}$	$\mathcal{T}_{24}$	$\mathcal{T}_{13}$	$\mathcal{T}_{24}$	t_{13}	t_{24}	t_{13}	t_{24}	f_{13}	f_{24}	f_{13}	f_{24}	$\mathcal{F}_{13}$	$\mathcal{F}_{24}$	$\mathcal{F}_{13}$	$\mathcal{F}_{24}$
$\mathbf{T_2}$	$\mathcal{T}_{13}$	$\mathcal{T}_{24}$	$\mathcal{T}_{13}$	$\mathcal{T}_{24}$	t_{13}	t_{24}	t_{13}	t_{24}	f_{13}	f_{24}	f_{13}	f_{24}	$\mathcal{F}_{13}$	$\mathcal{F}_{24}$	$\mathcal{F}_{13}$	$\mathcal{F}_{24}$
$\mathbf{T_3}$	$\mathcal{T}_{13}$	$\mathcal{T}_{24}$	$\mathcal{T}_{13}$	$\mathcal{T}_{24}$	t_{13}	t_{24}	t_{13}	t_{24}	f_{13}	f_{24}	f_{13}	f_{24}	$\mathcal{F}_{13}$	$\mathcal{F}_{24}$	$\mathcal{F}_{13}$	$\mathcal{F}_{24}$
$\mathbf{T_4}$	$\mathcal{T}_{13}$	$\mathcal{T}_{24}$	$\mathcal{T}_{13}$	$\mathcal{T}_{24}$	t_{13}	t_{24}	t_{13}	t_{24}	f_{13}	f_{24}	f_{13}	f_{24}	$\mathcal{F}_{13}$	$\mathcal{F}_{24}$	$\mathcal{F}_{13}$	$\mathcal{F}_{24}$
$\mathbf{t_1}$	$\mathcal{T}_{13}$	$\mathcal{T}_{13}$	$\mathcal{T}_{13}$	$\mathcal{T}_{13}$	$\mathcal{T}_{13}, t_{13}$	$\mathcal{T}_{13}, t_{13}$	$\mathcal{T}_{13}, t_{13}$	$\mathcal{T}_{13}, t_{13}$	f_{13}	f_{13}	f_{13}	f_{13}	$f_{13}, \mathcal{F}_{13}$	$f_{13}, \mathcal{F}_{13}$	$f_{13}, \mathcal{F}_{13}$	$f_{13}, \mathcal{F}_{13}$
$\mathbf{t_2}$	$\mathcal{T}_{13}$	$\mathcal{T}_{13}$	$\mathcal{T}_{13}$	$\mathcal{T}_{13}$	$\mathcal{T}_{13}, t_{13}$	$\mathcal{T}_{13}, t_{13}$	$\mathcal{T}_{13}, t_{13}$	$\mathcal{T}_{13}, t_{13}$	f_{13}	f_{13}	f_{13}	f_{13}	$f_{13}, \mathcal{F}_{13}$	$f_{13}, \mathcal{F}_{13}$	$f_{13}, \mathcal{F}_{13}$	$f_{13}, \mathcal{F}_{13}$
$\mathbf{t_3}$	$\mathcal{T}_{13}$	$\mathcal{T}_{13}$	$\mathcal{T}_{13}$	$\mathcal{T}_{13}$	$\mathcal{T}_{13}, t_{13}$	$\mathcal{T}_{13}, t_{13}$	$\mathcal{T}_{13}, t_{13}$	$\mathcal{T}_{13}, t_{13}$	f_{13}	f_{13}	f_{13}	f_{13}	$f_{13}, \mathcal{F}_{13}$	$f_{13}, \mathcal{F}_{13}$	$f_{13}, \mathcal{F}_{13}$	$f_{13}, \mathcal{F}_{13}$
$\mathbf{t_4}$	$\mathcal{T}_{13}$	$\mathcal{T}_{13}$	$\mathcal{T}_{13}$	$\mathcal{T}_{13}$	$\mathcal{T}_{13}, t_{13}$	$\mathcal{T}_{13}, t_{13}$	$\mathcal{T}_{13}, t_{13}$	$\mathcal{T}_{13}, t_{13}$	f_{13}	f_{13}	f_{13}	f_{13}	$f_{13}, \mathcal{F}_{13}$	$f_{13}, \mathcal{F}_{13}$	$f_{13}, \mathcal{F}_{13}$	$f_{13}, \mathcal{F}_{13}$
$\mathbf{f_1}$	$\mathcal{T}_{13}$	$\mathcal{T}_{24}$	$\mathcal{T}_{13}$	$\mathcal{T}_{24}$	t_{13}	t_{24}	t_{13}	t_{24}	$\mathcal{T}_{13}$	$\mathcal{T}_{24}$	$\mathcal{T}_{13}$	$\mathcal{T}_{24}$	t_{13}	t_{24}	t_{13}	t_{24}
$\mathbf{f_2}$	$\mathcal{T}_{13}$	$\mathcal{T}_{24}$	$\mathcal{T}_{13}$	$\mathcal{T}_{24}$	t_{13}	t_{24}	t_{13}	t_{24}	$\mathcal{T}_{13}$	$\mathcal{T}_{24}$	$\mathcal{T}_{13}$	$\mathcal{T}_{24}$	t_{13}	t_{24}	t_{13}	t_{24}
$\mathbf{f_3}$	$\mathcal{T}_{13}$	$\mathcal{T}_{24}$	$\mathcal{T}_{13}$	$\mathcal{T}_{24}$	t_{13}	t_{24}	t_{13}	t_{24}	$\mathcal{T}_{13}$	$\mathcal{T}_{24}$	$\mathcal{T}_{13}$	$\mathcal{T}_{24}$	t_{13}	t_{24}	t_{13}	t_{24}
$\mathbf{f_4}$	$\mathcal{T}_{13}$	$\mathcal{T}_{24}$	$\mathcal{T}_{13}$	$\mathcal{T}_{24}$	t_{13}	t_{24}	t_{13}	t_{24}	$\mathcal{T}_{13}$	$\mathcal{T}_{24}$	$\mathcal{T}_{13}$	$\mathcal{T}_{24}$	t_{13}	t_{24}	t_{13}	t_{24}
$\mathbf{F_1}$	$\mathcal{T}_{13}$	$\mathcal{T}_{13}$	$\mathcal{T}_{13}$	$\mathcal{T}_{13}$	$\mathcal{T}_{13}, t_{13}$	$\mathcal{T}_{13}, t_{13}$	$\mathcal{T}_{13}, t_{13}$	$\mathcal{T}_{13}, t_{13}$	$\mathcal{T}_{13}$	$\mathcal{T}_{13}$	$\mathcal{T}_{13}$	$\mathcal{T}_{13}$	$\mathcal{T}_{13}, t_{13}$	$\mathcal{T}_{13}, t_{13}$	$\mathcal{T}_{13}, t_{13}$	$\mathcal{T}_{13}, t_{13}$
$\mathbf{F_2}$	$\mathcal{T}_{13}$	$\mathcal{T}_{13}$	$\mathcal{T}_{13}$	$\mathcal{T}_{13}$	$\mathcal{T}_{13}, t_{13}$	$\mathcal{T}_{13}, t_{13}$	$\mathcal{T}_{13}, t_{13}$	$\mathcal{T}_{13}, t_{13}$	$\mathcal{T}_{13}$	$\mathcal{T}_{13}$	$\mathcal{T}_{13}$	$\mathcal{T}_{13}$	$\mathcal{T}_{13}, t_{13}$	$\mathcal{T}_{13}, t_{13}$	$\mathcal{T}_{13}, t_{13}$	$\mathcal{T}_{13}, t_{13}$
$\mathbf{F_3}$	$\mathcal{T}_{13}$	$\mathcal{T}_{13}$	$\mathcal{T}_{13}$	$\mathcal{T}_{13}$	$\mathcal{T}_{13}, t_{13}$	$\mathcal{T}_{13}, t_{13}$	$\mathcal{T}_{13}, t_{13}$	$\mathcal{T}_{13}, t_{13}$	$\mathcal{T}_{13}$	$\mathcal{T}_{13}$	$\mathcal{T}_{13}$	$\mathcal{T}_{13}$	$\mathcal{T}_{13}, t_{13}$	$\mathcal{T}_{13}, t_{13}$	$\mathcal{T}_{13}, t_{13}$	$\mathcal{T}_{13}, t_{13}$
$\mathbf{F_4}$	$\mathcal{T}_{13}$	$\mathcal{T}_{13}$	$\mathcal{T}_{13}$	$\mathcal{T}_{13}$	$\mathcal{T}_{13}, t_{13}$	$\mathcal{T}_{13}, t_{13}$	$\mathcal{T}_{13}, t_{13}$	$\mathcal{T}_{13}, t_{13}$	$\mathcal{T}_{13}$	$\mathcal{T}_{13}$	$\mathcal{T}_{13}$	$\mathcal{T}_{13}$	$\mathcal{T}_{13}, t_{13}$	$\mathcal{T}_{13}, t_{13}$	$\mathcal{T}_{13}, t_{13}$	$\mathcal{T}_{13}, t_{13}$

References

[1] Avron, A. and A. Zamansky, *Non-deterministic semantics for logical systems*, in: *Handbook of Philosophical Logic, vol.16*, Springer, 2011 pp. 227–304.

[2] Blackburn, P., M. d. Rijke and Y. Venema, "Modal Logic," Cambridge Tracts in Theoretical Computer Science, Cambridge University Press, 2001.

[3] Coniglio, M. E., L. Fariñas del Cerro and N. Peron, *Finite non-deterministic semantics for some modal systems*, Journal of Applied Non-Classical Logics **25** (2015), pp. 20–45.

[4] Coniglio, M. E., L. Fariñas del Cerro and N. Peron, *Errata and Addenda to 'Finite non-deterministic semantics for some modal systems'*, Journal of Applied Non-Classical Logics **26** (2016), pp. 336–345.

[5] Coniglio, M. E., L. Fariñas del Cerro and N. Peron, *Modal logic with non-deterministic semantics: Part I—Propositional case*, Logic Journal of the IGPL **28** (2020), pp. 281–315.

[6] Copeland, B. J., *On When a Semantics is Not a Semantics: Some Reasons for Disliking the Routley-Meyer Semantics for Relevance Logic*, Journal of Philosophical Logic **8** (1979), pp. 399–413.

[7] French, R., *Notational variance and its variants*, Topoi **38** (2019), pp. 321–331.

[8] Hiż, H., *A warning about translating axioms*, The American Mathematical Monthly **65** (1958), pp. 613–614.

[9] Humberstone, L., *Yet another "choice of primitives" warning: Normal modal logics*, Logique et Analyse **185–188** (2004), pp. 395–407.

[10] Ivlev, Y. V., *A semantics for modal calculi*, Bulletin of the Section of Logic **17** (1988), pp. 114–121.

[11] Ivlev, Y. V., "Modal logic. (in Russian)," Moskva: Moskovskij Gosudarstvennyj Universitet, 1991.

[12] Kearns, J., *Modal Semantics without Possible Worlds*, Journal of Symbolic Logic **46** (1981), pp. 77–86.

[13] Kearns, J., *Leśniewski's strategy and modal logic*, Notre Dame Journal of Formal Logic **30** (1989), pp. 77–86.

[14] Makinson, D., *A warning about the choice of primitive operators in modal logic*, Journal of Philosophical Logic **2** (1973), pp. 193–196.

[15] Marcos, J., *Modality and paraconsistency*, in: M. Bilková and L. Běhounek, editors, *The Logica Yearbook 2004*, Filosofia, 2005 pp. 213–222.

[16] Omori, H. and D. Skurt, *More modal semantics without possible worlds*, IfCoLog Journal of Logics and their Applications **3** (2016), pp. 815–846.

[17] Scroggs, S. J., *Extensions of the Lewis system S5*, Journal of Symbolic Logic **16** (1951), pp. 112–120.

[18] Shramko, Y. and H. Wansing, *Some useful 16-valued logics: How a computer network should think*, Journal of Philosophical Logic **34** (2005), pp. 121–153.

[19] Shramko, Y. and H. Wansing, "Truth and Falsehood - An Inquiry into Generalized Logical Values," Trends in Logic **36**, Springer Netherlands, 2012.

[20] Tarski, A. and J. Łukasiewicz, *Investigations into the sentential calculus*, in: *Logic, Semantics, Metamathematics; Collected Papers of A. Tarski*, Oxford University Press, 1956 pp. 38–59.

[21] Thomason, R., *Independence of the Dual Axiom in Modal* **K** *with Primitive* $\Diamond$, Notre Dame Journal of Formal Logic **59** (2018), pp. 381–385.

[22] Thomason, S. K., *Semantic analysis of tense logics*, Journal of Symbolic Logic **37** (1972), pp. 150–158.

[23] Wen, X., *Some Common Mistakes in the Teaching and Textbooks of Modal Logic (in Chinese)*, Studies in Logic **11** (2018), pp. 69–86, English translation is available at: `https://arxiv.org/abs/2005.10137`.

References

[1] Avron, A. and A. Zamansky, Non-deterministic semantics for logical systems, in Handbook of Philosophical Logic, vol. 16, Springer, 2011, pp. 227–304.

[2] Blackburn, P., M. de Rijke, and Y. Venema, Modal Logic, [illegible] Theoretical Computer Science, Cambridge University Press, 2001.

[3] Coniglio, M. E., L. Fariñas del Cerro, and N. Peron, Finite non-deterministic semantics for some modal systems, Journal of Applied Non-Classical Logics 25 (2015), pp. 20–45.

[4] Coniglio, M. E., L. Fariñas del Cerro, and N. Peron, [illegible] modal [illegible], Journal of Applied Non-Classical Logics 26 (2016), pp. 83–116.

[5] Coniglio, M. E., L. Fariñas del Cerro, and N. Peron, [illegible], Logic Journal of the IGPL [illegible]

[6] [illegible]

[7] [illegible]

[8] [illegible]

[9] [illegible]

An Extension of Connexive Logic C

Hitoshi Omori [1] Heinrich Wansing [2]

Department of Philosophy I, Ruhr University Bochum
Universitätsstraße 150, 44780, Bochum, Germany

Abstract

In [38], one of the present authors introduced a system of connexive logic, called **C**, as a simple variant of Nelson's Logic **N4**, obtained by making a small change in the falsification clause for the conditional. This was an important step marked in the field of connexive logic since **C** can be seen as the first system of connexive logic with an intuitively plausible semantics. The aim of this article is to consider an extension of **C** obtained by adding the law of excluded middle with respect to the strong negation. The extension of **C** is motivated by three questions. The first question comes from a system **CN** devised by John Cantwell. The second question concerns how many more connexive theses, beside the basic theses of Aristotle and Boethius, can be captured within the framework suggested in the above paper. The third question addresses the relation between constructivity and the law of excluded middle. We will show that the quantified version of our extension of **C** satisfies the Existence Property and its dual, but fails to satisfy the Disjunction Property and its dual when the law of excluded middle is restricted to atomic formulas. We will also mention some open problems related to the new system introduced in this article.

Keywords: Connexive logic, Constructive logic, Law of excluded middle, Nelson logic, Many-valued logic, **FDE**, Disjunction Property, Existence Property.

1 Introduction

Connexive logics are traditionally characterized as systems validating the theses of Aristotle and Boethius, namely the following theses (cf. [20,39] for surveys):

Aristotle $\sim(\sim A \to A)$, $\sim(A \to \sim A)$;

Boethius $(A \to B) \to \sim(A \to \sim B)$, $(A \to \sim B) \to \sim(A \to B)$.

Moreover, we require that $(A \to B) \to (B \to A)$ fails to be a theorem.

As one can easily observe, these characteristic theses are *not* valid in classical logic. In other words, connexive logics belong to a larger family of nonclassical logics known as contra-classical logics (cf. [15]). Thus it remained

---

[1] This research was supported by a Sofja Kovalevskaja Award of the Alexander von Humboldt-Foundation, funded by the German Ministry for Education and Research. We would like to thank the referees for their helpful comments. Email: Hitoshi.Omori@rub.de

[2] Email: Heinrich.Wansing@rub.de

as a non-trivial task to find an intuitive system of connexive logic. An important progress was marked when one of the present authors, in [38], suggested to capture connexive theses through a different falsity condition for the conditional by building on the elegant framework of the four-valued logic known as **FDE**, or Belnap-Dunn logic.[3] More specifically, a system of connexive logic called **C** was introduced as a variant of Nelson's logic **N4** (cf. [37,23,16] and references therein) by making a small change to the falsity condition for the conditional (cf. Remark 2.2).[4] As Graham Priest notes in [31, p. 178], the system **C** is most likely to be "one of the simplest and most natural."

The aim of this article is to consider an extension of **C** obtained by adding the law of excluded middle (LEM hereafter) with respect to the strong negation. The extension of **C** is motivated by the following three questions.

Question 1: Can we improve Cantwell's CN? In [6], John Cantwell addresses the question of how to negate indicative conditionals, and defends the following three-valued truth table, suggested by Nuel D. Belnap in [5]:

A	$\sim A$
t	**f**
–	–
f	**t**

$A \to B$	**t**	–	**f**
t	**t**	–	**f**
–	**t**	–	**f**
f	–	–	–

A notable feature here is that the third value is meant to be a gappy value, and that when a conditional has a false antecedent it lacks a truth value. As a byproduct, two formulas $\sim(A \to B)$ and $A \to \sim B$ receive exactly the same value for every assignment, and thus these two formulas are equivalent.[5]

Cantwell then defined the consequence relation by designating both **t** and –, which can be seen as preserving non-false values. As a consequence of this, Cantwell's logic **CN** includes both Aristotle's and Boethius' theses, and thus is connexive, though connexivity is not mentioned at all.

However, **CN** also has the feature of the pure $\to$-fragment of the resulting logic being classical. This implies that formulas such as $(A \to B) \vee (B \to C)$ hold for arbitrary A, B and C. We are fully aware that there are attempts aiming at making sense of the material conditional as an indicative conditional (cf. [32] and references therein). But, it also seems to be a natural question if we can replace the classical conditional by a better conditional, such as a constructive conditional. And this is precisely the first question that motivates us to explore the extension of **C** by LEM.

Question 2: How many desiderata, listed by Estrada-González and Ramírez-Cámara, can be met by connexive logics à la C? In [12], Luis Estrada-González and Elisángela Ramírez-Cámara offer a list of desiderata for connexive logics which contains more than two characteristic theses of

[3] For an overview of systems related to **FDE**, cf. [29].

[4] Tweaking the falsity condition for other connectives seems to be interesting from the perspective of contra-classical logics (cf. [30]).

[5] See [10,11] for a recent discussion on **CN**. See also [9,28] for negated indicative conditionals being equivalent to formulas involving a modality.

connexive logics. In particular, the list contains the following formulas:

Aristotle's second thesis $\sim((A \to B) \wedge (\sim A \to B))$;

Abelard's thesis $\sim((A \to B) \wedge (A \to \sim B))$.[6]

Note that on the one hand, in [40], it has been argued that the above theses should not be considered as defining principles of connexive logic because they are motivated by the idea of negation as cancellation, which is said to be unsuitable as a basis for any validity claims. On the other hand, however, the above formulas are included, for example, in probabilistic approaches to conditionals, and thus interests into these formulas go beyond technical curiosity. They rather seem to capture certain intuitions on conditionals.

Regardless of one's opinion on the above two theses, it is mentioned already in [40,41] that the system **C** fails to include Aristotle's second thesis and Abelard's thesis. But, this does not mean that variants of **C** will fail both or at least one of the above theses. And this is precisely the second question that motivates us to explore the extension of **C** by LEM.

Question 3: What is the relation between constructivity and LEM? As per Jeremy Avigad [1, p. 10], "the words "constructive" and "intuitionistic" are used today almost interchangeably." Yves Lafont [14, p. 149] explains that "[i]ntuitionistic logic is called constructive because of the correspondence between proofs and algorithms," so that intuitionistic implication is essential for regarding intuitonistic logic as constructive. According to Paul Gilmore [13, p. *xiv*], the Disjunction Property and the Existence Property together are "the hallmark property of an intuitionistic/constructive logic." This conception of constructivity (or constructiveness) has been challenged by David Nelson, who in addition to keeping the intuitionistic conditional and desiring the disjunction and existence properties suggested to require also the Constructible Falsity Property (if $\sim(A \wedge B)$ is provable, then $\sim A$ or $\sim B$ is provable) and the dual of the Existence Property, cf. Section 6.

Our third question does not make any sense if we assume intuitionistic logic as the logic in question since the addition of LEM will collapse the logic into classical logic, which is admittedly non-constructive. The situation is similar in the case of Nelson's constructive logics with strong negation. Indeed, for the case with his **N3**, the addition of LEM with respect to the strong negation will again collapse the logic into classical logic. Moreover, for the case with **N4**, the addition of LEM with respect to the strong negation will *not* collapse the logic into classical logic, but instead the resulting logic will be the three-valued logic, for example known as **CLuNs** (cf. [4]), obtained by expanding the well-known three-valued paraconsistent logic **LP**, by the following truth table:

$A \to B$	t	b	f
t	t	b	f
b	t	b	f
f	t	t	t

[6] This thesis is also known as Strawson's thesis.

In particular, the ~-free fragment of **CLuNs** is classical, and thus the addition of LEM indeed destroys the constructive features, including the intuitionistic conditional, the Disjunction Property, and the Existence Property.

Note, however, that the proof of the collapse of **N4** into **CLuNs** heavily relies on the falsity condition for the conditional, and thus motivates our third question to explore the effect of adding LEM to **C**, which has a falsity condition different from that of **N4**. As we will show, the addition of LEM to a logic with a constructive implication need not collapse into classical logic. Moreover, the addition of LEM for atomic formulas need not prevent the Existence Property and its dual form holding.

Based on these considerations, the paper is structured as follows. We first revisit connexive logic **C** in §2. This is followed by §3 in which we introduce the extension of **C** by LEM. We then discuss the first two questions in §4. After these discussions at the level of propositional logic, we introduce, in §5, the extension of **QC**, quantified **C**, obtained by adding LEM. We then discuss the proof theory for **QC** and its extension in §6, and conclude the paper by a summary and remarks on future directions in §7.

2 Revisiting C

The language $\mathcal{L}$ consists of a finite set $\{\sim, \wedge, \vee, \rightarrow\}$ of propositional connectives and a countable set Prop of propositional variables which we denote by p, q, etc. Furthermore, we denote the set of formulas defined as usual in $\mathcal{L}$ by Form, a formula of $\mathcal{L}$ by A, B, C, etc. and a set of formulas of $\mathcal{L}$ by Γ, Δ, Σ, etc.

2.1 Semantics

The following semantics, introduced in [38], is obtained by making a simple change to the standard semantics for Nelson's logic **N4**.

Definition 2.1 A **C**-model for the language $\mathcal{L}$ is a triple $\langle W, \leq, V\rangle$, where W is a non-empty set (of states); $\leq$ is a partial order on W; and $V : W \times \mathsf{Prop} \longrightarrow \{\varnothing, \{0\}, \{1\}, \{0,1\}\}$ is an assignment of truth values to state-variable pairs with the condition that $i \in V(w_1, p)$ and $w_1 \leq w_2$ only if $i \in V(w_2, p)$ for all $p \in \mathsf{Prop}$, all $w_1, w_2 \in W$ and $i \in \{0,1\}$. Valuations V are then extended to interpretations I of state-formula pairs by the following conditions:

- $I(w,p) = V(w,p)$,
- $1 \in I(w, \sim A)$ iff $0 \in I(w, A)$,
- $0 \in I(w, \sim A)$ iff $1 \in I(w, A)$,
- $1 \in I(w, A \wedge B)$ iff $1 \in I(w, A)$ and $1 \in I(w, B)$,
- $0 \in I(w, A \wedge B)$ iff $0 \in I(w, A)$ or $0 \in I(w, B)$,
- $1 \in I(w, A \vee B)$ iff $1 \in I(w, A)$ or $1 \in I(w, B)$,
- $0 \in I(w, A \vee B)$ iff $0 \in I(w, A)$ and $0 \in I(w, B)$,
- $1 \in I(w, A \rightarrow B)$ iff for all $w_1 \in W$: if $w \leq w_1$ and $1 \in I(w_1, A)$ then $1 \in I(w_1, B)$,
- $0 \in I(w, A \rightarrow B)$ iff for all $w_1 \in W$: if $w \leq w_1$ and $1 \in I(w_1, A)$ then $0 \in I(w_1, B)$.

Finally, the semantic consequence is now defined as follows: $\Gamma \models_{\mathbf{C}} A$ iff for all **C**-models $\langle W, \leq, I\rangle$, and for all $w \in W$: $1 \in I(w, A)$ if $1 \in I(w, B)$ for all $B \in \Gamma$.

Remark 2.2 Note that Nelson's logic **N4** is obtained by replacing the falsity condition for implication by the following condition:

$$0 \in I(w, A \to B) \text{ iff } 1 \in I(w, A) \text{ and } 0 \in I(w, B).$$

2.2 Proof System

We now turn to the proof system.

Definition 2.3 The axiomatic proof system **C** consists of the following axiom schemata and a rule of inference, where $A \leftrightarrow B$ abbreviates $(A \to B) \wedge (B \to A)$:

$$A \to (B \to A) \quad \text{(Ax1)}$$
$$(A \to (B \to C)) \to ((A \to B) \to (A \to C)) \quad \text{(Ax2)}$$
$$(A \wedge B) \to A \quad \text{(Ax3)}$$
$$(A \wedge B) \to B \quad \text{(Ax4)}$$
$$(C \to A) \to ((C \to B) \to (C \to (A \wedge B))) \quad \text{(Ax5)}$$
$$A \to (A \vee B) \quad \text{(Ax6)}$$
$$B \to (A \vee B) \quad \text{(Ax7)}$$
$$(A \to C) \to ((B \to C) \to ((A \vee B) \to C)) \quad \text{(Ax8)}$$
$$\sim\sim A \leftrightarrow A \quad \text{(Ax9)}$$
$$\sim(A \wedge B) \leftrightarrow (\sim A \vee \sim B) \quad \text{(Ax10)}$$
$$\sim(A \vee B) \leftrightarrow (\sim A \wedge \sim B) \quad \text{(Ax11)}$$
$$\sim(A \to B) \leftrightarrow (A \to \sim B) \quad \text{(Ax12)}$$
$$\frac{A \quad A \to B}{B} \quad \text{(MP)}$$

Finally, we write $\Gamma \vdash_{\mathbf{C}} A$ if there is a sequence of formulas $B_1, \ldots, B_n, A$, $n \geq 0$, such that every formula in the sequence $B_1, \ldots, B_n, A$ either (i) belongs to Γ; (ii) is an axiom of **C**; or (iii) is obtained by (MP) from formulas preceding it in sequence.

Remark 2.4 Note that if we replace (Ax12) by '$\sim(A \to B) \leftrightarrow (A \wedge \sim B)$', then we obtain an axiomatization of Nelson's logic **N4**.

We also note that the deduction theorem is provable.

Proposition 2.5 *For any* $\Gamma \cup \{A, B\} \subseteq \mathsf{Form}$, $\Gamma, A \vdash_{\mathbf{C}} B$ *iff* $\Gamma \vdash_{\mathbf{C}} A \to B$.

Proof. It can be proved in the usual manner in the presence of axioms (Ax1) and (Ax2), given that (MP) is the sole rule of inference. □

2.3 Basic results and an observation

As expected, we have a soundness and completeness result, established in [38].

Theorem 2.6 ([38]) *For any* $\Gamma \cup \{A\} \subseteq \mathsf{Form}$, $\Gamma \vdash_{\mathbf{C}} A$ *iff* $\Gamma \models_{\mathbf{C}} A$.

A highly unusual feature of **C** is non-trivial but inconsistent.

Proposition 2.7 *For any* $A \in \mathsf{Form}$, $\vdash_{\mathbf{C}} (A \wedge \sim A) \to A$ *and* $\vdash_{\mathbf{C}} \sim((A \wedge \sim A) \to A)$.

Proof. The first item is (Ax3). For the second item, it will suffice to establish $\vdash (A \wedge \sim A) \to \sim A$ in view of (Ax12) and (MP), but this is just an instance of (Ax4). □

Note that the proof of inconsistency relies on very weak assumptions. Indeed, even with an extremely weak relevant implication (cf. [26, p. 477]) and the very weak conditional considered in [41], the above inconsistency result will hold. This may motivate to address the question if the axiom (Ax12), or the falsity condition for implication from **C**, will *always* give rise to inconsistency of the system. The short answer is: No.

Theorem 2.8 *There is a consistent system with the axiom* (Ax12).

Proof. Take the propositional language only with $\sim$ and $\rightarrow$ (without $\wedge$), and consider an axiomatic proof system with (Ax9), (Ax12), (MP), and any axioms for $\rightarrow$ that only have an even number of occurrences of variables.

Then, we can see that this proof system is consistent. Indeed, we can take the two valued matrix for classical logic, and interpret $\sim$ and $\rightarrow$ as classical negation and classical biconditional, respectively. This clearly shows that the proof system is indeed consistent. □

In the rest of this paper, we will take the propositional language $\mathcal{L}$ that contains both conjunction and disjunction, and thus the inconsistency will be part of the features enjoyed by the system.

3 An extension of C

We now turn to the main extension of **C**. For the semantics, we simply close the gap.

Definition 3.1 A **C3**-model for the language $\mathcal{L}$ is a triple $\langle W, \leq, V\rangle$, where W is a non-empty set (of states); $\leq$ is a partial order on W; and $V : W \times \mathsf{Prop} \longrightarrow \{\{0\},\{1\},\{0,1\}\}$ is an assignment of truth values to state-variable pairs with the condition that $i \in V(w_1, p)$ and $w_1 \leq w_2$ only if $i \in V(w_2, p)$ for all $p \in \mathsf{Prop}$, all $w_1, w_2 \in W$ and $i \in \{0,1\}$. Valuations V are then extended to interpretations I of state-formula pairs by the same conditions as with **C**. The semantic consequence $\models_{\mathbf{C3}}$ is defined in a similar manner.

Remark 3.2 Note that the persistence condition carries over to all formulas.

For the proof system, we add the law of excluded middle with respect to the strong negation.

Definition 3.3 The axiomatic proof system for **C3** is obtained by adding $A \vee {\sim}A$, namely LEM, to the axiomatic proof system for **C**. We then define $\vdash_{\mathbf{C3}}$ in a similar manner.

As usual, the soundness part is rather straightforward.

Proposition 3.4 (Soundness) *For* $\Gamma \cup \{A\} \subseteq \mathsf{Form}$, *if* $\Gamma \vdash_{\mathbf{C3}} A$ *then* $\Gamma \models_{\mathbf{C3}} A$.

Proof. We only note that the elimination of the gappy value in the semantics guarantees that LEM is valid in all **C3**-models. □

For the completeness proof, we first introduce some standard notions.

Definition 3.5 $\Sigma \subseteq \mathsf{Form}$ is *deductively closed* iff if $\Sigma \vdash A$ then $A \in \Sigma$. The set Σ is *prime* iff $A \vee B \in \Sigma$ implies $A \in \Sigma$ or $B \in \Sigma$. Moreover, Σ is *prime deductively closed* (pdc) if it is both. Finally, Σ is *non-trivial* if $A \notin \Sigma$ for some A.

The following two lemmas are well-known, and thus the proofs are omitted.

Lemma 3.6 *If* $\Sigma \nvdash A$, *there is a non-trivial pdc* Δ *such that* $\Sigma \subseteq \Delta$ *and* $\Delta \nvdash A$.

Lemma 3.7 *If Σ is pdc and $A \to B \notin \Sigma$, there is a non-trivial pdc Θ such that $\Sigma \subseteq \Theta$, $A \in \Theta$ and $B \notin \Theta$.*

Now, we are ready to prove the completeness.

Theorem 3.8 (Completeness) *For $\Gamma \cup \{A\} \subseteq \mathsf{Form}$, if $\Gamma \vDash_{\mathbf{C3}} A$ then $\Gamma \vdash_{\mathbf{C3}} A$.*

Proof. Suppose that $\Gamma \nvdash_{\mathbf{C3}} A$. Then by Lemma 3.6, there is a $\Pi \supseteq \Gamma$ such that Π is a pdc and $A \notin \Pi$. Define the model $\mathfrak{A} = \langle X, \leq, I \rangle$, where $X = \{\Delta : \Delta \text{ is a non-trivial pdc}\}$, $\Delta \leq \Sigma$ iff $\Delta \subseteq \Sigma$ and I is defined thus. For every state Σ and propositional variable p:

$$1 \in I(\Sigma, p) \text{ iff } p \in \Sigma \text{ and } 0 \in I(\Sigma, p) \text{ iff } {\sim} p \in \Sigma$$

Note that $\mathfrak{A}$ is indeed a **C3**-model since $1 \in I(\Sigma, p)$ or $0 \in I(\Sigma, p)$ holds for all Σ and p in view of LEM.

We now show that the above definition holds for arbitrary formula, B:

$$1 \in I(\Sigma, B) \text{ iff } B \in \Sigma \text{ and } 0 \in I(\Sigma, B) \text{ iff } {\sim} B \in \Sigma$$

This can be proved by a simultaneous induction on the complexity of B with respect to the positive and the negative clause.[7] It then follows that $\mathfrak{A}$ is a counter-model for the inference, and hence that $\Gamma \nvDash_{\mathbf{C3}} A$. □

4 Questions 1 and 2 in view of C3

We now turn to address first two questions, raised in our introduction, in light of the new system **C3**.

4.1 An answer to Question 1: C3 is a generalization of CN

In brief, our first question concerned the system **CN** defended and explored by Cantwell in [6]. More specifically, we asked if we can replace the classical material conditional by a constructive conditional.

Note first that from a semantic perspective, it is easy to see that **C3** expands positive intuitionistic propositional logic conservatively. Moreover, we have the following result that justifies to claim that **C3** is a generalization of **CN**.

Proposition 4.1 *The extension of* **C3** *by Peirce's law is sound and complete with respect to the semantics induced by the following matrix with* **t** *and* **b** *as designated values. In other words, the extension is the system* **CN** *of Cantwell.*

A	${\sim}A$
t	f
b	b
f	t

$A \wedge B$	t	b	f
t	t	b	f
b	b	b	f
f	f	f	f

$A \vee B$	t	b	f
t	t	t	t
b	t	b	b
f	t	b	f

$A \to B$	t	b	f
t	t	b	f
b	t	b	f
f	b	b	b

Proof. For soundness, just note that every one-element model validates Peirce's law. For completeness, note first that the presence of Peirce's law makes the partial order on the canonical model trivial. More specifically, for two non-trivial pdcs Σ and Δ, we obtain that $\Sigma \subseteq \Delta$ only if $\Delta \subseteq \Sigma$. Indeed,

[7] The proof is the same as the one for [25, Theorem 2] with an obvious change to be made for the negative clause for the conditional.

suppose for reductio that $\Sigma \subseteq \Delta$ and that for some A_0, $A_0 \in \Delta$ but $A_0 \notin \Sigma$. Then, in view of Peirce's law, we also have $A \vee (A \to B)$ as a derivable formula, and thus we have $A_0 \vee (A_0 \to B) \in \Sigma$ for arbitrary B. In view of $A_0 \notin \Sigma$ and that Σ is prime, we obtain $(A_0 \to B) \in \Sigma$. This together with $\Sigma \subseteq \Delta$ implies $(A_0 \to B) \in \Delta$, and with $A_0 \in \Delta$, we obtain $B \in \Delta$. But since B is arbitrary, Δ will be trivial and this contradicts the assumption that Δ is non-trivial.

We can then consider the submodel of the canonical model with $X = \{\Pi\}$ where $\Pi \supseteq \Gamma$ such that Π is pdc and $A \notin \Pi$, obtained in view of Lemma 3.6. This completes the proof. □

Remark 4.2 Note that Grigory Olkhovikov, in [24], also introduced and discusses the above three-valued truth table for the conditional. Olkhovikov also motivates the truth table by considering conditionals in natural language, but with a different reading of the third value.

Remark 4.3 Compare the above result with **C** in which the addition of Peirce's law results in a four-valued logic, called **MC** (for material connexive logic) in [39], induced by the matrix obtained by adding the following truth table for implication in addition to the truth tables of **FDE**:

$A \to B$	**t**	**b**	**n**	**f**
t	**t**	**b**	**n**	**f**
b	**t**	**b**	**n**	**f**
n	**b**	**b**	**b**	**b**
f	**b**	**b**	**b**	**b**

An expansion of **MC** by the Boolean complement is explored in [27].

Remark 4.4 In view of the above result, there will be intermediate logics between **C3** and **CN**, as well as **C** and **MC**. Then, recalling that intermediate extensions of **N3** and **N4** (as well as $\mathbf{N4}^{\perp}$) are explored by Marcus Kracht in [18] and Sergei Odintsov in [23] respectively, describing the intermediate extensions of **C3** and **C** is an interesting open problem.

4.2 An answer to Question 2: C3 gives us all if we are careful

We now turn to the second question of how much connexivity can be captured through the present approach via a slightly different falsity condition. Our considerations build on a list provided by Estrada-González and Ramírez-Cámara in [12] which includes the following four theses that are contra-classical:

Aristotle $\sim(\sim A \to A)$, $\sim(A \to \sim A)$;
Boethius $(A \to B) \to \sim(A \to \sim B)$, $(A \to \sim B) \to \sim(A \to B)$;
Aristotle 2nd $\sim((A \to B) \wedge (\sim A \to B))$;
Abelard $\sim((A \to B) \wedge (A \to \sim B))$.

Then, we first observe that systems **C** and **C3** are not able to capture all of the above four theses.

Proposition 4.5 *Both Aristotle's and Boethius' theses are derivable in* **C** *and* **C3**. *However, (i)* **C** *fails to include Aristotle's second thesis and Abelard's thesis (this was mentioned already in [40,41]); (ii)* **C3** *includes Abelard's thesis* *but not Aristotle's second thesis.*

Proof. The first half, namely that **C** and **C3** include Aristotle's and Boethius' theses as derivable formulas is well-known and immediate in view of (Ax12). For the second half, we can make use of the truth tables.

- For the failures in **C**, it is enough to establish that the concerned theses are not valid in **MC**. Now, take the four-valued truth tables for **MC**. Then, if we assign values **b** and **t** to A and B respectively, Aristotle's second thesis receives the value **f**. Moreover, if we assign values **t** and **n** to A and B respectively, Abelard's second thesis receives the value **n**.
- For the cases of **C3**, we first observe that Aristotle's second thesis is not valid in **CN**. To this end, if we assign values **b** and **t** to A and B respectively, we obtain the desired result. For the derivability of Abelard's thesis, it is enough to establish that $(A{\to}B){\vee}(A{\to}{\sim}B)$ is derivable in **C3**, and this follows by LEM and (Ax1).

This completes the proof. □

The above results show that the move to **C3** from **C** will allow us to capture one more thesis, namely Abelard's thesis. However, Aristotle's second thesis is not captured, but this is not the end of the story. This is because we may consider another very natural conditional, $\Rightarrow$, in **C** and **C3** defined as follows:

$$A \Rightarrow B := (A \to B) \wedge ({\sim}B \to {\sim}A).$$

One of the obvious differences is that the contraposition rule holds for the "strong implication" $\Rightarrow$, which was not the case with $\to$. More importantly, we have the following result.

Theorem 4.6 *All four theses listed above, formulated in terms of $\Rightarrow$, are derivable in* **C3**. *However, only Aristotle's and Boethius' theses, formulated in terms of $\Rightarrow$, are derivable in* **C**.

Proof. We first check that Aristotle's and Boethius' theses, formulated in terms of $\Rightarrow$, are derivable in **C**, and thus also in **C3**. To this end, note that the following equivalences are derivable in view of the definition of $\Rightarrow$:

- $(A \Rightarrow {\sim}B) \leftrightarrow ((A \to {\sim}B) \wedge (B \to {\sim}A))$;
- ${\sim}(A \Rightarrow B) \leftrightarrow ((A \to {\sim}B) \vee ({\sim}B \to A))$;
- ${\sim}(A \Rightarrow {\sim}B) \leftrightarrow ((A \to B) \vee (B \to A))$.

Then, it is obvious that Aristotle's theses are derivable by the last equivalence. For Boethius' theses, we need to check that the following holds in **C**:

- $(A \Rightarrow B) \to {\sim}(A \Rightarrow {\sim}B)$;
- $(A \Rightarrow {\sim}B) \to {\sim}(A \Rightarrow B)$.

But these are obvious in view of the above equivalences, and thus we obtain Boethius' theses in **C** and also in **C3**.

We now turn to check that Abelard's thesis, formulated in terms of $\Rightarrow$, is derivable in **C3**. For this purpose, simply note that the thesis is equivalent to $((A{\to}{\sim}B){\vee}({\sim}B{\to}A)){\vee}((A{\to}B){\vee}(B{\to}A))$. Then, by looking at the first and the third disjuncts, we can see that the above formula is derivable since $(A \to B) \vee (A \to {\sim}B)$ is derivable in **C3**, as we observed in the previous proposition.

Similarly, for the case of Aristotle's second thesis, it suffices to derive $((A \to {\sim}B) \vee ({\sim}B \to A)) \vee (({\sim}A \to {\sim}B) \vee ({\sim}B \to {\sim}A))$. This time, by looking at the second and the fourth disjuncts, we can see that the above formula is derivable for the same reason.

Finally, in order to see that Aristotle's second thesis and Abelard's thesis are not derivable in **C**, it suffices to see that the above two formulas have a counter-model $\langle W, \leq, V\rangle$ defined as follows:

- $W := \{w_0, w_1, w_2\}$, $\leq := \{\langle w_0, w_0\rangle, \langle w_0, w_1\rangle, \langle w_0, w_2\rangle, \langle w_1, w_1\rangle, \langle w_2, w_2\rangle\}$,
- $V(w_0,p)=V(w_0,q)=V(w_1,p)=V(w_2,q)=\{\}$ and $V(w_1,q)=V(w_2,p)=\{1,0\}$.

This completes the proof. □

Remark 4.7 Note that all four theses, formulated in terms of $\Rightarrow$, hold in **CN** and **MC**. This is due to the fact that ${\sim}(A \Rightarrow B)$ is equivalent to $(A \to {\sim}B) \vee ({\sim}B \to A)$, an instance of the linearity axiom. Thus, for *arbitrary* formulas A and B, we have ${\sim}(A \Rightarrow B)$ as valid since $\to$ in both **CN** and **MC** is classical. This is in sharp contrast with **C3** since we have $\nvDash_{\mathbf{C3}} {\sim}(A \Rightarrow B)$. Indeed, we can consider a counter-model $\langle W, \leq, V\rangle$ defined as follows:

- $W := \{w_0, w_1, w_2\}$, $\leq := \{\langle w_0, w_0\rangle, \langle w_0, w_1\rangle, \langle w_0, w_2\rangle, \langle w_1, w_1\rangle, \langle w_2, w_2\rangle\}$,
- $V(w_0,p)=V(w_1,p)=\{0\}, V(w_0,q)=V(w_2,q)=\{1\}$, $V(w_1,q)=V(w_2,p)=\{1,0\}$.

We summarize our observations from this subsection in a table as follows:

	C		**C3**		**MC**		**CN**	
	$\to$	$\Rightarrow$	$\to$	$\Rightarrow$	$\to$	$\Rightarrow$	$\to$	$\Rightarrow$
Aristotle's theses	✓	✓	✓	✓	✓	✓	✓	✓
Boethius' theses	✓	✓	✓	✓	✓	✓	✓	✓
Aristotle's second theses	×	×	×	✓	×	✓	×	✓
Abelard's theses	×	×	✓	✓	✓	✓	✓	✓
Negated strong conditional	×	×	×	×	×	✓	×	✓

In short, the contraposible strong implication $\Rightarrow$ of **C3** seems to be a very interesting conditional, satisfying all four contra-classical theses, without having the problematic-looking formula, i.e. ${\sim}(A \Rightarrow B)$, as a valid formula.

4.3 Beyond Question 2: Totally connexive logics

The four theses we focused on in the previous subsection are part of a bigger list provided by Estrada-González and Ramírez-Cámara. Indeed, they also considered the following desiderata on top of the four theses.

Positive Paradox of Implication $\nvDash A \to (B \to A)$;
Negative Paradox of Implication $\nvDash A \to (\neg A \to B)$;
Paradox of Necessity $\nvDash A \to (B \to C)$ where A is a contingent truth and $B \to C$ is a logical truth;
Simplification $\vDash (A \wedge B) \to A$, $\vDash (A \wedge B) \to B$;
Idempotence $\vDash (A \wedge A) \to A$, $\vDash A \to (A \wedge A)$;
Kapsner-strong (i) $A \to {\sim}A$ is unsatisfiable and (ii) $A \to B$ and $A \to {\sim}B$ are

not simultaneously satisfiable.

Estrada-González and Ramírez-Cámara then introduced the notion of *totally connexive logics* as logics that satisfy all the desiderata, including the four theses. Moreover, they left as an open problem whether there are totally connexive logics, and if so then which is the minimal one (cf. [12, p. 348]).

In view of our observations in the previous section, there are three candidates for totally connexive logics, namely **C3**, **MC** and **CN**, by looking at the contraposible conditional $\Rightarrow$.[8] Since both **C3** and **MC** are subsystems of **CN**, and since the above desiderata involve some invalidities, let us focus on **CN**. Then, we obtain the following truth table for $\Rightarrow$ in **CN**:

$A \Rightarrow B$	**t**	**b**	**f**
t	**b**	**f**	**f**
b	**b**	**b**	**f**
f	**b**	**b**	**b**

This truth table is not new, but was already introduced and discussed by Chris Mortensen in [21] in a logic, later called **M3V** in [20]. Recall that **M3V** is an expansion of **LP** obtained by adding the above conditional. Before adding further remarks, we observe that **CN** and **M3V** are equivalent systems.

Proposition 4.8 **CN** *and* **M3V** *are equivalent.*

Proof. All connectives in **M3V** are definable in **CN** in view of the above discussion. For the other half, we need to show that the non-contraposible conditional of **CN** is definable in **M3V**. This can be checked by observing that $((A \Rightarrow B) \vee B) \wedge {\sim}(A \Rightarrow ({\sim}(A \Rightarrow B) \wedge (B \wedge {\sim}B)))$ defines $\rightarrow$ of **CN**. □

In fact, **M3V** is examined by Estrada-González and Ramírez-Cámara, and is shown to satisfy all the desiderata except Kapsner-strong. Building on this observation, we conclude that all three candidates enjoy the same status since Simplification and Idempotence are easy to check for the contraposible conditional of both **C** and **C3**.

For the Kapsner-strong condition, we wish to add a remark. We are fully aware of Andreas Kapsner's original motivation, argued in [17], as well as his intended way of spelling out the details in a rather classical manner.[9] However, if we are in the vicinity of **FDE**, then there are always at least two ways to formalize classical notions since truth and non-falsity (or, falsity and non-truth) are not necessarily equivalent.[10] In particular, the notion of satisfiability, which plays the crucial role in the Kapsner-strong condition, can be formalized as follows in **M3V**:

---

[8] Actually, **CN** is one of the systems examined by Estrada-González and Ramírez-Cámara, but with respect to the non-contraposible conditional $\rightarrow$.

[9] Recall that the original proposal of Kapsner in [17] was to call for *strong connexivity*, and that notion encompasses not only the theses of Aristotle and Boethius, but also the unsatisfiability clauses which we refer to as Kapsner-strong conditions following [12].

[10] For example, think of discussions on p– and q–consequence relations that have revived recently through a series of papers by Pablo Cobreros, Paul Egré, Dave Ripley, and Robert van Rooij (e.g. [7,8]).

- A is *positively satisfiable* iff for some **M3V**-valuation V, $1 \in I(A)$.
- A is *negatively satisfiable* iff for some **M3V**-valuation V, $0 \notin I(A)$.

Then, as observed by Estrada-González and Ramírez-Cámara, **M3V** is not Kapsner-strong, if satisfiability is understood as *positive* satisfiability. However, it *is* Kapsner-strong, if satisfiability is understood as *negative* satisfiability since $0 \in I(A \Rightarrow B)$ for all **M3V**-valuations V and for all A and B.

For **C3**, we may formulate two kinds of satisfiability as follows.

- A is *positively satisfiable* iff for some **C3**-model $\mathcal{M} = \langle W, \leq, V\rangle$, $1 \in I(w, A)$ for some $w \in W$.
- A is *negatively satisfiable* iff for some **C3**-model $\mathcal{M} = \langle W, \leq, V\rangle$, $0 \notin I(w, A)$ for some $w \in W$.

Then, **C3** is Kapsner-strong if satisfiability is understood as *negative* satisfiability. Indeed, for all **C**-models $\mathcal{M} = \langle W, \leq, V\rangle$ and for all $w \in W$, we have $0 \in I(w, A \Rightarrow {\sim}A)$ and $0 \in I(w, (A \Rightarrow B) \wedge (A \Rightarrow {\sim}B))$.

Based on these observations, we conclude that there are totally connexive logics, and examples include **C3**, **MC** and **CN** by looking at the contraposible conditional. Note also that **C3** enjoys the additional feature of being totally connexive *without* all negated conditionals being valid (cf. Remark 4.7).

Remark 4.9 The key idea in this section has been to consider the contraposible conditional in **C** and its extensions. Then, in view of results reported by Matthew Spinks and Robert Veroff, such as those in [33] and references therein, establishing clear and neat connections between Nelson logics and relevant logics, it is a natural and interesting question to explore the connections between extensions of **C** and relevant logics as well.

Moreover, a deeper understanding of these connections may also give us some new insights into the problem of finding sound and complete semantics for systems introduced by Everett Nelson. More specifically, Nelson, in his PhD thesis, introduced an axiomatic system of connexive logic, called **NL** by Edwin Mares and Francesco Paoli in [19]. Then, the open problem, noted in [39], is to find a sound and complete semantics for **NL**. Since one of the subsystems, called **NL**$^-$ in [19], is close to the relevant logic **DK**, we may seek for a suitable semantics via the contraposible conditional of one of the systems related to **C3**.

Before moving further, here is a table indicating the relations between extensions of **C** we discussed so far, with a comparison to extensions of **N4**.

$$
\begin{array}{ccc|ccc}
\mathbf{MC} & \xrightarrow{+\mathrm{LEM}} & \mathbf{CN}(=\mathbf{M3V}) & \mathbf{HBe} & \xrightarrow{+\mathrm{LEM}} & \mathbf{CLuNs}(=\mathbf{RM3}) \\
\uparrow{\scriptstyle +\mathrm{PL}} & & \uparrow{\scriptstyle +\mathrm{PL}} & \uparrow{\scriptstyle +\mathrm{PL}} & \nearrow{\scriptstyle +\mathrm{LEM}} & \\
\mathbf{C} & \xrightarrow{+\mathrm{LEM}} & \mathbf{C3} & \mathbf{N4} & &
\end{array}
$$

Note here that PL stands for Peirce's law. Moreover, **HBe** is an expansion of **FDE** explored by Arnon Avron in [3], and the equivalence of **CLuNs** and **RM3** is shown in [2, (a) of 2.10 Theorem] (Avron refers to **CLuNs** as $\mathbf{RM}_3^{\supset}$).

5 An extension of QC

We will now consider the expansion of **C** and **C3** to their quantified versions **QC** and **QC3** in a language without function symbols and equality.

We extend the propositional language $\mathcal{L}$ to a first-order language by adding denumerably many individual variables, x, y, z, ..., constants a, b, c, ..., and predicate symbols of finite arity. Terms (i.e, individual variables or constants) are denoted by $t, t_1, t_2, \ldots$, atomic formulas by $P, Q, \ldots$, and arbitrary formulas by A, B, etc.

5.1 Axiomatic proof systems

Definition 5.1 ([38, §4]) The schematic axioms and rules of **QC** are those of **C** together with:

$$\sim\exists xA \leftrightarrow \forall x{\sim}A; \qquad \sim\forall xA \leftrightarrow \exists x{\sim}A$$

$$A(t) \rightarrow \exists xA(x) \ (t \text{ is free for } x \text{ in } A); \qquad \forall xA(x) \rightarrow A(t) \ (t \text{ is free for } x \text{ in } A)$$

$$\frac{A \rightarrow B(x)}{A \rightarrow \forall xB(x)} \ (x \text{ not free in } A); \qquad \frac{A(x) \rightarrow B}{\exists xA(x) \rightarrow B} \ (x \text{ not free in } B)$$

Definition 5.2 The axiom schemata and rules of **QC3** and **QC3at** are those of **QC** together with LEM, resp. (LEMat): $P \vee {\sim}P$, for atomic formulas P.

Deducibility in **QC**, **QC3**, and **QC3at** and the consequence relations $\vdash_{\mathbf{QC}}$, $\vdash_{\mathbf{Q3C}}$, and $\vdash_{\mathbf{Q3Cat}}$ are defined in the usual way. In [38, Prop. 11] the axiomatic system **QC** is shown to be complete with respect to a suitable Kripke semantics by means of a faithful embedding into positive first-order intuitionistic logic.

Remark 5.3 Note that the embedding-based method is not available to us here since it is not clear into which system we can embed **QC3**. Thus, we leave this as an open problem.

Since the proof-theoretic aspect was not explored so far even for **QC**, we here focus on sequent calculi for **QC**, **QC3**, and **QC3at**.

5.2 Sequent calculi

In this section, we define cut-free sequent calculi for **QC**, **QC3**, and **QC3at**. The presentation is based on [22, §5.4] by Sara Negri and Jan von Plato, where the sequent calculus **G3i** for intuitionistic predicate logic (without equality) is extended by a rule, *Gem-at*, that captures LEMat. Whereas the addition of *Gem-at* to the sequent calculus **G3ip** for intuitionistic propositional logic results in a sequent system for classical propositional logic, the addition of *Gem-at* to **G3i** results in a proof system for an extension of classical propositional logic by the intuitionistic universal and particular quantifiers. Negri and von Plato [22, p. 121] remark that the proof of admissibility of excluded middle for arbitrary formulas for **G3ip** + *Gem-at* cannot be extended to quantified formulas. We shall therefore add an excluded middle rule for arbitrary formulas, *Gem*, to a sequent calculus **G3C** for **QC**. For the addition of *Gem-at* to **G3C** we prove the Existence Property and a Dual Existence Property.

We first present the sequent calculus **G3C** for **QC**. Uppercase Greek letters now stand for finite, possibly empty multisets of formulas, A, Γ stands for

$\{A\} \uplus \Gamma$, and Δ, Γ stands for $\Delta \uplus \Gamma$, where $\uplus$ is multiset union. Sequents are of the form $\Gamma \Rightarrow A$ ($\Rightarrow$ is used in this way, hereafter, not for strong implication).

Definition 5.4 The rules of the calculus **G3C** are the following:
Logical axioms:
$P, \Gamma \Rightarrow P \quad {\sim}P, \Gamma \Rightarrow {\sim}P$, for atomic formulas P
Logical rules:

$$\frac{A, B, \Gamma \Rightarrow C}{(A \wedge B), \Gamma \Rightarrow C}\ L\wedge \qquad \frac{\Gamma \Rightarrow A \quad \Gamma \Rightarrow B}{\Gamma \Rightarrow (A \wedge B)}\ R\wedge$$

$$\frac{A, \Gamma \Rightarrow C \quad B, \Gamma \Rightarrow C}{(A \vee B), \Gamma \Rightarrow C}\ L\vee \qquad \frac{\Gamma \Rightarrow A}{\Gamma \Rightarrow (A \vee B)}\ R\vee_1 \qquad \frac{\Gamma \Rightarrow B}{\Gamma \Rightarrow (A \vee B)}\ R\vee_2$$

$$\frac{(A \to B), \Gamma \Rightarrow A \quad B, \Gamma \Rightarrow C}{(A \to B), \Gamma \Rightarrow C}\ L\to \qquad \frac{A, \Gamma \Rightarrow B}{\Gamma \Rightarrow (A \to B)}\ R\to$$

$$\frac{A(t/x), \forall x A, \Gamma \Rightarrow B}{\forall x A, \Gamma \Rightarrow B}\ L\forall \qquad \frac{\Gamma \Rightarrow A(y/x)}{\Gamma \Rightarrow \forall x A}\ R\forall \qquad \frac{A(y/x), \Gamma \Rightarrow B}{\exists x A, \Gamma \Rightarrow B}\ L\exists$$

$$\frac{\Gamma \Rightarrow A(t/x)}{\Gamma \Rightarrow \exists x A}\ R\exists \qquad \frac{A, \Gamma \Rightarrow C}{{\sim}{\sim}A\Gamma \Rightarrow C}\ L{\sim}{\sim} \qquad \frac{\Gamma \Rightarrow A}{\Gamma \Rightarrow {\sim}{\sim}A}\ R{\sim}{\sim}$$

$$\frac{{\sim}A, {\sim}B, \Gamma \Rightarrow C}{{\sim}(A \vee B), \Gamma \Rightarrow C}\ L{\sim}\vee \qquad \frac{\Gamma \Rightarrow {\sim}A \quad \Gamma \Rightarrow {\sim}B}{\Gamma \Rightarrow {\sim}(A \vee B)}\ R{\sim}\vee$$

$$\frac{{\sim}A, \Gamma \Rightarrow C \quad {\sim}B, \Gamma \Rightarrow C}{{\sim}(A \wedge B), \Gamma \Rightarrow C}\ L{\sim}\wedge \qquad \frac{\Gamma \Rightarrow {\sim}A}{\Gamma \Rightarrow {\sim}(A \wedge B)}\ R{\sim}\wedge_1 \qquad \frac{\Gamma \Rightarrow {\sim}B}{\Gamma \Rightarrow {\sim}(A \wedge B)}\ R{\sim}\wedge_2$$

$$\frac{{\sim}(A \to B), \Gamma \Rightarrow A \quad {\sim}B, \Gamma \Rightarrow C}{{\sim}(A \to B), \Gamma \Rightarrow C}\ L{\sim}\to \qquad \frac{A, \Gamma \Rightarrow {\sim}B}{\Gamma \Rightarrow {\sim}(A \to B)}\ R{\sim}\to$$

$$\frac{{\sim}A(y/x), \Gamma \Rightarrow B}{{\sim}\forall x A, \Gamma \Rightarrow B}\ L{\sim}\forall \qquad \frac{\Gamma \Rightarrow {\sim}A(t/x)}{\Gamma \Rightarrow {\sim}\forall x A}\ R{\sim}\forall$$

$$\frac{{\sim}A(t/x), {\sim}\exists x A, \Gamma \Rightarrow B}{{\sim}\exists x A, \Gamma \Rightarrow B}\ L{\sim}\exists \qquad \frac{\Gamma \Rightarrow {\sim}A(y/x)}{\Gamma \Rightarrow {\sim}\exists x A}\ R{\sim}\exists$$

where (i) in $R\forall$ and in $R{\sim}\exists$, y must not occur free in $\Gamma, \forall x A$, resp. in $\Gamma, {\sim}\exists x A$ and (ii) in $L\exists$ and in $L{\sim}\forall$, y must not occur free in $\exists x A, \Gamma, B$, resp. in ${\sim}\forall x A, \Gamma, B$.

Definition 5.5 The rules of the calculus **G3C3**, respectively **G3C3at**, are those of **G3C** plus:

$$\frac{B, \Gamma \Rightarrow A \quad {\sim}B, \Gamma \Rightarrow A}{\Gamma \Rightarrow A}\ \mathit{Gem} \quad \text{resp.} \quad \frac{P, \Gamma \Rightarrow A \quad {\sim}P, \Gamma \Rightarrow A}{\Gamma \Rightarrow A}\ \mathit{Gem\text{-}at}$$

for atomic formulas P.

6 Some basic proof-theoretic results

As in [22, Lemmas 4.1 and 4.1.2], one can prove a formal version of the principle of renaming bound variables and a lemma showing that derivability of sequents is preserved with the same derivation height if a term t is substituted for a free variable x in a sequent $\Gamma \Rightarrow A$, provided that t is free for x in formulas from $\Gamma \Rightarrow A$. Also, one can easily show that for any formula A, sequents of the form $A, \Gamma \Rightarrow A$ are provable in **G3C** (and hence in **G3C3at** and **G3C3**). Moreover, the versions of the sequent rules $L\rightarrow$, $L{\sim}\rightarrow$, $L\forall$, and $L{\sim}\exists$ without repetitions of principal formulas are admissible in the calculi where they are present.

The proof of height-preserving admissibility of weakening and contraction

$$\frac{\Gamma \Rightarrow B}{\Gamma, A \Rightarrow B}\ Wk \qquad \frac{\Gamma, A \Rightarrow B}{\Gamma, A, A \Rightarrow B}\ Ctr$$

in [22] can be adapted to **G3C**, **G3C3at**, and **G3C3**. In particular, one has to show that the rules $L{\sim}\vee$, $L{\sim}\wedge$, $L{\sim}\forall$ are height-preserving invertible and that the rule $L{\sim}\rightarrow$ is height-preserving invertible for its second (right) premise.

Theorem 6.1 *The Cut rule*

$$\frac{\Gamma \Rightarrow A \quad A, \Delta \Rightarrow B}{\Gamma, \Delta \Rightarrow B}\ Cut$$

is an admissible rule of **G3C**, **G3C3at**, *and* **G3C3**.

Proof. The proof follows the standard pattern presented in [22]. In particular, if one of the premises of *Cut* is derived by an LEM rule, *Cut* with the same cut-formula is permuted upwards to applications of *Cut* with a smaller cut-height. The derivation

$$\frac{\Gamma \Rightarrow A \qquad \dfrac{B, A, \Delta \Rightarrow C \quad {\sim}B, A, \Delta \Rightarrow C}{A, \Delta \Rightarrow C}\ Gem}{\Gamma, \Delta \Rightarrow C}\ Cut$$

for example, is replaced by

$$\frac{\dfrac{\Gamma \Rightarrow A \quad B, A, \Delta \Rightarrow C}{B, \Gamma, \Delta \Rightarrow C}\ Cut \qquad \dfrac{\Gamma \Rightarrow A \quad {\sim}B, A, \Delta \Rightarrow C}{{\sim}B, \Gamma, \Delta \Rightarrow C}\ Cut}{\Gamma, \Delta \Rightarrow C}\ Gem$$

This completes the proof (sketch). □

To every finite set of formulas Γ, there corresponds a unique multiset (with no multiplicity of elements). If Γ is such a set, let $\bigwedge\Gamma$ be a conjunction of all formulas from the corresponding multiset. Conversely, to every finite multiset Γ, there corresponds a unique set, which we will also denote by Γ.

Theorem 6.2 (Equivalence of proof systems) *Let Γ be a finite set of formulas and let $\bigwedge\varnothing = (P \rightarrow P)$, for some fixed atomic formula P. Then $\Gamma \vdash_{\mathbf{QC3}} A$ iff $\Rightarrow \bigwedge\Gamma \rightarrow A$ is derivable in* **G3C3**.

Proof. Left-to-right: It is enough to show that $\Rightarrow A$ is derivable in **G3C3** for every theorem A of **QC3** and that the inference rules preserve derivability. We present two cases. For (LEM), we have

$$\frac{\dfrac{\dfrac{\vdots}{A \Rightarrow A}}{A \Rightarrow (A \vee {\sim}A)} \qquad \dfrac{\dfrac{\vdots}{{\sim}A \Rightarrow {\sim}A}}{{\sim}A \Rightarrow (A \vee {\sim}A)}}{\Rightarrow (A \vee {\sim}A)}$$

where the vertical dots indicate routine derivations. For the $\forall$-rule of **QC3** we have

$$\dfrac{\dfrac{\Rightarrow A \to B(x) \quad \overset{\vdots}{A, (A \to B(x)) \Rightarrow B(x)}}{A \Rightarrow B(x)}}{A \Rightarrow \forall x B(x)} R\forall$$

for x not free in A.
Right-to-left: By induction on the height of derivations in **G3C3** because $\Rightarrow \bigwedge\Gamma \to A$ is derivable in **G3C3** iff $\Gamma \Rightarrow A$ is. For axioms $P, \Delta \Rightarrow P$ and $\sim P, \Delta \Rightarrow \sim P$ we have $P \in \Delta \uplus \{P\}$, respectively $\sim P \in \Delta \uplus \{\sim P\}$. For the induction steps we consider the sequent rules. In the case of *Gem*, by the induction hypothesis we have $B, \Gamma \vdash_{\mathbf{QC3}} A$ and $\sim B, \Gamma \vdash_{\mathbf{QC3}} A$. By (LEM) and reasoning in positive intuitionistic propositional logic, as we can do in **QC3**, we obtain $\Gamma \vdash_{\mathbf{QC3}} A$. We present two more cases involving negation. Consider the rule $L{\sim}\forall$. By the induction hypothesis, we have $\sim A(y/x), \Gamma \vdash_{\mathbf{QC3}} B$, where y does not occur free in $\exists x A, \Gamma, B$. We easily obtain $\sim A(y/x) \vdash_{\mathbf{QC3}} \bigwedge\Gamma \to B$. By the $\exists$-rule of **QC3**, respecting its side-condition, we obtain $\exists x {\sim} A(x) \vdash_{\mathbf{QC3}} \bigwedge\Gamma \to B$. By axiom $\sim\forall x A(x) \leftrightarrow \exists x {\sim} A(x)$, we get $\sim\forall A(x) \vdash_{\mathbf{QC3}} \bigwedge\Gamma \to B$ and then $\sim\forall A(x), \Gamma \vdash_{\mathbf{QC3}} B$. Consider the rule $L{\sim}{\to}$. Since the version of the rule without repetition of the principal formula is admissible, we may assume by the induction hypothesis that $\Gamma \vdash_{\mathbf{QC3}} A$, and $\sim B, \Gamma \vdash_{\mathbf{QC3}} C$. Since $A, (A \to \sim B) \vdash_{\mathbf{QC3}} \sim B$, we successively obtain $(A \to \sim B), \Gamma \vdash_{\mathbf{QC3}} \sim B$ and $(A \to \sim B), \Gamma \vdash_{\mathbf{QC3}} C$. By (Ax12), we then get $\sim(A \to \sim B), \Gamma \vdash_{\mathbf{QC3}} C$. This completes the proof. □

Proposition 6.3 *The Disjunction Property and the Constructible Falsity Property fail for* **G3C3at**.

Proof. Both $\Rightarrow (P \vee \sim P)$ and $\Rightarrow \sim(P \wedge \sim P)$ are derivable in **G3C3at** for atomic formulas P. However, for no atomic formula P, *both* $\Rightarrow P$ and $\Rightarrow \sim P$ are derivable with the aid of *Gem-at.* □

Theorem 6.4 *The excluded middle rule Gem is admissible in* **G3C3at** *for arbitrary quantifier-free formulas.*

Proof. As in [22, Theorem 5.4.6], the proof is by induction on the length of a formula D. The rules shown to be admissible may be used, *Inv* indicates invertible rules, and *Ind* indicates applications of the induction hypothesis. For atomic formulas we have *Gem-at.*
D is a disjunction $(A \vee B)$. Apply the induction hypothesis to the following two derivations:

$$\dfrac{(A \vee B), \Gamma \Rightarrow C}{B, \Gamma \Rightarrow C} Inv$$

$$\dfrac{\dfrac{\dfrac{(A \vee B), \Gamma \Rightarrow C}{A, \Gamma \Rightarrow C} Inv}{A, \sim B, \Gamma \Rightarrow C} Wk \quad \dfrac{\dfrac{\overset{\vdots}{(\sim A \wedge \sim B) \Rightarrow \sim(A \vee B)} \quad \sim(A \vee B), \Gamma \Rightarrow C}{(\sim A \wedge \sim B), \Gamma \Rightarrow C} Cut}{\sim A, \sim B, \Gamma \Rightarrow C} Inv}{\sim B, \Gamma \Rightarrow C} Ind$$

D is a conjunction $(A \wedge B)$. Apply the induction hypothesis to the following two derivations:

$$\dfrac{\dfrac{(A \wedge B), \Gamma \Rightarrow C}{A, B, \Gamma \Rightarrow C}\, Inv \qquad \dfrac{\dfrac{\dfrac{\begin{matrix}\vdots \\ ({\sim}A \vee {\sim}B) \Rightarrow {\sim}(A \wedge B)\end{matrix} \quad {\sim}(A \wedge B), \Gamma \Rightarrow C}{({\sim}A \vee {\sim}B) \Rightarrow C}\, Cut}{{\sim}A, \Gamma \Rightarrow C}\, Inv}{{\sim}A, B, \Gamma \Rightarrow C}\, Wk}{B, \Gamma \Rightarrow C}$$

$$\dfrac{\dfrac{\begin{matrix}\vdots \\ ({\sim}A \vee {\sim}B) \Rightarrow {\sim}(A \wedge B)\end{matrix} \quad {\sim}(A \wedge B), \Gamma \Rightarrow C}{({\sim}A \vee {\sim}B), \Gamma \Rightarrow C}\, Cut}{{\sim}B, \Gamma \Rightarrow C}$$

D is an implication $(A \to B)$: Apply the induction hypothesis to the following two derivations

$$\dfrac{(A \to B), \Gamma \Rightarrow C}{B, \Gamma \Rightarrow C}\, Inv$$

$$\dfrac{\dfrac{\begin{matrix}\vdots \\ A, {\sim}B \Rightarrow {\sim}(A \to B)\end{matrix} \quad {\sim}(A \to B), \Gamma \Rightarrow C}{A, {\sim}B, \Gamma \Rightarrow C}\, Cut \qquad \dfrac{\dfrac{\dfrac{\begin{matrix}\vdots \\ (A \to {\sim}B) \Rightarrow {\sim}(A \to B)\end{matrix} \quad {\sim}(A \to B), \Gamma \Rightarrow C}{(A \to {\sim}B), \Gamma \Rightarrow C}\, Cut}{{\sim}B, \Gamma \Rightarrow C}\, Inv}{{\sim}A, {\sim}B, \Gamma \Rightarrow C}\, Wk}{{\sim}B, \Gamma \Rightarrow C}\, Ind$$

D is of the form ${\sim}(A \sharp B)$, $\sharp \in \{\vee, \wedge, \to\}$: Similar to the previous cases because $(A \sharp B), \Gamma \Rightarrow C$ is derivable from ${\sim}{\sim}(A \sharp B), \Gamma \Rightarrow C$.
D is a double negation ${\sim}{\sim}A$:

$$\dfrac{\dfrac{\dfrac{\begin{matrix}\vdots \\ A \Rightarrow A\end{matrix}}{A \Rightarrow {\sim}{\sim}A} \quad {\sim}{\sim}A, \Gamma \Rightarrow B}{A, \Gamma \Rightarrow B}\, Cut \qquad \dfrac{\dfrac{\begin{matrix}\vdots \\ {\sim}A \Rightarrow {\sim}A\end{matrix}}{{\sim}A \Rightarrow {\sim}{\sim}{\sim}A} \quad {\sim}{\sim}{\sim}A, \Gamma \Rightarrow A}{{\sim}A, \Gamma \Rightarrow B}\, Cut}{\Gamma \Rightarrow B}\, Ind$$

This completes the proof. □

Theorem 6.5 (Existence Properties) *If* $\Rightarrow \exists x A$ *is derivable in* **G3C3at**, *then so is* $\Rightarrow A(t/x)$ *for some term* t. *If* $\Rightarrow {\sim}\forall x A$ *is derivable in* **G3C3at**, *then so is* $\Rightarrow {\sim}A(t/x)$ *for some term* t.

Proof. We consider the first claim; the proof of the second claim is analogous. Suppose $\Rightarrow \exists x A$ is derivable in **G3C3at**. Then the last step in the derivation is either an application of $R\exists$, and we are done, or it is an application of *Gem-at*:

$$\dfrac{P \Rightarrow \exists x A \quad {\sim}P \Rightarrow \exists x A}{\Rightarrow \exists x A}$$

The same kind of case distinction applies to the derivations of $P \Rightarrow \exists A$ and $\sim P \Rightarrow \exists A$. Since every derivation is a finite tree, the derivation of $\Rightarrow \exists x A$ has the shape

$$\cfrac{\cfrac{\cfrac{\vdots}{\Delta_1 \Rightarrow A(t_1/x)}}{\Delta_1 \Rightarrow \exists x A}\,R\exists \quad \cdots \quad \cfrac{\cfrac{\vdots}{\Delta_n \Rightarrow A(t_n/x)}}{\Delta_n \Rightarrow \exists x A}\,R\exists}{\cfrac{P \Rightarrow \exists x A \quad \sim P \Rightarrow \exists x A}{\Rightarrow \exists x A}}$$

for some $n \in \mathbb{N}$ with $2 \leq n$, where every Δ_i $(1 \leq i \leq n)$ is a multiset of atomic formulas or negated atomic formulas that disappear in the derivation of $\Rightarrow \exists A$ by applications of *Gem-at* only. Suppose now for reductio that for every term t, the sequent $\Rightarrow A(t/x)$ is not derivable in **G3C3at**. Then for any term t, $(P \vee \sim P) \Rightarrow A(t/x)$ is not derivable because $\Rightarrow (P \vee \sim P)$ is derivable and *Cut* is admissible. Hence either $P \Rightarrow A(t/x)$ or $\sim P \Rightarrow A(t/x)$ is not derivable because otherwise $(P \vee \sim P) \Rightarrow A(t/x)$ is derivable by applying $L\vee$. Assume without loss of generality that $P \Rightarrow A(t/x)$ is not derivable. Then $P, (Q \vee \sim Q) \Rightarrow A(t/x)$ is not derivable for any atomic formula Q because $\Rightarrow (Q \vee \sim Q)$ is derivable and *Cut* is admissible. Therefore either $P, Q \Rightarrow A(t/x)$ or $P, \sim Q \Rightarrow A(t/x)$ is not derivable. Iterating this reasoning, we may conclude that for some i with $1 \leq i \leq n$, $\Delta_i \Rightarrow A(t_i/x)$ is not derivable, contrary to the assumption that we are considering a derivation of $\Rightarrow \exists x A$. □

Remark 6.6 It is known from work by Nobu-Yuki Suzuki [35] that the Disjunction Property and the Existence Property can come apart in intermediate predicate logics, in particular, that in general the Existence Property does not imply the Disjunction Property. The logic **QC3at** is an example of a naturally arising and independently motivated logic for which the Disjunction Property fails, whereas the Existence Property holds.

Remark 6.7 The proof of Theorem 6.5 uses classical logic in the metalanguage. In [36, p. 206 f.], Dirk van Dalen presents a constructive proof of the Existence Property for intuitionistic predicate logic (without identity). One may wonder whether a constructive proof of Theorem 6.5 is possible.

7 Concluding remarks

In this article, we introduced an extension of the connexive logic **C** from [38], with the following three questions as our motivations.

Q1 Can we improve John Cantwell's **CN**?

Q2 How much of the desiderata, listed by Estrada-González and Ramírez-Cámara, can be met by the approach to connexivity à la **C**?

Q3 What is the relation between constructivity and LEM?

Our answers to these questions, in view of the new extension **C3**, are as follows.

A1 Cantwell's classical conditional can be replaced by a constructive one.

A2 **C3** is a totally connexive logic with respect to the strong implication.

A3 LEM does not necessarily exclude properties that are usually regarded as indicating constructivity.

These answers give rise to additional questions such as:

Q1' Can we take other conditionals than the constructive conditional?
Q2' Which system is minimal among the family of totally connexive logics?
Q3' Are there any interesting variants of the Disjunction Property and the Existence Property, discussed in [34], that hold in **QC3** or related systems?

These questions, together with the open problems noted in Remarks 4.4, 4.9, 5.3 and 6.7 seem to show that there is a lot of room for further investigations. We hope some readers will be motivated to join the authors to continue with the development of connexive logics.

References

[1] Avigad, J., *Classical and constructive logic*, Available at `https://www.andrew.cmu.edu/user/avigad/Teaching/classical.pdf`(2000/09/19).

[2] Avron, A., *On an implication connective of RM*, Notre Dame Journal of Formal Logic **27** (1986), pp. 201–209.

[3] Avron, A., *Natural 3-valued logics–characterization and proof theory*, Journal of Symbolic Logic **56** (1991), pp. 276–294.

[4] Batens, D. and K. De Clercq, *A rich paraconsistent extension of full positive logic*, Logique et Analyse **185-188** (2004), pp. 227–257.

[5] Belnap, N. D., *Conditional assertion and restricted quantification*, Noûs **4** (1970), pp. 1–13.

[6] Cantwell, J., *The Logic of Conditional Negation*, Notre Dame Journal of Formal Logic **49** (2008), pp. 245–260.

[7] Cobreros, P., P. Egré, D. Ripley and R. van Rooij, *Tolerant, classical, strict*, Journal of Philosophical Logic **41** (2012), pp. 347–385.

[8] Cobreros, P., P. Égré, D. Ripley and R. Van Rooij, *Reaching transparent truth*, Mind **122** (2013), pp. 841–866.

[9] Egré, P. and G. Politzer, *On the negation of indicative conditionals*, in: M. F. M. Aloni and F. Roelofsen, editors, *Proceedings of the Amsterdam Colloquium*, 2013, pp. 10–18.

[10] Egré, P., L. Rossi and J. Sprenger, *De Finettian Logics of Indicative Conditionals. Part I: Trivalent Semantics and Validity*, Journal of Philosophical Logic (forthcoming).

[11] Egré, P., L. Rossi and J. Sprenger, *De Finettian Logics of Indicative Conditionals. Part II: Proof Theory and Algebraic Semantics*, Journal of Philosophical Logic (forthcoming).

[12] Estrada-González, L. and E. Ramírez-Cámara, *A comparison of connexive logics*, IfCoLog Journal of Logics and their Applications **3** (2016), pp. 341–355.

[13] Gilmore, P., "Logical Foundations for Mathematics And Computer Science," A.K. Peters, Wellesley, 2005.

[14] Girard, J., Y. Lafont and P. Taylor, "Proofs and Types," Cambridge University Press, Cambridge, 1989.

[15] Humberstone, L., *Contra-classical logics*, Australasian Journal of Philosophy **78** (2000), pp. 438–474.

[16] Kamide, N. and H. Wansing, "Proof Theory of N4-related Paraconsistent Logics," Studies in Logic, Vol. 54, College Publications, London, 2015.

[17] Kapsner, A., *Strong connexivity*, Thought **1** (2012), pp. 141–145.

[18] Kracht, M., *On extensions of intermediate logics by strong negation*, Journal of Philosophical Logic **27** (1998), pp. 49–73.

[19] Mares, E. and F. Paoli, *C.I. Lewis, E.J. Nelson, and the Modern Origins of Connexive Logic*, Organon F **26** (2019), pp. 405–426.

[20] McCall, S., *A history of connexivity*, in: *Handbook of the History of Logic, volume 11*, Elsevier, 2012 pp. 415–449.
[21] Mortensen, C., *Aristotle's Thesis in consistent and inconsistent logics*, Studia Logica **43** (1984), pp. 107–116.
[22] Negri, S. and J. von Plato, "Structural Proof Theory," Cambridge UP, Cambridge, 2001.
[23] Odintsov, S., "Constructive Negations and Paraconsistency," Trends in Logic **26**, Springer, 2008.
[24] Olkhovikov, G., *On a new three-valued paraconsistent logic (in Russian)*, in: *Logic of Law and Tolerance*, Ural State University Press, Yekaterinburg, 2001 pp. 96–113, English translation is available in *IfCoLog Journal of Logics and their Applications*, 3(3): 317–334, 2016.
[25] Omori, H., *A note on Francez' half-connexive formula*, IfCoLog Journal of Logics and their Applications **3** (2016), pp. 505–512.
[26] Omori, H., *A simple connexive extension of the basic relevant logic* **BD**, IfCoLog Journal of Logics and their Applications **3** (2016), pp. 467–478.
[27] Omori, H., *From paraconsistent logic to dialetheic logic*, in: H. Andreas and P. Verdée, editors, *Logical Studies of Paraconsistent Reasoning in Science and Mathematics*, Springer, 2016 pp. 111–134.
[28] Omori, H., *Towards a bridge over two approaches in connexive logic*, Logic and Logical Philosophy **28** (2019), pp. 553–566.
[29] Omori, H. and H. Wansing, *40 years of* **FDE***: An Introductory Overview*, Studia Logica **105** (2017), pp. 1021–1049.
[30] Omori, H. and H. Wansing, *On contra-classical variants of Nelson logic N4 and its classical extension*, The Review of Symbolic Logic **11** (2018), pp. 805–820.
[31] Priest, G., "An Introduction to Non-Classical Logic: From If to Is," Cambridge University Press, 2008, 2 edition.
[32] Rieger, A., *Conditionals are material: the positive arguments*, Synthese **190** (2013), pp. 3161–3174.
[33] Spinks, M. and R. Veroff, *Paraconsistent constructive logic with strong negation as a contraction-free relevant logic*, in: J. Czelakowski, editor, *Don Pigozzi on Abstract Algebraic Logic, Universal Algebra, and Computer Science*, Springer, 2018 pp. 323–379.
[34] Suzuki, N.-Y., *Some weak variants of the existence and disjunction properties in intermediate predicate logics*, Bulletin of the Section of Logic **46** (2017), pp. 93–109.
[35] Suzuki, N.-Y., *A negative solution to Ono's problem P52: Existence and disjunction properties in intermediate predicate logics*, in: N. Galatos and K. Terui, editors, *Hiroakira Ono on Residuated Lattices and Substructural Logics*, Springer, to appear .
[36] van Dalen, D., "Logic and Structure. Fourth Edition," Springer, Berlin, 2004.
[37] Wansing, H., *Negation*, in: L. Goble, editor, *The Blackwell Guide to Philosophical Logic*, Basil Blackwell Publishers, Cambridge/MA, 2001 pp. 415–436.
[38] Wansing, H., *Connexive modal logic*, in: R. Schmidt, I. Pratt-Hartmann, M. Reynolds and H. Wansing, editors, *Advances in Modal Logic. Volume 5*, King's College Publications, 2005 pp. 367–383.
[39] Wansing, H., *Connexive logic*, in: E. N. Zalta, editor, *The Stanford Encyclopedia of Philosophy*, https://plato.stanford.edu/archives/spr2020/entries/logic-connexive/, 2020, Spring 2020 edition.
[40] Wansing, H. and D. Skurt, *Negation as cancellation, connexive logic, and qLPm*, Australasian Journal of Logic **15** (2018), pp. 476–488.
[41] Wansing, H. and M. Unterhuber, *Connexive conditional logic. Part I*, Logic and Logical Philosophy **28** (2019), pp. 567–610.

Algorithmic properties of first-order modal logics of the natural number line in restricted languages

Mikhail Rybakov [1]

Institute for Information Transmission Problems, Russian Academy of Sciences, National Research University Higher School of Economics, Moscow, Russia, Tver State University, Tver, Russia

Dmitry Shkatov [2]

School of Computer Science and Applied Mathematics, University of the Witwatersrand, Johannesburg, South Africa

Abstract

We study algorithmic properties of first-order predicate monomodal logics of the frames $\langle \mathbb{N}, < \rangle$ and $\langle \mathbb{N}, \leqslant \rangle$ in languages with restrictions on the number of individual variables as well as the number and arity of predicate letters. The languages we consider have no constants, function symbols, or the equality symbol. We show that satisfiability for the logic of $\langle \mathbb{N}, < \rangle$ is Σ^1_1-hard in languages with two individual variables and two monadic predicate letters. We also show that satisfiability for the logic of $\langle \mathbb{N}, \leqslant \rangle$ is Σ^1_1-hard in languages with two individual variables, two monadic, and one 0-ary predicate letter. Thus, these logics are Π^1_1-hard, and therefore not recursively enumerable, in languages with the aforementioned restrictions. Similar results are obtained for the class of first-order predicate monomodal logics of frames $\langle \mathbb{N}, R \rangle$, where R is a binary relation between $<$ and $\leqslant$.

Keywords: first-order modal logic, predicate modal logic, restricted languages, decidability, undecidability, recursive enumerability, validity problem, satisfiability problem, Σ^1_1-hardness, Π^1_1-hardness, classification problem.

1 Introduction

The present paper aims to contribute to the understanding of the algorithmic properties of first-order predicate modal logics in languages with restrictions on

[1] m_rybakov@mail.ru

[2] shkatov@gmail.com

⋆ This work has been supported by the Russian Foundation for Basic Research, project 18-011-00869.

the number of individual variables, as well as the number and arity of predicate letters.

Interest in the algorithmic properties of non-classical, mostly modal and superintuitionistic (intermediate), predicate logics in restricted languages is a natural outgrowth of the extensive research into the Classical Decision Problem [5]. The study of the Classical Decision Problem aims, in light of undecidability [10] of the classical first-order predicate logic **QCl**, to identify maximal decidable and minimal undecidable fragments of **QCl**, i.e., the decidable fragments that become undecidable when slightly extended and the undecidable fragments that become decidable when slightly restricted. A similar effort has more recently been made to better understand the borderline between the decidable and the undecidable in predicate modal and superintuitionistic logics, mostly by looking at the fragments obtained by limiting the number of individual variables, as well as the number and arity of predicate letters, allowed in the construction of formulas [28], [31], [33], [34], [13], [3], [15], [47], [27], [41], [36].

In the present paper, we attempt to identify the minimal computationally hard fragments of the predicate monomodal logics of the frames $\langle \mathbb{N}, < \rangle$ and $\langle \mathbb{N}, \leqslant \rangle$, i.e., the natural numbers with a natural, respectively, strict and partial order. Interest in these logics is motivated by at least three considerations.

First, these logics are algorithmically quite hard: even thought the exact complexity seems to be unknown, they are, as follows from Lemmas 3.1 and 4.1 below, Π^1_1-hard. Most research into the algorithmic properties of non-classical predicate logics, as can be seen from the references above, has dealt with (un)decidability. While it is natural that (un)decidability is the main concern in the study of the Classical Decision Problem, it is to be expected that predicate modal logics are computationally harder than **QCl**; therefore, research into their algorithmic properties should involve identifying minimal, in the above sense, fragments that are hard in certain classes of the arithmetical, or the analytical, hierarchy. The only study to date, as far as we know, of algorithmic properties of the fragments of not recursively enumerable monomodal predicate logics has been the investigation [36] of the fragments of not recursively enumerable [43], [39, Lemma 3.3] monomodal predicate logics of finite Kripke frames (as discussed in [42], both the logics of finite frames and the logics considered here fall into the category of "awkward" predicate modal logics based on essentially second-order Kripke semantics).

Similar questions have, however, been studied in the context of richer predicate languages containing multiple modal operator—most recently by I. Hodkinson, F. Wolter, and M. Zakharyaschev [24], [47] (see also [14, Chapter 11]; for earlier work, see [2], [45], [46], [1], and [32]). The methods used in this paper are partially inspired by [47, Theorem 2.3], where a Σ^1_1-hard tiling problem is encoded using a predicate language with two modal operators, one corresponding to an atomic accessibility relation and the other to the reflexive transitive closure of that relation. A similar result [24, Theorem 2] has been obtained for the temporal predicate logic of $\langle \mathbb{N}, \leqslant \rangle$, i.e., a predicate logic with two modal operators, one for the "immediate successor" relation on $\mathbb{N}$, the other for its reflexive transitive closure, the partial order $\leqslant$ (of course, both of these can be expressed with a single binary temporal opertator "until"). The novelty of the present work lies, first, in obtaining a similar encoding for languages with a single unary modal operator and, second, similarly to [41] and [36], in further reductions to languages with only two monadic predicate letters—the encodings used in [24, Theorem 2] and [47, Theorem 2.3] require an unlimited supply of monadic predicate letters.

Second, the logics considered here are determined by linear frames, i.e., frames with a restriction on the branching factor in the sense that we cannot freely append to a world of a frame another frame without breaking the structure of the original frame. Modelling, in languages of such logics, of predicate letters with a limited number thereof presents certain difficulties: the methods used in [41] and [36]—which can be traced back to, and inherit the limitations of, the propositional-level techniques used in [20], [7], [38], [37] and [40]—are inapplicable in this setting. On the other hand, the methods used in [4] do not seem to be readily applicable to logics of transitive frames, such as $\langle \mathbb{N}, < \rangle$ and $\langle \mathbb{N}, \leqslant \rangle$. In this respect, the method used here should be of relevance in the study of the algorithmic properties of monomodal logics of various kinds of structures—such as reflexive and irreflexive trees with a limited branching factor—where similar restrictions apply.

Third, the structure $\langle \mathbb{N}, < \rangle$ has long been considered to be a natural model of the flow of time (see, e.g., [19], [17]), and so interest in the algorithmic properties of the predicate modal logics of this structure is partially motivated by applications of first-order temporal logics [9], [8], [17], [24], [25], [23], [14, Chapter 11], [29], [11], [21]. Clearly, the negative results, like those presented here, obtained for languages whose expressive power is weaker than those of predicate temporal logic are directly relevant to that area.

The paper is structured as follows. In Section 2, we introduce the necessary preliminaries on predicate modal logic. In Section 3, we present our results on the logic of $\langle \mathbb{N}, < \rangle$. In Section 4, we present similar results on the logic of $\langle \mathbb{N}, \leqslant \rangle$. We conclude by discussing problems for future research in Section 5.

2 Preliminaries

In this section, we recall the standard definitions related to predicate modal logic, our aim being mainly to fix the terminology and notation used throughout the paper; the reader wishing more background on predicate modal logic may consult [26], [12], [18], [6], and [16].

An unrestricted first-order predicate modal language—as considered in this paper—contains countably many individual variables; countably many predicate letters of every arity, including zero (0-ary predicate letters are propositional variables); the propositional constant $\bot$ (falsity), the binary propositional connective $\to$, the unary modal connective $\Box$, and the quantifier $\forall$. Formulas, as well as the symbols $\neg$, $\vee$, $\wedge$, $\leftrightarrow$, $\exists$, and $\Diamond$, are defined in the usual way. We also use the following abbreviations, where $n \in \mathbb{N}$:

$$\Box^0\varphi = \varphi, \quad \Box^{n+1}\varphi = \Box\Box^n\varphi, \quad \Diamond^n\varphi = \neg\Box^n\neg\varphi,$$
$$\Box^+\varphi = \varphi \wedge \Box\varphi, \quad \Diamond^+\varphi = \varphi \vee \Diamond\varphi.$$

When parentheses are omitted, $\neg$, $\Box$, $\forall$, and $\exists$ are assumed to bind tighter than $\wedge$ and $\vee$, which are assumed to bind tighter than $\to$ and $\leftrightarrow$. We usually write atomic formulas, or atoms, in prefix notation; for some predicate letters, however, we use infix.

A *normal predicate modal logic* is a set of predicate modal formulas containing the validities of the classical predicate logic **QCl**, as well as the formulas of the form $\Box(\varphi \to \psi) \to (\Box\varphi \to \Box\psi)$, and closed under predicate substitution, modus ponens, generalisation, and necessitation.[3]

We use the Kripke semantics to interpret predicate modal formulas.

A *Kripke frame* is a tuple $\mathfrak{F} = \langle W, R\rangle$, where W is a non-empty set of *worlds* and R is a binary *accessibility relation* on W. If wRv, we say that v *is accessible from* w or that w *sees* v. We say that v *is accessible from* w *in* k *steps*, for $k \geqslant 1$, if wR^kv, where R^k is the k-fold composition of R with itself.

A *predicate Kripke frame with expanding domains* is a tuple $\mathfrak{F}_D = \langle W, R, D\rangle$, where $\langle W, R\rangle$ is a Kripke frame and D is a function from W into the set of non-empty subsets of some set, *the domain of* $\mathfrak{F}_D$; the function D is required to satisfy the condition that wRw' implies $D(w) \subseteq D(w')$. We call the set $D(w)$ *the domain of* w. We often write D_w for $D(w)$. We also consider predicate frames satisfying the stronger condition that wRw' implies $D(w) = D(w')$; we call such frames *predicate frames with (locally) constant domains*. Whenever we say *predicate frame* simpliciter, we mean predicate frame with expanding domains.

A *Kripke model* is a tuple $\mathfrak{M} = \langle W, R, D, I\rangle$, where $\langle W, R, D\rangle$ is a predicate Kripke frame and I, called *the interpretation of predicate letters* with respect to worlds in W, is a function assigning to a world $w \in W$ and an n-ary predicate letter P an n-ary relation $I(w, P)$ on $D(w)$, i.e., $I(w, P) \subseteq D^n(w)$; in particular, if P is 0-ary, $I(w, P) \subseteq D^0(w) = \{\langle\rangle\}$. We often write $P^{I,w}$ for $I(w, P)$. We say that a model $\langle W, R, D, I\rangle$ is *based on* the frame $\langle W, R\rangle$ and is *based on* the predicate frame $\langle W, R, D\rangle$.

An *assignment* in a model is a function g associating with every individual variable x an element $g(x)$ of the domain of the underlying predicate frame. We write $g' \stackrel{x}{=} g$ to mean that assignment g' differs from assignment g in at most the value of x.

The truth of a formula φ at a world w of a model $\mathfrak{M}$ under an assignment g is defined inductively:

- $\mathfrak{M}, w \models^g P(x_1, \ldots, x_n)$ if $\langle g(x_1), \ldots, g(x_n)\rangle \in P^{I,w}$, where P is an n-ary predicate letter;

[3] The reader wishing a reminder of the definition of these closure conditions may consult [16, Definition 2.6.1]; for a detailed discussion of predicate substitution, see, e.g., [16, §2.3, §2.5].

- $\mathfrak{M}, w \not\models^g \bot$;
- $\mathfrak{M}, w \models^g \varphi_1 \to \varphi_2$ if $\mathfrak{M}, w \models^g \varphi_1$ implies $\mathfrak{M}, w \models^g \varphi_2$;
- $\mathfrak{M}, w \models^g \Box\varphi_1$ if wRw' implies $\mathfrak{M}, w' \models^g \varphi_1$;
- $\mathfrak{M}, w \models^g \forall x\, \varphi_1$ if $\mathfrak{M}, w \models^{g'} \varphi_1$, for every g' such that $g' \stackrel{x}{=} g$ and $g'(x) \in D_w$.

Notice that, given a Kripke model $\mathfrak{M} = \langle W, R, D, I\rangle$ and $w \in W$, the tuple $\mathfrak{M}_w = \langle D_w, I_w\rangle$, where $I_w(P) = I(w, P)$, is a classical predicate model.

Let $\mathfrak{M} = \langle W, R, D, I\rangle$ be a model, $w \in W$, and $a_1, \ldots, a_n \in D_w$; let also $\varphi(x_1, \ldots, x_n)$ be a formula whose free variables are among $x_1, \ldots, x_n$. We write $\mathfrak{M}, w \models \varphi(a_1, \ldots, a_n)$ to mean $\mathfrak{M}, w \models^g \varphi(x_1, \ldots, x_n)$, where $g(x_1) = a_1, \ldots, g(x_n) = a_n$. This notation is unambiguous since the languages we consider lack constants.

We say that a formula φ is *true at a world* w of a model $\mathfrak{M}$ (in symbols, $\mathfrak{M}, w \models \varphi$) if $\mathfrak{M}, w \models^g \varphi$, for every g assigning to the free variables of φ elements of D_w. We say that φ is *true in a model* $\mathfrak{M}$ (in symbols, $\mathfrak{M} \models \varphi$) if $\mathfrak{M}, w \models \varphi$, for every world w of $\mathfrak{M}$. We say that φ is *valid on a predicate frame* $\mathfrak{F}_D$ (in symbols, $\mathfrak{F}_D \models \varphi$) if φ is true in every model based on $\mathfrak{F}_D$. We say that φ is *valid on a frame* $\mathfrak{F}$ (in symbols, $\mathfrak{F} \models \varphi$) if φ is valid on every predicate frame $\langle \mathfrak{F}, D\rangle$. These notions, and the corresponding notation, can be extended to sets of formulas, in a natural way.

We write $w \models \varphi$, rather than $\mathfrak{M}, w \models \varphi$, when $\mathfrak{M}$ is clear from the context.

It is well known that the set of formulas valid on a class of frames is a normal predicate modal logic; this fact is sometimes referred to as soundness of Kripke semantics.

In this paper, we are mostly interested in the logics of frames $\langle \mathbb{N}, <\rangle$ and $\langle \mathbb{N}, \leqslant\rangle$; these logics are denoted, respectively, $L(\mathbb{N}, <)$ and $L(\mathbb{N}, \leqslant)$.

Observe that $L(\mathbb{N}, <) \not\subseteq L(\mathbb{N}, \leqslant)$ since $(\mathbb{N}, <) \models Z$ and $(\mathbb{N}, \leqslant) \not\models Z$, where $Z = \Box(\Box p \to p) \to (\Diamond\Box p \to \Box p)$. Also observe that $L(\mathbb{N}, \leqslant) \not\subseteq L(\mathbb{N}, <)$ since $(\mathbb{N}, \leqslant) \models \Box p \to p$ and $(\mathbb{N}, <) \not\models \Box p \to p$.

3 The first-order logic of $\langle \mathbb{N}, <\rangle$

In this section, we prove that satisfiability for the logic $L(\langle \mathbb{N}, <\rangle)$ is Σ^1_1-hard—hence, $L(\langle \mathbb{N}, <\rangle)$ is Π^1_1-hard, and therefore not recursively enumerable—in languages with two individual variables and two monadic predicate letters.

We do so by encoding the following recurrent tiling problem for $\mathbb{N} \times \mathbb{N}$, known to be Σ^1_1-complete [22].

We are given a set of tiles, a tile t being a 1×1 square, with a fixed orientation, whose edges are "colored" with *left*(t), *right*(t), *up*(t), and *down*(t). A tile type is fully determined by the edge colors. Every tile belongs to one of the finitely many types $T = \{t_0, \ldots, t_n\}$, there being an unlimited supply of tiles of each type. A *tiling* in an arrangement of tiles such that the edge colors of the adjacent tiles match, both horizontally and vertically. We are to determine whether there exists a tiling of an $\mathbb{N} \times \mathbb{N}$ grid, with tiles of the given types, such that a tile of type t_0 occurs infinitely often in the leftmost column. More precisely, we are to determine whether there exists a function $f : \mathbb{N} \times \mathbb{N} \to T$ such that, for every $n, m \in \mathbb{N}$,

(T_1) $\mathit{right}(f(n,m)) = \mathit{left}(f(n+1,m))$;

(T_2) $\mathit{up}(f(n,m)) = \mathit{down}(f(n,m+1))$;

(T_3) the set $\{n \in \mathbb{N} : f(0,n) = t_0\}$ is infinite.

The idea of the encoding we use is based on [14, Theorem 11.1] (see also [44], [30], [47], and [27]). To make the underlying idea clearer, we initially encode the recurrent tiling problem with predicate modal formulas of two individual variables, without regard for the number of predicate letters used; such a concern would complicate the formulas and, possibly, obfuscate their meaning. Subsequently, we eliminate all but two monadic predicate letters in the formulas obtained in the initial encoding.

Let $\lhd$ be a binary—while M and P_t, for every $t \in T$, monadic—predicate letters. Let (for brevity, we write l, r, u, and d rather than *left*, *right*, *up*, and *down*)

$$
\begin{aligned}
A_1 &= \forall x \exists y\, (x \lhd y);\\
A_2 &= \forall x \forall y\, [(x \lhd y \to \Box(x \lhd y)) \wedge (\neg(x \lhd y) \to \Box\neg(x \lhd y))];\\
A_3 &= \exists x\, M(x);\\
A_4 &= \forall x \forall y\, (x \lhd y \to \Box^+(M(x) \leftrightarrow \Diamond M(y) \wedge \neg\Diamond^2 M(y)));\\
A_5 &= \Box^+ \forall x\, [\bigvee_{t \in T} P_t(x) \wedge \bigwedge_{t' \neq t} (P_t(x) \to \neg P_{t'}(x))];\\
A_6 &= \forall x \forall y \bigwedge_{t \in T} [\Box^+(x \lhd y \wedge P_t(x) \to \bigvee_{r(t)=l(t')} P_{t'}(y))];\\
A_7 &= \forall x \forall y \bigwedge_{t \in T} [\Box^+(M(x) \wedge P_t(y) \to \Box(\exists y\, (x \lhd y \wedge M(y)) \to \bigvee_{u(t)=d(t')} P_{t'}(y)))];\\
A_8 &= \forall x\, (M(x) \to \Box\Diamond P_{t_0}(x)),
\end{aligned}
$$

Let A be a conjunction of formulas A_1 through A_8. Notice that A contains only two individual variables.

One may think of the relation represented by $x \lhd y$ as an "immediate successor" relation associated with a strict partial order. Then, A_2 says that this "order" is preserved throughout the frame. One may think of an element a of the domain D_w of the world w such that $w \models M(a)$ as "marking" the world w; so, we occasionally say that a is the mark of w. Then, formulas A_3 and A_4 can be understood as saying that every world in a model is "marked" with, as we shall see, a unique element of its domain and that the order of marks of successive worlds agrees with the relation $\lhd$. This, as we shall see, gives us an $\mathbb{N} \times \mathbb{N}$ grid whose rows correspond to the worlds of the frame $\langle \mathbb{N}, < \rangle$ and whose columns correspond to the (common) elements of the domains of the worlds. Building on this, formulas A_5 through A_8 describe a sought tiling of thus obtained grid. This is made precise in the following

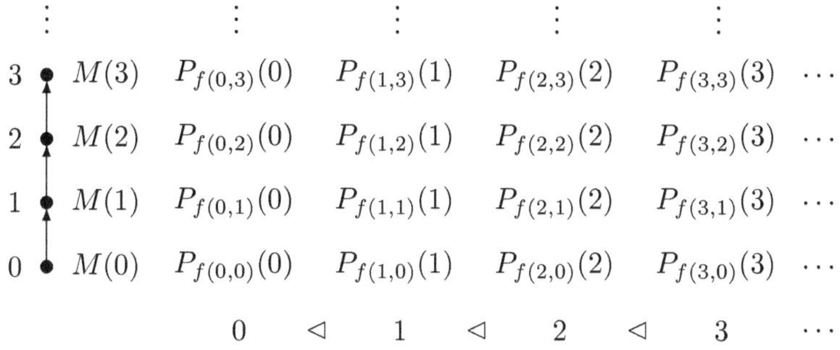

Fig. 1. Model $\mathfrak{M}_0$

Lemma 3.1 *There exists a recurrent tiling of* $\mathbb{N} \times \mathbb{N}$ *if, and only if,* $\langle \mathbb{N}, < \rangle \not\models \neg A$.

Proof. ("if") Suppose that $\mathfrak{M}, m \models A$, for some model $\mathfrak{M} = \langle \mathbb{N}, <, D, I \rangle$ and some $m \in \mathbb{N}$; we may assume without loss of generality that $m = 0$.

Since $0 \models A_3$, there exists $a_0 \in D_0$ such that $0 \models M(a_0)$.

Since $0 \models A_1$, there exists an infinite sequence $a_0, a_1, a_2, \dots$ of elements of D_0 such that $a_0 \lhd^{I,0} a_1 \lhd^{I,0} a_2 \lhd^{I,0} \dots$.

Since $0 \models A_2$, clearly, $a_0 \lhd^{I,n} a_1 \lhd^{I,n} a_2 \lhd^{I,n} \dots$, for every $n \in \mathbb{N}$.

Since $0 \models A_4$, clearly, $n \models M(a_n)$, for every $n \in \mathbb{N}$.

We next show that $a_0, a_1, a_2, \dots$ are pairwise distinct.

Suppose otherwise, i.e., let $a_i = a_{i+k}$, for some $i, k \in \mathbb{N}$. Then, as we have seen, $i \models M(a_i)$ and $i + k \models M(a_{i+k})$. Since $a_i = a_{i+k}$, we obtain $i + k \models M(a_i)$ and hence, by A_4, $i + k + 1 \models M(a_{i+1})$. Thus, $i \not\models M(a_i) \leftrightarrow \Diamond M(a_{i+1}) \wedge \neg \Diamond^2 M(a_{i+1})$, a contradiction.

Therefore, $w \models M(a_k)$ if, and only if, $w = k$.

Since $0 \models A_5$, for every $m, n \in \mathbb{N}$, there exists a unique $t \in T$ such that $m \models P_t(a_n)$. We can, therefore, define a function $f : \mathbb{N} \times \mathbb{N} \to T$ by

$$f(n, m) = t, \text{ where } t \text{ is such that } m \models P_t(a_n).$$

Since $0 \models A_6 \wedge A_7 \wedge A_8$, the function f satisfies (T_1) through (T_3). Observe that the subformula $\exists y\, (x \lhd y \wedge M(y))$ of A_7 ensures that a vertically matching tile t' is placed right on top of the current tile t.

Therefore, f is a recurrent tiling of $\mathbb{N} \times \mathbb{N}$ with T.

("only if") Suppose that f is a function satisfying (T_1) through (T_3). We define a model $\mathfrak{M}_0$, based on $\langle \mathbb{N}, < \rangle$, satisfying A.

To define $\mathfrak{M}_0$, let $D_n = \mathbb{N}$, for every $n \in \mathbb{N}$, and let I be an interpretation function such that, for every $n \in \mathbb{N}$,

- $n \models k \lhd l \;\leftrightharpoons\; l = k + 1$;
- $n \models M(k) \;\leftrightharpoons\; k = n$;
- $n \models P_t(k) \;\leftrightharpoons\; f(k, n) = t$.

Finally, let $\mathfrak{M}_0 = \langle \mathbb{N}, <, D, I \rangle$ (see Figure 1).

It is straightforward to check that $\mathfrak{M}_0, 0 \models A$. □

Thus, in the proof of the "if" part of Lemma 3.1, we obtained a grid for the tiling by treating the worlds of the model $\mathfrak{M}$ as rows and elements $a_0, a_1, a_2, \ldots$ of the domain D_0 of world 0 as columns.

We now make, in the following remarks, a few observations about those properties of the model $\mathfrak{M}_0$ defined in the proof of the "only if" part of Lemma 3.1 that we will rely on later on.

Remark 3.2 The model $\mathfrak{M}_0$ defined in the "only if" part of the proof of Lemma 3.1 is based on a predicate frame with a constant domain; even though this domain is $\mathbb{N}$, we denote it by $\mathcal{D}$ when we wish to emphasize that we are talking about the domain, rather than the set of worlds, of $\mathfrak{M}_0$.

Remark 3.3 In the model $\mathfrak{M}_0$ defined in "only if" part of the proof Lemma 3.1, the valuation of the binary predicate letter $\lhd$ is the same at every world.

Remark 3.4 In the model $\mathfrak{M}_0$ defined in "only if" part of the proof of Lemma 3.1, each world is marked with a unique element of the common domain, i.e., for every $w \in \mathbb{N}$, there exists a unique $a \in \mathcal{D}$ such that $w \models M(a)$.

We next eliminate, in a satisfiability-preserving way, all but two monadic predicate letters of the formula A, without increasing the number of individual variables in the resultant formula; we, thus, obtain a reduction of the $\mathbb{N} \times \mathbb{N}$ recurrent tiling problem to satisfiability in $L(\langle \mathbb{N}, < \rangle)$ in languages with two individual variables and two monadic predicate letters.

The elimination of predicate letters is carried out in two steps: first, we model the binary letter $\lhd$ with two monadic ones, obtaining formula A'; then, we model all the monadic letters of A' except M with a single monadic letter, thus obtaining a formula with only two monadic predicate letters and two individual variables.

From now on, we assume, for ease of notation, that A contains monadic predicate letters $P_0, \ldots, P_n$—rather than P_t, for $t \in \{t_0, \ldots, t_n\}$—to refer to the tile types.

First, following ideas of Kripke's [28], we eliminate, in a satisfiability-preserving way, the binary predicate letter $\lhd$ of A, without increasing the number of individual variables in the resultant formula.

Recall that Kripke's construction [28] transforms a model $\mathfrak{M}$ satisfying a formula containing a binary predicate letter, and no modal operators, at a world w in such a way that a sufficiently large number of worlds is added to $\mathfrak{M}$. More precisely, for every pair $\langle a, b \rangle$ of elements of the domain of w, a fresh world is introduced to $\mathfrak{M}$. This construction cannot be applied in a straightforward way in our setting, for two reasons.

Since we are restricted to the frame $\langle \mathbb{N}, < \rangle$, we may not introduce fresh worlds to a model satisfying A; we, rather, have to use the worlds of $\langle \mathbb{N}, < \rangle$ to simulate $\lhd$. Moreover, since $\lhd$ occurs within the scope of the modal operator $\Box$

in A, we need to simulate the valuation of $\lhd$ at every world of the model, not just at the world satisfying A.

We resolve these difficulties by using the fact that A is satisfied in the model $\mathfrak{M}_0$ defined in the "only if" part of the proof of Lemma 3.1 and drawing on the special properties of $\mathfrak{M}_0$—that, as noted in Remark 3.2, it is based on a predicate frame with a constant domain and that, as noted in Remark 3.3, the valuation of $\lhd$ is the same at every world of $\mathfrak{M}_0$.

Let P_{n+1} and P_{n+2} be monadic predicate letters distinct from $M, P_0, \dots, P_n$ and from each other, and let

$$\mu = \exists x\, M(x).$$

Lastly, let A' be the result of substituting $\Diamond(\mu \wedge P_{n+1}(x) \wedge P_{n+2}(y))$ for $x \lhd y$ in A.

Lemma 3.5 *There exists a recurrent tiling of* $\mathbb{N} \times \mathbb{N}$ *if, and only if,* $\langle \mathbb{N}, < \rangle \not\models \neg A'$.

Proof. ("if") This part is argued almost exactly as in the proof of Lemma 3.1, the only difference being that $\Diamond(\mu \wedge P_{n+1}(x) \wedge P_{n+2}(y))$ plays the role of the atom $x \lhd y$.

("only if") Suppose f is a function satisfying conditions (T_1) through (T_3). Let $\mathfrak{M}_0$ be the model defined in "if" part of the proof of Lemma 3.1. As we have seen there, $\mathfrak{M}_0, 0 \models A$. We use $\mathfrak{M}_0$ to define a model $\mathfrak{M}'_0$ satisfying A'.

Let $h : \mathbb{N} \to \mathbb{N} \times \mathbb{N}$ be a fixed enumeration of the pairs of natural numbers, thought of as elements of the domain $\mathcal{D}$ (i.e., we seek an enumeration of $\mathcal{D} \times \mathcal{D}$). Let α be the infinite sequence of natural numbers

$$0,\ 0,1,\ 0,1,2,\ 0,1,2,3,\ 0,1,2,3,4,\ \dots$$

and let α_k be the kth element of α.

To define $\mathfrak{M}'_0$, we use the predicate frame $\langle \mathbb{N}, <, D \rangle$ underlying the model $\mathfrak{M}_0$, together with the interpretation function I' defined as follows: for $w, a, b, c \in \mathbb{N}$,

$$\mathfrak{M}'_0, w \models P_{n+1}(c) \ \leftrightharpoons\ c = a \text{ and } \mathfrak{M}_0, 0 \models a \lhd b \text{ and } h(\alpha_w) = \langle a, b \rangle;$$

$$\mathfrak{M}'_0, w \models P_{n+2}(c) \ \leftrightharpoons\ c = b \text{ and } \mathfrak{M}_0, 0 \models a \lhd b \text{ and } h(\alpha_w) = \langle a, b \rangle;$$

and

$$\mathfrak{M}'_0, w \models S(c) \leftrightharpoons \mathfrak{M}_0, w \models S(c), \quad \text{for } S \in \{P_0, \dots, P_n, M\}.$$

Finally, let $\mathfrak{M}'_0 = \langle \mathbb{N}, <, D, I' \rangle$.

We prove that $\mathfrak{M}'_0, 0 \models A'$.

Since $\mathfrak{M}_0, 0 \models A$, if suffices to show that $\mathfrak{M}_0, w \models^g x \lhd y$ if, and only if, $\mathfrak{M}'_0, w \models^g \Diamond(\mu \wedge P_{n+1}(x) \wedge P_{n+2}(y))$, for every $w \in \mathbb{N}$ and every g.

Assume $\mathfrak{M}_0, w \models^g x \lhd y$. By definition of $\mathfrak{M}_0$ (see also Remark 3.3), $\mathfrak{M}_0, 0 \models^g x \lhd y$. Let $v \in \mathbb{N}$ be such that $w < v$ and $h(\alpha_v) = \langle g(x), g(y) \rangle$; it is evident from the definition of α that such a v exists. By definition,

$\mathfrak{M}'_0, v \models^g P_{n+1}(x) \wedge P_{n+2}(y)$. Since $w < v$ and since, as can be easily checked, $\mathfrak{M}'_0, w \models \mu$, we obtain $\mathfrak{M}'_0, w \models^g \Diamond(\mu \wedge P_{n+1}(x) \wedge P_{n+2}(y))$.

Conversely, assume $\mathfrak{M}'_0, w \models^g \Diamond(\mu \wedge P_{n+1}(x) \wedge P_{n+2}(y))$, and hence $\mathfrak{M}'_0, v \models^g P_{n+1}(x) \wedge P_{n+2}(y)$, for some v such that $w < v$. By definition, $\mathfrak{M}_0, 0 \models^g x \lhd y$. Thus, by definition of $\mathfrak{M}_0$ (see also Remark 3.3), $\mathfrak{M}_0, w \models^g x \lhd y$. □

We lastly model, in a satisfiability-preserving way, the occurrences of predicate letters $P_0, \ldots, P_{n+2}$ in A' with a single monadic letter P, without increasing the number of individual variables in the resultant formula. We, thus, obtain a reduction of the recurrent tiling problem using formulas with only two individual variables and only two monadic predicate letters, M and P.

Let P be a monadic predicate letter distinct from $P_0, \ldots, P_{n+2}, M$, and let, for $k \in \{0, \ldots, n+2\}$,

$$\begin{aligned}\beta_k(x) &= \mu \wedge \exists y\, [\Diamond^{n+4} M(y) \wedge \neg\Diamond^{n+5} M(y) \wedge \\ &\qquad \Diamond(\Diamond^{k+1} M(y) \wedge \neg\Diamond^{k+2} M(y) \wedge P(x))];\\ \beta_k(y) &= \mu \wedge \exists x\, [\Diamond^{n+4} M(x) \wedge \neg\Diamond^{n+5} M(x) \wedge \\ &\qquad \Diamond(\Diamond^{k+1} M(x) \wedge \neg\Diamond^{k+2} M(x) \wedge P(y))].\end{aligned}$$

Let $\cdot^*$ be the function replacing $P_k(x)$ with $\beta_k(x)$ and $P_k(y)$ with $\beta_k(y)$, for $k \in \{0, \ldots, n+2\}$.

Let A_i^*, where $1 \leqslant i \leqslant 8$ and $i \neq 4$, be the result of applying the function $\cdot^*$ to the formula A'_i and let

$$\begin{aligned}A_4^{\#} &= \forall x \forall y\, (\Diamond(\beta_{n+1}(x) \wedge \beta_{n+2}(y)) \rightarrow \\ &\qquad \Box(M(x) \leftrightarrow \Diamond^{n+4} M(y) \wedge \neg\Diamond^{n+5} M(y))).\end{aligned}$$

Finally, let

$$A^* = A_1^* \wedge A_2^* \wedge A_3^* \wedge A_4^{\#} \wedge A_5^* \wedge A_6^* \wedge A_7^* \wedge A_8^*.$$

To define a model satisfying A^*, provided a recurrent tiling of $\mathbb{N} \times \mathbb{N}$ exists, we take the model $\mathfrak{M}'_0$ defined in the "only if" part of the proof of Lemma 3.5 and, intuitively, stretch it out to include "additional" worlds whose sole purpose is to simulate the valuation of the predicate letters $P_0, \ldots, P_{n+2}$ at worlds of $\mathfrak{M}'_0$: $n+3$ worlds are "inserted" between m and $m+1$ to simulate the valuation of letters $P_0, \ldots, P_{n+2}$ at m. The valuation of P_k, where $k \in \{0, \ldots, n+2\}$, at m is simulated by the valuation of P at a "newly inserted" intermediate world k steps away from $m+1$ (see Figure 2, where $\beta_{f(a,b)}(x)$ stands for $\beta_k(x)$ such that $f(a,b) = t_k$). This is made precise in the following

Lemma 3.6 *There exists a recurrent tiling of* $\mathbb{N} \times \mathbb{N}$ *if, and only if,* $\langle \mathbb{N}, < \rangle \not\models \neg A^*$.

Proof. ("if") This part is argued as before, the only difference being that $\beta_k(x)$ and $\beta_k(y)$ are used instead of the atoms $P_k(x)$ and $P_k(y)$.

("only if") Suppose f is a function satisfying (T_1) through (T_3). Let $\mathfrak{M}'_0$ be the model defined in the "only if" part of the proof of Lemma 3.5. As we have seen there, $\mathfrak{M}'_0, 0 \models A'$. We use $\mathfrak{M}'_0$ to define a model $\mathfrak{M}^*_0$ satisfying A^*.

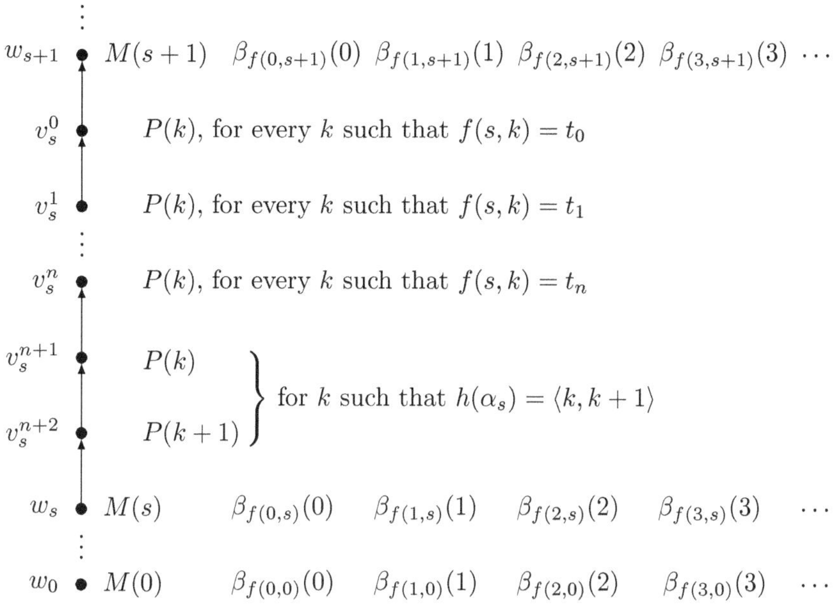

Fig. 2. Model $\mathfrak{M}_0^*$

Think of the worlds of $\mathfrak{M}_0^*$ as being labeled, in the ascending order,

$$w_0, v_0^{n+2}, \ldots, v_0^0, w_1, v_1^{n+2}, \ldots, v_1^0, w_2, v_2^{n+2}, \ldots, v_2^0, w_3, \ldots$$

(i.e., $w_0 = 0$, $v_0^{n+2} = 1$, etc.). Let, as before, $D_w = \mathcal{D} = \mathbb{N}$, for every $w \in \mathbb{N}$. Define the interpretation function I^* on the predicate frame $\langle \mathbb{N}, <, D\rangle$ underlying $\mathfrak{M}_0'$ by

$$\mathfrak{M}_0^*, x \models M(a) \leftrightharpoons x = w_m \text{ and } \mathfrak{M}_0', m \models M(a), \text{ for some } m \in \mathbb{N};$$

$$\mathfrak{M}_0^*, x \models P(a) \leftrightharpoons x = v_m^k \text{ and } \mathfrak{M}_0', m \models P_k(a), \text{ for some } m \in \mathbb{N} \text{ and } k \in \{0, \ldots, n+2\}.$$

We prove that $\mathfrak{M}_0^*, w_0 \models A^*$.

First, we show that, for every $s \in \mathbb{N}$, $k \in \{0, \ldots, n+2\}$, and g,

$$(1)\ \mathfrak{M}_0', s \models^g P_k(x) \iff \mathfrak{M}_0^*, w_s \models^g \beta_k(x);$$

$$(2)\ \mathfrak{M}_0', s \models^g P_k(y) \iff \mathfrak{M}_0^*, w_s \models^g \beta_k(y).$$

Assume $\mathfrak{M}_0', s \models^g P_k(x)$. As we have seen in the proof of Lemma 3.1 (see also Remark 3.4), for every world w in $\mathfrak{M}_0$, there exists a unique $a \in \mathcal{D}$ such that $\mathfrak{M}_0, w \models M(a)$.

Then, $\mathfrak{M}_0', s \models M(a)$, for some unique $a \in \mathcal{D}$; hence, by definition, $\mathfrak{M}_0^*, w_s \models M(a)$. Therefore, $\mathfrak{M}_0^*, w_s \models \mu$.

Similarly, $\mathfrak{M}'_0, s+1 \models M(b)$, for some unique $b \in \mathcal{D}$, and so, by definition, $\mathfrak{M}^*_0, w_{s+1} \models M(b)$. Observe that, due to uniqueness of b for $s+1$, if $t \neq s+1$, then $\mathfrak{M}^*_0, w_t \not\models M(b)$.

By definition, in $\mathfrak{M}^*_0$,

- w_{s+1} is accessible from w_s in $n+4$ steps;
- w_{s+1} is not accessible from w_s in $n+5$ steps;
- $w_s < v^k_s$;
- w_{s+1} is accessible from v^k_s in $k+1$ steps;
- w_{s+1} is not accessible from v^k_s in $k+2$ steps;
- $\mathfrak{M}^*_0, v^k_s \models^g P(x)$.

Therefore, $\mathfrak{M}^*_0, w_s \models^g \beta_k(x)$.

Conversely, assume $\mathfrak{M}^*_0, w_s \models^g \beta_k(x)$. Then, it is immediate from the definition of $\mathfrak{M}^*_0$ that $\mathfrak{M}'_0, s \models^g P_k(x)$.

This proves (1). The argument for (2) is analogous.

From (1) and (2) we immediately obtain $\mathfrak{M}^*_0, w_0 \models A^*_i$, where $1 \leqslant i \leqslant 8$ and $i \neq 4$. It is, moreover, straightforward to check, given (1) and (2), that $\mathfrak{M}^*_0, w_0 \models A^\#_4$. Therefore, $\mathfrak{M}^*_0, w_0 \models A^*$. □

We, thus, obtain (the reader wishing a reminder of the basic concepts of computability theory may consult [35])

Theorem 3.7 *Satisfiability for $L(\langle \mathbb{N}, < \rangle)$ is Σ^1_1-hard in languages with two individual variables and two monadic predicate letters.*

Proof. Immediate from Lemma 3.6. □

Thus, $L(\langle \mathbb{N}, < \rangle)$ is not recursively enumerable in such languages:

Theorem 3.8 *The logic $L(\langle \mathbb{N}, < \rangle)$ is Π^1_1-hard in languages with two individual variables and two monadic predicate letters.*

Proof. Immediate from Theorem 3.7. □

4 The first-order logic of $\langle \mathbb{N}, \leqslant \rangle$

We now modify the argument of the preceding section to prove Σ^1_1-hardness of satisfiability for the predicate monomodal logic of $\langle \mathbb{N}, \leqslant \rangle$ in languages with two individual variables, two monadic, and a single 0-ary predicate letter. It follows that the logic of $\langle \mathbb{N}, \leqslant \rangle$ is Π^1_1-hard, and therefore not recursively enumerable, in such languages. In comparison with languages considered in the previous section, we need an additional 0-ary predicate letter to deal with reflexivity.

As before, let $\lhd$ be a binary—while M and P_t, for every $t \in T$, monadic—predicate letters, and let p be a 0-ary predicate letter (i.e., a propositional variable). Given a formula φ in such a language, define

$$\blacklozenge\varphi = \Diamond(\neg p \wedge \Diamond(p \wedge \varphi));$$
$$\blacklozenge^0\varphi = \varphi, \quad \blacklozenge^{k+1}\varphi = \blacklozenge\blacklozenge^k\varphi.$$

Let

$$A_4^r = \forall x \forall y\, (x \lhd y \to \Box(M(x) \leftrightarrow p \wedge \diamondsuit M(y) \wedge \neg \diamondsuit^2 M(y)),$$
$$A_8^r = \forall x\, (M(x) \to \Box \diamondsuit P_{t_0}(x)),$$

and let $A^r = A_1 \wedge A_2 \wedge A_3 \wedge A_4^r \wedge A_5 \wedge A_6 \wedge A_7 \wedge A_8^r$.

The operator $\diamondsuit$ forces a transition to a different world when valuating a formula $\diamondsuit\varphi$, just as $\Diamond$ does in the absence of reflexivity.

Lemma 4.1 *There exists a recurrent tiling of* $\mathbb{N} \times \mathbb{N}$ *if, and only if,* $\langle \mathbb{N}, \leqslant \rangle \not\models \neg A^r$.

Proof. ("if") Suppose $\mathfrak{M}, m \models A^r$, for some model $\mathfrak{M} = \langle \mathbb{N}, \leqslant, D, I \rangle$ and some $m \in \mathbb{N}$; we may assume without a loss of generality that $m = 0$.

Since $0 \models A_3$, there exists $a_0 \in D_0$ such that $0 \models M(a_0)$.

Since $0 \models A_1$, there exists an infinite sequence $a_0, a_1, a_2, \ldots$ of elements of D_0 such that $a_0 \lhd^{I,0} a_1 \lhd^{I,0} a_2 \lhd^{I,0} \ldots$.

Since $0 \models A_2$, clearly, $a_0 \lhd^{I,n} a_1 \lhd^{I,n} a_2 \lhd^{I,n} \ldots$, for every $n \in \mathbb{N}$.

Since $0 \models A_4^r$, the following holds: $w \models M(a_k)$ if, and only if, $w \models p$ and there exist $w', w'' \in \mathbb{N}$ such that $w \leqslant w' \leqslant w''$ and $w' \not\models p$ and $w'' \models p \wedge M(a_{k+1})$. Observe that the valuation of p guarantees that $w < w' < w''$. Also observe that, if $w'' \leqslant v \leqslant v'$ and $v \not\models p$ and $v' \models p$, then $v' \not\models M(a_{k+1})$. Thus, a mark of the world changes once we have passed through a world refuting p.

For every $k \in \mathbb{N}$, let w_k be, for definiteness' sake, the least world (number) such that $w_k \models M(a_k)$. Observe that, since $0 \models A_5$, for every $m, n \in \mathbb{N}$ there exists a unique $t \in T$ such that $w_m \models P_t(a_n)$. Therefore, we can define a function $f : \mathbb{N} \times \mathbb{N} \to T$ by

$$f(n, m) = t, \text{ where } t \text{ is such that } w_m \models P_t(a_n).$$

Since $0 \models A_6 \wedge A_7 \wedge A_8^r$, conditions (T_1) through (T_3) are satisfied for f. Therefore, f is a recurrent tiling of $\mathbb{N} \times \mathbb{N}$ with T.

("only if") Suppose f is a function satisfying (T_1) through (T_3). We define a model $\mathfrak{M}_0$, based on $\langle \mathbb{N}, \leqslant \rangle$, satisfying A^r.

To define $\mathfrak{M}_0$, let $D_n = \mathbb{N}$, for every $n \in \mathbb{N}$, and let I be an interpretation function such that, for every $n \in \mathbb{N}$,

- $n \models k \lhd l \;\leftrightharpoons\; l = k + 1$;
- $n \models p \;\leftrightharpoons\; n = 2m$, for some m;
- $n \models M(k) \;\leftrightharpoons\; n = 2k$;
- $n \models P_t(k) \;\leftrightharpoons\; n = 2m$ and $f(k, m) = t$.

Finally, let $\mathfrak{M}_0 = \langle \mathbb{N}, <, D, I \rangle$.

It is straightforward to check that $\mathfrak{M}_0, 0 \models A^r$. □

Remark 4.2 Observe that Remarks 3.3 and 3.4 apply to those worlds of the model $\mathfrak{M}_0$ defined in the "only if" part of the proof of the Lemma 4.1 where p is true. Also observe that $\mathfrak{M}_0$ is based on a predicate frame with a constant domain.

We next eliminate, in a satisfiability-preserving way, all but two monadic predicate letters of the formula A^r, without increasing the number of individual variables in the resultant formula. As in the preceding section, this is done in two steps. We assume, for convenience, that A^r contains monadic predicate letters $P_0, \ldots, P_n$, rather than P_t, for $t \in \{t_0, \ldots, t_n\}$, to refer to the tile types.

Let P_{n+1} and P_{n+2} be monadic predicate letters distinct form $M, P_0, \ldots, P_n$ and from each other. Let formula μ be defined as before. Lastly, let $(A^r)'$ be the result of substituting $\Diamond(\mu \wedge P_{n+1}(x) \wedge P_{n+2}(y))$ for $x \lhd y$ in A^r.

Lemma 4.3 *There exists a recurrent tiling of* $\mathbb{N} \times \mathbb{N}$ *if, and only if,* $\langle \mathbb{N}, \leqslant \rangle \not\models \neg(A^r)'$.

Proof. Similar to the proof of Lemma 3.5. In the proof of the "only if" part, we only simulate the valuation of $\lhd$ at the worlds where p is true—or, equivalently, the worlds w such that $w \models M(a)$, for some a. Therefore, instead of the enumeration h, we use an enumeration of such worlds only. Once this change is made, we proceed as in the proof of Lemma 3.5. □

We, lastly, eliminate all but two monadic predicate letters of $(A^r)'$. Let P be a monadic predicate letter distinct from $P_0, \ldots, P_{n+2}, M$, and let, for $k \in \{0, \ldots, n+2\}$,

$$\begin{aligned}\gamma_k(x) &= \mu \wedge \exists y\, [\Diamond^{n+4} M(y) \wedge \neg\Diamond^{n+5} M(y) \wedge \\ &\qquad \Diamond(\Diamond^{k+1} M(y) \wedge \neg\Diamond^{k+2} M(y) \wedge P(x))];\\ \gamma_k(y) &= \mu \wedge \exists x\, [\Diamond^{n+4} M(x) \wedge \neg\Diamond^{n+5} M(x) \wedge \\ &\qquad \Diamond(\Diamond^{k+1} M(x) \wedge \neg\Diamond^{k+2} M(x) \wedge P(y))],\end{aligned}$$

Let $\cdot^*$ be the function replacing $P_k(x)$ with $\gamma_k(x)$ and $P_k(y)$ with $\gamma_k(y)$, for $k \in \{0, \ldots, n+2\}$, in $(A^r)'$.

Let $(A^r_i)^*$, where $1 \leqslant i \leqslant 8$ and $i \neq 4$, be the result of applying the function $\cdot^*$ to $(A^r_i)'$ and let

$$\begin{aligned}(A^r_4)^\# &= \forall x \forall y\, (\Diamond(\gamma_{n+1}(x) \wedge \gamma_{n+2}(y)) \to \\ &\qquad \Box(M(x) \leftrightarrow \Diamond^{n+4} M(y) \wedge \neg\Diamond^{n+5} M(y))).\end{aligned}$$

Finally, let

$$(A^r)^* = (A^r_1)^* \wedge (A^r_2)^* \wedge (A^r_3)^* \wedge (A^r_4)^\# \wedge (A^r_5)^* \wedge (A^r_6)^* \wedge (A^r_7)^* \wedge (A^r_8)^*.$$

Lemma 4.4 *There exists a recurrent tiling of* $\mathbb{N} \times \mathbb{N}$ *if, and only if,* $\langle \mathbb{N}, \leqslant \rangle \not\models \neg(A^r)^*$.

Proof. Similar to the proof of Lemma 3.6.

We take the model obtained in the "only if" part of the proof of Lemma 4.3 and, essentially, apply to it the construction used in the proof of Lemma 3.6, the only difference being that we make letter p true at the worlds that we "added" in Lemma 3.6 and "insert" an extra world refuting p in-between every pair of such worlds that are adjacent. □

We, thus, obtain

Theorem 4.5 *Satisfiability for $L(\langle \mathbb{N}, \leqslant \rangle)$ is Σ^1_1-hard in languages with two individual variables, two monadic predicate letters, and a single* 0*-ary predicate letter.*

Proof. Immediate from Lemma 4.4. □

Thus, $L(\langle \mathbb{N}, \leqslant \rangle)$ is not recursively enumerable in such languages:

Theorem 4.6 *The logic $L(\langle \mathbb{N}, \leqslant \rangle)$ is Π^1_1-hard in languages with two individual variables, two monadic predicate letters, and a single* 0*-ary predicate letter.*

Proof. Immediate from Theorem 4.5. □

5 Discussion

Observe that the results of Section 4 can be easily extended to predicate monomodal logics of frames $\langle \mathbb{N}, R \rangle$ where R is a binary relation between $<$ and $\leqslant$: given any such logic L, we reduce the recurrent tiling problem to satisfiability for L by applying to the formulas defined in Section 4 the translation replacing occurrences of the modal operator $\Box$ with those of $\Box^+$.

Also observe that we have never relied on the domains of the predicate frames we have been dealing with to be not equal; therefore, all of our results apply to the logics of predicate frames with constant domains.

Lastly, observe that our results apply to the first-order temporal logic of $\langle \mathbb{N}, \leqslant \rangle$, which is essentially the first-order linear time temporal logic **QLTL**.

The results presented here raise the following questions.

The first is whether the results presented here can be strengthened to languages with one fewer predicate letter. Both in [41] and in [36] we have been able to prove undecidability and Σ^0_1-hardness results for languages with a single monadic predicate letter. We conjecture that the logic of $\langle \mathbb{N}, < \rangle$ is Π^1_1-hard in languages with two individual variables and a single monadic predicate letter. If the conjecture is correct, an analogous result for $\langle \mathbb{N}, \leqslant \rangle$, at worst with an additional 0-ary predicate letter, should follow.

The second is whether analogous results can be obtained for the superintuitionistic logic of the frame $\langle \mathbb{N}, \leqslant \rangle$. Given that the accessibility relation in $\langle \mathbb{N}, \leqslant \rangle$ is reflexive and transitive, the only, by not means trivial, hurdle to clear is obtaining a model with a hereditary valuation. Whether this can be done is unclear to us, given the difficulty of modelling the changing values of the tile types on a linear frame with a hereditary valuation.

References

[1] Abadi, M., *The power of temporal proofs*, Theoretical Computer Science **65** (1989), pp. 35–83.

[2] Andréka, H., I. Németi and I. Sain, *Completeness problems in verification of programs and program schemes*, in: J. Bečvář, editor, *Mathematical Foundations of Computer Science 1979. MFCS 1979.*, Lecture Notes in Computer Science **74** (1979).

[3] Artemov, S. and G. Dzhaparidze, *Finite Kripke models and predicate logics of provability*, The Journal of Symbolic Logic **55** (1990), pp. 1090–1098.

[4] Blackburn, P. and E. Spaan, *A modal perspective on the computational complexity of attribute value grammar*, Journal of Logic, Language, and Information **2** (1993), pp. 129–169.

[5] Börger, E., E. Grädel and Y. Gurevich, "The Classical Decision Problem," Springer, 1997.

[6] Braüner, T. and S. Ghilardi, *First-order modal logic*, in: P. Blackburn, J. V. Benthem and F. Wolter, editors, *Handbook of Modal Logic*, Studies in Logic and Practical Reasoning **3**, Elsevier, 2007 pp. 549–620.

[7] Chagrov, A. and M. Rybakov, *How many variables does one need to prove PSPACE-hardness of modal logics?*, in: P. Balbiani, N.-Y. Suzuki, F. Wolter and M. Zakharyaschev, editors, *Advances in Modal Logic 4*, King's College Publications, 2003 pp. 71–82.

[8] Chomicki, J., *Temporal query languages: A survey*, in: D. M. Gabbay and H. Ohlbach, editors, *Temporal Logic. ICTL 1994*, Lecture Notes in Computer Science (Lecture Notes in Artificial Intelligence) **827** (1994).

[9] Chomicki, J. and D. Niwinski, *On the feasibility of checking temporal integrity constraints*, Journal of Computer and System Sciences **51** (1995), pp. 523–535.

[10] Church, A., *A note on the "Entscheidungsproblem"*, The Journal of Symbolic Logic **1** (1936), pp. 40–41.

[11] Figueira, D. and L. Segoufin, *Future-looking logics on data words and trees*, in: R. Královič and D. Niwinski, editors, *Mathematical Foundations of Computer Science 2009. MFCS 2009*, Lecture Notes in Computer Science **5734** (2009), pp. 331–343.

[12] Fitting, M. and R. L. Mendelsohn, "First-Order Modal Logic," Synthese Library **277**, Kluwer Academic Publishers, 1998.

[13] Gabbay, D., "Semantical Investigations in Heyting's Intuitionistic Logic," D. Reidel, 1981.

[14] Gabbay, D., A. Kurucz, F. Wolter and M. Zakharyaschev, "Many-Dimensional Modal Logics: Theory and Applications," Studies in Logic and the Foundations of Mathematics **148**, Elsevier, 2003.

[15] Gabbay, D. and V. Shehtman, *Undecidability of modal and intermediate first-order logics with two individual variables*, The Journal of Symbolic Logic **58** (1993), pp. 800–823.

[16] Gabbay, D., V. Shehtman and D. Skvortsov, "Quantification in Nonclassical Logic, Volume 1," Studies in Logic and the Foundations of Mathematics **153**, Elsevier, 2009.

[17] Gabbay, D. M., I. Hodkinson and M. Reynolds, "Temporal Logic: Mathematical Foundations and Computational Aspects, Volume 1," Oxford Logic Guides **28**, Oxford University Press, 1994.

[18] Garson, J. W., *Quantification in modal logic*, in: D. M. Gabbay and F. Guenthner, editors, *Handbook of Philosophical Logic, vol. 3*, Springer, Dordrecht, 2001 pp. 267–323.

[19] Goldblatt, R., "Logics of Time and Computation," CSLI Lecture Notes **7**, Center for the Study of Language and Information, 1992, second edition.

[20] Halpern, J. Y., *The effect of bounding the number of primitive propositions and the depth of nesting on the complexity of modal logic*, Aftificial Intelligence **75** (1995), pp. 361–372.

[21] Hampson, C. and A. Kurucz, *Undecidable propositional bimodal logics and one-variable first-order linear temporal logics with counting*, ACM Transactions on Computational Logic **16** (2015).

[22] Harel, D., *Effective transformations on infinite trees, with applications to high undecidability, dominoes, and fairness*, Journal of the ACM **33** (1986).

[23] Hodkinson, I., R. Kontchakov, A. Kurucz, F. Wolter and M. Zakharyaschev, *On the computational complexity of decidable fragments of first-order linear temporal logics*, in: *10th International Symposium on Temporal Representation and Reasoning 2003 and Fourth International Conference on Temporal Logic. Proceedings.* (2003), pp. 91–98.

[24] Hodkinson, I., F. Wolter and M. Zakharyaschev, *Decidable fragments of first-order temporal logics*, Annals of Pure and Applied Logic **106** (2000), pp. 85–134.

[25] Hodkinson, I., F. Wolter and M. Zakharyaschev, *Monodic fragments of first-order temporal logics: 2000–2001 A.D.*, in: R. Nieuwenhuis and A. Voronkov, editors, *Logic for Programming, Artificial Intelligence, and Reasoning. LPAR 2001.*, Lecture Notes in Computer Science **2250** (2001), pp. 1–23.

[26] Hughes, G. E. and M. J. Cresswell, "A New Introduction to Modal Logic," Routledge, 1996.

[27] Kontchakov, R., A. Kurucz and M. Zakharyaschev, *Undecidability of first-order intuitionistic and modal logics with two variables*, Bulletin of Symbolic Logic **11** (2005), pp. 428–438.

[28] Kripke, S., *The undecidability of monadic modal quantification theory*, Zeitschrift für Matematische Logik und Grundlagen der Mathematik **8** (1962), pp. 113–116.

[29] Lisitsa, A. and I. Potapov, *Temporal logic with predicate λ-abstraction*, in: *12th International Symposium on Temporal Representation and Reasoning (TIME'05)*, 2005, pp. 147–155.

[30] Marx, M., *Complexity of products of modal logics*, Journal of Logic and Computation **9** (1999), pp. 197–214.

[31] Maslov, S., G. Mints and V. Orevkov, *Unsolvability in the constructive predicate calculus of certain classes of formulas containing only monadic predicate variables*, Soviet Mathematics Doklady **6** (1965), pp. 918–920.

[32] Merz, S., *Decidability and incompleteness results for first-order temporal logics of linear time*, Journal of Applied Non-Classical Logics **2** (1992), pp. 139–156.

[33] Mints, G., *Some calculi of modal logic*, Trudy Matematicheskogo Instituta imeni V. A. Steklova **98** (1968), pp. 88–111.

[34] Ono, H., *On some intuitionistic modal logics*, Publications of the Research Institute for Mathematical Sciences **13** (1977), pp. 687–722.

[35] Rogers, H., "Theory of Recursive Functions and Effective Computability," McGraw-Hill, 1967.

[36] Rybakov, M. and D. Shkatov, *Algorithmic properties of first-order modal logics of finite Kripke frames in restricted languages*, To appear in Journal of Logic and Computation.

[37] Rybakov, M. and D. Shkatov, *Complexity and expressivity of branching- and alternating-time temporal logics with finitely many variables*, in: B. F. B. and T. Uustalu, editors, *Theoretical Aspects of Computing - ICTAC 2018*, Lecture Notes in Computer Science **11187**, 2018, pp. 396–414.

[38] Rybakov, M. and D. Shkatov, *Complexity and expressivity of propositional dynamic logics with finitely many variables*, Logic Journal of the IGPL **26** (2018), pp. 539–547.

[39] Rybakov, M. and D. Shkatov, *A recursively enumerable Kripke complete first-order logic not complete with respect to a first-order definable class of frames*, in: G. M. G. Bezhanishvili, G. D'Agostino and T. Studer, editors, *Advances in Modal Logic 12*, College Publications, 2018 pp. 531–540.

[40] Rybakov, M. and D. Shkatov, *Complexity of finite-variable fragments of propositional modal logics of symmetric frames*, Logic Journal of the IGPL **27** (2019), pp. 60–68.

[41] Rybakov, M. and D. Shkatov, *Undecidability of first-order modal and intuitionistic logics with two variables and one monadic predicate letter*, Studia Logica **107** (2019), pp. 695–717.

[42] Rybakov, M. and D. Shkatov, *Recursive enumerability and elementary frame definability in predicate modal logic*, Journal of Logic and Computation **30** (2020), pp. 549–560.

[43] Skvortsov, D., *On axiomatizability of some intermediate predicate logics*, Reports on Mathematical Logic **22** (1988), pp. 115–116.

[44] Spaan, E., "Complexity of Modal Logics," Ph.D. thesis, University of Amsterdam (1993).

[45] Szalas, A., *Concerning the semantic consequence relation in first-order temporal logic*, Theoretical Computer Science **47** (1986), pp. 329–334.

[46] Szalas, A. and L. Holenderski, *Incompleteness of first-order temporal logic with until*, Theoretical Computer Science **57** (1988), pp. 317–325.

[47] Wolter, F. and M. Zakharyaschev, *Decidable fragments of first-order modal logics*, The Journal of Symbolic Logic **66** (2001), pp. 1415–1438.

Goldblatt-Thomason-style Characterization for Intuitionistic Inquisitive Logic

Katsuhiko Sano [1]

Faculty of Humanities and Human Sciences, Hokkaido University
Nishi 7 Chome, Kita 10 Jo, Kita-ku
Sapporo, Hokkaido, 060-0810, Japan

Abstract

The purpose of this paper is to investigate a possible characterization of frame definability of intuitionistic inquisitive logic by Ciardelli et al. (2020) in terms of frame constructions such as generated subframes and bounded morphic images. Sano and Virtema (2015, 2019) provided a Goldblatt-Thomason-style characterization for (extended) modal dependence logic with the help of a normal form result for the logic. A key ingredient of establishing the characterization was to show that the ordinary modal logic expanded with positive occurrences of the universal modality and extended modal dependence logic have the same definability over Kripke models. This paper first reviews Goldblatt-Thomason-style characterization for intuitionistic logic from Rodenburg (1986)'s work on intuitionistic correspondence theory. Then we employ a similar strategy to Sano and Virtema (2015, 2019) and provide a Goldblatt-Thomason-style characterization for intuitionistic inquisitive logic.

Keywords: Intuitionistic Logic, Frame Definability, Universal Modality, Goldblatt-Thomason Theorem, Inquisitive Semantics, Inquisitive Logic

1 Introduction

Goldblatt-Thomason Theorem [11] for modal logic enables us to characterize elementary frame class definability in terms of frame construction. To be more specific, it states that an elementary (or first-order definable) frame class $\mathbb{F}$ is definable by a set of modal formulas iff $\mathbb{F}$ is closed under taking bounded morphic images, generated subframes, disjoint unions and $\mathbb{F}$ reflects ultrafilter

[1] v-sano@let.hokudai.ac.jp. I would like to thank three anonymous referees for their careful reading of the submission and helpful comments. This paper is based on my talk at Inquisitive Logic 2018 workshop, held at Amsterdam. In particular, I would like to thank Giuseppe Greco, Fan Yang, Vít Punčochář, Wesley Holliday, and Ivano Ciardelli for discussions at the workshop. I also wish to thank Jonni Virtema for our discussion at Sapporo in the spring of 2020. The work of this paper was partially supported by JSPS KAKENHI Grant-in-Aid for Scientific Research (C) Grant Number 19K12113, JSPS KAKENHI Grant-in-Aid for Scientific Research (B) Grant Number 17H02258, and JSPS Core-to-Core Program (A. Advanced Research Networks).

extensions (i.e., the complement of $\mathbb{F}$ is closed under taking ultrafilter extensions). Since then, Goldblatt-Thomason-style (GT-style, for short) characterization has been provided for a rich variety of logics: modal logic with the universal modality [9], hybrid logics [22], graded modal logic [17], modal logic over topological semantics [23], coalgebraic modal logic [12], intuitionistic logic [16], modal dependence logic [19], etc. Let us comment on intuitionistic logic. If we replace the reflection of ultrafilter extensions with the reflection of *prime filter extensions* in frame constructions for Goldblatt-Thomason Theorem for modal logic, we can obtain Rodenburg [16]'s characterization of intuitionistic elementary frame definability.

Inquisitive logic [3,2] (or inquisitive semantics [4]) is a recent theoretical framework for studying both declarative and interrogative sentences in one setting. It often assumes classical logic as a background logic. Then, on the top of classical logic, we add the *inquisitive disjunction* $\mathbin{\rotatebox[origin=c]{90}{$\geqslant$}}$, which allows us to formalize a question "Does Taro play tennis?" as $?p := p \mathbin{\rotatebox[origin=c]{90}{$\geqslant$}} \neg p$, where p denotes the declarative sentence "Taro plays tennis". Semantically, a formula is evaluated not by a single state but by a set of states (which is called a *team*). This semantic feature is also a core of (propositional) *dependence logic* (cf. [26]), where we can study the notion of *functional dependence* $\mathrm{dep}(q;p)$, "q truth-functionally determines p". In this sense, the semantics for dependence logic is called *team* semantics. Moreover, the recent interaction between the two communities reveal, e.g., that functional dependency $\mathrm{dep}(q;p)$ can be understood as an implication from the question $?q$ to the question $?p$ (see [2] for more detail).

Recently, the ideas of inquisitive logic and dependence logic are generalized also to non-classical logics, i.e., modal logic [24,8,7], (dynamic) epistemic logic [6], intuitionistic logic [13,14,5], substructural logic [15], etc. For modal dependence logic (modal logic extended with atoms for functional dependency), [19] provided a Goldblatt-Thomason-style characterization. A key ingredient for the characterization is that (extended) modal dependence logic (with team semantics) and modal logic expanded with the positive occurrences of the universal modality (with Kripke semantics) have the same definability for frame classes.

While modal dependence logic still assumes classical logic, intuitionistic inquisitive logic [5]'s background logic is intuitionistic. We add the inquisitive disjunction to the syntax of intuitionistic logic, and "lift" the ordinary state-based Kripke semantics for intuitionistic logic to the semantics based on teams (sets of states). Then, we can study questions and dependency also in the intuitionistic setting. As for frame definability, [5] raises the following research question ("[24]" and "[25]" in the citation correspond to [20] and [18] respectively):

> [...] it would also be interesting to look at the issue of frame definability in InqI. [...] Clearly, if a standard formula defines a certain frame class in IPL, then this formula still defines the same class in InqI. At the same time, however, some frame classes which cannot be characterized in IPL can

> now be characterized with the help of inquisitive formulas: for instance, $?p$ characterizes the class of singleton frames. Recent work on frame definability in the context of modal dependence logic (see, e.g., Sano and Virtema [24], [25]) might provide a handle on this question. [5, p.110]

This paper tackles this question and provide a GT-style characterization for intuitionistic inquisitive logic. For this purpose, we follow a similar strategy to [19]. That is, we first study intuitionistic logic with the universal modality A, which was, as far as the author knows, less studied in the literature (e.g., [21] studies the axiomatization of bi-intuitionistic tense logic expanded with the universal modality). Then, we provide GT-style characterizations for a special fragment of intuitionistic logic with the universal modality, which in turn gives us our intended GT-style characterization for intuitionistic inquisitive logic. An important insight is: we can mimic the behavior of inquisitive disjunction $\varphi \mathbin{\vee\!\!\!\vee} \psi$ by A-prefixed disjunction $\mathsf{A}\,\varphi \vee \mathsf{A}\,\psi$ where φ and ψ are intuitionistic formulas.

Our proof of Goldblatt-Thomason-type characterization is based on van Benthem's model-theoretic argument [25], though the original proof by Rodenburg [16] is based on the representation theorem of Heyting algebras. When we try to transfer the idea of Golcblatt-Thomason Theorems for modal dependence logic [19] to our current study, there is a tricky point on the negation. While we need to handle the intuitionistic negation for frame definability, we also need to deal with the classical negation when we use the standard translation to apply the first-order model theory. The results of this paper show that this tricky distinction can be overcome in applying van Benthem's model-theoretic argument [25].

We proceed as follows. Section 2 introduces Kripke semantics for the syntax of intuitionistic logic (the set of formulas is denoted by Form) and four frame constructions, and then reviews Rodenburg's Goldblatt-Thomason Theorem for intuitionistic logic. Section 3 adds the universal modality A to the syntax of intuitionistic logic (the resulting set of formulas is denoted by $\mathsf{Form}(\mathsf{A})$) and introduce the syntactic notion of disjunctive A-clauses, i.e., a formula of the form $\bigvee_{i \in I} \mathsf{A}\,\varphi_i$, where I is finite and φ_is does not contain any occurrences of A, i.e., an intuitionistic formula. We use $\bigvee \mathsf{A}\,\mathsf{Form}$ to denote the set of all disjunctive A-clauses. Section 4 provides two Goldblatt-Thomason-type characterizations of elementary frame definability in terms of $\mathsf{Form}(\mathsf{A})$ and $\bigvee \mathsf{A}\,\mathsf{Form}$ (Theorems 4.2 and 4.3, respectively). Section 5 introduces the *inquisitive disjunction* $\mathbin{\vee\!\!\!\vee}$ to Form (where the resulting set of formulas is denoted by $\mathsf{Form}(\mathbin{\vee\!\!\!\vee})$) and team semantics for it, and then establishes that $\mathsf{Form}(\mathbin{\vee\!\!\!\vee})$ and $\bigvee \mathsf{A}\,\mathsf{Form}$ have the same frame definability. This equi-definability result enables us to provide Goldblatt-Thomason-type Theorem for intuitionistic inquisitive logic (Theorem 5.12). Section 6 explains several directions of further research.

2 Goldblatt-Thomason Theorem for Intuitionistic Logic

2.1 Syntax and Kripke Semantics for Intuitionistic Logic

Let Prop be a set of propositional variables (we mostly assume that Prop is countably infinite). The set Form of all formulas for intuitionistic logic is defined inductively as follows:

$$\mathsf{Form} \ni \varphi ::= p \mid \bot \mid \varphi \wedge \varphi \mid \varphi \vee \varphi \mid \varphi \rightarrow \varphi \quad (p \in \mathsf{Prop}).$$

The negation is defined as $\neg\varphi := \varphi \rightarrow \bot$.

We move on to Kripke semantics. We say that $\mathfrak{F} = (W, R)$ is a *Kripke frame* (or simply *frame*) if W is a non-empty set of states and $R \subseteq W \times W$ is reflexive and transitive, i.e., (W, R) is a preorder or a quasi-order. We say that $\mathfrak{M} = (W, R, V)$ is a *Kripke model* (or simply *model*) if (W, R) is a frame and $V : \mathsf{Prop} \rightarrow \mathcal{P}(W)$ is a *valuation function* (or simply *valuation*) such that every $V(p)$ is *persistent* (or $V(p)$ is an *upset*) in the following sense: if $w \in V(p)$ and wRv then $v \in V(p)$, for all states $w, v \in W$. For a frame $\mathfrak{F}$ and a model $\mathfrak{M}$ we use $|\mathfrak{F}|$ and $|\mathfrak{M}|$ to mean the underlying domain.

Definition 2.1 Given a model $\mathfrak{M} = (W, R, V)$, a state $w \in W$ and a formula φ, the *satisfaction relation* $\mathfrak{M}, w \Vdash \varphi$ is defined inductively as follows:

$$\begin{array}{lll}
\mathfrak{M}, w \Vdash p & \text{iff} & w \in V(p), \\
\mathfrak{M}, w \nVdash \bot, & & \\
\mathfrak{M}, w \Vdash \varphi \wedge \psi & \text{iff} & \mathfrak{M}, w \Vdash \varphi \text{ and } \mathfrak{M}, w \Vdash \psi, \\
\mathfrak{M}, w \Vdash \varphi \vee \psi & \text{iff} & \mathfrak{M}, w \Vdash \varphi \text{ or } \mathfrak{M}, w \Vdash \psi, \\
\mathfrak{M}, w \Vdash \varphi \rightarrow \psi & \text{iff} & \forall v\, ((wRv \text{ and } \mathfrak{M}, v \Vdash \varphi) \text{ imply } \mathfrak{M}, v \Vdash \psi).
\end{array}$$

The truth set $[\![\varphi]\!]_{\mathfrak{M}}$ is defined as $\{\, w \in W \mid \mathfrak{M}, w \Vdash \varphi \,\}$. For a set Δ of formulas, we write $\mathfrak{M}, w \Vdash \Delta$ to mean $\mathfrak{M}, w \Vdash \varphi$ for all $\varphi \in \Delta$.

For the negation, we have the following satisfaction clause:

$$\mathfrak{M}, w \Vdash \neg\varphi \ \text{ iff } \ \forall v\, (wRv \text{ implies } \mathfrak{M}, v \nVdash \varphi).$$

Definition 2.2 Let $\mathfrak{F} = (W, R)$ be a frame and $X \subseteq W$. We define the upward closure $\uparrow X$ of X as the set $\{\, v \in W \mid \exists w \in X\, (wRv) \,\}$. We usually write $\uparrow w$ instead of $\uparrow \{\, w \,\}$ for $w \in W$. Given upsets $X, Y \subseteq W$, we define $X \Rightarrow Y := \{\, w \in W \mid \uparrow w \cap X \subseteq Y \,\}$.

For a model $\mathfrak{M}$, it is noted that $[\![\varphi \rightarrow \psi]\!]_{\mathfrak{M}} = [\![\varphi]\!]_{\mathfrak{M}} \Rightarrow [\![\psi]\!]_{\mathfrak{M}}$. Given a frame (W, R), it is remarked that X is an upset iff $\uparrow X = X$.

By induction on a formula, we can show that the persistency can be extended from propositional variables to formulas.

Proposition 2.3 *The set* $[\![\varphi]\!]_{\mathfrak{M}}$ *is an upset for all formulas* φ.

Definition 2.4 A formula φ is *valid* in a model $\mathfrak{M}$ (notation: $\mathfrak{M} \Vdash \varphi$) if $\mathfrak{M}, w \Vdash \varphi$ for all states $w \in W$, or equivalently, $[\![\varphi]\!]_{\mathfrak{M}} = W$. A set Γ of formulas is *valid* in a frame $\mathfrak{F} = (W, R)$ (notation: $\mathfrak{F} \Vdash \Gamma$) if, for every valuation V, $(\mathfrak{F}, V) \Vdash \varphi$ holds for all formulas $\varphi \in \Gamma$. When Γ is a singleton $\{\, \varphi \,\}$, we simply write $\mathfrak{F} \Vdash \varphi$ to mean $\mathfrak{F} \Vdash \{\varphi\}$. A set Γ of formulas *defines* a class $\mathbb{F}$ of frames if the following equivalence holds: $\mathfrak{F} \Vdash \Gamma$ iff $\mathfrak{F} \in \mathbb{F}$, for all frames $\mathfrak{F}$.

The following table demonstrates frame definability taken from [16].

Formula	Property of R
$p \vee \neg p$	$\forall w, v\,(wRv \text{ implies } vRw)$
$(p \to q) \vee (q \to p)$	$\forall w, v, u\,((wRv \text{ and } wRu) \text{ imply } (vRu \text{ or } uRv))$
$\neg p \vee \neg\neg p$	$\forall w, v, u\,((wRv \text{ and } wRu) \text{ imply } \exists z(vRz \text{ and } uRz))$

Definition 2.5 Let $\mathcal{L}^1_f$ be the first-order frame language (with equality) which has a binary predicate $x \leqslant y$ (corresponding to a relation R of a Kripke frame (W, R)). Let $\mathcal{L}^1_m$ be the first-order model language which expands $\mathcal{L}^1_f$ with a set $\{p(x) \mid p \in \mathsf{Prop}\}$ of unary predicates corresponding to Prop. Given any first-order variable x, we define the *standard translation* ST_x from Form to the set of first-order formulas in $\mathcal{L}^1_m$ as follows:

$$\begin{aligned}
\mathrm{ST}_x(p) &:= p(x),\\
\mathrm{ST}_x(\bot) &:= \bot,\\
\mathrm{ST}_x(\varphi \wedge \psi) &:= \mathrm{ST}_x(\varphi) \wedge \mathrm{ST}_x(\psi),\\
\mathrm{ST}_x(\varphi \vee \psi) &:= \mathrm{ST}_x(\varphi) \vee \mathrm{ST}_x(\psi),\\
\mathrm{ST}_x(\varphi \to \psi) &:= \forall y\,(x \leqslant y \wedge \mathrm{ST}_y(\varphi) \to \mathrm{ST}_y(\psi)),
\end{aligned}$$

where y is a fresh variable.

We note that $\mathfrak{M}$ and $\mathfrak{F}$ are regarded as first-order structures of $\mathcal{L}^1_m$ and $\mathcal{L}^1_f$, respectively. In what follows, we keep the symbol "$\models$" for the satisfaction relation for $\mathcal{L}^1_m$ or $\mathcal{L}^1_f$, while we keep "$\Vdash$" for Kripke semantics. By induction on φ, we get the following (see [16, p.7]).

Proposition 2.6 *Let $\mathfrak{M} = (W, R, V)$ be a model. For every formula $\varphi \in \mathsf{Form}$ and $w \in W$, $\mathfrak{M}, w \Vdash \varphi$ iff $\mathfrak{M} \models \mathrm{ST}_x(\varphi)[w]$.*

2.2 Frame Constructions and Rodenburg's Characterization of Intuitionistic Frame Definability

This subsection first introduces four frame constructions: bounded morphic images, generated subframes, disjoint unions, and prime filter extensions. Then, we review Rodenburg [16]'s Goldblatt-Thomason Theorem for intuitionistic logic in terms of the four frame constructions.

Definition 2.7 Let $\mathfrak{F} = (W, R)$ and $\mathfrak{F}' = (W', R')$. A mapping $f : W \to W'$ is a *bounded morphism* from $\mathfrak{F}$ to $\mathfrak{F}'$ if f satisfies the following:

(Forth) For every $w, v \in W$, wRv implies $f(w)R'f(v)$.

(Back) For every $w \in W$ and $b \in W'$, $f(w)R'b$ implies that $f(v) = b$ and wRv for some $v \in W$.

We say that $\mathfrak{F}'$ is a *bounded morphic images* of $\mathfrak{F}$ (notation: $\mathfrak{F} \twoheadrightarrow \mathfrak{F}'$) if there exists a *surjective* bounded morphism from $\mathfrak{F}$ onto $\mathfrak{F}'$. Given any models $\mathfrak{M} = (W, R, V)$ and $\mathfrak{M}' = (W', R', V')$, a mapping $f : W \to W'$ is a *bounded morphism* from $\mathfrak{M}$ to $\mathfrak{M}'$ if f is a bounded morphism from (W, R) to (W', R') and it also satisfies the following:

(Atom) $V(p) = f^{-1}[V'(p)]$ for all propositional variables p.

Definition 2.8 We say that $\mathfrak{F}' = (W', R')$ is a *generated subframe* of $\mathfrak{F} = (W, R)$ (notation: $\mathfrak{F}' \rightarrowtail \mathfrak{F}$) if the following conditions hold: (i) $W' \subseteq W$ is an upset with respect to R, and (ii) $R' = R \cap (W' \times W')$. A model $\mathfrak{M}' = (W', R', V')$ is a *generated submodel* of a model $\mathfrak{M} = (W, R, V)$ if (W', R') is a generated subframe of (W, R) and $V'(p) = V(p) \cap W'$ for all propositional variables p. Given a subset X of the domain of a frame $\mathfrak{F}$ (or a model $\mathfrak{M}$), $\mathfrak{F}_X$ (or $\mathfrak{M}_X$) is the smallest generated subframe (or submodel) whose domain contains X. When X is a singleton $\{ w \}$, we simply write $\mathfrak{F}_w$ and $\mathfrak{M}_w$ to mean $\mathfrak{F}_{\{ w \}}$ and $\mathfrak{M}_{\{ w \}}$, respectively.

By induction on φ, we can easily prove the following (cf. [16, p.5]).

Proposition 2.9 *Let $\mathfrak{M}' = (W, R, V)$ be a generated submodel of $\mathfrak{M}$. For every formula $\varphi \in \mathsf{Form}$ and $w \in W'$, $\mathfrak{M}', w \Vdash \varphi$ iff $\mathfrak{M}, w \Vdash \varphi$.*

Definition 2.10 Given a family $(\mathfrak{F}_i)_{i \in I}$ of frames where $\mathfrak{F}_i = (W_i, R_i)$, the *disjoint union* $\biguplus_{i \in I} \mathfrak{F}_i = (W, R)$ of $(\mathfrak{F}_i)_{i \in I}$ is defined as:

(i) $W := \bigcup_{i \in I}(W_i \times \{ i \})$ and

(ii) $(w, i)R(v, j)$ iff $i = j$ and wR_iv.

For a family $(\mathfrak{M}_i)_{i \in I}$ of models where $\mathfrak{M}_i = (W_i, R_i, V_i)$, the *disjoint union* $\biguplus_{i \in I} \mathfrak{M}_i = (W, R, V)$ of $(\mathfrak{M}_i)_{i \in I}$ is defined as follows: (W, R) is the disjoint union of $(W_i, R_i)_{i \in I}$ and $(w, i) \in V(p)$ iff $w \in V_i(p)$ for all $p \in \mathsf{Prop}$.

The following proposition has been already established in [16, Section 2.4].

Proposition 2.11 (i) *If $\mathfrak{F} \twoheadrightarrow \mathfrak{G}$, then $\mathfrak{F} \Vdash \varphi$ implies $\mathfrak{G} \Vdash \varphi$ for all $\varphi \in \mathsf{Form}$.*

(ii) *If $\mathfrak{F}' \rightarrowtail \mathfrak{F}$, then $\mathfrak{F} \Vdash \varphi$ implies $\mathfrak{F}' \Vdash \varphi$ for all $\varphi \in \mathsf{Form}$.*

(iii) *Given a family $(\mathfrak{F}_i)_{i \in I}$ of frames, if $\mathfrak{F}_i \Vdash \varphi$ for all $i \in I$, then $\biguplus_{i \in I} \mathfrak{F}_i \Vdash \varphi$, for all $\varphi \in \mathsf{Form}$.*

Now, we move to our final frame construction of *prime filter extensions.*

Definition 2.12 Let $\mathfrak{F} = (W, R)$ be a frame (or preorder) and define

$$\wp^{\uparrow}(W) := \{ X \subseteq W \mid X \text{ is an upset } \}.$$

We say that $\mathcal{F} \subseteq \wp^{\uparrow}(W)$ is a *filter* on W if $X \cap Y \in \mathcal{F}$ iff $X \in \mathcal{F}$ and $Y \in \mathcal{F}$, for every X, $Y \in \wp^{\uparrow}(W)$. A filter $\mathcal{F}$ is *prime* if the following two conditions hold: (i) $\varnothing \notin \mathcal{F}$ and $\mathcal{F} \neq \varnothing$, i.e., $\mathcal{F}$ is *proper*; (ii) $X \cup Y \in \mathcal{F}$ implies $X \in \mathcal{F}$ or $Y \in \mathcal{F}$, for every X, $Y \in \wp^{\uparrow}(W)$.

For a filter $\mathcal{F}$, $X \in \mathcal{F}$ and $X \subseteq Y$ imply $Y \in \mathcal{F}$ for all X, $Y \in \wp^{\uparrow}(W)$, i.e., $\mathcal{F}$ is *upward closed* (with respect to $\subseteq$).

Definition 2.13 The *prime filter extension* $\mathfrak{pe}\,\mathfrak{F} = (\mathrm{Pf}(W), R^{\mathfrak{pe}})$ of a frame $\mathfrak{F} = (W, R)$ is defined as follows: (i) $\mathrm{Pf}(W)$ is the set of all the prime filters on W; (ii) $\mathcal{F}_1 R^{\mathfrak{pe}} \mathcal{F}_2$ iff $\mathcal{F}_1 \subseteq \mathcal{F}_2$. We say that $\mathfrak{pe}\,\mathfrak{M} = (\mathrm{Pf}(W), R^{\mathfrak{pe}}, V^{\mathfrak{pe}})$ is the *prime filter extension* of a model $\mathfrak{M} = (W, R, V)$ if $(\mathrm{Pf}(W), R^{\mathfrak{pe}})$ is the prime filter extension of (W, R), and $\mathcal{F} \in V^{\mathfrak{pe}}(p)$ iff $V(p) \in \mathcal{F}$, for every propositional variable p.

It is noted that $V^{\mathfrak{pe}}(p)$ is clearly an upset with respect to $R^{\mathfrak{pe}}$.

Proposition 2.14 (Rodenburg [16]) (i) *Let* $\mathfrak{M} = (W, R, V)$ *be a model. Then, for any prime filter* $\mathcal{F}$ *on* W, $\mathfrak{pe}\,\mathfrak{M}, \mathcal{F} \Vdash \varphi$ iff $[\![\varphi]\!]_{\mathfrak{M}} \in \mathcal{F}$.

(ii) *Given any frame* $\mathfrak{F}$, *if* $\mathfrak{pe}\,\mathfrak{F} \Vdash \varphi$ *then* $\mathfrak{F} \Vdash \varphi$, *for every* $\varphi \in \mathsf{Form}$.

Item (ii) of Proposition 2.14 is from [16, Proposition 14.18.3], but there is no explicit proof of item (i) there, and so, we provide an outline of the argument for item (i).

Proof. (i) By induction on φ. We only deal with the case where φ is of the form $\psi \to \theta$. First, we prove the right-to-left direction. Assume that $[\![\psi \to \theta]\!]_{\mathfrak{M}} \in \mathcal{F}$. Fix any prime filter $\mathcal{F}' \in \mathrm{Pf}(W)$ such that $\mathcal{F} \subseteq \mathcal{F}'$ and $\mathfrak{pe}\,\mathfrak{M}, \mathcal{F}' \Vdash \psi$. Our goal is to show: $\mathfrak{pe}\,\mathfrak{M}, \mathcal{F}' \Vdash \theta$. It follows from $\mathfrak{pe}\,\mathfrak{M}, \mathcal{F}' \Vdash \psi$ and induction hypothesis that $[\![\psi]\!]_{\mathfrak{M}} \in \mathcal{F}'$. Thus, we have that $[\![\psi \to \theta]\!]_{\mathfrak{M}} \cap [\![\psi]\!]_{\mathfrak{M}} \in \mathcal{F}'$ hence $[\![\theta]\!]_{\mathfrak{M}} \in \mathcal{F}'$ since $[\![\psi \to \theta]\!]_{\mathfrak{M}} \cap [\![\psi]\!]_{\mathfrak{M}} \subseteq [\![\theta]\!]_{\mathfrak{M}}$. We can conclude $\mathfrak{pe}\,\mathfrak{M}, \mathcal{F}' \Vdash \theta$ by induction hypothesis. Second, we prove the left-to-right direction by the contrapositive implication and so assume that $[\![\psi \to \theta]\!]_{\mathfrak{M}} \notin \mathcal{F}$. Then, we can find a prime filter $\mathcal{F}'$ such that $\mathcal{F} \subseteq \mathcal{F}'$, $[\![\psi]\!]_{\mathfrak{M}} \in \mathcal{F}'$, and $[\![\theta]\!]_{\mathfrak{M}} \notin \mathcal{F}'$. By induction hypothesis, this implies that $\mathfrak{pe}\,\mathfrak{M}, \mathcal{F}' \nVdash \psi \to \theta$, as desired.

(ii) Fix any frame $\mathfrak{F} = (W, R)$ and formula φ. We prove the contrapositive implication and so assume that $\mathfrak{F} \nVdash \varphi$, i.e., there exists a valuation V and a state $w \in W$ such that $(\mathfrak{F}, V), w \nVdash \varphi$. Put $\mathfrak{M} := (\mathfrak{F}, V)$. Let $\mathcal{F}_w := \{\, X \in \wp^{\uparrow}(W) \mid w \in X \,\}$. It is easy to see that $\mathcal{F}_w$ is a prime filter. Since $w \notin [\![\varphi]\!]_{\mathfrak{M}}$, we get $[\![\varphi]\!]_{\mathfrak{M}} \notin \mathcal{F}_w$. It follows from item (i) that $\mathfrak{pe}\,\mathfrak{M}, \mathcal{F}_w \nVdash \varphi$, i.e., $\mathfrak{pe}\,\mathfrak{F} \nVdash \varphi$. □

Definition 2.15 Let $\mathbb{F}$ be a frame class. We say that $\mathbb{F}$ is *closed under taking bounded morphic images* if $\mathfrak{F} \in \mathbb{F}$ and $\mathfrak{F} \twoheadrightarrow \mathfrak{G}$ imply $\mathfrak{G} \in \mathbb{F}$, for all frames $\mathfrak{F}$ and $\mathfrak{G}$. The class $\mathbb{F}$ is *closed under taking generated subframes* if $\mathfrak{F} \in \mathbb{F}$ and $\mathfrak{G} \rightarrowtail \mathfrak{F}$ imply $\mathfrak{G} \in \mathbb{F}$, for all frames $\mathfrak{F}$ and $\mathfrak{G}$. The class $\mathbb{F}$ is *closed under taking disjoint unions* if, whenever $\mathfrak{F}_i \in \mathbb{F}$ for all $i \in I$, $\biguplus_{i \in I} \mathfrak{F}_i \in \mathbb{F}$ holds, for all families $(\mathfrak{F}_i)_{i \in I}$ of frames. A class $\mathbb{F}$ of frames *reflects prime filter extensions* if $\mathfrak{pe}\,\mathfrak{F} \in \mathbb{F}$ implies $\mathfrak{F} \in \mathbb{F}$, for all frames $\mathfrak{F}$. We say that a class $\mathbb{F}$ of frames is *elementary* (or *first-order definable*) if there exists a set Σ of sentences in $\mathcal{L}^1_f$ such that Σ defines $\mathbb{F}$ in the sense of first-order model theory.

Theorem 2.16 (Rodenburg [16]) *An elementary frame class* $\mathbb{F}$ *is definable by a set of intuitionistic formulas (i.e., a subset of* Form*) iff* $\mathbb{F}$ *is closed under taking bounded morphic images, generated subframes, and disjoint unions and* $\mathbb{F}$ *reflects prime filter extensions.*

It is noted that the left-to-right direction is shown by Propositions 2.11 and 2.14 where we do not need to use the assumption that $\mathbb{F}$ is elementary.

Rodenburg proved the right-to-left direction via the representation theorem of Heyting algebras (see a proof given for [16, Theorem 15.3]). We can also prove Theorem 2.16 by van Benthem's model-theoretic argument [25] (the reader can get an idea of it from our proof of Theorems 4.2 and 4.3).

Proposition 2.17 *The following frame properties are undefinable in the syntax of intuitionistic logic.*

(i) *Antisymmetry of* R, *i.e.*, $\forall\, x, y\, (xRy$ *and* yRx *imply* $x = y)$.

(ii) $\exists\, x, y\, (xRy$ *and* $x \neq y)$.

(iii) R *is a total relation, i.e.*, $\forall\, x, y\, (xRy)$.

(iv) $\forall\, x, y\, (xRy$ *or* $yRx)$.

(v) $\forall\, x, y\, \exists\, z\, (xRz$ *and* $yRz)$.

(vi) $\exists\, y\, \forall\, x\, (xRy)$, *i.e., the existence of the maximum element.*

(vii) $\exists\, y\, \forall\, x\, (yRx)$, *i.e., the existence of the minimum element.*

Proof. For (i), let us consider $\mathfrak{F} = (\mathbb{N}, \leqslant)$ with the ordinary partial order $\leqslant$ and $\mathfrak{G} = (\{0,1\}, \{0,1\} \times \{0,1\})$. Then the mapping sending even and odd numbers to 0 and 1 respectively is a surjective bounded morphism. While $\mathfrak{F}$ is anti-symmetric, $\mathfrak{G}$ is not. Then, Proposition 2.11 (i) implies the desired undefinability. The property (ii) is clearly not closed under generated subframes and we get the undefinability by Proposition 2.11 (ii). The remaining properties from (iii) to (vii) are not closed under disjoint unions. For example, the single point reflexive frame satisfies all the properties from (iii) to (vii) but two copies of it do not satisfy them. Then, Proposition 2.11 (iii) gives us the desired undefinability. □

3 Intuitionistic Logic with the Universal Modality

The set $\mathsf{Form}(\mathsf{A})$ of all formulas of the intuitionistic logic with the *universal modality* A is defined inductively as follows:

$$\mathsf{Form}(\mathsf{A}) \ni \varphi ::= p \mid \bot \mid \varphi \wedge \varphi \mid \varphi \vee \varphi \mid \varphi \to \varphi \mid \mathsf{A}\,\varphi, \quad (p \in \mathsf{Prop}).$$

The set $\bigvee \mathsf{A}\,\mathsf{Form}$ of *disjunctive* A-*clauses* is defined as follows.

$$\bigvee \mathsf{A}\,\mathsf{Form} \ni \rho ::= \bot \mid \mathsf{A}\,\varphi \mid \rho \vee \rho \quad (\varphi \in \mathsf{Form}).$$

where it is noted that $\varphi \in \mathsf{Form}$ is a formula of the intuitionistic logic and so it does not contain any occurrence of A. For example, $\mathsf{A}\,p \vee \mathsf{A}\,\neg p$ is a disjunctive A-clause. It is clear that $\bigvee \mathsf{A}\,\mathsf{Form} \subseteq \mathsf{Form}(\mathsf{A})$.

Given a model $\mathfrak{M} = (W, R, V)$, a state $w \in W$ and a formula $\varphi \in \mathsf{Form}(\mathsf{A})$, the satisfaction relation $\mathfrak{M}, w \Vdash \varphi$ is defined in the same way as in Definition 2.1 except

$$\mathfrak{M}, w \Vdash \mathsf{A}\,\varphi \;\text{ iff }\; \forall\, v \in W\, (\mathfrak{M}, v \Vdash \varphi).$$

It is easy to see that $[\![\mathsf{A}\,\varphi]\!]_{\mathfrak{M}} = W$ or $[\![\mathsf{A}\,\varphi]\!]_{\mathfrak{M}} = \varnothing$ for all models $\mathfrak{M} = (W, R, V)$. Since W and $\varnothing$ are upsets, we can easily obtain the following.

Proposition 3.1 *Given a model* $\mathfrak{M} = (W, R, V)$ *and a formula* $\varphi \in \mathsf{Form}(\mathsf{A})$, $[\![\varphi]\!]_{\mathfrak{M}}$ *is an upset.*

The reader may wonder if the existential dual E of A is defined as $\mathsf{E}\,\varphi := \neg\,\mathsf{A}\,\neg\varphi$. This is the case as shown in the following (we need to use reflexivity of R).

Proposition 3.2 *Given a model* $\mathfrak{M} = (W, R, V)$ *and a state* $w \in W$ *and a formula* $\varphi \in \mathsf{Form}(\mathsf{A})$, $\mathfrak{M}, w \Vdash \neg\,\mathsf{A}\,\neg\varphi$ *iff* $\mathfrak{M}, v \Vdash \varphi$ *for some* $v \in W$.

Proof. $\mathfrak{M}, w \Vdash \neg\,\mathsf{A}\,\neg\varphi$ iff $\forall\, v$ (wRv implies $\mathfrak{M}, v \nVdash \mathsf{A}\,\neg\varphi$) iff $\forall\, v$ (wRv implies $\exists\, u\,(\mathfrak{M}, u \nVdash \neg\varphi)$) iff $\forall\, v$ (wRv implies $\exists\, u\, \exists\, x\,(uRx$ and $\mathfrak{M}, x \Vdash \varphi)$) iff $\exists\, u\, \exists\, x$ (uRx and $\mathfrak{M}, x \Vdash \varphi$). The last statement implies $\mathfrak{M}, x \Vdash \varphi$ for some $x \in W$. Moreover, the converse direction of this is trivial by reflexivity of R. □

Therefore, we can define $\mathsf{E}\,\varphi := \neg\,\mathsf{A}\,\neg\varphi$ and obtain the following satisfaction clause:

$$\mathfrak{M}, w \Vdash \mathsf{E}\,\varphi \text{ iff } \mathfrak{M}, v \Vdash \varphi \text{ for some } v \in W.$$

Similarly to Form, we define the notions of validity, definability, etc. also for $\mathsf{Form}(\mathsf{A})$ hence also for $\bigvee \mathsf{A}\,\mathsf{Form}$. Some undefinable frame properties in the syntax of intuitionistic logic of Proposition 2.17 become definable with the help of A as follows.

Proposition 3.3 (i) $\mathsf{A}\,p \vee \mathsf{A}\,\neg p$ *defines* $\forall\, x, y\,(xRy)$.

(ii) $\mathsf{A}(p \to q) \vee \mathsf{A}(q \to p)$ *defines* $\forall\, x, y\,(xRy$ *or* $yRx)$.

(iii) $\mathsf{A}\,\neg p \vee \mathsf{A}\,\neg\neg p$ *defines* $\forall\, x, y\, \exists\, z\,(xRz$ *and* $yRz)$.

Proof.

(i) Fix any frame $\mathfrak{F} = (W, R)$. Suppose the frame property $\forall\, x, y\,(xRy)$. To show the validity of $\mathsf{A}\,p \vee \mathsf{A}\,\neg p$, fix any valuation V and any state $w \in W$ such that $\mathfrak{M}, w \nVdash \mathsf{A}\,\neg p$ where $\mathfrak{M} = (\mathfrak{F}, V)$. It follows that we can find states $v, u \in W$ such that vRu and $\mathfrak{M}, u \Vdash p$. To show $\mathfrak{M}, w \Vdash \mathsf{A}\,p$, fix any $a \in W$. Our goal is to show that $\mathfrak{M}, a \Vdash p$. By the supposed property, we get uRa. By $\mathfrak{M}, u \Vdash p$, we can conclude $\mathfrak{M}, a \Vdash p$.

Conversely, suppose that $\mathfrak{F} \Vdash \mathsf{A}\,p \vee \mathsf{A}\,\neg p$. Fix any $x, y \in W$. We show xRy. Define a valuation V such that $V(p) = \uparrow x$, which is an upset. By the supposition, we get 1) $V(p) = W$ or 2) $[\![\neg p]\!]_{(\mathfrak{F}, V)} = W$. But the case 2) is impossible by reflexivity of R and $V(p) = \uparrow x$. So, we get case 1), which implies $y \in \uparrow x$ hence xRy.

(ii) Fix any frame $\mathfrak{F} = (W, R)$. Suppose the property $\forall\, x, y\,(xRy$ or $yRx)$. To show $\mathfrak{F} \Vdash \mathsf{A}(p \to q) \vee \mathsf{A}(q \to p)$, fix any valuation V and any state $w \in W$. Put $\mathfrak{M} = (\mathfrak{F}, V)$ and assume that $\mathfrak{M}, w \nVdash \mathsf{A}(p \to q)$. This implies that we can find states v and u such that vRu, $\mathfrak{M}, u \Vdash p$ and $\mathfrak{M}, u \nVdash q$. We prove that $\mathfrak{M}, w \Vdash \mathsf{A}(q \to p)$. So, fix any a and b such that aRb and $\mathfrak{M}, b \Vdash q$. Our goal is to show $\mathfrak{M}, b \Vdash p$. By $\mathfrak{M}, b \Vdash q$ and $\mathfrak{M}, u \nVdash q$, bRu fails. By the supposed frame property, we get uRb. By $\mathfrak{M}, u \Vdash p$, we conclude $\mathfrak{M}, b \Vdash p$.

Conversely, suppose that $\mathfrak{F} \Vdash \mathsf{A}(p \to q) \vee \mathsf{A}(q \to p)$. Fix any $x, y \in W$. We show that xRy or yRx. Define a valuation V such that $V(p) = \uparrow x$ and $V(q) = \uparrow y$. By the supposition, we can derive $\uparrow x \cap V(p) \subseteq V(q)$ or $\uparrow y \cap V(q) \subseteq V(p)$, which implies xRy or yRx, as required.

(iii) Fix any frame $\mathfrak{F} = (W, R)$. Suppose that $\forall x, y \exists z\,(xRz$ and $yRz)$ and fix any valuation V and state $w \in W$ such that $\mathfrak{M}, w \nVdash \mathsf{A}\neg p$, where $\mathfrak{M} = (\mathfrak{F}, V)$. It follows that we can find states v and u such that vRu and $\mathfrak{M}, u \Vdash p$. We show $\mathfrak{M}, w \Vdash \mathsf{A}\neg\neg p$. So, fix any x. We show $\mathfrak{M}, x \Vdash \neg\neg p$. Moreover, fix any y such that xRy. Our goal now is $\mathfrak{M}, y \nVdash \neg p$. By applying the supposed frame property for states u and y, we can find a state z such that uRz and yRz. It follows from $\mathfrak{M}, u \Vdash p$ that $\mathfrak{M}, z \Vdash p$. Together with yRz, we can conclude that $\mathfrak{M}, y \nVdash \neg p$.

Conversely, suppose that $\mathfrak{F} \Vdash \mathsf{A}\neg p \vee \mathsf{A}\neg\neg p$. We show that $\forall x, y \exists z\,(xRz$ and $yRz)$. Fix any $x, y \in W$. Define a valuation V such that $V(p) := \uparrow x$. Put $\mathfrak{M} = (\mathfrak{F}, V)$. We show that $\mathfrak{M}, x \nVdash \mathsf{A}\neg p$, i.e., there exists v such that $\mathfrak{M}, v \Vdash \neg p$. This holds since xRx and $x \in V(p) = \uparrow x$. Buy the supposition, we get $\mathfrak{M}, x \Vdash \mathsf{A}\neg\neg p$. Thus, $\mathfrak{M}, y \Vdash \neg\neg p$ holds. It follows that $\mathfrak{M}, y \nVdash \neg p$ by yRy. Therefore, we can find a state z such that yRz and $z \in V(p)$, i.e., xRz, as desired.

□

We remark that $\bigvee\mathsf{A}\,\mathsf{Form}$-definable frame class is *not* closed under taking disjoint unions because $\mathsf{A}p \vee \mathsf{A}\neg p$ defines $\forall x, y\,(xRy)$ (it is remarked that, when we assume antisymmetry, the same formula defines $\#W = 1$, i.e., the cardinality of the domain is 1). Therefore, $\mathsf{Form}(\mathsf{A})$-definable frame class is also *not* closed under taking disjoint unions.

Proposition 3.4 (i) *Let f be a surjective bounded morphism from $\mathfrak{M} = (W, R, V)$ to $\mathfrak{M}' = (W', R', V')$. Then, for every formula $\varphi \in \mathsf{Form}(\mathsf{A})$ and $w \in W$, $\mathfrak{M}, w \Vdash \varphi$ iff $\mathfrak{M}', f(w) \Vdash \varphi$.*

(ii) *Given any model $\mathfrak{M} = (W, R, V)$, the following equivalence holds: for every formula $\varphi \in \mathsf{Form}(\mathsf{A})$, $\mathfrak{pe}\,\mathfrak{M}, \mathcal{F} \Vdash \varphi$ iff $[\![\varphi]\!]_{\mathfrak{M}} \in \mathcal{F}$.*

Proof. We only show the case where φ is of the form $\mathsf{A}\psi$ for both items.

(i) Since the right-to-left direction is easy, we focus on the converse direction. Suppose that $\mathfrak{M}, w \Vdash \mathsf{A}\psi$. To show $\mathfrak{M}', f(w) \Vdash \mathsf{A}\psi$, fix any state $v' \in W'$. Since f is surjective, there exists $v \in W$ such that $f(v) = v'$. By our supposition, $\mathfrak{M}, v \Vdash \psi$ hence $\mathfrak{M}', f(v) \Vdash \psi$, which is our goal.

(ii) Recall that $[\![\mathsf{A}\psi]\!]_{\mathcal{M}} = W$ or $\varnothing$. First, we prove the right-to-left direction and so assume that $[\![\mathsf{A}\psi]\!]_{\mathfrak{M}} \in \mathcal{F}$. Since $\varnothing \notin \mathcal{F}$, $[\![\mathsf{A}\psi]\!]_{\mathfrak{M}} = W$. It also follows that $[\![\psi]\!]_{\mathfrak{M}} = W \in \mathcal{F}'$ for all prime filters $\mathcal{F}'$. By induction hypothesis, we get $\mathfrak{pe}\,\mathfrak{M}, \mathcal{F} \Vdash \mathsf{A}\psi$, as required. Second, we prove the left-to-right direction. Suppose that $[\![\mathsf{A}\psi]\!]_{\mathfrak{M}} \notin \mathcal{F}$. Then $[\![\mathsf{A}\psi]\!]_{\mathfrak{M}} \neq W$ and so $[\![\mathsf{A}\psi]\!]_{\mathfrak{M}} = \varnothing$. It follows that $\mathfrak{M}, w \nVdash \psi$ for some $w \in W$. So, $W \nsubseteq [\![\psi]\!]_{\mathfrak{M}}$. Then we can find a prime filter $\mathcal{F}'$ such that $W \in \mathcal{F}'$ but $[\![\psi]\!]_{\mathfrak{M}} \notin \mathcal{F}'$. By induction hypothesis, we obtain $\mathfrak{pe}\,\mathfrak{M}, \mathcal{F}' \nVdash \psi$ hence $\mathfrak{pe}\,\mathfrak{M}, \mathcal{F} \nVdash \mathsf{A}\psi$, as desired. □

Proposition 3.5 (i) *If $\mathfrak{F} \twoheadrightarrow \mathfrak{G}$ and $\mathfrak{F} \Vdash \varphi$ then $\mathfrak{G} \Vdash \varphi$ for all $\varphi \in \mathsf{Form}(\mathsf{A})$.*

(ii) *If $\mathfrak{pe}\,\mathfrak{F} \Vdash \varphi$, then $\mathfrak{F} \Vdash \varphi$ for all $\varphi \in \mathsf{Form}(\mathsf{A})$.*

(iii) *If $\mathfrak{G} \rightarrowtail \mathfrak{F}$, then $\mathfrak{F} \Vdash \rho$ implies $\mathfrak{G} \Vdash \rho$ for all disjunctive A-clauses.*

Proof. Items (i) and (ii) follow from Proposition 3.4 (i) and (ii), respectively. Let us prove item (iii). Let $\rho = \bigvee_{i\in I} \mathsf{A}\,\varphi_i$ where $\varphi_i \in \mathsf{Form}$, i.e., an intuitionistic formula. Assume that $\mathfrak{G}$ is a generated subframe of $\mathfrak{F}$. Suppose also that $\mathfrak{F} \Vdash \bigvee_{i\in I} \mathsf{A}\,\varphi_i$. Our goal is to show $\mathfrak{G} \Vdash \bigvee_{i\in I} \mathsf{A}\,\varphi_i$. Fix any valuation V and $w \in |\mathfrak{G}|$. We show $(\mathfrak{G}, V), w \Vdash \bigvee_{i\in I} \mathsf{A}\,\varphi_i$, i.e., we show that there exists $i \in I$ such that $[\![\varphi_i]\!]_{(\mathfrak{G},V)} = |\mathfrak{G}|$. Because V is also a valuation on $|\mathfrak{F}|$, we get from the supposition that $[\![\varphi_i]\!]_{(\mathfrak{F},V)} = |\mathfrak{F}|$ for some $i \in I$. Fix such $i \in I$. Let us prove that $|\mathfrak{G}| \subseteq [\![\varphi_i]\!]_{(\mathfrak{G},V)}$. Fix any $v \in |\mathfrak{G}|$. Since $(\mathfrak{F}, V), v \Vdash \varphi_i$ iff $(\mathfrak{G}, V), v \Vdash \varphi_i$ by Proposition 2.9, we conclude from $[\![\varphi_i]\!]_{(\mathfrak{F},V)} = |\mathfrak{F}|$ that $v \in [\![\varphi_i]\!]_{(\mathfrak{G},V)}$. □

Proposition 3.6 *Let $\mathfrak{F} = (W, R)$ be a frame and ρ a disjunctive A-clause. If $\mathfrak{F}_X \Vdash \rho$ for all finite $X \subseteq W$ then $\mathfrak{F} \Vdash \rho$, where recall that $\mathfrak{F}_Z$ is the generated subframe of $\mathfrak{F}$ by Z.*

Proof. Let $\rho = \bigvee_{i\in I} \mathsf{A}\,\varphi_i$ where $\varphi_i \in \mathsf{Form}$. We prove the contrapositive implication and so assume that $(\mathfrak{F}, V), w \nVdash \bigvee_{i\in I} \mathsf{A}\,\varphi_i$ for some valuation V and state $w \in |\mathfrak{F}|$. Our goal is to show: there exists some finite $X \subseteq |\mathfrak{F}|$ such that $\mathfrak{F}_X \nVdash \bigvee_{i\in I} \mathsf{A}\,\varphi_i$. By assumption, for every choice $i \in I$, there exists $v_i \in |\mathfrak{F}|$ such that $(\mathfrak{F}, V), v_i \nVdash \varphi_i$. Put $X := \{\, v_i \mid i \in I \,\}$ and consider the finitely generated subframe $\mathfrak{F}_X$ of $\mathfrak{F}$ by the finite generator X. Let $V \upharpoonright |\mathfrak{F}_X|$ be a valuation V restricted to the domain $|\mathfrak{F}_X|$. For each $i \in I$, we have $(\mathfrak{F}_X, V \upharpoonright |\mathfrak{F}_X|), v_i \nVdash \varphi_i$ by $(\mathfrak{F}, V), w \nVdash \bigvee_{i\in I} \mathsf{A}\,\varphi_i$ (by Proposition 2.9). This allows us to conclude $\mathfrak{F}_X \nVdash \bigvee_{i\in I} \mathsf{A}\,\varphi_i$. □

Definition 3.7 We say that a class $\mathbb{F}$ of frames *reflects finitely generated subframes* if, for every frame $\mathfrak{F} = (W, R)$, whenever $\mathfrak{F}_X \in \mathbb{F}$ for all finite $X \subseteq W$, it holds that $\mathfrak{F} \in \mathbb{F}$.

Proposition 3.8 (i) *Each of antisymmetry and $\exists\, y\, \forall\, x\, (xRy)$ is not definable by any subset of $\mathsf{Form}(\mathsf{A})$.*

(ii) *Each of $\exists\, x, y\, (xRy$ and $x \neq y)$ and $\exists\, y\, \forall\, x\, (yRx)$ is not definable by any set of disjunctive A-clauses.*

Proof. For (i), it suffices to show that $\exists\, y\, \forall\, x\, (xRy)$ is not definable by any subset of $\mathsf{Form}(\mathsf{A})$ since we can use the same argument as in the proof of Proposition 2.17 for antisymmetry with the help of Proposition 3.5 (i). Consider $(\mathbb{N}, \leqslant)$, where $\leqslant$ is the ordinary partial ordering. Since $\wp^{\uparrow}(\mathbb{N}) = \{\, \varnothing \,\} \cup \{\, \uparrow n \mid n \in \mathbb{N} \,\}$, all the prime filters consist of $\{\, \uparrow n \mid n \in \mathbb{N} \,\}$ and $\mathcal{F}_n := \{\, X \in \wp^{\uparrow}(\mathbb{N}) \mid n \in X \,\} = \{\, \uparrow 0, \uparrow 1, \ldots, \uparrow n \,\}$ $(n \in \mathbb{N})$. Then, it is easy to see that $(\mathrm{Pf}(\mathbb{N}), \leqslant^{\mathfrak{pe}})$ satisfies $\exists\, y\, \forall\, x\, (xRy)$ ($\{\, \uparrow n \mid n \in \mathbb{N} \,\}$ is a maximum element) but $(\mathbb{N}, \leqslant)$ does not. Thus, Proposition 3.5 (ii) implies the intended undefinability.

For (ii), we only prove that $\exists\, y\, \forall\, x\, (yRx)$ is undefinable by any subset of $\bigvee \mathsf{A}\,\mathsf{Form}$, since the other property $\exists\, x, y\, (xRy$ and $x \neq y)$ is undefinable by the same argument given in the proof of Proposition 2.17 with the help of

Proposition 3.5 (iii). Consider the set of all integers $(\mathbb{Z}, \leqslant)$ with the ordinary partial ordering $\leqslant$. Then, all finitely generated subframes of $(\mathbb{Z}, \leqslant)$ satisfy $\exists y \forall x\,(yRx)$ but the original frame does not. Then Proposition 3.6 implies the undefinability in $\bigvee \mathsf{A}\,\mathsf{Form}$. □

4 Characterizing Elementary Frame Definability by Intuitionistic Logic with the Universal Modality

We employ van Benthem [25]'s purely model-theoretic argument for characterizing elementary frame definability of both $\mathsf{Form}(\mathsf{A})$ and $\bigvee \mathsf{A}\,\mathsf{Form}$.

Definition 4.1 Let Γ be a set of formulas, $\mathfrak{M}$ a model and $\mathbb{F}$ a frame class. We say that Γ is *satisfiable* in $\mathfrak{M}$ if there exists a state w in $\mathfrak{M}$ such that $\mathfrak{M}, w \Vdash \Gamma$ and that Γ is *finitely satisfiable* in $\mathfrak{M}$ if every finite subset of Γ is satisfiable in $\mathfrak{M}$. The set Γ is *satisfiable* in $\mathbb{F}$ if there exists a frame $\mathfrak{F} \in \mathbb{F}$ and a valuation V on $\mathfrak{F}$ such that Γ is satisfiable in $(\mathfrak{F}, V)$, and Γ is *finitely satisfiable* in $\mathbb{F}$ if every finite subset of Γ is satisfiable in $\mathbb{F}$.

In what follows in this section, we use some notions from first-order model theory such as (finite) satisfiability, compactness, elementary extension and ω-saturation, and so, the reader is unfamiliar with those is referred to [1].

4.1 Goldblatt-Thomason Theorem for $\mathsf{Form}(\mathsf{A})$

Theorem 4.2 *For any elementary frame class $\mathbb{F}$, the following are equivalent:*

(i) *$\mathbb{F}$ is definable by a subset of* $\mathsf{Form}(\mathsf{A})$,

(ii) *$\mathbb{F}$ is closed under taking bounded morphic images and it reflects prime filter extensions.*

If we replace "prime fitter extensions" with "ultrafilter extensions", we can obtain Gargov and Goranko [9]'s Goldblatt-Thomason-type characterization for modal logic with the universal modality. So, Theorem 4.2 can be regarded as the intuitionistic version of their result.

Proof. The direction from (i) to (ii) is due to Proposition 3.5. So, we focus on the direction from (ii) to (i) and so let us assume (ii). We show that $\mathbb{F}$ is defined by $\mathsf{Log}(\mathbb{F}) := \{\, \varphi \in \mathsf{Form}(\mathsf{A}) \mid \mathbb{F} \Vdash \varphi \,\}$. That is, we show that, for every frame $\mathfrak{F} = (W, R)$, $\mathfrak{F} \in \mathbb{F}$ iff $\mathfrak{F} \Vdash \mathsf{Log}(\mathbb{F})$. Let us fix any frame $\mathfrak{F} = (W, R)$. When $\mathfrak{F} \in \mathbb{F}$, it is easy to see that $\mathfrak{F} \Vdash \mathsf{Log}(\mathbb{F})$. Conversely, we suppose that $\mathfrak{F} \Vdash \mathsf{Log}(\mathbb{F})$. The rest of this proof is devoted to establishing $\mathfrak{F} \in \mathbb{F}$. Let us expand our syntax with a (possibly uncountably infinite) set $\{\, p_A \mid A \in \wp^{\uparrow}(W) \,\}$. Remark that we can still keep the supposition $\mathfrak{F} \Vdash \mathsf{Log}(\mathbb{F})$ even if we regard $\mathsf{Log}(\mathbb{F})$ as a set of formulas in the expanded language. Moreover, let us define $\Delta_{\mathfrak{F}}$ as the set of all the following formulas:

$$\mathsf{A}(p_{A\cap B} \leftrightarrow (p_A \wedge p_B)),\ \mathsf{A}(p_{A\cup B} \leftrightarrow (p_A \vee p_B)),\ \mathsf{A}(p_{A\Rightarrow B} \leftrightarrow (p_A \rightarrow p_B)),\ \mathsf{A}(p_{\varnothing} \leftrightarrow \bot),$$

where A, $B \in \wp^{\uparrow}(W)$ and recall that $A \Rightarrow B := \{\, w \in W \mid {\uparrow} w \cap A \subseteq B \,\}$. An underlying idea of $\Delta_{\mathfrak{F}}$ is to provide a complete enough description of the frame $\mathfrak{F}$ in terms of the propositional variables $\{\, p_A \mid A \in \wp^{\uparrow}(W) \,\}$.

We are going to show that $\Delta_{\mathfrak{F}}$ is finitely satisfiable in $\mathbb{F}$. So, let us fix any finite $\Delta' \subseteq \Delta_{\mathfrak{F}}$ and suppose for contradiction that $\bigwedge \Delta'$ is unsatisfiable in $\mathbb{F}$, i.e., for all frames $\mathfrak{G} \in \mathbb{F}$, valuations U and states v in $\mathfrak{G}$, we have $(\mathfrak{G}, U), v \nVdash \bigwedge \Delta'$. This implies $\mathbb{F} \Vdash \neg \bigwedge \Delta'$ (note that $\neg$ here is the intuitionistic negation). By our assumption of $\mathfrak{F} \Vdash \mathsf{Log}(\mathbb{F})$, we get $\mathfrak{F} \Vdash \neg \bigwedge \Delta'$. This means that $\bigwedge \Delta'$ is unsatisfiable in $\mathfrak{F}$. But Δ' is clearly satisfiable in $\mathfrak{F}$, a contradiction. Therefore, $\Delta_{\mathfrak{F}}$ is finitely satisfiable in $\mathbb{F}$.

Since $\mathbb{F}$ is elementary, we can deduce from finite satisfiability of $\Delta_{\mathfrak{F}}$ that $\Delta_{\mathfrak{F}}$ is satisfiable in $\mathbb{F}$ by compactness. Thus, there exists a frame $\mathfrak{G} \in \mathbb{F}$, a valuation U on $\mathfrak{G}$, and a state v in $\mathfrak{G}$ such that $(\mathfrak{G}, U), v \Vdash \Delta_{\mathfrak{F}}$. Since all the elements of $\Delta_{\mathfrak{F}}$ are A-prefixed, we get $(\mathfrak{G}, U) \Vdash \Delta_{\mathfrak{F}}$. For an ω-saturated elementary extension $(\mathfrak{G}^*, U^*)$ of $(\mathfrak{G}, U)$ such that $\mathfrak{G}^* \in \mathbb{F}$ (since $\mathbb{F}$ is elementary), we also have $(\mathfrak{G}^*, U^*) \Vdash \Delta_{\mathfrak{F}}$. Now we define a mapping $f : |\mathfrak{G}^*| \to |\mathfrak{pe}\,\mathfrak{F}|$ by: $f(s) := \{\, X \mid (\mathfrak{G}^*, U^*), s \Vdash p_X \,\}$. For this mapping f, the following claim holds.

Claim 1 f is a surjective bounded morphism from $\mathfrak{G}^*$ to $\mathfrak{pe}\,\mathfrak{F}$.

This claim implies $\mathfrak{pe}\,\mathfrak{F} \in \mathbb{F}$ because $\mathfrak{G}^* \in \mathbb{F}$ and $\mathfrak{G}^* \twoheadrightarrow \mathfrak{pe}\,\mathfrak{F}$. Moreover, since $\mathbb{F}$ reflects prime filter extensions, we obtain $\mathfrak{F} \in \mathbb{F}$, as desired.

So, let us provide a proof of the claim below (a basic idea of the proof is from [16, p.132, Lemma 15.2]) to finish the proof of this theorem.

(Proof of Claim) We show that $f : |\mathfrak{G}^*| \to |\mathfrak{pe}\,\mathfrak{F}|$ is a surjective bounded morphism. Let S be the binary relation of $\mathfrak{G}^*$.

ll-defined) We show that $f(s)$ is a prime filter. First, we check that $\varnothing \notin f(s)$ and $f(s) \neq \varnothing$. We have $\varnothing \notin f(s)$ because $p_\varnothing \leftrightarrow \bot$ is valid on $(\mathfrak{G}^*, U^*)$ and $\bot$ is unsatisfiable in $(\mathfrak{G}^*, U^*)$. As for $f(s) \neq \varnothing$, it suffices to note that we can derive from $(\mathfrak{G}^*, U^*) \Vdash \Delta_{\mathfrak{F}}$ that $(\mathfrak{G}^*, U^*), s \Vdash p_W$ hence $W \in f(s)$. The other conditions for prime filter are also established by $(\mathfrak{G}^*, U^*) \Vdash \Delta_{\mathfrak{F}}$.

(Forth) Suppose that sSs'. We prove that $f(s) R^{\mathfrak{pe}} f(s')$, i.e., $f(s) \subseteq f(s')$. Fix any $X \in f(s)$. Then we have $(\mathfrak{G}^*, U^*), s \Vdash p_X$. We want to show that $(\mathfrak{G}^*, U^*), s' \Vdash p_X$. Since the persistency is the first-order condition, the set $U^*(p_X)$ is an upset with respect to S. Therefore, we can conclude $(\mathfrak{G}^*, U^*), s' \Vdash p_X$.

(Back) Fix any $s \in |\mathfrak{G}^*|$ and $\mathcal{F} \in |\mathfrak{pe}\,\mathfrak{F}|$ such that $f(s) R^{\mathfrak{pe}} \mathcal{F}$, i.e., $f(s) \subseteq \mathcal{F}$. We establish that there exists $s' \in |\mathfrak{G}^*|$ such that sSs' and $f(s') = \mathcal{F}$. We need to use ω-saturation here. Let us put a type

$$\Gamma(x) := \{\, p_X(x) \mid X \in \mathcal{F} \,\} \cup \{\, \neg p_X(x) \mid X \notin \mathcal{F} \,\} \cup \{\, \underline{s} \leqslant x \,\}$$

of first-order formulas, where $\underline{s}$ denotes the corresponding constant symbol to s, "$\neg$" of $\neg p_X(x)$ is the classical negation since we are considering the first-order language $\mathcal{L}^1_m$. Now we show that $\Gamma(x)$ is finitely satisfiable in $(\mathfrak{G}^*, U^*)$ in the sense of the first-order model theory. Fix any $\Gamma'(x) := \{\, p_{X_1}(x), \ldots, p_{X_n}(x), \neg p_{Y_1}(x), \ldots, \neg p_{Y_n}(x), \underline{s} \leqslant x \,\}$ where $X_i \in \mathcal{F}$ and $Y_j \notin \mathcal{F}$, and suppose for contradiction that $\Gamma'(x)$ is not satisfiable in

$(\mathfrak{G}^*, U^*)$. It follows that, for every $s' \in |\mathfrak{G}^*|$,

$$(\mathfrak{G}^*, U^*) \models (\underline{s} \leqslant x \wedge \bigwedge_{1 \leqslant i \leqslant n} p_{X_i}(x)) \to \bigvee_{1 \leqslant j \leqslant m} p_{Y_j}(x))[s'].$$

Hence we get:

$$(\mathfrak{G}^*, U^*) \models \forall x\, ((\underline{s} \leqslant x \wedge \bigwedge_{1 \leqslant i \leqslant n} p_{X_i}(x)) \to \bigvee_{1 \leqslant j \leqslant m} p_{Y_j}(x)).$$

By shifting our semantics, this implies from Proposition 2.6 that

$$(\mathfrak{G}^*, U^*), s \Vdash \bigwedge_{1 \leqslant i \leqslant n} p_{X_i} \to \bigvee_{1 \leqslant j \leqslant m} p_{Y_j}$$

where "$\to$" is intuitionistic. It also follows from $(\mathfrak{G}^*, U^*) \Vdash \Delta_{\mathfrak{F}}$ that

$$(\mathfrak{G}^*, U^*), s \Vdash p_{\bigcap_{1 \leqslant i \leqslant n} X_i} \to p_{\bigcup_{1 \leqslant j \leqslant m} Y_j}.$$

Hence $(\mathfrak{G}^*, U^*), s \Vdash p_{X \Rightarrow Y}$, where $X := \bigcap_{1 \leqslant i \leqslant n} X_i$ and $Y := \bigcup_{1 \leqslant j \leqslant m} Y_j$. Thus, $X \Rightarrow Y \in f(s)$. Since $X \in f(s)$, we get $Y = \bigcup_{1 \leqslant j \leqslant m} Y_j \in f(s)$, which implies $Y_j \in f(s) \subseteq \mathcal{F}$ for some $1 \leqslant j \leqslant m$. This is a contradiction with $Y_j \notin \mathcal{F}$ for all indices j.

Therefore, we have shown that $\Gamma(x)$ is finitely satisfiable in $(\mathfrak{G}^*, U^*)$ in the sense of the first-order model theory. This implies that $\Gamma(x)$ is satisfiable in $(\mathfrak{G}^*, U^*)$ by ω-saturation. Thus, fix a solution s' of $\Gamma(x)$ at $(\mathfrak{G}^*, U^*)$. It is easy to see that sSs'. We can also establish $f(s') = \mathcal{F}$ as follows. By $(\mathfrak{G}^*, U^*) \models \Gamma(x)[s']$, it follows that $(\mathfrak{G}^*, U^*) \models p_X(x)[s']$ implies $X \in \mathcal{F}$ and that $(\mathfrak{G}^*, U^*) \not\models p_X(x)[s']$ implies $X \notin \mathcal{F}$. Therefore, $f(s') = \mathcal{F}$.

(Onto) Fix any prime filter $\mathcal{F} \in |\mathfrak{pe}\,\mathfrak{F}|$. Let us put a type

$$\Gamma(x) := \{\, p_X(x) \mid X \in \mathcal{F} \,\} \cup \{\, \neg p_X(x) \mid X \notin \mathcal{F} \,\}$$

of first-order formulas. Similarly to our argument for (Back), we can prove that $\Gamma(x)$ is satisfiable in $(\mathfrak{G}^*, U^*)$ hence $f(s') = \mathcal{F}$ for some $s' \in |\mathfrak{G}^*|$.

This finishes establishing that f is a surjective bounded morphism. ⊣

Therefore, we conclude that $\mathsf{Log}(\mathbb{F})$ defines $\mathbb{F}$. □

4.2 Goldblatt-Thomason Theorem for $\bigvee \mathsf{A\,Form}$

Theorem 4.3 *For any elementary frame class $\mathbb{F}$, the following are equivalent:*

(i) *$\mathbb{F}$ is definable by a set of disjunctive* A*-clauses, i.e., a subset of* $\bigvee \mathsf{A\,Form}$.

(ii) *$\mathbb{F}$ is closed under taking bounded morphic images and generated subframes and $\mathbb{F}$ reflects finitely generated subframes and prime filter extensions.*

If we replace "prime filter extensions" with "ultrafilter extension", we can obtain [20,19]'s Goldblatt-Thomason-style characterization for (extended) modal dependence logic. Therefore, Theorem 4.3 is an intuitionistic variant of the GT-style characterization in [20,19].

Proof. The direction from (i) to (ii) is established by Propositions 3.5 and 3.6 by $\bigvee \mathsf{A}\,\mathsf{Form} \subseteq \mathsf{Form}(\mathsf{A})$. So, we prove the converse direction. Assume (ii). Let us define $\mathsf{Log}_{\vee\mathsf{A}}(\mathbb{F}) := \{\, \rho \in \bigvee \mathsf{A}\,\mathsf{Form} \mid \mathbb{F} \Vdash \rho \,\}$. We show that $\mathsf{Log}_{\vee\mathsf{A}}(\mathbb{F})$ defines $\mathbb{F}$. Let us fix any frame $\mathfrak{F} = (W, R)$. We need to establish the following equivalence: $\mathfrak{F} \in \mathbb{F}$ iff $\mathfrak{F} \Vdash \mathsf{Log}_{\vee\mathsf{A}}(\mathbb{F})$. The left-to-right direction is easy to show, and so, we focus on showing the right-to-left direction. Suppose that $\mathfrak{F} \Vdash \mathsf{Log}_{\vee\mathsf{A}}(\mathbb{F})$. Our goal is to show $\mathfrak{F} \in \mathbb{F}$. The rest of the proof is devoted to showing it. Since $\mathbb{F}$ reflects finitely generated subframes, we can assume without loss of generality that $\mathfrak{F}$ is finitely generated, i.e., generated by a finite set $U \subseteq W$.

We expand our syntax with a set $\{\, p_A \mid A \in \wp^{\uparrow}(W) \,\}$ (which is possibly uncountably infinite). Similarly to the proof of Theorem 4.2, we can still keep $\mathfrak{F} \Vdash \mathsf{Log}_{\vee\mathsf{A}}(\mathbb{F})$ even if we regard $\mathsf{Log}_{\vee\mathsf{A}}(\mathbb{F})$ as a set of formulas in the expanded language. We define Δ as the set of all the following formulas:

$$p_{A\cap B} \leftrightarrow (p_A \wedge p_B),\ p_{A\cup B} \leftrightarrow (p_A \vee p_B),\ p_{A\Rightarrow B} \leftrightarrow (p_A \to p_B),\ p_{\varnothing} \leftrightarrow \bot,$$

where A, $B \in \wp^{\uparrow}(W)$. Moreover, we put $\Delta_{\mathfrak{F},u} := \{\, p_{\uparrow u} \wedge \varphi \mid \varphi \in \Delta \,\}$ for each $u \in U$. Since $\mathfrak{F}$ is finitely generated by U, $(\Delta_{\mathfrak{F},u})_{u\in U}$ encodes a complete enough description of $\mathfrak{F}$ in terms of the propositional variables $\{\, p_A \mid A \in \wp^{\uparrow}(W) \,\}$.

In what follows, we need to employ a different strategy from our proof of Theorem 4.2. Let us introduce a finite set $\{\, x_u \mid u \in U \,\}$ of mutually disjoint variables and ST_{x_u} be the standard translation from Form to all the formulas in the first-order correspondence model language $\mathcal{L}^1_m$.

We are going to show $\bigcup_{u\in U} \mathrm{ST}_{x_u}[\Delta_{\mathfrak{F},u}]$ is finitely satisfiable in $\mathbb{F}$, where $\mathrm{ST}_{x_u}[\Phi]$ is the direct image of Φ under the standard translation ST_{x_u}, i.e., $\{\, \mathrm{ST}_{x_u}(\varphi) \mid \varphi \in \Phi \,\}$. Let $\Gamma \subseteq \bigcup_{u\in U} \mathrm{ST}_{x_u}[\Delta_{\mathfrak{F},u}]$ be a finite set. Then we may write $\Gamma = \bigcup_{1\leqslant k\leqslant n} \mathrm{ST}_{u_k}[\Gamma_{u_k}]$ for some $u_1, \ldots, u_n \in U$ and some finite $\Gamma_{u_k} \subseteq \Delta_{\mathfrak{F},u_k}$ $(1 \leqslant k \leqslant n)$. Assume for contradiction that Γ is not satisfiable in $\mathbb{F}$, i.e.,

$$\forall\, \mathfrak{G} \in \mathbb{F}\, \forall V\, \forall \overrightarrow{a} \in |\mathfrak{F}|^n\, \exists\, 1 \leqslant k \leqslant n\, \left((\mathfrak{G}, V) \models \neg \mathrm{ST}_{x_{u_k}} \left(\bigwedge \Gamma_{u_k} \right) [\overrightarrow{a}] \right).$$

where $\overrightarrow{a} := (a_1, \ldots, a_n)$. This is also equivalent to:

$$\forall\, \mathfrak{G} \in \mathbb{F}\, \forall V\, \forall \overrightarrow{a} \in |\mathfrak{F}|^n\, \exists\, 1 \leqslant k \leqslant n\, \left((\mathfrak{G}, V) \models \neg \mathrm{ST}_{x_{u_k}} \left(\bigwedge \Gamma_{u_k} \right) [a_k] \right).$$

where we rewrite the assignment for variables. By first-order reasoning (in particular, we use the validity of $\forall x\, \forall y\, (P(x) \vee Q(y)) \to \forall x\, P(x) \vee \forall y\, P(y))$, we get:

$$\forall\, \mathfrak{G} \in \mathbb{F}\, \forall V\, \left((\mathfrak{G}, V) \models \bigvee_{1\leqslant k\leqslant n} \forall x_{u_k}\, \neg \mathrm{ST}_{x_{u_k}} \left(\bigwedge \Gamma_{u_k} \right) \right).$$

where "$\neg$" above is the classical negation. By changing our semantics to Kripke semantics, this also implies $\mathbb{F} \Vdash \bigvee_{1\leqslant k\leqslant n} \mathsf{A} \neg \bigwedge \Gamma_{u_k}$ by Proposition 2.6, where $\neg$ is the intuitionistic negation. It is noted that $\neg \bigwedge \Gamma_{u_k} \in \mathsf{Form}$, i.e., an intuitionistic formula and so $\bigvee_{1\leqslant k\leqslant n} \mathsf{A} \neg \bigwedge \Gamma_{u_k}$ is a disjunctive A-clause. Therefore,

$\bigvee_{1\leqslant k\leqslant n} \mathsf{A}\neg\bigwedge\Gamma_{u_k} \in \mathsf{Log}_{\vee\mathsf{A}}(\mathbb{F})$. Since we have assumed $\mathfrak{F} \Vdash \mathsf{Log}_{\vee\mathsf{A}}(\mathbb{F})$, we obtain $\mathfrak{F} \Vdash \bigvee_{1\leqslant k\leqslant n} \mathsf{A}\neg\bigwedge\Gamma_{u_k}$. This implies that Γ is not satisfiable in $\mathfrak{F}$ in the sense of first-order model theory. But Γ is clearly satisfiable in $\mathfrak{F}$, which implies a desired contradiction. We have shown that $\bigcup_{u\in U} \mathrm{ST}_{x_u}[\Delta_{\mathfrak{F},u}]$ is finitely satisfiable in $\mathbb{F}$.

Since $\mathbb{F}$ is elementary, $\bigcup_{u\in U} \mathrm{ST}_{x_u}[\Delta_{\mathfrak{F},u}]$ is satisfiable in $\mathbb{F}$ by compactness. We can find a frame $\mathfrak{G} \in \mathbb{F}$, a valuation U on $|\mathfrak{G}|$ and a finite sequence $\overrightarrow{w} = (w_u)_{u\in U}$ such that $(\mathfrak{G}, U) \models \bigcup_{u\in U} \mathrm{ST}_{x_u}[\Delta_{\mathfrak{F},u}][\overrightarrow{w}]$. By changing our semantics "($\models$)" to Kripke semantics ("$\Vdash$"), it follows that $(\mathfrak{G}, U), w_u \Vdash \Delta_{\mathfrak{F},u}$ by Proposition 2.6. Let us put $Z := \{ w_u \mid u \in U \}$. Let $(\mathfrak{G}^*_Z, V^*_Z)$ be an ω-saturated elementary extension of the Z-generated submodel $(\mathfrak{G}_Z, V_Z)$ of $(\mathfrak{G}, V)$. Because $\mathbb{F}$ is elementary and closed under taking generated subframes, we have $\mathfrak{G}_Z \in \mathbb{F}$ hence $\mathfrak{G}^*_Z \in \mathbb{F}$. It is also noted that $(\mathfrak{G}^*_Z, V^*_Z), w^*_u \Vdash \Delta_{\mathfrak{F},u}$ where w^*_u is the corresponding element in $\mathfrak{G}^*_Z$ to w_u in $|\mathfrak{G}_Z|$. Since $(\mathfrak{G}_Z, V_Z) \models \forall x\, \mathrm{ST}_x(\theta)$ for all $\theta \in \Delta$, we also get $(\mathfrak{G}^*_Z, V^*_Z) \models \forall x\, \mathrm{ST}_x(\theta)$ for all $\theta \in \Delta$, which implies $(\mathfrak{G}^*_Z, V^*_Z) \Vdash \Delta$.

Now we claim that $\mathfrak{G}^*_Z \twoheadrightarrow \mathfrak{pe}\,\mathfrak{F}$. By this claim and the closure and reflection properties of $\mathbb{F}$, we can conclude from $\mathfrak{G}^*_Z \in \mathbb{F}$ that $\mathfrak{F} \in \mathbb{F}$, as required. So, let us justify the claim. Define $f : |\mathfrak{G}^*_Z| \to |\mathfrak{pe}\,\mathfrak{F}|$ by: $f(s) := \{ X \subseteq W \mid (\mathfrak{G}^*_Z, V^*_Z), s \Vdash p_X \}$. We prove that f is a surjective bounded morphism. But the proof is almost the same as in the proof of Theorem 4.2, since $(\mathfrak{G}^*_Z, V^*_Z) \Vdash \Delta$. This finishes establishing the goal of $\mathfrak{F} \in \mathbb{F}$. Therefore, we conclude that $\mathsf{Log}_{\vee\mathsf{A}}(\mathbb{F})$ defines $\mathbb{F}$. □

5 Characterizing Elementary Frame Definability by Intuitionistic Inquisitive Logic

5.1 Team Semantics for Intuitionistic Logic

Definition 5.1 Let $\mathfrak{M} = (W, R, V)$ be a model. We say that $t \subseteq W$ is a *team*. Given a model $\mathfrak{M}$, a team $t \subseteq W$ and a formula $\varphi \in \mathsf{Form}$, the satisfaction relation $\mathfrak{M}, t \Vdash \varphi$ is defined inductively as follows:

$$\begin{array}{lll}
\mathfrak{M}, t \Vdash p & \text{iff} & t \subseteq V(p) \\
\mathfrak{M}, t \Vdash \bot & \text{iff} & t = \varnothing \\
\mathfrak{M}, t \Vdash \varphi \wedge \psi & \text{iff} & \mathfrak{M}, t \Vdash \varphi \text{ and } \mathfrak{M}, t \Vdash \psi \\
\mathfrak{M}, t \Vdash \varphi \vee \psi & \text{iff} & \exists t_1, t_2 (t = t_1 \cup t_2 \text{ and } \mathfrak{M}, t_1 \Vdash \varphi \text{ and } \mathfrak{M}, t_2 \Vdash \psi) \\
\mathfrak{M}, t \Vdash \varphi \to \psi & \text{iff} & \forall s \subseteq R[t]\, (\mathfrak{M}, s \Vdash \varphi \text{ implies } \mathfrak{M}, s \Vdash \psi),
\end{array}$$

where $R[t] := \{ v \in W \mid wRv \text{ for some } w \in t \}$.

For the negation, we can provide the following satisfaction clause:

$$\mathfrak{M}, t \Vdash \neg\varphi \text{ iff } \forall s \subseteq R[t]\, (s \neq \varnothing \text{ implies } \mathfrak{M}, s \nVdash \varphi).$$

By induction on φ, we can prove the following (see [5, Proposition 3.11]).

Proposition 5.2 *Let $\mathfrak{M}$ be a model. For all formulas $\varphi \in \mathsf{Form}$ and states w, $\mathfrak{M}, \{ w \} \Vdash \varphi$ iff $\mathfrak{M}, w \Vdash \varphi$.*

The following is from [5, Proposition 3.10].

Proposition 5.3 (Flatness) *Let $\mathfrak{M}$ be a model. For all formulas $\varphi \in \mathsf{Form}$ and teams $t \subseteq |\mathfrak{M}|$, $\mathfrak{M}, t \Vdash \varphi$ iff $\mathfrak{M}, \{w\} \Vdash \varphi$ for all $w \in t$.*

5.2 Intuitionistic Inquisitive Logic

We expand the syntax of intuitionistic logic with *inquisitive disjunction* $\mathbin{\rotatebox[origin=c]{-90}{$\geqslant$}}$.

Definition 5.4 The set $\mathsf{Form}(\mathbin{\rotatebox[origin=c]{-90}{$\geqslant$}})$ of all formulas for intuitionistic logic is defined inductively as:

$$\mathsf{Form}(\mathbin{\rotatebox[origin=c]{-90}{$\geqslant$}}) \ni \varphi ::= p \mid \bot \mid \varphi \wedge \varphi \mid \varphi \vee \varphi \mid \varphi \to \varphi \mid \varphi \mathbin{\rotatebox[origin=c]{-90}{$\geqslant$}} \varphi \quad (p \in \mathsf{Prop}).$$

Given any model $\mathfrak{M} = (W, R, V)$, the satisfaction relation in team semantics for the inquisitive disjunction is defined similarly to Definition 5.1 except:

$$\mathfrak{M}, t \Vdash \varphi \mathbin{\rotatebox[origin=c]{-90}{$\geqslant$}} \psi \text{ iff } \mathfrak{M}, t \Vdash \varphi \textit{ or } \mathfrak{M}, t \Vdash \psi$$

We note that flatness fails for $\mathsf{Form}(\mathbin{\rotatebox[origin=c]{-90}{$\geqslant$}})$, but we can still keep the persistency (see [5, Proposition 2.9]).

Proposition 5.5 (Persistency) *If $M, t \Vdash \varphi$ and $s \subseteq R[t]$ then $M, s \Vdash \varphi$, for all $\varphi \in \mathsf{Form}(\mathbin{\rotatebox[origin=c]{-90}{$\geqslant$}})$.*

Proposition 5.6 *Given any model $\mathfrak{M} = (W, R, V)$ and a formula $\varphi \in \mathsf{Form}(\mathbin{\rotatebox[origin=c]{-90}{$\geqslant$}})$, $\mathfrak{M}, t \Vdash \varphi$ for all teams $t \subseteq W$ iff $\mathfrak{M}, W \Vdash \varphi$.*

Proof. Since $R[W] = W$ (recall that R is reflexive and transitive), the statement follows from Proposition 5.5. □

Definition 5.7 We say that $\varphi \in \mathsf{Form}(\mathbin{\rotatebox[origin=c]{-90}{$\geqslant$}})$ is *valid* in a model $\mathfrak{M} = (W, R, V)$ (notation: $\mathfrak{M} \Vdash_T \varphi$, where the subscript "$T$" is used for emphasizing "Team semantics") if $\mathfrak{M}, W \Vdash \varphi$. A set $\Gamma \subseteq \mathsf{Form}(\mathbin{\rotatebox[origin=c]{-90}{$\geqslant$}})$ is *valid* in a frame $\mathfrak{M} = (W, R, V)$ (notation: $\mathfrak{F} \Vdash_T \Gamma$) if $(\mathfrak{F}, V) \Vdash_T \varphi$ for all formulas $\varphi \in \Gamma$ and valuations V.

Based on this notion of validity in a frame, we define the notion of frame definability as before. The following proposition is an immediate consequence from [5, Theorem 4.9].

Proposition 5.8 *For every $\varphi \in \mathsf{Form}(\mathbin{\rotatebox[origin=c]{-90}{$\geqslant$}})$, there are finitely many intuitionistic formulas $(\psi_i)_{i \in I} \subseteq \mathsf{Form}$ such that φ and $\mathbin{\rotatebox[origin=c]{-90}{$\geqslant$}}_{i \in I} \psi_i$ are equivalent, i.e., $\mathfrak{M}, t \Vdash \varphi$ iff $\mathfrak{M}, t \Vdash \mathbin{\rotatebox[origin=c]{-90}{$\geqslant$}}_{i \in I} \psi_i$ for every model $\mathfrak{M}$ and team $t \subseteq |\mathfrak{M}|$.*

5.3 Goldblatt-Thomason Theorem for Intuitionistic Inquisitive Logic

Proposition 5.9 *For any finite family $(\psi_i)_{i \in I} \subseteq \mathsf{Form}$ (i.e., I is finite) and model $\mathfrak{M} = (W, R, V)$, $\mathfrak{M} \Vdash \bigvee_{i \in I} \mathsf{A} \psi_i$ iff $\mathfrak{M} \Vdash_T \mathbin{\rotatebox[origin=c]{-90}{$\geqslant$}}_{i \in I} \psi_i$*

Proof. The equivalence is verified as follows: $\mathfrak{M} \Vdash \bigvee_{i \in I} \mathsf{A} \psi_i$ iff

there exists $i \in I$ such that $[\![\psi_i]\!]_{\mathfrak{M}} = W$
iff there exists $i \in I$ such that $\mathfrak{M}, w \Vdash \psi_i$ for all $w \in W$
iff there exists $i \in I$ such that $\mathfrak{M}, W \Vdash \psi_i$ (by Propositions 5.3 and 5.5)

and the last line is equivalent to $\mathfrak{M}, W \Vdash \mathbin{\rotatebox[origin=c]{-90}{$\geqslant$}}_{i \in I} \psi_i$ hence $\mathfrak{M} \Vdash_T \mathbin{\rotatebox[origin=c]{-90}{$\geqslant$}}_{i \in I} \psi_i$. □

Proposition 5.10 *For any class $\mathbb{F}$ of frames, the following are equivalent:*

(i) $\mathbb{F}$ *is definable by a set of disjunctive* A*-clauses.*

(ii) $\mathbb{F}$ *is definable by a set of formulas of intuitionistic inquisitive logic.*

Proof. We can establish the direction from (i) to (ii) by Proposition 5.9. The direction from (ii) to (i) follows from Propositions 5.8 and 5.9. □

By Propositions 5.10 and 3.3, we obtain the following frame definability results in the syntax of inquisitive intuitionistic logic.

Proposition 5.11 (i) $p \vee\!\!\!\vee \neg p$ *defines* $\forall x, y\,(xRy)$.

(ii) $(p \to q) \vee\!\!\!\vee (q \to p)$ *defines* $\forall x, y\,(xRy$ *or* $yRx)$.

(iii) $\neg p \vee\!\!\!\vee \neg\neg p$ *defines* $\forall x, y\, \exists z\,(xRz$ *and* $yRz)$.

By Proposition 5.10, we can also transfer the undefinability results from Proposition 3.8. For example, all the frame properties listed in Proposition 3.8 are also undefinable in the syntax of intuitionistic inquisitive logic.

Then, we can finally give GT-style characterization to intuitionistic inquisitive logic as follows.

Theorem 5.12 *An elementary frame class* $\mathbb{F}$ *is definable by a set of formulas of intuitionistic inquisitive logic iff* $\mathbb{F}$ *is closed under taking bounded morphic images, generated subframes and it reflects finitely generated subframes and prime filter extensions.*

Proof. By Proposition 5.10 and Theorem 4.3. □

6 Further Direction

There are several directions of further research. The first direction is that we may characterize relative frame definability of intuitionistic inquisitive logic within finite frames, as [19] did for modal dependence logic. For intuitionistic formulas, Rodenburg [16] provided a finitary version of Goldblatt-Thomason Theorem. The second direction is on model definability of both intuitionistic logic with the universal modality and intuitionistic inquisitive logic. Goldblatt [10] studied the characterization of intuitionistic definability of modal class. We may extend his result to this context.

As the final direction, we may define the notion of "normal form" of a formula of $\mathsf{Form}(\mathsf{A})$ in the spirit of [9]. Let us define $\mathsf{Form}(\mathsf{A}^+)$ as the set of all formulas φ in $\mathsf{Form}(\mathsf{A})$ such that all occurrences of A in φ are *positive*. For a formula in $\mathsf{Form}(\mathsf{A}^+)$, can we find an equivalent disjunctive A-clause via the normal form? A similar result held for modal logic with the universal modality as in [19]. This is ongoing work with Jonni Virtema.

References

[1] Chang, C. C. and H. J. Keisler, "Model Theory," North-Holland Publishing Company, Amsterdam, 1990, 3 edition.

[2] Ciardelli, I., *Dependency as question entailment*, in: S. Abramsky, J. Kontinen, J. Väänänen and H. Vollmer, editors, *Dependence Logic: Theory and Applications*, Springer, 2016 pp. 129–181.
[3] Ciardelli, I., *Questions as information types*, Synthese **195** (2018), pp. 321–365.
[4] Ciardelli, I., J. Groenendijk and F. Roelofsen, "Inquisitive Semantics," Oxford University Press, 2017.
[5] Ciardelli, I., R. Iemhoff and F. Yang, *Questions and dependency in intuitionistic logic*, Notre Dame Journal of Formal Logic **61** (2020), pp. 75–115.
[6] Ciardelli, I. A. and F. Roelofsen, *Inquisitive dynamic epistemic logic*, Synthese **192** (2015), pp. 1643–1687.
[7] Ebbing, J., L. Hella, A. Meier, J.-S. Müller, J. Virtema and H. Vollmer, *Extended modal dependence logic* $\mathcal{EMDL}$, in: L. Libkin, U. Kohlenbach and R. de Queiroz, editors, *Logic, Language, Information, and Computation. WoLLIC 2013*, Lecture Notes in Computer Science **8071** (2013), pp. 126–137.
[8] Ebbing, J. and P. Lohmann, *Complexity of model checking for modal dependence logic*, in: M. Bieliková, G. Friedrich, G. Gottlob, S. Katzenbeisser and G. Turán, editors, *SOFSEM 2012: Theory and Practice of Computer Science*, Lecture Notes in Computer Science **7147** (2012), pp. 226–237.
[9] Gargov, G. and V. Goranko, *Modal logic with names*, Journal of Philosophical Logic **22** (1993), pp. 607–636.
[10] Goldblatt, R., *Axiomatic classes of intuitionistic models*, Journal of Universal Computer Science **11** (2005), pp. 1945–1962.
[11] Goldblatt, R. I. and S. K. Thomason, *Axiomatic classes in propositional modal logic*, in: J. N. Crossley, editor, *Algebra and Logic*, Springer-Verlag, 1975 pp. 163–173.
[12] Kurz, A. and J. Rosický, *The Goldblatt-Thomason theorem for coalgebras*, in: T. Mossakowski, U. Montanari and M. Haveraaen, editors, *Algebra and Coalgebra in Computer Science. CALCO 2007*, Lecture Notes in Computer Science, **4624** (2007), pp. 342–355.
[13] Punčochář, V., *A generalization of inquisitive semantics*, Journal of Philosophical Logic **45** (2016), pp. 399–428.
[14] Punčochář, V., *Algebras of information states*, Journal of Logic and Computation **27** (2017), pp. 1643–1675.
[15] Punčochář, V., *Substructural inquisitive logics*, The Review of Symbolic Logic **12** (2019), pp. 296–330.
[16] Rodenburg, P. H., "Intuitionistic Correspondence Theory," Ph.D. thesis, Universiteit van Amsterdam (1986).
[17] Sano, K. and M. Ma, *Goldblatt-Thomason-style theorems for graded modal language*, in: L. Beklemishev, V. Goranko and V. Shehtman, editors, *Advances in Modal Logic 2010* (2010), pp. 330–349.
[18] Sano, K. and J. Virtema, *Characterizing relative frame definability in team semantics via the universal modality*, in: J. Väänänen, Åsa Hirvonen and R. de Queiroz, editors, *Logic, Language, Information, and Computation. WoLLIC 2016*, Lecture Notes in Computer Science **9803** (2016), pp. 392–409.
[19] Sano, K. and J. Virtema, *Characterising modal definability of team-based logics via the universal modality*, Annals of Pure and Applied Logic **170** (2019), pp. 1100–1127.
[20] Sano, K. and V. Virtema, *Characterizing frame definability in team semantics via the universal modality*, in: V. de Paiva, R. J. de Queiroz and L. S. Moss, editors, *Logic, Language, Information, and Computation. WoLLIC 2015*, Lecture Notes in Computer Science **9160** (2015), pp. 140–155.
[21] Sindoni, G., K. Sano and J. G. Stell, *Axiomatizing discrete spatial relations*, in: J. Desharnais, W. Guttmann and S. Joosten, editors, *Relational and Algebraic Methods in Computer Science. RAMiCS 2018.*, Lecture Notes in Computer Science **11194**, 2018, pp. 1–18.
[22] ten Cate, B., "Model theory for extended modal languages," Ph.D. thesis, University of Amsterdam, Institute for Logic, Language and Computation (2005).
[23] ten Cate, B., D. Gabelaia and D. Sustretov, *Modal languages for topology: expressivity and definability*, Annals of Pure and Applied Logic **159** (2008), pp. 146–170.

[24] Väänänen, J., *Modal dependence logic*, in: K. R. Apt and R. van Rooij, editors, *New Perspectives on Games and Interaction*, Texts in Logic and Games **4**, Amsterdam University Press, 2008 pp. 237–254.
[25] van Benthem, J., *Modal frame classes revisited*, Fundamenta Informaticae **18** (1993), pp. 303–317.
[26] Yang, F., "On Extensions and Variants of Dependence Logic," Ph.D. thesis, University of Helsinki (2014).

Finitely-valued Propositional Dynamic Logic

Igor Sedlár [1]

The Czech Academy of Sciences, Institute of Computer Science
Pod Vodárenskou věží 271/2
Prague, The Czech Republic

Abstract

We study a many-valued generalization of Propositional Dynamic Logic where formulas in states and accessibility relations between states of a Kripke model are evaluated in a finite FL-algebra. One natural interpretation of this framework is related to reasoning about costs of performing structured actions. We prove that PDL over any finite FL-algebra is decidable. We also establish a general completeness result for a class of PDLs based on commutative integral FL-algebras with canonical constants.

Keywords: FL-algebras, Many-valued modal logic, Propositional Dynamic Logic, Residuated lattices, Substructural logics, Weighted structures.

1 Introduction

Propositional dynamic logic, PDL, is a well-known modal logic formalizing reasoning about structured actions, e.g. computer programs or actions performed by physical agents, and their correctness properties [10,17]. PDL is subject to two limiting design features. First, being based on classical propositional logic, it formalizes actions that modify values of Boolean variables. A more general setting, one where variables take values from an arbitrary set (integers, characters, trees etc.), is offered by variants of first-order Dynamic Logic, DL [16,17]; these variants, however, are mostly undecidable. Second, PDL can express the fact that one action is guaranteed to attain a certain goal while another action is not, but it is not able to express that one action is a more efficient way of attaining the goal than another action. In other words, accessibility between states mediated by actions is modelled as a crisp rather than a graded relation; the former approach is a convenient idealization, but the latter one is more realistic and often also practically required.

[1] Email: sedlar@cs.cas.cz. This work was supported by the Czech Science Foundation grant GJ18-19162Y for the project *Non-Classical Logical Models of Information Dynamics*. Some results presented here were obtained while the author was visiting Manuel A. Martins and Alexandre Madeira at the University of Aveiro; our discussions helped to initiate this work and shaped the paper. The author is also grateful to Petr Cintula and Libor Běhounek for discussions on their earlier work on many-valued dynamic logic and the audience of the Seminar of Applied Mathematical Logic at the Institute of Computer Science CAS. The valuable comments of the anonymous reviewers are gratefully acknowledged.

Both of these limitations of "classical" PDL are avoided in a *many-valued* setting. In such a setting, values of formulas in states of a Kripke model are taken from an algebra that is typically distinct from the two-element Boolean algebra used in classical PDL. In a many-valued setting, accessibility between states can also be evaluated in such an algebra, naturally leading to a representation of "costs" or other "weights" associated with performing actions under specific circumstances.

Research into many-valued modal logics dates back to the 1960s, see the pioneering [25] and the later [23]. Fitting [11,12] was the first to study modal logics where both formulas in states and accessibility relations between states in the Kripke model take values from a non-Boolean algebra. Fitting considers finite Heyting algebras; generalizations studied for example in [4,7,6,15,28] focus on various kinds of finite or infinite *residuated lattices* [13]. Residuated lattices are algebraic structures related to *substructural logics*, with many important special cases such as Boolean and Heyting algebras, relation algebras, lattice-ordered groups, powersets of monoids, various algebras on the $[0,1]$-interval and so on.

Investigations of PDL based on residuated lattices are relatively scarce. The work in [5,18,19] focuses on expressivity of PDL with many-valued accessibility, but technical results such as decidability or completeness are not provided. Teheux [27] establishes decidability and completeness of PDLs based on finite Łukasiewicz chains and the present author [24] establishes decidability and completeness of PDL extending the paraconsistent modal logic of [22]; both papers, however, deal with crisp acessibility relations. As an attempt to sytematize the work in many-valued PDL, Madeira et al. [20,21] put forward a general method of producing many-valued versions of PDL, based on the matrix representation of Kleene algebras; their method, however, applies only if models are defined to be finite.

In this paper we add to this literature by studying PDLs based on finite Full Lambek algebras, that is, residuated lattices with a distinguished, though arbitrary, 0 element. We assume that both evaluations of formulas in states and accessibility between states are many-valued. Our main technical results are general completeness and decidability proofs for logics in the family. To the best of our knowledge, our results are the first decidability and completeness results concerning non-crisp many-valued PDL. To be more specific, we work with versions of test-free PDL based on finite Full Lambek algebras with canonical constants; we prove that any PDL based on a finite FL-algebra with canonical constants is decidable; we also establish a completeness result for PDLs based on finite commutative integral FL-algebras with canonical constants.

The paper is structured as follows. Section 2 introduces the general framework of PDL based on finite FL-algebras. We note that, for technical reasons discussed in §6, our version of PDL uses the transitive closure operator, or Kleene plus, as primitive instead of the more standard reflexive transitive closure operator, the Kleene star. An informal interpretation of the framework is discussed in §3. Section 4 establishes our decidability result using a generalization of the smallest filtration technique. Section 5 establishes the completeness result for PDLs based on finite integral commutative FL-algebras with canonical constants. Our work there builds on the results of [4], but the canonical model construction used in our proof is novel to this paper (it is a suitable generalization of the greatest filtration construction, though the model itself is infinite).

2 Preliminaries

In this section we briefly recall two-valued PDL (§2.1), and we define FL-algebras and many-valued models for the language of PDL based on them (§2.2). We point out some basic facts that we will use later on.

2.1 Two-valued PDL

We begin by recalling some well-known facts about two-valued test-free PDL; see [17]. Fix $Ac = \{\mathrm{a}_i \mid i \in \omega\}$, a countable set of atomic action expressions. The set of *standard action expressions*, STA, is the closure of Ac under applying binary operators ; ("composition"), $\cup$ ("choice") and unary $*$ ("Kleene star"). That is, STA are regular expressions over Ac without the empty expression. For example, $(\mathrm{a}_0;\mathrm{a}_1)^* \cup \mathrm{a}_0$ is in STA. Let $Pr = \{\mathrm{p}_i \mid i \in \omega\}$ be a countable set of propositional variables. Take $\mathbf{2}$, the two-element Boolean algebra on the set $\{0,1\}$ with meet $\sqcap$, join $\sqcup$ and complement $-$; the binary operation $\Rightarrow$ is defined as usual: $a \Rightarrow b := -a \sqcup b$. Formulas of the *standard language for* $\mathbf{2}$, $Fm(\mathcal{L}_{\mathbf{2}}^{STA})$, are defined by

$$\varphi := \mathrm{p} \mid \bar{c} \mid \varphi \wedge \varphi \mid \varphi \vee \varphi \mid \varphi \to \varphi \mid [\alpha]\varphi$$

where $\mathrm{p} \in Pr$, $c \in \mathbf{2}$ and $\alpha \in STA$. For example, $\mathrm{p}_0 \to [\mathrm{a}_0;(\mathrm{a}_1)^*](\mathrm{p}_1 \to \bar{0})$ is a formula of $\mathcal{L}_{\mathbf{2}}^{STA}$.

A $\mathbf{2}$-*valued frame for* STA is $\mathfrak{F} = (S, \{R_\alpha\}_{\alpha \in STA})$ where S is a non-empty set and, for each $\alpha \in STA$, R_α is a function from $S \times S$ to $\mathbf{2}$. We denote $R(\alpha) := \{(s,t) \mid R_\alpha(s,t) = 1\}$; and the functions in $\{R_\alpha\}_{\alpha \in STA}$ are required to satisfy the following: (i) $R(\alpha \cup \beta) = R(\alpha) \cup R(\beta)$; (ii) $R(\alpha;\beta) = R(\alpha) \circ R(\beta)$, the composition of $R(\alpha)$ and $R(\beta)$; (iii) $R(\alpha^*) = R(\alpha)^*$, the reflexive transitive closure of $R(\alpha)$.

Let $\mathfrak{F} = (S, \{R_\alpha\}_{\alpha \in STA})$ be a $\mathbf{2}$-valued frame. A $\mathbf{2}$-*valued model based on* $\mathfrak{F}$ is $\mathfrak{M} = (S, \{R_\alpha\}_{\alpha \in STA}, V)$ where $V : Fm(\mathcal{L}_{\mathbf{2}}^{STA}) \times S \to \mathbf{2}$ such that

- $V(\bar{c}, s) = c$;
- $V(\varphi \wedge \psi, s) = V(\varphi, s) \sqcap V(\psi, s)$, $V(\varphi \vee \psi, s) = V(\varphi, s) \sqcup V(\psi, s)$, and $V(\varphi \to \psi, s) = V(\varphi, s) \Rightarrow V(\psi, s)$;
- $V([\alpha]\varphi, s) = \bigsqcap_{t \in S}\big(R_\alpha(s,t) \Rightarrow V(\varphi, t)\big)$.

Note that $V([\alpha]\varphi, s) = \bigsqcap_{R_\alpha(s,t)=1} V(\varphi, t)$. A formula φ is *valid in* $\mathfrak{M}$ iff $V(\varphi, s) = 1$ for all s; validity in frames and classes of frames is defined as expected.

This is the standard presentation of test-free PDL, phrased in a way that invites generalizations obtained by replacing $\mathbf{2}$ by another algebra. We will study some such generalizations in this paper but, as we discuss in more detail below, the story is somewhat more complicated. For reasons discussed in §6, our generalizations will use a different primitive iteration operator instead of the Kleene star. The operator we will use, however, is conveniently related to the Kleene star.

The set of *action expressions over Ac*, ACT, is the closure of Ac under composition, choice and the unary operator $+$ ("Kleene plus"). Formulas of the *language* $\mathcal{L}_\mathbf{2}$ are defined as expected (we omit reference to ACT), with $\alpha \in ACT$; for example, $\mathrm{p}_0 \to [\mathrm{a}_0; (\mathrm{a}_1)^+](\mathrm{p}_1 \to \bar{0})$ is a formula of $\mathcal{L}_\mathbf{2}$. The definition of **2**-*valued frames for* ACT is the same as the definition of **2**-valued frames for STA, with an obvious exception, namely, the requirement that $R(\alpha^+)$ be the *transitive closure* of $R(\alpha)$, i.e. $R(\alpha^+) = \bigcup_{n>0} R^n(\alpha)$, where $R^1(\alpha) = R(\alpha)$ and $R^{n+1}(\alpha) = R^n(\alpha) \circ R(\alpha)$. Compare this with the *reflexive* transitive closure $R(\alpha)^* = \bigcup_{n\geq 0} R^n(\alpha)$, where $R^0(\alpha) = \{(s,s) \mid s \in S\}$. Models based on frames for ACT are defined as before.

Proposition 2.1 *Let For each* $\alpha \in ACT$ *and* $\varphi \in Fm(\mathcal{L}_\mathbf{2})$,

$$V(\varphi \wedge [\alpha^+]\varphi, s) = 1 \quad \textit{iff} \quad \forall t((s,t) \in R(\alpha)^* \implies V(\varphi, t) = 1)\,.$$

Proposition 2.1 implies that $\varphi \wedge [\alpha^+]\varphi$ "simulates" $[\alpha^*]\varphi$ in $\mathcal{L}_\mathbf{2}$. This provides a justification for our using languages based on ACT rather than on STA in what follows. However, we admit that this choice is related to the technical issues discussed in §6.

2.2 FL-algebras and finitely-valued PDL

In this section we generalize two-valued PDL by replacing the two-element Boolean algebra **2** by a more general structure, namely, a finite FL-algebra. FL-algebras provide semantics for a wide class of *substructural logics* [13].

Definition 2.2 An *FL-algebra* ("full Lambek algebra", [13]) is a set X with binary operations $\sqcap, \sqcup, \backslash, \cdot, /$ and two distinguished elements $1, 0$ such that

- $(X, \sqcap, \sqcup)$ is a lattice (let $a \sqsubseteq b$ iff $a \sqcup b = b$);
- $(X, \cdot, 1)$ is a monoid;
- $(\backslash, \cdot, /)$ are residuated over $(X, \sqsubseteq)$, i.e.

$$a \cdot b \sqsubseteq c \quad \text{iff} \quad b \sqsubseteq a\backslash c \quad \text{iff} \quad a \sqsubseteq c/b;$$

- 0 is an arbitrary element of X.

Residuated lattices are 0-free reducts of FL-algebras. Each finite FL-algebra $\boldsymbol{X}$ contains a least element $\bot^{\boldsymbol{X}}$ (for all $a \in X$, $\bot^{\boldsymbol{X}} \sqsubseteq a$) and a greatest element $\top^{\boldsymbol{X}}$ (for all $a \in X$, $a \sqsubseteq \top^{\boldsymbol{X}}$).

We usually write ab instead of $a \cdot b$ and $a \Rightarrow b$ instead of b/a. Two varieties of FL-algebras will be important in this paper:

- *commutative* FL-algebras satisfy $ab = ba$ for all $a, b \in X$;

- *integral* FL-algebras satisfy $a \sqsubseteq 1$ for all $a \in X$.

Note that in commutative FL-algebras $a\backslash b = b/a$.

Example 2.3 The two-element Boolean algebra **2** is a commutative integral FL-algebra, where $\cdot$ is $\sqcap$ and $\backslash$ (identical to $/$) is $\Rightarrow$.

Example 2.4 Let $N > 0$ and define $\boldsymbol{N} = (N, max, min, +_N, \rightarrow_N)$ where

$$a +_N b = min(a + b, N - 1) \quad \text{and} \quad a \rightarrow_N b = max(b - a, 0)\,.$$

$\boldsymbol{N}$ is a finite commutative integral FL-algebra, with 0 as the monoid identity with respect to $+_N$ and the greatest element under the $\geq$-ordering induced by taking min as join. We note that $\boldsymbol{N}$ is isomorphic to the N-element Łukasiewicz lattice $\boldsymbol{Ł}_N$ over $\{\frac{k}{N-1} \mid k \in N\}$.

Example 2.5 As an example of a non-commutative, non-integral infinite FL-algebra, take the power set of the free monoid over some set Σ, i.e. the set of languages over Σ, with intersection as meet, union as join, $L \cdot L' := \{xx' \mid x \in L \;\;\&\;\; x' \in L'\}$, $\{\varepsilon\}$ as the monoid identity (ε is the empty word) and $L\backslash L' := \{x \in \Sigma \mid L \cdot \{x\} \subseteq L'\}$, $L'/L := \{x \in \Sigma \mid \{x\} \cdot L \subseteq L'\}$.

The following lemma summarizes some of the properties of FL-algebras we will rely on in this paper (we will often say that something holds "by the properties FL-algebras" in our proofs).

Lemma 2.6 *Let $\boldsymbol{X}$ be an arbitrary FL-algebra. Then* (i) $a \sqsubseteq b$ *iff* $1 \sqsubseteq a \Rightarrow b$*;* (ii) *If $a \sqsubseteq b$ and $c \sqsubseteq d$, then $b \Rightarrow c \sqsubseteq a \Rightarrow d$, $b\backslash c \sqsubseteq a\backslash d$ and $ac \sqsubseteq bd$;* (iii) $(a \sqcup b)c = ac \sqcup ab$ *and* $c(a \sqcup b) = ca \sqcup cb$*;* (iv) $a \Rightarrow (b \sqcap c) = (a \Rightarrow b) \sqcap (a \Rightarrow c)$*;* (v) $a \sqcup b \Rightarrow c = (a \Rightarrow c) \sqcap (b \Rightarrow c)$*;* (vi) $a \Rightarrow (b \Rightarrow c) = ab \Rightarrow c$*;* (vii) $(a \Rightarrow b)(b \Rightarrow c) \sqsubseteq a \Rightarrow c$*;* (viii) $(1 \Rightarrow a) = a$

If S is a non-empty set, then $\Pi(S)$ is the set of all finite sequences of elements of S; that is, $\pi \in \Pi(S)$ iff π is a function from some $n \in \omega$, called the length of π, to S. The unique sequence of length 0 is $\emptyset$. If π is a sequence of length n and $s \in S$, then $\pi^\frown s$ is the unique sequence of length $n+1$ such that $(\pi^\frown s)(k) = \pi(k)$ for all $k < n$ and $(\pi^\frown s)(n) = s$. Note that each sequence π of length $n > 0$ can be expressed as $(\ldots(\emptyset^\frown \pi(0))^\frown \ldots)^\frown \pi(n-1)$.

Definition 2.7 Let $\boldsymbol{X}$ be a finite FL-algebra and S a non-empty set. A *binary $\boldsymbol{X}$-valued relation on S* is any function from $S \times S$ to $\boldsymbol{X}$. Let R, Q be binary $\boldsymbol{X}$-valued relations on a set S; then

- the *union* of R and Q is the function $R \cup Q$ defined by $(R \cup Q)(s,t) := R(s,t) \sqcup Q(s,t)$;
- the *composition* of R and Q is the function $R \circ Q$ defined by $(R \circ Q)(s,t) = \bigsqcup_{x \in S} \big(R(s,x) \cdot Q(x,t)\big)$;
- the *transitive closure* of R is the function R^+ defined by $R^+(s,t) = \bigsqcup_{\pi \in \Pi(S)} Rs\pi t$ where $Rs\pi t$ is defined as follows:
  - $Rs\emptyset t = R(s,t)$ and
  - $Rs(\pi^\frown u)t = Rs\pi u \cdot R(u,t)$.

We say that Q *extends* R, notation $R \sqsubseteq Q$, iff $R(s,t) \sqsubseteq Q(s,t)$ for all $s,t \in S$; R is the *smallest* relation in a set $\{R_i\}_{i \in I}$ if $R = R_i$ for some $i \in I$ and each R_i extends R. R is *transitive* if $R(s,t) \cdot R(t,u) \sqsubseteq R(s,u)$ for all $s,t,u \in S$; and R is *reflexive* if $1 \sqsubseteq R(s,s)$ for all $s \in S$.

Note that we need to assume that all the required joins exist in $\boldsymbol{X}$; hence the restriction to finite FL-algebras (however, a restriction to complete $\boldsymbol{X}$ is sufficient, as is the assumption that R, Q are "$\boldsymbol{X}$-safe" [14, ch. 5]).

Proposition 2.8 *Let $\boldsymbol{X}$ be a finite FL-algebra and R a binary $\boldsymbol{X}$-valued relation on a set S. Then R^+ is the smallest transitive relation extending R. For any R, define R^* as follows:*

$$R^*(s,t) = \begin{cases} 1 & \textit{if } s = t \\ R^+(s,t) & \textit{otherwise.} \end{cases}$$

Then R^* is the smallest reflexive transitive relation extending R.

Proof. It is clear that R^+ is a transitive relation extending R. Now assume that so is Q. The conclusion that $R^+ \sqsubseteq Q$ follows from two facts that are easily established by induction on the length of π: (a) For all $s,t \in S$ and $\pi \in \Pi(S)$, $Rs\pi t \sqsubseteq Qs\pi t$ (the assumption that $R \sqsubseteq Q$ is used here); (b) For all $s,t \in S$ and $\pi \in \Pi(S)$, $Qs\pi t \sqsubseteq Q(s,t)$ (the assumption that Q is transitive is used). Since $\boldsymbol{X}$ is finite, the two claims imply that, for any given s and t, $\bigsqcup_\pi Rs\pi t \sqsubseteq \bigsqcup_\pi Qs\pi t \sqsubseteq Q(s,t)$.

It is clear that R^* is a reflexive transitive relation extending R. If so is Q, then we reason for any given s and t by cases as follows. If $s = t$, then $R^*(s,t) \sqsubseteq Q(s,t)$ is equivalent to $1 \sqsubseteq Q(s,s)$, which holds by reflexivity of Q. If $s \neq t$, then $R^*(s,t) \sqsubseteq Q(s,t)$ is equivalent to $R^+(s,t) \sqsubseteq Q(s,t)$, which follows from the assumption that Q is a transitive relation extending R. Hence, $R^*(s,t) \sqsubseteq Q(s,t)$ for any s and t. □

Lemma 2.9 *Let $\boldsymbol{X}$ be a finite FL-algebra and S a set; the $\boldsymbol{X}$-valued identity relation on S is defined as follows:*

$$Id_{\boldsymbol{X}}(s,t) := \begin{cases} 1 & \textit{if } s = t \\ \bot^{\boldsymbol{X}} & \textit{otherwise.} \end{cases}$$

If $\boldsymbol{X}$ is integral, then $R^* = Id_{\boldsymbol{X}} \cup R^+$ for any binary $\boldsymbol{X}$-valued relation on S.

Proof. We omit the proof; we just note that if $s = t$, then $R^*(s,t) = Id_{\boldsymbol{X}}(s,t) \sqcup R^+(s,t)$ is equivalent to $R^+(s,t) \sqsubseteq 1$, which is guaranteed to hold only if $\boldsymbol{X}$ is integral. □

Definition 2.10 Let $\boldsymbol{X}$ be a finite FL-algebra. An *$\boldsymbol{X}$-valued frame for ACT* is a pair $\mathfrak{F} = (S, \{R_\alpha\}_{\alpha \in ACT})$ where S is a non-empty set and, for all $\alpha \in ACT$, R_α is an $\boldsymbol{X}$-valued binary relation on S such that (i) $R_{\alpha \cup \beta} = R_\alpha \cup R_\beta$; (ii) $R_{\alpha;\beta} = R_\alpha \circ R_\beta$; and (iii) $R_{\alpha^+} = R_\alpha^+$.

$\boldsymbol{X}$-valued frames will also be referred to as $\boldsymbol{X}$-frames or simply frames if $\boldsymbol{X}$ is clear from the context or immaterial. We will sometimes write $R_\alpha st$ instead of $R_\alpha(s,t)$.

Definition 2.11 Formulas of the *language* $\mathcal{L}_{\boldsymbol{X}}$ are defined as follows:

$$\varphi := \mathrm{p} \mid \bar{c} \mid \varphi \wedge \varphi \mid \varphi \vee \varphi \mid \varphi \backslash \varphi \mid \varphi \cdot \varphi \mid \varphi / \varphi \mid [\alpha]\varphi\,,$$

where $\mathrm{p} \in Pr$, $c \in \boldsymbol{X}$ and $\alpha \in ACT$. We use $\bot, \top$ instead of $\overline{\bot^{\boldsymbol{X}}}$ and $\overline{\top^{\boldsymbol{X}}}$, respectively. We often write $\varphi\psi$ instead of $\varphi \cdot \psi$, $\varphi \to \psi$ instead of ψ/φ, m instead of a_m, and $\alpha\beta$ instead of $\alpha;\beta$. We define $\varphi \leftrightarrow \psi := (\varphi \to \psi) \wedge (\psi \to \varphi)$, $\neg\varphi := \varphi \to \bot$ and $\langle\alpha\rangle\varphi := \neg[\alpha]\neg\varphi$.

Note that we use the same symbol $\otimes \in \{\backslash, \cdot, /\}$ for the implication and fusion connectives of the language and for the residuated operations on FL-algebras. We will denote the operations on a given $\boldsymbol{X}$ as $\otimes^{\boldsymbol{X}}$ in contexts where it is convenient for the reader to distinguish the connectives of the language from the operations on the algebra. (However, $\Rightarrow$ denotes the operation $/^{\boldsymbol{X}}$ and $\to$ denotes the connective $/$ throughout.)

Definition 2.12 A *model* based on an $\boldsymbol{X}$-frame $(S, \{R_\alpha\}_{\alpha \in ACT})$ is $\mathfrak{M} = (S, \{R_\alpha\}_{\alpha \in ACT}, V)$, where V is a function from $Fm(\mathcal{L}_{\boldsymbol{X}}) \times S$ to $\boldsymbol{X}$ such that

- $V(\bar{c}, s) = c$;
- $V(\varphi \wedge \psi, s) = V(\varphi, s) \sqcap V(\psi, s)$ and $V(\varphi \vee \psi, s) = V(\varphi, s) \sqcup V(\psi, s)$;
- $V(\varphi \otimes \psi, s) = V(\varphi, s) \otimes^{\boldsymbol{X}} V(\psi, s)$ for $\otimes \in \{\backslash, \cdot, /\}$;
- $V([\alpha]\varphi, s) = \bigsqcap_{t \in S}\big(R_\alpha st \Rightarrow V(\varphi, t)\big)$.

A formula φ is *valid in* $\mathfrak{M}$ iff $1 \sqsubseteq V(\varphi, s)$ for all s in $\mathfrak{M}$. Validity in frames and classes of frames is defined as expected. The *theory* of a frame is the set of formulas valid in the frame; the theory of a class of frames is the set of formulas valid in each frame in the class. $Th(\boldsymbol{X})$ is the theory of the class of all $\boldsymbol{X}$-frames.

The following addendum to Proposition 2.1 suggests that integral FL-algebras are particularly suitable for us.

Proposition 2.13 *Take an arbitrary $\boldsymbol{X}$-frame for a finite integral $\boldsymbol{X}$. Then* $V(\varphi \wedge [\alpha^+]\varphi, s) = \bigsqcap_{t \in S}(R^*_\alpha st \Rightarrow V(\varphi, t))$.

Proof. The $\sqsubseteq$-inequality is straightforward and the $\sqsupseteq$-inequality follows from Lemma 2.9. □

It is clear that two-valued PDL is a special case of the present framework for $\boldsymbol{X} = \mathbf{2}$.

Lemma 2.14 *The following are valid in each $\boldsymbol{X}$-frame:*

(a) $[\alpha](\varphi \wedge \psi) \leftrightarrow ([\alpha]\varphi \wedge [\alpha]\psi)$ (c) $[\alpha\beta]\varphi \leftrightarrow [\alpha][\beta]\varphi$
(b) $[\alpha \cup \beta]\varphi \leftrightarrow ([\alpha]\varphi \wedge [\beta]\varphi)$ (d) $[\alpha^+]\varphi \leftrightarrow [\alpha](\varphi \wedge [\alpha^+]\varphi)$

Proof. To prove that $\varphi \leftrightarrow \psi$ is valid if suffices to show that $V(\varphi, s) = V(\psi, s)$ for all s in all models. (a) The proof relies on the fact that $a \Rightarrow (b \sqcap c) = (a \Rightarrow b) \sqcap (a \Rightarrow c)$ in all FL-algebras. (b) The proof relies on the fact that $(a \sqcup b) \Rightarrow c = (a \Rightarrow c) \sqcap (b \Rightarrow c)$ in all FL-algebras. (c) The proof relies on the fact that $a \Rightarrow (b \Rightarrow c) = ab \Rightarrow c$ in all FL-algebras. (Note that composition of relations needs to be defined using monoid multiplication $\cdot$, not lattice meet.) (d) The proof relies on the fact that $R_\alpha st \sqsubseteq R_{\alpha^+} st$, it also uses simple composition of paths. □

We will discuss an informal interpretation of a special case of the many-valued framework in the next section. Speaking generally, however, we may adapt the slogan characterizing modal logic as providing languages for talking about relational structures [3, p. viii] and say that many-valued modal logics provide *simple yet expressive languages for talking about many-valued relational structures.* Examples of many-valued relational structures include weighted structures such as weighted graphs etc. Choosing an FL-algebra as the algebra of weights brings the framework closer to substructural logics that include well-known formalisms for reasoning about resources (variants of linear logic) or graded properties and relations (fuzzy logics). Many-valued PDL adds to this the capacity to articulate reasoning about *structured* many-valued relations using the PDL relational operations of choice, composition and iteration. An intriguing connection here is the relation of finitely-valued PDL to weighted automata over finite semirings [9], but a more thorough investigation of this connection is left for another occasion.

3 Motivation

This section discusses the informal interpretation of finitely-valued PDL. We give two general interpretations of the framework first and then we zoom in to PDLs over a specific class of FL-algebras. Our overview is cursory; the present paper is focused more on basic technical results than on informal interpretations and applications. A more thorough exploration of the latter is left for another occasion. We only note here that we consider many-valued PDL to be sufficiently mathematically interesting to be studied independently of informal interpretations and applications.

We have mentioned before the slogan that modal logics provide simple yet expressive languages for talking about relational structures [3, p. viii]; by the same token, many-valued modal logics can be seen as providing means of talking about "weighted" relational structures. Two-valued PDL has been applied to at least two kinds of relational structures which have very natural weighted generalizations. We discuss these in turn.

First, take the interpretation of modal logic that relates it to *description logics* [2]. Simply put, formulas of a modal language can be seen as expressing "concepts", i.e. properties of objects, and indices of modal operators as expressing various "roles", i.e. relations between objects. On this reading, "states" in a Kripke model represent arbitrary objects and "accessibility relations" between them represent relations between these objects. Structured modal indices that come with PDL (i.e. "action expressions" as we call them) can be seen as expressing structured relations between objects; union, composition and transitive closure have been found particularly suitable for expressing various important concepts and roles [1]. Many-valued description logics (see [26] for instance) are a generalization of description logics designed for management of *uncertain and imprecise information.* These logics can express the fact that an object is subsumed under a given concept (e.g. "tall" if the reader will forgive the platitudinous example) only to some degree or that only imprecise information about a relation holding between two objects is available. Finitely-valued PDL as presented here can be seen as a family of many-valued description logics with transitive closure of roles.

Second, the original motivation of PDL was reasoning about the behaviour of computer programs [10]. From a more general perspective, PDL can be seen as a logic formalising reasoning about types of *structured actions*, represented by "action expressions". On this reading, a Kripke frame consists of states and transitions between states labelled by types of action; for instance $R_\alpha st$ means that action of type α can be used to get from state s to state t. States can be thought of as physical locations, states of a complex system such as a database or states of a computer during the run of a program; but states can also be thought of as "states of the world" that can be modified by actions of intelligent agents. PDL can be used to formalize reasoning about properties of actions that modify these kinds of states. One important example is correctness, related to the question if a specific kind of action is guaranteed to lead to a specific outcome when performed under specific circumstances. (This more general perspective makes PDL relevant to automated planning, for example.) Many-valued Kripke models can be seen as transition systems where transitions carry *weights*; these can be costs or resources needed to perform a transition using the given action type. Běhounek [5] suggested a many-valued version of PDL for reasoning about costs of program runs that is close to our framework, but he did not establish completeness or decidability results.

Let us now discuss a special case of the finitely-valued PDL framework giving rise to a natural class of weighted relational structures; we show that formulas of the PDL language are able to express interesting features of these structures. Let $\boldsymbol{N}$ be the FL-algebra of Example 2.4, that is, $\boldsymbol{N} = (N, max, min, +_N, \rightarrow_N)$ where

$$a +_N b = min(a + b, N - 1) \quad \text{and} \quad a \rightarrow_N b = max(b - a, 0)\,,$$

where $N \in \omega$ is non-empty. The set N is seen as a *weight scale* with 0 representing zero weight ("for free") and $N - 1$ representing the maximal weight (considered "infeasible"). The operation $+_N$, namely, sum bounded by the maximal weight, represents *weight addition.* N is given a (distributive) lattice structure by including *max* as meet and *min* as join; the associated lattice order $\sqsubseteq$ is defined as usual, $a \sqsubseteq b$ iff $min(a, b) = b$. Hence, $a \sqsubseteq b$ (i.e. $b \leq a$) means that weight b is at most as big as weight a. The choice of *max* as meet and *min* as join—not the other way around—may seem unintuitive at first, but it yields the result that $a \sqsubseteq 0$ for all $a \in N$. It is important to note in this respect that 0 is the identity element with respect to $+_N$. (Hence, choosing the natural ordering on N as our lattice ordering would mean that each element of the lattice would be above the monoid identity, which is problematic given our definition of validity.) It is clear that $a +_N b = b +_N a$. The residual $\rightarrow_N$ of $+_N$ is truncated subtraction or monus; the crucial feature of $\rightarrow_N$ is that $a \rightarrow_N b = 0$ iff $a \sqsubseteq b$ (iff $b \sqsubseteq a$). We note that $\boldsymbol{N}$ is isomorphic to the N-element Łukasiewicz lattice $\boldsymbol{Ł}_N$ over $\{\frac{k}{N-1} \mid k \in N\}$, but we prefer $\boldsymbol{N}$ to $\boldsymbol{Ł}_N$ as a representation of an N-element weight scale.

$\boldsymbol{N}$-frames are weighted relational structures that can be informally interpreted in a number of ways. On the "description reading", for instance, states $s \in S$ are objects and R_α represent structured weighted relations between these objects. On the "transition cost reading", states can be seen as physical locations or states of a system and $R_\alpha st \in N$ is the cost of accessing state t from s by performing action α (hence, frames are weighted labelled transition systems). If $R_\alpha st = N - 1$, then we say that t is not in relation α with s, or that t cannot be accessed from s by performing α; if $R_\alpha st = 0$, then t is "clearly" in relation α with s, or t can be accessed from s by α for free. Let us now discuss some properties of weighted relational structures that can be expressed by PDL formulas.

Since $\boldsymbol{N}$ is $(N-1)$-involutive, i.e. $(a \Rightarrow (N-1)) \Rightarrow (N-1) = a$ for all $a \in \boldsymbol{N}$, we have

$$V(\langle\alpha\rangle\bar{0}, s) = \bigsqcup_{t \in S} (R_\alpha st +_N 0) = min\{R_\alpha st \mid t \in S\}\,.$$

In other words, $V(\langle\alpha\rangle\bar{0}, s)$ is the minimal guaranteed cost of performing α at s (on the transition cost reading) or the maximal degree to which s is α-related to any object (on the description reading). Let us write simply α instead of $\langle\alpha\rangle\bar{0}$ if the context clears up any possible confusion. Note that $a \Rightarrow b$ is the difference between b and a if $a < b$ and 0 otherwise. The following features of weighted relation structures can be expressed (we use the transition cost reading and the reader is invited to translate to the description reading):

- the minimal cost of performing α is at most m (this is true in state s if $V(\bar{m} \rightarrow \alpha, s) = 0$); the "at least" direction is expressed dually;
- performing α is at least as costly as performing β (this is true in state s if $V(\alpha \rightarrow \beta, s) = 0$); the "at most" direction is expressed dually;
- the difference between the minimal guaranteed cost of β and α is at most m (this is true in state s if $V(\bar{m} \rightarrow (\alpha \rightarrow \beta), s) = 0$).

On the transition cost reading, atomic formulas in Pr can be seen as representing various items that can be obtained at states for a given cost, with

$V(\mathrm{p}, s)$ representing the cost of item p at s (e.g. time needed to charge the battery at the charger location). Observe that $V(\langle\alpha\rangle\mathrm{p}, s) = \bigsqcup_{t\in S}(R_\alpha st +_N V(\mathrm{p}, t))$ is the minimal cost of getting from s to a state t by performing α and obtaining p at t; we may also say that this is the minimal guaranteed cost of obtaining p by α. On the description reading, atomic formulas can be seen as expressing graded, imprecise or vague properties of objects; thus the value of $\langle\alpha\rangle\mathrm{p}$ at s is the "grade of truth" of the statement that s is α-related to an object with property p. The interesting case obtains where both the relation and the property are graded or vague; think of "Alice was in contact with a person displaying symptoms of COVID-19". We write φ^α instead of $\langle\alpha\rangle\varphi$. The following features of weighted relation structures can be expressed (we use the transition cost reading and the reader is again invited to translate to the description reading):

- the minimal cost of obtaining p by α is at most m (this is true in state s if $V(\bar{m} \to \mathrm{p}^\alpha, s) = 0$); the "at least" direction is expressed dually;
- obtaining p by α is at least as costly as obtaining q by β (this is true in state s if $V(\mathrm{p}^\alpha \to \mathrm{q}^\beta, s) = 0$); the "at most" direction is expressed dually;
- the difference between the minimal guaranteed cost of obtaining q by β and obtaining p by α is at most m (this is true in state s if $V(\bar{m} \to (\mathrm{p}^\alpha \to \mathrm{q}^\beta), s) = 0$).

This cursory overview shows that the PDL language provides means to expressing a variety of features of weighted relational structures and so finitely-valued PDL can be used to formalize reasoning about these features. A more thorough exploration of expressivitiy and applications is left for another occasion.

4 Finite model property and decidability

In this section we prove that $Th(\boldsymbol{X})$ is decidable for all finite $\boldsymbol{X}$. We prove this by showing that each such $Th(\boldsymbol{X})$ has the bounded finite model property. The result is established using a many-valued generalization of the smallest filtration construction; see [8], where the construction is applied to some many-valued modal logics with $\Box$ and $\Diamond$.[2] Even though the decidability result is not surprising, we consider it to be a "sanity check" for the many-valued dynamic framework. We note that presence of canonical constants is not necessary for the decidability result (in contrast to the completeness result of §5).

Definition 4.1 The *closure* of a set of formulas Ψ is the smallest $\Phi \supseteq \Psi$ such that

- Φ is closed under subformulas (that is, if $\varphi \in \Phi$ and ψ is a subformula of φ, then $\psi \in \Phi$);
- $[\alpha \cup \beta]\varphi \in \Phi$ implies $[\alpha]\varphi \in \Phi$ and $[\beta]\varphi \in \Phi$;

[2] We are grateful to an anonymous reviewer for pointing the reference out.

- $[\alpha\beta]\varphi \in \Phi$ implies $[\alpha][\beta]\varphi \in \Phi$;
- $[\alpha^+]\varphi \in \Phi$ implies $[\alpha][\alpha^+]\varphi \in \Phi$ and $[\alpha]\varphi \in \Phi$.

Φ is *closed* iff Φ is the closure of Φ.

Definition 4.2 For each set of formulas Φ and each model $\mathfrak{M}$, we define the binary two-valued equivalence relation $\approx_\Phi$ on states of $\mathfrak{M}$ by

$$s \approx_\Phi t \iff (\forall\varphi \in \Phi)\big(V(\varphi, s) = V(\varphi, t)\big).$$

The equivalence class of s under $\approx_\Phi$ will be denoted as $[s]_\Phi$ or just as $[s]$ if Φ is clear from the context.

Definition 4.3 Take an $\boldsymbol{X}$-valued model $\mathfrak{M}$ and a finite closed set Φ. The *filtration of* $\mathfrak{M}$ *through* Φ is the $\boldsymbol{X}$-valued model $\mathfrak{M}^\Phi = (S^\Phi, R^\Phi, V^\Phi)$ such that

- $S^\Phi = \{[s] \mid s \in S\}$;
- $R^\Phi_{\mathrm{a}_m}([s],[t]) = \bigsqcup\{R_{\mathrm{a}_m}(u,v) \mid s \approx_\Phi u \;\;\&\;\; t \approx_\Phi v\}$; R^Φ_α for $\alpha \notin Ac$ is defined as in models;
- $V^\Phi(\mathrm{p},[s]) = V(\mathrm{p}, s)$ for $\mathrm{p} \in \Phi$; $V^\Phi(\mathrm{p},[s]) = 0^{\boldsymbol{X}}$ for $\mathrm{p} \notin \Phi$; $V^\Phi(\varphi,[s])$ for $\varphi \notin Pr$ is defined as in models.

It is clear that if Φ is the closure of a finite set Ψ, then Φ is finite. If Φ is finite, then so is $\mathfrak{M}^\Phi$; in fact, $|S^\Phi| \leq |\boldsymbol{X}|^{|\Phi|}$. We usually omit reference to Φ while discussing accessibility relations on S^Φ and we also write $\approx$ instead of $\approx_\Phi$. We will write R_m instead of R_{a_m}. In the rest of the section, we fix an $\boldsymbol{X}$-model $\mathfrak{M}$ and a finite closed set Φ.

Lemma 4.4 *For all $\alpha \in ACT$ and all $x, y \in S$,*

(a) $R_\alpha xy \sqsubseteq R_\alpha[x][y]$;

(b) For all $[\alpha]\varphi \in \Phi$, $V([\alpha]\varphi, x) \sqsubseteq R_\alpha[x][y] \Rightarrow V(\varphi, y)$.

Proof. Both claims are established by induction on the complexity of α. The base case of (a) holds by definition and the rest is established easily using the induction hypothesis. In the case of $\alpha = \beta^+$, we define for each $\pi \in \Pi(S)$ of length n the sequence $[\pi] \in \Pi(S^\Phi)$ of length n by $[\pi](k) := [\pi(k)]$ for all $k < n$; it is then easy to establish by induction on n that $R_\beta x\pi y \sqsubseteq R_\beta[x][\pi][y]$.)

The base case of (b) is follows from the fact that, for all $x' \in [x]$ and $y' \in [y]$, $\bigsqcap_{z\in S}\big(R_m x'z \Rightarrow V(\varphi, z)\big) \cdot R_m x'y' \sqsubseteq V(\varphi, y')$ using the definition of $\approx_\Phi$, closure of Φ under subformulas and properties of FL-algebras. The fact itself follows easily from properties of FL-algebras. The induction step uses Lemma 2.14 and is easy; for instance, in the case $\alpha = \beta^+$ we may use the fact that, for all x and y, $V([\beta](\varphi \wedge [\beta^+]\varphi, x) \sqsubseteq R_\beta[x][y] \Rightarrow V([\beta^+]\varphi, y)$ and hence, for all s, t and $\pi \in \Pi(S)$, $V([\beta^+]\varphi, s) \sqsubseteq R_\beta[s][\pi][t] \Rightarrow V(\varphi, t)$ as required. □

Lemma 4.5 *For all models $\mathfrak{M}$, all $\varphi \in \Phi$ and $s \in \mathfrak{M}$, $V(\varphi, s) = V^\Phi(\varphi, [s])$.*

Proof. The proof is by induction on the complexity of φ. The base case $\varphi \in Pr$ holds by definition, the cases for constants and propositional connectives are trivial and the case $\varphi = [\alpha]\psi$ is established using Lemma 4.4. □

Theorem 4.6 *$Th(\boldsymbol{X})$ is decidable for each finite $\boldsymbol{X}$.*

Proof. Lemma 4.5 implies $\varphi \in Th(\boldsymbol{X})$ iff φ is valid in all frames where $|S| \leq |\boldsymbol{X}|^{|\Phi|}$ where Φ is the closure of $\{\varphi\}$. Now $m := |\boldsymbol{X}| = m$, $n := m^{|\Phi|}$ and let n-frames be the frames with $|S| \leq n$. There are at most

$$n \times m^{n^2}$$

n-frames. On each n-frame, there are $n \times m^{\omega}$ models, but there are at most $n \times |\Phi| \times m$ possible ways to evaluate elements of $|\Phi|$ on an n-frame. Hence, there are at most

$$m^{n^2+1} \times n^2 \times |\Phi|$$

models to check. It is not hard to show that there is an algorithm checking validity of formulas in finite models. □

5 Completeness

Bou et al. [4] establish a general weak completeness result for modal logics based on finite commutative integral FL-algebras with canonical constants where 0 is the bottom element. In this section we build on their work to show how a Hilbert-style axiomatic presentation of any finite commutative integral FL-algebra $\boldsymbol{X}$ with canonical constants can be extended to a sound and weakly complete axiomatization of PDL based on $\boldsymbol{X}$. The restriction to commutative FL-algebras seems to be necessary for our style of argument to go through and we discuss this at appropriate places in more detail; the restriction to integral FL-algebras is convenient. We leave generalizations of our result as an open problem.

Fix a finite commutative integral FL-algebra $\boldsymbol{X}$ with canonical constants denoting elements of $\boldsymbol{X}$, together with a Hilbert-style axiomatic presentation $\mathsf{Log}(\boldsymbol{X})$ in the language $\mathcal{L}_{\boldsymbol{X}}$ that is *strongly complete with respect to* $\boldsymbol{X}$. That is, we assume that $\varphi \in \mathcal{L}_{\boldsymbol{X}}$ is derivable from $\Gamma \subseteq \mathcal{L}_{\boldsymbol{X}}$ in $\mathsf{Log}(\boldsymbol{X})$, in symbols $\Gamma \vdash_{\mathsf{Log}(\boldsymbol{X})} \varphi$, iff each non-modal homomorphism $u : \mathcal{L}_{\boldsymbol{X}} \to \boldsymbol{X}$ such that $1 \sqsubseteq \bigsqcap u[\Gamma]$ satisfies $1 \sqsubseteq u(\varphi)$ (values $u([\alpha]\psi)$ of modal formulas under u are arbitrary, so u "treats" modal formulas as propositional atoms).[3] For the details on how $\mathsf{Log}(\boldsymbol{X})$ looks like, see [4]. Since $\boldsymbol{X}$ is finite, $\vdash_{\mathsf{Log}(\boldsymbol{X})}$ is *finitary* in the sense that if $\Gamma \vdash_{\mathsf{Log}(\boldsymbol{X})} \varphi$, then there is a finite $\Delta \subseteq \Gamma$ such that $\Delta \vdash_{\mathsf{Log}(\boldsymbol{X})} \varphi$. We note that $\vdash_{\mathsf{Log}(\boldsymbol{X})}$ is also *monotonic* in the sense that if $\Gamma \vdash_{\mathsf{Log}(\boldsymbol{X})} \varphi$ and $\Gamma \subseteq \Delta$, then $\Delta \vdash_{\mathsf{Log}(\boldsymbol{X})} \varphi$.

Since $\boldsymbol{X}$ is commutative, we have $a \backslash b = b/a$ and so we use only a single "official" implication operator $\to$; see [13, p. 95]. Recall that $\varphi \leftrightarrow \psi := (\varphi \to \psi) \wedge (\psi \to \varphi)$; we define similarly $a \Leftrightarrow b := (a \Rightarrow b) \sqcap (b \Rightarrow a)$.

[3] A function $f : \mathcal{L}_{\boldsymbol{X}} \to \boldsymbol{X}$ is a non-modal homomorphism iff $f(\bar{c}) = c$ and f commutes with the propositional connectives $\oplus$ of $\mathcal{L}_{\boldsymbol{X}}$ and the corresponding operations $\oplus^{\boldsymbol{X}}$ on $\boldsymbol{X}$; we assume that $\wedge^{\boldsymbol{X}}$ is $\sqcap$ and $\vee^{\boldsymbol{X}}$ is $\sqcup$.

Definition 5.1 $\mathsf{PDL}(\boldsymbol{X})$ is the Hilbert-style axiom system extending $\mathsf{Log}(\boldsymbol{X})$ with the following axioms and rules (for all formulas φ, ψ, all action expressions $\alpha, \beta \in ACT$ and all canonical constants $\bar{c}$):

(A-1)	$[\alpha]\bar{1}$	(A-∪)	$[\alpha \cup \beta]\varphi \leftrightarrow ([\alpha]\varphi \wedge [\beta]\varphi)$
(A-reg)	$[\alpha]\varphi \wedge [\alpha]\psi \to [\alpha](\varphi \wedge \psi)$	(A-;)	$[\alpha\beta]\varphi \leftrightarrow [\alpha][\beta]\varphi$
(A-$\bar{c}$)	$[\alpha](\bar{c} \to \varphi) \leftrightarrow (\bar{c} \to [\alpha]\varphi)$	(A-+)	$[\alpha^+]\varphi \leftrightarrow [\alpha](\varphi \wedge [\alpha^+]\varphi)$
(R-mon)	$\dfrac{\varphi \to \psi}{[\alpha]\varphi \to [\alpha]\psi}$	(R-+)	$\dfrac{\varphi \to [\alpha]\varphi}{\varphi \to [\alpha^+]\varphi}$

The notions of proof, derivability, theorem and a formula derivable from a set of formulas are defined as usual (see [4]). $\mathsf{Thm}(\mathsf{PDL}(\boldsymbol{X}))$ is the set of theorems of $\mathsf{PDL}(\boldsymbol{X})$.

Since $\boldsymbol{X}$ is fixed, we write L instead of $\mathsf{Log}(\boldsymbol{X})$, PDL instead of $\mathsf{PDL}(\boldsymbol{X})$, Thm instead of $\mathsf{Thm}(\mathsf{PDL}(\boldsymbol{X}))$ and $\mathcal{L}$ instead of $\mathcal{L}_{\boldsymbol{X}}$ for the rest of this section.

Theorem 5.2 *If φ is a theorem of* PDL, *then φ is valid in the class of all $\boldsymbol{X}$-frames.*

Proof. The axioms and the rule in the left column are taken from [4]. Validity of the axioms in the right column in all FL-algebras was established in Lemma 2.14. To show that the rule (R-+) preserves validity in models, assume that $V(\varphi, s) \sqsubseteq V([\alpha]\varphi, s)$ for all s in an arbitrary model. Take some t and assume that $a \sqsubseteq V(\varphi, t)$; we prove that $a \sqsubseteq R_{\alpha^+}tu \Rightarrow V(\varphi, u)$ for all u. The claim to be proved is equivalent to $(\forall \pi \in \Pi(S))(a \sqsubseteq R_\alpha t\pi u \Rightarrow V(\varphi, u))$. This claim is easily established by induction on the length of π. □

We note that, without the assumption of commutativity, versions of (A-$\bar{c}$) are not sound; the axiom is used in the proof of Lemma 5.6 which is in turn applied in most of our arguments below.

From now on, let S be the set of non-modal homomorphisms $s : \mathcal{L} \to \boldsymbol{X}$ such that $s[\mathsf{Thm}] = \{1\}$ and let Φ be a fixed finite closed set.

Definition 5.3 The *Φ-equivalence relation* on S is an $\boldsymbol{X}$-valued binary relation $\sim_\Phi$ on S defined by

$$s \sim_\Phi t \quad := \quad \textstyle\bigsqcap_{\varphi \in \Phi} \big(s(\varphi) \Leftrightarrow t(\varphi)\big)\,.$$

If Φ is clear from the context, we will write $s \sim t$ or just st instead of $s \sim_\Phi t$.

Lemma 5.4 *The relation $\sim_\Phi$ is an $\boldsymbol{X}$-valued equivalence relation, that is, (a) $1 \sqsubseteq s \sim s$, (b) $s \sim t = t \sim s$ and (c) $(s \sim t)(t \sim u) \sqsubseteq s \sim u$, for all $s, t, u \in S$.*

Proof. Claims (a) and (b) are clear; claim (c) follows from Lemma 2.6. □

Completeness proofs for two-valued PDL typically use a filtration-like construction of the canonical model, where states are (or boil down to) equivalence classes of states taken from some other structure. A natural approach in our case would be to take "equivalence classes" of non-modal homomorphisms under $\sim$, where $s \sim t$ expresses "how much equivalent" s and t are with respect to Φ. However, in our case a simpler approach is available. We take S itself as the set of states of the canonical model and we refer to Φ only in the definition of the canonical R_α, which is a generalization of the definition of accessibility relations in the greatest filtration of a Kripke model.

Definition 5.5 The *canonical model modulo* Φ is $\mathfrak{M} = (S, R, V)$ where

- S is the set of non-modal homomorphisms $s : \mathcal{L} \to \boldsymbol{X}$ such that $s[\mathsf{Thm}] = \{1\}$;
- $R_m st := \bigsqcap_{[m]\varphi \in \Phi} \big(s([m]\varphi) \Rightarrow t(\varphi)\big)$ for all $\mathrm{a}_m \in Ac$ and $R_\alpha st$ for $\alpha \notin Ac$ is defined as in models;
- $V(\mathrm{p}, s) := s(\mathrm{p})$ and $V(\varphi, s)$ for $\varphi \notin Pr$ is defined as in models.

We define for each α the relation $R^{\mathcal{L}}_\alpha$ on S by $R^{\mathcal{L}}_\alpha st := \bigsqcap_{\varphi \in \mathcal{L}} \big(s([\alpha]\varphi) \Rightarrow t(\varphi)\big)$.

Note that $R^{\mathcal{L}}_n st \sqsubseteq R_n st$ for all $\mathrm{a}_n \in Ac$ and all s, t since $R^{\mathcal{L}}_n$ "cares" about more formulas. $R^{\mathcal{L}}_\alpha$ is the usual canonical many-valued accessibility relation, see [4], but we cannot use it here because of the presence of the Kleene plus iteration operator in ACT, similarly as in the case of two-valued PDL.

The following lemma states some properties of $R^{\mathcal{L}}_\alpha$ that will be useful in our proofs; the proof of the lemma can be found in [4] (the logics studied there are mono-modal, but the same approach applies here).

Lemma 5.6 *The following holds for all $\alpha \in ACT$ and all $s \in S$ of the canonical model:*

(a) For all t, $R^{\mathcal{L}}_\alpha st = \bigsqcap_{\varphi \in \mathcal{L}} \{t(\varphi) \mid 1 \sqsubseteq s([\alpha]\varphi)\}$ ([4], Proposition 4.1.);

(b) For all $\varphi \in \mathcal{L}$, $s([\alpha]\varphi) = \bigsqcap_{u \in S} \{R^{\mathcal{L}}_\alpha su \Rightarrow u(\varphi)\}$ ([4], Lemma 4.8.).

Lemma 5.7 *For all $[\alpha]\varphi \in \Phi$ and all $s, t \in S$, $s([\alpha]\varphi) \sqsubseteq R_\alpha st \Rightarrow t(\varphi)$.*

Proof. The claim is proved by induction on the complexity of α. The base case is established as follows. We know that $s([n]\varphi) \cdot (s([n]\varphi) \Rightarrow t(\varphi)) \sqsubseteq t(\varphi)$; from this $s([n]\varphi) \cdot R_n st \sqsubseteq t(\varphi)$ follows by the definition of R_n.

The cases of choice and composition in the induction step are straightforward. The case $\alpha = \beta^+$ is established by showing that, for all $\pi \in \Pi(S)$, all $s; t$, and all φ such that $[\beta^+]\varphi \in \Phi$, $s([\beta^+]\varphi) \sqsubseteq R_\beta s\pi t \Rightarrow t(\varphi)$. This claim, call it (A), follows from the claims (s, t and $[\beta^+]\varphi \in \Phi$ are fixed)

(B) $s([\beta]\varphi) \sqsubseteq R_\beta st \Rightarrow t(\varphi)$;

(C) for all $\sigma \in \Pi(S)$ and all u, $s([\beta^+]\varphi) \sqsubseteq R_\beta s\sigma u \Rightarrow u([\beta^+]\varphi)$.

The proof of (C) is left to the reader; (B) holds by the induction hypothesis. □

Lemma 5.8 *For all α and s, t, u, $R_\alpha su(ut) \sqsubseteq R_\alpha st$.*

Proof. We argue by induction on the complexity of α. The base case is established as follows. If $a \sqsubseteq R_n su(ut)$, then, by definition, $a \sqsubseteq \bigsqcap_{[n]\varphi \in \Phi} \big(s([n]\varphi) \Rightarrow u(\varphi)\big)(ut)$. Hence, for all $[n]\varphi \in \Phi$, $a \sqsubseteq \big(s([n]\varphi) \Rightarrow u(\varphi)\big)\big(u(\varphi) \Rightarrow t(\varphi)\big)$ by the definition of $u \sim t$ and monotonicity of monoid multiplication (also, $[n]\varphi \in \Phi$ implies $\varphi \in \Phi$). It follows by the properties of FL-algebras that

$a \sqsubseteq (s([n]\varphi) \Rightarrow t(\varphi))$. Since $[n]\varphi \in \Phi$ was arbitrary, we obtain $a \sqsubseteq R_n st$. All cases of the induction step are easy. □

Definition 5.9 For all α and s, we define the following formula:

$$R_\alpha s \quad := \quad \bigvee_{x \in S} \left(\overline{R_\alpha sx} \cdot \bigwedge_{\varphi \in \Phi} \overline{(x(\varphi) \leftrightarrow \varphi)} \right)$$

Note that $R_\alpha s$ is well defined even though S is infinite – there are only finitely many possible values of $R_\alpha sx$ for $x \in S$, as $\boldsymbol{X}$ is finite. Note also that $t(R_\alpha s) = \bigsqcup_{x \in S} (R_\alpha sx(xt))$.

Lemma 5.10 *For all s, t and α, $t(R_\alpha s) = R_\alpha st$.*

Proof. First, $R_\alpha st \sqsubseteq R_\alpha st(tt)$ by Lemma 5.4(a), and $R_\alpha st(tt) \sqsubseteq \bigsqcup_{x \in S} (R_\alpha sx(xt)) = t(R_\alpha s)$. Second, $R_\alpha sx(xt) \sqsubseteq R_\alpha st$ for all $x \in S$ by Lemma 5.8. Hence, $\bigsqcup_{x \in S} R_\alpha sx(xt)$ and so $t(R_\alpha s) \sqsubseteq R_\alpha st$. □

Lemma 5.11 *For all $s, t \in S$ and all $\alpha \in ACT$, $R^{\mathcal{L}}_\alpha st \sqsubseteq R_\alpha st$.*

Proof. Induction on the complexity of α. The base case follows from definition. To establish the induction step, we reason by cases. Note that the induction hypothesis is equivalent to the claim that, for all α, β and x, $1 \sqsubseteq x([\alpha]R_\alpha x)$ and $1 \sqsubseteq x([\beta]R_\beta x)$ by Lemmas 5.6(b) and 5.10.

If $a \sqsubseteq R^{\mathcal{L}}_{\alpha \cup \beta} st$, then $a \sqsubseteq \bigsqcap_{\varphi \in \mathcal{L}} \{t(\varphi) \mid 1 \sqsubseteq s([\alpha \cup \beta]\varphi)\}$ by Lemma 5.6(a). By the definition of S, this entails $a \sqsubseteq \bigsqcap \{t(\varphi) \mid 1 \sqsubseteq s([\alpha]\varphi) \sqcap s([\beta]\varphi)\}$. By the induction hypothesis, $1 \sqsubseteq s([\alpha]R_\alpha s)$ and $1 \sqsubseteq s([\beta]R_\beta s)$. Hence, $1 \sqsubseteq s([\alpha](R_\alpha s \vee R_\beta s))$ and $1 \sqsubseteq s([\beta](R_\alpha s \vee R_\beta s))$ by the definition of S. It follows that $a \sqsubseteq t(R_\alpha s) \sqcup t(R_\beta s)$. By Lemma 5.10, $a \sqsubseteq R_\alpha st \sqcup R_\beta st$ and so $a \sqsubseteq R_{\alpha \cup \beta} st$.

If $a \sqsubseteq R^{\mathcal{L}}_{\alpha\beta} st$, then $a \sqsubseteq \bigsqcap_{\varphi \in \mathcal{L}} \{t(\varphi) \mid 1 \sqsubseteq s([\alpha\beta]\varphi)\}$ by Lemma 5.6(a) and so $a \sqsubseteq \bigsqcap_{\varphi \in \mathcal{L}} \{t(\varphi) \mid 1 \sqsubseteq s([\alpha][\beta]\varphi)\}$ by the definition of S. For all x and y, $R^{\mathcal{L}}_\alpha sx R^{\mathcal{L}}_\beta xy \sqsubseteq y(R_{\alpha\beta} s)$ by the induction hypothesis, Lemma 5.10 and the definition of $R_{\alpha\beta}$. Hence, for all x, $R^{\mathcal{L}}_\alpha sx \sqsubseteq x([\beta]R_{\alpha\beta} s)$ by residuation and Lemma 5.6(b); from this is follows that $1 \sqsubseteq s([\alpha][\beta]R_{\alpha\beta} s)$ by another application of residuation and Lemma 5.6(b). Therefore, $a \sqsubseteq t(R_{\alpha\beta} s)$ and so $a \sqsubseteq R_{\alpha\beta} st$ by Lemma 5.10.

Finally, we discuss the case of α^+. Fix s; we write F instead of $R_{\alpha^+} s$. Note that R_{α^+} is a transitive relation extending $R^{\mathcal{L}}_\alpha$. Hence, for all $t, u \in S$, $u(F) \cdot R^{\mathcal{L}}_\alpha ut \sqsubseteq t(F)$ by Lemma 5.10 and the induction hypothesis applied to $R^{\mathcal{L}}_\alpha$; we obtain from this $u(F) \sqsubseteq u([\alpha]F)$ for all $u \in S$ by Lemma 5.6(b). Hence, by definition of S, we have $F \to [\alpha]F \in \mathsf{Thm}$. Hence, using (R-+), we have $F \to [\alpha^+]F \in \mathsf{Thm}$ and, using (R-mon) and (A-+), we obtain $[\alpha]F \to [\alpha^+]F \in \mathsf{Thm}$. By the induction hypothesis we have $R^{\mathcal{L}}_\alpha st \sqsubseteq R_\alpha st \sqsubseteq R_{\alpha^+} st$ for all t and so $1 \sqsubseteq R^{\mathcal{L}}_\alpha st \Rightarrow t(F)$ for all t by Lemma 5.10. This means that $1 \sqsubseteq s([\alpha]F)$ and so $1 \sqsubseteq s([\alpha^+]F)$ which means that $R^{\mathcal{L}}_{\alpha^+} st \sqsubseteq t(F)$ for all t by Lemma 5.6(b). Hence, $R^{\mathcal{L}}_{\alpha^+} st \sqsubseteq R_{\alpha^+} st$ by Lemma 5.10. □

Lemma 5.12 *For all $\varphi \in \Phi$ and $s \in S$, $s(\varphi) = V(\varphi, s)$.*

Proof. Induction on the complexity of φ. The base case holds by definition and the cases for non-modal formulas and canonical constants are straightforward. Finally, $s([\alpha]\varphi) \sqsubseteq V([\alpha]\varphi, s)$ holds thanks to Lemma 5.7 and $V([\alpha]\varphi, s) \sqsubseteq s([\alpha]\varphi)$ holds thanks to Lemma 5.6(b) and Lemma 5.11. □

Theorem 5.13 *For all finite commutative integral* $\boldsymbol{X}$ *with canonical constants,* φ *is valid in all* $\boldsymbol{X}$*-frames iff* φ *is a theorem of* $\mathsf{PDL}(\boldsymbol{X})$.

Proof. Soundness is established by Theorem 5.2. Completeness is established as usual. If φ is not in Thm, then $\mathsf{Thm} \nvdash_{\mathsf{L}} \varphi$ since Thm is obviously closed under $\vdash_{\mathsf{L}}$. By strong completeness of L, there is a non-modal homomorphism from $\mathcal{L}$ to $\boldsymbol{X}$ such that $s[\mathsf{Thm}] = \{1\}$ and $s(\varphi) \neq 1$. Let Φ be the closure of $\{\varphi\}$; φ is not valid in the canonical model modulo Φ by Lemma 5.12. □

6 On Kleene star and test

Our syntactic presentation of propositional dynamic logic differs from the standard presentation in two important respects, namely, (i) our action operators do not include the *Kleene star*, but rather the Kleene plus operator; (ii) we do not include the *test operator*. Kleene star and test are instrumental in the ability of classical PDL to express standard programming constructs such as while loops and conditionals (test suffices for the latter). In this section we discuss these omissions.

Concerning the Kleene star, Proposition 2.13 suggests that, working with frames based on finite integral FL-algebras, we can define, for all $\alpha \in ACT$ and $\varphi \in Fm(\mathcal{L}_{\boldsymbol{X}})$,

$$[\alpha^*]\varphi := [\alpha^+]\varphi \wedge \varphi$$

as a semantically equivalent surrogate for formulas with the Kleene star. For instance, $[(\mathrm{a} \cup \mathrm{b})^*; \mathrm{a}^*]\mathrm{p}$ is short for $[(\mathrm{a} \cup \mathrm{b})^+]([\mathrm{a}^+]\mathrm{p} \wedge \mathrm{p}) \wedge ([\mathrm{a}^+]\mathrm{p} \wedge \mathrm{p})$. However, it is clear that not all action expressions in STA can be expressed by action expressions in ACT. Therefore, for example, $[(\mathrm{a}^*; \mathrm{b})^*]\mathrm{p}$ is not a well-formed formula since $\mathrm{a}^* \notin ACT$.

The technical problem that precluded us from working with Kleene star as a primitive operator is related to Lemma 5.8. Take the reflexive transitive closure R^*_α of R_α, defined as in Proposition 2.8. The issue is that Lemma 5.8 fails if Kleene star is a primitive operator and we define $R_{\alpha^*} := R^*_\alpha$. In particular, if $s = u \neq t$, then $R^*_\alpha su(ut) \sqsubseteq R^*_\alpha st$ boils down to $s \sim t \sqsubseteq R_{\alpha^+} st$, which does not hold in all canonical models. (Take the canonical $\mathbf{2}$-model modulo the closure Φ of $\Psi = \{[\mathrm{a}]\bot\}$. As both $\Psi \cup \{\mathrm{p}_0\}$ and $\Psi \cup \{\mathrm{p}_1\}$ are consistent, there are two distinct s, t such that $s \sim_\Phi t$ equals 1, but $R_{\mathrm{a}^+} st$ equals 0.)

Concerning test, a natural semantic interpretation of $\varphi?$, endorsed also in [18,19], is

$$R_{\varphi?}(s,t) = \begin{cases} V(\varphi, s) & \text{if } s = t \\ \bot^{\boldsymbol{X}} & \text{otherwise.} \end{cases}$$

However, Lemma 5.8 turns out to be problematic for such a relation as well. (Take the model from the previous paragraph and let $\varphi = [\mathrm{a}]\bot$; clearly

$R_{\varphi?}ss(st)$ equals 1, but $R_{\varphi?}st$ equals 0.)

It is clear that a more substantial modification of our completeness argument is needed to accommodate logics with Kleene star and test. This is an interesting problem we leave open here.

7 Conclusion

We have studied a general framework for many-valued versions of Propositional Dynamic Logic where both formulas in states and accessibility relations between states of a Kripke model are evaluated in a finite FL-algebra. We established a general decidability result and we provided a general completeness argument for PDLs based on commutative integral FL-algebras with canonical constants. We build on previous work on many-valued modal logic and our techniques are generalizations of the arguments used in the two-valued case; however, to the best of our knowledge, the technical results presented here are the first decidability and completeness results on PDL with many-valued accessibility relations. As our discussion of the informal interpretations of the framework suggests, many-valued PDL has links to existing research in description logics and potential applications in reasoning about weighted labelled transition systems.

Our paper also suggests a number of topics for future research. We would like to mention especially the addition of test and further work on the standard version of PDL with primitive Kleene star in the many-valued setting. Another topic are generalizations of our results beyond finite (commutative integral) FL-algebras with canonical constants; in many cases the work here would require modifications of existing techniques used in completeness arguments for many-valued modal logics without "structured" modal operators. Finally, informal interpretations and applications of our framework need to be explored in more detail.

References

[1] Baader, F., *Augmenting concept languages by transitive closure of roles: An alternative to terminological cycles*, in: *IJCAI'91*, 1991, pp. 446–451.

[2] Baader, F., D. Calvanese, D. L. McGuiness, D. Nardi and P. F. Patel-Schneider, editors, "The Description Logic Handbook: Theory, Implementation, and Applications," Cambridge University Press, 2007, 2nd edition edition.

[3] Blackburn, P., M. de Rijke and Y. Venema, "Modal Logic," Cambridge University Press, 2001.

[4] Bou, F., F. Esteva, L. Godo and R. O. Rodríguez, *On the minimum many-valued modal logic over a finite residuated lattice.*, Journal of Logic and Computation **21** (2011), pp. 739–790.

[5] Běhounek, L., *Modeling costs of program runs in fuzzified propositional dynamic logic*, in: F. Hakl, editor, *Doktorandské dny '08* (2008), pp. 6 – 14.

[6] Caicedo, X. and R. O. Rodríguez, *Standard Gödel modal logics*, Studia Logica **94** (2010), pp. 189–214.

[7] Caicedo, X. and R. O. Rodríguez, *Bi-modal Gödel logic over [0,1]-valued Kripke frames*, Journal of Logic and Computation **25** (2015), pp. 37–55.

[8] Conradie, W., W. Morton and C. Robinson, *Filtrations for many-valued modal logic with applications* (2017), presentation at TACL 2017, Prague.
[9] Droste, M., W. Kuich and H. Vogler, editors, "Handbook of Weighted Automata," Springer, 2009.
[10] Fischer, M. J. and R. E. Ladner, *Propositional dynamic logic of regular programs*, Journal of Computer and System Sciences **18** (1979), pp. 194–211.
[11] Fitting, M., *Many-valued modal logics*, Fundamenta Informaticae **15** (1991), pp. 235–254.
[12] Fitting, M., *Many-valued modal logics II*, Fundamenta Informaticae **17** (1992), pp. 55–73.
[13] Galatos, N., P. Jipsen, T. Kowalski and H. Ono, "Residuated Lattices: An Algebraic Glimpse at Substructural Logics," Elsevier, 2007.
[14] Hájek, P., "Metamathematics of Fuzzy Logic," Kluwer, 1998.
[15] Hansoul, G. and B. Teheux, *Extending Łukasiewicz logics with a modality: Algebraic approach to relational semantics*, Studia Logica **101** (2013), pp. 505–545.
[16] Harel, D., "First-Order Dynamic Logic," Lecture Notes in Computer Science **68**, Springer, 1979.
[17] Harel, D., D. Kozen and J. Tiuryn, "Dynamic Logic," MIT Press, 2000.
[18] Hughes, J., A. Esterline and B. Kimiaghalam, *Means-end relations and a measure of efficacy*, Journal of Logic, Language and Information **15** (2006), pp. 83–108.
[19] Liau, C.-J., *Many-valued dynamic logic for qualitative decision theory*, in: N. Zhong, A. Skowron and S. Ohsuga, editors, *New Directions in Rough Sets, Data Mining, and Granular-Soft Computing* (1999), pp. 294–303.
[20] Madeira, A., R. Neves and M. A. Martins, *An exercise on the generation of many-valued dynamic logics*, Journal of Logical and Algebraic Methods in Programming **85** (2016), pp. 1011–1037.
[21] Madeira, A., R. Neves, M. A. Martins and L. S. Barbosa, *A dynamic logic for every season*, in: C. Braga and N. Martí-Oliet, editors, *Formal Methods: Foundations and Applications* (2015), pp. 130–145.
[22] Odintsov, S. and H. Wansing, *Modal logics with Belnapian truth values*, Journal of Applied Non-Classical Logics **20** (2010), pp. 279–301.
[23] Ostermann, P., *Many-valued modal propositional calculi*, Mathematical Logic Quarterly **34** (1988), pp. 343–354.
[24] Sedlár, I., *Propositional dynamic logic with Belnapian truth values*, in: *Advances in Modal Logic. Vol. 11* (2016).
[25] Segerberg, K., *Some modal logics based on a three-valued logic*, Theoria **33** (1967), pp. 53–71.
[26] Straccia, U., *Description logics over lattices*, International Journal of Uncertainty, Fuzziness and Knowledge-Based Systems **14** (2006), pp. 1–16.
[27] Teheux, B., *Propositional dynamic logic for searching games with errors*, Journal of Applied Logic **12** (2014), pp. 377–394.
[28] Vidal, A., F. Esteva and L. Godo, *On modal extensions of product fuzzy logic*, Journal of Logic and Computation **27** (2017), p. 299.

[9] [illegible] and O. [illegible] modal logic and [illegible] presentation at [illegible] 2017 [illegible]
[9] [illegible] and [illegible] Handbook [illegible] Springer, 2009.
[10] Fischer, M. J. and R. E. Ladner, Propositional dynamic logic of regular programs, Journal of Computer and System Sciences 18 (1979), pp. [illegible]–211.
[11] [illegible] Foundations of [illegible] (1981), pp. 23[illegible]
[12] [illegible] Fundamenta Mathematicae [illegible] (1969), pp. [illegible]
[13] [illegible], L. Kovács and R. Ortiz [illegible] Applied Non-classical Logics [illegible], 2007.
[14] Hájek, P., Metamathematics of fuzzy logic, Kluwer, 1998.
[15] Hansoul, G. and B. Teheux, Extending Łukasiewicz logics with a modality: Algebraic approach to relational semantics, Studia Logica 101 (2013), pp. 505–545.
[16] Harel, D., First-Order Dynamic Logic, Lecture Notes in Computer Science 68, Springer, 1979.
[17] [illegible], D. Kozen and J. Tiuryn, Dynamic logic, MIT Press, 2000.
[18] [illegible] Kaminski and [illegible] Information 15 (2006), pp. [illegible]
[19] [illegible] theory [illegible] Mathematical Logic [illegible] 29–305.
[20] [illegible]
[21] [illegible]
[22] [illegible]
[23] [illegible] logic with [illegible] Applied Non-classical Logics 20 (2010), pp. 2[illegible]
[24] [illegible]
14 (2004), pp. [illegible]

Global neighbourhood completeness of the provability logic GLP

Daniyar Shamkanov [1]

Steklov Mathematical Institute of Russian Academy of Sciences
Gubkina str. 8, 119991, Moscow, Russia

National Research University Higher School of Economics
Moscow, Russia

In memoriam Sergei Ivanovich Adian
(01.01.1931 —- 05.05.2020)

Abstract

The provability logic GLP introduced by G. Japaridze is a propositional polymodal logic with important applications in proof theory, specificially, in ordinal analysis of arithmetic. Though being incomplete with respect to any class of Kripke frames, the logic GLP is complete for its neighbourhood interpretation. This completeness result, established by L. Beklemishev and D. Gabelaia, implies strong neighbourhood completeness of this system for the case of the so-called local semantic consequence relation. In the given article, we consider Hilbert-style non-well-founded derivations in the provability logic GLP and establish that GLP with the obtained derivability relation is strongly neighbourhood complete in the case of the global semantic consequence relation.

Keywords: provability logic, algebraic semantics, neighbourhood semantics, global consequence relations, non-well-founded derivations.

1 Introduction

The provability logic GLP introduced by G. Japaridze [6] is a propositional modal logic in a language with infinitely many modal connectives $\Box_0, \Box_1, \ldots$. It is sound and complete with respect to a natural provability semantics, where the modal connective $\Box_n$ corresponds to the provability predicate "*... is provable from the axioms of Peano arithmetic together with all true arithmetical Π^0_n-sentences*". This system has important applications in proof theory, specificially, in ordinal analysis of arithmetic [1]. In the given article, we consider non-well-founded derivations in the provability logic GLP and study algebraic

[1] *E-mail: daniyar.shamkanov@gmail.com*

and neighbourhood semantics of the system GLP with the obtained derivability relation.

Neighbourhood semantics is an interesting generalization of Kripke semantics independently developed by D. Scott and R. Montague in [9] and [7]. The logic GLP is incomplete with respect to any class of Kripke frames. At the same time GLP is complete for its neighbourhood interpretation [3]. Notice that this completeness result implies strong neighbourhood completeness of this system for the case of the so-called local semantic consequence relation. Over neighbourhood GLP-models, a formula φ is a local semantic consequence of Γ if for any neighbourhood GLP-model $\mathcal{M}$ and any world x of $\mathcal{M}$

$$(\forall \psi \in \Gamma \ \mathcal{M}, x \vDash \psi) \Rightarrow \mathcal{M}, x \vDash \varphi.$$

A formula φ is a global semantic consequence of Γ if for any neighbourhood GLP-model $\mathcal{M}$

$$(\forall \psi \in \Gamma \ \mathcal{M} \vDash \psi) \Rightarrow \mathcal{M} \vDash \varphi.$$

Recently, global neighbourhood completeness of the Gödel-Löb provability logic GL with non-well-founded derivations was established in [10,11]. In the given article, we obtain an analogous result for the provability logic GLP.

2 Non-well-founded derivations in GLP

In this section we recall the provability logic GLP and define Hilbert-style non-well-founded derivations for the given system.

The provability logic GLP is a propositional modal logic in a language with infinitely many modal connectives $\Box_0, \Box_1, \dots$. In other words, formulas of the logic are built from the countable set of variables $PV = \{p, q, \dots\}$ and the constant $\bot$ using propositional connectives $\to$ and $\Box_i$ for each $i \in \mathbb{N}$. We treat other Boolean connectives and modal connectives $\Diamond_i$ as abbreviations:

$$\neg\varphi := \varphi \to \bot, \qquad \top := \neg\bot, \qquad \varphi \wedge \psi := \neg(\varphi \to \neg\psi),$$
$$\varphi \vee \psi := \neg\varphi \to \psi, \qquad \varphi \leftrightarrow \psi := (\varphi \to \psi) \wedge (\psi \to \varphi), \qquad \Diamond_i\varphi := \neg \Box_i \neg\varphi.$$

By Fm, we denote the set of formulas of GLP.

The provability logic GLP is defined by the following axiom schemes and inference rules.

Axiom schemes:

(i) the tautologies of classical propositional logic;

(ii) $\Box_i(\varphi \to \psi) \to (\Box_i\varphi \to \Box_i\psi)$;

(iii) $\Box_i(\Box_i\varphi \to \varphi) \to \Box_i\varphi$;

(iv) $\Diamond_i\varphi \to \Box_{i+1} \Diamond_i \varphi$;

(v) $\Box_i\varphi \to \Box_{i+1}\varphi$.

Inference rules:

$$\mathsf{mp}\ \frac{\varphi \qquad \varphi \to \psi}{\psi}, \qquad \mathsf{nec}\ \frac{\varphi}{\Box_0\varphi}.$$

We remark that transitivity of the modal connectives $\Box_i$ is provable in GLP, i.e. $\mathsf{GLP} \vdash \Box_i\psi \to \Box_i\Box_i\psi$ for any formula ψ and any $i \in \mathbb{N}$.

Now we define non-well-founded derivations in GLP. An ∞-*derivation* is a (possibly infinite) tree whose nodes are marked by formulas of GLP and that is constructed according to the rules (mp) and (nec). In addition, any infinite branch in an ∞-derivation must contain infinitely many applications of the rule (nec). An *assumption leaf* of an ∞-derivation is a leaf that is not marked by an axiom of GLP.

The *main fragment* of an ∞-derivation is a finite tree obtained from the ∞-derivation by cutting every infinite branch at the nearest to the root application of the rule (nec). The *local height* $|\pi|$ *of an* ∞-*derivation* π is the length of the longest branch in its main fragment. An ∞-derivation consisting of a single formula only has height 0.

For example, consider the following ∞-derivation

$$\dfrac{\dfrac{\dfrac{\dfrac{\dfrac{\vdots}{\Box_0 p_3} \qquad \Box_0 p_3 \to p_2}{p_2}\,\mathsf{mp}}{\Box_0 p_2}\,\mathsf{nec} \qquad \Box_0 p_2 \to p_1}{p_1}\,\mathsf{mp}}{\Box_0 p_1}\,\mathsf{nec} \qquad \Box_0 p_1 \to p_0}{p_0}\,\mathsf{mp} \;,$$

where assumption leaves are marked by formulas of the form $\Box_0 p_{i+1} \to p_i$. The local height of this ∞-derivation equals to 1 and its main fragment has the form

$$\mathsf{mp}\,\dfrac{\Box_0 p_1 \qquad \Box_0 p_1 \to p_0}{p_0} \;.$$

Definition 2.1 We set $\Gamma \vdash_g \varphi$ if there is an ∞-derivation with the root marked by φ in which all assumption leaves are marked by some elements of Γ.

Proposition 2.2 *For any formula* φ, *we have*

$$\mathsf{GLP} \vdash \varphi \Longleftrightarrow \varnothing \vdash_g \varphi.$$

We give a proof of this proposition in the Appendix since this statement is not essential for the global neighbourhood completeness result of the final section.

3 Algebraic semantics

In this section we consider algebraic semantics for the provability logic GLP enriched with non-well-founded derivations.

A *Magari algebra* (or a *diagonalizable algebra*) $\mathcal{A} = (A, \wedge, \vee, \to, 0, 1, \Box)$ is a Boolean algebra $(A, \wedge, \vee, \to, 0, 1)$ together with a unary map $\Box\colon A \to A$ satisfying the identities:

$$\Box 1 = 1, \quad \Box(x \wedge y) = \Box x \wedge \Box y, \quad \Box(\Box x \to x) = \Box x.$$

For any Magari algebra $\mathcal{A}$, the mapping $\Box$ is monotone with respect to the order (of the Boolean part) of $\mathcal{A}$. Indeed, if $a \leqslant b$, then $a \wedge b = a$, $\Box a \wedge \Box b =$

$\Box(a \wedge b) = \Box a$, and $\Box a \leqslant \Box b$. In addition, we remark that an inequality $\Box x \leqslant \Box\Box x$ holds in any Magari algebra.

We call a Magari algebra *$\Box$-founded (or Pakhomov-Walsh-founded)*[2] if, for every sequence of its elements $(a_i)_{i \in \mathbb{N}}$ such that $\Box a_{i+1} \leqslant a_i$, we have $a_0 = 1$. Note that, for any such sequence $(a_i)_{i \in \mathbb{N}}$, all elements a_i are equal to 1 in any $\Box$-founded Magari algebra.

We give a series of examples of $\Box$-founded Magari algebras. A Magari algebra is called *σ-complete* if its underlying Boolean algebra is σ-complete, that is, each of its countable subsets S has the least upper bound $\bigvee S$. An equivalent condition is that every countable subset S has the greatest lower bound $\bigwedge S$.

Proposition 3.1 *Any σ-complete Magari algebra is $\Box$-founded.*

Proof. Assume we have a σ-complete Magari algebra $\mathcal{A}$ and a sequence of its elements $(a_i)_{i \in \mathbb{N}}$ such that $\Box a_{i+1} \leqslant a_i$. We shall prove that $a_0 = 1$.

Put $b = \bigwedge\limits_{i \in \mathbb{N}} a_i$. For any $i \in \mathbb{N}$, we have $b \leqslant a_{i+1}$ and $\Box b \leqslant \Box a_{i+1} \leqslant a_i$. Hence,

$$\Box b \leqslant b, \qquad \Box b \to b = 1, \qquad \Box b = \Box(\Box b \to b) = \Box 1 = 1, \qquad b = 1.$$

We obtain that $a_0 = 1$. □

Remark 3.2 Let us additionally mention an arithmetical example of $\Box$-founded Magari algebra without going into details. If we consider the second-order arithmetical theory $\Sigma^1_1 - \mathsf{AC}_0$ extended with all true Σ^1_1-sentences, then its provability algebra forms a $\Box$-founded Magari algebra. This observation can be obtained following the lines of Theorem 3.2 from [8].

The notion of $\Box$-founded Magari algebra $\mathcal{A}$ can be defined in terms of the binary relation $<_{\mathcal{A}}$ on $\mathcal{A}$:

$$a <_{\mathcal{A}} b \Longleftrightarrow \Box a \leqslant b.$$

Proposition 3.3 (see Proposition 3.1 from [11]) *For any Magari algebra $\mathcal{A} = (A, \wedge, \vee, \to, 0, 1, \Box)$, the relation $<_{\mathcal{A}}$ is a strict partial order on $A \setminus \{1\}$.*

Proposition 3.4 (see Proposition 3.2 from [11]) *For any Magari algebra $\mathcal{A} = (A, \wedge, \vee, \to, 0, 1, \Box)$, the algebra $\mathcal{A}$ is $\Box$-founded if and only if the partial order $<_{\mathcal{A}}$ on $A \setminus \{1\}$ is well-founded.*

A Boolean algebra $(A, \wedge, \vee, \to, 0, 1)$ together with a sequence of unary mappings $\Box_0, \Box_1, \ldots$ is called a GLP-*algebra* if it satisfies the following conditions for each $i \in \mathbb{N}$:

(i) $(A, \wedge, \vee, \to, 0, 1, \Box_i)$ is a Magari algebra;

(ii) $\Diamond_i a \leqslant \Box_{i+1} \Diamond_i a$ for any $a \in A$;

(iii) $\Box_i a \leqslant \Box_{i+1} a$ for any $a \in A$.

[2] This notion has been inspired by an article of F. Pakhomov and J. Walsh [8].

A GLP-algebra $\mathcal{A} = (A, \wedge, \vee, \to, 0, 1, \Box_0, \Box_1, \dots)$ is called *$\Box$-founded* if the Magari algebra $\mathcal{A}_0 = (A, \wedge, \vee, \to, 0, 1, \Box_0)$ is $\Box$-founded. In the same way, we apply notions defined for the Magari algebra $\mathcal{A}_0$ to $\mathcal{A}$. From Proposition 3.1, we immediately see that any σ-complete GLP-algebra is $\Box$-founded.

Now we define a semantic consequence relation over $\Box$-founded GLP-algebras, which, we shall see, corresponds to the derivability relation $\vdash_g$. A *valuation in a* GLP-*algebra* $\mathcal{A} = (A, \wedge, \vee, \to, 0, 1, \Box_0, \Box_1, \dots)$ is a function $v\colon Fm \to A$ such that $v(\bot) = 0$, $v(\varphi \to \psi) = v(\varphi) \to v(\psi)$, and $v(\Box_i \varphi) = \Box_i v(\varphi)$.

Definition 3.5 Given a set of formulas Γ and a formula φ, we set $\Gamma \Vdash_g \varphi$ if for any $\Box$-founded GLP-algebra $\mathcal{A}$ and any valuation v in $\mathcal{A}$

$$(\forall \psi \in \Gamma \;\; v(\psi) = 1) \Rightarrow v(\varphi) = 1.$$

Lemma 3.6 *For any set of formulas Γ and any formula φ, we have*

$$\Gamma \vdash_g \varphi \Longrightarrow \Gamma \Vdash_g \varphi.$$

Proof. Assume π is an ∞-derivation with the root marked by φ in which all assumption leaves are marked by some elements of Γ. In addition, assume we have a $\Box$-founded GLP-algebra $\mathcal{A} = (X, \wedge, \vee, \to, 0, 1, \Box_0, \Box_1, \dots)$ together with a valuation v in $\mathcal{A}$ such that $v(\psi) = 1$ for any $\psi \in \Gamma$. We shall prove that $v(\varphi) = 1$.

For any node w of the ∞-derivation π, let π_w be the subtree of π with the root w. Also, put $r(w) = |\pi_w|$. In addition, let φ_w be the formula of the node w. A node w belongs to the *$(n+1)$-th slice of* π if there are precisely n applications of the rule (nec) on the path from this node to the root of π. By c_n, we denote the element $\bigwedge\{v(\varphi_w) \mid w \text{ belongs to the } (n+1)\text{-th slice of } \pi\}$.

We claim that $\Box_0 c_{n+1} \leqslant c_n$ for any $n \in \mathbb{N}$. It is sufficient to prove that $\Box_0 c_{n+1} \leqslant v(\varphi_w)$ whenever w belongs to the $(n+1)$-th slice of π. The proof is by induction on $r(w)$.

If φ_w is an axiom of GLP or an element of Γ, then we immediately obtain the required statement. Otherwise, φ_w is obtained by an application of an inference rule in π.

If φ_w is obtained by the rule (nec), then this formula has the form $\Box_0 \varphi_u$, where u is the premise of w. We see that u belongs to the $(n+2)$-th slice of π. Consequently $c_{n+1} \leqslant v(\varphi_u)$ and $\Box_0 c_{n+1} \leqslant v(\varphi_w)$.

Suppose φ_w is obtained by the rule (mp). Consider the premises u_1 and u_2 of w. We have $r(u_1) < r(w)$ and $r(u_2) < r(w)$. By our induction hypotheses, we obtain $\Box_0 c_{n+1} \leqslant v(\varphi_{u_1}) \wedge v(\varphi_{u_2}) \leqslant v(\varphi_w)$.

This proves the claim that $\Box_0 c_{n+1} \leqslant c_n$ for any $n \in \mathbb{N}$. Applying $\Box$-foundedness of $\mathcal{A}$, we note that $c_0 = 1$. Since the root of the ∞-derivation π belongs to the first slice of π, we conclude that $c_0 \leqslant v(\varphi)$ and $v(\varphi) = 1$. □

Theorem 3.7 (Algebraic completeness) *For any set of formulas Γ and any formula φ, we have*

$$\Gamma \vdash_g \varphi \Longleftrightarrow \Gamma \Vdash_g \varphi.$$

Proof. The left-to-right implication follows from Lemma 3.6. We prove the converse. Assume $\Gamma \Vdash_g \varphi$. Consider the theory $T = \{\theta \in Fm \mid \Gamma \vdash_g \theta\}$. We see that T contains all axioms of GLP and is closed under the rules (mp) and (nec). We define an equivalence relation $\sim_T$ on the set of formulas Fm by putting $\mu \sim_T \rho$ if and only of $(\mu \leftrightarrow \rho) \in T$. Let us denote the equivalence class of μ by $[\mu]_T$. Applying the Lindenbaum-Tarski construction, we obtain a GLP-algebra $\mathcal{L}_T$ on the set of equivalence classes of formulas, where $[\mu]_T \wedge [\rho]_T = [\mu \wedge \rho]_T$, $[\mu]_T \vee [\rho]_T = [\mu \vee \rho]_T$, $[\mu]_T \to [\rho]_T = [\mu \to \rho]_T$, $0 = [\bot]_T$, $1 = [\top]_T$ and $\Box_i[\mu] = [\Box_i\mu]$.

Let us check that the algebra $\mathcal{L}_T$ is $\Box$-founded. Assume we have a sequence of formulas $(\mu_i)_{i\in\mathbb{N}}$ such that $\Box_0[\mu_{i+1}]_T \leqslant [\mu_i]_T$. We have $[\Box_0\mu_{i+1} \to \mu_i]_T = 1$ and $(\Box_0\mu_{i+1} \to \mu_i) \in T$. For every $i \in \mathbb{N}$, there exists an ∞-derivation π_i for the formula $\Box_0\mu_{i+1} \to \mu_i$ such that all assumption leaves of π_i are marked by some elements of Γ. We obtain the following ∞-derivation for the formula μ_0:

$$\mathrm{mp}\,\dfrac{\mathrm{nec}\,\dfrac{\mathrm{mp}\,\dfrac{\mathrm{nec}\,\dfrac{\mathrm{mp}\,\dfrac{\begin{matrix}\vdots\\ \Box_0\mu_3\end{matrix} \qquad \begin{matrix}\pi_2\\ \vdots\\ \Box_0\mu_3 \to \mu_2\end{matrix}}{\mu_2}}{\Box_0\mu_2} \qquad \begin{matrix}\pi_1\\ \vdots\\ \Box_0\mu_2 \to \mu_1\end{matrix}}{\mu_1}}{\Box_0\mu_1} \qquad \begin{matrix}\pi_0\\ \vdots\\ \Box_0\mu_1 \to \mu_0\end{matrix}}{\mu_0}\ ,$$

where all assumption leaves are marked by some elements of Γ. Hence, $\mu_0 \in T$ and $[\mu_0]_T = [\top]_T = 1$. We conclude that the GLP-algebra $\mathcal{L}_T$ is $\Box$-founded.

Consider the valuation $v{:}\theta \mapsto [\theta]_T$ in the GLP-algebra $\mathcal{L}_T$. Since $\Gamma \subset T$, we have $v(\psi) = 1$ for any $\psi \in \Gamma$. From the assumption $\Gamma \Vdash \varphi$, we obtain that $v(\varphi) = 1$. Consequently $\varphi \in T$ and $\Gamma \vdash_g \varphi$. $\Box$

4 Neighbourhood semantics

In this section we recall neighbourhood semantics of the provability logic GLP.

An *Esakia frame* (or a *Magari frame*) $\mathcal{X} = (X, \Box)$ is a set X together with a mapping $\Box : \mathcal{P}(X) \to \mathcal{P}(X)$ such that the powerset Boolean algebra $\mathcal{P}(X)$ with the mapping $\Box$ forms a Magari algebra.

We briefly recall a connection between scattered topological spaces and Esakia frames (cf. [4]). Note that we allow Esakia frames and topological spaces to be empty.

In a topological space, an open set U containing a point x is called a *neighbourhood* of x. A set U is a *punctured neighbourhood* of x if $x \notin U$ and $U \cup \{x\}$ is open. For a topological space (X, τ) and a subset V the *derived set* $d_\tau(V)$ *of* V is the set of limit points of V:

$$x \in d_\tau(V) \Longleftrightarrow \forall U \in \tau\ (x \in U \Rightarrow \exists y \neq x\ (y \in U \cap V))\,.$$

The *co-derived set* $cd_\tau(V)$ *of* V is defined as $X \setminus d_\tau(X \setminus V)$. By definition, $x \in cd_\tau(V)$ if and only if there is a punctured neighbourhood of x entirely contained in V.

In a topological space, a point having an empty punctured neighbourhood is called *isolated.* A topological space (X,τ) is *scattered* if each non-empty subset of X (as a topological space with the inherited topology) has an isolated point.

Proposition 4.1 (L. Esakia [5]) *If $(X,\Box)$ is an Esakia frame, then X bears a unique topology τ for which $\Box = cd_\tau$. Moreover, the space (X,τ) is scattered.*

Proposition 4.2 (H. Simmons [12], L. Esakia [5]) *If (X,τ) is a scattered topological space, then (X, cd_τ) is an Esakia frame.*

A *neighbourhood* GLP*-frame* $\mathcal{X} = (X,\Box_0,\Box_1,\dots)$ is a set X together with a sequence of unary mappings $\Box_0,\Box_1,\dots$ on $\mathcal{P}(X)$ such that the powerset Boolean algebra $\mathcal{P}(X)$ with the given mappings forms a GLP-algebra. Elements of X are called *worlds* of the frame $\mathcal{X}$. A *neighbourhood* GLP*-model* is a pair $\mathcal{M} = (\mathcal{X},v)$, where $\mathcal{X}$ is a neighbourhood GLP-frame and v is a valuation in the powerset GLP-algebra of $\mathcal{X}$. A formula φ is *true at a world x of a model* $\mathcal{M}$, written as $\mathcal{M},x \vDash \varphi$, if $x \in v(\varphi)$. A formula φ is called *true in* $\mathcal{M}$, written as $\mathcal{M} \vDash \varphi$, if φ is true at all worlds of $\mathcal{M}$.

A GLP*-space* is a polytopological space $(X,\tau_0,\tau_1,\dots)$, where, for each $i \in \mathbb{N}$, τ_i is scattered, $\tau_i \subset \tau_{i+1}$, and $d_{\tau_i}(V) \in \tau_{i+1}$ for any $V \in \mathcal{P}(X)$.

Proposition 4.3 (see Proposition 4 from [4])

(i) *If $(X,\Box_0,\Box_1,\dots)$ is a* GLP*-frame, then X bears a unique series of topologies $\tau_0,\tau_1,\dots$ such that $\Box_i = cd_{\tau_i}$ for every $i \in \mathbb{N}$. Moreover, the polytopological space $(X,\tau_0,\tau_1,\dots)$ is a* GLP*-space.*

(ii) *If $(X,\tau_0,\tau_1,\dots)$ is a* GLP*-space, then $(X, cd_{\tau_0}, cd_{\tau_1},\dots)$ is a* GLP*-frame.*

In the sequel, we don't distinguish GLP-frames and corresponding polytopological spaces so that we use the topological terminology referring to $(X,\tau_0,\tau_1,\dots)$ for the frame $(X, cd_{\tau_0}, cd_{\tau_1},\dots)$. For example, we say that a subset U is *n-open* in $(X,\Box_0,\Box_1,\dots)$ if it belongs to the corresponding n-th topology on X (which is equivalent to $U \subset \Box_n U$).

Now we define a global semantic consequence relation over GLP-frames.

Definition 4.4 Given a set of formulas Γ and a formula φ, we set $\Gamma \vDash_g \varphi$ if for any GLP-model $\mathcal{M}$

$$(\forall \psi \in \Gamma \;\; \mathcal{M} \vDash \psi) \Rightarrow \mathcal{M} \vDash \varphi.$$

Let us recall the following neighbourhood completeness result obtained by L. Beklemishev and D. Gabelaia in [3].

Theorem 4.5 *For any formula φ, if* GLP $\nvdash \varphi$*, then there is a* GLP*-model $\mathcal{M}$ and a world x such that $\mathcal{M},x \nvDash \varphi$.*

We notice that, for any GLP-frame $\mathcal{X}$, the powerset GLP-algebra of $\mathcal{X}$ is σ-complete. Consequently this algebra is $\Box$-founded by Proposition 3.1. Hence we immediately obtain the following proposition.

Proposition 4.6 *For any set of formulas Γ and any formula φ, we have*

$$\Gamma \Vdash_g \varphi \Longrightarrow \Gamma \vDash_g \varphi.$$

5 Representation of $\Box$-founded Magari algebras

In this section we prove that any $\Box$-founded Magari algebra can be embedded into the powerset Magari algebra of an Esakia frame. We also obtain some related results, which will be applied in the next section.

From Proposition 3.4, we know that a Magari algebra $\mathcal{A} = (A, \wedge, \vee, \to, 0, 1, \Box)$ is $\Box$-founded if and only if the binary relation $<_{\mathcal{A}}$ is well-founded on $A \smallsetminus \{1\}$, where

$$a <_{\mathcal{A}} b \Longleftrightarrow \Box a \leqslant b.$$

Let us recall some basic properties of well-founded relations.

A *well-founded set* is a pair $\mathcal{S} = (S, <)$, where $<$ is a well-founded relation on S. For any element a of $\mathcal{S}$, its ordinal height in $\mathcal{S}$ is denoted by $ht_{\mathcal{S}}(a)$. Recall that $ht_{\mathcal{S}}$ is defined by transfinite recursion on $<$ as follows:

$$ht_{\mathcal{S}}(a) = \sup\{ht_{\mathcal{S}}(b) + 1 \mid b < a\}.$$

A *homomorphism* from $\mathcal{S}_1 = (S_1, <_1)$ to $\mathcal{S}_2 = (S_2, <_2)$ is a function $f\colon S_1 \to S_2$ such that $f(b) <_2 f(c)$ whenever $b <_1 c$.

Proposition 5.1 *Suppose $f\colon \mathcal{S}_1 \to \mathcal{S}_2$ is a homomorphism of well-founded sets and a is an element of $\mathcal{S}_1$. Then $ht_{\mathcal{S}_1}(a) \leqslant ht_{\mathcal{S}_2}(f(a))$.*

For well-founded sets $\mathcal{S}_1 = (S_1, <_1)$ and $\mathcal{S}_2 = (S_2, <_2)$, their product $\mathcal{S}_1 \times \mathcal{S}_2$ is defined as the set $S_1 \times S_2$ together with the following relation

$$(b_1, b_2) < (c_1, c_2) \Longleftrightarrow b_1 <_1 c_1 \text{ and } b_2 <_2 c_2.$$

Clearly, $<$ is a well-founded relation on $S_1 \times S_2$.

Proposition 5.2 *Suppose a and b are elements of well-founded sets $\mathcal{S}_1$ and $\mathcal{S}_2$ respectively. Then $ht_{\mathcal{S}_1 \times \mathcal{S}_2}((a, b)) = \min\{ht_{\mathcal{S}_1}(a), ht_{\mathcal{S}_2}(b)\}$.*

For an element a of a $\Box$-founded Magari algebra $\mathcal{A}$, define $ht_{\mathcal{A}}(a)$ as the ordinal height of a with respect to $<_{\mathcal{A}}$. We put $ht_{\mathcal{A}}(a) = \infty$ if $a = 1$.

Lemma 5.3 *Suppose a and b are elements of a $\Box$-founded Magari algebra $\mathcal{A}$. Then $ht_{\mathcal{A}}(a \wedge b) = \min\{ht_{\mathcal{A}}(a), ht_{\mathcal{A}}(b)\}$ and $ht_{\mathcal{A}}(a) + 1 \leqslant ht_{\mathcal{A}}(\Box a)$, where we define $\infty + 1 := \infty$.*

Proof. Assume we have a $\Box$-founded Magari algebra $\mathcal{A} = (A, \wedge, \vee, \to, 0, 1, \Box)$ and two elements a and b of $\mathcal{A}$.

First, we prove that $ht_{\mathcal{A}}(a \wedge b) = \min\{ht_{\mathcal{A}}(a), ht_{\mathcal{A}}(b)\}$. If $a = 1$ or $b = 1$, then the equality immediately holds. Suppose $a \neq 1$ and $b \neq 1$. Let $\mathcal{S}$ be the set $A \smallsetminus \{1\}$ together with the well-founded relation $<_{\mathcal{A}}$. We have $a \wedge b \neq 1$, $ht_{\mathcal{A}}(a) = ht_{\mathcal{S}}(a)$, $ht_{\mathcal{A}}(b) = ht_{\mathcal{S}}(b)$ and $ht_{\mathcal{A}}(a \wedge b) = ht_{\mathcal{S}}(a \wedge b)$. The mapping

$$f\colon (c, d) \mapsto c \wedge d$$

is a homomorphism from $\mathcal{S} \times \mathcal{S}$ to $\mathcal{S}$. From Proposition 5.2 and Proposition 5.1, we have

$$\min\{ht_{\mathcal{S}}(a), ht_{\mathcal{S}}(b)\} = ht_{\mathcal{S} \times \mathcal{S}}((a, b)) \leqslant ht_{\mathcal{S}}(a \wedge b).$$

Consequently,

$$\min\{ht_{\mathcal{A}}(a), ht_{\mathcal{A}}(b)\} \leqslant ht_{\mathcal{A}}(a \wedge b).$$

On the other hand, $ht_{\mathcal{A}}(a \wedge b) \leqslant ht_{\mathcal{A}}(a)$ since

$$\{e \in A \setminus \{1\} \mid e <_{\mathcal{A}} (a \wedge b)\} \subset \{e \in A \setminus \{1\} \mid e <_{\mathcal{A}} a\}.$$

Analogously, we have $ht_{\mathcal{A}}(a \wedge b) \leqslant ht_{\mathcal{A}}(b)$. It follows that

$$ht_{\mathcal{A}}(a \wedge b) = \min\{ht_{\mathcal{A}}(a), ht_{\mathcal{A}}(b)\}.$$

Now we prove $ht_{\mathcal{A}}(a) + 1 \leqslant ht_{\mathcal{A}}(\Box a)$. If $\Box a = 1$, then the inequality immediately holds. Suppose $\Box a \neq 1$. Then $a \neq 1$. We see $a <_{\mathcal{A}} \Box a$. The required inequality holds from the definition of $ht_{\mathcal{A}}$. □

For a □-founded Magari algebra $\mathcal{A} = (A, \wedge, \vee, \to, 0, 1, \Box)$ and an ordinal γ, put $M_{\mathcal{A}}(\gamma) = \{a \in A \mid \gamma \leqslant ht_{\mathcal{A}}(a)\}$. We see that $M_{\mathcal{A}}(0) = A$ and $M_{\mathcal{A}}(\delta) \supset M_{\mathcal{A}}(\gamma)$ whenever $\delta \leqslant \gamma$.

Lemma 5.4 *For any □-founded Magari algebra $\mathcal{A}$ and any ordinal γ, the set $M_{\mathcal{A}}(\gamma)$ is a filter in $\mathcal{A}$.*

Proof. Suppose a and b belong to $M_{\mathcal{A}}(\gamma)$. Then $\gamma \leqslant ht_{\mathcal{A}}(a)$ and $\gamma \leqslant ht_{\mathcal{A}}(b)$. We have $\gamma \leqslant \min\{ht_{\mathcal{A}}(a), ht_{\mathcal{A}}(b)\} = ht_{\mathcal{A}}(a \wedge b)$ by Lemma 5.3. Consequently $a \wedge b$ belongs to $M_{\mathcal{A}}(\gamma)$.

Now suppose c belongs to $M_{\mathcal{A}}(\gamma)$ and $c \leqslant d$. We shall show that $d \in M_{\mathcal{A}}(\gamma)$. We have $\gamma \leqslant ht_{\mathcal{A}}(c) = ht_{\mathcal{A}}(c \wedge d) = \min\{ht_{\mathcal{A}}(c), ht_{\mathcal{A}}(d)\} \leqslant ht_{\mathcal{A}}(d)$ by Lemma 5.3. Hence $d \in M_{\mathcal{A}}(\gamma)$. □

Let $Ult\,\mathcal{A}$ be the set of all ultrafilters of (the Boolean part of) a Magari algebra $\mathcal{A} = (A, \wedge, \vee, \to, 0, 1, \Box)$. Put $\widehat{a} = \{u \in Ult\,\mathcal{A} \mid a \in u\}$ for $a \in A$. We recall that the mapping $\widehat{\cdot}\colon a \mapsto \widehat{a}$ is an embedding of the Boolean algebra $(A, \wedge, \vee, \to, 0, 1)$ into the powerset Boolean algebra $\mathcal{P}(Ult\,\mathcal{A})$ by Stone's representation theorem.

Lemma 5.5 *For any □-founded Magari algebra $\mathcal{A}$, there exists a scattered topology τ on $Ult\,\mathcal{A}$ such that $\widehat{\Box a} = cd_{\tau}(\widehat{a})$ for any element a of $\mathcal{A}$.*

Proof. Assume we have a □-founded Magari algebra $\mathcal{A} = (A, \wedge, \vee, \to, 0, 1, \Box)$. Let $ht(\mathcal{A}) = \sup\{ht_{\mathcal{A}}(a) + 1 \mid a \in A \setminus \{1\}\}$. We see that $M_{\mathcal{A}}(ht(\mathcal{A})) = \{1\}$. For an ultrafilter u of $\mathcal{A}$, set $rk(u) := \min\{\delta \leqslant ht(\mathcal{A}) \mid M_{\mathcal{A}}(\delta) \subset u\}$. Also, for an ordinal γ, put $I(\gamma) := \{u \in Ult\,\mathcal{A} \mid rk(u) < \gamma\}$.

Set $\tau = \{V \subset Ult\,\mathcal{A} \mid \forall u \in V\ \exists a \in A\ \ (\Box a \in u) \wedge (\widehat{\boxdot a} \cap I(rk(u)) \subset V)\}$, where $\boxdot a = a \wedge \Box a$.

Let us check that τ is a topology on $Ult\,\mathcal{A}$. Trivially, $\varnothing \in \tau$ and τ is closed under arbitrary unions. For any $u \in Ult\,\mathcal{A}$, we see that $\Box 1 = 1 \in u$ and $\widehat{\boxdot 1} \cap I(rk(u)) \subset Ult\,\mathcal{A}$. Consequently $Ult\,\mathcal{A} \in \tau$. Assume $S_0 \in \tau$ and $S_1 \in \tau$. Consider an arbitrary $u \in S_0 \cap S_1$. By definition of τ, there exist elements b and c of A such that $\Box b \in u$, $\Box c \in u$, $\widehat{\boxdot b} \cap I(rk(u)) \subset S_0$ and $\widehat{\boxdot c} \cap I(rk(u)) \subset S_1$. We have $\Box(b \wedge c) = (\Box b \wedge \Box c) \in u$ and $\widehat{\boxdot(b \wedge c)} \cap I(rk(u)) = \widehat{\boxdot a} \cap \widehat{\boxdot c} \cap I(rk(u)) \subset S_0 \cap S_1$. Therefore $S_0 \cap S_1 \in \tau$. This shows that τ is a topology on $Ult\,\mathcal{A}$.

It easily follows from the definition of τ that $\widehat{\boxdot a} \in \tau$, for any $a \in A$, and $I(\gamma) \in \tau$, for any ordinal γ. Now we claim that τ is scattered. Consider any non-empty subset S of $Ult\,\mathcal{A}$. There is an ultrafilter $h \in S$ such that $rk(h) = \min\{rk(u) \mid u \in S\}$. We see that a set $\{h\} \cup I(rk(h))$ is a τ-neighbourhood of h and $S \cap (\{h\} \cup I(rk(h))) = \{h\}$. Hence the ultrafilter h is an isolated point in S. This proves that τ is a scattered topology.

It remains to show that $\widehat{\Box a} = cd_\tau(\widehat{a})$ for any $a \in A$. First, we check that $\widehat{\Box a} \subset cd_\tau(\widehat{a})$. For any ultrafilter d, if $d \in \widehat{\Box a}$, then $\widehat{\boxdot a} \cap I(rk(d))$ is a punctured neighbourhood of d. Also, $\widehat{\boxdot a} \cap I(rk(d)) \subset \widehat{a}$. By definition of the co-derived-set operator, $d \in cd_\tau(\widehat{a})$. Consequently $\widehat{\Box a} \subset cd_\tau(\widehat{a})$.

Now we claim that $cd_\tau(\widehat{a}) \subset \widehat{\Box a}$. Consider any ultrafilter d such that $d \notin \widehat{\Box a}$. Let W be an arbitrary punctured neighbourhood of d. It is sufficient to show that W is not included in $\widehat{a}$.

By definition of τ, there exists an element e of A such that $\Box e \in d$ and $\widehat{\boxdot e} \cap I(rk(d)) \subset W$. From the conditions $\Box e \in d$ and $\Box a \notin d$, it follows that $\Box(\boxdot e \to a) \notin d$. Hence $\Box(\boxdot e \to a) \notin M_{\mathcal{A}}(rk(d)) \subset d$ and $ht_{\mathcal{A}}(\Box(\boxdot e \to a)) < rk(d)$. Note that $(\boxdot e \to a) \notin M_{\mathcal{A}}(ht_{\mathcal{A}}(\boxdot e \to a)+1)$. By the Boolean ultrafilter theorem, there exists an ultrafilter w of $\mathcal{A}$ such that $(\boxdot e \to a) \notin w$ and $M_{\mathcal{A}}(ht_{\mathcal{A}}(\boxdot e \to a)+1) \subset w$. We see that $\boxdot e \in w$, $a \notin w$ and $rk(w) \leqslant ht_{\mathcal{A}}(\boxdot e \to a) + 1$. From Lemma 5.3, we have $ht_{\mathcal{A}}(\boxdot e \to a) + 1 \leqslant ht_{\mathcal{A}}(\Box(\boxdot e \to a)) < rk(d)$. Thus $rk(w) < rk(d)$, $w \in \widehat{\boxdot e} \cap I(rk(d))$ and $w \notin \widehat{a}$. Consequently w is an element of W, which does not belong to $\widehat{a}$.

We obtain that none of the punctured neighbourhoods of d are included in $\widehat{a}$. In other words, $d \notin cd_\tau(\widehat{a})$ for any $d \notin \widehat{\Box a}$. We conclude that $cd_\tau(\widehat{a}) \subset \widehat{\Box a}$. Hence $\widehat{\Box a} = cd_\tau(\widehat{a})$. □

Theorem 5.6 *A Magari algebra is $\Box$-founded if and only if it is embeddable into the powerset Magari algebra of an Esakia frame.*

Proof. (if) Suppose a Magari algebra $\mathcal{A}$ is isomorphic to a subalgebra of the powerset Magari algebra of an Esakia frame $\mathcal{X}$. The powerset Magari algebra of $\mathcal{X}$ is σ-complete. Hence, by Proposition 3.1, it is $\Box$-founded. Since any subalgebra of a $\Box$-founded Magari algebra is $\Box$-founded, the algebra $\mathcal{A}$ is $\Box$-founded.

(only if) Suppose a Magari algebra $\mathcal{A}$ is $\Box$-founded. By Lemma 5.5, there exists a scattered topology τ on $Ult\,\mathcal{A}$ such that $\widehat{\Box a} = cd_\tau(\widehat{a})$ for any element a of $\mathcal{A}$. We know that $\mathcal{X} = (Ult\,\mathcal{A}, cd_\tau)$ is an Esakia frame by Proposition 4.2. We see that the mapping $\widehat{\cdot} : a \mapsto \widehat{a}$ is an injective homomorphism from $\mathcal{A}$ to the powerset Magari algebra of the frame $\mathcal{X}$. Therefore the algebra $\mathcal{A}$ is embeddable into the powerset Magari algebra of an Esakia frame. □

For a Magari algebra $\mathcal{A}$, by $Top\,\mathcal{A}$, we denote the set of all scattered topologies τ on $Ult\,\mathcal{A}$ such that $\widehat{\Box a} = cd_\tau(\widehat{a})$ for any element a of $\mathcal{A}$.

Lemma 5.7 *Suppose $\mathcal{A}$ is a Magari algebra and $\tau \in Top\,\mathcal{A}$. Then there is a maximal with respect to inclusion element of $Top\,\mathcal{A}$ that extends τ.*

Proof. Consider the set $P = \{\sigma \in Top\,\mathcal{A} \mid \tau \subset \sigma\}$, which is a partially ordered set with respect to inclusion. We claim that any chain in P has an upper bound.

Assume C is a chain in P. Let ν be the coarsest topology containing τ and $\bigcup C$. Note that the topology ν is scattered as an extension of a scattered topology. For any element a of $\mathcal{A}$, we have $\widehat{\Box a} = cd_\tau(\widehat{a}) \subset cd_\nu(\widehat{a})$, because ν is an extension of τ.

Now assume c is an arbitrary element of $\mathcal{A}$ and $u \in cd_\nu(\widehat{c})$. We check that $u \in \widehat{\Box c}$. By definition of the co-derived-set operator, there is a punctured ν-neighbourhood V of u such that $V \subset \widehat{c}$. Since the set $\tau \cup \bigcup C$ is closed under finite intersections, it is a basis of ν. Consequently there is a subset W of V with $W \cup \{u\} \in \tau \cup \bigcup C$. We see that $W \subset \widehat{c}$ and W is a punctured neighbourhood of u with respect to a topology $\kappa \in \{\tau\} \cup C \subset Top\,\mathcal{A}$. Hence $u \in cd_\kappa(\widehat{c}) = \widehat{\Box c}$.

We obtain that $\widehat{\Box a} = cd_\nu(\widehat{a})$ for any element a of $\mathcal{A}$. Therefore $\nu \in Top\,\mathcal{A}$ and ν is an upper bound for C in P.

We see that any chain in P has an upper bound. By Zorn's lemma, there is a maximal element in P, which is the required maximal extension of τ. □

The following lemma was inspired by Lemma 4.5 from [3].

Lemma 5.8 *Suppose $\mathcal{A}$ is a Magari algebra and τ is a maximal element of $Top\,\mathcal{A}$. Then, for any $u \in Ult\,\mathcal{A}$ and any $V \in \tau$, we have $V \cup \{u\} \in \tau$ or there are a τ-open set W and an element a of $\mathcal{A}$ such that $u \in W$, $\Box a \notin u$ and $V \cap W \subset \widehat{a}$.*

Proof. Assume $u \in Ult\,\mathcal{A}$ and $V \in \tau$. It is sufficient to consider the case when $V \cup \{u\} \notin \tau$. Let σ be the coarsest topology containing τ and the set $V \cup \{u\}$. The topology σ is scattered as an extension of a scattered topology. Since τ is a maximal element of $Top\,\mathcal{A}$, the topology σ does not belong to $Top\,\mathcal{A}$ and there exists an element a of $\mathcal{A}$ such that $\widehat{\Box a} \neq cd_\sigma(\widehat{a})$. Notice that $\widehat{\Box a} = cd_\tau(\widehat{a}) \subset cd_\sigma(\widehat{a})$, because $\tau \subset \sigma$. Thus there is an ultrafilter h such that $h \in cd_\sigma(\widehat{a})$ and $h \notin cd_\tau(\widehat{a}) = \widehat{\Box a}$. Hence there is a punctured σ-neighbourhood of h that is included in $\widehat{a}$. In addition, note that $\tau \cup \{W \cap (V \cup \{u\}) \mid W \in \tau\}$ is a basis of σ. We see that $h \in B$ and $B \setminus \{h\} \subset \widehat{a}$ for some $B \in \tau \cup \{W \cap (V \cup \{u\}) \mid W \in \tau\}$. If $B \in \tau$, then $h \in cd_\tau(\widehat{a})$. This is a contradiction with the condition $h \notin cd_\tau(\widehat{a})$. Therefore B has the form $W \cap (V \cup \{u\})$ for some $W \in \tau$. Since $h \in B = W \cap (V \cup \{u\})$, we have $h \in V$ or $h = u$. If $h \in V$, then $h \in W \cap V$ and $(W \cap V) \setminus \{h\} \subset \widehat{a}$. In this case, we obtain $h \in cd_\tau(\widehat{a})$, which is a contradiction. Consequently $h \notin V$ and $h = u$. It follows that $\Box a \notin u$, $u \in W$ and $W \cap V = (W \cap (V \cup \{u\})) \setminus \{h\} \subset \widehat{a}$.

□

For a scattered topological space (X, τ), the *derivative topology τ^+ on X* is defined as the coarsest topology including τ and $\{d_\tau(Y) \mid Y \subset X\}$. The next lemma was inspired by Lemma 5.1 from [3].

Lemma 5.9 *Suppose $\mathcal{A} = (A, \wedge, \vee, \rightarrow, 0, 1, \Box)$ is a Magari algebra and τ is a maximal element of $Top\,\mathcal{A}$. Then the topology τ^+ is generated by τ and the sets $d_\tau(\widehat{a})$ for $a \in A$.*

Proof. Assume τ is a maximal element of $Top\,\mathcal{A}$. Let τ' be the topology generated by τ and the sets $d_\tau(\widehat{a})$ for $a \in A$. It is clear that $\tau' \subset \tau^+$. We prove the converse. We shall check that $d_\tau(Y)$ is τ'-open for any $Y \subset Ult\,\mathcal{A}$.

Consider any $Y \subset Ult\,\mathcal{A}$ and any $u \in d_\tau(Y)$. We claim that there is a τ'-neighbourhood of u entirely contained in $d_\tau(Y)$. Suppose $\Box\Box a \in u$ and $\Box a \notin u$ for some $a \in A$. In this case, we see $u \notin \widehat{\Box a}$ and

$$\{u\} \cup \widehat{\Box a} \subset \widehat{\Box\Box a} = cd_\tau(\widehat{\Box a}) \subset cd_\tau(\{u\} \cup \widehat{\Box a}).$$

Hence the set $\{u\} \cup \widehat{\Box a}$ is τ-open. In addition, we see

$$u \in (\,Ult\,\mathcal{A} \smallsetminus \widehat{\Box a}) = (\,Ult\,\mathcal{A} \smallsetminus cd_\tau(\widehat{a})) = d_\tau(\widehat{\neg a}) \in \tau'.$$

It implies that

$$\{u\} = (\{u\} \cup \widehat{\Box a}) \cap (\,Ult\,\mathcal{A} \smallsetminus \widehat{\Box a}) \in \tau'.$$

In other words, the ultrafilter u is a τ'-isolated point of $Ult\,\mathcal{A}$.

Now consider the case when, for any $a \in A$, we have $\Box\Box a \notin u$ whenever $\Box a \notin u$. By $int_\tau(X)$, we denote the τ-interior of a set X. Recall that $cd_\tau(X) = cd_\tau(int_\tau(X))$ for any set X in any topological space. Put $X = Ult\,\mathcal{A} \smallsetminus Y$. Since $u \in d_\tau(Y)$ and $u \notin cd_\tau(X) = cd_\tau(int_\tau(X))$, the set $\{u\} \cup int_\tau(X) \notin \tau$. By Lemma 5.8, there are a τ-open set W and an element c of A such that $u \in W$, $\Box c \notin u$ and $int_\tau(X) \cap W \subset \widehat{c}$. Since, for any $a \in A$, $\Box\Box a \notin u$ whenever $\Box a \notin u$, we obtain $\Box\Box c \notin u$. It follows that

$$u \in W \cap (\,Ult\,\mathcal{A} \smallsetminus \widehat{\Box\Box c}) = W \cap d_\tau(\widehat{\neg\Box c}) \in \tau'.$$

Thus $W \cap (\,Ult\,\mathcal{A} \smallsetminus \widehat{\Box\Box c})$ is a τ'-neighbourhood of u. It remains to show that

$$W \cap (\,Ult\,\mathcal{A} \smallsetminus \widehat{\Box\Box c}) \subset d_\tau(Y).$$

Indeed, we have

$$\begin{aligned} cd_\tau(X) \cap W &\subset cd_\tau(int_\tau(X)) \cap cd_\tau(W) = \\ &= cd_\tau(int_\tau(X) \cap W) \subset cd_\tau(\widehat{c}) \subset cd_\tau(cd_\tau(\widehat{c})) = \widehat{\Box\Box c}, \end{aligned} \quad (1)$$

because W is a τ-open set and $int_\tau(X) \cap W \subset \widehat{c}$. Hence,

$$\begin{aligned} W \cap (\,Ult\,\mathcal{A} \smallsetminus \widehat{\Box\Box c}) &\subset W \cap (\,Ult\,\mathcal{A} \smallsetminus (cd_\tau(X) \cap W)) \quad \text{(from 1)} \\ &= W \cap ((\,Ult\,\mathcal{A} \smallsetminus cd_\tau(X)) \cup (\,Ult\,\mathcal{A} \smallsetminus W)) \\ &= W \cap (d_\tau(Y) \cup (\,Ult\,\mathcal{A} \smallsetminus W)) \\ &= (W \cap d_\tau(Y)) \cup (W \cap (\,Ult\,\mathcal{A} \smallsetminus W)) \\ &= W \cap d_\tau(Y) \\ &\subset d_\tau(Y). \end{aligned}$$

This argument shows that any element of $d_\tau(Y)$ belongs to this set together with a τ'-neighbourhood. We conclude that $d_\tau(Y)$ is τ'-open and $\tau' = \tau^+$.

□

6 Global neighbourhood completeness

In this section we prove that any $\Box$-founded GLP-algebra can be embedded into the powerset algebra of a GLP-frame. As a corollary, we obtain global neighbourhood completeness for GLP w.r.t. non-well-founded derivations.

Analogously to the case of Magari algebras, by *Ult* $\mathcal{A}$, we denote the set of ultrafilters of a GLP-algebra $\mathcal{A}$. For a GLP-algebra $\mathcal{A} = (A, \wedge, \vee, \to, 0, 1, \Box_0, \Box_1, \dots)$, we denote the Magari algebra $(A, \wedge, \vee, \to, 0, 1, \Box_i)$ by $\mathcal{A}_i$. We see $\mathit{Ult}\,\mathcal{A} = \mathit{Ult}\,\mathcal{A}_i$ for any $i \in \mathbb{N}$. We call (maximal with respect to inclusion) elements of $\mathit{Top}\,\mathcal{A}_i$ *(maximal) i-topologies* on *Ult* $\mathcal{A}$.

Lemma 6.1 *For any* GLP*-algebra* $\mathcal{A}$ *and any maximal i-topology* τ *on* $\mathit{Ult}\,\mathcal{A}$, *there exists a maximal* $(i+1)$*-topology* ν *on* $\mathit{Ult}\,\mathcal{A}$ *such that* $\tau \subset \nu$ *and* $d_\tau(Y)$ *is* ν*-open for each* $Y \subset \mathit{Ult}\,\mathcal{A}$.

Proof. Assume we have a GLP-algebra $\mathcal{A}$ and a maximal i-topology τ on $\mathit{Ult}\,\mathcal{A}$. Consider the coarsest topology τ' containing τ^+ and all sets of the form $\{u\} \cup \widehat{\boxdot_{i+1} a}$, where $u \in \mathit{Ult}\,\mathcal{A}$, $\Box_{i+1} a \in u$ and $\boxdot_{i+1} a = a \wedge \Box_{i+1} a$. We see that $\tau \subset \tau'$ and $d_\tau(Y)$ is τ'-open for each $Y \subset \mathit{Ult}\,\mathcal{A}$. Trivially, the topology τ' is scattered as an extension of a scattered topology. We claim that $\tau' \in \mathit{Top}\,\mathcal{A}_{i+1}$.

We shall show that $\widehat{\Box_{i+1} a} = cd_{\tau'}(\widehat{a})$ for any element a of $\mathcal{A}$. First, we check that $\widehat{\Box_{i+1} a} \subset cd_{\tau'}(\widehat{a})$. For any ultrafilter d, if $d \in \widehat{\Box_{i+1} a}$, then $\widehat{\boxdot_{i+1} a}$ is a punctured τ'-neighbourhood of d. Also, $\widehat{\boxdot_{i+1} a} \subset \widehat{a}$. By definition of the co-derived-set operator, $d \in cd_{\tau'}(\widehat{a})$. Consequently $\widehat{\Box_{i+1} a} \subset cd_{\tau'}(\widehat{a})$.

Now we check that $cd_{\tau'}(\widehat{a}) \subset \widehat{\Box_{i+1} a}$. Consider any ultrafilter d such that $d \notin \widehat{\Box_{i+1} a}$. In addition, let W be an arbitrary punctured τ'- neighbourhood of d. It is sufficient to show that W is not included in $\widehat{a}$.

We have $\Box_{i+1} a \notin d$, $d \notin W$ and $W \cup \{d\} \in \tau'$. From Lemma 5.9, there is a basis of τ' consisting of alls sets of the form

$$V \cap d_\tau(\widehat{b_1}) \cap \dots \cap d_\tau(\widehat{b_n}) \cap (\{u_1\} \cup \widehat{\boxdot_{i+1} c_1}) \cap \dots \cap (\{u_m\} \cup \widehat{\boxdot_{i+1} c_m}),$$

where $V \in \tau$, $\{b_1, \dots, b_n\}$ and $\{c_1, \dots, c_m\}$ are (possibly empty) subsets of A, $\{u_1, \dots, u_m\}$ is a subset of $\mathit{Ult}\,\mathcal{A}$. In addition, $\Box_{i+1} c_k \in u_k$ for $k \in \{1, \dots, m\}$. Hence we have

$$d \in \big(V \cap d_\tau(\widehat{b_1}) \cap \dots \cap d_\tau(\widehat{b_n}) \cap (\{u_1\} \cup \widehat{\boxdot_{i+1} c_1}) \cap \dots \cap (\{u_m\} \cup \widehat{\boxdot_{i+1} c_m})\big) \subset W \cup \{d\}$$

for some element of the basis of τ'. We see that the ultrafilter d contains $\Diamond_i b_1, \dots, \Diamond_i b_n$ and $\Box_{i+1} c_1, \dots, \Box_{i+1} c_m$. Also, $\Diamond_{i+1} \neg a \in d$. In any GLP-algebra, we have

$$\bigwedge\{\Diamond_i b_1, \dots, \Diamond_i b_n\} \leqslant \Box_{i+1} \bigwedge\{\Diamond_i b_1, \dots, \Diamond_i b_n\},$$
$$\bigwedge\{\Box_{i+1} c_1, \dots, \Box_{i+1} c_m\} \leqslant \Box_{i+1} \bigwedge\{\boxdot_{i+1} c_1, \dots, \boxdot_{i+1} c_m\}.$$

Further, we have

$$\begin{aligned}
&(\Diamond_{i+1}\neg a) \wedge \bigwedge\{\Diamond_i b_1,\dots,\Diamond_i b_n,\Box_{i+1}c_1,\dots,\Box_{i+1}c_m\} \leqslant\\
&\quad\leqslant (\Diamond_{i+1}\neg a) \wedge \Box_{i+1}\bigwedge\{\Diamond_i b_1,\dots,\Diamond_i b_n\} \wedge \Box_{i+1}\bigwedge\{\boxdot_{i+1}c_1,\dots,\boxdot_{i+1}c_m\} \leqslant\\
&\qquad\leqslant \Diamond_{i+1}\big((\neg a) \wedge \bigwedge\{\Diamond_i b_1,\dots,\Diamond_i b_n,\boxdot_{i+1}c_1,\dots,\boxdot_{i+1}c_m\}\big) \leqslant\\
&\qquad\quad\leqslant \Diamond_{i}\big((\neg a) \wedge \bigwedge\{\Diamond_i b_1,\dots,\Diamond_i b_n,\boxdot_{i+1}c_1,\dots,\boxdot_{i+1}c_m\}\big)
\end{aligned}$$

We obtain $\Diamond_i\big((\neg a) \wedge \bigwedge\{\Diamond_i b_1,\dots,\Diamond_i b_n,\boxdot_{i+1}c_1,\dots,\boxdot_{i+1}c_m\}\big) \in d$ and

$$d \in d_\tau\big(\widehat{\neg a} \cap d_\tau(\widehat{b_1}) \cap \dots \cap d_\tau(\widehat{b_n}) \cap \widehat{\boxdot_{i+1}c_1} \cap \dots \cap \widehat{\boxdot_{i+1}c_m}\big).$$

Since V is a τ-neighbourhood of d, there exists an ultrafilter w such that

$$w \in (V \smallsetminus \{d\}) \cap \widehat{\neg a} \cap d_\tau(\widehat{b_1}) \cap \dots \cap d_\tau(\widehat{b_n}) \cap \widehat{\boxdot_{i+1}c_1} \cap \dots \cap \widehat{\boxdot_{i+1}c_m} \subset W.$$

Consequently w is an element of W, which does not belong to $\widehat{a}$.

We obtain that none of the punctured τ'-neighbourhoods of d are included in $\widehat{a}$. In other words, $d \notin cd_{\tau'}(\widehat{a})$ for any $d \notin \widehat{\Box_{i+1}a}$. This argument shows that $cd_{\tau'}(\widehat{a}) \subset \widehat{\Box_{i+1}a}$. Hence $\widehat{\Box_{i+1}a} = cd_{\tau'}(\widehat{a})$. We see $\tau' \in Top\,\mathcal{A}_{i+1}$.

Now we extend the topology τ' applying Lemma 5.7 and obtain the required maximal $(i+1)$-topology ν on *Ult* $\mathcal{A}$.

□

Lemma 6.2 *For any* □*-founded* GLP*-algebra* $\mathcal{A}$*, there exists a series of topologies* $\tau_0, \tau_1, \dots$ *on* Ult $\mathcal{A}$ *such that* $(\mathit{Ult}\,\mathcal{A}, \tau_0, \tau_1, \dots)$ *is a* GLP*-space and* $\tau_i \in Top\,\mathcal{A}_i$ *for any* $i \in \mathbb{N}$.

Proof. From Lemma 5.5, there exists a topology $\tau \in Top\,\mathcal{A}_0$. By Lemma 5.7, the topology τ can be extended to a maximal 0-topology τ_0. Applying Lemma 6.1, we obtain a series of topologies $\tau_1, \tau_2, \dots$ on *Ult* $\mathcal{A}$ such that $(\mathit{Ult}\,\mathcal{A}, \tau_0, \tau_1, \dots)$ is a GLP-space and $\tau_i \in Top\,\mathcal{A}_i$ for any $i \in \mathbb{N}$. □

The following theorem is analogous to Theorem 5.6 and is obtained by a similar argument. So we omit the proof.

Theorem 6.3 *A* GLP*-algebra is* □*-founded if and only if it is embeddable into the powerset* GLP*-algebra of a* GLP*-frame.*

Theorem 6.4 *For any set of formulas* Γ *and any formula* φ*, we have*

$$\Gamma \vdash_g \varphi \Longleftrightarrow \Gamma \Vdash_g \varphi \Longleftrightarrow \Gamma \vDash_g \varphi.$$

Proof. From Theorem 3.7 and Proposition 4.6, it remains to show that $\Gamma \Vdash \varphi$ whenever $\Gamma \vDash \varphi$. Assume $\Gamma \vDash \varphi$. Also assume we have a □-founded GLP-algebra $\mathcal{A} = (A, \wedge, \vee, \to, 0, 1, \Box_0, \Box_1, \dots)$ and a valuation v in $\mathcal{A}$ such that $v(\psi) = 1$ for any $\psi \in \Gamma$. We shall prove $v(\varphi) = 1$.

By the previous theorem, there exist a GLP-frame $\mathcal{X} = (X, \Box_0, \Box_1, \dots)$ and a mapping $f\colon A \to \mathcal{P}(X)$ such that f is an embedding of $\mathcal{A}$ into the powerset GLP-algebra of $\mathcal{X}$. We see that $w = f \circ v$ is valuation over $\mathcal{X}$, where $(\mathcal{X}, w) \vDash \psi$ for any $\psi \in \Gamma$. From the assumption $\Gamma \vDash \varphi$, we obtain $(\mathcal{X}, w) \vDash \varphi$. Since f is an embedding, we conclude that $v(\varphi) = 1$. □

Acknowledgements. I thank anonymous reviewers for their constructive comments and attention to this work. SDG.

Appendix

Proof. [Proof of Proposition 2.2] First, we recall an important result from [2]. The logic J is obtained from GLP by replacing axiom schemes (iv-v) with the following ones all of which are provable in GLP:

(vi) $\Diamond_i\psi \to \Box_j \Diamond_i \psi$ for $i < j$;

(vii) $\Box_i\psi \to \Box_j \Box_i \psi$ for $i < j$;

(viii) $\Box_i\psi \to \Box_i \Box_j \psi$ for $i < j$.

A Kripke J-frame $(W, R_0, R_1, \dots)$ is a set W together with a sequence of binary relations on W such that

- R_i are transitive and conversely well-founded relations;
- xR_iy and yR_jz implies xR_iz, for $i < j$;
- xR_jy and xR_iz implies yR_iz, for $i < j$;
- xR_jy and yR_iz implies xR_iz, for $i < j$.

A notion of Kripke J-model is defined in the standard way.

L. Beklemishev showed in [2] that the logic J is Kripke complete, i.e. it is complete for its relational interpretation over the class of Kripke J-frames. In addition, he proved the following result: *if* GLP $\nvdash \psi$, *then there is a* J*-model* $\mathcal{K}$ *such that all theorems of* GLP *are true in* $\mathcal{K}$ *and* $\mathcal{K} \nvDash \psi$ (see Theorem 4 from [2]).

Now we prove that for any formula ξ

$$\mathsf{GLP} \vdash \xi \Longleftrightarrow \varnothing \vdash_g \xi.$$

The left-to-right implication trivially holds. We prove the converse by *reductio ad absurdum.* Assume GLP $\nvdash \xi$ and there is an ∞-derivation π with the root marked by ξ in which all leaves are marked by some axioms of GLP. Then there exist a J-model $\mathcal{K}$ and its world w such that $\mathcal{K}, w \nvDash \xi$ and all theorems of GLP are true at all worlds of $\mathcal{K}$. For a node x of the ∞-derivation π, let ψ_x be the formula of the node x. We define a sequence of pairs (x_n, w_n), where x_n is a node of π and w_n is a world of $\mathcal{K}$, such that $\mathcal{K}, w_n \nvDash \psi_{x_n}$. Let x_0 be the root of π and $w_0 = w$.

Given a pair (x_n, w_n) such that $\mathcal{K}, w_n \nvDash \psi_{x_n}$, we define (x_{n+1}, w_{n+1}). We see that x_n is not a leaf of π. Indeed, if x_n is a leaf of π, then the formula ψ_{x_n} is an axiom of GLP, which is a contradiction with the assertion that all theorems of GLP are true at all worlds of $\mathcal{K}$. We have that x_n is not a leaf of π and ψ_{x_n} is obtained by an application of an inference rule in π.

Suppose ψ_{x_n} is obtained by the rule (nec). Let x_{n+1} be the premise of x_n. We have $\psi_{x_n} = \Box_0\psi_{x_{n+1}}$ and $\mathcal{K}, w_n \nvDash \Box_0\psi_{x_{n+1}}$. Then there is a world w_{n+1} such that $w_nR_0w_{n+1}$ and $\mathcal{K}, w_{n+1} \nvDash \psi_{x_{n+1}}$.

If ψ_{x_n} is obtained by the rule (mp), then there is a node y such that y is a premise of x_n and $\mathcal{K}, w_n \nVdash \psi_y$. Set $x_{n+1} = y$ and $w_{n+1} = w_n$.

The sequence (x_n, w_n) is well-defined. We see that $x_0, x_1, \dots$ is an infinite branch in π. In addition, the sequence $w_0, w_1, \dots$ satisfies the condition: $w_n R_0 w_{n+1}$ if x_n is a conclusion of the rule (nec) in π, and $w_n = w_{n+1}$, otherwise. Since π is an ∞-derivation, the branch $x_0, x_1, \dots$ contains infinitely many applications of the rule (nec). Consequently, there is an infinite ascending sequence of worlds in $\mathcal{K}$ with respect to the relation R_0, which is a contradiction with the assertion that $\mathcal{K}$ is a J-model. This contradiction concluds the proof. □

References

[1] Beklemishev, L., *Reflection principles and provability algebras in formal arithmetic*, Russian Mathematical Surveys **60** (2005), pp. 197–268.

[2] Beklemishev, L., *Kripke semantics for provability logic GLP*, Annals of Pure and Applied Logic **161** (2010), pp. 756–774.

[3] Beklemishev, L. and D. Gabelaia, *Topological completeness of the provability logic GLP*, Annals of Pure and Applied Logic **164** (2013), pp. 1201–1223.

[4] Beklemishev, L. and D. Gabelaia, *Topological interpretations of provability logic*, in: G. Bezhanishvili, editor, *Outstanding Contributions to Logic. Leo Esakia on duality in modal and intuitionistic logics*, Springer, Dordrecht, 2014 pp. 257–290.

[5] Esakia, L., *Diagonal constructions, Löb's formula and Cantor's scattered space*, Studies in Logic and Semantics **132** (1981), pp. 128–143, in Russian.

[6] Japaridze, G., "The modal logical means of investigation of provability," Candidate dissertation in philosophy, Moscow State University, Moscow (1986), in Russian.

[7] Montague, R., *Universal grammar*, Theoria **36** (1970), pp. 373–398.

[8] Pakhomov, F. and J. Walsh, *Reflection ranks and ordinal analysis* (2018), arXiv:1805.02095.

[9] Scott, D., *Advice on modal logic*, in: K. Lambert, editor, *Philosophical problems in Logic*, Reidel, Dordrecht, 1970 pp. 143–173.

[10] Shamkanov, D., *Global neighbourhood completeness of the Gödel-Löb provability logic*, in: J. Kennedy and R. de Queiroz, editors, *Logic, Language, Information, and Computation. 24th International Workshop, WoLLIC 2017 (London, UK, July 18-21, 2017)*, number 103888 in Lecture Notes in Computer Science (2017), pp. 358–371.

[11] Shamkanov, D., *Non-well-founded derivations in the Gödel-Löb provability logic*, The Review of symbolic Logic (2019), https://doi.org/10.107/S1755020319000613.

[12] Simmons, H., *Topological aspects of suitable theories*, Proceedings of the Edinburgh Mathematical Society **19** (1975), pp. 383–391.

William of Sherwood on Necessity and Contingency

Sara L. Uckelman

Department of Philosophy, Durham University

Abstract

In [14], I presented three 13th-century approaches to modality and modal logic, focusing on the well-developed and clearly articulated views of William of Sherwood (fl. 1250), and contrasting them with the more nascent and brief views found in Pseudo-Aquinas and Aquinas. That paper focused on Sherwood's modal theory as found in his *Introductiones ad Logicam* [10], without attempting to integrate it with what he has to say about modes, modality, and modal reasoning in his other main treatise, the *Syncategoremata* [11]. This paper extends [14] by doing this integration.

Keywords: 13th century, contingency, modal logic, mode, necessity, syncategorematic terms, William of Sherwood

1 Introduction

In [14], I presented three 13th-century approaches to modality and modal logic, focusing on the well-developed and clearly articulated view found in William of Sherwood's (fl. 1250) *Introductiones ad Logicam* [10], and contrasting them with the more nascent and brief views found in Pseudo-Aquinas's *Summa totius logicae Aristotelis* and Thomas Aquinas's short, early treatise *De propositionibus modalibus*. There, we considered the three authors' definitions of *mode* and *modal proposition*; the ways in which modal propositions can be constructed and classified according to their quality, quantity, and whether they are *de re* or *de dicto* (or adverbial or nominal); their truth conditions; the inferential relations that hold between these propositions; and the treatments of modal syllogisms.

Of the three accounts considered in that paper, Sherwood's was by far the most sophisticated and consistent; this despite the fact that he does not discuss modal syllogisms in his *Introductiones* (or indeed in any other known extant text). But at the time, I focused on Sherwood's modal theory as found in his *Introductiones* without attempting to integrate it with what he has to say about modes, modality, and modal logic in his other main treatise, the *Syncategoremata* [11], preferring instead to compare his analysis with two contemporary texts. As a contribution to our understanding of modality and modal logic in the 13th century, this paper extends the previous analysis of Sherwood on modality in [14] by doing this integration. This paper has four main parts: First, in §2, I introduce William of Sherwood and discuss his importance for the study of the history of logic generally and modal logic specifically. In §3 we provide the context for his discussion of modal terms in the *Syncategoremata* by explaining the subgenre of medieval logic it is situated in, and why one would expect to find modal terms in it. With this background in place, the main contribution of the paper is in §4, an exposition and analysis of Sherwood's chapter on modal terms in the *Syncategoremata*. This material is supplemented in §5 by consideration of what Sherwood has to say about necessity and contingency in the other chapters. We conclude in §6, outlining scope for future work and some of the limitations of the present study.

2 Who is William of Sherwood, and why should we care about him?

This is not the place to rehearse the medieval history of logic more generally nor indeed of modal logic more specifically; the reader interested in such a comprehensive overview is directed to [16]. What is important to know is that the 13th century was a period of both consolidation—as the recently translated texts of Aristotle and Avicenna circulated amongst the newly-birthed universities—and of invention—as these texts provided European logicians with new sources for innovation and development. By the middle of the 13th century, the establishment of the universities of Paris and Oxford fifty years earlier and the foundation of their curriculum upon the *trivium* (the disciplines of logic, grammar, and rhetoric) created a need for textbooks on these topics. Between 1240 and 1260, four influential textbooks were produced by authors whose names and identities we know (albeit some to a lesser degree). These are the *Introductiones ad Logicam* by William of Sherwood [10,12]; Roger Bacon's *Art and Science of Logic* [2]; Lambert of Auxerre's *Summa Lamberti* [9,4]; and Peter of Spain's *Summulae Logicales* [3].

Of these books, Sherwood's is the most interesting because it is one of the earliest and was directly influential on the succeeding books—Bacon even says, in his *Opus tertium* (1267), that Sherwood was "much wiser than Albert [the Great]; for in *philosophia communis*, no one is greater than he" [10, p. 6]. Sherwood was born in the early 13th century, probably between 1200 and 1205, and died sometime between 1266 and 1272. Though records of his early life are uncertain, from references by other scholars (not just Bacon) to him and his works, it seems likely that Sherwood was teaching logic at the University of Paris between 1235 and 1250, and then became a master at Oxford in 1250 [11, p. 3]. As a result, Sherwood is one of the earliest named writers we know of in the *logica nova* tradition, the tradition of logic that built upon Aristotle but extended it with the introduction of two new areas of study: the study of the properties of terms (*proprietates terminorum*), and of syncategorematic words (*syncategoremata*). The topic of the properties of terms, which include signification, supposition, copulation, and appellation, makes up one chapter of his *Introductiones*, which also covers such basic logical notions as propositions, predicables, syllogisms, different types of non-syllogistic arguments, and sophisms and sophistical reasoning. But syncategorematic words were important enough to get a treatise of their own.

There is no clear evidence as to when the *Syncategoremata*, or *Treatise on Syncategorematic Words*, was written. Kretzmann argues that "although Sherwood exhibits a higher level of logical sophistication" in the *Syncategoremata* than in the *Introductiones*, he "regularly omits details and ignores technical distinctions he had laid down in the earlier book"; as a result, Kretzmann concludes that the *Syncategoremata* was likely written quite awhile after the *Introductiones* [11, p. 6]. The text was first edited by O'Donnell [8] and translated into English by Kretzmann [11]. A more recent Latin edition, along with a German translation, was produced by Kann and Kirchhoff [13]. All Latin references are taken from [8] because I did not have access to [13] while completing the paper (see §6 for a further discussion of this).

Kretzmann describes this text as "an advanced treatise", designed for students who have already mastered the basics found in the *Introductiones.* After a short introduction where Sherwood introduces the topic and provides foundational definitions, the text is divided up into chapters each covering a specific syncategorematic term or group of related syncategorematic terms. These include: *omnis* ('every'/'all'); *totum* ('whole'); *dictiones numerales* (number words); *infinita in plurali* ('infinitely many'); *uterque* ('both'); *quaelelibet* ('of every sort'); *nullus* ('no'); *nihil* ('nothing'); *neutrum* ('neither'); *praeter* ('but'); *solus* ('alone'); *tantum* ('only'); *est* ('is'); *non* ('not'); *necessario* ('necessarily') and *contingenter* ('contingently'); *incipit* ('begin') and *desinit* ('ceases'); *si* ('if'); *nisi* ('unless'); *quin* ('but that'); *et* ('and'); *vel* ('or'); *an* ('whether'/'or'); *ne* (an enclitic negating particle); and *sive* ('whether... or'). Naturally, our interest here is the chapter on *necessario* and *contingenter*, though some relevant material is also found in other chapters.

3 What are syncategorematic words?

Sherwood opens his treatise with the following claim:

> In order to understand anything one must understand its parts; thus in order that the statement may be fully understood one must understand the parts of it [11, p. 13].[1]

Understanding the parts that make up a statement is the central focus of medieval treatises on the properties of terms and on syncategorematic terms. Sherwood's introduction to the *Syncategoremata* proceeds to a definition of 'syncategoremata' or 'syncategorematic term' via a series of binary divisions, resulting in a complete classification of the parts of statements. This classification is represented in Figure 1.

The first division is between the *principal* parts of the statement, that is,

[1] *Quid ad cognitionem alicujus oportet cognoscere suas partes; ideo ut plene cognoscatur enuntiatio oportet ejus partes cognoscere* [8, p. 48].

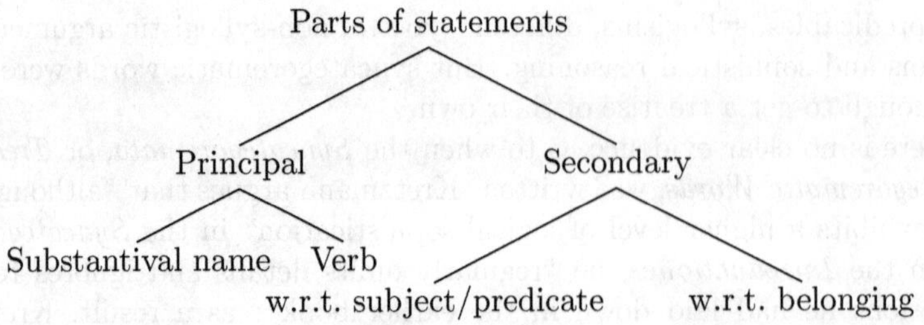

Fig. 1. Sherwood's classification of parts of statements.

substantival names (*nomen substantivum*) and verbs (*verbum*), and the *secondary* parts of a statement, which are 'determinations' of the principal parts. As Kretzmann notes, "*determinatio* is a technical notion in medieval logic" [11, p. 14], deriving from the notion of secondary substance outlined in Aristotle's *Categories* 3b10ff, where secondary substances are distinguished from primary substances by means of certain qualifications. While this section of the *Categories* is discussing metaphysics rather than language or logic, the parallel with the parts of speech is clear: Substances are either primary (without qualification) or secondary (with qualification); the parts of speech are either principal (the unqualified or undetermined parts without which a statement or sentence cannot exist) or secondary (the parts that qualify or determine the principal parts).

These principal parts, substantive names (or nouns) and verbs, are parts of speech that can be used as subjects and predicates of sentences; furthermore, they are principal because a complete statement can be made with these parts and no others. A name (or noun or *nomen*) is, per Sherwood's *Introductiones*, "an utterance significant by convention, apart from time, finite[2] and direct[3], no part of which taken by itself signifies anything" [10, p. 23][4]; this definition distinguishes names from (a) non-significative words, (b) words which signify but not by convention, (c) sentences and phrases, and (d) verbs. The substantive names include general names like 'human' and 'animal', proper names like 'Socrates' and 'Sara', and substantivised adjectives such as 'the white [thing]'. A verb is "an utterance significant by convention, together with time, finite and direct, no part of which taken by itself signifies anything" [10, p. 24]; that is, verbs are distinguished from names by being tensed.[5] What ties these words

[2] A *finite* term, in medieval parlance, is one which signifies a determinate or definite number of things. Infinite terms are the complements of finite terms. For instance, 'man' is a finite term (and hence a noun), and 'non-man' is an infinite term.

[3] A term is *direct* (Lat. *recte*) if it is in the nominative case (or indicative mood, for verbs). Sherwood notes that this is the logician's definition of a noun, under which oblique cases of terms are not nouns.

[4] At the time of writing, I did not have access to a Latin edition of the *Introductiones*; see the conclusion of this paper.

[5] In giving these definitions, Sherwood is drawing upon the earlier grammatical tradition due to Abelard via Priscian and going all the way back to Dionysius of Thrax [1,5].

together is that they are significative, that is, they have meaning in isolation from other words. It is the properties of *these* words, (signification, supposition, copulation, and appellation), that form the topic of the chapter on the properties of terms in Sherwood's *Introductiones.* As a result, they are *not* our present focus. However, as Kretzmann points out, "some understanding of these notions is essential for a thorough understanding of Sherwood's treatment of syncategorematic words" [11, p. 5], so we will provide the necessary background information as required.

Secondary parts "are not necessary for the statement's being", and include things such as "the adjectival name, the adverb, and conjunctions and prepositions" [11, p. 13].[6] They are divided into two types, those which are "determinations of principal parts in respect of the things belonging to them" and those which are "determinations of principal parts insofar as they are subjects or predicates" [11, pp. 14–15].[7] The distinction here is between words that limit the scope of a noun or a verb and those that affect the way the noun or verb functions as a grammatical or logical subject or predicate in a statement. For instance (to use the examples Sherwood provides), in 'white man' (*albus homo*), 'white' is a secondary part of the first type; it provides a qualification of something that belongs to the word 'human'[8], while in 'every human', 'every' does not provide a qualification of a thing or things which are human, but instead says something about the relationship between the subject of the sentence and the predicate. What this something is will depend not merely on the phrase alone, but also where that phrase occurs in the sentence: For in "Every human is an animal", 'every human' will not be given the same analysis as it will in "An animal is every human".

It is for this reason—that the signification of a phrase like 'every human' is only knowable in a complete grammatical context, and not in isolation—that secondary parts of the second type are called *syncategorematic*, that is "'*sin-*'—i.e., 'con-'—and '*categoreuma*'—i.e., 'significative' or 'predicative', for a syncategorematic word is always joined with something else in discourse" [11, p. 16][9], where a *categoreuma* or *categorematic word* is one that is either a primary part, or a secondary part of the first type.

So here we have our final definition:

Definition 3.1 (Syncategoremata) *A* syncategorematic word *or* term *is a*

[6] *non sunt necessaria ad esse enuntiationis... nomen adjectivum et adverbium et conjunctiones et praepositiones* [8, p. 48].

[7] *quaedam sunt determinationes partium principalium ratione suarum rerum... quaedam sunt determinationes partium principalium in quantum sunt subjecta vel praedicata* [8, p. 48].

[8] Latin *homo* refers to humans of any gender; the addition of the word *albus*, with masculine grammatical gender, not only restricts *homo* to those humans which are white, but also to those humans which are male. Because English is not as strongly gendered as Latin is, it is sometimes hard to reproduce these subtleties in translation. Nevertheless, I will in general use 'human' for unmodified *homo*, but 'man' or 'woman' where appropriate for modified *homo*.

[9] *'sin' quod est 'con' et 'categoreuma' quod est 'significativum' vel 'praedicativum' quasi conpraedicativum; semper enim cum alio jungitur in sermone* [8, p. 48].

secondary part of a statement which is a determination of the principal parts of the statement with respect to their being subjects and predicates.

From the preceding definition, it should be clear that 'necessarily', 'contingently', and many other modal adverbs are syncategoremata: They are determinations of the principal parts of a statement in so far as those parts are subjects or predicates.

Here it is worth noting that Sherwood *only* considers modal *adverbs* in his discussion of syncategorematic terms. This hearkens back to his discussion of modes and modality in the *Introductiones*, in which he admits only adverbs as modes, unlike Aquinas and Pseudo-Aquinas who also allow modal adjectives (e.g., "That Socrates is a man is necessary"), cf. [14, p. 391].

4 Necessity and contingency as syncategoremata

In this section, we work through the chapter on modal syncategorematic terms, providing an analysis of and commentary on Sherwood's views.

First, Sherwood notes that such words can be used either categorematically or syncategorematically, which brings to light a point we have not yet made and so should make now: While we speak of 'categorematic terms' and 'syncategorematic terms', this is somewhat sloppy usage. Instead, we should speak of the *uses* of terms: For some terms are sometimes used categorematically—for instance, when we speak of 'the whole man' (*omnis homo*)—and sometimes used syncategorematically—for instance, when we speak of 'every man' (*omnis homo*)—while other terms can be used only categorematically and others only syncategorematically.[10] The focus in this chapter is, naturally, the syncategorematic use of the terms, and we will continue to sometimes speak of 'syncategorematic terms' as opposed to 'terms used syncategorematically'.

Sherwood argues that modal adverbs such as 'necessarily' can be used both categorematically (that is, determining the verb it modifies "in respect of the thing belonging to it" [11, p. 101][11]) and syncategorematically (that is, determining it "in respect of the composition belonging to it, or insofar as it is a predicate" [11, p. 101][12]). In support of this he gives the following example [11, p. 101][13]:

The heaven moves necessarily. (1)

There are two ways that (1) can be understood. In the first case, "it signifies... that the motion of the heaven is necessary" [11, p. 101].[14] On this understanding, 'necessarily' modifies the motion of the heaven, which is a thing that belongs to the term 'moves', and it is an answer to the question "How does the heaven move?"—it moves necessarily—or "What kind of movement does

[10] For more on this point, and the consequences it has in terms of logical analysis, see [15].

[11] *ratione suae rei* [8, p. 74].

[12] *ratione compositionis suae vel in quantum est praedicatum* [8, p. 74].

[13] *Caelum movetur necessario* [8, p. 74].

[14] *significat quod motus caeli sit necessarius* [8, p. 74].

the heaven have?"—necessary movement. The sentence itself, though, is not modal; it is a simple assertoric sentence, which can be either true (if the heavens in fact do move necessarily) or false (if they either do not move at all or their movement is contingent).

In the second way, the sentence signifies "that the composition of the verb with the subject is necessary" [11, p. 101] [15], that is, "the heaven moves" is a necessary statement. (An analogous sophism, involving Socrates, running, and moving, is discussed in the chapter on conditionals [11, p. 125].)

The preliminaries being rehearsed, the primary focus of this chapter is an analysis of possible sophisms (logical puzzles or puzzles in analysis) that can arise from either conflating the syncategorematic and categorematic uses of a word, or from ambiguities resulting from combining the words with other syncategorematic words. The procedure is to raise a particular sophism and then solve it, and from this deduce certain rules governing the use of modal adverbs.

What is a sophism? Briefly, it is a sentence which has two seemingly equally plausible analyses that lead to opposite conclusions. (An example of a sophism familiar to modern readers is the Liar sentence: Both the analysis from which one concludes that it is true and the analysis from which one concludes that it is false seem equally plausible.) Medieval logicians used these sentences, and their opposing analyses, to distinguish good logical inference from sophistical inference. In the case of many of the sophisms discussed in this chapter, the existence of opposing analyses trades on a conflation of the syncategorematic and categorematic use of the same term. Other analyses involve scope ambiguities introduced by distributives (including quantifiers) and exceptives. In each case, Sherwood presents a sophism sentence, and then gives both a *probatio* 'proof' and a *contra* '[proof] contra'. We will follow suit in presenting the sophisms, in what follows.

4.1 The sophisms

The first sophism is this:

Sophism 4.1 *The soul of the Antichrist will be necessarily [11, p. 101].* [16]

Proof. Proof: The soul of Antichrist will have necessary being because at some time it will have unceasing, incorruptible being.

On the contrary, [the soul of Antichrist] will be contingently because it is possible that it will not be [11, p. 101]. [17] □

This sophism is solved by distinguishing the categorematic use of 'necessarily' and the syncategorematic use, as in the analysis of (1). If 'necessarily' is

[15] *quod compositio hujus verbi cum hoc subjecto sit necessaria* [8, p. 74].

[16] *Anima antichristi erit necessario* [8, p. 74].

[17] *Probatio: anima antichristi habebit esse necessarium quia aliquando habebit esse non cessans incorruptibile.*
Contra: contingenter erit quia possibile est ipsum non fore [8, p. 74].

taken categorematically, then it determines what type of being Antichrist's soul will have. For when Antichrist exists, their soul will have its existence necessarily, following traditional 13th-century thought that souls exist eternally, necessarily, and incorruptibly (cf., e.g., [7]). Thus, the *probatio* is correct under the categorematic analysis.

If, however, 'necessarily' is taken syncategorematically, then the *contra* is correct: "The soul of the Antichrist will be" is not necessarily true, because Antichrist's existence is contingent (and if they don't exist, then there is no soul will be 'the soul of the Antichrist').

In this sophism, we see a sentence that is true when the modal adverb modifies the predicate, but false when it modifies the sentence as a whole. In the next sophism, we see the opposite:

Sophism 4.2 *Contingents necessarily are true [11, p. 102].* [18]

Proof. Proof: 'Contingents are true' is necessary; therefore it will be true when it has been modified by the mode of necessity; therefore 'contingents necessarily are true' is true.

On the contrary, no contingents are necessarily true [11, p. 102]. [19] □

Both Sherwood's example and his argumentation is substantially compressed here, so let us unpick it. First, the fact that

$$\text{Contingents are true.} \tag{2}$$

is an indefinite sentence is important for its analysis; for Sherwood, such indefinite (unquantified) sentences "do not determine whether the discourse is about the whole [of the subject] or about a part" [10, p. 29]. Second, though Sherwood does not state this explicitly anywhere in either the *Introductiones* or the *Syncategoremata*, he takes it as given that contingent sentences are sometimes true and sometimes false. Thus (2) is not only a true statement, it is also necessary, for if a contingent sentence was never true, then it would not be a contingent sentence, and this is true of any contingent sentence. Since the statement is necessarily true, we can add the modal adverb 'necessarily' to it, scoping over the entire sentence, and maintain truth. This is the syncategorematic use of the term. However, if we take 'necessarily' categorematically, to modify the predicate 'true' only, then it is clear why the statement would be false: For no contingent sentence is necessarily-true. [20]

At this point, Sherwood introduces some new vocabulary to describe what's going on: He says that in the first case (when the term is interpreted syn-

[18] *Contingentia necessaria [sic] sunt vera* [8, p. 74].

[19] *Probatio: contingentia sunt vera; haec est necessaria; ergo modificato modo necessitatis erit vert; ergo haec est vera: contingentia necessario sunt vera.*
Contra: nulla contingentia necessario sunt vera [8, p. 74].

[20] Sentence structure in English is less flexible, and hence more ambiguous, than in Latin. When we intend the categorematic reading of 'necessarily' as a modifier of a subject or predicate term, we will hyphenate it with that term. When it is not so hyphenated, it should be read syncategorematically, as an adverb modifying the verb.

categorematically), the modal adverb 'necessarily' is being used as a *note of coherence*, because it modifies the coherence of the subject and the predicate. In the second case (when the term is used categorematically), it functions as a *note of inherence*, because it expresses something about how the predicate inheres in some subject (namely, necessarily).

The third sophism illustrates how modal adverbs interact with exclusive words such as 'only' (*solus*) or 'alone' (*tantum*).

Sophism 4.3 *Suppose that Socrates, Plato, and Cicero are running necessarily and that a fourth [man is running] contingently, and that there are no more [men]. Then only three men are running necessarily [11, p. 103].*[21]

Proof. Proof: Three men necessarily are running, [and no others necessarily are running;] therefore only three [men are running necessarily].

On the contrary, 'only three men are running' is contingent, because when the fourth is running it will be false and when he is not running it will be true; therefore it will be false when it has been modified by the mode of necessity [11, p. 103].[22] □

In analysing this sophism, Sherwood points out that the inclusion of 'only' or 'alone' introduces an ambiguity depending on whether the modal adverb scopes over it or not. The distinction highlighted in the *probatio* and *contra* is between the categorematic usage, where 'only' modifies 'three men' and 'necessarily' modifies 'running':

$$\text{Three men and no more than three men are necessarily-running.} \tag{3}$$

and the syncategorematic usage, where 'only' still modifies 'three men' but 'necessarily' modifies the entire sentence:

$$\text{Necessarily: Three men and no more than three men are running.} \tag{4}$$

In (3), neither the exclusive nor the modal adverb are within the scope of each other; because each of the three men is individually necessarily-running, and no other man is necessarily-running, it is true that only three men are necessarily-running. However, in (4), the wide scope of the modal adverb makes the claim false, because there is no reason why the fourth man couldn't start running.

A similar analysis is given of the next sophism, which involves the exclusive 'alone' instead of 'only':

Sophism 4.4 *Necessaries alone are necessarily true [11, p. 103].*[23]

---

[21] *Currant Sortes et Plato et Cicero necessario et quartus contingenter et non sint plures. Deinde: tantum tres homines currunt necessario* [8, p. 74].

[22] *Probatio: tres homines necessario currunt, ergo tantum tres.*
Contra: tantum tres homines currunt. Hoc est contingens, quia quarto currente erit falsa, et illo non currente erit vera: ergo modificato modo necessitatis erit falsa [8, p. 74].

[23] *Sola necessaria necessario sunt vera* [8, p. 74].

Proof. Proof: Necessaries necessarily are true, and no others [necessarily are true] (which is proved inductively); therefore necessaries alone necessarily [are true].

On the contrary, 'necessaries alone are true' is false; therefore, it will be false when the mode of necessity has been added. Alternatively: on the contrary, contingents necessarily are true since 'contingents are true' is necessary; therefore not necessaries alone [necessarily are true] [11, pp. 103–104].[24] □

To see how this sophism is analogous to Sophism 4.3, it is sufficient to observe that the sophism sentence is equivalent to:

$$\text{Only necessaries are necessarily true.} \tag{5}$$

The analysis of this sophism then can proceed entirely analogously to the previous one.

The final two sophisms that Sherwood considers in this chapter involve the interaction of modal terms with distributive ones. The most familiar distributive term is *omnis* ('all' or 'every'), and it is the topic of the first chapter of virtually every treatise on syncategorematic words (Sherwood's included), and both of the next two sophisms we look at involve this distributive term.

Sophism 4.5 *Suppose that all men who exist now are running necessarily as long as they exist, and similarly with respect to future men. Thus every man necessarily is running [11, p. 104].*[25]

Proof. Proof: 'Every man is running' is necessary; therefore it will be true when it has been modified with the mode of necessity. Then if Socrates is a man, Socrates necessarily is running [11, p. 104].[26] □

Note that in O'Donnell's edition and in Kretzmann's translation, there is no proof *contra*. In the discussion of this sophism at [6, p. 480], an alternative reading of the Latin text is provided: *Contra: sed Sortes est homo, ergo Sortes necessario currit* ("Contra: But Socrates is a man, therefore Socrates is necessarily running").[27] This alternative removes that curiosity in the O'Donnell-Kretzmann versions.

This sophism is solved by introducing a distinction between whether the necessity ties to the universal statement that every man is running or whether it ties to all of the singular statements that are implied by this universal statement

[24] *Probatio: necessaria necessario sunt vera et nulla alia: quod probatur inductive; ergo sola necessaria necessario etc.*
Contra: sola necessaria sunt vera. Haec est falsa; ergo addito modo necessitatis erit falsa. Vel sic contra: contingentia necessario sunt vera quia haec est necessaria: contingentia sunt vera; ergo non sola necessaria [8, p. 74].

[25] *Verbi gratia, currant omnes homines qui nunc sunt necessario dum sunt, et similiter de futuribus hominibus; inde omnis homo necessario currit* [8, p. 75].

[26] *Probatio: haec est necessaria 'omnis homo currit'; ergo modificato modo necessitatis erit vera. Deinde: si Sortes est homo; ergo Sortes necessario currit* [8, p. 75].

[27] I would also like to thank one of the anonymous referees who provided me with this text and reference.

(e.g., "Socrates is running", "Sara is running", etc.), and Sherwood says that the same distinction applies to the next sophism as well:

Sophism 4.6 *Every man of necessity is an animal, but Socrates is a man; therefore Socrates of necessity is an animal [11, p. 105].*[28]

Here is a bit of a surprise: While these two sophisms involve the same distributive term, they don't both involve modal *adverbs*—this despite the fact that we noted above that Sherwood, both here and in the *Introductiones*, restricted the definition of 'mode' to include only adverbs. The second of the two sophisms uses the phrase *de necessitate* 'of necessity' instead of the adverbial form *necessario* 'necessarily'. What we should conclude here is that these two phrases, despite their grammatical differences, do not differ in their logical import.

4.2 The Rules

From consideration of these six sophisms, Sherwood gives the following rules that govern the use of these adverbs, both on their own and in conjunction with other syncategorematic words:

Rule 4.1 *The word 'necessarily' can sometimes be a note of coherence and at other times a note of inherence [11, p. 102].*[29]

This rule is illustrated by Sophism 4.2.

Sophisms 4.3 and 4.4 form the basis for the next rule:

Rule 4.2 *Sometimes there is an ambiguity in that the word 'necessarily' can include the word 'alone' or 'only', or vice versa [11, p. 103].*[30]

Finally, sophisms 4.5 and 4.6 give rise to the following rule:

Rule 4.3 *Sometimes ambiguity occurs in that the word 'necessarily' can either include a division or be included by it [11, p. 104].*[31]

These rules are disappointingly banal, especially in the context of Kretzmann describing the *Syncategoremata* as "an advanced treatise". They certainly don't seem to be very advanced principles, or have the feeling of something being discovered through the analysis of these sophisms: Surely anyone after a modicum of reflection could tell you that syntax (whether Latin or English) is in many cases inherently ambiguous. What do we gain from identifying these principles and classifying them as logical rules, elevating them above other principles?

It's hard to give a satisfactory answer to this question. When one looks at a historical logic text, one always hopes to gain insight not only into the history of the field but into the field itself; but it is not clear what we can learn from

---

[28] *Omnis homo de necessitate est animal; sed Sortes est homo; ergo Sortes de necessitate est animal* [8, p. 75].

[29] *Item regula quod haec dictio 'necessario' quandoque potest esse nota cohaerentiae, quandoque nota inhaerentiae* [8, p. 74].

[30] *Item quandoque est multiplicitas eo quod haec dictio 'necessario' possit includere hanc dictionem 'solus' sive 'tantum', vel e converso* [8, p. 74].

[31] *Item quandoque accidit multiplicitas eo quod haec dictio 'necessario' possit includere divisionem vel includi ab ea* [8, p. 75].

Sherwood here. The chapter on necessity and contingency does not, itself, tell us much more than what we already knew, or could have easily discovered on our own through reflection on the material presented in the *Introductiones.*

Perhaps an alternative question to ask is not what *we*, as modern logicians, can learn from looking at this text, but rather what Sherwood and his contemporaries could learn from going through these exercises. Here the answer is clearer: Analysing these various sophismata makes explicit the need to be precise about the interaction between modality and quantification, so that propositions involving both can be properly disambiguated. In this respect, one could compare the developments of Sherwood and his contemporaries with the modern-day developments in quantified modal logic due to Barcan Marcus.[32]

5 Necessity and contingency elsewhere in the *Syncategoremata*

But the chapter of necessity and contingency is not the only place in the *Syncategoremata* where modal terms are discussed. Just as this chapter included rules governing how modal adverbs interact with other syncategorematic terms such as 'alone' or 'every', chapters covering other terms sometimes include sophisms involving modal adverbs or other modal terms. In this section, we outline these sophisms and use their analyses to augment the overall picture we're developing.

In the chapter on conditionals, Sherwood distinguishes natural and nonnatural consequences and this distinction relies on modal notions. Natural consequences are those where "the consequent follows from the antecedent in respect of some state of the one relative to the other" and in this case "it notes an ordering of things in reality" [11, p. 123].[33] Nonnatural consequences, on the other hand, are those where

> the consequent follows from the antecedent not in respect of a state of the one relative to the other but solely because of the impossibility of the antecedent or the necessity of the consequent... it notes an ordering of things in discourse [11, p. 123].[34]

That is, nonnatural consequences are ones that rely on the principles of *ex impossibili quodlibet sequitur* and *necessarium ex quolibet sequitur.*

In this chapter, he also offers a distinction between conditional propositions and categorical statements with conditional predicates [11, ch. XVII, §§16, 17]. He considers the following sophism:

---

[32] Thanks to one of the anonymous referees for suggesting this more positive way of looking at Sherwood's contribution.

[33] *consequens sequi ad antecedens ratione alicujus habitudinis unius ad aliud... notat ordinem rerum secundum rem* [8, p. 80].

[34] *consequens sequi ad antecedens, non ratione habitudinis unius ad aliud, sed solum propter impossibilitatem antecedentis vel necessitatem consequentis... notat ordinem rerum secundum sermonem* [8, p. 80].

Sophism 5.1 *What is true is false if Antichrist exists [11, p. 122].*[35]

The proof *contra* relies on blocking the modal inference from a contingent statement to an impossible one: "the antecedent [Antichrist exists] is contingent, and what follows [what is true is false] is impossible; therefore the conditional is false" [11, p. 122].[36] This same principle is appealed to in solving a sophism in §13 of the same chapter; here we also find another modal principle, related to the one used in the definition of nonnatural consequences, invoked: "the antecedent is impossible, therefore the conditional is necessary" [11, p. 124].[37] These two are just examples of true nonnatural consequences, as discussed above.

In the chapter on 'unless' (*nisi*)[38], we find the correlate of the principle just described: a conditional is false when "the antecedent is necessary and what follows it is contingent" [11, p. 129].[39]

Finally, in the chapter on disjunction, Sherwood notes that disjunctions combined with modal terms are ambiguous between scoping over the whole disjunction or the individual disjuncts; a sentence where the modality has wide scope can be true without the corresponding sentence with narrow scope being true. That is, he rejects the (obviously invalid) inference:

$$\models \Box(p \vee \neg p) \qquad \Rightarrow \qquad \models \Box p \vee \Box \neg p \tag{6}$$

The example he gives is [11, p. 147][40]:

That Socrates is running or not running is necessary. (7)

What is interesting here is that he follows this example up with two more example combinations of a modal term with a disjunction, but this time (one of the only times in the text) the modal term is not alethic, but epistemic [11, pp. 147–148][41]:

You know that the stars are even or uneven [in number]. (8)

and

That the stars are even [in number]
or that the stars are odd [in number] is known to you. (9)

---

[35] *Verum est falsum si antichristus est* [8, p. 80].

[36] *antecedit contingens et sequitur impossibile; ergo conditionalis falsa* [8, p. 80].

[37] *antecedens est impossibile; ergo conditionalis necessaria* [8, p. 81].

[38] An interesting chapter in itself, for anyone who has had to motivate to undergraduates why 'unless' in English can be translated into a conditional with a negated antecedent; in Latin, *nisi* is literally a compound of the negative particle *non* 'not' plus the conditional marker *si* 'if'.

[39] *antecedit necessarium et sequitur contingens* [8, p. 83].

[40] *Sortem currere vel non currere est necessarium* [8, p. 89].

[41] *Tu scis astra esse paria vel imparia... astra esse paria vel astra esse imparia scitur a te* [8, p. 89].

These are classic examples of sentences where one of the disjuncts must be true, but which one is true is not known. Unfortunately, Sherwood does not otherwise discuss epistemic modalities, in this text or in the *Introductiones.*

To sum up: In other places in the *Syncategormata* where modal terms occur, we can see Sherwood relying on the following modal inferential principles:

Rule 5.1 *Impossibility never follows from contingency.*

Rule 5.2 *Contingency never follows from necessity.*

Rule 5.3 *Any conditional with an impossible antecedent is necessary.*

As with the case of the rules deduced at the end of the previous section, these are neither especially interesting nor especially novel rules to pose: They are quite basic and quite orthodox.

What we saw both in this section and in §4 is not a systematic approach: For the rules are derived from the analyses as consequences of them, rather than the rules being stipulated in advance and then used to analyse the sophisms. Additionally, there is a lack of systematicity in terms of completeness: There is no guarantee that the sophisms considered in this text exhaust all of the possible sophisms that arise from the use of modal adverbs.

This lack of systematicity may cause a contemporary logician to bristle: For logicians are, if anything, systematic. One could even use this lack of systematicity to dismiss Sherwood as a worthwhile object of study. In my final remarks below, I would like to briefly argue that even if what we have found in this analysis seems straightforward and banal, there is still value to be gained from having undertaken this study.

6 Some final remarks

In this paper, we have revisited William of Sherwood's modal theory through the lens of what he has to say about necessity and contingency in his later treatise, the *Syncategoremata.* This deepens our understanding of Sherwood's account of modality which we originally discussed in [14]. The primary features of Sherwood's views on modal terms in the *Syncategoremata* are a distinction between the categorematic use and the syncategorematic use of modal adverbs, which is used to solve various sophisms, and rules that govern the interaction of modal adverbs with distributive and exclusive terms. From this, we can see Sherwood's close attention to the ways in which modal terms are used in actual discourse and where sophistical reasoning can arise from ambiguity or equivocation. This highlights a possible explanation for why Sherwood's approach lacks the systematicity that modern logicians strive for in their theories: Sherwood is fundamentally interested in analysing language in discourse, and language is inherently unsystematic. There is no way to survey all possible sophisms involving modal terms; but it is possible to highlight common problems and errors that people can make, and to provide rules for recognizing and avoiding those problems. In fact, the lack of systematicity and completeness can be seen as a virtue: By identifying types of sophisms and types of problems, and rules to deal with these, Sherwood makes it possible for us to extrapolate fro these rules to novel situations. What results may not be terribly interesting modal logic, but *is* of straightforward use and application in ordinary everyday modal reasoning.

In an ideal world, I would have included a comparison, at the end of this paper, of what Sherwood has to say on these topics with his colleagues in the Big Four, Peter of Spain, Lambert of Auxerre, and Roger Bacon. All four men covered similar topics in their introductory textbooks and treatises on syncategorematic terms, and a comparison of the other three with Sherwood is often in valuable because there are clear paths of influence from Sherwood's *Introductiones* to the other texts and because Sherwood's views, as some of the earliest, are often both more nascent and more interesting. Furthermore, Kretzmann puts forth an argument that Sherwood's *Syncategormata* post-dates Peter of Spain's *Tractatus Syncategorematum*; but whichever is earlier, Kretzmann says it is clear that "the study of either man's work on syncategorematic words is greatly aided by a close comparison with the work of the other" [11, p. 7].)

Unfortunately, this research was not produced in an ideal world. This paper was completed while I was in isolation due to Covid-19 lockdown measures in the United Kingdom, with the books that I would have needed to reference inaccessible in my office.[42] I believe, nevertheless, that an analysis of Sherwood's views alone is still of research value and of historical interest, and that noting the shortcomings of the present paper can serve as a reminder to future readers that research is not produced in a vacuum, but depends on so many vital factors all coming together in the right way at the same moment—time to work without constant interruption from a child, access to the right books, space to work without wondering when you or one of your family members will be the next person you know who is sick. When all these factors are lacking at the same time, the result is papers whose scope must perforce be more modest.

References

[1] Abaelardus, P., "Glossae Super Peri Hermeneias," Brepols Publishers, 2010, K. Jacobi and C. Strub, eds.

[2] Bacon, R., "The Art and Science of Logic," Pontifical Institute of Medieval Studies, 2009, T. S. Maloney, trans.

[3] Copenhaver, B. P., editor, "Peter of Spain: Summaries of Logic, Text, Translation, Introduction, and Notes," Oxford University Press, 2014, with C. Normore and T. Parsons.

[4] d'Auxerre, L., "Logica (Summa Lamberti)," La Nuova Italia, 1971, F. Alessio, ed.

[5] Hall, R. and C. Lejewski, *Symposium: Parts of speech*, Proceedings of the Aristotelian Society, Supplementary Volumes **39** (1963), pp. 173–204.

[6] Kirchhoff, R., "Die Syncategoremata des Wilhelm von Sherwood: Kommentierung und historische Einordnung," Brill, 2008.

---

[42] My sincere thanks go out to two internet friends who helped me source electronic versions of texts or looked up relevant information in their copies of books for me, Mark Thakkar and Justin Vlasits.

[7] Nauta, L., *The preexistence of the soul in medieval thought*, Recherches de Théologie Ancienne et Médiévale **63** (1996), pp. 93–135.
[8] O'Donnell, J. R., *The Syncategoremata of William of Sherwood*, Mediaeval Studies **3** (1941), pp. 46–93.
[9] of Auxerre, L., "Logica or Summa Lamberti," University of Notre Dame Press, 2015, T. S. Maloney, trans.
[10] of Sherwood, W., "William of Sherwood's *Introduction to Logic*," University of Minnesota Press, 1966, N. Kretzmann, trans.
[11] of Sherwood, W., "William of Sherwood's *Treatise on Syncategorematic Words*," University of Minnesota Press, 1968, N. Kretzmann, trans.
[12] of Sherwood, W., "Introductiones in logicam / Einführung in die Logik," Hamburg: Felix Meiner Verlag, 1995, H. Brands & C. Kann, eds. and trans.
[13] of Sherwood, W., "Syncategoremata," Felix Meiner Verlag, 2012, Latin-German edition, C. Kann and R. Kirchhoff, eds. and trans.
[14] Uckelman, S. L., *Three 13th-century views of quantified modal logic*, in: C. Areces and R. Goldblatt, editors, *Advances in Modal Logic*, Advances in Modal Logic **7**, 2008, pp. 389–406.
[15] Uckelman, S. L., *The logic of categorematic and syncategorematic infinity*, Synthese **192** (2015), pp. 2361–2377.
[16] Uckelman, S. L. and H. Lagerlund, *Logic in the 13th century*, in: C. D. Novaes and S. Read, editors, *Cambridge Companion to Medieval Logic*, Cambridge University Press, 2016 pp. 119–141.

www.ingramcontent.com/pod-product-compliance
Lightning Source LLC
Chambersburg PA
CBHW070712160426
43192CB00009B/1160
* 9 7 8 1 8 4 8 9 0 3 4 1 8 *